JACARANDA PHYSICS 2

VCE UNITS 3 AND 4 | FIFTH EDITION

MURRAY ANDERSON

DAN O'KEEFFE

BARBARA MCKINNON

MICHAEL ROSENBROCK

GRAEME LOFTS

ROSS PHILLIPS

PETER PENTLAND

jacaranda
A Wiley Brand

Fifth edition published 2024 by
John Wiley & Sons Australia, Ltd
155 Cremorne Street, Cremorne, Vic 3121

First edition published 1997
Second edition published 2003
Third edition published 2008
eBookPLUS edition published 2016
Fourth edition published 2020

Typeset in 10.5/13 pt TimesLTStd

ISBN: 978-1-119-88833-8

Front cover image: © Kristian Danu wijanarko/Shutterstock

Illustrated by various artists, diacriTech and Wiley Composition Services

Typeset in India by diacriTech

A catalogue record for this book is available from the National Library of Australia

Printed in Singapore
M WEP220017 180823

The Publishers of this series acknowledge and pay their respects to Aboriginal Peoples and Torres Strait Islander Peoples as the traditional custodians of the land on which this resource was produced.

This suite of resources may include references to (including names, images, footage or voices of) people of Aboriginal and/or Torres Strait Islander heritage who are deceased. These images and references have been included to help Australian students from all cultural backgrounds develop a better understanding of Aboriginal and Torres Strait Islander Peoples' history, culture and lived experience.

It is strongly recommended that teachers examine resources on topics related to Aboriginal and/or Torres Strait Islander Cultures and Peoples to assess their suitability for their own specific class and school context. It is also recommended that teachers know and follow the guidelines laid down by the relevant educational authorities and local Elders or community advisors regarding content about all First Nations Peoples.

All activities in this resource have been written with the safety of both teacher and student in mind. Some, however, involve physical activity or the use of equipment or tools. **All due care should be taken when performing such activities.** To the maximum extent permitted by law, the author and publisher disclaim all responsibility and liability for any injury or loss that may be sustained when completing activities described in this resource.

The Publisher acknowledges ongoing discussions related to gender-based population data. At the time of publishing, there was insufficient data available to allow for the meaningful analysis of trends and patterns to broaden our discussion of demographics beyond male and female gender identification.

Contents

UNIT 4 HOW HAVE CREATIVE IDEAS AND INVESTIGATION REVOLUTIONISED THINKING IN PHYSICS? 415

AREA OF STUDY 1 HOW HAS UNDERSTANDING ABOUT THE PHYSICAL WORLD CHANGED?

About this resource

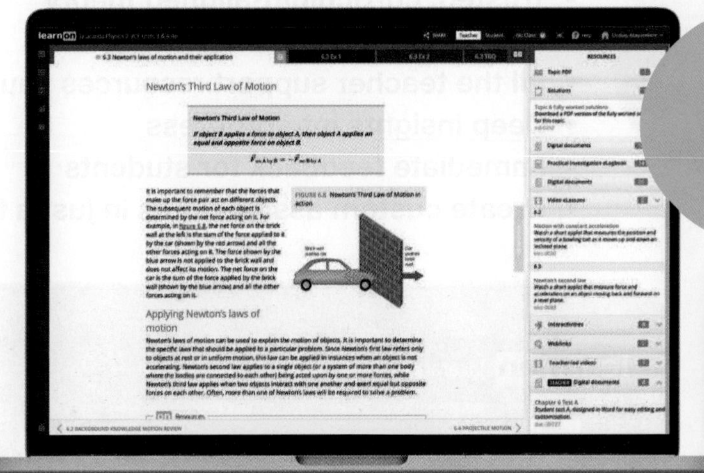

Aligned to the VCE Physics Study Design (2023–2027)

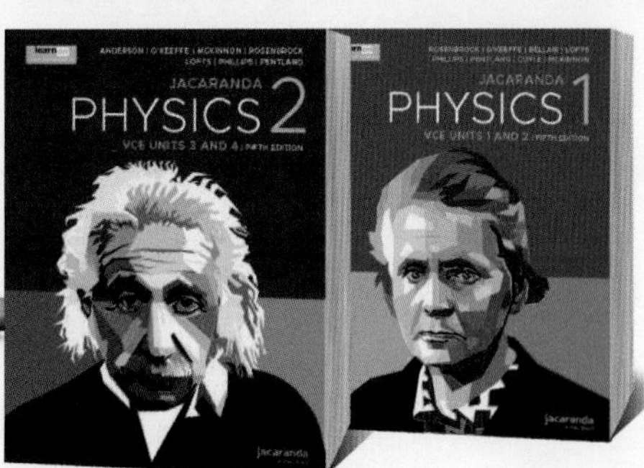

JACARANDA

PHYSICS 2 VCE UNITS 3 AND 4 FIFTH EDITION

Developed by expert Victorian teachers for VCE students

Tried, tested and trusted. The NEW Jacaranda VCE Physics series continues to deliver curriculum-aligned material that caters to students of all abilities.

Completely aligned to the VCE Physics Study Design

Our expert author team of practising teachers and assessors ensures 100% coverage of the new VCE Physics Study Design (2023–2027).

Everything you need for your students to succeed, including:

- NEW! Access targeted questions sets including exam-style questions and all relevant past VCAA exam questions since 2013. Ensure assessment preparedness with practice SACs.
- NEW! Enhanced practical investigation support including practical investigation videos, and eLogbook with fully customisable practical investigations — including teacher advice and risk assessments.
- NEW! Teacher-led videos to unpack challenging concepts, VCAA exam questions, exam-style questions, investigations and sample problems to fill learning gaps after COVID-19 disruptions.

Learn online with Australia's most

Everything you need for each of your lessons in one simple view

- **Trusted, curriculum-aligned theory**
- **Engaging, rich multimedia**
- **All the teacher support resources you need**
- **Deep insights into progress**
- **Immediate feedback for students**
- **Create custom assignments in just a few clicks.**

Practical teaching advice and ideas for each lesson provided in teachON

Each lesson linked to the Key Knowledge (and Key Science Skills) from the VCE Physics Study Design

Reading content and rich media including embedded videos and interactivities

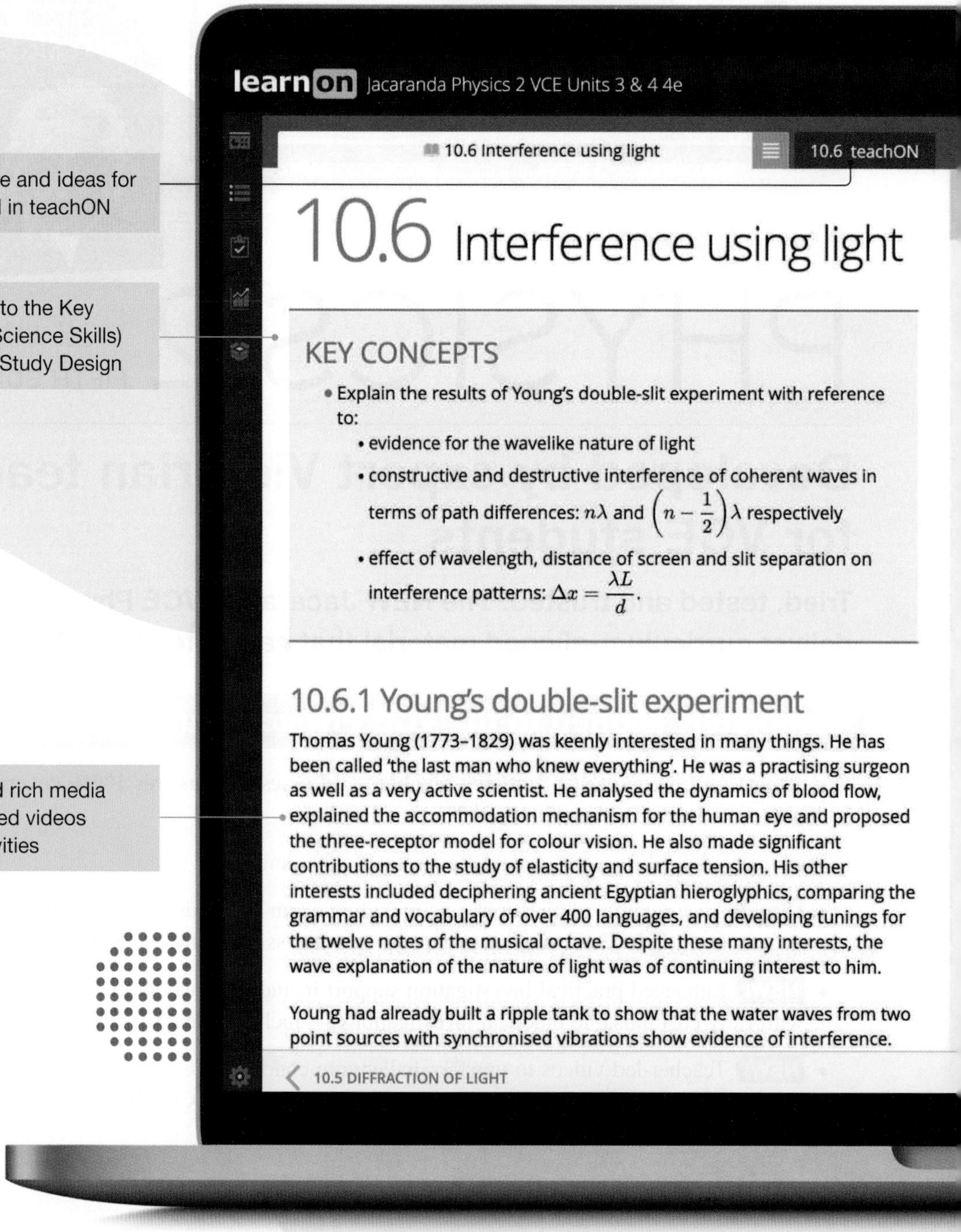

learnON Jacaranda Physics 2 VCE Units 3 & 4 4e

10.6 Interference using light | 10.6 teachON

10.6 Interference using light

KEY CONCEPTS

- Explain the results of Young's double-slit experiment with reference to:
  - evidence for the wavelike nature of light
  - constructive and destructive interference of coherent waves in terms of path differences: $n\lambda$ and $\left(n - \frac{1}{2}\right)\lambda$ respectively
  - effect of wavelength, distance of screen and slit separation on interference patterns: $\Delta x = \frac{\lambda L}{d}$.

10.6.1 Young's double-slit experiment

Thomas Young (1773–1829) was keenly interested in many things. He has been called 'the last man who knew everything'. He was a practising surgeon as well as a very active scientist. He analysed the dynamics of blood flow, explained the accommodation mechanism for the human eye and proposed the three-receptor model for colour vision. He also made significant contributions to the study of elasticity and surface tension. His other interests included deciphering ancient Egyptian hieroglyphics, comparing the grammar and vocabulary of over 400 languages, and developing tunings for the twelve notes of the musical octave. Despite these many interests, the wave explanation of the nature of light was of continuing interest to him.

Young had already built a ripple tank to show that the water waves from two point sources with synchronised vibrations show evidence of interference.

< 10.5 DIFFRACTION OF LIGHT

powerful learning tool, learnON

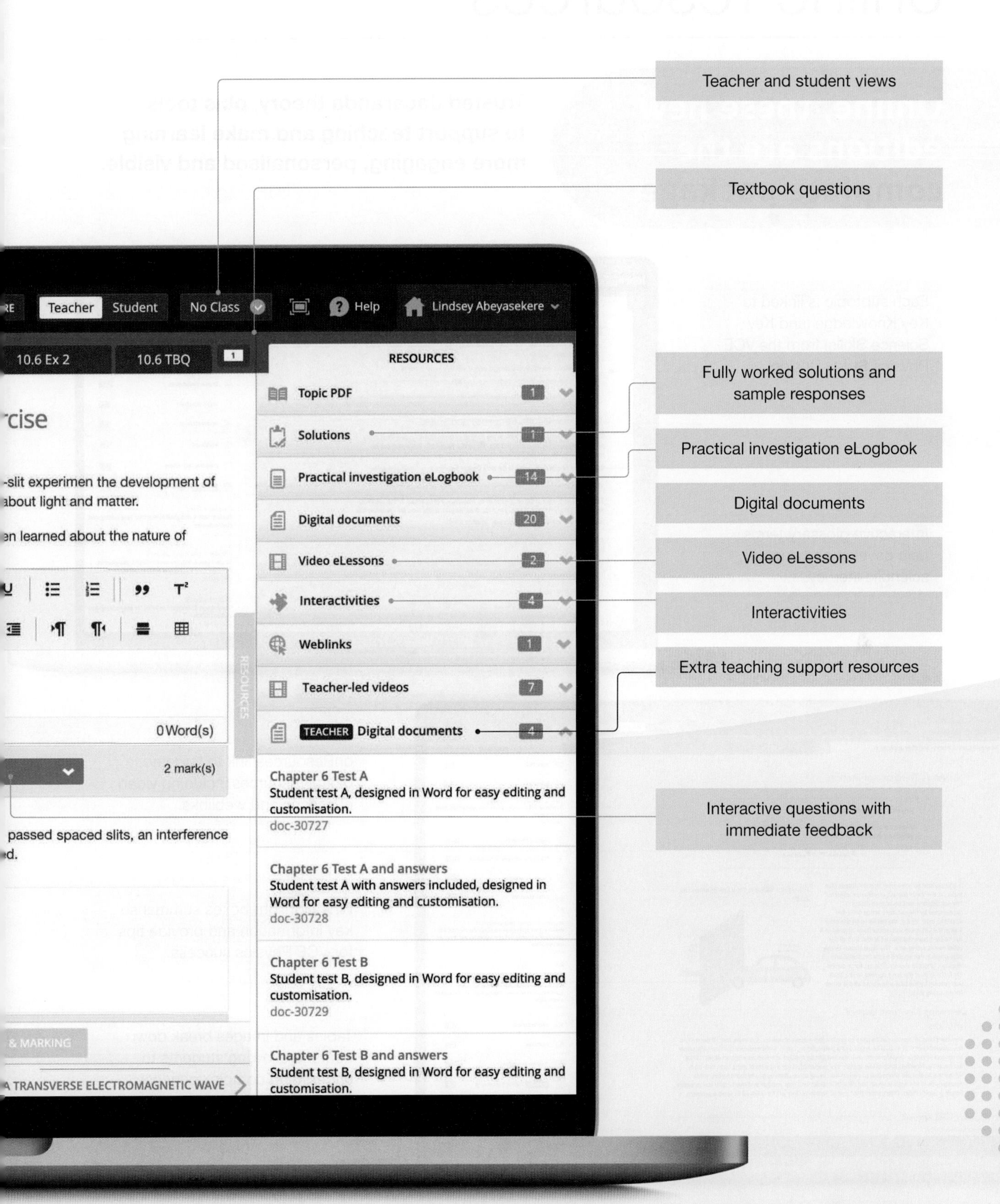

Teacher and student views
Textbook questions
Fully worked solutions and sample responses
Practical investigation eLogbook
Digital documents
Video eLessons
Interactivities
Extra teaching support resources
Interactive questions with immediate feedback
Teacher
Student
No Class
Help
Lindsey Abeyasekere
10.6 Ex 2
10.6 TBQ
RESOURCES
Topic PDF
Solutions
Practical investigation eLogbook
Digital documents
Video eLessons
Interactivities
Weblinks
Teacher-led videos
TEACHER Digital documents
Chapter 6 Test A
Student test A, designed in Word for easy editing and customisation.
doc-30727
Chapter 6 Test A and answers
Student test A with answers included, designed in Word for easy editing and customisation.
doc-30728
Chapter 6 Test B
Student test B, designed in Word for easy editing and customisation.
doc-30729
Chapter 6 Test B and answers
Student test B, designed in Word for easy editing and customisation.
0 Word(s)
2 mark(s)

Get the most from your online resources

Online, these new editions are the **complete package**

Trusted Jacaranda theory, plus tools to support teaching and make learning more engaging, personalised and visible.

Each subtopic is linked to Key Knowledge (and Key Science Skills) from the VCE Physics Study Design.

Interactive glossary terms help develop and support scientific literacy.

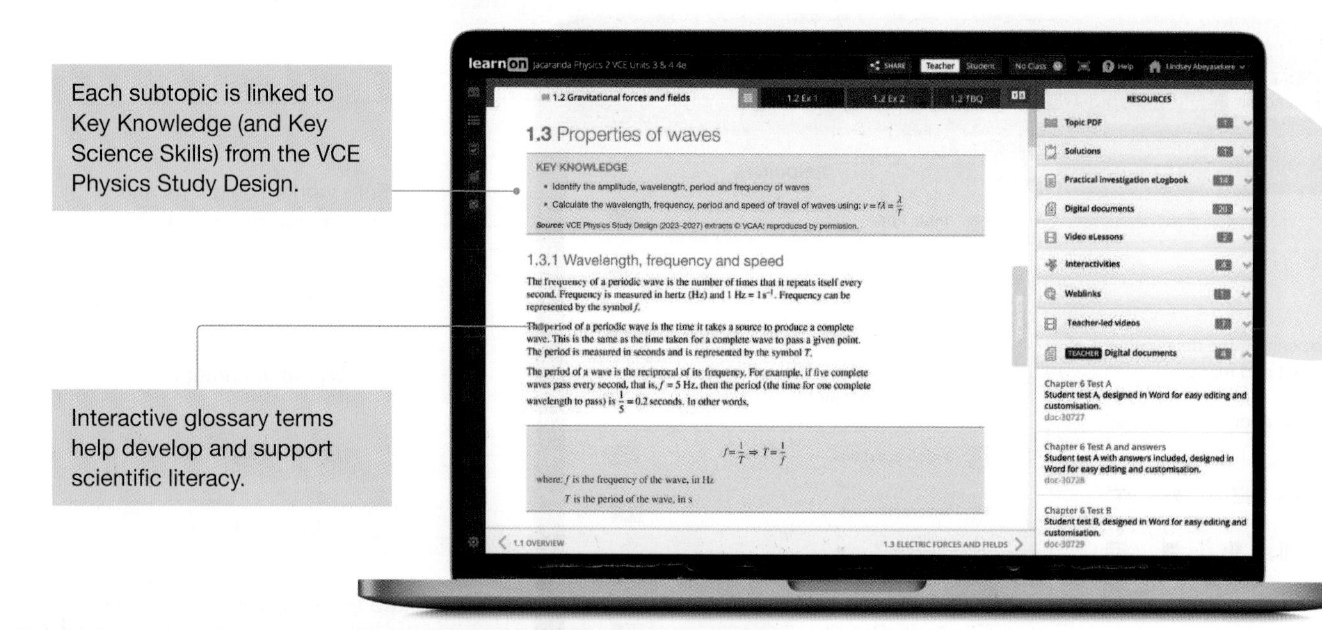

onResources link to targeted digital resources including video eLessons and weblinks.

Pink highlight boxes summarise key information and provide tips for VCE Physics success.

Tables and images break down content, allowing students to understand complex concepts.

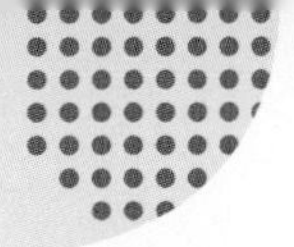

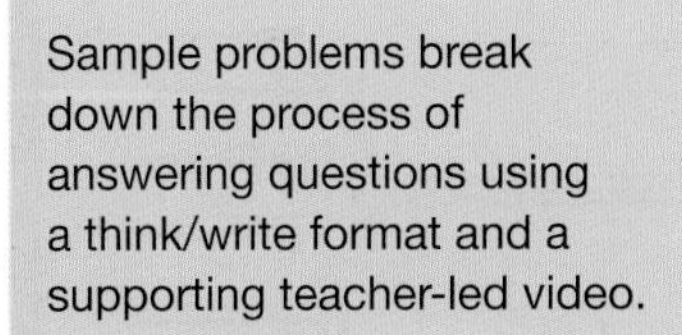

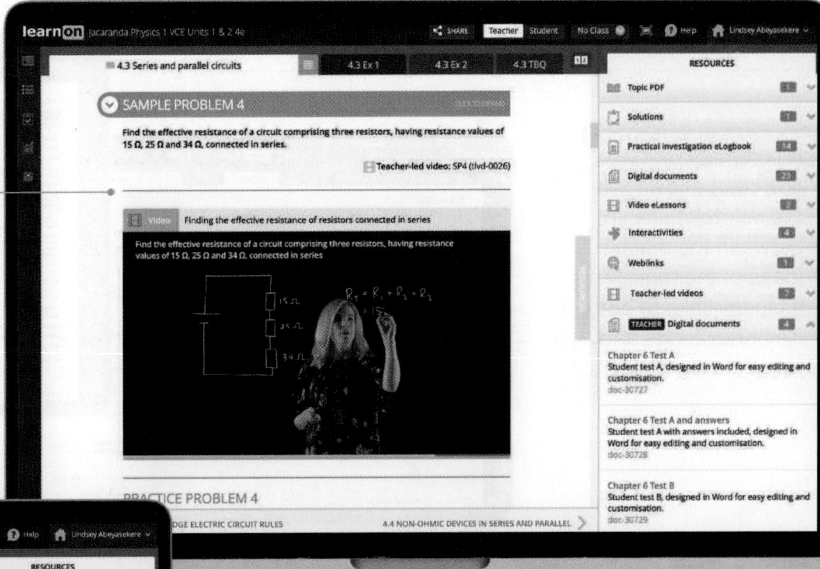

Sample problems break down the process of answering questions using a think/write format and a supporting teacher-led video.

Practical investigations are highlighted throughout topics, and are supported by teacher-led videos and downloadable student and teacher version eLogbooks.

- Online and offline question sets contain practice questions and past VCAA exam questions with exemplary responses and marking guides.
- Every question has immediate, corrective feedback to help students to overcome misconceptions as they occur and to study independently — in class and at home.

Topic reviews

A summary flowchart shows the interrelationship between the main ideas of the topic. This includes links to both Key Knowledge and Key Science Skills.

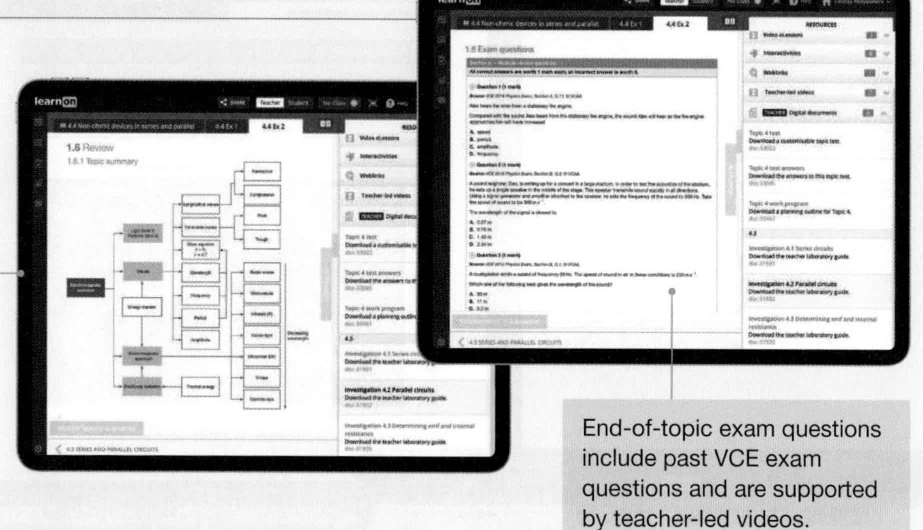

End-of-topic exam questions include past VCE exam questions and are supported by teacher-led videos.

Area of Study reviews

Areas of study reviews include practice examinations and practice SACs with worked solutions and sample responses. Teachers have access to customisable quarantined SACs with sample responses and marking rubrics.

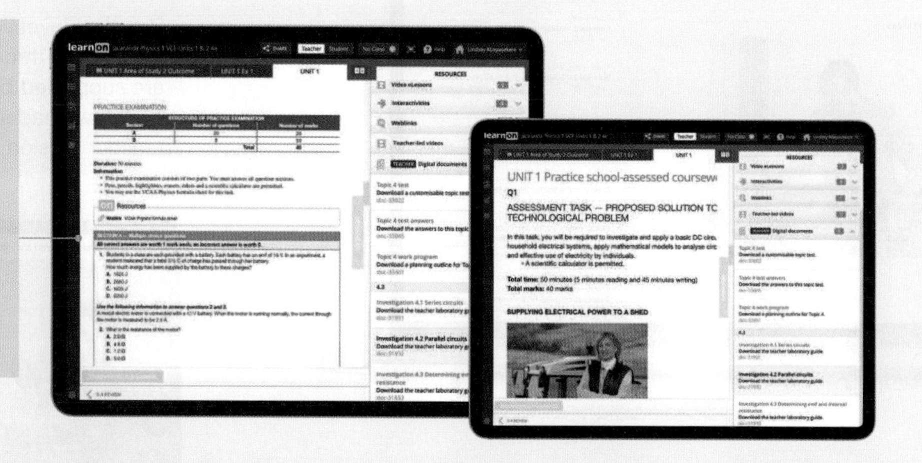

Practical investigation eLogbook

Enhanced practical investigation support includes practical investigation videos and an eLogbook with fully customisable practical investigations — including teacher advice and risk assessments.

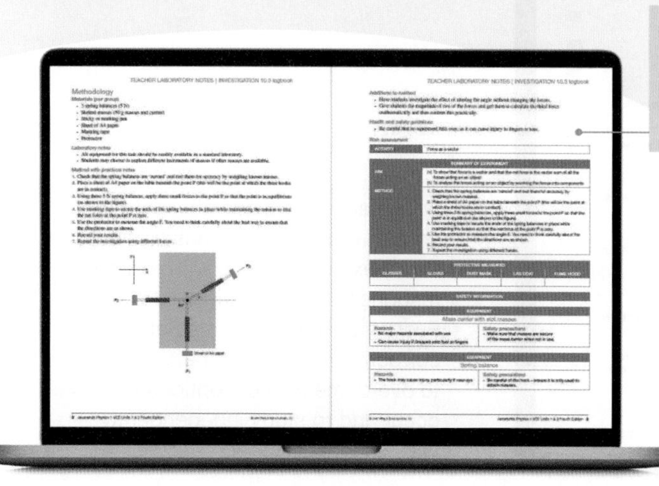

A wealth of teacher resources

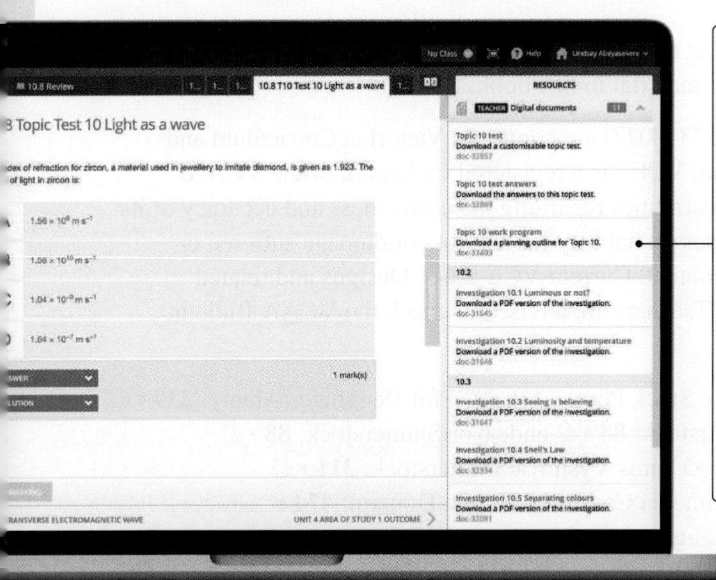

Enhanced teacher support resources, including:

- work programs and curriculum grids
- teaching advice
- additional activities
- teacher laboratory eLogbook, complete with solution and risk assessments
- quarantined topic tests (with solutions)
- quarantined SACs (with worked solutions and marking rubrics).

Customise and assign

A testmaker enables you to create custom tests from the complete bank of thousands of questions (including past VCAA exam questions).

Reports and results

Data analytics and instant reports provide data-driven insights into progress and performance within each lesson and across the entire course.

Show students (and their parents or carers) their own assessment data in fine detail. You can filter their results to identify areas of strength and weakness.

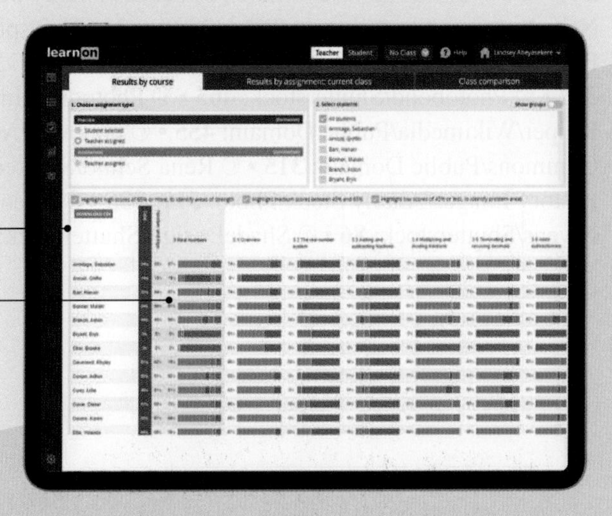

Acknowledgements

The authors and publisher would like to thank the following copyright holders, organisations and individuals for their assistance and for permission to reproduce copyright material in this book.

Selected extracts from the VCE Physics Study Design (2023–2027) are copyright Victorian Curriculum and Assessment Authority (VCAA), reproduced by permission. VCE® is a registered trademark of the VCAA. The VCAA does not endorse this product and makes no warranties regarding the correctness and accuracy of its content. To the extent permitted by law, the VCAA excludes all liability for any loss or damage suffered or incurred as a result of accessing, using or relying on the content. Current VCE Study Designs and related content can be accessed directly at www.vcaa.vic.edu.au. Teachers are advised to check the VCAA Bulletin for updates.

• © Anders93/Adobe Stock Photos: **417** • © Mopic/Alamy Stock Photo: **236** • © Phil Degginger/Alamy: **239** • © lightpoet/Adobe Stock Photos: **627** • © CHEN WS/Shutterstock: **88** • © endeavor/Shutterstock: **88** • © NikoNomad: **143** • © StockPhotosArt/Shutterstock: **107** • © Taras Vyshnya/Shutterstock: **311** • © yelantsevv/Shutterstock: **365** • Ferdinand Schmutzer/Wikimedia Commons/Public Domain: **474** • Thongsuk7824/Shutterstock: **473** • © 3Dsculptor: **144** • © agsandrew/Shutterstock: **518** • © Alan Porritt/AAP Image: **562** • © Alejo Miranda/Shutterstock: **346** • © arogant/Shutterstock: **337** • © Australian Synchrotron: **602** • © Bahadir Yeniceri/Shutterstock: **20** • © Ben Jeayes/Shutterstock: **562** • © By Kyle Senior - Own work: CC BY-SA 4.0, https://commons.wikimedia.org/w/index.php?curid=118428178: **196** • © By Miropa01 - Own work: CC BY-SA 4.0, https://commons.wikimedia.org/w/index.php?curid=81735397: **209** • © By Teravolt (talk): CC BY 3.0, https://commons.wikimedia.org/w/index.php?curid=24784084: **239** • © By wdwd - Own work: map data based on this locator map: File:AUS locator map.svg, CC BY 3.0, https://commons.wikimedia.org/w/index.php?curid=15312815: **390** • © By Wisky - Own work: CC BY-SA 3.0, https://commons.wikimedia.org/w/index.php?curid=15113998: **625** • © Sakurambo/Wikimedia Commons/Public Domain, **625** • © christianto soning/EyeEm/Adobe Stock Photos: **1** • © dennizn/Shutterstock: **595** • © Designua/Shutterstock: **344** • © Diagram courtesy of Paul G. Hewitt: **544** • © Dragon Images/Shutterstock: **261** • © dvande/Shutterstock: **249** • © Edward R. Degginger/Alamy Stock Photo: **448** • © Evgeny Murtola/Shutterstock: **115** • © Fineart1/Shutterstock: **261** • © Georgios Kollidas/Shutterstock: **563** • © GIPHOTOSTOCK/Science Photo Library (SPL): **452** • © haryigit/Shutterstock: **279** • © imagIN.gr photography/Shutterstock: **560** • © irin-k/Shutterstock: **418** • © jabiru/Shutterstock: **4, 124** • © JackF/Adobe Stock Photos: **415** • © Jari Sokka/Adobe Stock Photos: **205** • © JASPERIMAGE/Shutterstock: **47** • © jennyt/Shutterstock: **52** • © KGBobo/Shutterstock: **3** • © Kim Christensen/Shutterstock: **250** • © Marccophoto/iStock/Getty Images: **378** • © Mark Garlick/Science Photo Library/Alamy Stock Photo: **559** • © mironov/Shutterstock: **85** • © mosufoto/Shutterstock: **276** • © muratart/Shutterstock: **37, 124** • © NASA: **167, 260** • © NASA/Dembinsky Photo Associates/Alamy Stock Photo: **167** • © Natursports/Shutterstock: **53** • © New Anawach/Shutterstock: **383** • © News Ltd/Newspix: **565** • © Olga Besnard/Shutterstock: **52** • © Pannawish/Shutterstock: **347** • © PATTARAWIT CHOMPIPAT/Alamy Stock Photo: **345** • © Peter Pentland: **347** • © Photobond/Shutterstock: **465** • © Photodisc: Inc., **44** • © PhysicsOpenLab: **235** • © Pieter Kuiper/Wikimedia/Public Domain: **455** • © popov48/Adobe Stock Photos: **570** • © RadRafe/Wikimedia Commons/Public Domain: **315** • © Rena Schild/Shutterstock: **53** • © Roger Bruce/Shutterstock: **383** • © Science History Images/Alamy Stock Photo: **519, 525** • © sciencephotos/Alamy Stock Photo: **195** • © Sergey Nivens/Shutterstock: **86** • © ShadeDesign/Shutterstock: **240**, **241** • © Spencer Bliven/Wikimedia

Commons/Public Domain: **443** • © Springer Nature: **458** • © STELR Solar Cars. STELR: Australian Academy of Technological Sciences & Engineering: https://www.atse.org.au/stelr: **346**, **348** • © Teun van den Dries/Shutterstock: **308** • © The National High Magnetic Field Laboratory: **464** • © thka/Shutterstock: **45** • © Triff/Shutterstock: **604** • © udaix/Shutterstock: **250** • © University of Cambridge/CC BY-NC-SA 2.0 UK: **441** • © Vereshchagin Dmitry/Shutterstock: **117** • © Volkmar Holzwarth: **250** • © Wirestock Creators/Shutterstock: **517** • © World History Archive/Alamy Stock Photo: **561** • © Yiorgos GR/Shutterstock: **533** • © sciencephotos/Alamy Stock Photo: **533** • © Physics Dept.: Imperial College/Science Photo: **533** • © Phil Degginger/Alamy Stock Photo: **533** • © Zoonar GmbH/Alamy Stock Photo: **366** • © Dn Br/Shutterstock: **284**

Every effort has been made to trace the ownership of copyright material. Information that will enable the publisher to rectify any error or omission in subsequent reprints will be welcome. In such cases, please contact the Permissions Section of John Wiley & Sons Australia, Ltd.

UNIT

3 How do fields explain motion and electricity?

AREA OF STUDY 1

How do physicists explain motion in two dimensions?

OUTCOME 1

Investigate motion and related energy transformations experimentally, and analyse motion using Newton's laws of motion in one and two dimensions.

AREA OF STUDY 2

How do things move without contact?

OUTCOME 2

Analyse gravitational, electric and magnetic fields, and apply these to explain the operation of motors and particle accelerators, and the orbits of satellites.

AREA OF STUDY 3

How are fields used in electricity generation?

OUTCOME 3

Analyse and evaluate an electricity generation and distribution system.

1 Newton's laws of motion

KEY KNOWLEDGE

In this topic, you will:

- investigate and apply theoretically and practically Newton's three laws of motion in situations where two or more coplanar forces act along a straight line and in two dimensions
- investigate and analyse theoretically and practically the motion of projectiles near Earth's surface, including a qualitative description of the effects of air resistance
- investigate and analyse theoretically and practically the uniform circular motion of an object moving in a horizontal plane: $\left(F_{\text{net}} = \frac{mv^2}{r}\right)$, including:
  - a vehicle moving around a circular road
  - a vehicle moving around a banked track
  - an object on the end of a string
- model natural and artificial satellite motion as uniform circular motion
- investigate and apply theoretically Newton's second law to circular motion in a vertical plane (forces at the highest and lowest positions only).

Source: VCE Physics Study Design (2024–2027) extracts © VCAA; reproduced by permission.

PRACTICAL WORK AND INVESTIGATIONS

Practical work is a central component of VCE Physics. Experiments and investigations, supported by a **practical investigation eLogbook** and **teacher-led video**, are included in this topic to provide opportunities to undertake investigations and communicate findings.

EXAM PREPARATION

Access past VCAA questions and exam-style questions and their video solutions in every lesson, to ensure you are ready.

1.1 Overview

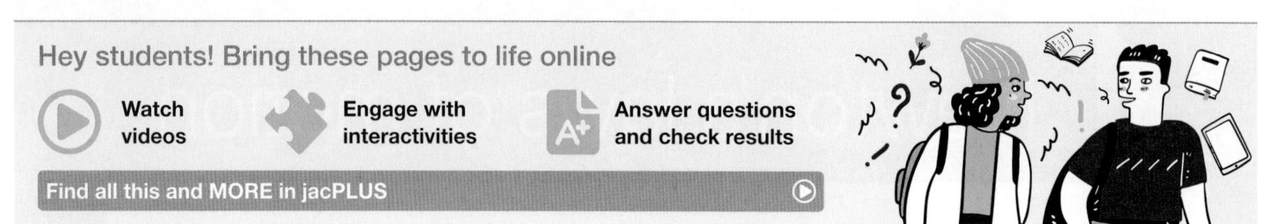

1.1.1 Introduction

Backwards and forwards. Faster and slower. Why do objects in motion behave the way they do, and how can such behaviour be consistently described?

Before Sir Isaac Newton published *Principia Mathematica* in 1687, there was no plausible theory that could clearly describe objects, also called bodies, the forces acting upon them, and their movements in response to those forces. Newton's laws of motion changed humanity's understanding of the entire universe by providing a mathematical description of the universe. They describe everything from a ball rolling down a hill to the positions of the planets.

To this day, more than 300 years later, Newton's laws of motion and gravity remain the foundations on which mechanics and engineering are based. They gave humanity the knowledge needed to send astronauts to the Moon and put satellites into orbit. Everyday motion, from driving a car and riding a bike, to enjoying a ride on a roller-coaster, are all governed by forces that can be described by Newton's laws.

FIGURE 1.1 Whether you are driving a car, riding a bike or riding a roller-coaster, your motion is controlled by the forces acting on the vehicle.

LEARNING SEQUENCE

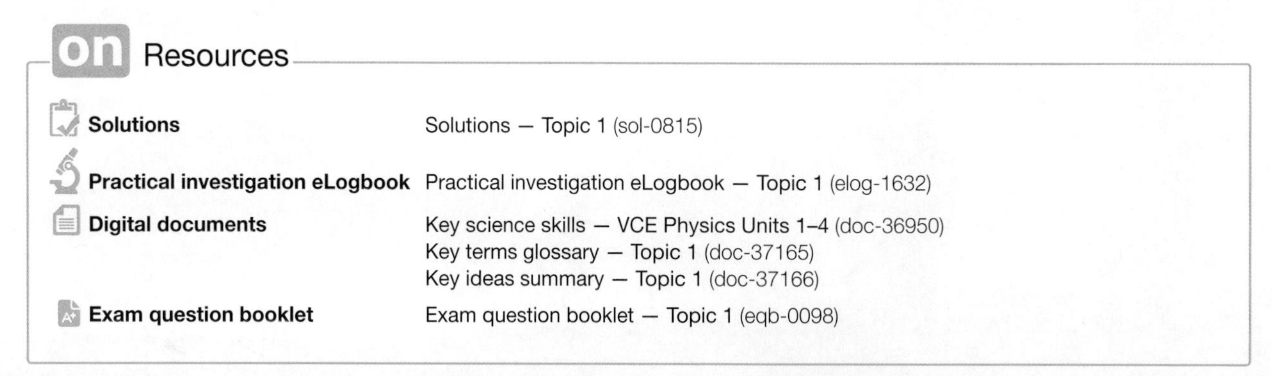
on Resources

Solutions	Solutions — Topic 1 (sol-0815)
Practical investigation eLogbook	Practical investigation eLogbook — Topic 1 (elog-1632)
Digital documents	Key science skills — VCE Physics Units 1–4 (doc-36950) Key terms glossary — Topic 1 (doc-37165) Key ideas summary — Topic 1 (doc-37166)
Exam question booklet	Exam question booklet — Topic 1 (eqb-0098)

1.2 BACKGROUND KNOWLEDGE Motion review

BACKGROUND KNOWLEDGE

- identify parameters of motion as vectors or scalars
- analyse graphically, numerically and algebraically, straight-line motion under constant acceleration:

$$v = u + at \qquad v^2 = u^2 + 2as \qquad s = \frac{1}{2}(u + v)t \qquad s = ut + \frac{1}{2}at^2 \qquad s = vt - \frac{1}{2}at^2$$

- graphically analyse non-uniform motion in a straight line
- apply concepts of momentum to linear motion: $\boldsymbol{p} = m\boldsymbol{v}$

1.2.1 Describing motion

To explain the motion of objects, it is important to be able to measure and describe the motion clearly. The language used to describe motion must therefore be precise and unambiguous.

The language of motion

The physical quantities used to describe and explain motion fall into two distinct groups: scalar quantities and vector quantities.

Scalar quantities are fully described by magnitude (size) only. Mass, energy, time, power and temperature are all examples of scalar quantities.

Vector quantities are fully described by specifying both a direction and a magnitude. Force, displacement, velocity and acceleration are all examples of vector quantities.

Note: Vector quantities are bolded in this resource but other notations are common, such as an arrow above or below the variable.

Distance is a measure of the length of the path taken during the change in position of an object. Distance is a scalar quantity. It does not specify a direction.

Displacement is a measure of the length of the change in position of an object. To fully describe a displacement, a direction must be specified as well as a magnitude. Displacement is therefore a vector quantity.

Speed is a measure of the rate at which an object moves over a distance. Because distance is a scalar quantity, speed is also a scalar quantity. The average speed of an object can be calculated by dividing the distance travelled by the time taken:

$$\text{average speed} = \frac{\text{distance travelled}}{\text{time interval}}$$

Velocity is a measure of the rate of displacement of an object. Because displacement (change in position) is a vector quantity, velocity is also a vector quantity. The velocity has the same direction as the displacement. The symbol $\boldsymbol{v}$ is used to denote velocity. (The symbol $\boldsymbol{v}$ is often used to represent speed as well.)

The average velocity of an object, $\boldsymbol{v}_{\text{av}}$, during a time interval can be expressed as:

$$\boldsymbol{v}_{\text{av}} = \frac{\Delta \boldsymbol{s}}{\Delta t}$$

where: $\boldsymbol{s}$ represents the displacement

Δt represents the time interval

scalar quantity quantity with only a magnitude (size)

vector quantity quantity requiring both a direction and a magnitude

distance measure of the full length of the path taken when an object changes position, a scalar quantity

displacement measure of the change in position of an object, a vector quantity

speed the rate at which distance is covered per unit time; a scalar quantity

velocity the rate of change of position of an object; a vector quantity

Neither the average speed nor the average velocity provide information about movement at any particular instant of time. The speed at any particular instant of time is called the **instantaneous speed**. The velocity at any particular instant of time is called the **instantaneous velocity**. It is only if an object moves with a constant velocity during a time interval that its instantaneous velocity throughout the interval is the same as its average velocity.

The rate at which an object changes its velocity is called its **acceleration**. Because velocity is a vector quantity, it follows that acceleration is also a vector quantity. The average acceleration of an object, $\boldsymbol{a}_{av}$, can be expressed as:

$$\boldsymbol{a}_{av} = \frac{\Delta \boldsymbol{v}}{\Delta t} = \frac{\boldsymbol{v} - \boldsymbol{u}}{t_f - t_i}$$

where: $\Delta \boldsymbol{v}$ represents the change in velocity $(\boldsymbol{v} - \boldsymbol{u})$, $\boldsymbol{v}$ is the final velocity and $\boldsymbol{u}$ the initial velocity

Δt represents the time interval, where t_i is the initial time and t_f is the final time, where $\Delta t = t_f - t_i$

The direction of the average acceleration is the same as the direction of the change in velocity. The instantaneous acceleration of an object is its acceleration at a particular instant of time.

A non-zero acceleration is not always caused by a change in speed. The vector nature of acceleration means that the object can be accelerating if it has a constant speed but is changing direction. Hence, acceleration is the rate of change of velocity.

Reviewing vectors

A vector quantity is one that has both direction and magnitude. These are represented diagrammatically using arrows, in which the length of the arrow reflects the magnitude of the quantity and the arrowhead allows the direction to be shown. Vectors are used constantly in physics, particularly in the study of motion, in which many variables are vector quantities.

Adding vectors

When vector quantities are added together, both direction and magnitude need to be taken into account. The example of forces in figure 1.2 illustrates this. The labelled arrows that represent vectors can be used to perform the addition by placing them 'head to tail'. When adding pairs of vectors, the labelled arrows are redrawn so that the 'tail' of the second arrow abuts the 'head' of the first arrow. The sum of the vectors is represented by the arrow drawn between the tail of the first vector and the head of the second. Figure 1.2 illustrates how this method has been used to determine the net force in the three examples shown. The sum of the vectors ($\boldsymbol{F}_{net}$) is represented in each case.

instantaneous speed speed at a particular instant of time

instantaneous velocity velocity at a particular instant of time

acceleration rate of change of velocity; a vector quantity

FIGURE 1.2 When adding vectors, both magnitude and direction need to be considered.

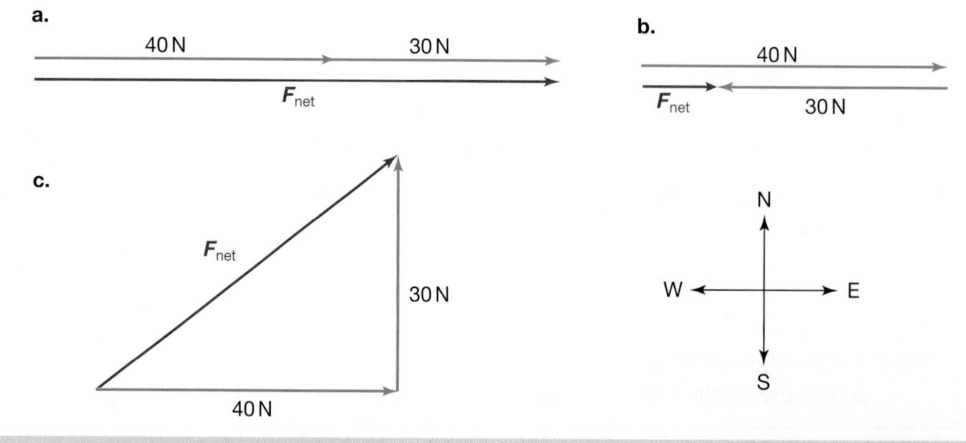

Determining the magnitude of a vector sum (resultant vector)

The vectors in figure 1.2 have been drawn to scale. This means that the length of the arrow representing the vector sum (resultant vector) can be measured. The magnitude of the vector sum can then be calculated. The direction of the vector sum is given by the direction in which the third arrow points. If the vectors have been drawn to scale, the direction can be determined by measuring the appropriate angle with a protractor.

The vector addition shown in figure 1.3 results in a right-angled triangle. The magnitude of the vector sum can be determined by using Pythagoras's theorem. The hypothenuse arrow of the triangle is the vector sum and its length represents the magnitude.

$$\begin{aligned} c^2 &= a^2 + b^2 \\ &= (40)^2 + (30)^2 \\ &= 2500 \text{ (calculating the sum of the squares of both sides)} \\ \Rightarrow c &= 50 \text{ N (taking the positive square root of the sum of the squares)} \end{aligned}$$

The direction of the net force can be found using trigonometric ratios. In this case, we can use $\tan \text{B} = \dfrac{\text{O}}{\text{A}}$.

$$\begin{aligned} \tan \text{B} &= \frac{30}{40} \\ &= 0.75 \\ \text{B} &= \tan^{-1}(0.75) \\ &= 37° \end{aligned}$$

The vector sum, and net force, is 50 N at an angle of N53°E (53° clockwise from north as A + B = 90°).

Knowing the various trigonometric ratios is important when finding unknown angles (when at least 2 side lengths are known) or unknown sides (when at least one angle and one side length is known) for right-angled triangles. Alternatively, the sine or cosine rule could be used.

FIGURE 1.3 The magnitude and direction of vector addition can be determined using Pythagoras's theorem. $\boldsymbol{F}_{net}$ is the resultant vector, or vector sum.

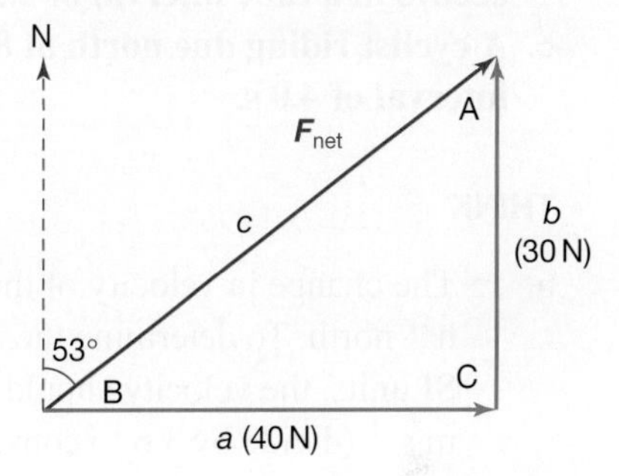

Subtracting vectors

One vector can be subtracted from another simply by adding its negative. It works because subtracting a vector is the same as adding the negative vector (just as subtracting a positive number is the same as adding the negative of that number). Another way to *subtract* vectors is to place them tail to tail, as shown in figure 1.4. The difference between the vectors ***a*** and ***b*** (***b*** – ***a***) is given by the vector that begins at the head of vector ***a*** and ends at the head of vector ***b***.

FIGURE 1.4 Subtracting vectors can be done either by placing them tail to tail, or through adding the negative vector.

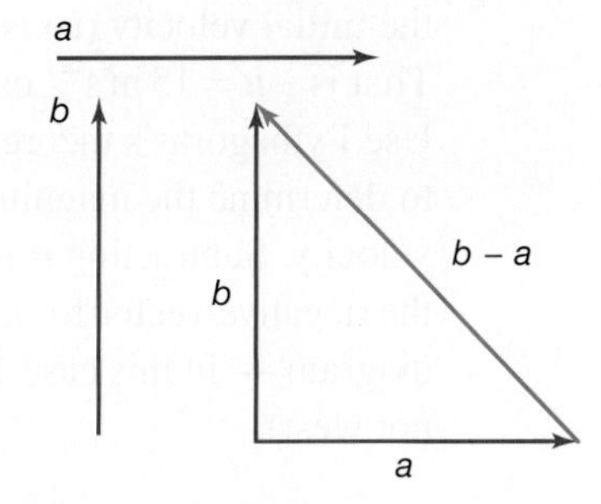

Finding vector components

The magnitude of vector components can be determined using trigonometric ratios. The vector ***P*** in figure 1.5 can be resolved into vertical and horizontal components.

The magnitude of the horizontal component, labelled $\boldsymbol{P}_\text{H}$, is given by:

$$\boldsymbol{P}_\text{H} = \boldsymbol{P}\cos 40° \left(\text{since } \cos 40° = \frac{\boldsymbol{P}_\text{H}}{\boldsymbol{P}}\right)$$
$$\Rightarrow P_\text{H} = 500 \text{ units} \times \cos 40°$$
$$\Rightarrow P_\text{H} = 500 \text{ units} \times 0.7660$$
$$= 383 \text{ units.}$$

The magnitude of the vertical component, labelled as $\boldsymbol{P}_\text{V}$, is given by:

$$\boldsymbol{P}_\text{V} = \boldsymbol{P}\sin 40° \left(\text{since } \sin 40° = \frac{\boldsymbol{P}_\text{V}}{\boldsymbol{P}}\right)$$
$$\Rightarrow P_\text{V} = 500 \text{ units} \times \sin 40°$$
$$\Rightarrow P_\text{V} = 500 \text{ units} \times 0.6428$$
$$= 321 \text{ units.}$$

FIGURE 1.5 A vector can be split into vertical and horizontal components.

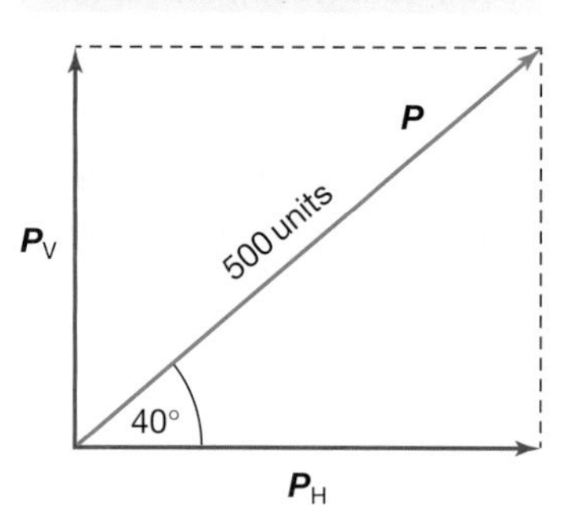

tlvd-8940

SAMPLE PROBLEM 1 Determining the average acceleration

Determine the average acceleration of each of the following objects.

a. A car starting from rest reaches a velocity of 60 km h^{-1} due north in 5.0 s.

b. A car travelling due west at a speed of 15 m s^{-1} turns due north at a speed of 20 m s^{-1}. The change occurs in a time interval of 2.5 s.

c. A cyclist riding due north at 8.0 m s^{-1} turns right to ride due east without changing speed in a time interval of 4.0 s.

THINK	WRITE
a. 1. The change in velocity of the car is 60 km h^{-1} north. To determine the acceleration in SI units, the velocity should be expressed in m s^{-1} (divide by 3.6 to convert from km h^{-1} to m s^{-1}).	**a.** $60 \text{ km h}^{-1} = \frac{60}{3.6} \text{ m s}^{-1} = 16.7 \text{ m s}^{-1}$
2. Use the formula $\boldsymbol{a}_\text{av} = \frac{\Delta \boldsymbol{v}}{\Delta t}$ to calculate the average acceleration.	$\boldsymbol{a}_\text{av} = \frac{\Delta \boldsymbol{v}}{\Delta t} = \frac{16.7}{5.0}$ $= 3.3 \text{ m s}^{-2}$ north
b. 1. The change in velocity must first be found by subtracting vectors because $\Delta \boldsymbol{v} = \boldsymbol{v} - \boldsymbol{u}$. The final velocity ($\boldsymbol{v}$) is 20 m s^{-1} north and the initial velocity ($\boldsymbol{u}$) is 15 m s^{-1} west. That is $-\boldsymbol{u} = 15 \text{ m s}^{-1}$ east. Use Pythagoras's theorem or trigonometry to determine the magnitude of the change of velocity. Subtracting $\boldsymbol{u}$ is the same as adding the negative vector for $\boldsymbol{u}$ (as shown in the diagram — in this case 15 m s^{-1} east, not west).	**b.** 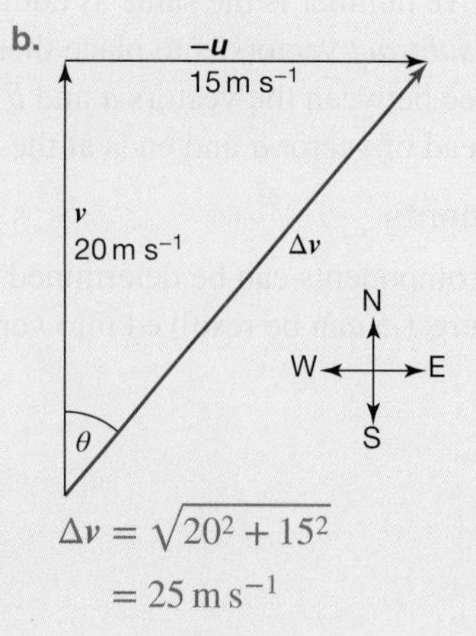 $\Delta \boldsymbol{v} = \sqrt{20^2 + 15^2}$ $= 25 \text{ m s}^{-1}$

2. The direction of a vector is usually given as an angle of rotation of the vector about its *tail*. Then look at where the *head* of the vector sum (resultant vector) is pointing. The direction can be found by calculating the value of the angle θ.

$$\tan\theta = \frac{15}{20} = 0.75 \Rightarrow \theta = 37°$$

The direction of the change in velocity is therefore N37°E.

3. Use the formula $\boldsymbol{a}_{av} = \frac{\Delta \boldsymbol{v}}{\Delta t}$ to calculate the average acceleration, where $\Delta \boldsymbol{v}$ = 25 m s^{-1} N37°E and t = 2.5 s.

$$\boldsymbol{a}_{av} = \frac{\Delta \boldsymbol{v}}{\Delta t} = \frac{25}{2.5} = 10\,\text{m s}^{-2}\ \text{N}\,37°\text{E}$$

c. 1. The change in velocity must first be found by subtracting vectors because $\Delta \boldsymbol{v} = \boldsymbol{v} - \boldsymbol{u}$. The final velocity ($\boldsymbol{v}$) is 8.0 m s^{-1} east and the initial velocity ($\boldsymbol{u}$) is 8.0 m s^{-1} north. That is $-\boldsymbol{u}$ = 8.0 m s^{-1} south. Use Pythagoras's theorem or trigonometry to determine the magnitude of the change of velocity. Subtracting $\boldsymbol{u}$ is the same as adding the negative vector for $\boldsymbol{u}$ (as shown in the diagram as going 8.0 m s^{-1} south, not north).

c.

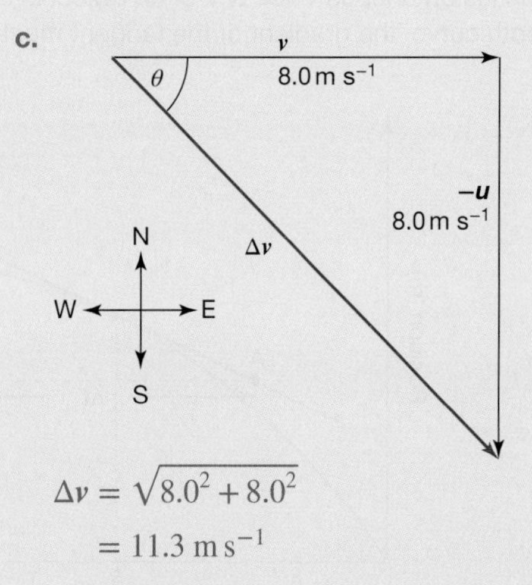

$$\Delta \boldsymbol{v} = \sqrt{8.0^2 + 8.0^2} = 11.3\ \text{m s}^{-1}$$

2. The direction can be found by calculating the value of the angle θ.

The triangle formed by the vector diagram shown is a right-angled isosceles triangle. The angle θ is therefore 45° and the direction of the change in velocity is south-east.

3. Use the formula $\boldsymbol{a}_{av} = \frac{\Delta \boldsymbol{v}}{\Delta t}$ to calculate the average acceleration, where $\Delta \boldsymbol{v}$ = 11.3 m s^{-1} S45°E and Δt = 4.0 s.

$$\boldsymbol{a}_{av} = \frac{\Delta \boldsymbol{v}}{\Delta t} = \frac{11.3}{4.0} = 2.8\ \text{m s}^{-2}\ \text{south-east (or S45°E)}$$

PRACTICE PROBLEM 1

Determine the average acceleration (in m s^{-2}) of:

a. **a rocket launched from rest that reaches a velocity of 15 m s^{-1} during the first 5.0 s after lift-off**

b. **a roller-coaster cart travelling due north at 20 m s^{-1} that turns 90 degrees to the left during an interval of 4.0 s without changing speed**

c. **a rally car travelling west at 100 km h^{-1} that turns 90 degrees to the left and slows to a speed of 80 km h^{-1} south. The turn takes 5.0 s to complete.**

1.2.2 Graphical analysis of motion

A description of motion in terms of displacement, average velocity and average acceleration is not complete. These quantities provide a 'summary' of motion but do not provide detailed information. By describing the motion of an object in graphical form, it is possible to estimate its displacement, velocity or acceleration at any instant during a chosen time interval.

Position–time graphs

A graph of position versus time provides information about the displacement and velocity at any instant of time during the interval described by the graph. If the graph is a straight line or smooth curve, it is also possible to estimate the displacement and velocity outside the time interval described by the data displayed in the graph using extrapolation.

FIGURE 1.6 The instantaneous velocity ***v*** of an object is equal to the gradient of the position–time graph. If the graph is a smooth curve, the gradient of the tangent must be determined.

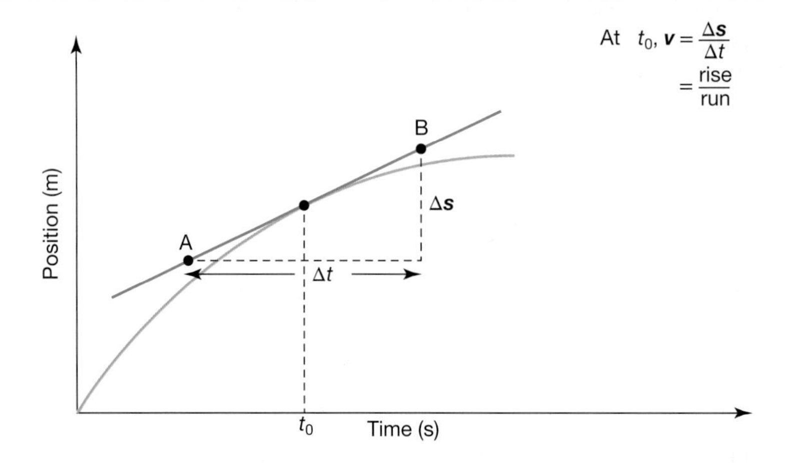

The velocity of an object at an instant of time can be obtained from a position–time graph by determining the gradient of the line or curve at the point representing that instant. This method is a direct consequence of the fact that velocity is a measure of the rate of change in position. If the graph is a smooth curve, the gradient at an instant of time is the same as the gradient of the tangent to the curve at that instant.

Similarly, the speed of an object at an instant of time can be obtained by determining the gradient of a graph of the object's distance from a reference point versus time.

Velocity–time graphs

A graph of velocity versus time provides information about the velocity and acceleration at any instant of time during the interval described by the graph. It also provides information about the displacement between any two instants.

The instantaneous acceleration of an object can be obtained from a velocity–time graph by determining the gradient of the line or tangent to a curve at the point representing that instant in time. This method is a direct consequence of the fact that acceleration is defined as the rate of change of velocity.

The displacement of an object during a time interval can be obtained by determining the area under the velocity–time graph representing that time interval. The actual position of an object at any instant during the time interval can be found only if the starting position is known.

Similarly, the distance travelled by an object during a time interval can be obtained by determining the corresponding area under the speed versus time graph for the object.

Acceleration–time graphs

A graph of acceleration versus time provides information about the acceleration at any instant of time during the time interval described by the graph. It also provides information about the change in velocity between any two instants.

The change in velocity of an object during a time interval can be obtained by determining the area under the acceleration–time graph representing that time interval. The actual velocity of the object can be found at any instant during the time interval only if the initial velocity is known.

The relationship between position, velocity and acceleration–time graphs

Position–time, velocity–time and acceleration–time graphs are all related. As velocity is the change of position over time, it is equivalent to the gradient of the position–time graph. Acceleration is the change of velocity over time, and is the gradient of the velocity–time graph, as seen in figure 1.7.

FIGURE 1.7 The position–time, velocity–time and acceleration–time graphs for an object thrown vertically into the air (assume negligible air resistance). As long as one graph is given, the other two can be deduced. Some extra information may be needed.

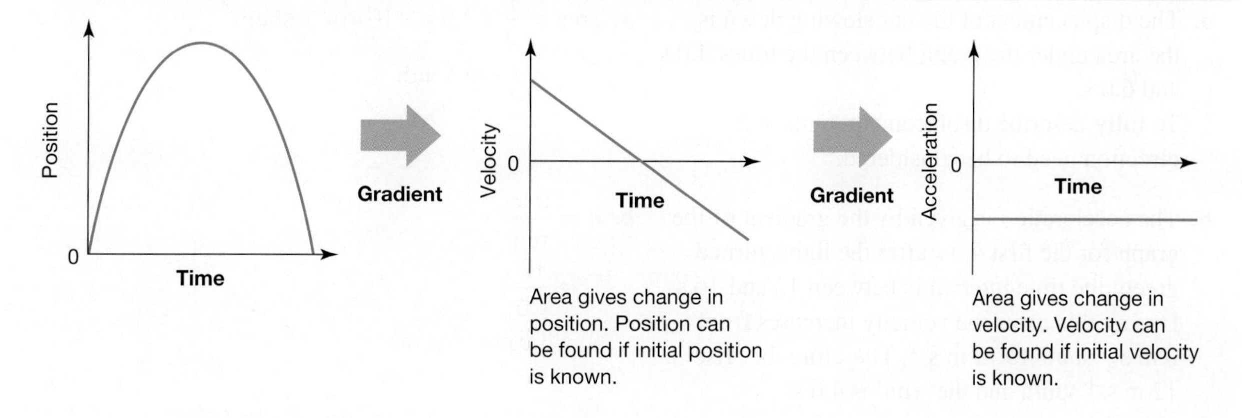

TABLE 1.1 Summary of motion graphs

		Position versus time graphs	Velocity versus time graphs	Acceleration versus time graphs
Quantities that can be read directly from the graph	Horizontal axis	Time	Time	Time
	Vertical axis	Position	Velocity	Acceleration
Quantities that can be calculated from the graph	Gradient of tangent	Instantaneous velocity	Instantaneous acceleration	
	Area under the graph		Change in position (displacement)	Change in velocity

lvd-8941

SAMPLE PROBLEM 2 Using a velocity-time graph

The following velocity–time graph describes the motion of a car travelling south through an intersection. The car was stationary for 6.0 s while the traffic lights were red.

a. **What was the displacement of the car during the interval in which it was slowing down?**
b. **What was the acceleration of the car during the first 4.0 s after the lights turned green?**
c. **Determine the average velocity of the car during the interval described by the graph.**

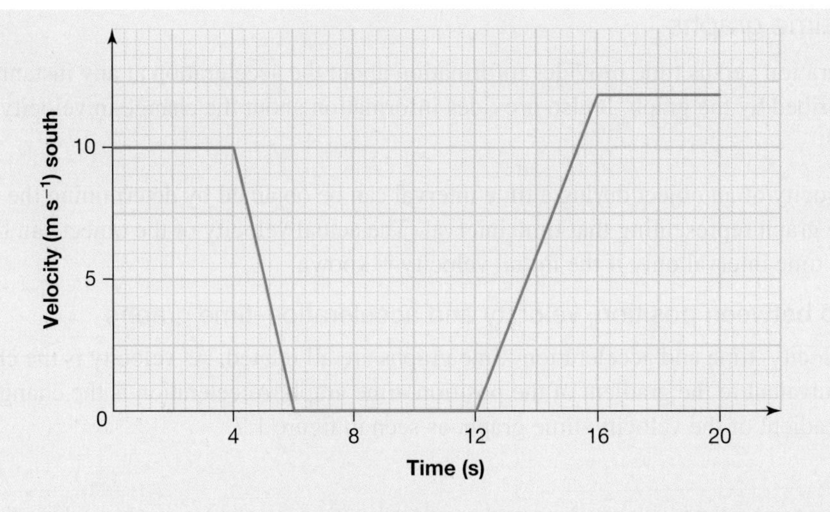

THINK	WRITE
a. The displacement of the car slowing down is the area under the graph between the times 4.0 s and 6.0 s. To fully describe displacement, units and direction need to be considered.	**a.** $\text{area} = \frac{1}{2} \times 2.0\,\text{s} \times 10\,\text{m}\,\text{s}^{-1}$ south $= 10$ m south
b. The acceleration is given by the gradient of the graph for the first 4.0 s after the lights turned green; the time interval is between 12 and 16 s. During this time, the velocity increases from 0 m s^{-1} south to 12 m s^{-1}. Therefore the 'rise' is 12 m s^{-1} south and the 'run' is 4.0 s.	**b.** $a = \frac{\text{rise}}{\text{run}} = \frac{12}{4.0} = 3.0\,\text{m}\,\text{s}^{-2}$ south
c. 1. The displacement during the whole time interval described by the graph is given by the total area under the graph. In this case, the area is split into four shapes (two rectangles and two triangles).	**c.** $\text{area} = (4.0 \times 10) + \left(\frac{1}{2} \times 2.0 \times 10\right) + \left(\frac{1}{2} \times 4.0 \times 12\right) + (4.0 \times 12) = 40 + 10 + 24 + 48 = 122$ m south
2. The average velocity is determined by the formula: $v_{\text{av}} = \frac{\Delta s}{\Delta t}$, where $\Delta s = 122$ m south and $\Delta t = 20$ s	$v_{\text{av}} = \frac{\Delta s}{\Delta t} = \frac{122}{20} = 6.1\,\text{m}\,\text{s}^{-1}$ south

PRACTICE PROBLEM 2

Use the velocity–time graph in sample problem 2 to answer the following questions.

a. **Determine the acceleration of the car while it had a positive southerly acceleration.**

b. **Determine the acceleration of the car during the 2.0 s before it came to a stop at the traffic lights.**

c. **Determine the average velocity of the car during the 6.0 s before it stopped at the traffic lights.**

Resources

Video eLesson Motion with constant acceleration (eles-0030)

1.2.3 Algebraic analysis of motion

The motion of an object moving in a straight line with a constant acceleration can be described by a number of formulas. The formulas can be used to determine unknown quantities of straight-line motion with constant acceleration.

$$v = u + at \quad s = \frac{1}{2}(u + v)t \quad s = ut + \frac{1}{2}at^2 \quad v^2 = u^2 + 2as \quad s = vt - \frac{1}{2}at^2$$

where: ***u*** is the initial velocity

v is the final velocity

s is the displacement

a is the acceleration

t is the time interval

Because the formulas describe motion along a straight line, vector notation is not necessary. The displacement, velocity and acceleration can be expressed as positive or negative quantities.

It is possible to rearrange each of the equations to make different variables the subject. Table 1.2 summarises all possible versions of the equations and may be useful when solving problems. Note that each formula uses 4 out of the 5 possible variables (***s u v a t***).

TABLE 1.2 Equations for solving problems

		Variables that are involved in the problem				
		u v a t	*u v a s*	*u v t s*	*u a t s*	*v a t s*
Variable that is to be calculated	*u*	$u = v - at$	$u^2 = v^2 - 2as$	$u = \frac{2s}{t} - v$	$u = \frac{s}{t} - \frac{at}{2}$	—
	v	$v = u + at$	$v^2 = u^2 + 2as$	$v = \frac{2s}{t} - u$	—	$v = \frac{s}{t} + \frac{at}{2}$
	a	$a = \frac{v - u}{t}$	$a = \frac{v^2 - u^2}{2s}$	—	$a = \frac{2(s - ut)}{t^2}$	$a = \frac{2(vt - s)}{t^2}$
	t	$t = \frac{v - u}{a}$	—	$t = \frac{2s}{(u + v)}$	Determine ***v*** then solve	Determine ***u*** then solve
	s	—	$s = \frac{v^2 - u^2}{2a}$	$s = \frac{1}{2}(u + v)t$	$s = ut + \frac{1}{2}at^2$	$s = vt - \frac{1}{2}at^2$

tlvd-8942

SAMPLE PROBLEM 3 Algebraic analysis of straight line motion with constant acceleration

Amy rides a toboggan down a steep snow-covered slope. Starting from rest, she reaches a speed of 12 m s^{-1} as she passes her brother, who is standing 19 m further down the slope from her starting position. Assume that Amy's acceleration is constant.

a. **Determine Amy's acceleration.**
b. **How long did she take to reach her brother?**
c. **How far had Amy travelled when she reached an instantaneous velocity equal to her average velocity?**
d. **At what instant was Amy travelling at an instantaneous velocity equal to her average velocity?**

THINK	WRITE
a. 1. List the given information. Let's consider down the slope as the positive direction.	a. $u = 0$, $v = 12$ m s^{-1}, $s = 19$ m, $a = ?$
2. The appropriate formula is: $v^2 = u^2 + 2as$, because it includes the three known quantities and the unknown quantity a.	$v^2 = u^2 + 2as$ $\Rightarrow 12^2 = 0 + 2a \times 19$ $144 = 38 \times a$ $\Rightarrow a = 3.8$ m s^{-2} down the slope
b. 1. List the information. (Note that it is better to use the data given rather than data calculated in the previous part of the question. That way, rounding or errors in an earlier part of the question will not affect the answer.)	b. $u = 0$, $v = 12$ m s^{-1}, $s = 19$ m, $t = ?$
2. The appropriate formula is: $s = \frac{1}{2}(u + v)t$.	$s = \frac{1}{2}(u + v)t$ $\Rightarrow 19 = \frac{1}{2}(0 + 12)t$ $19 = 6.0 \times t$ $\Rightarrow t = \frac{19}{6.0}$ $= 3.2$ s
c. 1. The magnitude of the average velocity during a period of constant acceleration is given by: $v_{av} = \frac{u + v}{2}$	c. $v_{av} = \frac{u + v}{2}$ $= \frac{0 + 12}{2}$ $= 6.0$ m s^{-1}
2. The distance travelled when Amy reaches an instantaneous velocity of this magnitude can now be calculated. List the information.	$u = 0$, $v = 6.0$ m s^{-1}, $a = 3.8$ m s^{-2}, $s = ?$

3. The appropriate formula is $v^2 = u^2 + 2as$.

$$v^2 = u^2 + 2as$$
$$\Rightarrow (6.0)^2 = 0 + 2 \times 3.8 \times s$$
$$36 = 7.6 \times s$$
$$\Rightarrow s = \frac{36}{7.6}$$
$$= 4.7 \text{ m}$$

d. 1. List the information.

d. $u = 0$, $v = 6.0$ m s^{-1}, $a = 3.8$ m s^{-2}, $t = ?$

2. The appropriate formula is $v = u + at$.

$$v = u + at$$
$$\Rightarrow 6.0 = 0 + 3.8 \times t$$
$$6.0 = 3.8 \times t$$
$$\Rightarrow t = \frac{6.0}{3.8}$$
$$= 1.6 \text{ s}$$

This is the midpoint of the entire time interval. In fact, during any motion in which the acceleration is constant, the instantaneous velocity halfway (in time) through the interval is equal to the average velocity during the interval.

PRACTICE PROBLEM 3

A car initially travelling at a speed of 20 m s^{-1} on a straight road accelerates at a constant rate for 16 s over a distance of 400 m.

a. **Calculate the final speed of the car.**
b. **Determine the car's acceleration without using your answer to part (a).**
c. **What is the average speed of the car?**
d. **What is the instantaneous speed of the car after:**
   i. **2.0 s**
   ii. **8.0 s?**

1.2.4 Momentum

When explaining changes in motion, it is necessary to consider another property of the object: its mass. Consider how much more difficult it is to stop a truck moving at 20 m s^{-1} than it is to stop a tennis ball moving at the same speed. The physical quantity **momentum** is useful in explaining changes in motion, because it takes into account the mass as well as the velocity of a moving object.

momentum the product of the mass of an object and its velocity; a vector quantity

Newton described momentum as 'quantity of motion' and understood the special nature of mass in motion.

The momentum $\boldsymbol{p}$ of an object is defined as the product of its mass m and its velocity $\boldsymbol{v}$.

$$p = mv$$

where: p is the momentum, in kg m s^{-1}, or N s
m is the mass, in kg
v is the velocity, in m s^{-1}

Momentum is a vector quantity that has the same direction as that of the velocity. The SI unit of momentum is kg m s^{-1}. Momentum is also sometimes expressed in N s.

1.2 Activities

learnon

Students, these questions are even better in jacPLUS

Receive immediate feedback and access sample responses

Access additional questions

Track your results and progress

Find all this and MORE in jacPLUS

1.2 Quick quiz on

1.2 Exercise

1.2 Exercise

1. Two Physics students are trying to determine the instantaneous speed of a bicycle 5.0 m from the start of a 1000-m sprint. They use a stopwatch to measure the time taken for the bicycle to cover the first 10 m. If the acceleration was constant, and the measured time was 4.0 s, what was the instantaneous speed of the bicycle at the 5.0 m mark?
2. A car travelling north at a speed of 40 km h^{-1} turns right to head east at a speed of 30 km h^{-1}. This change in direction and speed takes 2.0 s. Calculate the average acceleration of the car in:
   a. km h^{-1} s^{-1}
   b. m s^{-2}.
3. An aeroplane approaches Melbourne Airport and touches down on the runway while travelling at 70 m s^{-1}. This speed is maintained for 8.0 s. Following this, the brakes are engaged, and the aeroplane comes to a stop with a uniform deceleration of 4.0 m s^{-2}.
   a. Calculate how long it takes the aeroplane to stop after landing.
   b. Draw a velocity–time graph to describe the motion of the aeroplane from landing to the moment it comes to a stop. Ensure your graph is fully labelled.
   c. Use your graph to determine the length of runway used in the landing process.
4. A 65.0-kg student runs at a velocity of 35.0 km h^{-1}. Determine the student's momentum.

1.3 Newton's laws of motion and their application

KEY KNOWLEDGE

- Investigate and apply theoretically and practically Newton's three laws of motion in situations where two or more coplanar forces act along a straight line and in two dimensions.

1.3.1 Newton's three laws of motion

Sir Isaac Newton's three laws of motion, first published in 1687, explain changes in the motion of objects in terms of the forces acting on them. However, Einstein and others have since shown that Newton's laws have limitations. Newton's laws fail, for example, to explain successfully the motion of objects travelling at speeds close to the speed of light. They do not explain the bending of light by the gravitational forces exerted by stars, planets and other large bodies in the universe. However, they do successfully explain the motion of most objects at Earth's surface, the motion of satellites and the orbits of the planets that make up the solar system. In fact, it was Newton's laws that enabled NASA to plan the trajectories of artificial satellites.

Newton's First Law of Motion

Every object continues in its state of rest or uniform motion unless made to change by a non-zero net force.

net force the vector sum of all the forces acting on an object

Newton's First Law of Motion explains why things move. For example, you need to strike a golf ball with the club before it will soar through the air. Without a **net force** acting on the golf ball, it will remain in its state of rest on the tee or grass. (Recall that the vector sum of the forces acting on an object is called the net force.) The law explains why seatbelts should be worn in a moving car and why you should never leave loose objects (like books, luggage or pets) in the back of a moving car. When a car stops suddenly, it does so because there is a large net force acting on it — as a result of braking or a collision. However, the large force does not act on you or the loose objects in the car. The loose objects continue their motion until they are stopped by a non-zero net force. Without a properly fitted seatbelt, you would move forward until stopped by the airbag in the steering wheel, the windscreen or even the road. The loose objects in the car will also continue moving forward, posing a danger to anyone in the car. This is why Newton's First Law is sometimes referred to as the Law of Inertia. Inertia is a property of mass and describes the tendency of a body to remain at rest or to move with a constant velocity in a straight line unless acted upon by a net force.

Newton's First Law of Motion can also be expressed in terms of momentum by stating that the momentum of an object does not change unless the object is acted upon by a non-zero net force.

Newton's Second Law of Motion

The rate of change in momentum is directly proportional to the magnitude of the net force and is in the direction of the net force.

$$\boldsymbol{F}_{net} = \frac{\Delta \boldsymbol{p}}{\Delta t} = \frac{m\Delta \boldsymbol{v}}{\Delta t} = m\boldsymbol{a}$$

(provided the mass is constant)

where: $\Delta \boldsymbol{p}$ is the change in momentum, in kg m s^{-1} (or N s)

$\Delta \boldsymbol{v}$ is the change in velocity, in m s^{-1}

Δt is the time interval, in s

m is the mass, in kg

$\boldsymbol{a}$ is the acceleration, in m s^{-2}

This expression of Newton's Second Law of Motion is especially useful because it relates the net force to a description of the motion of objects. An acceleration of 1 m s^{-2} results when a net force of 1 N acts on an object of mass 1 kg. Newton's Second Law gives the relationship between the net force acting upon a body, its mass and the resulting acceleration of that body. Acceleration is directly proportional to the net force applied to a body but inversely proportional to its mass. The derived equation above, $\boldsymbol{F}_{\text{net}} = m\boldsymbol{a}$ is commonly associated with the second law.

Newton's Third Law of Motion

Newton's Third Law of Motion

If object B applies a force on object A, then object A applies an equal and opposite force on object B:

$$\boldsymbol{F}_{\text{on A by B}} = -\boldsymbol{F}_{\text{on B by A}}$$

It is important to remember that the forces that make up the force pair act on different objects. The subsequent motion of each object is determined by the net force acting on it. For example, in figure 1.8, the net force on the brick wall at the left is the sum of the force applied to it by the car (shown by the red arrow) and all the other forces acting on it. The force shown by the orange arrow is applied to the car and does not affect the state of motion of the brick wall. The net force on the car is the sum of the force applied by the brick wall (shown by the orange arrow) and all the other forces acting on it. The 'Newton pairs' of forces are of the same type but act on different bodies. They are equal and opposite in nature.

FIGURE 1.8 Newton's Third Law of Motion in action

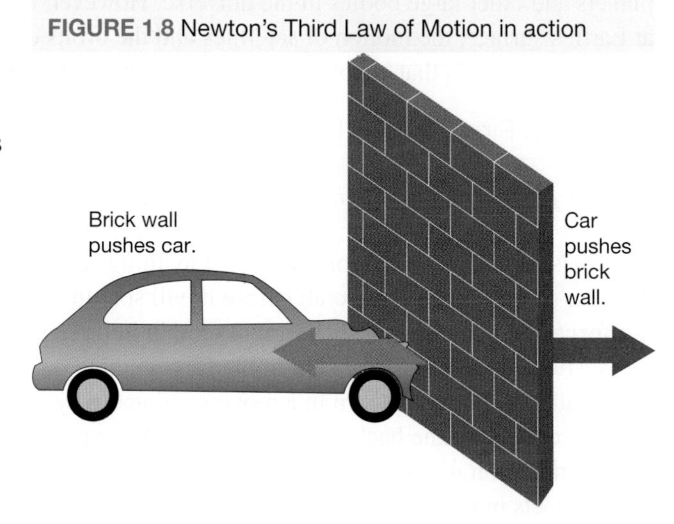

Applying Newton's laws of motion

Newton's laws of motion can be used to explain the motion of objects. It is important to determine the specific laws that should be applied to a particular problem:

- Newton's First Law refers only to objects at rest or in uniform motion and can be applied in instances when an object is *not* accelerating
- Newton's Second Law applies to a *single* object (or a system of more than one body where the bodies are connected to each other) being acted upon by one or more forces
- Newton's Third Law applies when *two* objects interact with one another and exert equal but opposite forces on each other

Often, more than one of Newton's laws will be required to solve a problem.

on Resources

Video eLesson Newton's Second Law (eles-0033)

1.3.2 On level ground

Whether you are walking on level ground, driving a car, riding in a roller-coaster or flying in an aeroplane, your motion is determined by the net force acting on you.

Figure 1.9 shows the forces acting on a car moving at a constant velocity on a level surface. The net force on the car is zero as the forces are balanced (in equilibrium).

FIGURE 1.9 Forces acting on a car moving on a level surface. The car's engine is making the front wheels turn.

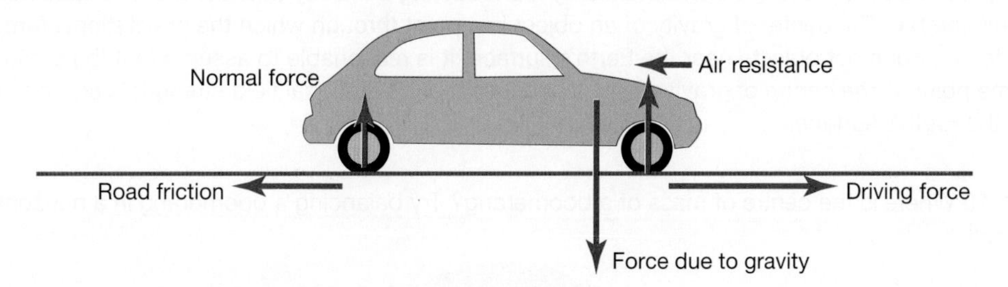

The forces acting on the car are described as follows.

- *The force due to gravity*. The force due to gravity of an object is given by:

$$\boldsymbol{F}_g = m\boldsymbol{g}$$

  where m is mass and $\boldsymbol{g}$ is the gravitational field strength.
  Throughout this text, the magnitude of $\boldsymbol{g}$ at the Earth's surface will be taken as 9.8 N kg^{-1}. The force due to gravity of a medium-sized sedan carrying a driver and passenger is about 15 000 N.
- *Normal force*. The normal force is the upward push of the supporting surface on the car. A normal force acts on all four wheels of the car. It is described as a normal force because it acts at right angles to the surface. Considering that the road surface is not accelerating up or down, the force applied to the surface by the object is the same as the force due to gravity acting on the object. The total normal force is therefore equal and opposite in direction to the force due to gravity.
- *Driving force*. The force that pushes the car forward is provided at the driving wheels — the wheels that are turned by the motor. In most cars, either the front wheels or the rear wheels are the driving wheels. The motor of a four-wheel drive pushes all four wheels. As the tyres push back on the road, the road pushes forward on the tyres, propelling the car forward. The forward push of the road on the tyres is a frictional force, as it is the resistance to movement of one surface across another. In this case, it is the force that prevents the tyres from sliding across the road. If the tyres or the road is too smooth, the driving force is reduced, the tyres slide backwards and the wheels spin. Note that if a car is braking, the wheels are not being driven by the engine and the driving force is not present.
- *Resistance forces*. As the car moves, it applies a force to the air in front of it. The air applies an equal force opposite to its direction of motion. This force is called **air resistance**. The air resistance on an object increases as its speed increases. The other resistance force acting on the car is **road friction**. It opposes the forward motion of the non-driving wheels, rotating them in the same direction as the driving wheels. In the car in figure 1.9, the front wheels are the driving wheels. Road friction opposes the motion of the rear wheels along the road and, therefore, the forward motion of the car. This road friction is an example of static friction, which is considerably smaller than the kinetic or sliding friction that acts if a wheel slips on the surface and spins.

The centre of mass

The forces on a moving car do not all act at the same point on the car. When analysing the translational motion of an object (its movement across space without considering rotational motion), all of the forces applied to an object can be considered to be acting at one particular point. That point is the **centre of mass**. The centre of mass of a symmetrical object with uniform mass distribution is at the centre of the object. For example, the centre of mass of a ruler, a solid ball or an ice cube is at the centre, unlike the centres of mass of asymmetrical objects, such as a person or a car. Note that the centre of mass can also lie outside of an object.

air resistance the force applied to an object opposite to its direction of motion, by the air through which it is moving

road friction the force applied by the road surface to the wheels of a vehicle in a direction opposite to the direction of motion of the vehicle

centre of mass the point at which all of the mass of an object can be considered to be positioned when modelling the external forces acting on the object

AS A MATTER OF FACT

If you hold an object such as a ruler at its centre of mass, it will perfectly balance. However, the centre of mass does not have to be within the object. For example, the centre of mass of a donut is in the middle of its hole. A high jumper can improve her performance by manoeuvring her body over the bar so that her centre of mass is below the bar. The centre of gravity of an object is a point through which the gravitational force can be considered to act. For most objects near the Earth's surface, it is reasonable to assume that the centre of mass is at the same point as the centre of gravity. This is because the gravitational field strength is approximately constant at the Earth's surface.

FIGURE 1.10 Where is the centre of mass of a boomerang? Try balancing a boomerang in a horizontal plane on one finger.

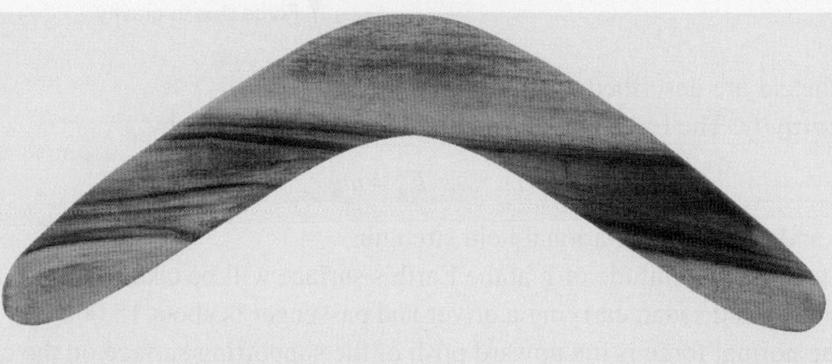

1.3.3 Applying Newton's Second Law of Motion

Sample problem 4 shows how Newton's Second Law of Motion can be applied to single objects or to a system of two objects, often referred to as connected bodies.

Remember that when the problem involves connected bodies, the whole system, as well as each individual part of the system, will have the same acceleration.

Tips for using Newton's Second Law of Motion

1. Draw a simplified diagram of the system.
2. Clearly label the diagrams to represent the forces acting on each object in the system. Draw all the forces as though they were acting through the centre of mass.
3. Apply Newton's Second Law to the system and/or each individual object as required.

tlvd-8943

SAMPLE PROBLEM 4 Applying Newton's Second Law of Motion

A car of mass 1600 kg starts from rest on a horizontal road with a forward driving force of 5400 N east. The sum of the forces resisting the motion of the car is 600 N.

a. **Determine the acceleration of the car.**

b. **The same car is used to tow a 400 kg trailer with the same driving force as before. The sum of the forces resisting the motion of the trailer is 200 N.**
   i. **Determine the acceleration of the system of the car and the trailer.**
   ii. **What is the magnitude of the force, F_{ct}, exerted on the car by the trailer?**

THINK	WRITE
a. 1. Determine the net force acting on the car. The forward driving force is 5400 N and the sum of the forces acting in the negative direction is 600 N.	**a.** F_{net} = Driving forces F_D − Resisting forces on car F_{rc} $F_{net} = 5400 - 600$ $= 4.80 \times 10^3$ N
2. Apply Newton's Second Law to determine the acceleration of the car, where the net force is 4800 N and the mass is 1600 kg.	$F_{net} = ma$ $a = \dfrac{F_{net}}{m}$ $= \dfrac{4.80 \times 10^3}{1600}$ $= 3.00\ \text{m s}^{-2}$ east
b. i. • A diagram must be drawn to show the forces acting on the car and trailer. The vertical forces can be omitted because their sum is zero. (If not, there would be a vertical component of acceleration.) Assign east as positive.	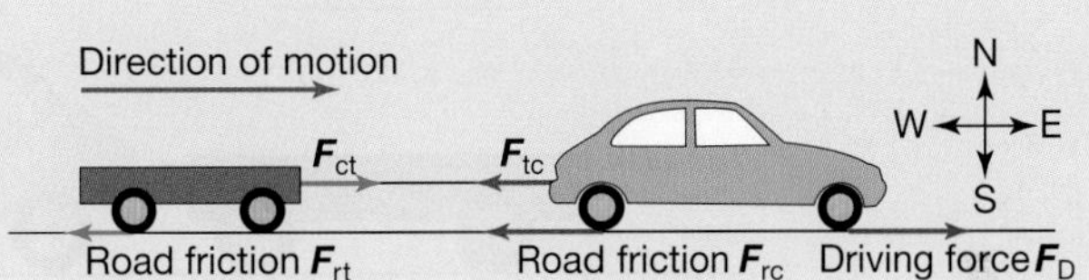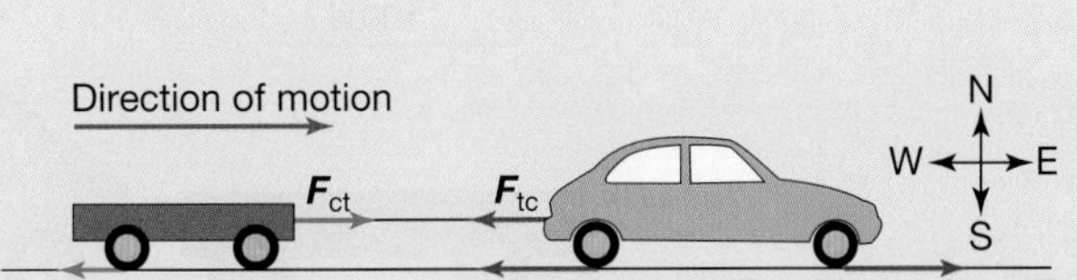
• Determine the net force acting on the entire system.	**b. i.** $F_{net} = F_D - F_{rc} - F_{rt}$ $F_{net} = 5400 - 600 - 200$ $= 4600$ N
• Apply Newton's Second Law to determine the acceleration of the system, where the net force is 4600 N and the mass of the system is 2000 kg.	$F_{net} = ma$ $a = \dfrac{F_{net}}{m}$ $= \dfrac{4600}{2000}$ $= 2.30\ \text{m s}^{-2}$ east
ii. Newton's Second Law can be applied to either the car or the trailer to answer this question.	**ii.** $F_{net} = ma$ $= 1600 \times 2.30$ $= 3.68 \times 10^3$ N
• *Method 1: Applying Newton's Second Law to the car*	
• Write an expression for the net force acting on the car, and use it to determine the magnitude of the force exerted on the car by the trailer.	$F_{net} = 5400 - 600 - F_{ct}$ where F_{ct} is the magnitude of the force exerted by the trailer on the car. $3680 = 5400 - 600 - F_{ct}$ $\Rightarrow F_{ct} = 5400 - 600 - 3680$ $= 1.12 \times 10^3$ N
or	
• *Method 2: Applying Newton's Second Law to the trailer*	$F_{net} = ma$ $= 400 \times 2.30$ $= 920$ N

- Write an expression for the net force acting on the trailer, and use it to determine the magnitude of the force exerted on the trailer by the car.

Note: $F_{ct} = F_{tc}$

$F_{net} = F_{tc} - 200$

where F_{tc} is the magnitude of the force exerted on the trailer by the car.

$920 = F_{tc} - 200$

$\Rightarrow F_{tc} = 920 + 200$

$= 1.12 \times 10^3$ N

PRACTICE PROBLEM 4

a. **A car of mass 1400 kg tows a trailer of mass 600 kg north along a level road at constant speed. The forces resisting the motion of the car and trailer are 400 N and 100 N respectively.**

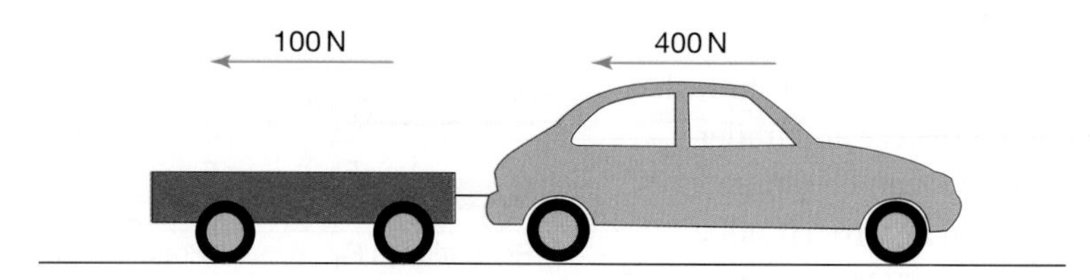

i. **Determine the forward driving force applied to the car.**

ii. **What is the magnitude of the tension in the bar between the car and the trailer?**

b. **If the car and trailer in part (a), with the same resistance forces, have a northerly acceleration of 2.0 m s^{-2}, what is:**

i. **the net force applied to the trailer**

ii. **the magnitude of the tension in the bar between the car and the trailer**

iii. **the forward driving force applied to the car?**

1.3.4 Feeling lighter — feeling heavier

As you sit reading this, the force due to gravity on you by the Earth $(F_g = mg)$ is pulling you down towards the centre of the Earth, but the chair is in the way. The material in the chair is being compressed and pushes up on you. This force is called the normal force (F_N) because it is perpendicular or normal to the surface. If these two forces, the force due to gravity and the normal force, balance, then the net force on you is zero.

The normal force is responsible for the feeling of 'heaviness'.

The greater the normal force, the 'heavier' you will feel.

You 'feel' the Earth's pull on you because of Newton's Third Law. The upward compressive force on you by the chair is paired with the downward force on the chair by you. You sense this upward force through the compression in the bones in your pelvis.

$$F_{\text{on you by chair}} = -F_{\text{on chair by you}}$$

What happens to these forces when you are in a lift? A lift going up initially accelerates upwards, then travels at a constant speed (no acceleration) and finally slows down (the direction of acceleration is downwards). You experience each of these stages differently.

Accelerating upwards

When you are accelerating upwards, the net force on you is upwards. The only forces acting on you are the force due to gravity downwards and the normal force by the floor acting upwards. The force due to gravity is not going to change. So, if the net force on you is up, then the normal force on you must be greater than the force due to gravity: $\boldsymbol{F}_N > m\boldsymbol{g}$.

You 'feel heavier'.

Accelerating downwards

When you are accelerating downwards, the net force on you is downwards. So, the normal force on you must be less than the force on you due to gravity: $\boldsymbol{F}_N < m\boldsymbol{g}$. You 'feel lighter'. Note that if the lift were in free-fall, the person would not experience a normal force from the floor of the lift, and she would experience 'apparent weightlessness'. This concept of absence of normal force will be revisited in subtopic 3.4, to explain astronauts floating in space.

Note that 'apparent weightlessness' is no longer part of the study design.

FIGURE 1.11 The magnitude of the normal force determines how 'heavy' you feel.

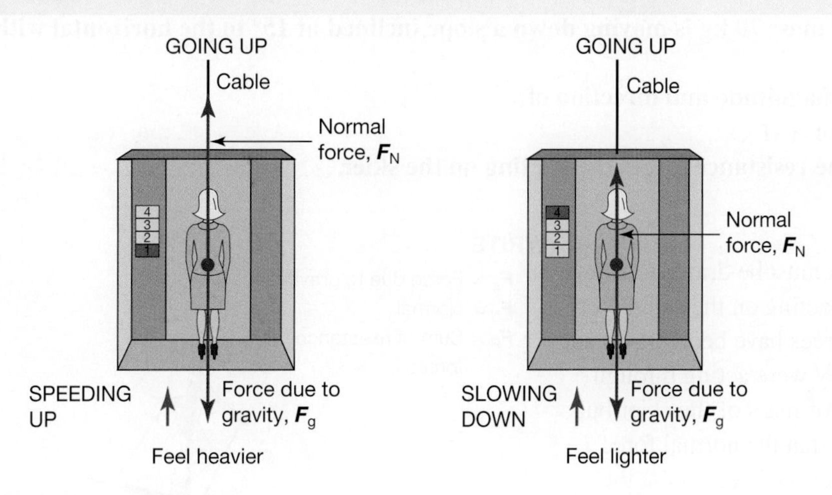

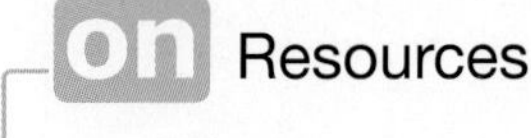

Resources

Interactivity Going up? (int-6606)

1.3.5 Motion on an incline

The forces acting on objects on an inclined plane are similar to those acting on the same objects on a level surface.

The forces acting on a car rolling down an inclined plane are shown in figure 1.12. The car is considered to behave like a single particle and the rotational motion of the wheels is ignored.

FIGURE 1.12 The forces acting on a car rolling down an inclined plane

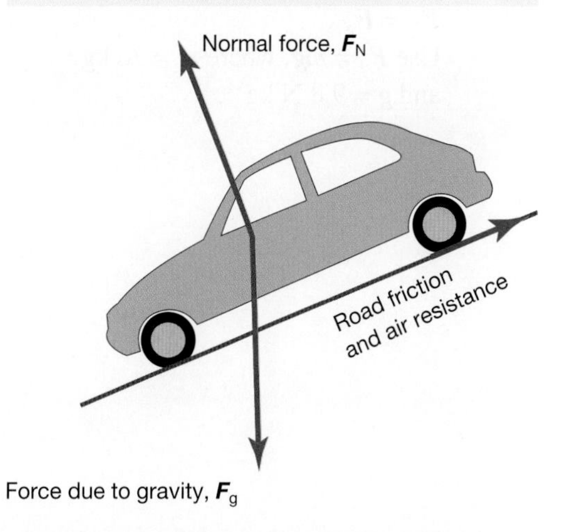

Resolving forces into components

The net force on a car can be found by finding the vector sum of the forces acting on it. Figure 1.13 shows how the force due to gravity can be resolved (divided) into two components — one parallel to the surface and one perpendicular to the surface.

By resolving the force due to gravity into these components, the motion of the car can be analysed. Consider the forces perpendicular to the inclined plane in figure 1.13. The magnitude of the normal force is equal to the component of the force due to gravity that is perpendicular to the surface. These forces balance each other out; therefore, the net force has no perpendicular component. (Imagine what would happen if this wasn't the case!)

Now consider the forces parallel to the inclined plane. It can be seen that the horizontal component of the force due to gravity is greater than the sum of road friction and air resistance. The net force is therefore parallel to the surface. The car will accelerate down the slope, as the downslope force exceeds the smaller upslope frictional force in the example.

FIGURE 1.13 Forces can be resolved into components. In this case, the force due to gravity has been resolved into two components.

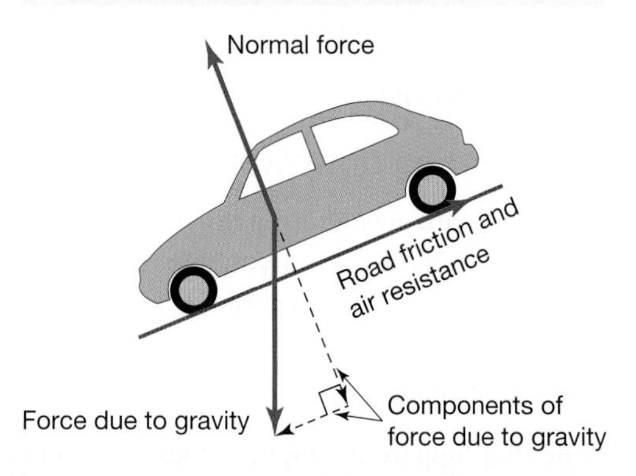

tlvd-8944

SAMPLE PROBLEM 5 Calculating the normal reaction force and sum of resistance forces

A snow skier of mass 70 kg is moving down a slope inclined at 15° to the horizontal with a constant velocity.
Determine the magnitude and direction of:
a. **the normal force (F_N)**
b. **the sum of the resistance forces (F_R) acting on the skier.**

THINK

a. 1. A diagram must be drawn to show the forces acting on the skier. Note that the forces have been drawn as though they were acting through the centre of mass of the skier but remember that the normal force and the friction forces act at the surface between the slope and the skis.

WRITE

a. F_g = Force due to gravity
F_N = Normal
F_R = Sum of resistance forces

THINK

2. The perpendicular net force is zero, so $F_N = F_{gy}$.
Use $F_g = mg$, where $m = 70$ kg and $g = 9.8$ N kg^{-1}.

WRITE

$$F_N = F_{gy}$$
$$= F_g \cos 15° \left(\text{since } \cos 15° = \frac{F_{gy}}{F_g} \Rightarrow F_{gy} = F_g \cos 15° \right)$$
$$= mg \cos 15°$$
$$= 70 \times 9.8 \times \cos 15°$$
$$= 663 \text{ N, rounded to } 6.6 \times 10^2 \text{ N}$$

The normal force is therefore 6.6×10^2 N in the direction perpendicular to the surface as shown (as we are specifying the direction, we do not need to use a negative sign).

b. The skier has a constant velocity, so the net force on the skier in the direction parallel to the surface is zero. Therefore, the magnitude of the sum of resistance forces must be equal to the component of the force due to gravity that is parallel to the surface, $F_R = F_{gx}$.

b. $F_R = F_{gx}$

$$= F_g \sin 15° \left(\text{since } \sin 15° = \frac{F_{gx}}{F_g} \Rightarrow F_{gx} = F_g \sin 15° \right)$$

$$= mg \sin 15°$$

$$= 70 \times 9.8 \times \sin 15°$$

$$= 178 \text{ N, rounded to } 1.8 \times 10^2 \text{ N}$$

The sum of the resistance forces (air resistance and friction) acting on the skier is 1.8×10^2 N opposite to the direction of motion.

PRACTICE PROBLEM 5

a. A cyclist rides at constant velocity up a hill that is inclined at 15° to the horizontal. The total mass of the cyclist and bicycle is 90 kg. The sum of the resistive forces on the cyclist and bicycle is 20 N. Determine:

i. the forward driving force provided by the road on the bicycle

ii. the normal force.

b. If the cyclist in part (a) coasts down the same hill with a constant total resistance of 50 N, what is the cyclist's acceleration?

1.3 Activities

learnon

1.3 Quick quiz on	1.3 Exercise	1.3 Exam questions

1.3 Exercise

1. Draw a sketch (the length of the arrows should give a rough indication of relative size of the force) showing all the forces acting on a tennis ball while it is:
   a. falling to the ground
   b. in contact with the ground just before rebounding upwards
   c. on its upward path after bouncing on the ground.
2. A coin is allowed to slide with a constant velocity down an inclined plane as shown in the diagram. Which of the arrows A to G on the diagram represents the direction of each of the following? Write X if no direction is correct.
   a. The force due to gravity on the coin
   b. The normal force
   c. The net force

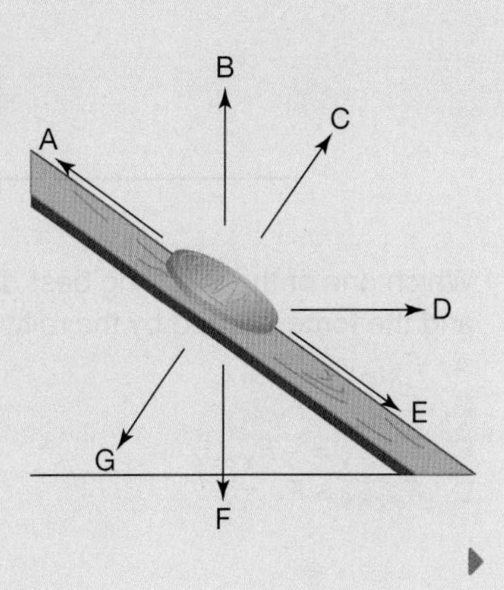

3. A child pulls a 4.0-kg toy cart along a horizontal path with a rope so that the rope makes an angle of 30° with the horizontal. The tension in the rope is 12 N.
   a. What is the force due to gravity acting on the toy cart?
   b. What is the component of tension in the direction of motion?
   c. What is the magnitude of the normal force?

4. A 200-kg dodgem car is driven due south into a rigid barrier at an initial speed of 5.0 m s^{-1}. The dodgem rebounds at a speed of 2.0 m s^{-1}. It is in contact with the barrier for 0.20 s. Calculate:
   a. the average acceleration of the car during its interaction with the barrier
   b. the average net force applied to the car during its interaction with the barrier.
5. A 1500-kg car is resting on a slope inclined at 20° to the horizontal. It has been left out of gear, so the only reason it doesn't roll down the hill is that the handbrake is on.
   a. Draw a labelled diagram showing the forces acting on the car.
   b. Calculate the magnitude of the normal force. Give your answer to 2 significant figures.
   c. What is the net force acting on the car?
   d. What is the magnitude of the frictional force acting on the car? Give your answer to 2 significant figures.
6. An experienced downhill skier with a mass of 60 kg (including skis) moves in a straight line down a slope inclined at 30° to the horizontal with a constant speed of 15 m s^{-1}.
   a. What is the direction of the net force acting on the skier?
   b. What is the magnitude of the resistive forces opposing the skier's motion? Give your answer to 2 significant figures.
7. A 70.0-kg waterskier is towed in a northerly direction by a 350-kg speedboat. The frictional forces opposing the forward motion of the waterskier total 240 N.
   a. If the waterskier has an acceleration of 2.0 m s^{-2} due north, what is the tension in the rope towing the waterskier?
   b. If the frictional forces opposing the forward motion of the speedboat are increased to total 600 N, what is the thrust force applied to the boat due to the action of the motor?
8. A 0.3-kg magpie flies towards a very tight plastic wire on a clothes line. The wire is perfectly horizontal and is stretched between poles 4.0 m apart. The magpie lands on the centre of the wire, depressing it by a vertical distance of 4.0 cm. What is the magnitude of the tension in the wire?
9. An old light globe hangs by a wire from the roof of a train. What angle does the globe make with the vertical when the train is accelerating at 1.5 m s^{-2}?

1.3 Exam questions

Question 1 (1 mark)

Source: *VCE 2022 Physics Exam, Section A, Q.7;* © VCAA

MC A railway truck (X) of mass 10 tonnes, moving at 3.0 m s^{-1}, collides with a stationary railway truck (Y), as shown in the diagram below.

After the collision, they are joined together and move off at speed v = 2.0 m s^{-1}.

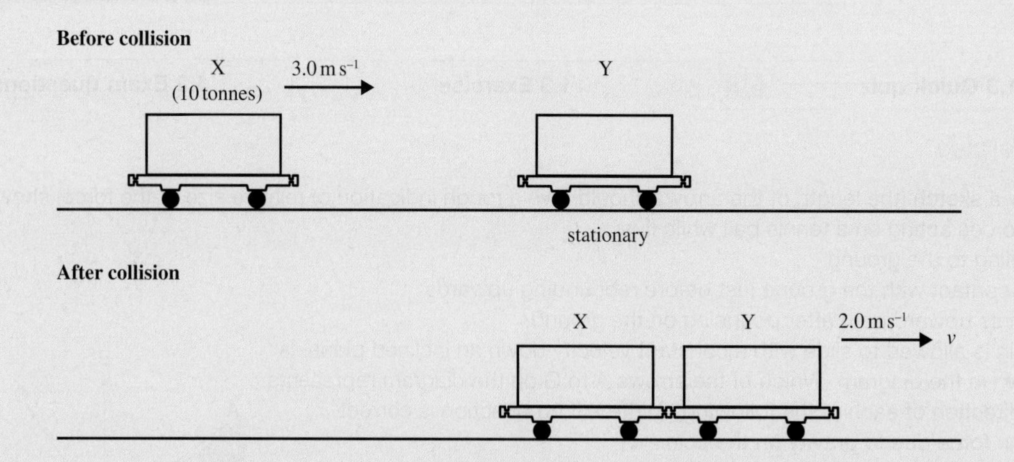

Which one of the following best describes the force exerted by the railway truck X on the railway truck Y ($F_{X \text{ on } Y}$) and the force exerted by the railway truck Y on the railway truck X (F_Y on X) at the instant of collision?

A. $F_{X \text{ on } Y} < F_{Y \text{ on } X}$
B. $F_{X \text{ on } Y} = F_{Y \text{ on } X}$
C. $F_{X \text{ on } Y} = -F_{Y \text{ on } X}$
D. $F_{X \text{ on } Y} > F_{Y \text{ on } X}$

Question 2 (1 mark)

***Source:** VCE 2022 Physics Exam, Section A, Q.9; © VCAA*

MC Two students pull on opposite ends of a rope, as shown in the diagram below. Each student pulls with a force of 400 N.

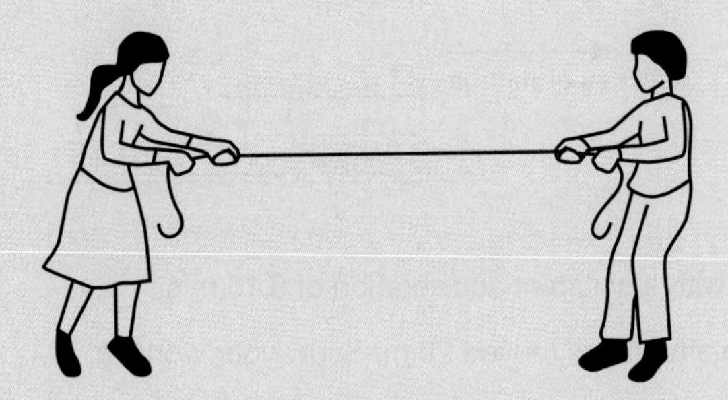

Which one of the following is closest to the magnitude of the force of the rope on each student?

A. 0 N
B. 400 N
C. 600 N
D. 800 N

Question 3 (5 marks)

***Source:** VCE 2021 Physics Exam, NHT, Section B, Q.8; © VCAA*

A car is driving up a uniform slope with a trailer attached, as shown in Figure 11. The slope is angled at 15° to the horizontal. The trailer has a mass of 200 kg and the car has a mass of 750 kg. Ignore all retarding friction forces down the slope.

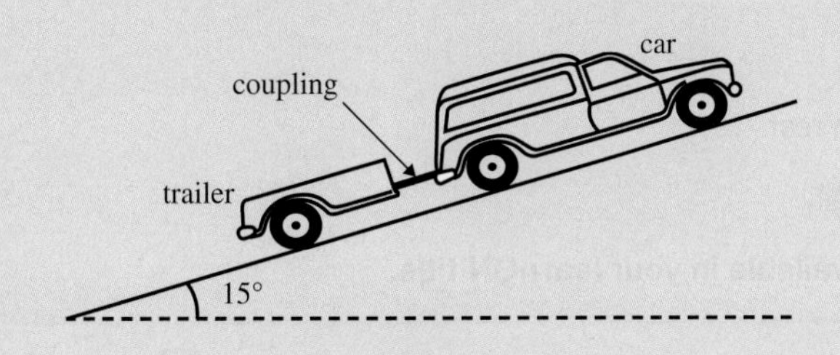

Figure 11

a. On Figure 12 below, draw labelled arrows to indicate the direction of the forces acting on the trailer. The labels should also indicate the kind of force that each arrow represents. **(3 marks)**

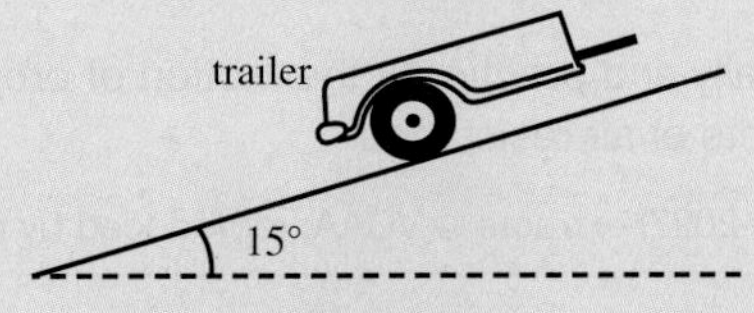

Figure 12

b. The car and trailer are travelling at a constant speed of 8 m s^{-1} up the slope.
Calculate the magnitude of the force that the car exerts on the trailer. Show your working. **(2 marks)**

Question 4 (2 marks)

Source: *VCE 2016, Physics Exam, Q.1.a; © VCAA*

A train consists of an engine of mass 20 tonnes (20 000 kg) towing one wagon of mass 10 tonnes (10 000 kg), as shown in the figure.

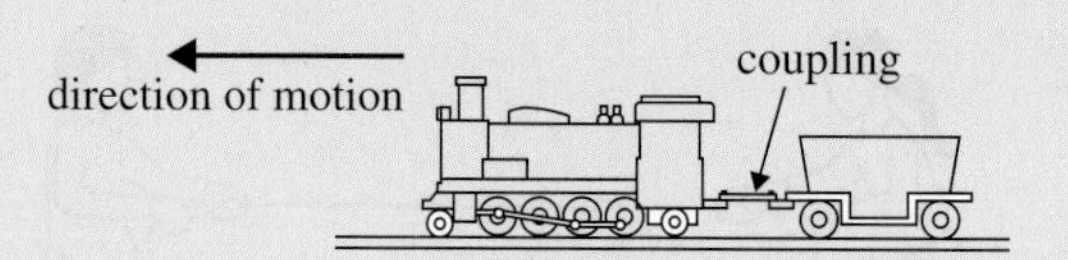

The train accelerates from rest with a constant acceleration of 0.10 m s^{-2}.

Calculate the speed of the train after it has moved 20 m. Show your working.

Question 5 (2 marks)

Source: *VCE 2015, Physics Exam, Q.2.a; © VCAA*

Students set up an experiment as shown in the figure.

M_1, of mass 4.0 kg, is connected by a light string (assume it has no mass) to a hanging mass, M_2, of 1.0 kg.

The system is initially at rest. Ignore mass of string and friction.

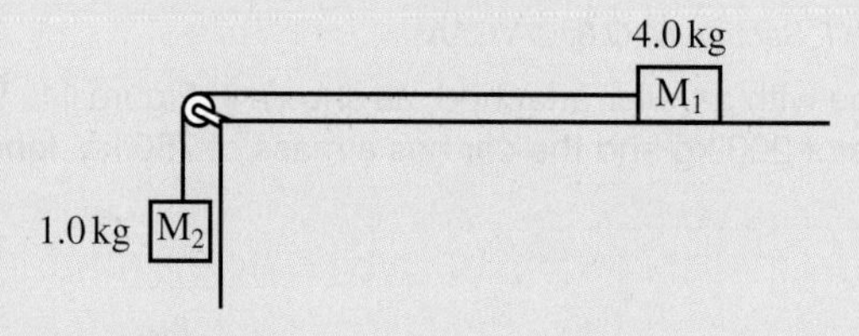

The masses are released from rest.

Calculate the acceleration of M_1.

More exam questions are available in your learnON title.

1.4 Projectile motion

KEY KNOWLEDGE

- Investigate and analyse theoretically and practically the motion of projectiles near Earth's surface, including a qualitative description of the effects of air resistance.

Source: VCE Physics Study Design (2024–2027) extracts © VCAA; reproduced by permission.

Any object that is launched into the air is a projectile. A basketball thrown towards a goal, a trapeze artist soaring through the air, and a package dropped from a helicopter are all examples of projectiles.

Except for those projectiles whose motion is initially straight up or down, or those that have their own power source (such as a guided missile), projectiles generally follow a parabolic path. Deviations from this path can be caused either by air resistance, by the spinning of the object or by wind. These effects are often small and can be ignored in many cases. A major exception, however, is the use of spin in many ball sports, but this effect will not be dealt with in this title.

1.4.1 Falling down

Imagine a ball that has been released some distance above the ground. Once the ball is set in motion, the only forces acting on it are the force due to gravity (straight down) and air resistance (straight up).

After the ball is released, the projection device (hand, gun, slingshot or whatever) stops exerting a force on the ball.

The net force on the ball in figure 1.14 is downwards. As a result, the ball accelerates downwards. If the size of the forces and the mass of the ball are known, the acceleration can be calculated using Newton's Second Law of Motion.

Often the force exerted on the ball by air resistance is very small in comparison to the force of gravity, and so can be ignored. This makes it possible to model projectile motion by assuming gravity is the only force on it so its acceleration is 9.8 m s^{-2} downwards.

FIGURE 1.14 The forces acting on a ball falling downwards

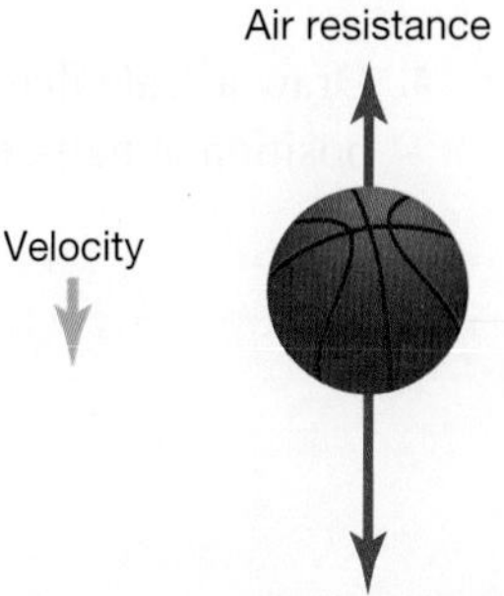

tlvd-8945

SAMPLE PROBLEM 6 Calculating the time and distance for an object to fall from a stationary object

A helicopter delivering supplies to a flood-stricken farm hovers 100 m above the ground. A package of supplies is dropped from rest, just outside the door of the helicopter. Air resistance can be ignored.

a. Calculate how long it takes the package to reach the ground.

b. Calculate how far from its original position the package has fallen after 0.50 s, 1.0 s, 1.5 s, 2.0 s and so on until it hits the ground. (A spreadsheet could be used here.) Draw a scale diagram of the package's position at half-second intervals.

THINK	WRITE
a. 1. List the known information.	**a.** $u = 0$ m s^{-1}, $s = 100$ m, $a = 9.8$ m s^{-2}
2. Use the rule $s = ut + \frac{1}{2}at^2$ to determine the time taken for the package to reach the ground. (*Note*: The negative square root can be ignored as time will be positive.)	$s = ut + \frac{1}{2}at^2$ $100 = 0 \times t + \frac{1}{2}(9.8)t^2$ $\frac{100}{4.9} = t^2$ $t = \sqrt{\frac{100}{4.9}}$ $t = 4.5$ s
b. 1. Look at the position of the package after 0.50 s. List the known information.	**b.** $t = 0.50$ s, $u = 0$ m s^{-1}, $a = 9.8$ m s^{-2}
2. Use the rule $s = ut + \frac{1}{2}at^2$ to determine the distance the package has travelled between $t = 0.00$ s and $t = 0.50$ s.	$s = ut + \frac{1}{2}at^2$ $= 0 \times 0.5 + \frac{1}{2}(9.8)(0.5)^2$ $= 1.2$m

3. Repeat this for $t = 1.0$ s, 1.5 s, 2.0 s and so on until the package hits the ground, and list the results in a table.

Time (s)	0.50	1.0	1.5	2.0
Vertical distance (m)	1.2	4.9	11	20

Time (s)	2.5	3.0	3.5	4.0	4.5
Vertical distance (m)	31	44	60	78	99

4. Draw a scale diagram of the package's position at half-second intervals.

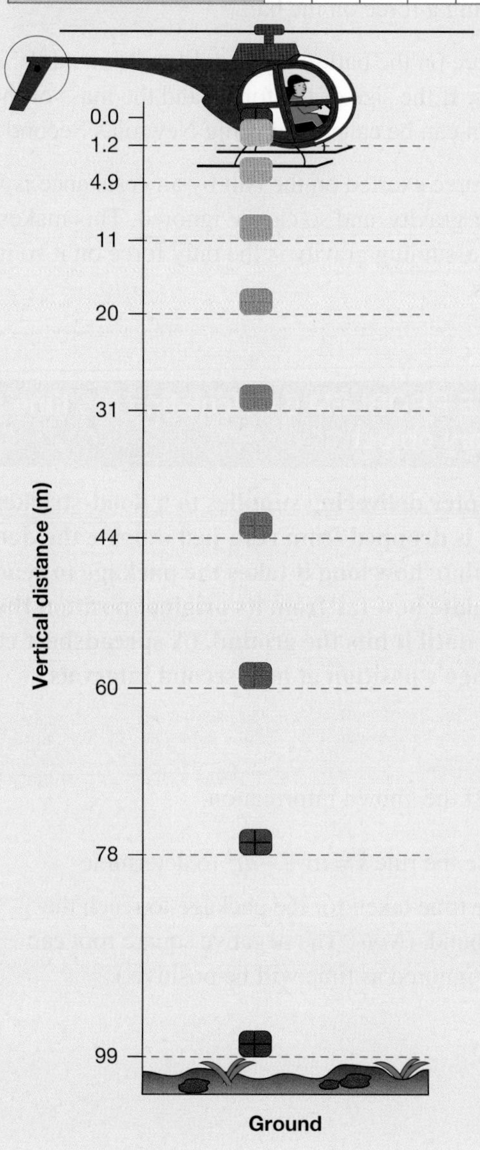

PRACTICE PROBLEM 6

A camera is dropped by a tourist from a lookout and falls vertically to the ground. The thud of the camera hitting the hard ground below is heard by the tourist 3.0 seconds later. Air resistance and the time taken for the sound to reach the tourist can be ignored.

a. **How far did the camera fall?**

b. **What was the velocity of the camera when it hit the ground?**

Terminal velocity

The air resistance on a falling object increases as its velocity increases. An object falling from rest initially experiences no air resistance. Eventually, if the object doesn't hit a surface first, the air resistance will become as large as the force due to gravity acting on the object. The net force on it becomes zero and the object continues to fall with a constant velocity, referred to as its **terminal velocity**.

terminal velocity velocity reached by a falling object when the upward air resistance becomes equal to the downward force of gravity

 Resources

 **Video eLessons** Ball toss (eles-0031)
Air resistance (eles-0035)

1.4.2 Moving and falling

Moving and falling

If a ball is thrown horizontally, as in figure 1.15, the only force acting on the ball once it has been released is the force due to gravity (ignoring air resistance). As the force of gravity is the same regardless of the motion of the ball, the ball will still accelerate downwards at the same rate as if it were dropped. There will not be any horizontal acceleration as there is no net force acting horizontally. This means that, while the ball's vertical velocity will change, its horizontal velocity remains the same throughout its motion.

It is the constant horizontal velocity and changing vertical velocity that give projectiles their characteristic parabolic motion.

FIGURE 1.15 Position of a ball at constant time intervals

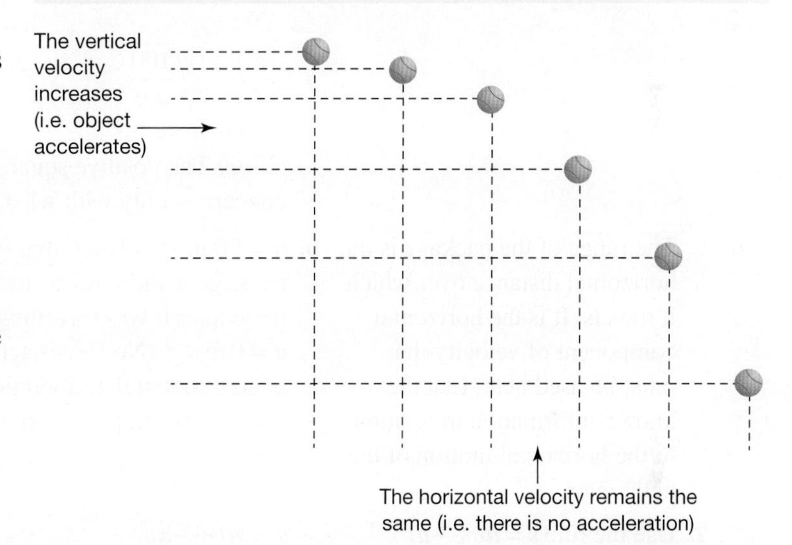
The vertical velocity increases (i.e. object accelerates)
The horizontal velocity remains the same (i.e. there is no acceleration)

Notice that the vertical distance travelled by the ball in each time period increases, but that the horizontal distance is constant.

In modelling projectile motion, the vertical and horizontal components of the motion are treated separately.

1. The total time taken for the projectile motion is determined by the vertical part of the motion as the projectile cannot continue to move horizontally once it has hit the ground, the target or whatever else it might collide with.
2. This total time can then be used to calculate the horizontal distance, or range, over which the projectile travels.

tlvd-8946

SAMPLE PROBLEM 7 Calculating the time and distance for an object to fall from a moving object

Imagine that the helicopter described in sample problem 6 is not stationary but is flying at a slow and steady speed of 20 m s^{-1} and is 100 m above the ground when the package is dropped.

a. **Calculate how long it takes the package to hit the ground.**

b. **What is the range of the package?**

c. Calculate the vertical distance the package has fallen after 0.50 s, 1.0 s, 1.5 s, 2.0 s and so on until the package has reached the ground. (You may like to use a spreadsheet here.) Then calculate the corresponding horizontal distance, and hence draw a scale diagram of the package's position at half-second intervals.

THINK	WRITE
a. 1. Remember, the horizontal and vertical components of the package's motion must be considered separately. In this part of the question, the vertical component is important.	**a.** $u = 0$ m s^{-1}, $s = 100$ m, $a = 9.8$ m s^{-2}
2. Use the rule $s = ut + \frac{1}{2}at^2$ to determine the time taken for the package to reach the ground.	$s = ut + \frac{1}{2}at^2$ $100 = 0 \times t + \frac{1}{2}(9.8)\,t^2$ $\frac{100}{4.9} = t^2$ $t = \sqrt{\frac{100}{4.9}}$ $t = 4.5$ s (*Note:* The positive square root is taken, as the concern is only with what happens after $t = 0$.)
b. 1. The range of the package is the horizontal distance over which it travels. It is the horizontal component of velocity that must be used here. List the known information in relation to the horizontal motion of the helicopter.	**b.** $u = 20$ m s^{-1} (The initial velocity of the package is the same as the velocity of the helicopter it was travelling in.) $a = 0$ m s^{-2} (No forces act horizontally so there is no horizontal acceleration.) $t = 4.5$ s (from part (a) of this example)
2. Use the rule $s = ut + \frac{1}{2}at^2$ to determine the horizontal distance over which the package travels.	$s = ut + \frac{1}{2}at^2$ $= 20 \times 4.5 + 0$ $= 90$ m
c. 1. For $t = 0.50$ s, list the known information regarding the vertical and horizontal motions separately. Use the rule $s = ut + \frac{1}{2}at^2$ to determine the vertical and horizontal distance over which the package travels.	**c. Vertical component** $u = 0$ m s^{-1}, $t = 0.50$ s, $a = 9.8$ m s^{-2} $s = ut + \frac{1}{2}at^2$ $= 0 \times 0.5 + \frac{1}{2}(9.8)(0.50)^2$ $= 1.2$ m **Horizontal component** $u = 20$ m s^{-1}, $t = 0.50$ s, $a = 0$ m s^{-2} $s = ut + \frac{1}{2}at^2$ $= 20 \times 0.50 + 0$ $= 10$ m

2. Repeat the calculations for t = 1.0 s, 1.5 s, 2.0 s and so on until the package reaches the ground.

Time (s)	Vertical distance (m)	Horizontal distance (m)
0.5	1.2	10
1.0	4.9	20
1.5	11.0	30
2.0	20.0	40
2.5	31.0	50
3.0	44.0	60
3.5	60.0	70
4.0	78.0	80
4.5	99.0	90

3. Draw a scale diagram of the package's position at half-second intervals.

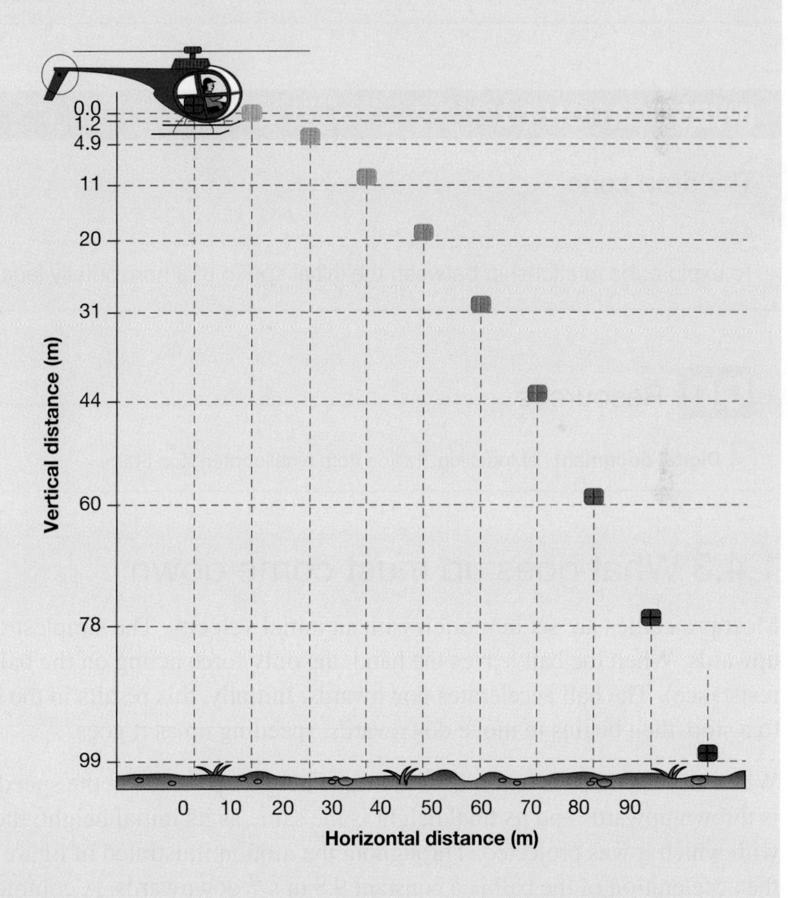

PRACTICE PROBLEM 7

A ball is thrown horizontally at a speed of 40 m s^{-1} from the top of a cliff into the ocean below and takes 4.0 seconds to land in the water. Air resistance can be ignored.

a. **What is the height of the cliff above sea level if the thrower's hand releases the ball from a height of 2.0 metres above the ground?**
b. **What horizontal distance did the ball cover?**
c. **Calculate the vertical component of the velocity at which the ball hits the water.**
d. **At what angle to the horizontal does the ball strike the water?**

elog-1694

tlvd-10809

INVESTIGATION 1.1 online only

Modelling projectile motion

Aim

To model projectile motion by studying the motion of a ball bearing projected onto an inclined plane

elog-1695
tlvd-10810

INVESTIGATION 1.2 online only

The drop zone

Aim

To explore the relationship between the initial speed of a horizontally launched object and its range

Resources

Digital document eModelling: Falling from a helicopter (doc-0005)

1.4.3 What goes up must come down

Most projectiles are set in motion with an initial velocity. The simplest case is that of a ball thrown vertically upwards. When the ball leaves the hand, the only force acting on the ball is the force due to gravity (ignoring air resistance). The ball accelerates downwards. Initially, this results in the ball slowing down. Eventually, it comes to a stop, then begins to move downwards, speeding up as it goes.

When air resistance is ignored, at the same height up or down, the speed will be the same. Therefore, if a ball is thrown upwards and its final height is the same as its initial height, the ball will return with the same speed with which it was projected. Throughout the motion illustrated in figure 1.16 (graphs are shown in figure 1.17), the acceleration of the ball is a constant 9.8 m s^{-2} downwards. A common error made by physics students is to suggest that the acceleration of the ball is zero at the top of its flight. If this were true, would the ball ever come down? The velocity is zero but not the acceleration. Remember, acceleration is the rate of change of velocity.

FIGURE 1.16 The motion of a ball projected vertically upwards

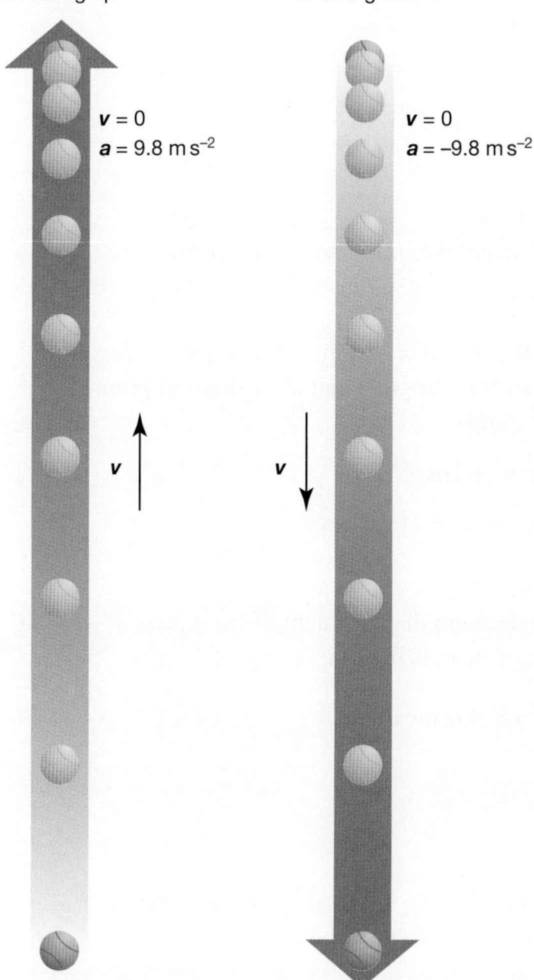

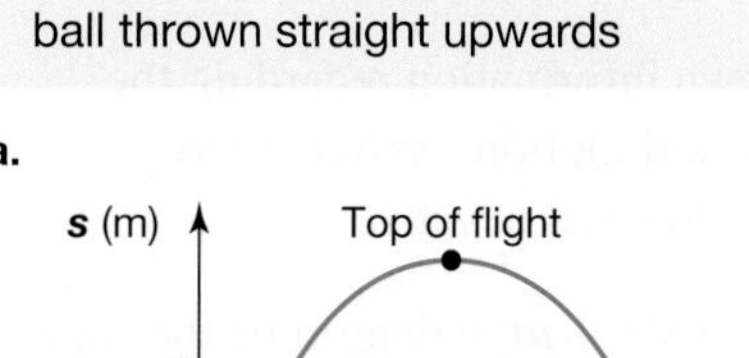

FIGURE 1.17 Graphs of motion for a ball thrown straight upwards

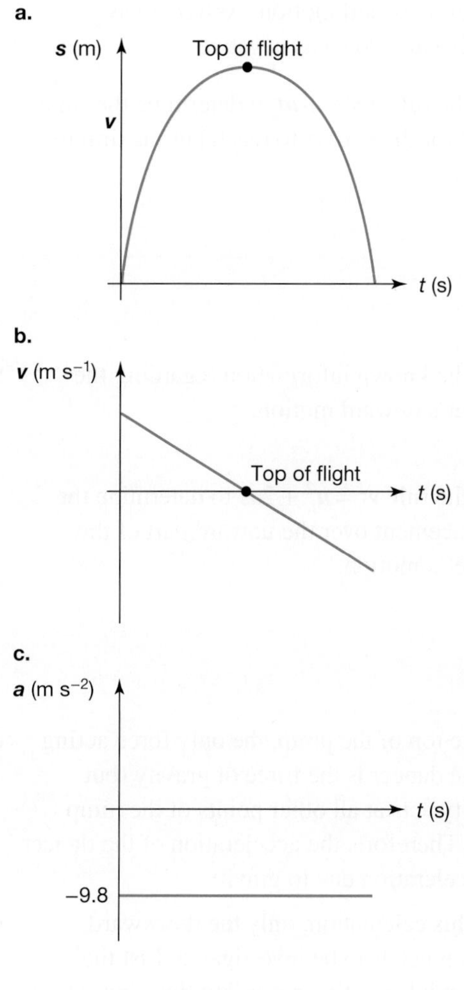

AS A MATTER OF FACT

The axiom 'what goes up must come down' applies equally so to bullets as it does to balls. Unfortunately, this means that people sometimes get killed when they shoot guns straight up into the air. If the bullet left the gun at a speed of 60 m s^{-1}, it will return to Earth at that speed. This speed is fast enough to kill a person who is hit by the returning bullet.

tlvd-8947

SAMPLE PROBLEM 8 Examining the displacement, velocity and acceleration of a dancer

A dancer jumps vertically upwards with an initial velocity of 4.0 m s^{-1}. Assume the dancer's centre of mass was initially 1.0 m above the ground, and ignore air resistance.

a. **How long did the dancer take to reach her maximum height?**
b. **What was the maximum displacement of the dancer's centre of mass?**
c. **What is the acceleration of the dancer at the top of her jump?**
d. **Calculate the velocity of the dancer's centre of mass when it returns to its original height above the ground.**

THINK	WRITE
a. 1. List the known information regarding the dancer's upward motion. Assign up as positive and down as negative.	**a.** $\boldsymbol{u} = 4.0\ \text{m s}^{-1}$, $\boldsymbol{a} = -9.8\ \text{m s}^{-2}$, $\boldsymbol{v} = 0\ \text{m s}^{-1}$ (as the dancer comes to a halt at the highest point of the jump)
2. Use the rule $\boldsymbol{v = u + at}$ to determine the time taken for the dancer to reach her maximum height.	$v = u + at$ $0 = 4.0 + (-9.8) \times t$ $\Rightarrow t = \dfrac{4.0}{9.8}$ $= 0.41\ \text{s}$ The dancer takes 0.41 s to reach her highest point.
b. 1. List the known information regarding the dancer's upward motion.	**b.** $\boldsymbol{u} = 4.0\ \text{m s}^{-1}$, $\boldsymbol{a} = -9.8\ \text{m s}^{-2}$, $\boldsymbol{v} = 0\ \text{m s}^{-1}$ (as the dancer comes to a halt at the highest point of the jump)
2. Use the rule $\boldsymbol{v^2 = u^2 + 2as}$ to determine the displacement over the upward part of the dancer's motion.	$v^2 = u^2 + 2as$ $(0)^2 = (4.0)^2 + 2(-9.8)\,s$ $16 = 19.6s$ $\Rightarrow s = 0.82\ \text{m}$ The maximum displacement of the dancer's centre of mass is 0.82 m.
c. At the top of the jump, the only force acting on the dancer is the force of gravity (but gravity acts at all other points of the jump too). Therefore, the acceleration of the dancer is acceleration due to gravity.	**c.** $9.8\ \text{m s}^{-2}$ downwards
d. 1. For this calculation, only the downward motion needs to be investigated. List the known information regarding the dancer's downward motion. (Alternatively, you could look at the whole motion rather than using previously calculated values.)	**d.** $\boldsymbol{u} = 0\ \text{m s}^{-1}$, $\boldsymbol{a} = -9.8\ \text{m s}^{-2}$, $\boldsymbol{s} = -0.82\ \text{m}$ (as the motion is downwards)
2. Use the rule $\boldsymbol{v^2 = u^2 + 2as}$ to determine the dancer's velocity when she returns to the ground. (*Note:* Here, the negative square root is used, as the dancer is moving downwards. Remember, the positive and negative signs show direction only.)	$v^2 = u^2 + 2as$ $v^2 = (0)^2 + 2(-9.8)(-0.82)$ $v^2 = 16.072$ $\Rightarrow v = -4.0\ \text{m s}^{-1}$ The velocity of the dancer's centre of mass when it returns to its original height is $4.0\ \text{m s}^{-1}$ downwards.

PRACTICE PROBLEM 8

A basketball player jumps vertically upwards so that his centre of mass reaches a maximum displacement of 50 cm.

a. **What is the velocity of the basketballer's centre of mass when it returns to its original height above the ground?**

b. **For how long was the basketballer's centre of mass above its original height?**

HANGING IN MID AIR

Sometimes dancers, basketballers and high jumpers seem to hang in mid air. It is as though the force of gravity had temporarily stopped acting on them. Of course this is not so! It is only the person's centre of mass that moves in a parabolic path. The arrangement of the person's body can change the position of the centre of mass, causing the body to appear to be hanging in mid air even though the centre of mass is still following its original path.

High jumpers can use this effect to increase the height of their jumps. By bending her body as she passes over the bar, a high jumper can cause her centre of mass to be outside her body! This allows her body to pass over the bar, while her centre of mass passes under it. The amount of energy available to raise the high jumper's centre of mass is limited, so she can raise her centre of mass only by a certain amount. This technique allows her to clear a higher bar than other techniques for the same amount of energy.

FIGURE 1.18 Croatian high jumper Ana Simic's centre of mass passes under the bar, while her body passes over the bar!

1.4.4 Shooting at an angle

Generally, projectiles are shot, thrown or driven at some angle to the horizontal. In these cases, the initial velocity may be resolved into its horizontal and vertical components to help simplify the analysis of the motion.

If the velocity and the angle to the horizontal are known, the size of the components can be calculated using trigonometry.

The motion of projectiles with an initial velocity at an angle to the horizontal can be dealt with in exactly the same manner as those with a velocity straight up or straight across. Once the initial velocity has been separated into its vertical and horizontal components, the vertical and horizontal motion can be analysed separately. The time of flight must be the same for both the vertical and horizontal motion and this is often used to link them when solving problems.

FIGURE 1.19 The velocity can be resolved into a vertical and a horizontal component.

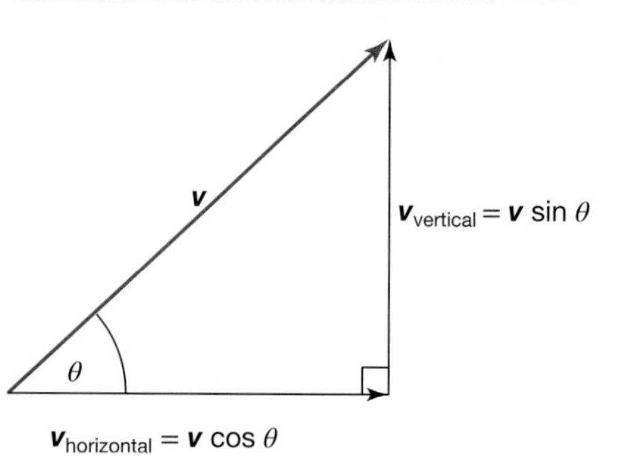

Projectile motion calculations

Tips for projectile motion calculations

- Draw a diagram and to write down all known information that you can identify.
- Always separate the motion into vertical and horizontal components.
- Remember to resolve the initial velocity into its components if necessary.
- The time in flight is the link between the separate vertical and horizontal components of the motion.
- At the end of any calculation, check to see if the quantities you have calculated are reasonable.

tlvd-8948

SAMPLE PROBLEM 9 Calculating a ramp jump

A stunt driver is trying to drive a car over a small river. The car will travel up a ramp (at an angle of 40°) and leave the ramp at 22 m s^{-1} before landing back at its initial height. The river is 50 m wide. Will the car make it?

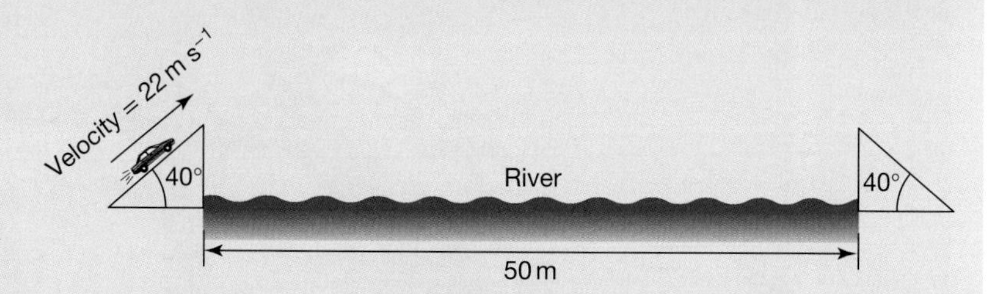

THINK

1. Before either part of the motion can be examined, it is important to calculate the vertical and horizontal components of the initial velocity. Assign up as positive and down as negative.

WRITE

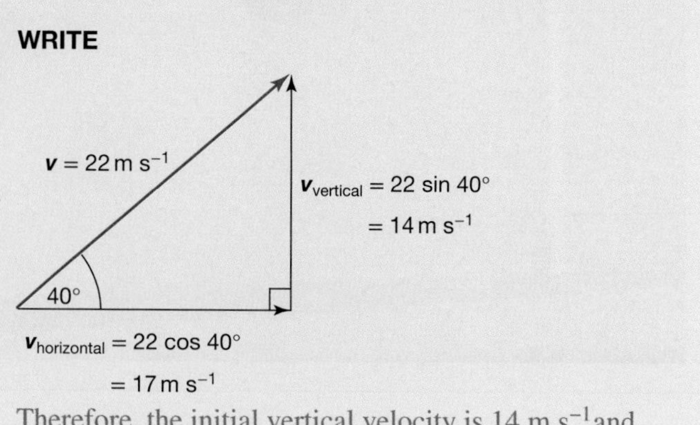

$v_{vertical} = 22 \sin 40°$
$= 14 \text{ m s}^{-1}$

$v_{horizontal} = 22 \cos 40°$
$= 17 \text{ m s}^{-1}$

Therefore, the initial vertical velocity is 14 m s^{-1}and the initial horizontal velocity is 17 m s^{-1}.

2. The vertical motion is used to calculate the time in the air. Use the first half of the motion — from take-off until the car has reached its highest point.
(It is possible to double the time in this situation because air resistance has been ignored. The two parts of the motion are symmetrical.)

Vertical component

$u = 14 \text{ m s}^{-1}$, $a = -9.8 \text{ m s}^{-2}$, $v = 0 \text{ m s}^{-1}$ (as the car comes to a vertical halt at its highest point):

$$v = u + at$$
$$0 = 14 + (-9.8) \times t$$
$$\Rightarrow t = \frac{14}{-9.8}$$
$$= 1.4 \text{ s}$$

As this is only half the motion, the total time in the air is 2.8 s.

3. The horizontal component is used to calculate the range.

Horizontal component

$u = 17 \text{ m s}^{-1}$, $t = 2.8 \text{ s}$, $a = 0 \text{ m s}^{-2}$:

$$s = ut$$
$$= 17 \times 2.8$$
$$= 48 \text{ m}$$

Therefore, the unlucky stunt driver will fall short of the second ramp and will land in the river. Perhaps the study of physics should be a prerequisite for all stunt drivers!

PRACTICE PROBLEM 9

A hockey ball is hit towards the goal at an angle of 25° to the ground with an initial speed of 32 km h^{-1}.

a. **What are the horizontal and vertical components of the initial velocity of the ball?**
b. **How long does the ball spend in flight?**
c. **What is the range of the hockey ball?**

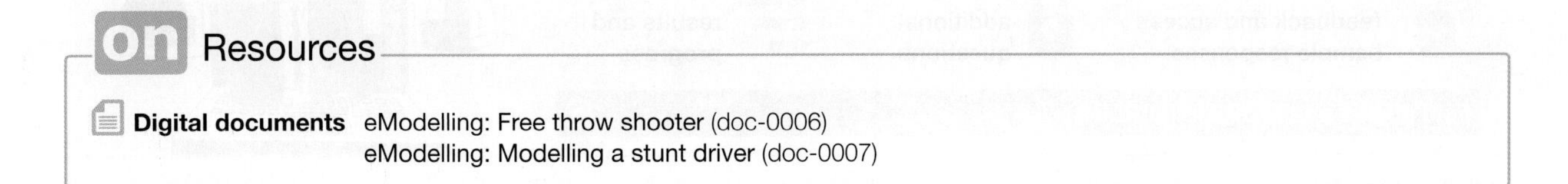

on Resources
Digital documents eModelling: Free throw shooter (doc-0006)
eModelling: Modelling a stunt driver (doc-0007)

1.4.5 The real world — including air resistance

So far in this topic, the effects of air resistance have been ignored so that projectile motion can easily be modelled. The reason the force of air resistance complicates matters so much is that it is not constant throughout the motion. It depends on a number of factors. Consider what differences you might expect when you throw a crumpled-up piece of paper versus when you throw a cricket ball. Or, the difference between throwing a cricket ball in humid air conditions and dry air conditions.

The impact of air resistance can be influenced by the velocity (v) of the object — the faster an object moves, the greater the air resistance. The size of the object in cross-section to the direction it is being thrown also has an impact — the greater the area, the greater the air resistance. Related to the size of the object is the shape of the object — objects that are more streamlined will experience less air resistance. Finally, the density of the air can impact air resistance — the more dense the air, the greater the air resistance.

FIGURE 1.20 While the magnitude of air resistance changes throughout the motion, it always opposes the direction of the motion. Note that the projectile has a steeper descent than its initial ascent when air resistance is taken into consideration.

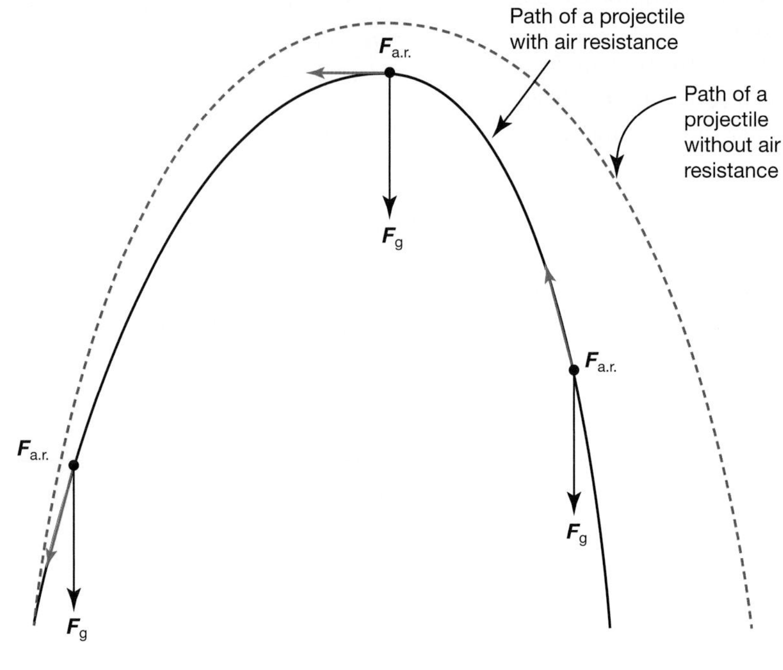

Resources

Weblink Projectile motion applet

1.4 Activities

learnon

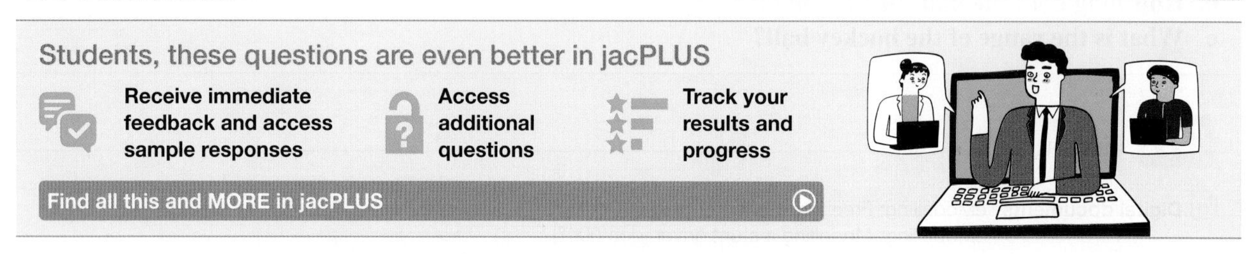

1.4 Quick quiz | 1.4 Exercise | 1.4 Exam questions

1.4 Exercise

1. A ball has been thrown directly upwards. Draw the ball at three points during its flight (going up, at the top and going down) and mark on the diagram(s) all the forces acting on the ball at each stage.
2. Ignoring air resistance, the acceleration of a projectile in flight is always the same, whether it is going up or down. Use graphs of motion to show why this is true.
3. In each of the following cases, calculate the magnitude of the vertical and horizontal components of the velocity.

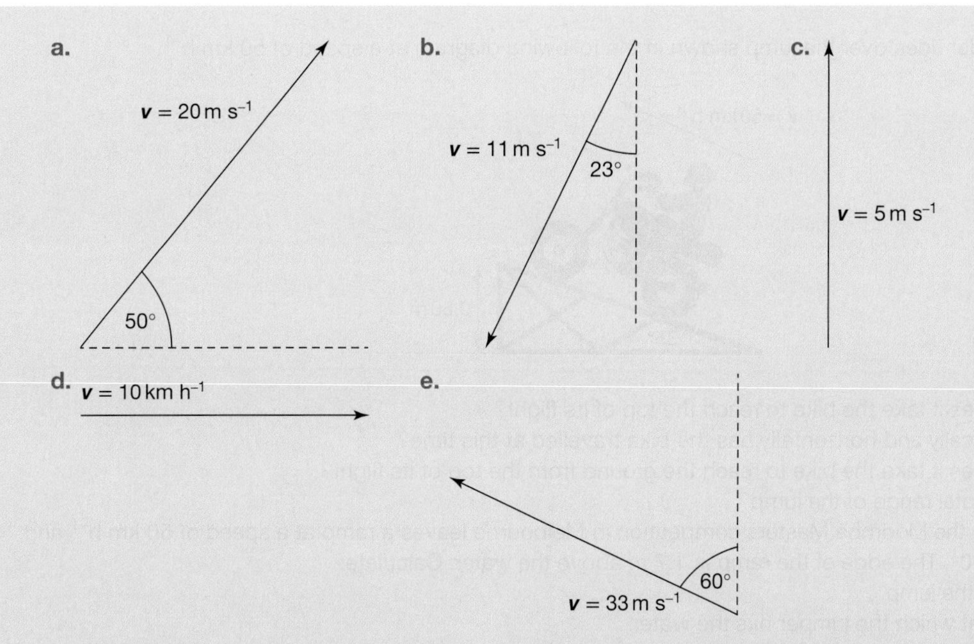

4. After taking a catch, a cricketer throws the cricket ball up into the air in jubilation.
   a. The vertical velocity of the ball as it leaves her hands is 18.0 m s^{-1}. How long will the ball take to return to its original position?
   b. What was the ball's maximum vertical displacement?
   c. Draw vectors to indicate the net force on the ball (ignoring air resistance),
      i. the instant it left her hands
      ii. at the top of its flight
      iii. as it returns to its original position.
5. A friend wants to get into the *Guinness Book of Records* by jumping over 11 people on his pushbike and landing on the other side at the same height he jumped from. He has set up two ramps as shown in the following figure, and has allowed a space of 0.5 m within which each person can lie down. In practice attempts, he has averaged a speed of 7.0 m s^{-1} at the top of the ramp. Will you lie down as the eleventh person between the ramps? Justify your answer using physics calculations.

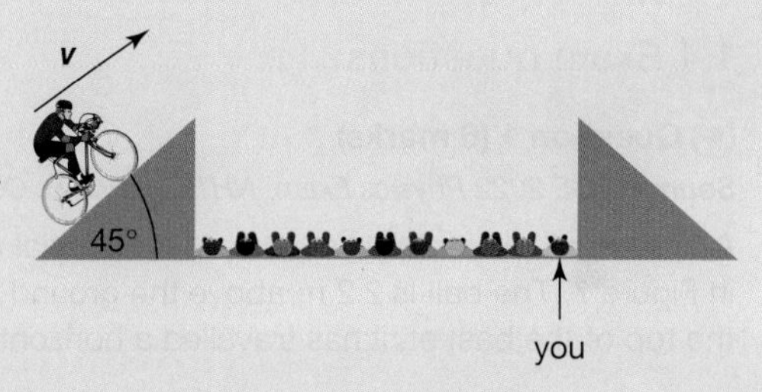

6. A skateboarder jumps a horizontal distance of 2.0 m (starting from the ground), taking off at a speed of 5.0 m s^{-1}. The jump takes 0.42 s to complete.
   a. What was the skateboarder's initial horizontal velocity?
   b. What was the angle of take-off?
   c. What was the maximum height above the ground reached during the jump?
7. During practice, a soccer player shoots for goal. The goalkeeper is able to stop the ball only if it is more than 30 cm beneath the crossbar. The ball is kicked at an angle of 45° with a speed of 9.8 m s^{-1}. The arrangement of the players is shown in the following diagram.
   a. How long does it take the ball to reach the top of its flight?
   b. How far vertically and horizontally has the ball travelled at this time?
   c. How long does it take the ball to reach the soccer net from the top of its flight?
   d. Will the ball go into the soccer net, over it, or will the goalkeeper stop it?

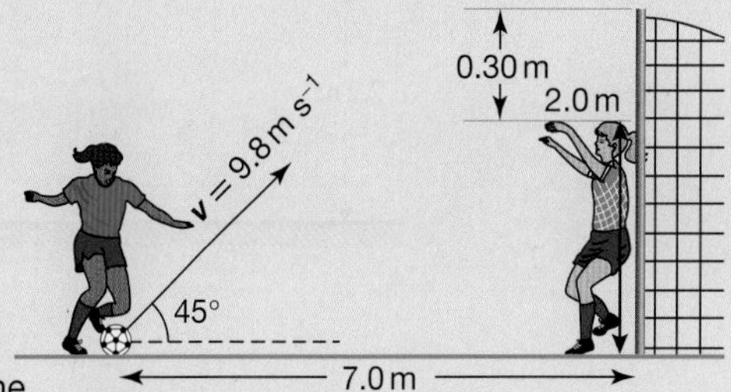

8. A motocross rider rides over the jump shown in the following diagram at a speed of 50 km h^{-1}.

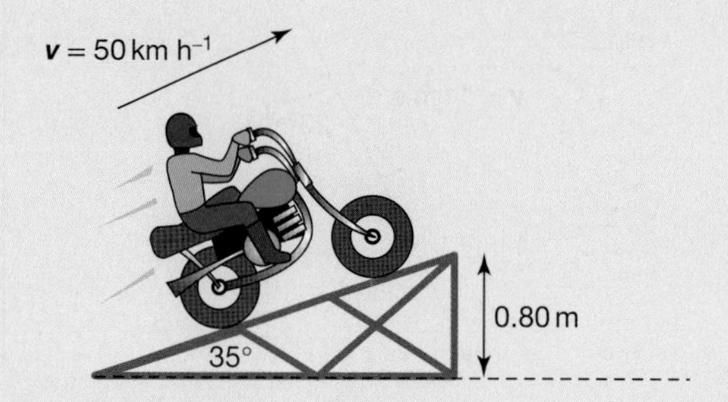

   a. How long does it take the bike to reach the top of its flight?
   b. How far vertically and horizontally has the bike travelled at this time?
   c. How long does it take the bike to reach the ground from the top of its flight?
   d. What is the total range of the jump?
9. A water skier at the Moomba Masters competition in Melbourne leaves a ramp at a speed of 50 km h^{-1} and at an angle of 30°. The edge of the ramp is 1.7 m above the water. Calculate:
   a. the range of the jump
   b. the velocity at which the jumper hits the water.
   (*Hint*: Split the waterskier's motion into two sections, before the highest point and after the highest point, to avoid solving a quadratic equation.)
10. A gymnast wants to jump a horizontal distance of 2.5 m, leaving the ground at an angle of 28°. With what speed must the gymnast take off?
11. A horse rider wants to jump a stream that is 3.0 metres wide. The horse can approach the stream with a speed of 7.0 m s^{-1}. At what angle must the horse take off? (This question is a challenge. *Hint:* You will need to use trigonometric ratios from mathematics, or model the situation using a spreadsheet to solve this problem.)

1.4 Exam questions

Question 1 (6 marks)

Source: *VCE 2022 Physics Exam, NHT, Section B, Q.10;* © VCAA

A basketball player throws a ball with an initial velocity of 7.0 m s^{-1} at an angle of 50° to the horizontal, as shown in Figure 7. The ball is 2.2 m above the ground when it is released. By the time the ball passes through the ring at the top of the basket, it has travelled a horizontal distance of 3.2 m. Ignore air resistance.

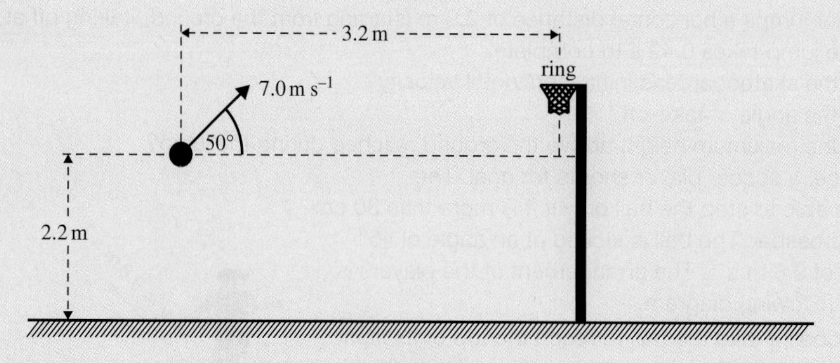

Figure 7

a. Show that the time taken for the ball's flight from launch to passing through the ring is 0.71 s. Show your working. **(2 marks)**
b. How far above the ground is the ring at the top of the basket? Show your working. **(4 marks)**

Question 2 (6 marks)

***Source:** VCE 2018 Physics Exam, Section B, Q.7; © VCAA*

A small ball of mass 0.20 kg rolls on a horizontal table at 3.0 m s^{-1}, as shown in Figure 9.

The ball hits the floor 0.40 s after rolling off the edge of the table. The radius of the ball may be ignored. In this question, take the value of g to be 10 m s^{-2}.

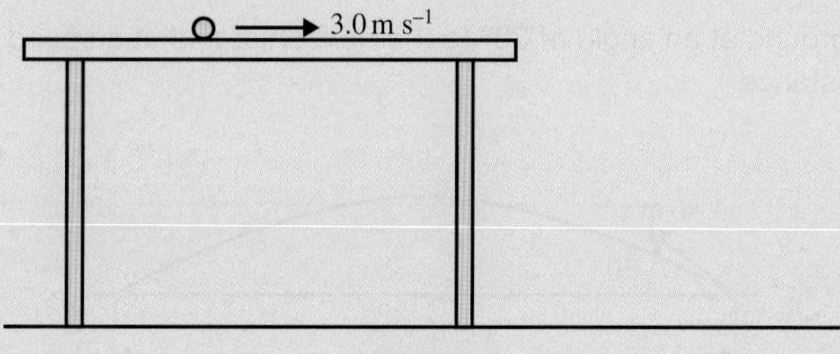

Figure 9

a. Calculate the horizontal distance from the right-hand edge of the table to the point where the ball hits the floor. **(1 mark)**

b. Calculate the height of the table. Show your working. **(2 marks)**

c. Calculate the speed at which the ball hits the floor. Show your working. **(3 marks)**

Question 3 (3 marks)

***Source:** VCE 2017 Physics Exam, Section B, Q.9a; © VCAA*

Students use a catapult to investigate projectile motion. In their first experiment, a ball of mass 0.10 kg is fired from the catapult at an angle of 30° to the horizontal. Ignore air resistance. In this first experiment, the ball leaves the catapult at ground level with a speed of 20 m s^{-1}.

However, instead of reaching the ground, the ball strikes a wall 26 m from the launching point, as shown in Figure 8a. Figure 8b shows an enlarged view of the catapult.

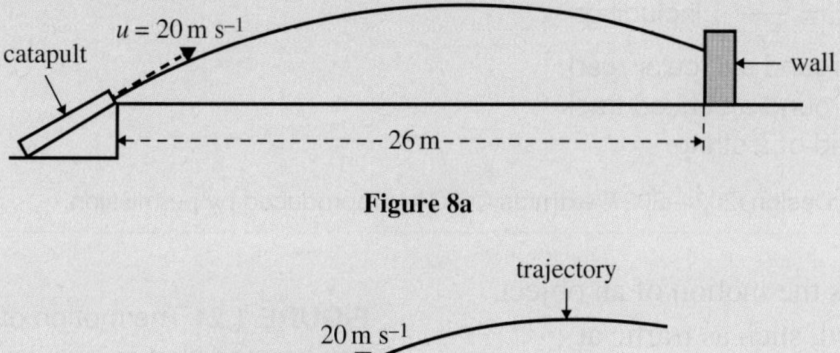

Figure 8a

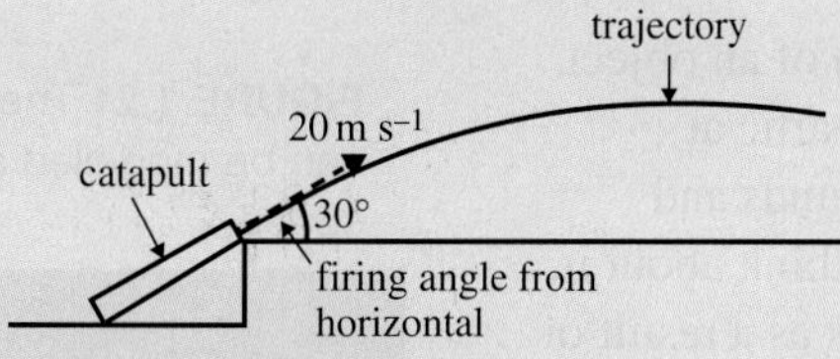

Figure 8b

Calculate the height of the ball above the ground when it strikes the wall. Show your working

Question 4 (7 marks)

***Source:** VCE 2018 Physics Exam, NHT, Section B, Q.6; © VCAA*

A rock of mass 2.0 kg is thrown horizontally from the top of a vertical cliff 20 m high with an initial speed of 25 m s^{-1}, as shown in Figure 3.

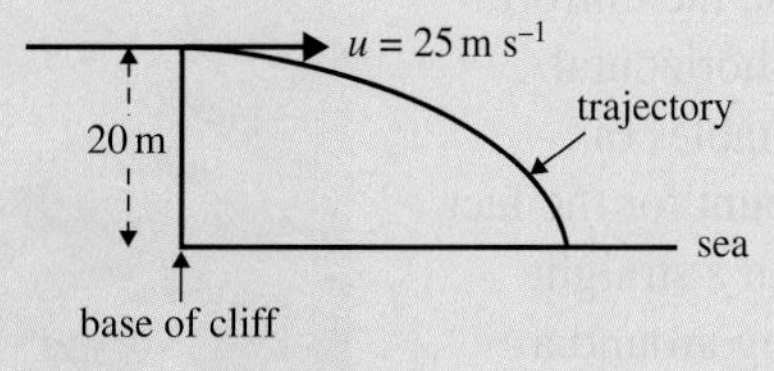

Figure 3

a. Calculate the time taken for the rock to reach the sea. Show your working. **(3 marks)**
b. Calculate the horizontal distance from the base of the cliff to the point where the rock reaches the sea. Show your working. **(2 marks)**
c. Calculate the kinetic energy of the rock as it reaches the surface of the sea. Show your working. **(2 marks)**

Question 5 (3 marks)

Source: *VCE 2016, Physics Exam, Q.5.a; © VCAA*

A ball is projected from the ground at an angle of 30° to the horizontal and at a speed of 40 m s^{-1}, as shown in the figure. Ignore any air resistance.

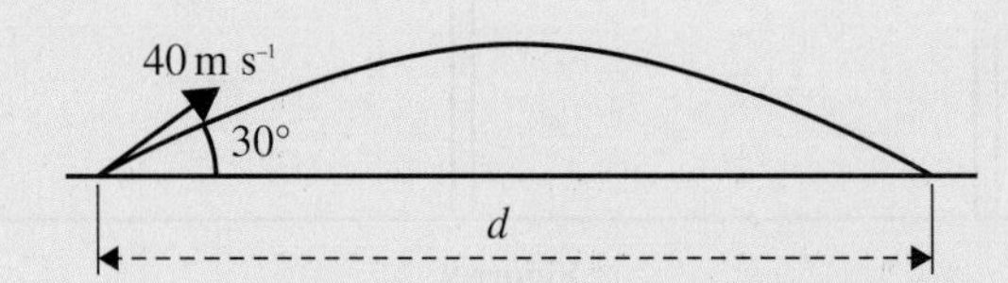

Calculate the distance, d, to the point where the ball hits the ground. Show your working.

More exam questions are available in your learnON title.

1.5 Uniform circular motion

KEY KNOWLEDGE

- Investigate and analyse theoretically and practically the uniform circular motion of an object moving in a horizontal plane $\left(F_{\text{net}} = \frac{mv^2}{r}\right)$ including:
  - a vehicle moving around a circular road
  - a vehicle moving around a banked track
  - an object on the end of a string.

Source: VCE Physics Study Design (2024–2027) extracts © VCAA; reproduced by permission.

Uniform circular motion is the motion of an object in a circle at constant speed, such as traffic at roundabouts, children on merry-go-rounds and cyclists in velodromes. If you stop to think about it, you are always going around in circles as a result of Earth's rotation.

The satellites orbiting Earth, including the Moon, travel in ellipses. However, their orbits can be modelled as circular motion. This motion is covered in subtopic 3.4, when gravitational forces are closely explored.

This section will investigate and analyse the uniform circular motion of objects moving in a horizontal plane. In doing so, the way that the variables of motion are calculated will need to account for the fact that the motion is circular, rather than in a straight line. Examples include a vehicle moving around a circular road, a vehicle moving around a banked track and an object at the end of a string.

FIGURE 1.21 The motion of satellites around Earth can be modelled as circular motion with a constant speed.

1.5.1 Period and frequency

The time taken for an object, moving in a circular path and at a constant speed, to complete one revolution is called the **period**, *T*. The number of revolutions the object completes each second is called the **frequency**, *f*.

period time taken for an object, moving in a circular path and at a constant speed, to complete one revolution

frequency number of revolutions that an object completes each second

$$f = \frac{1}{T} \text{ and } T = \frac{1}{f}$$

where: f is the frequency in Hertz (Hz)

T is the period in seconds (s)

1.5.2 Instantaneous velocity

Imagine this scenario: Ralph the dog is chained to pole in the backyard while his owner does the gardening (don't worry — the chain is long enough so that he can still move around). A neighbourhood cat likes to tease him and makes Ralph run around in circles at a constant speed. Ralph's owner, Julie, is a physics teacher. She knows that no matter how great Ralph's average speed, if he always ends up in the same place his average velocity is always zero.

FIGURE 1.22 A dog running in circles at a constant speed will have a constantly changing instantaneous velocity but an average velocity of zero, assuming the dog's starting and stopping positions are the same.

Although Ralph's average velocity for a single lap is zero, his instantaneous velocity is continually changing. Velocity is a vector and has a magnitude and direction. While the magnitude of Ralph's velocity may be constant, the direction is continually changing. At one point, Ralph is travelling east, so his instantaneous velocity is in an easterly direction. A short time later, he will be travelling south, so his instantaneous velocity is in a southerly direction.

If Ralph could maintain a constant speed, the magnitude of his velocity would not change, but the direction would be continually changing.

The speed is therefore constant and can be calculated using the formula speed $= \frac{d}{t}$, where d is the distance travelled, and t is the time interval. It is most convenient to use the period of the object travelling in a circle. Thus:

FIGURE 1.23 An object moving in circular motion

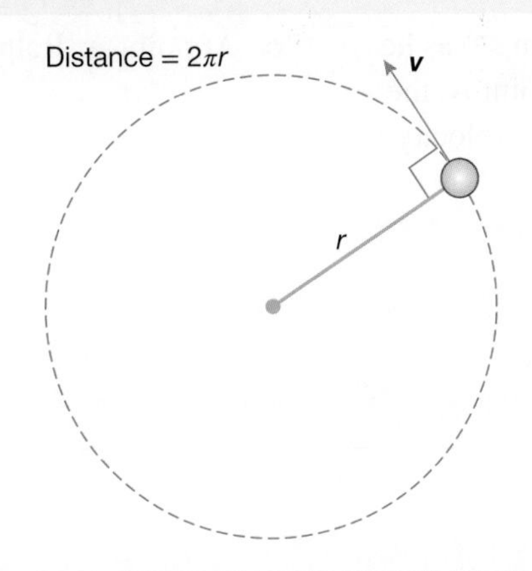

$$\text{speed} = \frac{d}{t}$$
$$= \frac{\text{circumference}}{\text{period}}$$
$$= \frac{2\pi r}{T}$$

where: r = radius of the circle
T = period

tlvd-8949

SAMPLE PROBLEM 10 Determining the average speed and instantaneous velocity in a uniform circular motion

Ralph's chain is 7.0-m long and attached to a small post in the middle of the garden. It takes an average of 9 s to complete one lap.

a. What is Ralph's average speed?

b. What is Ralph's average velocity after three laps?

c. What is Ralph's instantaneous velocity at point A? (Assume he travels at a constant speed around the circle.)

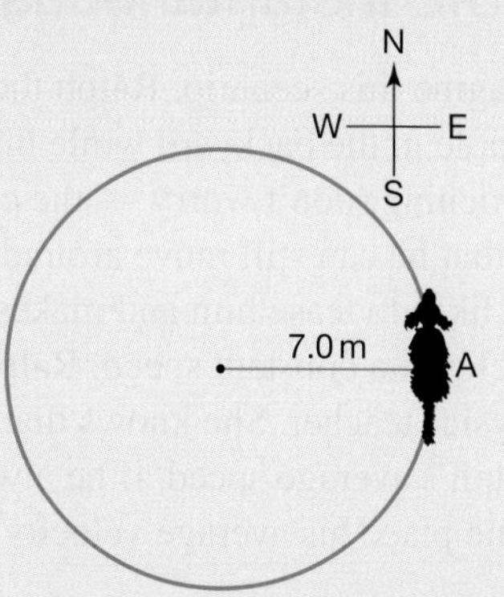

THINK	WRITE
a. 1. To calculate Ralph's speed, the distance he has travelled is required. Use the formula for the circumference of a circle: distance $= 2\pi r$.	**a.** distance $= 2\pi r$ $= 2\pi \times 7.0$ m $= 44$ m
2. Now the average speed can be calculated.	$\text{speed} = \frac{d}{t}$ $= \frac{44}{9}$ $= 5\text{ m s}^{-1}$ The average speed is 5 m s^{-1}.
b. After three laps, Ralph is in exactly the same place as he started, so his displacement is zero. No matter how long he took to run these laps, his average velocity would still be zero, as $v_{av} = \frac{\Delta x}{\Delta t}$.	**b.** $v_{av} = \frac{\Delta x}{\Delta t}$ $= \frac{0}{3 \times 9}$ $= 0\text{ m s}^{-1}$
c. Ralph's speed is a constant 5 m s^{-1} as he travels around the circle. At point A, the magnitude of his instantaneous velocity is also 5 m s^{-1}.	**c.** At point A, Ralph's velocity is 5 m s^{-1} north.

PRACTICE PROBLEM 10

A battery-operated toy car completes a single lap of a horizontal circular track in 15 s with an average speed of 1.3 m s^{-1}. Assume that the speed of the toy car is constant.

a. **What is the radius of the track?**
b. **What is the magnitude of the toy car's instantaneous velocity halfway through the lap?**
c. **What is the average velocity of the toy car after half of the lap has been completed?**
d. **What is the average velocity of the toy car over the entire lap?**

1.5.3 Changing velocities and accelerations

As all objects with changing velocities are experiencing an acceleration, this means all objects that are moving in a circle are accelerating.

An acceleration can be caused only by an unbalanced force, so non-zero net force is needed to move an object in a circle. For example, a hammer thrower must apply a force to the hammer to keep it moving in a circle. When the hammer is released, this force is no longer applied and the hammer moves off with the velocity it had when released. The hammer will then experience projectile motion.

FIGURE 1.24 The hammer is always accelerating while it moves in a circle.

In which direction is the force?

Figure 1.26 shows diagrammatically the head of the hammer moving in a circle at two different times. It takes t seconds to move from A to B. (This movement is also covered in subtopic 3.4, in which the motion of satellites is explored.)

FIGURE 1.25 The hammer moves in a circle while the thrower turns. When the hammer is released, it moves in a straight line.

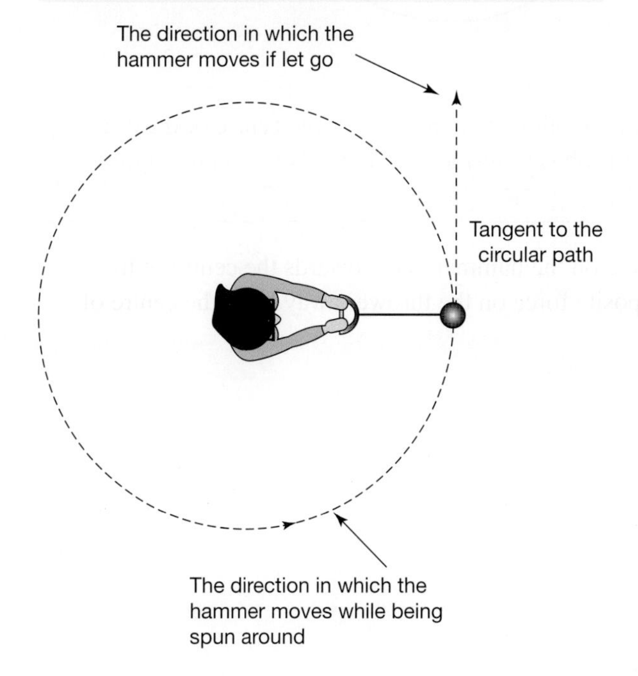

FIGURE 1.26 Velocity vectors for a hammer moving in an anticlockwise circle

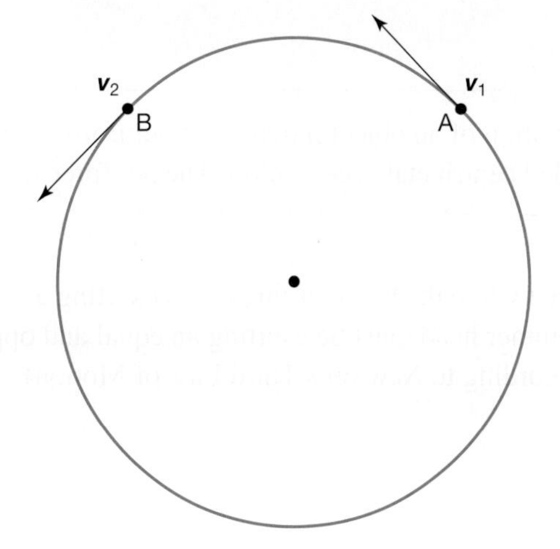

To determine the acceleration, the change in velocity between these two points must be known. Vector addition needs to be used:

$$\Delta \boldsymbol{v} = \boldsymbol{v}_2 - \boldsymbol{v}_1$$

$$\Delta \boldsymbol{v} = \boldsymbol{v}_2 + (-\boldsymbol{v}_1)$$

Notice that when the $\Delta \boldsymbol{v}$ vector is transferred back to the original circle halfway between the two points in time, it is pointing towards the centre of the circle. (See figure 1.27b. Such calculations become more accurate when very small time intervals are used; however, a large time interval has been used here to make the diagram clear.)

As $\boldsymbol{a} = \frac{\Delta \boldsymbol{v}}{t}$, the acceleration vector is in the same direction as $\Delta \boldsymbol{v}$, but has a different magnitude and different units.

No matter which time interval is chosen, the acceleration vector always points towards the centre of the circle. So, for an object to have uniform circular motion, the acceleration of the object *must be towards the centre* of the circle. Such an acceleration is called **centripetal acceleration**. The word *centripetal* literally means 'centre-seeking'. As stated in Newton's Second Law of Motion, the net force on an object is in the same direction as the acceleration ($\boldsymbol{F}_{\text{net}} = m\boldsymbol{a}$). Therefore, the net force on an object moving with uniform circular motion is towards the centre of the circle.

centripetal acceleration the acceleration towards the centre of a circle experienced by an objec moving in a circular motion

FIGURE 1.27 a. Vector addition b. The change in velocity is towards the centre of the circle.

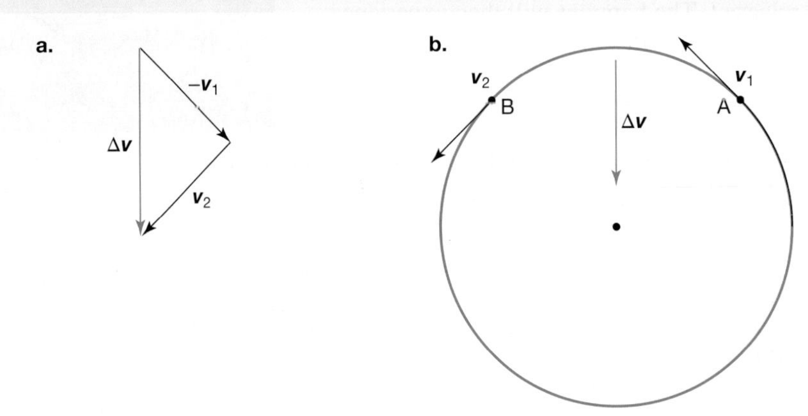

The acceleration of an object moving with uniform circular motion must be towards the centre of the circle. This is called centripetal acceleration. The net force on this object must also be towards the centre of the circle.

Remember that while the hammer thrower is exerting a force on the hammer head towards the centre of the circle, the hammer head must be exerting an equal and opposite force on the thrower away from the centre of the circle (according to Newton's Third Law of Motion).

1.5.4 Calculating accelerations and forces

Using vector diagrams and the formulas $a = \frac{\Delta v}{t}$ and $F_{net} = ma$, it is possible to calculate the accelerations and forces involved in circular motion. However, doing calculations this way is tedious, and results can be inaccurate if the vector diagrams are not drawn carefully. It is simpler to use a formula that will avoid these difficulties. The derivation of such a formula is a little challenging, but it is worth the effort!

FIGURE 1.28 The triangles shown in parts (a) and (b) are both isosceles triangles.

a.

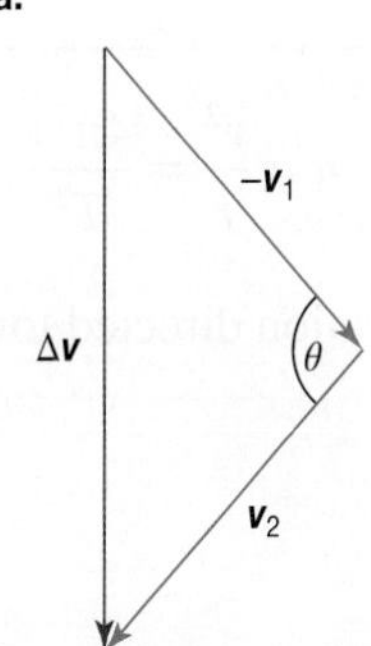

b.

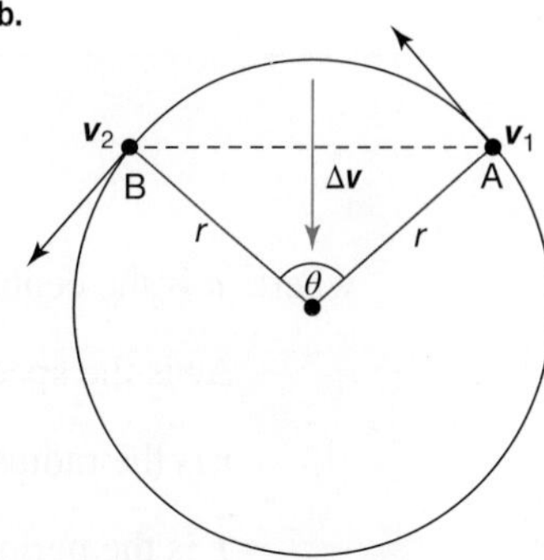

The circular motion formula explained

By re-examining the two previous figures, it is possible to see that they both 'contain' isosceles triangles. These are shown in figure 1.29.

In figure 1.29, diagram (a) shows distances. It has the radius of the circle marked in twice. These radii form two sides of an isosceles triangle. The third side is formed by a line, or chord, joining point A with point B. It is the distance between the two points. When the angle θ is very small, the length of the chord is virtually the same as the length of the arc that also joins these two points. As this is a distance, its length can be calculated using $\boldsymbol{s} = \boldsymbol{v}t$.

In figure 1.29, diagram (b) shows velocities. As the object was moving with *uniform* circular motion, the length of the vectors $\boldsymbol{v}_2$ and $-\boldsymbol{v}_1$ are identical and form two sides of an isosceles triangle. As both diagrams (a) and (b) are derived from figure 1.27, both of the angles marked as θ are the same size. Therefore, the triangles are both isosceles triangles, containing the same angle, θ. This means they are similar triangles — they can be thought of as the same triangle drawn on two different scales. Figure 1.29 shows these triangles redrawn to make this more obvious.

FIGURE 1.29 The two triangles are *similar* triangles.

a.

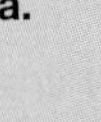

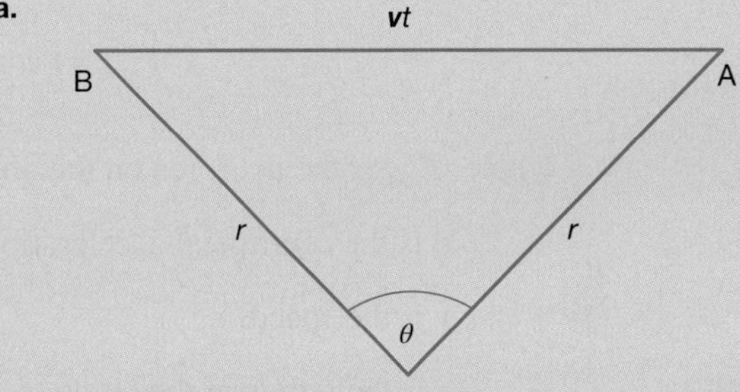

b.

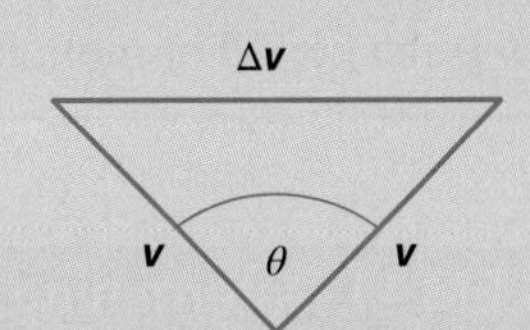

As the triangles are similar, the ratio of their sides must be constant, so:

$$\frac{\Delta \boldsymbol{v}}{\boldsymbol{v}t} = \frac{\boldsymbol{v}}{r}$$

Multiplying both sides by $\boldsymbol{v}$:

$$\frac{\Delta \boldsymbol{v}}{t} = \frac{\boldsymbol{v}^2}{r}$$

As $\boldsymbol{a} = \frac{\Delta \boldsymbol{v}}{t}$:

$$\boldsymbol{a} = \frac{\boldsymbol{v}^2}{r}$$

Sometimes it is not easy to measure the velocity of the object undergoing circular motion. However, this can be calculated from the radius of the circle and the time taken to complete one circuit using the equation $\boldsymbol{v} = \frac{2\pi r}{T}$. It can also be found using the equation $\boldsymbol{a} = \frac{v^2}{r}$. Combining these two equations yields the following relationship:

$$\boldsymbol{a} = \frac{v^2}{r} = \frac{4\pi^2 r}{T^2}$$

where: $\boldsymbol{a}$ is the centripetal acceleration directed towards the centre of the circle

$\Delta \boldsymbol{v}$ is the speed

r is the radius of the circle

T is the period of motion

This formula provides a way of calculating the centripetal acceleration of a mass moving with uniform circular motion having speed $\boldsymbol{v}$ and radius r.

If the acceleration of a known mass moving in a circle with constant speed has been calculated, the net force can be determined by applying $\boldsymbol{F}_{\text{net}} = m\boldsymbol{a}$. Note that because in this scenario the net force is causing the centripetal acceleration, you may sometimes see it referred to as the centripetal force, $\boldsymbol{F}_{\text{c}}$.

The magnitude of the net force can also be calculated using:

$$\boldsymbol{F}_{\text{net}} = m\boldsymbol{a} = \frac{mv^2}{r} = \frac{4\pi^2 rm}{T^2} = \boldsymbol{F}_{\text{c}}$$

where: $\boldsymbol{F}_{\text{net}}$ is the net force on the object

$\boldsymbol{a}$ is the centripetal acceleration directed towards the centre of the circle

$\boldsymbol{v}$ is the speed

r is the radius of the circle

T is the period of motion

tlvd-8950

SAMPLE PROBLEM 11 Determining the magnitude and direction of the acceleration and the force of an object moving in a circular motion

A car is driven around a roundabout at a constant speed of 20 km h^{-1} (5.6 m s^{-1}). The roundabout has a radius of 3.5 m and the car has a mass of 1200 kg.

a. **What is the magnitude and direction of the acceleration of the car?**

b. **What is the magnitude and direction of the net force on the car?**

THINK	WRITE
a. 1. List the known information.	a. $v = 5.6\ \text{m s}^{-1}$, $r = 3.5\ \text{m}$
2. Calculate the acceleration.	$a = \frac{v^2}{r}$ $= \frac{(5.6)^2}{3.5}$ $= 9.0\ \text{m s}^{-2}$ The car accelerates at $9.0\ \text{m s}^{-2}$ towards the centre of the roundabout.
b. 1. There are two different formulas that can be used to calculate this answer. i. Use the answer to (a) and substitute into $F_{net} = ma$.	b. $a = 9.0\ \text{m s}^{-2}$, $m = 1200\ \text{kg}$ $F_{net} = ma$ $= 1200 \times 9.0$ $= 1.1 \times 10^4\ \text{N}$
ii. Use the formula $F_{net} = \frac{mv^2}{r}$.	$v = 5.6\ \text{m s}^{-1}$, $r = 3.5\ \text{m}$, $m = 1200\ \text{kg}$ $F_{net} = \frac{mv^2}{r}$ $= \frac{1200\,(5.6)^2}{3.5}$ $= 1.1 \times 10^4\ \text{N}$ Both methods give the force on the car as 1.1×10^4 N towards the centre of the roundabout.

PRACTICE PROBLEM 11

Kwong (mass 60 kg) rides the Gravitron at the amusement park. This ride moves Kwong in a circle of radius 3.5 m, at a rate of one rotation every 2.5 s.

a. **What is Kwong's acceleration?**
b. **What is the net force acting on Kwong? (Include a magnitude and a direction.)**
c. **Draw a labelled diagram showing all the forces acting on Kwong.**

INVESTIGATION 1.3

eog-1696

evd-0238

Exploring circular motion

Aim

To examine some of the factors affecting the motion of an object undergoing uniform circular motion, and then to determine the quantitative relationship between the variables of force, velocity and radius

1.5.5 Forces that produce centripetal acceleration

Whenever an object is in uniform circular motion, the net force on that object must be towards the centre of the circle. Some examples of situations involving forces producing centripetal acceleration follow.

Tension

The force applied by an object that is being pulled or stretched can be referred to as a tension force.

FIGURE 1.30 a. Tension contributes to the net force in many amusement park rides. **b.** The net force acting on a compartment in the ride

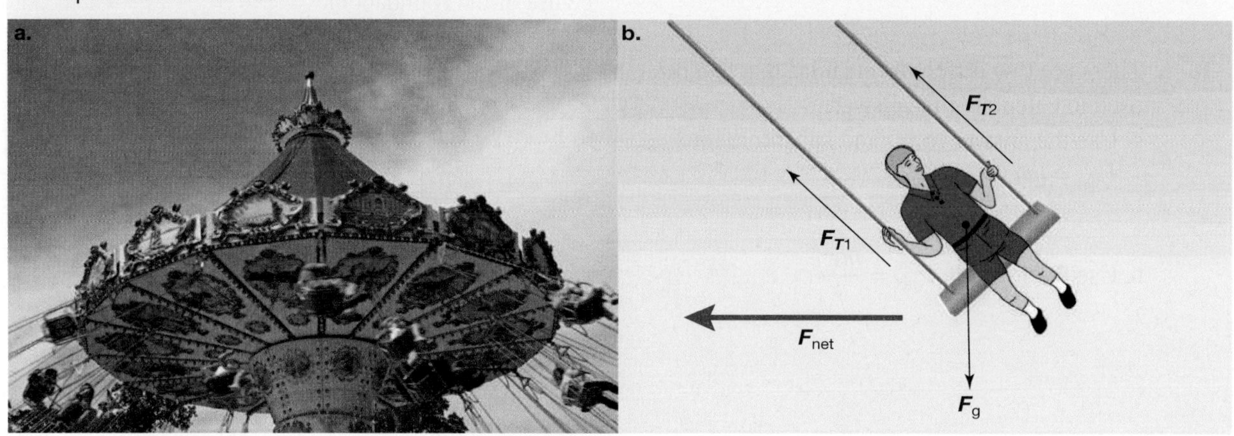

FIGURE 1.31 A component of the tension is the net force acting on the female skater when she is performing a 'death roll'.

F_T

F_N

F_{net}

F_g

Friction

When a car rounds a corner, the sideways frictional forces contribute to the net force. The forward frictional forces by the ground on the tyres keep the car moving, but if the sideways frictional forces are not sufficient, the net force on the car will not be towards the centre of the curve. Since the net force is less than the force required to keep the car moving in a circle at this radius, the car will not make it around the corner but move sideways!

FIGURE 1.32 The sideways frictional forces of the ground on the tyres enable a car to move around a corner.

The formula $F_{net} = \frac{mv^2}{r}$ shows that as the velocity increases, the force needed to move in a circle greatly increases $(F_{net} \propto v^2)$. This is why it is vital that cars do not attempt to corner while travelling too fast.

Track athletes, cyclists and motorcyclists also rely on sideways frictional forces to enable them to manoeuvre around corners. They often lean into corners to increase the size of the sideways frictional forces, to turn more quickly. Leaning also means that they are pushing on the surface at an angle, so the reaction force is no longer normal to the ground (Newton's Third Law). It has an upward component, the normal force F_N, which balances the force due to gravity, F_g, and a horizontal component towards the centre of their circular motion due to the frictional force F_f.

FIGURE 1.33 Leaning into a corner increases the size of the net force, allowing a higher speed while cornering. Leaning sideways induces a sideways frictional force, resulting in the horizontal net force experienced by the bicycle.

In velodromes, the track is banked so that a component of the normal force acts towards the centre of the velodrome, thus increasing the net force in this direction. As the centripetal force is larger, the cyclists can move around the corners faster than if they had to rely on friction alone.

Going around a bend

When a vehicle travels around a bend, or curve, at constant speed, its motion can be considered to be part of a circular motion. The curve makes up the arc of a circle. For a car to travel around a corner safely, the net force acting on it must be towards the centre of the circle.

Figure 1.34a shows the forces acting on a vehicle of mass m travelling around a curve with a radius r at a constant speed $\boldsymbol{v}$. The forces acting on the car are the force due to gravity, $\boldsymbol{F}_g$, friction, $\boldsymbol{F}_f$, and the normal force, $\boldsymbol{F}_N$.

FIGURE 1.34 a. For the vehicle to take the corner safely, the net force must be towards the centre of the circle. **b.** Banking the road allows a component of the normal force to contribute to the centripetal force. Note that the forces have been drawn as though they were acting through the centre of mass.

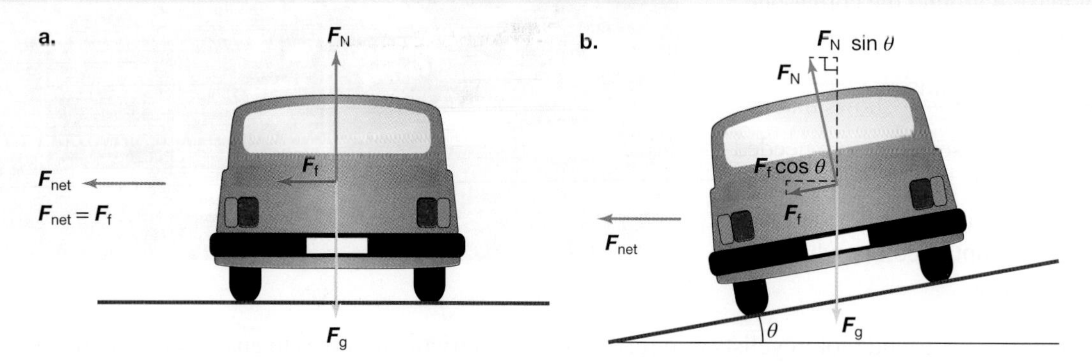

On a level road, the only force with a component towards the centre of the circle is the 'sideways' friction. This sideways friction makes up the whole of the magnitude of the net force on the vehicle. That is:

$$\boldsymbol{F}_{net} = \text{sideways friction} = \frac{mv^2}{r}$$

If you drive the vehicle around the curve with a speed so that $\frac{mv^2}{r}$ is greater than the sideways friction, the motion is no longer circular and the vehicle will skid off the road. If the road is wet, sideways friction is less and a lower speed is necessary to drive safely around the curve.

CASE STUDY: Calculating the net force on a banked road

If the road is banked at an angle θ towards the centre of the circle, a component of the normal force, $\boldsymbol{F}_N \sin\theta$, can also contribute to the net force, which acts in the horizontal direction. This is shown in figure 1.34b.

$$\boldsymbol{F}_{net} = \boldsymbol{F}_f \cos\theta + \boldsymbol{F}_N \sin\theta$$

The larger net force means that, for a given curve, banking the road makes a higher speed possible.

Loose gravel on bends in roads is dangerous because it reduces the sideways friction force. At low speeds this is not a problem, but a vehicle travelling at high speed is likely to lose control and run off the road in a straight line.

Cycling velodromes are steeply banked (often up to 40°), allowing cyclists to achieve very fast speeds. When engineers design velodromes (and other banked tracks, such as banked roadways) they need to consider the speed at which the force due to friction becomes zero. This is dependent on the angle at which the track is banked. The speed at which the force due to friction becomes zero is called the **design speed** and means that frictional force is not required to keep the vehicle on the track. As shown in figure 1.35, only the horizontal component of the normal force is contributing to the net force and there is no frictional force acting sideways.

design speed speed at which the force due to friction becomes zero as seen on a banked track

The equation for net force can be simplified as follows:

$$\begin{aligned} \boldsymbol{F}_{\text{net}} &= \boldsymbol{F}_{\text{f}}\cos\theta + \boldsymbol{F}_{\text{N}}\sin\theta \\ &= 0 + \boldsymbol{F}_{\text{N}}\sin\theta \\ &= \boldsymbol{F}_{\text{N}}\sin\theta \end{aligned}$$

From figure 1.35, it can also be shown that:

$$\boldsymbol{F}_{\text{g}} = \boldsymbol{F}_{\text{N}}\cos\theta$$

$$\boldsymbol{F}_{\text{net}} = \boldsymbol{F}_{\text{g}}\tan\theta$$

FIGURE 1.35 **a.** When travelling in circular motion at the design speed, the force due to friction is zero, so the only forces acting on the object are the force due to gravity and the normal force. **b.** The vector addition of the force due to gravity and the normal force gives the net force as acting horizontally towards the centre of the circle.

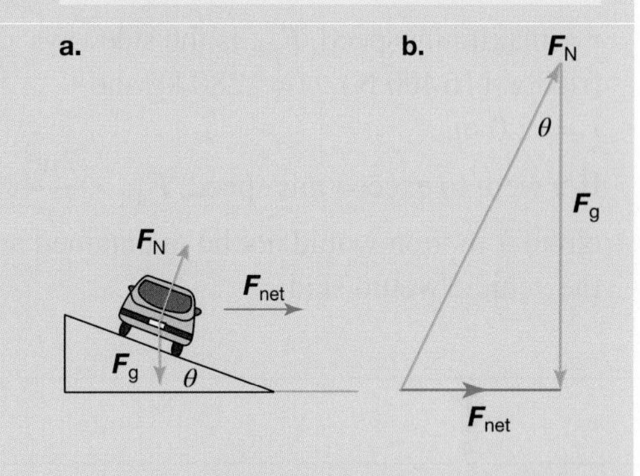

These equations can be used to determine the angle a road needs to be banked at to achieve a certain design speed, or if the angle is known, the design speed can be determined.

Recall that $\boldsymbol{F}_{\text{net}} = \frac{m\boldsymbol{v}^2}{r}$ and $\boldsymbol{F}_{\text{g}} = m\boldsymbol{g}$. If the following is used in substitution, $\tan\theta = \frac{\frac{mv^2}{r}}{m\boldsymbol{g}}$, then:

$$\tan\theta = \frac{\boldsymbol{v}^2}{r\boldsymbol{g}}$$

$$\Rightarrow \theta = \tan^{-1}\left(\frac{\boldsymbol{v}^2}{r\boldsymbol{g}}\right)$$

When travelling at the design speed, where the frictional force will be zero, the angle of the bank can be found by using:

$$\tan\theta = \frac{\boldsymbol{F}_{\text{net}}}{\boldsymbol{F}_{\text{g}}} = \frac{v^2}{r\boldsymbol{g}}$$

$$\theta = \tan^{-1}\left(\frac{v^2}{r\boldsymbol{g}}\right)$$

where: θ is the angle of the bank

$\boldsymbol{F}_{\text{net}}$ is the net force

$\boldsymbol{F}_{\text{g}}$ is the force due to gravity

v is the speed

r is the radius of the track

$\boldsymbol{g}$ is the acceleration due to gravity

The equation can be arranged so that the design speed can be determined if the bank angle is known.

The design speed can be found by using:

$$v^2 = r\boldsymbol{g}\tan\theta$$

$$v = \sqrt{r\boldsymbol{g}\tan\theta}$$

tlvd-8951

SAMPLE PROBLEM 12 Calculating the maximum constant speed of a car with sideways wheel friction

A car of mass 1280 kg travels around a bend with a radius of 12.0 m. The total sideways friction on the wheels is 16 400 N. The road is not banked. Calculate the maximum constant speed at which the car can be driven around the bend without skidding off the road.

THINK	WRITE
The car will maintain the circular motion around the bend if: $F_{net} = \frac{mv^2}{r}$ where v = maximum speed, F_{net} is the sideways friction (16 400 N), m = 1280 kg and r = 12.0 m. If v were to exceed this speed, $F_{net} < \frac{mv^2}{r}$, the circular motion would not be maintained and the vehicle would skid.	$F_{net} = 1280 \times \frac{v^2}{12.0}$ $v^2 = 16\,400 \times \frac{12.0}{1280}$ $= 153.75\ \text{m}^2\ \text{s}^{-2}$ $\Rightarrow v = 12.4\ \text{m s}^{-1}$ The maximum constant speed at which the vehicle can be driven around the bend is $12.4\ \text{m s}^{-1}$.

PRACTICE PROBLEM 12

A motorcyclist is travelling around a circular track at a constant speed of 30 m s^{-1}. The surface is flat and horizontal. The radius of the track is 100 m. The mass of the cyclist with her motorbike is 200 kg. What is the net force experienced by the rider and her bike?

tlvd-8952

SAMPLE PROBLEM 13 Calculating the maximum constant speed of a car on a banked road

Calculate the maximum speed of the car in sample problem 12 (without skidding off the road) if the road is banked at an angle of 10° to the horizontal.

THINK	WRITE
1. Draw a diagram to represent the known information.	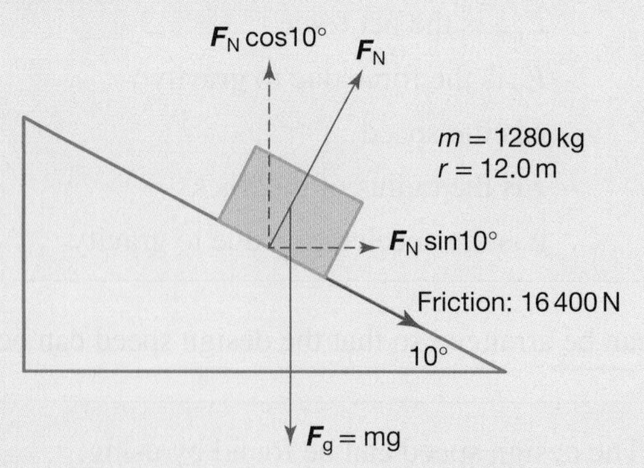

2. The vertical forces are balanced.

$$F_N \cos 10° = 16\,400 \sin 10° + 1280 \times 9.8$$
$$F_N = \frac{15\,392}{\cos 10°}$$
$$= 15\,629 \text{ N}$$

3. The net force is equal to the sum of the horizontal forces.

$$F_{net} = \frac{mv^2}{r}$$
$$F_N \sin 10° + 16\,400 \cos 10° = \frac{1280 \times v^2}{12}$$
$$15\,629 \sin 10° + 16\,400 \cos 10° = \frac{1280 \times v^2}{12}$$
$$\Rightarrow v = 13.3 \text{ m s}^{-1}$$

PRACTICE PROBLEM 13

A cyclist is training at her local velodrome. The velodrome has a radius of 25 m and she is travelling at 8.0 m s^{-1}. The total mass of the cyclist and her bike is 80 kg. The velodrome track is banked at an angle that results in there being no sideways frictional force on the bike's wheels by the track. Calculate the angle at which the track is banked for there to be no sideways frictional force.

1.5.6 Inside circular motion

What happens to people and objects inside larger objects that are travelling in circles? The answer to this question depends on several factors.

Consider passengers inside a bus. The sideways frictional forces by the road on the bus tyres act towards the centre of the circle, which increases the net force on the bus and keeps the bus moving around the circle. If the passengers are also to move in a circle with the bus (therefore keeping the same position in the bus) they must also have a net force towards the centre of the circle. Without such a force, they would continue to move in a straight line and probably hit the side of the bus! Usually the friction between the seat and a passenger's body is sufficient to prevent this happening.

However, if the bus is moving quickly, friction alone may not be adequate. In such cases, passengers may grab hold of the seat in front, thus adding a force through their arms. Hopefully, the sum of the frictional force of the seat on a passenger's legs and the horizontal component of force through the passenger's arms will provide a large enough force to keep that person moving in the same circle as the bus!

tlvd-8953

SAMPLE PROBLEM 14 Calculating the angle of an object inside circular motion

When travelling around a roundabout, John notices that the fluffy dice suspended from his rear-vision mirror swing out. If John is travelling at 8.0 m s^{-1} and the roundabout has a radius of 5.0 m, what angle will the string connected to the fluffy dice (mass 100 g) make with the vertical?

THINK

WRITE

1. When John enters the roundabout, the dice, which are hanging straight down, will begin to move outwards. As long as John maintains a constant speed, they will reach a point at which they become stationary at some angle to the vertical. At this point, the net force on the dice is the centripetal force. Because the dice appear stationary to John, they must be moving in the same circle, with the same speed, as John and his car.

$v = 8.0$ m s^{-1}, $r = 5.0$ m, $m = 0.100$ kg

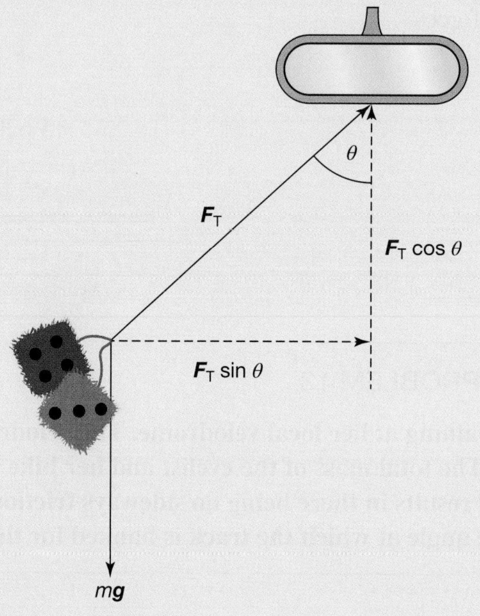

2. Consider the vertical components of the forces. The acceleration has no vertical component.

$$mg = F_T \cos\theta$$
$$\Rightarrow F_T = \frac{mg}{\cos\theta} \ldots (1)$$

3. Consider the horizontal components of the forces.

$$F_{net} = \frac{mv^2}{r} = F_T \sin\theta$$
$$\Rightarrow \frac{mv^2}{r} = F_T \sin\theta \ldots (2)$$

4. To solve the simultaneous equations, substitute for T (from equation (1) into equation (2)).

$$\frac{mv^2}{r} = \frac{mg}{\cos\theta} \times \sin\theta$$
$$= mg\tan\theta$$
$$\Rightarrow \tan\theta = \frac{v^2}{rg}$$
$$= \frac{(8.0)^2}{5.0 \times 9.8}$$
$$\Rightarrow \theta = 53°$$

PRACTICE PROBLEM 14

A 50 kg circus performer grips a vertical rope with her teeth and sets herself moving in a circle with a radius of 5.0 m at a constant horizontal speed of 3.0 m s^{-1}.

a. What angle does the rope make with the vertical?

b. What is the magnitude of the tension in the rope?

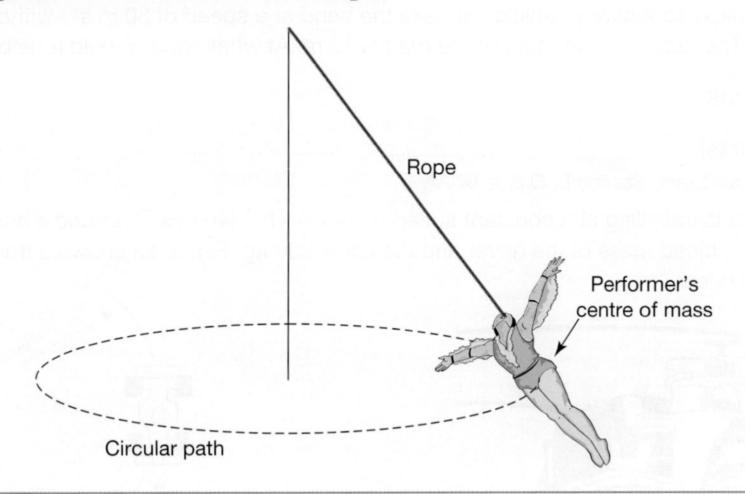

1.5 Activities

learnon

Students, these questions are even better in jacPLUS

Receive immediate feedback and access sample responses

Access additional questions

Track your results and progress

Find all this and MORE in jacPLUS

1.5 Quick quiz | 1.5 Exercise | 1.5 Exam questions

1.5 Exercise

1. A 65-kg jogger runs around a circular track of radius 120 m with an average speed of 6.0 km h^{-1}.
   a. What is the centripetal acceleration of the jogger?
   b. What is the net force acting on the jogger?
2. At a children's amusement park, the miniature train ride completes a circuit of radius 350 m, maintaining a constant speed of 15 km h^{-1}.
   a. What is the centripetal acceleration of the train?
   b. What is the net force acting on a 35-kg child riding on the train?
   c. What is the net force acting on the 1500-kg train?
   d. Explain why the net forces acting on the child and the train are different and yet the train and the child are moving along the same path.
3. Explain why motorcyclists lean into bends.
4. A rubber stopper of mass 50.0 g is whirled in a horizontal circle on the end of a 1.50-m length of string. The time taken for ten complete revolutions of the stopper is 8.00 s. The string makes an angle of 6.03 with the horizontal. Calculate the following:
   a. the speed of the stopper
   b. the centripetal acceleration of the stopper
   c. the net force acting on the stopper
   d. the magnitude of the tension in the string.

5. Carl is riding around a corner on his bike at a constant speed of 15 km h^{-1}. The corner approximates part of a circle of radius 4.5 m. The combined mass of Carl and his bike is 90 kg. Carl keeps the bike in a vertical plane.
   a. What is the net force acting on Carl and his bike?
   b. What is the sideways frictional force acting on the tyres of the bike?
   c. Carl rides onto a patch of oil on the road; the sideways frictional forces are now 90% of their original amount. If Carl maintains a constant speed, what will happen to the radius of the circular path he is taking?
6. A road is to be banked so that any vehicle can take the bend at a speed of 30 m s^{-1} without having to rely on sideways friction. The radius of curvature of the road is 12 m. At what angle should it be banked?

1.5 Exam questions

Question 1 (5 marks)

Source: *VCE 2022 Physics Exam, Section B, Q.8;* © VCAA

A Formula 1 racing car is travelling at a constant speed of 144 km h^{-1} (40 m s^{-1}) around a horizontal corner of radius 80.0 m. The combined mass of the driver and the car is 800 kg. Figure 8a shows a front view and Figure 8b shows a top view.

Figure 8a – Front view

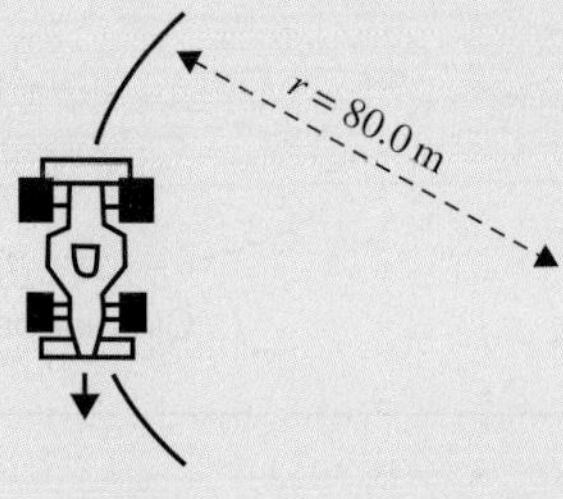

Figure 8b – Top view

a. Calculate the magnitude of the net force acting on the racing car and driver as they go around the corner. **(2 marks)**
b. On Figure 8b, draw the direction of the net force acting on the racing car using an arrow. **(1 mark)**
c. Explain why the racing car needs a net horizontal force to travel around the corner and state what exerts this horizontal force. **(2 marks)**

Question 2 (4 marks)

Source: *VCE 2018 Physics Exam, Section B, Q.10;* © VCAA

Members of the public can now pay to take zero gravity flights in specially modified jet aeroplanes that fly at an altitude of 8000 m above Earth's surface. A typical trajectory is shown in Figure 12. At the top of the flight, the trajectory can be modelled as an arc of a circle.

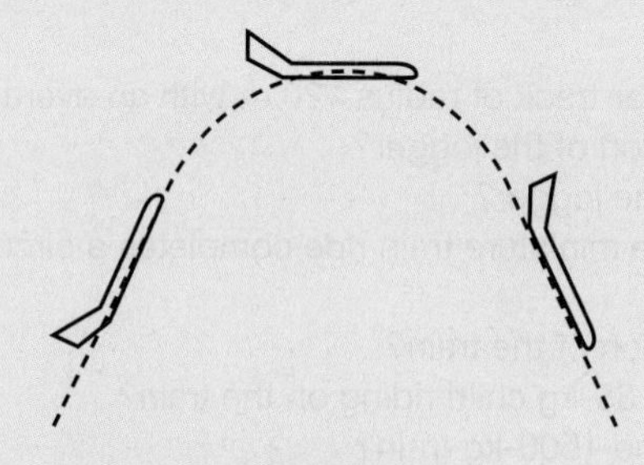

Figure 12

a. Calculate the radius of the arc that would give passengers zero gravity at the top of the flight if the jet is travelling at 180 m s^{-1}. Show your working. **(2 marks)**
b. Is the force of gravity on a passenger zero at the top of the flight? Explain what 'zero gravity experience' means. **(2 marks)**

Question 3 (7 marks)

Source: *VCE 2022 Physics Exam, NHT, Section B, Q.7*; © VCAA

A spherical mass of 2.0 kg is attached to a piece of string with a length of 2.0 m. The spherical mass is pulled back until it makes an angle of 60° with the vertical, as shown in Figure 4. The spherical mass is then released. Ignore the mass of the string.

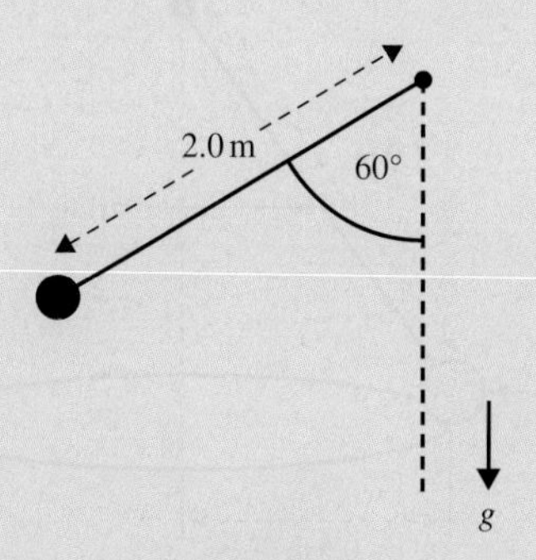

Figure 4

a. Show that the maximum speed of the spherical mass is 4.4 m s^{-1}. **(2 marks)**

b. At what part of its path is the spherical mass at its maximum speed? Explain your reasoning. **(2 marks)**

c. Calculate the maximum tension in the string. **(3 marks)**

Question 4 (4 marks)

Source: *VCE 2017, Physics Exam, Section B, Q.7*; © VCAA

A bicycle and its rider have a total mass of 100 kg and travel around a circular banked track at a radius of 20 m and at a constant speed of 10 m s^{-1}, as shown in the figure. The track is banked so that there is no sideways friction force applied by the track on the wheels.

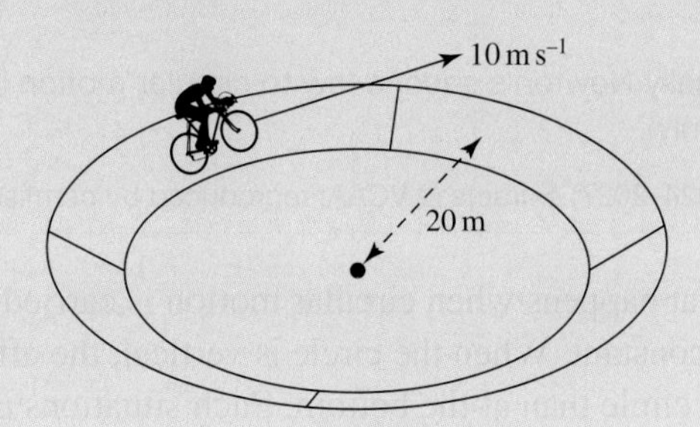

a. On the diagram below, draw all of the forces on the rider and the bicycle, considered as a single object, as arrows. Draw the net resultant force as a dashed arrow labelled F_{net}. **(2 marks)**

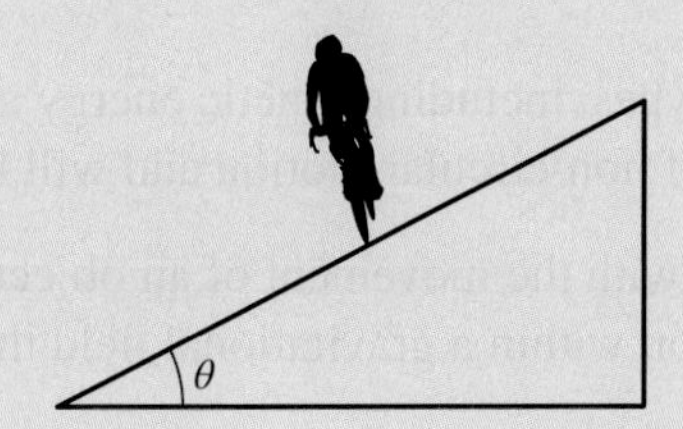

b. Calculate the correct angle of bank for there to be no sideways friction force applied by the track on the wheels. Show your working. **(2 marks)**

Question 5 (5 marks)

Source: *Adapted from VCE 2016, Physics Exam, Section A, Q.2; © VCAA*

A steel ball of mass 2.0 kg is swinging in a circle of radius 0.50 m at a constant speed of 1.7 m s^{-1} at the end of a string of length 1.0 m, as shown in the figure.

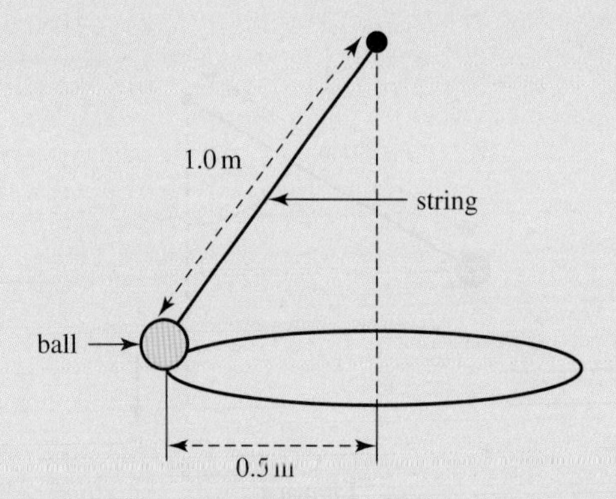

a. On the figure, draw all the forces **on the ball**. Label all forces. Draw the resultant force as a dotted line labelled $\boldsymbol{F}_R$. **(2 marks)**

b. Calculate the tension in the string. Show your working. **(3 marks)**

More exam questions are available in your learnON title.

1.6 Non-uniform circular motion

KEY KNOWLEDGE

- Investigate and apply theoretically Newton's second law to circular motion in a vertical plane (forces at the highest and lowest positions only).

Source: VCE Physics Study Design (2024–2027) extracts © VCAA; reproduced by permission.

So far, we have considered only what happens when circular motion is carried out at a constant speed. However, in many situations the speed is not constant. When the circle is vertical, the effects of gravity can cause the object to go slower at the top of the circle than at the bottom. Such situations can be examined either by analysing the energy transformations that take place or by applying Newton's laws of motion.

1.6.1 Energy review

Energy can be classified into different types, including kinetic energy and gravitational potential energy. These concepts will help in the investigation of non-circular motion and will be explored further in topic 2.

Kinetic energy is the energy associated with the movement of an object. Gravitational potential energy is the energy an object has based on its position within a gravitational field that can cause work to be done on it.

The Law of Conservation of Energy states that energy cannot be created or destroyed, only converted from one form to another. When motion is in a vertical circle you will need to consider that energy is converted from gravitational potential energy to kinetic energy.

$$E_g = mg\Delta h$$

$$E_k = \frac{1}{2}mv^2$$

where: E_g is the gravitational potential energy

E_k is the kinetic energy

m is the mass

g is the acceleration due to gravity

Δh is the change in height, or the height above a reference point

v is the velocity

1.6.2 Uniform horizontal motion

If a person is sitting in a car moving in a straight line at a constant speed, the force due to gravity on them by the Earth balances the normal force from the seat.

FIGURE 1.37 The forces acting on an object in uniform horizontal motion

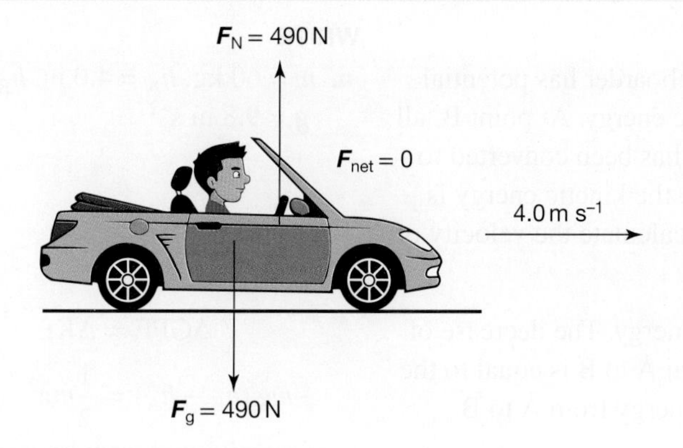

1.6.3 Travelling through dips

When a skateboarder enters a half-pipe from the top, that person has a certain amount of gravitational potential energy, but a velocity, and hence kinetic energy, close to zero. At the bottom of the half-pipe, most of the gravitational potential energy of the skateboarder has been transformed into kinetic energy. As long as the person's change in height is known, it is possible to calculate the speed at that point.

FIGURE 1.38 Forces acting on the skateboarder at the bottom of a dip. The normal force is greater than the force due to gravity; the skateboarder feels 'heavier'.

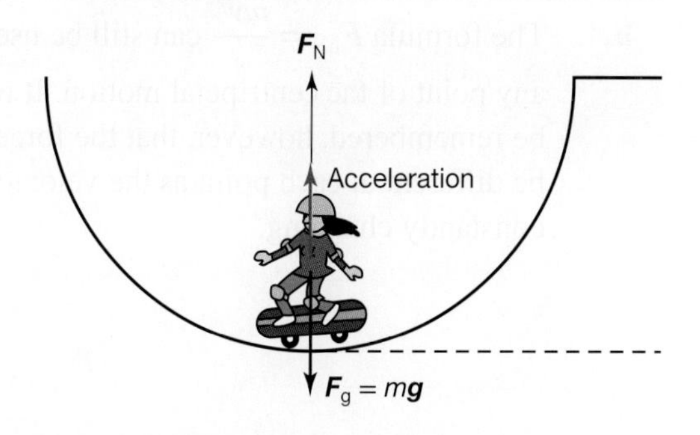

At the bottom of the half-pipe, the normal force acting on the skateboarder is greater than the force due to gravity, causing the skateboarder to feel 'heavier' than usual. The net force acting on the skateboarder is given by $\boldsymbol{F}_{net} = m\boldsymbol{a} = \boldsymbol{F}_N + \boldsymbol{F}_g = \boldsymbol{F}_N - mg$ (taking the upward direction as positive). In this case, the normal force is greater than the force due to gravity. The net force, and hence the acceleration, is directed upwards towards the centre of the circle.

For circular motion, the acceleration is centripetal and is given by the expression $\frac{v^2}{r}$:

$$\frac{mv^2}{r} = F_N - mg$$

tlvd-8954

SAMPLE PROBLEM 15 Calculating the speed and forces acting on an object travelling through a dip

A skateboarder, with an initial velocity of 0 m s^{-1} and a mass of 60 kg, enters the half-pipe at point A, as shown in the figure. Assume the frictional forces are negligible.

a. **What is the skateboarder's speed at point B?**

b. **What is the net force on the skateboarder at B?**

c. **What is the normal force on the skateboarder at B? Explain whether the skateboarder feel lighter or heavier than usual.**

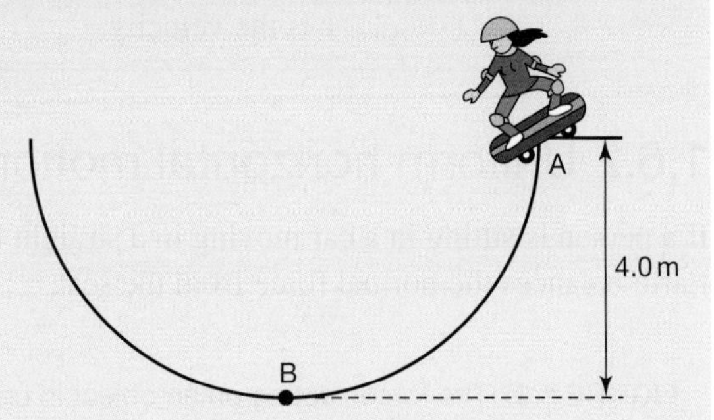

THINK	WRITE
a. 1. At point A, the skateboarder has potential energy but no kinetic energy. At point B, all the potential energy has been converted to kinetic energy. Once the kinetic energy is known, it is easy to calculate the velocity of the skateboarder.	a. $m = 60$ kg, $h_A = 4.0$ m, $h_B = 0.0$ m, $g = 9.8$ m s^{-2}
2. Find the change in energy. The decrease of potential energy from A to B is equal to the increase of kinetic energy from A to B.	$\Delta GPE = \Delta KE$ $-mg\,(h_B - h_A) = \frac{1}{2}mv^2$ Cancelling m from both sides: $-g\,(h_B - h_A) = \frac{1}{2}v^2$ $-9.8\,(0 - 4.0) = \frac{1}{2}v^2$ $\Rightarrow v^2 = 78.4$ $\Rightarrow v = 8.9$ m s^{-1} The skateboarder's speed at B is 8.9 m s^{-1}.
b. The formula $F_{net} = \frac{mv^2}{r}$ can still be used for any point of the centripetal motion. It must be remembered, however, that the force will be different at each point as the velocity is constantly changing.	b. $m = 60$ kg, $r = 4.0$ m, $v = 8.9$ m s^{-1} $F_{net} = \frac{mv^2}{r}$ $= \frac{60 \times (8.9)^2}{4.0}$ $\approx 1.2 \times 10^3$ N The net force acting on the skateboarder at point B is 1200 N upwards.

c. 1. As there is more than one force acting on the skateboarder, it helps to draw a diagram.

c.

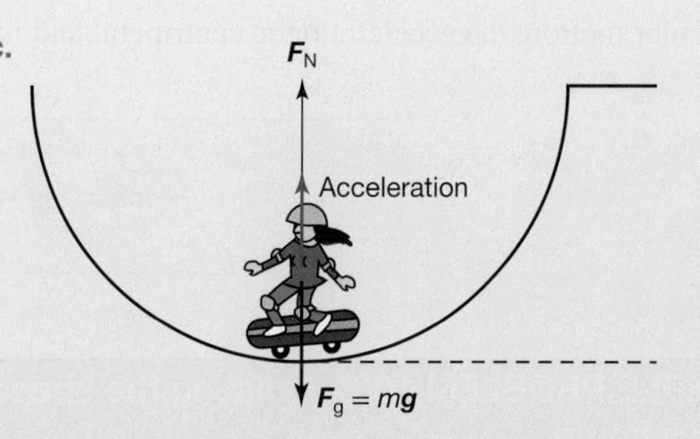

2. The net force is determined by adding together all the forces acting on the skateboarder.

$$F_{net} = F_N + F_g$$

3. Taking the upward direction as positive.

$$F_{net} = F_N - mg$$
$$\Rightarrow F_N = F_{net} + mg$$
$$= 1.2 \times 10^3 + 60 \times 9.8$$
$$\approx 1.8 \times 10^3 \text{ N}$$

The normal force acting on the skateboarder at point B is 1.8×10^3 N upwards. This is larger than the normal force if the skateboarder was stationary. This causes the skateboarder to experience a sensation of heaviness.

PRACTICE PROBLEM 15

A roller-coaster car travels through the bottom of a dip of radius 9.0 m at a speed of 13 m s^{-1}.

a. **What is the net force on a passenger of mass 60 kg?**

b. **What is the normal force on the passenger by the seat?**

c. **Compare the size of the normal force to the force due to gravity and comment on how the passenger would feel.**

1.6.4 Travelling over humps

The experience of 'heaviness' described in the previous section, when the normal force is greater than the force due to gravity, occurs on a roller-coaster when the roller-coaster car travels through a dip at the bottom of a vertical arc. When the car is at the top of a vertical arc, the passengers experience a feeling of being 'lighter'. How can this be explained?

FIGURE 1.39 The forces acting on a roller-coaster car at the top of a hump

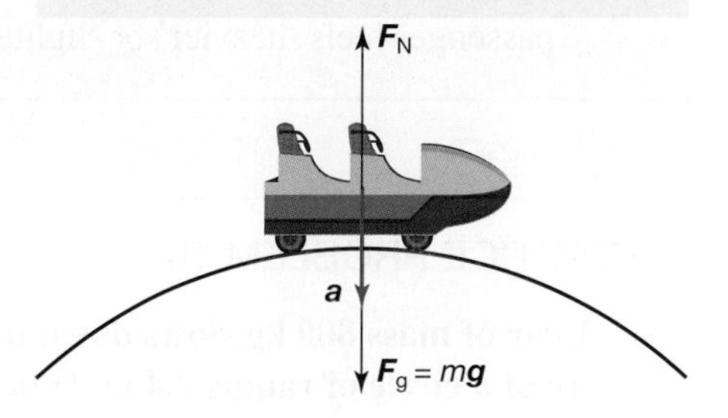

When the roller-coaster car is on the top of the track, the normal force is upwards, and the force due to gravity and the net force are downwards. Taking the upward direction as positive, $F_{net} = ma = \frac{mv^2}{r} = F_g - F_N$. This clearly shows that F_g is larger than F_N, and hence the passenger will feel 'lighter'.

For circular motion, the acceleration is centripetal and is given by the expression $\frac{v^2}{r}$.

$$ma = mg - F_N = \frac{mv^2}{r}$$

tlvd-8955

SAMPLE PROBLEM 16 Calculating the speed and forces acting on an object travelling over a hump

A passenger is in a roller-coaster car at the top of a circular arc of radius 9.0 m.

a. At what speed would the normal force on the passenger equal half the force due to gravity?

b. What happens to the normal force as the speed increases?

c. What would the passenger experience?

THINK	WRITE
a. 1. Write the known information.	a. $F_N = \frac{mg}{2}$, $r = 9.0$ m
2. Calculate the speed using $\frac{mv^2}{r} = mg - F_N$.	$\frac{mv^2}{r} = mg - \frac{mg}{2}$ $\frac{v^2}{r} = \frac{g}{2}$ $\Rightarrow v = \sqrt{\frac{gr}{2}}$ $= \sqrt{\frac{9.8 \times 9.0}{2}}$ $= 6.6 \text{ m s}^{-1}$
b. 1. Rearrange $\frac{mv^2}{r} = mg - F_N$ to make F_N the subject of the equation.	b. Rearranging $\frac{mv^2}{r} = mg - F_N$ gives $F_N = mg - \frac{mv^2}{r}$
2. Comment on the effect of increasing v on F_N.	The force due to gravity, mg, is constant, so as the speed, v, increases, the normal force, F_N, gets smaller.
c. The normal force determines whether the passenger feels 'heavier' or 'lighter'.	c. The normal force is less than the force due to gravity, so the passenger will feel lighter.

PRACTICE PROBLEM 16

a. A car of mass 800 kg slows down to a speed of 4.0 m s^{-1} to travel over a speed hump that forms the arc of a circle of radius 2.4 m. What normal force acts on the car at the top of the speed hump?

b. At what minimum speed would a car of mass 1000 kg have to travel to momentarily leave the road at the top of the speed hump described in part (a)? (To leave the road, the normal force would have to decrease to zero.)

The normal force is a push by the track on the wheels of the roller-coaster car. The track can only push up on the wheels; it cannot pull down on the wheels to provide a downward force. So as the speed increases, there is a limit on how small the normal force can be. The smallest value is zero. What would the passenger feel? And what is happening to the roller-coaster car?

When the normal force is zero, the passenger will feel as if they are floating just above the seat. They will feel no compression in the bones of their backside. At this point the car has lost contact with the track. Any attempt to put on the brakes will not slow down the car, as the frictional contact with the track depends on the size of the normal force. No normal force means no friction.

Modern roller-coaster cars have two sets of wheels, one set above the track and one set below the track, so that if the car is moving too fast, the track can supply a downward normal force on the lower set of wheels.

The safety features of roller-coasters cannot be applied to cars on the road. If a car goes too fast over a hump on the road, the situation is potentially very dangerous. Loss of contact with the road means that turning the steering wheel to avoid an obstacle or an oncoming car will have no effect whatsoever. The car will continue on in the same direction.

tlvd-8956

SAMPLE PROBLEM 17 Determining the speed and forces acting on a toy car inside a loop

A toy car travels through a vertical loop of radius 15.0 cm on a racetrack. The toy car, of mass 200 g, is released from rest at point A, which is 2.00 m above the lowest point on the track. The car rolls down the track and travels inside the loop. Friction can be ignored.

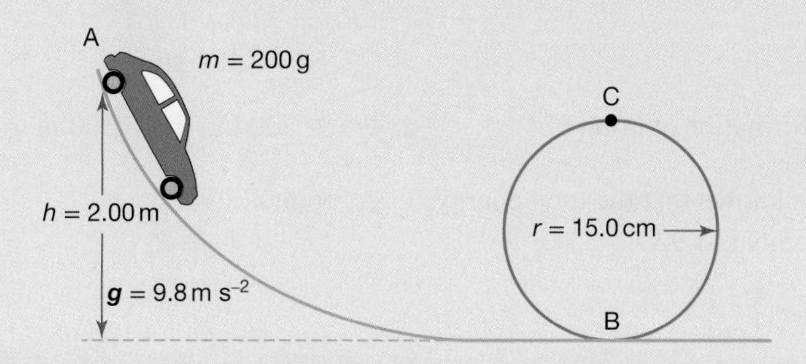

a. **Calculate the speed of the car at point B, the bottom of the loop.**
b. **What is the net force on the toy car at point B?**
c. **What is the normal force on the car at point B?**
d. **What is the speed of the car when it reaches point C?**
e. **What is the normal force on the car at point C?**

THINK	WRITE
a. 1. List all known information at points A and B.	a. At point A, $m = 0.200$ kg, $h = 2.00$ m, $v = 0$ m s^{-1}, $g = 9.8$ m s^{-2} At point B, $m = 0.2$ kg, $h = 0.00$ m, $g = 9.8$ m s^{-2}
2. Calculate the total energy of the car at point A by adding the car's gravitational potential energy and kinetic energy.	$E_A = E_k + E_g$ $= \frac{1}{2}mv^2 + mg\Delta h$ $= 0 + (0.200 \times 9.8 \times 2.00)$ $= 3.92$ J

3. The total energy of the car at point B is equal to the total energy of the car at point A. This is the Law of Conservation of Energy.

$$E_B = E_k + E_g$$
$$= \frac{1}{2}mv^2 + mg\Delta h$$
$$3.92 = \left(\frac{1}{2} \times 0.200 \times v^2\right) + 0$$
$$3.92 = 0.100v^2$$
$$v^2 = \frac{3.92}{0.100}$$
$$\Rightarrow v = \sqrt{39.2}$$
$$= 6.26 \text{ m s}^{-1}$$

b. During circular motion, the net force is always directed towards the centre of the circle. So, at point B, the net force will be directed upwards.

b. $$F_{net} = \frac{mv^2}{r}$$
$$= \frac{0.200 \times 6.26^2}{0.15}$$
$$= 78.4 \text{ N up}$$

c. The net force is equal to the sum of the normal force and the force due to gravity. Take the upward direction to be positive.

c. $$F_{net} = F_N + F_g$$
$$F_{net} = F_N - mg$$
$$78.4 = F_N - (0.200 \times 9.8)$$
$$78.4 = F_N - 1.96$$
$$F_N = 78.4 + 1.96$$
$$= 80.4 \text{ N up}$$

d. 1. List all known information at point C.

d. $m = 0.200$ kg, $h = 0.300$ m, $g = 9.8$ m s^{-2}

2. From part (a), it is known that the total energy of the car at all points is 3.92 J.

At point C,

$$E_C = E_k + E_g$$
$$= \frac{1}{2}mv^2 + mg\Delta h$$
$$3.92 = \left(\frac{1}{2} \times 0.200 \times v^2\right) + (0.200 \times 9.8 \times 0.300)$$
$$3.92 = 0.100v^2 + 0.588$$
$$0.100v^2 = 3.332$$
$$v^2 = \frac{3.332}{0.100}$$
$$\Rightarrow v = \sqrt{33.32}$$
$$= 5.77 \text{ m s}^{-1}$$

e. 1. Determine the net force acting on the car at point C. The net force on the car at point C will be directed downwards towards the centre of the circle.

e. $F_{net} = \frac{mv^2}{r}$

$= \frac{0.200 \times 5.77^2}{0.15}$

$= 44.4$ N down

2. The net force is equal to the sum of the normal force and the force due to gravity. Take the downward direction to be positive. *Note:* In this case, both the normal force and the force due to gravity act in the downward direction.

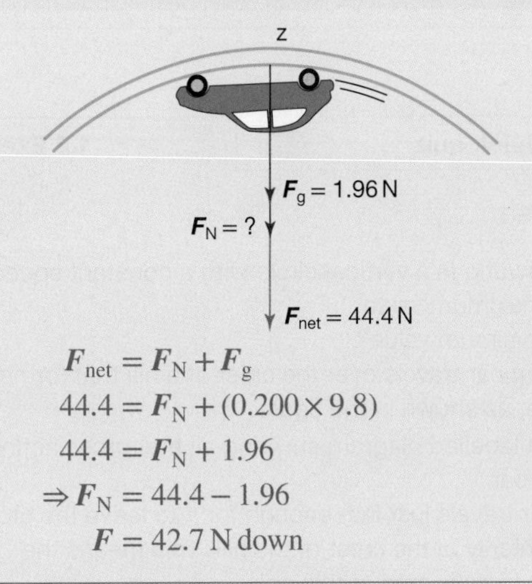

$$\begin{aligned} F_{net} &= F_N + F_g \\ 44.4 &= F_N + (0.200 \times 9.8) \\ 44.4 &= F_N + 1.96 \\ \Rightarrow F_N &= 44.4 - 1.96 \\ F &= 42.7 \text{ N down} \end{aligned}$$

PRACTICE PROBLEM 17

A toy car travels through a vertical loop of radius 20.0 cm on a racetrack. The toy car, of mass 100 g, is released from rest at point A, which is 1.30 m above the lowest point on the track. The car rolls down the track and travels inside the loop. Friction can be ignored.

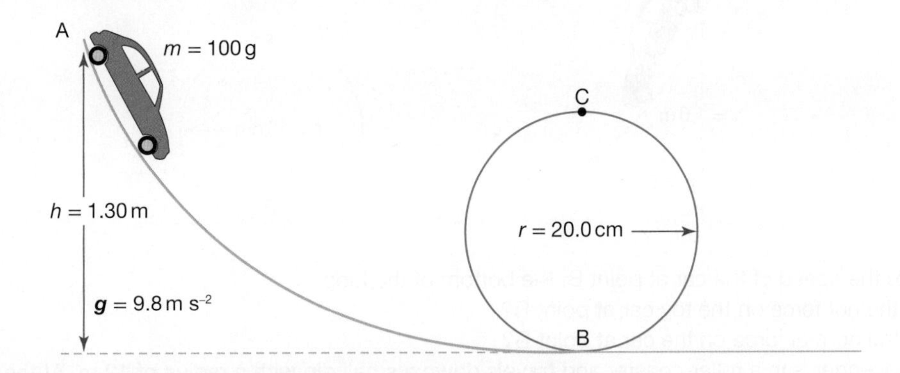

a. **Calculate the speed of the car at point B, the bottom of the loop.**
b. **What is the net force on the toy car at point B?**
c. **What is the normal force on the car at point B?**
d. **What is the speed of the car when it reaches point C?**
e. **What is the normal force on the car at point C?**

1.6 Activities

learn**on**

Students, these questions are even better in jacPLUS

Receive immediate feedback and access sample responses

Access additional questions

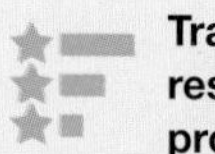
Track your results and progress

Find all this and MORE in jacPLUS

1.6 Quick quiz on	1.6 Exercise	1.6 Exam questions

1.6 Exercise

1. A ball is swung in a vertical circle with a constant speed. At which point is the tension force:
   a. at its maximum value
   b. at its minimum value?
2. An 800-kg car travels over the crest of a hill that forms the arc of a circle, as shown in the figure.
   a. Draw a labelled diagram showing all the forces acting on the car.
   b. The car travels just fast enough for it to leave the ground momentarily at the crest of the hill. This means the normal force is zero at this point.
      i. What is the net force acting on the car at this point?
      ii. What is the speed of the car at this point?

v

r = 4.0 m

3. A 120-g toy car travels through a vertical loop on a racetrack. The loop has a radius of 10 cm.
   The car is released from the start of the track, which is at a height of 1.0 m (position A), and travels inside the loop. Assume ***g*** is 9.8 m s^{-1} downwards and ignore friction.

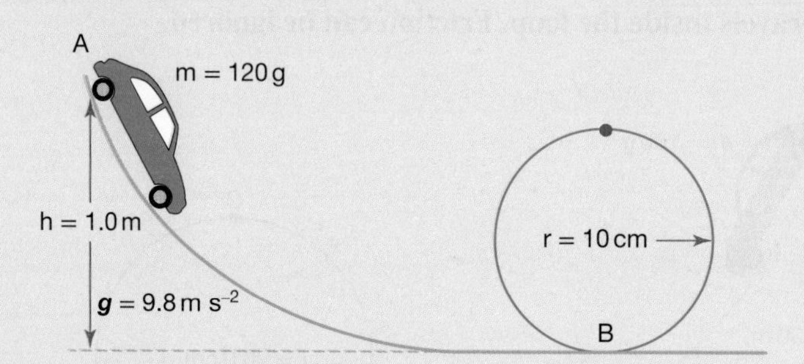

   a. Calculate the speed of the car at point B, the bottom of the loop.
   b. What is the net force on the toy car at point B?
   c. What is the normal force on the car at point B?
4. A 60-kg passenger is in a roller-coaster and travels down a small dip with a radius of 12 m. At the bottom of the dip, the passenger is travelling with a speed of 14 m s^{-1} and is feeling a larger than normal force. Use Newton's Second Law to calculate the normal force acting upon their body.
5. A 75-kg BMX rider is riding in a half-pipe with a radius of 2.5 m. At the lowest point of the half-pipe, the rider attains a speed of 7.0 m s^{-1}. Assume there is no air resistance or friction.
   a. What is the acceleration of the rider at the lowest point of the half-pipe?
   b. Determine the magnitude of the normal force acting on the rider at the lowest point of the half-pipe.

1.6 Exam questions

Question 1 (5 marks)

Source: *VCE 2022 Physics Exam, NHT, Section B, Q.9;* © VCAA

A small ball of mass 0.30 kg travels horizontally at a speed of 6 m s^{-1}. It enters a vertical circular loop of diameter 0.80 m, as shown in Figure 6. Assume that the radius of the ball and that the frictional forces are negligible.

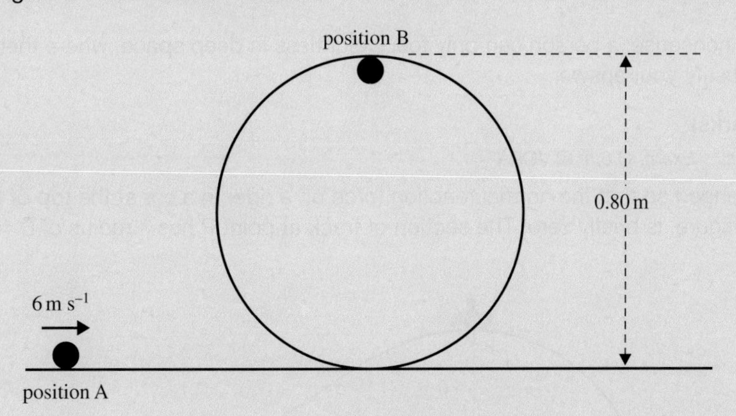

Figure 6

a. Show that the kinetic energy of the ball at position A in Figure 6 is 5.4 J. **(1 mark)**

b. Will the ball remain on the track at the top of the loop (position B in Figure 6)? Give your reasoning. **(4 marks)**

Question 2 (5 marks)

Source: *VCE 2018 Physics Exam, NHT, Section B, Q.8;* © VCAA

In an experiment, a ball of mass 2.5 kg is moving in a vertical circle at the end of a string, as shown in Figure 5. The string has a length of 1.5 m.

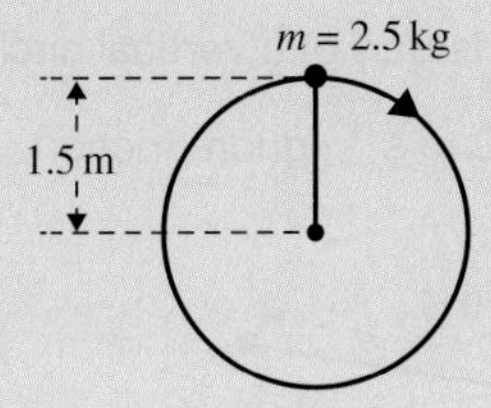

Figure 5

a. Calculate the minimum speed the ball must have at the top of its arc for the string to remain tight (under tension). **(2 marks)**

b. In another experiment, the ball is moving at 6.0 m s^{-1} at the top of its arc. Calculate the speed of the ball at the lowest point. **(3 marks)**

Question 3 (7 marks)

Source: *VCE 2017 Physics Exam, NHT, Section B, Q.3;* © VCAA

An amusement park has a car ride consisting of vertical partial circular tracks, as shown in Figure 4a. The track is arranged so that the car remains upright at both the top and bottom positions. The track has a radius of 12.0 m and its lowest point is point P.

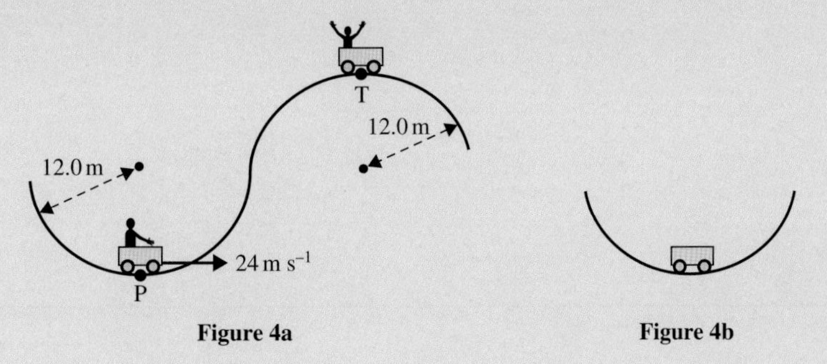

Figure 4a **Figure 4b**

a. On the diagram in Figure 4b, draw labelled arrows showing all of the forces on the car at point P and draw the resultant force with a dotted arrow labelled F_R. **(3 marks)**

b. At point P, the car is moving at 24 m s^{-1}.
Calculate the force of the car seat on a passenger of mass 50 kg as the car passes point P. Show your working. **(2 marks)**

c. Emily says that if the car moves at the correct speed at the top, point T, a person can feel weightless at that point.
Roger says this is nonsense; a person can only feel weightless in deep space, where there is no gravity.
Who is correct? Justify your answer. **(2 marks)**

Question 4 (2 marks)

Source: *VCE 2017, Physics Exam, Q.8.a; © VCAA*

A roller-coaster is arranged so that the normal reaction force on a rider in a car at the top of the circular arc at point P, shown in the figure, is briefly zero. The section of track at point P has a radius of 6.4 m.

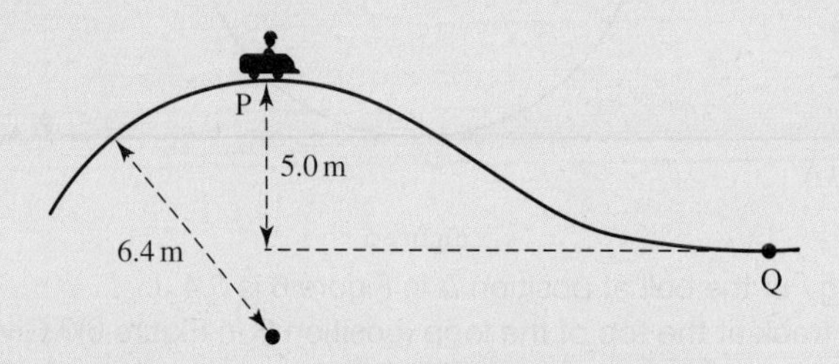

Calculate the speed that the car needs to have to achieve a zero normal reaction force on the rider at point P.

Question 5 (2 marks)

Source: *VCE 2015, Physics Exam, Q.3.b; © VCAA*

A model car of mass 2.0 kg is on a track that is part of a vertical circle of radius 4.0 m, as shown in the figure.

At the lowest point, L, the car is moving at 6.0 m s^{-1}. Ignore friction.

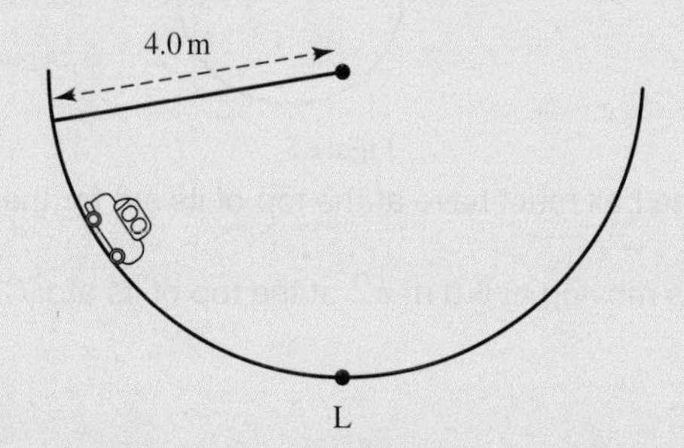

Calculate the magnitude of the force exerted by the track on the car at its lowest point (L). Show your working.

More exam questions are available in your learnON title.

1.7 Review

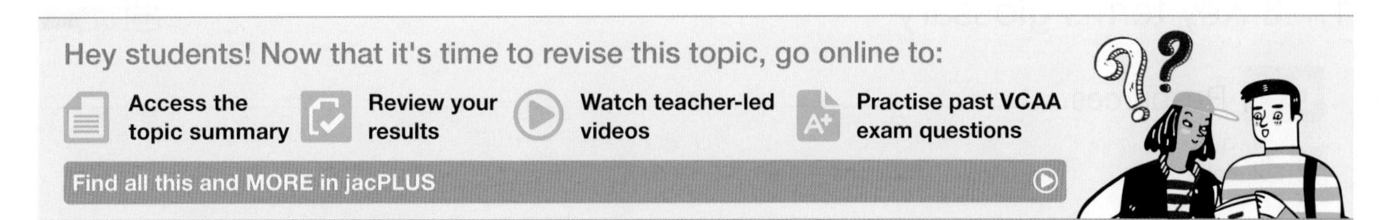

1.7.1 Topic summary

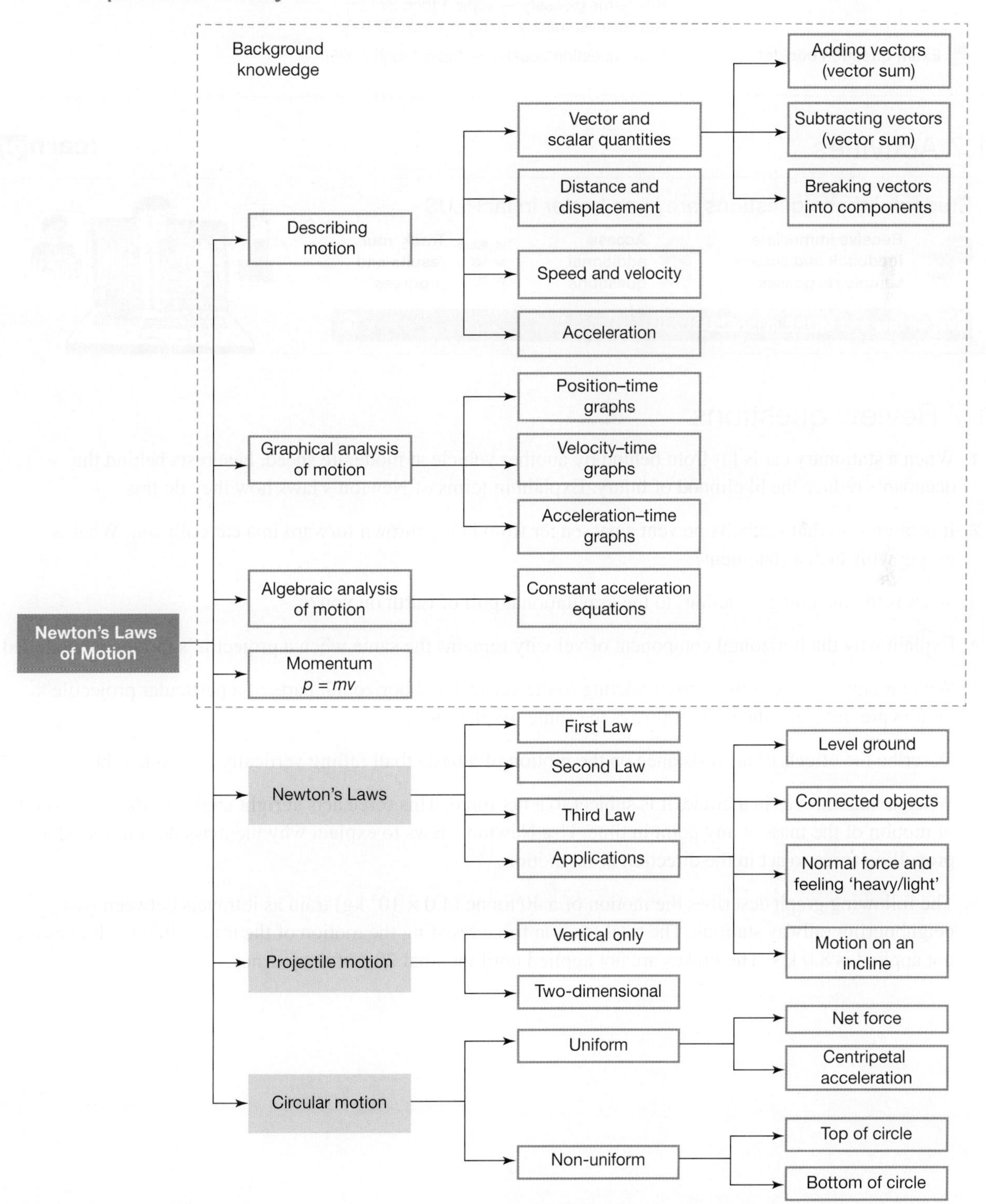

1.7.2 Key ideas summary

1.7.3 Key terms glossary

online only

Solutions	Solutions — Topic 1 (sol-0815)
Practical investigation eLogbook	Practical investigation eLogbook — Topic 1 (elog-1632)
Digital documents	Key science skills — VCE Physics Units 1–4 (doc-36950) Key terms glossary — Topic 1 (doc-37165) Key ideas summary — Topic 1 (doc-37166)
Exam question booklet	Exam question booklet — Topic 1 (eqb-0098)

1.7 Activities

learnon

Students, these questions are even better in jacPLUS

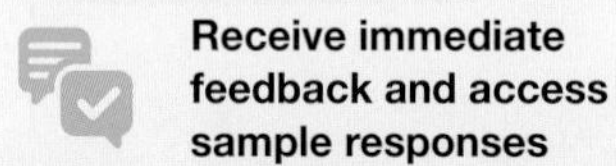

Access additional questions

Track your results and progress

Find all this and MORE in jacPLUS

1.7 Review questions

1. When a stationary car is hit from behind by another vehicle at moderate speed, headrests behind the occupants reduce the likelihood of injury. Explain in terms of Newton's laws how they do this.
2. It is often said that seatbelts prevent a passenger from being thrown forward in a car collision. What is wrong with such a statement?
3. What is the matching 'reaction' to the gravitational pull of Earth on you?
4. Explain why the horizontal component of velocity remains the same when a projectile's motion is modelled.
5. While many pieces of information relating to the vertical and horizontal parts of a particular projectile's motion are different, the time is always the same. Explain why.
6. Describe the effects of air resistance on the motion of a basketball falling vertically from a height.
7. When a mass moves in a circle, it is subject to a net force. This force acts at right angles to the direction of motion of the mass at any point in time. Use Newton's laws to explain why the mass does not need a propelling force to act in the direction of its motion.
8. The following graph describes the motion of a 40 tonne (4.0×10^4 kg) train as it travels between two neighbouring railway stations. The total friction force resisting the motion of the train while the brakes are not applied is 8.0 kN. The brakes are not applied until the final 20 s of the journey.

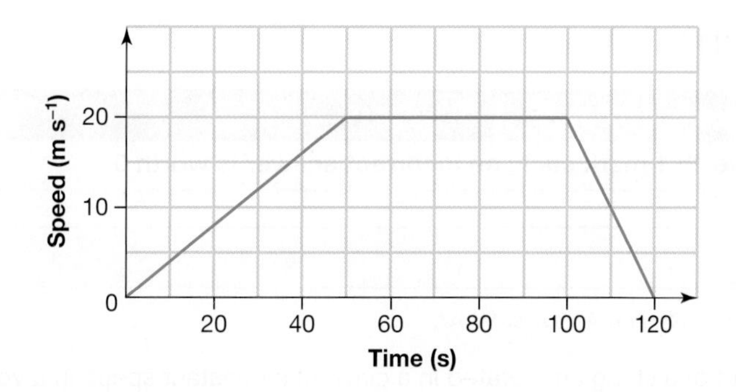

a. What is the braking distance of the train?
b. A cyclist travels between the stations at a constant speed, leaving the first station and arriving at the second station at the same time as the train. What is the constant speed of the cyclist?
c. What forward force is applied to the train by the tracks while it is accelerating?
d. What additional frictional force is applied to the train while it is braking?

9. At a children's amusement park, the miniature train ride completes a circuit of radius 300 m, maintaining a constant speed of 12 km h^{-1}.
a. What is the centripetal acceleration of the train?
b. What is the net force acting on a 45-kg child riding on the train?
c. What is the net force acting on the 1250-kg train?
d. Explain why the net forces acting on the child and the train are different and yet the train and the child are moving along the same path.

10. During a game of totem tennis, a 100-g ball is whirled in a horizontal circle on the end of a 1.30-m length of string. The time taken for ten complete revolutions of the ball is 12.0 s. The string makes an angle of 30.0° with the horizontal. Calculate:
a. the speed of the ball
b. the centripetal acceleration of the ball
c. the net force acting on the ball
d. the magnitude of the tension in the string

11. A road is to be banked so that any vehicle can take the bend at a speed of 40.0 m s^{-1} without having to rely on sideways friction. The radius of curvature of the road is 15.0 m. At what angle should it be banked?

12. A 65-kg gymnast, who is swinging on the rings, follows the path shown in the following figure.
a. What is the speed of the gymnast at point B, if he is at rest at point A?
b. What is the centripetal force acting on the gymnast at point B?
c. Draw a labelled diagram of the forces acting on the gymnast at point B. Include the magnitude of all forces.

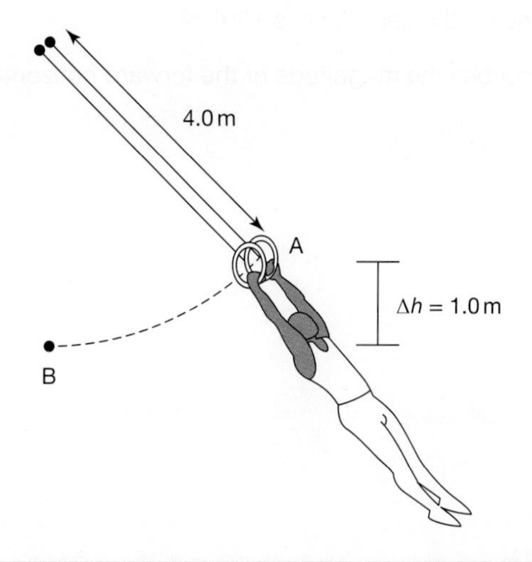

1.7 Exam questions

Section A — Multiple choice questions

All correct answers are worth 1 mark each; an incorrect answer is worth 0.

Question 1

Source: *VCE 2020 Physics Exam, Section A, Q.8;* © VCAA

A ball is attached to the end of a string and rotated in a circle at a constant speed in a vertical plane, as shown in the diagram below.

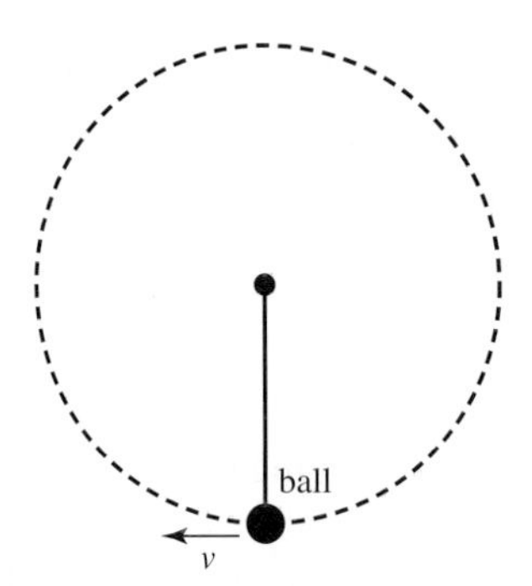

The arrows in options A. to D. below indicate the direction and the size of the forces acting on the ball.

Ignoring air resistance, which one of the following best represents the forces acting on the ball when it is at the bottom of the circular path and moving to the left?

A.

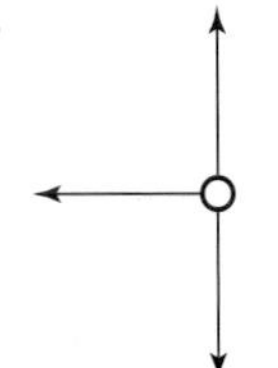

B.

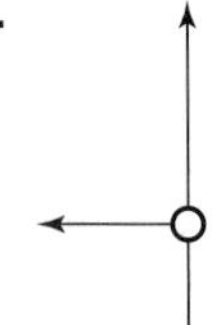

C.

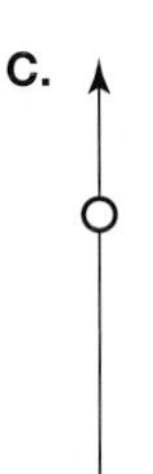

D.

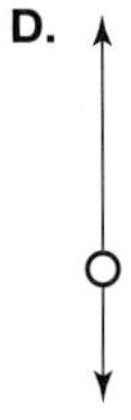

Question 2

Source: *VCE 2019, Physics Exam, Section A, Q.11;* © VCAA

An ultralight aeroplane of mass 500 kg flies in a horizontal straight line at a constant speed of 100 m s^{-1}.

The horizontal resistance force acting on the aeroplane is 1500 N.

Which one of the following best describes the magnitude of the forward horizontal thrust on the aeroplane?

A. 1500 N
B. slightly less than 1500 N
C. slightly more than 1500 N
D. 5000 N

Question 3

Source: *VCE 2018, Physics Exam, Section A, Q.5;* © VCAA

Four students are pulling on ropes in a four-person tug of war. The relative sizes of the forces acting on the various ropes are $F_W = 200$ N, $F_X = 240$ N, $F_Y = 180$ N and $F_Z = 210$ N. The situation is shown in the diagram below.

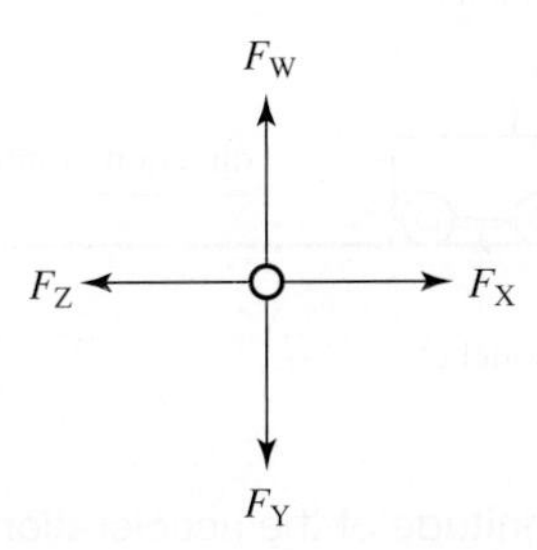

Which one of the following **best** gives the magnitude of the resultant force acting at the centre of the tug-of-war ropes?

A. 28.3 N
B. 30.0 N
C. 36.1 N
D. 50.0 N

Question 4

Source: *VCE 2018, Physics Exam, Section A, Q.6;* © VCAA

Lisa is driving a car of mass 1000 kg at 20 m s^{-1} when she sees a dog in the middle of the road ahead of her. She takes 0.50 s to react and then brakes to a stop with a constant braking force. Her speed is shown in the graph below. Lisa stops before she hits the dog.

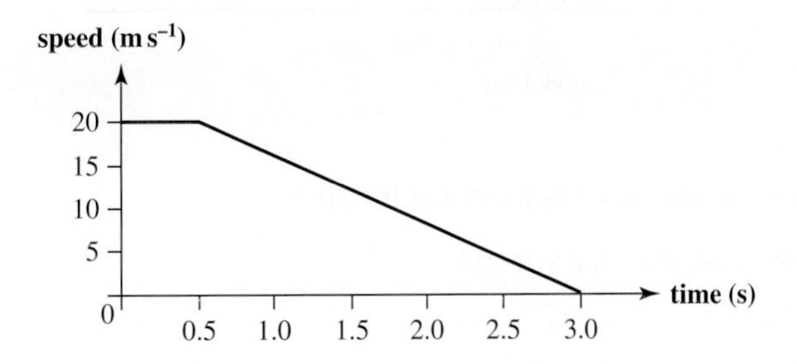

Which one of the following is closest to the magnitude of the braking force acting on Lisa's car during her braking time?

A. 6.7 N
B. 6.7 kN
C. 8.0 kN
D. 20.0 kN

Question 5

Source: *VCE 2017, Physics Exam, Section A, Q.7;* © VCAA

A model car of mass 2.0 kg is propelled from rest by a rocket motor that applies a constant horizontal force of 4.0 N, as shown below. Assume that friction is negligible.

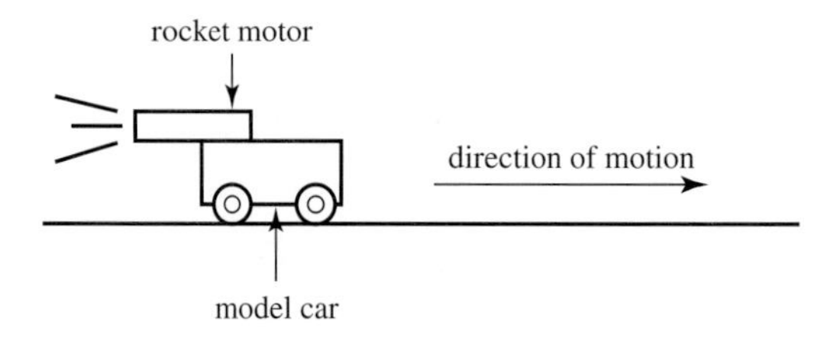

Which one of the following best gives the magnitude of the acceleration of the model car?

A. 0.50 m s^{-2}
B. 1.0 m s^{-2}
C. 2.0 m s^{-2}
D. 4.0 m s^{-2}

Question 6

Source: *VCE 2017, Physics Exam, Section A, Q.9;* © VCAA

A model car of mass 2.0 kg is propelled from rest by a rocket motor that applies a constant horizontal force of 4.0 N, as shown below. Assume that friction is negligible.

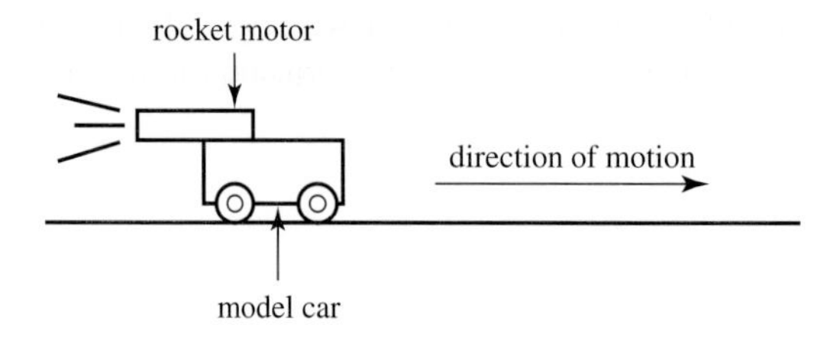

With the same rocket motor, the car accelerates from rest for 10 s.

Which one of the following best gives the final speed?

A. 6.3 m s^{-1}
B. 10 m s^{-1}
C. 20 m s^{-1}
D. 40 m s^{-1}

Question 7

Source: *VCE 2019, Physics Exam, Section A, Q.12;* © VCAA

A small ball is rolling at constant speed along a horizontal table. It rolls off the edge of the table and follows the parabolic path shown in the diagram below. Ignore air resistance.

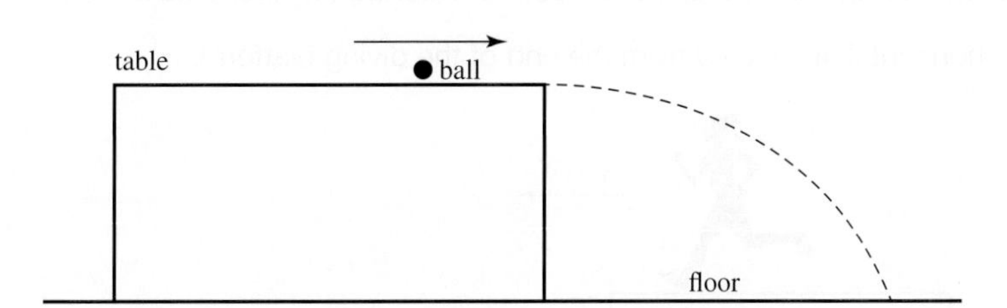

Which one of the following statements about the motion of the ball as it falls is correct?

A. The ball's speed increases at a constant rate.
B. The momentum of the ball is conserved.
C. The acceleration of the ball is constant.
D. The ball travels at constant speed.

Question 8

Source: *VCE 2020 Physics Exam, Section A, Q.11;* © VCAA

The International Space Station (ISS) is travelling around Earth in a stable circular orbit, as shown in the diagram below.

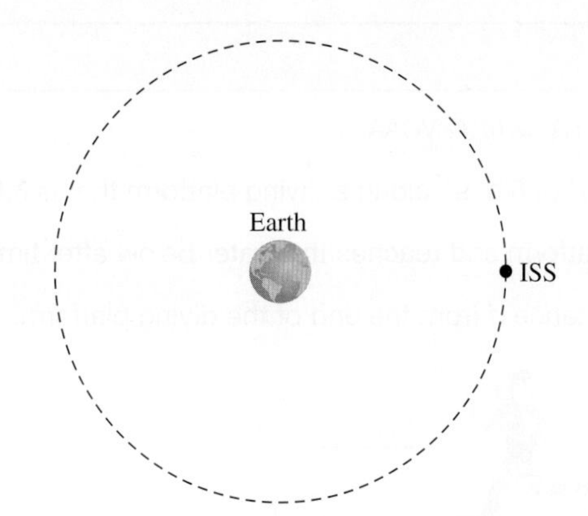

Which one of the following statements concerning the momentum and the kinetic energy of the ISS is correct?

A. Both the momentum and the kinetic energy vary along the orbital path.
B. Both the momentum and the kinetic energy are constant along the orbital path.
C. The momentum is constant, but the kinetic energy changes throughout the orbital path.
D. The momentum changes, but the kinetic energy remains constant throughout the orbital path.

Question 9

Source: *VCE 2021 Physics Exam, Section A, Q.9;* © VCAA

Lucy is running horizontally at a speed of 6 m s^{-1} along a diving platform that is 8.0 m vertically above the water.

Lucy runs off the end of the diving platform and reaches the water below after time *t*.

She lands feet first at a horizontal distance *d* from the end of the diving platform.

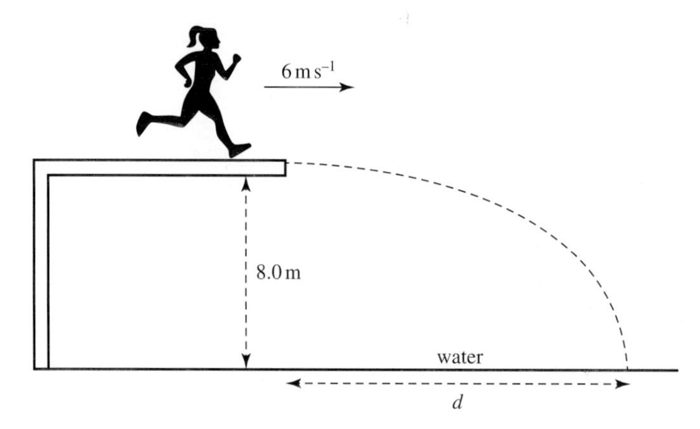

Which one of the following expressions correctly gives the distance *d*?

A. $0.8t$
B. $6t$
C. $5t^2$
D. $6t + 5t^2$

Question 10

Source: *VCE 2021, Physics Exam, Section A, Q.10;* © VCAA

Lucy is running horizontally at a speed of 6 m s^{-1} along a diving platform that is 8.0 m vertically above the water.

Lucy runs off the end of the diving platform and reaches the water below after time *t*.

She lands feet first at a horizontal distance *d* from the end of the diving platform.

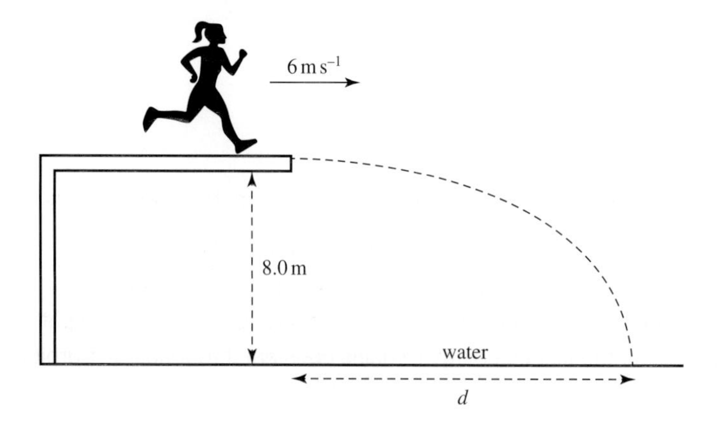

Which one of the following is closest to the time taken, *t*, for Lucy to reach the water below?

A. 0.8 s
B. 1.1 s
C. 1.3 s
D. 1.6 s

Section B – Short answer questions

Question 11 (2 marks)

Source: *VCE 2021, Physics Exam, Section B, Q.4;* © VCAA

Liesel, a student of yoga, sits on the floor in the lotus pose, as shown in the figure. The action force, F_g, on Liesel due to gravity is 500 N down.

Identify and explain what the reaction force is to the action force, F_g, shown in the figure.

Question 12 (3 marks)

Source: *VCE 2021 Physics Exam, Section B, Q.8a;* © VCAA

On 30 July 2020, the National Aeronautics and Space Administration (NASA) launched an Atlas rocket containing the Perseverance rover space capsule on a scientific mission to explore the geology and climate of Mars, and search for signs of ancient microbial life.

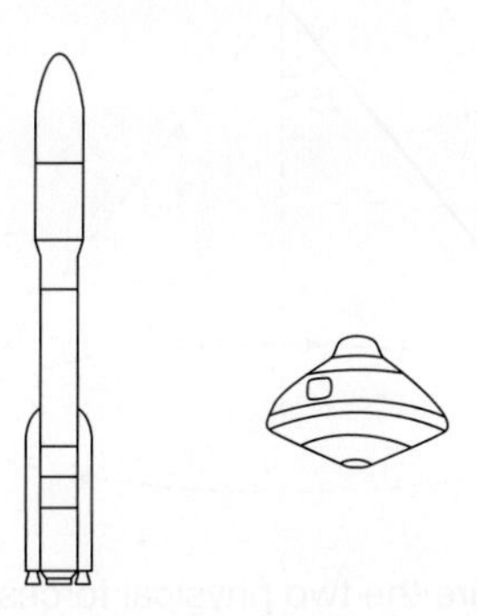

At lift-off from launch, the acceleration of the rocket was 7.20 m s^{-2}. The total mass of the rocket and capsule at launch was 531 tonnes.

Calculate the magnitude and the direction of the thrust force on the rocket at launch. Take the gravitational field strength at the launch site to be g = 9.80 N kg^{-1}. Give your answer in meganewtons. Show your working.

Question 13 (2 marks)

Source: *VCE 2021 Physics Exam, Section B, Q.9b;* © VCAA

Abbie and Brian are about to go on their first loop-the-loop roller-coaster ride. As competent Physics students, they are working out if they will have enough speed at the top of the loop to remain in contact with the track while they are upside down at point C, shown in the figure. The radius of the loop CB is *r*.

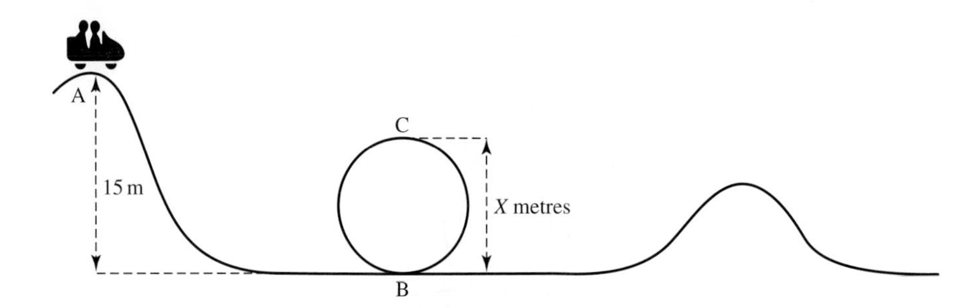

The highest point of the roller-coaster (point A) is 15 m above point B and the car starts at rest from point A. Assume that there is negligible friction between the car and the track.

By considering the forces acting on the car, show that the condition for the car to just remain in contact with the track at point C is given by $\frac{v^2}{r} = g$. Show your working.

Question 14 (6 marks)

Source: *VCE 2020 Physics Exam, Section B, Q.8;* © VCAA

The figure below shows a small ball of mass 1.8 kg travelling in a horizontal circular path at a constant speed while suspended from the ceiling by a 0.75-m long string.

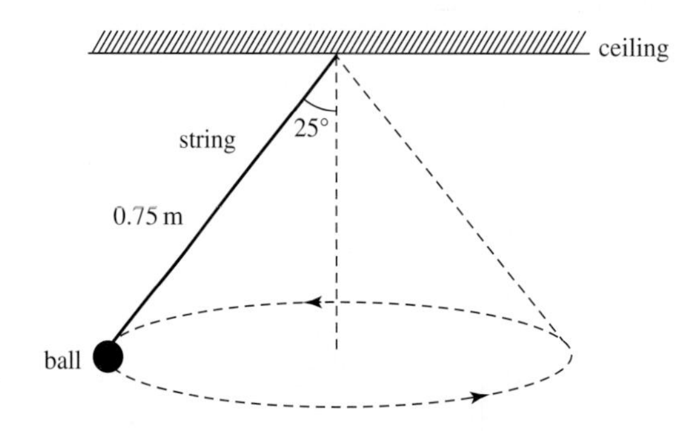

a. Use labelled arrows to indicate on the figure the two physical forces acting on the ball. **(2 marks)**

b. Calculate the speed of the ball. Show your working. **(4 marks)**

Question 15 (4 marks)

Source: *VCE 2019, Physics Exam, Section B, Q.10*; © VCAA

A projectile is launched from the ground at an angle of 39° and at a speed of 25 m s^{-1}, as shown in the figure. The maximum height that the projectile reaches above the ground is labelled h.

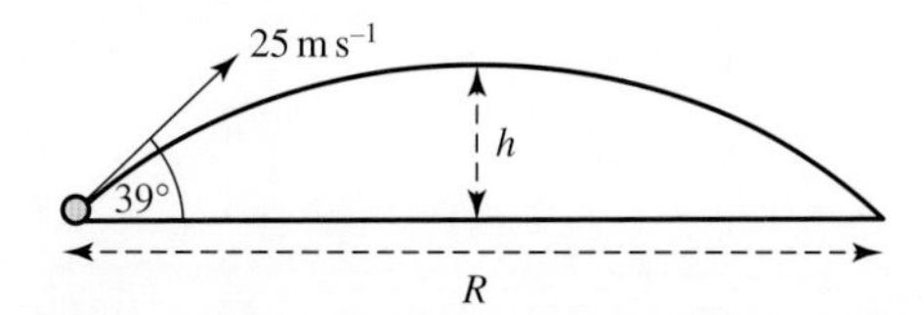

a. Ignoring air resistance, show that the projectile's time of flight from the launch to the highest point is equal to 1.6 s. Give your answer to two significant figures. Show your working and indicate your reasoning. **(2 marks)**

b. Calculate the range, R, of the projectile. Show your working. **(2 marks)**

2 Relationships between force, energy and mass

KEY KNOWLEDGE

In this topic, you will:

- investigate and analyse theoretically and practically impulse in an isolated system for collisions between objects moving in a straight line: $\boldsymbol{F}\Delta t = m\Delta \boldsymbol{v}$
- investigate and apply theoretically and practically the laws of energy and momentum conservation in isolated systems in one dimension
- investigate and apply theoretically and practically the concept of work done by a force using:
  - work done = force × displacement
  - work done = area under force vs distance graph (one dimensional only)
- analyse transformations of energy between kinetic energy, elastic potential energy, gravitational potential energy and energy dissipated to the environment (considered as a combination of heat, sound and deformation of material):
  - kinetic energy at low speeds: $E_k = \frac{1}{2}mv^2$; elastic and inelastic collisions with reference to conservation of kinetic energy
  - strain potential energy: area under force-distance graph including ideal springs obeying Hooke's Law: $E_s = \frac{1}{2}kx^2$
  - gravitational potential energy: $E_g = mg\Delta h$ or from area under a force-distance graph and area under a field-distance graph multiplied by mass.

PRACTICAL WORK AND INVESTIGATIONS

Practical work is a central component of VCE Physics. Experiments and investigations, supported by a **practical investigation eLogbook** and **teacher-led video**, are included in this topic to provide opportunities to undertake investigations and communicate findings.

EXAM PREPARATION

Access past VCAA questions and exam-style questions and their video solutions in every lesson, to ensure you are ready.

2.1 Overview

Hey students! Bring these pages to life online

Watch videos

Engage with interactivities

Answer questions and check results

Find all this and MORE in jacPLUS

2.1.1 Introduction

Have you been in a collision today? It is possible that you have, without even realising it. Collisions feature more often than you might expect in your day-to-day life. Perhaps you knocked into a fellow student as you walked down the corridor, or you had a boxing match and took a hit to the head; or maybe you are a basketball player and you hit the floor after being fouled. Whatever happened in your day, it is likely that you were involved in some sort of collision. How significant that collision was depends on the forces involved, the energy of the interaction and the mass of the objects colliding.

FIGURE 2.1 Collisions occur often in sports such as boxing.

LEARNING SEQUENCE

on Resources

Solutions	Solutions — Topic 2 (sol-0816)
Practical investigation eLogbook	Practical investigation eLogbook — Topic 2 (elog-1633)
Digital documents	Key science skills — VCE Physics Units 1–4 (doc-36950) Key terms glossary — Topic 2 (doc-37167) Key ideas summary — Topic 2 (doc-37168)
Exam question booklet	Exam question booklet — Topic 2 (eqb-0099)

2.2 Momentum and impulse

KEY KNOWLEDGE

- Investigate and analyse theoretically and practically impulse in an isolated system for collisions between objects moving in a straight line: $\boldsymbol{F}\Delta t = m\Delta\boldsymbol{v}$.
- Investigate and apply theoretically and practically the laws of energy and momentum conservation in isolated systems in one dimension.

2.2.1 Momentum and impulse in a collision

Calculating momentum and impulse algebraically

Momentum was introduced in sections 1.2.4 and 1.3.1 as one way to describe the motion of an object. Momentum is useful in explaining changes in motion, because it takes into account the mass as well as the velocity of the moving object. Momentum is a vector quantity with the same direction as the velocity; it is expressed in the units kg m s^{-1}. Newton's First Law of Motion can be explained in terms of momentum. The rate of change in momentum is directly proportional to the magnitude of the net force. This means momentum and impulse are useful quantities for understanding what happens in a collision.

The momentum of an object is the product of the mass and velocity of the object:

$$\boldsymbol{p} = m \times \boldsymbol{v}$$

where: $\boldsymbol{p}$ is the momentum, in kg m s^{-1}
m is the mass, in kg
$\boldsymbol{v}$ is the velocity, in m s^{-1}

Newton's Second Law of Motion describes how the effect of the average net force on an object depends on its mass ($\boldsymbol{F} = m\boldsymbol{a}$). In previous topics, it was useful to express Newton's Second Law in terms of acceleration. Here, it is useful to express it in terms of the change in momentum of an object. That is:

$$\begin{aligned} \boldsymbol{F}_{\text{av}} &= \frac{\Delta \boldsymbol{p}}{\Delta t} \\ \Rightarrow \boldsymbol{F}_{\text{av}}\Delta t &= \Delta \boldsymbol{p} \\ &= m\Delta \boldsymbol{v} \end{aligned}$$

If the force is constant, it can be written as $\boldsymbol{F}_{\text{net}}$.

The product $\boldsymbol{F}_{\text{av}} \times \Delta t$ is called the **impulse** of the average net force. Impulse is a vector quantity that has SI units of N s. Calculations can be carried out to show that:

$$1 \text{ N s} = 1 \text{ kg m s}^{-1}$$

impulse product of a force and the time interval during which it acts. Impulse is a vector quantity with SI units of N s.

Thus, the effect of an average net force on the motion of an object can be summarised by:

$$\text{impulse } (\boldsymbol{I}) = \text{change in momentum } (\Delta \boldsymbol{p})$$
$$\boldsymbol{I} = \boldsymbol{F}\Delta t = m\Delta \boldsymbol{v}$$

When two or more objects collide, the change in the motion of each object can be described by Newton's Second Law of Motion.

When a car collides with an 'immovable' object such as a large tree, its change in momentum is fixed. It is determined by the mass of the car and its initial velocity at the instant of impact. The final momentum is zero because its final velocity is zero. Since the impulse is equal to the change in momentum, the impulse $\boldsymbol{F}_{av} \times \Delta t$ is also fixed. By designing the car so that Δt is as large as possible, the magnitude of the average net force on the car (and hence its deceleration) can be reduced. The smaller the deceleration of the car, the safer it is for the occupants.

Airbags, collapsible steering wheels and padded dashboards are all designed to increase the time interval during which the momentum of a human body changes during a collision — the bigger the stopping time, the smaller the impact force.

Likewise, the polystyrene liner of bicycle helmets is designed to crush during a collision. This increases the time interval during which the skull accelerates (or decelerates), thus decreasing the average net force applied to the head.

FIGURE 2.2 Cars are designed to crumple in collisions. This increases the time interval over which the momentum changes, decreasing the average net force on the car.

FIGURE 2.3 Bicycle helmets: Newton's Second Law provides an explanation for their life-saving function.

tlvd-8957

SAMPLE PROBLEM 1 Calculating the impulse, change in momentum and magnitude of a force during a collision

A 1200-kg car collides with a concrete wall at a speed of 15 m s^{-1} and takes 0.060 s to come to rest.

a. What is the change in momentum of the car?

b. What is the impulse on the car?

c. What is the average magnitude of the force exerted by the wall on the car?

d. What would be the average magnitude of the force exerted by the wall on the car if the car bounced back from the wall with a speed of 3.0 m s^{-1} after being in contact for 0.060 s?

THINK	WRITE
a. 1. Assign the initial direction of the car as positive. Calculate the change in momentum.	a. $m = 1200$ kg; $u = 15$ m s^{-1}; $v = 0$ m s^{-1}; $\Delta t = 0.060$ s $\Delta p = mv - mu$ $= m(v - u)$ $= 1200 \times (0 - 15)$ $= 1200 \times (-15)$ $= -1.8 \times 10^4$ kg m s^{-1}
2. State the change in momentum of the car; this should be a positive number.	The change in momentum is 1.8×10^4 kg m s^{-1} in a direction opposite to the original direction of the car.

b. 1. Determine the impulse of the car using the change in momentum.

b. Impulse on car = change in momentum of car
$= -1.8 \times 10^4 \text{ kg m s}^{-1}$

2. State the impulse on the car.

The impulse on the car is 1.8×10^4 N s in a direction opposite to the original direction of the car.

c. Determine the magnitude of force using $\Delta p = F\Delta t$.

c.
$$\Delta p = F\Delta t$$
$$1.8 \times 10^4 = F \times 0.06$$
$$\Rightarrow F = \frac{1.8 \times 10^4}{0.060}$$
$$= 3.0 \times 10^5 \text{ N}$$

d. 1. Determine the impulse of the car. In this case, the final velocity is $v = -3.0 \text{ m s}^{-1}$ and the initial velocity is $u = 15 \text{ m s}^{-1}$ (remember the change in velocity or Δv is equal to the initial velocity subtracted from the final velocity).

d.
$$\text{Impulse} = m\Delta v$$
$$= 1200\,(-3 - 15)$$
$$= 1200 \times (-18)$$
$$= -2.16 \times 10^4 \text{ N s or kg m s}^{-1}$$

2. Determine the magnitude of force using $\Delta p = F\Delta t$, where $\Delta p = 2.16 \times 10^4$ N s and $\Delta t = 0.060$ s.

$$\Delta p = F\Delta t$$
$$2.16 \times 10^4 = F \times 0.060$$
$$\Rightarrow F = \frac{2.16 \times 10^4}{0.060}$$
$$= 3.6 \times 10^5 \text{ N}$$

PRACTICE PROBLEM 1

A dodgem car of mass 200 kg strikes a barrier head-on at a speed of 8.0 m s^{-1} due west and rebounds in the opposite direction with a speed of 2.0 m s^{-1}.

a. **What is the impulse delivered to the dodgem car?**

b. **If the dodgem car is in contact with the barrier for 0.80 s, what average force does the barrier apply to the car?**

c. **What average force does the car apply to the barrier?**

Calculating impulse from a force–time graph

The impulse delivered by a changing force is given by $F_{av}\Delta t$.

If a graph of force versus time is plotted, the impulse can be determined from the area under the graph.

The area under the graph can be calculated using a variety of methods:

- If a grid is provided, the area can be determined by finding the area of each 'square' and multiplying it by the number of squares (found by counting the 'squares' between the graph and the horizontal axis). You can also add sections of partial squares together to make full squares (for example, two half squares make up one full square).
- Another way to calculate areas is by drawing a regular shape (or shapes) that have approximately the same area as the area under the graph. For example, the graph shown in sample problem 2 may be divided into triangles and rectangles, and the areas may be calculated and added together. (This is much easier for graphs with straight-line sections as opposed to curved sections.)

tlvd-8958

SAMPLE PROBLEM 2 Using a force versus time graph to calculate speed and magnitude of impulse

The following graph describes the changing horizontal force on a 40-kg rollerskater as she begins to move from rest. Estimate her speed after 2.0 seconds.

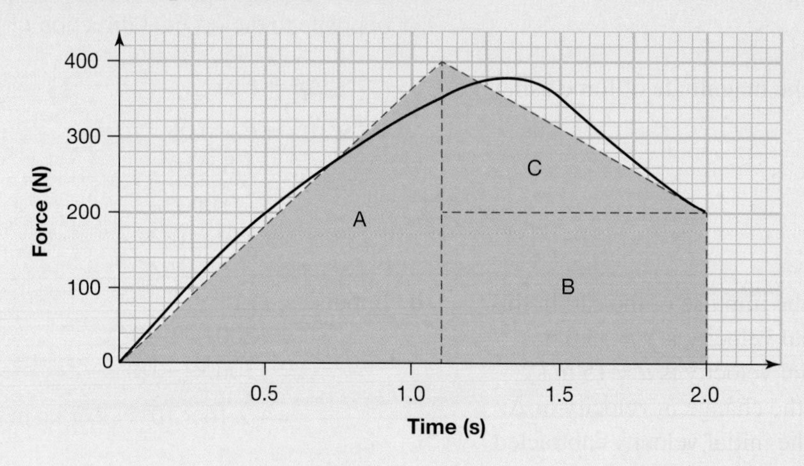

THINK	WRITE
1. The magnitude of the impulse on the skater can be determined by calculating the area under the graph. This can be determined by either counting squares or by determining the shaded area. In this example, the areas have been provided.	Magnitude of impulse = area A + area B + area C $= \left(\frac{1}{2} \times 1.1 \times 400 + 0.9 \times 200 + \frac{1}{2} \times 0.9 \times 200\right)$ $= (220 + 180 + 90)$ $= 490\text{ N s}$
2. Determine the magnitude of impulse using the change in momentum, $\Delta \boldsymbol{p} = m\Delta \boldsymbol{v}$, where $\Delta \boldsymbol{p} = 490$ N s, and $m = 40$ kg.	Magnitude of impulse = magnitude of change in momentum $= m\Delta v$ $490\text{ N s} = 40\text{ kg} \times \Delta v$ $\Rightarrow \Delta v = \frac{490}{40}$ $= 12\text{ m s}^{-1}$
3. Determine the speed.	As her initial speed is zero (she started from rest), her speed after 2.0 seconds is 12 m s^{-1}.

PRACTICE PROBLEM 2

Estimate the speed of the rollerskater in sample problem 2 after 1.0 s.

2.2.2 Conservation of momentum

Newton's Second Law of Motion can be applied to the system of two objects just as it can be applied to each object. By applying the formula $\boldsymbol{F}_{\text{av}} = \frac{\Delta \boldsymbol{p}}{\Delta t}$ to a system of one or more objects, another expression of Newton's Second Law can be written: if the net force acting on a system is zero, the total momentum of the system does not change.

This statement is an expression of the Law of Conservation of Momentum. It is also expressed as follows: if there are no external forces acting on a system, the total momentum of the system remains constant.

> The Law of Conservation of Momentum can be written as:
>
> $$p_{\text{before}} = p_{\text{after}}$$

A system on which no external forces act is called an **isolated system**. The only forces acting on objects in the system are those applied by other objects in the system. In practice, collisions at the surface of Earth do not take place within isolated systems. Consider a system comprising two cars that collide. This is not isolated because forces are applied to the cars by objects outside the system, such as road friction and the gravitational pull of Earth.

However, if the cars collide on an icy horizontal road, the collision can be considered to take place in an isolated system. The sum of external forces (including the force of gravity and the normal force) acting on the system of the cars would be negligible compared with the forces that each car applies to the other. A system comprising a car and a tree struck by the car could not be considered an isolated system because Earth exerts a large external force on the tree in the opposite direction to that applied to the tree by the car.

2.2.3 Modelling a collision

Consider the system of the two blocks labelled A and B in figure 2.4. The blocks are on a smooth horizontal surface. The system can be treated as isolated because the gravitational force and normal force on each of the blocks have no effect on their horizontal motion. Because the surface is described as smooth, the frictional force can be assumed to be negligible. Thus, the net force on the *system* is zero and the total momentum of the system remains constant. The momentum of the centre of mass of the system also remains constant. However, the momentum of each of the blocks changes during the collision because each block has a non-zero net force acting on it.

isolated system system where no external forces act; the only forces acting on objects in the system are those applied by other objects within the system.

FIGURE 2.4 The net force on this system of two blocks is zero. Its total momentum therefore remains constant.

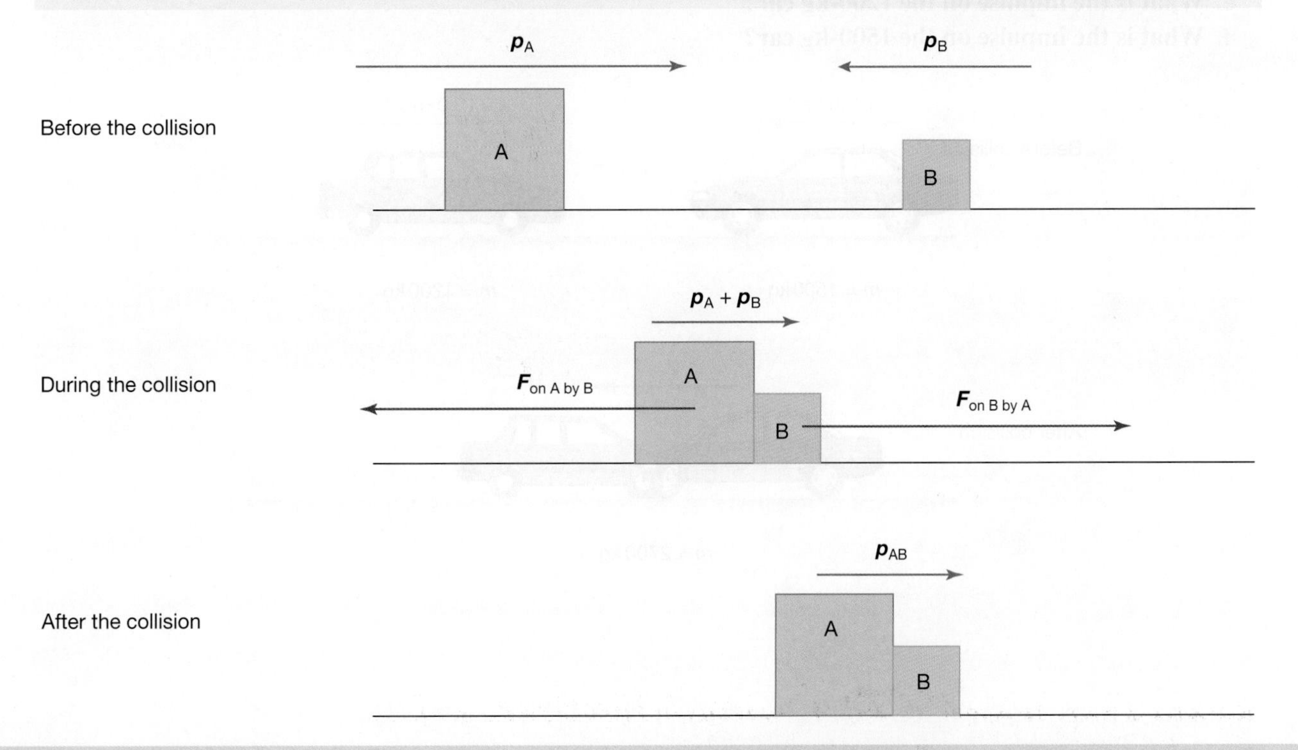

The force exerted on block A by block B ($\boldsymbol{F}_{\text{on A by B}}$) during the collision is equal in magnitude and opposite in direction to the force exerted on block B by block A ($\boldsymbol{F}_{\text{on B by A}}$). Therefore, the change in momentum of block A ($\Delta \boldsymbol{p}_{\text{A}}$) is equal and opposite to the change in momentum of block B ($\Delta \boldsymbol{p}_{\text{B}}$). That is:

$$\boldsymbol{F}_{\text{on A by B}} = \boldsymbol{F}_{\text{on B by A}}$$
$$\Rightarrow \boldsymbol{F}_{\text{on A by B}} \Delta t = -\boldsymbol{F}_{\text{on B by A}} \Delta t$$

where: Δt = time duration of interaction

$$\Rightarrow \quad \Delta \boldsymbol{p}_{\text{A}} = -\Delta \boldsymbol{p}_{\text{B}}$$
$$\Rightarrow \Delta \boldsymbol{p}_{\text{A}} + \Delta \boldsymbol{p}_{\text{B}} = 0$$

This result should be no surprise as, in order for the total momentum of the system consisting of the two blocks to be constant, the total change in momentum must be zero.

The interaction between blocks A and B can be summarised as follows:

- The total momentum of the system remains constant.
- The change in momentum of the system is zero.
- The momentum of the centre of mass of the system remains constant.
- The force that block A exerts on block B is equal and opposite to the force that block B exerts on block A.
- The change in momentum of block A is equal and opposite to the change in momentum of block B.

tlvd-8959

SAMPLE PROBLEM 3 Calculating the momentum and impulse of vehicles before and after a collision

A 1500-kg car travelling at 12.0 m s^{-1} on an icy road collides with a 1200-kg car travelling at the same speed, but in the opposite direction. The cars lock together after impact.

a. **What is the momentum of each car before the collision?**
b. **What is the total momentum before the collision?**
c. **What is the total momentum after the collision?**
d. **With what speed is the tangled wreck moving *immediately* after the collision?**
e. **What is the impulse on the 1200-kg car?**
f. **What is the impulse on the 1500-kg car?**

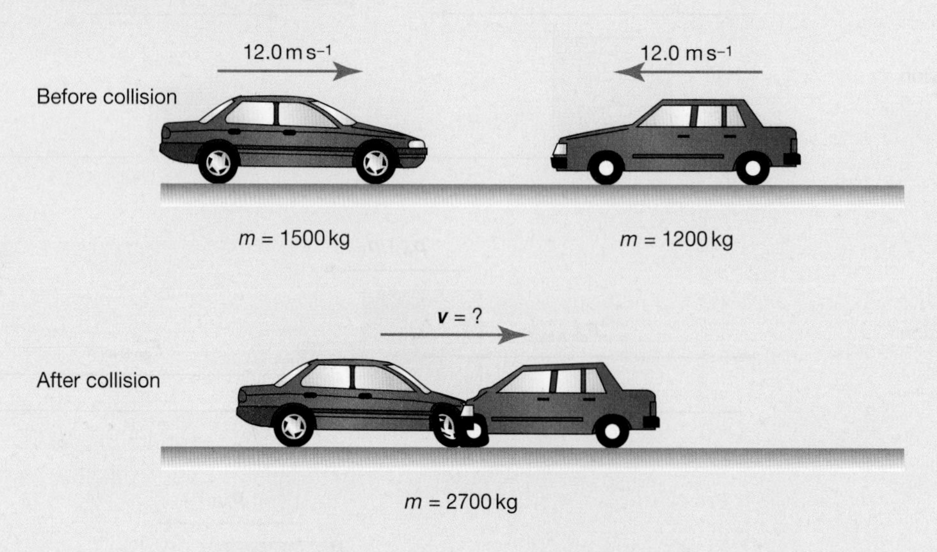

THINK	WRITE
a. 1. Assign the direction in which the first car is moving as positive.	**a.** The 1500-kg car will have a positive velocity, and the 1200-kg car will have a negative velocity.
2. Determine the momentum of the first car. (Remember, momentum is in the units kg m s^{-1} or N s.)	$m = 1500$ kg; $v = 12.0$ m s^{-1} $p = mv$ $= 1500 \times 12.0$ $= 18\,000$ kg m s^{-1} $= 1.80 \times 10^4$ kg m s^{-1}
3. Determine the momentum of the second car.	$m = 1200$ kg; $v = -12.0$ m s^{-1} $p = mv$ $= 1200 \times (-12.0)$ $= -14\,400$ kg m s^{-1} $= -1.44 \times 10^4$ kg m s^{-1}
b. Determine the total momentum before the collision by adding the initial momentum of the 1500-kg car to the initial momentum of the 1200-kg car that was calculated in part **a**.	**b.** $p_i = 1.80 \times 10^4 + \left(-1.44 \times 10^4\right)$ $= 3.60 \times 10^3$ kg m s^{-1}
c. The description of the road suggests that friction is insignificant. It can be assumed that there are no external forces acting on the system.	**c.** $p_f = p_i$ $= 3.60 \times 10^3$ kg m s^{-1}
d. The tangled wreck can be considered as a single mass of 2700 kg.	**d.** $p_f = mv$ $3.60 \times 10^3 = 2700 \times v$ $v = \dfrac{3.60 \times 10^3}{2700}$ $\Rightarrow v = 1.33$ m s^{-1} in the direction of the initial velocity of the first car
e. The impulse on the 1200-kg car is equal to its change in momentum. The initial momentum was calculated in part **a**, and the final momentum can be calculated using $p = mv$, where the mass = 1200 kg, and the final velocity = 1.33 m s^{-1} (as calculated in part **d**).	**e.** $\Delta p = p_f - p_i$ $= (1200 \times 1.33) - \left(-1.44 \times 10^4\right)$ $= 1600 + 1.44 \times 10^4$ $= 1.60 \times 10^4$ kg m s^{-1} (or N s) in the direction of motion of the tangled wreck
f. The impulse or change in momentum on the 1500-kg car is equal to the impulse on the 1200-kg car. This can be verified by calculating the change in momentum of the 1500-kg car. The initial momentum was calculated in part **a**, and the final momentum can be calculated using $p = mv$, where the mass = 1500 kg, and the final velocity = 1.33 m s^{-1} (as calculated in part **d**).	**f.** $\Delta p = p_f - p_i$ $= (1500 \times 1.33) - \left(1.80 \times 10^4\right)$ $= 2000 - 1.80 \times 10^4$ $= -1.60 \times 10^4$ kg m s^{-1} (or N s) in the direction opposite that of the 1200-kg car

PRACTICE PROBLEM 3

A 1000-kg car travelling north at 30 m s^{-1} (108 km h^{-1}) collides with a stationary delivery van of mass 2000 kg on an icy road. The two vehicles lock together after impact.

a. **What is the velocity of the tangled wreck immediately after the collision?**
b. **What is the impulse on the delivery van?**
c. **What is the impulse on the speeding car?**
d. **After the collision, if — instead of locking together — the delivery van moved forward separately at a speed of 12 m s^{-1}, what velocity would the car have?**

EXTENSION: Feeling Earth move

Can you feel Earth move when you bounce a basketball on the court? If Earth and your basketball were an isolated system, Earth *would* move! Its change in speed can be calculated by applying the Law of Conservation of Momentum.

The mass of Earth is 6.0×10^{24} kg. If the mass of a basketball is 600 g and it strikes the ground with a velocity of 12 m s^{-1} downwards in an isolated system, estimate the velocity of Earth after impact.

elog-1876

INVESTIGATION 2.1

online only

Who's pulling whom?

Aim

(a) To demonstrate that the 'action' and 'reaction' described in Newton's Third Law are equal in magnitude, opposite in direction and act on different objects
(b) To demonstrate that the total momentum of a system remains constant if there are no unbalanced external forces acting on the system

2.2 Activities

learnon

Students, these questions are even better in jacPLUS

Receive immediate feedback and access sample responses

Access additional questions

Track your results and progress

Find all this and MORE in jacPLUS

2.2 Quick quiz on	2.2 Exercise	2.2 Exam questions

2.2 Exercise

1. Summarise the relationship between impulse and momentum in eight words or fewer.
2. In a real collision between two cars on a bitumen road on a dry day, is it reasonable to assume that the total momentum of the two cars is conserved? Explain your answer.

3. An empty railway cart of mass 500 kg is moving along a horizontal low-friction track at a velocity of 3.0 m s^{-1} south when a 250-kg load of coal is dropped into it from a stationary container directly above it.
   a. Calculate the velocity of the railway cart immediately after the load has been emptied into it.
   b. What happens to the vertical momentum of the falling coal as it lands in the railway cart?
   c. If the fully loaded railway cart is travelling along the track at the velocity calculated in part **a** and the entire load of coal falls out through a large hole in its floor, what is the final velocity of the cart?
4. Explain whether you are generally safer in a big car or a small car. To do so, consider the following questions by making some estimates and applying Newtons' laws to each car. What assumptions have you made?
   a. How do the forces on each car compare?
   b. How do the masses of the cars compare with each other?
   c. What is the subsequent change in velocity of each car as a result of the collision?
   d. How does your body move during a collision and what does it collide with?
5. A 400-g rubber ball hits a wall at 15 m s^{-1} and bounces back after being in contact with the wall for 0.10 seconds. The momentum of the ball changes by 10 kg m s^{-1}. (Air resistance and the gravitational force can be ignored.) Take the initial velocity as positive.
   a. Calculate the magnitude and direction of the impulse that the wall exerts on the ball.
   b. What is the magnitude of the average force that is exerted on the ball?
   c. Calculate the velocity with which the ball bounces back.

2.2 Exam questions

Question 1 (10 marks)

Source: *VCE 2022 Physics Exam, Section B, Q.7*; © VCAA

Kym and Kelly are experimenting with trolleys on a ramp inclined at 25°, as shown in Figure 7. They release a trolley with a mass of 2.0 kg from the top of the ramp. The trolley moves down the ramp, through two light gates and onto a horizontal, frictionless surface. Kym and Kelly calculate the acceleration of the trolley to be 3.2 m s^{-2} using the information from the light gates.

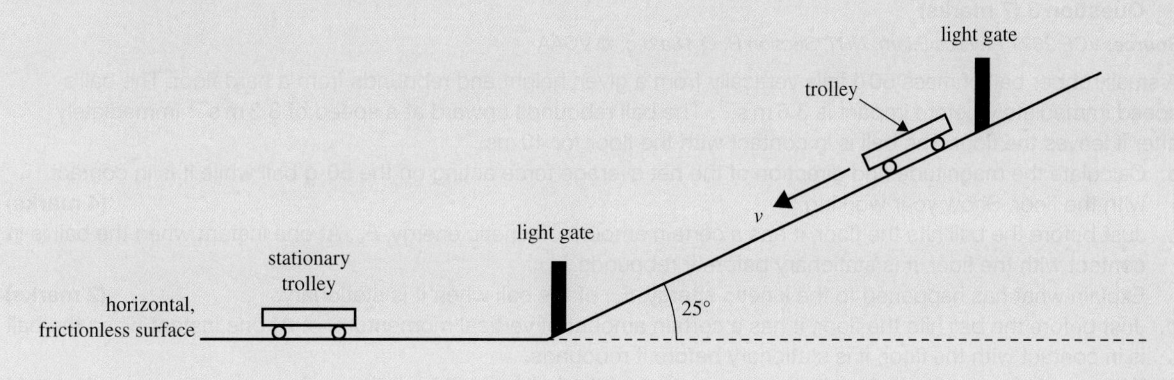

Figure 7

a. i. Show that the component of the gravitational force of the trolley down the slope is 8.3 N. Use $g = 9.8$ m s^{-2}. **(2 marks)**
   ii. Assume that on the ramp there is a constant frictional force acting on the trolley and opposing its motion. Calculate the magnitude of the constant frictional force acting on the trolley. **(2 marks)**

b. When it reaches the bottom of the ramp, the trolley travels along the horizontal, frictionless surface at a speed of 4.0 m s^{-1} until it collides with a stationary identical trolley. The two trolleys stick together and continue in the same direction as the first trolley.
   i. Calculate the speed of the two trolleys after the collision. Show your working and clearly state the physics principle that you have used. **(3 marks)**
   ii. Determine, with calculations, whether this collision is an elastic or inelastic collision. Show your working. **(3 marks)**

Question 2 (10 marks)

Source: *VCE 2021 Physics Exam, NHT, Section B, Q.9; © VCAA*

In a model of a proposed ride at a theme park, a 5.0-kg smooth block slides down a ramp from point W and into an ideal spring bumper without any friction or air resistance, as shown in Figure 13. The final section of the ramp, between points X and Y, is horizontal. The block comes to an instantaneous stop at point Y.

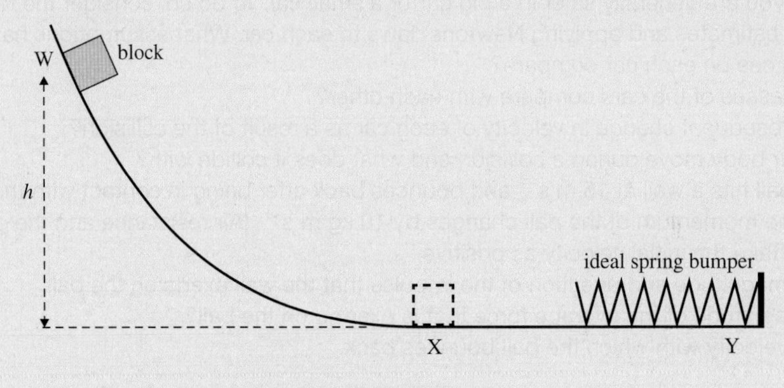

Figure 13

a. Describe the acceleration of the block at points W, X and Y. **(4 marks)**

b. The maximum compression of the spring is measured as 3.0 m and its spring constant, k, is 100 N m^{-1}.
Calculate the release height, h. Show your working. **(3 marks)**

c. Calculate the magnitude of the maximum momentum of the block. Show your working. **(2 marks)**

d. When the block comes to rest, its momentum is zero. In terms of the principle of conservation of momentum, state what has happened to the momentum of the block as it comes to rest. **(1 mark)**

Question 3 (7 marks)

Source: *VCE 2021 Physics Exam, NHT, Section B, Q.18a,b,c; © VCAA*

A small rubber ball of mass 50 g falls vertically from a given height and rebounds from a hard floor. The ball's speed immediately before impact is 3.6 m s^{-1}. The ball rebounds upward at a speed of 3.3 m s^{-1} immediately after it leaves the floor. The ball is in contact with the floor for 40 ms.

a. Calculate the magnitude and direction of the net average force acting on the 50-g ball while it is in contact with the floor. Show your working. **(4 marks)**

b. Just before the ball hits the floor, it has a certain amount of kinetic energy, E_k. At one instant when the ball is in contact with the floor, it is stationary before it rebounds.
Explain what has happened to the kinetic energy, E_k, of the ball when it is stationary. **(2 marks)**

c. Just before the ball hits the floor, it has a certain amount of vertical momentum, p. At one instant when the ball is in contact with the floor, it is stationary before it rebounds.
What has happened to the vertical momentum, p, of the ball when it is stationary? **(1 mark)**

Question 4 (7 marks)

Source: *VCE 2019 Physics Exam, NHT, Section B, Q.7; © VCAA*

Students are using high-speed photography to analyse the collision between a bat and a ball. The experiment is arranged so that the bat and the ball are both moving horizontally just before and just after the collision, as shown in Figure 8. Assume that the bat and the ball are point masses. The students record the following measurements.

mass of bat	2.0 kg
mass of ball	0.20 kg
speed of bat immediately before collision	10 m s^{-1} (bat is stationary after collision)
speed of ball immediately before collision	60 m s^{-1} (towards bat)
speed of ball immediately after collision	40 m s^{-1} (away from bat)
time ball is in contact with bat	0.010 s

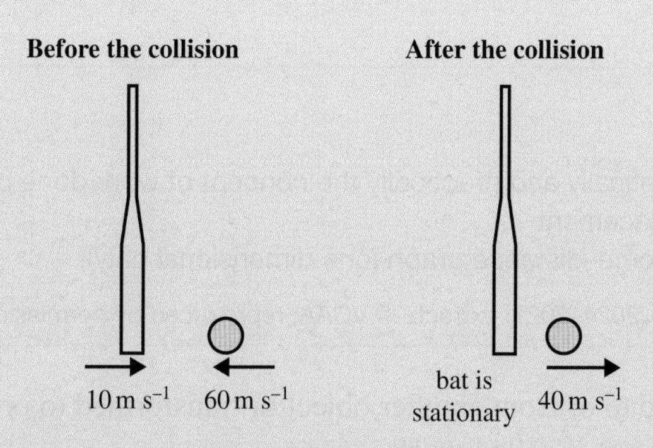

Figure 8

a. Calculate the magnitude of the impulse given by the bat to the ball. Include an appropriate unit. Show your working. **(3 marks)**

b. Calculate the average force of the bat on the ball during the collision. Show your working. **(2 marks)**

c. Use calculations to determine whether the collision between the bat and the ball is elastic or inelastic. Show your working. **(2 marks)**

Question 5 (2 marks)

Source: *VCE 2016, Physics Exam, Section A, Q.4.c; © VCAA*

In a test, an unpowered toy car of mass 4.0 kg is held against a spring, compressing the spring by 0.50 m, and then released, as shown in the figure.

There is negligible friction while the car is in contact with the spring.

The figure also shows the force–extension graph for the spring.

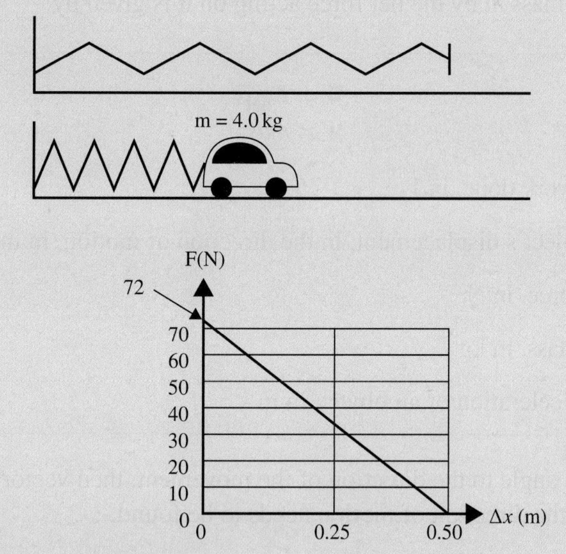

A second test is done, where the spring is not compressed as far, and the car moves off at a speed of 2.0 m s^{-1}.

Calculate the impulse given to the car by the spring. Include an appropriate unit.

More exam questions are available in your learnON title.

2.3 Work done

KEY KNOWLEDGE

- Investigate and apply theoretically and practically the concept of work done by a constant force using:
  - work done = force × displacement
  - work done = area under force–distance graph (one dimensional only).

The amount of energy transferred to or from another object or transformed to or from another form by the action of a force is called **work**.

work energy transferred to or from another object by the action of a force. Work is a scalar quantity.

The change in energy, ΔE, caused by a force acting on the object in the same plane as the motion is the work being done.

2.3.1 Calculating work done by a constant force

The work, W, done when a force, $\boldsymbol{F}$, causes a movement along a displacement, $\boldsymbol{s}$, in the direction of the force is defined as:

$$\text{work} = \text{force} \times \text{displacement along the direction of the force}$$
$$W = \boldsymbol{F} \times \boldsymbol{s}$$

Work is a scalar quantity. The SI unit of work is the joule. One joule of work is done when a force of 1 newton causes a displacement of 1 metre in the same direction as the force.

The work done on an object of mass m by the net force acting on it is given by:

$$W = \boldsymbol{F}_{\text{net}}\boldsymbol{s}$$
$$W = m\boldsymbol{a}\boldsymbol{s}$$

where: W is the work done, in J

$\boldsymbol{s}$ is the object's displacement, in the direction of motion, in m

$\boldsymbol{F}$ is the force, in N

m is the mass, in kg

$\boldsymbol{a}$ is the acceleration of an object, in m s^{-2}

When the force is applied at an angle to the direction of the movement, then vectors need to be used. The component of force parallel to the direction of motion needs to be found.

Consider a student pulling a box along the ground using a piece of rope. The angle between the applied force and the direction of displacement is θ. It is this angle that allows the separation of the force in the direction of the displacement from the total force applied.

FIGURE 2.5 The angle between the applied force and direction changes the work done.

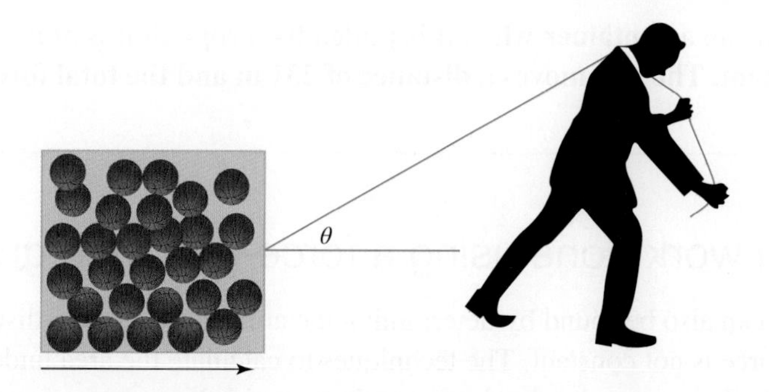

$$W = Fs\cos\theta$$

where: $\cos\theta$ is the cosine of angle of the force to the distance moved of the object

W is the work done, in J

s is the magnitude of the object's displacement, in m

F is the magnitude of the force, in N

When the force and the movement of the object are in the same direction, $\theta = 0°$. Since $\cos 0° = 1$, the formula for work done can be simplified to $W = Fs$.

When the force and the displacement of the object are in opposite directions, $\theta = 180°$. Since $\cos 180° = -1$, it can be concluded that the work leads to a loss of energy.

Since work is only considered to be done when a force (or a component of a force) is in the direction of the movement of the object, it can be seen that if an object is being carried at 90° to the direction of movement, no work is being done on the object. This is because $\cos 90° = 0$. Sometimes, rather than s, you may see Δx or d used to describe the movement in the horizontal direction. In this resource, the displacement s is being used. However, as work is a scalar quantity, we often use distance, d, for the magnitude of displacement.

vd-8960

SAMPLE PROBLEM 4 Calculating the work done when an object is pulled at an angle

Calculate the work done on a box when it is pulled by a rope that is at an angle of 30° to the direction of the movement. The box moves a distance of 585 m and the total force exerted on the box is 500 N.

THINK	WRITE
1. Use the formula for work when the direction of movement is angled.	$W = Fs\cos\theta$
2. Substitute in the values F = 500 N; s = 585 m; $\theta = 30°$	$W = 500 \times 585\cos 30°$
3. Determine the work done on a box in joules.	$W = 2.53 \times 10^5$ J

PRACTICE PROBLEM 4

Calculate the work done on a container when it is pulled by a rope that is at an angle of 52° to the direction of the movement. The box moves a distance of 231 m and the total force exerted on the box is 440 N.

2.3.2 Calculating work done using a force–distance graph

The work done by a force can also be found by determining the area under a force–distance graph. This is particularly useful if the force is not constant. The techniques to calculate the area under a graph were outlined in section 2.2.1 under 'Calculating impulse from a force–time graph'. Note that where the nature of the motion described is such that distance and displacement are the same, you can also calculate work from the area under a force–displacement graph.

tlvd-8961

SAMPLE PROBLEM 5 Using a force–distance graph to determine the amount of the work done by a force

Using the following graph, determine the amount of work done by the force when the object moves from 0 to 5 m.

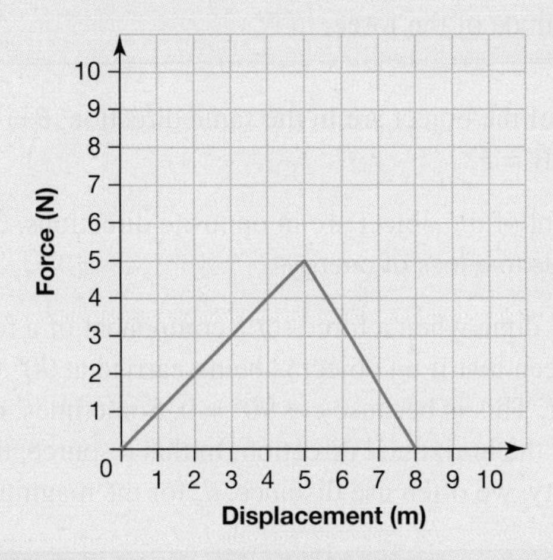

THINK

1. The work done is the area under the line between 0 and 5 m.
   Determine the dimensions of this area.

WRITE

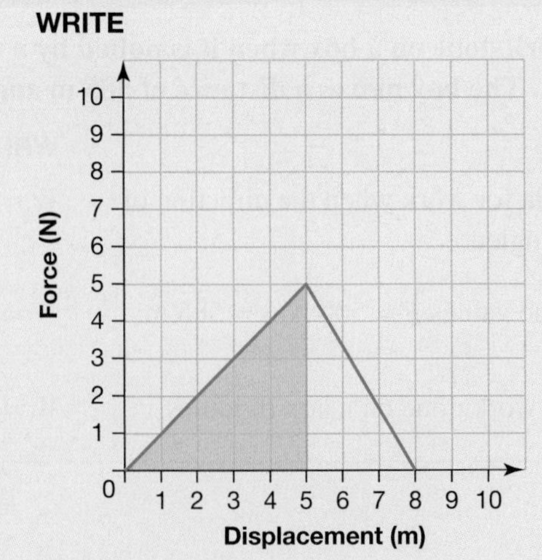

The base of the triangle is 5 and the height is 5.

2. Calculate the area of the triangle to determine the amount of work done using the following formula: $A_{triangle} = \frac{1}{2} \times \text{base} \times \text{height}$

$$A_{triangle} = \frac{1}{2} \times 5 \times 5$$
$$= 12.5 \text{ J}$$
$$\simeq 1 \times 10^1, \text{ to 1 s.f.}$$

PRACTICE PROBLEM 5

Determine the amount of work done by the force when the object moves from 5 to 8 m (using the graph from sample problem 5).

2.3 Activities

learnon

Students, these questions are even better in jacPLUS

Receive immediate feedback and access sample responses

Access additional questions

Track your results and progress

Find all this and MORE in jacPLUS

2.3 Quick quiz  | 2.3 Exercise | 2.3 Exam questions

2.3 Exercise

1. A total of 10 000 J of energy is used to move a ute a distance of 50 m. What is the size of the force applied?
2. You use a rope, at 30° to the horizontal, to pull a box of your old physics textbooks along a brick-paved pathway. The box has a mass of 8.0 kg and you drag it for 5.0 m towards the shed. The tension in the rope, F_T, is 40 N, and the frictional force, F_R, is 20 N. Determine the total work done on the box.
3. A weightlifter lifts a 125-kg barbell to a height of 1.1 m, at constant speed. Calculate the work done by the weightlifter.
4. An elastic band is stretched 40 cm and then released. Using the graph calculate the work done when the elastic band is released.

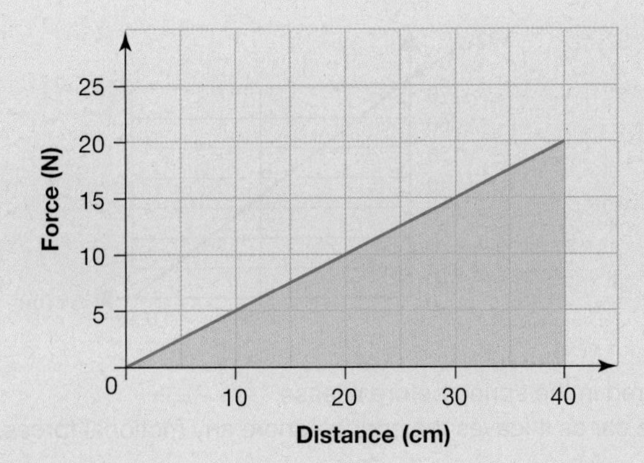

5. The science laboratory is on the second floor of the school building. Jo, a fellow science student, carries her 5.1-kg backpack up the two flights of stairs. The two flights of stairs have a height of 7.0 m. How much work has Jo done on her backpack while carrying it up the stairs?

2.3 Exam questions

Question 1 (1 mark)

Source: *VCE 2022 Physics Exam, Section A, Q.8; © VCAA*

MC The graph below shows force versus compression for a spring used in a Physics investigation.

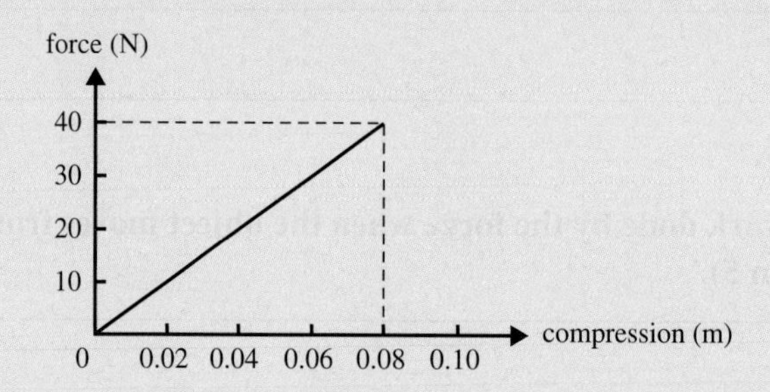

Which one of the following is closest to the compression required to store 0.9 J of potential energy in the spring?

A. 0.05 m
B. 0.06 m
C. 0.07 m
D. 0.08 m

Question 2 (4 marks)

Source: *VCE 2016, Physics Exam, Section A, Q.4a, b; ©VCAA*

In a test, an unpowered toy car of mass 4.0 kg is held against a spring, compressing the spring by 0.50 m, and then released, as shown in the figure.

There is negligible friction while the car is in contact with the spring.

The figure also shows the force–extension graph for the spring.

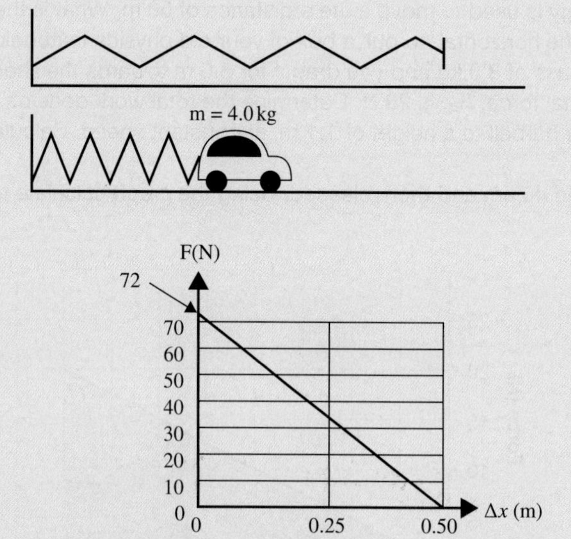

a. Determine the energy stored in the spring before release. **(2 marks)**
b Calculate the speed of the car as it leaves the spring. Ignore any frictional forces. **(2 marks)**

Question 3 (2 marks)

A satellite moves in uniform circular motion under the influence of the gravitational force of the planet at the centre of its orbit.

A student claims that the kinetic energy of the satellite should change because of the work done on the satellite by the gravitational pull of the planet.

Explain whether this claim is correct or incorrect.

Question 4 (2 marks)

The following graph shows the force applied by a bulldozer moving a large boulder a total distance of 4 metres. Calculate the work done by the bulldozer. Assume both the force and the displacement are in the same direction.

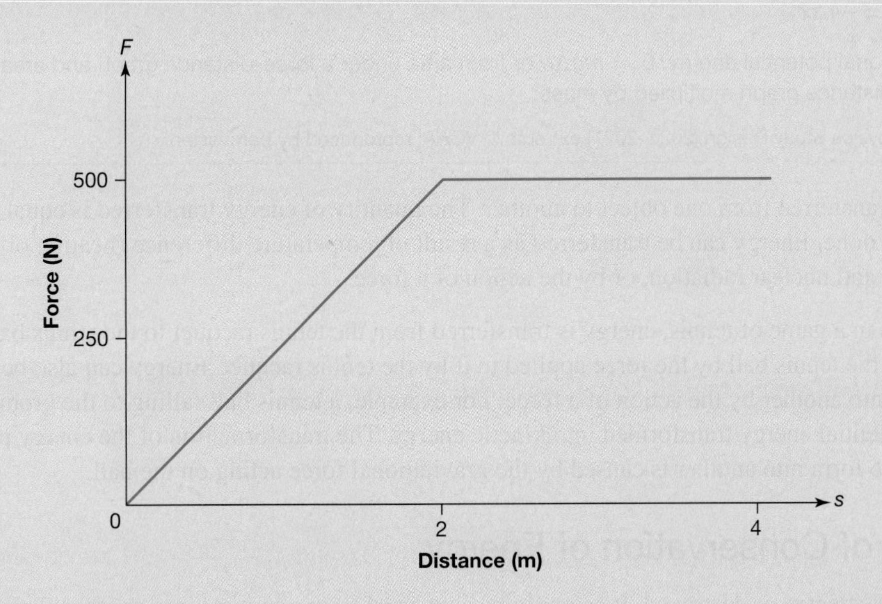

Question 5 (2 marks)

A student applies a force with their foot to crush an aluminium can. The variation of the force exerted by their foot on the can is shown in the following graph. Calculate the work done by their foot on the can.

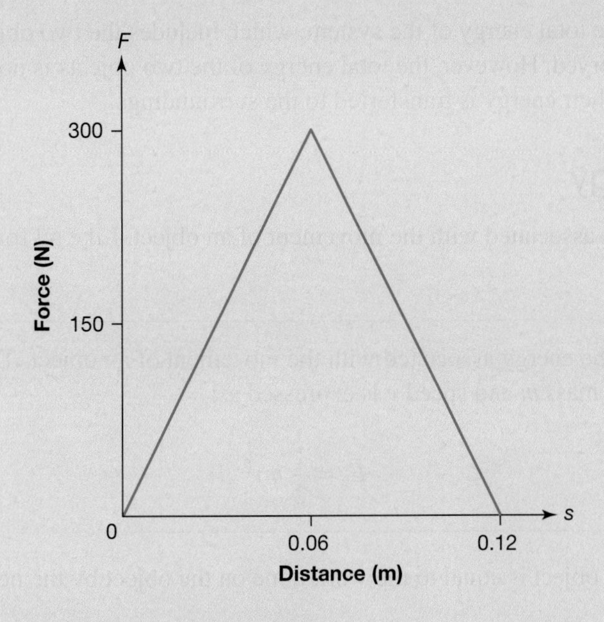

More exam questions are available in your learnON title.

2.4 Kinetic and potential energy

KEY KNOWLEDGE

- Analyse transformations of energy between kinetic energy, strain potential energy, gravitational potential energy and energy dissipated to the environment (considered as a combination of heat, sound and deformation of material):
  - kinetic energy at low speeds: $E_k = \frac{1}{2}mv^2$; elastic and inelastic collisions with reference to conservation of kinetic energy
  - strain potential energy: area under force–distance graph including ideal springs obeying Hooke's Law: $E_s = \frac{1}{2}k\Delta x^2$
  - gravitational potential energy: $E_g = mg\Delta h$ or from area under a force-distance graph and area under a field-distance graph multiplied by mass.

Energy can be transferred from one object to another. The quantity of energy transferred is equal to the amount of work done. Energy can be transferred as a result of temperature difference (heating or cooling), by electromagnetic and nuclear radiation, or by the action of a force.

When you serve in a game of tennis, energy is transferred from the tennis racquet to the tennis ball. The energy is transferred to the tennis ball by the force applied to it by the tennis racquet. Energy can also be transformed from one form into another by the action of a force. For example, a tennis ball falling to the ground has its gravitational potential energy transformed into kinetic energy. The transformation of the energy possessed by the ball from one form into another is caused by the gravitational force acting on the ball.

2.4.1 Law of Conservation of Energy

Energy cannot be created or destroyed. It can only be converted from one form into another. This is the Law of Conservation of Energy. During most energy transformations, some energy is degraded into less useful forms, heating the surroundings and causing noise. If air resistance and other types of friction are small, the amount of energy degraded can be considered negligible.

When two objects collide, the total energy of the system, which includes the two objects and the surroundings (the air and ground), is conserved. However, the total energy of the two objects is not conserved because, when they make contact, some of their energy is transferred to the surroundings.

2.4.2 Kinetic energy

Kinetic energy is the energy associated with the movement of an object. Like all forms of energy, kinetic energy is a scalar quantity.

Kinetic energy is the energy associated with the movement of an object. The kinetic energy, E_k, of an object of mass m and speed v is expressed as:

$$E_k = \frac{1}{2}mv^2$$

The kinetic energy, E_k, of an object is equal to the work done on the object by the net force.

Recall from subtopic 2.3 that the work done on an object of mass m by the net force acting on it is given by:

$$W = F_{net}s$$
$$= mas$$

where: s = the displacement of an object.

If we use the constant acceleration formulas (where the displacement, s, can be equated to the distance, d), work can also be written as:

$$W = \frac{ma\left(v^2 - u^2\right)}{2a}$$
$$= \frac{1}{2}mv^2 - \frac{1}{2}mu^2$$
$$= \Delta E_k$$

If the initial kinetic energy of the object is zero, the work done by the net force is equal to the final kinetic energy. If work is done to stop an object, the work done is equal to the initial kinetic energy.

kinetic energy energy associated with the movement of an object. Like all forms of energy, it is a scalar quantity.

tlvd-8962

SAMPLE PROBLEM 6 Determining the work and magnitude of a force to stop an object

A car of mass 600 kg travelling at 12.0 m s^{-1} collides with a concrete wall and comes to a complete stop over a distance of 30.0 cm. Assume that the frictional forces acting on the car are negligible.

a. **How much work was done by the concrete wall to stop the car?**

b. **What was the magnitude of net force acting on the car as it came to a halt?**

THINK	WRITE
a. As it is known that the car comes to a complete stop, the final kinetic energy is 0. As the change in kinetic energy is being calculated, you only need to focus on the initial kinetic energy.	a. The net force on the car is equal to the force applied by the wall. The work done by the wall, W, is given by: $W = \Delta E_k$ $\Delta E_k = \frac{1}{2}mv^2$, where m = 600 kg and v = 12.0 m s^{-1}. $W = \frac{1}{2}mv^2$ $= \frac{1}{2} \times 600 \times (12.0)^2$ $= 4.32 \times 10^4$ J
b. The magnitude is determined by: $W = F_{net} \times d$	b. $W = F_{av} \times d$ $4.32 \times 10^4 = F_{net} \times 0.300$ ($F_{av} = F_{net}$ in this case) $\Rightarrow F_{net} = 1.44 \times 10^5$ N

PRACTICE PROBLEM 6

A car travelling at 15 m s^{-1} brakes heavily before colliding with another vehicle. The total mass of the car is 800 kg. The car skids for a distance of 20 m before making contact with the other vehicle at a speed of 5.0 m s^{-1}.

a. **How much work is done on the car by road friction during braking?**

b. **Calculate the average road friction during braking.**

Elastic and inelastic collisions

The Law of Conservation of Momentum states that when a collision between two objects occurs, the total momentum of the two objects remains constant.

This statement is valid as long as the two objects comprise an isolated system — that is, as long as there are no external forces acting on each of the objects.

Consider the differences between the two collisions shown in figure 2.6: a collision between two billiard balls on a smooth, level billiard table, and a head-on collision between two cars travelling in opposite directions on a level, icy road.

The two billiard balls can be considered to be an isolated system. The total momentum of the two billiard balls immediately after the collision is the same as it was immediately before the collision. (It is also the same during the collision. Momentum, unlike energy, cannot be stored.) The two cars can also be considered to be an isolated system, because the frictional forces on the cars are relatively small. Therefore, the total momentum of the cars immediately after the collision is the same as it was immediately before the collision.

FIGURE 2.6 Two collisions — momentum is conserved in both of them.

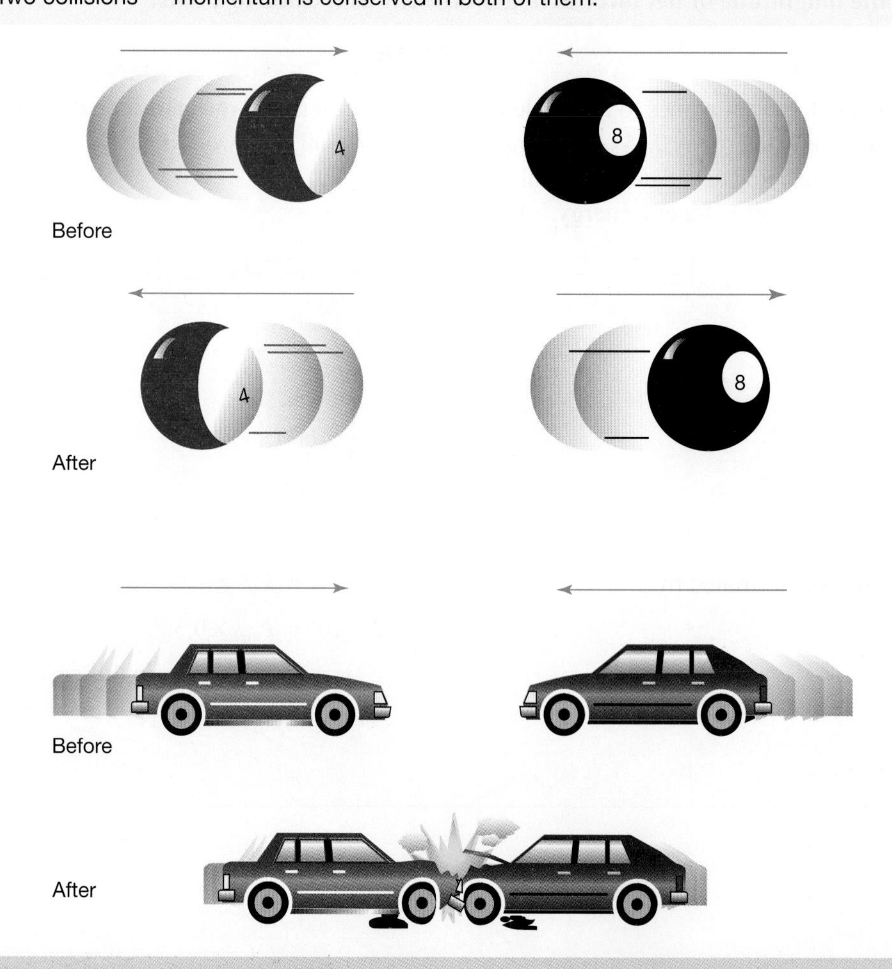

What's the difference?

Apart from the difference between the masses of the objects involved in the collisions, there is one obvious difference:

- The collision between the two billiard balls is an almost perfect elastic collision. An **elastic collision** is one in which the total kinetic energy after the collision is the same as it was before the collision. The sound made when the balls collide provides evidence that the collision is not quite perfectly elastic. Some of the initial kinetic energy of the system is transferred to particles in the surrounding air (and within the balls themselves). However, when making predictions about the outcome of such a collision, it would be quite reasonable to treat the collision as a perfectly elastic one. In fact, a perfectly elastic 'collision' can only take place if the interacting objects do not actually make contact with each other. A perfectly elastic interaction can take place when two electrons move towards each other in a vacuum.
- The collision between the two cars is an **inelastic collision**. Even though momentum is conserved, the total kinetic energy of the cars after the collision is considerably less than it was before the collision. A significant proportion of the initial kinetic energy of the system is transferred to the bodies of both cars, changing their shapes and heating them. Some of the initial kinetic energy is also transformed to sound energy.

elastic collision collision in which the total kinetic energy is conserved

inelastic collision collision in which the total kinetic energy is not conserved

strain potential energy energy stored in an object as a result of a reversible change in shape

on Resources

 **Interactivity** Colliding dodgems (int-6610)

2.4.3 Strain potential energy

The energy stored in an object by changing its length or shape is usually called **strain potential energy** if the object can return naturally to its original shape. Work must be done on an object by a force in order to store energy as strain potential energy. However, when objects are compressed, stretched, bent or twisted, the force needed to change their shape is not constant. For example, the more you stretch a rubber band, the harder it is to stretch it further. The more you compress the sole of a running shoe, the harder it is to compress it further.

The strain potential energy of an object can be determined by calculating the amount of work done on it by the force. This can be found by calculating the area under a graph of force versus change in length (which is referred to as the distance, compression or extension).

Strain potential energy stored in an object can do work on other objects, transforming the energy into another form. When you close the lid of a jack-in-the-box, you do work on the spring to increase its strain potential energy, transferring energy from your body to the spring. When the lid is opened and the spring is released, the spring does work on the 'jack', transforming strain potential energy into kinetic energy and gravitational potential energy.

FIGURE 2.7 A jack-in-the-box. When the lid opens, the spring does work on the 'jack', transforming strain potential energy into kinetic energy.

tlvd-8963

SAMPLE PROBLEM 7 Calculating the strain potential energy stored in a spring using a force versus compression graph

The following graph shows how the force required to compress a jack-in-the-box spring changes as the compression of the spring increases.
How much energy is stored in the spring when it is compressed by 25 cm?

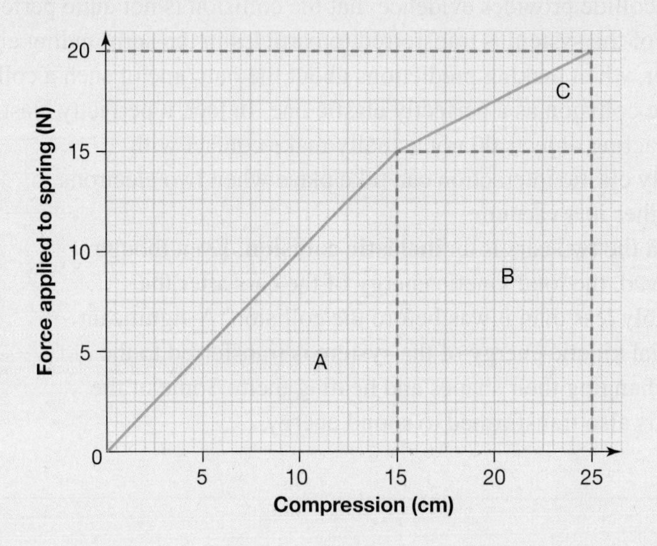

THINK	WRITE
1. The energy stored in the spring is equal to the amount of work done on it. This is the area under the line in the graph.	$W =$ area under graph $=$ area A + area B + area C
2. Determine the sections under the graph to calculate the area. As work is in m, convert the horizontal units to m from cm.	Area A $= \frac{1}{2} \times 0.15 \times 15 = 1.125$ N m Area B $= 0.10 \times 15 = 1.5$ N m Area C $= \frac{1}{2} \times 0.10 \times 5.0 = 0.25$ N m
3. Calculate the total area.	A + B + C $= 1.125 + 1.5 + 0.25$ $= 2.9$ N m
4. Determine the stored energy (remembering that J is the same as N m).	The stored energy is 2.9 J.

PRACTICE PROBLEM 7

The length of the spring represented by the graph in sample problem 7 is 35 cm.

a. **How much strain potential energy is stored in the spring when its length is 15 cm?**

b. **What is the length of the spring when 0.50 J of strain potential energy is stored in it due to compression? (This question is a little harder.)**

Hooke's Law springs to mind

Robert Hooke (1635–1703) investigated the behaviour of elastic springs and found that the **restoring force** exerted by the spring was directly proportional to its displacement. The force is called a restoring force because it acts in a direction that would restore the spring to its natural length.

In vector notation, Hooke's Law states:

$$\boldsymbol{F} = -k\boldsymbol{x}$$

where: $\boldsymbol{F}$ = restoring force, in N

$\boldsymbol{x}$ = displacement (stretch or compression) of the end of the spring from its natural position, in m

k = spring constant (known as the force constant), in N m^{-1}

The negative sign is necessary because the restoring force is always in the opposite direction to the displacement.

It is usually more convenient to express Hooke's Law in terms of magnitude so that the negative sign is not necessary. That is:

$$F = kx$$

where: F = magnitude of the restoring force

x = compression or extension of the spring

k = spring constant

restoring force force applied by a spring to resist compression or extension

Important points to remember include:

- Hooke's Law applies to springs within certain limits. If a spring is compressed or extended so much that it is permanently deformed — unable to return to its original natural length — Hooke's Law no longer applies.
  - The magnitude of the restoring force is equal to the force that is compressing or extending the spring (Newton's Third Law).
- The measure x is not the length of the spring. It is a measure of its compression or extension — the change in length of the spring.
- The spring constant has SI units of N m^{-1}.
- A graph of F versus x produces a straight line with a gradient of k.

The strain potential energy of a spring that obeys Hooke's Law can be expressed as:

$$E_s = \frac{1}{2}kx^2$$

This can be verified by calculating the work done in extending the spring described in figure 2.8. If the spring obeys Hooke's Law, then calculating the area under the graph of force versus extension gives:

$$\begin{aligned} \text{strain potential energy} &= \text{work done on spring} \\ &= \text{area under graph} \\ &= \frac{1}{2}\Delta x \times kx \\ E_s &= \frac{1}{2}kx^2 \end{aligned}$$

FIGURE 2.8 The strain potential energy of a spring is equal to the area under the graph.

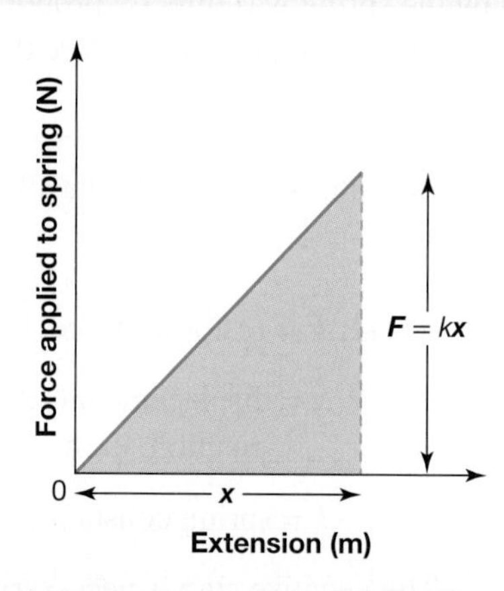

tlvd-8964

SAMPLE PROBLEM 8 Calculating the spring constant and strain potential energy using Hooke's Law

The following graph describes the behaviour of two springs that obey Hooke's Law. Both springs are extended by 20 cm.

a. What is the spring constant of spring A?

b. Which spring has the greatest spring constant?

c. What is the strain potential energy of spring B when its extension is 0.20 m?

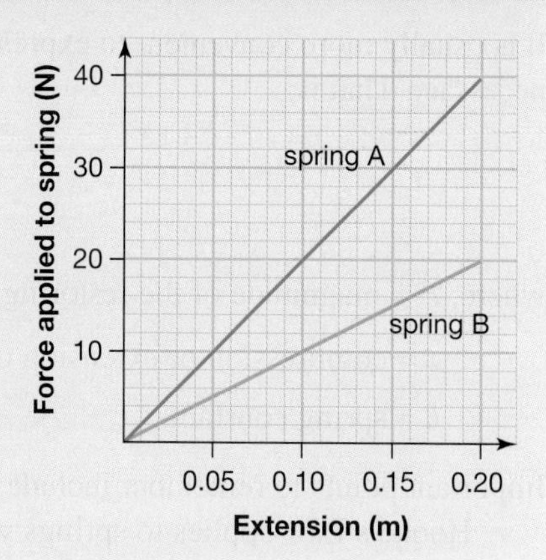

THINK	WRITE
a. The spring constant k is equal to the gradient of the graph. Looking at spring A, the 'rise' of the graph is 40 N and the 'run' is 0.20 m.	a. $k = \frac{40}{0.20}$ $= 2.0 \times 10^2\ \text{N m}^{-1}$
b. The gradient of the graph for spring A is greater than that for spring B.	b. Therefore, spring A has a greater spring constant than spring B — in fact, it is twice the size.
c. 1. Since the spring obeys Hooke's Law, the strain potential energy of spring B can be calculated using the formula.	c. Strain potential energy $= \frac{1}{2}kx^2$
2. Determine k (the gradient). Looking at spring B, the 'rise' of the graph is 20 N and the 'run' is 0.20 m.	k = gradient $= \frac{20}{0.20}$ $= 1.0 \times 10^2\ \text{N m}^{-1}$
3. Use the formula to determine the strain potential energy.	Strain potential energy $= \frac{1}{2} \times 1.0 \times 10^2 \times 0.20^2$ $= 2.0\ \text{J}$

PRACTICE PROBLEM 8

a. **What is the spring constant of spring B, described in sample problem 8?**

b. **How much strain potential energy is stored in spring A when it is extended by 20 cm?**

tlvd-8965

SAMPLE PROBLEM 9 Calculating the speed of an object using Hooke's Law

A toy car of mass 0.50 kg is pushed against a spring so that it is compressed by 0.10 m. The spring obeys Hooke's Law and has a spring constant of 50 N m^{-1}. When the toy car is released, what will its speed be at the instant that the spring returns to its natural length? Assume there is no friction within the spring and no frictional force resisting the motion of the toy car.

THINK	WRITE
1. Determine the formula for strain potential energy.	Strain potential energy $= \frac{1}{2}kx^2$
2. Substitute in the provided values: $k = 50$ N m^{-1}; $x = 0.10$ m	$\frac{1}{2}kx^2 = \frac{1}{2} \times 50 \times (0.10)^2$ $= 0.25$ J
3. Determine the speed when the spring returns to its natural length using $\frac{1}{2}mv^2 =$ energy transformed where: $m = 0.50$ kg and the energy transformed $= 0.25$ J (from step 2)	$\frac{1}{2}mv^2 = \text{energy transformed}$ $\frac{1}{2} \times 0.50 \times v^2 = 0.25$ $\Rightarrow v^2 = \frac{0.25}{\frac{1}{2} \times 0.50}$ $\Rightarrow v = 1.0\,\text{m s}^{-1}$
4. Respond to the question.	The speed of the toy car is 1.0 m s^{-1}.

PRACTICE PROBLEM 9

A model car of mass 0.40 kg travels along a frictionless horizontal surface at a speed of 0.80 m s^{-1}. It collides with the free end of a spring that obeys Hooke's Law. The spring constant is 100 N m^{-1}.

a. **How much strain potential energy is stored in the spring when the car comes to a stop?**

b. **What is the maximum compression of the spring?**

elog-1878

tlvd-8745

INVESTIGATION 2.2

online only

The properties of a coil spring

Aim

To determine whether the force applied by a coil spring extending a suspended mass is directly proportional to the extension of the coil spring and thereby verify Hooke's Law

2.4.4 Gravitational potential energy

Gravitational potential energy is the energy stored in an object as a result of its position relative to another object to which it is attracted by the force of gravity. The gravitational potential energy of an object increases the further it moves away from the object it is attracted to and decreases the closer it gets to the attracted object.

When you drop an object, the gravitational force does work on it, transforming gravitational potential energy to kinetic energy as it falls. When you lift an object, you do work on the object to increase its gravitational potential energy. (Energy is transferred from your body to the object.)

Calculating gravitational potential energy

A quantitative definition of gravitational potential energy can be stated by determining how much work is done in lifting an object of mass m through a height Δh. To lift an object without changing its kinetic energy, a force, $\boldsymbol{F}$, equal to the force due to gravity acting on the object is needed. The work done is:

gravitational potential energy energy stored in an object as a result of its position relative to another object to which it is attracted by the force of gravity

$$W = \boldsymbol{F}\boldsymbol{s}$$
$$= mg\Delta h$$
$$\Rightarrow \Delta E_g = mg\Delta h$$

where: ΔE_g is the change in gravitational potential energy, in J

m is the mass of the object, in kg

$\boldsymbol{g}$ is the acceleration due to gravity, in m s^{-2}

Δh is the change in height of an object, in m

It is important to remember that the change in gravitational potential energy as a result of a particular change in height is independent of the path taken. The change in gravitational potential energy of the diver in figure 2.9 is the same whether she falls from rest, jumps upwards first or completes a complicated dive with twists and somersaults.

FIGURE 2.9 The change in gravitational potential energy of the diver is independent of the path taken.

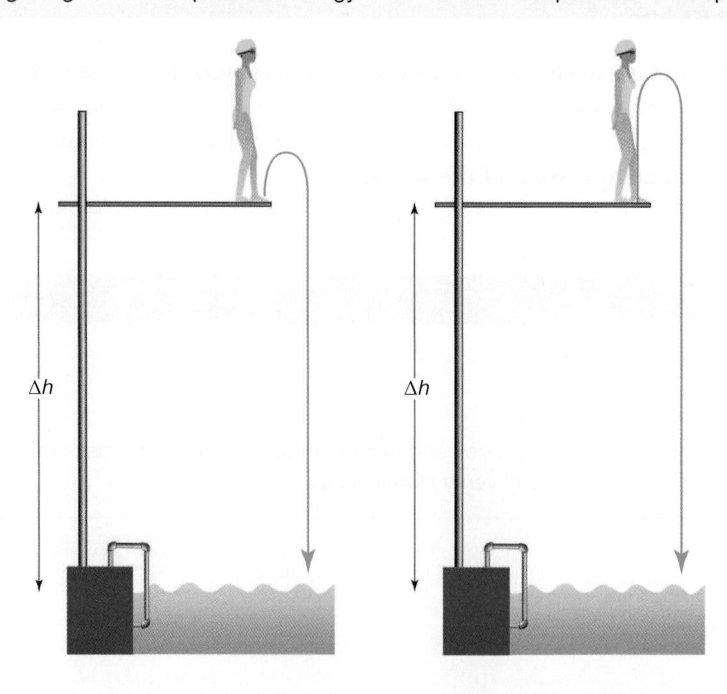

tlvd-8966

SAMPLE PROBLEM 10 Calculating gravitational potential energy and velocity

A water slide has a drop of 9.0 m. A child of mass 35 kg sits at the top.
a. What is the child's gravitational potential energy?
b. How fast will the child be travelling when they hit the water? Ignore any frictional losses.

THINK	WRITE
a. Determine the gravitational energy of the child using $\Delta E_g = mg\Delta h$, where $m = 35$ kg, $g = 9.8$ m s^{-2} and $\Delta h = 9.0$ m.	a. $\Delta E_g = mg\Delta h$ $= 35 \times 9.8 \times 9.0$ $= 3.1 \times 10^3$ J
b. Determine the gravitational energy of the child using $\frac{1}{2}mv^2 = mg\Delta h$.	b. $\frac{1}{2}mv^2 = mg\Delta h$ $\Rightarrow v^2 = 2g\Delta h$ $v = \sqrt{2 \times 9.8 \text{ m s}^{-2} \times 9.0 \text{ m}}$ $= 13 \text{ m s}^{-1}$

PRACTICE PROBLEM 10

The maximum height of a roller-coaster ride is 30 m above the ground. The lowest height of the ride is 5.0 m.
a. What is the change in gravitational potential energy of a 60-kg passenger?
b. If the passenger was travelling at 0.50 m s^{-1} at the top, what would be their maximum speed at the lowest point?

Calculating gravitational potential energy using a graph of force versus height

The change in gravitational potential energy is equal to the work done on an object. It can be found by calculating the area under a graph of force versus height.

FIGURE 2.10 The area under a graph of force versus height can be used to calculate the change in gravitational potential energy.

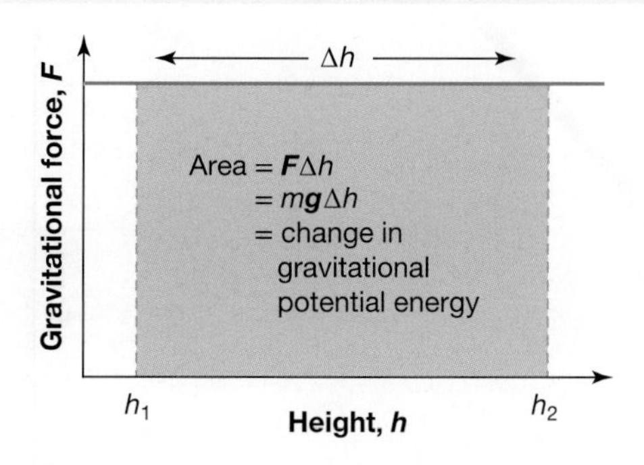

The quantity **g** is known as the gravitational field strength (sometimes referred to as gravitational field).

The change in gravitational potential energy of an object can be determined by calculating the area under a graph of gravitational field strength versus height (equal to $\mathbf{g}\Delta h$) and multiplied by its mass.

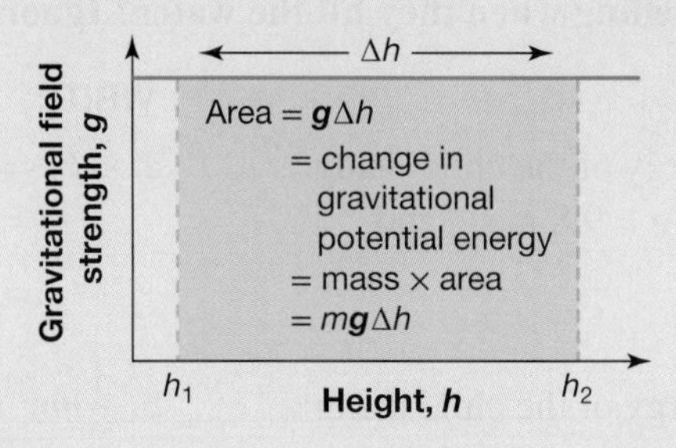

2.4.5 Energy transformations in collisions

Whether or not a collision is elastic depends on what happens to the colliding objects during the collision. When two objects collide, each object is deformed. Each object applies a force on the other — the forces are equal and opposite! The size of the applied force increases as the deformation increases (just like a compressed spring). If each object behaves elastically, all the energy stored as strain potential energy during deformation is returned to the other object as kinetic energy. The collision is therefore elastic.

In the collision between the two billiard balls discussed earlier, the work done on each ball as it returns to its original shape is almost as much as the work done during deformation. Therefore, almost all the strain potential energy stored in each ball while they are in contact with each other is returned as kinetic energy. This is an example of nearly perfect elastic collision.

The graph in figure 2.11 shows that, in an elastic collision, the work done on an object during deformation (the area under the force versus deformation graph) is equal to the work that the object does on the other object as it returns to its original shape. The graph in figure 2.12 illustrates a collision between two electrons. The work done to slow down the approaching electron is the same as the work done to increase its speed during separation.

FIGURE 2.11 A graph of force versus deformation for an object involved in an elastic collision

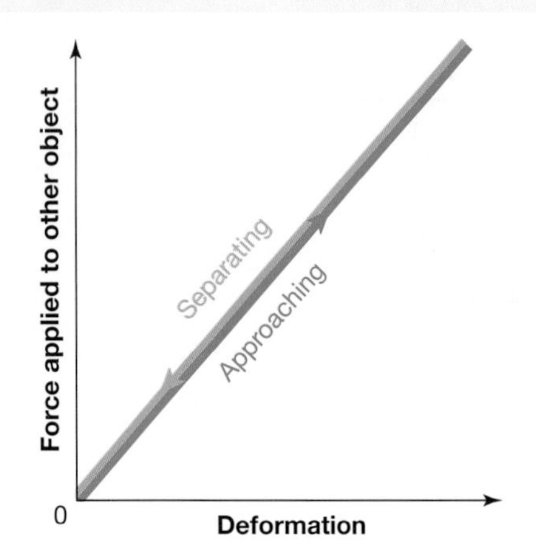

FIGURE 2.12 A graph of force versus separation for an electron approaching another electron

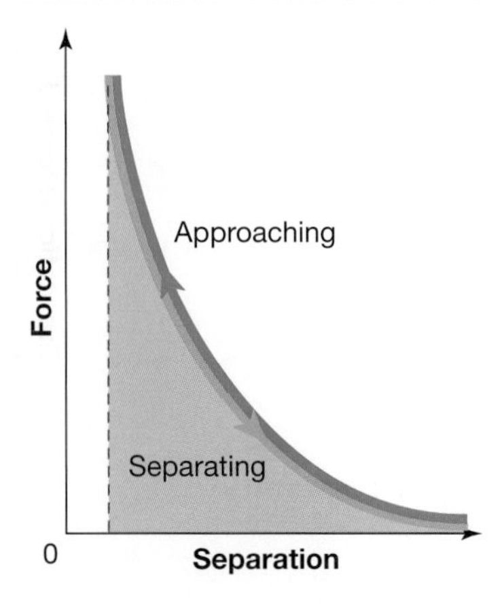

Figure 2.13 shows that, even though the total kinetic energy and total strain potential energy change during an elastic collision, the sum of the kinetic energy and strain potential energy is constant. In an inelastic collision, the sum of the kinetic energy and strain potential energy decreases because energy is dissipated from the system of objects to the environment as heat, permanent deformation of the objects and sound.

FIGURE 2.13 Energy transformations during an elastic collision

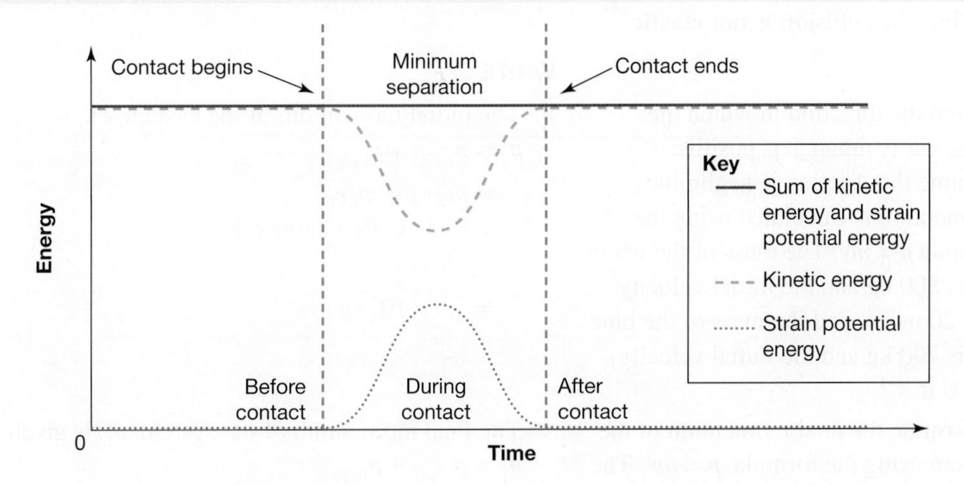

EXTENSION: Crumple zones

The crumple zones at the front and rear of cars are designed to reduce injuries by ensuring that the collisions are not elastic. Between these crumple zones is the more rigid passenger 'cell'. This is designed to protect occupants from the intrusion of the engine or other solid objects that could injure or even kill them.

According to Newton's Second Law, the car's crumple zone increases the time during which the velocity changes. The result is a decrease in the deceleration of the occupants, reducing the severity of injury.

The reason that crumple zones work can be also understood by analysing a collision's energy transformations. When a car collides with a rigid object, the object does work on the car, transforming its kinetic energy into other forms of energy and some to its surroundings. Most of the kinetic energy of the car is used to deform the body of the car while some heats the surrounding air. Without the crumple zone, the distance over which the force acts lessens, and the car would likely rebound. The result would be a greater acceleration (in magnitude) of occupants, and therefore a greater chance of serious injury or death.

The effectiveness of gloves in baseball and cricket relies on this same principle. Like the crumple zones of cars, they are designed to ensure that collisions are inelastic.

FIGURE 2.14 Crumple zones at the front and rear of cars absorb energy and reduce the magnitude of acceleration during an accident.

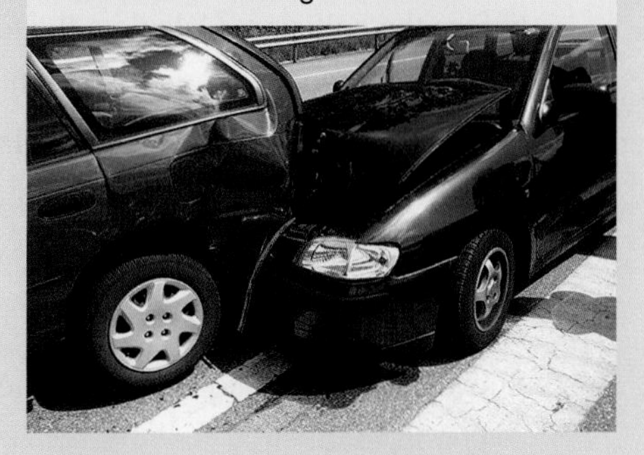

tlvd-8967

SAMPLE PROBLEM 11 Calculating speed after a collision and determining whether a collision is elastic

A white car of mass 800 kg is driven along a slippery straight road at a speed of 20 m s^{-1} (72 km h^{-1}). It collides at a stationary blue car of mass 700 kg. During the collision, the blue car is pushed forward at a speed of 12 m s^{-1}.

a. What is the speed of the white car after the collision?

b. Show that the collision is not elastic.

THINK

a. 1. Assign the direction in which the white car is moving as positive. Assume that friction is negligible. Momentum is calculated using the formula $\boldsymbol{p} = m\boldsymbol{v}$. The mass of the white car is 800 kg and the initial velocity was 20 m s^{-1}, and the mass of the blue car is 700 kg and the initial velocity was 0 m s^{-1}.

WRITE

a. The initial momentum of the system, $\boldsymbol{p}_\text{i}$, is given by:

$$\begin{aligned}\boldsymbol{p}_\text{i} &= \boldsymbol{p}_\text{white} + \boldsymbol{p}_\text{blue}\\ &= m_\text{w}\boldsymbol{v}_\text{w} + m_\text{b}\boldsymbol{v}_\text{b}\\ &= (800 \times 20) + (700 \times 0)\\ &= 16\,000 + 0\\ &= 1.6 \times 10^4 \text{ kg m s}^{-1}\end{aligned}$$

2. Determine the final momentum of the system using the formula, $\boldsymbol{p} = m\boldsymbol{v}$. The mass of the white car is 800 kg and the mass of the blue car is 700 kg and the final velocity was 12 m s^{-1}.

The final momentum of the system, $\boldsymbol{p}_\text{f}$, is given by:

$$\begin{aligned}\boldsymbol{p}_\text{f} &= \boldsymbol{p}_\text{white} + \boldsymbol{p}_\text{blue}\\ &= (800 \times \boldsymbol{v}_\text{white}) + (700 \times 12)\\ &= 800 \times \boldsymbol{v}_\text{white} + 8.4 \times 10^3\end{aligned}$$

$\boldsymbol{v}_\text{white}$ = velocity of the white car after the collision

3. Conservation of momentum states $\boldsymbol{p}_\text{i} = \boldsymbol{p}_\text{f}$.

$$\begin{aligned}800\,\boldsymbol{v}_\text{white} + 8.4 \times 10^3 &= 1.6 \times 10^4\\ 800\,\boldsymbol{v}_\text{white} &= 7.6 \times 10^3\\ \Rightarrow \boldsymbol{v}_\text{white} &= 9.5 \text{ m s}^{-1}\end{aligned}$$

The speed of the white car after the collision is 9.5 m s^{-1}.

b. 1. If the collision is elastic, the total kinetic energy after the collision will be the same as the total kinetic energy before the collision. Calculate the total kinetic energy using $\frac{1}{2}mv^2$ for both the white and the blue car and adding these together.

m_white = 800 kg, $\boldsymbol{v}_\text{white initial}$ = 20 m s^{-1}, $\boldsymbol{v}_\text{white final}$ = 9.5 m s^{-1}

m_blue = 700 kg, $\boldsymbol{v}_\text{blue initial}$ = 0 m s^{-1}, $\boldsymbol{v}_\text{blue final}$ = 12 m s^{-1}

Total kinetic energy before the collision is given by:

$$\frac{1}{2} \times 800 \times 20^2 + \frac{1}{2} \times 700 \times 0^2 = 1.6 \times 10^5 \text{ J}$$

Total kinetic energy after the collision is given by:

$$\begin{aligned}&\frac{1}{2} \times 800 \times 9.5^2 + \frac{1}{2} \times 700 \times 12^2\\ &= 8.7 \times 10^4 \text{ J}\end{aligned}$$

2. Show that the collision is not elastic.

Kinetic energy is not conserved. The collision is not elastic.

PRACTICE PROBLEM 11

a. **A green dodgem car of mass 400 kg has a head-on collision with a red dodgem car of mass 300 kg. Both dodgem cars were travelling at a speed of 2.0 m s^{-1} before the collision. What is the rebound speed of the green dodgem car if the red dodgem car rebounds at a speed of:**
   i. **1.0 m s^{-1}**
   ii. **2.0 m s^{-1}?**

b. **Are either of the collisions in part (a) elastic? If so, which one?**

EXTENSION: The importance of airbags

Most deaths and injuries in car crashes are caused by collisions between occupants and the interior of the car. Driver front airbags are designed to reduce the injuries caused by impact with the steering wheel and should inflate only in head-on collisions.

FIGURE 2.15 Testing airbags

Airbags inflate when the crash sensors in the car detect a large deceleration. When the sensors are activated, an electric current is used to ignite a chemical called sodium azide (NaN_3). The sodium azide stored in a metal container at the opening of the airbag burns rapidly, producing sodium compounds and nitrogen gas. The reaction is explosive, causing a noise like the sound of gunfire. The nitrogen gas inflates the airbag to a volume of about 45 L in only 30 ms. When the driver's head makes contact with the airbag, the airbag deflates as the nitrogen gas escapes through vents in the bag. The dust produced when an airbag is activated is a mixture of the talcum powder used to lubricate the bags and the sodium compounds produced by the chemical reaction. Deflation must be rapid enough to allow the driver to see ahead after the accident. The collision of the driver with the airbag is inelastic. Most of the kinetic energy of the driver's body is transferred to the nitrogen gas molecules, as kinetic energy.

INVESTIGATION 2.3

online only

elog-1880

vd-10811

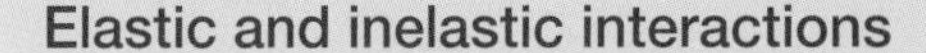

Elastic and inelastic interactions

(a) To record the motion of two objects during an interaction between them and determine the velocity of each object before and after the interaction
(b) To determine the momentum of two objects before and after three different types of interaction and, subsequently, determine whether the total momentum is conserved
(c) To determine the kinetic energy of two objects before and after three different types of interaction and, subsequently, determine whether the total kinetic energy is conserved
(d) To distinguish between elastic and inelastic collisions

on Resources

Weblink Car safety systems

2.4 Activities

learn**on**

Students, these questions are even better in jacPLUS

Receive immediate feedback and access sample responses

Access additional questions

Track your results and progress

Find all this and MORE in jacPLUS

2.4 Quick quiz on	2.4 Exercise	2.4 Exam questions

2.4 Exercise

1. A tennis ball drops vertically onto a hard surface.
   a. Is the collision of the falling ball with the ground elastic?
   b. How do you know?
   c. Is momentum conserved during this collision?
2. Two cars of equal mass and travelling in opposite directions on a wet and slippery road collide and lock together after impact. Neither car brakes before the collision. The tangled wreck moves off in an easterly direction at 5.0 m s^{-1} immediately after the collision. One car was travelling west at 20 m s^{-1} immediately before the collision.
   a. What was the initial velocity of the other car?
   b. What fraction of the initial kinetic energy was 'conserved' during the collision?
3. The graph shown describes the behaviour of three springs as known masses are suspended from one end.
   a. What is the force applied by spring A to a 1.0-kg mass suspended from one end?
   b. What is the spring constant of spring B?
   c. Which spring has the greatest stiffness?
   d. How much work is done by a 500-g mass on spring C to extend it fully?
   e. Which spring has the greatest strain energy at maximum extension?
4. A weightlifter raises a 150-kg barbell vertically through a height of 1.20 m.
   a. Sketch a graph of gravitational field strength versus height of the barbell.
   b. Use the graph to determine the change in gravitational potential energy of the barbell.
   c. How much work did the weightlifter do on the barbell?
5. Calculate the gravitational potential energy of the following objects.
   a. A 70-kg pole vaulter 6.0 m above the ground
   b. An 80-kg pile driver raised 7.0 m above the pile
   c. A 400-kg lift at the bottom of an 80-m mine shaft relative to the ground
6. A 900-kg car travelling at 20 m s^{-1} on an icy road collides with a stationary truck. The car comes to rest over a distance of 40 cm.
   a. What is the initial kinetic energy of the car?
   b. How much work is done by the truck to stop the car?
   c. What average force does the car apply to the truck during the collision?
7. A rock is dropped from a height into mud and penetrates. If it was dropped from twice the height, compare the original penetration depth to the second penetration depth.

2.4 Exam questions

Question 1 (7 marks)

Source: *VCE 2021 Physics Exam, Section B, Q.9a; © VCAA*

Abbie and Brian are about to go on their first loop-the-loop roller-coaster ride. As competent Physics students, they are working out if they will have enough speed at the top of the loop to remain in contact with the track while they are upside down at point C, shown in Figure 9. The radius of the loop CB is r.

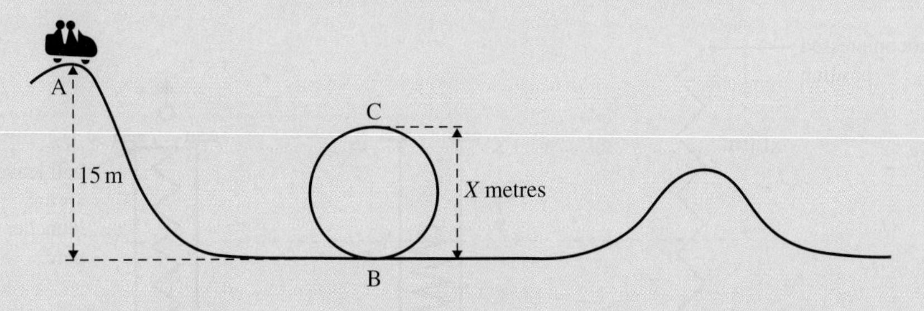

Figure 9

The highest point of the roller-coaster (point A) is 15 m above point B and the car starts at rest from point A. Assume that there is negligible friction between the car and the track.

a. What is the speed of the car at point B at the bottom of the loop? Show your working. **(2 marks)**

b. What is the maximum height of the loop (X metres) that will ensure that the car stays in contact with track at point C? Show your working. **(2 marks)**

c. If friction is taken into account, will Abbie and Brian need to increase or decrease their predicted value for the radius of the loop? Explain your answer. **(3 marks)**

Question 2 (8 marks)

Source: *VCE 2019 Physics Exam, NHT, Section B, Q.5; © VCAA*

Students conduct an experiment in which a mass of 2.0 kg is suspended from a spring with spring constant $k = 100\ \text{N m}^{-1}$.

Ignore the mass of the spring.

Take the gravitational field, g, to be $10\ \text{N kg}^{-1}$

Take the zero of gravitational potential energy when the mass is at its lowest point.

The experimental arrangement is shown in Figure 6.

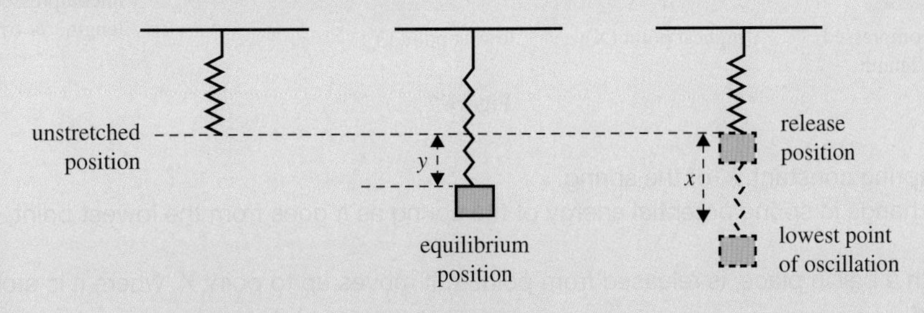

Figure 6

a. The mass is attached to the spring and slowly lowered to its equilibrium position. Calculate the extension, y, of the spring from its unstretched position to its equilibrium position. Show your working. **(2 marks)**

b. The mass is now raised to the unstretched length of the spring and released so that it oscillates vertically.

i. Determine the distance, x, from the release position to the point at which the mass momentarily comes to rest at the lowest point of oscillation. Ignore frictional losses. Show your working. **(2 marks)**

ii. Calculate the maximum speed of the mass. Show your working. **(4 marks)**

Question 3 (9 marks)

Source: *VCE 2018 Physics Exam, NHT, Section B, Q.9; © VCAA*

A spring launcher is used to project a rubber ball of mass 2.0 kg vertically upwards. The arrangement is shown in Figure 6.

The ball is driven by a spring, which is compressed and released. When the spring reaches the top, point X, it is held stationary, but is still partly compressed as the ball leaves the launcher. Assume that the spring has no mass.

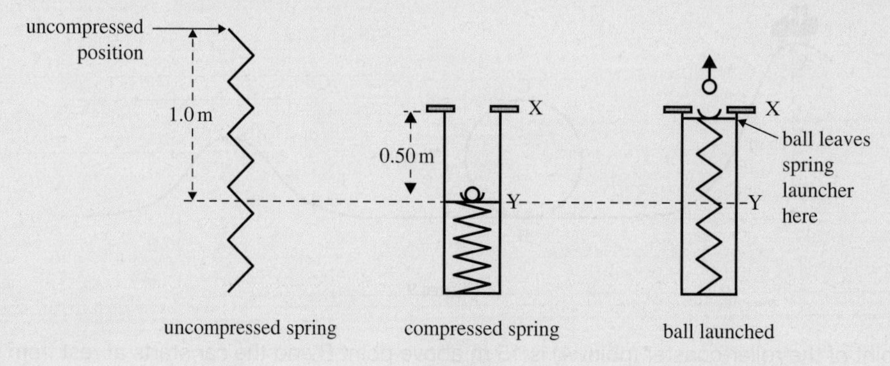

Figure 6

The force–distance graph of the spring is shown in Figure 7, on which the lower and upper positions of the spring in the spring launcher are marked.

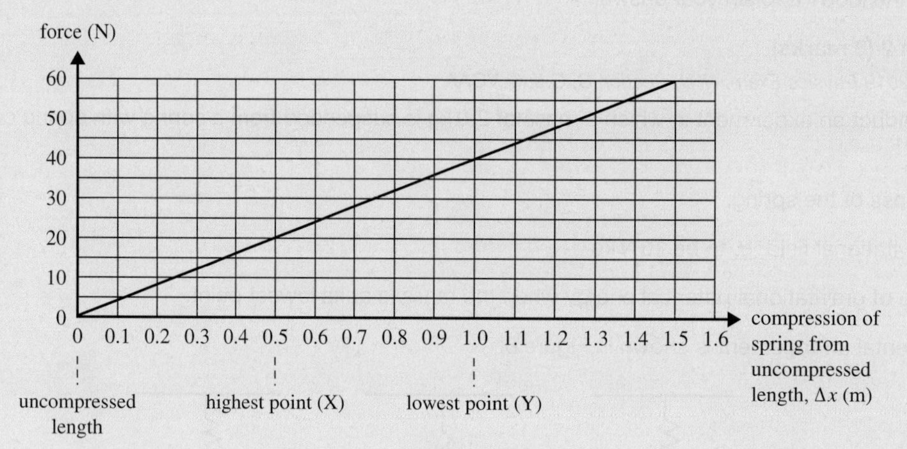

Figure 7

a. Calculate the spring constant, k, of the spring. **(2 marks)**

b. Calculate the change in spring potential energy of the spring as it goes from the lowest point, Y, to the highest point, X. **(3 marks)**

c. The spring, with a ball in place, is released from point Y. It moves up to point X, where it is stopped and the ball is launched.
Calculate the speed of the ball when it leaves the spring launcher. Show the steps involved in your working. **(4 marks)**

Question 4 (2 marks)

Source: *VCE 2017, Physics Exam, Section B, Q.8.b; © VCAA*

A roller-coaster is arranged so that the normal reaction force on a rider in a car at the top of the circular arc at point P, shown in Figure 7, is briefly zero. The section of track at point P has a radius of 6.4 m.

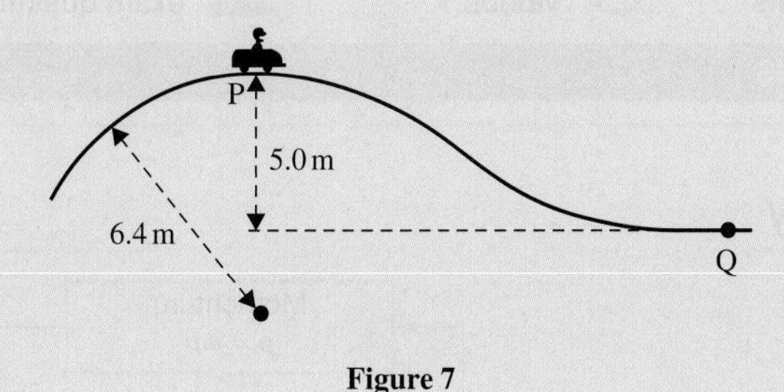

Figure 7

The car is faulty and only achieves a speed of 4.0 m s^{-1} at the top of the arc at point P.

Calculate how fast this car would be moving when it reaches the bottom at point Q, 5.0 m below point P. Assume that there is no friction and no driving force on the car.

Question 5 (3 marks)

Source: *VCE 2017, Physics Exam, Section B, Q.12; © VCAA*

Students are using two trolleys, Trolley A of mass 4.0 kg and Trolley B of mass 2.0 kg, to investigate kinetic energy and momentum in collisions.

Before the collision, Trolley A is moving to the right at 5.0 m s^{-1} and Trolley B is moving to the right at 2.0 m s^{-1}, as shown in Figure 10a. The trolleys collide and lock together, as shown in Figure 10b.

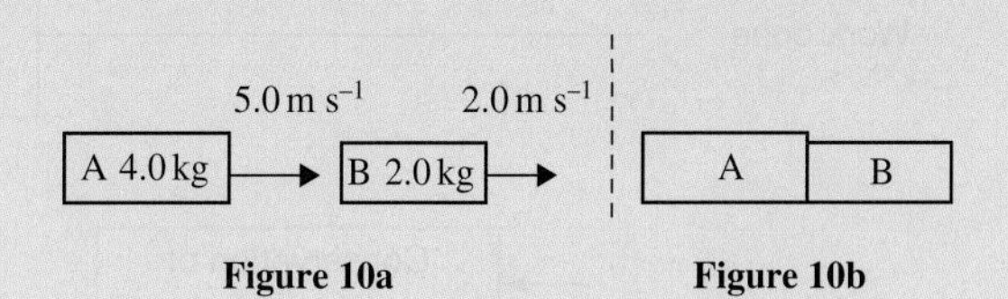

Figure 10a **Figure 10b**

Determine, using calculations, whether the collision is elastic or inelastic. Show your working and justify your answer.

More exam questions are available in your learnON title.

2.5 Review

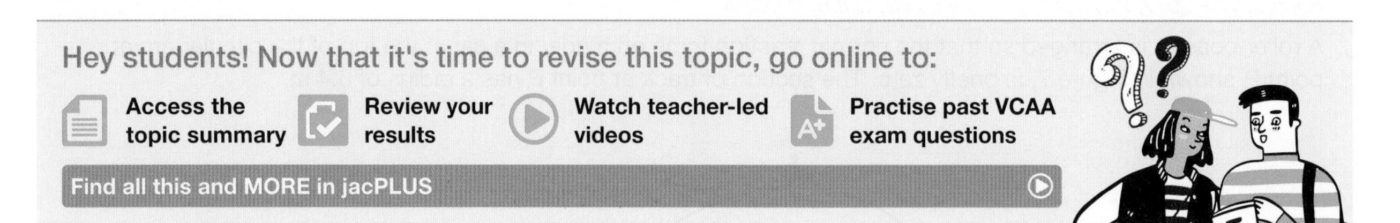

2.5.1 Topic summary

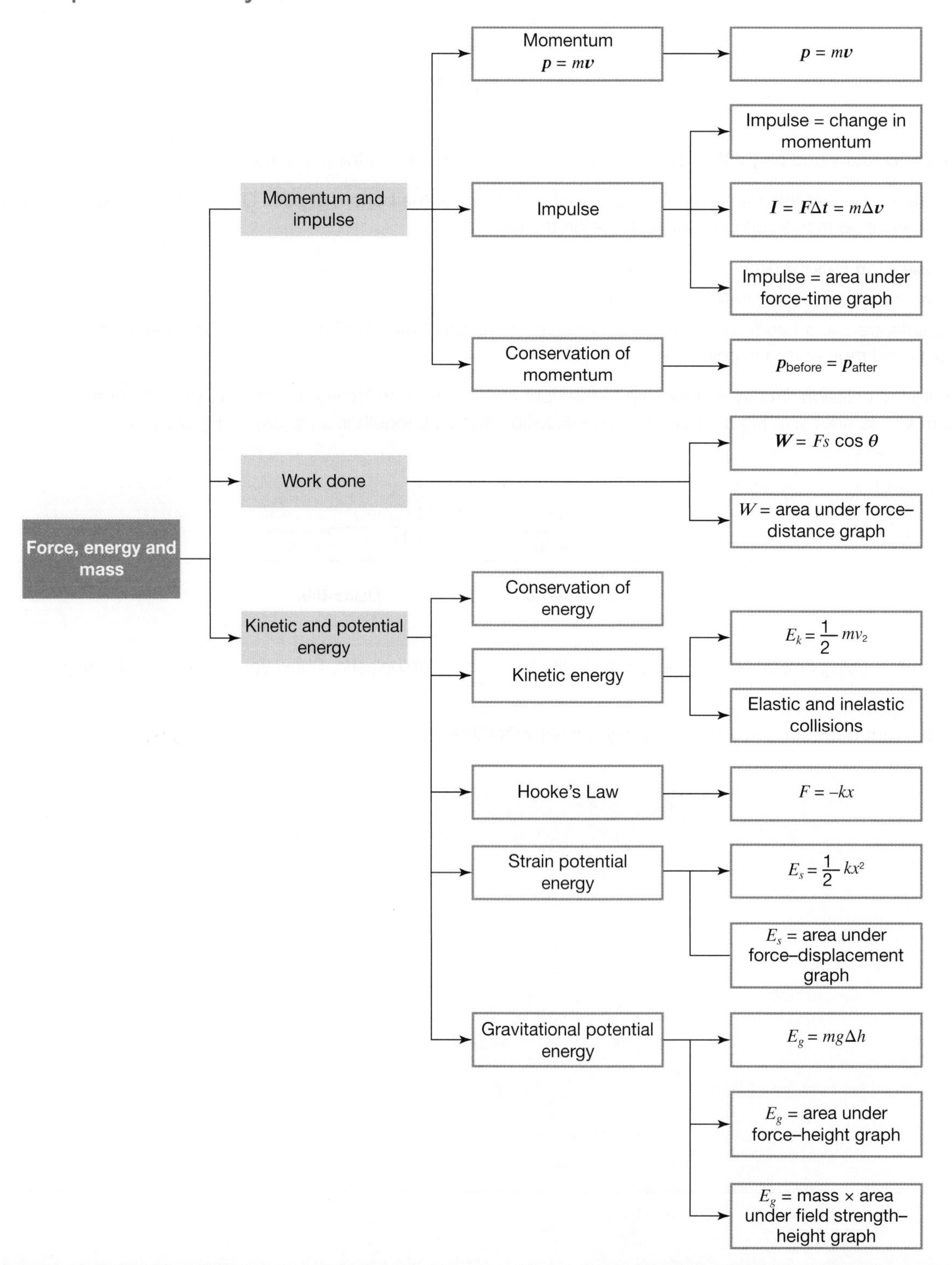

2.5.2 Key ideas summary

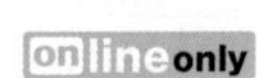

2.5.3 Key terms glossary

online only

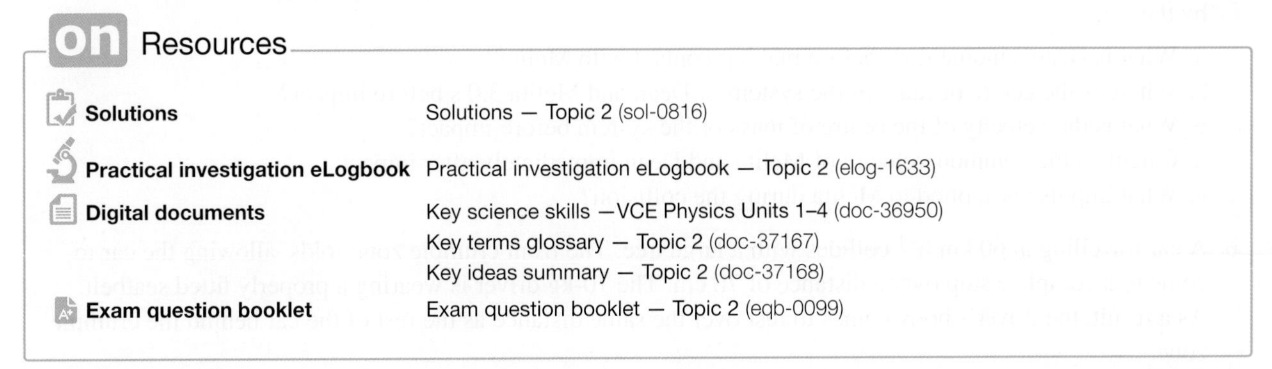

on Resources

Solutions	Solutions — Topic 2 (sol-0816)
Practical investigation eLogbook	Practical investigation eLogbook — Topic 2 (elog-1633)
Digital documents	Key science skills —VCE Physics Units 1–4 (doc-36950) Key terms glossary — Topic 2 (doc-37167) Key ideas summary — Topic 2 (doc-37168)
Exam question booklet	Exam question booklet — Topic 2 (eqb-0099)

2.5 Activities

learn on

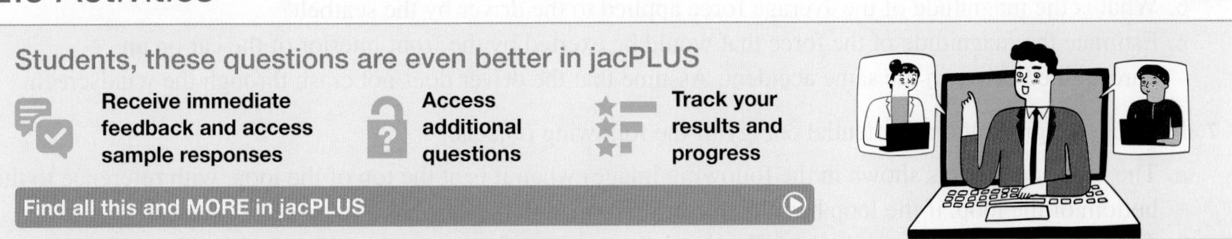

2.5 Review questions

1. Two billiard balls with identical mass, *m*, collide with one another. Ball 1 has an initial velocity of 5 m s^{-1} to the right. After the collision, ball 1 continues to move to the right, now with a velocity of 0.8 m s^{-1}. Calculate the velocity of ball 2 before the collision if the collision is completely elastic. Assume the system is isolated and has no external forces acting on it.

2. A contractor lifts a 10-kg tool box up 0.40 m to hand to a colleague, at constant speed. Calculate the work done by the contractor.

3. The ancient Egyptians relied on knowledge of the physics of energy transformations to build the Great Pyramids at Giza. They used ramps to push limestone blocks with an average mass of 2300 kg to heights of almost 150 m. The ramps were sloped at about 10° to the horizontal. Friction was reduced by pumping water onto the ramps.
   a. How much work would have been done to lift an average limestone block vertically through a height of 150 m?
   b. How much work would have been done to push an average limestone block to the same height along a ramp inclined at 10° to the horizontal? Assume that friction is negligible.

4. A 60-kg bungee-jumper falls from a bridge 50 m above a deep river. The length of the bungee cord when it is not under tension is 30 m. Calculate the:
   a. kinetic energy of the bungee-jumper at the instant that the cord begins to stretch beyond its natural length
   b. strain energy of the bungee cord at the instant that the tip of the jumper's head touches the water. (Her head just makes contact with the water before she is pulled upwards by the cord.) The height of the bungee-jumper is 170 cm.

5. Two ice skaters, Melita and Dean, are performing an ice dancing routine. Dean (mass of 70 kg) glides smoothly at a velocity of 2.0 m s^{-1} east towards a stationary Melita (mass of 50 kg), holds her around the waist, and then they both move off together. During the whole move, no significant frictional force is applied by the ice.
   a. What is Dean's momentum before making contact with Melita?
   b. Where is the centre of mass of the system of Dean and Melita 3.0 s before impact?
   c. What is the velocity of the centre of mass of the system before impact?
   d. Calculate the common velocity of Melita and Dean immediately after impact.
   e. What impulse is applied to Melita during the collision?
6. A car travelling at 60 km h^{-1} collides with a large tree. The front crumple zone folds, allowing the car to come to a complete stop over a distance of 70 cm. The 70-kg driver is wearing a properly fitted seatbelt. As a result, the driver's body comes to rest over the same distance as the rest of the car behind the crumple zone.
   a. Determine the amount of work done by the seatbelt in stopping the driver
   b. What is the magnitude of the average force applied to the driver by the seatbelt?
   c. Estimate the magnitude of the force that would be exerted by the front interior of the car on an unrestrained driver in the same accident. Assume that the driver does not crash through the windscreen.
7. Estimate the gravitational potential energy of the following objects.
   a. The roller-coaster (as shown in the following image) when it is at the top of the loop, with reference to the bottom of the loop, if the loop has a diameter of 10 m, and the cart has a mass of 4.5 tonnes
   b. The high jumper (as shown in the following image) with reference to the ground, if the jumper has a mass of 60 kg, and is attempting a 1.7-m jump
   c. A 1.3-kg textbook on a 1.2-m high table, with reference to the floor
   d. A 58-g tennis ball about to be hit, at a height of 3 m, during a serve with reference to the ground

8. Angela rides a toboggan down a slope inclined at 30° to the horizontal. She starts from rest and rides a distance of 25 m down the slope. Angela and her toboggan have a combined mass of 60 kg.
   a. How much work is done on Angela by the force of gravity?
   b. If friction is negligible, what would her speed be at the end of her ride?
   c. How much work is done on Angela by the normal reaction?
   d. In reality, the frictional force on Angela is not negligible. Her speed at the end of her ride is measured to be 7.2 m s^{-1}. What is the magnitude of the frictional force?
9. A 1500-kg car travelling west at a speed of 20 m s^{-1} on an icy road collides with a 2000-kg truck travelling at the same speed in the opposite direction. The vehicles lock together at impact.
   a. What is the velocity of the tangled wreck immediately after the collision?
   b. Use your answer to part a to determine what impulse is applied to the truck during the collision.
   c. Which vehicle experiences the greatest change in velocity, in magnitude?
   d. Which vehicle experiences the greatest change in momentum?
   e. Which vehicle experiences the greatest force?

10. The graph shows how the force applied by the rubber bumper at the front of a 450-kg dodgem car changes as it is compressed during factory testing.

  a. If the dodgem car collides head-on with a solid wall at a speed of 2.0 m s^{-1}, what is the maximum compression of the front rubber bumper?
  b. How much work is done on the dodgem car by the rubber bumper as it is compressed?
  c. If the rubber bumper obeys Hooke's Law, with what speed will the dodgem car rebound from the wall?

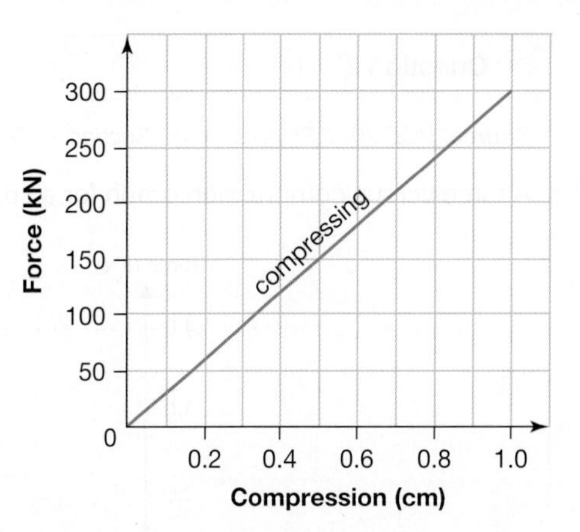

2.5 Exam questions

Section A — Multiple choice questions

All correct answers are worth 1 mark each; an incorrect answer is worth 0.

Question 1

Source: *VCE 2021 Physics Exam, Section A, Q.11; © VCAA*

A force versus compression graph for a suspension spring is shown below.

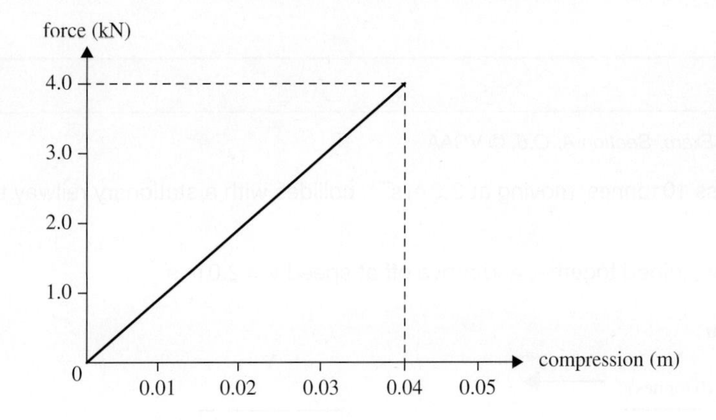

Which one of the following is closest to the spring constant of the spring?

A. 0.16 N m^{-1}
B. 1.0×10^2 N m^{-1}
C. 1.6×10^2 N m^{-1}
D. 1.0×10^5 N m^{-1}

Question 2

Source: *VCE 2021 Physics Exam, Section A, Q.12; © VCAA*

A force versus compression graph for a suspension spring is shown below.

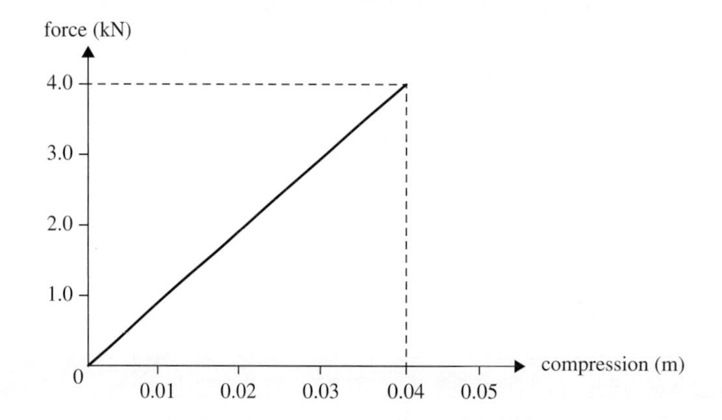

The spring is compressed to 0.02 m.

Which one of the following is closest to the potential energy stored in the spring?

A. 0.04 J
B. 0.20 J
C. 20 J
D. 40 J

Question 3

Source: *VCE 2022 Physics Exam, Section A, Q.6*; © VCAA

A railway truck (X) of mass 10 tonnes, moving at 3.0 m s^{-1}, collides with a stationary railway truck (Y), as shown in the diagram below.

After the collision, they are joined together and move off at speed $v = 2.0 \text{ m s}^{-1}$.

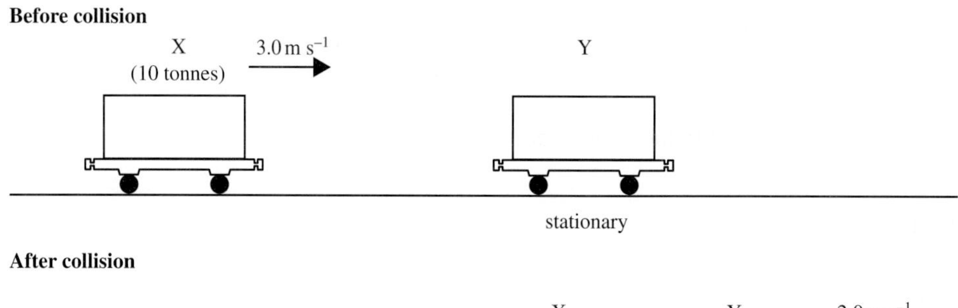

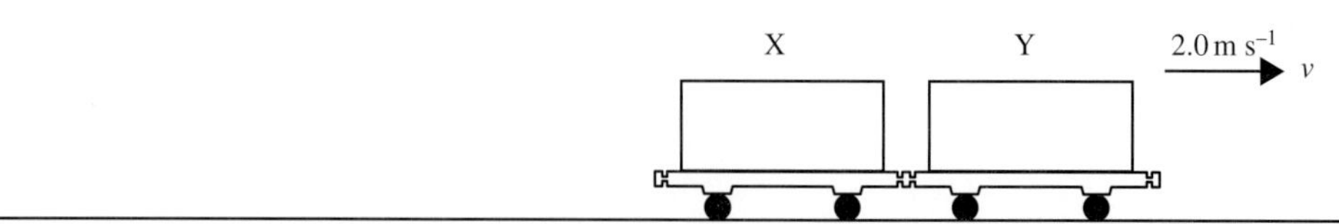

Which one of the following is closest to the mass of railway truck Y?

A. 3 tonnes
B. 5 tonnes
C. 6.7 tonnes
D. 15 tonnes

Question 4

Source: *VCE 2020 Physics Exam, Section A, Q.9;* © VCAA

Two blocks of mass 5 kg and 10 kg are placed in contact on a frictionless horizontal surface, as shown in the diagram below. A constant horizontal force, *F*, is applied to the 5-kg block.

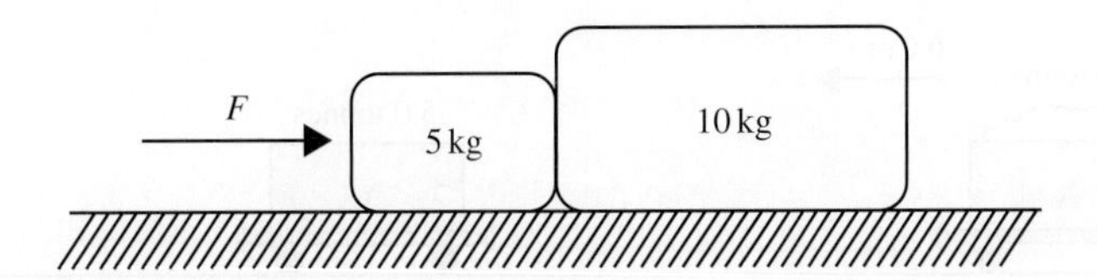

Which one of the following statements is correct?

A. The net force on each block is the same.
B. The acceleration experienced by the 5 kg block is twice the acceleration experienced by the 10 kg block.
C. The magnitude of the net force on the 5 kg block is half the magnitude of the net force on the 10 kg block.
D. The magnitude of the net force on the 5 kg block is twice the magnitude of the net force on the 10 kg block.

Question 5

Source: *VCE 2020 Physics Exam, Section A, Q.10;* © VCAA

Two blocks of mass 5 kg and 10 kg are placed in contact on a frictionless horizontal surface, as shown in the diagram below. A constant horizontal force, *F*, is applied to the 5-kg block.

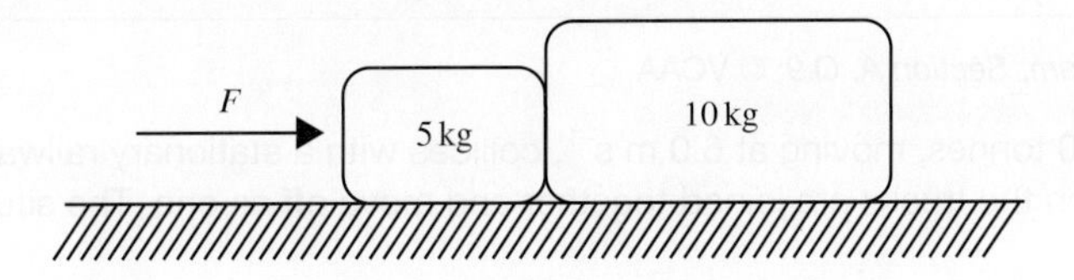

If the force *F* has a magnitude of 250 N, what is the work done by the force in moving the blocks in a straight line for a distance of 20 m?

A. 5 kJ
B. 25 kJ
C. 50 kJ
D. 500 kJ

Question 6

Source: *VCE 2018 Physics Exam, NHT, Section A, Q.12;* © VCAA

A golf club strikes a stationary golf ball of mass 0.040 kg. The golf club is in contact with the ball for one millisecond. The ball moves off at 50 m s^{-1}.

The average force exerted by the club on the ball is closest to

A. 2.0 N
B. 1.0×10^3 N
C. 2.0×10^3 N
D. 1.0×10^6 N

Question 7

Source: *VCE 2018, Physics Exam, Section A, Q.8*; © VCAA

A railway truck X of mass 10 tonnes, moving at 6.0 m s^{-1}, collides with a stationary railway truck Y of mass 5.0 tonnes. After the collision the trucks are joined together and move off as one. The situation is shown below.

Before the collision

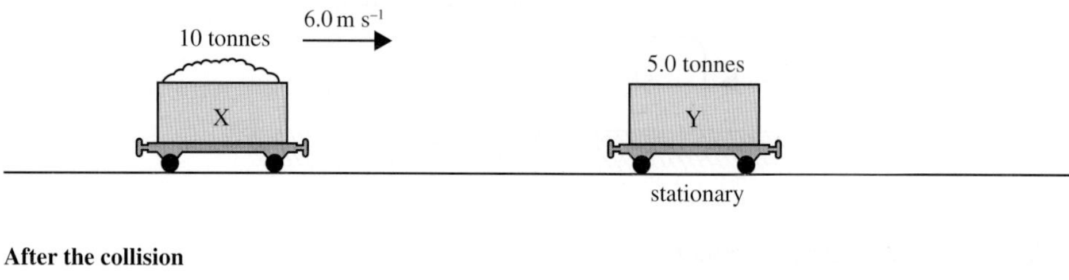

After the collision

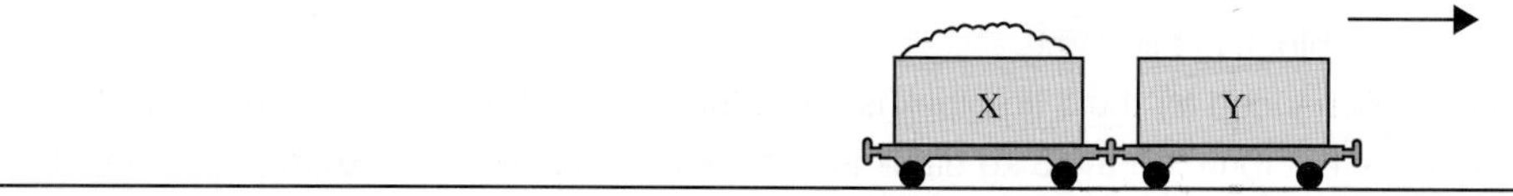

The final speed of the joined railway trucks after the collision is closest to

A. 2.0 m s^{-1}
B. 3.0 m s^{-1}
C. 4.0 m s^{-1}
D. 6.0 m s^{-1}

Question 8

Source: *VCE 2018, Physics Exam, Section A, Q.9*; © VCAA

A railway truck X of mass 10 tonnes, moving at 6.0 m s^{-1}, collides with a stationary railway truck Y of mass 5.0 tonnes. After the collision the trucks are joined together and move off as one. The situation is shown below.

Before the collision

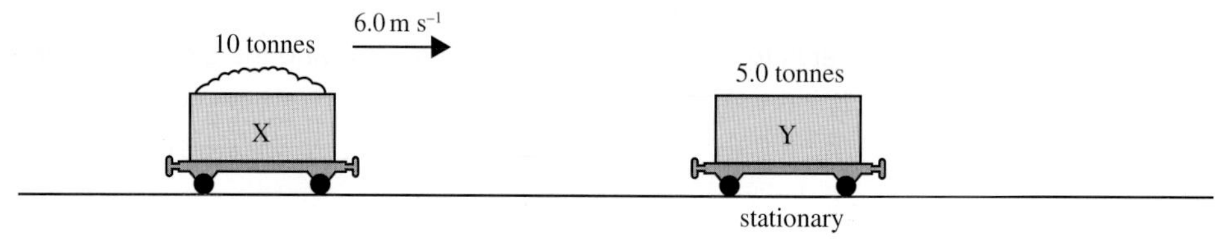

After the collision

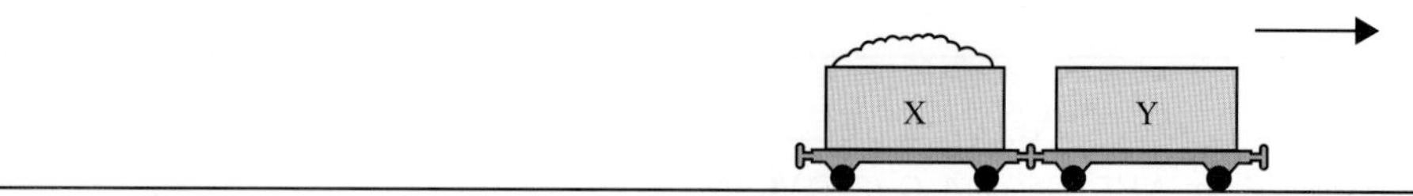

The collision of the railway trucks is best described as one where

A. kinetic energy is conserved but momentum is not conserved.
B. kinetic energy is not conserved but momentum is conserved.
C. neither kinetic energy nor momentum is conserved.
D. both kinetic energy and momentum are conserved.

Question 9

Source: *VCE 2017, Physics Exam, Section A, Q.8;* © VCAA

A model car of mass 2.0 kg is propelled from rest by a rocket motor that applies a constant horizontal force of 4.0 N, as shown below. Assume that friction is negligible.

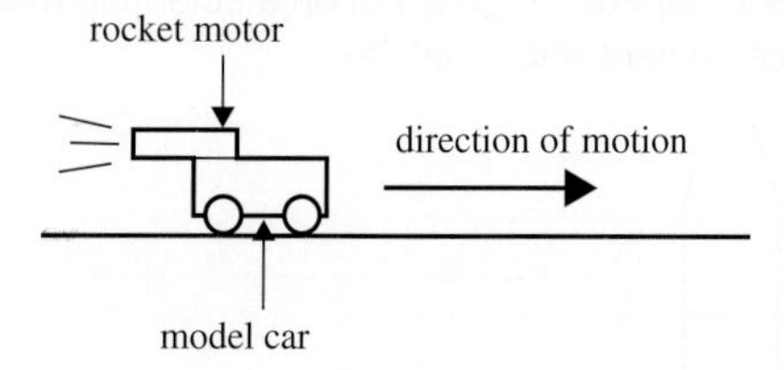

Which one of the following best gives the magnitude of the impulse given to the car by the rocket motor in the first 5.0 s?

A. 4.0 N s
B. 8.0 N s
C. 20 N s
D. 40 N s

Question 10

Source: *VCE 2017, Physics Exam, Section A, Q.13;* © VCAA

A model car is on a track and moving to the right. It collides with and compresses a spring that is considered ideal, as shown in the diagram below.

The car compresses the spring to 0.50 m when the car comes to rest. The force–distance graph for the spring is also shown below.

Assume that friction is negligible.

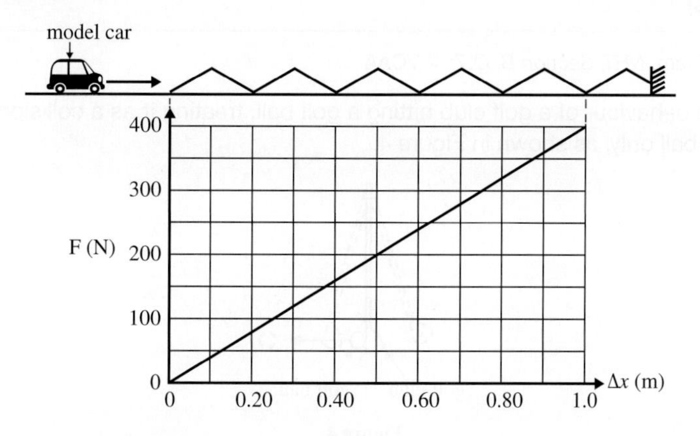

What is the initial kinetic energy of the car?

A. 25 J
B. 50 J
C. 100 J
D. 200 J

Section B — Short answer questions

Question 11 (2 marks)

Source: *VCE 2021 Physics Exam, Section B, Q.8b; © VCAA*

On 30 July 2020, the National Aeronautics and Space Administration (NASA) launched an Atlas rocket (Figure 7a) containing the Perseverance rover space capsule (Figure 7b) on a scientific mission to explore the geology and climate of Mars, and search for signs of ancient microbial life.

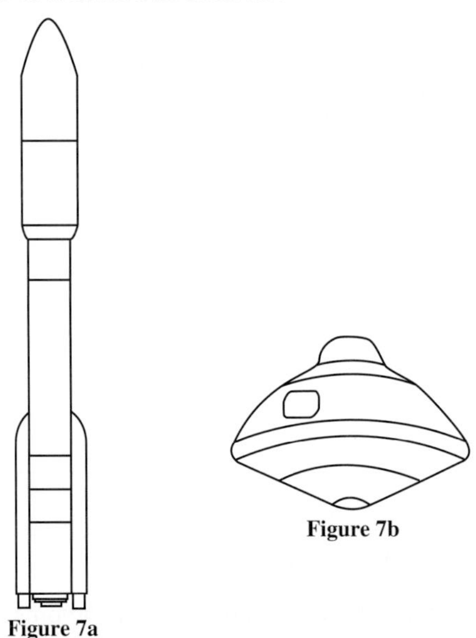

Figure 7b

Figure 7a

On 18 February 2021, the Perseverance rover space capsule, travelling at 20 000 km h^{-1}, entered Mars's atmosphere at an altitude of 300 km above the surface of Mars. The mass of the capsule was 1000 kg.

Calculate the kinetic energy of the capsule at this point. Show your working.

Question 12 (3 marks)

Source: *VCE 2018 Physics Exam, NHT, Section B, Q.7; © VCAA*

Students are studying the behaviour of a golf club hitting a golf ball, treating it as a collision between the head of the golf club and the golf ball only, as shown in Figure 4.

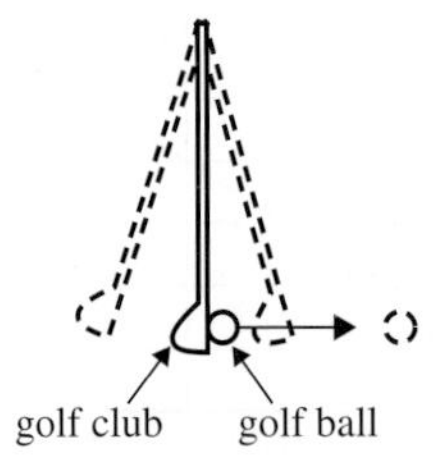

Figure 4

The students take the following measurements.

mass of head of golf club	0.50 kg
mass of golf ball	0.040 kg
initial speed of golf club	45 m s^{-1}
final speed of golf club after hitting golf ball	40 m s^{-1}

The golf ball is stationary before being hit. The ball's speed immediately after being hit is 63 m s^{-1}.

Use calculations to determine whether the collision is elastic or inelastic. Show your working.

Question 13 (5 marks)

Source: *VCE 2020 Physics Exam, Section B, Q.9;* © VCAA

An ideal spring is compressed by 0.15 m. A ball of mass 0.20 Kg is placed in contact with the compressed spring. The spring is then released, causing the ball to move horizontally, with a velocity of v, across a smooth surface, as shown in the figure.

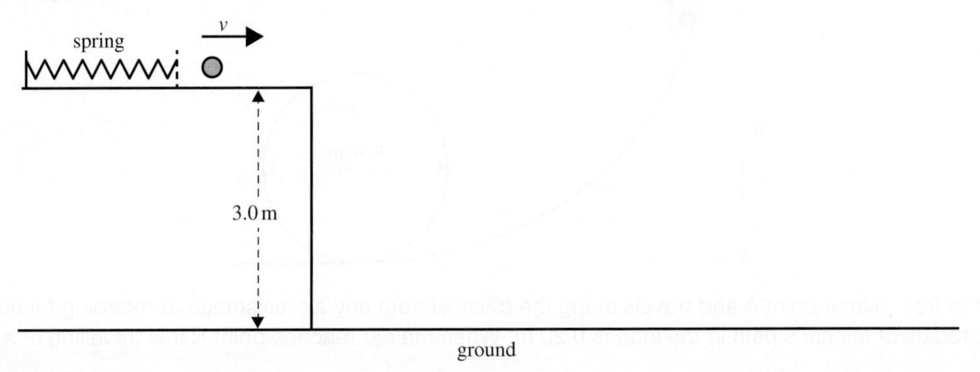

a. If the spring constant is 1250 N m^{-1}, show that the magnitude of the initial velocity, v, of the ball is 12 m s^{-1}, correct to two significant figures. Show your working. **(2 marks)**

b. Calculate the speed of the ball after it has fallen a vertical distance of 2.5 m. Show your working. **(3 marks)**

Question 14 (12 marks)

Source: *VCE 2020 Physics Exam, Section B, Q.10;* © VCAA

Jacinda designs a computer simulation program as part of her practical investigation into the physics of vehicle collisions. She simulates colliding a car of mass 1200 kg, moving at 10 m s^{-1}, into a stationary van of mass 2200 kg. After the collision, the van moves to the right at 6.5 m s^{-1}. This situation is shown in the figure.

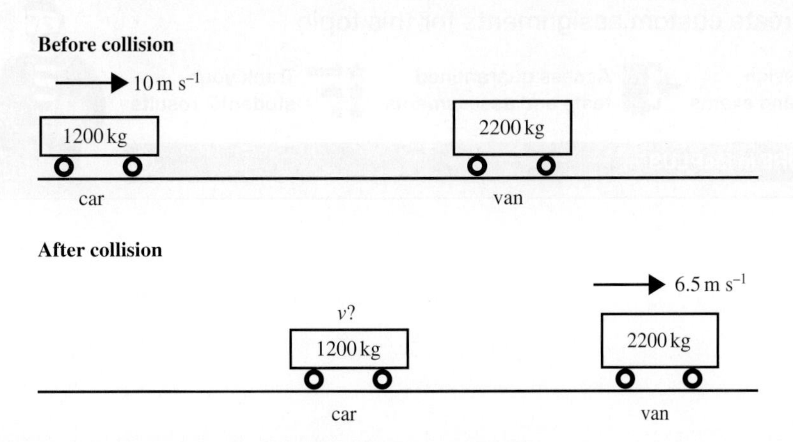

a. Calculate the speed of the car after the collision and indicate the direction it would be travelling in. Show your working. **(4 marks)**

b. Explain, using appropriate physics, why this collision represents an example of either an elastic or an inelastic collision. **(3 marks)**

c. The collision between the car and the van takes 40 ms.

i. Calculate the magnitude and indicate the direction of the average force on the van by the car. **(3 marks)**

ii. Calculate the magnitude and indicate the direction of the average force on the car by the van. **(2 marks)**

Question 15 (3 marks)

Source: *VCE 2019, Physics Exam, Section B, Q.8;* © VCAA

A 250 g toy car performs a loop in the apparatus shown in the figure.

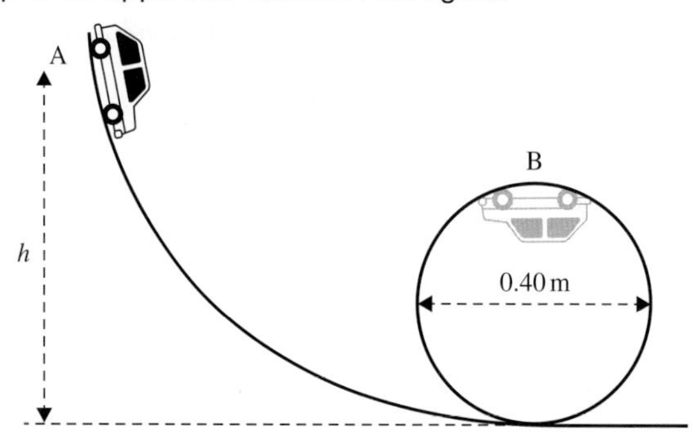

The car starts from rest at point A and travels along the track without any air resistance or retarding frictional forces. The radius of the car's path in the loop is 0.20 m. When the car reaches point B it is travelling at a speed of 3.0 m s^{-1}.

Calculate the value of h. Show your working.

UNIT 3 | AREA OF STUDY 1 REVIEW

AREA OF STUDY 1 How do physicists explain motion in two dimensions?

OUTCOME 1

Investigate motion and related energy transformations experimentally, and analyse motion using Newton's laws of motion in one and two dimensions.

PRACTICE EXAMINATION

STRUCTURE OF PRACTICE EXAMINATION		
Section	**Number of questions**	**Number of marks**
A	20	20
B	5	20
	Total	**40**

Duration: 50 minutes

Information:

- This practice examination consists of two parts. You must answer all question sections.
- Pens, pencils, highlighters, erasers, rulers and a scientific calculator are permitted.
- You may use the VCAA Physics formula sheet for this task.

Resources

Weblink VCAA Physics formula sheet

SECTION A — Multiple choice questions

All correct answers are worth 1 mark each; an incorrect answer is worth 0.

Use the following information to answer questions 1, 2 and 3:

A small van, with a mass of 2500 kg, was travelling at 25 m s^{-1} south when it collided with the rear of a car travelling in the same direction. The velocity of the van is reduced to 15 m s^{-1} south over a time interval of 0.80 s.

1. Which of the following best describes the average acceleration of the van?

A. 12.5 m s^{-2} north

B. 12.5 m s^{-2} south

C. 8.0 m s^{-2} north

D. 8.0 m s^{-2} south

2. What is the impulse on the car by the van? You may assume the collision is isolated.

A. 12 500 N s north

B. 12 500 N s south

C. 25 000 N s north

D. 25 000 N s south

3. How much kinetic energy was lost by the van during the collision?
   A. 1.3×10^4 J
   B. 2.8×10^5 J
   C. 5.0×10^5 J
   D. 7.8×10^5 J

Use the following information to answer questions 4, 5 and 6:
A golfer tees off by hitting a golf ball with a velocity of 40 m s^{-1} at an angle of 34° to the horizontal. You may ignore the effects of air resistance.

4. Assuming the ball lands at the same height as the height from which it was hit, which of the following is the closest to the expected range of the golf ball?
   A. 61 m
   B. 99 m
   C. 134 m
   D. 151 m
5. What is the maximum height the ball attains above the point where it was hit?
   A. 26 m
   B. 46 m
   C. 56 m
   D. 68 m
6. If the effects of air resistance are taken into consideration, what would happen to the range and the maximum height attained by the golf ball?
   A. The range would increase and the maximum height would decrease.
   B. The range would increase and the maximum height would increase.
   C. The range would decrease and the maximum height would increase.
   D. The range would decrease and the maximum height would decrease.

Use the following information to answer questions 7 and 8:
A cyclist and her bicycle, total mass of 78 kg, negotiates a roundabout at a constant speed of 12 m s^{-1}. The diameter of her path around the roundabout is 18 m.

7. What is the nearest value for the net force on the tyres of the bicycle?
   A. 50 N
   B. 100 N
   C. 620 N
   D. 1250 N
8. The cyclist increases her speed to 14 m s^{-1}; however, she decides to change the diameter of her path around the roundabout so that the net force on the bicycle tyres remains the same. What should she do?
   A. Increase the diameter of her path
   B. Decrease the diameter of her path
   C. Retain the same diameter of her path
   D. There is insufficient information to determine the change required.

Use the following information to answer questions 9 and 10:
A car is travelling over a crest with a radius of 9.0 m at a constant speed, $\boldsymbol{v}$, as shown in the following diagram.

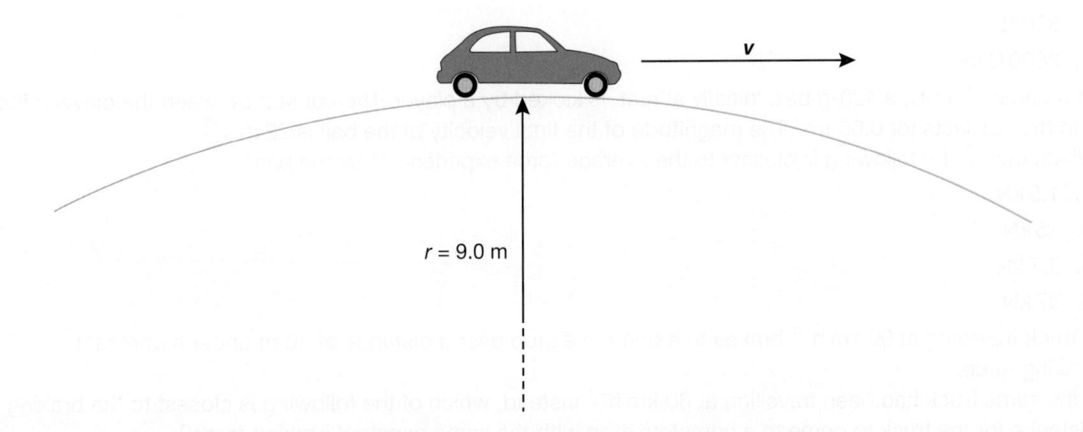

9. The normal force on the car when it is at the top of the crest is zero ($\boldsymbol{F}_N = 0$ N). At what speed is the car travelling?

A. 4.9 m s^{-1}
B. 9.4 m s^{-1}
C. 77 m s^{-1}
D. 96 m s^{-1}

10. The car travels over the crest again but this time at a lower speed. Which of the following is the expected normal force on the car?

A. $\boldsymbol{F}_N = 0$ N
B. $\boldsymbol{F}_N < 0$ N
C. $\boldsymbol{F}_N > 0$ N
D. Unknown, since there is insufficient information

11. Two large dogs are playing tug-of-war. Each dog pulls with a force of magnitude 600 N. The magnitude of the force exerted by the rope on each dog is closest to:

A. 0 N.
B. 300 N.
C. 600 N.
D. 1200 N.

12. Devi and Helen want to use a spring for a practical experiment but the one they found does not have its spring constant labelled. They used slotted masses and a ruler to create the following force–distance plot with a line of best fit.

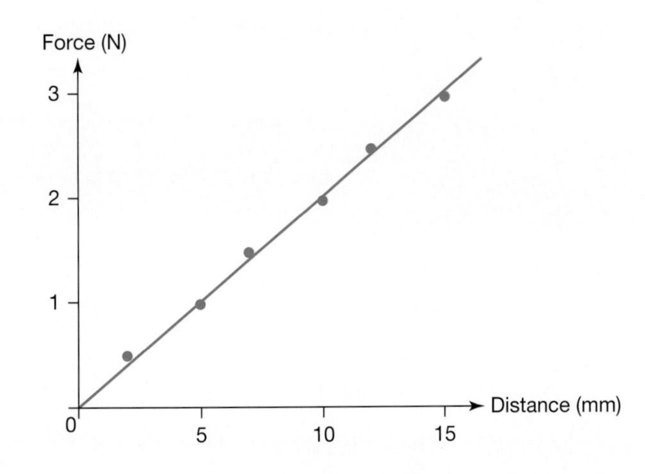

The spring constant of the spring they found is closest to:

A. $0.2\,\mathrm{N\,m^{-1}}$.

B. $200\,\mathrm{N\,m^{-1}}$.

C. $500\,\mathrm{N\,m^{-1}}$.

D. $2000\,\mathrm{N\,m^{-1}}$.

13. At a game of footy, a 420-g ball, initially at rest, is kicked by a player. The contact between the player's foot and the ball lasts for 0.60 ms. The magnitude of the final velocity of the ball is $22\,\mathrm{m\,s^{-1}}$.
Which one of the following is closest to the average force experienced by the ball?

A. 1.5 kN

B. 15 kN

C. 3.7 kN

D. 37 kN

14. A truck travelling at $90\,\mathrm{km\,h^{-1}}$ brakes to a complete stop over a distance of 40 m under a constant braking force.
If the same truck had been travelling at $60\,\mathrm{km\,h^{-1}}$ instead, which of the following is closest to the braking distance for the truck to come to a complete stop with the same constant braking force?

A. 40 m

B. 30 m

C. 27 m

D. 18 m

15. The force–distance graph of a bow being drawn is shown below.

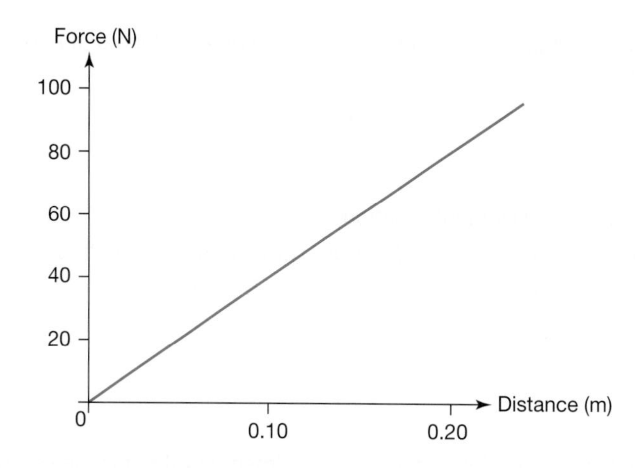

The amount of work done to draw the bow back 20 cm is closest to:

A. 8 J.

B. 16 J.

C. 8 kJ.

D. 16 kJ.

Use the following information to answer questions 16 and 17:

A 3.0-kg mass is resting on a frictionless surface. It is connected via a frictionless pulley to a 2.0-kg mass by a massless cable. The 2.0-kg mass is then allowed to fall due to gravity.

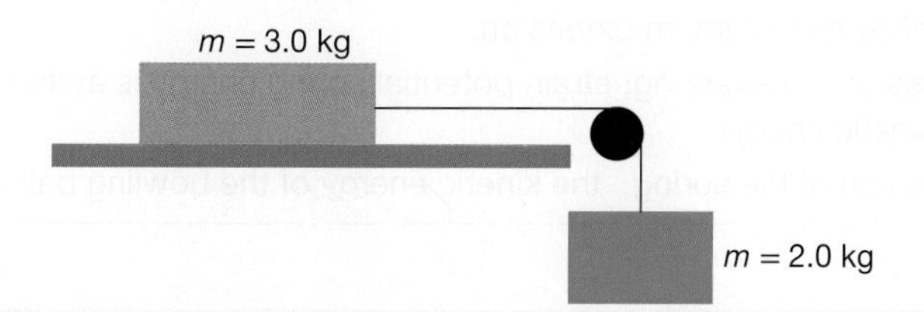

16. What is the magnitude of the acceleration of both masses?

A. 1.9 m s^{-2}

B. 3.9 m s^{-2}

C. 5.9 m s^{-2}

D. 7.9 m s^{-2}

17. If the two masses were swapped, what would happen to the magnitude of the acceleration when the 3.0-kg falling mass is allowed to fall, compared to the magnitude of the acceleration in the case represented in the diagram?

A. It would increase.

B. It would decrease.

C. It would stay the same.

D. There is insufficient information to determine this.

Use the following information to answer questions 18, 19 and 20:

A bowling ball is allowed to fall, due to gravity, onto a spring, as shown in the following diagram. As it comes into contact with the spring, it begins to compress the spring. The ball continues to compress the spring until it stops momentarily. After this moment, the spring begins to rebound and the bowling ball begins to move upwards. Air resistance may be ignored.

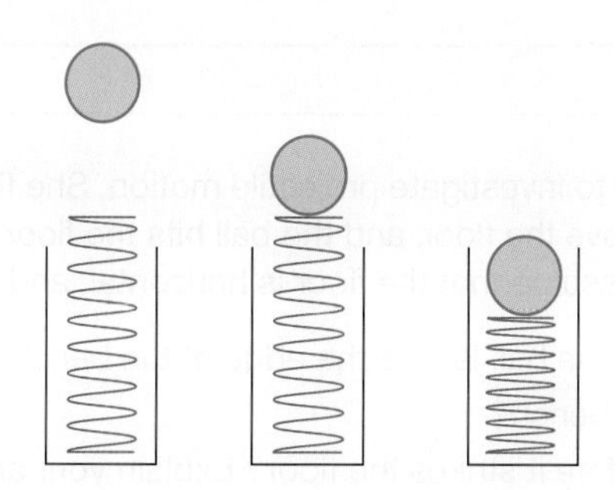

18. What is the magnitude of the acceleration of the bowling ball just after it touches the spring?

A. Equal to g

B. Less than g

C. Greater than g

D. Equal to 0

19. What is the direction of the acceleration of the bowling ball just after it touches the spring?

A. Upwards

B. Downwards

C. Neither upwards nor downwards

D. Unknown, since there is insufficient information

20. Which of the following statements is **correct**?

   A. When the spring is extending, the strain potential energy of the spring increases, and when the spring is compressing, its strain potential energy decreases.

   B. The direction of the acceleration of the bowling ball changes from downwards to upwards when the extending spring reaches its maximum extension.

   C. At maximum compression of the spring, strain potential spring energy is at its maximum, and is transformed into kinetic energy.

   D. At maximum compression of the spring, the kinetic energy of the bowling ball is maximal.

SECTION B — Short answer questions

Question 21 (3 marks)

A minibus of mass 1700 kg travelling with a velocity of 25 m s^{-1} west collides head-on with a solid wall. The minibus rebounds with a velocity of 5.0 m s^{-1} east. The minibus is in contact with the wall for 0.80 s. Assume that no external force acts on this system.

Calculate the average force that the wall exerts on the minibus during the collision. Show your working.

Question 22 (5 marks)

A 3.0-kg model car starts from rest on an inclined plane. The car is initially 1.7 m above ground level. After descending the ramp, the car comes to rest along a rough carpet surface, taking 6.0 m to completely stop. Ignore any friction effects on the ramp.

a. Determine the initial gravitational potential energy of the car. **(1 mark)**

b. What is the kinetic energy of the car as it first leaves the ramp and meets the carpet? Explain your answer using energy considerations. **(1 mark)**

c. Calculate the maximum speed of the car. **(1 mark)**

d. What is the work done on the car by the carpet? Explain your answer. **(1 mark)**

e. Determine the average force exerted on the car by the carpet. **(1 mark)**

Question 23 (4 marks)

A student is carrying out an experiment to investigate projectile motion. She fires a metal ball horizontally from a bench top. The bench top is 0.90 m above the floor, and the ball hits the floor at a distance of 1.6 m from a point directly below the edge of the bench. Assume that the floor is horizontal, and air resistance is negligible.

a. What is the elapsed time from when the ball leaves the edge of the bench to when it hits the floor? **(1 mark)**

b. At what speed did the ball leave the bench? **(1 mark)**

c. What is the speed of the ball just before it strikes the floor? Explain your answer. **(2 marks)**

Question 24 (3 marks)

A car of mass 1100 kg is travelling at a constant speed of 12 m s^{-1} on a flat section of road that forms an arc of a circle of radius 30 m. This situation is shown in the following diagram.

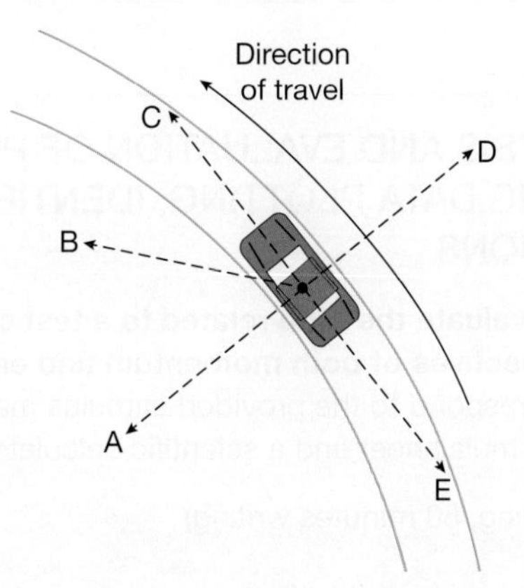

a. What is the magnitude of the net force acting on the car? **(1 mark)**

b. Which of the arrows A to E gives the direction of the net force acting on the car at the instant shown? **(1 mark)**

A road engineer has proposed replacing this flat section of road with a banked curve. The angle of the curve is such that the same 1100-kg car could travel at the same speed of 12 m s^{-1} around the banked curve, without any sideways friction on the tyres.

c. What should be the angle of the banked curve to the horizontal? **(1 mark)**

Question 25 (5 marks)

A 28.0-kg spherical mass is attached to a 67.0-m wire of negligible mass. The spherical mass is pulled back until it makes an angle of 60.0° with the vertical, as illustrated in the figure below.

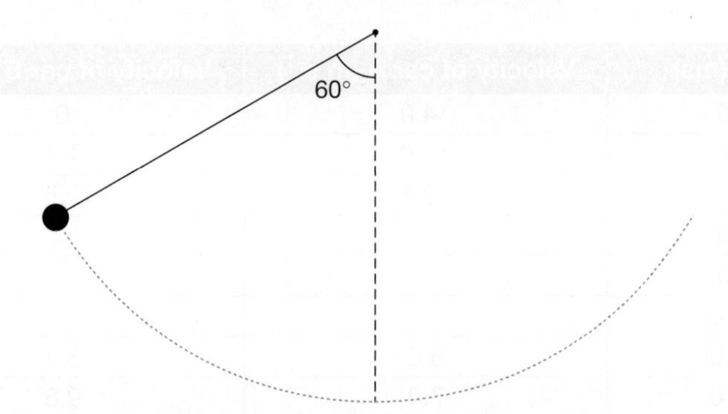

The spherical mass is then released with no initial speed. Assume that air resistance is negligible.

a. Explain what happens to the gravitational potential energy and the kinetic energy of the pendulum as it swings back and forth. **(2 marks)**

b. Determine the maximum speed of the spherical mass. Explain your answer. **(2 marks)**

c. Determine the maximum tension in the wire. **(1 mark)**

UNIT 3 | AREA OF STUDY 1

PRACTICE SCHOOL-ASSESSED COURSEWORK

ASSESSMENT TASK — ANALYSIS AND EVALUATION OF PRIMARY AND/OR SECONDARY DATA, INCLUDING DATA PLOTTING, IDENTIFIED ASSUMPTIONS OR DATA LIMITATIONS, AND CONCLUSIONS

In this task, you will be required to evaluate the data related to a test collision between two cars and evaluate the collision from the perspectives of both momentum and energy.

- This practice SAC requires you to respond to the provided stimulus material and analyse data.
- You may use the VCAA Physics formula sheet and a scientific calculator for this task.

Total time: 55 minutes (5 minutes reading, 50 minutes writing)

Total marks: 36 marks

INVESTIGATING COLLISIONS BETWEEN CARS

At a vehicle testing facility, collisions between cars are used to determine safety measures for passengers. In one test, car A, of mass 1400 kg, collides with stationary car B, of mass 800 kg.

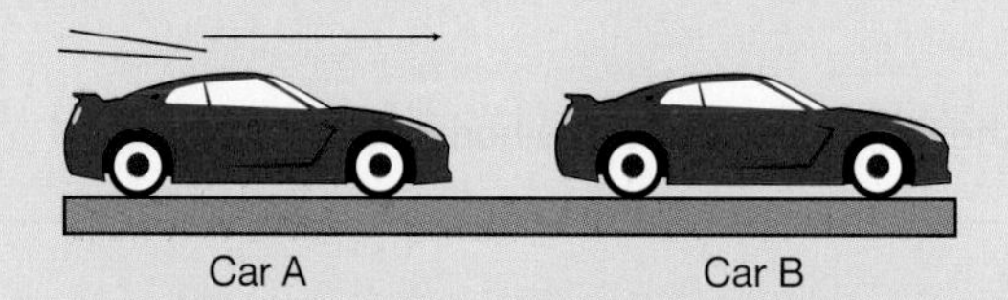

In the following table, the velocities of the two cars A and B are given for each 2.0-millisecond interval during which the collision is occurring. The table is incomplete.

Time (ms)	Velocity of car A (m s^{-1})	Velocity of car B (m s^{-1})
0.0	14.0	0
2.0	13.0	1.4
4.0	12.0	2.8
6.0		
8.0		
10.0		
12.0	8.0	8.4
14.0	7.0	9.8
16.0	7.0	9.8

You must answer all questions in both part A and part B.

PART A: Analysing the collision from the perspective of momentum

1. Calculate the initial momentum for car A and the initial momentum for car B and hence the total momentum of the two cars before the collision.
2. Assuming the acceleration of both cars A and B is uniform, complete the missing entries for velocity in the table.

3. Graph the velocities (using a line graph) of both car A and car B versus time on the one set of axes. Label the axes and each graph clearly, using a well-chosen scale for your graph.
4. Use your graph to estimate the time when the cars are travelling at the same speed.
5. Calculate the final momentum for car A and the final momentum for car B and hence the total momentum of the two cars after the collision.
6. Does it appear that the collision between the two cars is an isolated one? Is it likely that the road surface exerted a force on both cars during the collision? Give an explanation to support your choice.
7. Use your graph to calculate the average acceleration of each of the cars during the collision.
8. Determine the average force acting on car A and the average force acting on car B during the collision. Are the two forces the same or different in value? Explain your answer in relation to Newton's Third Law.
9. Calculate the distance travelled by car B during the collision.

PART B: Analysing the collision from the perspective of energy

10. Calculate the initial kinetic energy for each of car A and car B and hence the total kinetic energy of the two cars before the collision.
11. Calculate the final kinetic energy for each of car A and car B and hence the total kinetic energy of the two cars after the collision.
12. Is the collision an elastic collision? Explain your choice using your results for questions **10** and **11**.
13. Calculate the amount of energy that has been transformed from the initial kinetic energy of car A into other forms. State the likeliest form that most of the kinetic energy has transformed into.

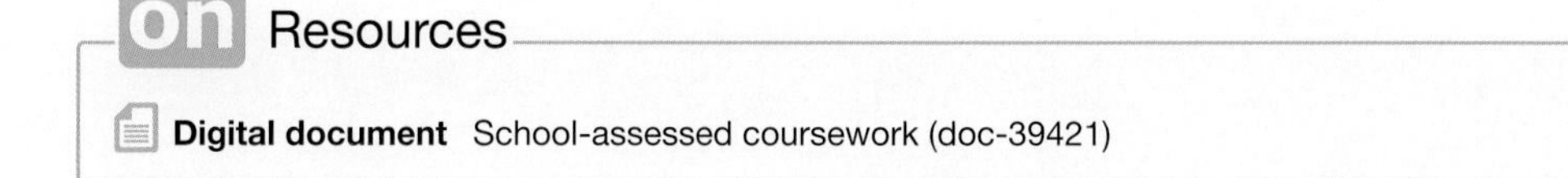

on Resources
Digital document School-assessed coursework (doc-39421)

3 Gravitational fields and their applications

KEY KNOWLEDGE

In this topic, you will:

- describe gravitation using a field model
- investigate theoretically and practically gravitational fields, including directions and shapes of fields, attractive and repulsive effects, and the existence of dipoles and monopoles
- investigate theoretically and practically gravitational fields about a point mass or charge (positive or negative) with reference to:
  - the direction of the field
  - the shape of the field
  - the use of the inverse square law to determine the magnitude of the field
  - potential energy changes (qualitative) associated with a point mass moving in the field
- identify fields as static or changing, and as uniform or non-uniform
- analyse the use of gravitational fields to accelerate mass, including:
  - gravitational field and gravitational force concepts: $g = G\frac{M}{r^2}$ and $F_g = G\frac{m_1 m_2}{r^2}$
  - potential energy changes in a uniform gravitational field: $E_g = mg\Delta h$
- analyse the change in gravitational potential energy from area under a force vs distance graph and area under a field vs distance graph multiplied by mass
- apply the concepts of force due to gravity and normal force including in relation to satellites in orbit where the orbits are assumed to be uniform and circular
- model satellite motion (artificial, Moon, planet) as uniform circular orbital motion: $a = \frac{v^2}{r} = \frac{4\pi^2 r}{T^2}$
- describe the interaction of two fields, allowing that masses only attract each other.

PRACTICAL WORK AND INVESTIGATIONS

Practical work is a central component of VCE Physics. Experiments and investigations, supported by a **practical investigation eLogbook** and **teacher-led video,** are included in this topic to provide opportunities to undertake investigations and communicate findings.

EXAM PREPARATION

Access past VCAA questions and exam-style questions and their video solutions in every lesson, to ensure you are ready.

3.1 Overview

Hey students! Bring these pages to life online

Watch videos | Engage with interactivities | Answer questions and check results

Find all this and MORE in jacPLUS

3.1.1 Introduction

Our lives are profoundly affected by gravity. Our muscles are constantly working against the gravitational pull on our bodies towards the centre of Earth. Our heart has to pump our blood so that it can circulate around our body. When we throw a ball, the pull on the ball by Earth dictates the path of the ball. Upon observing the night sky, we see the Moon, which orbits Earth every 29 days. The gravitational pull on the Moon towards the centre of Earth causes the Moon to orbit Earth. Gravitational attractions between masses explain the motion of the whole solar system.

FIGURE 3.1 Understanding gravitational forces has allowed humans to put satellites in orbit around Earth.

The existence of an attractive force between masses was a revolutionary idea articulated by Isaac Newton in his Law of Universal Gravitation. In this topic we also introduce the concept of a gravitational field. Imagine a map showing the strength and direction of the attractive force on a 1-kg mass at all points in space. A gravitational field is such a map. Knowledge of the gravitational field allows calculation of the energy changes experienced by objects moving through space, enabling humans to successfully launch rockets and send probes to other planets.

LEARNING SEQUENCE

on Resources

Solutions	Solutions — Topic 3 (sol-0817)
Practical investigation eLogbook	Practical investigation eLogbook — Topic 3 (elog-1634)
Digital documents	Key science skills — VCE Physics Units 1–4 (doc-36950) Key terms glossary — Topic 3 (doc-37169) Key ideas summary — Topic 3 (doc-37170)
Exam question booklet	Exam question booklet — Topic 3 (eqb-0100)

3.2 Newton's Universal Law of Gravitation and the inverse square law

KEY KNOWLEDGE

- Investigate theoretically and practically gravitational forces between point masses with reference to:
  - the direction of the force
  - the use of the inverse square law to determine the magnitude of the force

Source: Adapted from VCE Physics Study Design (2024–2027) extracts © VCAA; reproduced by permission.

3.2.1 Mass and gravitation

Nearly three hundred and fifty years ago, Isaac Newton was trying to explain the motion of the planets and the Moon. He had already established his three laws of motion. Newton realised if all masses are attracted to each other, there is an attractive force that can explain the motion of Earth around the Sun, the motion of the Moon around Earth and even the motion of a dropped ball.

Newton's Law of Universal Gravitation says that all masses attract each other and that the strength of the gravitational force of attraction between any two masses is proportional to the magnitude of each of the masses.

3.2.2 The inverse square law

The force due to gravity between two masses gets weaker as the distance between them increases. It is inversely proportional to the square of the distance between the centres of the two masses. This is an **inverse square law**.

inverse square law relationship in which one variable is proportional to the reciprocal of the square of another variable

EXTENSION: Standing on the shoulders of giants

Newton was preceded by the brilliant Johannes Kepler (1571–1630). Kepler's study of astronomical data obtained by his teacher and master, the remarkable Danish astronomer Tycho Brahe, led him to discover Kepler's first law, namely that planets move in elliptical orbits with the Sun at one focus. Newton's laws of motion show that an object moving in a circle, or an ellipse, experiences an overall force directed towards the focus, a force whose magnitude is inversely proportional to the square of the radius from the focus. Newton's genius was to realise that the force came from the attraction between the masses of the planets and the mass of the Sun, a force which he called gravitation.

The inverse square law explained

Imagine Earth as a perfect sphere. A ball placed on the surface of Earth is approximately 6400 km from Earth's centre (R_E is the radius of Earth) and is attracted to Earth by all of Earth's mass. The ball could be placed at any point on the surface and experience the same size of attraction towards Earth's centre.

At Earth's surface, the attraction towards the Earth's mass is spread over a surface area of $4\pi R_E^2$. Imagine now that the ball is moved to a position R_E above Earth's surface, $2R_E$ from Earth's centre. The amount of mass attracting the ball has not changed; however, the area that the attractiveness is spread over is now $4\pi(2R_E)^2 = 16\pi R_E^2$. As a result, the attractive force on a ball placed a distance of $2R_E$ from the centre is four times weaker than the force experienced at a distance of R_E from the centre. At a distance of $3R_E$ from Earth's centre, the attractiveness is spread over an area of $4\pi(3R_E)^2 = 36\pi R_E^2$, so the force on ball is nine times weaker than the force experienced at a distance of R_E from Earth's centre.

The attraction is not affected by the radius of either mass, so the masses can be modelled by point particles located at the centre of each mass.

FIGURE 3.2 The inverse square law illustrated. As the distance from a mass increases, the attraction is spread over an increasingly large area. The strength of attraction to the mass is inversely proportional to the square of the distance from the centre of the mass.

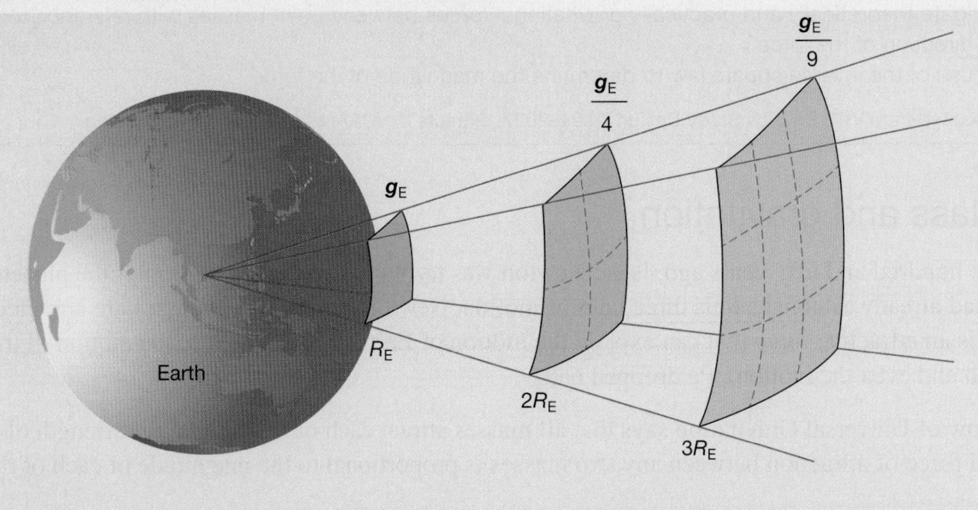

Putting it all together, Newton's Law of Universal Gravitation states that the magnitude of the force on a point mass, m_1, by another point mass, m_2, where the centres of the masses are separated by a distance r is given by:

$$\boldsymbol{F}_g = G\frac{m_1 m_2}{r^2}$$

where: $\boldsymbol{F}_g$ is the gravitational force on m_1 by m_2, in N

G is the gravitational constant (6.67×10^{-11} N kg^{-2} m^2)

m_1 is the mass of object 1, in kg

m_2 is the mass of object 2, in kg

r is the distance between the masses, in m

Note that the direction of the attractive force experienced by m_1 points towards m_2, along an imaginary line connecting the centres of the two masses.

The value of G is very small, and over short distances, gravitational forces are generally weaker than electrical and magnetic forces.

EXTENSION: Measuring the gravitational constant

The value of G could not be determined at the time of Newton because the mass of Earth was not known. Another 130 years passed before Henry Cavendish was able to measure the gravitational attraction between two known masses and calculate the value of G.

tlvd-8968

SAMPLE PROBLEM 1 Calculating the force due to gravity between two objects

Calculate the force due to gravity of the following:

a. **Earth (with a mass of 5.97×10^{24} kg) on a 70-kg person standing on the equator**

b. **a 70-kg person standing on Earth's equator**

THINK	WRITE
a. Recall the formula for Newton's Law of Universal Gravitation. The distance r is the separation of the centres of the two masses.	a. Assume that the direction towards Earth is positive. $F_{\text{on person by Earth}} = G\frac{m_{\text{Earth}}m_{\text{person}}}{r^2}$ $= \frac{6.67 \times 10^{-11} \times 5.97 \times 10^{24} \times 70}{(6.37 \times 10^6)^2}$ $= 6.9 \times 10^2$ N to two significant figures. The force due to gravity of Earth on a 70-kg person standing on the equator is 6.9×10^2 N towards the centre of Earth.
b. Remember Newton's Third Law: The force on A by B is equal in magnitude and opposite in direction to the force on B by A.	b. $F_{\text{on Earth by the person}} = -F_{\text{on person by Earth}}$ $= -6.9 \times 10^2$ N The negative value means that the direction of the force is in the opposite direction to the force on the person by Earth. The force due to gravity of a 70-kg person standing on Earth's equator is 6.9×10^2 N towards the person.

PRACTICE PROBLEM 1

Use the following data to calculate the force due to gravity by Earth on the Moon and the force due to gravity by the Moon on Earth:

mass of Earth = 5.97×10^{24} kg

mass of Moon = 7.35×10^{22} kg

distance between Earth and the Moon = 3.84×10^8 m.

3.2 Activities

Students, these questions are even better in jacPLUS

Receive immediate feedback and access sample responses

Access additional questions

Track your results and progress

Find all this and MORE in jacPLUS

3.2 Quick quiz | on | 3.2 Exercise | 3.2 Exam questions

3.2 Exercise

1. The gravitational force experienced by an object is known as the force due to gravity. Using the data in the table below, calculate the magnitude of the force due to gravity F_g experienced by a 70.0-kg person at the surface of Mars and at the surface of Jupiter.
   Give your answers to 3 significant figures.

Planet	Mass(kg)	Radius (m)
Mars	6.39×10^{23}	3.39×10^{6}
Jupiter	1.90×10^{27}	6.69×10^{7}

   $\left(G = 6.67 \times 10^{-11}\ \text{N m}^2\ \text{kg}^{-2}\right)$

2. Using the data below, calculate the magnitude of the force of attraction between Earth and the Sun.
   Give your answer to 3 significant figures.
   $\text{Mass}_{\text{Earth}} = 5.97 \times 10^{24}$ kg
   $\text{Mass}_{\text{Sun}} = 1.99 \times 10^{30}$ kg
   $\text{Distance}_{\text{Earth-Sun}} = 1.49 \times 10^{11}$ m
   $\left(G = 6.67 \times 10^{-11}\ \text{N m}^2\ \text{kg}^{-2}\right)$

3. If Earth expanded so that its radius is multiplied by 3, without any change in its mass, determine what would happen to the magnitude of the force due to gravity on an individual on the surface of Earth and thus copy and complete the following sentence.
   The magnitude of the force due to gravity would be ___ of the initial force.

4. Using the data below, calculate the ratio of the magnitude of the gravitational force on a ball at the surface of Mercury to the magnitude of the gravitational force on the same ball at the surface of Earth. Give your answer to 3 significant figures.

Planet	Mass (kg)	Radius (m)
Mercury	3.29×10^{23}	2.44×10^{6}
Earth	5.97×10^{24}	6.37×10^{6}

5. A 100-kg satellite S_1 orbits Earth at a distance of one Earth radius (r_E) above Earth's surface.
   A 200-kg satellite S_2 orbits Earth at a distance of two Earth radii above Earth's surface.
   a. Calculate the ratio of the magnitude of the gravitational force experienced by satellite S_1 to S_2.
   b. How far above Earth should S_2 orbit so that the magnitude of the gravitational force experienced by satellite S_1 is two times larger than that experienced by S_2? Give your answer in terms of Earth radii (r_E) above Earth's radius.

6. Using the data below, determine how many Earth radii (r_E) from the centre of Earth an object must be positioned for the magnitude of the gravitational force by Earth on the object to equal the magnitude of the gravitational force that would be exerted by the Moon on the same object if the object was on the Moon's surface. Give your answer to 3 significant figures.

Celestial object	Mass (kg)	Radius (m)
The Moon	7.35×10^{22}	1.74×10^{6}
Earth	5.97×10^{24}	6.37×10^{6}

3.2 Exam questions

Question 1 (1 mark)

Source: *VCE 2017 Physics Sample Exam, Section A, Q.3;* © VCAA

MC Students measure the gravitational force between two masses of 1.0 kg and 100 kg, placed 10 cm apart. The universal gravitational constant, G, is 6.67×10^{-11} N m^2 kg^{-2}. Which one of the following best gives the gravitational force of attraction between the two masses?

A. 1.0×10^{-3} N
B. 6.7×10^{-5} N
C. 6.7×10^{-7} N
D. 1.0×10^{6} N

Question 2 (1 mark)

MC An astronaut arriving on the Moon feels lighter than on Earth. Which of the following gives the correct reason for this?

A. The gravitational field strength of the Moon is smaller than that of Earth; therefore, the force experienced by the astronaut is less than on Earth.
B. The astronaut is a long way from Earth, so the force due to gravity is smaller.
C. The diameter of the Moon is smaller than the diameter of Earth, so the gravitational field strength is smaller.
D. The gravitational field strength of the Moon is smaller than that of Earth; therefore, the masses are smaller on the Moon.

Question 3 (1 mark)

MC Mass A and Mass B are placed in a gravitational field, at exactly the same distance from a central body. Mass A is 1000 times larger than Mass B. There are no other forces acting on the masses. Which of the following is correct?

A. The force on mass A is the same as the force on mass B.
B. The acceleration of mass A is 1000 times larger than the acceleration of mass B.
C. The gravitational field experienced by mass A is 1000 times larger than the gravitational field strength experienced by mass B.
D. The acceleration of mass A is the same as the acceleration of mass B.

Question 4 (2 marks)

A 300-kg terrestrial satellite has a circular orbit 10 000 km above the surface of Earth.

Using the information below, calculate the magnitude of the gravitational force experienced by the satellite. Give your answer to 3 significant figures.

$G = 6.67 \times 10^{-11}$ N m^2 kg^{-2}

$m_E = 5.97 \times 10^{24}$ kg

$r_E = 6.37 \times 10^{6}$ m

Question 5 (2 marks)

Using the information below, calculate the magnitude of the gravitational force of Earth on a 50-g hailstone formed at an altitude of 6000 m. Give your answer to 3 significant figures.

$G = 6.67 \times 10^{-11}$ N m^2 kg^{-2}

$m_E = 5.97 \times 10^{24}$ kg

$r_E = 6.37 \times 10^{6}$ m

More exam questions are available in your learnON title.

3.3 The field model

KEY KNOWLEDGE

- Describe gravitation using a field model
- Investigate theoretically and practically gravitational fields about a point mass with reference to:
  - the direction of the field
  - the shape of the field
  - the use of the inverse square law to determine the magnitude of the field
- Describe the interaction of two fields, allowing that masses only attract each other

3.3.1 The field model

Imagine a probe travelling through our solar system. At any point, the probe is attracted towards the Sun, planets and even asteroids, each of which is much more massive than the probe. The combined attraction by the Sun, planets and other massive bodies can be described as a **gravitational field**.

gravitational field vector field describing the property of space that causes an object with mass to experience a force in a particular direction

The gravitational field, $\boldsymbol{g}$, is defined as the gravitational force experienced by the probe, $\boldsymbol{F}_g$, divided by the mass of the probe, m. The force, field and probe mass are related by the equation:

$$\boldsymbol{F}_g = m\boldsymbol{g}$$

where: $\boldsymbol{F}_g$ is the force on an object due to gravity, its magnitude is expressed in N

m is the mass of the object, in kg

$\boldsymbol{g}$ is the gravitational field, and its magnitude, the gravitational field strength, is in N kg^{-1} (or m s^{-2})

Note that for a body on Earth, the gravitational force $\boldsymbol{F}_g$ is also denoted as $\boldsymbol{F}_{\text{on body by Earth}}$.

A field is defined as a physical quantity that has a value at each point in space. There are different types of fields. The gravitational field is a vector field as it also has a direction at every point in space. A field is often represented by a diagram.

In a gravitational field diagram:

- arrows show the direction of force that a mass at that point would experience
- the spacing of the lines indicates field strength: more closely spaced lines indicate a stronger field
- field lines never touch or intersect, because the force on an object cannot have multiple magnitudes or directions at the same time.

In a diagram of Earth's gravitational field (see figure 3.3) the arrows point towards Earth, indicating that any mass placed in Earth's gravitational field will experience a force towards Earth. The field lines become further apart as the distance from Earth increases, showing that the field is weakening and is non-uniform (in a uniform field, the value of the field strength remains the same at all points). In addition, Earth's gravitational field does not change with time, it is static.

FIGURE 3.3 Diagram of Earth's gravitational field

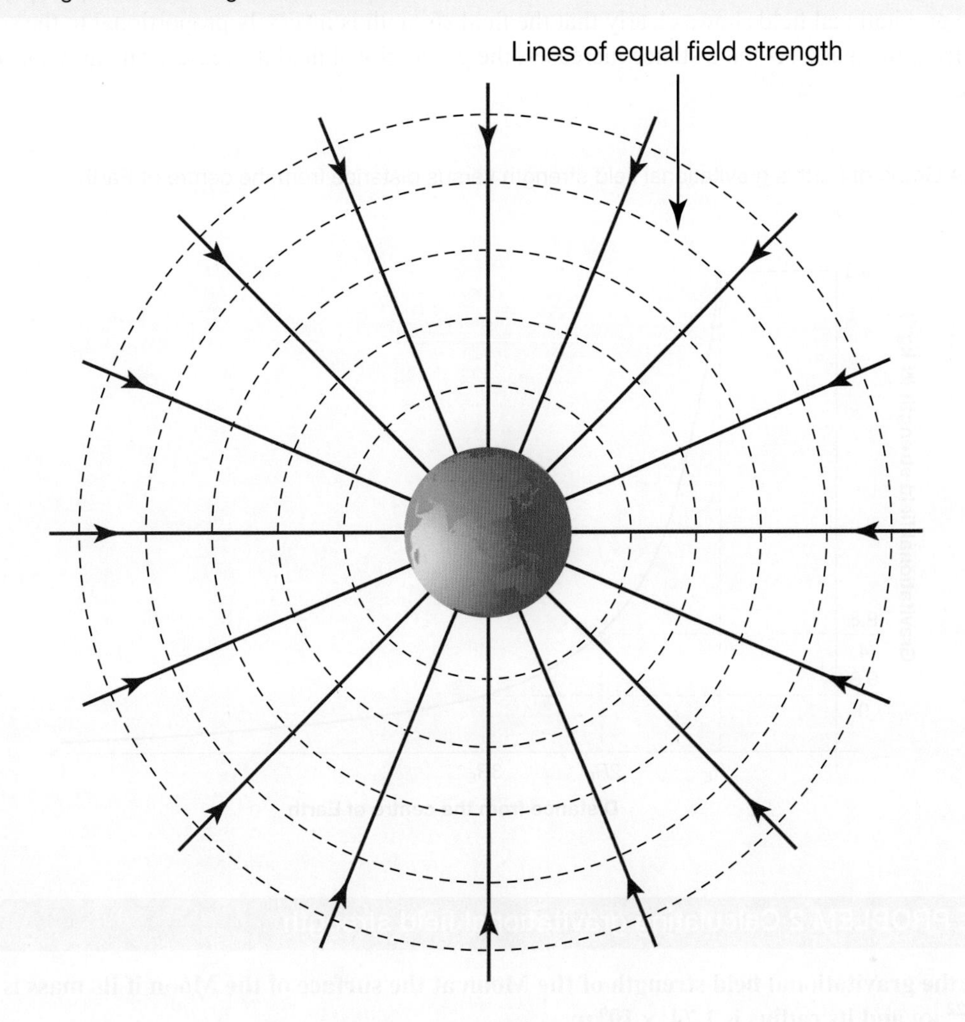

3.3.2 The gravitational field from a point mass

We have seen that the gravitational force on a mass m due to the attraction to a point mass $\boldsymbol{M}$ is given by $\boldsymbol{F}_g = G\frac{Mm}{r^2}$ towards the point mass M. This means that, as $\boldsymbol{F}_g = m\boldsymbol{g}$ near Earth's surface, the gravitational field strength, g, of the point mass $\boldsymbol{M}$ is simply $g = G\frac{M}{r^2}$.

$$g = G\frac{M}{r^2}$$

where: g is the gravitational field strength, in N kg^{-1}

G is the gravitational constant (6.67×10^{-11} N m^2 kg^{-2})

M is the mass of the point source, in kg

r is the distance to the point source, in m

The direction of the gravitational field is radially inwards, towards the point mass $\boldsymbol{M}$. This can be used as a model for the gravitational field from objects such as Earth, the Moon, other planets and the Sun. The larger the mass source, the greater the magnitude of the gravitational field.

Graphing the gravitational field

Graphing the gravitational field shows clearly that the field strength is inversely proportional to the square of the distance from the mass. As the distance increases, the gravitational field decreases. The gravitational field is non-uniform.

FIGURE 3.4 Graph of Earth's gravitational field strength versus distance from the centre of Earth

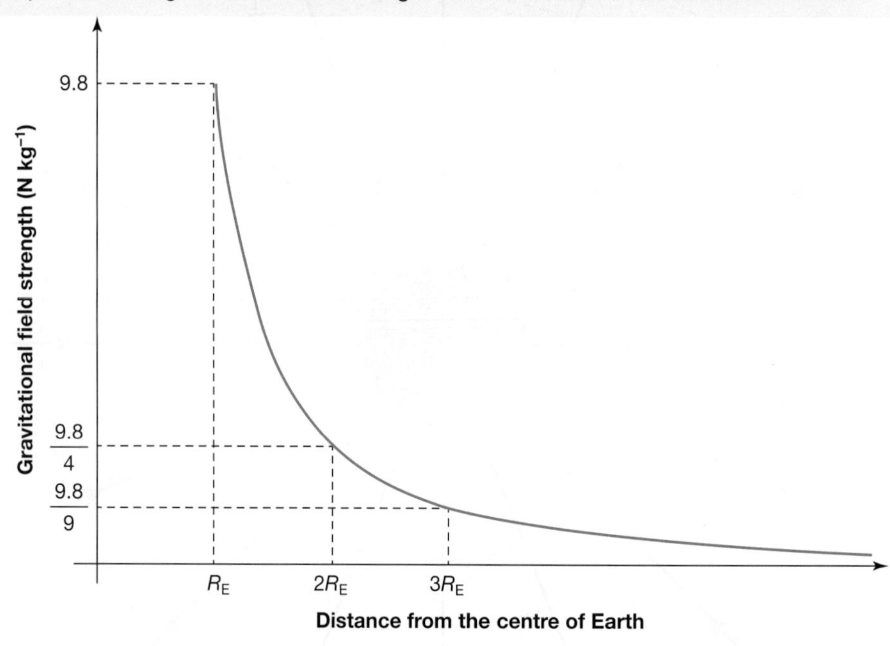

tlvd-8969

SAMPLE PROBLEM 2 Calculating gravitational field strength

Calculate the gravitational field strength of the Moon at the surface of the Moon if its mass is 7.35×10^{22} kg and its radius is 1.74×10^6 m.

THINK	WRITE
Recall the formula $g = G\frac{M}{r^2}$, where: $G = 6.67 \times 10^{-11}\,\text{N m}^2\,\text{kg}^{-2}$, $M = 7.35 \times 10^{22}$ kg and $r = 1.74 \times 10^6$ m	$g = G\frac{M}{r^2}$ $= \frac{6.67 \times 10^{-11} \times 7.35 \times 10^{22}}{(1.74 \times 10^6)^2}$ $= 1.62\,\text{N kg}^{-1}$

PRACTICE PROBLEM 2

Calculate the gravitational field strength of the Sun at the centre of Earth, given the following:

$M_{Sun} = 1.99 \times 10^{30}$ kg

distance from the centre of Earth to the centre of the Sun = 1.50×10^{11} m.

EXTENSION: The gravitational field inside Earth

What does the gravitational field look like inside Earth? Newton showed that the gravitational field inside a hollow spherical shell is in fact zero. This means that, closer to the centre of Earth, only the mass inside that radius contributes to the gravitational field at that point. The overall effect is that the gravitational field decreases linearly from the outside of Earth towards the centre.

FIGURE 3.5 Graph of Earth's gravitational field strength, including the strength inside Earth

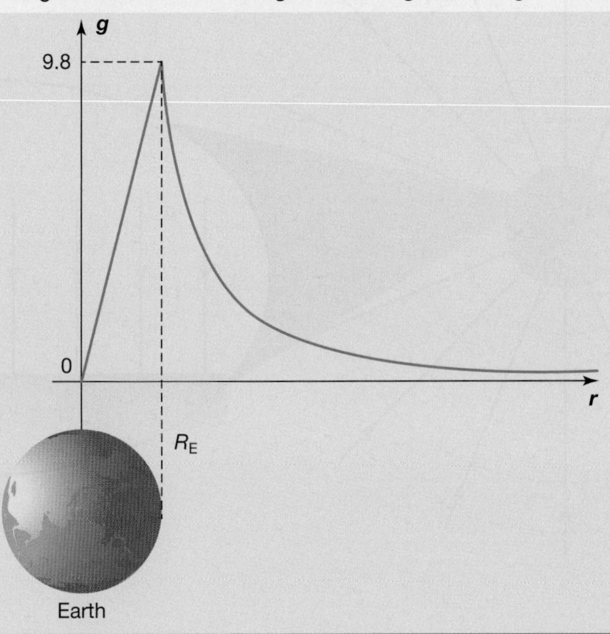

3.3.3 The gravitational field close to the surface of Earth

You have just learned that the gravitational field of Earth decreases as the square of the distance from the centre of Earth. The strength of the gravitational field at Earth's surface can be estimated using the values for the mass and radius of Earth, $M = 5.97 \times 10^{24}$ kg; $R_E = 6.38 \times 10^6$ m:

$$\begin{aligned} g &= \frac{6.67 \times 10^{-11} \times 5.97 \times 10^{24}}{(6.38 \times 10^6)^2} \\ &= 9.80 \text{ N Kg}^{-1} \end{aligned}$$

However, Earth is huge compared to the scale of ordinary human actions. For a distance of 1000 m from the surface of Earth, the gravitational field has decreased by a mere 0.03%, from 9.805 N kg^{-1} to 9.802 N kg^{-1}. The gravitational field strength experienced by a ball dropping from an initial height of a couple of metres above the surface of Earth is effectively constant or uniform with strength 9.8 N kg^{-1} vertically downwards towards the surface of Earth. A field that has the same magnitude and same direction everywhere in a given space is called a uniform field.

FIGURE 3.6 Zooming in on the gravitational field very close to the surface of Earth, from a human perspective, the gravitational field appears to be constant and uniform. A mass placed at any point in this field would experience the same magnitude and direction of force: $\boldsymbol{F} = m\boldsymbol{g}$.

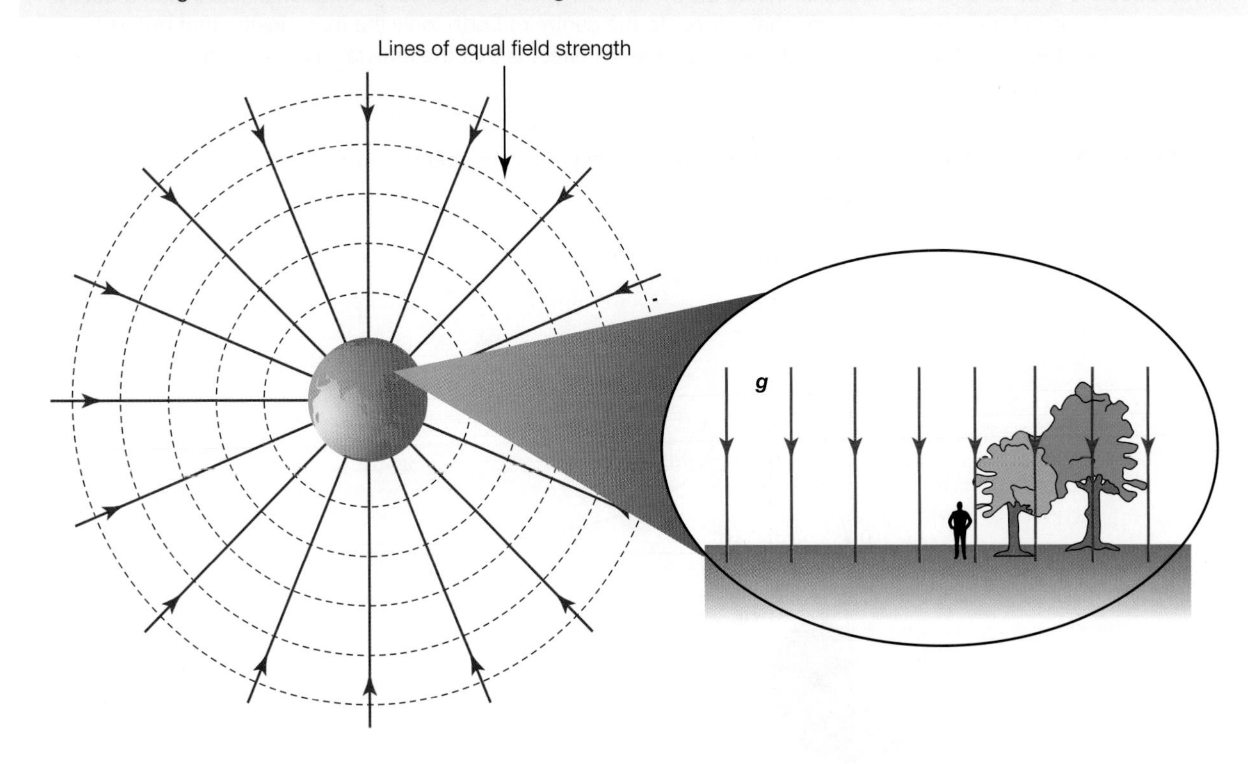

3.3.4 Gravitational fields from two masses

When a rocket travels from Earth to the Moon, we need to take into account both the gravitational attraction to Earth and to the Moon. The gravitational field, $\boldsymbol{g}$, experienced by the rocket is the vector sum of the field from Earth and the field from the Moon.

$$\boldsymbol{g} = \boldsymbol{g}_{\text{Earth}} + \boldsymbol{g}_{\text{Moon}}$$

Because the two fields are in opposite directions, to find the magnitude of the total field we subtract one field from the other. Taking the positive direction towards the Moon, as illustrated in figure 3.7, the magnitude of the gravitational field experienced by the rocket is given by:

$$g = -G\frac{M_{\text{Earth}}}{r^2_{\text{rocket - earth}}} + G\frac{M_{\text{Moon}}}{r^2_{\text{rocket - Moon}}}$$

Initially, the overall field is towards Earth. As the rocket approaches the Moon, the size of Earth's field decreases and the size of the Moon's field increases, so the overall field will become smaller. The rocket will pass through a point where the field is zero. From there, the overall field will be towards the Moon and its size increases as the rocket approaches the Moon's surface.

FIGURE 3.7 A rocket travelling from Earth to the Moon experiences the gravitational pull from both Earth and the Moon.

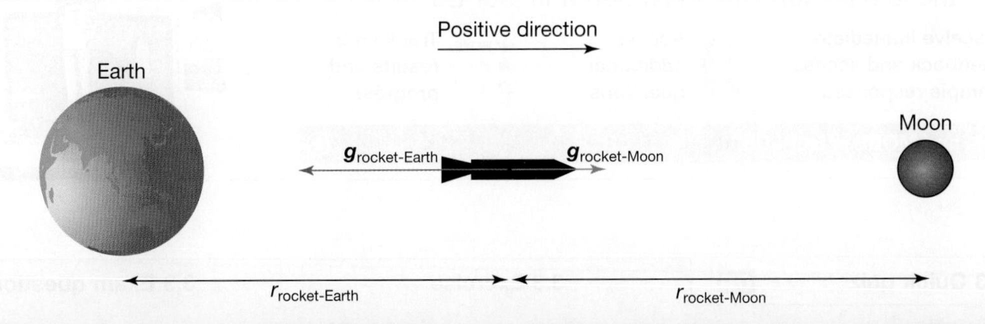

tlvd-8970

SAMPLE PROBLEM 3 Calculating the magnitude of the gravitational field on a rocket halfway between Earth and the Moon

A rocket travels in a straight line from Earth to the Moon. What is the magnitude and direction of the gravitational field experienced by the rocket at the halfway point?
Use the following information:
Earth's mass is $M_{\text{Earth}} = 5.97 \times 10^{24}$ kg
The Moon's mass is $M_{\text{Moon}} = 7.35 \times 10^{22}$ kg
and the distance between Earth and the Moon is 384 400 km.

THINK

- Earth and the Moon can be modelled as point masses.
- The gravitational field from a point mass is directed towards the centre of the mass.
- This means the gravitational field from Earth and the Moon act in opposite directions on the rocket.
- At the halfway point, the distance to centre of Earth and the distance to the centre of the Moon is $r = 1.922 \times 10^8$ m.

WRITE

- Take towards the Moon as the positive direction.

$$g = -G\frac{M_{\text{Earth}}}{r^2} + G\frac{M_{\text{Moon}}}{r^2}$$

$$= -6.67 \times 10^{-11}\frac{5.97 \times 10^{24}}{\left(1.922 \times 10^8\right)^2} + 6.67 \times 10^{-11}\frac{7.35 \times 10^{22}}{\left(1.922 \times 10^8\right)^2}$$

$= -1.06 \times 10^2 \text{ N kg}^{-1}$ to three significant figures

- The negative value for g means that at the point halfway between Earth and the Moon, the direction of the overall gravitational field is towards Earth.

PRACTICE PROBLEM 3

Consider the situation where Mercury is directly on a line between the Sun and Venus. How much does Venus's gravitational field affect the gravitational field experienced by Mercury?

Use the following information:

- **The Sun's mass is $M_{\text{Sun}} = 1.99 \times 10^{30}$ kg**
- **Venus's mass is $M_{\text{Venus}} = 4.87 \times 10^{24}$ kg**
- **The distance Mercury–Sun is $r_{\text{Sun–Mercury}} = 5.79 \times 10^{10}$ m**
- **The distance Mercury–Venus is $r_{\text{Venus–Mercury}} = 5.01 \times 10^{10}$ m**

3.3 Activities

learn on

Students, these questions are even better in jacPLUS

Receive immediate feedback and access sample responses

Access additional questions

Track your results and progress

Find all this and MORE in jacPLUS

3.3 Quick quiz on	3.3 Exercise	3.3 Exam questions

3.3 Exercise

1. a. Calculate the gravitational field strength on a 1.50-kg ball held at a height of 2.00 metres above the surface of Earth (the radius of Earth is $r_E = 6.37 \times 10^6$ m; its mass is $m_E = 5.97 \times 10^{24}$ kg). Give your answer to 3 significant figures. $(G = 6.67 \times 10^{-11}\ \text{N m}^2\ \text{kg}^{-2})$
   b. The ball is dropped. Calculate the magnitude of the net force on the 1.50-kg ball as it falls to the surface of Earth. Give your answer to 3 significant figures
2. Answer the following.
   a. Compare the magnitude of the force on an object that is positioned one Earth radius above the surface of Earth to the magnitude of the force on the same object if it was positioned two Earth radii above the surface of Earth.
   b. Copy and complete the diagram below to sketch the gravitational field of Earth: draw a few vectors, every 30° for instance, indicating the force on an object that is positioned one Earth radius above the surface of Earth, and a few vectors indicating the force on the same object if it was positioned two Earth radii above the surface of Earth.

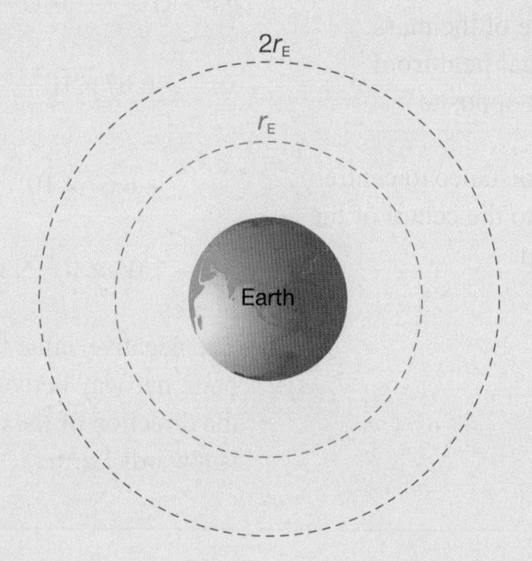

   c. Is the gravitational field uniform or non-uniform? Explain how this is indicated in your sketch.
3. Using the data provided in the table below, and $G = 6.67 \times 10^{-11}\ \text{N m}^2\ \text{kg}^{-2}$, calculate the gravitational field strength at the surface of Earth, Mars, Venus and Pluto. Give your answers to 3 significant figures.

Planet/dwarf planet	Mass (kg)	Radius (m)
Earth	5.97×10^{24}	6.37×10^6
Mars	6.39×10^{23}	3.39×10^6
Venus	4.87×10^{24}	6.05×10^6
Pluto	1.31×10^{22}	1.19×10^6

4. Explain why the gravitational field lines for an isolated mass always point radially inwards towards the centre of the mass.
5. Answer the following.
   a. Using the data provided below, calculate the gravitational field strength at the distances d of 10 000 km, 20 000 km and 30 000 km from the centre of Earth.
      Give your answers to 3 significant figures.
      $G = 6.67 \times 10^{-11}\ \text{N m}^2\ \text{kg}^{-2}$
      Earth's mass: $m_E = 5.97 \times 10^{24}$ kg
   b. Calculate the following ratios of the field strengths at 10 000 km, 20 000 km and 30 000 km from the centre of Earth.
      i. Field strength at 20 000 km : Field strength at 10 000 km
      ii. Field strength at 30 000 km : Field strength at 10 000 km
      iii. Field strength at 30 000 km : Field strength at 20 000 km
   c. Compare the ratios and explain the pattern you observe.
6. A gravitational field strength detector is released into Earth's atmosphere and reports back a reading of $9.73\ \text{N kg}^{-1}$.
   a. If the detector has a mass of 10.0 kg, calculate the magnitude of the force of gravity acting on it.
   b. If the detector is to remain stationary at this height, what is the magnitude of the upward force that must be exerted on the detector?
   c. Use the information below to calculate how far the detector is from the centre of Earth. Give your answer to 3 significant figures.
      $G = 6.67 \times 10^{-11}\ \text{N m}^2\ \text{kg}^{-2}$
      Earth's mass: $m_E = 5.97 \times 10^{24}$ kg
7. Answer the following.
   a. Use the information below to calculate g_S, the gravitational field strength experienced by the Moon from the Sun, and g_E, the gravitational field strength experienced by the Moon from Earth. Give your answers to 3 significant figures.
      $G = 6.67 \times 10^{-11}\ \text{N m}^2\ \text{kg}^{-2}$
      Earth's mass: $m_E = 5.97 \times 10^{24}$ kg
      Sun's mass: $m_S = 1.989 \times 10^{30}$ kg
      Average distance Earth–Moon (centre to centre): $r_{E\text{-}M} = 3.84 \times 10^5$ km
      Average distance Sun–Moon (centre to centre): $r_{S\text{-}M} = 1.50 \times 10^8$ km
   b. Compare the gravitational field strength experienced by the Moon from the Sun with that from Earth and discuss whether this result is surprising. Comment on its significance in regard to the motion of the Moon.
8. A spacecraft leaves Earth to travel to the Moon.
   a. Use the following data to determine how far from the centre of the Earth the spacecraft is when it experiences a net force of zero. Give your answer 3 significant figures.
      Earth's mass: $m_E = 5.97 \times 10^{24}$ kg
      Moon's mass: $m_M = 7.35 \times 10^{22}$ kg
      Average distance Earth-Moon (centre to centre): $r_{E\text{-}M} = 3.84 \times 10^5$ km
   b. Copy and complete the diagram below to show the location (to scale) of the spacecraft when the net force it experiences is zero.

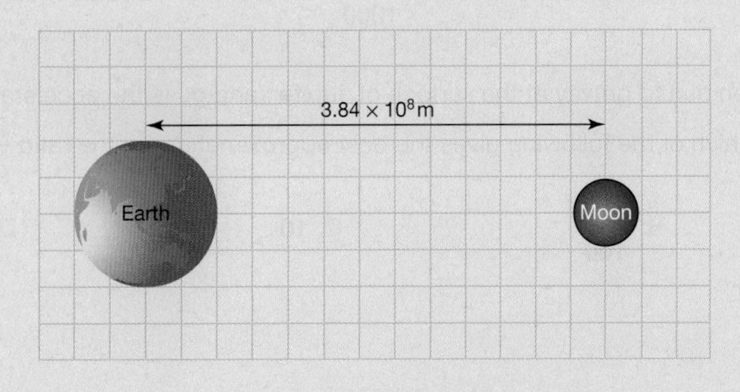

9. Meredith and Julian are interested in determining whether the alignment of the Sun and the Moon on the same side of Earth, or on opposite side of Earth, has a noticeable effect on the gravitational field strength on opposite points on Earth's surface.
They gathered the information below.
$G = 6.6743 \times 10^{-11}\ \text{N m}^2\ \text{kg}^{-2}$
Earth's radius: $r_E = 6.3781 \times 10^6$ m
Moon's mass: $m_M = 7.3460 \times 10^{22}$ kg
Sun's mass: $m_S = 1.98847 \times 10^{30}$ kg
Average distance Earth–Moon (centre to centre): $r_{E\text{-}M} = 3.84400 \times 10^5$ km
Average distance Sun–Earth (centre to centre): $r_{S\text{-}M} = 1.4742 \times 10^8$ km
They consider the following two scenarios (not drawn to scale):

(a)

(b)

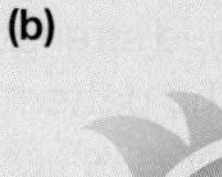

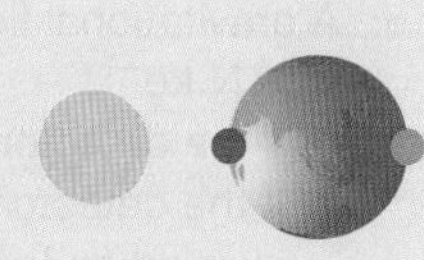

Meredith makes the hypothesis that the gravitational field would be significantly greater at the red dot location in scenario **(b)** than in scenario **(a)** while Julian makes the hypothesis that the gravitational field would be approximately the same at both dot locations in both scenario **(a)** and scenario **(b)**.

a. Using the information gathered by Meredith, calculate the combined gravitational field strength $\boldsymbol{g}_c$ from the Sun and the Moon at the point on the surface of the Earth indicated by the red and orange dots for both scenarios illustrated on the figure and complete the table below.
Give your answer to 5 significant figures and take towards the centre of the Sun to be the positive direction.

Scenario	dot	g_c (N kg^{-1})
(a)	red	
(a)	orange	
(b)	red	
(b)	orange	

b. Using your results to part **a**, evaluate the two students' hypotheses.

3.3 Exam questions

Question 1 (1 mark)

MC The planet Jupiter has a mass approximately $\frac{1}{1000}$ that of the Sun and its radius is approximately $\frac{1}{10}$ that of the Sun.

If $\boldsymbol{g}_J$ is the acceleration due to gravity at the surface of Jupiter, and $\boldsymbol{g}_S$ is the acceleration due to gravity at the surface of the Sun, which of the following gives the best approximation for the ratio $\frac{g_J}{g_S}$?

A. $\frac{1}{10}$

B. $\frac{1}{100}$

C. 10

D. $\frac{1}{1000}$

Question 2 (2 marks)

The gravitational field strength on Jupiter's second largest moon, Europa, is $1.31\ \text{N kg}^{-1}$. Its radius is 1.56×10^6 m. Calculate its mass. Give your answer to 3 significant figures.

$(G = 6.67 \times 10^{-11}\ \text{N m}^2\ \text{kg}^{-2})$

Question 3 (2 marks)

The asteroid Icarus has a mass of 10^{12} kg and a radius of 1400 m. Calculate the gravitational field strength on its surface. Give your answer to 3 significant figures.

$(G = 6.67 \times 10^{-11}\ \text{N m}^2\ \text{kg}^{-2})$

Question 4 (2 marks)

A new planet has been discovered that has exactly the same mass as Earth, but a radius half that of Earth. What would be the gravitational field strength on the surface of this new planet? Assume that the gravitational field strength on the surface of Earth can be taken as $\boldsymbol{g} = 9.8\ \text{N kg}^{-1}$.

Question 5 (2 marks)

The force due to gravity on a space capsule before launch has a magnitude of 110 000 N. The capsule reaches an altitude 3 times the radius of Earth. Calculate magnitude of the force due to gravity on it at this altitude.

More exam questions are available in your learnON title.

3.4 Motion in gravitational fields, from projectiles to satellites in space

KEY KNOWLEDGE

- Apply the concepts of force due to gravity and normal force including in relation to satellites in orbit where the orbits are assumed to be uniform and circular
- Analyse the use of gravitational fields to accelerate mass, including:
  - gravitational field and gravitational force concepts: $g = G\frac{M}{r^2}$ and $F_g = G\frac{m_1 m_2}{r^2}$
- Model satellite motion (artificial, Moon, planet) as uniform circular orbital motion: $a = \frac{v^2}{r} = \frac{4\pi^2 r}{T^2}$

Source: Adapted from VCE Physics Study Design (2024–2027) extracts © VCAA; reproduced by permission.

How do apples, satellites and planets move in gravitational fields?

When the gravitational force, $\boldsymbol{F}_g$, is the only force acting on a mass, the mass is in 'free fall'. Applying Newton's Second Law, $\boldsymbol{F}_{net} = m\boldsymbol{g} = m\boldsymbol{a}$, it is clear that in free fall, the acceleration $\boldsymbol{a}$, experienced by the mass, is equal to the gravitational field $\boldsymbol{g}$ at that point in space.

You will now look at what this means for masses close to the surface of the Earth and what it means for those further away.

3.4.1 Motion close to Earth's surface

Close to Earth's surface the gravitational field is constant with a value of 9.8 N kg^{-1} (see calculation in section 3.3.3). Things that fall with little air resistance experience an acceleration of 9.8 m s^{-2}, increasing their vertical velocity by 9.8 m s^{-1} every second.

Using the familiar constant acceleration equations, you can find how high a ball will go when launched with a particular velocity, or how long it will take for an object dropped from a height to hit the ground.

FIGURE 3.8 Position of an object in free fall in Earth's gravitational field, **a.** with no initial velocity and **b.** with an initial velocity

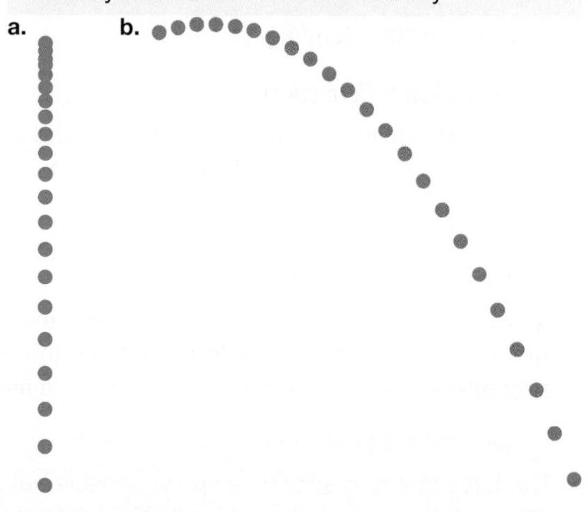

tlvd-8971

SAMPLE PROBLEM 4 Determining the height of an object thrown close to Earth's surface

A ball is thrown vertically into the air with a speed of 20.0 m s^{-1}. How high does it go? (Ignore air resistance.)

THINK

After leaving the hand of the person throwing the ball, the only force on the ball is the gravitational attraction of the ball to Earth:

$$F_g = mg$$

This means that the net force, $F_{nat} = ma = mg$ and the acceleration of the ball is simply $g = 9.8$ m s^{-2} downwards. This is motion under constant acceleration in a straight line, so the equation $v^2 = u^2 + 2as$ can be used to determine the maximum height reached by the ball.

$$s = \frac{v^2 - u^2}{2a}$$

WRITE

Take upwards as the positive direction.

$a = g = -9.8$, $u = 20.0$ m s^{-1}, $v = 0$ m s^{-1}

$$v^2 = u^2 + 2as$$

$$0^2 = 20^2 + (2 \times -9.8 \times s)$$

$$0^2 - 20^2 = 2 \times -9.8 \times s$$

$$s = \frac{0^2 - 20^2}{2 \times -9.8}$$

$$= 20 \text{ m}$$

The maximum height reached by the ball is 20 m.

PRACTICE PROBLEM 4

A ball is dropped from a bridge. The initial speed of the ball is 0 m s^{-1}. How fast is the ball travelling after falling 20 metres?

elog-1916

tlvd-10438

INVESTIGATION 3.1

Using a pendulum to determine the strength of the gravitational field

AIM

To determine the rate of acceleration due to gravity using the motion of a pendulum

3.4.2 Orbital motion

Not all objects affected by Earth's gravitational field fall to Earth's surface. The Moon and other artificial **satellites** such as the International Space Station circle Earth at constant speeds.

satellite an object that is orbiting a larger central mass. Satellites can be natural (such as the Moon) or man-made (such as the International Space Station)

centripetal acceleration the acceleration towards the centre of a circle experienced by an object moving in a circular motion

Figure 3.9 shows how the constantly changing direction of the velocity of an object moving in a circle is due to an acceleration towards the centre of the circle. This kind of acceleration is called **centripetal acceleration**. When an object moves along a circular path with a constant speed, the direction of the change in the velocity always points along a line towards the centre of the circle, causing the acceleration to point in the same direction.

Determining the orbital acceleration

In figure 3.9, it can be deduced from the similar triangles that:

$$\frac{\Delta \mathbf{v}}{\mathbf{v}} = \frac{\mathbf{v}\Delta t}{r}$$

$$\Rightarrow \Delta \mathbf{v} = \frac{\mathbf{v}^2 \Delta t}{r} n$$

The acceleration of the object towards the centre of the circle (centripetal acceleration) is:

$$\mathbf{a} = \frac{\Delta \mathbf{v}}{\Delta t} = \frac{\mathbf{v}^2 \Delta t}{r\Delta t} = \frac{\mathbf{v}^2}{r}$$

The speed of the satellite, **v**, is constant and can be expressed in terms of the radius, r, and period, T, of the orbit:

$$\mathbf{v} = \frac{\text{distance travelled}}{\text{time taken}} = \frac{2\pi r}{T}$$

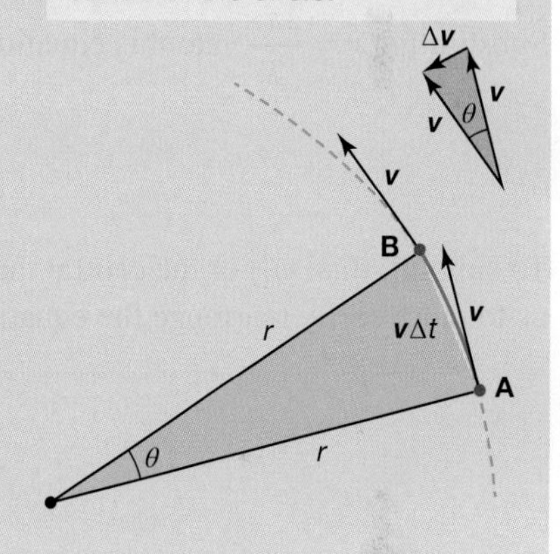

FIGURE 3.9 The acceleration of an object moving in a circle at a constant speed always points inwards towards the centre of the circle.

The magnitude of the centripetal acceleration of a satellite is given by:

$$a = \frac{v^2}{r} = \frac{4\pi^2 r}{T^2}$$

where: v is the speed of the satellite, in m s^{-1}

r is the radius of the circular orbit, in m

T is the period of orbit, in s

The magnitude of the net (centripetal) force on the satellite then becomes:

$$F_{\text{net}} \text{ (or } \Sigma F) = ma = \frac{mv^2}{r}$$

where: m is the mass of the satellite, in kg

v is the speed of the satellite, in m s^{-1}

r is the radius of the circular orbit, in m

Newton had already shown that objects moving in a circle with a constant speed experience a net force pointing radially inwards towards the centre of the circle. His great achievement with the Law of Universal Gravitation was to realise that the gravitational field from a large mass, such as that of Earth, could be the source of the radially inward net force on a satellite, such as the Moon.

Newton equated the gravitational force on a satellite of mass m_2 orbiting a central mass m_1 with the net force on a satellite experiencing circular motion.

$$F_g = G\frac{m_1 m_2}{r^2} = \frac{m_s v^2}{r}$$

where: r is the distance between the centres of the central mass and the satellite's mass

m_1 is the central mass, in kg

m_2 (or m_s) is the mass of the satellite, in kg

G is the gravitational constant, 6.67×10^{-11} N m^2 kg^{-2}

v is the speed of the satellite, in m s^{-1}

Substituting $v = \frac{2\pi r}{T}$ into this equation, the following is obtained:

$$G\frac{m_1 m_2}{r^2} = \frac{4 m_s \pi^2 r}{T^2}$$

To simplify this, m_1, or the central mass can be shown as M, and the mass of the satellite m_2 as m_s. This allows us to more easily rearrange the equation and show the relationship:

$$G\frac{M}{4\pi^2} = \frac{r^3}{T^2}$$

where: r is the distance between the centres of the two bodies (orbital radius), in m

M is the central mass, in kg

G is the gravitational constant, 6.67×10^{-11} N m^2 kg^{-2}

T is the orbital period, in s

This expression reveals the profound insight that the ratio of the cube of the radius of orbit of a satellite to the square of the period of the orbit has a constant value and does not depend on the value of the mass of the satellite. This had already been discovered by the exceptional astronomer Johannes Kepler in his work on explaining the motion of the planets around the Sun and is known as Kepler's Third Law.

tlvd-8972

SAMPLE PROBLEM 5 Calculating speed, mass and acceleration of satellites in orbit

The Moon orbits Earth with a period of 2.36×10^6 s. The distance from the centre of Earth to the centre of the Moon is 3.84×10^8 m.

a. **Calculate the speed of the Moon as it orbits Earth.**
b. **Calculate the acceleration experienced by the Moon.**
c. **Use the provided data to calculate the mass of Earth.**

THINK	WRITE
a. Recall that, for an object moving in a circle at a constant speed, $v = \frac{2\pi r}{T}$.	a. $v = \frac{2\pi \times 3.84 \times 10^8}{2.36 \times 10^6}$ $= 1020 \text{ m s}^{-1}$ (to 3 significant figures)
b. Recall that, for an object moving in a circle at a constant speed, $a = \frac{v^2}{r} = \frac{4\pi^2 r}{T^2}$. Ensure you use the non-rounded value if using the velocity from part **a**.	b. $a = \frac{4\pi^2 \times 3.84 \times 10^8}{(2.36 \times 10^6)^2}$ $= 0.002\,72 \text{ m s}^{-2}$ or $a = \frac{1020^2}{3.84 \times 10^8}$ $= 0.002\,72 \text{ m s}^{-2}$
c. Use the relationship $G\frac{M}{4\pi^2} = \frac{r^3}{T^2}$ to determine the mass of Earth.	c. $G\frac{M}{4\pi^2} = \frac{r^3}{T^2}$ $\frac{6.67 \times 10^{-11} M_{\text{Earth}}}{4\pi^2} = \frac{(3.84 \times 10^8)^3}{(2.36 \times 10^6)^2}$ $\Rightarrow M_{\text{Earth}} = \frac{4\pi^2 \times (3.84 \times 10^8)^3}{(6.67 \times 10^{-11}) \times (2.36 \times 10^6)^2}$ $= 6.02 \times 10^{24} \text{ kg}$

PRACTICE PROBLEM 5

Mars's moon, Phobos, orbits Mars with a period of 2.76×10^4 s. The distance from the centre of Mars to the centre of Phobos is 9.38×10^6 m.

a. **What is the speed of Phobos as it orbits Mars?**
b. **What is the acceleration experienced by Phobos?**
c. **Use the provided data to calculate the mass of Mars.**

EXTENSION: Period of the Moon

The orbit of the Moon is slightly elliptical, but the average radius of the Moon's orbit is about 384 000 km or about 60 Earth radii.

The period of the Moon in relation to the stars is called the sidereal or **orbital period** and has been measured at 27.321 582 days (or approximately 2.36×10^6 seconds). As a close approximation, 27.3 days can be used. The period of the Moon in relation to the Sun, that is the time between full moons, is 29.5 days; this is longer than the sidereal period because during this time Earth has moved further around the Sun.

orbital period the time it takes for a satellite to complete one orbit around a central object

TABLE 3.1 The solar system: some useful data

	Mean radius of orbit		Orbital period			
	(au)	(m)	(years)	(seconds)	Equatorial radius (m)	Mass (kg)
Sun					6.96×10^{8}	1.99×10^{30}
Mercury	0.387	5.79×10^{10}	0.241	7.60×10^{6}	2.44×10^{6}	3.30×10^{23}
Venus	0.723	1.08×10^{11}	0.615	1.94×10^{7}	6.05×10^{6}	4.87×10^{24}
Earth	1.00	1.50×10^{11}	1.00	3.16×10^{7}	6.37×10^{6}	5.97×10^{24}
Moon	2.57×10^{-3}	3.84×10^{8}	27.3 days*	2.36×10^{6}*	1.74×10^{6}	7.35×10^{22}
Mars	1.52	2.28×10^{11}	1.88	5.94×10^{7}	3.39×10^{6}	6.42×10^{23}
Jupiter	5.20	7.78×10^{11}	11.9	3.74×10^{8}	6.99×10^{7}	1.90×10^{27}
Saturn	9.58	1.43×10^{12}	29.5	9.30×10^{8}	5.82×10^{7}	5.68×10^{26}
Titan	8.20×10^{-3}	1.22×10^{9}	15.9 days*	1.37×10^{6}*	2.57×10^{6}	1.35×10^{23}
Uranus	19.2	2.87×10^{12}	84.0	2.65×10^{9}	2.54×10^{7}	8.68×10^{25}
Neptune	30.1	4.50×10^{12}	165	5.21×10^{9}	2.46×10^{7}	1.02×10^{26}
Pluto	39.48	5.91×10^{12}	248	7.82×10^{9}	1.15×10^{6}	1.31×10^{22}

*The orbital period for the Moon and Titan is the time it takes to complete one orbit around Earth and Saturn respectively. All other listed measurements is the time to orbit the Sun.

tlvd-8973

SAMPLE PROBLEM 6 Confirming Kepler's Third Law

Use the data in table 3.1 to calculate the value of $\frac{r^3}{T^2}$ for Mercury, Venus and Mars, and therefore confirm Kepler's Third Law.

THINK	WRITE
1. For Mercury, substitute the values $r = 5.79 \times 10^{10}$ m and $T = 7.60 \times 10^{6}$ s into the expression $\frac{r^3}{T^2}$.	$\frac{r^3}{T^2} = \frac{(5.79 \times 10^{10})^3}{(7.60 \times 10^{6})^2}$ $= 3.36 \times 10^{18}$
2. For Venus, substitute the values $r = 1.08 \times 10^{11}$ m and $T = 1.94 \times 10^{7}$ s into the expression $\frac{r^3}{T^2}$.	$\frac{r^3}{T^2} = \frac{(1.08 \times 10^{11})^3}{(1.94 \times 10^{7})^2}$ $= 3.35 \times 10^{18}$
3. For Mars, substitute the values $r = 2.28 \times 10^{11}$ m and $T = 5.94 \times 10^{7}$ s into the expression $\frac{r^3}{T^2}$.	$\frac{r^3}{T^2} = \frac{(2.28 \times 10^{11})^3}{(5.94 \times 10^{7})^2}$ $= 3.36 \times 10^{18}$
4. Compare the values of $\frac{r^3}{T^2}$ for the three planets.	The values of $\frac{r^3}{T^2}$ for the three planets are approximately the same, confirming Kepler's Third Law.

PRACTICE PROBLEM 6

Use the data in table 3.1 to calculate the value of $\frac{r^3}{T^2}$ for Saturn, Uranus and Neptune, and therefore confirm Kepler's Third Law.

3.4.3 Method of ratios

Newton used his Law of Universal Gravitation to show that Kepler's Third Law applies to all satellites going around the same central mass.

In the context of Earth, this means that the ratio $\frac{r^3}{T^2}$ is the same for every single artificial satellite, regardless of the orientation of its orbit, as well as for the Moon itself. If the period and the radius of one satellite's orbit is known, the method of ratios can be used to calculate the characteristics of any other satellite:

$$\frac{r_1^3}{T_1^2} = \frac{r_2^3}{T_2^2}$$

$$\left(\frac{r_1}{r_2}\right)^3 = \left(\frac{T_1}{T_2}\right)^2$$

where: r_1 is the distance between the central mass and satellite 1

r_2 is the distance between the central mass and satellite 2

T_1 is the period of orbit of satellite 1

T_2 is the period of orbit of satellite 2

The benefit of this method is that, because you are working with ratios, you don't need to use metres and seconds for your data. However, it is important that you are consistent. (If you are using metres for the distance for one satellite, you must use metres for the other satellite.) Earth radii (1 Earth radius = 6.37×10^6 m) and days can be used, making for simpler calculations.

tlvd-8974

SAMPLE PROBLEM 7 Determining the orbital radius using ratios

The communications satellite Intelsat has an orbital period of 24.54 hours and an orbital radius of 4.25×10^6 m. The International Space Station (ISS) has an orbital period of 92.75 minutes. Use the method of ratios to find the orbital radius of the ISS.

THINK	WRITE
1. Convert all time periods to the same unit.	$T_{ISS} = 92.75 \text{ minutes} = 5565 \text{ seconds}$ $T_{Intelsat} = 24.54 \text{ hours} = 88\,344 \text{ seconds}$
2. Apply Kepler's Third Law: $\frac{r_{ISS}^3}{T_{ISS}^2} = \frac{r_{Intelsat}^3}{T_{Intelsat}^2}$	$\frac{r_{ISS}^3}{T_{ISS}^2} = \frac{r_{Intelat}^3}{T_{Innelat}^2}$ $\frac{r_{ISS}^3}{5565^2} = \frac{(4.25 \times 10^6)^3}{88\,344^2}$ $\Rightarrow r_{ISS}^3 = \frac{5565^2 \times (4.25 \times 10^6)^3}{88\,344^2}$ $= 3.046 \times 10^{17}$ $r_{ISS} = \sqrt[3]{3.046 \times 10^{17}}$ $= 6.73 \times 10^5 \text{ m}$

PRACTICE PROBLEM 7

Mars has two moons, Phobos and Deimos. The orbital period of Phobos is 7.66 hours, whereas the orbital period of Deimos is 36.75 hours. If the orbital radius of Phobos is 9.38×10^6 m, calculate the orbital radius of Deimos using the method of ratios.

3.4.4 Artificial satellites

Artificial satellites are used for communication and exploration. Some transmit telephone and television signals around the world, some photograph cloud patterns to help weather forecasters, some are fitted with scientific equipment that enables them to collect data about X-ray sources in outer space, whereas others spy on neighbours! The motion of an artificial satellite depends on what it is designed to do.

Those satellites that are required to stay constantly above one place on Earth's surface are called **geostationary** satellites and they are said to be in geostationary orbit. In order to stay in position, a geostationary satellite must have an orbital period that is the same as the time for the central mass to complete one rotation about its axis. Therefore, geostationary satellites have a period of 24 hours, or 1 day, with the movement of the geostationary satellite corresponding to the rotation of Earth about its axis.

geostationary stationary relative to a point directly below it on Earth's surface. A geostationary orbit has the same period as the rotation of Earth

tlvd-8975

SAMPLE PROBLEM 8 Determining the altitude at which a geostationary satellite orbits Earth

A geostationary satellite has a period of 24.0 hours. The mass of Earth is 5.97×10^{24} kg. The radius of Earth is 6.37×10^6 m. Determine the altitude at which the geostationary satellite orbits Earth.

THINK

1. Convert the orbital period of the satellite to seconds.

2. Apply Kepler's Third Law: $G\frac{M}{4\pi^2} = \frac{r^3}{T^2}$

3. The question asks for the altitude of the satellite, which is the height **above** the surface of Earth. Subtract the radius of Earth from the radius of the satellite's orbit to determine the altitude.

WRITE

1. $T = 24.0 \text{ hours} = 86\,400 \text{ seconds}$

2. $$G\frac{M_{\text{Earth}}}{4\pi^2} = \frac{r^3}{T^2}$$

$$\frac{6.67 \times 10^{-11} \times 5.97 \times 10^{24}}{4\pi^2} = \frac{r_3}{86\,400^2}$$

$$\Rightarrow r^3 = \frac{6.67 \times 10^{-11} \times 5.97 \times 10^{24} \times 86\,400^2}{4\pi^2}$$

$$\Rightarrow r = \sqrt[3]{\frac{6.67 \times 10^{-11} \times 5.97 \times 10^{24} \times 86\,400^2}{4\pi^2}}$$

$$= 4.22 \times 10^7 \text{ m}$$

3. $$\text{altitude} = \text{radius of orbit} - \text{radius of Earth}$$

$$= 4.22 \times 10^7 - 6.37 \times 10^6$$

$$= 3.58 \times 10^7 \text{ m}$$

PRACTICE PROBLEM 8

A satellite is placed into a geostationary orbit above Mars. A day on Mars lasts for 24 hours and 37 minutes. The mass of Mars is 6.42×10^{23} kg and the radius is 3.39×10^6 m. Calculate the altitude of the satellite.

3.4.5 'Floating' in a spacecraft

The appearance of an astronaut floating around inside a spacecraft suggests that there is no force acting on them, leading some people to mistakenly think that there is no gravity in space. In fact, both the astronaut and the spacecraft are in a circular orbit about Earth.

FIGURE 3.10 An astronaut inside the International Space Station. Both the astronaut and the station are in orbit around Earth.

However, you also know that if an object is moving in a curved path, therefore changing its direction, there must be acceleration. If the path is circular, the acceleration is directed towards the centre of that path. The astronaut and the spacecraft are in the same gravitational field. They are at the same distance from the centre of Earth. They are travelling at the same speed, taking the same time to orbit Earth. Therefore, their centripetal accelerations provided by the gravitational field are the same.

TABLE 3.2 Comparing the gravitational fields experienced by an astronaut and spacecraft

For the spacecraft:	For the astronaut:
$G\frac{M_{\text{Earth}}M_{\text{spacecraft}}}{r^2} = M_{\text{spacecraft}} \times \frac{4\pi^2 r}{T^2}$ $G\frac{M_{\text{Earth}}}{r^2} = \frac{4\pi^2 r}{T^2}$	$G\frac{M_{\text{Earth}}m_{\text{astronaut}}}{r^2} = m_{\text{astronaut}} \times \frac{4\pi^2 r}{T^2}$ $G\frac{M_{\text{Earth}}}{r^2} = \frac{4\pi^2 r}{T^2}$

There is no need for a normal force by the spacecraft on the astronaut to explain the astronaut's motion. The astronaut inside the spacecraft circles Earth as if the spacecraft was not there. Indeed, if the astronaut is outside the spacecraft doing a spacewalk, the astronaut's speed and acceleration around Earth will be unchanged as they 'float' beside the spacecraft. Once back inside, their speed and acceleration are still unchanged, and this time they are 'floating' inside the spacecraft.

If the astronaut steps onto a set of bathroom scales, they will give a reading of zero. As shown in figure 3.11, an astronaut running on a treadmill needs stretched springs attached to his waist to pull him down to the treadmill.

FIGURE 3.11 The cloth-covered stretched springs are pulling the astronaut down so he can exercise on the treadmill.

tlvd-8976

SAMPLE PROBLEM 9 Explaining the reading on scales in a spacecraft

An astronaut on the International Space Station stands on a set of bathroom scales. Explain what the expected reading would be.

THINK	WRITE
The only force acting on the astronaut is the gravitational force from Earth. The only force acting on the bathroom scales is the gravitational force from Earth.	The net force on the astronaut is equal to the force due to gravity on the astronaut, because the astronaut is in circular motion around Earth with acceleration $\boldsymbol{g}$. The bathroom scales are also in circular motion around Earth with acceleration $\boldsymbol{g}$. This means that the normal force experienced by the astronaut from the bathroom scales is 0 N.

PRACTICE PROBLEM 9

In an attempt to increase the reading on the bathroom scales, the astronaut on the International Space Station holds a 10-kg mass. What difference will this make to the reading on the bathroom scales?

on Resources

Interactivity How the speed and period changes with the radius of a satellite's orbit (int-0062)

Digital document Kepler's Laws (doc-39320)

3.4 Activities

learnon

Students, these questions are even better in jacPLUS

Receive immediate feedback and access sample responses

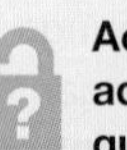
Access additional questions

Track your results and progress

Find all this and MORE in jacPLUS

3.4 Quick quiz on	3.4 Exercise	3.4 Exam questions

3.4 Exercise

1. The Moon orbits the Earth with a period of 27.32 days. The orbital radius is 3.84×10^8 m. Calculate the orbital speed of the Moon. Give your answer to 3 significant figures.
2. The asteroid 243 Ida was discovered in 1884. The Galileo spacecraft, on its way to Jupiter, visited the asteroid in 1993. The asteroid was the first to be found to have a natural satellite, that is, its own moon, now called Dactyl. Dactyl orbits Ida at a radius of 100 km and with a period of 27.0 hours. Calculate the mass of the asteroid.
$(G = 6.67 \times 10^{-11}\ \text{N m}^2\ \text{kg}^{-2})$
3. Answer the following.
   a. Using the information below, calculate the magnitude of the centripetal acceleration of a person standing on Earth's equator due to Earth's rotation about its axis. Give your answer to 3 significant figures.
   $$r_E = 6.37 \times 10^6\ \text{m}$$
   $$T = 24.0\ \text{hours}$$
   b. Would the centripetal acceleration be greater or less for a person standing in Victoria? Justify your answer.

4. A space station orbits at a height of 355 km above Earth and completes one orbit every 92.0 minutes. (Earth's radius is $r_E = 6.37 \times 10^6$ m and Earth's mass is $m_E = 5.97 \times 10^{24}$ kg.)
   a. Calculate the orbital speed of the space station. Give your answer to 3 significant figures.
   b. Calculate the magnitude of centripetal acceleration of the space station. Give your answer to 3 significant figures.
   c. What gravitational field strength does the space station experience? Give your answer to 3 significant figures.
      ($G = 6.67 \times 10^{-11}$ N m^2 kg^{-2})
   d. Compare your answers to parts **b** and **c**; should they be the same? Explain your reasoning and account for discrepancies.
   e. If the mass of the space station is $m_{SS} = 1200$ tonnes, calculate the magnitude of the centripetal force on the space station. Give your answer to 3 significant figures.
   f. The mass of an astronaut and the special spacesuit they wear when outside the space station is 270 kg. If they are a distance of 10 m from the centre of mass of the space station, what is the force exerted on the astronaut by the floor of the space station?
5. A spacecraft of mass 470 kg is orbiting Earth with a period of 2.50 hours. Using the information below, calculate its altitude. Give your answer to 3 significant figures.
   $G = 6.67 \times 10^{-11}$ N m^2 kg^{-2}
   $m_E = 5.97 \times 10^{24}$ kg
   $r_E = 6.37 \times 10^6$ m
6. A geostationary satellite remains above the same position on Earth's surface. Once in orbit, the only force acting on the satellite is that of gravity towards the centre of the Earth. Explain why the satellite doesn't fall straight back down to Earth.
7. A spacecraft orbits Earth with an orbital radius of $r_A = 20\,000$ km. If its orbital radius were doubled, calculate the expected change in the period of orbit.
8. Venus and Saturn both orbit the Sun. Using the information about the Sun and the periods of the two planets from table below, calculate the value of the following ratio:

$$\frac{\text{distance of Saturn from the Sun}}{\text{distance of Venus from the Sun}}$$

   Give your answer to 3 significant figures.

	Mean radius of orbit		Orbital period			
	(au)	(m)	(years)	(seconds)	Equatorial radius (m)	Mass (kg)
Sun					6.96×10^8	1.989×10^{30}
Venus	0.723	1.08×10^{11}	0.615	1.94×10^7	6.05×10^6	4.87×10^{24}
Saturn	9.58	1.43×10^{12}	29.5	9.30×10^8	5.82×10^7	5.68×10^{26}

9. An astronaut is living on the International Space Station. When the astronaut sits down, they need to strap themselves in so that they do not float away.
   Explain why they need to do this.
10. In the future, it is predicted that space stations may rotate to simulate the gravitational field of Earth and therefore make life more normal for the occupants.
    a. Draw a diagram of such a space station. Include on your diagram:
       - the axis of rotation
       - the distance of the occupants from the axis
       - arrows indicating the direction the occupants would consider as 'down'.

       (Remember to consider the frame of reference of the occupants!)
    b. Make an estimate of the period of rotation your space station would need to simulate Earth's gravitational field.

3.4 Exam questions

Question 1 (7 marks)

Source: *VCE 2022 Physics Exam, NHT, Section B, Q.2; © VCAA*

The speed of a satellite in a circular orbit around a planet is given by $v = \sqrt{\frac{GM}{r}}$, where G is the universal gravitational constant, M is the mass of the planet and r is the orbital radius of the satellite.

Titan is the largest moon of Saturn and has an orbital radius of 1.2×10^9 m. The mass of Saturn is 5.7×10^{26} kg. Assume that Titan's orbit is circular.

a. Calculate Titan's orbital speed. **(2 marks)**

Another moon of Saturn is Rhea. Rhea is in a circular orbit of radius 5.3×10^8 m.

b. Does Rhea travel faster than, at the same speed as or slower than Titan? Explain your answer. **(2 marks)**

c. Titan's period around Saturn is 16 days.
Calculate Rhea's period around Saturn. Show your working. **(3 marks)**

Question 2 (6 marks)

Source: *VCE 2022 Physics Exam, NHT, Section B, Q.8; © VCAA*

A satellite is moving in a stable circular orbit 25 Earth radii from the centre of Earth, as shown in Figure 5. The period of the satellite is T.

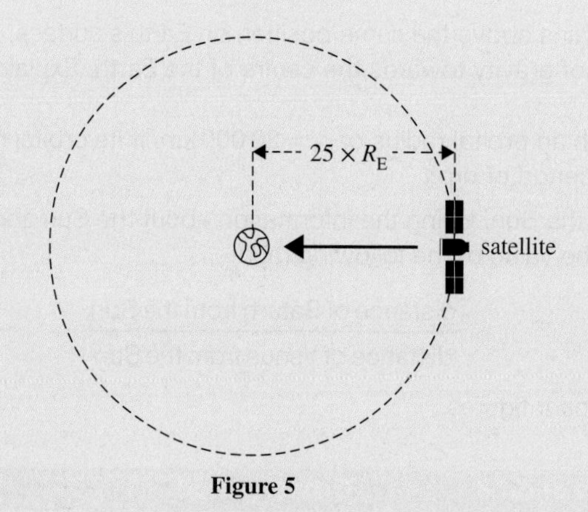

Figure 5

a. Calculate the magnitude of the acceleration of the satellite. Show your working. **(2 marks)**

b. Indicate the direction of the acceleration of the satellite by drawing an arrow on the satellite shown in Figure 5. **(1 mark)**

c. Another identical satellite is placed in a stable circular orbit 30 Earth radii from the centre of Earth.
Using the terms 'less than', 'same as' or 'more than', copy and complete the table below to describe the magnitude of the acceleration, the kinetic energy and the period of this satellite compared to those of the satellite that is orbiting at 25 Earth radii. **(3 marks)**

Magnitude of acceleration	
Kinetic energy	
Period	

Question 3 (3 marks)

Source: *VCE 2021 Physics Exam, Section B, Q.3; © VCAA*

To calculate the mass of distant pulsars, physicists use Newton's law of universal gravitation and the equations of circular motion.

The planet Phobetor orbits pulsar PSR B1257+12 at an orbital radius of 6.9×10^{10} m and with a period of 8.47×10^6 s.

Assuming that Phobetor follows a circular orbit, calculate the mass of the pulsar. Show all your working.

Question 4 (6 marks)

Source: *VCE 2021 Physics Exam, NHT, Section B, Q.3;* © VCAA

The motion of Earth's moon can be modelled as a circular orbit around Earth, as shown in Figure 3.

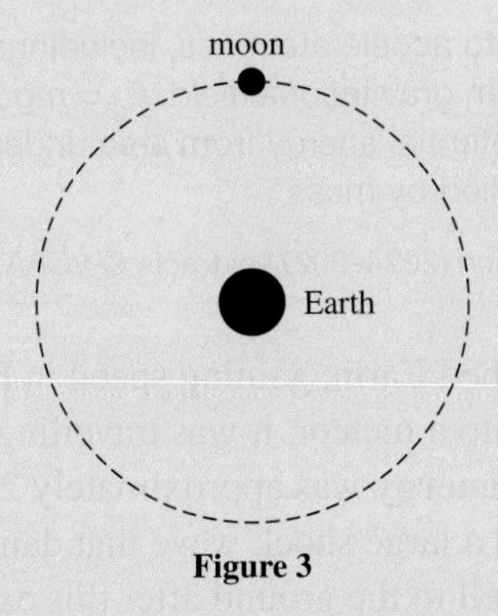

Figure 3

Data

Mass of Earth	5.98×10^{24} kg
Mass of the Moon	7.35×10^{22} kg
Radius of the Moon's orbit around Earth	3.84×10^{8} m
Universal gravitational constant (G)	6.67×10^{-11} N m^{2} kg^{-2}

a. Calculate the magnitude of the gravitational force that Earth exerts on the orbiting moon. Give your answer correct to three significant figures. Show your working. **(3 marks)**

b. The average orbital period of Earth's moon is 27.32 days. The moon is moving slightly further away from Earth at an average rate of 4 cm per year. **(3 marks)**

Given this information, will the average orbital period of Earth's moon decrease, stay the same or increase? Explain your answer.

Question 5 (4 marks)

Source: *VCE 2018, Physics Exam, Section B, Q.10;* © VCAA

Members of the public can now pay to take zero gravity flights in specially modified jet aeroplanes that fly at an altitude of 8000 m above Earth's surface. A typical trajectory is shown in Figure 12. At the top of the flight, the trajectory can be modelled as an arc of a circle.

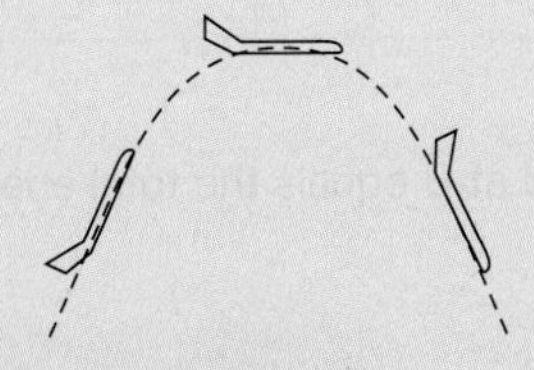

Figure 12

a. Calculate the radius of the arc that would give passengers zero gravity at the top of the flight if the jet is travelling at 180 m s^{-1}. Show your working. **(2 marks)**

b. Is the force of gravity on a passenger zero at the top of the flight? Explain what 'zero gravity experience' means. **(2 marks)**

More exam questions are available in your learnON title.

3.5 Energy changes in gravitational fields

KEY KNOWLEDGE

- Analyse the use of gravitational fields to accelerate mass, including:
  - potential energy changes in a uniform gravitational field: $E_g = mg\Delta h$
- Analyse the change in gravitational potential energy from area under a force vs distance graph and area under a field vs distance graph multiplied by mass

Source: Adapted from VCE Physics Study Design (2024–2027) extracts © VCAA; reproduced by permission.

On 15 February 2013, an asteroid approached Earth, gaining speed in Earth's gravitational field. By the time it reached the atmosphere and vaporised into a meteor, it was travelling at a speed of 19 km s^{-1}. With a mass of approximately 1.2×10^7 kg, its **kinetic energy** was approximately 2.2×10^{15} J. It exploded about 30 km above Chelyabinsk, in Russia. This caused a large shock wave that damaged buildings and injured over 1000 individuals. Smaller meteorite fragments fell to the ground after this explosion (with one fragment being found to be over 600 kg). The Chelyabinsk asteroid gains speed as it approaches Earth. Its kinetic energy increases. Where does this energy come from? The asteroid is experiencing a force from Earth's gravitational field, which causes the asteroid to accelerate. The work done by the gravitational force is transferred into kinetic energy of the asteroid.

However, the total energy of the asteroid–Earth system must remain the same. The kinetic energy of the asteroid increases but the **gravitational potential energy** of the asteroid–Earth system decreases.

In a similar way, as Halley's Comet orbits the Sun, it has a speed of 38 km s^{-1} at the point of closest approach, but slows to 0.64 km s^{-1} at the point furthest away.

kinetic energy the energy associated with the movement of an object. Like all forms of energy, i is a scalar quantity.

gravitational potential energy energy stored in an object as a result of its position relative to another object to which it is attracted by the force of gravity

3.5.1 Force–distance graphs

The asteroid and Halley's Comet both experience gravitational forces that change in size. The change in gravitational potential energy in each case is due to work done by and against the gravitational force and can be found by calculating the area under the graph of gravitational force versus the displacement in the direction of the force.

The change in gravitational potential energy can be obtained from the area under a force–distance graph.

FIGURE 3.12 The total energy of the asteroid at B equals the total energy of the asteroid at A.

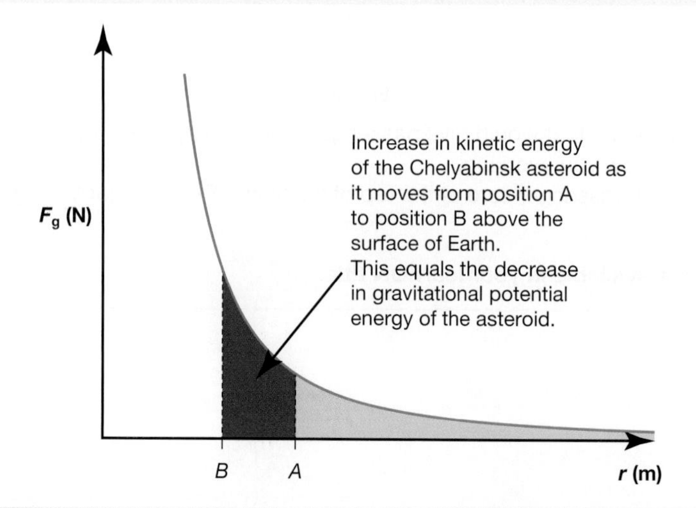

Note: Sometimes you may be given a negative value for the gravitational force rather than a positive value, and so the graph will be inverted vertically. It is important that you are calculating the area between the graph and the horizontal axis in this case when calculating the change in kinetic energy.

tlvd-8977

SAMPLE PROBLEM 10 Calculating the change in gravitational potential energy of a falling mass using force–distance graphs

A mass of 10 kg falls to the surface of Earth from an altitude equal to two Earth radii (or 1.27×10^7 m). What is the change in gravitational potential energy?

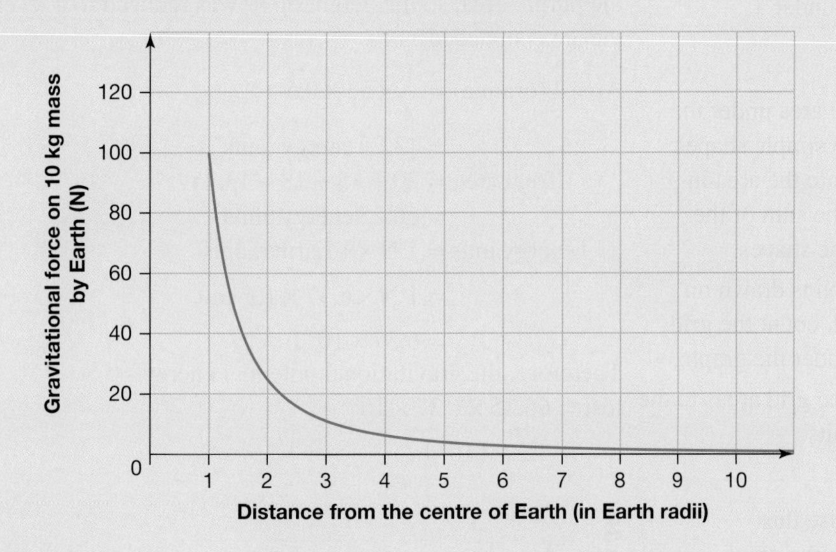

THINK	WRITE
1. The gravitational field does the work on the mass, and the change in gravitational potential energy can be found from the area under the force–displacement graph. (Remember that the surface of Earth is 1 Earth radius away from the centre of Earth. So an altitude of 2 radii from the surface is in fact 3 Earth radii from the centre, as given in the graph.)	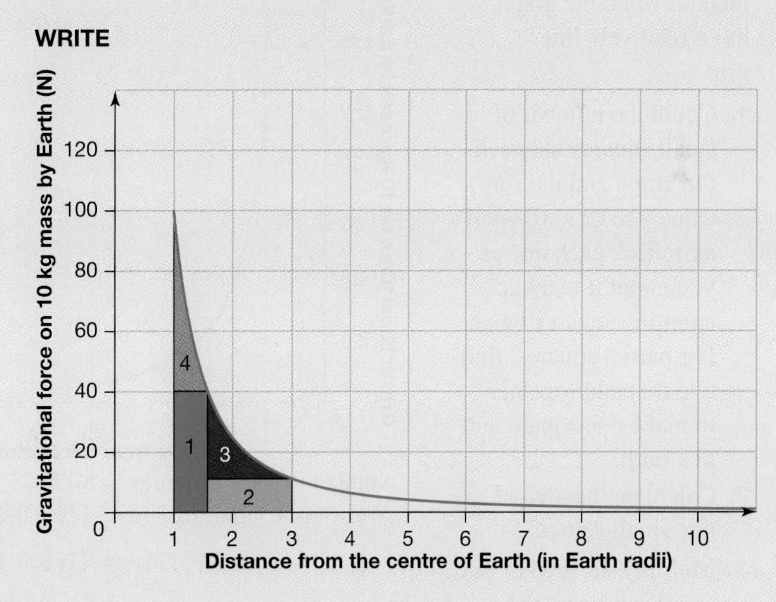

There are two methods you can use to determine the area. For these questions, VCAA will accept a range of answers, so either technique may be used, depending on the grid you are given. How to find the area under a curved graph:

Method 1:

1. Divide the area under the graph into simple shapes and estimate the area in terms of the sum of the areas of the shapes.
2. If the graph is drawn on a fine grid, count the grid squares under the graph.
3. Convert the grid areas to the correct units.

$$\text{Area 1 (blue)} = 40 \times 0.5 = 20 \text{ energy units}$$

$$\text{Area 2 (green)} = 10 \times 1.5 = 15 \text{ energy units}$$

$$\text{Area 3 (purple)} = \frac{1}{2} \times 24 \times 1.5 = 18 \text{ energy units}$$

(*Note:* The triangle with area $\frac{1}{2} \times 30 \times 1.5$ would be larger than the purple area, so the height of 30 was reduced to a level where the areas matched.)

$$\text{Area 4 (orange)} = \frac{1}{2} \times 53 \times 0.5 = 13.25 \text{ energy units}$$

$$\text{Total area} = 20 + 15 + 18 + 13.25 = 66.25 \text{ energy units}$$

$$1 \text{ energy unit} = 1 \text{ N} \times 1 \text{ Earth radius} = 1 \text{ N} \times 6.37 \times 10^6 \text{ m} = 6.37 \times 10^6 \text{ J}$$

Therefore, the gravitational potential energy lost $= 66.25 \times 6.37 \times 10^6 = 4.22 \times 10^8$ J

Method 2: Use this method when the graph has a relatively fine grid.

1. Count the number of small squares between the graph and the zero-value line or horizontal axis. Tick each one as you count it to avoid counting squares twice. For partial squares, find two that add together to make one square and tick both.
2. Calculate the area of one small square.
3. Multiply the area of one small square by the number of small squares.

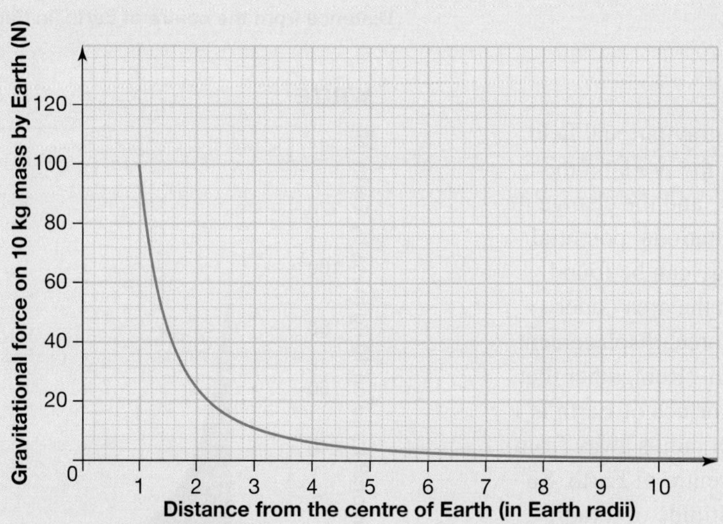

Number of small squares = 80.5

$$\text{Area of one small square} = 4 \text{ N} \times 0.2 \times 1 \text{ Earth radius} = 4 \text{ N} \times 0.2 \times 6.37 \times 10^6 \text{ m} = 5.1 \times 10^6 \text{ J}$$

Therefore, the loss in gravitational potential $= 80.5 \times 5.1 \times 10^6$ J $= 4.11 \times 10^8$ J

PRACTICE PROBLEM 10

a. **Use the following graph of the gravitational force on the Chelyabinsk asteroid to show that, in moving from an altitude of two Earth radii to an altitude of one Earth radius, the asteroid lost 1.25×10^{14} joules of gravitational potential energy.**

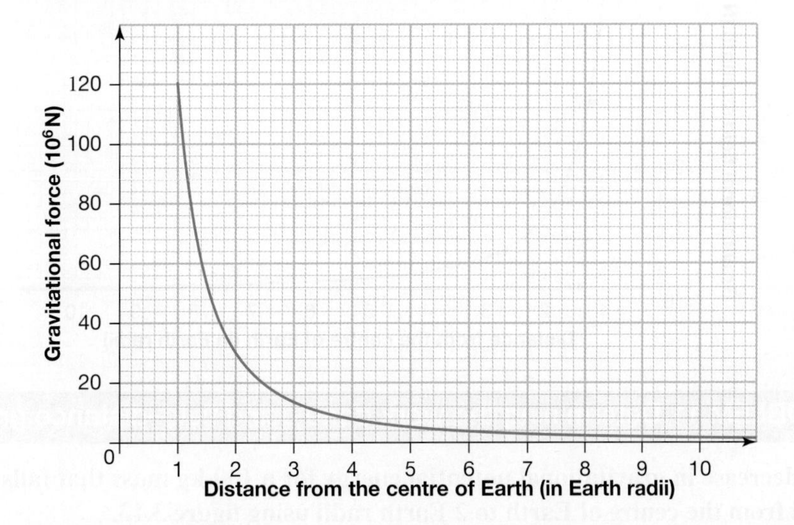

b. **Use the graph to find, to the nearest whole number, approximately how much gravitational potential energy was lost by falling from an altitude of one Earth radius to Earth's surface. Compare this value with your answer to part a.**

Note: **Sometimes you may be given a negative value for the gravitational force rather than a positive value, and so the graph will be inverted vertically. It is important that you are calculating the area between the graph and the horizontal axis in this case when calculating the change in energy.**

3.5.2 Field–distance graphs

The graph of the force versus distance experienced by a 10-kg mass falling to Earth's surface and the Chelyabinsk asteroid have the same shape, but very different scales. However, the gravitational fields experienced by the 10-kg mass and the Chelyabinsk asteroid are identical. An alternative approach to finding the change in gravitational potential energy of an object is to find the area under the gravitational field versus distance graph and multiply that value by the mass of the object.

The change in gravitational potential energy for an object that moves from one point to another can be obtained by multiplying the area under the graph of the gravitational field against distance by its mass.

The unit for gravitational field is newtons per kilogram. The unit for the area under a graph of gravitational field against distance is (newtons per kilogram) × metre, hence newton metre per kilogram or simply joule per kilogram. The change in energy can be obtained from this area by multiplying by the mass of the object. Sometimes, instead of being given the distance in metres, you may be given the distance in Earth radii (as seen in figure 3.13). In this case, you need to ensure you convert into metres before multiplying by newtons per kilogram.

FIGURE 3.13 Graph of the Earth's gravitational field strength

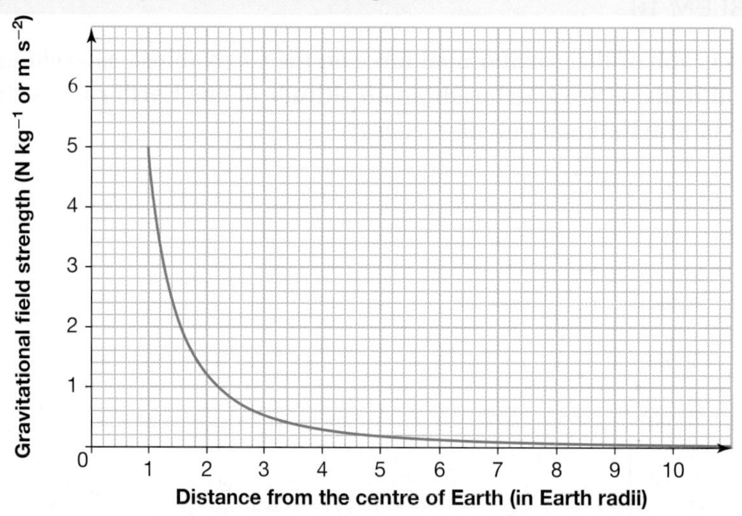

tlvd-8978

SAMPLE PROBLEM 11 Determining the work done using field–distance graphs

Determine the decrease in gravitational potential energy for a 100-kg mass that falls from a distance of 4 Earth radii from the centre of Earth to 2 Earth radii using figure 3.13.

THINK	WRITE
1. Determine the area between the curve and the axis between the distance of 2 and 4 Earth radii and convert the units to SI units. (Remember, there are two methods you can use for these questions. You may count the squares or split them into smaller areas. Due to the provision of small squares, the method of counting squares has been used.)	The area under the curve between 2 Earth radii and 4 Earth radii is 29 squares, with each square having an area of $0.2\ \text{N kg}^{-1} \times 0.2$ Earth radii. One Earth radius $= 6.37 \times 10^6$ m Area under curve $= 0.2 \times 0.2 \times 6.37 \times 10^6 \times 29$ $= 7.39 \times 10^6\ \text{N m kg}^{-1}$
2. Remember that $F_g = mg$. Multiply the area by the mass to obtain the change in energy.	Change in energy $= 100\ \text{kg} \times 7.39 \times 10^6\ \text{N m kg}^{-1}$ $= 7.39 \times 10^8$ N m $= 7.39 \times 10^8$ J

PRACTICE PROBLEM 11

Calculate the decrease in gravitational potential energy for a 20-kg mass that falls from a distance of 3 Earth radii from the centre of Earth to 2 Earth radii from the centre of Earth.

3.5.3 Energy changes close to Earth's surface

You have learned that the gravitational field of Earth decreases as the square of distance from the centre of Earth. However, Earth is huge compared to the scale of ordinary human actions. For a distance of 1000 m from the surface of Earth, the gravitational field has decreased by a mere 0.03%, from 9.805 N kg^{-1} to 9.802 N kg^{-1}. The gravitational field strength experienced by a ball dropped from an initial height of a couple of metres above the surface of Earth is effectively constant.

What about the energy transformations experienced by the ball? As the ball falls, work is done on the ball by the gravitational field, and the ball speeds up. Energy is transferred from the gravitational field into the kinetic energy of the ball.

Because the force of gravity, $\boldsymbol{F}_g$, is constant ($\boldsymbol{F}_g = mg$), the area under the force versus distance graph is simply the product of the force and the change in height of the object (Δh).

The change in gravitational potential energy is:

$$\Delta E_g = mg \times \text{change in height} = mg\,\Delta h$$

where: ΔE_g is the change in gravitational potential energy, in J

m is the mass of the object, in kg

g is the gravitational field strength, in N kg^{-1}

Δh is the change in height, in m

$$\Delta E = W = Fs$$

where: ΔE is the change in gravitational potential energy, in J

W is the work done, in J

F is the force of gravity, in N

s is the displacement in the gravitational field, in m

tlvd-8979

SAMPLE PROBLEM 12 Calculating the work done by gravitational fields (extending)

a. Determine the work done by the gravitational field on a 30-kg ball that is dropped from the top of a 150-m building.

b. Calculate the change in kinetic energy of a 30-kg ball that is dropped from the top of 150-m building. Assume that air resistance is negligible.

THINK	WRITE
a. 1. Determine the magnitude of the force due to gravity acting on the ball.	a. $F = mg$ $= 30 \times 9.8$ $= 294\text{ N}$
2. The work done can be calculated using $W = Fs$.	$W = Fs$ $= 294 \times 150$ $= 44\,100\text{ J}$
b. The change in kinetic energy is equal to the work done by the field.	b. ΔE_k = work done by field $= 44\,100\text{ J}$ $= 4.4 \times 10^4\text{ J}$

PRACTICE PROBLEM 12

A 2.0-kg ball is thrown vertically upwards from a height of 1.0 m, with an initial speed of 5.0 m s^{-1}. What is the kinetic energy of the ball just before it hits the ground?

elog-1917

INVESTIGATION 3.2

Exploring the relationship between gravitational potential energy and kinetic energy

AIM

To calculate the gravitational potential energy and kinetic energy of a falling ball and determine if the change in gravitational potential energy is equal to the kinetic energy

3.5 Activities

learnon

Students, these questions are even better in jacPLUS

Receive immediate feedback and access sample responses

Access additional questions

Track your results and progress

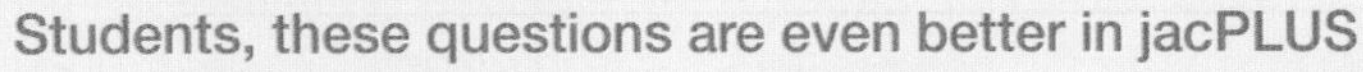

Find all this and MORE in jacPLUS

3.5 Quick quiz on | 3.5 Exercise | 3.5 Exam questions

3.5 Exercise

1. A 2000-kg satellite falls from a distance of 3.5×10^7 m to 1.0×10^7 m.
   Using the force–displacement graph below, estimate the change in the kinetic energy of the satellite. Give your estimate to 1 significant figure.

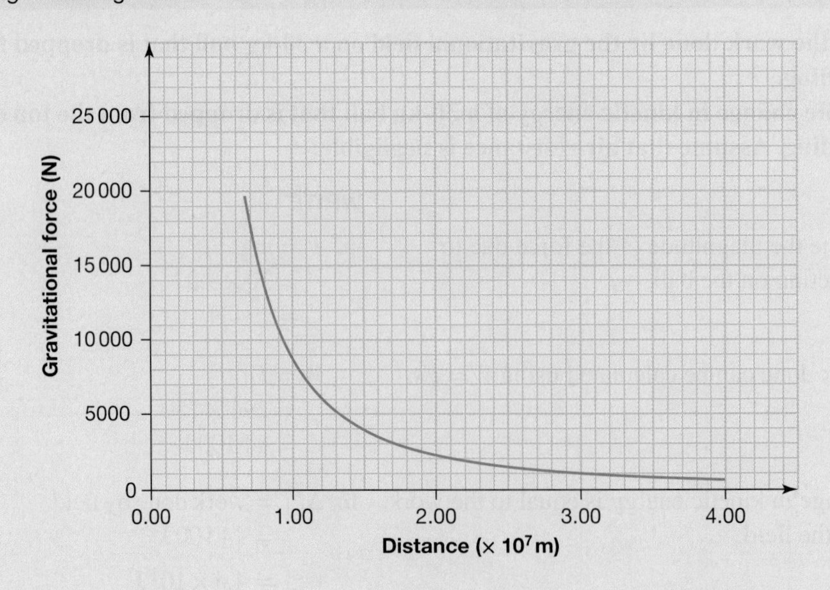

2. Using the force–displacement graph below, estimate the change in gravitational potential energy in moving a 150-kg mass from a distance of 1 Earth radius to 3 Earth radii from the centre of Earth. Give your estimate to 1 significant figure.

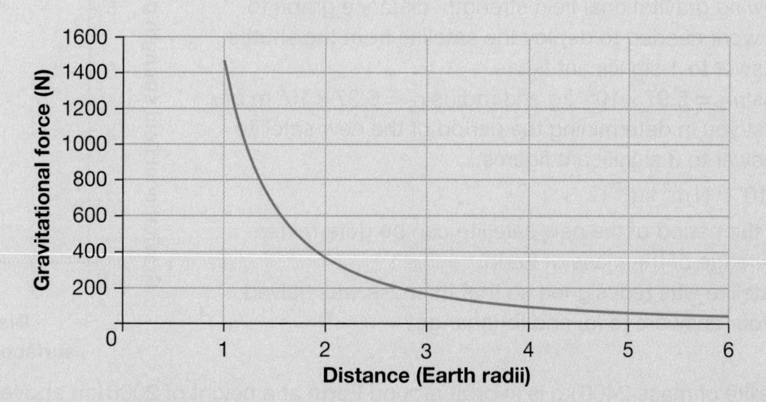

The change in in gravitational potential energy is approximately $\Delta E_g =$ ___J.

3. Explain why the area under a gravitational force–distance graph gives the energy needed to launch a satellite, but the area under a gravitational field strength–distance graph gives the energy *per kilogram* needed to launch a satellite.
4. A space probe traverses an elliptical orbit as it passes around a distant unknown planet, Planet Q. Explain how the kinetic energy of the space probe changes as it moves from X to Y and how the total energy of the space probe changes as it moves from X to Y.

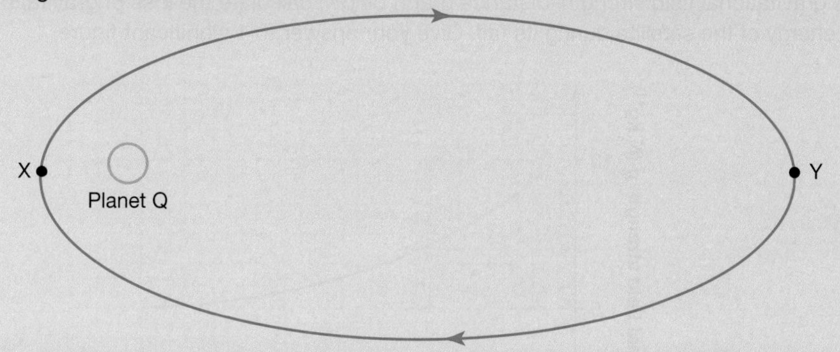

5. Using the gravitational field strength–distance graph below, estimate the change in gravitational potential energy of a 300-kg satellite carried from the surface of Earth (radius 6.37×10^6 m) to an orbit at an altitude of 1.4×10^7 m. Give your estimate to 1 significant figure.

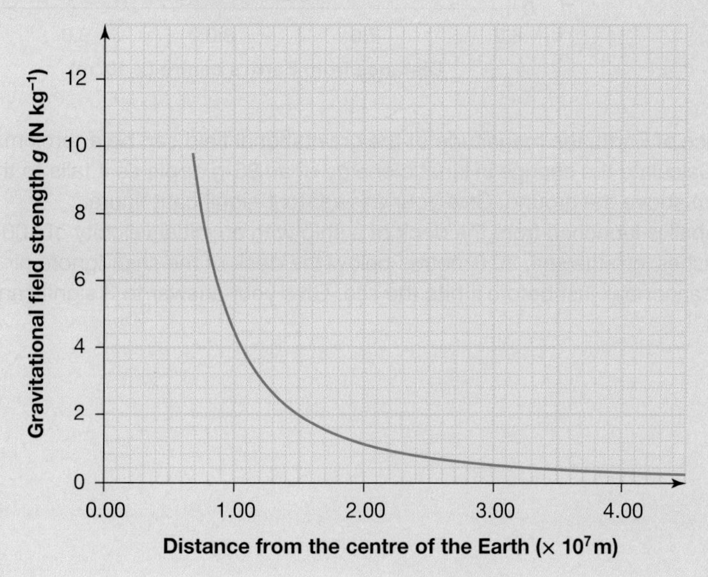

6. A space shuttle, orbiting Earth once every 93 minutes at a height of 400 km above the surface, deploys a new 800-kg satellite that is to orbit a further 200 km away from Earth.
   a. Use the following gravitational field strength–distance graph to estimate the work needed to deploy the satellite from the shuttle. Give your answer to 1 significant figure.
   b. Use the mass $m_E = 5.97 \times 10^{24}$ kg and radius $r_E = 6.37 \times 10^6$ m of Earth to assist you in determining the period of the new satellite. Give your answer to 3 significant figures. $(G = 6.67 \times 10^{-11}\ \text{N m}^2\ \text{kg}^{-2})$
   c. Explain how the period of the new satellite can be determined without knowledge of the mass of Earth.
   d. If the new satellite was redesigned so that its mass was halved, how would your answers to (**a**) and (**b**) change?

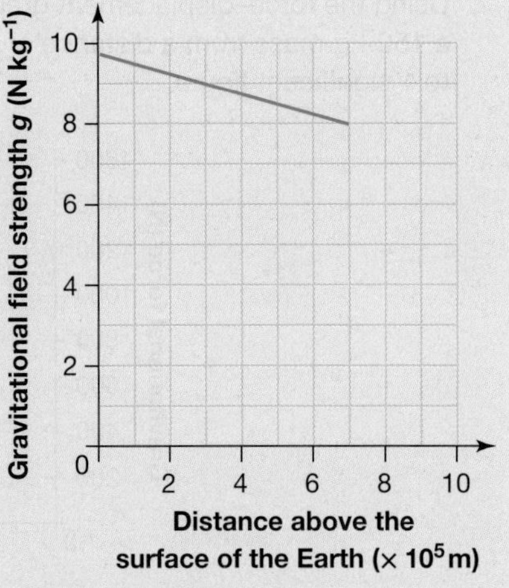

7. A disabled satellite of mass 2400 kg is in orbit around Earth at a height of 2000 km above sea level. It falls to a height of 800 km before its built-in rocket system can be activated to stop the fall.
   a. Calculate the magnitude of the gravitational force on the satellite while it is in its initial orbit. Give your answer to 3 significant figures.
      $G = 6.67 \times 10^{-11}\ \text{N m}^2\ \text{kg}^{-2}$
      $m_E = 5.97 \times 10^{24}$ kg
      $r_E = 6.37 \times 10^6$ m
   b. Using the gravitational field strength–distance graph below, calculate the loss of gravitational potential energy of the satellite during its fall. Give your answer to 1 significant figure.

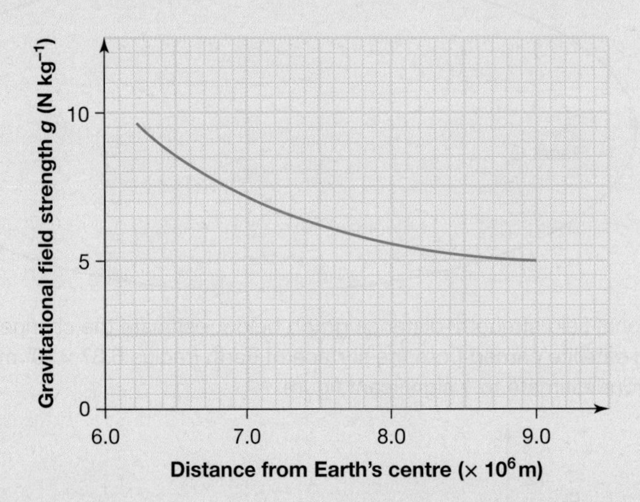

8. Close to the surface of Earth, the magnitude of the gravitational field can be approximated as a constant (**g** $= 9.8\ \text{N kg}^{-1}$). Calculate the change in kinetic energy of an 85-*g* apple as it falls to the ground from a branch at a height of 3.0 m above the ground. Give your answer to 1 significant figure.
9. A 4.50-kg cannonball is launched from the deck of a ship with an initial velocity of $100\ \text{m s}^{-1}$. It misses its target and splashes into the sea, 30.0 metres below the deck of the ship. Ignore air resistance and estimate the speed of the cannonball just before it hits the sea. Give your answer to 3 significant figures. Use $g = 9.81\ \text{N kg}^{-1}$.

3.5 Exam questions

Question 1 (1 mark)

MC The change in kinetic energy of a space probe in a gravitational field as it moves further away from a planet can be calculated from:

A. the area under a gravitational force versus distance graph.
B. the area under a gravitational field strength versus distance graph.
C. the area under a gravitational force versus distance graph multiplied by the mass of the space probe.
D. the area under a gravitational field strength versus distance graph divided by the mass of the space probe.

Question 2 (1 mark)

MC A space probe can use the gravitational fields of planets that it is passing to change its speed. Which of the following is **not** possible?

A. Gravitational potential energy can be converted into kinetic energy.
B. Work can be done on the space probe by the gravitational field, increasing its kinetic energy.
C. Kinetic energy can be converted into gravitational potential energy.
D. The gravitational field can do work on the space probe, increasing its gravitational potential and kinetic energy.

Question 3 (11 marks)

Source: *VCE 2021 Physics Exam, Section B, Q.8; © VCAA*

On 30 July 2020, the National Aeronautics and Space Administration (NASA) launched an Atlas rocket (Figure 7a) containing the Perseverance rover space capsule (Figure 7b) on a scientific mission to explore the geology and climate of Mars, and search for signs of ancient microbial life.

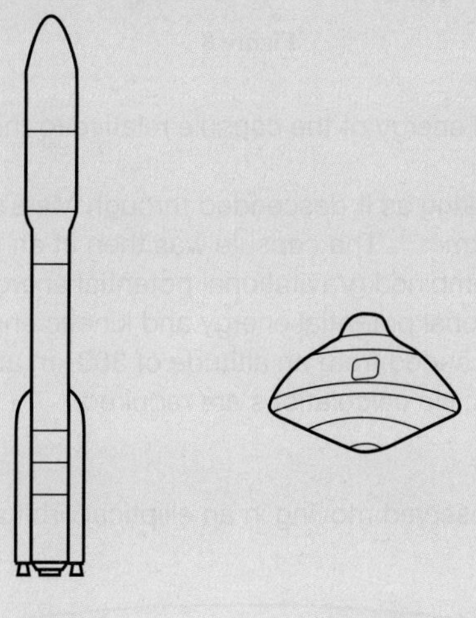

Figure 7a **Figure 7b**

a. At lift-off from launch, the acceleration of the rocket was $7.20\ \text{m s}^{-2}$. The total mass of the rocket and capsule at launch was 531 tonnes.
Calculate the magnitude and the direction of the thrust force on the rocket at launch. Take the gravitational field strength at the launch site to be $g = 9.80\ \text{N kg}^{-1}$. Give your answer in meganewtons. Show your working. **(3 marks)**

On 18 February 2021, the Perseverance rover space capsule, travelling at $20\,000\ \text{km h}^{-1}$, entered Mars's atmosphere at an altitude of 300 km above the surface of Mars. The mass of the capsule was 1000 kg.

b. Calculate the kinetic energy of the capsule at this point. Show your working. **(2 marks)**

Figure 8 shows the gravitational field strength of Mars g versus altitude h.

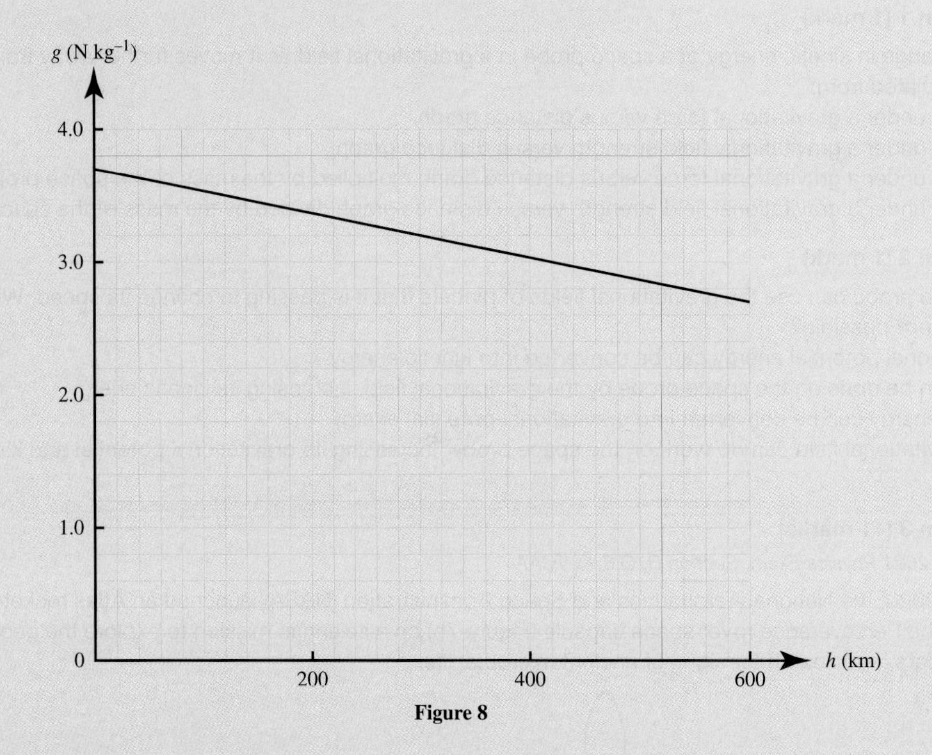

Figure 8

c. Calculate the gravitational potential energy of the capsule relative to the surface of Mars at an altitude of 300 km. Show your working. **(3 marks)**

d. The capsule used aerodynamic braking as it descended through Mars's atmosphere to reduce its speed from 20 000 km h^{-1} to 1600 km h^{-1}. The capsule was then at an altitude of 10 km above the surface of Mars and had ~1% of its original combined gravitational potential energy and kinetic energy remaining. Describe how ~99% of the gravitational potential energy and kinetic energy of the capsule was transformed and dissipated as the capsule descended from an altitude of 300 km above the surface of Mars to an altitude of 10 km above the surface of Mars. No calculations are required. **(3 marks)**

Question 4 (3 marks)

Planet P in a distant solar system is observed moving in an elliptical orbit around the central star, Peres.

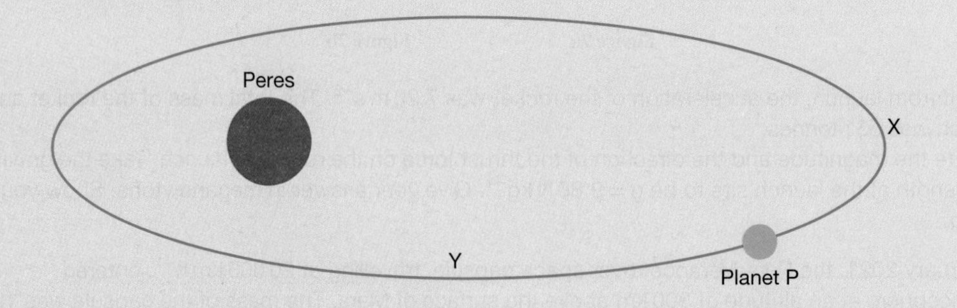

The gravitational force experienced by the planet is shown in the following graph.

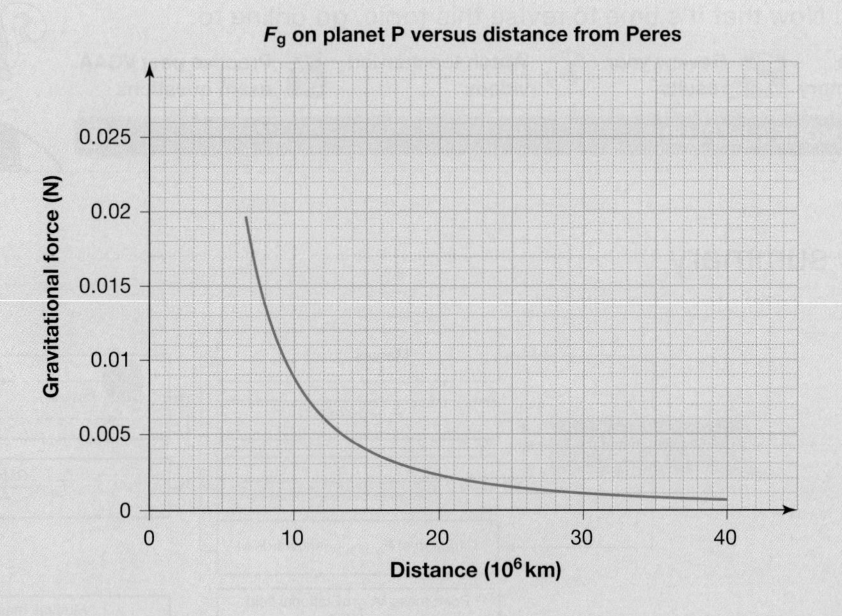

Calculate the change in kinetic energy as the planet moves from the point X, 30×10^6 km from Peres, to point Y, 10×10^6 km from Peres.

Question 5 (3 marks)

A 1.5-kg weather balloon is placed 50 km above the surface of the Earth. Use the following graph to calculate the energy required to move the balloon from the Earth's surface to its new position.

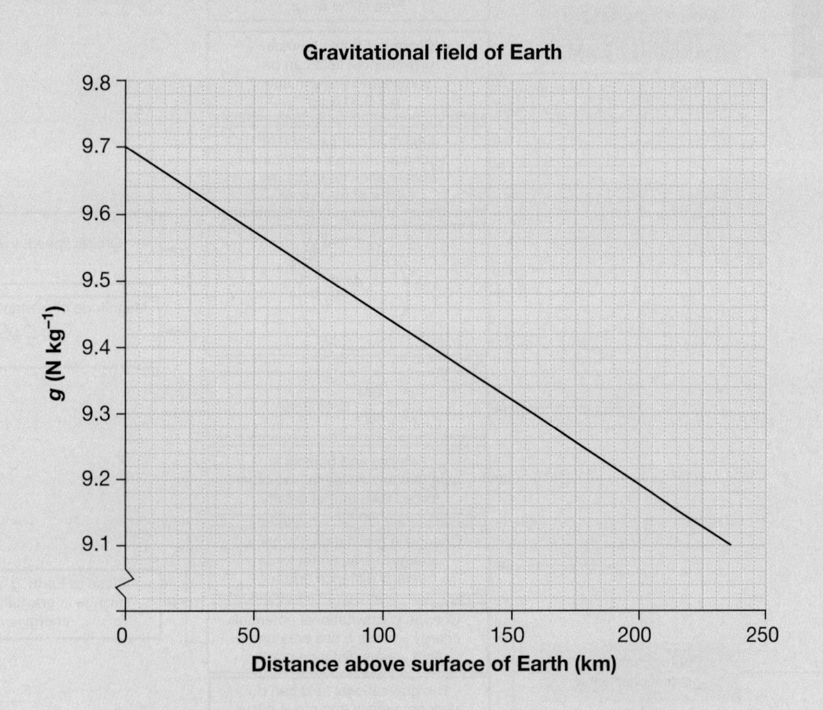

More exam questions are available in your learnON title.

3.6 Review

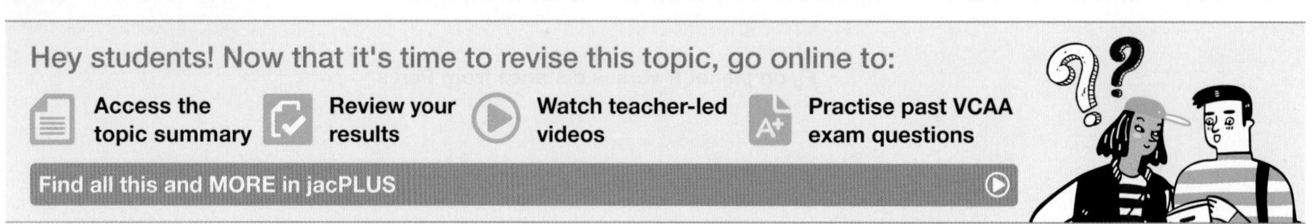

3.6.1 Topic summary

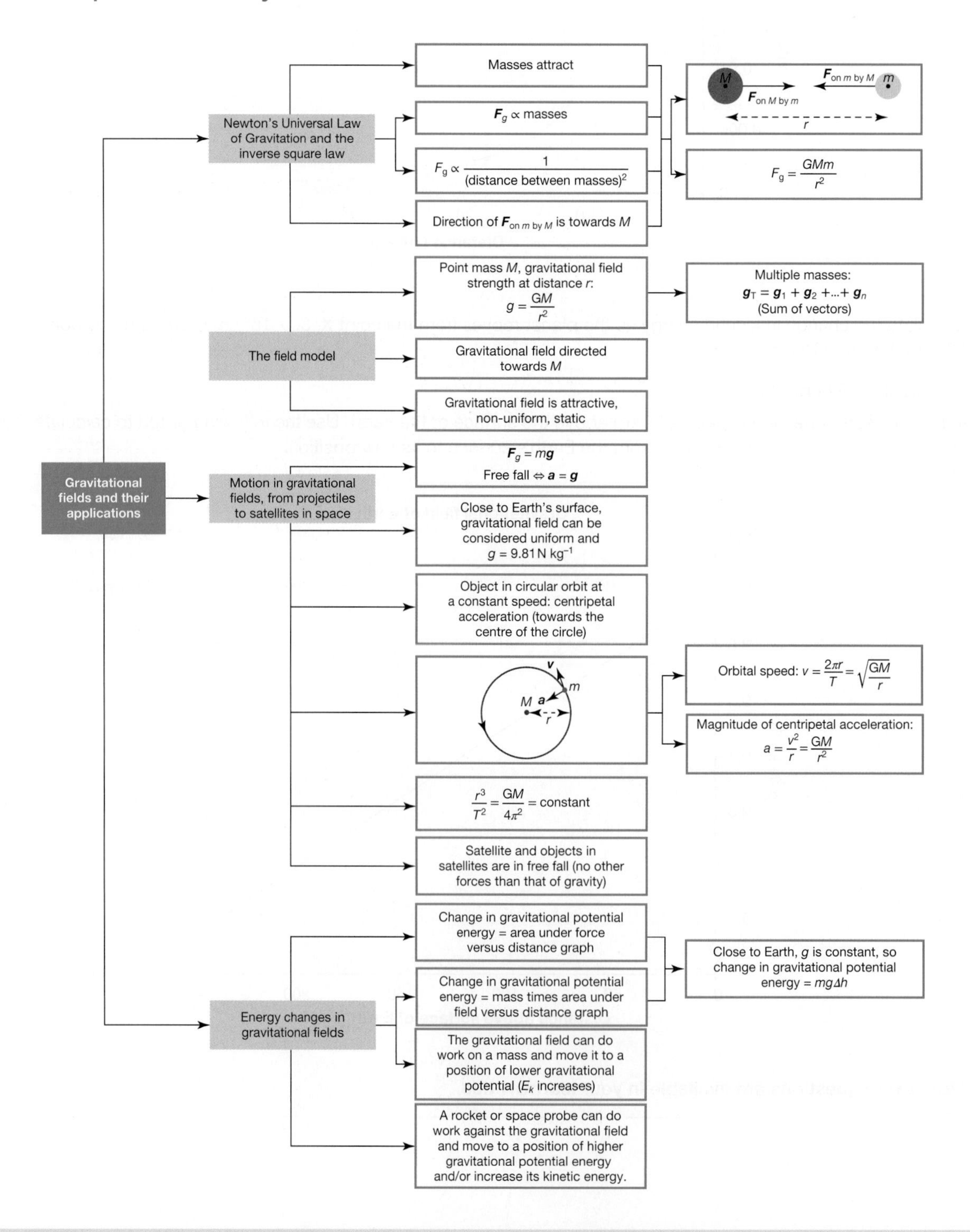

3.6.2 Key ideas summary

3.6.3 Key terms glossary

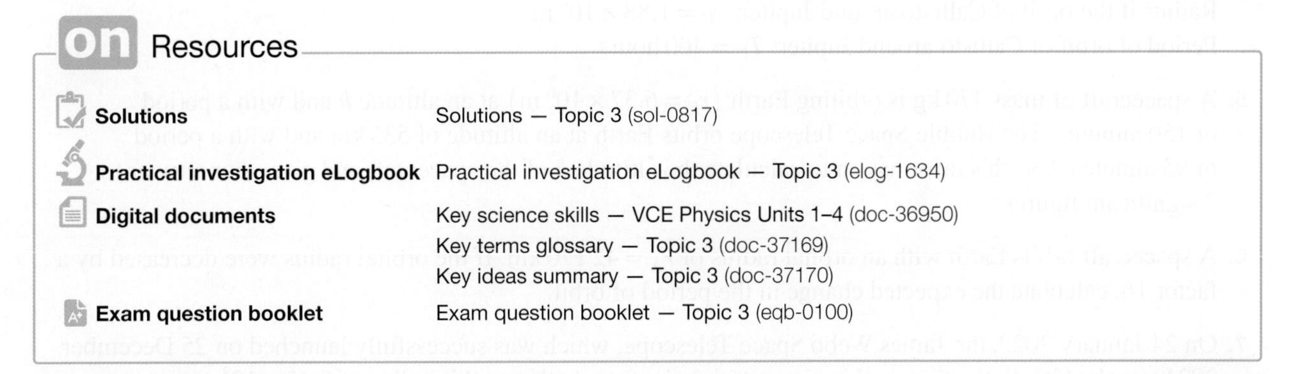

on Resources
Solutions
Solutions — Topic 3 (sol-0817)
Practical investigation eLogbook
Practical investigation eLogbook — Topic 3 (elog-1634)
Digital documents
Key science skills — VCE Physics Units 1–4 (doc-36950)
Key terms glossary — Topic 3 (doc-37169)
Key ideas summary — Topic 3 (doc-37170)
Exam question booklet
Exam question booklet — Topic 3 (eqb-0100)

3.6 Activities

learnon

Students, these questions are even better in jacPLUS

Receive immediate feedback and access sample responses

Access additional questions

Track your results and progress

Find all this and MORE in jacPLUS

3.6 Review questions

1. Which of the following correctly describes the gravitational field of Earth at large distances from Earth's surface?
   A. The shape of the field is uniform, it attracts other masses, and the strength of the field is inversely proportional to the square of the distance from the surface of Earth.
   B. The shape of the field is non-uniform, it attracts other masses, and the strength of the field is inversely proportional to the square of the distance from the surface of Earth.
   C. The shape of the field is uniform, it attracts other masses, and the strength of the field is inversely proportional to the square of the distance from the centre of Earth.
   D. The shape of the field is non-uniform, it attracts other masses, and the strength of the field is inversely proportional to the square of the distance from the centre of Earth.

2. Using the data below, calculate the ratio of the magnitude of the gravitational force on a ball at the surface of Mars to the magnitude of the gravitational force on the same ball at the surface of the Moon. Give your answer to 3 significant figures.

	Mass (kg)	Radius (m)
Mars	6.39×10^{23}	3.39×10^{6}
Moon	7.35×10^{22}	1.74×10^{6}

3. If a rocket travels in a straight line from Earth to the Moon, the gravitational field experienced by the rocket is initially pointing towards Earth. As the rocket journeys, the field strength becomes weaker, until it is finally zero, before increasing again, but with a direction pointing towards the Moon. Explain the change in magnitude and direction of the gravitational field.

4. Callisto is one of the many moons of Jupiter. Use the data provided below to calculate the mass of Jupiter. Give your answer to 3 significant figures.
   $G = 6.67 \times 10^{-11}\ \text{N m}^2\ \text{kg}^{-2}$
   Radius if the orbit of Callisto around Jupiter: $r_C = 1.88 \times 10^9$ m
   Period of orbit of Callisto around Jupiter: $T_C = 400$ hours
5. A spacecraft of mass 470 kg is orbiting Earth $\left(r_E = 6.37 \times 10^6\ \text{m}\right)$ at an altitude h and with a period of 150 minutes. The Hubble Space Telescope orbits Earth at an altitude of 535 km and with a period of 95 minutes. Use this information to calculate the altitude h of the spacecraft and give your answer to 2 significant figures.
6. A spacecraft orbits Earth with an orbital radius of $r_A = 42\,170$ km. If the orbital radius were decreased by a factor 16, calculate the expected change in the period of orbit.
7. On 24 January 2022, the James Webb Space Telescope, which was successfully launched on 25 December 2021, reached its destination, and is now orbiting the Sun, with an orbit radius of 9.45×10^8 m. Earth also orbits the Sun, with an orbit radius of 1.50×10^9 m.
   a. It takes 365 days for Earth to complete a revolution around the Sun. Determine the orbital period of the James Webb Space Telescope. Give your answer to 3 significant figures.
   b. Calculate the orbital speed of the James Webb space Telescope. Give your answer to 3 significant figures.
8. Lakshmi has read in her VCE Physics textbook that near Earth's surface, the gravitational field is uniform. She decides to check this information by calculating the gravitational field strength at different altitudes above Earth's surface.
   a. Using the information below and a calculator or a spreadsheet, calculate the gravitational field strength for the different altitudes h listed in the table. Give your answers to 3 significant figures.

$$G = 6.67 \times 10^{-11}\ \text{N m}^2\ \text{kg}^{-2}$$
$$m_E = 5.97 \times 10^{24}\ \text{kg}$$
$$r_E = 6.37 \times 10^6\ \text{m}$$

Altitude h (m)	g (N kg)$^{-1}$
0	
10^3	
10^4	
10^5	
10^6	
6.37×10^6	

   b. Based on those results, Lakshmi considers that using $g = 9.8\ \text{N kg}^{-1}$ for any altitude between sea level and 10 km is acceptable. Do you agree with her? Justify your answer.
9. Using the information below, calculate the change in gravitational potential energy of a 25-kg object dropped from 10 metres above the surface of the Moon. Give your answer to 3 significant figures.
   Mass of the Moon: $m_M = 7.35 \times 10^{22}$ kg.
   Radius of the Moon: $r_M = 1.74 \times 10^6$ m.
   $G = 6.67 \times 10^{-11}\ \text{N m}^2\ \text{kg}^{-2}$

10. The following graph shows how the gravitational field strength, g, varies with distance from the centre of the Moon. An artificial satellite of mass 150 kg orbits the Moon at a distance of 2.0×10^6 m.

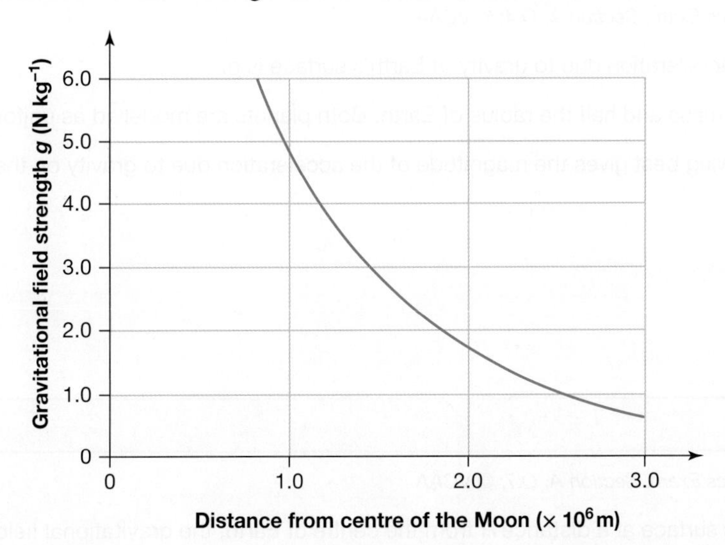

Estimate the change in the satellite's gravitational potential energy if it is moved to a new orbit with a radius of 3.0×10^6 m. Give your answer to 1 significant figure.

3.6 Exam questions

Section A — Multiple choice questions

All correct answers are worth 1 mark each; an incorrect answer is worth 0.

Question 1

Source: *VCE 2021 Physics Exam, Section A, Q.4;* © VCAA

The planet Phobetor has a mass four times that of Earth. Acceleration due to gravity on the surface of Phobetor is $18\,\text{m s}^{-2}$.

If Earth has a radius ***R***, which one of the following is closest to the radius of Phobetor?

A. R
B. $1.5R$
C. $2R$
D. $4R$

Question 2

Source: *VCE 2020 Physics Exam, Section A, Q.2;* © VCAA

Jupiter's moon Ganymede is its largest satellite.

Ganymede has a mass of 1.5×10^{23} kg and a radius of 2.6×10^6 m.

Which one of the following is closest to the magnitude of Ganymede's surface gravity?

A. $0.8\,\text{m s}^{-2}$
B. $1.5\,\text{m s}^{-2}$
C. $3.8\,\text{m s}^{-2}$
D. $9.8\,\text{m s}^{-2}$

Question 3

Source: *VCE 2019, Physics Exam, Section A, Q.4;* © VCAA

The magnitude of the acceleration due to gravity at Earth's surface is g.

Planet Y has twice the mass and half the radius of Earth. Both planets are modelled as uniform spheres.

Which one of the following best gives the magnitude of the acceleration due to gravity on the surface of Planet Y?

A. $\frac{1}{2}g$

B. $1\,g$

C. $4\,g$

D. $8\,g$

Question 4

Source: *VCE 2018, Physics Exam, Section A, Q.7;* © VCAA

At one point on Earth's surface at a distance R from the centre of Earth, the gravitational field strength is measured as 9.76 N kg^{-1}.

Which one of the following is closest to Earth's gravitational field strength at a distance $2R$ **above** the surface of Earth at that point?

A. 1.08 N kg^{-1}

B. 2.44 N kg^{-1}

C. 3.25 N kg^{-1}

D. 4.88 N kg^{-1}

Question 5

Source: *VCE 2020 Physics Exam, Section A, Q.11;* © VCAA

The International Space Station (ISS) is travelling around Earth in a stable circular orbit, as shown in the diagram below.

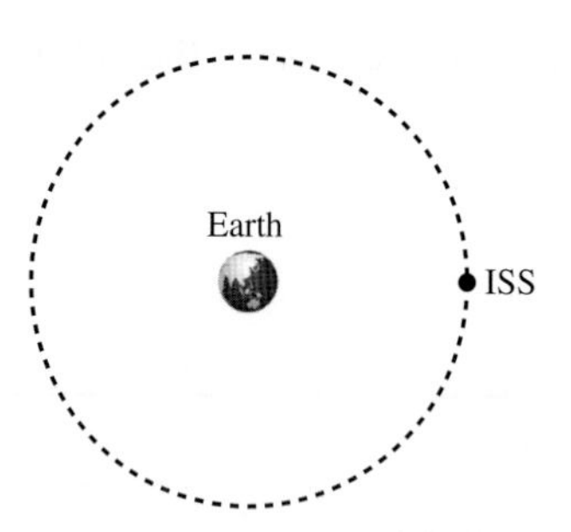

Which one of the following statements concerning the momentum and the kinetic energy of the ISS is correct?

A. Both the momentum and the kinetic energy vary along the orbital path.

B. Both the momentum and the kinetic energy are constant along the orbital path.

C. The momentum is constant, but the kinetic energy changes throughout the orbital path.

D. The momentum changes, but the kinetic energy remains constant throughout the orbital path.

Question 6

Source: *VCE 2022 Physics Exam, NHT, Section A, Q.4;* © VCAA

The Mars *Odyssey* spacecraft was launched from Earth to explore Mars. The graph below shows the gravitational force acting on the 700 kg Mars *Odyssey* spacecraft plotted against its height above Earth's surface.

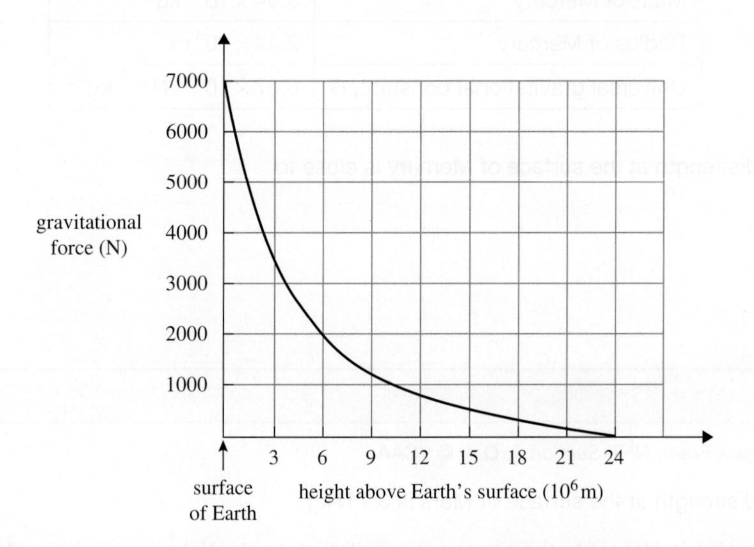

Which one of the following is closest to the minimum launch energy needed for the Mars *Odyssey* spacecraft to 'escape' Earth's gravitational attraction?

A. 4.0×10^4 J
B. 1.5×10^5 J
C. 4.0×10^{10} J
D. 1.5×10^{11} J

Question 7

Source: *VCE 2021 Physics Exam, NHT, Section A, Q.4;* © VCAA

A person has a mass of 60.0 kg.

Which one of the following is closest to the weight of this person on Earth's surface?

A. 60.0 kg
B. 60.0 N
C. 588 kg
D. 588 N

Question 8

Source: *VCE 2022 Physics Exam, NHT, Section B, Q.5;* © VCAA

When a spacecraft orbits Earth, its orbital period is **not** a function of the

A. mass of Earth.
B. mass of the spacecraft.
C. velocity of the spacecraft.
D. height of the spacecraft above Earth.

Question 9

***Source:** VCE 2018 Physics Exam, NHT, Section A, Q.2; © VCAA*

Data

Mass of Mercury	3.34×10^{23} kg
Radius of Mercury	2.44×10^{6} m
Universal gravitational constant, G	6.67×10^{-11} N m^2 kg^{-2}

The gravitational field strength at the surface of Mercury is close to

A. 9.00×10^{6} N kg^{-1}
B. 9.81×10^{6} N kg^{-1}
C. 3.74×10^{6} N kg^{-1}
D. 3.74×10^{-2} N kg^{-1}

Question 10

***Source:** VCE 2019 Physics Exam, NHT, Section A, Q.4; © VCAA*

The gravitational field strength at the surface of Mars is 3.7 N kg^{-1}.

Which one of the following is closest to the change in gravitational potential energy when a 10-kg mass falls from 2.0 m above Mars's surface to Mars's surface?

A. 3.7 J
B. 7.4 J
C. 37 J
D. 74 J

Section B — Short answer questions

Question 11 (9 marks)

***Source:** VCE 2022 Physics Exam, Section B, Q.2; © VCAA*

There are over 400 geostationary satellites above Earth in circular orbits. The period of orbit is one day (86400 s). Each geostationary satellite remains stationary in relation to a fixed point on the equator. Figure 2 shows an example of a geostationary satellite that is in orbit relative to a fixed point, X, on the equator.

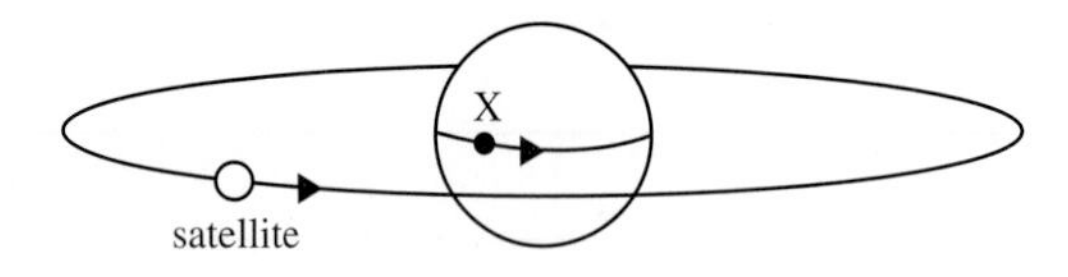

Figure 2

a. Explain why geostationary satellites must be vertically above the equator to remain stationary relative to Earth's surface. **(2 marks)**
b. Using $G = 6.67 \times 10^{-11}$ N m^2 kg^{-2}, $M_E = 5.97 \times 10^{24}$ kg and $R_E = 6.37 \times 10^{6}$ m, show that the altitude of a geostationary satellite must be equal to 3.59×10^{7} m. **(4 marks)**
c. Calculate the speed of an orbiting geostationary satellite. **(3 marks)**

Question 12 (10 marks)

Source: *VCE 2020 Physics Exam, Section B, Q.4; © VCAA*

The Ionospheric Connection Explorer (ICON) space weather satellite, constructed to study Earth's ionosphere, was launched in October 2019. ICON will study the link between space weather and Earth's weather at its orbital altitude of 600 km above Earth's surface. Assume that ICON's orbit is a circular orbit. Use $R_E = 6.37 \times 10^6$ m.

a. Calculate the orbital radius of the ICON satellite. **(1 mark)**

b. Calculate the orbital period of the ICON satellite correct to three significant figures. Show your working. **(4 marks)**

c. Explain how the ICON satellite maintains a stable circular orbit without the use of propulsion engines. **(2 marks)**

d. Figure 3 shows the strength of Earth's gravitational field, g, as a function of orbital altitude, h, above the surface of Earth. **(3 marks)**

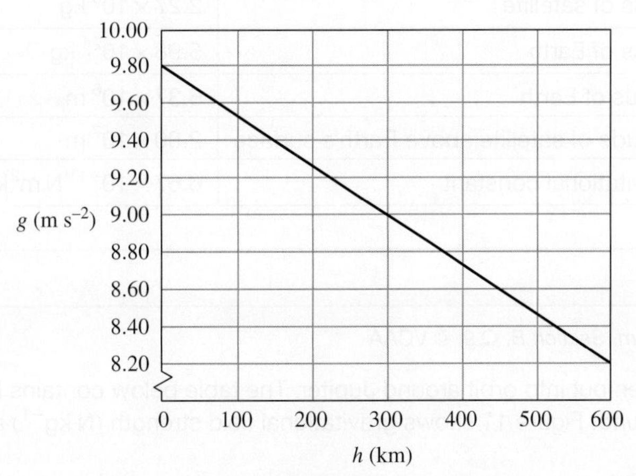

Figure 3

Determine the change in gravitational potential energy of the ICON satellite as it travels from Earth's surface to its orbital altitude of 600 km above Earth's surface. The mass of the ICON satellite is 288 kg.

Question 13 (5 marks)

Source: *VCE 2019, Physics Exam, Section B, Q.4; © VCAA*

Assume that a journey from approximately 2 Earth radii ($2R_E$) down to the centre of Earth is possible. The radius of Earth R_E is 6.37×10^6 m. Assume that Earth is a sphere of constant density.

A graph of gravitational field strength versus distance from the centre of Earth is shown in Figure 4.

a. What is the numerical value of Y? **(1 mark)**

b. Explain why gravitational field strength is 0 N kg^{-1} at the centre of Earth. **(2 marks)**

c. Calculate the increase in potential energy for a 75-kg person hypothetically moving from the centre of Earth to the surface of Earth. Show your working. **(2 marks)**

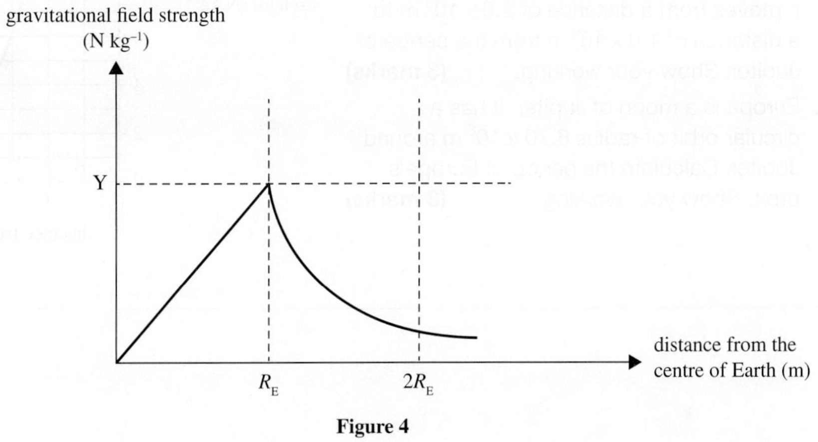

gravitational field strength (N kg^{-1})

Y

distance from the centre of Earth (m)

R_E

$2R_E$

Figure 4

Question 14 (5 marks)

***Source:** VCE 2019, Physics Exam, Section B, Q.5; © VCAA*

Navigation in vehicles or on mobile phones uses a network of global positioning system (GPS) satellites. The GPS consists of 31 satellites that orbit Earth.

In December 2018, one satellite of mass 2270 kg, from the GPS Block IIIA series, was launched into a circular orbit at an altitude of 20 000 km above Earth's surface.

a. Identify the type(s) of force(s) acting on the satellite and the direction(s) in which the force(s) must act to keep the satellite orbiting Earth. **(2 marks)**

b. Calculate the period of the satellite to three significant figures. You may use data from the table below in your calculations. Show your working. **(3 marks)**

Data

mass of satellite	2.27×10^{3} kg
mass of Earth	5.98×10^{24} kg
radius of Earth	6.37×10^{6} m
altitude of satellite above Earth's surface	2.00×10^{7} m
gravitational constant	6.67×10^{-11} N m^{2}kg^{-2}

Question 15 (8 marks)

***Source:** VCE 2018, Physics Exam, Section B, Q.9; © VCAA*

The spacecraft *Juno* has been put into orbit around Jupiter. The table below contains information about the planet Jupiter and the spacecraft *Juno*. Figure 11 shows gravitational field strength (N kg^{-1}) as a function of distance from the centre of Jupiter.

Data

mass of Jupiter	1.90×10^{27} kg
radius of Jupiter	7.00×10^{7} m
mass of spacecraft Juno	1500 kg

a. Calculate the gravitational force acting on *Juno* by Jupiter when *Juno* is at a distance of 2.0×10^{8} m from the centre of Jupiter. Show your working. **(2 marks)**

b. Use the graph in Figure 11 to estimate the magnitude of the change in gravitational potential energy of the spacecraft *Juno* as it moves from a distance of 2.0×10^{8} m to a distance of 1.0×10^{8} m from the centre of Jupiter. Show your working. **(3 marks)**

c. Europa is a moon of Jupiter. It has a circular orbit of radius 6.70×10^{8} m around Jupiter. Calculate the period of Europa's orbit. Show your working. **(3 marks)**

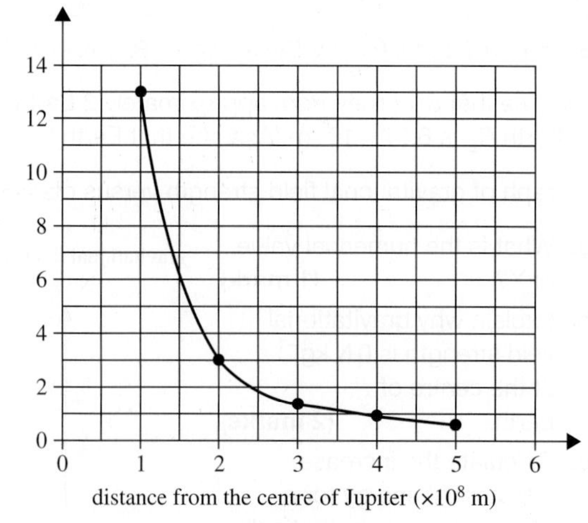

Figure 11

4 Electric fields and their applications

KEY KNOWLEDGE

In this topic, you will:

- investigate theoretically and practically electric fields, including directions and shapes of fields, attractive and repulsive effects, and the existence of dipoles and monopoles
- investigate theoretically and practically electric fields about a point charge (positive or negative) with reference to:
  - the direction of the field
  - the shape of the field
  - the use of the inverse square law to determine the magnitude of the field
  - potential energy changes (qualitative) associated with a point mass or charge moving in the field
- identify fields as static or changing, and as uniform or non-uniform
- describe the interaction of two fields, allowing that electric charges can either attract or repel
- analyse the use of electric fields to accelerate a charge, including:
  - electric field and electric force concepts: $E = k\frac{Q}{r^2}$ and $F = k\frac{q_1q_2}{r^2}$
  - potential energy changes in a uniform electric field: $W = qV$, $E = \frac{V}{d}$
  - the magnitude of the force on a charged particle due to a uniform electric field: $F = qE$.

Source: Adapted from VCE Physics Study Design (2024–2027) extracts © VCAA; reproduced by permission.

PRACTICAL WORK AND INVESTIGATIONS

Practical work is a central component of VCE Physics. Experiments and investigations, supported by a **practical investigation eLogbook** and **teacher-led video,** are included in this topic to provide opportunities to undertake investigations and communicate findings.

EXAM PREPARATION

Access past VCAA questions and exam-style questions and their video solutions in every lesson, to ensure you are ready.

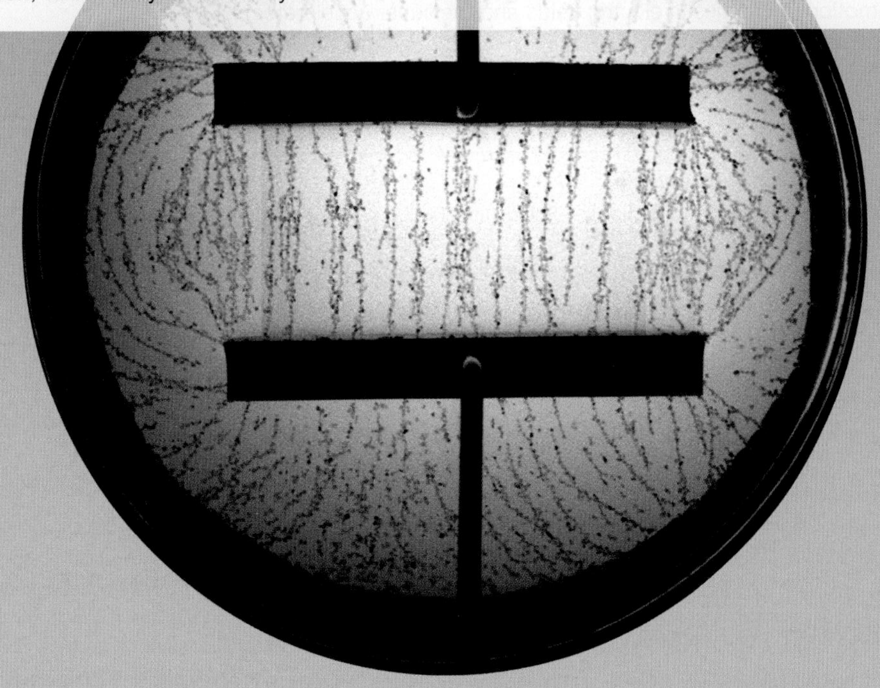

4.1 Overview

Hey students! Bring these pages to life online

Watch videos

Engage with interactivities

Answer questions and check results

Find all this and MORE in jacPLUS

4.1.1 Introduction

Electric forces explain the form and function of much of our everyday environment, from nerve cell responses in the human body to the structure of the atom. Electric forces exist between all electrically charged objects. Like gravitational forces between masses, charged objects experience electric forces when they are not touching each other. Unlike masses, charged objects can attract or repel each other, depending on whether they are positive or negative. This topic explores electric fields produced by one or more charged objects, the motion of charged particles affected by electric fields and the use of electric fields to solve technological problems. An example is the electron gun, a key component of an X-ray imaging system. In an electron gun, electrons are accelerated by a uniform electric field and collided with a target metal such as tungsten. This produces X-rays, which can be used in medical imaging.

FIGURE 4.1 A dental X-ray tube, with an electron gun on the left-hand side and a tungsten target in the copper sleeve on the right-hand side.

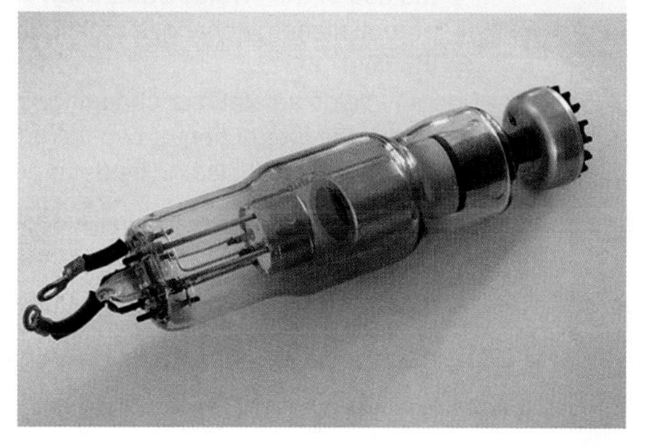

LEARNING SEQUENCE

on Resources

Solutions	Solutions — Topic 4 (sol-0818)
Practical investigation eLogbook	Practical investigation eLogbook — Topic 4 (elog-1635)
Digital documents	Key science skills — VCE Physics Units 1–4 (doc-36950) Key terms glossary — Topic 4 (doc-37171) Key ideas summary — Topic 4 (doc-37172)
Exam question booklet	Exam question booklet — Topic 4 (eqb-0101)

4.2 Coulomb's Law and electric force

KEY KNOWLEDGE

- Investigate theoretically and practically electric forces about a point charge (positive or negative) with reference to:
  - the direction of the force
  - the use of the inverse square law to determine the magnitude of the force

Source: Adapted from VCE Physics Study Design (2024–2027) extracts © VCAA; reproduced by permission.

4.2.1 Coulomb's Law

Many scientists in the 1700s thought that the force between charged objects would follow an inverse square law, like the gravitational force. The first to measure it definitively was Charles-Augustin de Coulomb, who published his findings in 1785.

Coulomb's Law states that the force between two point charges is directly proportional to the product of the magnitude of each charge and inversely proportional to the square of the distance between them. This relationship is represented by the equation:

$$F = \frac{kq_1q_2}{r^2}$$

where: F is the force value on each charged particle, in newtons (N)

k is Coulomb's constant (8.99×10^9 N m^2 C^{-2})

q_1 and q_2 are the signed magnitudes of the two charges, in coulombs (C)

r is the distance between the two charges, in metres (m)

EXTENSION: The Coulomb's constant

The value of k is affected by the medium that the charges are in. If the charged particles are in a vacuum, k has a value of 8.99×10^9 N m^2 C^{-2}.The value of k is smaller than 8.99×10^9 N m^2 C^{-2} in all other media. In water, k is 80 times smaller; however, in air the reduction is negligible.

This constant is referred to by various names, such as 'the electric force constant' and 'Coulomb's constant'.

Because the signs of the charges are included in the formula expressing Coulomb's Law, the force value can be positive or negative. A positive force value indicates that the two charges repel each other; a negative force value indicates that the two charges attract one another.

tlvd-8980

SAMPLE PROBLEM 1 Using Coulomb's Law to calculate the force between two charges

Two balloons are rubbed on a woollen jumper and each gain a charge of 2.0 nC. What force does one balloon exert on the other when they are 10.0 cm apart? Is this an attractive or repulsive force?

THINK	WRITE
Recall Coulomb's Law: $F = k\dfrac{q_1 q_2}{r^2}$, where: $k = 8.99 \times 10^9$ N m^2 C^{-2} q_1 and $q_2 = 2.0 \times 10^{-9}$ C and $r = 0.100$ m	$F_{\text{B1 on B2}} = \dfrac{8.99 \times 10^9 \times 2.0 \times 10^{-9} \times 2.0 \times 10^{-9}}{(0.100)^2}$ $= 3.6 \times 10^{-6}$ N. The force has a positive sign, so the two balloons repel each other. (This can also be determined by the fact both charges are the same, and will therefore repel.)

PRACTICE PROBLEM 1

In the hydrogen atom, the electron and proton are on average 5.3×10^{-11} m apart. An electron has a charge of -1.6×10^{-19}C, and a proton has a charge of $+1.6 \times 10^{-19}$C. Find the magnitude of the electrical force between the electron and proton. Is it attractive or repulsive?

Repulsive and attractive electric forces

There are two types of charges, positive and negative.

Consider the three scenarios represented in figure 4.2.

In these three scenarios, the magnitude of the force is $|F| = k\dfrac{|q_1| \times |q_2|}{r^2}$.

In the first scenario, in which one object is negatively charged while the other is positively charged, $q_1 q_2 < 0$, and thus the force value is negative and the force is attractive.

In the other two scenarios, $q_1 q_2 > 0$, and thus the force value is positive and the force is repulsive.

Like charges repel and **unlike charges** attract.

FIGURE 4.2 Unlike the gravitational force, the force between charged objects can be attractive or repulsive.

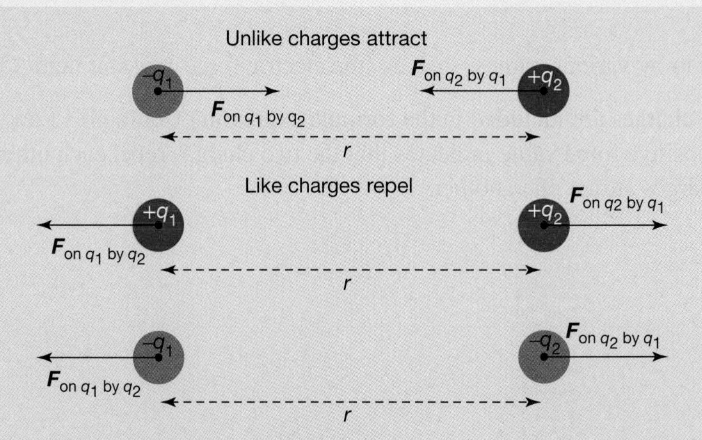

like charges charges with the sam[e] type (both positive, or both negativ[e])

unlike charges charges with opposite type (one negative, one positive)

4.2 Activities

learnon

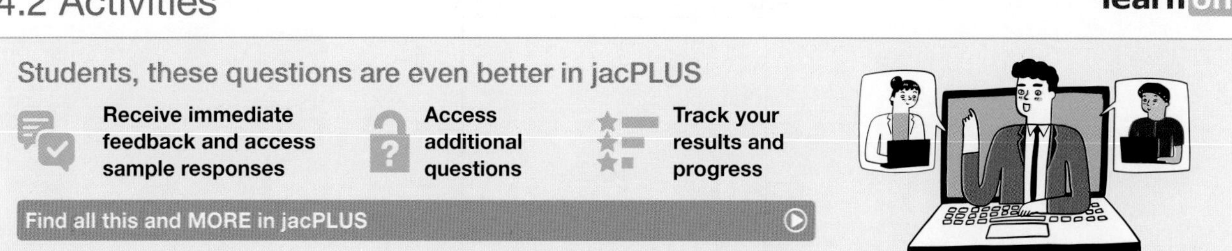

4.2 Quick quiz | 4.2 Exercise | 4.2 Exam questions

4.2 Exercise

1. A 10-μC point charge is placed 30 cm to the right of a 120-μC point charge.
   a. Determine the magnitude of the electric force on the 10-μC charge. Give your answer to 2 significant figures.
   b. Determine the direction of the electric force on the 10-μC charge.
2. A 5.0-μC point charge is placed 20 cm to the left of a −7.0-μC point charge.
   a. Determine the magnitude of the electric force on the 5.0-μC charge. Give your answer to 2 significant figures.
   b. Determine the direction of the electric force on the 5.0-μC charge.
3. A and B are two point charges placed x m apart. Initially, each point has a charge of $+q$ C and the force they exert on each other has a magnitude of 1.20×10^{-3} N. Determine the magnitude of the electric force in each of the following situations.
   (Consider the situations separately.)
   a. The distance between A and B is doubled, to $2 \times$ m.
   b. A charge of $+2q$ C is added to B.
   c. A charge of $-3q$ C is added to A.
   d. The distance between the point spheres is halved and the charges are changed to +0.5 C on A and $4q$ C on B.
4. If the magnitude of the electric force between two electrons is equal to 2.40×10^{-18} N, determine how far apart they are. Give your answer to 2 significant figures.
5. Two point charges A and B are placed 20 cm apart. The charges on A and B are $+4.0 \times 10^{-8}$ C and $+9.0 \times 10^{-8}$ C respectively. Determine the distance from A where a test charge would experience zero net force.
   (Consider the gravitational force to be negligible.)
6. MC A particle is moved along an imaginary line between two masses, m_1 and m_2. Points A and B have a charge q_1 and q_2 respectively. The charge on the particle is +1 nC. As the particle is moved along the line, the force experienced by the particle changes.
   Which of the following statements is **not** true?
   A. If q_1 and q_2 are both positive, there will be a point somewhere on the line where the net electric force experienced by the +1-nC charge will be zero.
   B. If q_1 and q_2 are both negative, there will be a point somewhere on the line where the net electric force experienced by the +1-nC charge will be zero.
   C. If q_1 is positive and q_2 is negative, there will be a point on the line where the net electric force experienced by the +1-nC charge will be zero.
   D. There will be a point on the line where the gravitational attraction of the +1-nC charge to m_1 is equal and opposite to the gravitational attraction of the +1-nC charge to m_2.

4.2 Exam questions

Question 1 (1 mark)

MC Four charges (−2*Q*, −3*Q*, +2*Q*, −3*Q*) are placed at the vertices of a square and a test charge +*Q* is placed at the centre of the square, as shown below.

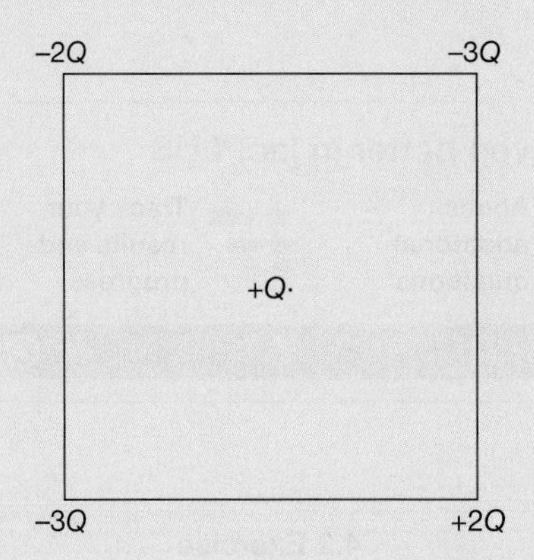

Which of the following arrows best represent the direction of the net electric force on the charge +*Q*?

A. ↗
B. ↘
C. ↙
D. ↖

Question 2 (2 marks)

A point charge of 2.0 C is1.5 m from another point charge of 5.0 C. Calculate the magnitude of the electric force acting between them.

Question 3 (2 marks)

Two point charges, each of 1.0 C, experience an electric force between them of 2.0 N. Calculate the distance between the two charges.

Question 4 (2 marks)

An electron experiences an electric force of 100 N to the right from an unknown charge placed 3.2 m to its left. Calculate the value of the unknown charge.

Question 5 (4 marks)

In a hydrogen atom, the distance between the proton and the electron is about 5.2×10^{-11} m.

Using the data in the table below, determine how far apart an electron and a proton should be so that the magnitude of the electrical force between them is equal to the magnitude of the gravitational force exerted by Earth on a hydrogen atom on its surface.

Data

mass of a hydrogen atom	1.6735×10^{-27} kg
charge of an electron/proton (e)	$(\pm)1.6 \times 10^{-19}$ C
Coulomb's constant (k)	9.0×10^{9} N m^{2} C^{-2}
acceleration due to gravity on Earth (g)	9.8 m s^{-2}

Give your answers to the appropriate number of significant figures.

More exam questions are available in your learnON title.

4.3 The field model for point-like charges

KEY KNOWLEDGE

- Investigate theoretically and practically electric fields about a point charge (positive or negative) with reference to:
  - the direction of the field
  - the shape of the field
  - the use of the inverse square law to determine the magnitude of the field
- Investigate theoretically and practically electric fields, including directions and shapes of fields, attractive and repulsive effects, and the existence of dipoles and monopoles
- Identify fields as static or changing, and as uniform or non-uniform
- Analyse the use of electric fields to accelerate a charge, including:
  - electric field and electric force concepts: $E = k\frac{Q}{r^2}$ and $F = k\frac{q_1 q_2}{r^2}$

4.3.1 The electric field from a charged particle

Attraction and repulsion between charged objects occurs even when they are not touching. To explain such interactions, Michael Faraday (1791–1867) introduced the concept of a 'field'. He proposed that an **electric field** exists in the space around a charged object. If another charged object is placed in that space, it experiences an **electric force** due to the field from the other object.

electric field vector field describing the property of the space around a charge that causes a second charge in that space to experience a force due only to the presence of the first charge

electric force force experienced by a charged particle if it is placed within the electric field of a second charged particle

We have seen in subtopic 4.2 that the force on the test charge q by Q is given by $F = k\frac{Qq}{r^2}$.

Thus, the strength of the electric field E generated at a distance r from an object with a charge Q is given by:

$$E = \frac{kQ}{r^2}$$

where: E is the strength of the electric field due to a point charge, in N C^{-1}

Q is the charge of the point charge (that is causing the electric field), in C

r is the distance from the point charge, in m

k is Coulomb's constant (8.99×10^9 N m^2 C^{-2})

The unit for the electric field is newtons per coulomb, or N C^{-1}.

The relation between the electric force and the electric field is given by:

$$\boldsymbol{F} = q\boldsymbol{E}$$

where: $\boldsymbol{F}$ is the force on the charged particle, in newtons (N)

q is the charge of the particle experiencing the force, in coulombs (C)

$\boldsymbol{E}$ is the strength of the electric field, measured in N C^{-1}

Notice the similarity of the expressions for the electric force and field from point charges to the gravitational force and field expression for point masses. Both are inverse square laws.

TABLE 4.1 Comparison between expressions for electric and gravitational fields

Force and field between masses	Force and field between charges
$F_g = G\frac{mM}{r^2}$	$F = \frac{qQ}{r^2}$
$F_g = mg$	$F = qE$
$g = G\frac{M}{r^2}$	$E = \frac{kQ}{r^2}$

However, electrical interactions are different from gravitational interactions in that electric charges can attract and repel whereas gravitational interactions are always attractive.

tlvd-8981

SAMPLE PROBLEM 2 Determining the magnitude of an electric field from a point charge

What is the magnitude of the electric field at a point 50 cm to the left of a point charge of $+2.0 \times 10^{-7}$ C?

THINK	WRITE
Recall the formula for an electric field:	$E = \frac{kQ}{r^2}$
Ensure that the distance is in metres and the charge is in coulombs. In this case, the distance of 50 cm needs to be converted to 0.5 m.	$E = \frac{8.99 \times 10^9 \times 2.0 \times 10^{-7}}{(0.5)^2}$ $= 7.2 \times 10^3\ \text{N}\,\text{C}^{-1}$ The magnitude of the electric field is $7.2 \times 10^3\ \text{N}\,\text{C}^{-1}$.

PRACTICE PROBLEM 2

What is the magnitude of the electric field at a point 30 cm to the right of a point charge of -3.0×10^{-6} C?

Drawing an electric field diagram

When you draw a gravitational field, the field lines indicate the direction of gravitational force on a mass in the field. But for an electric field, it is ambiguous because the direction of the force experienced by a charge in a field depends on whether the charge is positive or negative. A convention is needed.

> In an electric field diagram, the direction of the field is the direction in which a positive charge would experience the electrical force.

This means that the lines are directed away from positively charged sources, and towards negatively charged sources (see figure 4.3).

FIGURE 4.3 Fields around a. a positive charge and b. a negative charge

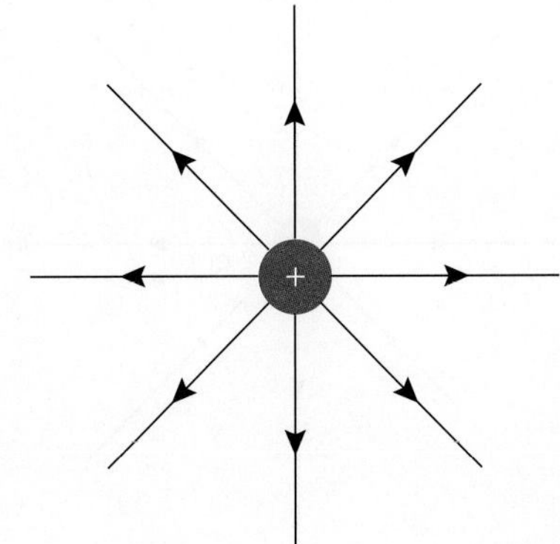

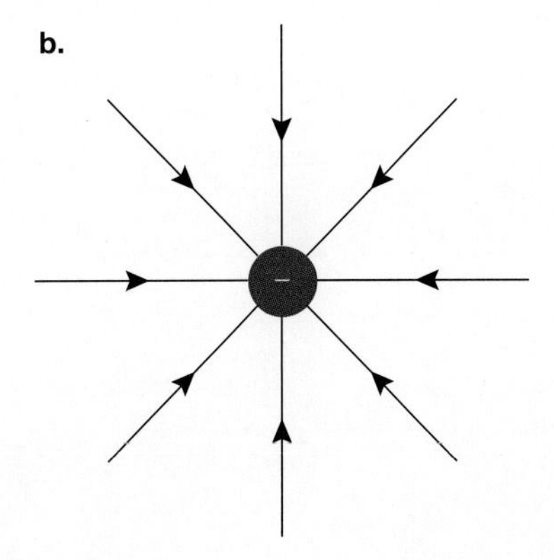

Remember that the changes in the spacing of the field lines indicate changes in the strength of the field. For the **electric monopoles** in figure 4.3, the field lines spread out from the centre. The increased spacing between field lines, as the distance from the point charge increases, shows that the field is weakening. The field is not uniform everywhere in space. However, the field is unchanging or static over time.

electric monopole single electric point charge, in which all the electric field lines point inward for a net negative electric charge or away for a net positive electric charge

tlvd-8982

SAMPLE PROBLEM 3 Drawing forces between charged particles

Consider a particle with a charge of 1.0 μC.

a. **Draw a diagram of the field of the particle.**
b. **Draw a vector showing the force on a 1.0-nC charge placed 2.0 cm to the right of the particle.**
c. **Draw a vector showing the force on a −1.0-nC charge placed 2.0 cm to the right of the particle.**
d. **Draw a vector showing the force on a 1.0-nC charge placed 4.0 cm to the right of the particle.**

THINK

a. Recall the shape of the electric field around a point charge, remembering that the direction of the field is the direction in which a positive charge would experience a force (away from the charge).

WRITE

a.

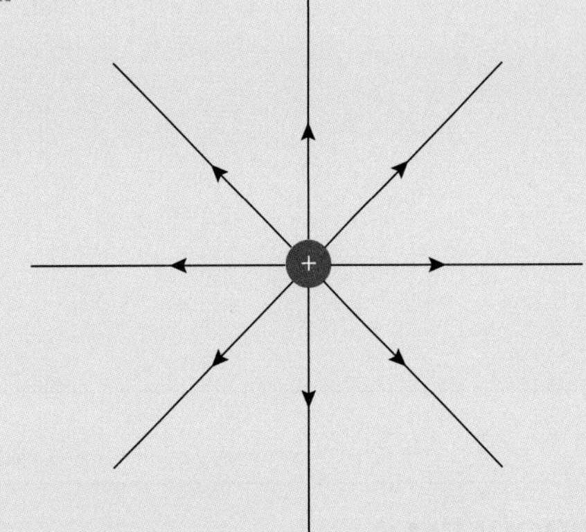

b. As the charges are both positive, they will repel each other.

b.

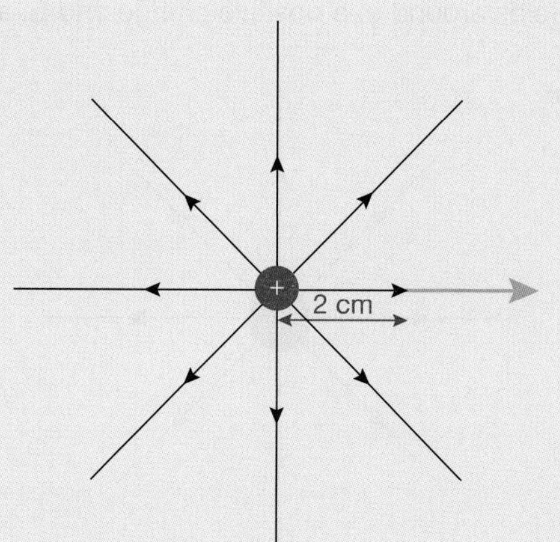

c. As the charges are opposite, they will attract each other.

c.

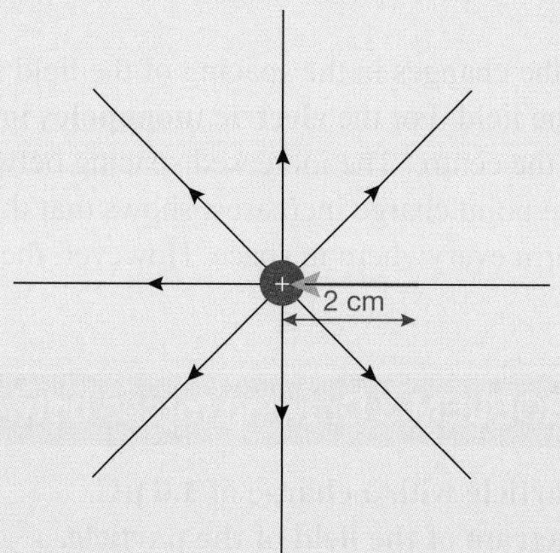

d. As the charges are both positive, they will repel each other. As the distance between the charges has doubled, the force between them has reduced to $\frac{1}{4}$ of the original force.

d.

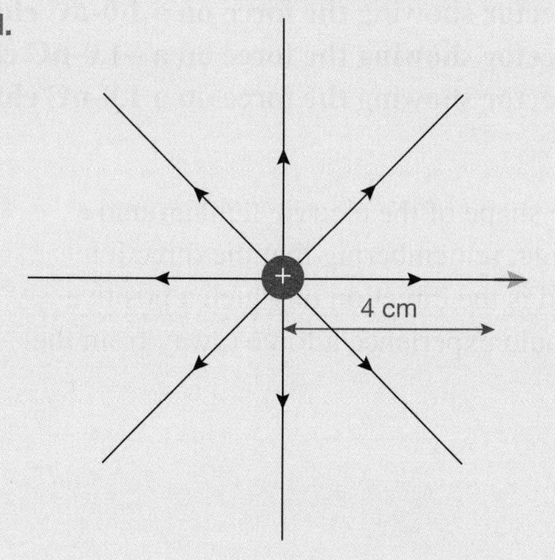

PRACTICE PROBLEM 3

Consider a particle with a charge of −1.0 μC.

a. **Draw a diagram of the field of the particle.**

b. **Draw a vector showing the force on a 1.0-nC charge placed 2.0 cm to the left of the particle.**

c. **Draw a vector showing the force on a –1.0-nC charge placed 2.0 cm south-east of the particle.**

d. **Draw a vector showing the force on a 1.0-nC charge placed 4.0 cm north-east of the particle.**

Graphing the electric field

The direction of the electric field is the direction in which a positive test charge would experience a force.

FIGURE 4.4 The electric field at an increasing distance from **a.** a positive charge, **b.** a negative charge

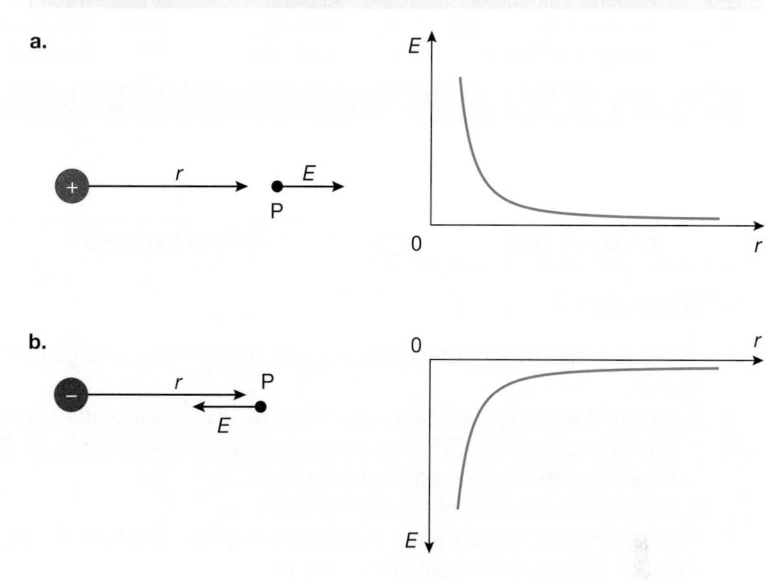

For a central positive charge, the direction of the electric field vector at a point P is in the same direction as the position vector of P. This means that the graph of the electric field is above the distance axis. This can be further understood when you consider the formula for an electric field, where $E=\frac{kQ}{r^2}$. If the value of Q (the charge) is positive, the strength of the electric field (E) is also positive.

For a central negative charge, the direction of the electric field vector is in the opposite direction to the position vector, so the graph of the electric field around a negative charge will be below the distance axis. This can be further understood when you consider the formula for an electric field, where $E=\frac{kQ}{r^2}$. If the value of Q (the charge) is negative, the direction of the electric field $\boldsymbol{E}$ is in the opposite direction from the position of any test charge. A positive test charge would experience an attractive force towards the negative charge.

EXTENSION: Electric attraction

Recent research has shown that electric attractions between bees and flowers have an important role in bees locating flowers with pollen. Bees acquire a slight positive charge as they fly through the air. Like the surface of Earth, flowers tend to be negatively charged. Bees are very sensitive to the shape of the electric field from the flower, using the electric field to distinguish between different flower types.

FIGURE 4.5 Bumblebees can sense the electric field surrounding a flower.

4.3 Activities

Students, these questions are even better in jacPLUS

Receive immediate feedback and access sample responses

Access additional questions

Track your results and progress

Find all this and MORE in jacPLUS

4.3 Quick quiz	4.3 Exercise	4.3 Exam questions

4.3 Exercise

1. Determine the strength of the electric field 30 cm from a charge of 120 μC. Give your answer to 2 significant figures.
2. Calculate the strength of the electric field 5.2 μC from a proton. Give your answer to 2 significant figures.
3. a. An electric force of 1.5 N acts upwards on a charge of +3.0 μC. Calculate the strength of the electric field. Give your answer to 2 significant figures.
   b. What is the direction of the electric field?
4. A negative test charge is placed at a point in an electric field. It experiences a force to the right. What is the direction of the electric field at that point?
5. a. An electric force of 3.0 N acts downwards on a charge of −1.5 μC. Calculate the strength of the electric field. Give your answer to 2 significant figures.
   b. What is the direction of the electric field?
6. MC Which of the following statements is **incorrect**?
   A. Electric fields can attract or repel charges, depending on the type of charge and nature of the field.
   B. Electric fields can be uniform or non-uniform.
   C. The source of an electric field can be a monopole.
   D. Electric field lines must always form closed loops.
7. Answer the following
   a. Explain why electric field lines can never cross.
   b. If a particle is free to move, will it move along a field line?

4.3 Exam questions

Question 1 (1 mark)

Source: *VCE 2018 Physics Exam, NHT, Section A, Q.3;* © VCAA

MC A Van de Graaff generator, which is a piece of electric field demonstration equipment, consists of a small sphere that is electrically charged, as shown in the diagram below.

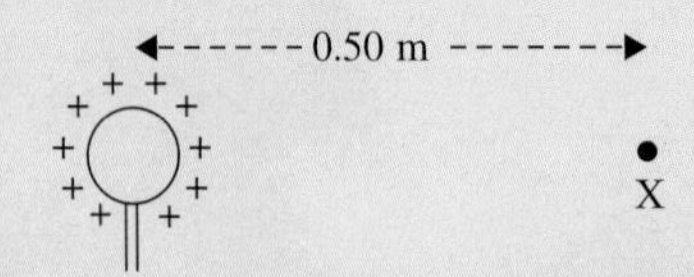

A particular Van de Graaff generator has a sphere that has a charge of 5.0×10^{-7} coulombs on it. Take the Coulomb's law constant to be $k = 9.0 \times 10^{9}$ N m^2 C^{-2}.

Which one of the following best gives the magnitude of the electric field at point X in the diagram above, 0.50 m from the sphere?

A. 1.8×10^{-2} V m^{-1}
B. 3.6×10^{-2} V m^{-1}
C. 1.8×10^{4} V m^{-1}
D. 3.6×10^{4} V m^{-1}

Question 2 (1 mark)

An electron is placed in an electric field and experiences a force due to the field that is in an upward direction. What is the direction of the field at that location?

Question 3 (2 marks)

Calculate the strength of the electric field 12 cm away from a point charge of 6.0×10^{-4} C.

Question 4 (2 marks)

At a distance of 15 cm from an unknown charge, the strength of the electric field is $80.5\,\text{N}\,\text{C}^{-1}$. Determine the unknown charge in coulombs.

Question 5 (2 marks)

At what distance from a proton, of charge 1.6×10^{-19} C, is the strength of the electric field $5.4\,\text{N}\,\text{C}^{-1}$?

More exam questions are available in your learnON title.

4.4 Electric fields from more than one point-like charge

KEY KNOWLEDGE

- Describe the interaction of two electric fields, allowing that electric charges can either attract or repel

Source: Adapted from VCE Physics Study Design (2024–2027) extracts © VCAA; reproduced by permission.

4.4.1 Electric fields from two or more charges: dipoles

When two or more charged sources are present, the fields from each of the individual sources are added together to give the total field. Remember that electric fields are vectors and obey vector addition laws.

Consider the following diagram of the electric field for two positive charges:

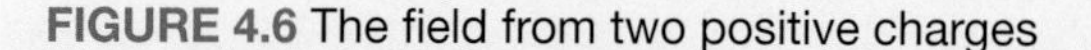

FIGURE 4.6 The field from two positive charges

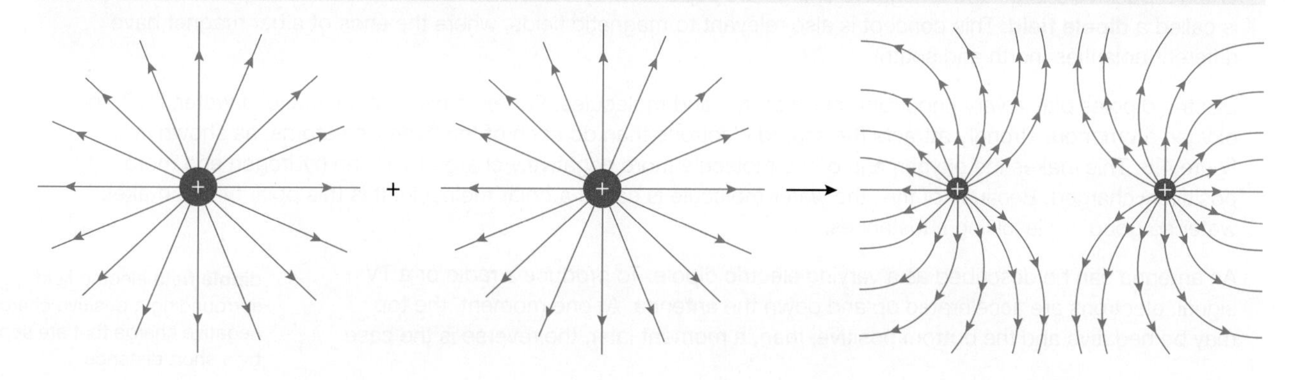

At the point halfway between the two charges, the fields from the two charges completely cancel, resulting in a field of zero strength. Notice that at large distances from the charges, the field lines start to look more like field lines from a point with twice as much charge.

When a positive charge and negative charge of equal size get near to each other, they create an **electric dipole**, and you obtain a very different pattern:

FIGURE 4.7 The field from a positive and a negative charge, known as an electric dipole

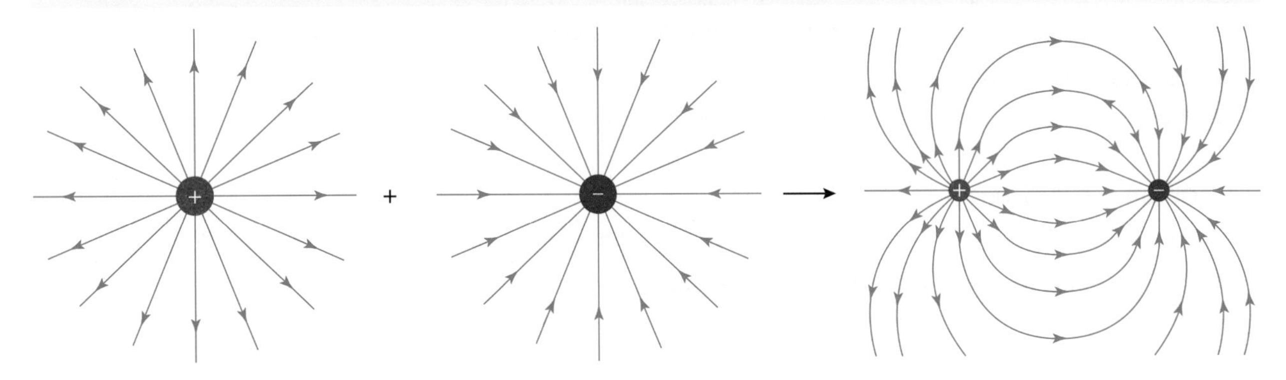

The field pattern from an electric dipole cannot be confused with the field from a single charge. The dipole field pattern has a definite attractive end and a definite repulsive end. It is also clearly non-uniform.

electric dipole a positive charge and a negative charge that are separated by a short distance

elog-1924

tlvd-10486

INVESTIGATION 4.1

Investigating the electric field using an electric compass

AIM

To investigate the electric field from positive and negative sources of charge, both singly and in combination

EXTENSION: Dipole fields

When a positive charge and a negative charge are separated by a short distance, the electric field around them is called a **dipole field**. This concept is also relevant to magnetic fields, where the ends of a bar magnet have different polarities (north and south).

Electric dipoles play a very important role in atoms and molecules. For example, in a molecule of water, H_2O, the oxygen atom more strongly attracts the shared electrons than do each of the hydrogen atoms, as shown in figure 4.8. This makes the oxygen end of the molecule more negatively charged and the hydrogen end more positively charged. Because of this, the water molecule is called a polar molecule. It is this polarity that makes water so good at dissolving substances.

An antenna can be described as a varying electric dipole. To produce a radio or a TV signal, electrons are accelerated up and down the antenna. At one moment, the top may be negative and the bottom positive, then, a moment later, the reverse is the case.

dipole field electric field surrounding a positive charge and negative charge that are separated by a short distance

FIGURE 4.8 A water molecule (H_2O) displays polarity.

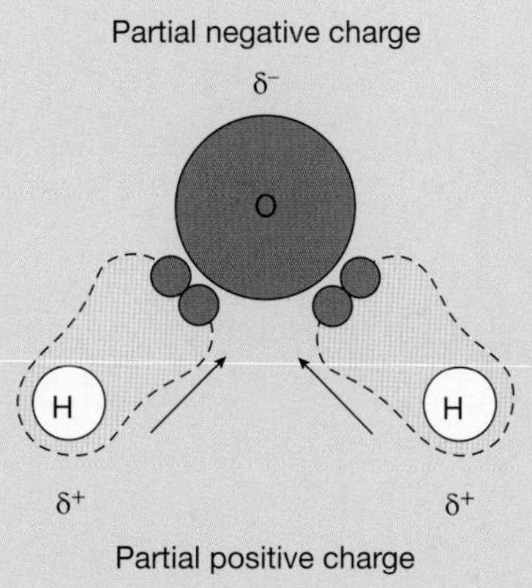

Partial negative charge

δ⁻

O

H H

δ⁺ δ⁺

Partial positive charge

FIGURE 4.9 Partial circuit diagram of an antenna

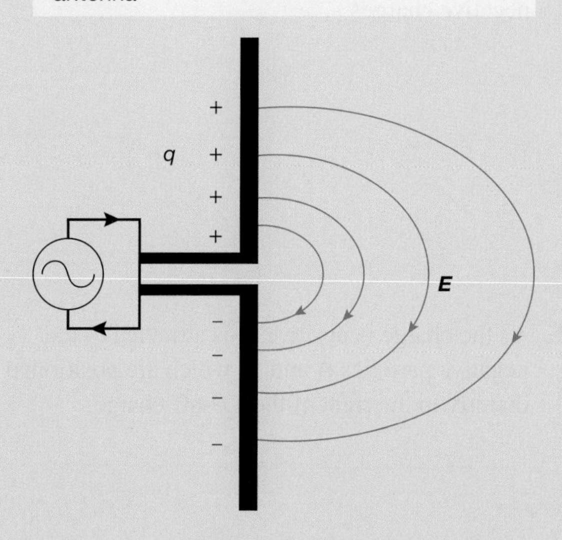

Many fish generate and sense electric fields. Some use them to communicate, others to sense prey, and others like electric eels can stun their prey with an electric shock.

FIGURE 4.10 An electric eel produces an electric dipole field.

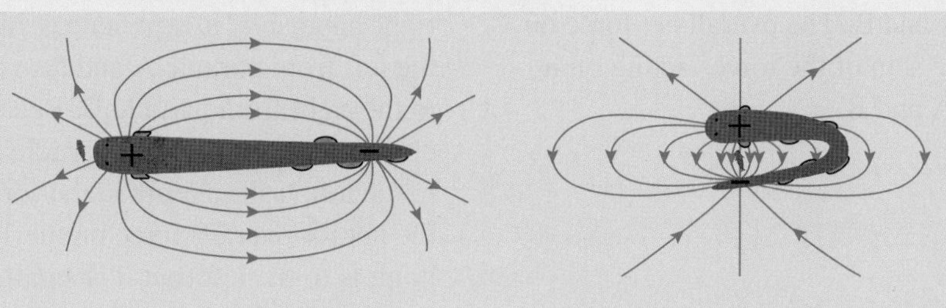

tlvd-8983

SAMPLE PROBLEM 4 Sketching electric fields due to point charges

a. **Sketch the field due to the two particles shown in the following diagram, each with a charge of −1.0 μC separated by 4.0 cm.**

b. **Draw a vector indicating the direction of the force on a 1.0-nC charge placed to the left of particle A.**

c. **Describe how the direction of the force on the 1.0-nC charge changes as it is moved along an imaginary line connecting particles A and B.**

THINK	WRITE
a. Recall the shape of the electric field around two negative charges.	a. 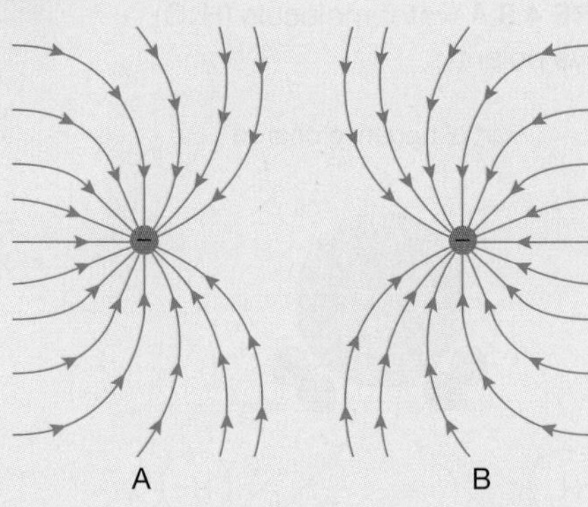
b. As the charge is positive, it is attracted to the negative particles A and B, which are positioned directly to the right of the 1.0-nC charge.	b. 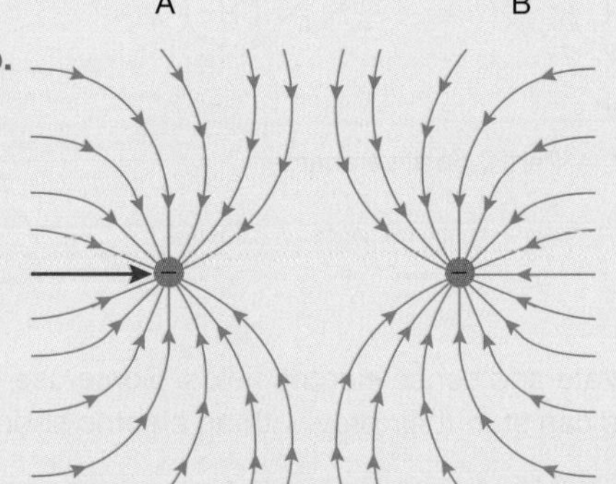
c. The positive charge is attracted to both negative point charges A and B. The overall net force on the charge is the sum of the forces acting on it from particles A and B.	c. When the 1.0-nC charge moves past particle A to the right-hand side, it experiences an attractive force to the left from particle A and also an attractive force to the right from particle B. Because the charge is closer to particle A than particle B, the force acting to the left is stronger than the force to the right, so the net (or overall) force on the 1.0-nC charge at this point is to the left (but it is smaller in magnitude than the force it experienced in part b). As the charge is moved closer to particle B, the magnitude of the leftward net force decreases to zero at the point halfway between the two particles. When the charge moves closer to particle B than particle A, the net force acts to the right and becomes stronger as the charge moves closer to particle B.

PRACTICE PROBLEM 4

Sketch the electric field that would result from particle A having a negative charge of 1 μC and particle B (which is placed to the right of particle A) having a positive charge of 1 μC.

a. **Compare the force on a 1-nC charge placed to the left of particle A with one placed to the right of particle A.**
b. **Describe how the force changes on the 1-nC charge as it is moved along an imaginary line between A and B.**

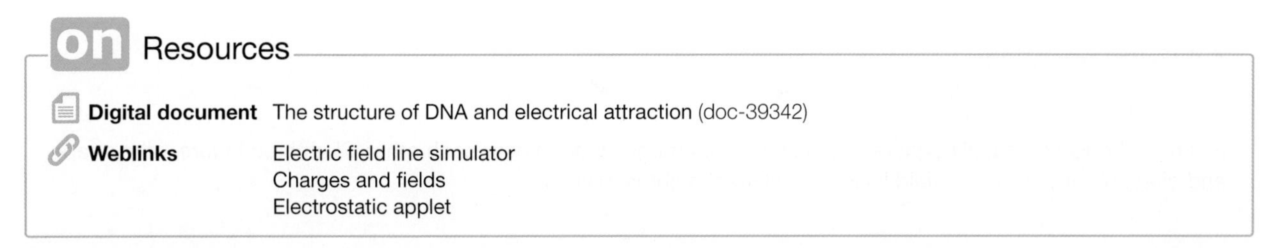

on Resources

Digital document The structure of DNA and electrical attraction (doc-39342)

Weblinks Electric field line simulator
Charges and fields
Electrostatic applet

4.4 Activities

learn**on**

Students, these questions are even better in jacPLUS

Receive immediate feedback and access sample responses

Access additional questions

Track your results and progress

Find all this and MORE in jacPLUS

4.4 Quick quiz **on**	4.4 Exercise	4.4 Exam questions

4.4 Exercise

1. Two small spheres, A and B, are placed with their centres 10 cm apart. P is placed between A and B, at a position 2.5 cm from A. Determine the direction of the electric field at P in the following situations.
   a. A and B have the same positive charge.
   b. A has a positive charge, B has a negative charge and the magnitudes are the same.
2. A +2-nC charge is placed to the right of a −2-nC charge. Sketch the electric field around those two charges.

3. Sketch the electric field around two positive charges, A and B, where the charge on A is twice that on B.
4. **MC** For the dipole field shown, which of the following statements are true?

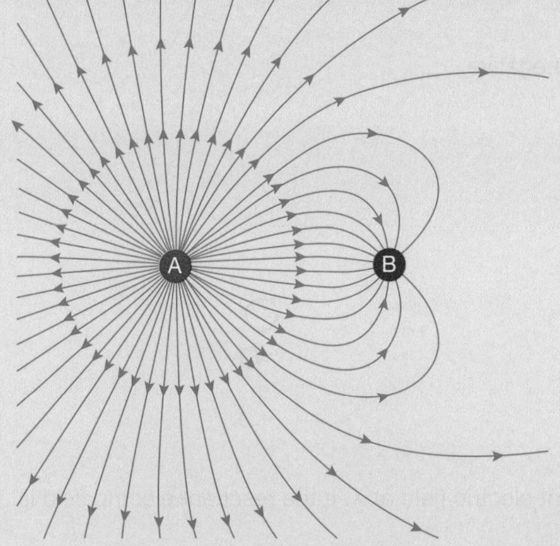

   A. The point charges at A and B have the same size.
   B. The size of the charge at point B is smaller than the one at point A.
   C. Both charges are positive.
   D. The charge at point A is positive, and the one at point B is negative.
5. A +2.0-nC charge is placed 15 cm to the right of a +3.0-nC charge. Josh predicts there will be a point between the two particles at which the combined electric field strength will be zero.
   a. Determine the value of the combined electric field strength E at the following distances x to the right of the +3.0-nC charge. Consider the positive direction to be to the right.
   Give your answers to 3 significant figures.
   b. Explain your results from part **a** and decide whether Josh is correct.

4.4 Exam questions

Question 1 (2 marks)

Figure 1 shows two equal negative stationary point charges placed near each other. Sketch on Figure 1 the shape and direction of the electric field lines. Use at least eight field lines.

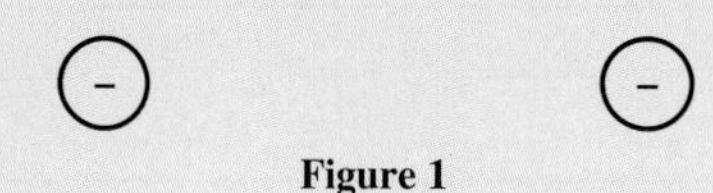

Figure 1

Question 2 (1 mark)

Source: *Adapted from VCE 2020 Physics Exam, Section A, Q.1; © VCAA*

MC The diagram below shows the electric field lines between two charges of equal magnitude.

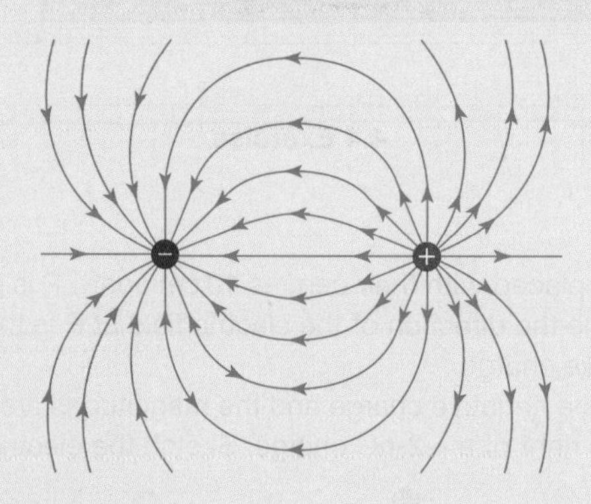

The best description of the two charges is that the

A. charges are both positive.
B. charges are both negative.
C. charges can be either both positive or both negative.
D. left-hand charge is positive and the right-hand charge is negative.

Question 3 (1 mark)

Source: *VCE 2017, Physics Exam, Section B, Q.1; © VCAA*

Three charges are arranged in a line, as shown in Figure 1.

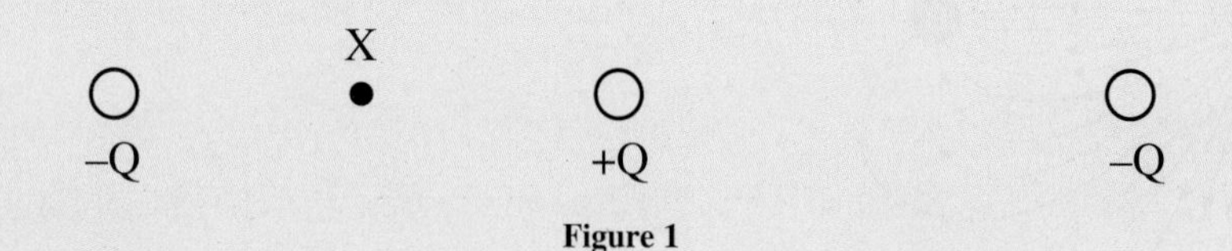

Figure 1

Draw an arrow at point X to show the direction of the resultant electric field at X. If the resultant electric field is zero, write the letter 'N' at X.

Question 4 (2 marks)

Two charges, $-2Q$ and $+Q$, are separated by a distance of 1.0 cm. A point X is 1.0 cm to the right of the positive charge. Draw an arrow at X to show the direction of the resultant field at X. If the resultant field is zero, write the letter N at X.

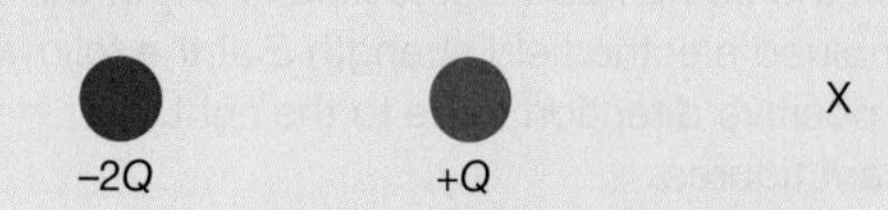

Question 5 (8 marks)

A particle acquires a −3.0-nC charge.

a. Sketch the electric field due to the −3.0-nC charge. **(2 marks)**

A second particle with a charge of +5.0 nC is placed 4.0 cm to the left of the −3.0-nC charge.

b. Calculate the magnitude of the force on the +5.0-nC charge due to the −3.0-nC charge. **(2 marks)**

c. What is the direction of the electric force exerted by the −3.0-nC charge on the +5.0-nC charge? **(1 mark)**

d. Is there any point at which the combined electric field from the two charges will have a value of zero? Explain your response without using calculations. **(3 marks)**

More exam questions are available in your learnON title.

4.5 Uniform electric fields

KEY KNOWLEDGE

- Analyse the use of electric fields to accelerate a charge, including:
  - the magnitude of the force on a charged particle due to a uniform electric field: $F = qE$
- Identify fields as static or changing, and as uniform or non-uniform

4.5.1 The electric field between parallel charged plates

If a set of positive and negative charges were lined up in two rows facing each other, the lines of electric field in the space between the rows would be evenly spaced, that is, the value of the strength of the field would be constant. This is called a **uniform electric field**. Note that the electric field vector (and thus the field lines) are always perpendicular to conducting surfaces.

uniform electric field electrical field in which the strength and direction are constant at every point

It is also very easy to set up. Just set two metal plates a few centimetres apart, then connect one plate to the positive terminal of a battery and connect the other plate to the negative terminal of the battery. The battery will transfer electrons from one plate, making it positive, and put them on the other, making that one negative. The battery will keep on doing this until the positive plate is so positive that the battery's voltage, or the energy it gives to each coulomb of electrons, is insufficient to overcome the attraction of the positively charged plate. Similarly, the negatively charged plate will become so negative that the repulsion from this plate prevents further electrons being added.

FIGURE 4.11 A uniform electric field

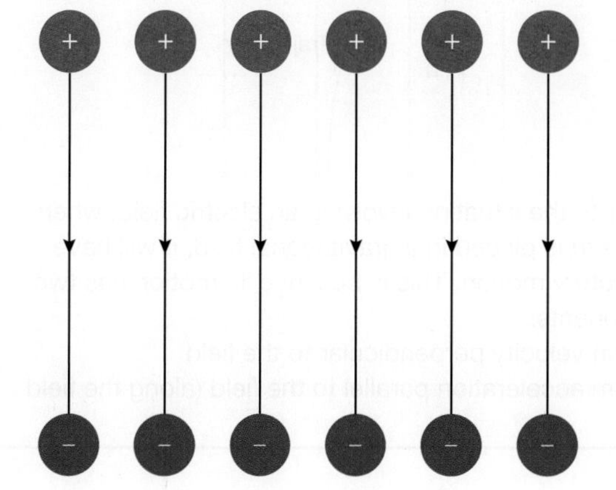

FIGURE 4.12 An electric field between two plates

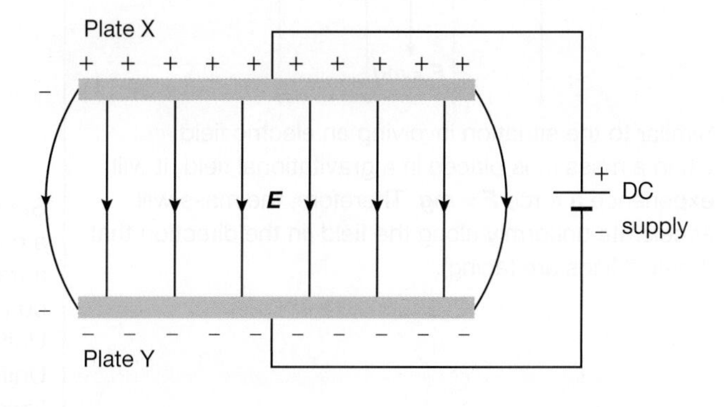

The uniform field between the plates means that the force experienced by a charged particle placed at any position between the plates is the same: $\boldsymbol{F} = q\boldsymbol{E}$.

The field remains static, or unchanging, for as long as the parallel plates remain charged.

The force on a charge placed between the plates is the same at all places between the plates. This has many important applications and is the basic building block of a particle accelerator.

If there is no other force acting on the charged particle q, then the force due to the interaction with the electric field is the net force, and, applying Newton's Second Law, the particle experiences an acceleration parallel to the field, $\boldsymbol{a}$, where $\boldsymbol{a} = \frac{\boldsymbol{F}}{m} = \frac{q\boldsymbol{E}}{m}$. This can be observed in table 4.2. It is clear that this motion is very similar to that experienced in gravitational fields.

Hence, the motion of a charged particle has two components:

1. uniform velocity perpendicular to the field
2. uniform acceleration parallel to the field.

The combination of these two components produces a trajectory very similar to that of projectile motion.

TABLE 4.2 Comparing trajectories in electric fields to those in gravitational fields

Electric fields	
Charge placed at a point	Charge with velocity
q; $\boldsymbol{E}$; $\boldsymbol{F} = q\boldsymbol{E}$	q; $\boldsymbol{v}$; $\boldsymbol{E}$; Trajectory; $\boldsymbol{F}$, $\boldsymbol{a}$
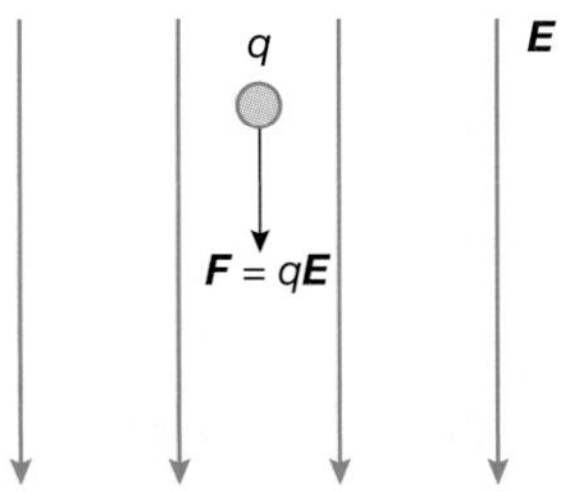 When a charge q is placed in a uniform electric field, it will experience a force $\boldsymbol{F} = q\boldsymbol{E}$. Therefore, the charge will accelerate uniformly along the field (in the direction that the field lines are facing).	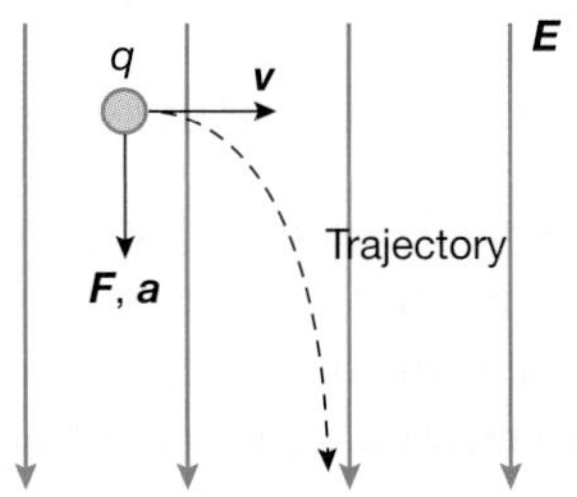 When a charge q travelling with a velocity $\boldsymbol{v}$ across the field is placed in an electric field, it will have a trajectory motion. This is because its motion has two components: Uniform velocity perpendicular to the field Uniform acceleration parallel to the field (along the field lines).
Gravitational fields	
Mass placed at a point	Mass with velocity
m; $\boldsymbol{g}$; $\boldsymbol{F} = m\boldsymbol{g}$	m; $\boldsymbol{v}$; $\boldsymbol{g}$; Trajectory; $\boldsymbol{F}$, $\boldsymbol{a}$
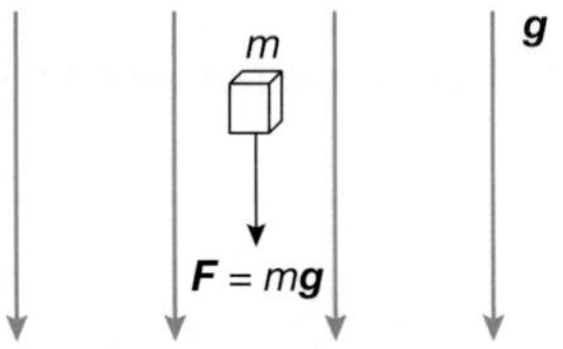 Similar to the situation involving an electric field, when a mass m is placed in a gravitational field, it will experience a force $\boldsymbol{F} = m\boldsymbol{g}$. Therefore, the mass will accelerate uniformly along the field (in the direction that the field lines are facing).	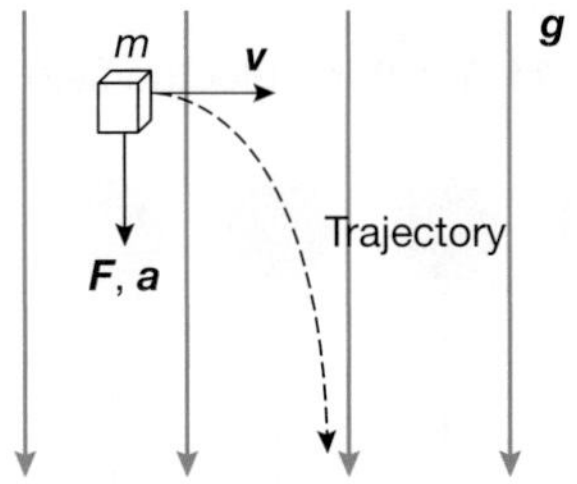 Similar to the situation involving an electric field, when a mass m is placed in a gravitational field, it will have a trajectory motion. This is because its motion has two components: Uniform velocity perpendicular to the field Uniform acceleration parallel to the field (along the field lines).

tlvd-8984

SAMPLE PROBLEM 5 Calculating forces on a free charge in a uniform electric field

The electric field strength between two parallel plates separated by 2.0 cm is 10 N C^{-1}. An electron (with a charge of -1.6×10^{-19} C) enters the space between the plates.

a. **Calculate the magnitude of the force on the electron.**

b. **Describe the direction of the force on the electron.**

c. **Calculate the change in velocity of an electron that starts at rest halfway between the two plates.**

THINK	WRITE
a. Recall that $F = qE$ and that the charge on the electron is -1.6×10^{-19} C.	a. $F = qE$ $= -1.6 \times 10^{-19} \times 10$ Hence, the magnitude of the force is 1.6×10^{-18} N.
b. Remember that the electron has a negative charge.	b. The sign of the force is opposite to the sign of the field, so the electron experiences a force in the opposite direction to the electric field and will be accelerated towards the positive plate.
c. 1. The electron experiences a constant force, so it undergoes constant acceleration. Apply Newton's Second Law, $F = ma$, to determine the acceleration. Use your formula sheet to find that the mass of an electron is equal to 9.1×10^{-31} kg.	c. $a = \frac{F_{net}}{m}$ $= \frac{1.6 \times 10^{-18}}{9.1 \times 10^{-31}}$ $= 1.8 \times 10^{12}$ m s^{-2}
2. The values of the initial velocity, acceleration and distance travelled are known, so these can be used in the equation $v^2 = u^2 + 2as$ to calculate the final velocity. For the distance, remember that the electron is halfway between the two plates, so the distance is 1.0 cm (converted to 0.01 m for use in the formula). You also need to ensure that you use your non-rounded value for acceleration during your calculations.	$v^2 = u^2 + 2as$ $v^2 = 0 + 2 \times 1.8 \times 10^{12} \times 0.01$ $= 3.5 \times 10^{10}$ $\Rightarrow v = 1.9 \times 10^5$ m s^{-1} As the initial velocity was 0 m s^{-1} and the final velocity was 1.9×10^5 m s^{-1}, the change in velocity was 1.9×10^5 m s^{-1}.

PRACTICE PROBLEM 5

The electric field strength between two parallel plates separated by 2.0 cm is 20 N C^{-1}. A proton is placed halfway between the plates. The proton has a charge of 1.6×10^{-19} C and a mass of 1.67×10^{-27} kg.

a. **What is the magnitude of the force on the proton?**

b. **Describe the direction of the force on the proton.**

c. **Calculate the final velocity of a proton that starts at rest halfway between the two plates.**

4.5 Activities

Students, these questions are even better in jacPLUS

Receive immediate feedback and access sample responses

Access additional questions

Track your results and progress

Find all this and MORE in jacPLUS

4.5 Quick quiz  | 4.5 Exercise | 4.5 Exam questions

4.5 Exercise

1. The electric field between two charged plates is represented below. Determine which plate is positively charged.

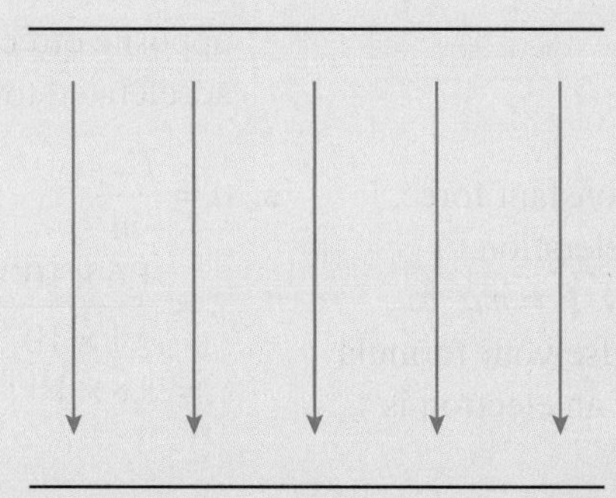

2. An electron is moving between two charged plates, and its trajectory is represented below.
Draw an arrow to represent the direction of the electric field, an arrow to represent the direction of the electric force on the electron at a random point of its trajectory and label the plates with their respective charges.

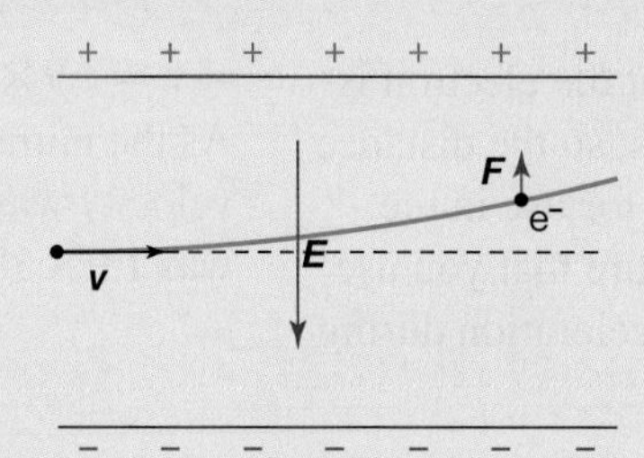

3. Answer the following:
   a. Calculate the magnitude of the acceleration of an electron of mass 9.1×10^{-31} kg in a uniform electric field of strength 1.0×10^{6} N C^{-1}. Give your answer to 2 significant figures.
   b. Starting from rest, determine how long it would take for the speed of the electron to reach 10% of the speed of light in a vacuum ($c = 3.0 \times 10^{8}$ m s^{-1})? Ignore relativistic effects and give your answer to 2 significant figures.
   c. Calculate the distance the electron would travel in that time. Give your answer to 2 significant figures.
4. In an inkjet printer, small drops of ink are given a controlled charge and fired between two charged plates. The electric field deflects each drop and thus controls where the drop lands on the page, as outlined in the diagram.

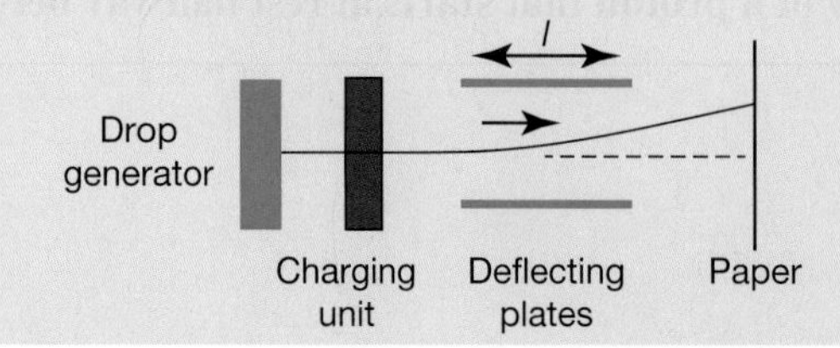

Let m be the mass of the drop, q be the charge of the drop, $\boldsymbol{v}$ be the speed of the drop, l be the horizontal length of the plate crossed by the drop, and E be the electric field strength.

a. Develop an expression for the deflection of the drop.
Hint: This is like a projectile motion question.

b. With the values $m = 1.0 \times 10^{-20}$ kg, $\boldsymbol{v} = 20\text{ m s}^{-1}$, $l = 1.0$ cm and $E = 1.2 \times 10^{6}\text{ N C}^{-1}$, calculate the charge required on the drop to produce a deflection of 1.2 mm. Give your answer to 2 significant figures.

5. **MC** A charged particle is placed between two oppositely charged conducting plates. If the strength of the electric field is doubled (and ignoring gravitational effects), which of the following is true? (Select all that apply.)
A. The net force experienced by the charge doubles.
B. The acceleration experienced by the charge doubles.
C. The final velocity of the particle doubles.
D. The charged particle does not move.

4.5 Exam questions

Question 1 (1 mark)

Source: *VCE 2017, Physics Exam, Section A, Q.2; © VCAA*

MC Millikan, a famous scientist, measured the size of the electron charge by balancing an upwards electric force with a gravitational force on a small oil drop. In a repeat of this experiment, an oil drop with a charge of 9.6×10^{-19} C was placed in an electric field of 10^{4} V m^{-1}.

Which one of the following is closest to the electrical force on the oil drop?
A. 9.6×10^{-14} N
B. 9.6×10^{-15} N
C. 9.6×10^{-22} N
D. 9.6×10^{-23} N

Question 2 (2 marks)

Calculate the magnitude of the electric force acting on an electron of charge 1.6×10^{-19} C in a uniform electric field of strength $2.5 \times 10^{4}\text{ N C}^{-1}$.

Question 3 (2 marks)

Calculate the acceleration of an electron of charge 1.6×10^{-19} C and mass 9.1×10^{-31} kg due to the force acting on it in a uniform electric field of strength $2.0 \times 10^{8}\text{ N C}^{-1}$.

Question 4 (1 mark)

An electron gun is used to inject electrons into the LINAC of a synchrotron.

MC The charge of an electron is 1.6×10^{-19} C and the mass of an electron is 9.1×10^{-31} kg. If the acceleration of the electrons is $1.2 \times 10^{14}\text{ m s}^{-2}$, then the electric field strength is closest to:
A. $1.0 \times 10^{5}\text{ V C}^{-1}$.
B. $1.0 \times 10^{8}\text{ V C}^{-1}$.
C. $1.0 \times 10^{3}\text{ V C}^{-1}$.
D. $1.0 \times 10^{-2}\text{ V C}^{-1}$.

Question 5 (6 marks)

Answer the following.

a. Calculate the magnitude of the acceleration of an alpha particle of mass 6.64×10^{-27} kg, and of charge 2e, in a uniform electric field of strength $4.0 \times 10^{6}\text{ N C}^{-1}$. **(2 marks)**

b. Starting from rest, the alpha particle travels for 1.0 ns in the uniform electric field. Determine how far the alpha particle travelled. (Ignore relativistic effects.) **(2 marks)**

c. Calculate the speed of the alpha particle after 1.0 ns. **(2 marks)**

More exam questions are available in your learnON title.

4.6 Energy and motion of charges in electric fields and the linear accelerator

KEY KNOWLEDGE

- Investigate theoretically and practically electrical fields about a point mass or charge (positive or negative) with reference to:
  - potential energy changes (qualitative) associated with a point mass or charge moving in the field
- Analyse the use of an electric field to accelerate a charge, including:
  - potential energy changes in a uniform electric field: $W = qV$, $E = \frac{V}{d}$
- Model the acceleration of particles in a particle accelerator by a uniform electric field

4.6.1 Potential energy changes for charges moving in electric fields

A particle placed in an electric field experiences an electric force. If the particle is displaced in a direction parallel to the force, the electric force does work on the particle. There is a corresponding decrease in electrical potential energy of the particle in the field. An example of this occurs in an electron gun, when an electron is released near a positively charged object. The electron accelerates towards the positive object. The electron's kinetic energy increases and the electrical potential energy in the field decreases.

If a force is used to move a particle in a direction opposing the electric force on the particle, then the potential energy of the charged particle will increase. For example, if two positively charged balloons are pushed closer together, the electrical potential energy will increase.

4.6.2 Energy changes for a charged particle in a uniform electric field

A source of EMF, such as a battery, supplies energy to a circuit. Each amount of charge, q, that passes from one side of the battery to the other is supplied with qV joules of energy, where V is the value of the battery EMF or voltage, in volts (V).

When two parallel plates are connected to a battery, a total amount of charge, Q, passes through the battery until the battery is no longer able to supply an electron with enough energy to overcome its attraction to the now positively charged plate. At this point, the electrical potential difference between the plates is equal to the EMF, V, of the battery.

The electric field strength between the plates, E, can be related to the electrical potential difference between the plates, V, and the separation of the plates, d.

If a charge q is placed on the positive plate, it experiences an attractive force, $q\mathbf{E}$, due to the field between the charged plates. It will be accelerated across the gap to the negatively charged plate. The work done on the charge by the field equals qEd, so the kinetic energy of the charge will increase by qEd. It is equivalent to an excess of charge on the plate, q, moving back through the battery, which would be a decrease in electrical potential energy of qV.

$$\Delta E_k = W = qV$$

where: ΔE_k is the change in kinetic energy, in J

W is the work done on q coulombs of charge, in J (or N m)

q is the quantity of charge, in C

V is the potential difference of the voltage drop, in V

Equating the work done on the charge by the field with the change in electrical potential energy of the battery, we find the following relationship:

$$E = \frac{V}{d}$$

where: E is the electric field strength, in V m^{-1}

V is the electric potential difference, in V

d is the distance between two points in the electric field, in m

This provides an alternative unit for electric field of volts per metre or V m^{-1}. So, like gravitational field strength, electric field strength has two equivalent units: either newtons per coulomb $\left(\text{N C}^{-1}\right)$ or volts per metre. Using volts per metre makes it very easy to determine the strength of a uniform electric field.

tlvd-8985

SAMPLE PROBLEM 6 Determining the strength of a uniform electric field between two plates

What is the strength of the electric field between two plates 5.0 cm apart connected to a 100 V DC supply?

THINK	WRITE
Recall the formula for the electric field between two parallel plates with a potential difference V across the plates: $E = \frac{V}{d}$ where V = 100 V and d = 5.0 cm = 0.05 m.	$E = \frac{V}{d}$ $= \frac{100}{0.05}$ $= 2.0 \times 10^3 \text{ V m}^{-1}$

PRACTICE PROBLEM 6

Calculate the strength of the electric field between a storm cloud 1.5 km above ground and the ground itself if the electric potential difference is 30 000 000 V. Assume a uniform field.

4.6.3 The linear accelerator

An electric field can be used to increase the speed and kinetic energy of charged particles. This principle is the basis of a **linear particle accelerator**.

linear particle accelerator type of particle accelerator based on the work done by the field in moving a charge from one plate to the other

The world's longest linear accelerator is at Stanford University in California. It is 3.2 km long and can accelerate electrons to energies of 50 billion electron volts.

Figure 4.13 shows two metal plates with a small hole cut in the middle of each plate. The plates have been connected to a DC power supply. In the hole of the negative plate is a filament of wire, like the filament in an incandescent light globe, connected to a low voltage. When the current flows in this circuit, the filament glows red hot. The electrons are, in a sense, 'boiling at the surface' of the filament. The electric field can easily pull the electrons off the surface of the filament.

FIGURE 4.13 The electrons on the hot filament are attracted across to the positive plate and pass through the hole that is in line with the beam.

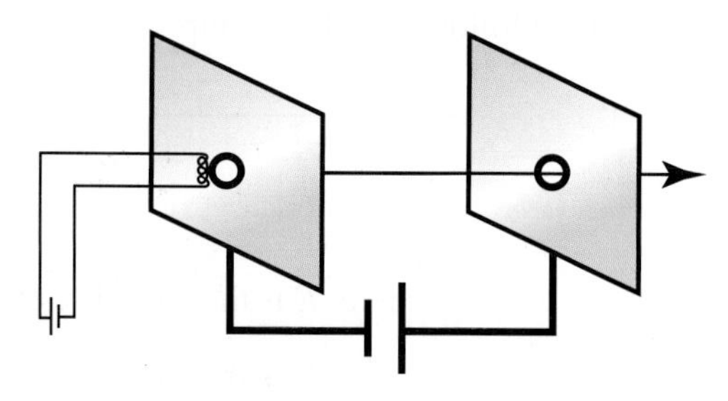

The hole in the positive plate is in a direct line with the filament, so as the electrons are accelerated across the space between the plates, they go straight through the hole to the next part of the machine. This design is called an electron gun. It produces the electrons required for many devices, such as for a synchrotron.

tlvd-8986

SAMPLE PROBLEM 7 Calculations involving an electron field acting as a particle accelerator

An electron is accelerated from one plate to another. The electrical potential difference between the plates is 100 V.

a. How much energy does the electron gain as it moves from the negative plate to the positive plate?

b. How fast will the electron be travelling when it hits the positive plate, if it left the negative plate with a speed of 0 m s^{-1}?

mass of electron = 9.1 × 10^{-31} kg; charge on electron = 1.6 × 10^{-19} C.

THINK

a. The change in kinetic energy can be found by the work done by the electric field:

$$\Delta E_k = W = qV$$

where $q = 1.6 \times 10^{-19}$ C and $V = 100$ V.

WRITE

a.
$$\Delta E_k = qV$$
$$= 1.6 \times 10^{-19} \times 100$$
$$= 1.6 \times 10^{-17} \text{ J}$$

THINK

b. The electron began with a speed of 0 m s^{-1}, so the final speed can be found from the final kinetic energy, $E_k = \frac{1}{2}mv^2$.

WRITE

b.
$$E_k = \frac{1}{2}mv^2$$
$$1.6 \times 10^{-17} = \frac{1}{2} \times 9.1 \times 10^{-31} \times v^2$$
$$v = \sqrt{\frac{2 \times 1.6 \times 10^{-17}}{9.1 \times 10^{-31}}}$$
$$= 5.9 \times 10^6 \text{ m s}^{-1}$$

Note that the final speed of the electron is about 2% of the speed of light!

PRACTICE PROBLEM 7

Repeat the calculation in sample problem 7 for an electrical potential difference across the plates of 1000 V.

tlvd-8987

SAMPLE PROBLEM 8 Calculating the acceleration and velocity of electrons in a cathode ray tube

The first televisions relied on cathode ray tube technology. In a cathode ray tube, electrons are emitted from a metal cathode, accelerated and then passed between parallel plates that allow the electron beam to be deflected horizontally and vertically by the television signal, striking the screen in different places with different intensities, creating an image.
The following diagram shows the trajectory of a particular electron, of charge -1.6×10^{-19} C and mass 9.1×10^{-31} kg, with an initial horizontal velocity v as it passes between horizontal charged parallel plates:

Electron beam passing between parallel plates

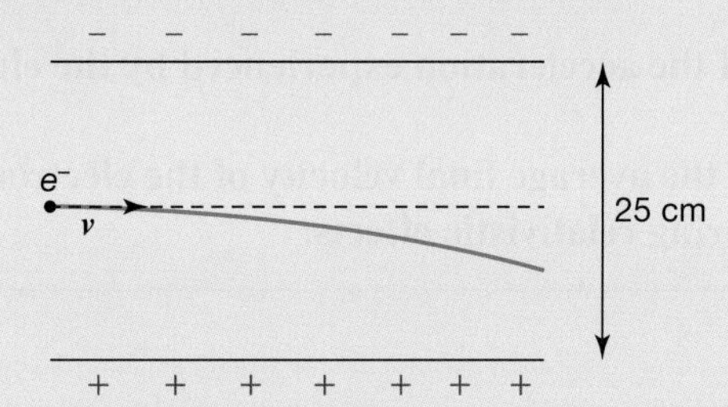

The electric field strength is 10^4 V m^{-1} (or N C^{-1}). The distance between the plates is 25.0 cm.
a. **What is the magnitude and direction of the acceleration experienced by the electron?**
b. **What is the vertical velocity of the electron beam when it leaves the plates?**
Note that the gravitational force on the electrons is negligible compared to the electric force and can be ignored in this case.

THINK

a. Let's take upwards as the positive direction. Recall that the size of the force on the electron in the field is given by $F = qE$, where E is the electric field strength.

WRITE

$$F = qE$$
$$= -1.6 \times 10^{-19}\text{C} \times 10^4\,\text{N C}^{-1}$$
$$= -1.6 \times 10^{-15}\,\text{N}$$

In the diagram above, the electrons (negatively charged) are attracted to the positively charged plate.
The electric force is in the downward direction.

The acceleration can be found by applying Newton's Second Law, $F = ma$.

$$a = \frac{F}{m}$$
$$= \frac{-1.6 \times 10^{-15}\,\text{N}}{9.1 \times 10^{-31}\,\text{kg}}$$
$$= 1.758 \times 10^{15}\,\text{m s}^{-2}$$
$$= -1.8 \times 10^{15}\,\text{m s}^{-2} \text{ to 2 s. f.}$$

The acceleration is in the downward direction. Its magnitude is 1.8×10^{15} m s^{-2} to 2 s. f.

b. The constant acceleration formulae can be used to find other kinematic variables, such as the change in vertical velocity.
The vertical displacement of the beam above is half the distance separating the plates, $s = -12.5$ cm.
The initial vertical velocity of the electrons is $0\ \text{m s}^{-1}$.
Using $v^2 = u^2 + 2as$, we can find the final vertical velocity of the electrons.

$$v^2 = u^2 + 2as$$
$$v^2 = 0 + 2 \times (-1.758 \times 10^{15}\ \text{m s}^{-2}) \times (-0.125\ \text{m})$$
$$\Rightarrow v = 2.1 \times 10^7\ \text{m s}^{-1}$$

The vertical velocity is downwards.
Its magnitude is $2.1 \times 10^7\ \text{m s}^{-1}$.

PRACTICE PROBLEM 8

An electric field strength of $2.0 \times 10^6\ \text{N C}^{-1}$ is used to accelerate electrons emitted from a cathode over a distance of 25 mm.

a. **Calculate the magnitude of the acceleration experienced by the electrons, ignoring the gravitational force.**
b. **Estimate the magnitude of the average final velocity of the electrons, assuming that their initial velocity is $0\ \text{m s}^{-1}$ and ignoring relativistic effects.**

4.6 Activities

learnon

Students, these questions are even better in jacPLUS

Receive immediate feedback and access sample responses

Access additional questions

Track your results and progress

Find all this and MORE in jacPLUS

4.6 Quick quiz	4.6 Exercise	4.6 Exam questions

4.6 Exercise

1. The electric field strength between two oppositely charged conducting plates is $15.0\ \text{V m}^{-1}$. If the electrical potential difference across the plates is 100 V, determine the distance d between the plates. Give your answer to 2 significant figures.
2. Two metal plates, X and Y, are set up 20 cm apart. The X plate is connected to the positive terminal of a 120-V battery and the Y plate is connected to the negative terminal. A small positively charged sphere is midway between the plates and it experiences a force of 4.0×10^{-3} N.
   a. If it was placed 7.5 cm from plate X, what would be the magnitude of the force on the sphere?
   b. The sphere is placed back in the middle and the plates are moved closer so that the distance between them is now 15 cm. What is the magnitude of the electric force now? Give your answer to 2 significant figures.
   c. The plates are moved apart so that the distance between them is 20 cm once again, but the battery is changed. With the new battery, the magnitude of the force is now 6.0×10^{-3} N. What is the voltage of the new battery? Give your answer to 3 significant figures.

3. The electric field strength between two oppositely charged conducting plates 15 mm apart is $200\ \text{N C}^{-1}$. Give your answer to 2 significant figures. Calculate the work done by the electric field on a 2.0-nC charge as it moves from one plate to the other.
4. A 10-nC charge travels 1.0 cm parallel to a $300\ \text{V m}^{-1}$ electric field. Calculate the change in kinetic energy of the 10-nC charge. Give your answer to 2 significant figures.
5. Murali finds that, in dry weather, the rubbing of his clothes when walking causes static electricity to build up. When he goes to touch a metal door handle, he observes a brief spark and feels a small, sharp shock. The discharging occurs when the electric field strength between himself and the door handle exceeds $3.0 \times 10^{6}\ \text{V m}^{-1}$. His finger is 5.0 mm from the door handle when the discharge occurs. Calculate the electrical potential difference between himself and the door handle. Give your answer to 2 significant figures.
6. Two oppositely charged conducting plates have an electrical potential difference of 1000 V. An electron is accelerated from the negative plate to the positive plate.
   a. Determine the electrical potential energy E_{PE} of the electron at the following points:
   Give your answer to 2 significant figures.

Position of the electron	E_{PE} (J)
i. at the negative plate	
ii. halfway between the plates	
iii. at the positive plate	

   b. Determine the change in kinetic energy, in eV, of the electron at the following points:

Position of the electron	ΔE_k (eV)
i. halfway between the plates	
ii. at the positive plate	

7. Jacinta applies a uniform electric field to slow down over-energetic electrons. She wants to reduce the kinetic energy of the electrons from 1000 eV to 100 eV as they pass between a pair of parallel plates. The spacing of the plates is 1.0 cm. Calculate what the electric field strength should be between the plates. Give your answer to 2 significant figures.
8. Electrons from a hot filament are emitted into the space between two parallel plates, as shown in the following diagram, and are accelerated across the space between them.

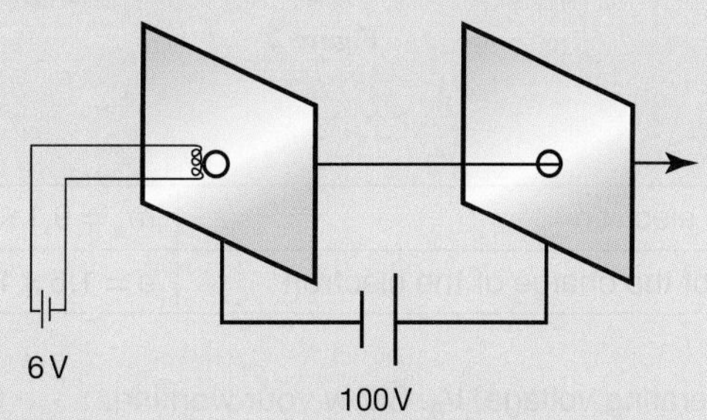

   a. Which battery supplies the field to accelerate the electrons?
   b. How much energy would be gained by an electron in crossing the space between the plates? Give your answer to 2 significant figures.
   c. How would your answer to part **b** change if the plate separation was halved?
   d. How would your answer to part **b** change if the terminals of the 6-V battery were reversed?
   e. How would your answer to part **b** change if the terminals of the 100-V battery were reversed?
   f. How would the size of the electric field between the plates, and thus the electric force on the electron, change if the plate separation was halved?
   g. Explain how your answers to parts **c** and **f** are connected.
9. A student wants to increase the energy of particles emitted from an electron gun that has a design essentially the same as the diagram in question 8. He proposes to do this by doubling the strength of the electric field through halving the separation between the charged conducting plates. Explain whether or not this change will achieve the aim of increasing the energy of the emitted particles.

10. **MC** A proton is placed between two conducting plates separated by 1 cm, with a potential difference of 500 V across the plates. Which of the following statements are correct? (Select all that apply.)
   A. The strength of the electric field varies from 50 000 $V m^{-1}$ to 0 $V m^{-1}$ between the plates.
   B. The potential difference halfway between the plates is 250 V.
   C. The proton is accelerated towards the plate at the lowest potential difference.
   D. The field between the plates is uniform, so the particle moves with a constant speed.

4.6 Exam questions

Question 1 (4 marks)

Source: *VCE 2022 Physics Exam, NHT, Section B, Q.1;* © VCAA

A particle with mass m and charge q is accelerated from rest by a potential difference, V. The only force acting on the particle is due to the electric field associated with this potential difference.

a. Show that the speed of the particle is given by $v = \sqrt{\frac{2qV}{m}}$ and state the principle of physics used in your answer. **(2 marks)**

b. Calculate the speed of an electron accelerated from rest by a potential difference of 200 V. **(2 marks)**

Data

mass of the electron	$m_e = 9.1 \times 10^{-31}$ kg
magnitude of the charge of the electron	$e = 1.6 \times 10^{-19}$ C

Question 2 (3 marks)

Source: *Adapted from VCE 2021 Physics Exam, NHT, Section B, Q.1a;* © VCAA

An electron is accelerated from rest by a potential difference of V_0. It emerges at a speed of 2.0×10^7 m s^{-1} as shown in Figure 2.

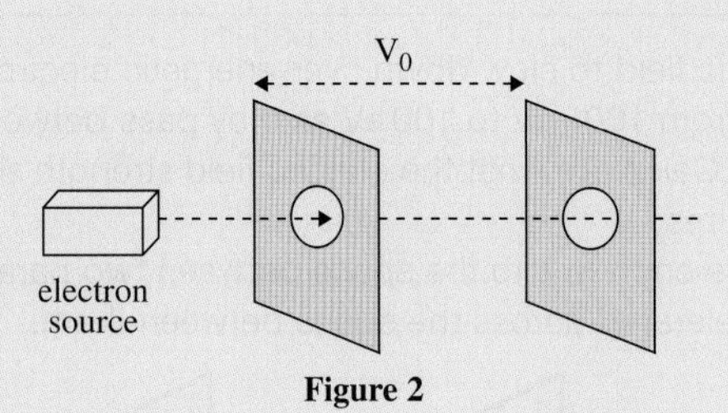

Figure 2

Data

mass of the electron	$m_e = 9.1 \times 10^{-31}$ kg
magnitude of the charge of the electron	$e = 1.6 \times 10^{-19}$ C

Calculate the value of the accelerating voltage, V_0. Show your working.

Question 3 (1 mark)

Source: *VCE 2017, Physics Exam, Section A, Q.3;* © VCAA

MC Two large charged plates with equal and opposite charges are placed close together, as shown in the diagram below. A distance of 5.0 mm separates the plates. The electric field between the plates is equal to 1000 N C^{-1}.

Which one of the following is closest to the voltage difference between the plates?

A. 5.0 V
B. 200 V
C. 5000 V
D. 5 000 000 V

Question 4 (1 mark)

Source: *VCE 2013 Physics Exam, Section B, DS4, Q.1; © VCAA*

MC Figure 1 shows a simplified diagram of the electron gun in the Australian Synchrotron.

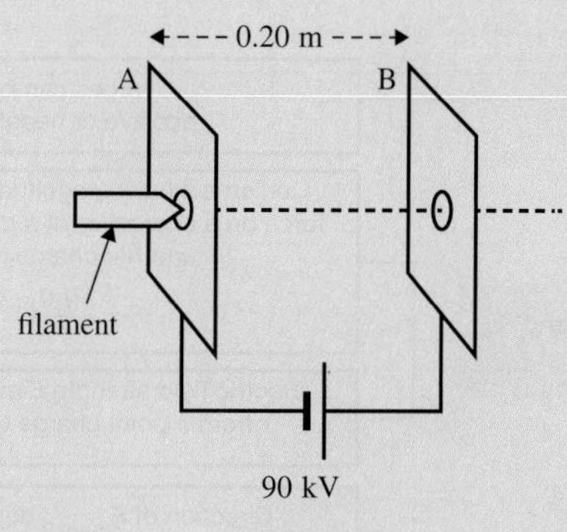

Figure 1

The potential difference between the plates is equal to 90 kV and the separation of the plates is 0.20 m.

Which one of the following best gives the magnitude of the force acting on electrons that enter the space between the plates?

A. 7.2×10^{14} N
B. 7.2×10^{15} N
C. 4.5×10^{-4} N
D. 4.5×10^{4} N

Question 5 (4 marks)

Source: *Adapted from VCE 2017 Physics sample Exam, Section B, Q.2 a&b; © VCAA*

Figure 1 shows part of a particle accelerator. Electrons are accelerated by a voltage of 10 000 V in an electron gun consisting of two plates that are 0.10 m apart. Ignore relativistic effects.

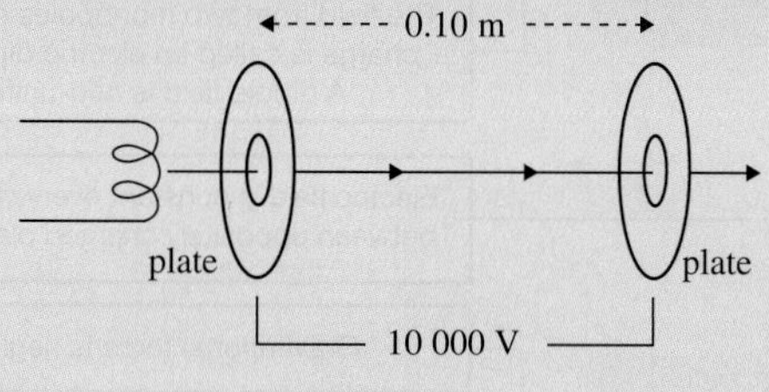

Figure 1

Data

mass of the electron	$m_e = 9.1 \times 10^{-31}$ kg
charge of the electron	$e = 1.6 \times 10^{-19}$ C

a. Calculate the strength of the electric field between the plates. **(2 marks)**
b. Calculate the speed of the electrons as they exit the electron gun. **(2 marks)**

More exam questions are available in your learnON title.

4.7 Review

Hey students! Now that it's time to revise this topic, go online to:

Access the topic summary | Review your results | Watch teacher-led videos | Practise past VCAA exam questions

Find all this and MORE in jacPLUS

4.7.1 Topic summary

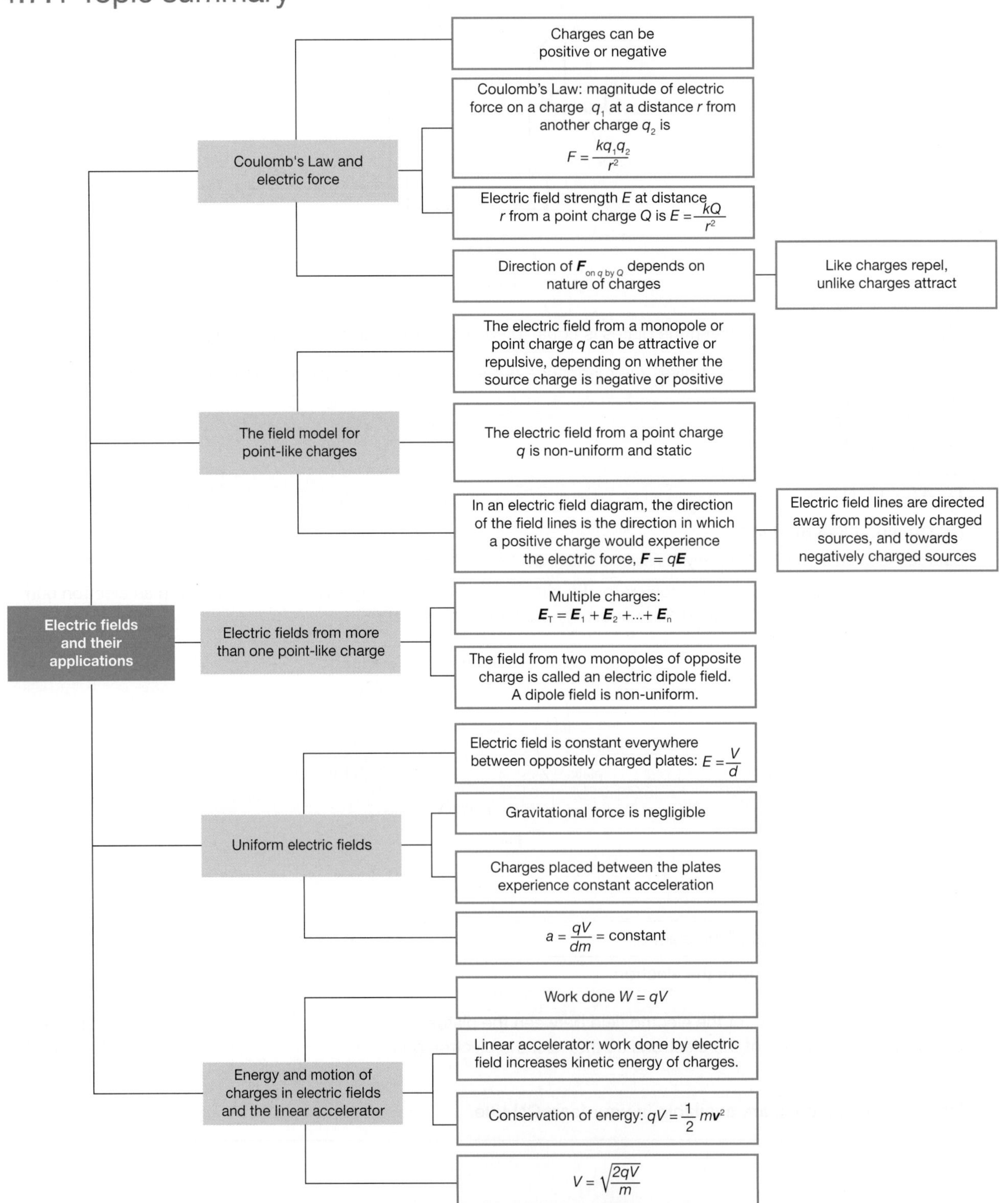

Electric fields and their applications
Coulomb's Law and electric force
Charges can be positive or negative
Coulomb's Law: magnitude of electric force on a charge q_1 at a distance r from another charge q_2 is
$F = \frac{kq_1q_2}{r^2}$
Electric field strength E at distance r from a point charge Q is $E = \frac{kQ}{r^2}$
Direction of $\boldsymbol{F}_{\text{on } q \text{ by } Q}$ depends on nature of charges
Like charges repel, unlike charges attract
The field model for point-like charges
The electric field from a monopole or point charge q can be attractive or repulsive, depending on whether the source charge is negative or positive
The electric field from a point charge q is non-uniform and static
In an electric field diagram, the direction of the field lines is the direction in which a positive charge would experience the electric force, $\boldsymbol{F} = q\boldsymbol{E}$
Electric field lines are directed away from positively charged sources, and towards negatively charged sources
Electric fields from more than one point-like charge
Multiple charges: $\boldsymbol{E}_T = \boldsymbol{E}_1 + \boldsymbol{E}_2 + ... + \boldsymbol{E}_n$
The field from two monopoles of opposite charge is called an electric dipole field. A dipole field is non-uniform.
Uniform electric fields
Electric field is constant everywhere between oppositely charged plates: $E = \frac{V}{d}$
Gravitational force is negligible
Charges placed between the plates experience constant acceleration
$a = \frac{qV}{dm}$ = constant
Energy and motion of charges in electric fields and the linear accelerator
Work done $W = qV$
Linear accelerator: work done by electric field increases kinetic energy of charges.
Conservation of energy: $qV = \frac{1}{2}mv^2$
$v = \sqrt{\frac{2qV}{m}}$

4.7.2 Key ideas summary

4.7.3 Key terms glossary

online only

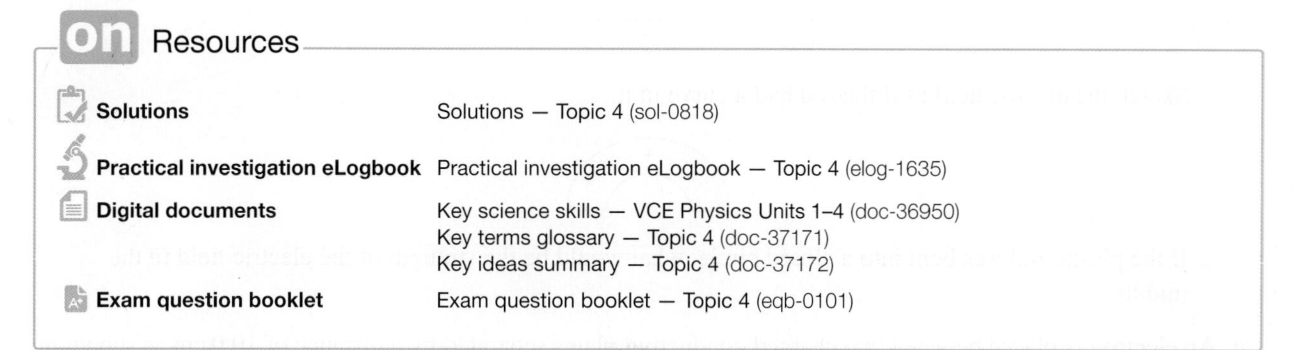

Resources

Solutions	Solutions — Topic 4 (sol-0818)
Practical investigation eLogbook	Practical investigation eLogbook — Topic 4 (elog-1635)
Digital documents	Key science skills — VCE Physics Units 1–4 (doc-36950) Key terms glossary — Topic 4 (doc-37171) Key ideas summary — Topic 4 (doc-37172)
Exam question booklet	Exam question booklet — Topic 4 (eqb-0101)

4.7 Activities

learnon

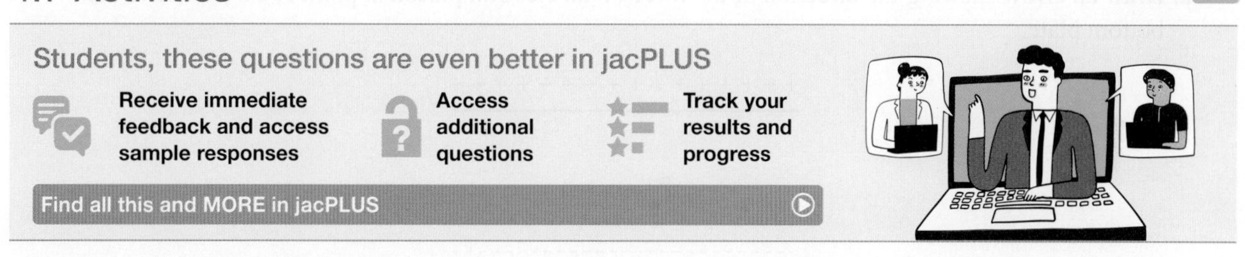

4.7 Review questions

1. Calculate the magnitude of the force of repulsion between two point charges with charges of 5.0 μC and 7.0 μC if they are 20 cm apart. Give your answer to 2 significant figures.
2. An electric force of 3.0 N acts downwards on a charge of −1.5 μC.
   a. Calculate the strength of the electric field to 2 significant figures.
   b. Determine the direction of the electric field.
3. Calculate the strength of the electric field 1.0 mm from a proton. Give your answer to 2 significant figures.
4. If the magnitude of the force between two charges separated by a distance *r* is 400 mN, express as a function of *r* how far apart the charges would need to be moved for the magnitude of the force between them to become 50 mN.
5. A particle acquires a 15.0-nC charge.
   a. Sketch the electric field due to the 15.0-nC charge.

   A second particle with a charge of –10.0 nC is placed 2.0 cm to the right of the 15.0-nC charge.

   b. Calculate the magnitude of the force on the −10.0-nC charge due to the +15.0-nC charge. Give your answer to 2 significant figures.
   c. Is there any point at which the combined electric field from the two charges will have a value of zero? Explain your response without using calculations.
6. Sketch the electric field around a proton and an electron separated by 0.1 nanometres.

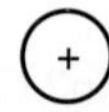

7. Answer the following:
   a. Sketch the electric field around a positively charged straight plastic rod. Assume the charge is distributed evenly.

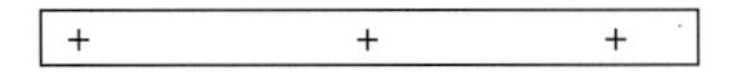

   b. Sketch the electric field as if the rod had a curve in it.

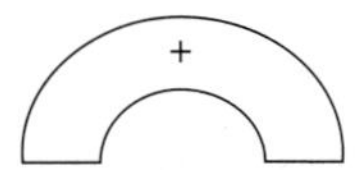

   c. If the plastic rod was bent into a closed circle, what would be the strength of the electric field in the middle?

8. An electron is placed between two charged conducting plates separated by a distance of 10.0 cm as shown in the following diagram. The electric field strength between the plates is $100\ \text{N C}^{-1}$.
   a. Draw an arrow showing the direction of the force on an electron placed at point P, 2.5 cm above the bottom plate.

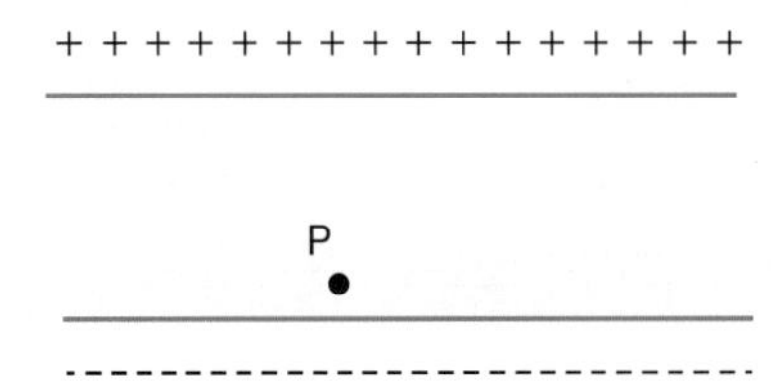

   b. Calculate the potential difference between the top and bottom plate.
   c. Calculate the change in kinetic energy of the electron as it moves from point P to the top plate. Give your answer to 2 significant figures.

9. An electron passes between two charged conducting plates separated by 3.0 cm, with a potential difference of 2000 V.

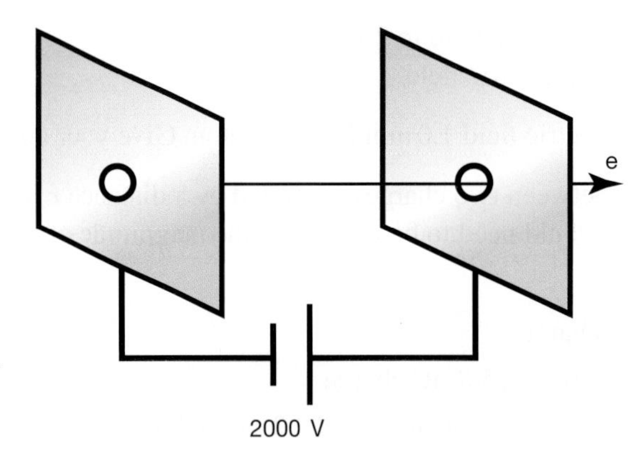

   a. What is the direction of the electric field between the plates?
   b. Calculate the electric field strength between the plates. Give your answer to 2 significant figures.
   c. Assuming that the initial speed of the electron, of mass 9.1×10^{-31} kg, is $0\ \text{m s}^{-1}$, calculate how long it takes the electron to pass from the first to the second plate. Give your answer to 2 significant figures. (Consider the gravitational force to be negligible.)
   d. What is the work done by the electric field on the electron as it passes from the first to the second plate? Give your answer to 2 significant figures.
   e. What is the kinetic energy change experienced by the electron? Give your answer to 2 significant figures.
   f. What is the kinetic energy change experienced by the electron when it is halfway between the plates? Give your answer to 2 significant figures.
   g. What would be the effect of reversing the applied potential difference?

10. A student wants to increase the energy of particles emitted from an electron gun that has a design essentially the same as the diagram below. He proposes to do this by doubling the strength of the electric field through halving the separation between the charged conducting plates. Explain whether or not this change will achieve the aim of increasing the energy of the emitted particles.

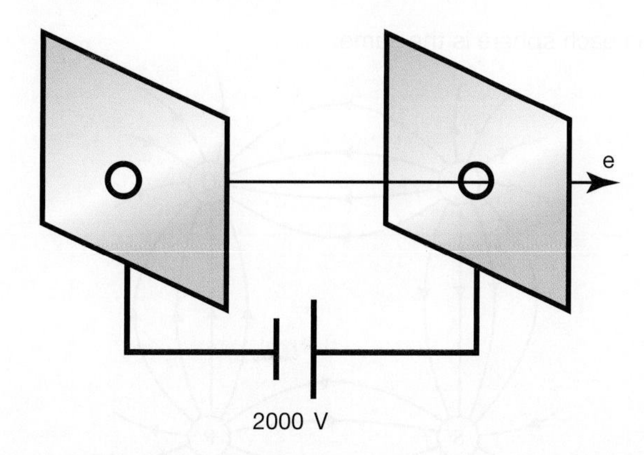

4.7 Exam questions

Section A — Multiple choice questions

All correct answers are worth 1 mark each; an incorrect answer is worth 0.

Question 1

Source: *VCE 2022 Physics Exam, Section A, Q.4;* © VCAA

Two point charges, Q and $4Q$, are placed 12 cm apart, as shown in the diagram below.

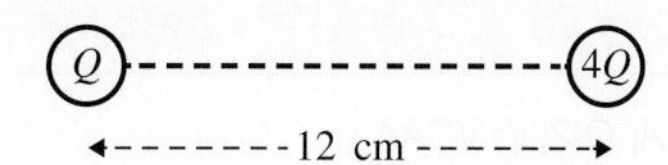

On the straight line between the charges Q and $4Q$, the electric field is

A. non-zero everywhere.
B. zero at a point 2.4 cm from Q.
C. zero at a point 3 cm from Q.
D. zero at a point 4 cm from Q.

Question 2

Source: *VCE 2022, Physics Exam, NHT, Section A, Q.1;* © VCAA

Two parallel plates are 10 mm apart and have a potential difference of 5.0 kV between them.

Which one of the following best gives the strength of the electric field between the plates?

A. 5.0×10^{-1} V m^{-1}
B. 5.0×10^{1} V m^{-1}
C. 5.0×10^{2} V m^{-1}
D. 5.0×10^{5} V m^{-1}

Question 3

Source: *VCE 2021 Physics Exam, Section A, Q.2;* © VCAA

The diagram below shows the electric field lines between four charged spheres: P, Q, R and S.

The magnitude of the charge on each sphere is the same.

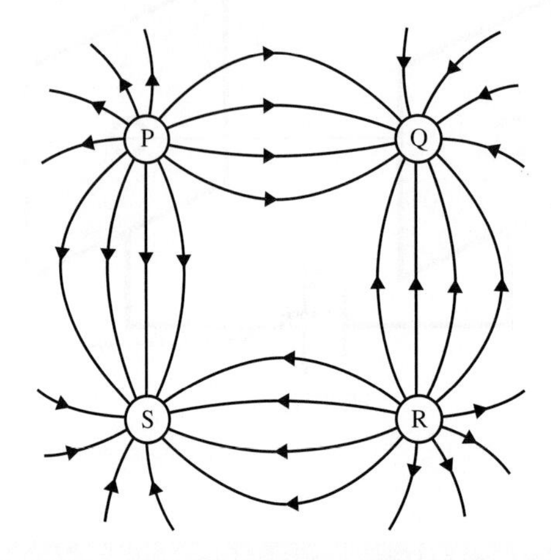

Which of the following correctly identifies the type of charge (+ positive or – negative) that resides on each of the spheres P, Q, R and S?

A.

P	Q	R	S
–	+	–	+

B.

P	Q	R	S
+	–	+	–

C.

P	Q	R	S
–	–	+	+

D.

P	Q	R	S
+	+	–	–

Question 4

Source: *VCE 2021 Physics Exam, NHT, Section A, Q.2;* © VCAA

Three charges, $-Q$, $+2Q$ and $+2Q$, are placed at the vertices of an equilateral triangle, as shown in the diagram below.

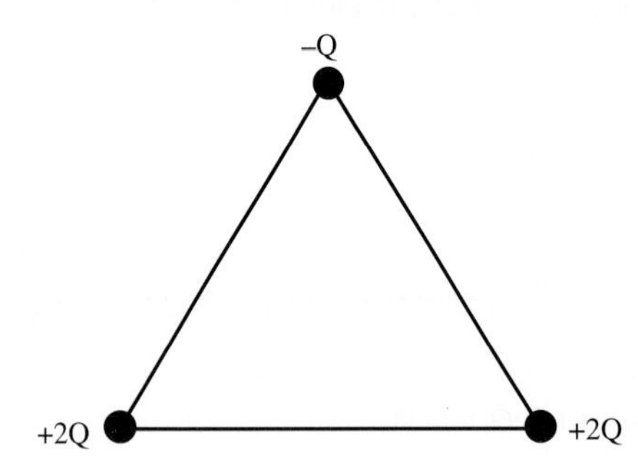

Which one of the following arrows best represents the direction of the net force on the charge $-Q$?

A. ↑

B. ↓

C.

D.

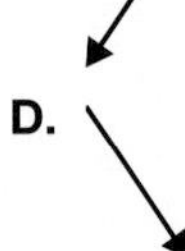

Question 5

Source: *VCE 2019, Physics Exam, Section A, Q.2;* © VCAA

The electric field between two parallel plates that are 1.0×10^{-2} m apart is 2.0×10^{-4} N C^{-1}.

Which one of the following is closest to the voltage between the plates?

A. 2.0×10^{-8} V
B. 2.0×10^{-6} V
C. 2.0×10^{-4} V
D. 1.0×10^{-2} V

Question 6

Source: *VCE 2018, Physics Exam, Section A, Q.4;* © VCAA

A small sphere has a charge of 2.0×10^{-6} C on it. Take $k = 8.99 \times 10^{9}$ N m^2 C^{-2}.

The strength of the electric field due to this charge at a point 3.0 m from the sphere is best given by

A. 2.0×10^{-3} V m^{-1}
B. 6.0×10^{-3} V m^{-1}
C. 9.0×10^{-3} V m^{-1}
D. 2.0×10^{3} V m^{-1}

Question 7

Source: *VCE 2016, Physics Exam, Section B, DS4, Q.2;* © VCAA

In the electron gun of a synchrotron, electrons are accelerated from rest over a distance of 12 cm to reach a final speed of 8.0×10^{7} m s^{-1}.

What is the accelerating voltage of the electron gun in kilovolts? (Ignore any relativistic effects.)

A. 2.67 kV
B. 5.30 kV
C. 6.67 kV
D. 18.2 kV

Question 8

Source: *VCE 2015 Physics Exam, Section B, DS4, Q.4;* © VCAA

Electrons are accelerated from rest in an electron gun by a potential difference of 50 kV.

What is their final speed? (Ignore relativistic effects.)

A. 1.0×10^{5} m s^{-1}
B. 4.7×10^{7} m s^{-1}
C. 1.3×10^{8} m s^{-1}
D. 2.9×10^{8} m s^{-1}

Question 9

Source: *VCE 2013 Physics Exam, Section B, DS4, Q.2; © VCAA*

Figure 1 shows a simplified diagram of the electron gun in the Australian Synchrotron.

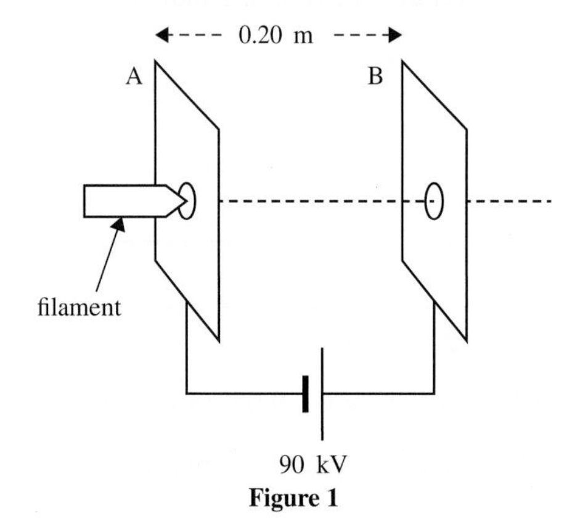

Figure 1

The potential difference between the plates is equal to 90 kV and the separation of the plates is 0.20 m.

Which one of the following is closest to the kinetic energy of an electron that reaches the positive plate?

A. 90 kJ
B. 90 keV
C. 1.44×10^{15} kJ
D. 1.44×10^{-17} keV

Question 10

Which of the following diagrams correctly describes the electric field due to +1 nC and −1 nC charges separated by 10 nm?

A.

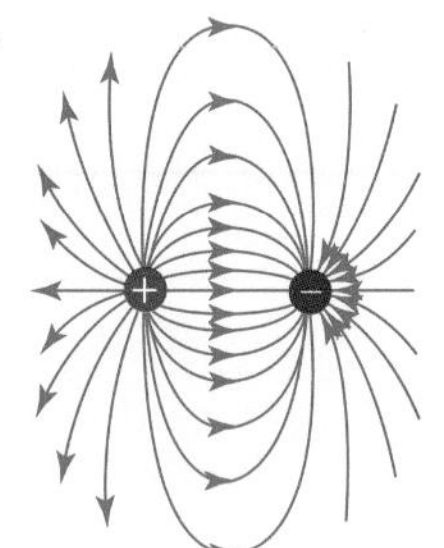

B.

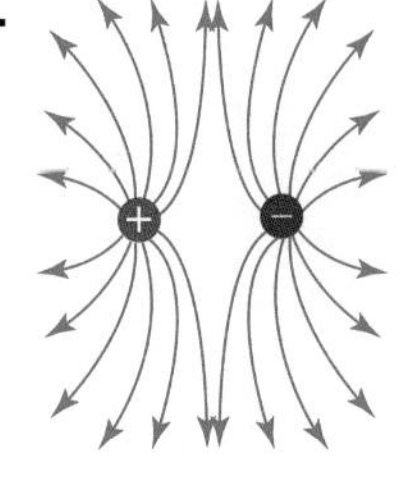

C.

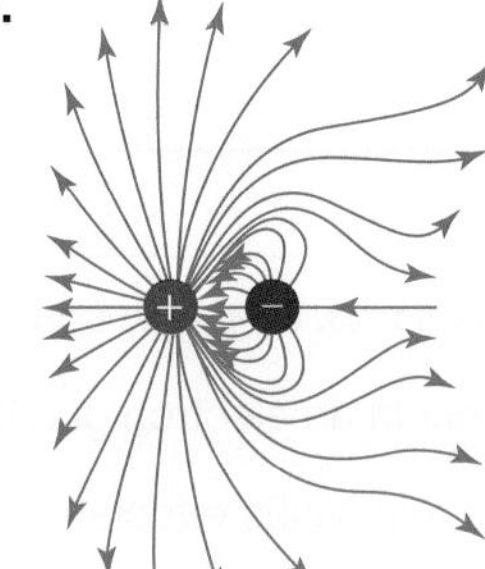

D.

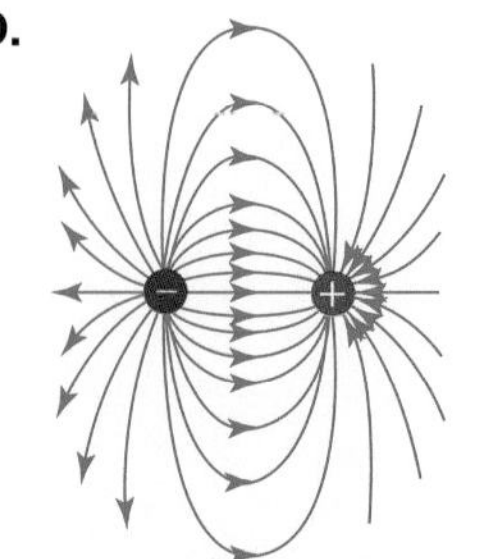

Section B — Short answer questions

Question 11 (3 marks)

Source: *VCE 2021 Physics Exam, NHT, Section B, Q.1;* © VCAA

A small sphere carrying a charge of $-2.7\ \mu C$ is placed between charged parallel plates, as shown in Figure 1. The potential difference between the plates is set at 15.5 V, which just holds the sphere stationary. The electric field between the plates is uniform.

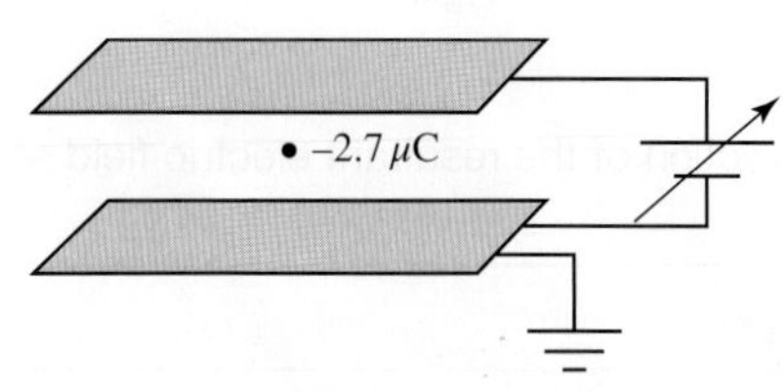

Figure 1

a. In which direction (up, down, right, left) will the sphere move if the voltage is increased? **(1 mark)**

b. Calculate the value of the electric force that is holding the sphere stationary if the plates are 2.0 mm apart. Show your working. **(2 marks)**

Question 12 (2 marks)

Source: *VCE 2019, Physics Exam, Section B, Q.2;* © VCAA

Figure 2 shows two equal positive stationary point charges placed near each other.

Figure 2

Sketch on Figure 2 the shape and direction of the electric field lines. Use at least **eight** field lines.

Question 13 (6 marks)

Source: *VCE 2018 Physics Exam, NHT, Section B, Q.2;* © VCAA

The electron gun section of a particle accelerator accelerates electrons between two plates that are 10 cm apart and have a potential difference of 5000 V between them.

Data

mass of electron	9.1×10^{-31} kg
charge of electron	$(-)1.6 \times 10^{-19}$ C

a. Calculate the electric field between the plates. Include an appropriate unit. **(2 marks)**

b. Calculate the magnitude of the force on an electron between the plates. **(2 marks)**

c. Calculate the speed of the electrons as they exit the electron gun. Ignore any relativistic effects. Assume that the initial speed of the electrons is zero. **(2 marks)**

Question 14 (1 mark)

Source: Adapted from *VCE 2017, Physics Exam, Section B, Q.1; © VCAA*

Three charges are arranged in a line, as shown in Figure 1.

X

○ • ○ ○

–2Q +3Q –2Q

Figure 1

Draw an arrow at point X to show the direction of the resultant electric field at X. If the resultant electric field is zero, write the letter 'N' at X.

Question 15 (5 marks)

Source: *VCE 2017, Physics Exam, Section B, Q.2; © VCAA*

According to one model of the atom, the electron in the ground state of a hydrogen atom moves around the stationary proton in a circular orbit with a radius of 53 pm $(53 \times 10^{-12}\ \text{m})$.

a. Show that the magnitude of the force acting between the proton and the electron at this separation is equal to 8.2×10^{-8} N. Take $k = 9.0 \times 10^{9}\ \text{N m}^2\ \text{C}^{-2}$ and the magnitude of the electron and proton charges as 1.6×10^{-19} C. Show all the steps of your working. **(2 marks)**

b. Using 8.2×10^{-8} N as the value of the magnitude of the force given in part a, calculate the speed of the electron in its circular path. Take the mass of the electron to be 9.1×10^{-31} kg. Show your working. **(3 marks)**

5 Magnetic fields and their applications

KEY KNOWLEDGE

In this topic, you will:

- describe magnetism using a field model
- investigate theoretically and practically magnetic fields, including directions and shapes of fields, and attractive and repulsive fields
- investigate and apply theoretically and practically a vector model to magnetic phenomena, including directions and shapes of fields produced by bar magnets
- investigate and apply theoretically and practically a vector field model to magnetic phenomena, including shapes and directions of fields produced by current carrying wires, loops and solenoids
- analyse the use of a magnetic field to change the path of a charged particle, including:
  - the magnitude and direction of the force applied to an electron beam by a magnetic field: $F = qvB$, in cases where the directions of v and B are perpendicular or parallel
  - the radius of the path followed by an electron in a magnetic field: $qvB = \frac{mv^2}{r}$, where $v \ll c$
- investigate and analyse theoretically and practically the force on a current carrying conductor due to an external magnetic field, $F = nIlB$, where the directions of I and B are either perpendicular or parallel to each other
- investigate and analyse theoretically and practically the operation of simple DC motors consisting of one coil, containing a number of loops of wire, which is free to rotate about an axis in a uniform magnetic field and including the use of a split ring commutator
- investigate, qualitatively, the effect of current, external magnetic field and the number of loops of wire on the torque of a simple motor
- model the acceleration of particles in a particle accelerator (limited to linear acceleration by a uniform electric field and direction change by a uniform magnetic field)
- compare theoretically and practically gravitational, magnetic and electric fields, including directions and shapes of fields, attractive and repulsive fields, and the existence of dipoles and monopoles
- identify fields as static or changing, and as uniform or non-uniform.

PRACTICAL WORK AND INVESTIGATIONS

Practical work is a central component of VCE Physics. Experiments and investigations, supported by a **practical investigation eLogbook** and **teacher-led video,** are included in this topic to provide opportunities to undertake investigations and communicate findings.

EXAM PREPARATION

Access past VCAA questions and exam-style questions and their video solutions in every lesson, to ensure you are ready.

5.1 Overview

Hey students! Bring these pages to life online

Watch videos

Engage with interactivities

Answer questions and check results

Find all this and MORE in jacPLUS

5.1.1 Introduction

Magnetism has been known since the beginning of recorded history. The ancient Athenians (600 BCE) observed that a particular stone could attract pieces of iron. They called this stone 'magnet' because it was found in an area that was then called Magnesia. The tendency of freely spinning small magnetic particles to always point in the same direction was known by 800 CE and magnets became an essential tool for navigation and exploration. Magnets were found to affect other magnetic materials even when they were not touching each other, leading Michael Faraday to suggest the concept of a magnetic field. The magnetic field is a property of the space around the magnet and causes a magnetic particle placed in the field to experience a force. The discovery that moving electric charges were also sources of magnetic fields revolutionised the understanding of light and the relationship between electric and magnetic fields. The interplay between charged particles and magnetic fields underlies many important technologies from the simple DC electric motor to analytical chemistry tools such as the mass spectrometer.

FIGURE 5.1 It has been proposed that the motion of the molten iron in Earth's core is what generates the magnetic field protecting Earth from cosmic radiation and solar winds.

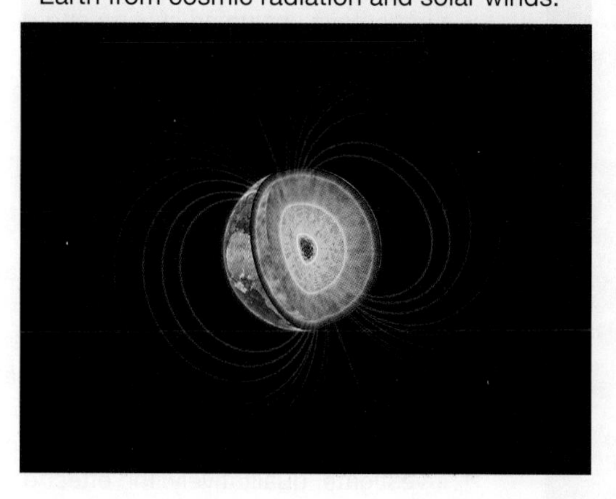

LEARNING SEQUENCE

on Resources

Solutions	Solutions — Topic 5 (sol-0819)
Practical investigation eLogbook	Practical investigation eLogbook — Topic 5 (elog-1636)
Digital documents	Key science skills — VCE Physics Units 1–4 (doc-36950) Key terms glossary — Topic 5 (doc-37173) Key ideas summary — Topic 5 (doc-37174)
Exam question booklet	Exam question booklet — Topic 5 (eqb-0102)

5.2 Magnets and magnetic fields

KEY KNOWLEDGE

- Describe magnetism using a field model
- Investigate theoretically and practically magnetic fields, including directions and shapes of fields, and attractive and repulsive fields
- Investigate and apply theoretically and practically a vector model to magnetic phenomena, including directions and shapes of fields produced by bar magnets

Source: Adapted from VCE Physics Study Design (2024–2027) extracts © VCAA; reproduced by permission.

Like masses and charged particles, which can interact without contact through, respectively, the effect of gravitational fields and electric fields, **magnets** can interact without contact through the effect of **magnetic fields**.

Unlike the gravitational force between masses, which is always attractive, the direction of the forces between two magnets depends on the orientation of each magnet and can be attractive or repulsive.

As observed by Peter Peregrinus in the thirteenth century, magnets have two ends, named poles. A freely spinning magnet, such as the iron needle in a compass, will orient itself with one end, the north pole, pointing towards geographic north and the other, the south pole, towards geographic south.

magnet material or object capable of producing a magnetic field and attracting unlike poles and repelling like poles

magnetic field vector field describing the property of the space in which a magnetic object experiences a force

FIGURE 5.2 A magnet will line up with a line from north to south if it is allowed to spin freely.

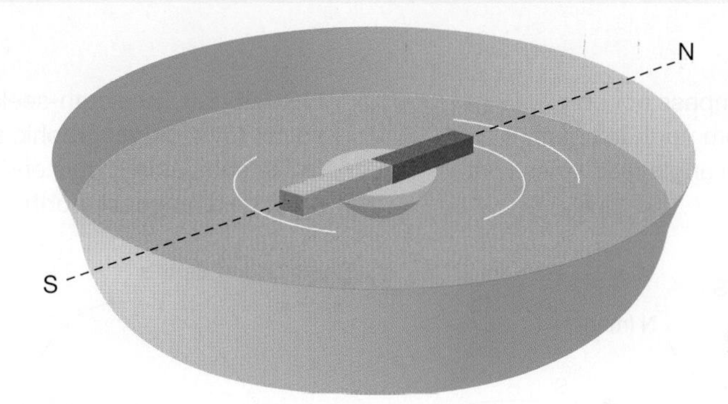

The molten iron in the centre of Earth behaves like a giant bar magnet, with a north pole and a south pole, and a small compass magnet moves to align itself with Earth's magnetic field.

FIGURE 5.3 The south end of the magnet points generally towards geographic south.

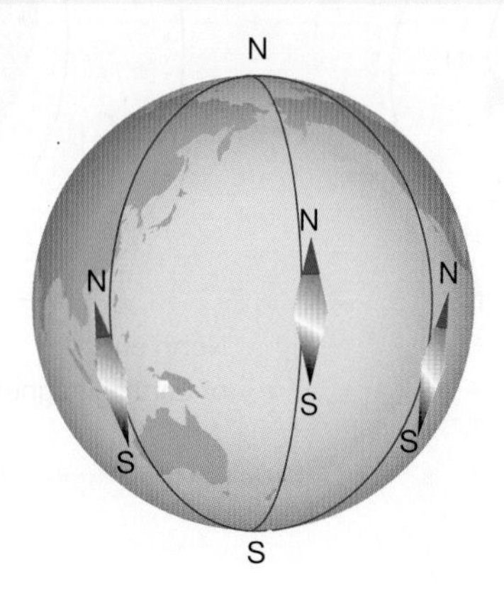

The poles of a magnet cannot be separated. Instead, breaking a magnet in half creates two new magnets, each with a north (N) and south (S) pole. The like ends of two magnets repel each other and the unlike ends of two magnets attract each other.

FIGURE 5.4 a. Like poles repel and **b.** unlike poles attract.

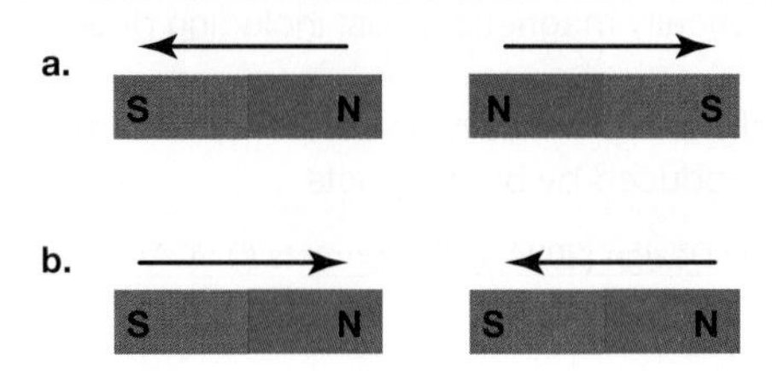

5.2.1 Magnetic fields and forces

A magnetic particle placed in the space around a magnet experiences a force. The force is due to the magnetic field from the magnet, which exists everywhere in the space around the magnet. Magnetic field lines have the following characteristics:

- Magnetic field lines form continuous loops, leaving the north end of the magnet and entering the south end of the magnet.
- Field lines do not intersect.
- The direction of the magnetic field at any point is along the tangent to the field line.
- The strength of the magnetic field is represented by the spacing of the field lines.
- In regions where the field lines are more closely spaced, the field is stronger. If the spacing of the field lines remains constant, the field is uniform. If the spacing between the field lines changes, then the field is non-uniform.

FIGURE 5.5 A magnet compass not only aligns itself along a line from north to south, it also dips downwards at an angle that varies with latitude.

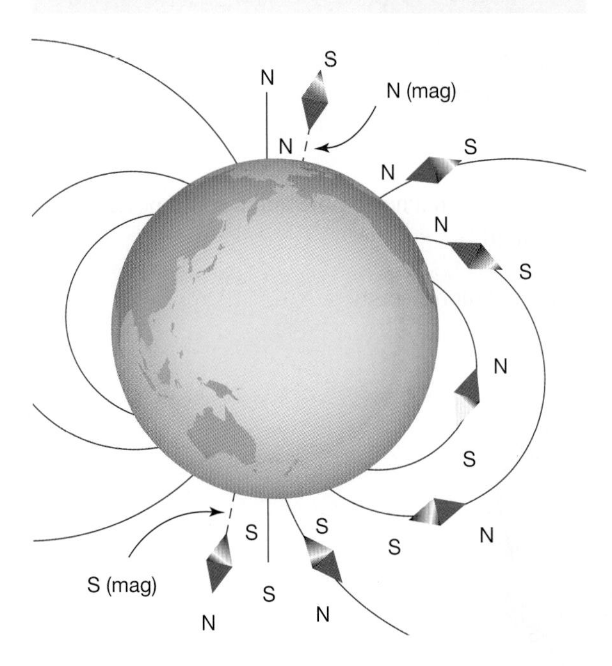

FIGURE 5.6 The south-seeking end of the needle points towards geographic south. But because unlike ends attract, this end of Earth's magnet must be a magnetic north end.

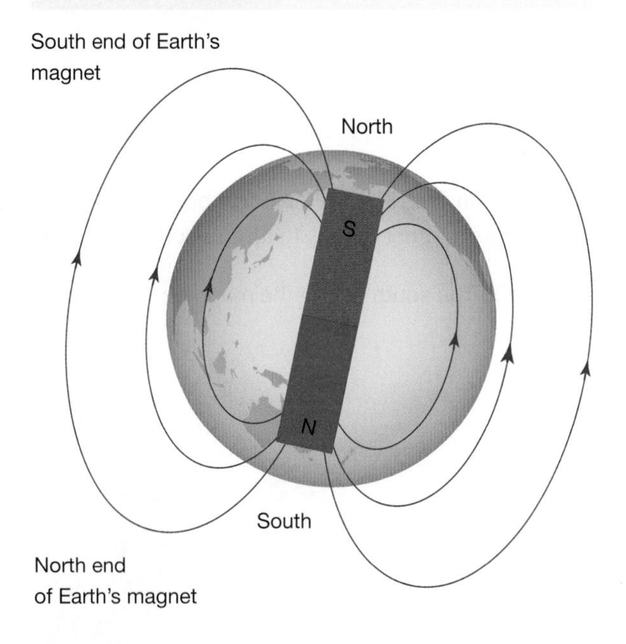

5.2.2 Magnets of various shapes

A popular activity to visualise magnetic field lines is to sprinkle iron filings around a **bar magnet**. The iron filings are **ferromagnetic** and align themselves to the field around the magnet as shown in figure 5.7.

bar magnet object with a rectangular shape, generally made up of iron or other ferromagnetic substance, showing permanent magnetic properties

ferromagnetic property of materials, such as iron, cobalt and nickel, that can be easily magnetised (act like a magnet)

magnetic induction process by which a substance, such as iron, becomes magnetised by a magnetic field

FIGURE 5.7 Using a bar magnet and iron filings is an easy way to visualise magnetic field lines.

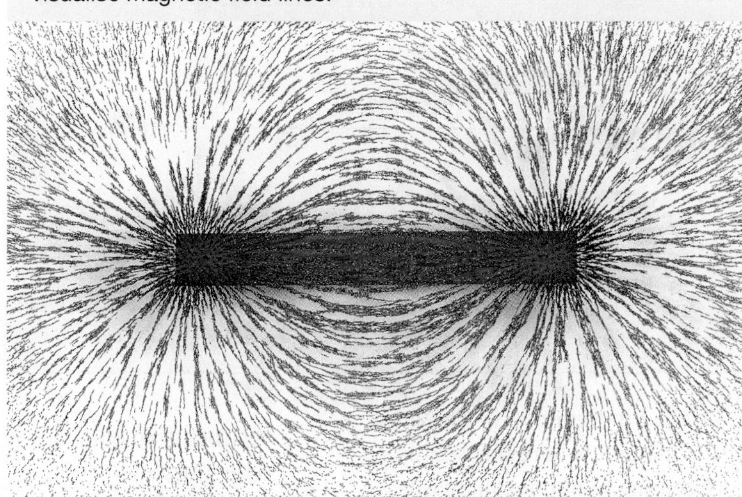

EXTENSION: Naturally occurring magnets

Magnetite is the most common iron oxide mineral. It has the chemical formula Fe_3O_4. In appearance, it is black, metallic (shiny) and quite hard. It is also a magnetic substance. Lodestone, which comes from 'leading stone', is a naturally occurring piece of magnetite that humans have used for at least 2600 years.

FIGURE 5.8 Lodestone attracting small bits of iron

But how do natural magnets form?

Ferromagnetic materials have magnetic domains, which are regions within a material in which the magnetisation of its atoms is in a uniform direction. Generally, magnetic domains are oriented randomly, and there is no net magnetic field. However, when a ferromagnetic material is placed within an external magnetic field, its domain will rotate and align with that external field, creating a magnet.

When molten ferromagnetic materials in lava cool down, the magnetic domains within the material are aligned by Earth's magnetic field, which form natural permanent magnets.

You can experience something similar to this by rubbing a piece of iron along a magnet. This will align the magnetic domains in your piece of iron, which has now become a magnet. The magnetic field from the magnet has 'magnetised' the iron it was in contact with. This process is called **magnetic induction**.

Magnetic domains

In a piece of iron or other ferromagnetic material, groups of nearby atoms line up together throughout the metal into regions called magnetic domains. When the iron is placed in a magnetic field, the domains that are already lined up with the external field increase in size as other domains shrink.

FIGURE 5.9 a. The magnetic fields of adjacent iron atoms align themselves in local areas called domains. b. Domains in a piece of iron exposed to a magnetic field, acting to the right.

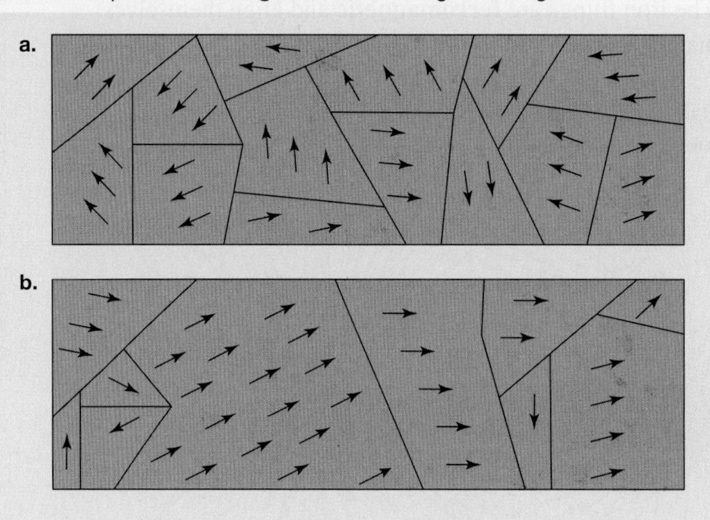

The magnetic field of a bar magnet is sketched in figure 5.10. The field lines are closest together at the ends of the magnet, where the field is strongest. Further away, the spacing between the field lines increases, indicating that the field is getting weaker. This is similar to what happens with gravitational fields and electric fields, for which the field strength decreases as the distance to the mass or charge source increases.

Observe that all the field lines leave the north end of the magnet and loop around to return to the south end.

Magnets can be designed to produce fields of different shapes. A horseshoe magnet (see figure 5.11a) produces a strong and fairly uniform field between the ends.

Figure 5.11b shows a circular ring magnet, oriented south side up, surrounding a cylindrical magnet oriented north side up. This produces a radial field between the two magnets, pointing outwards all the way around the circle. This design is used in loudspeakers.

FIGURE 5.10 Magnetic field of a bar magnet

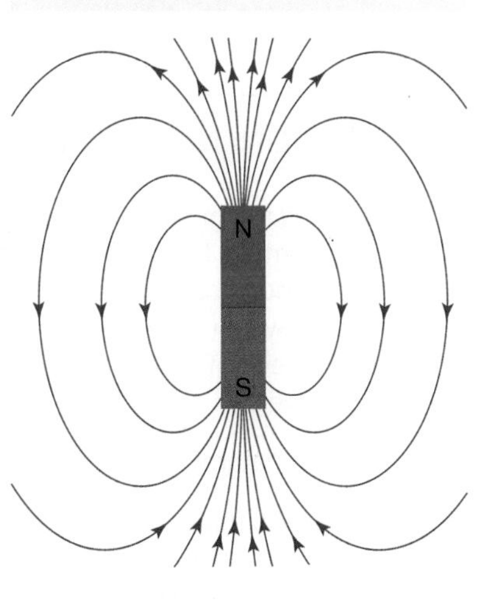

FIGURE 5.11 Differently shaped magnetic fields can be created by arranging the north and south ends of the magnet, as shown by a. a horseshoe magnet and b. a circular magnet.

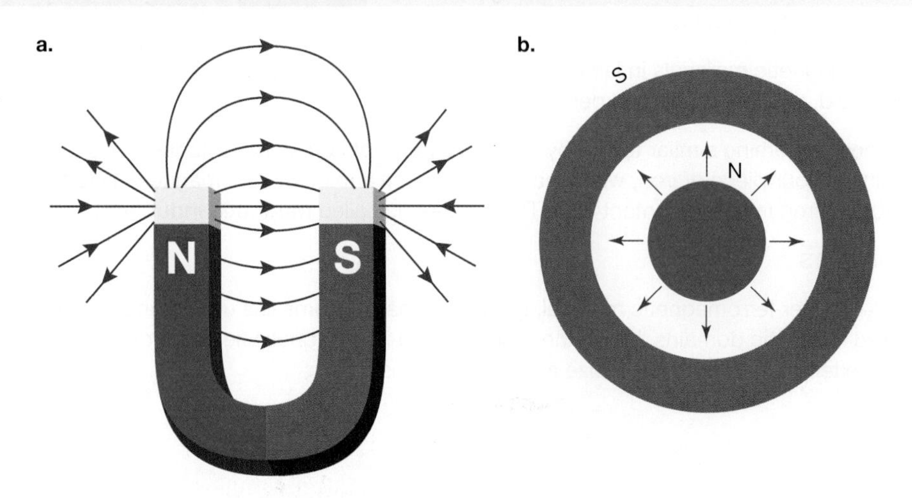

Some magnets have stronger fields than others. The strength of a magnetic field is measured in tesla (T). The strength of Earth's magnetic field at its surface is quite small, about 10^{-4} T or 0.1 mT (0.1 millitesla).

The strength of a typical school magnet is about 0.1 T. A fridge magnet is about 30 mT. The strongest permanent magnetic fields typically produced have field strengths of about 1.0 T.

SAMPLE PROBLEM 1 Exploring the magnetic field around a bar magnet

Consider the following diagram of the magnetic field of a bar magnet.

a. In what direction would the north end of a compass point if placed at points X, Y and Z?

b. Rank the points in order of increasing field strength.

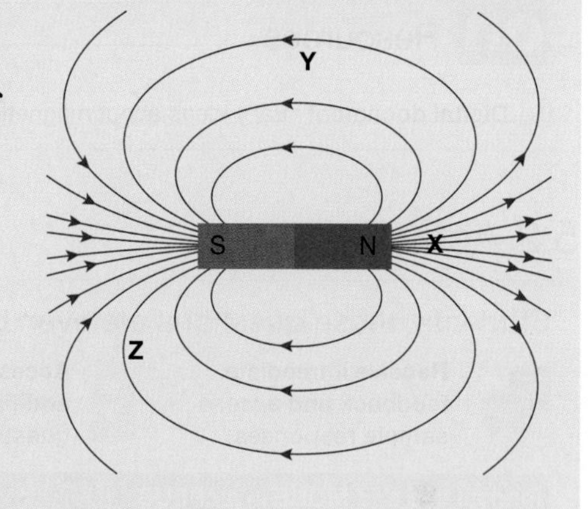

THINK	WRITE
a. Remember that the field lines indicate the direction that a magnetic compass would point.	a. At X, the field lines point away from the north pole of the bar magnet, so a compass placed there would point towards the right, away from the north pole. At Y, the field lines point towards the left, so the compass needle would point towards the left. At Z, the field lines curve, so the compass would follow the tangent to the curve and would point to the right at an angle of approximately 45° upwards.
b. Remember that the density of the field lines indicates the relative strength of the field.	b. The field lines are most dense at X and least dense at Y, so the points in order of increasing field strength would be Y, Z, X.

PRACTICE PROBLEM 1

Consider the magnetic field of a horseshoe magnet.

a. In what direction would the north end of a compass point if placed just to the left of the north pole?

b. In what direction would the north end of a compass point if placed halfway between the north and south pole?

c. In which position would the field be strongest?

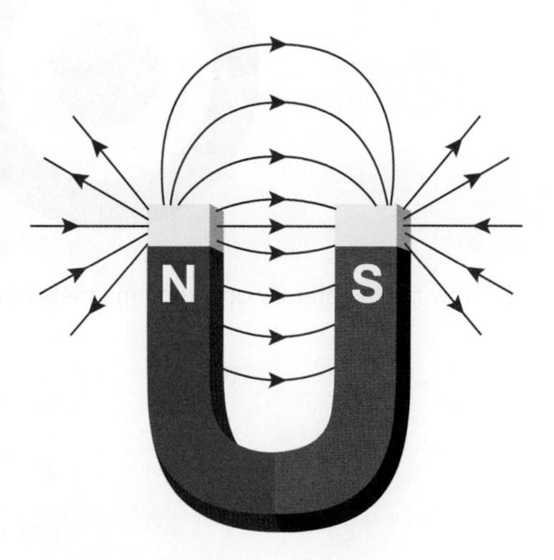

elog-1944

tlvd-10602

INVESTIGATION 5.1

online only

Magnetic field mapping

AIM

To explore the magnetic fields of different arrangements of permanent magnets

Resources

Digital document Early ideas about magnetism (doc-39375)

5.2 Activities

learnon

Students, these questions are even better in jacPLUS

Receive immediate feedback and access sample responses

Access additional questions

Track your results and progress

Find all this and MORE in jacPLUS

5.2 Quick quiz  | **5.2 Exercise** | **5.2 Exam questions**

5.2 Exercise

1. How would you use a magnet to test whether a piece of metal was magnetic?
2. Why do both ends of a magnet attract an iron nail?
3. What is the polarity of Earth's magnetic field at the magnetic pole in the southern hemisphere?
4. Draw the magnetic field lines for the following items:
   a. a loudspeaker magnet
   b. a horseshoe magnet.

a.

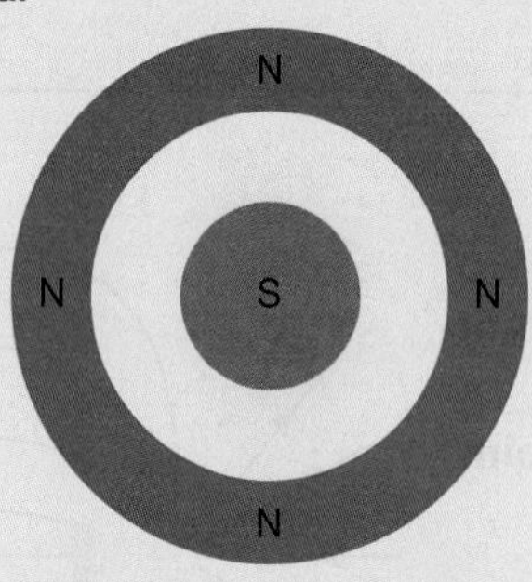

b.

5. How could naturally occurring magnets have been formed?

5.2 Exam questions

Question 1 (1 mark)

Earth's geomagnetic field can be modelled as a bar magnet. Using the field lines in the illustration, identify the south (S) and north (N) poles.

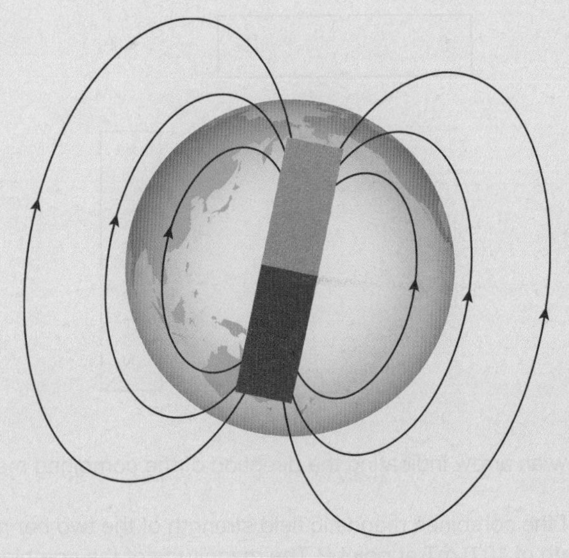

Question 2 (1 mark)

MC Consider the magnetic field of a horseshoe magnet between the arms of the magnet. What would be the orientation of a compass needle at point P?

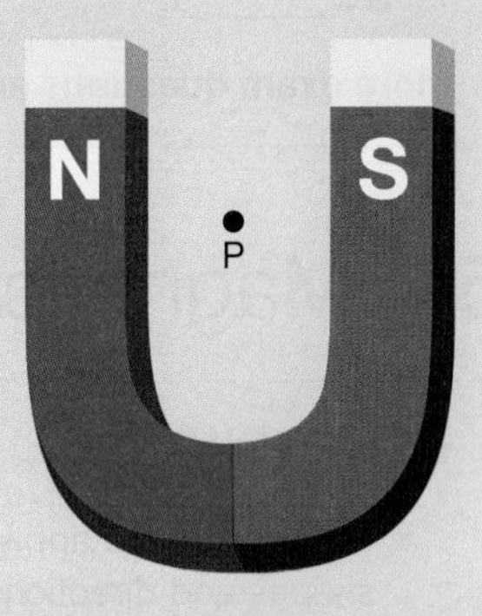

A.

B.  

C. 

D.

Question 3 (3 marks)

Consider two bar magnets placed as illustrated.

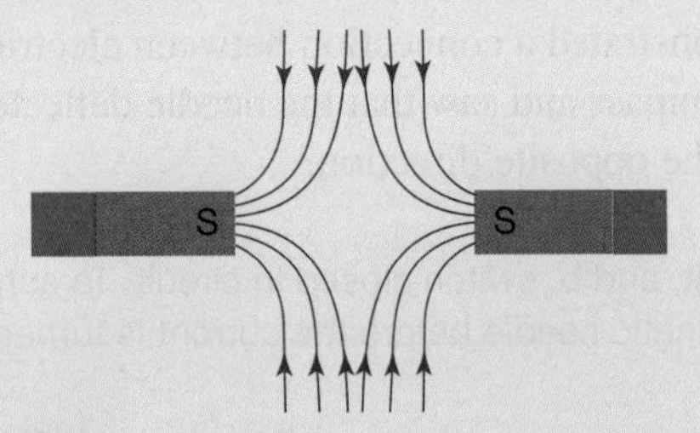

a. Is the magnetic force between the magnets attractive or repulsive? **(1 mark)**
b. Draw the field lines between the poles. **(2 marks)**

Question 4 (1 mark)

MC Which of the following statements is incorrect?

A. Magnetic fields can attract or repel other magnets, depending on the type of magnet and nature of the field.
B. Magnetic fields can be uniform or non-uniform.
C. The source of a magnetic field can be a monopole.
D. Magnetic field lines must always form closed loops.

Question 5 (3 marks)

Source: *VCE 2021 Physics Exam, Section B, Q.1;* © VCAA

Two identical bar magnets of the same magnetic field strength are arranged at right angles to each other and at the same distance from point P, as shown in Figure 1.

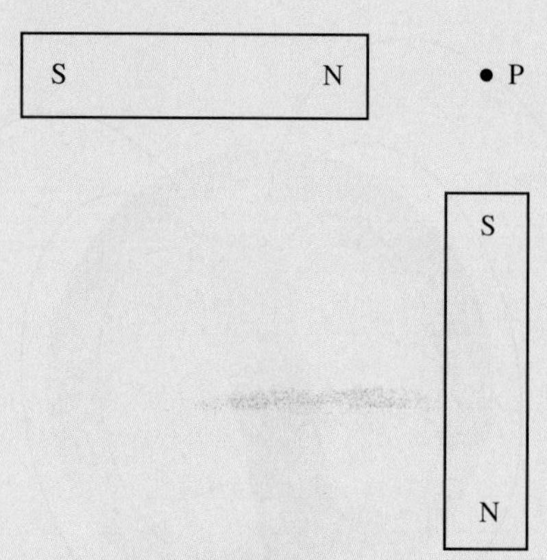

Figure 1

a. At point P on Figure 1, draw an arrow indicating the direction of the combined magnetic field of the two bar magnets. **(1 mark)**

b. Calculate the magnitude of the combined magnetic field strength of the two bar magnets if each bar magnet has a magnetic field strength of 10.0 mT at point P. The magnitude of the combined magnetic field strength is ***B*** = _______ T. **(2 marks)**

More exam questions are available in your learnON title.

5.3 Magnetic fields from moving charged particles

KEY KNOWLEDGE

- Investigate and apply theoretically and practically a vector field model to magnetic phenomena, including shapes and directions of fields produced by current carrying wires, loops and solenoids

Source: Adapted from VCE Physics Study Design (2024–2027) extracts © VCAA; reproduced by permission.

In 1820 Hans Christian Oersted demonstrated a connection between electricity and magnetism. He placed a wire carrying a current over a magnetic compass and saw that the needle deflected. He then placed the wire under the compass and the needle deflected in the opposite direction.

FIGURE 5.12 a. Switch open in circuit, and **b.** switch closed in circuit. To achieve maximum deflection, the wire should be placed in line with the magnetic needle before the current is turned on.

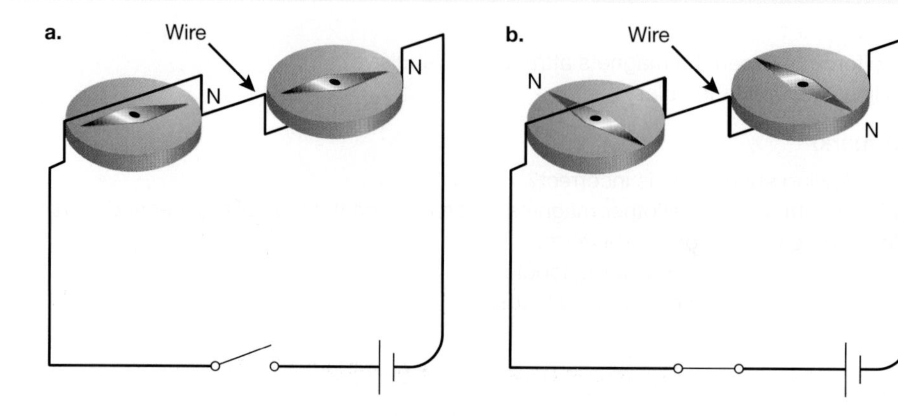

The deflection of the compass needle is evidence that the current is a source of a magnetic field. The moving charged particles, electrons, produce a magnetic field as they move along the wire. A compass can be used to map the field around a current-carrying wire. The field is non-uniform, becoming weaker as the distance from the wire increases.

Representing current and its magnetic field often requires a three-dimensional view. To achieve this on a flat two-dimensional page, a convention is adopted.

The symbol of a circle with a dot in the middle is used to represent a magnetic field coming *out* of a page. A circle with a diagonal cross is used to represent a magnetic field going *into* the page.

Figure 5.13 shows the direction of the magnetic field from the perspective of observers A and B. The dot symbol can be thought of as the point of an arrow coming towards the reader, while the cross symbol represents the feathers of an arrow going away from the reader.

FIGURE 5.13 Magnetic fields going into the page (from B) and coming out of the page (from A)

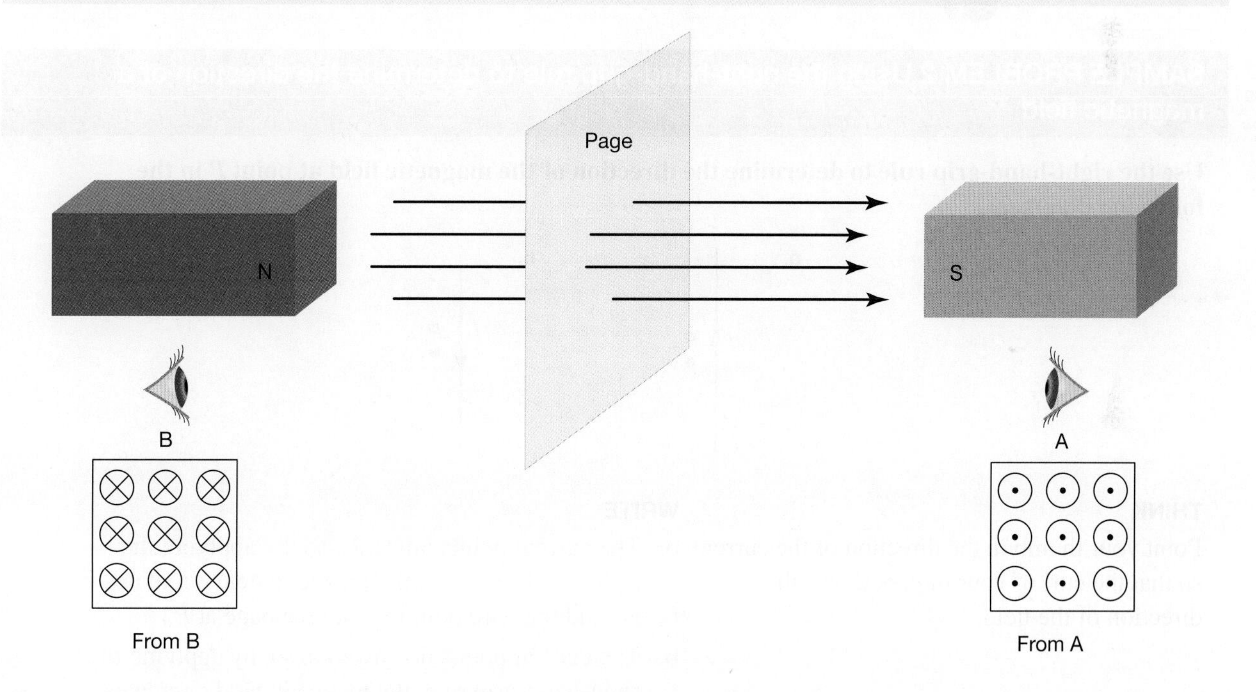

5.3.1 Fields from a current-carrying wire

Oersted's experiments show that the direction of the magnetic field around a current-carrying wire depends on the direction of the current flow in the wire. The right-hand-grip rule relates the direction of the current flow to the direction of the magnetic field from the wire.

The right-hand-grip rule: The wire carrying the current is gripped by the right hand, with the thumb pointing in the direction that conventional current (the direction a positive charge would move) flows. The fingers wrap around the wire in the direction of the magnetic field.

FIGURE 5.14 The compasses around the wire show a circular magnetic field.

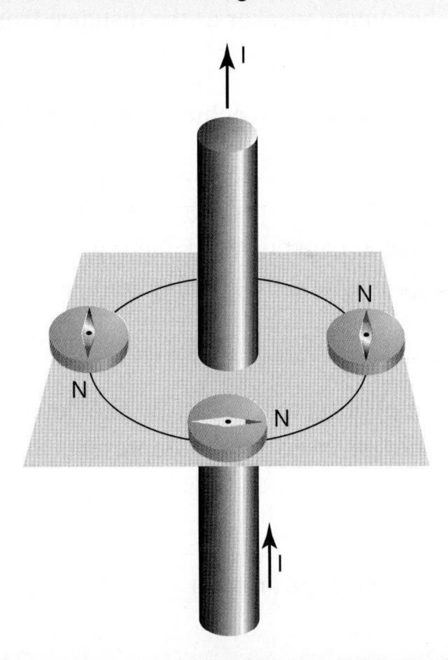

FIGURE 5.15 If a person's right hand holds the wire with the thumb pointing in the direction of the conventional current, the fingers curl around the wire in the direction of the magnetic field.

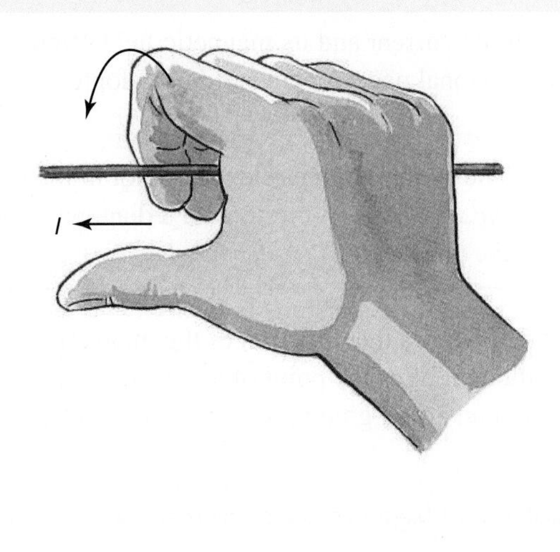

tlvd-8989

SAMPLE PROBLEM 2 Using the right-hand-grip rule to determine the direction of a magnetic field

Use the right-hand-grip rule to determine the direction of the magnetic field at point *P* in the following diagrams.

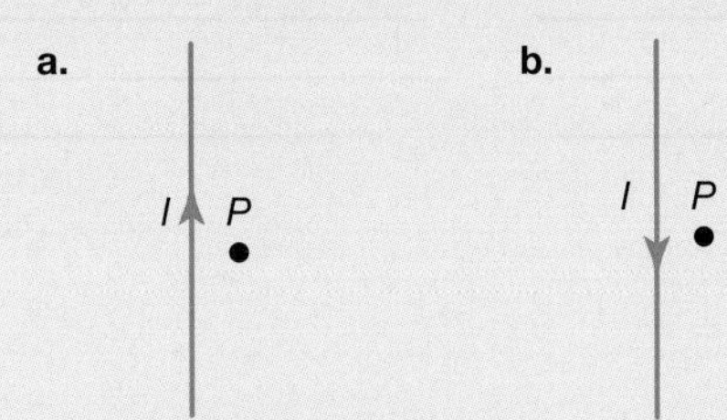

THINK	WRITE
Point your thumb in the direction of the current so that the curl of your fingers shows the direction of the field.	**a.** The current points upwards, so, by applying the right-hand-grip rule, the magnetic field circulates around the wire pointing into the page at *P*. **b.** The current points downwards, so, by applying the right-hand-grip rule, the magnetic field circulates around the wire, pointing out of the page at *P*.

PRACTICE PROBLEM 2

Use the right-hand-grip rule to determine the direction of the magnetic field at point Z in the following diagrams.

5.3.2 Fields from current-carrying loops and solenoids

solenoid coil of wire wound into a cylindrical shape

Applying the right-hand-grip rule to a loop of wire shows that the magnetic field enters at one side of the loop and exits from the other side, all the way around the loop. Joining loops together results in a **solenoid**. The magnetic fields from each loop add together to produce a stronger magnetic field.

If the loops are very close together, the field lines within the coil are parallel to the axis of the coil. Inside the coil, the field is approximately uniform. The field lines then emerge from one end of the solenoid, curve around and enter the other end of the solenoid, completing the path for the field lines. The shape of this field is similar to that of a bar magnet and is non-uniform. The ends of the solenoid can be labelled north and south. The field emerges from the north end.

FIGURE 5.16 Applying the right-hand-grip rule to each part of the loop reveals that, at all points of the loop, the magnetic field is curving in the same direction.

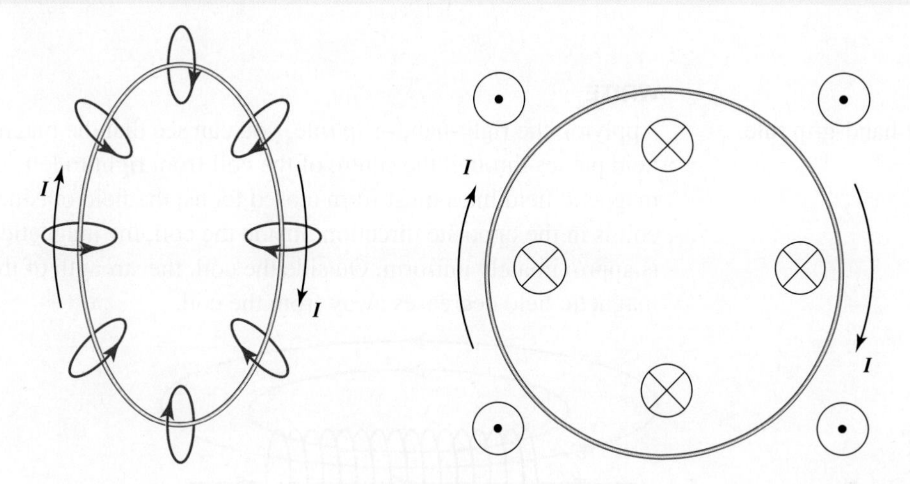

The direction of current flow in a loop can be described using words such as 'clockwise' and 'anticlockwise'. If you use these words, you also need to describe the direction that you are looking in. Looking through the solenoid from the north end, the current in the loop flows anticlockwise. However, as shown in figure 5.17, if you are looking at the solenoid from the south end, the current in the loop flows clockwise.

FIGURE 5.17 Using the right-hand-grip rule with a solenoid

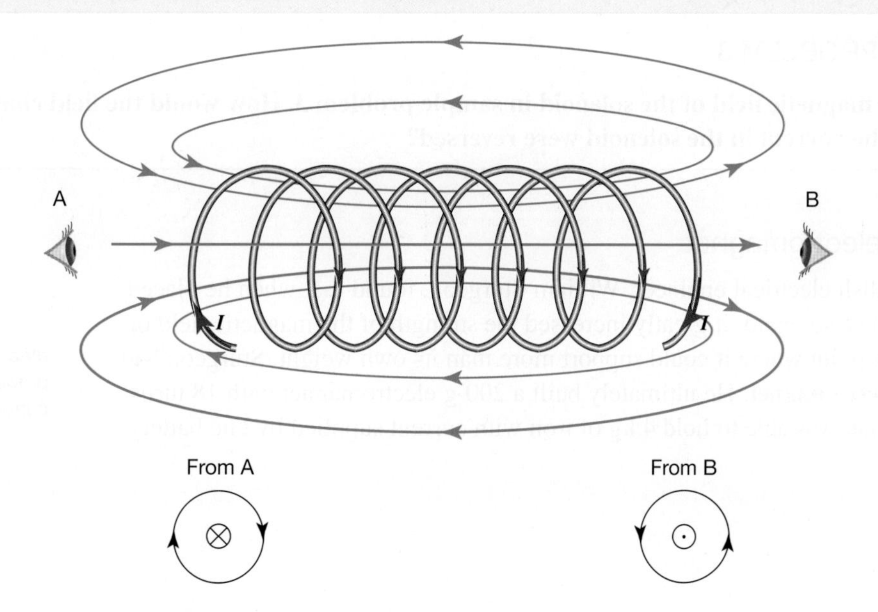

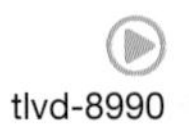
tlvd-8990

SAMPLE PROBLEM 3 Drawing magnetic field lines around a solenoid

The following solenoid has a current passing through it in the direction shown. Draw five field lines representing the resultant magnetic field.

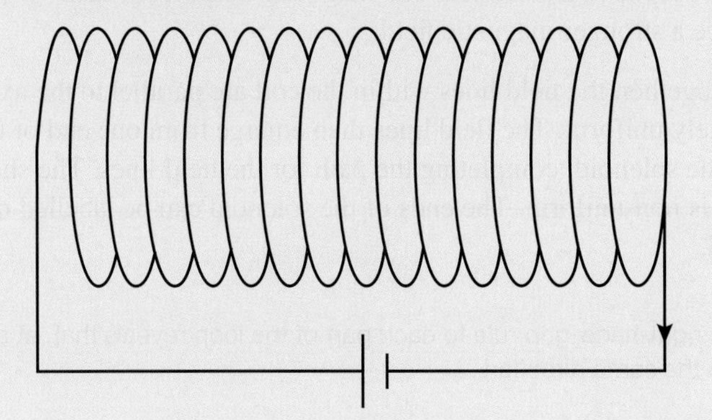

THINK	WRITE
Use the right-hand-grip rule.	Applying the right-hand-grip rule, you can see that the magnetic field passes through the centre of the coil from right to left. Because magnetic field lines must form closed loops, the field outside the coil is in the opposite direction. Inside the coil, the magnetic field is approximately uniform. Outside the coil, the strength of the magnetic field decreases away from the coil.

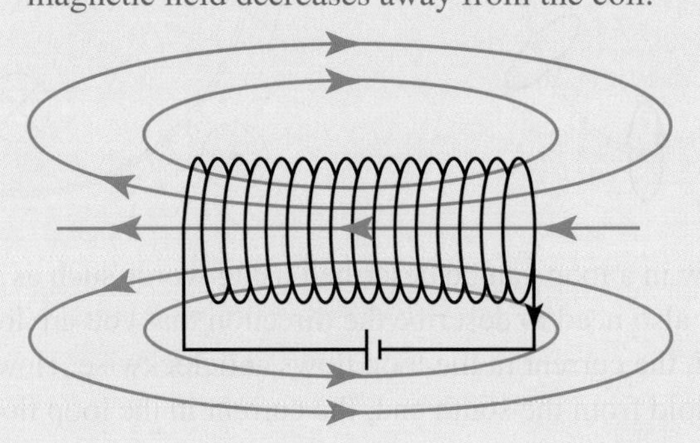

PRACTICE PROBLEM 3

Consider the magnetic field of the solenoid in sample problem 3. How would the field change if the direction of the current in the solenoid were reversed?

Creating an electromagnet

In 1823, an English electrical engineer, William Sturgeon, found that when he placed an iron rod inside a solenoid, it greatly increased the strength of the magnetic field of the device to the point where it could support more than its own weight. Sturgeon had invented the **electromagnet**. He ultimately built a 200-g electromagnet with 18 turns of copper wire that was able to hold 4 kg of iron with current supplied by one battery.

electromagnet temporary magnet produced in the presence of a current-carrying wire

By placing an iron core inside a solenoid, Sturgeon had made a magnet that could be turned on and off at the flick of a switch, and made stronger by increasing the current. His invention has many applications. In a wrecking yard, for example, electromagnets are used to separate metals containing iron from other metals.

FIGURE 5.18 Car parts being lifted by an electromagnet in a car wrecking yard

The difficulty with using iron in an electromagnet is that when the current is turned off, the iron loses its magnetism. However, by adding carbon to the iron to produce an alloy, the magnetism is not lost when the current is turned off — a permanent magnet has been made. Stronger and more long-lasting magnets are made with different combinations of elements. The common 'alnico' magnets in schools are made from iron (54%), nickel (18%), cobalt (12%), aluminium (10%) and copper (6%).

EXTENSION: Explaining magnetism in atoms

The solenoid provides a model for the magnetism in a magnet and the iron rod. The shapes of the magnetic fields around a solenoid and of a magnet are identical. The magnetic field of the solenoid is produced by a current travelling in a circle, and the magnetic field is at right angles to the plane of the circle.

If electrons are visualised as orbiting the nucleus, then, like electrons in a current-carrying loop, they produce a magnetic field. In most atoms, the paths of the electrons are randomly oriented, so their magnetic fields cancel out. However, the paths of a few electrons in an iron atom always line up. These are shielded by outer electrons, so they are not disturbed by other atoms. In this way, each iron atom can act as a little magnet.

When there is a current flowing through a solenoid with an iron core, the magnetic field lines up all the atoms in the iron core so their magnetic fields all point in the same direction. This creates a very strong field. However, when the current is turned off, the motion of the atoms rapidly produces a random rearrangement due to their temperature.

In artificial magnets (e.g. fridge magnets), other elements are added to iron to hold the iron atoms in place while they are lined up by another magnetic field so they stay lined up. This produces a permanent magnet. The crystal structure of magnetite (or the iron oxide in the magnet) forces the iron atoms to stay lined up after the magnetite is formed.

elog-1945

:lvd-10603

INVESTIGATION 5.2

online only

Making an electromagnet

AIM

To investigate the properties of the magnetic field produced by a solenoid, in particular the dependence of the magnitude and direction of the magnetic field on the current carried by the solenoid, the number of turns in the solenoid and the presence of an iron core

5.3.3 Combining magnetic fields

Like the gravitational field and electric field, the magnetic field is also a vector field. When more than one magnetic source is present, the total magnetic field is the vector sum of the original fields. By drawing the vectors of each of the original fields head to tail, the direction of the resultant field can be determined by drawing a vector from the start of the first vector to the end of the second vector. Pythagoras's theorem may be used to determine the magnitude of this vector if a right-angled triangle is formed.

FIGURE 5.19 The magnetic field from two bar magnets

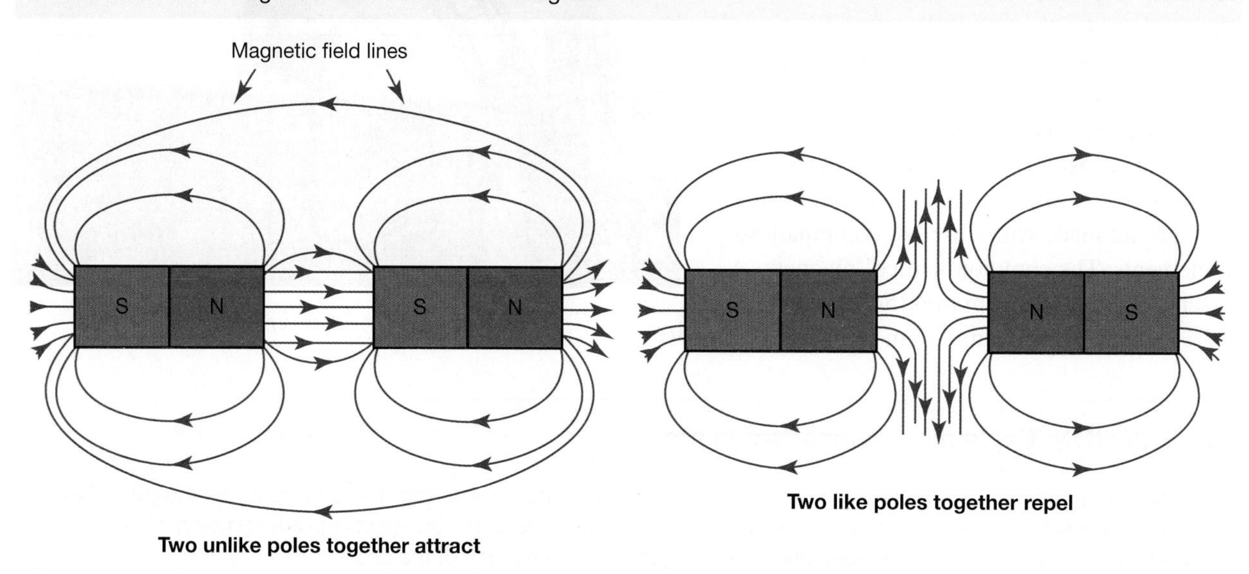

The effect of the Earth's magnetic field can be seen in the orientation of the compass needles in figure 5.21, particularly in those to the left and right of the wire. The compass needles are not perfectly perpendicular to the wire, but slightly tilted to the left-hand side of the page, indicating that the Earth's magnetic field points towards the left-hand side of the page in this experiment.

FIGURE 5.20 The magnetic field from a binary star system in which each star has a magnetic field

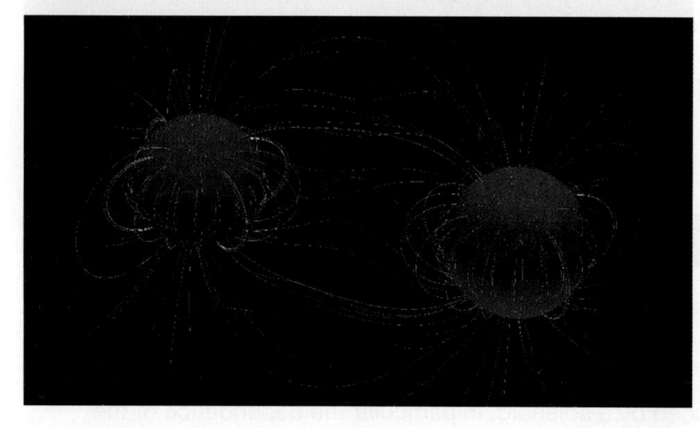

FIGURE 5.21 Magnetic field measurements around a current-carrying wire

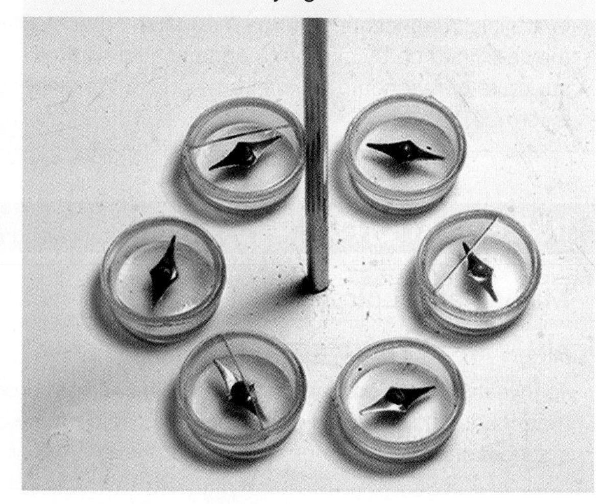

tlvd-8991

SAMPLE PROBLEM 4 Determining the strength and direction of the magnetic field around a solenoid

The magnetic field at point X from the current-carrying solenoid shown in the following illustration is 25 μT. The Earth's magnetic field is 25 μT.

a. **What direction would a compass point if it were placed at point X?**

b. **Calculate the strength of the resultant magnetic field at X.**

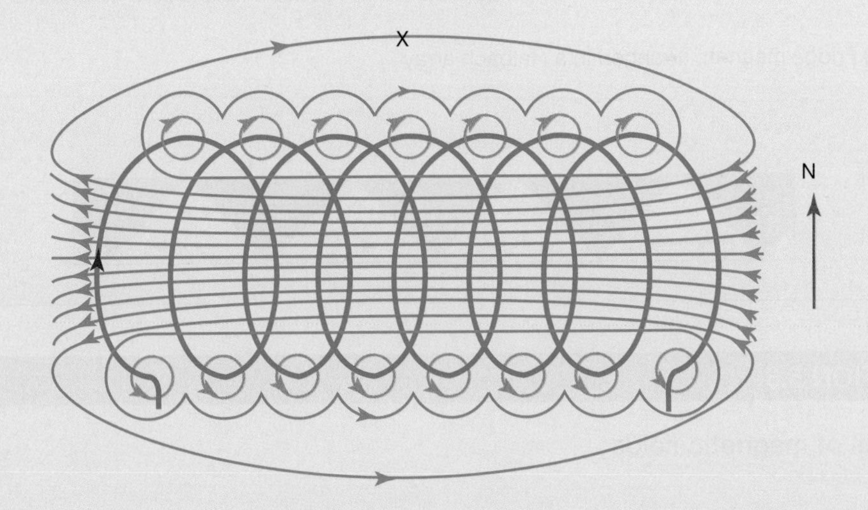

THINK

a. At point X, the compass would point along the direction of the sum of the magnetic fields of the solenoid and the Earth. The Earth's magnetic field is pointed up and, at point X, the magnetic field in the solenoid is to the right.

WRITE

a. The combined vector sum is

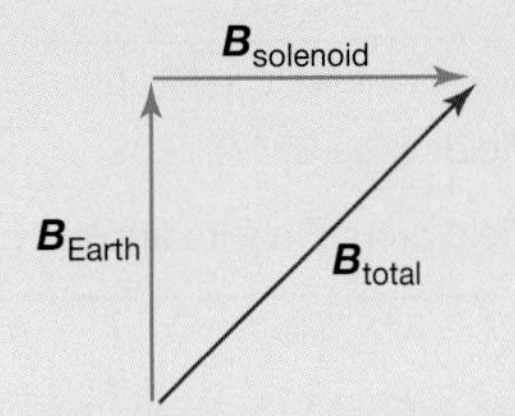

The diagram shows that the resultant would point in a north-east direction.

THINK

b. The strength can be calculated using Pythagoras's theorem.

WRITE

b. $B = \sqrt{\left(B^2_{\text{solenoid}} + B^2_{\text{Earth}}\right)}$
$= \sqrt{(25^2 + 25^2)}\ \mu\text{T}$
$= 35\ \mu\text{T}$

PRACTICE PROBLEM 4

A neodymium iron boron magnet is placed 1.0 cm to the left of a magnetic probe (located at point X). The field strength due to the magnet is 0.01 T. A second identical magnet is also placed 1.0 cm below a magnetic probe, but at right angles to the first magnet.

a. **What is the resultant strength of the magnetic field at point X?**

b. **What are the possible directions of the magnetic field at point X?**

EXTENSION: The curious case of the fridge magnets

You may have noticed that fridge magnets are magnetic on one side only, which seems to contradict our understanding of magnetic field lines. This is because a fridge magnet is a composition of magnets arranged in a Halbach array. The superposition of the Halbach array of alternating horizontal and vertical magnetic fields results in a magnetic field on one side of the array only. During manufacture, a rubber sheet embedded with ferromagnetic particles is exposed to an alternating magnetic field so that the sheet acquires a magnetic field that alternates vertically and horizontally, as shown in figure 5.22. The resultant field is zero on one side of the sheet.

FIGURE 5.22 Fridge magnets arranged in a Halbach array

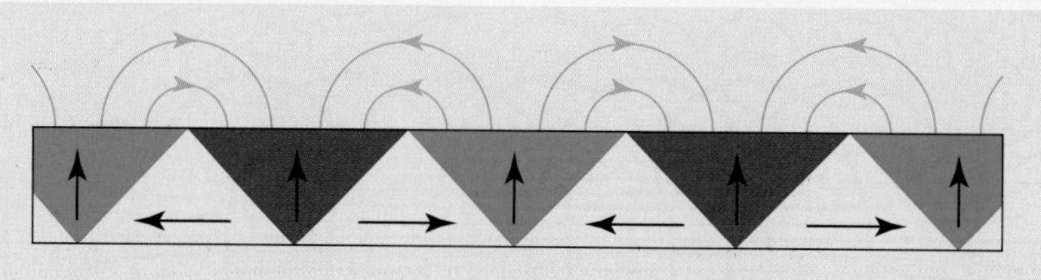

INVESTIGATION 5.3

online only

Vector model of magnetic fields

AIM

To observe the effect of Earth's magnetic field upon the magnetic field of a current-carrying wire

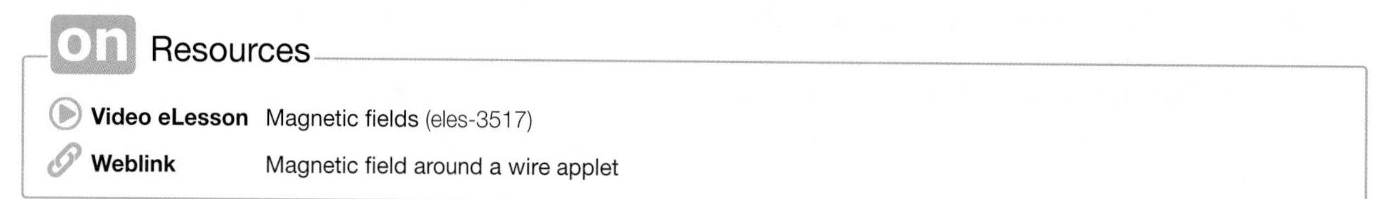

on Resources

Video eLesson Magnetic fields (eles-3517)

Weblink Magnetic field around a wire applet

5.3 Activities

learnon

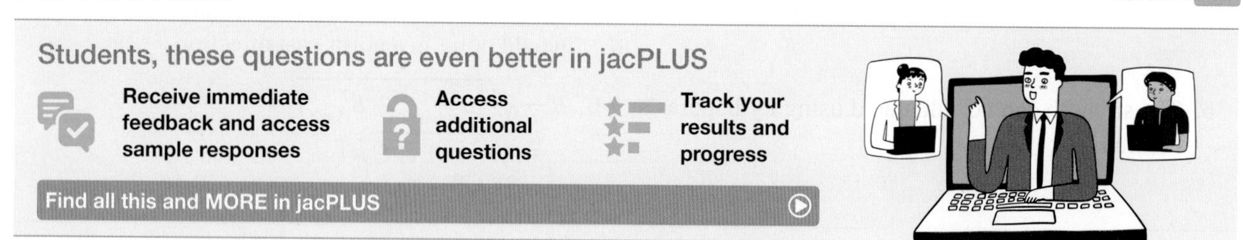

5.3 Quick quiz on | 5.3 Exercise | 5.3 Exam questions

5.3 Exercise

1. Use the right-hand-grip rule to determine the direction of the magnetic field at point X in the following diagrams.

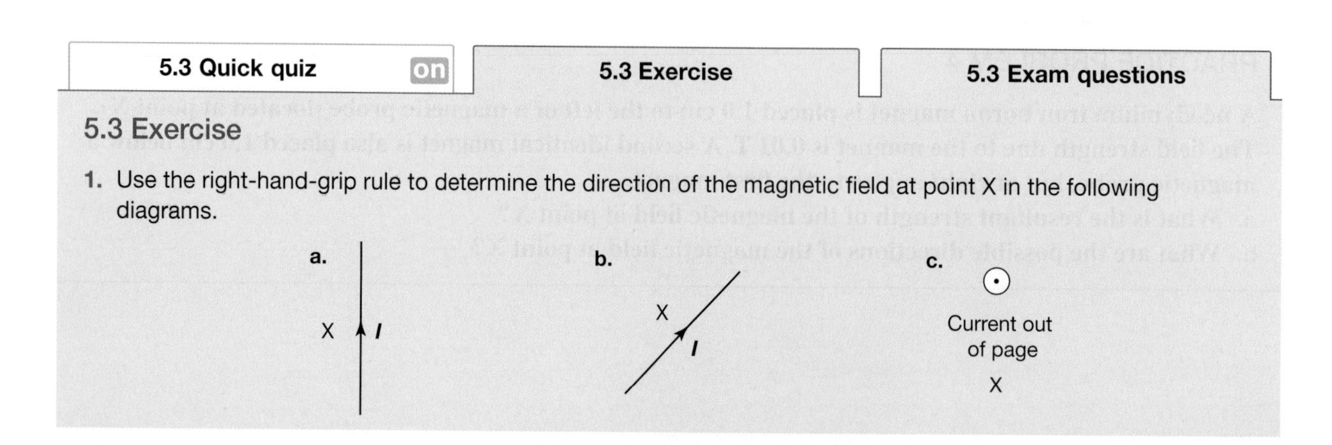

2. Copy the following diagrams and use the right-hand-grip rule and the direction of the magnetic field at X to determine the direction of the current in the wire in each case.

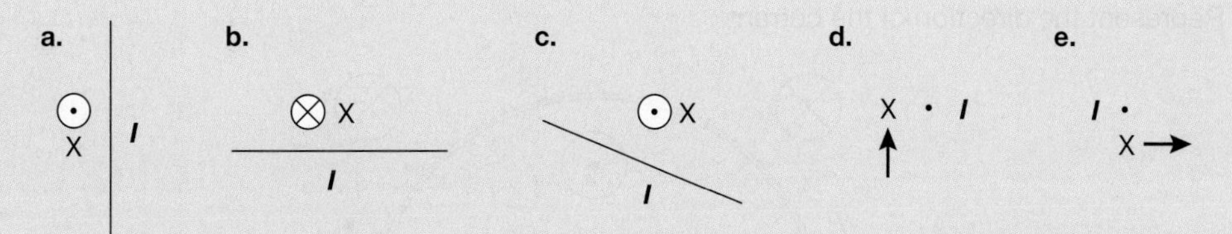

3. Use the right-hand-grip rule to determine the direction of the magnetic field at W, X, Y and Z in the following diagrams. Figure (a) represents a circular loop of wire with a current and figure (b) represents a solenoid.

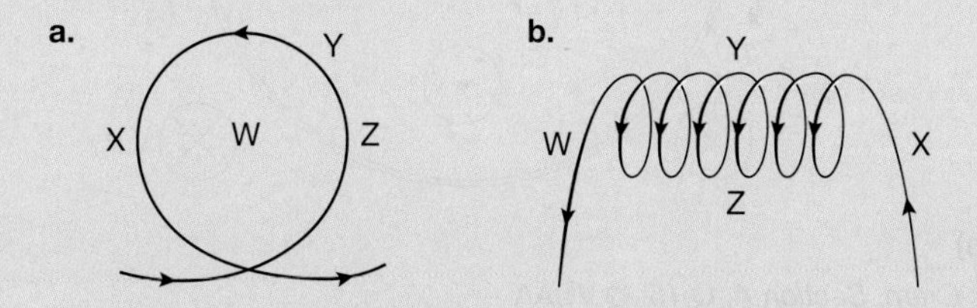

4. When current is connected to a solenoid that is wrapped around two iron rods that have been placed in a line, the two rods move apart. Explain why this happens.
5. In Oersted's experiment, the compass needle initially points north–south. What would happen if the current in the wire above the needle ran:
   a. west–east
   b. east–west?

5.3 Exam questions

Question 1 (1 mark)

Source: *VCE 2022 Physics Exam, Section A, Q.1; © VCAA*

MC A single loop of wire carries a current, I, as shown in the diagram below.

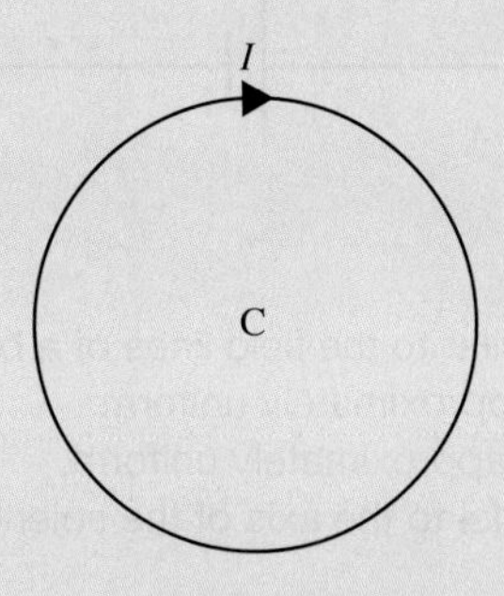

Which one of the following best describes the direction of the magnetic field at the centre of the circle, C, which is produced by the current carrying wire?

A. to the left
B. to the right
C. into the page
D. out of the page

Question 2 (1 mark)

Consider the loop of current-carrying wire represented with the direction of the magnetic field produced by the current I. Represent the direction of the current.

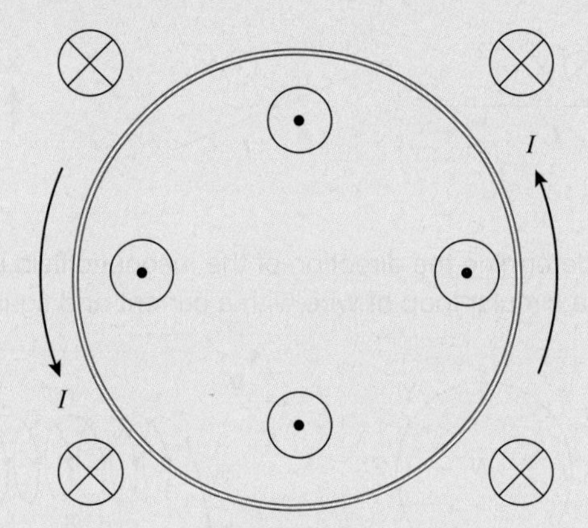

Question 3 (2 marks)

Source: *VCE 2015, Physics Exam, Section A, Q.15;* © VCAA

The diagram below shows a solenoid.

Draw five lines with arrows to show the magnetic field of the solenoid.

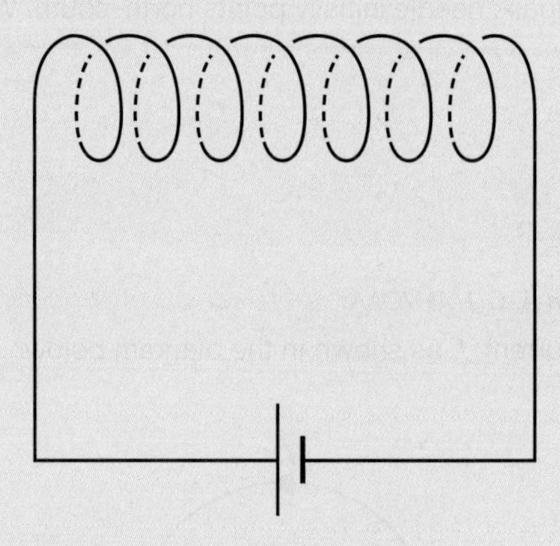

Question 4 (1 mark)

MC Which of the following is *incorrect*?

A. The field lines outside a solenoid are similar to the field lines of a bar magnet.
B. The magnetic field inside a solenoid is approximately uniform.
C. The magnetic field outside a solenoid is approximately uniform.
D. The field lines inside a solenoid are parallel to the axis of the solenoid.

Question 5 (3 marks)

A student has come up with the following mnemonic illustration, using the direction of the current and the letter S for south and N for north, to identify the nature of the face of a solenoid when looking at it.

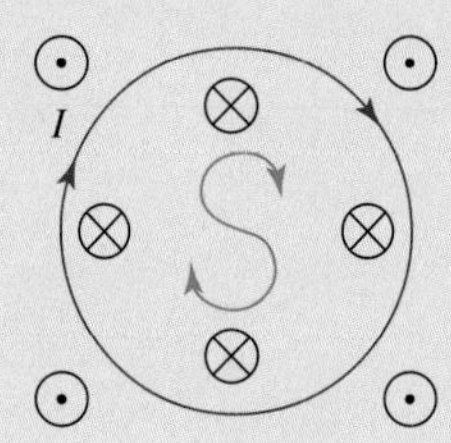

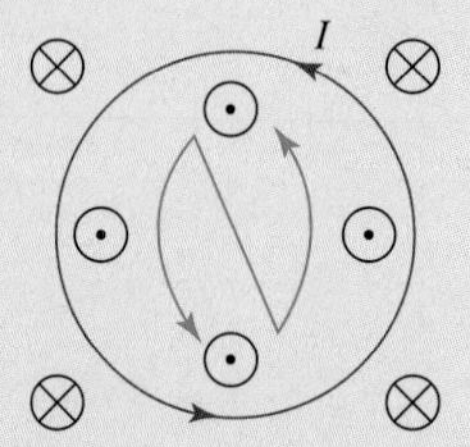

Discuss whether this is a correct tool to identify the north and south face of a solenoid.

More exam questions are available in your learnON title.

5.4 Using magnetic fields to control charged particles, cyclotrons and mass spectrometers

KEY KNOWLEDGE

- Analyse the use of a uniform magnetic field to change the path of a charged particle, including:
  - the magnitude and direction of the force applied to an electron beam by a magnetic field: $\boldsymbol{F} = q\boldsymbol{vB}$, in cases where the directions of $\boldsymbol{v}$ and $\boldsymbol{B}$ are perpendicular or parallel
  - the radius of the path followed by an electron in a magnetic field: $q\boldsymbol{vB} = \frac{mv^2}{r}$ where $v \ll c$
- Model the acceleration of particles in a particle accelerator (limited to linear acceleration by a uniform electric field and direction change by a uniform magnetic field)

We have seen that moving charges produce magnetic fields. So, can moving charges be affected by other magnetic fields? The answer is yes!

5.4.1 Forces on moving charges in magnetic fields

A charged particle moving in a magnetic field experiences a force that is at right angles to *both* its velocity and the magnetic field. This behaviour was initially identified by J.J. Thomson in the late 1800s and described, using an equation, by Hendrik Lorentz in 1895.

The magnitude of the magnetic force of a moving particle for a charge, q, moving with a velocity, $\boldsymbol{v}$, perpendicular to a magnetic field, $\boldsymbol{B}$, is:

$$F = qvB$$

where: F is the magnitude of the force on the particle, in newtons (N)

q is the charge of the particle, in coulombs (C)

v is the component of the velocity of the particle that is perpendicular to the magnetic field, in m s^{-1}

B is the strength of the magnetic field, in tesla (T)

The direction of the force on the charged particle can be determined using a helpful hand rule. There are two alternative hand rules commonly used: the right-hand-slap and left-hand rules. The rules apply both to moving single positive charges and to currents. Either rule will give you the same result for the direction of the force, so you can choose the one which works best for you. It is important to note that both the right-hand-slap and left-hand rules assume a positive charge and positive current. If the particle has a negative charge, or the direction of the current is reversed, then the direction of the force on the particle or current is reversed.

Left-hand rule

The **left-hand rule** applies as follows:

- the index finger, pointing straight ahead, represents the magnetic field (***B***)
- the middle finger, at right angles to the index finger, represents the current (*I*) or particle velocity (***v***) that is perpendicular to the magnetic field
- the thumb, upright at right angles to both fingers, represents the force (***F***).

Lock the three fingers in place so they are at right angles to one another. Now rotate your hand so that the field and current (or particle velocity) line up with the directions in your problem. The thumb will now point in the direction of the force.

FIGURE 5.23 Left-hand rule for determining the direction of the magnetic force of a magnetic field on a current or moving positive charge

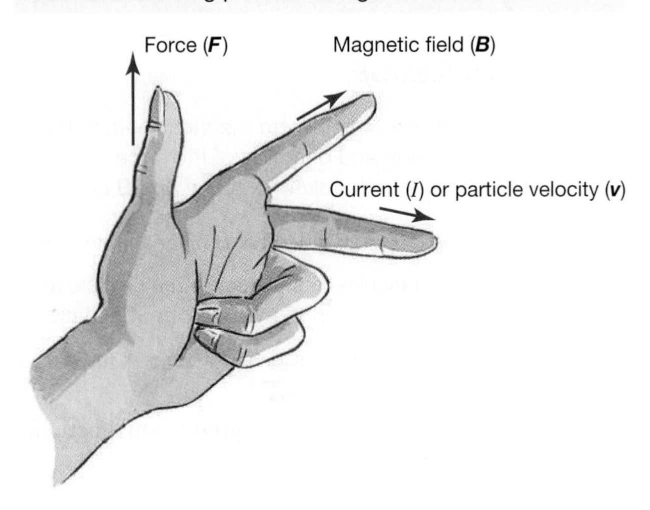

Right-hand-slap rule

The **right-hand-slap rule** applies as follows:

- the fingers (out straight) represent the magnetic field (***B***)
- the thumb (out to the side of the hand) represents the current (*I*) or particle velocity (***v***) that is perpendicular to the magnetic field
- the palm of the hand represents the force (***F***).

Hold your hand flat with the fingers outstretched and the thumb out to the side, at right angles to your fingers. Now rotate your hand so that the field and current or particle velocity line up with the direction in your problem. The palm of your hand now gives the direction of the force, hence the name.

FIGURE 5.24 Right-hand-slap rule for determining the direction of the magnetic force of a magnetic field on a current or moving positive charge

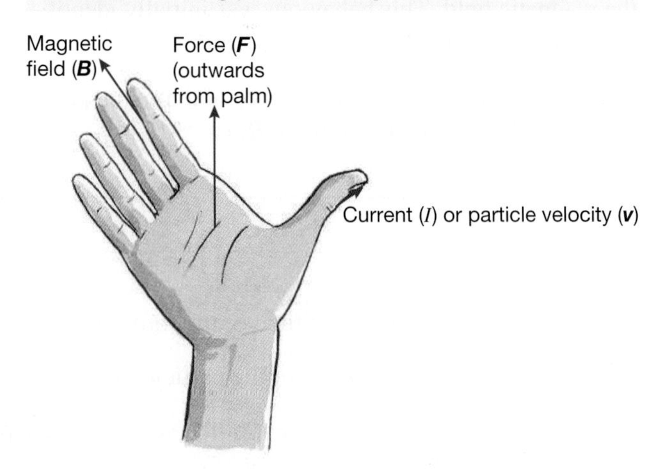

5.4.2 Charged particle moving parallel to the magnetic field

When the particle's velocity is parallel to the magnetic field, the magnetic force (***F***) is zero. Because there is no net force, the particle continues to move in a straight line and the velocity will remain constant.

Imagine a magnetic field that is pointing out of the page. If we had an electron moving either out of or into the page, this velocity would be parallel to the field. There would be no force on the electron and it would not change its motion.

left-hand rule rule used to determine the direction of the magnetic force of a magnetic field on a current or moving positive charge
right-hand-slap rule rule used to determine the direction of the magnetic force of a magnetic field on a current or moving positive charge

These different scenarios can be summarised in table 5.1:

TABLE 5.1 Summarising the effect on charged particles with different velocities

$\boldsymbol{B}$, $+q$, $\boldsymbol{v} = 0$	If velocity is equal to 0, $\boldsymbol{F} = q\boldsymbol{vB}$ is equal to 0. Therefore the motionless charged particles in a magnetic field experience no force.
$\boldsymbol{B}$, $+q$, $\boldsymbol{v}$, $\theta = 0$	If the velocity is parallel to the field lines, charged particles in a magnetic field experience no force and will continue moving at the same velocity.
$\boldsymbol{B}$, $+q$, $\boldsymbol{v}$, θ	If velocity is perpendicular to the field lines (90°), then there will be a force, the direction of which can be determined by the left-hand or right-hand-slap rule. This allows for the magnetic force to be at its maximal value.
$\boldsymbol{B}$, $+q$, $\boldsymbol{v}$, θ	If velocity is at an angle between 0 and 90°, there will be a force that will be greater than 0 N, but less than the force if the velocity was perpendicular to the magnetic field.

tlvd-8992

SAMPLE PROBLEM 5 Calculating the magnitude and direction of a magnetic force on an electron

An electron moving towards the right of the page at 1.5×10^5 m s^{-1} enters a uniform 0.060 T magnetic field that is coming out of the page.

a. Calculate the magnitude of the magnetic force on the electron.

b. Determine the direction of the force on the electron.

THINK	WRITE
a. Recall the formula $F = qvB$, where $q = -1.6 \times 10^{19}$ C, $v = 1.5 \times 10^5$ m s^{-1} and $B = 0.060$ T (as we are being asked only for magnitude of force here, vector notation is not required for F).	a. $F = qvB$ $= -1.6 \times 10^{-19} \times 1.5 \times 10^5 \times 0.060$ $= 1.4 \times 10^{-15}$ N

b. Apply the right-hand-slap rule and reverse the direction of the current, as the electron has a negative charge. The electron moves to the right of the page, so the conventional current flows to the left of the page.

b. 1. v is to the left and will be along the thumb.
2. B is out of the page and follows the direction of the fingers.
3. The force on the electron is in the direction of the palm, namely pointing up the page.

PRACTICE PROBLEM 5

An alpha particle (charge +2e) moving towards the right of the page at 1.0×10^7 m s^{-1} enters a uniform 0.060 T magnetic field pointing down the page.

a. **Calculate the magnitude of the magnetic force on the alpha particle.**
b. **Determine the direction of the force.**
c. **Calculate the magnitude of the magnetic field on a positron (charge +e) moving up the page at 3.0×10^7 m s^{-1} in the same magnetic field.**

on Resources

Digital document Hand-rules (doc-18541)

5.4.3 Motion of charged particles in magnetic fields

We have seen that satellites move in circular orbits because they constantly experience a net force towards the centre of their orbit that is perpendicular to their velocity. Similarly, the force on charged particles moving at right angles to a magnetic field causes them to constantly change the direction of their motion so that they follow a circular path perpendicular to the magnetic field.

The magnitude of the force on the moving charged particle due to the magnetic field is $F = qvB$. The magnetic force is always at right angles to the direction of the charge's motion, which means that the magnetic force does no work on the charge and it cannot increase the speed of the charge — it can only change its direction at a constant rate.

The mass spectrometer, the electron microscope and the synchrotron pass charged particles through a magnetic field causing the charged particles to follow a circular path.

So what is the radius of the circle? How does it depend on the strength of the magnetic field, the speed of the charge and size of the charge?

TABLE 5.2 Observations concerning the strength of magnetic fields, and speed and size of the charge

What is observed?	What does the formula predict?	Do the observations match the formula's prediction?
If the charge is stationary, the current is zero, so there is no force.	If $v = 0$, then $F = 0$.	Yes
A stronger magnetic field will produce a larger force and deflect the charge more.	Force is proportional to the field.	Yes

The net force on the charged particle as it moves in the magnetic field is $\boldsymbol{F}_{\text{net}} = m\boldsymbol{a}$. When the only significant force is the magnetic force:

$$\begin{aligned} \boldsymbol{F} &= q\boldsymbol{v}\boldsymbol{B} \\ \Rightarrow q\boldsymbol{v}\boldsymbol{B} &= m\boldsymbol{a} \end{aligned}$$

The acceleration of an object moving in a circle is $a = \frac{v^2}{r}$, where r is the radius of the circular motion. Combining this with the net force on the particle, we see that:

$$qvB = \frac{mv^2}{r}$$

where: q is the charge on the particle, in C

v is the component of the velocity of the particle perpendicular to the magnetic field, in m s^{-1}

B is the strength of the magnetic field, in T

m is the mass of the particle, in kg

r is the radius of the path followed by the particle, in m

The expression for the radius is therefore:

$$r = \frac{mv}{Bq}$$

Does this relationship make sense?

TABLE 5.3 Observations concerning the radius of circular motion

What is observed?	What does the formula predict?	Do the observations match the formula's prediction?
Hard to turn heavy objects	The heavier the mass, the larger the radius	Yes
Hard to turn fast objects	The faster the object, the larger the radius	Yes
The larger the force, the smaller the radius	The stronger the field, the smaller the radius; the larger the charge, the smaller the radius	Yes

tlvd-8993

SAMPLE PROBLEM 6 Calculating the radius of the circular motion of charged particles in a magnetic field

An electron travelling at 5.9 × 10⁶ m s⁻¹ enters a magnetic field in a direction perpendicular to its motion with a magnitude of 6.0 mT. What is the radius of its path in the magnetic field?

THINK	WRITE
Recall the expression for the radius that comes from equating the net force on objects moving in a circle with the magnetic force. Recall the mass of an electron is 9.1×10^{-31} kg and the charge of an electron is 1.6×10^{-19} C.	$r = \frac{mv}{qB}$ $= \frac{9.1 \times 10^{-31} \times 5.9 \times 10^{6}}{1.6 \times 10^{-19} \times 6.0 \times 10^{-3}}$ $= 5.6 \times 10^{-3}$ m

PRACTICE PROBLEM 6

Calculate the speed of an electron moving perpendicular to a 3.0 mT magnetic field in a circle with radius 1.0 mm.

EXTENSION: Aurorae

The aurorae at the North Pole and South Pole are glorious displays of waves of coloured light, high in the atmosphere. They are produced when charged particles ejected by the Sun enter Earth's magnetic field.

The particles spiral down to the poles, producing an amazing display of light as they move in smaller and smaller circles from the increasing magnetic field, exciting mainly oxygen and nitrogen molecules in the atmosphere in high-speed collisions.

FIGURE 5.25 Charged particles entering Earth's magnetic field

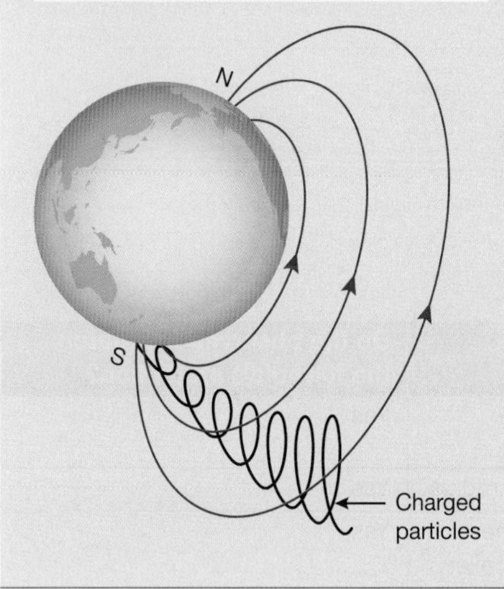

FIGURE 5.26 Aurora Australis, seen from the International Space Station

elog-1947

INVESTIGATION 5.4

online only

Electrons in a magnetic field

AIM

To use a cathode-ray oscilloscope (CRO) to display variations in voltage applied to the input terminals

Resources

Interactivity Electrons in a magnetic field (int-0124)

5.4.2 Controlling charged particle motion with magnetic fields

Magnetic fields can be used to control charged particles in a number of important applications.

The mass spectrometer is able to separate ions with different ratios of mass to charge because, as shown previously, the radius of the trajectory of an ion passing through a magnetic field depends on the ratio of mass to charge. The sample is initially vapourised and ionised, accelerated by an electric field and then passed through a magnetic field, which deflects the ions. The radius of this deflection depends on the mass to charge ratio of the particle being examined. The mass spectrometer is an essential analytical tool for determining the composition of a chemical sample and is often used to determine the mass of various atomic particles, measuring the radius of their deflection in a magnetic field.

FIGURE 5.27 A mass spectrometer

FIGURE 5.28 Mass spectrometers deflect particles based on their mass to charge ratios. When the magnetic field is pointing into the page, lighter positive ions move through a circle of smaller radius than heavier positive ions, causing them to separate and to arrive at different points on the detector.

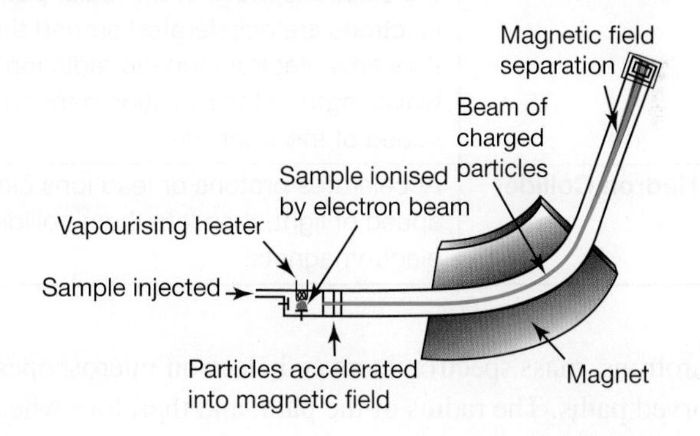

In the electron microscope, magnetic fields focus the electron beam after it is scattered from a sample, in the same way that optical lenses focus the light scattered from a sample under an ordinary microscope.

FIGURE 5.29 Electron microscopes use magnetic fields to focus the electron beam, after it is scattered when passing through a specimen, allowing an image to be observed.

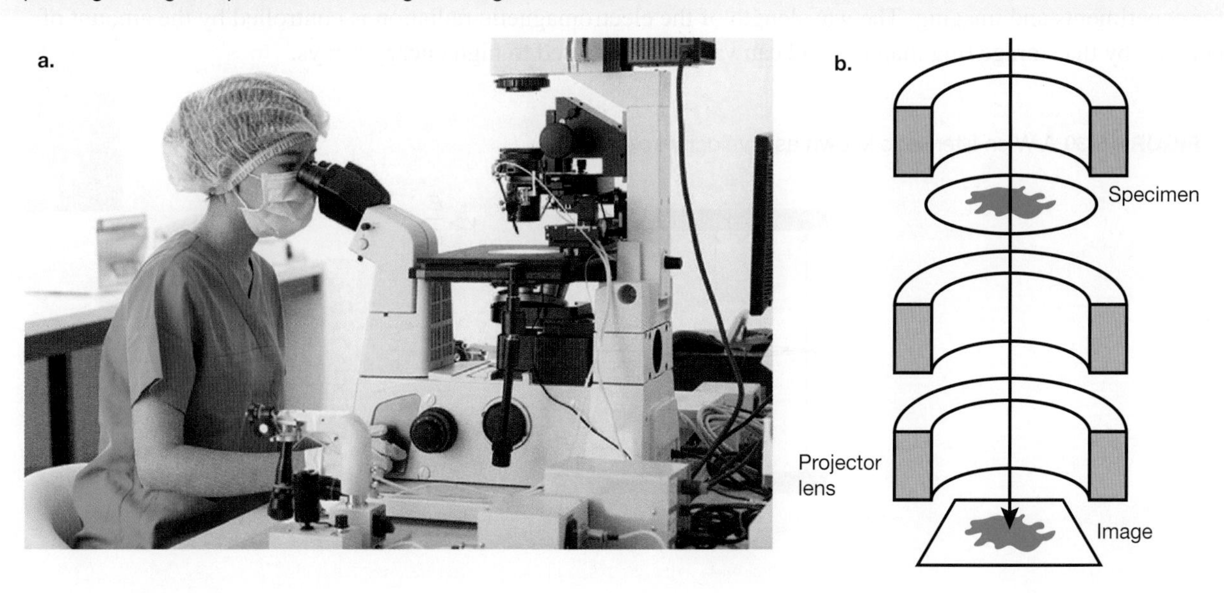

5.4.3 Combining electric and magnetic fields

Many devices relies on combining magnetic and electric fields to achieve their function. Examples are given in Table 5.4.

TABLE 5.4 Devices that use electric fields and magnetic fields

Device	Operation	Purpose
Mass spectrometer	Accelerates positive ions of different mass, using an electric field. Uses a magnetic field to cause ions of different masses to deflect by different amounts.	To measure the abundance of different elements and isotopes in a sample
Electron microscope	Accelerates electrons, which are scattered by a sample and then passed through electric and magnetic lenses to form an image of the sample.	To use an electron beam to examine very small objects
Synchrotron	Accelerates electrons, using an electric field, close to the speed of light, then feeds them into a storage ring. Magnetic fields make the electrons move in a circular path. As the electrons are accelerated around the loop, they emit electromagnetic radiation. The wavelength of the radiation depends on the speed of the electrons.	To produce intense and very narrow beams of mainly X-rays to examine the fine structure of substances such as proteins
Large Hadron Collider	Accelerates protons or lead ions close to the speed of light, then lets them collide, using electromagnets.	To test the predictions of theories of particle physics, for example, the existence of the Higgs boson

In synchrotrons, mass spectrometers and electron microscopes, magnetic fields cause charged particles to move along curved paths. The radius of the path, and therefore where the particle ends up, depends on the particle velocity, so it can be important to filter out particles that are not moving with the desired velocity.

At the Australian Synchrotron, electrons are first accelerated by an electric field and injected at high speeds ($0.999997c$) into a booster ring. In the booster ring, a combination of increasing electric and magnetic fields is used to further accelerate the electrons, increasing their kinetic energy by a factor of thirty. The electrons are then passed to the storage ring, which uses magnetic fields to keep the electrons moving around a circle. The accelerated electrons emit high intensity, coherent, and polarized electromagnetic radiation, which is used for experiments and imaging. The wavelength of the electromagnetic radiation is controlled by the amount of bending by the storage ring magnets and can vary from infrared to high energy x-rays.

FIGURE 5.30 A Wien filter (also known as a velocity selector)

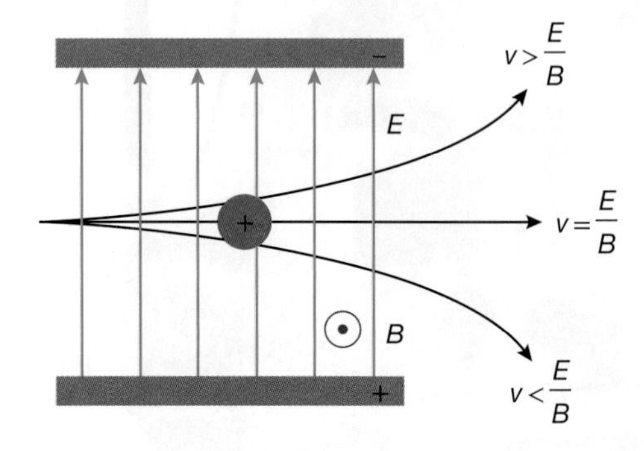

In 1898, Wilhelm Wien (after whom Wien's Law in thermodynamics is named) was investigating the charged particles that are produced when electricity is passed through gases. To investigate their speed and their charge, he set up a magnetic field to deflect the beam of charged particles in one direction, and an electric field to deflect the beam in the opposite direction. For the charged particles that were undeflected, the magnetic force must have been balanced by the electric force. This configuration is now called a **Wien filter**.

Wien filter device that can be used as a velocity filter using a magnetic field to deflect the beam of charged particles in one direction, and an electric field to deflect the beam in the opposite direction

The electric force on a charge in an electric field is $\boldsymbol{F} = q\boldsymbol{E}$, and the magnetic force on a moving charge is $\boldsymbol{F} = qv\boldsymbol{B}$. If the magnetic and electric forces are balanced, then:

$$q\boldsymbol{E} = qv\boldsymbol{B}$$

and cancelling q gives:

$$v = \frac{E}{B}$$

where: v is the velocity of an undeflected particle, in m s^{-1}

E is the strength of the electric field, in N C^{-1}

B is the strength of the magnetic field, in T

This formula implies that by controlling the strength of the magnetic and electric fields, one can choose which particles will be undeflected by the fields. Particles travelling faster or slower than the undeflected particles will be deflected out of the beam. This means that a Wien filter can essentially be used to filter out particles that are travelling faster or slower than a desired velocity and select only those particles travelling at the desired velocity. For this reason, a Wien filter is often referred to as a **velocity selector**.

velocity selector device that can be used as a velocity filter using a magnetic field to deflect the beam of charged particles in one direction, and an electric field to deflect the beam in the opposite direction

tlvd-8996

SAMPLE PROBLEM 7 Calculating the velocity of an electron travelling through a Wien filter

An electron travels undeflected through a Wien filter, where the magnetic field is 0.0030 T. The electric field strength is 500 N C^{-1}. Calculate the velocity of the electron.

THINK	WRITE
For the undeflected electron, the magnetic and electric forces are balanced. Use the formula $v = \frac{E}{B}$ to calculate the velocity of an electron.	$v = \frac{500}{0.0030}$ $= 1.7 \times 10^5$ m s^{-1}

PRACTICE PROBLEM 7

An electron beam requires electrons moving at 6 × 10^6 m s^{-1}. A Wien filter, or velocity selector, is set up with a magnetic field of 0.02 T. How large must the electric field be to select electrons travelling at 6 × 10^6 m s^{-1}?

5.4 Quick quiz  | 5.4 Exercise | 5.4 Exam questions

5.4 Exercise

1. A proton travelling vertically at 3.0×10^5 m s^{-1} enters a uniform horizontal magnetic field with a field strength of 4.0 mT. Calculate the magnitude of the force experienced by the proton.
2. An alpha particle enters a uniform magnetic field of 4.0 mT with a speed of 1.5×10^7 m s^{-1} as shown in the diagram.

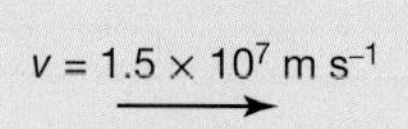

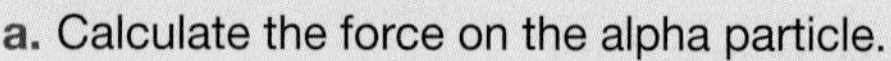

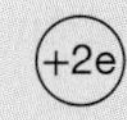

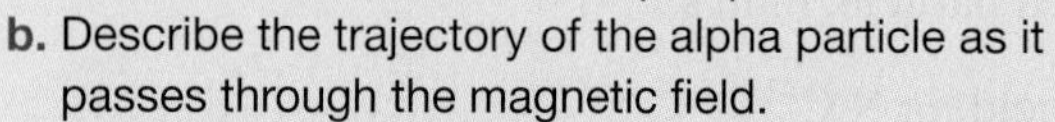

   a. Calculate the force on the alpha particle.
   b. Describe the trajectory of the alpha particle as it passes through the magnetic field.
3. An electron travelling east at 1.2×10^5 m s^{-1} enters a region of uniform magnetic field of strength 2.4 T pointing north.
   a. Calculate the magnitude of the magnetic force acting on the electron.
   b. Describe the path taken by the electron, giving a reason for your answer.
   c. Calculate the magnitude of the acceleration of the electron.
4. In the diagram, a tauon (tau particle) enters the magnetic field from the left and executes the trajectory shown.

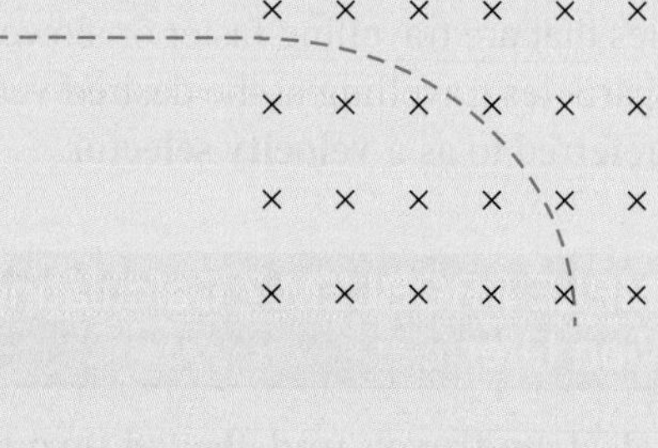

   a. Does the tauon carry a positive or negative charge? Give reasons for your choice.
   b. The tauon exits the magnetic field at the point shown. Describe its trajectory after it leaves the field.
5. Determine the direction of the magnetic force in the following situations, using your preferred hand rule. Use the following terminology in your answers: up the page, down the page, left, right, into the page, out of the page.
   a. Magnetic field into the page, electron entering from left

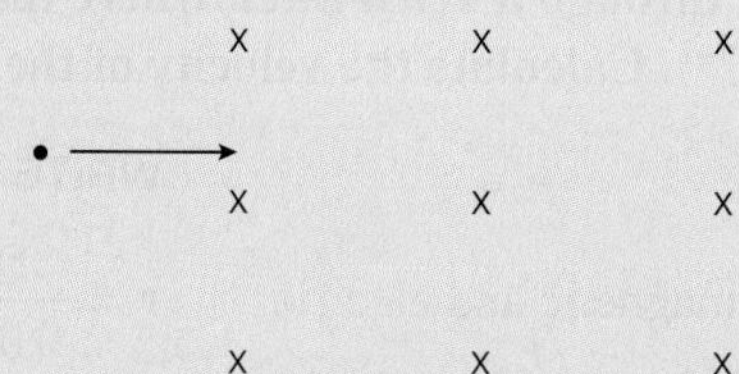

   b. Magnetic field down the page, electron entering from left

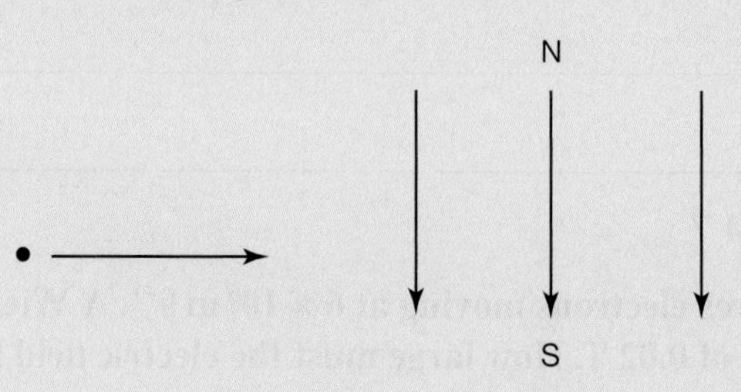

c. Magnetic field out of the page, proton entering obliquely from left

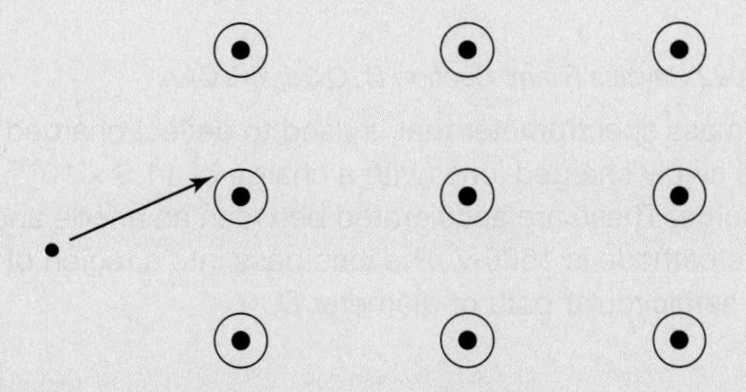

6. How could a moving electron remain undeflected in a magnetic field?
7. An ion beam consisting of three different types of charged particles is directed eastwards into a region having a uniform magnetic field, B, directed out of the page. The particles making up the beam are (i) an electron, (ii) a proton and (iii) a helium nucleus or alpha particle. Copy the following diagram and draw the paths that the electron, proton and helium nucleus could take.

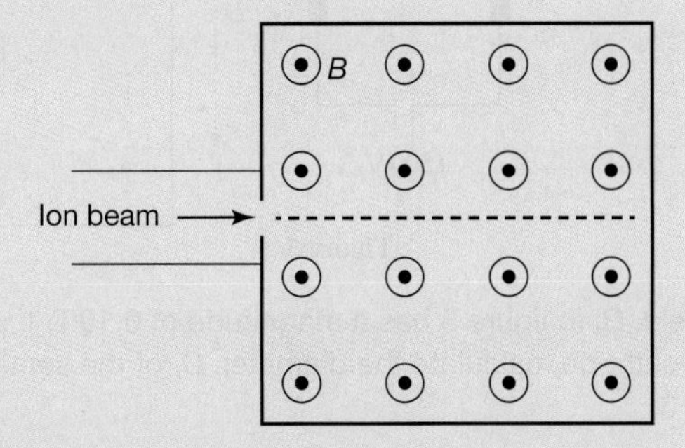

8. Calculate the radius of curvature of the following particles travelling at 10% of the speed of light in a magnetic field of 4.0 T.
   a. an electron
   b. a proton
   c. a helium nucleus.
9. Explain how magnetic fields are used in the functioning of mass spectrometers.
10. A positron enters the magnetic field shown here, with a velocity in the direction indicated.

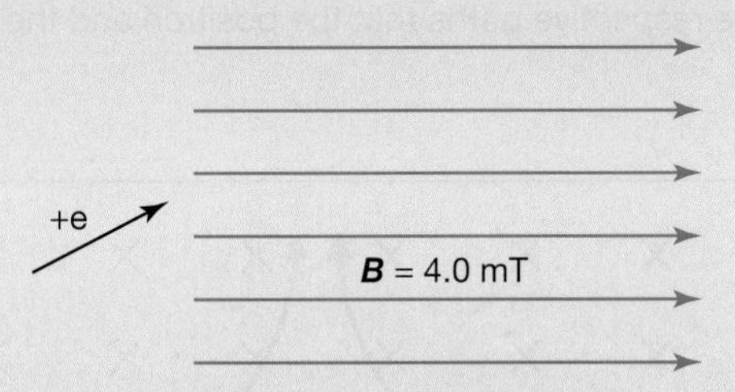

    Describe the trajectory of the positron as it traverses the magnetic field.
11. An electron beam requires electrons moving at 8.0×10^6 m s^{-1}. A velocity selector is set up with an electric field of 5000 V m^{-1}. How large must the magnetic field be in order to select electrons travelling at 8.0×10^6 m s^{-1} perpendicular to the field?
12. In a mass spectrometer, the sample is vapourised and ionised, resulting in ions with a random distribution of velocities that are accelerated by a uniform electric field. The ions are then passed through a velocity selector with a magnetic field strength of 500 mT and an electric field provided by an electrical potential difference of 1000 V across conducting plates separated by 1.5 mm. Calculate the velocity of ions that pass through undeflected.

5.4 Exam questions

Question 1 (3 marks)

Source: *Adapted from VCE 2022 Physics Exam, Section B, Q.3d; © VCAA*

A schematic diagram of a mass spectrometer that is used to deflect charged particles to determine their mass is shown in Figure 3. Positive singly charged ions (with a charge of $+1.6 \times 10^{-19}$ C and a mass of 4.80×10^{-27} kg) are produced at the ion source. These are accelerated between an anode and a cathode. The potential difference between the anode and the cathode is 1500 V. The ions pass into a region of uniform magnetic field, *B*, and are directed by the field into a semicircular path of diameter *D*.

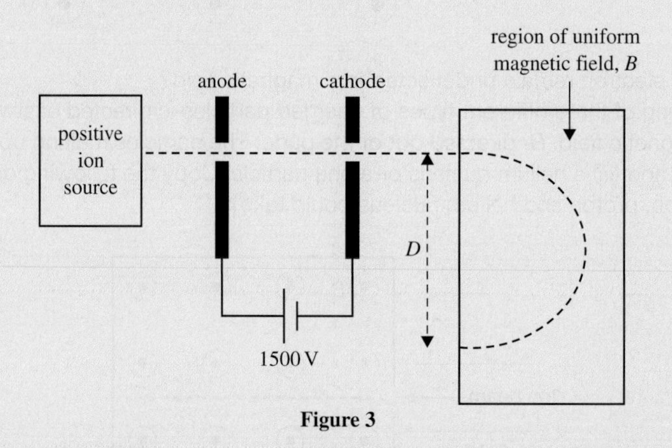

Figure 3

The region of uniform magnetic field, *B*, in figure 3 has a magnitude of 0.10 T. If each ion has a speed of 3.16×10^5 m s^{-1} when it exits the cathode, calculate the diameter, *D*, of the semicircular path followed by ions within the magnetic field in figure 3.

Question 2 (4 marks)

Source: *VCE 2022 Physics Exam, NHT, Section B, Q.3; © VCAA*

A positron and an electron are fired one at a time into a strong uniform magnetic field in an evacuated chamber. They are fired at the same speed but from opposite sides of the chamber. Their initial velocities are initially perpendicular to the magnetic field and opposite in direction to each other, as shown in Figure 1. A positron has the same mass as an electron (9.1×10^{-31} kg) and has the same magnitude of electric charge as an electron (-1.6×10^{-19} C) but is positively charged ($+1.6 \times 10^{-19}$ C).

On Figure 1, sketch and label the respective paths that the positron and the electron will take while in the uniform magnetic field.

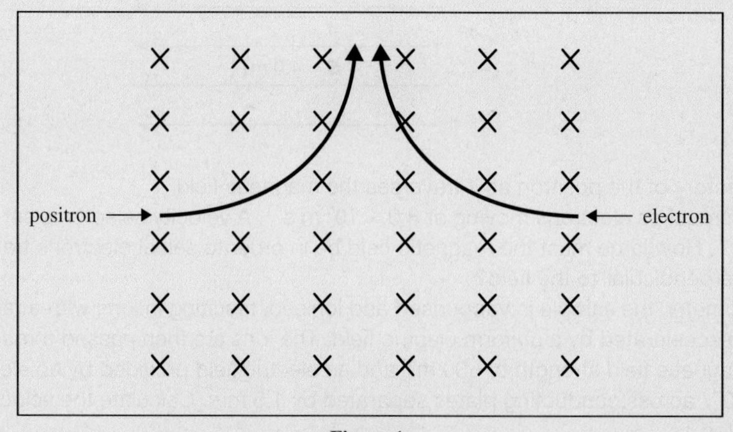

Figure 1

Question 3 (1 mark)

Source: *VCE 2022 Physics Exam, NHT, Section A, Q.2;* © VCAA

MC A loudspeaker consists of a current carrying coil within a radial magnetic field, as shown in the diagram below. The direction of the current in the coil is also shown.

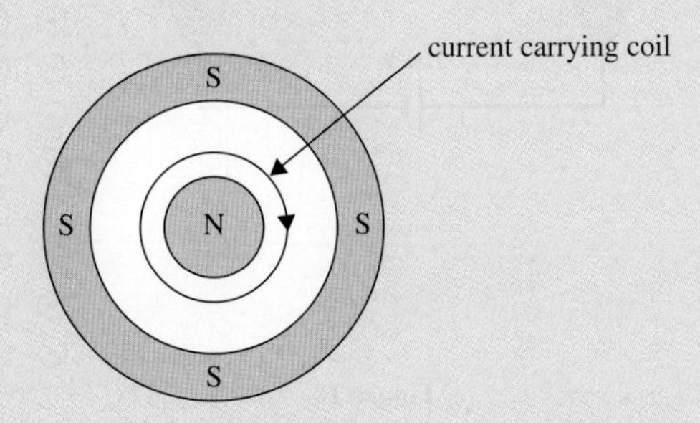

Which one of the following best describes the direction of the force on the coil?

A. out of the page
B. down the page
C. into the page
D. up the page

Question 4 (3 marks)

Source: *VCE 2019, Physics Exam, Section B, Q.1;* © VCAA

A particle of mass m and charge q travelling at velocity v enters a uniform magnetic field B, as shown in Figure 1.

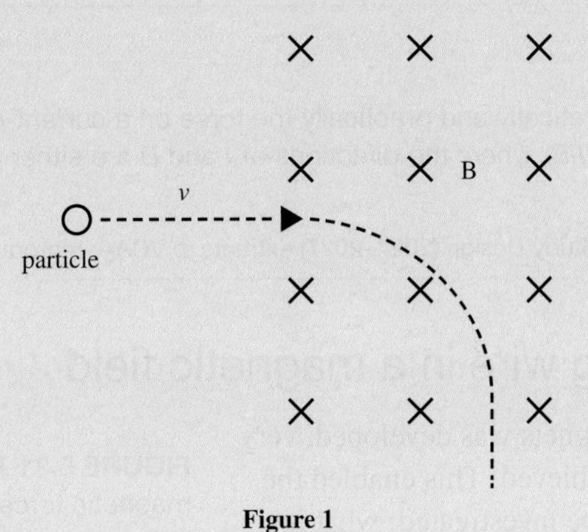

Figure 1

a. Is the charge q positive or negative? Give a reason for your answer. **(1 mark)**
b. Explain why the path of the particle is an arc of a circle while the particle is in the magnetic field. **(2 marks)**

Question 5 (2 marks)

Source: *Adapted from VCE 2018, Physics Exam, Section B, Q.1c;* © VCAA

An electric field accelerates a proton between two plates. The proton exits into a region of uniform magnetic field at right angles to its path, directed out of the page, as shown in Figure 1.

Data

mass of proton	1.7×10^{-27} kg
charge on proton	$+1.6 \times 10^{-19}$ C
accelerating voltage	10 kV
distance between plates	20 cm
strength of magnetic field	2.0×10^{-2} T

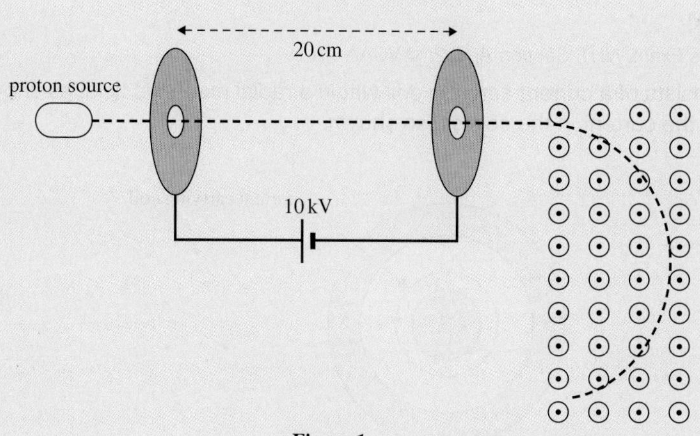

Figure 1

With a different accelerating voltage, the proton exits the electric field at a speed of 1.0×10^6 m s^{-1}.

Calculate the radius of the path of this proton in the magnetic field.

More exam questions are available in your learnON title.

5.5 Magnetic forces on current-carrying wires

KEY KNOWLEDGE

- Investigate and analyse theoretically and practically the force on a current-carrying conductor due to an external magnetic field, $F = nIlB$, where the directions of I and B are either perpendicular or parallel to each other

Source: Adapted from VCE Physics Study Design (2024–2027) extracts © VCAA; reproduced by permission.

5.5.1 Current-carrying wire in a magnetic field

Once the technology of electromagnets was developed, very strong magnetic fields could be achieved. This enabled the reverse of Oersted's discovery to be investigated: what is the effect of a magnetic field on a current in a wire?

Observations of the force experienced by a wire carrying a current in a magnetic field show that the force increases if

- the magnetic field increases
- the current in the wire increases
- the length of the wire increases
- there is more than one wire carrying current in the same direction.

It is also observed that only the component of the magnetic field perpendicular to the current causes there to be a force on the wire.

FIGURE 5.31 The magnetic field exerts a magnetic force on a current-carrying wire.

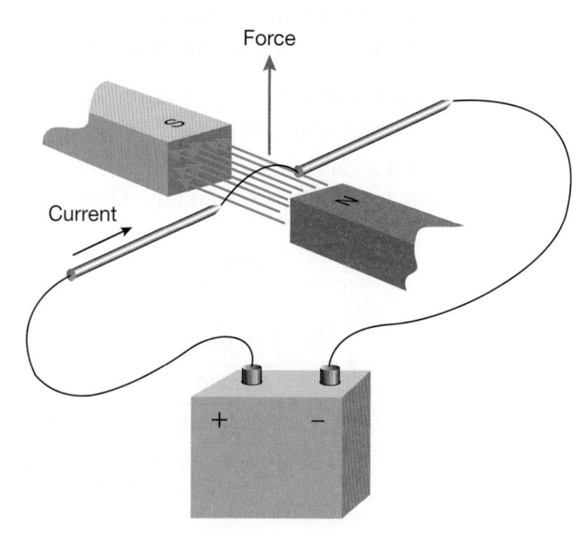

$$F = nIlB$$

where: F is the force on the conductor perpendicular to the magnetic field, in N

n is the number of conductors (or loops)

I is the current in the conductor, in A

l is the length of the conductor, in m

B is the component of the magnetic field that is perpendicular to the current, in T

When the magnetic field is not perpendicular to the direction of the current, it is important to remember that the force on the wire is less. In fact, if the magnetic field is parallel to the direction of the current, the force on the wire is zero. That is because there is no component of magnetic field perpendicular to the current.

Note that the same result can also be obtained starting from Lorentz's expression for the magnetic force on a charged particle.

EXTENSION: Using Lorentz's expression

The current, I, is the motion of charges along a length of wire, l. If a charge q travels a distance l in a time Δt, then the velocity of the charge is $v = \frac{l}{\Delta t}$. Rewriting the force in terms of the current and length of wire, the following is obtained:

$$F = qvB = q\frac{l}{\Delta t}B = \frac{q}{\Delta t}lB = IlB$$

tlvd-8994

SAMPLE PROBLEM 8 Calculating the magnitude force on a current-carrying wire

If a straight wire of length 8.0 cm carries a current of 300 mA, calculate the magnitude of the force acting on it when it is in a magnetic field of strength 0.25 T if:

a. the wire is at right angles to the field

b. the wire is parallel with the field.

THINK	WRITE
a. Recall that $F = nIlB$ and that B is the component of the field that is perpendicular to the length of the wire, l. Ensure your units are correct. In this case, you need to convert mA to A and cm to m.	a. $F = nIlB$ $= 1 \times 300 \times 10^{-3} \times 0.080 \times 0.25$ $= 6.0 \times 10^{-3}$ N
b. The wire is parallel to the field so there is no component of the field perpendicular to the wire.	b. The force on the wire is 0 N.

PRACTICE PROBLEM 8

a. Calculate the force on a 100-m length of wire carrying a current of 250 A when the strength of Earth's magnetic field at right angles to the wire is 5.00×10^{-5} T.

b. The force on a 10-cm wire carrying a current of 15 A when placed in a magnetic field perpendicular to B has a maximum value of 3.5 N. What is the strength of the magnetic field?

Forces between current-carrying parallel wires

Two current-carrying wires each have their own magnetic field, the direction of which is determined using the right-hand grip rule. When these wires are brought together in close proximity and placed parallel to each other, the wires are observed to either repel each other or be attracted to each other, depending on the direction of current flow in each wire. This can be explained by understanding that each wire experiences a force on the current that it is carrying due to the magnetic field that is being produced by the other wire. If the current in the left-hand wire is up the page, the magnetic field experienced by the right-hand wire is into the page. If the current in the right-hand wire is also up the page, then applying either the left-hand rule or right-hand-slap rule shows that the magnetic force on the right-hand wire is towards the left-hand wire.

Similarly, the magnetic force on the left-hand wire due to the magnetic field from the right-hand wire acting on the current in the left-hand wire is towards the right-hand wire. The two wires attract one another!

If, however, the current in the right-hand wire is down the page, then the right-hand wire experiences a magnetic force away from the left-hand wire and the left-hand wire experiences a magnetic force away from the right-hand wire. Hence, parallel wires carrying currents in opposite directions repel each other. The effects can be observed in figure 5.32.

FIGURE 5.32 Magnetic fields interact between parallel wires, leading to either **a.** attraction when the current runs in the same direction or **b.** repulsion when the current runs in opposite directions.

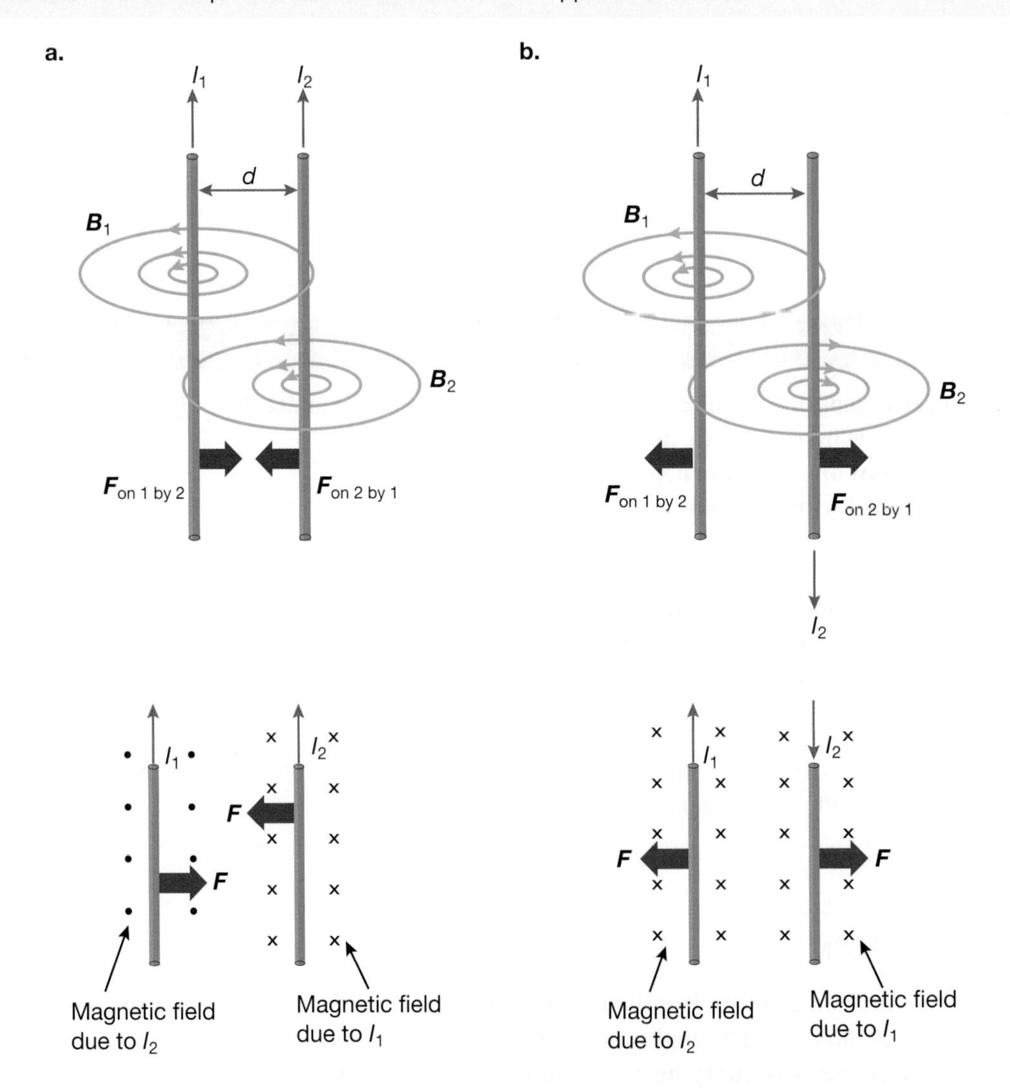

elog-1948

vd-10604

INVESTIGATION 5.5

online only

Measuring the force on a current-carrying wire in a magnetic field

AIM

To investigate the relationship $F = IlB$ and determine strength of the magnetic field experienced by the wire loop

on Resources

Interactivity Changing magnetic force (int-0115)

Weblink Lorentz force applet

5.5 Activities

learnon

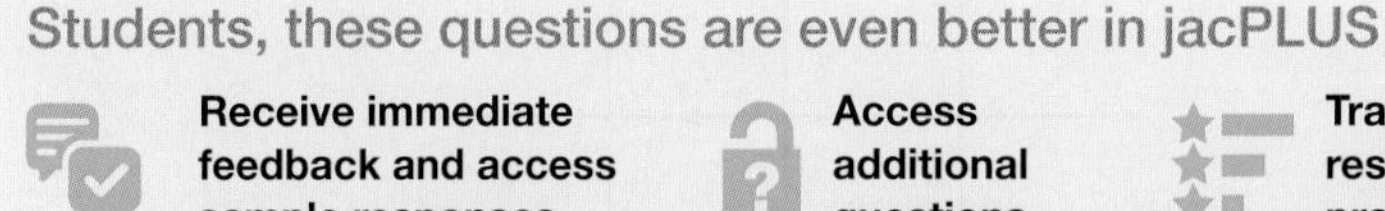
Students, these questions are even better in jacPLUS

Receive immediate feedback and access sample responses

Access additional questions

Track your results and progress

Find all this and MORE in jacPLUS

5.5 Quick quiz on	5.5 Exercise	5.5 Exam questions

5.5 Exercise

1. Calculate the size of the force on a wire of length 0.05 m in a magnetic field of strength 0.30 T if the wire is at right angles to the field and it carries a current of 4.5 A.
2. Calculate the size of the force on an 8.0-cm wire carrying a current of 1.8 A at an angle parallel to a magnetic field with strength 40 mT.
3. Calculate the size of the force exerted on a loudspeaker coil of radius 1.5 cm and 500 turns that carries a current of 15 mA in a radial magnetic field of 2.0 T. (*Hint:* Consider what aspect of the circle takes the place of *l* in this question.)
4. Use the answer key provided to indicate the direction of the force of the magnetic field on the current-carrying wire in the following diagrams (a) to (h).

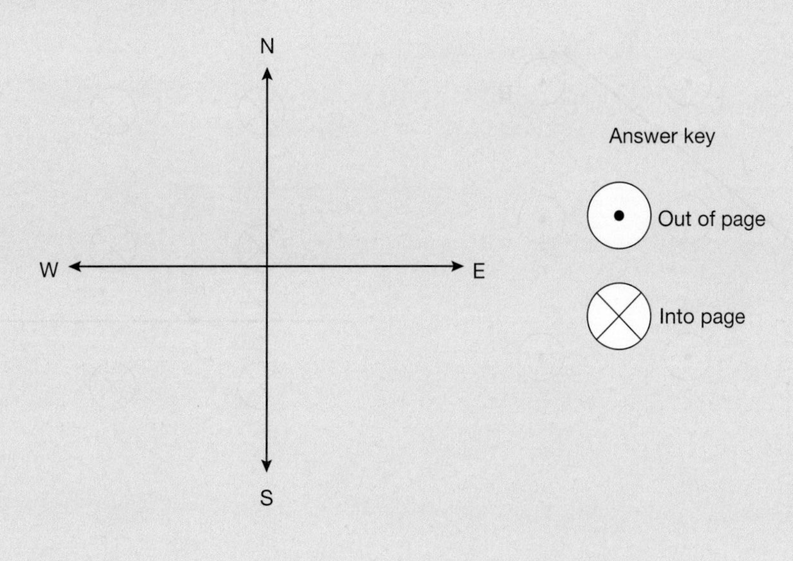

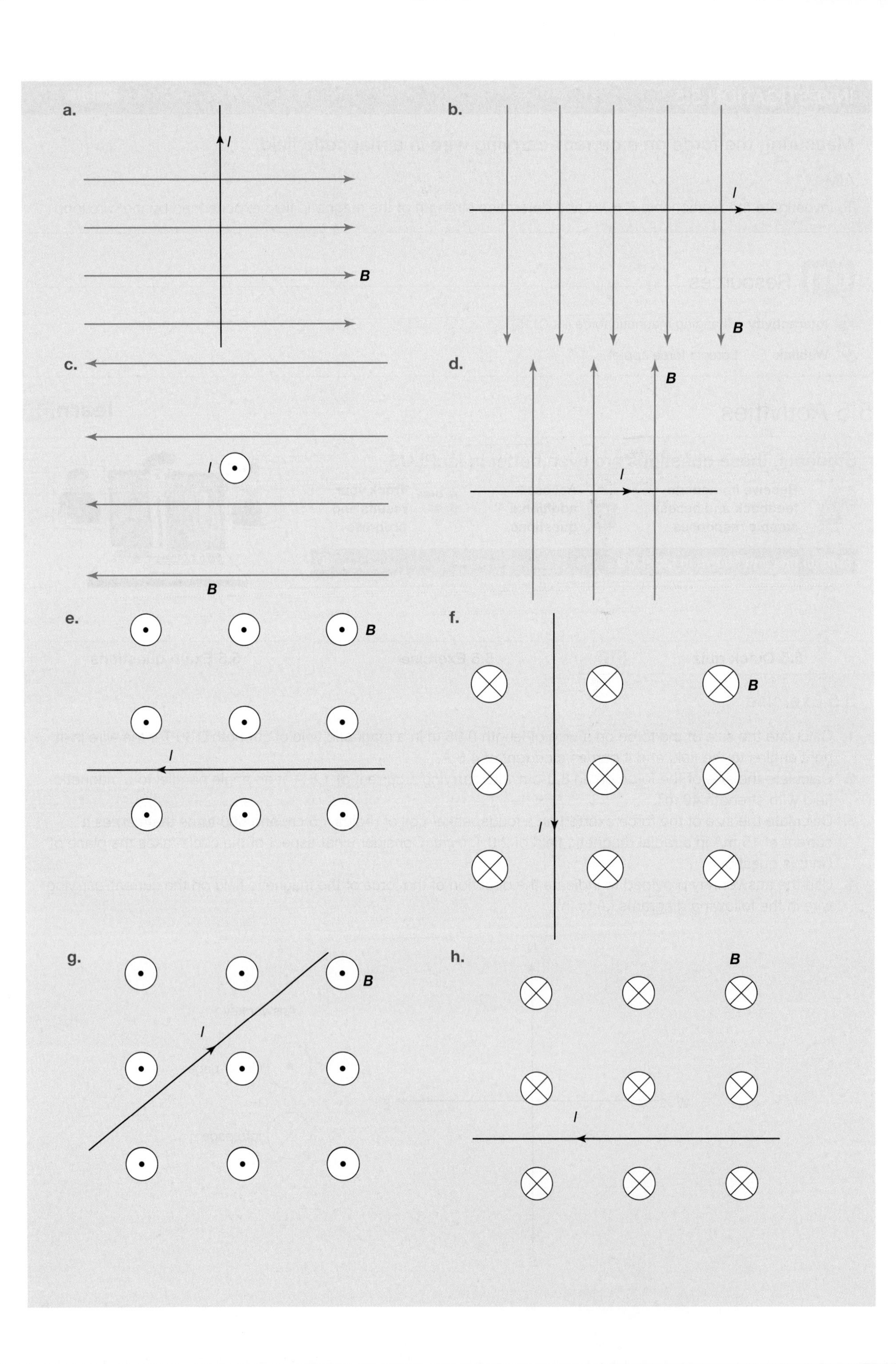

a.
I
B
b.
I
B
c.
I
B
d.
B
I
e.
B
I
f.
B
I
g.
B
I
h.
B
I

5. Wires A and B are parallel to each other and carry current in the same direction.
   a. Draw a diagram to represent this situation, and determine the direction of the magnetic field at B due to wire A.
   b. This magnetic force will act on the current in wire B. What is the direction of the force by wire A on wire B?
   c. Now draw a diagram and determine the direction of the magnetic field at A due to wire B and the direction of the force by wire B on wire A.
   d. Is the answer to (c) what you expected? Why? (*Hint:* Consider Newton's laws of motion.)

5.5 Exam questions

Question 1 (1 mark)

Source: *VCE 2019 Physics Exam, NHT, Section A, Q.2;* © VCAA

MC A powerline carries a current of 1000 A DC in the direction east to west. At the point of measurement, Earth's magnetic field is horizontally north and its strength is 5.0×10^{-5} T.

Which one of the following best gives the direction of the electromagnetic force on the powerline?

A. horizontally west
B. horizontally north
C. vertically upwards
D. vertically downwards

Question 2 (1 mark)

Source: *VCE 2019 Physics Exam, NHT, Section A, Q.3;* © VCAA

MC A powerline carries a current of 1000 A DC in the direction east to west. At the point of measurement, Earth's magnetic field is horizontally north and its strength is 5.0×10^{-5} T.

The magnitude of the force on **each metre** of the powerline is best given by

A. 5.0×10^{3} N
B. 5.0×10^{2} N
C. 5.0×10^{-2} N
D. 5.0×10^{-5} N

Question 3 (3 marks)

Source: *Adapted from VCE 2019, Physics Exam, Section B, Q.3c&d;* © VCAA

Figure 3 shows a schematic diagram of a DC motor. The motor has a coil, JKLM, consisting of 100 turns. The permanent magnets provide a uniform magnetic field of 0.45 T. The commutator connectors, X and Y, provide a constant DC current, I, to the coil. The length of the side JK is 5.0 cm.

The current I flows in the direction shown in the diagram.

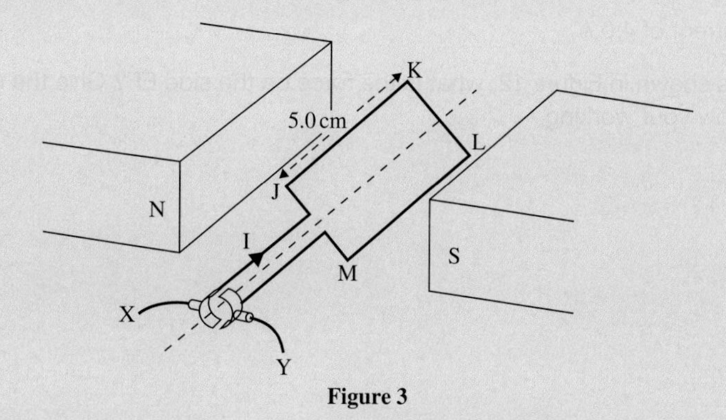

Figure 3

a. Draw an arrow on Figure 3 to indicate the direction of the magnetic force acting on the side JK. **(1 mark)**

b. A current of 6.0 A flows through the 100 turns of the coil JKLM.
The side JK is 5.0 cm in length.
Calculate the size of the magnetic force on the side JK in the orientation shown in Figure 3. Show your working. **(2 marks)**

Question 4 (4 marks)

***Source**: VCE 2016, Physics Exam, Section A, Q.13; © VCAA*

A 3.0 m long, vertical, copper lightning conductor is located in a region where Earth's magnetic field is horizontal and pointing north. A current of 2000 A flows down the conductor to Earth during an electrical storm. Force detectors measure a force on the lightning conductor of 0.32 N.

a. Calculate the magnitude of Earth's magnetic field acting on the lightning conductor. **(2 marks)**

b. **MC** Which one of the following (A.–F.) is the best description of the direction of the magnetic force acting on the lightning conductor? Explain your answer. **(2 marks)**

A. north
B. south
C. east
D. west
E. vertically up
F. vertically down

Question 5 (3 marks)

***Source**: VCE 2015, Physics Exam, Section A, Q.12.a; © VCAA*

Students have a model that can be used as a motor or generator, depending on the connections used.

The magnets provide a uniform magnetic field of 2.0×10^{-3} tesla.

EFGH is a square coil of each side length 4.0 cm with 10 turns.

A 6.0 V battery and an ammeter are connected to the shaft through a commutator.

This is shown in Figure 12.

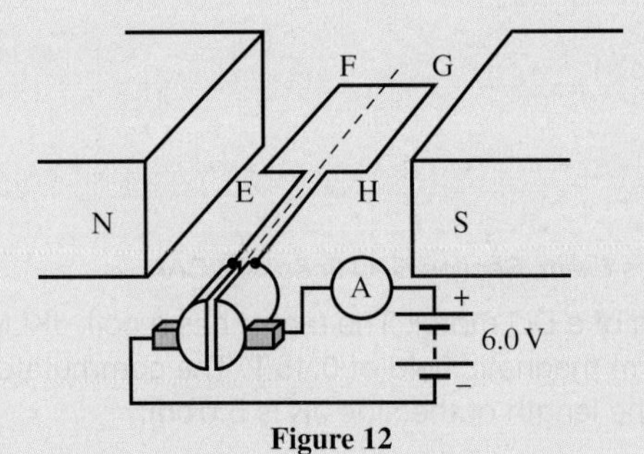

Figure 12

The ammeter shows a current of 4.0 A.

With the coil horizontal as shown in Figure 12, what is the force on the side EF? Give the magnitude and direction (up, down, left, right). Show your working.

More exam questions are available in your learnON title.

5.6 Applying magnetic forces — the DC motor

KEY KNOWLEDGE

- Investigate and analyse theoretically and practically the operation of simple DC motors consisting of one coil, containing a number of loops of wire, which is free to rotate about an axis in a uniform magnetic field and including the use of a split ring commutator
- Investigate, qualitatively, the effect of current, external magnetic field and the number of loops of wire on the torque of a simple motor

Source: Adapted from VCE Physics Study Design (2024–2027) extracts © VCAA; reproduced by permission.

5.6.1 The DC motor

An electric motor is a device that transforms electrical potential energy into rotational kinetic energy. A DC motor uses the current from a battery flowing through a coil in a magnetic field to produce continuous rotation of a shaft. The coil carries a direct current. Figure 5.33 shows a simplified example of a DC motor, with the coil illustrated by a single rectangular loop of wire. The straight sides make it easier to visualise how forces on the sides come about. The coil is usually wound around a frame known as an **armature**, which is made of ferromagnetic material and can rotate on an axle. The armature and coil together are known as the rotor. How do forces allow the coil to rotate?

FIGURE 5.33 A simplified DC motor

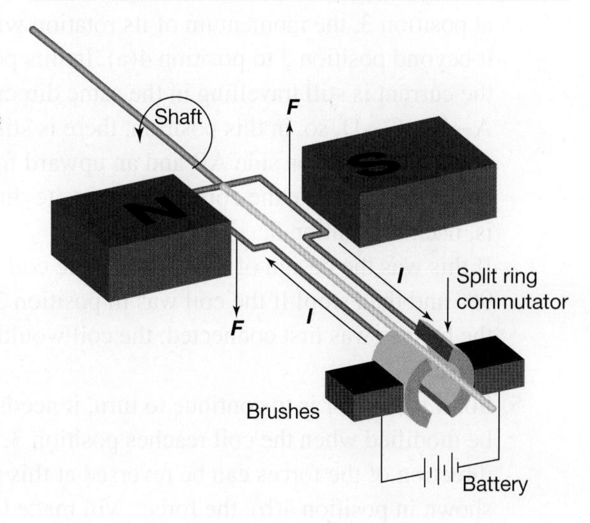

1. In figure 5.34 when the coil is in position 1 with the coil parallel to the magnetic field and the current flowing A→B→C→D, there will be a downward force on side AB and an upward force on side CD, causing the coil to rotate.

2. As the coil rotates, as shown in position 2, the forces remain unchanged in size and direction. This is because the magnetic field and the current in the wire are still the same size and in the same direction. However, the turning effect or torque is greatest when the forces are perpendicular to the plane of the loop. The forces are no longer perpendicular to the plane of the loop, so the turning effect is smaller.

FIGURE 5.34 Force on a coil in a DC motor

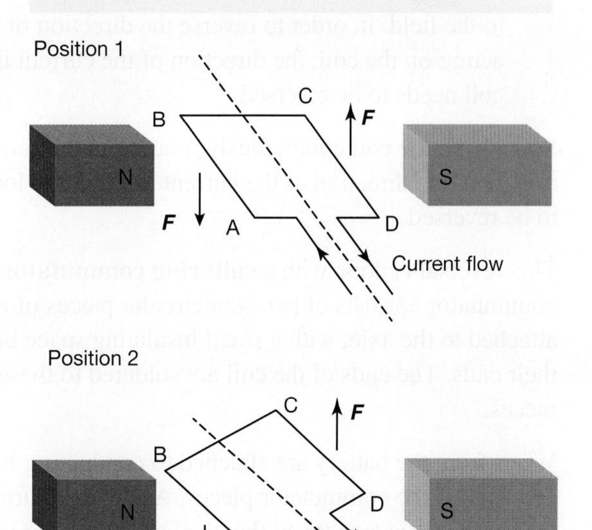

armature frame around which a coil of wire is wound, which rotates in a motor's magnetic field

3. When the coil reaches position 3, at right angles to the magnetic field, the forces are still unchanged in size and direction but, in this case, the forces are completely parallel to the plane of the loop, so there is no turning effect.

FIGURE 5.34 Force on a coil in a DC motor *(continued)*

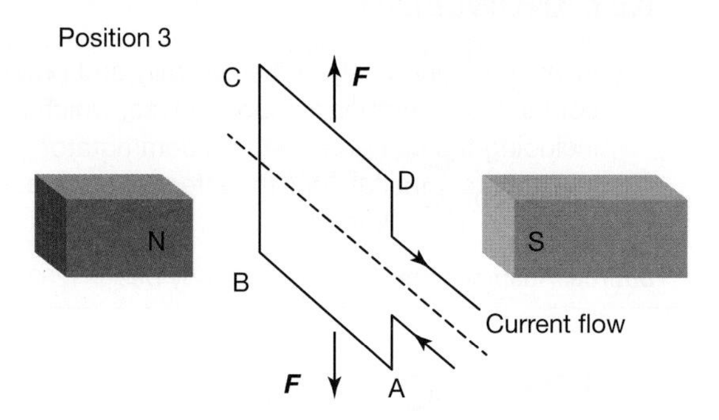

4. Since the coil was already moving before it arrived at position 3, the momentum of its rotation will carry it beyond position 3 to position 4(a). In this position, the current is still travelling in the same direction A→B→C→D, so, in this position, there is still a downward force on side AB and an upward force on side CD acting to rotate the coil in the opposite direction, that is, back to position 3.
   If this was the design of a DC motor, the coil would turn 90° and then stop! If the coil was in position 3 when the battery was first connected, the coil would not even move.
5. So, if the motor is to continue to turn, it needs to be modified when the coil reaches position 3. If the direction of the forces can be reversed at this point, as shown in position 4(b), the forces will make the coil continue to turn for another 180°. The forces need to be reversed again at this point to complete the rotation, and reversed every half turn when the coil is perpendicular to the field. In order to reverse the direction of the forces acting on the coil, the direction of the current through the coil needs to be reversed.

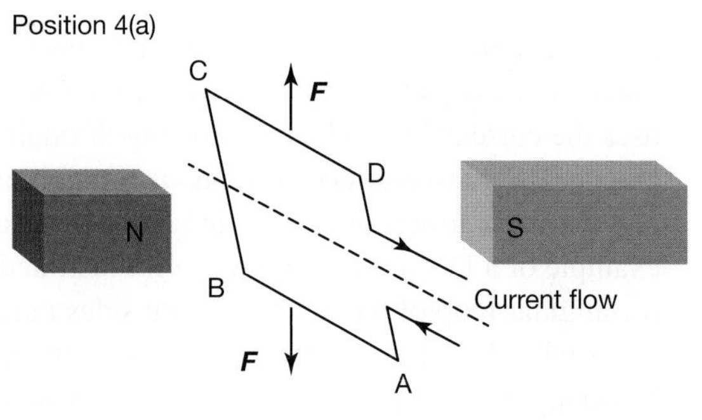

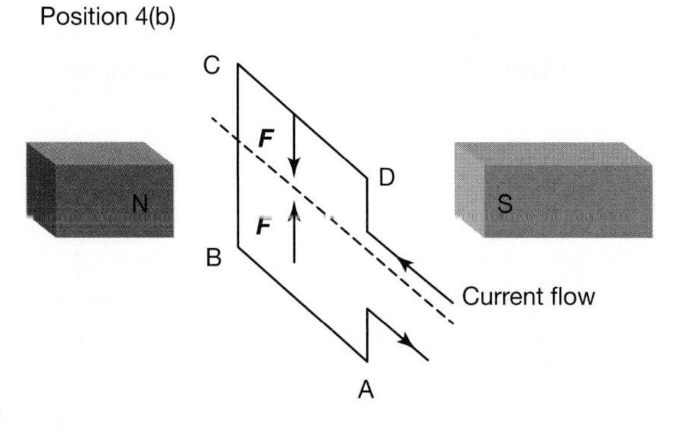

So, to keep the coil continuously rotating in the same direction, the direction of the current through the loop needs to be reversed.

This reversal is done with a **split ring commutator**. The commutator consists of two semicircular pieces of metal attached to the axle, with a small insulating space between their ends. The ends of the coil are soldered to these metal pieces.

FIGURE 5.35 Commutator and coil from a hair dryer

Wires from the battery are attached to conducting **brushes** that rest against the commutator pieces. As the axle turns, these commutator pieces turn so that they are no longer in contact with the brushes, stopping the flow of electrical current through the coil. As the commutators continue to rotate, they again make contact with the brushes, but are now in contact with the opposite brush, causing current to flow through the coil in the opposite direction. This enables the current through the coil to change direction every time the insulating spaces pass the brushes.

split ring commutator a device th reverses the direction of the curren flowing through an electric circuit

brushes conductors that make electrical contact with the moving split ring commutator in a DC motc

Brushes are often small carbon blocks that allow charge to flow and the axle to turn smoothly. The brushes are often made of graphite, which conducts electricity and acts as a lubricant.

As an energy transfer device of some industrial significance, there are some important questions to be asked about the design for a DC motor. Are there some starting positions of the coil that won't produce rotation? How can this be overcome? Can it run backwards and forwards? Can it run at different speeds?

elog-1949

INVESTIGATION 5.6 online only

The motor effect

AIM

To observe the direction of the force on a current-carrying conductor in an external magnetic field

elog-1950

vd-10605

INVESTIGATION 5.7 online only

Designing an electric DC motor

AIM

To construct a working electric motor and analyse and explain the physics principles behind its operation

Resources

Weblink DC motor applet

5.6.2 Torque

The turning effect of the forces on the coil in an electric motor is called a **torque**. The magnitude of the torque on a coil is the product of the component of the force perpendicular to the plane of the coil and the perpendicular distance between the point of application of the force and the shaft or axle (the axis of rotation).

torque the turning effect of the forces on the coil in an electric motor

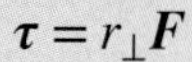

$$\tau = r_{\perp} F$$

where: τ is the torque, in N m

$r_{\perp}$ is the perpendicular distance between the point of application of the force and the axis of rotation, in m

F is the component of the force perpendicular to the axis of rotation, in N

Consider the case of the coil in figure 5.34. The sides of the coil perpendicular to the magnetic field, sides AB and CD, experience a force that is constant in size and magnitude no matter which position they are in; however, only the component of this force that is perpendicular to the plane of the coil contributes to the torque. When the coil is in the vertical position with its plane perpendicular to the field, the force on AB and CD is parallel to the plane of the coil, so the component of the force perpendicular to the plane is zero and hence the torque on the coil is zero. When the coil is parallel to the field, the force acting on AB and CD is perpendicular to the plane of the coil, so the coil experiences maximum torque in this position.

Each of the two sides that are perpendicular to the magnetic field, sides AB and CD, experience a force that contributes to the overall torque on the coil, so the total torque is equal to twice the torque acting on one of these sides.

To summarise, increasing the maximum torque will increase the speed of a DC motor. This can be achieved by:

- Increasing the force acting on the side (by either increasing the current in the coil, increasing the number of loops in the coil, producing a stronger magnetic field or adding an iron core in the centre of the loop as part of the armature)
- Increasing the width of the coil
- Using multiple coils mounted on the armature.

tlvd-8995

SAMPLE PROBLEM 9 Explaining DC motors and the split ring commutator

A student constructs a simple DC motor, exactly like the motor in figure 5.33, with a square loop having side lengths 2.0 cm and a uniform 0.0050 T magnetic field. A current of 1.5 A passes through the coil.

a. **Calculate the magnitude and direction of the force on the wire in the following sections, referring to figure 5.34 for the positions of these sections:**
   i. **AB**
   ii. **BC**
   iii. **CD**

b. **Explain why the motor is most easily started when the coil is parallel to the magnetic field.**

c. **Explain why the commutator is essential if the coil is to rotate continuously in one direction.**

THINK	WRITE
a. Recall that the force on a current-carrying wire in a magnetic field is $F = nIlB$ when the current and magnetic field are perpendicular to each other. Use the right-hand-slap rule or left-hand rule to find the direction of the force on each side of the wire loop.	a. $F = nIlB$ i. $F = 1 \times 1.5 \times 0.020 \times 0.0050$ $= 1.5 \times 10^4$ N, down ii. $F = 0$ N, because I and B are parallel to each other iii. $F = 1 \times 1.5 \times 0.020 \times 0.0050$ $= 1.5 \times 10^4$ N, up
b. Recall that the turning force, or torque, is given by $\tau = r_{\perp}F$.	b. Torque is a maximum when the axis of rotation and the line of action of the force are perpendicular to each other, and the distance between them is at a maximum. This occurs when the coil is parallel to the field.
c. Review the operation of the commutator.	c. Without the commutator, after the loop passes through the vertical position, the direction of the forces on the coil would reverse, causing the loop to slow down and then reverse the direction of rotation. Every time the loop passes through the vertical position, the direction of the forces reverse again, causing the loop to slow and reverse the direction of rotation. The loop would just oscillate around the vertical position. The commutator enables the direction of current through the loop to be reversed every time the loop passes through the vertical position, reversing the direction of the forces so that the loop can rotate continuously in the one direction.

PRACTICE PROBLEM 9

The terminals of the battery supplying the motor in sample problem 9 are reversed.

a. **Calculate the magnitude and direction of the force on the wire in the following sections:**
   i. **AB**
   ii. **BC**
   iii. **CD**

b. **Explain the differences, if any, between the two motors.**

on Resources

Video eLessons How maglev trains work (eles-2556)
Torque (eles-0025)

5.6 Activities

learnon

Students, these questions are even better in jacPLUS

Receive immediate feedback and access sample responses

Access additional questions

Track your results and progress

Find all this and MORE in jacPLUS

5.6 Quick quiz on	5.6 Exercise	5.6 Exam questions

5.6 Exercise

1. What is the purpose of each of the following in a DC motor?
   a. The magnet
   b. The brushes
   c. The commutator (mention three aspects)
   d. The large number of turns in the wire
2. Look at the simplified DC motor in figure 5.34.
   a. Are there some starting positions of the coil that won't produce rotation? How can this be overcome?
   b. Can the DC motor run backwards and forwards?
   c. Can it run at different speeds? If so, how?
3. Would a DC motor work if it was connected to an alternating current (AC) power source?
4. An example of a DC motor without a commutator is the homopolar motor shown in the following image. The upper section of the copper loop is in continuous contact with the positive terminal of the battery, and the lower section of the loop is in continuous contact with the negative terminal of the battery via the surface of the magnet. Explain how this arrangement results in a continuously rotating loop.

5. Stronger magnetic fields can be obtained with an electromagnet. The same DC power source can supply current to the electromagnet as well as to the rotating coil. The two components of the circuit, the electromagnet and the rotating coil, can be connected to the power source in two different ways.
   a. What are these ways?
   b. How do you think the starting and operating characteristics of these two types will differ?

6. **MC** Consider the simple DC motor shown in the diagram. Which of the following actions would increase the average torque on the loop?

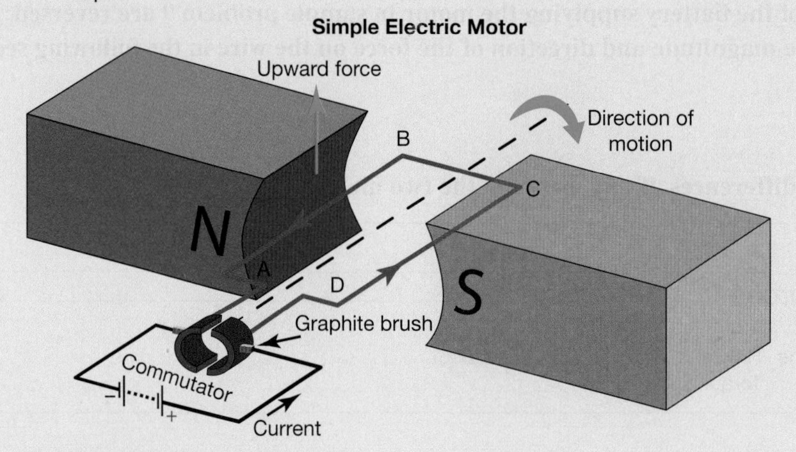

A. Decreasing the side length BC
B. Halving the voltage supplied to the circuit
C. Decreasing the size of the magnetic field
D. Increasing the number of loop windings.

7. **MC** Which of the following changes would stop the motor in question **6** from rotating continuously?
A. Reversing the direction of the magnetic field
B. Reversing the direction of the voltage
C. Rotating the direction of the magnetic field to be perpendicular to the direction shown in the diagram
D. Reversing the direction of the current in the circuit.

5.6 Exam questions

Question 1 (5 marks)

Source: *VCE 2022 Physics Exam, Section B, Q.1;* © VCAA

Figure 1 shows four positions (1, 2, 3 and 4) of the coil of a single-turn, simple DC motor. The coil is turning in a uniform magnetic field that is parallel to the plane of the coil when the coil is in Position 1, as shown.

When the motor is operating, the coil rotates about the axis through the middle of sides *LM* and *NK* in the direction indicated. The coil is attached to a commutator. Current for the motor is passed to the commutator by brushes that are not shown in Figure 1.

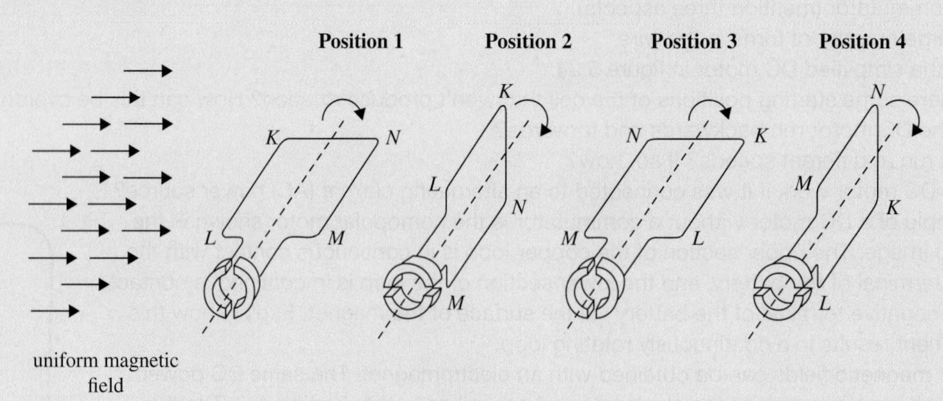

Figure 1

a. When the coil is in Position 1, in which direction is the current flowing in the side *KL* — from *K* to *L* or from *L* to *K* ? Justify your answer. **(2 marks)**

b. When the coil is in Position 3, in which direction is the current flowing in the side *KL* — from *K* to *L* or from *L* to *K* ? **(1 mark)**

c. The side *KL* of the coil has a length of 0.10 m and experiences a magnetic force of 0.15 N due to the magnetic field, which has a magnitude of 0.5 T. Calculate the magnitude of the current in the coil. **(2 marks)**

Question 2 (4 marks)

Source: *VCE 2019, Physics Exam, NHT, Section B, Q.3; © VCAA*

Figure 3 shows a simple DC motor consisting of a square loop of wire of side 10 cm and 10 turns, a magnetic field of strength 2.0×10^{-3} T, and a commutator connected to a 12 V battery. The current in the loop is 2.0 A.

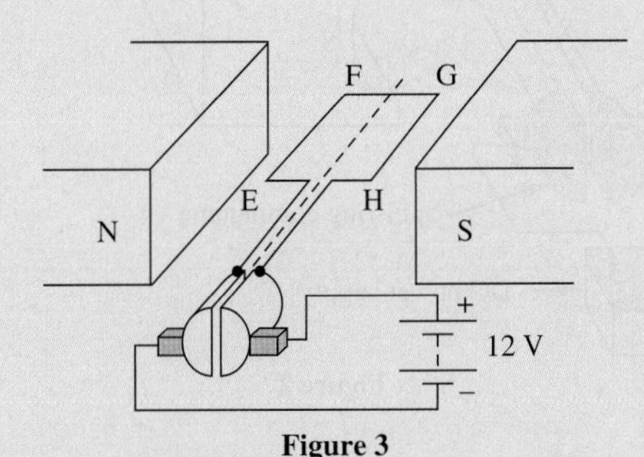

Figure 3

a. Calculate the magnitude of the total force acting on the side EF when the loop is in the position shown in Figure 3. Show your working. **(2 marks)**

b. Explain the role of the commutator in the operation of the DC motor. **(2 marks)**

Question 3 (3 marks)

Source: *VCE 2018, Physics Exam, Section B, Q.3a; © VCAA*

Students build a model of a simple DC motor, as shown in Figure 3.

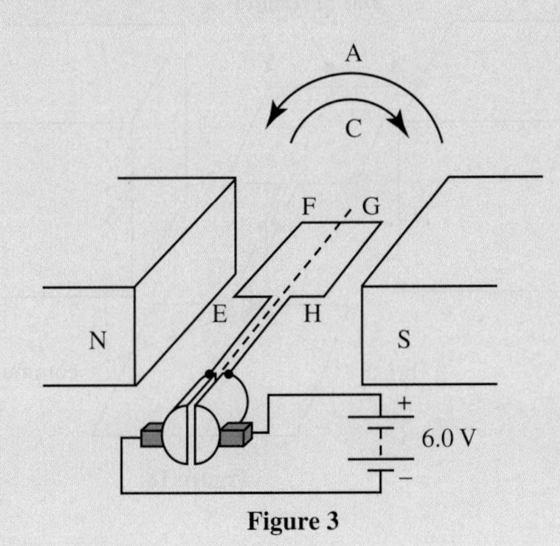

Figure 3

The motor is set with the coil horizontal, as shown, and the power source is applied.

Will the motor rotate in a clockwise (C) or anticlockwise (A) direction? Explain your answer.

Question 4 (5 marks)

Source: *VCE 2017, Physics Exam, Section B, Q.3; © VCAA*

Figure 2 shows a schematic diagram of a simple DC motor.

It consists of two magnets, a single 9.0 V DC power supply, a split-ring commutator and a rectangular coil of wire consisting of 10 loops.

The total resistance of the coil of wire is 6.0 Ω.

The length of the side JK is 12 cm and the length of the side KL is 6.0 cm.

The strength of the uniform magnetic field is 0.50 T.

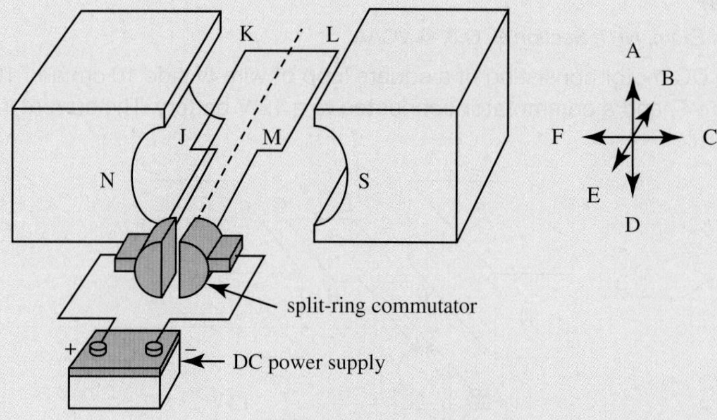

Figure 2

a. Determine the size and the direction (A-F) of the force acting on the side JK. **(3 marks)**
b. What is the size of the force acting on the side KL in the orientation shown in Figure 2? Explain your answer. **(2 marks)**

Question 5 (5 marks)

Source: *VCE 2016 Physics Exam, NHT, Section A, Q.14; © VCAA*

Students build a simple electric motor, as shown in Figure 18.

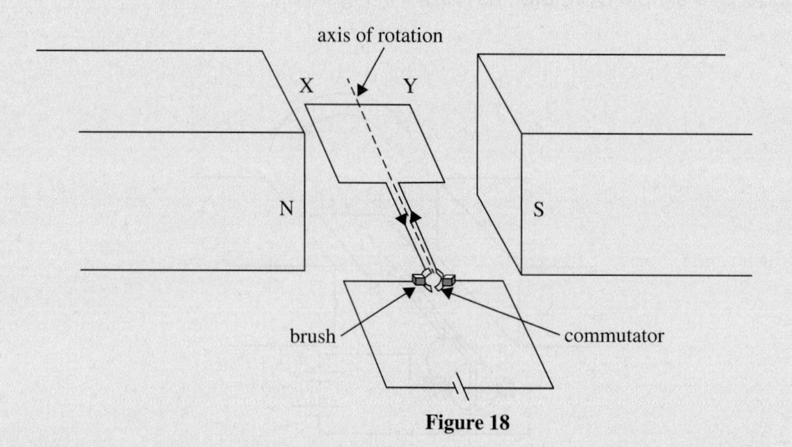

Figure 18

a. **MC** At what position(s) (A.–D.) of the rotating coil is the magnetic force on the side XY zero? One or more answers may be selected. **(1 mark)**
   A. horizontal with the current as shown in Figure 18
   B. horizontal with the current in the opposite direction to that shown in Figure 18
   C. vertical
   D. at all orientations of the coil
b. The students discover that the motor starts moving more easily with the coil in some orientations than in others. Explain the best orientation(s) for starting the motor to move from rest. **(2 marks)**
c. **MC** To increase the speed of rotation of the motor, the students suggest a number of improvements. Which suggested improvement(s) (A.–D.) is likely to increase the speed of rotation of the coil? One or more answers may be selected. Explain your answer. **(2 marks)**
   A. increase the battery voltage
   B. replace the single coil of the motor with several turns
   C. increase the resistance of the coil
   D. reverse one of the poles of the permanent magnets

More exam questions are available in your learnON title.

5.7 Similarities and differences between gravitational, electric and magnetic fields

KEY KNOWLEDGE

- Compare theoretically and practically gravitational, magnetic and electric fields, including directions and shapes of fields, attractive and repulsive fields, and the existence of dipoles and monopoles
- Identify fields as static or changing, and as uniform or non-uniform.

Source: Adapted from VCE Physics Study Design (2024–2027) extracts © VCAA; reproduced by permission.

5.7.1 Comparing gravitational, magnetic and electric fields

Gravitational, electric and magnetic fields are all properties of the space around an object, whether the object is a mass, a charge or a magnet. Lines are used to show the direction of the field, that is, the direction of force experienced by a test object; the strength of the field is shown by the density of the lines. For some field diagrams, it is not possible to tell the type of field simply by looking at the diagram.

FIGURE 5.36 Field diagrams

a.

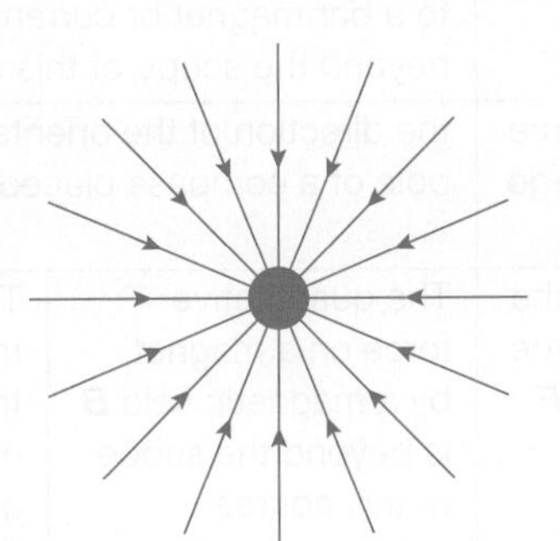

b.

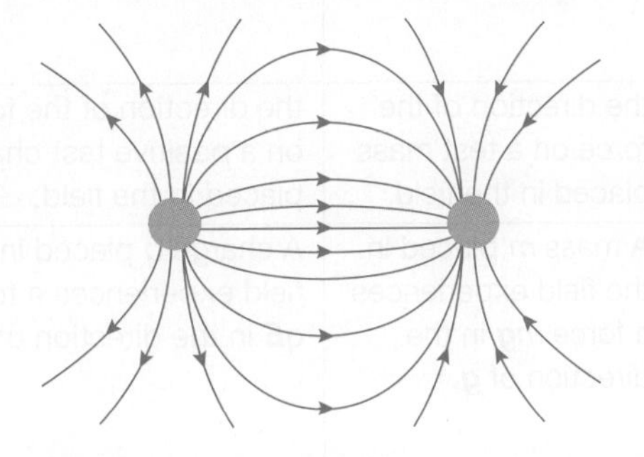

For example, field diagram (a) in figure 5.36 could show either a gravitational field around a mass or an electric field around a negative point charge. It could not be a magnetic field as, even though it might look like the field near the south pole of a magnet, further away the field lines would start to curve away towards the north pole.

Similarly, field diagram (b) in figure 5.36 could show either an electric field around two opposite charges or a magnetic field around north and south poles. However, it could not be a gravitational field, because mass does not come in two opposite versions.

The fields described in this topic are all vector fields, and they share the property that fields from multiple sources can be added to find the total field at any point.

The essential features of gravitational, electric and magnetic fields are summarised in table 5.5.

TABLE 5.5 Summary of gravitational, electric and magnetic fields

	Gravitational field	Electric field	Magnetic field	
Surrounds	a mass	a charge	(i) a magnet	(ii) a moving charge
Diagram			N S	Current-carrying wire
Affects:	other masses	other charges	other magnetic materials and/or moving charges	
Field strength symbol and units	$\boldsymbol{g}$ N kg^{-1} (or m s^{-2})	$\boldsymbol{E}$ N C^{-1} (or V m^{-1})	$\boldsymbol{B}$ T (or N m A^{-1})	
Magnitude of field from a point source	$g = G\frac{M}{r^2}$	$E = \frac{kq}{r^2}$	A magnetic north or south pole cannot be isolated to make a point source (monopole). Calculating the magnetic field strength due to a bar magnet or current-carrying wire is beyond the scope of this course.	
Direction of the field is defined by:	the direction of the force on a test mass placed in the field.	the direction of the force on a positive test charge placed in the field.	the direction of the orientation of the north pole of a compass placed in the field.	
Force in the field	A mass *m* placed in the field experiences a force *m***g** in the direction of **g**.	A charge *q* placed in the field experiences a force *q***E** in the direction of **E**.	The quantitative force on a magnet by a magnetic field **B** is beyond the scope of this course.	The force on a moving charge in the presence of a magnetic field **B** is *qv***B** and the force on a current-carrying wire in the presence of a magnetic field **B** is *nIl***B**.
Polarity of the field	Unipolar	Positive or negative monopoles. A dipole field is the field from a separated positive and negative charge.	The north or south pole cannot be isolated to create a magnetic monopole.	
A uniform field exists:	on the Earth's surface.	between two oppositely charged parallel conducting plates.	between the poles of a large horseshoe magnet.	inside a current-carrying coil.
A static field exists:	around a mass.	around charges that are not moving.	around magnetised materials.	around a wire carrying a constant current.

tlvd-8997

SAMPLE PROBLEM 10 Comparing gravitational fields from a point mass to electric fields from a positive charge

Identify similarities and differences between the gravitational field generated from a point mass and the electric field generated from a positive charge.

THINK

Recall that fields are described by whether they are uniform or non-uniform, their shape and direction, and the nature of their source.

WRITE

Similarities:

1. The gravitational field generated from a point mass and the electric field generated from a positive charge both obey an inverse square law.
2. The gravitational field and electric field are both non-uniform and static.
3. The point mass and positive charge can both be described as monopoles.

Differences:

1. The direction of the gravitational field is inwards towards the point mass, whereas the electric field generated from a positive charge is outwards, away from the point charge.
2. The gravitational field is attractive for all other masses, whereas the electric field generated from a positive charge repels other positive charges and attracts negative charges.

PRACTICE PROBLEM 10

Identify similarities and differences between the electric field of a dipole and the magnetic field of a bar magnet.

You can see in table 5.5 that there are circumstances in which each type of field can be uniform. There are also circumstances in which each type of field is non-uniform, particularly as the distance away from the source (such as an object with mass, a magnet or a point charge) increases. Gravitational sources can only attract, whereas electric and magnetic sources can attract or repel. Gravitational and electric monopoles exist, which can be seen by the way that field lines can begin from a single point, whereas magnetic field lines can never have a beginning and an end but must instead form closed loops. The intriguing similarity between the electric dipole field and the magnetic field from a bar magnet or solenoid hints at the fundamental connection between electric and magnetic phenomena, a connection that was elegantly and brilliantly captured by James Clerk Maxwell at the end of the nineteenth century.

Ultimately, the field model is a pathway to describing the forces experienced by objects and to understanding and predicting the motion of objects and the transfer of energy in gravitational, electric and magnetic fields.

5.7 Activities

learnon

Students, these questions are even better in jacPLUS

Receive immediate feedback and access sample responses

Access additional questions

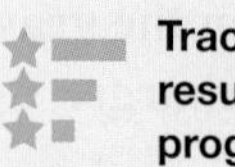

Track your results and progress

Find all this and MORE in jacPLUS

5.7 Quick quiz	5.7 Exercise	5.7 Exam questions

5.7 Exercise

1. What is the experimental evidence for there being two types of charge?
2. Coulomb's Law is very similar to Newton's Law of Universal Gravitation. How do these two laws differ? Compare electric charge and gravitational mass.
3. Give an example of a source of a uniform and a non-uniform magnetic field. Sketch the field in each case.
4. Identify similarities and differences between the electric field due to two positive charges separated by 1.0 cm and the magnetic field due to two bar magnets with the north poles facing each other, but separated by 1.0 cm.
5. Compare the magnitude of the electric and gravitational forces of attraction between an electron and a proton that are a distance 5.3×10^{-11} m apart in a hydrogen atom. (Given data: $m_e = 9.11 \times 10^{-31}$ kg; $m_p = 1.67 \times 10^{-27}$ kg)
6. A proton is suspended so that it is stationary in an electric field. Using the value of $g = 9.8$ m s^{-2}, determine the strength of the electric field.
7. Millikan observed charged oil drops falling at a constant speed in an electric field provided by two charged plates. In a simplified version of the experiment, the oil drops are in a vacuum, so that when they fall at a constant speed, the electric force on the drop is equal and opposite to the gravitational force on the drop. If the mass of an oil drop is 10 μg and the charge of the oil drop is $10e^-$, how large does the electric field provided by the plates need to be for the gravitational and electric forces to balance?
8. What equal positive charge would the Earth and the Moon need to have for the electrical repulsion to balance the gravitational attraction? Why don't you need to know the distance of separation of the two objects? The mass of the Moon is 7.35×10^{22} kg.
9. The strength of the magnetic field can be determined using the following formula:

$$B = \frac{\mu_0 I}{2\pi r},$$

where r is the distance from a wire, and μ_0 is the vacuum permittivity constant ($4\pi \times 10^{-7}$ T m A^{-1}). The strength of the Earth's magnetic field at the surface is approximately 50 μT. Calculate how much current needs to flow through the wire for the magnetic field at a distance of 2.0 cm from the wire to be equal in magnitude to the Earth's magnetic field at that point.

5.7 Exam questions

Question 1 (1 mark)

MC Which of the following is ***incorrect***?

A. The gravitational field around a mass is static.
B. The field strength decreases when the distance to the source of the field increases.
C. No static magnetic field exists
D. The electric field around a charge at rest is static.

Question 2 (1 mark)

MC Observe the following field shape.

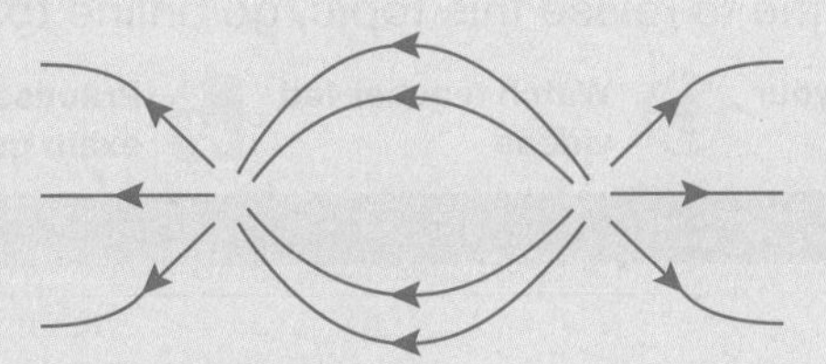

Among the following, select the possible sources for it.

A. Two point masses separated by a distance r

B. A bar magnet

C. Two identical point charges separated by a distance r

D. An electric dipole.

Question 3 (2 marks)

When making notes on field shapes, a student recorded the following unlabelled diagram.

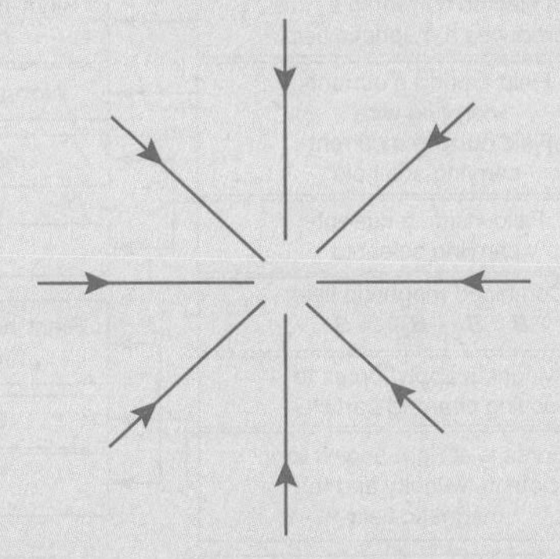

Help the student identify the type of fields and possible sources of field for this diagram.

Question 4 (3 marks)

Millie and Noah are discussing the similarities between gravitational, electric and magnetic fields. Noah has drawn the following field shape and affirms that it could correspond to a gravitational field, to an electric field or even to a magnetic field.

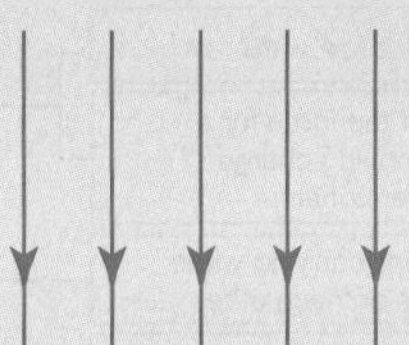

Justify whether Noah is correct, and if so, give an example of potential source of this field shape for each of the field types.

Question 5 (3 marks)

When making notes to summarise the characteristics of gravitational, electric and magnetic fields, Lara created the following table.

Source of a gravitational field	Mass
Source of an electric field	Charge
Source of a magnetic field	Magnetic material

Their teacher told them that they forgot to include another possible source for one of their fields. Explain what Lara forgot to include.

More exam questions are available in your learnON title.

5.8 Review

Hey students! Now that it's time to revise this topic, go online to:

Access the topic summary  **Review your results**  **Watch teacher-led videos**  **Practise past VCAA exam questions**

Find all this and MORE in jacPLUS

5.8.1 Topic summary

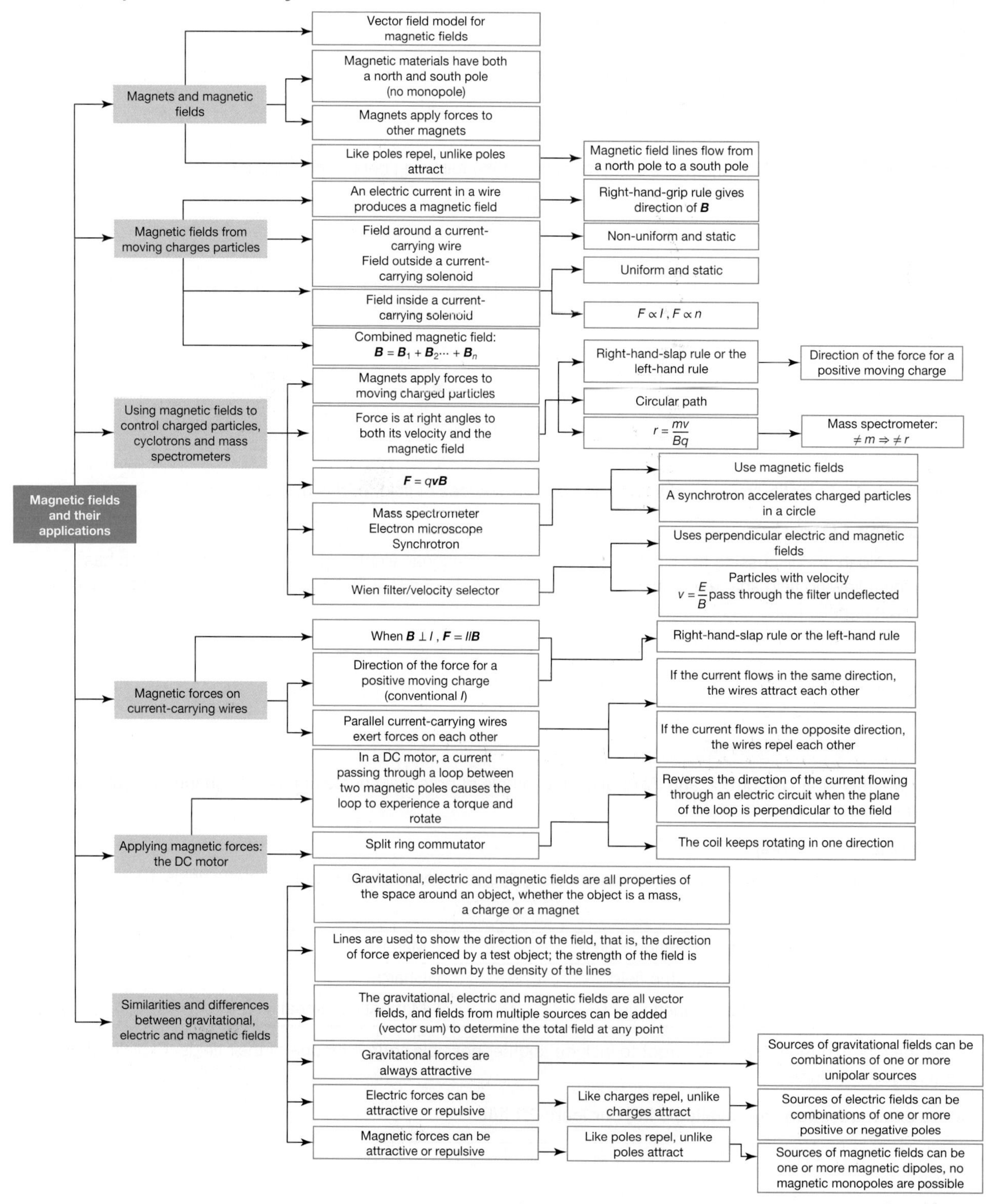

5.8.2 Key ideas summary

5.8.3 Key terms glossary

online only

Solutions — Solutions — Topic 5 (sol-0819)

Practical investigation eLogbook — Practical investigation eLogbook — Topic 5 (elog-1636)

Digital documents — Key science skills — VCE Physics Units 1–4 (doc-36950)
Key terms glossary — Topic 5 (doc-37173)
Key ideas summary — Topic 5 (doc-37174)

Exam question booklet — Exam question booklet — Topic 5 (eqb-0102)

5.8 Activities

learnon

Students, these questions are even better in jacPLUS

Receive immediate feedback and access sample responses

Access additional questions

Track your results and progress

Find all this and MORE in jacPLUS

5.8 Review questions

1. A wire pointing into the page carries a 5.0-A current.
   a. Sketch the magnetic field from the wire, ensuring that the magnetic field passes through points P and Q.
   b. The Earth's magnetic field is in the direction shown in the following diagram and has a uniform magnitude of 10 μT. At the points P and Q, the value of the magnetic field from the wire is 20 μT. Calculate the resulting magnitude and direction of the total magnetic field at P and Q.

2. Sketch the magnetic field from a solenoid and the electric field from an electric dipole. Identify similarities and differences between the two fields.
3. When making notes on field shapes, a student records the following unlabelled sets of field lines in their notebook:

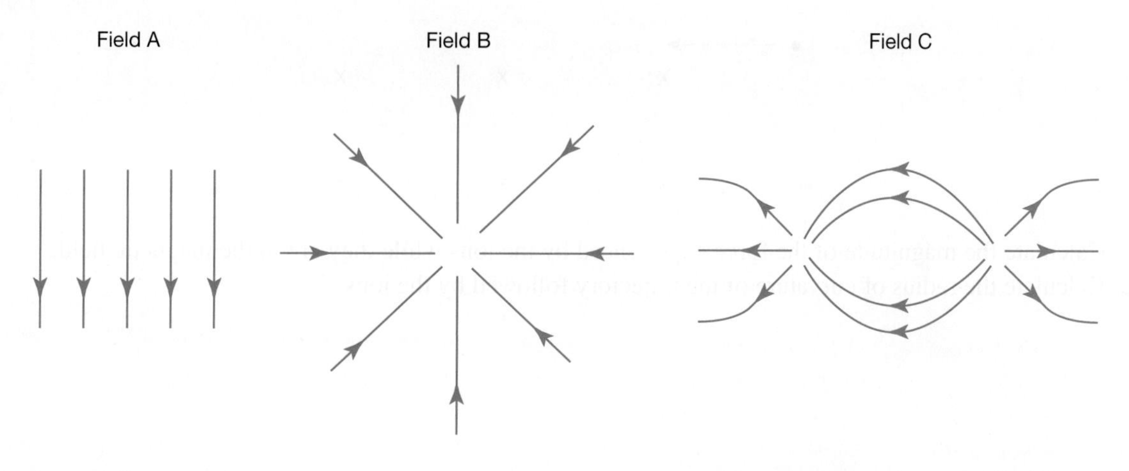

The student returns to their work and tries to identify the type of field and a possible source of field for each set of data.

Data Set	Gravitational	Electric	Magnetic
Field A			
Field B			
Field C	Not possible	−ve and +ve charge, electric dipole	Not possible (similar to electric dipole, but not identical)

Complete the table for the remaining data sets.

4. a. What is the size of the magnetic force on an electron entering perpendicularly to a magnetic field of 250 mT at a speed of 5.0×10^6 m s^{-1}?
   b. Use the mass of the electron (mass = 9.1×10^{-31} kg) to determine its centripetal acceleration.
   c. If a proton entered the same field with the same speed, what would its centripetal acceleration be?

5. After a radioactive decay, a beta particle is observed to execute the following trajectory in a cloud chamber. What is the direction of the magnetic field experienced by the beta particle?

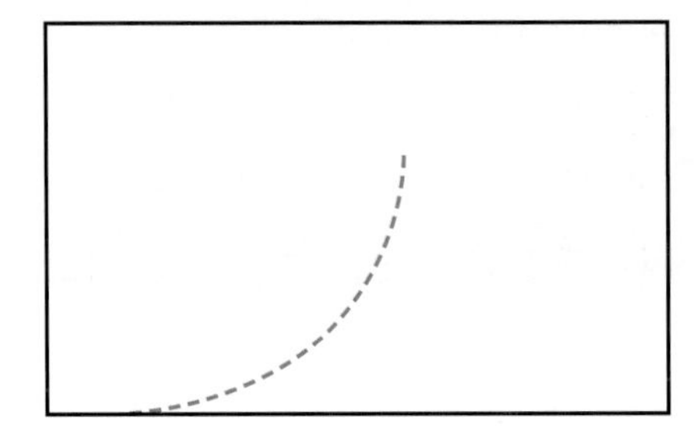

6. What strength of magnetic field would be needed to obtain a radius of 1000 m if an electron has momentum of 1.0×10^{-18} kg m s^{-1}? (Assume the direction of the momentum of the electrons is perpendicular to the direction of the magnetic field.)

7. A mass spectrometer is designed to separate particles based on the ratio of their mass to charge. The spectrometer consists of three stages: a linear accelerator, a velocity selector and a uniform magnetic field. Ions with a mass of 1.2×10^{-26} kg and a charge of +2e are passed through a linear accelerator with a potential difference of 2000 V.
   a. Assume that the ions are initially at rest. Show that, after passing through the linear accelerator, the velocity of the ions is 3.3×10^5 m s^{-1}.

   The ions pass through the velocity selector undeflected into a uniform 0.50 T magnetic field pointing into the page, as shown in the diagram.

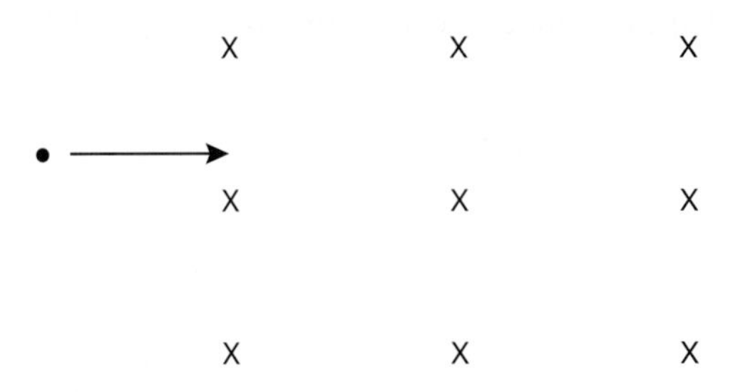

   b. Calculate the magnitude of the force experienced by the ions while they are in the magnetic field.
   c. Calculate the radius of curvature of the trajectory followed by the ions.

8. Two sturdy wire cables connect a power supply to a piece of heavy machinery as shown in the diagram.
   a. Draw an arrow at P indicating the direction of the magnetic field from the right-hand wire.
   b. Draw an arrow at Q indicating the direction of the force on the right-hand wire.
   c. Do the two wires attract or repel each other? Explain your response.

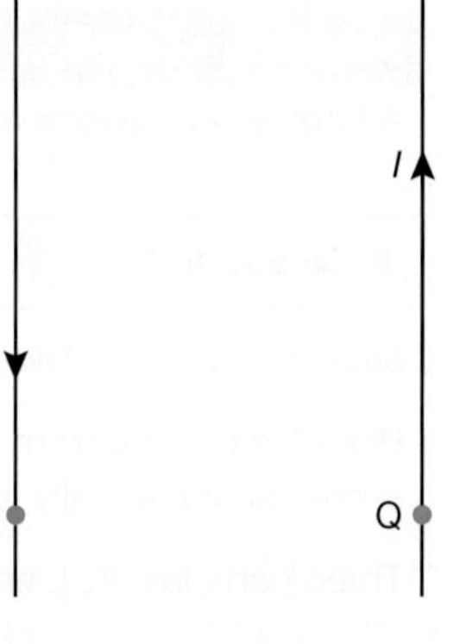

9. Consider the DC motor shown in the following diagram. The strength of the magnetic field is 0.030 T. The loop consists of 50 turns of wire. The loop is a rectangle with side lengths AB and CD of 15 cm and side lengths BC and DA of 10 cm. The current in the wire is 3.0 A.

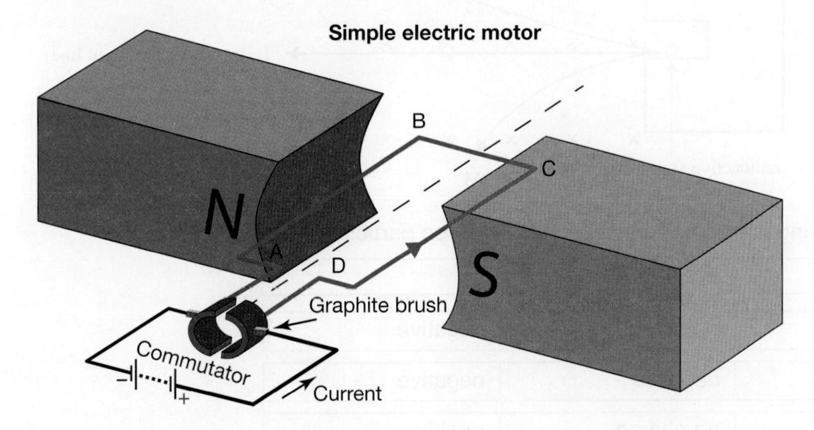

   a. Calculate the magnitude of the force on the side length AB.
   b. Draw an arrow on side AB indicating the direction of the force on AB at that moment.
   c. The polarity of the DC supply is reversed. Explain how this will affect the operation of the motor.
   d. A student attempts to make their own motor but has difficulty attaching the commutator. The commutator is attached with the split in the ring at an angle of 45° from the vertical. Explain how this will affect the operation of the motor.

10. Consider the velocity selector shown in the following diagram:

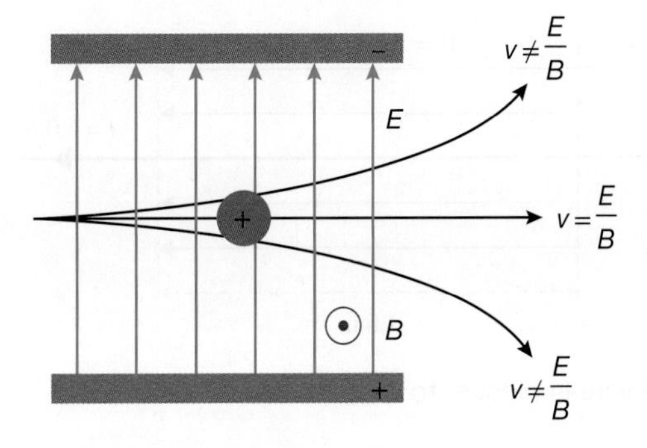

Would the same configuration of electric and magnetic fields work for negatively charged particles?

5.8 Exam questions

Section A — Multiple choice questions

All correct answers are worth 1 mark each; an incorrect answer is worth 0.

Question 1

Source: *VCE 2022, Physics Exam, Section A, Q.3*; © VCAA

Particles emitted from a radioactive source travel through a magnetic field, B_{in}, directed into the page, as shown schematically in the diagram below.

Three particles, K, L and M, follow the paths indicated by the arrows.

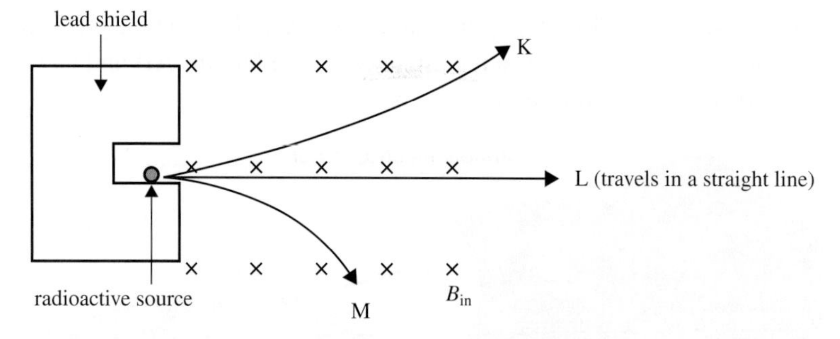

Which of the following correctly identifies the charges on particles K, L and M?

	K	L	M
A.	positive	no charge	negative
B.	positive	negative	negative
C.	negative	no charge	positive
D.	no charge	no charge	no charge

Question 2

Source: *VCE 2021 Physics Exam, NHT, Section A, Q.1*; © VCAA

A wire carrying a current, I, of 6.0 A passes through a magnetic field, B, of strength 1.4×10^{-5} T, as shown below. The magnetic field is exactly 1.0 m wide.

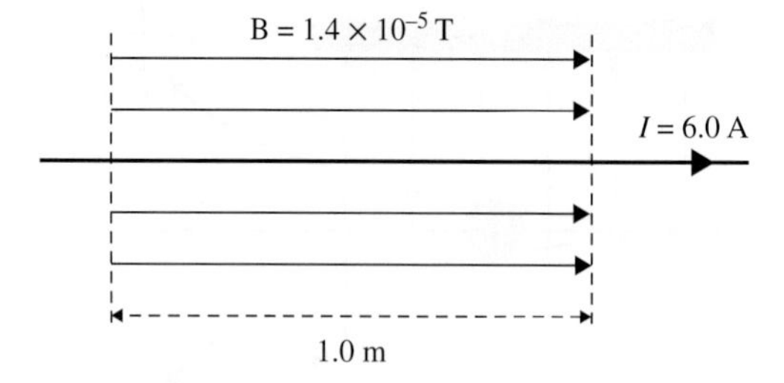

The magnitude of the force on the wire is closest to

A. 0 N
B. 2.3×10^{-6} N
C. 8.4×10^{-5} N
D. 4.3×10^{5} N

Question 3

Source: *VCE 2020, Physics Exam, Section A, Q.3; © VCAA*

A positron with a velocity of 1.4×10^6 m s^{-1} is injected into a uniform magnetic field of 4.0×10^{-2} T, directed into the page, as shown in the diagram below. It moves in a vacuum in a semicircle of radius *r*. The mass of the positron is 9.1×10^{-31} kg and the charge on the positron is 1.6×10^{-19} C. Ignore relativistic effects.

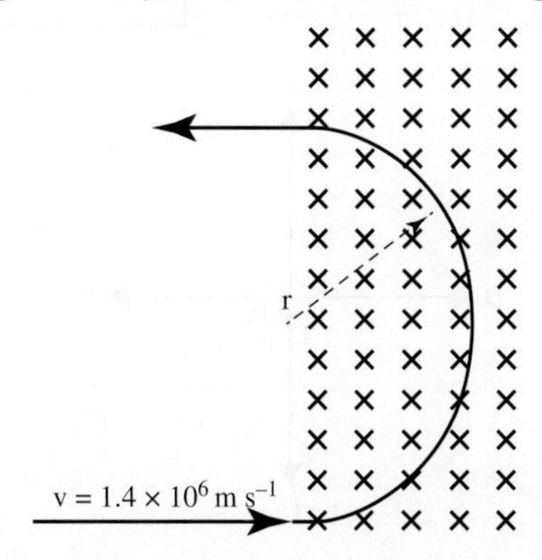

Which one of the following best gives the speed of the positron as it exits the magnetic field?

A. $0\,\text{m s}^{-1}$
B. much less than $1.4 \times 10^6\,\text{m s}^{-1}$
C. $1.4 \times 10^6\,\text{m s}^{-1}$
D. greater than $1.4 \times 10^6\,\text{m s}^{-1}$

Question 4

The speed of the positron is changed to 7.0×10^5 m s^{-1}.
Which one of the following best gives the value of the radius *r* for this speed?

A. $\frac{r}{4}$
B. $\frac{r}{2}$
C. r
D. $2r$

Question 5

Source: *VCE 2019, Physics Exam, Section A, Q.1; © VCAA*

Magnetic and gravitational forces have a variety of properties.

Which of the following best describes the attraction/repulsion properties of magnetic and gravitational forces?

	Magnetic forces	Gravitational forces
A.	either attract or repel	only attract
B.	only repel	neither attract nor repel
C.	only attract	only attract
D.	either attract or repel	either attract or repel

A. Option A
B. Option B
C. Option C
D. Option D

Question 6

Source: *VCE 2019, Physics Exam, NHT, Section A, Q.1; © VCAA*

Two identical bar magnets are placed end to end, as shown below. Point X is midway between the bar magnets.

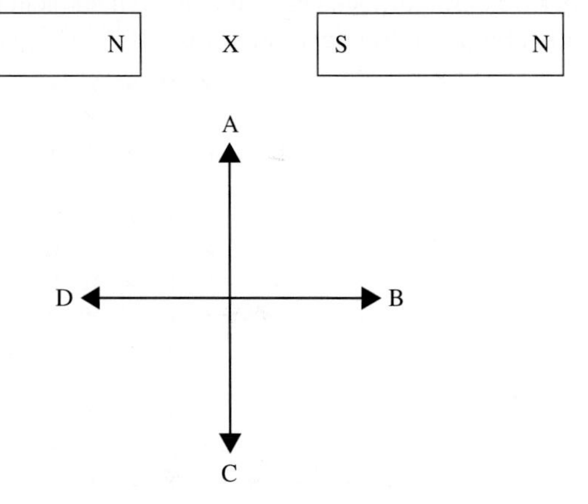

Which direction best shows the direction of the magnetic field at point X?

A. A
B. B
C. C
D. D

Question 7

Source: *VCE 2018, Physics Exam, Section A, Q.2; © VCAA*

A wire carrying a current of 10 A is placed in a uniform magnetic field of $B = 4.0 \times 10^{-4}$ T, as shown below. 10 cm of the wire is in the field.

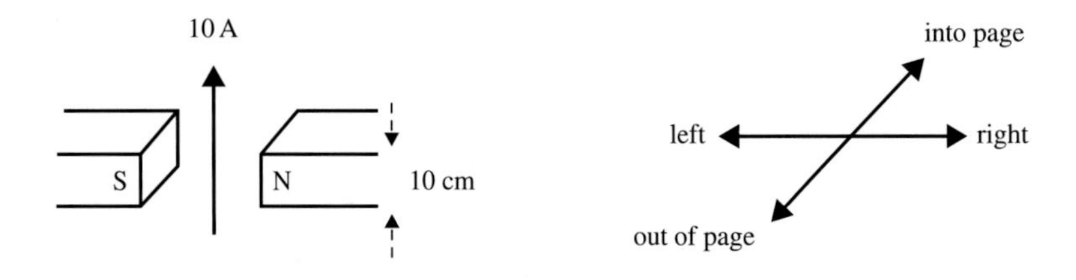

Which one of the following best gives the direction of the force acting on the wire?

A. out of page
B. into page
C. right
D. left

Question 8

Source: *VCE 2018, Physics Exam, Section A, Q.3;* © VCAA

A wire carries a current of 10 A.

Which one of the following diagrams best shows the magnetic field associated with this current?

A.

10 A

B.

10 A

C.

10 A

D.

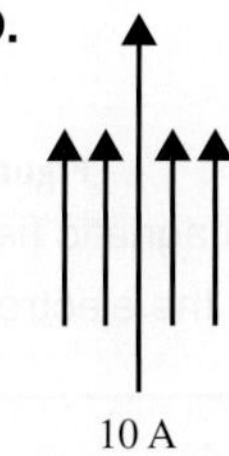

Question 9

Source: *VCE 2018 Physics Exam, NHT, Section A, Q.1;* © VCAA

Engineers are measuring the force due to Earth's magnetic field on the supply wire of a railway line. The wire runs east–west and carries a current of 2000 A. Earth's magnetic field is horizontal and due north at the place where measurements are taken.

The engineers measure the force on a 10 m length of the wire to be 1.0 N.

Which one of the following best gives the strength of Earth's magnetic field at this point?

A. 2.0×10^{-8} T
B. 5.0×10^{-5} T
C. 5.0×10^{-4} T
D. 200 T

Question 10

Source: *VCE 2017, Physics Exam, Section A, Q.1; © VCAA*

A group of students is considering how to create a magnetic monopole.

Which one of the following is correct?

A. Break a bar magnet in half.
B. Pass a current through a long solenoid.
C. Pass a current through a circular loop of wire.
D. It is not known how to create a magnetic monopole.

Section B — Short answer questions

Question 11 (5 marks)

Source: *Adapted from VCE 2021 Physics Exam, NHT, Section B, Q.2b&c*; © VCAA

An electron is accelerated from rest by a potential difference of V_0. It emerges at a speed of 2.0×10^7 m s^{-1} into a magnetic field, B, of strength 2.5×10^{-3} T and follows a circular arc, as shown in Figure 2.

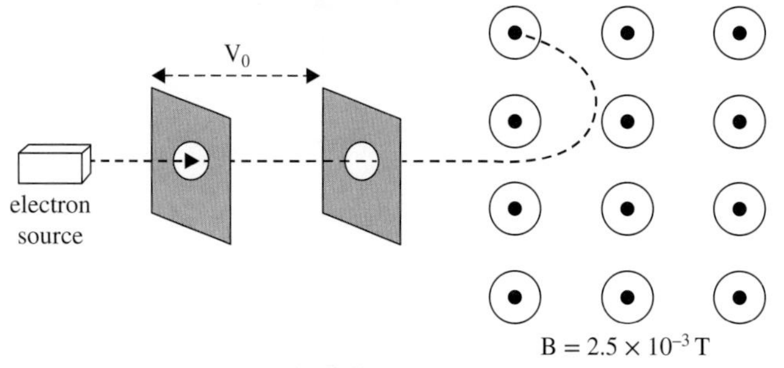

Figure 2

a. Explain why the path of the electron in the magnetic field follows a circular arc. **(2 marks)**

b. Calculate the radius of the path travelled by the electron. Show your working. **(3 marks)**

Question 12 (3 marks)

Source: *VCE 2021 Physics Exam, NHT, Section B, Q.4;* © VCAA

Figure 4 shows a schematic diagram of a simple one-coil DC motor. A current is flowing through the coil.

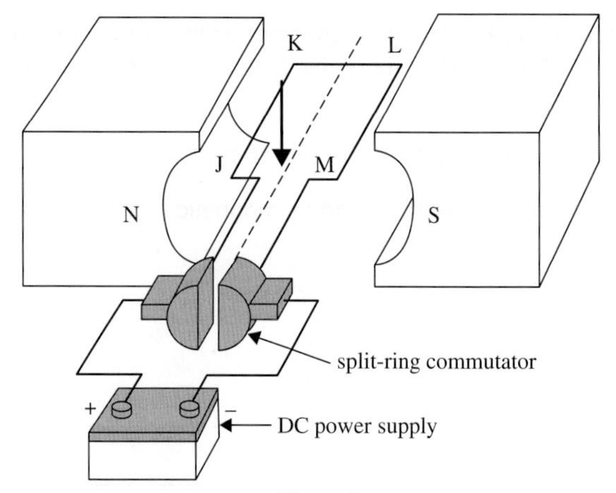

Figure 4

a. Draw an arrow on Figure 4 to indicate the direction of the force acting on the side JK of the coil. **(1 mark)**

b. Explain the purpose of the split-ring commutator. **(2 marks)**

Question 13 (2 marks)

Source: *VCE 2020, Physics Exam, Section B, Q.1; © VCAA*

Two bar magnets are placed close to each other, as shown in Figure 1.

Sketch the shape and the direction of **at least four** magnetic field lines between the two poles within the dashed border shown in Figure 1.

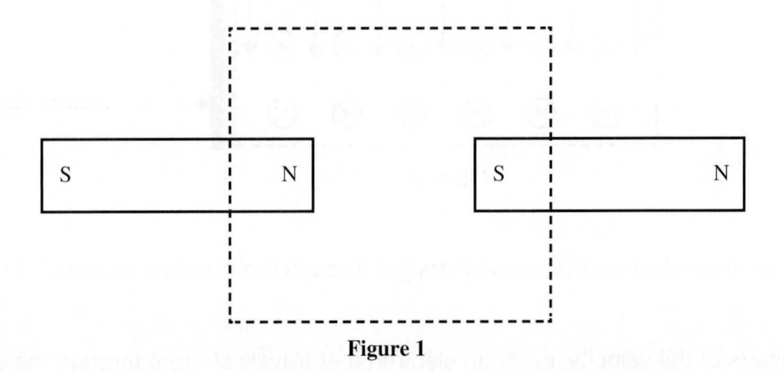

Figure 1

Question 14 (3 marks)

Source: *VCE 2020 Physics Exam, Section B, Q.2; © VCAA*

Gravitation, magnetism and electricity can be explained using a field model. According to our understanding of physics and current experimental evidence, these three field types can be associated with only monopoles, only dipoles or both monopoles and dipoles.

In the table below, indicate whether each field type can be associated with only monopoles, only dipoles or both monopoles and dipoles by ticking (✓) the appropriate box.

Field type	Only monopoles	Only dipoles	Both monopoles and dipoles
gravitation			
magnetism			
electricity			

Question 15 (6 marks)

Source: *VCE 2020 Physics Exam, Section B, Q.3; © VCAA*

Electron microscopes use a high-precision electron velocity selector consisting of an electric field, E, perpendicular to a magnetic field, B.

Electrons travelling at the required velocity, v_0, exit the aperture at point Y, while electrons travelling slower or faster than the required velocity, v_0, hit the aperture plate, as shown in Figure 2.

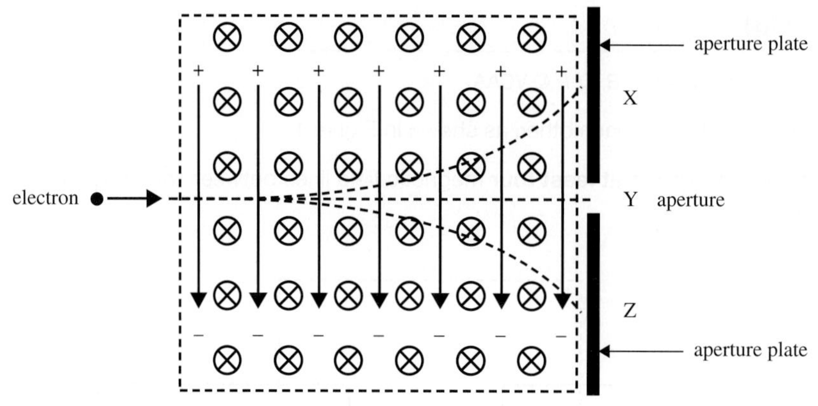

Figure 2

a. Show that the velocity of an electron that travels straight through the aperture to point Y is given by $v_0 = \frac{E}{B}$. **(1 mark)**

b. Calculate the magnitude of the velocity, v_0, of an electron that travels straight through the aperture to point Y if $E = 500\ \text{kV m}^{-1}$ and $B = 0.25$ T. Show your working. **(2 marks)**

c. i. At which of the points – X, Y or Z – in Figure 2 could electrons travelling faster than v_0 arrive? **(1 mark)**

ii. Explain your answer to **part c. i.** **(2 marks)**

UNIT 3 | AREA OF STUDY 2 REVIEW

AREA OF STUDY 2 How do things move without contact?

OUTCOME 2

Analyse gravitational, electric and magnetic fields, and apply these to explain the operation of motors and particle accelerators and the orbits of satellites.

PRACTICE EXAMINATION

STRUCTURE OF PRACTICE EXAMINATION		
Section	**Number of questions**	**Number of marks**
A	20	20
B	3	25
	Total	**45**

Duration: 50 minutes
Information:

- This practice examination consists of two parts. You must answer all question sections.
- Pens, pencils, highlighters, erasers, rulers and a scientific calculator are permitted.
- You may use the VCAA Physics formula sheet for this task.

Resources

Weblink VCAA Physics formula sheet

SECTION A — Multiple choice questions

All correct answers are worth 1 mark each; an incorrect answer is worth 0.

1. The planet Jupiter has a mass of 1.9×10^{27} kg and a radius of 7.0×10^{7} m. What is the gravitational field strength on the surface of Jupiter?
   A. 26 N kg^{-1}
   B. 59 N kg^{-1}
   C. 62 N kg^{-1}
   D. 95 N kg^{-1}
2. The gravitational field strength on the surface of the Moon, a distance of r_{Moon} from the centre of the Moon, is approximately 1.6 N kg^{-1}. Which of the following is the best estimate of the gravitational field strength at an altitude of $2 \times r_{\text{Moon}}$ above the Moon?
   A. 0.18 N kg^{-1}
   B. 0.40 N kg^{-1}
   C. 0.53 N kg^{-1}
   D. 0.80 N kg^{-1}

3. Two spherical bodies, each with a mass of 4.5 kg and radius of 4.5 m, are touching each other in deep space. They exert an attractive force on each other. The magnitude of this attractive force would be closest to:
   A. 1.5×10^{-11} N.
   B. 1.7×10^{-11} N.
   C. 6.7×10^{-11} N.
   D. 3.0×10^{-10} N.
4. The gravitational field around Earth may be described as:
   A. uniform and static.
   B. non-uniform and changing in time.
   C. uniform and changing in time.
   D. non-uniform and static.

Use the following information to answer questions 5 and 6:
Two satellites, Rhodium-45 and Palladium-64, are orbiting Earth at the same altitude. Palladium-64 is twice the mass of Rhodium-45.

5. Which of the following statements best describes the orbital period of each satellite and the gravitational force by Earth on each satellite?
   A. The orbital period of each satellite is the same and the gravitational force acting on each satellite is the same.
   B. The orbital period of each satellite is different and the gravitational force acting on each satellite is the same.
   C. The orbital period of each satellite is the same and the gravitational force acting on each satellite is different.
   D. The orbital period of each satellite is different and the gravitational force acting on each satellite is different.
6. Rhodium-45 is moved by a booster rocket to a higher orbit. Which of the following statements best describes the orbital period and orbital speed of Rhodium-45 in the new orbit?
   A. The orbital period has increased and the orbital speed has increased.
   B. The orbital period has increased and the orbital speed has decreased.
   C. The orbital period has decreased and the orbital speed has increased.
   D. The orbital period has decreased and the orbital speed has decreased.
7. The force acting on an electric charge of $+9.6 \times 10^{-19}$C in a uniform electric field with a magnitude of 250 V m^{-1} is closest to:
   A. 2.4×10^{-16} N.
   B. 2.6×10^{-16} N.
   C. 2.4×10^{-19} N.
   D. 2.6×10^{-19} N.

Use the following information to answer questions 8 and 9:
A small sphere is fixed in place with a charge of +4.5 μC. Another sphere, with a charge of –3.2 μC and suspended on a string, is brought close to the fixed sphere. The two spheres may be considered as point charges.

8. The distance between the centres of the two spheres is 7.5 cm. What is the magnitude of the electric force acting on the fixed sphere?
   A. 2.3×10^{-4} N
   B. 2.3×10^{-3} N
   C. 23 N
   D. 2.3×10^{2} N

9. What is the size of the electric field strength due to the fixed sphere at the position of the suspended sphere?
   A. 5.1×10^6 N C^{-1}
   B. 7.2×10^6 N C^{-1}
   C. 8.1×10^6 N C^{-1}
   D. 9.2×10^6 N C^{-1}
10. Two metal plates are placed parallel to each other and connected to a 120-V DC power supply. A uniform electric field is created between the two plates, which are separated by a distance of 32 mm. The strength of the uniform electric field between the two plates is approximately:
   A. 38 N C^{-1}.
   B. 3.8×10^2 N C^{-1}.
   C. 3.8×10^3 N C^{-1}.
   D. 3.8×10^4 N C^{-1}.
11. An electron is travelling in a straight line in a vacuum tube. Its path may be deflected by a force field. Which of the following fields could cause the deflection?
   A. An electric field only
   B. Either an electric field or a magnetic field only
   C. Either an electric field, a magnetic field or a gravitational field
   D. A gravitational field only
12. Two metal plates are used to create a uniform electric field between them with a field strength of 2300 N C^{-1}. The distance between the plates is 3.0 cm. An electron is held at rest on the negatively charged plate. It is released and accelerated towards the positively charged plate. What is the gain in kinetic energy by the electron when it just reaches the positively charged plate?
   A. 1.1×10^{-24} J
   B. 1.1×10^{-21} J
   C. 1.1×10^{-19} J
   D. 1.1×10^{-17} J
13. The diagram shows the electric field lines between two charges q_1 and q_2 of different magnitude.
   The best description of the two charges is that the:
   A. charges are both positive.
   B. charges are both negative.
   C. left-hand charge is positive and the right-hand charge is negative.
   D. left-hand charge is negative and the right-hand charge is positive.

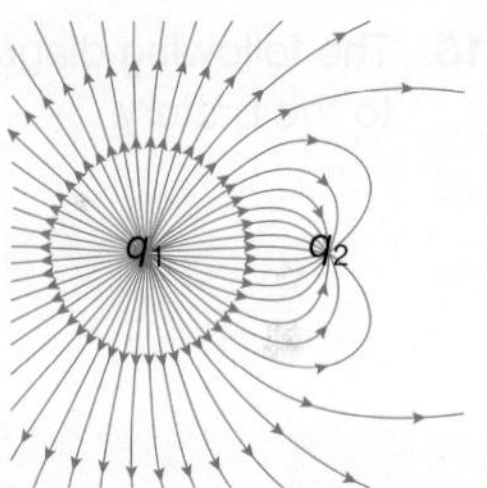

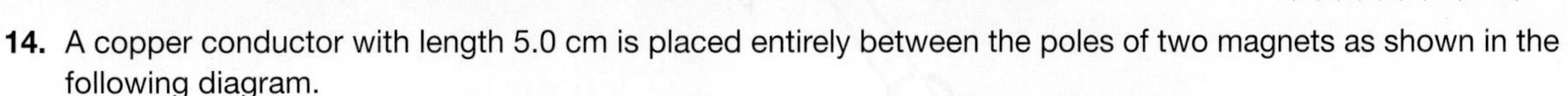

14. A copper conductor with length 5.0 cm is placed entirely between the poles of two magnets as shown in the following diagram.

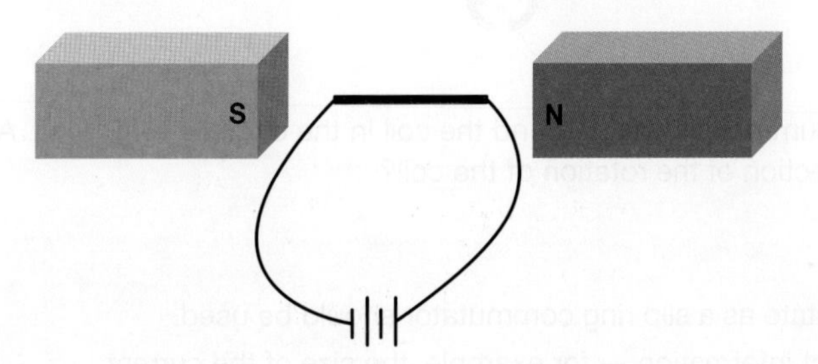

   The magnetic field strength is 3.0 mT and the current in the conductor is 4.0 A. What is the magnitude of the magnetic force on the conductor?
   A. 0.0 N
   B. 2.0×10^{-4} N
   C. 4.0×10^{-4} N
   D. 6.0×10^{-4} N

15. The field lines of Earth's magnetic field run from south to north on the surface of the Earth. A power line carrying a DC current runs from east to west. What is the direction of the magnetic force acting on the power line?

    A. Up
    B. Down
    C. North
    D. South

16. A straight conductor is carrying a current of 5.0 A. Which of the following diagrams best represents the direction of the magnetic field, in blue, around the conductor?

    A.
    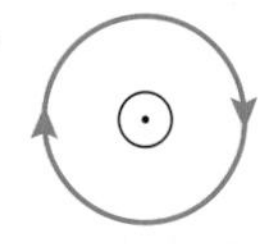

    B.
    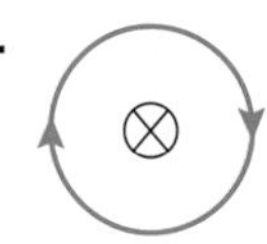

    C.
    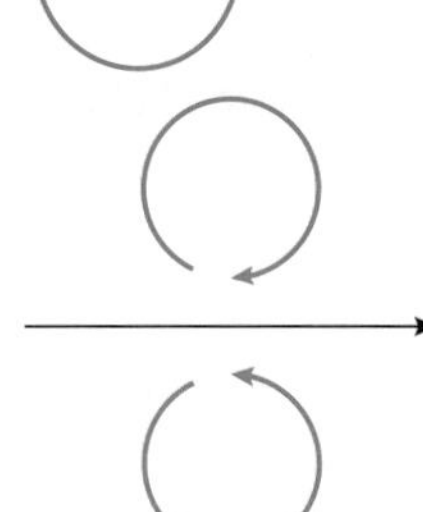

    D.
    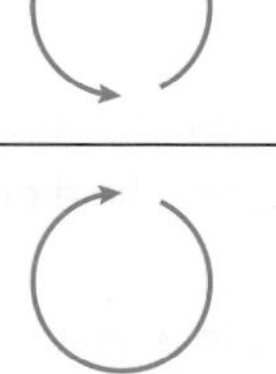

17. A straight length of wire carries a current of 3.5 A and passes perpendicularly within a magnetic field of unknown field strength. A force meter measures the magnetic force on the wire to be 0.35 N. The length of the portion of wire within the magnetic field is 15 cm. The magnetic field strength is approximately:

    A. 6.7 T.
    B. 0.67 T.
    C. 0.067 T.
    D. 0.0067 T.

18. The following diagram shows a model DC motor with a split ring commutator that connects the rotating coil to the battery.

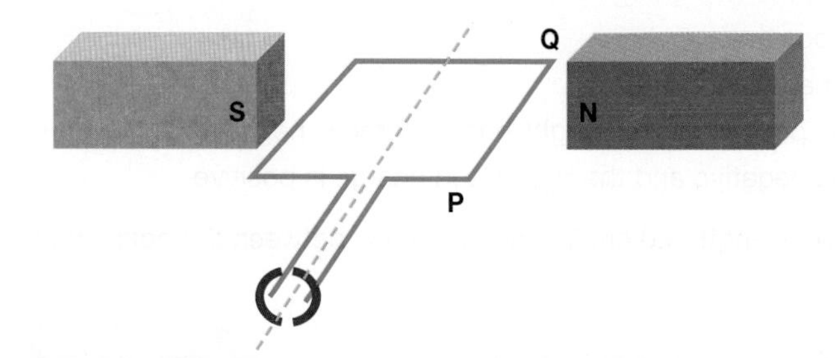

    At one instance, the current is flowing around the coil in the direction of P to Q. As seen from the front of this motor, what is the direction of the rotation of the coil?

    A. Clockwise
    B. Anticlockwise
    C. The coil will not rotate as a slip ring commutator should be used.
    D. There is insufficient information — for example, the size of the current.

19. A stream of electrons enters a region of magnetic field with a field strength of 8.0 mT that is oriented perpendicularly to the direction of electron motion. The electrons are travelling at a velocity of 8.0×10^6 m s^{-1}. What is the radius of the electron path within the magnetic field?

    A. 5.7 μm
    B. 5.7 mm
    C. 5.7 cm
    D. 5.7 m

20. A stream of electrons may be deflected by a magnetic field or by an electric field. Which of the following options best describes the path taken by the electrons after being deflected?

	Uniform magnetic field	Uniform electric field
A.	Circular path	Circular path
B.	Parabolic path	Parabolic path
C.	Circular path	Parabolic path
D.	Parabolic path	Circular path

SECTION B — Short answer questions

Question 21 (10 marks)

Saturn has at least 18 natural satellites, two of which are Titan and Tethys. Data for Titan and Tethys is shown below:

Mass of Titan	1.35×10^{23} kg
Radius of Titan	2.57×10^{6} m
Period of Titan's orbit	1.38×10^{6} s
Radius of Titan's orbit	1.22×10^{9} m
Mass of Tethys	6.18×10^{20} kg
Radius of Tethys	5.31×10^{5} m
Radius of Tethys's orbit	2.95×10^{8} m

a. Calculate the gravitational field strength on the surface of Titan. **(2 marks)**

b. Determine the period of Tethys's orbit. **(3 marks)**

c. Determine the mass of Saturn. **(3 marks)**

d. What is the orbital speed of Titan? **(2 marks)**

Question 22 (8 marks)

Electrons are 'liberated' from a hot filament, and are then accelerated across a vacuum by a potential difference of 120 V applied between two plates that are 15 cm apart. The electric field between the two plates may be considered to be uniform in strength.

a. Determine the electric field strength between the plates. **(2 marks)**

b. Calculate the acceleration of the electrons between the plates. **(2 marks)**

c. What is the gain in kinetic energy by the electrons as they cross between the two plates? **(2 marks)**

d. Assuming that the initial velocity of the electrons is zero, what is the final velocity attained by the electrons? **(2 marks)**

Question 23 (7 marks)

A student builds a model DC motor with 40 turns of wire, and places the coil entirely within a magnetic field as shown in the following diagram. She connects the commutator to a battery, and the coil rotates clockwise when viewed from the position of the commutator.

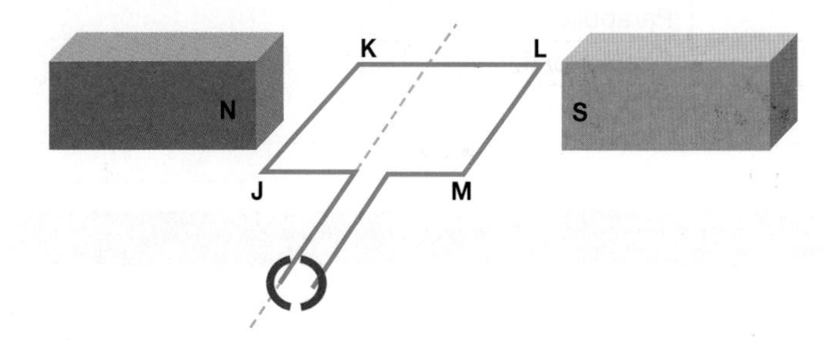

a. At one moment as the coil is rotating, the coil is in the position shown in the diagram. What is the direction of the current in the coil? **(1 mark)**

b. The current in the coil is 3.2 A, and the length of the side JK is 2.9 cm. The force on the side JK is measured to be 1.2 N. What is the magnitude of the magnetic field? **(2 marks)**

c. The motor is stopped and the coil is positioned again as shown in the diagram. The student replaces the commutator with split rings. She then reconnects the battery. Carefully explain what happens to the motion of the coil. **(2 marks)**

d. Another student decides to alter his own motor and instead uses only 17 turns of wire. He wants to ensure that the current and the force stay the same as the first student's motor. What would the length of the side JK need to be to enable this? **(2 marks)**

UNIT 3 | AREA OF STUDY 2

PRACTICE SCHOOL-ASSESSED COURSEWORK

ASSESSMENT TASK — APPLICATION OF PHYSICS CONCEPTS TO EXPLAIN A MODEL, THEORY, DEVICE, DESIGN OR INNOVATION

In this task, you will be required to analyse gravitational, electric and magnetic fields to explain the operation of a type of particle accelerator, the Pelletron.

- This practice SAC requires you to describe the operation of a device; a structured set of questions is supplied to assist you to write your description. You may choose to present your explanation as a report or a multimodal or oral presentation, or break it down and answer question by question.
- You may use the VCAA Physics formula sheet and a scientific calculator for this task.

Total time: 55 minutes (5 minutes reading, 50 minutes writing)

Total marks: 46 marks

EXPLORING THE OPERATION OF 5U PELLETRON

The University of Melbourne houses a particle accelerator, the 5U Pelletron, which is capable of accelerating positively charged particles. These charged particles can be directed at targets used for research purposes ranging from fundamental nuclear physics through to medical research projects and the analysis of materials.

The Pelletron consists of an ion source that provides particles capable of being accelerated, and a device similar to a Van de Graaff generator to provide a variable potential difference V. At the base of the accelerator, a large magnet is used to direct beams of charged particles towards samples to be analysed. In essence, the machine is capable of directing positive ions of a specific kinetic energy at a target. A schematic of the machine is shown in the following diagram.

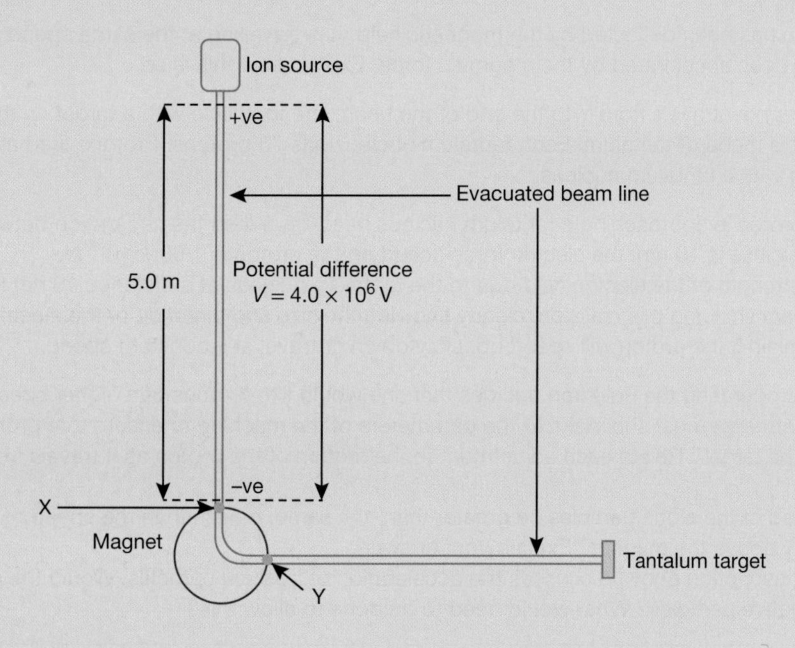

Relevant data:

- mass of a proton: 1.67×10^{-27} kg
- charge of a proton: 1.60×10^{-19} C
- mass of an alpha particle: 6.68×10^{-26} kg
- charge of an alpha particle: 3.20×10^{-19} C

In one experiment, protons are produced by the ion source and accelerated by the potential difference, V.

1. Why is it necessary for the ions produced by the ion source to be positively charged and not negatively charged?

The accelerated protons travel along an evacuated beam line from the ion source to the target.

2. Why is it important for the beam line to be evacuated — that is, for the beam line to be a vacuum?
3. Determine the work done on these protons and hence their kinetic energy when they arrive at point X after leaving the ion source and being accelerated.
4. Use your value for the kinetic energy of the proton to determine its speed at point X directly above the magnet.

Consider now if the ion source produced alpha particles instead of protons.

5. Calculate the strength of the electric field in the region between the positive and negative plates of the Pelletron. Give the direction of the electric field in this region (up or down) and explain why you have chosen it.
6. Calculate the size of the electric force acting on a proton moving from the ion source to point X and hence its acceleration down the beam line.

Protons moving from the ion source of the Pelletron to point X also undergo a loss of gravitational potential energy.

7. Explain why this transformation of energy can be ignored for protons moving from the ion source to point X. A calculation would assist in your answer. Use $\boldsymbol{g}$ = 9.8 N kg^{-1} down.

At point X, the protons enter a region where there is a uniform magnetic field. This field is used to exert a force on the protons so they move in the arc of a circle. They exit the magnetic field at point Y from which they are directed towards a target situated at the end of a beam line.

8. Explain why the magnetic field interacting with the protons causes them to move in the arc of a circle as they move from point X to point Y and why this could not be achieved with a uniform electric field.
9. Determine the direction of the magnetic field, clearly explaining your method.

The radius of the circle that the protons travel is 0.80 m.

10. Calculate the magnitude of the magnetic field of the magnet.
11. Use your results for questions 4 and 11 to determine the size of the magnetic force acting on a proton as it moves from X to Y.
12. After the proton has been deflected by the magnetic field, it is travelling at the same speed as at point X, yet the proton has been accelerated by the magnetic force. Explain how this is so.

The beam of protons now travels from Y to the end of the beam line to collide with a target. In this particular instance, the target is made of tantalum. Each tantalum nucleus has 73 protons. Protons fired at the target are capable of colliding with a tantalum nucleus.

In one collision, a proton is approaching a tantalum nucleus head-on. When the separation between the proton and the tantalum nucleus is 10 nm, the electric force acting on the proton is 1.68×10^{-8} N.

13. Calculate the strength of the electric field due to the tantalum nucleus at a distance 10 nm from its centre.
14. Sketch a diagram showing this collision, clearly showing the size and direction of the electric force acting on the proton. Explain if the proton will speed up, slow down or travel at a constant speed.

A university student operating the Pelletron decides that she would like protons of a higher speed.

15. What two adjustments must she make to the parameters of the machine to ensure that protons of higher speed still hit the target? Relate each adjustment to the motion of the proton as it travels from the ion source to the target.
16. Would the speed of the alpha particles be greater than, the same, or less than the speed of the protons at point X, directly above the magnet? Explain your answer.
17. The Pelletron description above discusses the acceleration of positive particles. Would the device be able to accelerate negative particles? What would need to be done to allow this?

Digital document School-assessed coursework (doc-39422)

6 Generation of electricity

KEY KNOWLEDGE

In this topic, you will:

- calculate magnetic flux when the magnetic field is perpendicular to the area, and describe the qualitative effect of differing angles between the area and the field: $\Phi_B = B_{\perp}A$
- investigate and analyse theoretically and practically the generation of electromotive force (emf) including AC voltage and calculations using induced emf: $\varepsilon = -N\frac{\Delta\Phi_B}{\Delta t}$, with reference to:
  - rate of change of magnetic flux
  - number of loops through which the flux passes
  - direction of induced emf in a coil
- explain the production of DC voltage in DC generators and AC voltage in alternators, including the use of split ring commutators and slip rings respectively.
- describe the production of electricity using photovoltaic cells and the need for an inverter to convert power from DC to AC for use in the home (not including details of semiconductors action or inverter circuitry).

PRACTICAL WORK AND INVESTIGATIONS

Practical work is a central component of VCE Physics. Experiments and investigations, supported by a **practical investigation eLogbook** and **teacher-led video**, are included in this topic to provide opportunities to undertake investigations and communicate findings.

EXAM PREPARATION

Access past VCAA questions and exam-style questions and their video solutions in every lesson, to ensure you are ready.

6.1 Overview

Hey students! Bring these pages to life online

Watch videos

Engage with interactivities

Answer questions and check results

Find all this and MORE in jacPLUS

6.1.1 Introduction

Hans Christian Oersted's discovery that a current-carrying wire induces a magnetic field was the first indication of the relationship between electricity and magnetism. Following this, scientists started to investigate the relationship in more detail. In 1831, Michael Faraday completed the link between electricity and magnetism when he found that electricity could be produced from magnetism. This discovery was monumental, allowing the generation of the massive amounts of electricity that society now relies on.

In this topic, you will be introduced to the concept of magnetic flux and apply Faraday's findings to create devices that generate electricity.

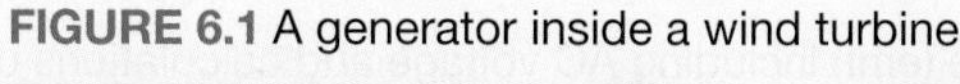
FIGURE 6.1 A generator inside a wind turbine

LEARNING SEQUENCE

on Resources

Solutions — Solutions — Topic 6 (sol-0820)

Practical investigation eLogbook — Practical investigation eLogbook — Topic 6 (elog-1637)

Digital documents — Key science skills — VCE Physics Units 1–4 (doc-36950)
Key terms glossary — Topic 6 (doc-37175)
Key ideas summary — Topic 6 (doc-37176)

Exam question booklet — Exam question booklet — Topic 6 (eqb-0103)

6.2 BACKGROUND KNOWLEDGE Generating voltage and current with a magnetic field

BACKGROUND KNOWLEDGE

- Explain how voltage and current is generated with a magnetic field.
- Investigate and analyse the electromotive force (emf) induced in a moving conductor: $\varepsilon = Blv$.

6.2.1 Generating voltage

What should happen when a metal rod moves through a magnetic field? Imagine a horizontal rod falling down through a magnetic field as shown in figure 6.2.

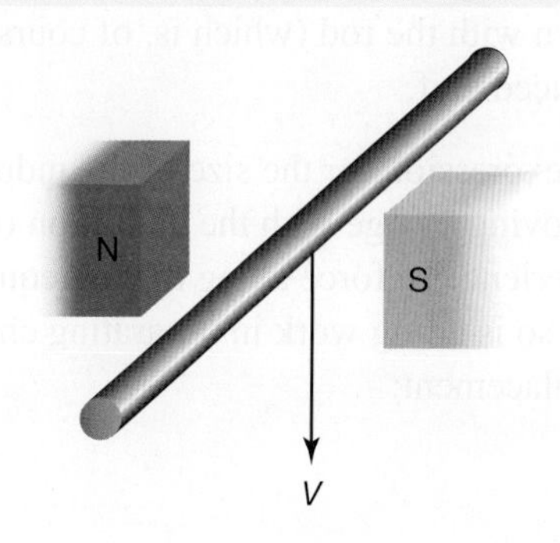

FIGURE 6.2 A metal rod falling down through a magnetic field

As the rod falls, the electrons and the positively charged nuclei in the rod are both moving down through the magnetic field. (It is important to recall that velocity, $\boldsymbol{v}$, is perpendicular to the direction of the magnetic field. If the velocity is parallel to the magnetic field, the force will be zero.) The magnetic field will exert a magnetic force on the electrons, and on the nuclei. The magnitude of this force, $\boldsymbol{F}$, is equal to the charge, q, multiplied by the velocity of the charge, $\boldsymbol{v}$, multiplied by the strength of the magnetic field, $\boldsymbol{B}$.

In which direction will the magnetic force act on the electrons and the nuclei?

The hand rules from topic 5 can be used for both the electrons and the nuclei, keeping in mind that the hand rules use conventional current, so electrons moving down are equivalent to positive charges moving up.

The force on the electrons will be towards the far end of the rod (into the page), while the force on the nuclei will be to the near end of the rod (out of the page), as is shown in figure 6.3.

FIGURE 6.3 The magnetic field forces electrons to the far end of the falling rod.

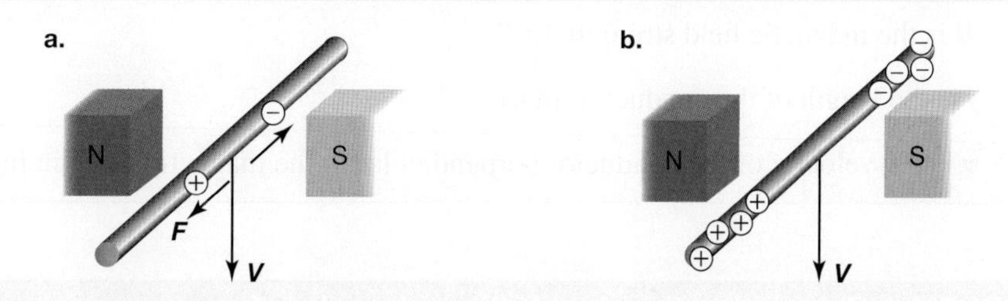

The atomic structure of the metal prevents the movement of the positively charged nuclei. The negatively charged electrons, on the other hand, are free to move. The electrons move towards the far end of the rod, leaving the near end deficient in electrons and thus positively charged.

Not all electrons move to the far end. As the far end becomes more negative, there will be an increasingly repulsive force on any extra electrons at this end of the rod. Similarly, there will be an increasingly attractive force from the positively charged near end, attempting to keep the remaining electrons at that end.

The movement of the metal rod through the magnetic field has resulted in the separation of charge, causing a potential difference, or voltage drop, between the ends. This is called **induced voltage**. As long as the rod keeps moving, the charges will remain separated. As soon as the rod stops falling (at which point the velocity of the charges equals zero), the magnetic force on the charges is reduced to zero; electrons are then attracted back to the positive end and the electrons return to being distributed evenly in the rod.

The charge in the moving rod is separated by the magnetic field, but the charge has nowhere to go. A source of voltage, an **emf** (electromotive force), has been produced. It is like a DC battery with one end positive and the other negative.

The size of the induced emf depends on the number of electrons shifted to one end. The electrons are shifted by the magnetic force until their own repulsion balances this force. So, the larger the magnetic force pushing the electrons, the more electrons are shifted to one end of the rod.

The size of this pushing magnetic force depends on the size of the magnetic field and the velocity of the charge. Assuming the magnetic field is constant, the size of the force depends on how fast the electrons are moving down with the rod (which is, of course, how fast the rod is falling). So the faster the rod falls, the larger the induced emf.

An expression for the size of the induced emf can be obtained by combining the expression for the force on a moving charge with the definition of voltage. When the rod is moving down with velocity $\boldsymbol{v}$, each electron experiences a force along the rod equal to $q\boldsymbol{vB}$. This force pushes the electron along the length l of the rod and so is doing work in separating charge. The amount of work done, in joules, is equal to the force times the displacement:

$$W = \text{force} \times \text{displacement} = q\boldsymbol{vB} \times l$$

induced voltage voltage caused by the separation of charge due to the presence of a magnetic field

emf source of voltage that can cause an electric current to flow

However, the definition of emf (or the voltage drop across the rod) is energy supplied per unit of charge, measured in joules per coulomb, or volts. So the induced emf (ε) is given by $\varepsilon = \frac{W}{q} = \frac{Blqv}{q}$, which gives:

$$\varepsilon = \boldsymbol{B}l\boldsymbol{v}$$

where: ε is the induced emf, in V

$\boldsymbol{B}$ is the magnetic field strength, in T

l is the length of the conductor, in m

$\boldsymbol{v}$ is the velocity of the conductor perpendicular to the magnetic field, in m s^{-1}

tlvd-8998

SAMPLE PROBLEM 1 Calculating the size of the induced emf across a rod

A 5.0-cm metal rod moves at right angles across a magnetic field of strength 0.25 T at a speed of 40 cm s^{-1}. What is the size of the induced emf across the ends of the rod?

THINK	WRITE
Recall the formula for the induced emf of a conductor: $\varepsilon = \boldsymbol{B}l\boldsymbol{v}$ where: $l = 0.05$ m, $\boldsymbol{v} = 0.4$ m s^{-1} and $\boldsymbol{B} = 0.25$ T. Ensure that the units have been converted to those required in the formula.	$\varepsilon = \boldsymbol{B}l\boldsymbol{v}$ $= 0.25 \times 5.0 \times 10^{-2} \times 0.4$ $= 5.0 \times 10^{-3}$ V $= 5.0$ mV

PRACTICE PROBLEM 1

At what speed would the rod need to move to induce an emf of 1.0 V?

6.2.2 Generating a current

Emfs can be used to produce a current by attaching a wire to each end of the metal rod and connecting these wires outside the magnetic field, producing a closed circuit. Now the electrons have the path of a low-resistance conductor to move to the positively charged end of the rod.

Once the electrons reach the positive end, they will be back in the magnetic field, falling down through the magnetic field with the metal rod, and will again experience a magnetic force pushing them to the far end of the rod. The electrons will then move around the circuit for a second time.

The electrons will continue to flow around the circuit as long as the wire is moving through the magnetic field. An electric current has been generated (see figure 6.4).

FIGURE 6.4 The accumulated electrons at the far end of a rod move to the positive end through the connecting wire.

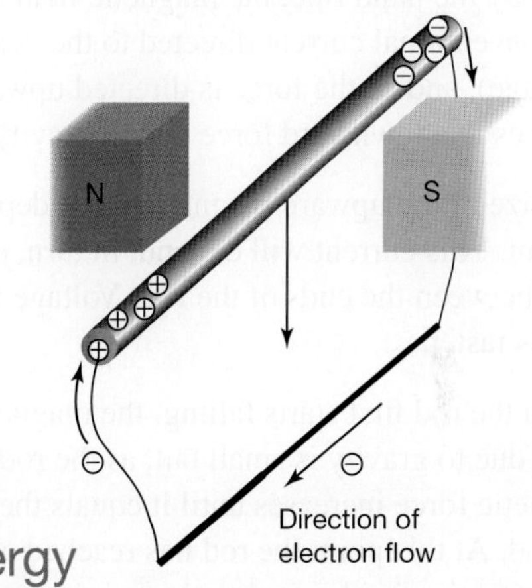

6.2.3 The source of a current's electrical energy

Electric current has electrical energy. Where did this energy come from? Before the rod was released, it had gravitational potential energy. If it is dropped outside the magnetic field (see figure 6.5b), this gravitational potential energy is transformed into kinetic energy. If it is dropped inside the magnetic field (see figure 6.5a), the potential energy is transformed into kinetic energy and electrical energy. Since energy is conserved (i.e. it cannot be created or destroyed), there must be less kinetic energy in the rod falling in the magnetic field than the rod falling outside of the magnetic field. That is, the rod in the magnetic field is falling at a slower speed. Why?

FIGURE 6.5 (a) Inside the magnetic field, the gravitational potential energy of the falling rod is converted into both kinetic energy and electrical energy, whereas (b) outside the magnetic field, it is converted only into kinetic energy.

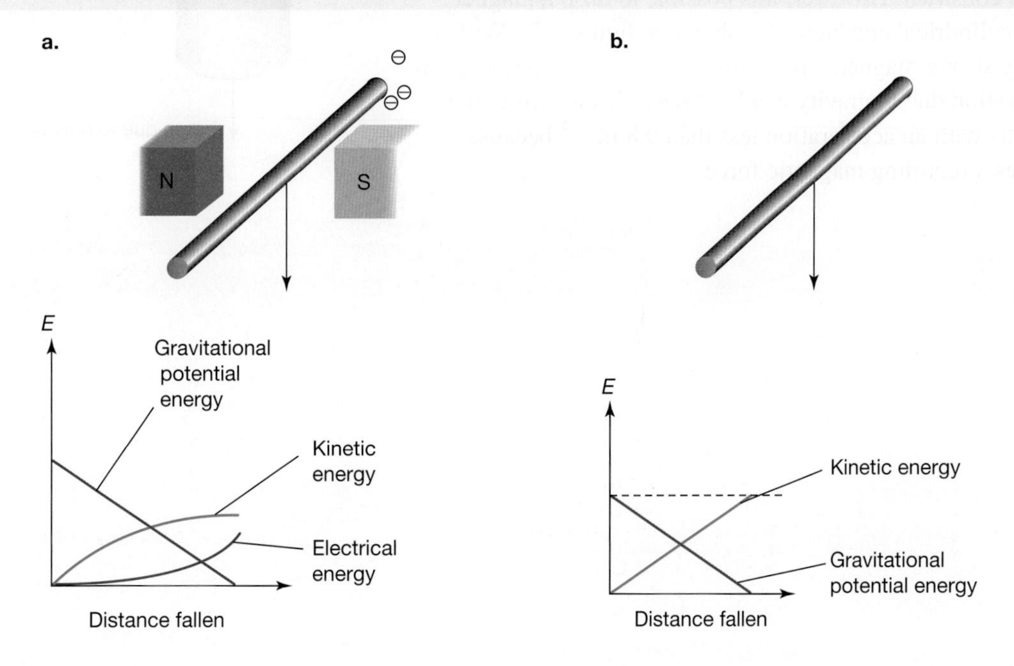

The induced current in the falling rod means that, when the electrons are in the rod, they are moving in two directions — downwards with the rod and along the rod.

The downward movement of the rod through the magnetic field produces the sideways force along the rod that keeps the current flowing. But if the electrons are moving along the rod, what is the effect of the magnetic field?

The movement of electrons along the rod means there is a current-carrying conductor in the magnetic field, causing the field to exert a second force on the rod. The direction of this force is once again given by the hand rule; the magnetic field is directed to the right, the conventional current directed to the near end of the rod (out of the page), and so the force is directed upwards. This magnetic force opposes the downward force due to gravity on the rod.

The size of the upward magnetic force depends on the size of the current. This current will depend, in turn, on the size of the voltage drop between the ends of the rod. Voltage will increase as the rod moves faster.

When the rod first starts falling, the magnetic force opposing the force due to gravity is small but, as the rod falls faster, the opposing magnetic force increases until it equals the force due to gravity on the rod. At this point the rod has reached a maximum steady speed. This situation is identical to the terminal velocity experienced by objects falling through the air.

As the metal rod falls through the magnetic field at constant speed, its kinetic energy remains constant, so the loss in gravitational potential energy is converted to electrical energy as the generated emf drives the current through the circuit.

This effect is difficult to demonstrate in practice; a magnetic field large enough for the rod to achieve terminal velocity is too difficult to construct. However, it is possible to drop a magnet through a cylindrical conductor (as shown in figure 6.7). With a sufficiently strong magnet, a measurable decrease in speed against the acceleration due to gravity can be observed. Therefore, the magnet falls with an acceleration less than 9.8 m s^{-2} because it experiences a retarding magnetic force.

FIGURE 6.6 The magnetic force opposes the weight of the rod.

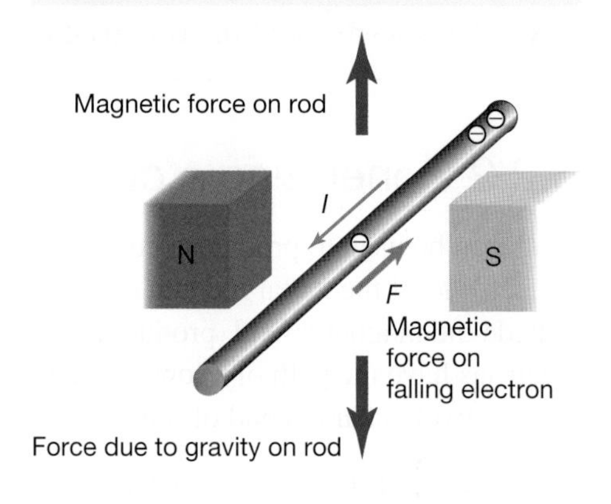

FIGURE 6.7 A magnet falling through a metal tube

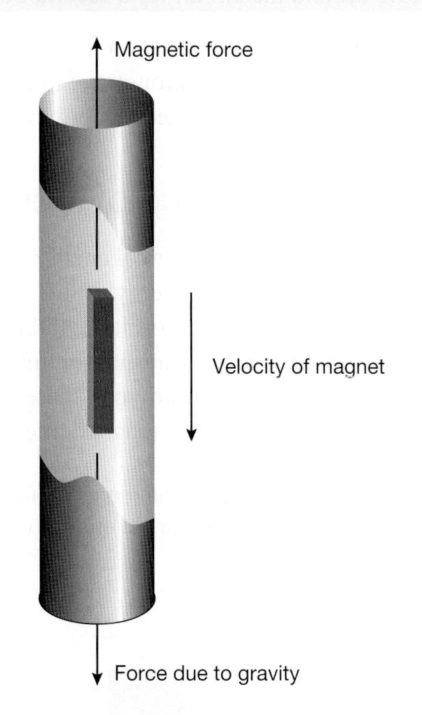

6.2 Activities

Students, these questions are even better in jacPLUS

Receive immediate feedback and access sample responses

Access additional questions

Track your results and progress

Find all this and MORE in jacPLUS

6.2 Quick quiz	6.2 Exercise	6.2 Exam questions

6.2 Exercise

1. A rod of length 10 cm is being moved at a speed of 0.20 m s^{-1} perpendicularly to a magnetic field of strength 0.70 T. Calculate the induced emf in the rod.
2. A rod of length 14.3 cm is being moved perpendicularly to a magnetic field of strength 0.850 T. It generates an emf of 100 mV. Determine the speed at which the rod is moving.
3. A rod is being moved at a speed of 0.18 m s^{-1} perpendicularly to a magnetic field of strength 0.60 T. It generates an emf of 90 mV. Determine the length of the rod.
4. A rod of length 15 cm is held stationary in a vertical position and then dropped downwards through a magnetic field of strength 100 T. The magnetic field is directed vertically downwards. Determine the induced emf in the rod 4.0 s after it was dropped.
5. A rod of length 70 mm is being moved at a speed of 5.0 cm s^{-1} perpendicularly to a magnetic field. It generates an emf of 35 mV. Determine the strength of the magnetic field.
6. An orbiting satellite has a small module tethered to it by a 5.0-km conducting cable. As the satellite and its module orbit Earth, they cut across Earth's magnetic field at right angles.
   a. If the pair are travelling at a speed of 6000 m s^{-1}, how far do they travel in 1.0 s?
   b. What area does the conducting cable cross during the 1.0-s period?

6.2 Exam questions

Question 1 (3 marks)

An aeroplane with a wingspan of 30 m is flying at right angles to the Earth's magnetic field at a speed of 2200 km h^{-1}. If the strength of the Earth's magnetic field is 5.0×10^{-5} T, determine the size of the induced emf between the tips of the aeroplane's wings.

Question 2 (2 marks)

As the metal rod in the diagram falls through the magnetic field, charge is separated and a voltage is established between the two ends of the rod. This requires energy. Where did the energy come from?

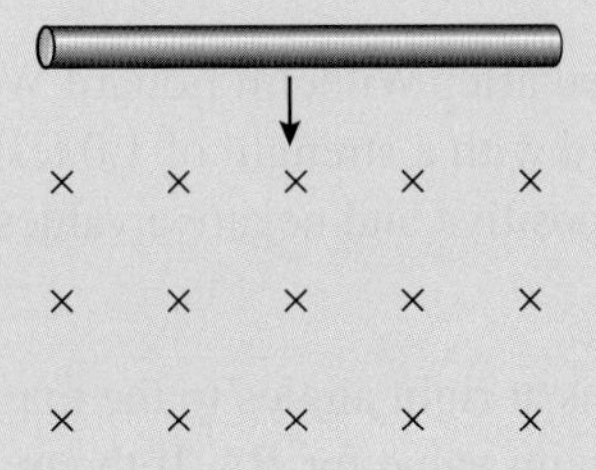

Question 3 (3 marks)

Explain what happens to the voltage between the ends of the rod in question 2 as the rod falls faster.

Question 4 (2 marks)

Calculate the average emf in the axle of a car travelling at 120 km h^{-1} if the vertical component of the Earth's magnetic field is 40 μT and the length of the axle is 1.5 m.

Question 5 (2 marks)

A boat sails at 2.5 m s^{-1} due west, where the Earth's magnetic field is 5.0×10^{-5} T due north and horizontal. The boat carries a vertical aerial 4.0 m long. Calculate the magnitude of the emf induced in the aerial.

More exam questions are available in your learnON title.

6.3 Magnetic flux

KEY KNOWLEDGE

- Calculate magnetic flux when the magnetic field is perpendicular to the area, and describe the qualitative effect of differing angles between the area and the field: $\Phi_B = B_{\perp}A$.

6.3.1 What is magnetic flux?

Magnetic flux is the amount of magnetic field passing through an area, such as a coil.

The stronger the magnetic field going through an area, the larger the magnetic flux. Similarly, the larger the area that the magnetic field is going through, the larger the magnetic flux. Due to this definition, the magnetic field strength is sometimes referred to as the **magnetic flux density**.

Magnetic flux can be calculated as the product of the strength of the magnetic field and the area that is perpendicular to the field lines.

$$\Phi_B = B_{\perp}A$$

where: Φ_B is the magnetic flux, in Wb

B is the strength of the magnetic field, in T

A is the area perpendicular to the magnetic field, in m^2

$\perp$ indicates that the area referred to in the formula is the area perpendicular to the magnetic field.

Magnetic flux is measured in webers (named after Wilhelm Eduard Weber). One weber (Wb) is the amount of magnetic flux from a uniform magnetic field with a strength of 1.0 tesla passing through an area of 1.0 square metre. The magnetic flux can also take on positive and negative values, depending on which side of the area the magnetic field is coming from.

This description has assumed that the area is at right angles to the magnetic field, as shown in figure 6.8a. This leads to a maximum value for ***BA***. If the magnetic field went through the area at an angle less than 90°, as shown in figure 6.8b, the amount of magnetic flux passing through the area would be less. In fact, if the magnetic field is parallel to the area, the amount of magnetic flux will be zero, as shown in figure 6.9. None of the magnetic field lines pass through the area from one side to the other.

magnetic flux measure of the amount of magnetic field passing through an area; measured in webers (Wb)

magnetic flux density (B) strength of a magnetic field; measured in tesla (T) or weber per square metre (Wb m^{-2})

FIGURE 6.8 Magnetic flux is the amount of magnetic field passing through an area. In **a.** it is the maximum ***BA***; in **b.** the value is less, as fewer field lines pass through the coil.

a.

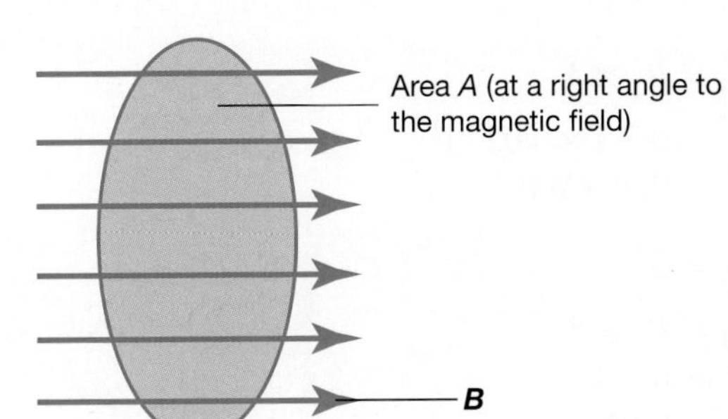

b.

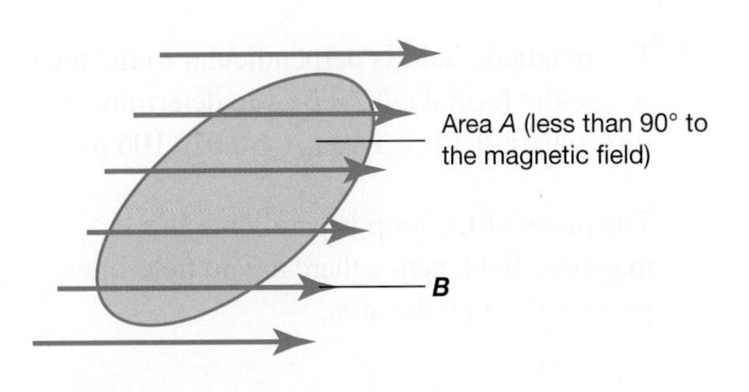

FIGURE 6.9 Zero magnetic flux, as no field lines 'thread' the loop

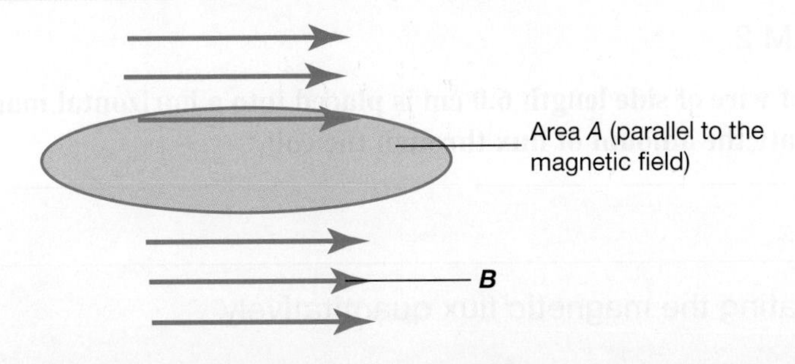

tlvd-8999

SAMPLE PROBLEM 2 Calculating the magnetic flux through a loop

Calculate the magnetic flux in each of the following situations.

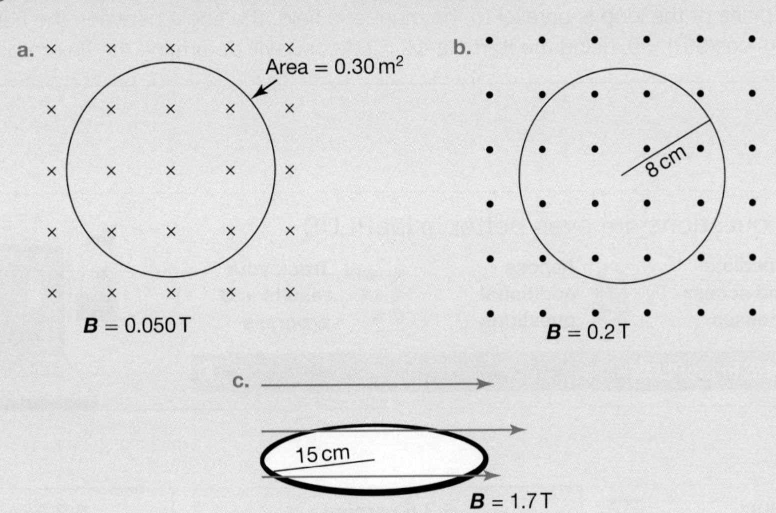

THINK	WRITE
a. The magnetic field is perpendicular to the area, so use the formula $\Phi_B = B_\perp A$ to determine the flux, where $B = 0.05$ T and $A = 0.30$ m^2.	**a.** $\Phi_B = B_\perp \times A$ $= 0.050\ \text{T} \times 0.30\ \text{m}^2$ $= 0.015$ Wb

b. 1. First calculate area A. (Don't forget to convert the radius to metres.)	**b.** $A = \pi r^2$ $= \pi \times 0.08^2$ $= 0.02 \text{ m}^2$
2. The magnetic field is perpendicular to the area, so use the formula $\Phi_B = B_\perp A$ to determine the flux, where $B = 0.2$ T and $A = 0.020\,106 \text{ m}^2$.	$\Phi_B = B_\perp \times A$ $= 0.2 \text{ T} \times 0.02 \text{ m}^2$ $= 0.004 \text{ Wb}$
c. The plane of the loop is parallel to the magnetic field, hence there are no field lines passing through the area.	**c.** $B_\perp = 0$ $\Phi_B = B_\perp \times A$ $= 0 \times A$ $= 0 \text{ Wb}$

PRACTICE PROBLEM 2

A vertical square coil of wire of side length 6.0 cm is placed into a horizontal magnetic field of strength 0.15 T. Calculate the amount of flux through the coil.

EXTENSION: Calculating the magnetic flux quantitatively

In general, the magnetic flux can be expressed in terms of the magnetic field strength, the area and the angle (θ) between the magnetic field and a normal to the area:

$$\Phi_B = BA\cos\theta$$

For example, if the plane of the loop is parallel to the magnetic field, the angle between the field and the normal to the area is 90°. As cos(90°) = 0, using the formula $\Phi_B = BA\cos\theta$ will determine the flux to be 0 Wb.

6.3 Activities

learnon

Students, these questions are even better in jacPLUS

Receive immediate feedback and access sample responses

Access additional questions

Track your results and progress

Find all this and MORE in jacPLUS

6.3 Quick quiz on	6.3 Exercise	6.3 Exam questions

6.3 Exercise

1. What is the difference between magnetic flux and magnetic field strength?
2. Calculate the maximum magnetic flux passing through:
   a. a single coil of area 0.050 m^2 in a magnetic field of strength 3.0 T
   b. a single coil of area 4.5 cm^2 in a magnetic field of strength 0.4 T
   c. a single coil with a radius of 0.3 m in a magnetic field of strength 0.025 T.

3. A square loop of wire of side length 5.0 cm is placed in a magnetic field of strength 0.030 T, as shown in the following diagram. The loop is free to rotate about its horizontal axis.

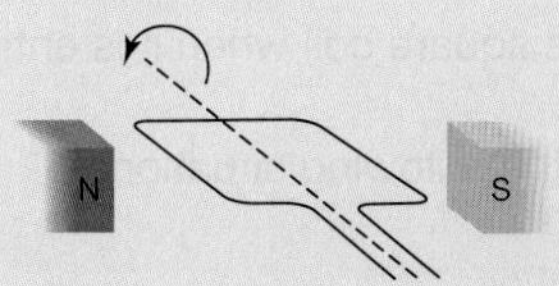

   a. What is the magnetic flux passing through the loop when it is in the position shown in the diagram?
   b. The loop is now rotated 90° about its axis from its original position. What is the magnetic flux passing through the loop when it is in its new position?
   c. Draw a graph of the magnetic flux passing through the loop as it is rotated anticlockwise through one revolution from the position shown in the diagram.
4. In the situation represented in the diagram, the loop has an area of 2.5 m^2 and the magnetic field strength is 7.0 T.

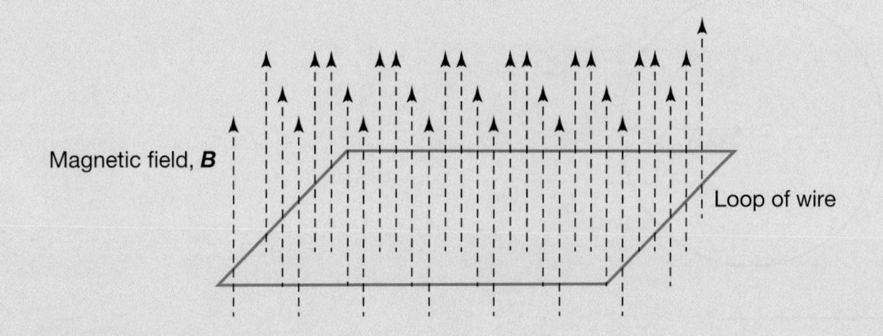

   Calculate the magnetic flux through the loop and state the units used in your answer.
5. A circular conducting coil with 80 loops is placed inside a magnetic field with a field strength of 0.09 T. The area of the circular coil is 38 cm^2.
   Calculate the magnetic flux through the loop.
6. a. In the situation in the diagram shown, the loop has an area of 2.0 m^2 and the magnetic flux through the loop is 6.0 Wb. Calculate the magnetic field strength.
   b. In the situation in the diagram shown, the magnetic flux density is 9 mT and the magnetic flux through the loop is 7.2×10^{-2} Wb. Calculate the area of the loop.

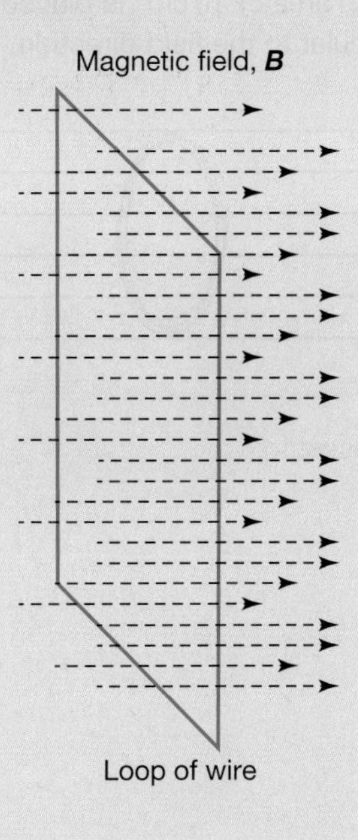

7. Grace and Joshua are investigating electromagnetic induction using a square coil, with side length of 2 cm, which they place between the poles of a magnet. The uniform magnetic field between the poles of the magnet is 5.0×10^{-2} T, and zero elsewhere.
   Calculate the magnetic flux through the square coil when it is entirely within and perpendicular to the magnetic field.
8. Calculate the magnetic flux in each of the following situations.
   a.
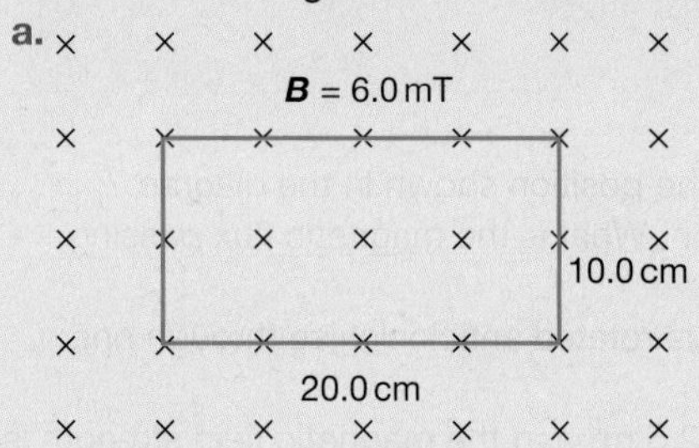

   b.
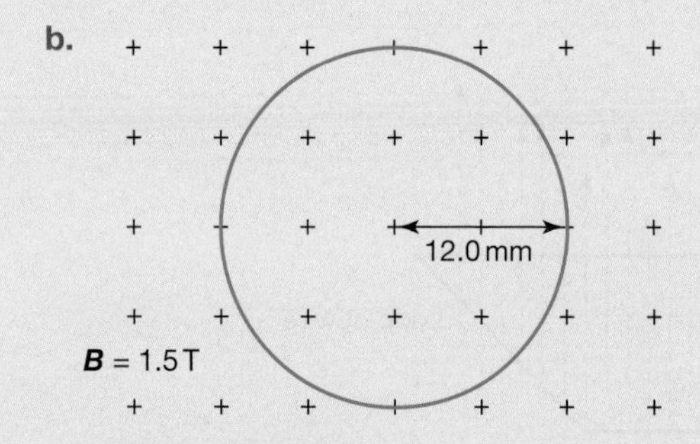

   c.
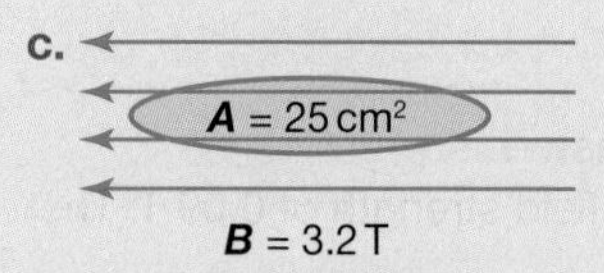

6.3 Exam questions

Question 1 (1 mark)

***Source:** VCE 2020 Physics Exam, Section A, Q5; © VCAA*

MC A coil consisting of 20 loops with an area of 10 cm^2 is placed in a uniform magnetic field B of strength 0.03 T so that the plane of the coil is perpendicular to the field direction, as shown in the diagram below.

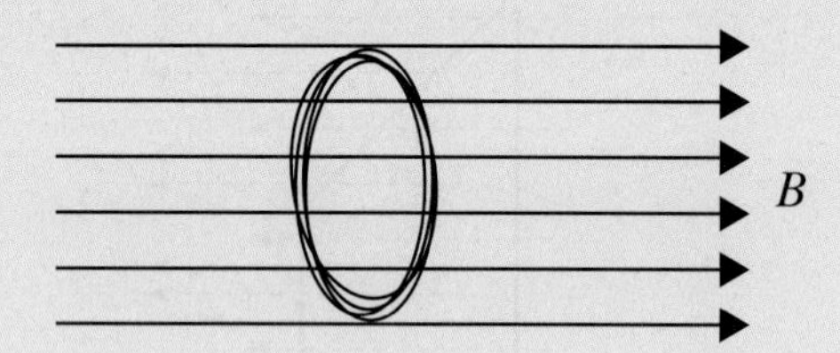

The magnetic flux through the coil is closest to

A. 0 Wb
B. 3.0×10^{-5} Wb
C. 6.0×10^{-4} Wb
D. 3.0×10^{-1} Wb

Question 2 (3 marks)

Source: *VCE 2017, Physics Exam, Section B, Q.5.a; © VCAA*

The alternator in the figure has a rectangular coil with sides of 0.30 m × 0.40 m and 10 turns. The coil rotates four times a second in a uniform magnetic field. The magnetic flux through the coil in the position shown is 0.20 Wb.

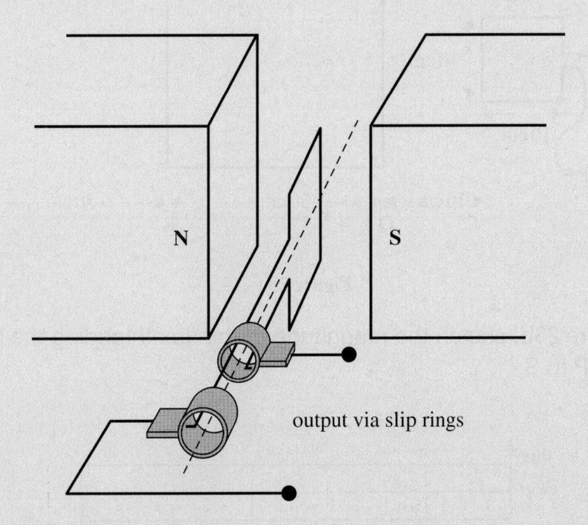

Calculate the magnitude of the magnetic field. Include an appropriate unit.

Question 3 (2 marks)

Source: *VCE 2016, Physics Exam, Section A, Q.15.a; © VCAA*

A coil is wound around a cardboard cylinder, as shown in the figure. The cross-sectional area of the coil is 0.0060 m^2. There are 1000 turns in the coil (not all are shown in the diagram).

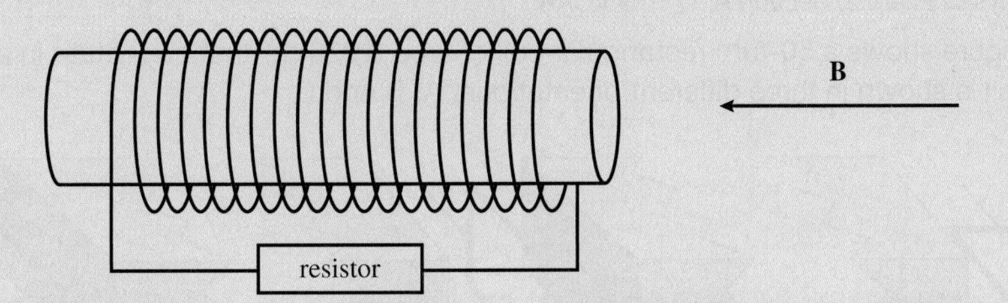

The axis of the coil is immersed in a uniform external magnetic field of strength 0.0050 T and its direction is shown by the arrow labelled B in the figure.

Calculate the magnitude of the flux through the first turn of the coil. Include an appropriate unit.

Question 4 (1 mark)

Source: *VCE 2017 Physics Exam, NHT, Section A, Q.14a; © VCAA*

A square loop of side 10 cm is allowed to move horizontally through a region of a magnetic field. This is shown in Figure 22.

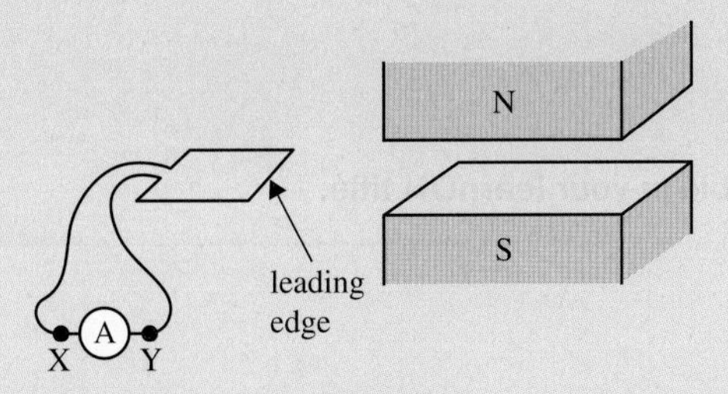

Figure 22

Assume the magnetic field is uniform and does not extend beyond the limits of the magnets.

The arrangement is shown as viewed from above in Figure 23a.

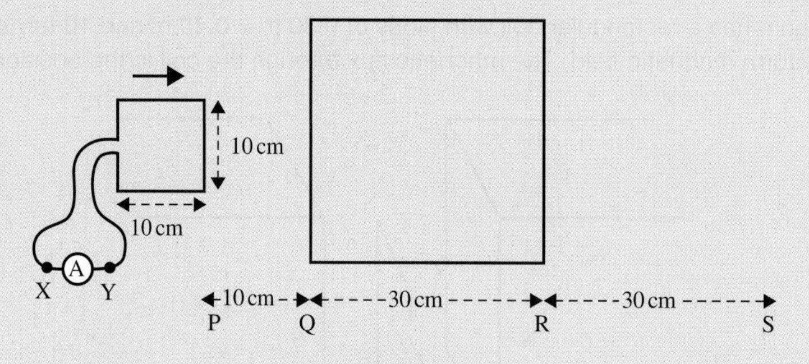

Figure 23a

On the graph provided in Figure 23b, sketch the magnitude of the flux threading the loop as the loop moves with its leading edge moving from P to S.

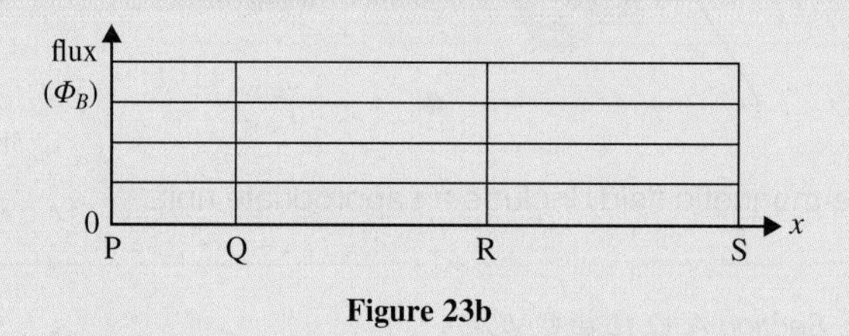

Figure 23b

Question 5 (1 mark)

Source: *VCE 2011, Physics Exam 2, Section A, Q.8; © VCAA*

MC The following figure shows a 50-turn rectangular coil of area 0.020 m^2 that can rotate in a uniform magnetic field of 2.0 T. The coil is shown in three different orientations, A, B and C.

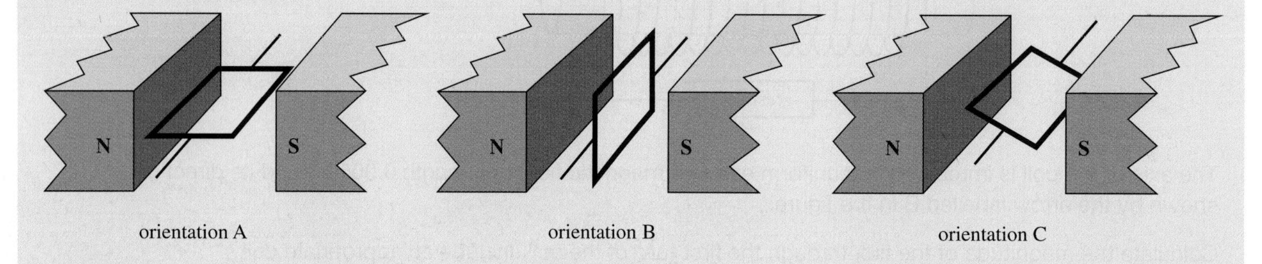

In orientation A the coil is horizontal; in orientation B it is vertical; and in orientation C it is inclined at 45° to the vertical.

Which of the following is the closest to the value of the magnetic flux through the coil when it is at orientation C?

A. 0 Wb
B. 0.03 Wb
C. 0.04 Wb
D. 1.5 Wb

More exam questions are available in your learnON title.

6.4 Generating emf from a changing magnetic flux

KEY KNOWLEDGE

- Investigate and analyse theoretically and practically the generation of electromotive force (emf) including AC voltage and calculations using induced emf, $\varepsilon = -N\frac{\Delta\Phi_B}{\Delta t}$, with reference to:
  - rate of change of magnetic flux
  - number of loops through which the flux passes
  - direction of induced emf in a coil.

Source: VCE Physics Study Design (2024–2027) extracts © VCAA; reproduced by permission.

6.4.1 Faraday's discovery of electromagnetic induction

Michael Faraday was aware of the magnetic effect of a current and he spent six years searching for the reverse effect — that is, the electrical effect of magnetism.

His equipment consisted of two coils of insulated wire, wrapped around a wooden ring. One coil was connected to a battery, the other to a **galvanometer**, a sensitive current detector. Faraday observed that the galvanometer needle gave a little kick as the battery switch was opened and a little kick the opposite way as the switch was closed. The rest of the time, when the switch was either open or closed, the needle was stationary, reading zero. The current was momentary, not the constant current he was looking for. What Faraday had observed came to be called **electromagnetic induction.**

galvanometer instrument used to detect small electric currents, or to detect the direction of current (such as in AC)

electromagnetic induction generation of an electromotive force (emf) in a coil (an electrical conductor) as a result of a changing magnetic field

FIGURE 6.10 When the switch in the battery circuit is opened or closed, there is a momentary current through the galvanometer.

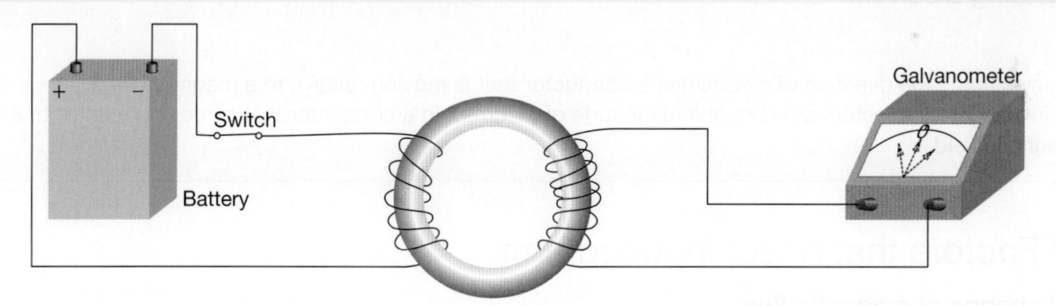

Investigating further, Faraday found that using an iron ring instead of a wooden ring increased the size of the current. He concluded that when the magnetic field of the battery coil was changing, there was a current induced in the other coil.

He therefore replaced the battery coil with a magnet. Moving the magnet through the other coil changed the magnetic field and produced a current. The faster the magnet moved, the larger the current. When the magnet was moved back away from the coil, current flowed in the opposite direction.

If there was an induced current, there must have been an induced emf. An emf, ε, gives energy to a charge to move it through the wire, and the resistance of the wire limits the size of the current. So it is more correct to say that the changing magnetic field induced an emf.

FIGURE 6.11 Magnet a. moving into a coil and b. away again

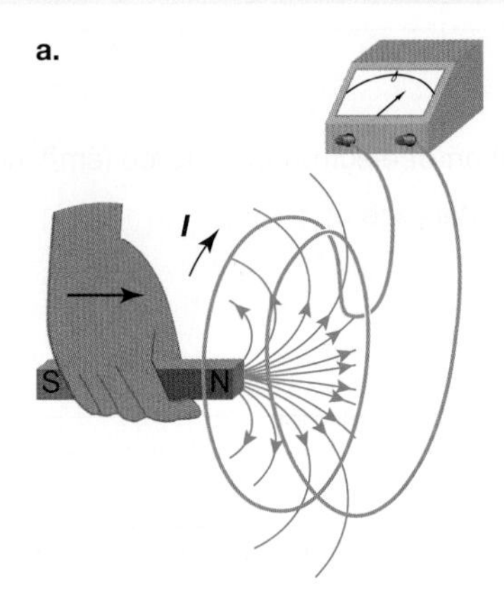

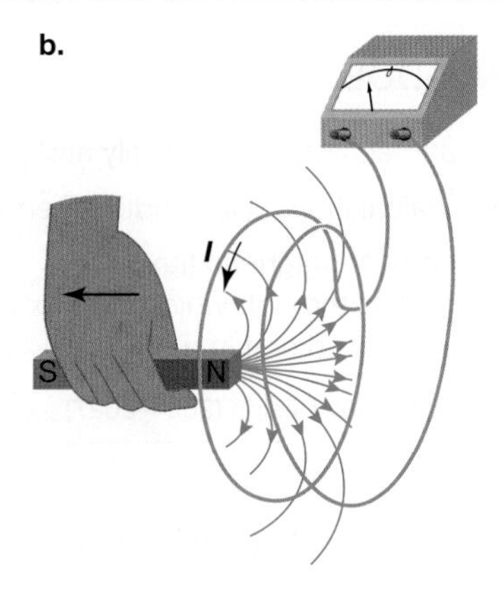

Changing magnetic flux and induced emf

Whenever the magnetic flux passing through a coil changes, an emf is induced in the coil.

A change in magnetic flux occurs when one of the following happens:
- the angle between the magnetic field and the coil changes
- the magnetic field strength changes
- the area of the coil changes.

elog-1882

INVESTIGATION 6.1

Generating a current

Aim

a. To investigate the direction of a current in a conductor that is moving relative to a magnetic field
b. To investigate the factors affecting the magnitude of a current in a conductor that is moving relative to a magnetic field

6.4.2 Factors that affect induced emf

Rate of change of magnetic flux

The concept of magnetic flux can be used to explain the induced emf. The two principles are described here.
1. An emf is induced in a coil when the amount of magnetic flux passing through the coil changes.
2. The size of the emf depends on how quickly the amount of magnetic flux changes.

These two statements can be written formally as follows:

$$\varepsilon_{\text{average}} = \frac{\Delta\Phi_B}{\Delta t}$$

This statement is known as Faraday's Law. The word 'average' is included because the change in magnetic flux takes place over a finite interval of time.

Direction of induced emf in a coil

There is no magnetic flux passing through the loop in figure 6.12a. When the magnet approaches the coil from the left (figure 6.12b), there is an increase in the amount of magnetic field passing through the coil from left to right. The loop has experienced a change in the magnetic flux passing through it (figure 6.12c), and the direction of this change is from left to right. The change in magnetic flux induces a current in the coil, and the current in the coil then induces a magnetic field. The direction of the induced magnetic field due to the induced current in the loop must be such that it opposes the change in the magnetic flux. This means its direction will be from right to left (figure 6.12d). To achieve an induced magnetic field from right to left, the induced current, using the right-hand-grip rule, must be travelling up the front of the loop (figure 6.12e).

The coil responds in such a way as to keep its magnetic environment constant. In the example shown in figure 6.12, there an increase in flux from left to right, so the induced magnetic field opposes this increase by going from right to left. When the magnet is pulled back, the flux that is still going from left to right is decreasing, so the induced magnetic field adds to the existing flux to compensate for the loss; therefore, this induced field points from left to right.

FIGURE 6.12 The loop **a.** before and **b.** after; **c.** change in flux, **d.** direction of induced field and **e.** direction of current

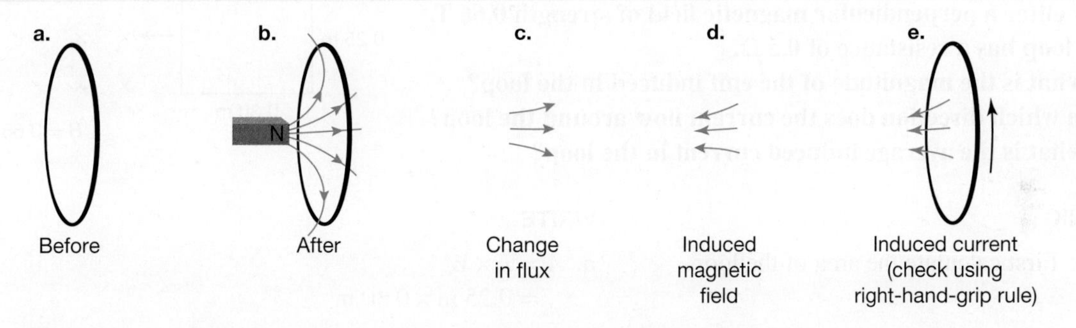

> To determine the direction of the induced current in a coil according to Lenz's Law, the problem needs to be broken down into several steps:
> 1. Determine the change in flux: is it increasing or decreasing (and in which direction)?
> 2. Determine the direction of the induced magnetic field such that it compensates for the increase or decrease in flux.
> 3. Determine the direction that the induced current must flow in order to produce the induced magnetic field in step 2.

Lenz's Law states:

> The direction of the induced current is such that its magnetic field is in the opposite direction to the change in magnetic flux.

In other words, a coil responds to a change in magnetic flux in such a way as to keep its magnetic environment constant. If the coil experiences an increase in flux, it will compensate by inducing a magnetic field in the opposite direction to the flux. If the coil experiences a decrease in flux, it will compensate for the loss by inducing a magnetic field in the same direction as the flux.

This can be incorporated in the equation as a minus sign:

$$\varepsilon = \frac{-\Delta\Phi_B}{\Delta t}$$

Number of loops

Examples concerning emf have been previously discussed when there is a single coil of wire. When there are multiple turns (or loops) in the coil, the emf produced will be the sum of all the individual turns. Therefore, if the coil consists of several turns of wire, the equation can be generalised further:

$$\varepsilon = -N\frac{\Delta\Phi_B}{\Delta t}$$

where: ε is the induced emf, in V

N is the number of turns (or loops) in the coil

$\Delta\Phi_B$ is the change in magnetic flux, in Wb

Δt is the time interval, in s

tlvd-9000

SAMPLE PROBLEM 3 Determining the induced emf and induced current in a loop

The rectangular loop shown in the diagram takes 2.0 s to fully enter a perpendicular magnetic field of strength 0.66 T. The loop has a resistance of 0.5 Ω.

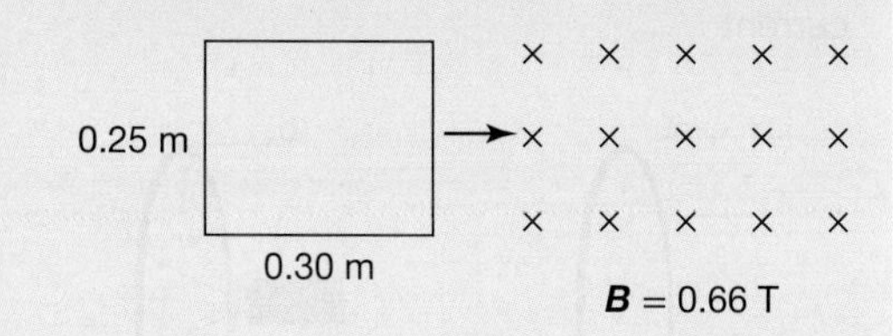

a. What is the magnitude of the emf induced in the loop?

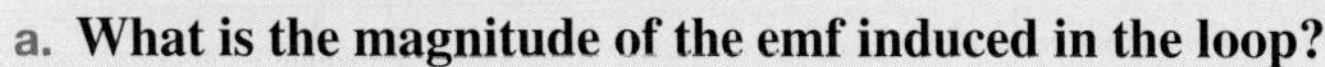

b. In which direction does the current flow around the loop?

c. What is the average induced current in the loop?

THINK

WRITE

a. 1. First calculate the area of the loop.

a. $A = L \times W$
$= 0.25 \text{ m} \times 0.30 \text{ m}$
$= 0.075 \text{ m}^2$

2. Now find the change in flux.

$\Delta\Phi_B = \Phi_{B \text{ final}} - \Phi_{B \text{ initial}}$
$= (BA)_{\text{final}} - (BA)_{\text{initial}}$
$= \left(0.66 \text{ T} \times 0.075 \text{ m}^2\right) - \left(0 \text{ T} \times 0.075 \text{ m}^2\right)$
(The initial field strength through the coil is zero.)
$= \left(0.050 \text{ T m}^2\right) - \left(0 \text{ T m}^2\right)$
$= 0.050$ Wb into the page

3. Use Faraday's Law to determine the induced emf.

$\varepsilon = -N\frac{\Delta\Phi_B}{\Delta t}$
$= -1 \times \frac{0.050 \text{ Wb}}{2.0 \text{ s}}$
$= -0.025 \text{ V}$

The magnitude of the induced voltage is 0.025 V.
(The minus sign indicates that the induced emf opposes the change in magnetic flux.)

b. 1. Determine the direction of the flux, and whether the flux is increasing or decreasing.

b. The flux increases from 0 Wb to 0.050 Wb into the page as the loop moves into the field.

2. Use Lenz's Law to determine the direction of the induced magnetic field.

Lenz's Law states that the induced magnetic field opposes the change in flux. Because the flux is increasing, the direction of the induced magnetic field will be in the opposite direction to the flux. Hence the direction of the induced field will be out of the page.

3. Use the right-hand-grip rule to determine the direction of the induced current.	In order for the induced magnetic field to be directed out of the page, the induced current must flow in an anticlockwise direction around the loop.
c. Use Ohm's Law, $V = IR$, to calculate the average current in the loop, where $V = 0.025$ V and $R = 0.5\ \Omega$.	c. $V = IR$ $0.025\ \text{V} = I \times 0.5\ \Omega$ $I = \dfrac{0.025\ \text{V}}{0.5\ \Omega}$ $= 0.05\ \text{A}$

PRACTICE PROBLEM 3

A spring is bent into a circle and stretched out to a radius of 5.0 cm. It is then placed in a magnetic field of strength 0.55 T. The spring is released and contracts down to a circle of radius 3.0 cm. This happens in 0.15 seconds. The spring has a resistance of 0.4 Ω.

a. **What is magnitude of the induced emf?**
b. **In what direction does the current move?**
c. **What is the average induced current in the spring?**

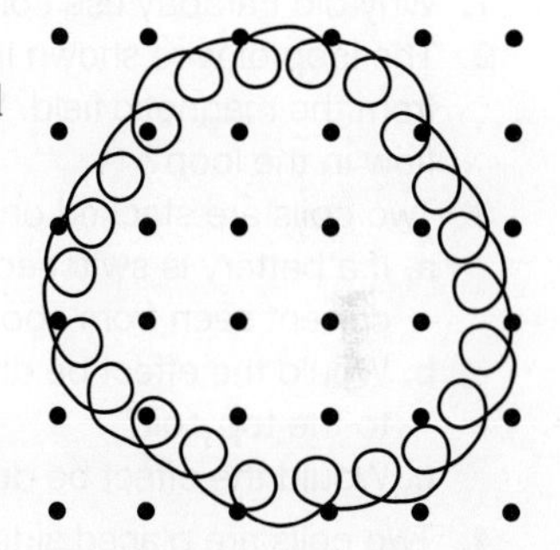

elog-1883

vd-10812

INVESTIGATION 6.2

online only

Inducing a current in a coiled conductor

Aim

To investigate how a current can be induced in a coiled conductor, and to determine factors that affect the size of the induced current

elog-1884

vd-10606

INVESTIGATION 6.3

online only

Direction of induced currents

Aim

To investigate the direction of an induced current in a conducting coil and verify Lenz's Law

Resources

Video eLesson Magnetic flux and Lenz's Law (eles-0026)

Interactivities Magnetic flux and Lenz's Law (int-0050)
Generating an emf (int-0116)

6.4 Activities

Students, these questions are even better in jacPLUS

 Receive immediate feedback and access sample responses

 Access additional questions

 Track your results and progress

Find all this and MORE in jacPLUS 

6.4 Quick quiz on | 6.4 Exercise | 6.4 Exam questions

6.4 Exercise

1. Why did Faraday use coils with many turns of copper wire?
2. The loop of wire shown in the diagram is quickly withdrawn from the magnetic field. Which way does the current flow in the loop?
3. Two coils are stacked one on top of the other with their centres in line.
   a. If a battery is switched on in the bottom coil, producing a clockwise current seen from above, what happens in the top coil?
   b. Would the effect be different if the battery was connected to the top coil?
   c. Would the effect be different if the battery was switched off?
4. Two coils are placed side by side on a page with their centres in line, as in the diagram.
   a. If a battery is switched on in the left coil, producing a clockwise current (seen from above the page), what happens in the right coil?
   b. What would be the effect if the current was anticlockwise?

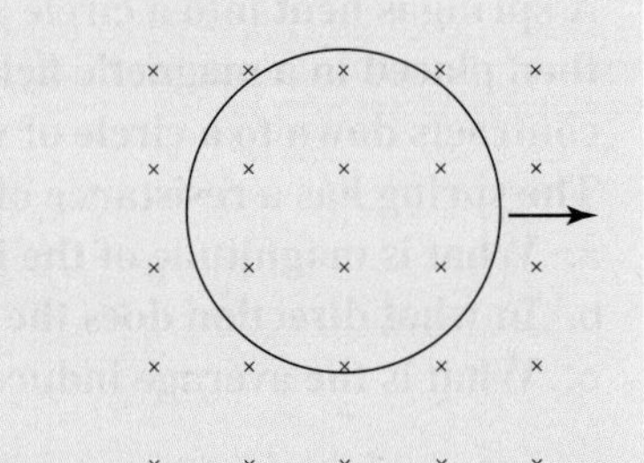

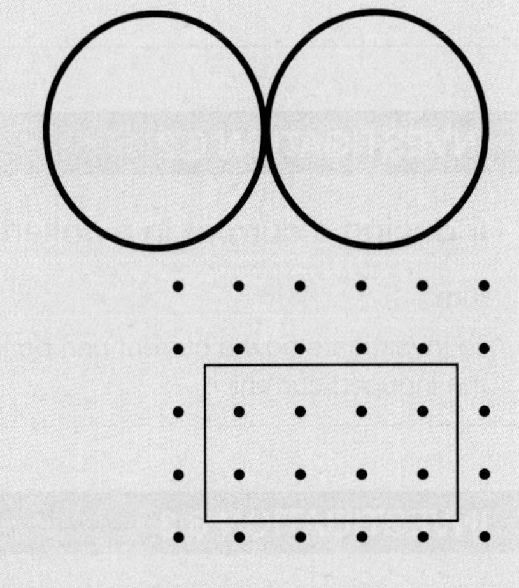

5. The diagram shows a confined uniform magnetic field coming out of the page with a wire coil in the plane of the page. Is there an induced current in the coil as it is moved:
   a. to the right
   b. upwards
   c. into the page
   d. out of the page?

   Give a reason for each answer. If there is a current, indicate the direction.
6. Calculate the average induced emf in each of the following situations.
   a. A circular loop of wire with a 5.0-cm radius is removed from a magnetic field of strength 0.40 T in a time of 0.20 s.
   b. The magnetic flux through a coil changes from 60 Wb to 35 Wb in 1.5 s.
   c. The magnetic flux through a coil changes from 60 Wb to −35 Wb in 2.5 s.
7. Calculate the average induced current in each of the following situations.
   a. A circular loop of wire, 10 cm long with a resistance 0.40 Ω, is removed from a magnetic field of strength 0.60 T in a time of 0.30 s.
   b. The magnetic field strength perpendicular to a square loop, of side length 0.26 m and resistance 2.5 Ω, is increased from 0.20 T to 1.20 T in 0.50 s.
   c. A stretched circular spring coil with an 8-cm radius and resistance 0.2 Ω is threaded by a perpendicular magnetic field of strength 2.0 T. The coil shrinks back to a radius of 4 cm in 0.8 s.
8. An orbiting satellite has a small module tethered to it by a 5.0-km conducting cable. As the satellite and its module orbit Earth, they cut across Earth's magnetic field at right angles.
   If the strength of Earth's magnetic field at this distance is 0.1 mT, what is the size of the induced emf?

9. A bar magnet, with its north end down, is dropped through a horizontal wire loop.
   a. What is the direction of the induced current when the magnet is:
      i. just above the loop
      ii. halfway through the loop (so the north end is below the loop and south end is above the loop)
      iii. just below the loop?
   b. Draw the graph of the induced current against time.
   c. Where did the electrical energy of the induced current come from?
   d. If the magnet falls from a very long distance above the loop to a very long distance below the loop, what is the overall change in magnetic flux through the loop? What does this imply about the area under the current–time graph?
10. How much charge, in coulombs, flows in a loop of wire of area 1.6×10^{-3} m^2 and resistance 0.20 Ω when it is totally withdrawn from a magnetic field of strength 3.0 T over 1 second?
11. A magnet passes through two loops, one wire and the other plastic. Compare the induced emfs and the induced currents of the two loops.
12. Lenz's Law is an illustration of the conservation of energy. Explain why the reverse of Lenz's Law (the direction of the induced current reinforces the change in magnetic flux) contravenes the law about the conservation of energy. Use the example of a north end of a magnet approaching a loop of conducting wire (as shown in the diagram).

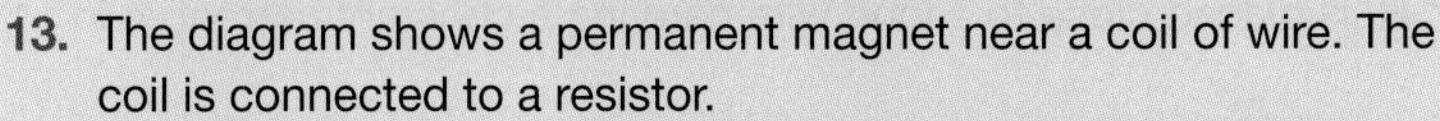

13. The diagram shows a permanent magnet near a coil of wire. The coil is connected to a resistor.
   a. Which way will a current flow through the resistor when the magnet moves towards the coil from the position shown? Explain your answer.
   b. When the magnet is held stationary inside the coil, no current is detected through the resistor. Explain why this is the case.
   c. Describe two ways to increase the magnitude of the induced current in the resistor.

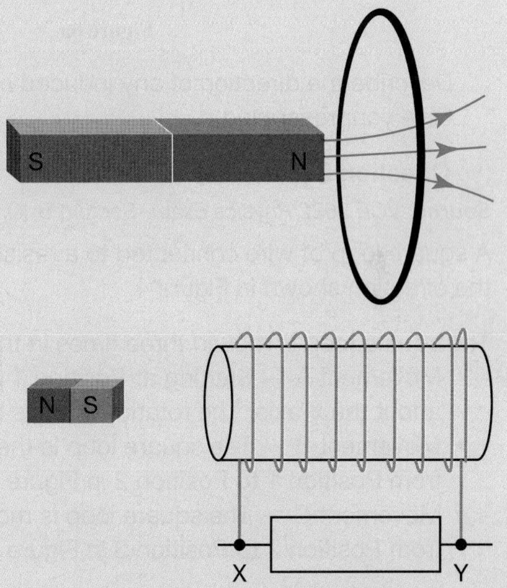

6.4 Exam questions

Question 1 (6 marks)

Source: *VCE 2020 Physics Exam, Section B, Q.6; © VCAA*

Two Physics students hold a coil of wire in a constant uniform magnetic field, as shown in Figure 5a. The ends of the wire are connected to a sensitive ammeter. The students then change the shape of the coil by pulling each side of the coil in the horizontal direction, as shown in Figure 5b. They notice a current register on the ammeter.

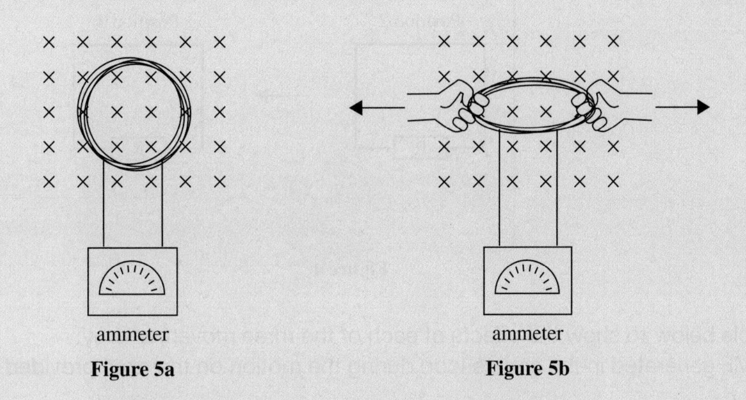

Figure 5a **Figure 5b**

a. Will the magnetic flux through the coil increase, decrease or stay the same as the students change the shape of the coil? **(2 marks)**

b. Explain, using physics principles, why the ammeter registered a current in the coil and determine the direction of the induced current. **(2 marks)**

c. The students then push each side of the coil together, as shown in Figure 6a, so that the coil returns to its original circular shape, as shown in Figure 6b, and then changes to the shape shown in Figure 6c.

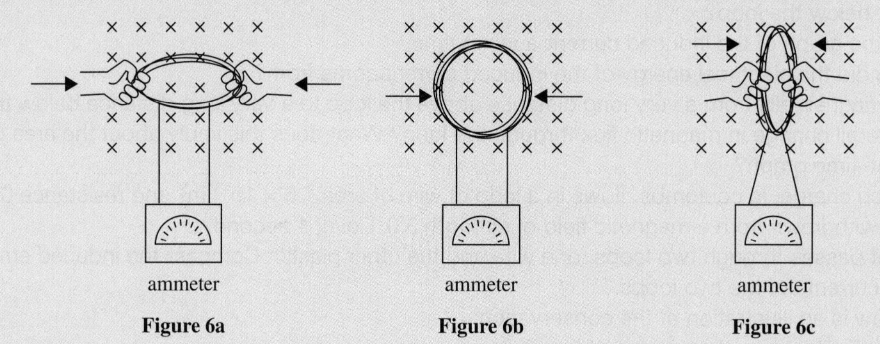

Figure 6a **Figure 6b** **Figure 6c**

Describe the direction of any induced currents in the coil during these changes.
Give your reasoning. **(2 marks)**

Question 2 (6 marks)

Source: *VCE 2022 Physics Exam, Section B, Q.4;* © VCAA

A square loop of wire connected to a resistor, R, is placed close to a long wire carrying a constant current, I, in the direction shown in Figure 4.

The square loop is moved three times in the following order:

- Movement A — Starting at Position 1 in Figure 4, the square loop rotates one full rotation at a steady speed about the x-axis. The rotation causes the resistor, R, to first move out of the page.
- Movement B — The square loop is then moved at a constant speed, parallel to the current-carrying wire, from Position 1 to Position 2 in Figure 4.
- Movement C — The square loop is moved at a constant speed, perpendicular to the current-carrying wire, from Position 2 to Position 3 in Figure 4.

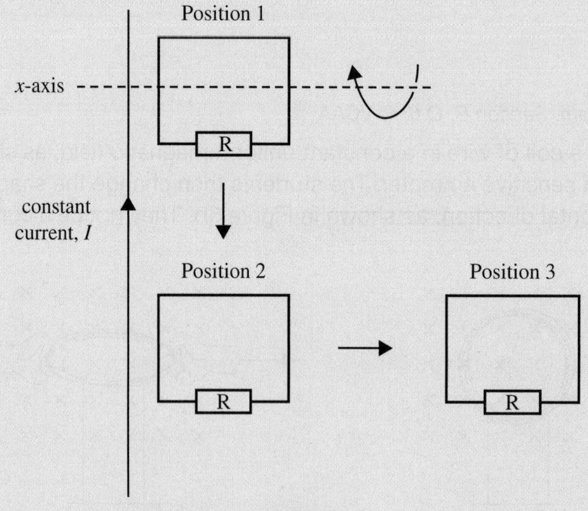

Figure 4

Complete the table below to show the effects of each of the three movements by:

- sketching any EMF generated in the square loop during the motion on the axes provided (scales and values are not required)
- stating whether any induced current in the square loop is 'alternating', 'clockwise', 'anticlockwise' or has 'no current'.

Movement	Possible EMF	Direction of any induced current (alternating/clockwise/anticlockwise/no current)
A rotation about *x*-axis	EMF (V) time (s)	
B moving from Position 1 to Position 2	EMF (V) time (s)	
C moving from Position 2 to Position 3	EMF (V) time (s)	

Question 3 (6 marks)

Source: *VCE 2022 Physics Exam, NHT, Section B, Q.4;* © VCAA

Figure 2 shows a schematic diagram of a simple DC generator with the output voltage connected to a cathode-ray oscilloscope (CRO).

The DC generator consists of a rectangular wire coil of 200 turns placed in a uniform magnetic field of strength 5.0 mT. The coil is rotated with a frequency of 60 Hz in the direction shown in Figure 2. The average EMF generated in the coil for the first quarter turn is 35 mV. The coil is initially in the position shown in Figure 2.

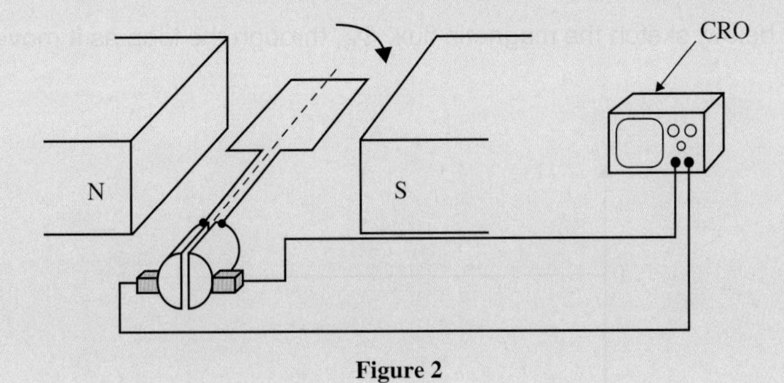

Figure 2

a. When viewed from above, will the induced current in the coil be clockwise or anticlockwise during the first quarter turn? **(1 mark)**

b. Calculate the area of one loop of the rectangular wire coil. Show your working. **(3 marks)**

c. The graph below shows the EMF induced in the coil over two full turns.
On the same axes, sketch the output EMF that would result if the number of turns in the coil is changed to 100 turns and the frequency of rotation is changed to 30 Hz. **(2 marks)**

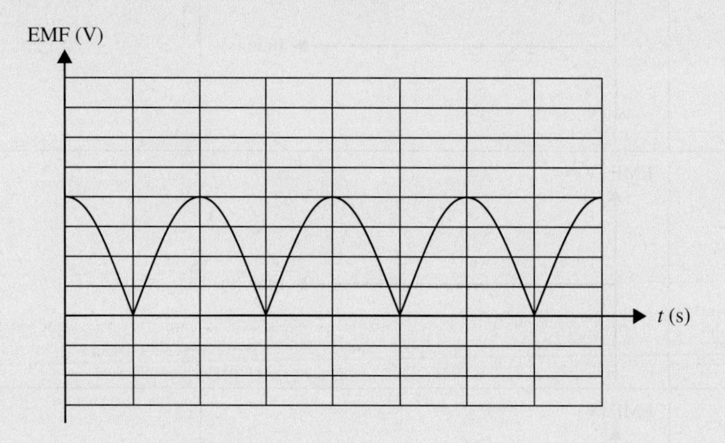

Question 4 (9 marks)

Source: *VCE 2018 Physics Exam, NHT, Section B, Q.4;* © VCAA

Students move a square loop of wire of 100 turns and of cross-sectional area 4.0×10^{-4} m^2. The loop moves at constant speed from outside left, into, through and out of a magnetic field, as shown in Figure 1a. The area between the poles has a uniform magnetic field of magnitude 2.0×10^{-3} T. Figure 1b shows the view from above.

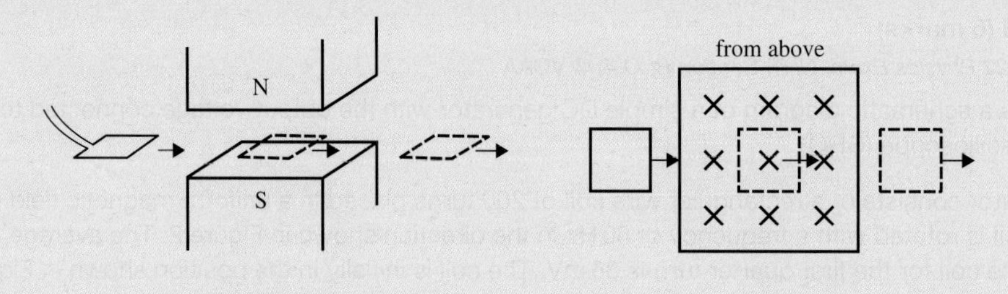

Figure 1a **Figure 1b**

a. On the axes provided below, sketch the magnetic flux, Φ_B, through the loop as it moves into, through and out of the magnetic field. **(2 marks)**

b. On the axes provided below, sketch the EMF induced through the loop as it moves into, through and out of the magnetic field. **(2 marks)**

c. The loop takes 2.0 s to move from completely outside to completely inside the magnetic field. Calculate the magnitude of the induced EMF in the loop as it moves into the magnetic field. **(2 marks)**

d. Determine the direction of the induced current in the loop as it moves into the magnetic field as viewed from above (clockwise or anti-clockwise). Justify your answer. **(3 marks)**

Question 5 (6 marks)

Source: *VCE 2018 Physics Exam, Section B, Q.2;* © VCAA

A square loop of wire of 10 turns with a cross-sectional area of 1.6×10^{-3} m^2 passes at a constant speed into, through and out of a magnetic field of magnitude 2.0×10^{-2} T, as shown in Figure 2. The loop takes 0.50 s to go from position X to position Y.

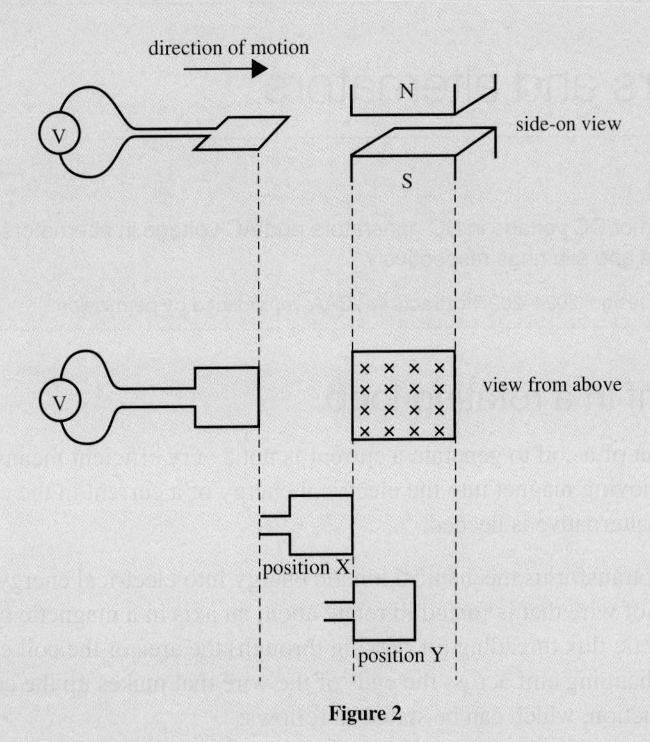

Figure 2

a. Calculate the average EMF induced in the loop as it passes from just outside the magnetic field at position X to just inside the magnetic field at position Y. Show your working. **(3 marks)**

b. Sketch the EMF induced in the loop as it passes into, through and out of the magnetic field. You do not need to include values on the axes. **(3 marks)**

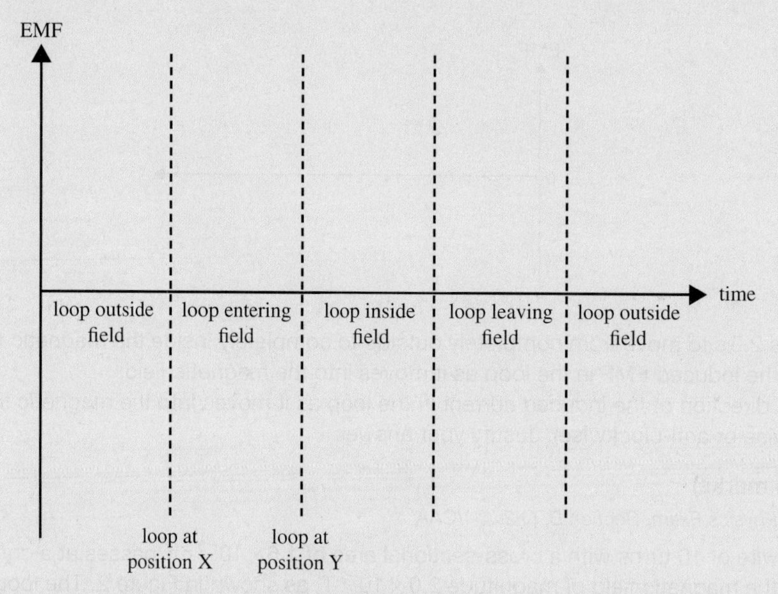

More exam questions are available in your learnON title.

6.5 Generators and alternators

KEY KNOWLEDGE

- Explain the production of DC voltage in DC generators and AC voltage in alternators, including the use of split ring commutators and slip rings respectively

Source: VCE Physics Study Design (2024–2027) extracts © VCAA; reproduced by permission.

6.5.1 Induced emf in a rotating loop

A magnet moving in and out of a coil to generate a current is not a very efficient means of converting the mechanical energy of the moving magnet into the electrical energy of a current in the coil. It does not have much technological potential; an alternative is needed.

A generator is a device that transforms mechanical kinetic energy into electrical energy. In its simplest form, a generator consists of a coil of wire that is forced to rotate about an axis in a magnetic field. As the coil rotates, the magnitude of the magnetic flux threading (or passing through) the area of the coil changes. This changing magnetic flux produces a changing emf across the ends of the wire that makes up the coil. This is in accordance with Faraday's Law of Induction, which can be stated as follows:

The induced emf in a coil is proportional in magnitude to the rate at which the magnetic flux through the coil is changing with time.

Rotating a coil in a magnetic field continually changes the rate at which the magnetic flux changes through the coil and therefore induces a changing voltage.

This way of changing the amount of magnetic flux passing through a loop is shown in figure 6.13.

When the loop is 'face on' to the magnetic field, the maximum amount of magnetic flux is passing through the loop: $\Phi_B = BA$. As the loop turns, the amount of flux decreases. When it has turned 90° and is parallel to the field, there is no flux passing through it at all: $\Phi_B = 0$. As the loop continues to turn between 90° and 180°, the magnetic field passes through the loop from the other side: a negative amount of flux, from the point of view of the loop.

As seen in figure 6.14, the magnitude of the magnetic flux passing through the loop increases to a maximum again, but in the opposite direction: $\Phi_B = -BA$. Then it decreases back to zero when the loop has rotated 270°, and finally passes through the original face of the loop.

FIGURE 6.13 A loop 'face on' to a magnetic field has maximum magnetic flux.

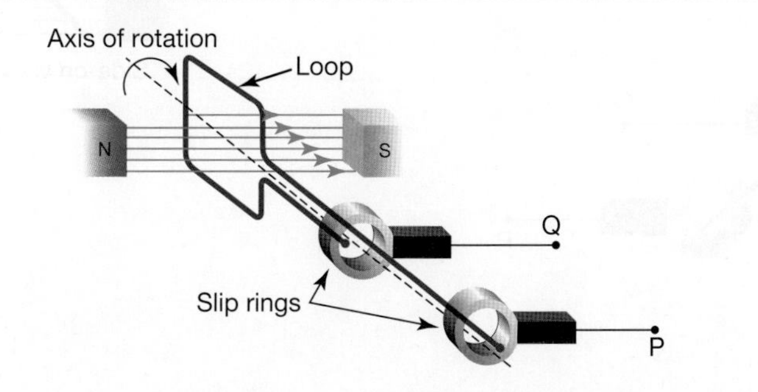

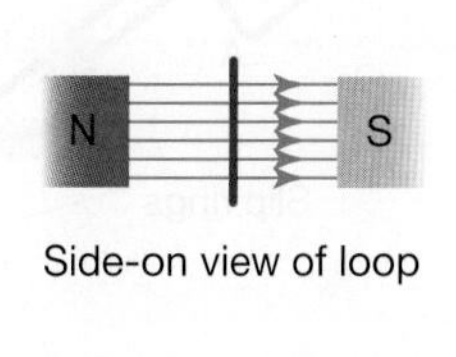

The amount of magnetic flux passing through the loop varies in the form of a sine wave, as shown in figure 6.14. The induced emf across the ends of the loop is proportional to the change of magnetic flux with time. Since $\varepsilon \propto -\dfrac{\Delta \Phi_B}{\Delta t}$, the induced emf is shown on the graph as the negative gradient function of the magnetic flux–time graph, and hence is also a sine wave.

FIGURE 6.14 Flux–time graph

The emf graph is the same shape as the flux graph but shifted sideways, so that when the flux is a maximum, the emf is zero. When the flux is at a maximum (or minimum) the flux–time graph is flat, so the gradient is zero and hence the emf is zero.

Similarly, when the flux is zero, the flux–time graph is steepest, so the gradient is a maximum (or minimum) and hence the emf is a minimum (or maximum).

6.5.2 AC alternators

Which way does the current travel when a rotating loop is placed in a magnetic field? Consider again figure 6.13. From which connection, P or Q, does the current leave the loop to go around the external circuit? It can be worked out using Lenz's Law or using the magnetic force of electrons in the loop. Both methods are explained as follows. (Note — in VCE Physics you only need to understand one explanation, so use the one in which you feel most confident.)

Using Lenz's Law

As the loop rotates, the magnetic flux changes from passing through one side to passing through the other.

In figure 6.15a, the magnetic flux is entering the loop from left to right above the edge AB. As the coil rotates clockwise to a horizontal position, the flux decreases to zero, inducing a current through the coil. The induced current induces a magnetic field from left to right to restore the decreasing flux.

In figure 6.15b, the loop continues to rotate in a clockwise direction from the horizontal position to the vertical position. The magnetic flux increases from zero, entering the loop from the other side below the edge AB. The induced magnetic field will now be from right to left to counteract the increase in flux.

To produce this field, the right-hand-grip rule shows that conventional current needs to run in the direction A→B→C→D.

FIGURE 6.15 Direction of current flow as loop passes through the horizontal position

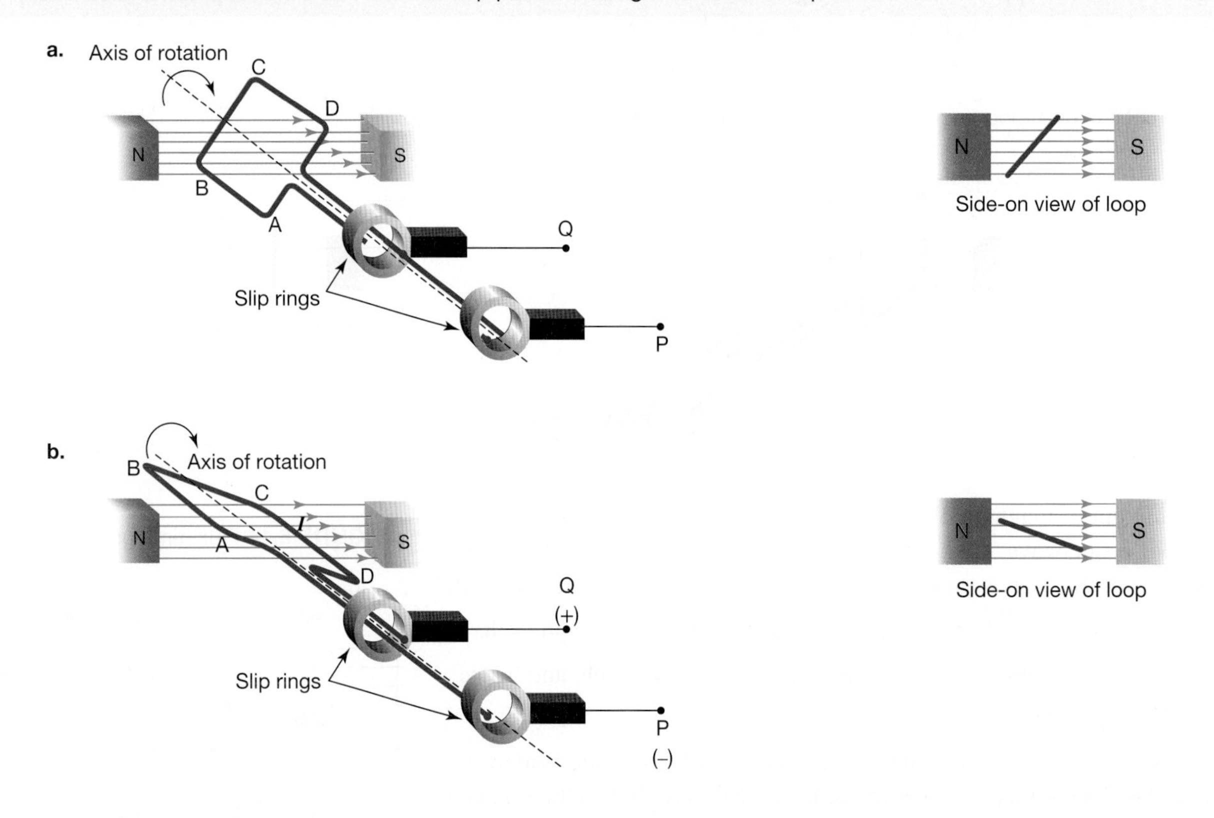

At this point in the rotation, the current will enter the external circuit from the **slip ring** at Q and return to the loop by the slip ring at P. So, for the time being, the current flows through the external circuit from Q to P, with Q being the positive terminal and P the negative.

In the diagrams in figure 6.15, the wire from A is attached to the front metal ring, the one connected to P, and the wire from D is attached to the back ring, the one connected to Q. These connections are fixed. When the loop rotates about its axis, the two slip rings also rotate about the same axis. The black blocks are made of graphite. They are being held in place against the spinning slip rings by the springs. Graphite is used because it not only conducts electricity but is also a lubricant. The spinning slip rings easily slide past the fixed block. The blocks are also called 'brushes' because early designs used thin metal strips that brushed against the slip rings.

Consider the loop as it continues to rotate clockwise from the position shown in figure 6.15b through the vertical position. The flux is increasing from left to right through the loop as it moves to the vertical position. The induced current flows through A→B→C→D in order to induce a magnetic field from right to left through the loop to counteract the increasing flux. When the loop rotates past the vertical position and begins to move again towards the horizontal position, the flux from left to right decreases, inducing a magnetic field from left to right through the loop, below the edge AB. For this to be possible, the induced current must now move in the direction D→C→B→A. The current in the external circuit will now flow from P to Q, that is, in the opposite direction! The direction of the induced current will reverse every time the loop passes through the vertical position (when the plane of the loop is perpendicular to the field), or every half-turn.

The sinusoidal emf drives current through the external circuit first one way, then the opposite way, and is thus called **alternating current** (AC).

slip ring an electromechanical device carrying current from a stationary to a rotating structure

alternating current (AC) electric current that reverses direction at short, regular intervals

This design of a rotating coil in a magnetic field is called a **generator**. If the ends of the coil are connected to slip rings, then the voltage across the external connections is alternating in direction, producing an alternating current. The device can also be called an **alternator**.

Slip rings

Slip rings maintain a continuous electrical connection with the spinning loop and are used when an AC output is required.

Using magnetic force on the charges in the wire

As the loop rotates through the horizontal plane, the left side of the loop, AB (see figure 6.16), is moving up, and the right side, CD, is moving down. Using the right-hand-slap rule, the force of the magnetic field on the positive charges in AB will be towards B, while the force on the electrons in AB will be towards A.

Similarly, the positive charges in CD will be pushed to D, while the electrons will be pushed towards C.

This means that conventional current will flow in the direction A→B→C→D, while the electrons will travel around the loop in the order D→C→B→A. The conventional current will enter the external circuit from the slip ring at Q and return to the loop by the slip ring at P.

FIGURE 6.16 Using the left-hand rule to determine current direction in a rotating loop

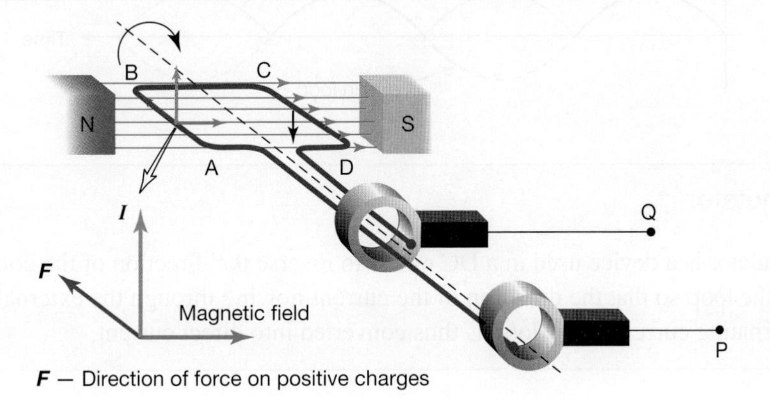

Once the loop passes through the vertical plane, AB will move downwards and CD will move upwards. The force of the magnetic field on the positive charges in AB will be towards A, while the force on the electrons in AB will be towards B, and the conventional current will now flow in the opposite direction: D→C→B→A, entering the external circuit at P and returning to the loop at Q. This is the same result obtained as with the previous method.

generator device in which a rotating coil in a magnetic field is used to induce a voltage

alternator device in which the ends of the coil are connected to slip rings, causing the voltage to alternate in direction, inducing an alternating current

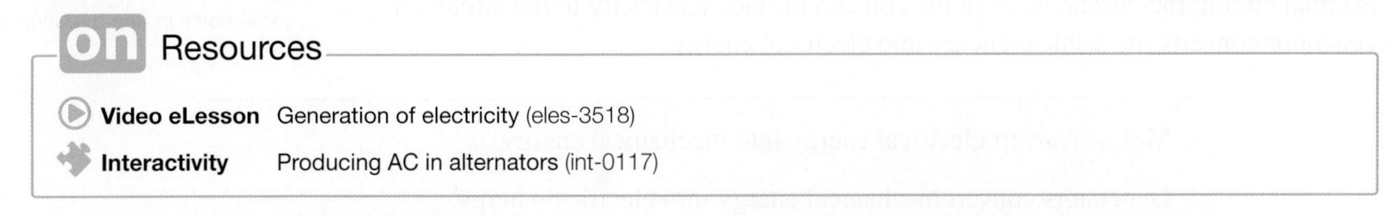

on Resources

Video eLesson Generation of electricity (eles-3518)

Interactivity Producing AC in alternators (int-0117)

6.5.3 DC generators

What happens if the slip rings are replaced with split rings like those used in a DC motor? When the current in the loop reverses every half-cycle as the loop rotates through the vertical position, the ends of the coil swap to the other side of the split ring so that the direction of the current flowing through the external circuit remains the same. Essentially, the direction of the output to the external circuit is changed by the **split ring commutator**, and so the alternating current in the loop is converted into pulsating **direct current** (DC) in the external circuit. The device is now called a DC generator.

FIGURE 6.17 AC voltage coming from loop, and DC coming from commutator

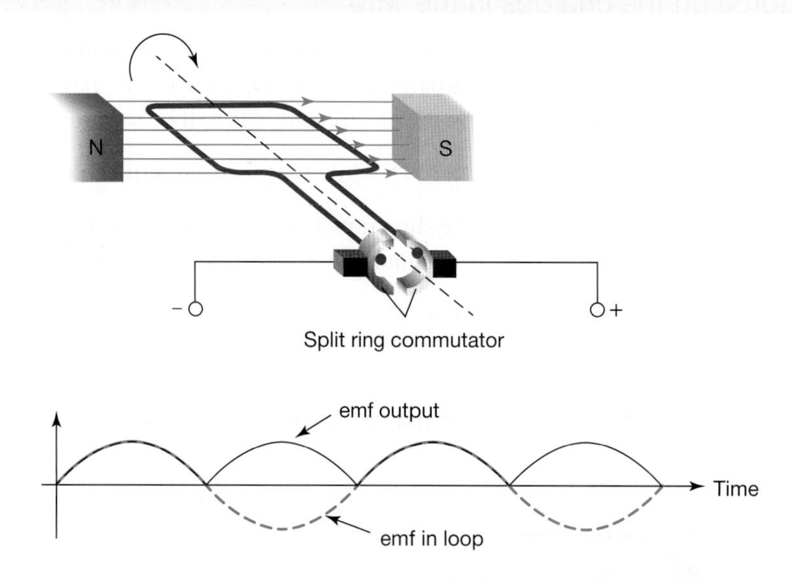

Split ring commutator

A split ring commutator is a device used in a DC motor to reverse the direction of the coil's current every half revolution of the loop so that the direction of the current flowing through the external circuit remains the same. The alternating current in the loop is thus converted into direct current.

6.5.4 Comparison of motors and generators

Recall from topic 5 that a motor consists of a coil connected to a power supply in the presence of a magnetic field. The magnetic field induces a force on the electrons moving through the coil, causing the coil to turn. A motor converts electrical energy into mechanical energy.

A generator consists of a coil in the presence of a magnetic field. By manually rotating the coil, the magnetic flux through the coil changes, inducing an emf in the coil, which subsequently induces a current through the coil. If connected to an external circuit, the current through the coil can provide electricity to the circuit. A generator converts mechanical energy into electrical energy.

split ring commutator a device tha reverses the direction of the curren flowing through an electric circuit

direct current (DC) electric current that flows in one direction only

Motors convert electrical energy into mechanical energy.

Generators convert mechanical energy into electrical energy.

EXTENSION: Back emf

Considering what you have learned in this topic, when the shaft of a motor is turning (the motor is doing work), an emf is generated. Lenz's Law informs us that this emf opposes the emf powering the motor. This self-generated emf is known as **back emf**. This results in the net emf in the motor being less than the supplied emf when the motor is rotating.

back emf electromagnetic force that opposes the main current flow in a circuit. When the coil of a motor rotates, a back emf is induced in the coil due to its motion in the external magnetic field.

Resources

Weblink Generator applet

6.5.5 Producing a greater emf

The AC voltage produced by a generator has a substantial technological application because it is easy to make things spin. Hydroelectricity is produced when water falls under gravity through pipes and hits the vanes of a propeller connected to a generator. In coal and gas-fired turbines, the burning fuel heats up water to a high temperature, producing steam at high pressure. The steam is directed against the vanes of the turbine.

FIGURE 6.18 Vanes of a turbine

The emf that is produced by a generator has a frequency that is the same as the frequency of the rotation of a coil in a magnetic field.

Using the Faraday equation for average emf:

$$\varepsilon = \frac{-N\Delta\Phi_B}{\Delta t} = \frac{-N\Delta(B_{\perp}A)}{\Delta t}$$

and ignoring the – sign (which relates to direction), the ways to produce a larger emf can be deduced, as follows:

- increase the number of turns or coils
- increase the strength of the magnetic field
- increase the area of each coil
- decrease the time for one turn (that is, increase the frequency of rotation).
  (Note that turning the coil twice as fast doubles both the induced emf and the frequency — that is, it halves the period.)

FIGURE 6.19 Doubling the frequency doubles the induced emf.

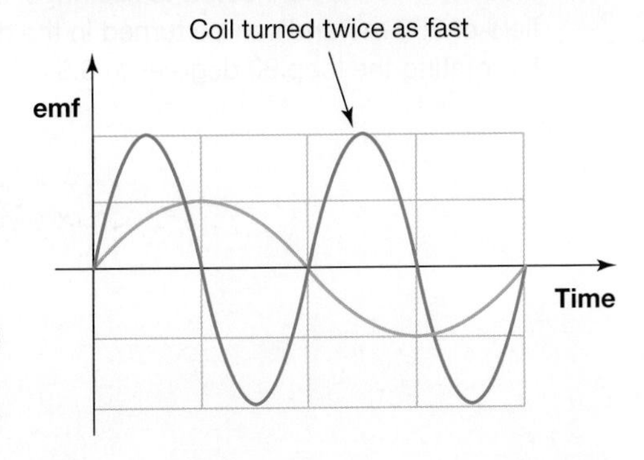

EXTENSION: Other technological strategies can also increase the emf

- The pole ends of the magnet can be curved so that the coils are close to the magnets for more time during the rotation.
- An iron core can be placed inside the coils to strengthen the magnetic field.
- The coils can be wound onto the iron core in grooves cut into the outer surface so that the iron core is as close as possible to the magnetic poles to increase the magnetic field.

FIGURE 6.20 Improvements to the design of a DC motor and an alternator

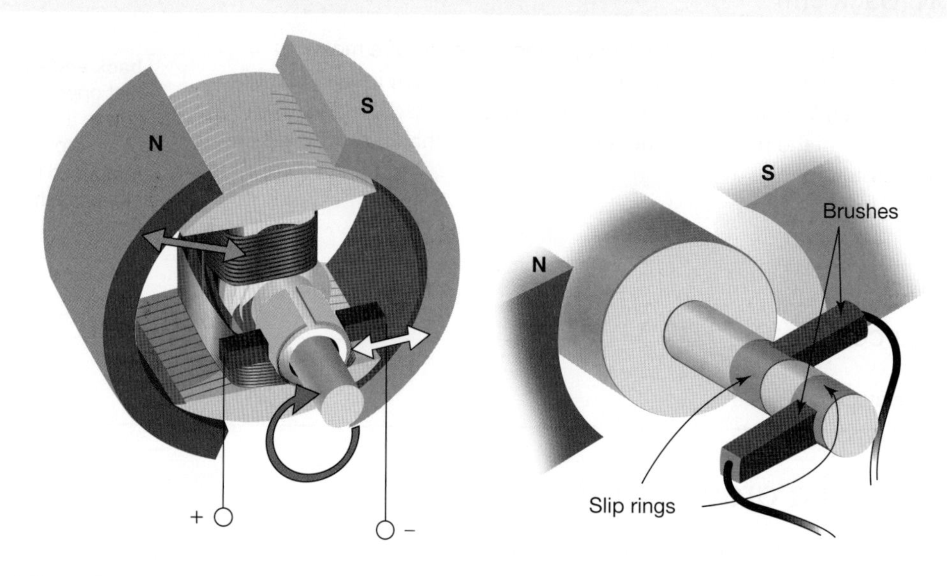

6.5 Activities

learnon

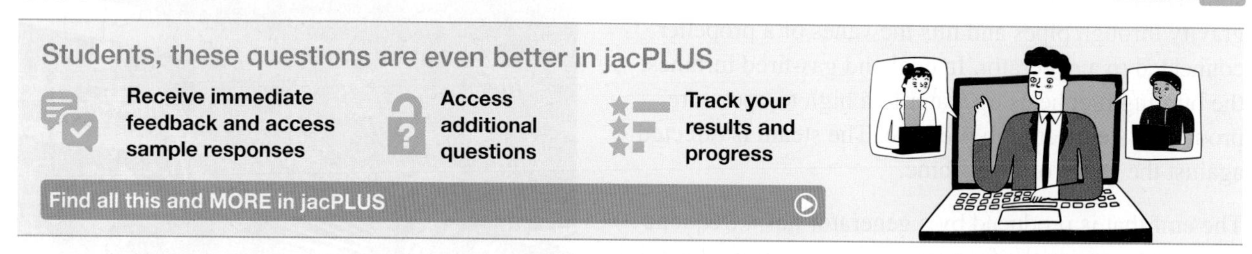

6.5 Quick quiz	6.5 Exercise	6.5 Exam questions

6.5 Exercise

1. Jaidev and Ahn have constructed a simple alternator. It consists of a single rectangular coil of wire, 0.60 m × 0.50 m, which is connected to slip rings, as shown in the following diagram. The coil is in a uniform magnetic field of 250 mT and can be turned in the direction as shown in the diagram. Jaidev and Ahn test this alternator by rotating the loop 90 degrees in 0.5 s.

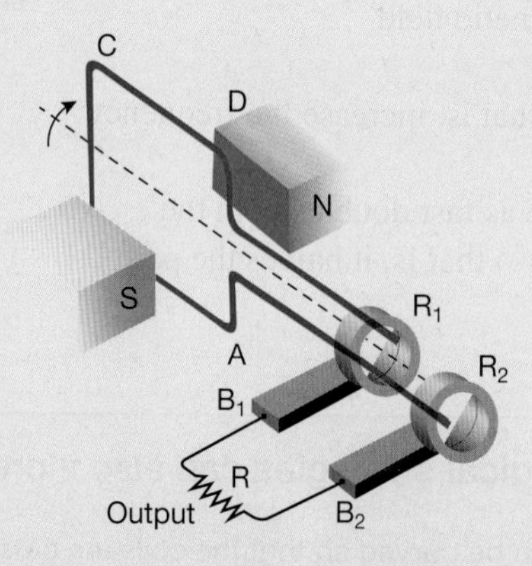

   a. Calculate the magnitude of flux through the coil when orientated as shown in the diagram.
   b. Calculate the average voltage measured through the slip rings.

c. Ahn now rotates the coil at a constant rate of 6 revolutions per second, and the students observe that the voltage between the slip rings varies with time as shown in the following graph.

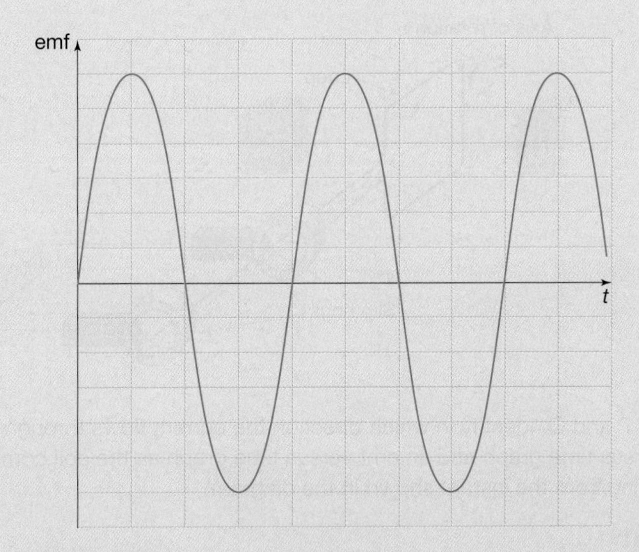

Jaidev decides to test the effect of rotation speed and turns the coil at a rate of 3 revolutions per second. Copy the graph shown and sketch the effect this would now have on the variation of voltage with time.

2. A model electric generator, similar to the one shown in the following diagram, is being tested by some students. The coil consists of 50 turns of wire and has an area of 6.0×10^{-3} m^2. The magnetic field between the poles of the magnets is measured to be 0.15 T.

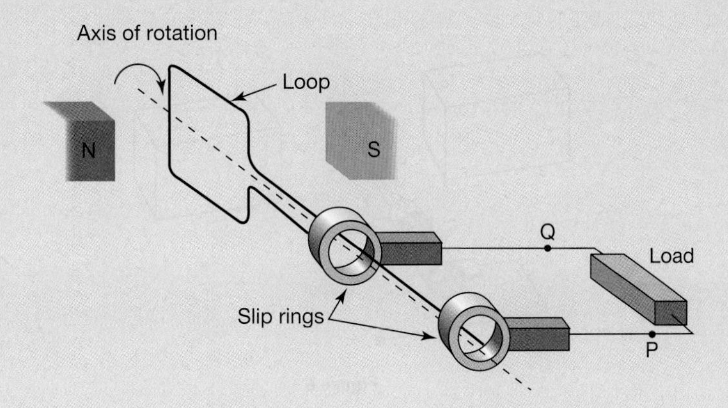

a. Calculate the magnitude of the magnetic flux threading the coil when it is in the position shown.
b. As the students rotate the coil through an angle of 90°, the magnitude of the average induced emf is 40 mV. What time interval is required for this rotation to produce an average induced emf of 40 mV during a quarter-turn?

3. Describe the effect of the following changes on the size of the current produced by a generator.
a. The number of loops of the coil is increased.
b. The rate of rotation of the coil is decreased.
c. The strength of the magnetic field is increased.

4. A student builds a model electric generator, similar to that shown below. The coil consists of 50 turns of wire. The student rotates the coil in the direction indicated on the diagram. The coil is rotated a quarter turn (90°).

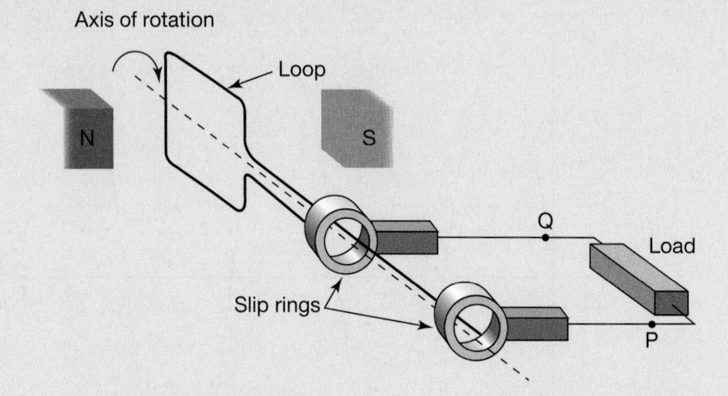

a. Using the points P and Q, identify in which direction the current flows through the external resistor.
b. Sketch a flux versus time graph and an emf versus time graph as the coil completes one rotation at a steady rate starting from the instant shown in the diagram.

6.5 Exam questions

Question 1 (5 marks)

Source: *VCE 2022 Physics Exam, Section B, Q.6; © VCAA*

Figure 6 shows a simple alternator consisting of a rectangular coil of area 0.060 m^2 and 200 turns, rotating in a uniform magnetic field. The magnetic flux through the coil in the vertical position shown in Figure 6 is 1.2×10^{-3} Wb.

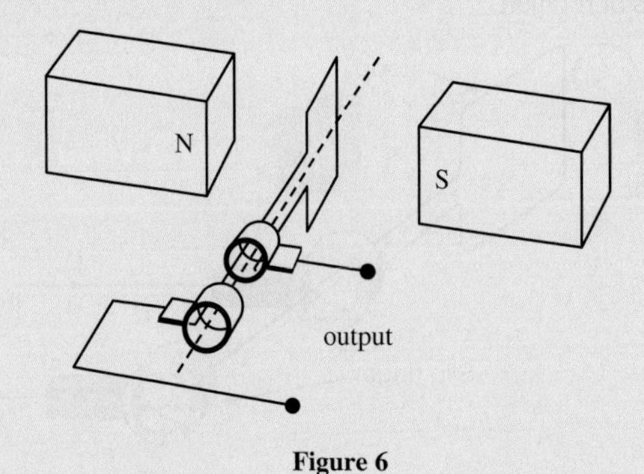

Figure 6

a. Calculate the strength of the magnetic field in Figure 6. Show your working. **(2 marks)**
b. The rectangular coil rotates at a frequency of 2.5 Hz.
Calculate the average induced EMF produced in the first quarter of a turn. Begin the quarter with the coil in the vertical position shown in Figure 6. **(3 marks)**

Question 2 (3 marks)

Source: *VCE 2021 Physics Exam, NHT, Section B, Q.4; © VCAA*

Figure 4 shows a schematic diagram of a simple one-coil DC motor. A current is flowing through the coil.
a. Draw an arrow on Figure 4 to indicate the direction of the force acting on the side JK of the coil. **(1 mark)**

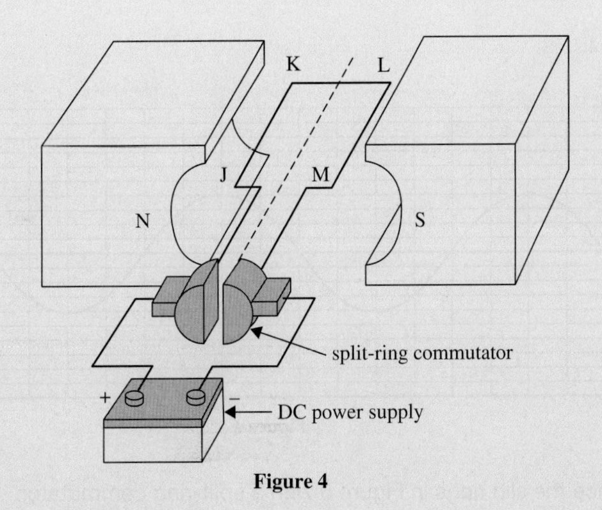

Figure 4

b. Explain the purpose of the split-ring commutator. **(2 marks)**

Question 3 (11 marks)

Source: *VCE 2019, Physics Exam, Section B, Q7; © VCAA*

Students in a Physics practical class investigate the piece of electrical equipment shown in Figure 5. It consists of a single rectangular loop of wire that can be rotated within a uniform magnetic field. The loop has dimensions 0.50 m × 0.25 m and is connected to the output terminals with slip rings. The loop is in a uniform magnetic field of strength 0.40 T.

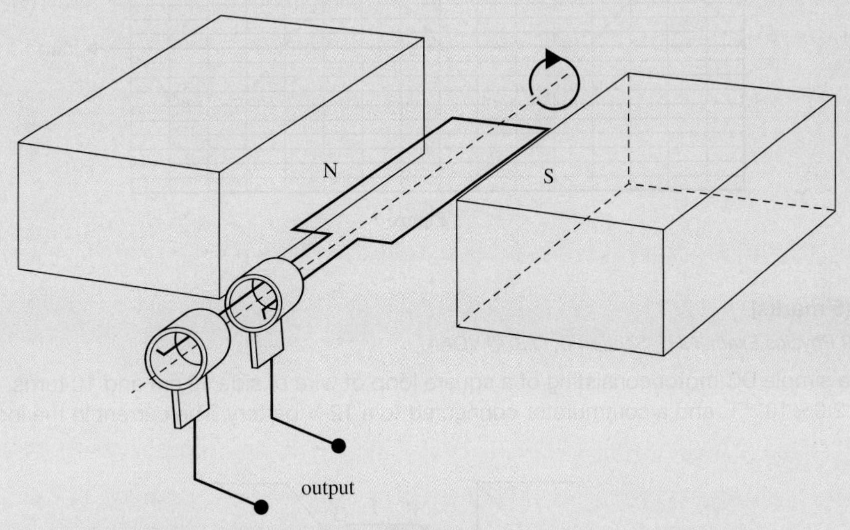

Figure 5

a. Circle the name that best describes the piece of electrical equipment shown in Figure 5. **(1 mark)**

alternator   DC generator   DC motor   AC motor

b. i. What is the magnitude of the flux through the loop when it is in the position shown in Figure 5? **(1 mark)**

ii. Explain your answer to part **b.i.** **(1 mark)**

The students connect the output terminals of the piece of electrical equipment to an oscilloscope. One student rotates the loop at a constant rate of 20 revolutions per second.

c. Calculate the period of rotation of the loop. **(1 mark)**

d. Calculate the maximum flux through the loop. Show your working. **(2 marks)**

e. The loop starts in the position shown in Figure 5.
What is the average voltage measured across the output terminals for the first quarter turn? Show your working. **(2 marks)**

f. State **two** ways that the amplitude of the voltage across the output terminals can be increased. **(2 marks)**

g. Figure 6 shows the output voltage graph shown on the oscilloscope for two cycles.

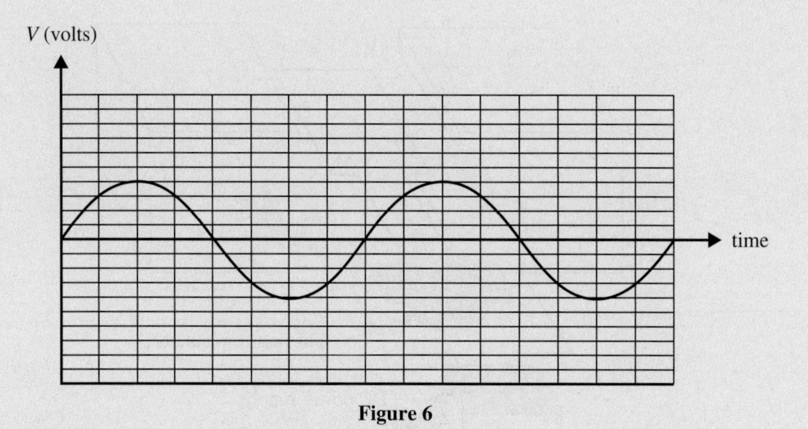

Figure 6

The students now replace the slip rings in Figure 5 with a split-ring commutator.

On Figure 7, sketch with a solid line the output that the students will now observe on the oscilloscope. Show **two** complete revolutions. The original output is shown with a dashed line. **(1 mark)**

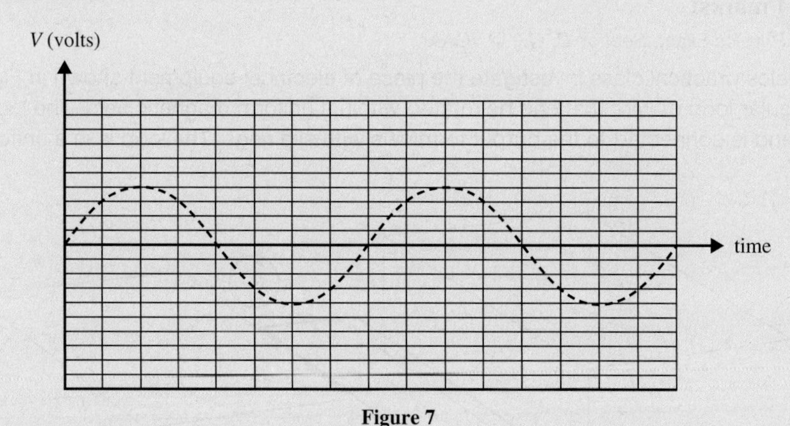

Figure 7

Question 4 (5 marks)

Source: *VCE 2019 Physics Exam, NHT, Section B, Q.3; © VCAA*

Figure 3 shows a simple DC motor consisting of a square loop of wire of side 10 cm and 10 turns, a magnetic field of strength 2.0×10^{-3} T, and a commutator connected to a 12-V battery. The current in the loop is 2.0 A.

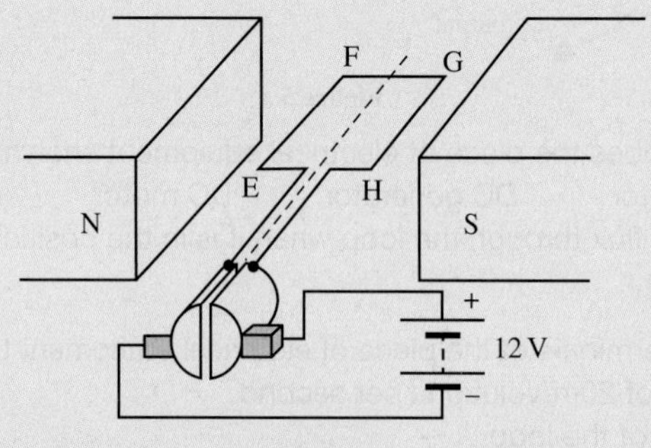

Figure 3

a. Calculate the magnitude of the total force acting on the side EF when the loop is in the position shown in Figure 3. Show your working. **(2 marks)**

b. Explain the role of the commutator in a DC motor. **(3 marks)**

Question 5 (5 marks)

Source: *VCE 2018 Physics Exam, Section B, Q.3;* © VCAA

Students build a model of a simple DC motor, as shown in Figure 3.

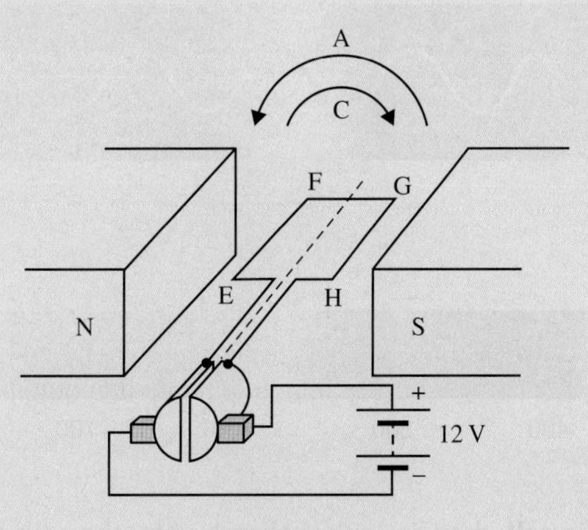

Figure 3

a. The motor is set with the coil horizontal, as shown, and the power source is applied. Will the motor rotate in a clockwise (C) or anticlockwise (A) direction? Explain your answer. **(3 marks)**

b. One student suggests that slip rings would be easier to make than a commutator and that they should use slip rings instead.
Explain the effect that replacing the commutator with slip rings would have on the operation of the motor, if no other change was made. **(2 marks)**

More exam questions are available in your learnON title.

6.6 Photovoltaic cells

KEY KNOWLEDGE

- Describe the production of electricity using photovoltaic cells and the need for an inverter to convert power from DC to AC for use in the home (not including details of semiconductors action or inverter circuitry)

Source: VCE Physics Study Design (2024–2027) extracts © VCAA; reproduced by permission.

6.6.1 The production of electricity using photovoltaic cells

Visible light is a part of the electromagnetic spectrum. Light behaves like waves, and it is often described in terms of its wavelength. The wavelength of light depends on the energy of the light: high energy light has a short wavelength, and low energy light has a long wavelength. Red light is low energy and has wavelengths greater than 700 nm (nanometres). Violet light is high energy and has wavelengths less than 400 nm.

Light also behaves like particles or discrete packets of energy. These particles are called photons.

You will learn more about this dual behaviour of light in Unit 4.

Solar energy is the energy we receive from the Sun. The Sun emits the full range of radiation of the electromagnetic spectrum.

FIGURE 6.21 The electromagnetic spectrum

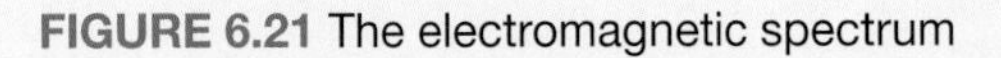

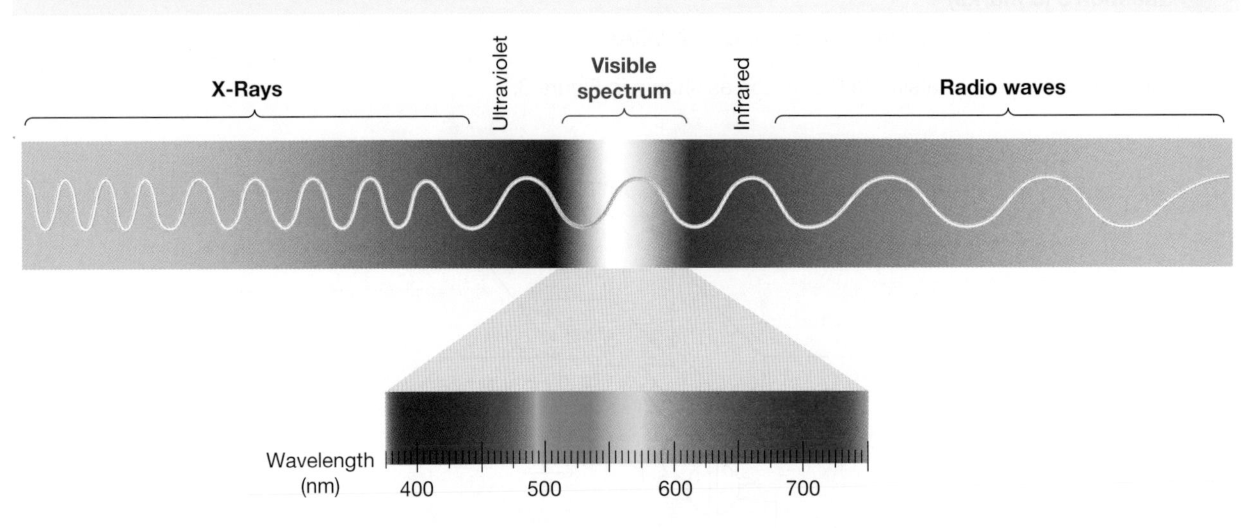

When photons strike a **photovoltaic cell**, they are either absorbed by the cell or reflected from the surface.

If photons have enough energy, they cause electrons to be removed from atoms in the cell.

If the photons do not have enough energy, their energy is transformed into thermal energy and the solar cell heats up. Solar cells are not very efficient because most of the photons in light from the Sun do not have enough energy to release electrons in the cells.

Released electrons are collected and travel around an electrical circuit. The resulting electrical current is a direct current (DC). It flows in one direction. As the intensity of the light increases, it has more photons that release more electrons, creating a bigger electric current.

The voltage produced by a solar cell depends on the materials used to manufacture the cell. The voltage produced by the cell is not affected by the intensity of the light.

Silicon-based solar cells

Most household solar cells are silicon-based. Silicon-based solar cells are p-n junction **diodes**.

A silicon-based solar cell has two thin layers of silicon sandwiched together, as shown in figure 6.22. The two silicon layers in the figure are both made from highly purified silicon. In the silicon layer on the right, phosphorus atoms are inserted among the silicon atoms, in a process called 'doping'. A phosphorus atom has one more electron in the outer shell than does a silicon atom. This extra electron is held quite loosely, which is why a phosphorus atom releases an electron when it absorbs energy from a photon that has enough energy. This layer is called an n-type layer because it is a source of negatively charged electrons.

The silicon layer on the left side has been 'doped' with boron atoms. A boron atom has one less negatively charged electron in the outer shell than does a silicon atom. The presence of boron atoms in the silicon lattice therefore creates 'positive holes', and is thus called a p-type layer.

photovoltaic cell device that transforms electromagnetic energy, such as light from the Sun, directly into electrical energy. Also known as a PV cell or solar cell.

diode a two-terminal semiconductor device that allows current to pass through it in one direction but not the other

Where the two layers meet is termed a p-n junction. Electrons from the n-type material drift across the junction to fill holes, forming a 'depletion layer'.

FIGURE 6.22 A silicon-based solar cell, or junction diode

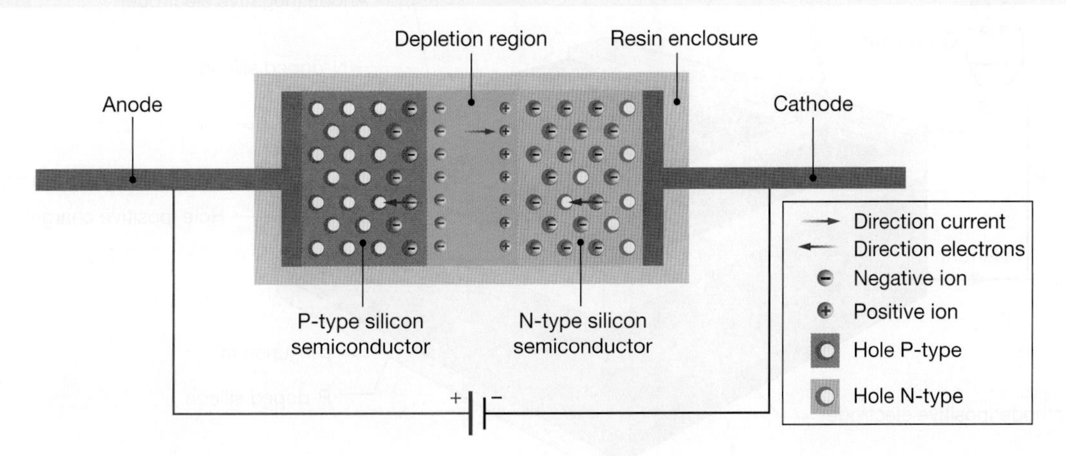

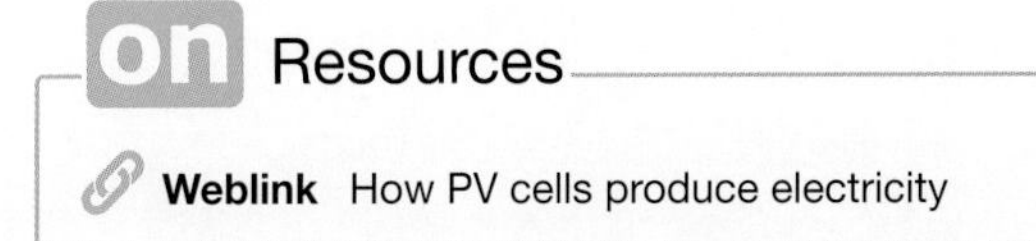

Resources

Weblink How PV cells produce electricity

EXTENSION: Semiconductors in photovoltaic cells and electricity generation

Semiconductors (generally silicon) in photovoltaic cells exposed to light will absorb the light's energy.

Electrons in the silicon valence band are promoted to the conduction band, and holes are formed in the valence band.

The charge carriers can travel in the form of an electrical current through the semiconductor material and an external circuit.

They can also recombine, with their energy being dissipated as heat.

When the n-type semiconductor and the p-type semiconductor are put together in a p-n junction, an electric field is formed at the p-n junction, moving the flow of electrons and holes in opposite directions, which reduces electron-hole recombination.

Thus, when light is shone on the top layer of a photovoltaic cell, the top layer is covered with a grid of silver to collect the released electrons. The result is the generation of an electric current when it is connected to a closed electric circuit.

Silicon-based solar cells produce a maximum voltage of about 0.65 V (DC). The structure of silicon-based solar cells is shown in figure 6.23. The top is a thin layer of n-type material. This allows photons to reach the junction.

FIGURE 6.23 Cross-section of a silicon-based solar cell

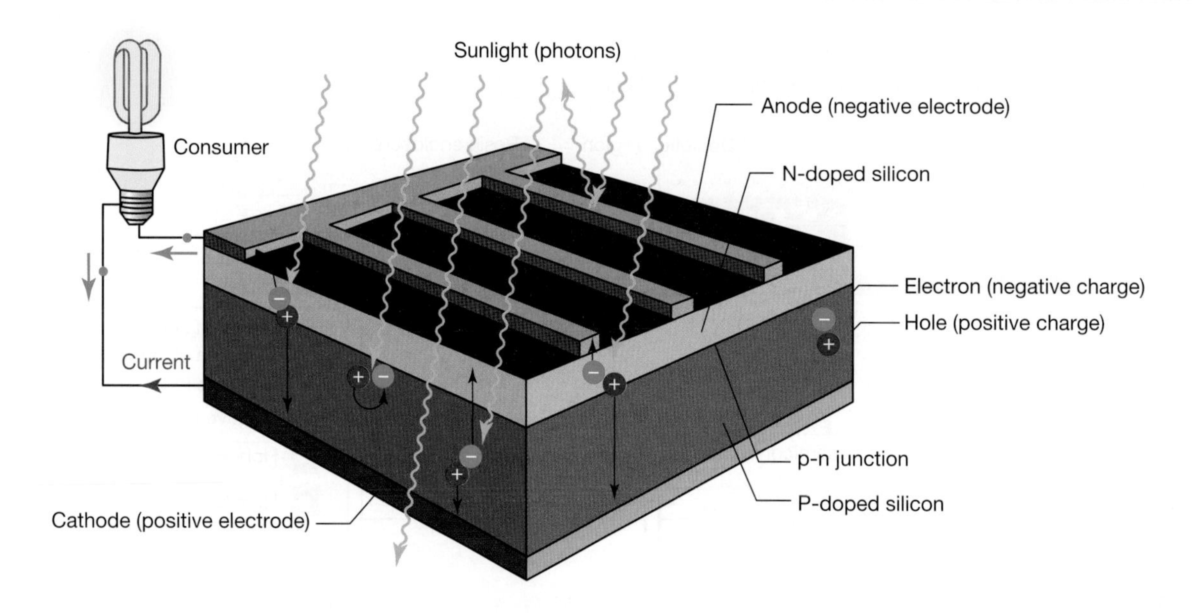

6.6.2 Solar panels and solar arrays

A solar panel consists of a set of solar cells connected in series and/or in parallel to produce a desired voltage and current. The solar cells are set into a watertight frame, as shown in figure 6.24.

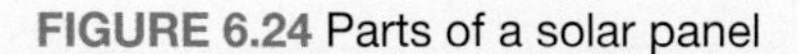

FIGURE 6.24 Parts of a solar panel

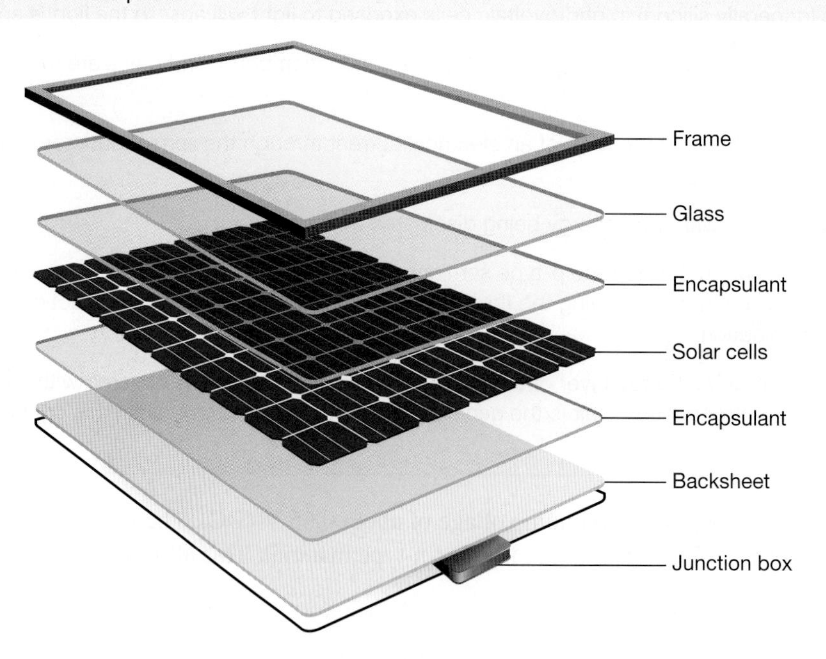

A single solar cell has a maximum output voltage of about 0.65 V DC. The solar panel in the illustration has 60 solar cells connected in series. This gives a nominal output voltage of 24 V DC. The maximum voltage can be greater than 36 V.

A solar array is a set of solar panels connected in a grid. Solar arrays are often used on the rooftops of buildings. If the solar array generates more electricity than is required in the building, the excess electrical energy is fed back into the electricity grid or stored in a battery.

inverter device that converts direct current electricity produced by solar panels into alternating current electricity, usable in the home

6.6.3 Solar electricity for buildings

A rooftop solar system is the name given to the solar panels together with the electrical circuit that must be set up to link the solar array to the electrical circuitry in the building.

The electrical current produced by a solar cell is a direct current (DC). In Victoria, household electrical appliances operate on a 230 V 50 Hz AC supply. For this reason, a device known as an **inverter** must be inserted into a rooftop solar system to convert the direct current into an alternating current and change the voltage to 230 V_{rms}. The electrical energy generated that is not needed at the time can be stored in batteries and/or fed back into the power grid. When electrical energy is fed back into power grid, a meter measures the electrical energy that has been supplied and the owners are paid for the power they supply.

FIGURE 6.25 Solar panels are connected together in a rooftop array.

FIGURE 6.26 Schematic diagram of a photovoltaic array connected to an inverter before the electricity is fed into the household wiring.

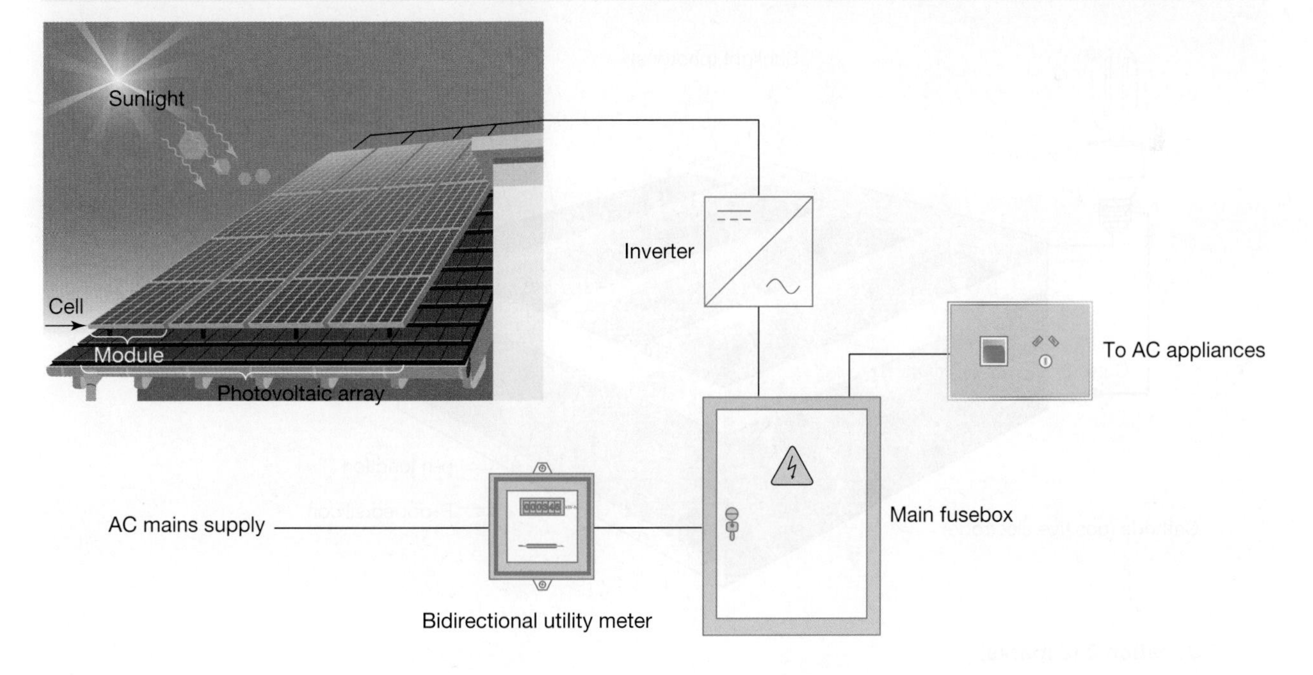

6.6 Activities

learnon

Students, these questions are even better in jacPLUS

Receive immediate feedback and access sample responses

Access additional questions

Track your results and progress

Find all this and MORE in jacPLUS

6.6 Quick quiz	6.6 Exercise	6.6 Exam questions

6.6 Exercise

1. Explain what a photovoltaic cell is.
2. Why are inverters needed in connecting solar panel arrays to the electric circuits of a building?
3. If ____________ have enough energy, they cause ____________ to be removed from atoms in the cell. If the photons striking a solar cell ____________ have enough energy, their energy is transformed into ____________ energy and the solar cell heats up.
4. Why is it important that one side of the silicon layer is an n-type layer and the other a p-type layer?
5. **MC** Photovoltaic cells can convert all sunlight into electricity. True or false?
   A. True
   B. False

6.6 Exam questions

Question 1 (4 marks)

Explain why the light bulb in the figure will be able to work.

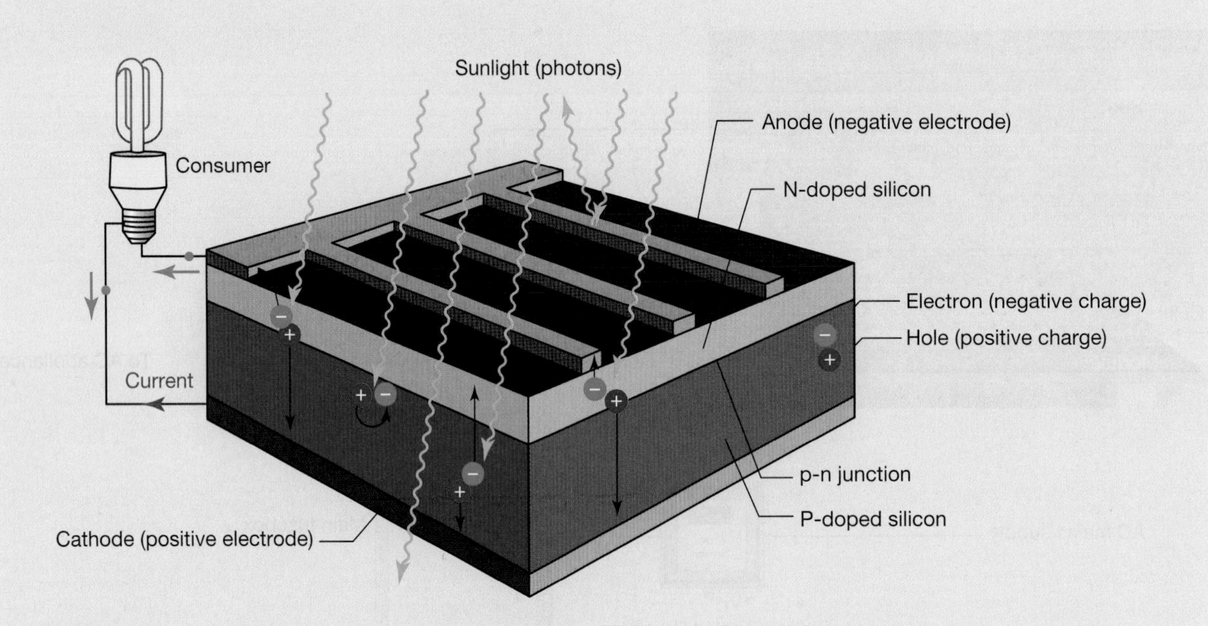

Question 2 (2 marks)

Silicon is used as a base for photovoltaic cells. The silicon is doped with elements from Group III and V. Why is this doping process important for photovoltaic cells?

Question 3 (1 mark)

MC Even though photovoltaic cells have a low efficiency, the government has been offering rebates and incentives to residences that install them. Why is the government supportive of this energy source?

A. To provide people with more energy options
B. To help reduce greenhouse gas emissions as part of the Renewable Energy Target scheme
C. To allow people to reduce their electricity bills
D. To reduce the number of households reliant on fossil fuels

Question 4 (2 marks)

What is the advantage of connecting two solar cells with a rated voltage of 40 V and a rated amperage of 5 A in series, rather than in parallel?

Question 5 (2 marks)

If no inverter was used for a rooftop solar system, would you still be able to use electrical appliances in your home? Explain.

More exam questions are available in your learnON title.

6.7 Review

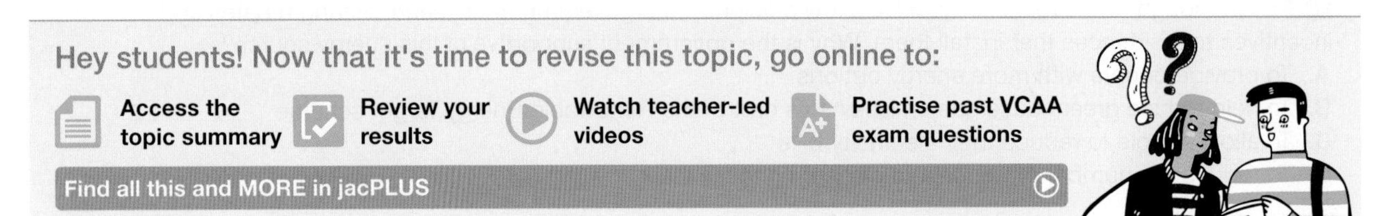

6.7.1 Topic summary

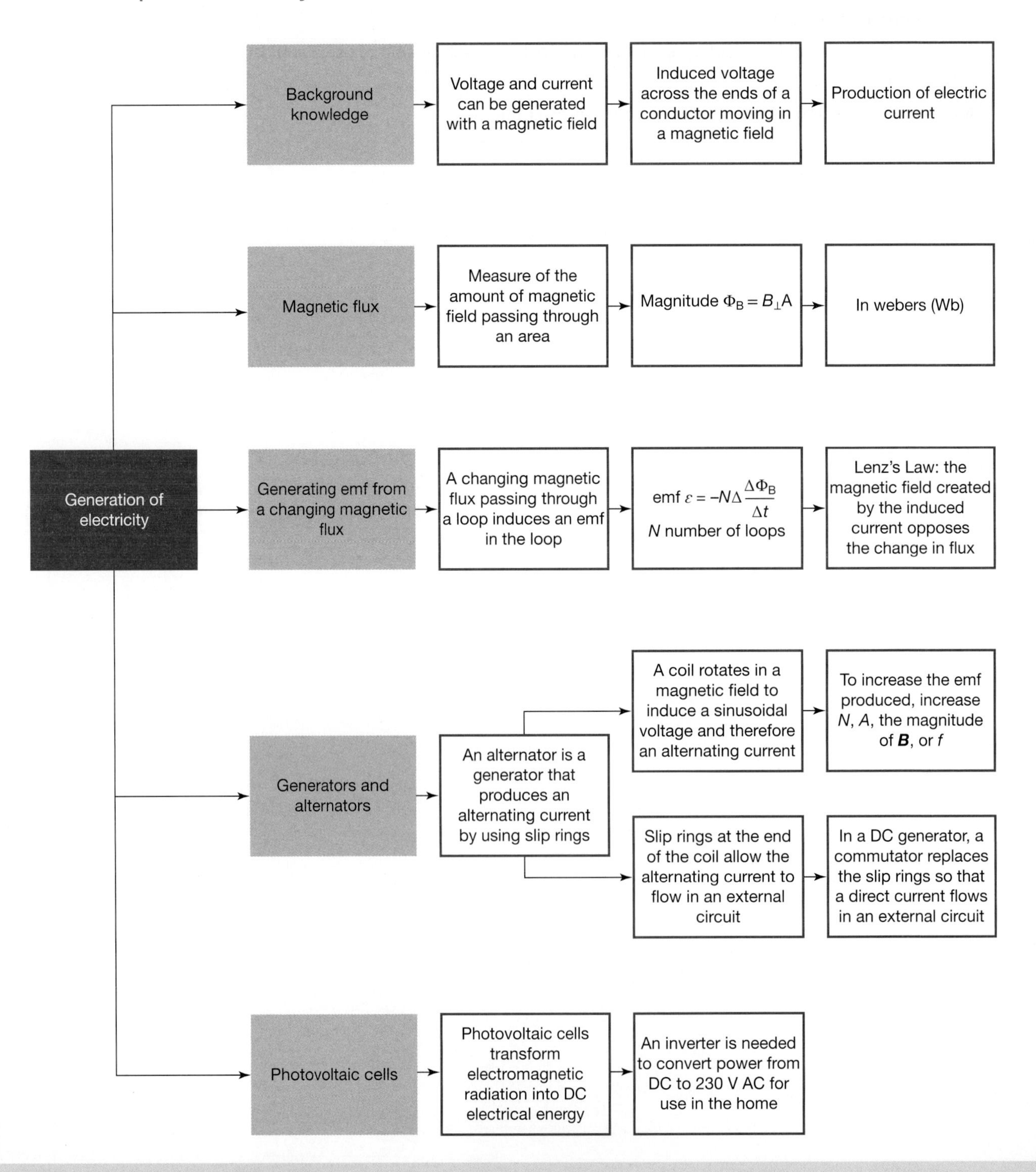

6.7.2 Key ideas summary

online only

6.7.3 Key terms glossary

online only

Resources

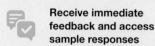

Solutions	Solutions — Topic 6 (sol-0820)
Practical investigation eLogbook	Practical investigation eLogbook — Topic 6 (elog-1637)
Digital documents	Key science skills — VCE Physics Units 1–4 (doc-36950) Key terms glossary — Topic 6 (doc-37175) Key ideas summary — Topic 6 (doc-37176)
Exam question booklet	Exam question booklet — Topic 6 (eqb-0103)

6.7 Activities

learnon

6.7 Review questions

1. A magnet passes through a copper tube at constant velocity along the path shown. A current is induced in the tube by the motion of the magnet.

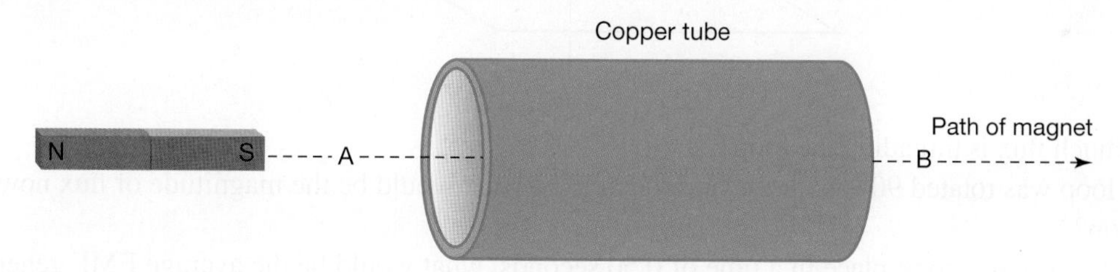

   a. Describe the force acting between the tube and the magnet at point A and justify your answer.
   b. Describe the force acting between the tube and the magnet at point B and justify your answer.

2. This figure shows a single loop of wire placed completely within a region of uniform magnetic field ($\boldsymbol{B} = 3.2 \times 10^{-2}$ T) directed into the page. The loop of wire has an area of 8.0 cm^2 and is mounted so that it can be turned about the vertical axis as shown. The loop rotates at a constant rate of 10 turns each second.
   What is the value of the emf induced in the loop at the instant when the plane of the loop is first perpendicular to the direction of the magnetic field if the coil begins parallel to the field?

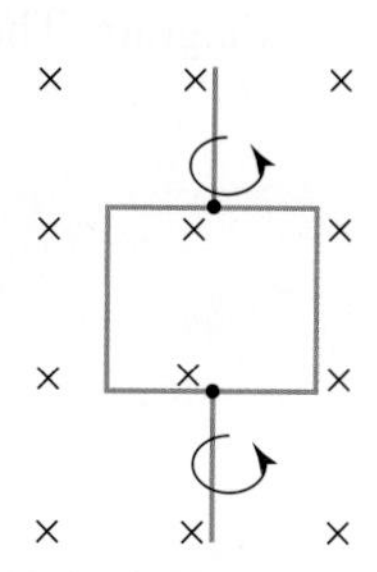

3. Veronica and Ron are investigating electromagnetic induction using a square coil. They place the coil between the poles of a magnet. The sides of the coil are 0.020 m long. The uniform magnetic field between the poles is 5.0×10^{-2} T, and elsewhere in air it is assumed to be zero. Calculate the magnetic flux through the coil when it is entirely within and perpendicular to the magnetic field.

4. A coil consisting of 10 turns of wire is held at right angles to, and completely within, a uniform magnetic field.
The area enclosed by the coil is A. The magnetic field is decreased from $\boldsymbol{B}$ to $\frac{\boldsymbol{B}}{3}$ in time t. Write an expression for the induced emf in the 10-turn coil in terms of the quantities given.

5. A boat sails at 3.0 m s^{-1} due east, where the Earth's magnetic field is 5.0×10^{-5} T due north and horizontal. The boat carries a vertical aerial 5.0 m long.

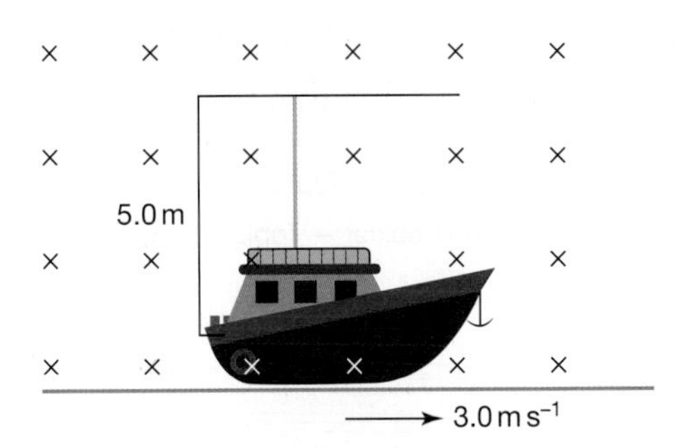

What will be the magnitude of the emf induced in the aerial?

6. State an advantage and a disadvantage of using photovoltaic cells as an energy source.

7. The following diagram shows a square loop of metal in a horizontal plane, situated in a uniform magnetic field $\boldsymbol{B}$ that is directed vertically downwards. The side length of the square is 0.25 m. The magnitude of $\boldsymbol{B}$ is 9.0×10^{-5} T.

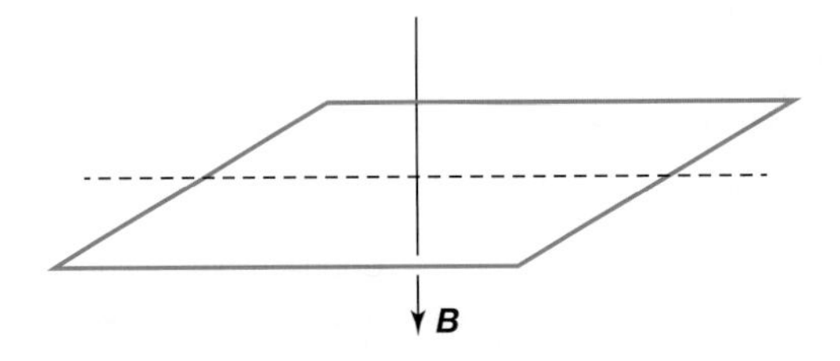

a. How much flux is threading the loop?
b. If the loop was rotated 90° about the horizontal axis, what would be the magnitude of flux now threading the loop?
c. If this 90° rotation took place in a time of 0.50 seconds, what would be the average EMF generated during this period of time?

8. A single, circular coil of wire is placed perpendicularly to a magnetic field as shown in the following diagram. The magnetic field strength is 0.70 T and the flux threading the coil is 4.0×10^{-3} Wb.

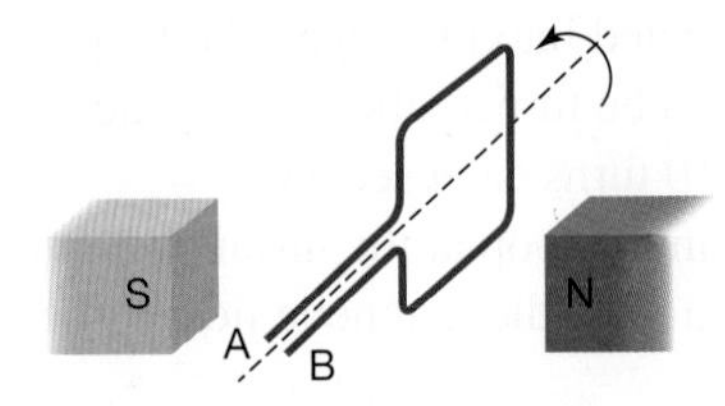

a. Calculate the area of the loop.
b. The loop is now rotated 90° anticlockwise as shown in the diagram. If it is rotated at a rate of 10 revolutions per second, calculate the average emf generated in the loop.
c. On one set of axes, sketch the graphs of flux and emf generated over one complete rotation of the coil. A vertical scale is not required.
d. State two ways to increase the size of the emf generated in the loop.

9. A wire spring is made into a circular loop of area 0.30 m^2 and is placed in a uniform magnetic field, ***B***, of 0.82 T, perpendicular to the plane of the loop, as in part (a) of the following diagram.

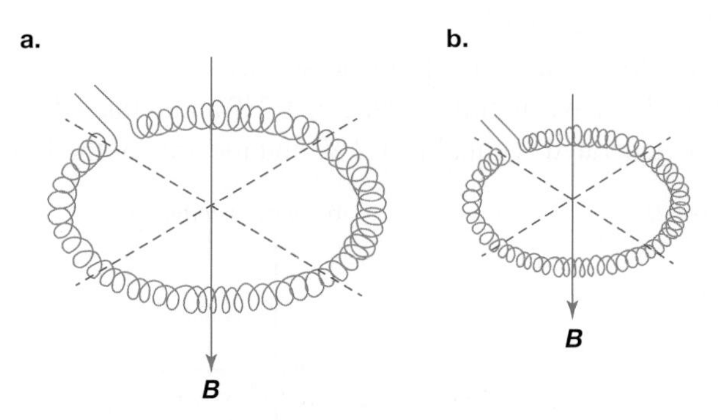

a. What is the magnetic flux through the loop shown in diagram (a)?
b. The area enclosed by the loop shrinks from 0.30 m^2 in diagram (a) to 0.10 m^2 in diagram (b) in a time of 0.10 seconds. What is the average emf induced?

At time t_0, the loop has an area of 0.10 m^2, as in diagram (b). At time t_1, the loop is made to expand back to its original size, which it reaches at time t_2, in such a way that its area grows at a steady rate, as shown in the following graph:

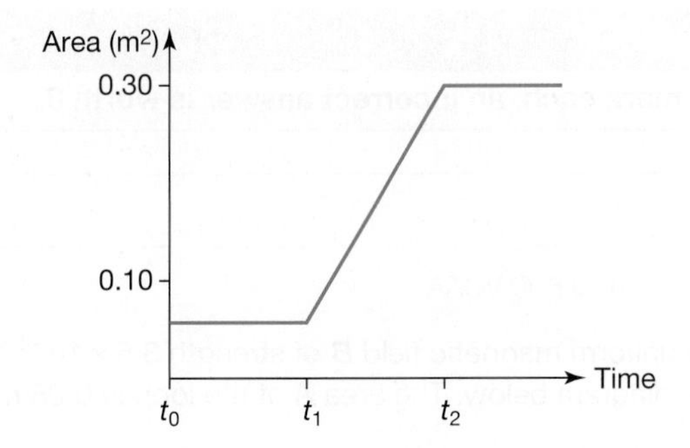

c. Sketch a graph that represents the emf in the loop as a function of time.

10. A student investigates electromagnetic induction using a single loop of wire and an electromagnet. As shown in the following diagram, the loop is placed between the pole pieces of the electromagnet, perpendicular to the magnetic field, and connected to an oscilloscope so that any induced voltage can be measured.

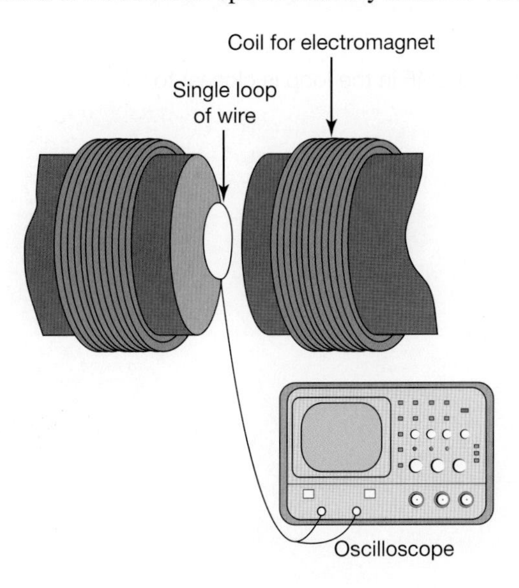

The current in the electromagnet's coil is reduced so that the magnetic field, $\boldsymbol{B}$, decreases to zero at a constant rate over a time, t, as shown in part (a) of the following diagram. The induced voltage measured on the oscilloscope is shown in diagram (b).

a. Explain why the induced voltage varies with time as shown.
b. The magnetic field is found to have an initial value of 1.7 T and it takes 5.0×10^{-3} s to reduce to 0 T. If the magnitude of voltage generated is equal to 1.2 V, find the area of the loop of wire.

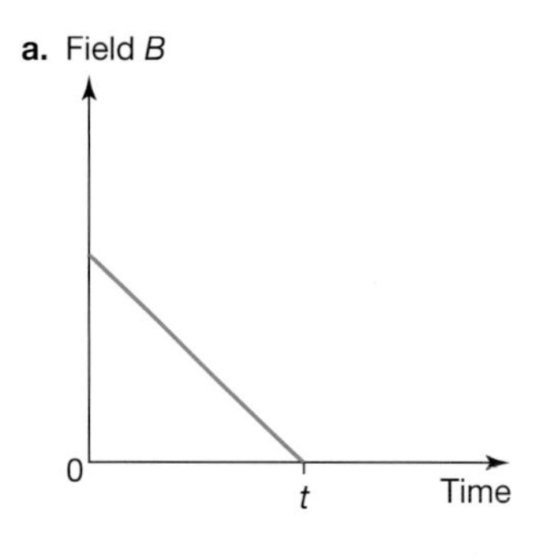

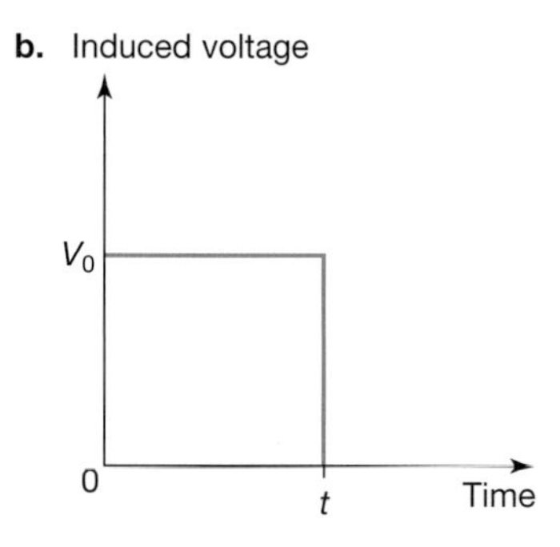

6.7 Exam questions

Section A — Multiple choice questions

All correct answers are worth 1 mark each; an incorrect answer is worth 0.

Question 1

***Source:** VCE 2020 Physics Exam, Section A, Q.6; © VCAA*

A single loop of wire moves into a uniform magnetic field $\boldsymbol{B}$ of strength 3.5×10^{-4} T over time $t = 0.20$ s from point X to point Y, as shown in the diagram below. The area A of the loop is 0.05 m^2.

The magnitude of the average induced EMF in the loop is closest to

A. 0 V
B. 3.5×10^{-6} V
C. 8.8×10^{-5} V
D. 8.8×10^{3} V

Question 2

Source: *VCE 2021 Physics Exam, Section A, Q.6; © VCAA*

A magnet approaches a coil with six turns, as shown in the diagram below. During time interval Δt, the magnetic flux changes by 0.05 Wb and the average induced EMF is 1.2 V.

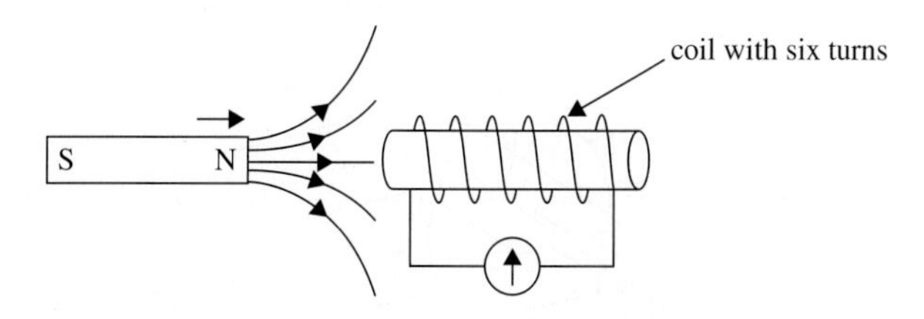

Which one of the following is closest to the time interval Δt?

A. 0.04 s
B. 0.01 s
C. 0.25 s
D. 0.50 s

Question 3

Source: *VCE 2017, Physics Exam, Section A, Q.6; © VCAA*

The graph below shows the change in magnetic flux (Φ) through a coil of wire as a function of time (t).

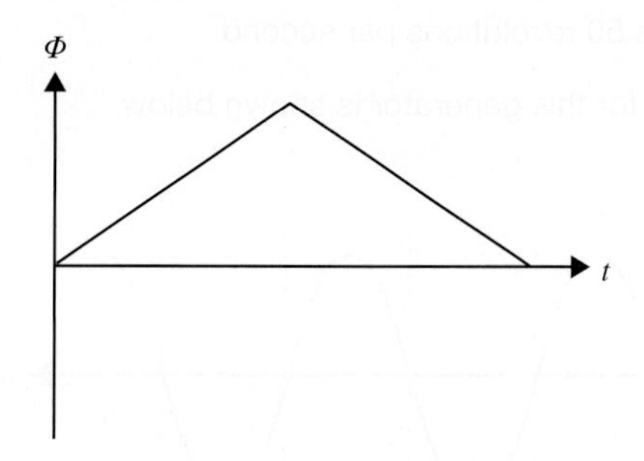

Which one of the following graphs best represents the induced EMF (ε) across the coil of wire as a function of time (t)?

A.

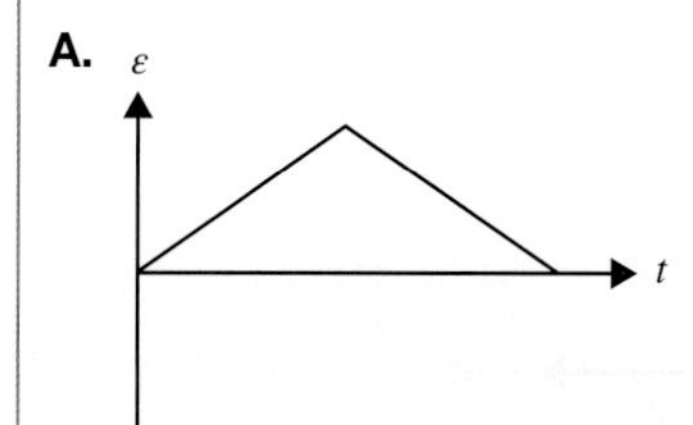

B.

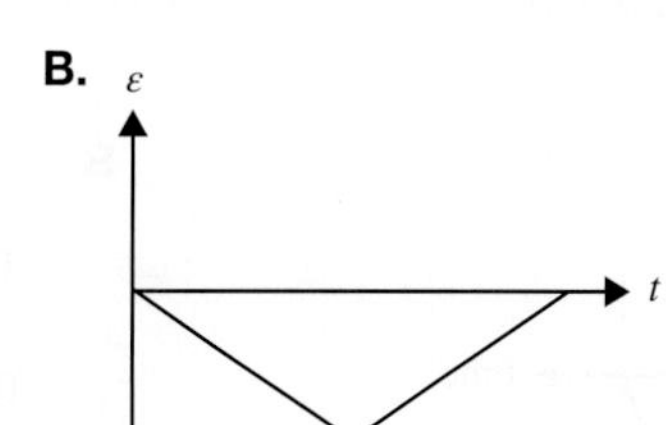

C.

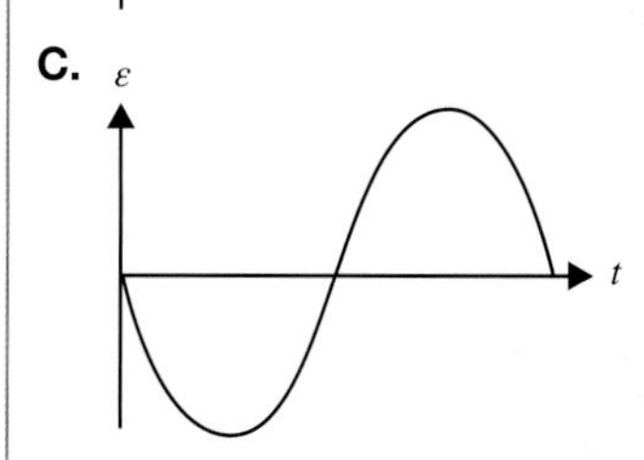

D.

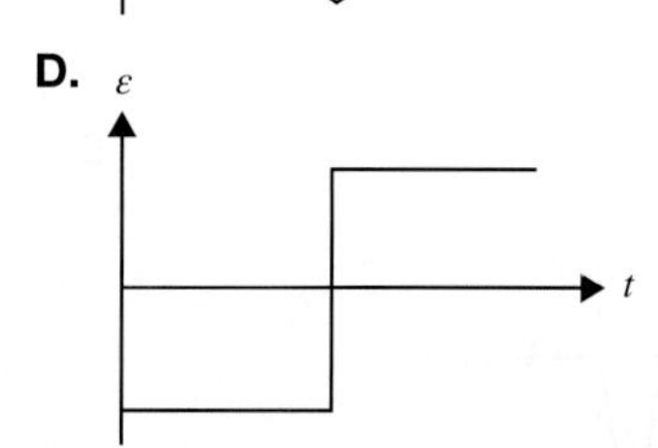

Question 4

Source: *VCE 2021, Physics Exam, Section A, Q.8;* © VCAA

The diagram below shows a simple electrical generator consisting of a rotating wire loop in a magnetic field, connected to an external circuit with a light globe, a split-ring commutator and brushes. The direction of rotation is shown by the arrow.

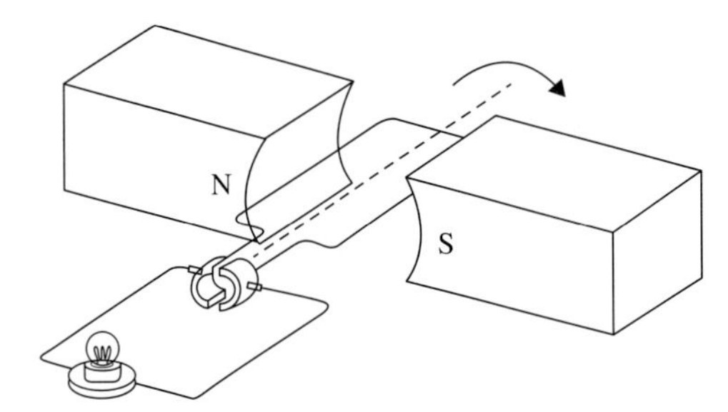

Which one of the following best describes the function of the split-ring commutator in the external circuit?

A. It delivers a DC current to the light globe.
B. It delivers an AC current to the light globe.
C. It ensures the force on the side of the loop nearest the north pole is always up.
D. It ensures the force on the side of the loop nearest the north pole is always down.

Question 5

Source: *VCE 2019, Physics Exam, Section A, Q.7;* © VCAA

The coil of an AC generator completes 50 revolutions per second.

A graph of output voltage versus time for this generator is shown below.

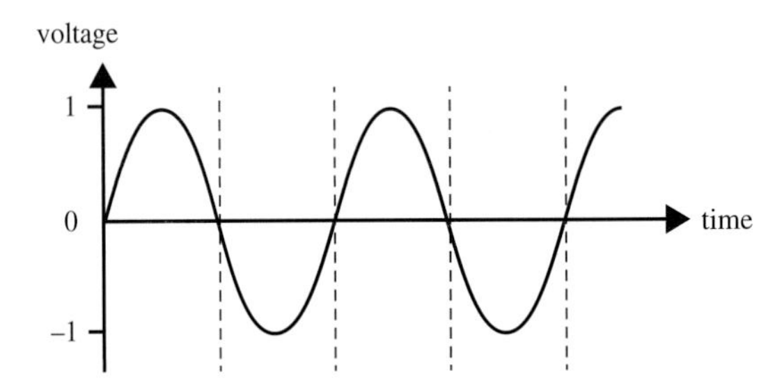

Which one of the following graphs best represents the output voltage if the rate of rotation is changed to 25 revolutions per second?

A.

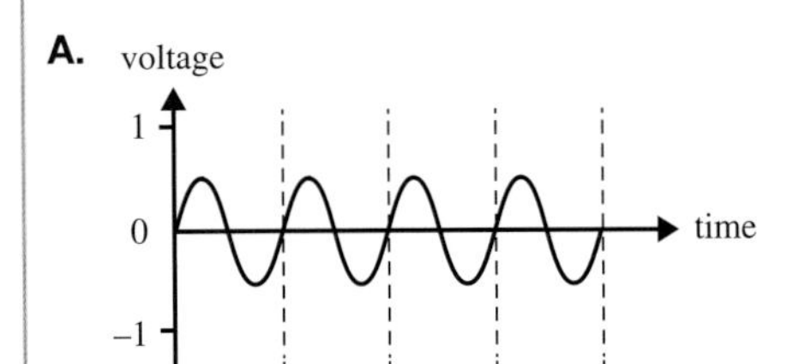

B.

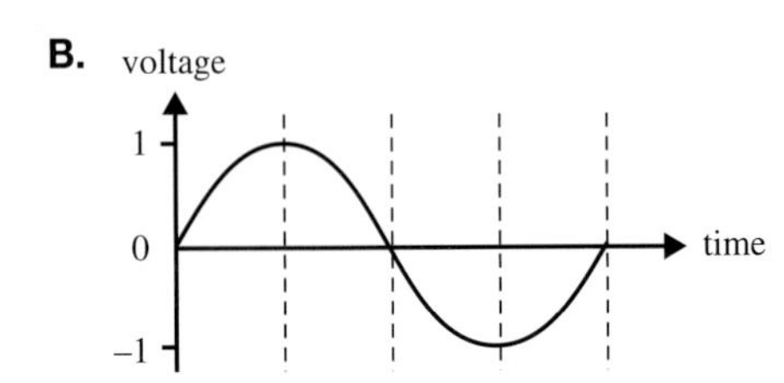

C.

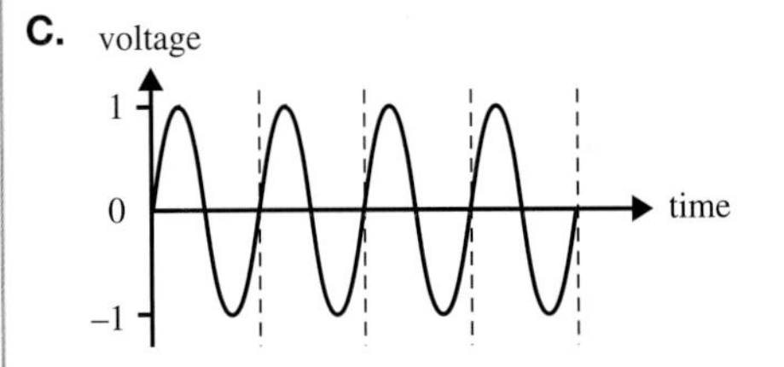

D.

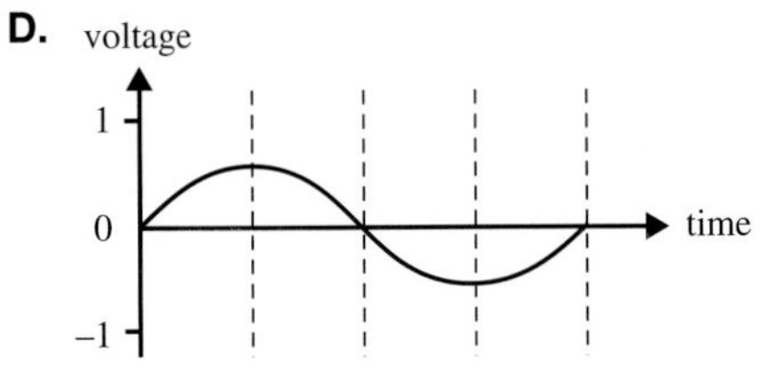

Question 6

Source: *VCE 2022 Physics Exam, Section A, Q.2;* © VCAA

The diagram below shows the magnetic flux variation through the coil of an AC generator.

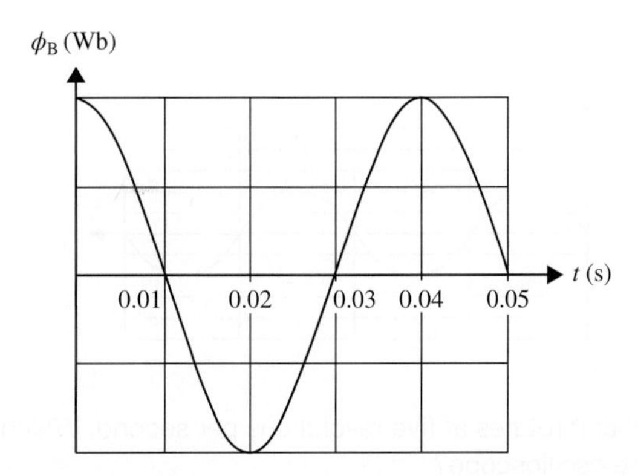

Which one of the following is closest to the frequency of the magnetic flux variation through the coil of the AC generator?

A. 0.04 Hz
B. 10 Hz
C. 20 Hz
D. 25 Hz

Question 7

Source: *VCE 2021 Physics Exam, NHT, Section A, Q.8;* © VCAA

In the diagram below, the solid line represents the graph of output EMF, ε, versus time produced by an AC generator. A single change is made to the AC generator and its operation, and the new graph of output EMF, ε, versus time is shown as a dashed line.

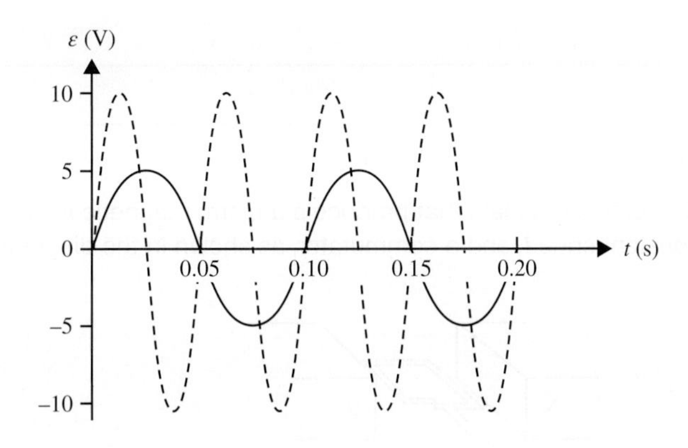

Which one of the following best describes the change made to the AC generator?

A. The area of the coil was doubled.
B. The speed of rotation was halved.
C. The speed of rotation was doubled.
D. The number of turns of the wire in the coil was doubled.

Question 8

Source: *VCE 2019 Physics Exam, NHT, Section A, Q.7*; © VCAA

An alternator is rotating at 10 revolutions per second. Its output is measured by an oscilloscope. The signal produced is shown below.

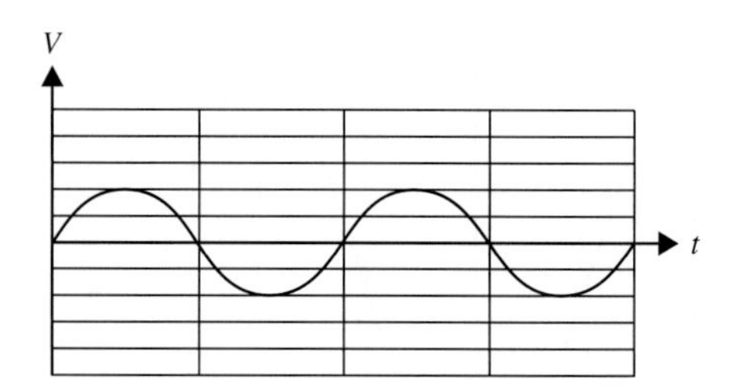

The alternator is then slowed so that it rotates at five revolutions per second. Which one of the following best shows the display observed on the oscilloscope?

A.

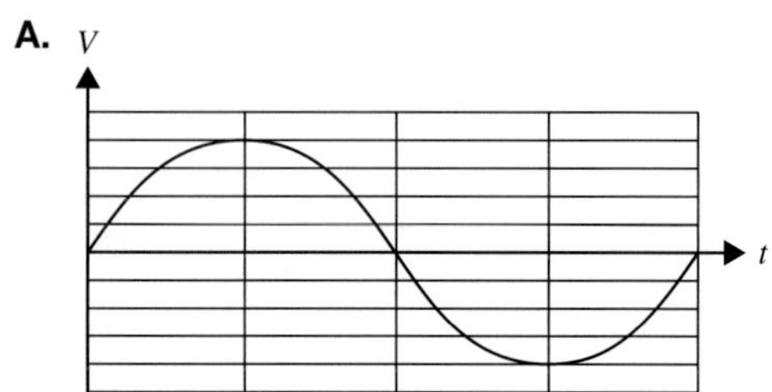

B.

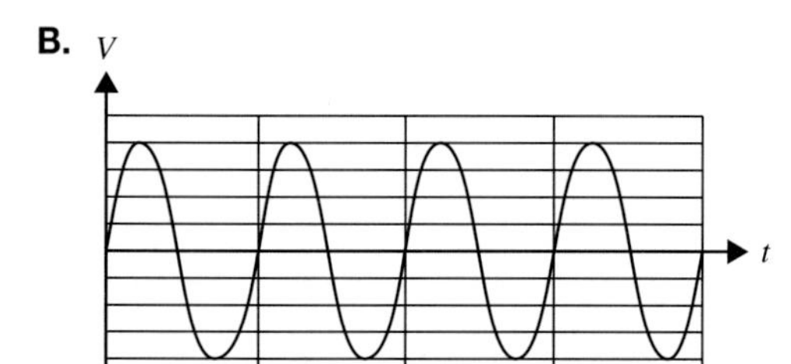

C.

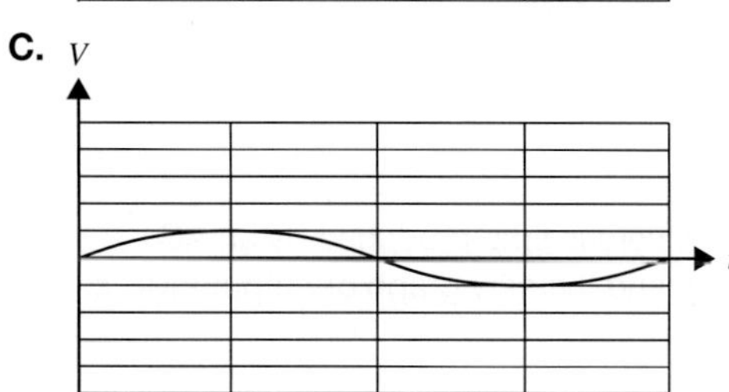

D.

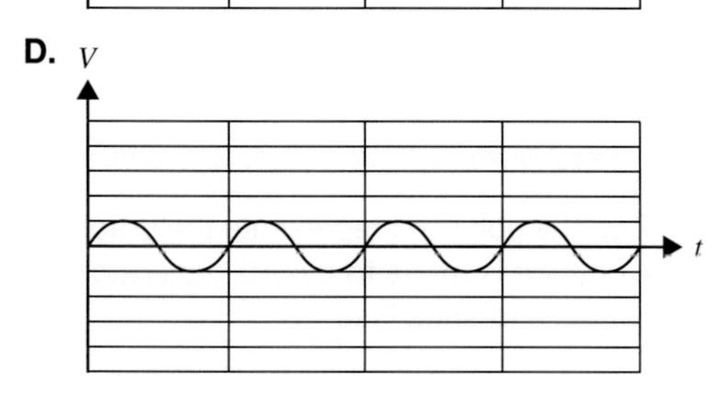

Question 9

Source: *VCE 2018 Physics Exam, NHT, Section A, Q.4*; © VCAA

A simple DC generator consists of two magnets that produce a uniform magnetic field, in which a square loop of wire of 100 turns rotates at constant speed, and a commutator, as shown in the diagram below.

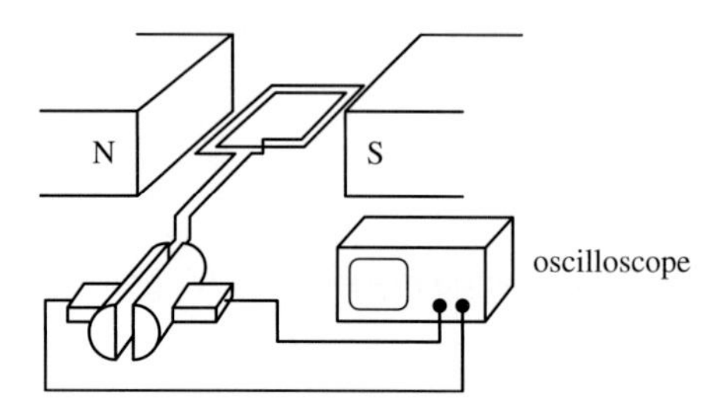

Which one of the following best shows the display observed on the oscilloscope?

A.

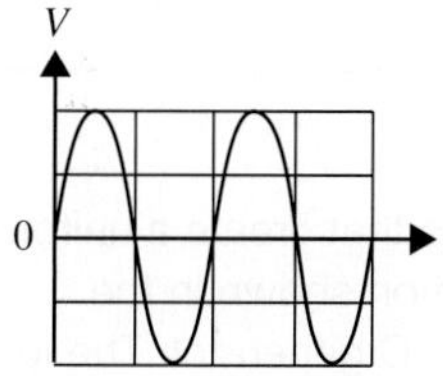

B.

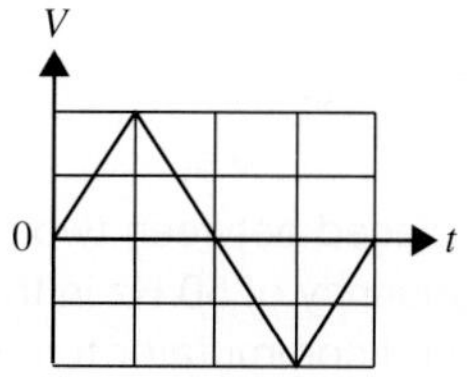

C.

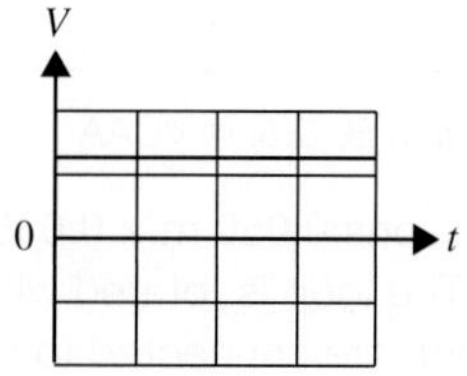

D.

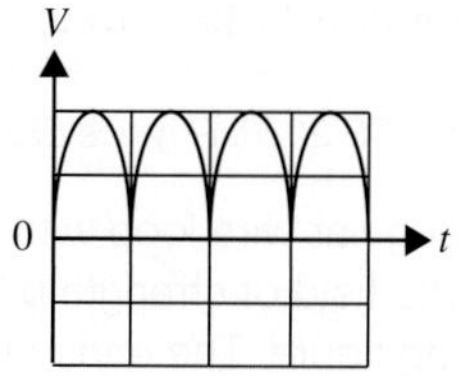

Question 10

Source: *VCE 2019, Physics Exam, Section A, Q.8; © VCAA*

An electrical generator is shown in the diagram below. The generator is turning clockwise.

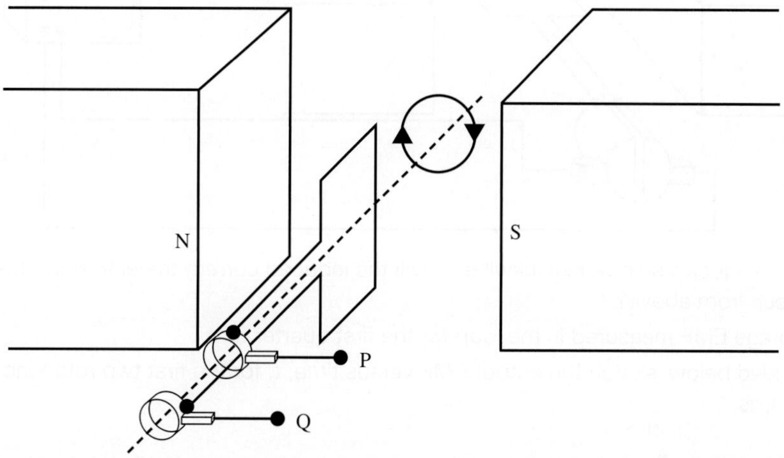

The voltage between P and Q and the magnetic flux through the loop are both graphed as a function of time, with voltage versus time shown as a solid line and magnetic flux versus time shown as a dashed line.

Which one of the following graphs best shows the relationships for this electrical generator?

A.

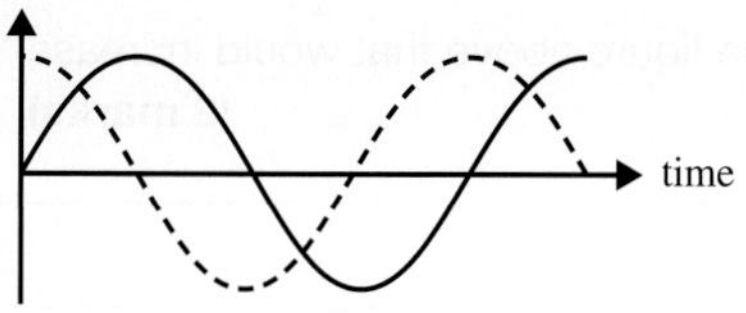

Key

— voltage

- - - magnetic flux

B.

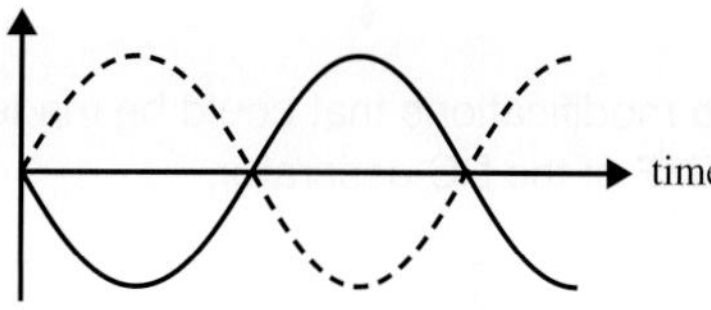

C.

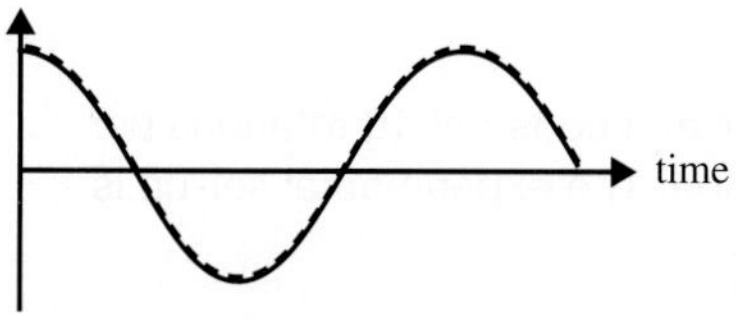

D.

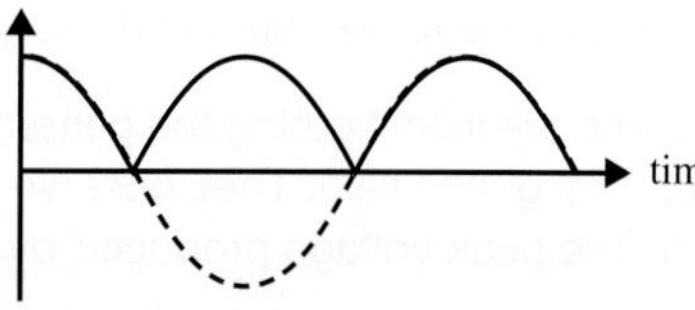

Section B — Short answer questions

Question 11 (9 marks)

Source: *VCE 2020, Physics Exam, Section B, Q.5;* © VCAA

A rectangular wire loop with dimensions 0.050 m × 0.035 m is placed between two magnets that create a uniform magnetic field of strength 0.2 mT. The loop is rotated with a frequency of 50 Hz in the direction shown in the following figure. The ends of the loop are connected to a split-ring commutator to create a DC generator. The loop is initially in the position shown in the figure below.

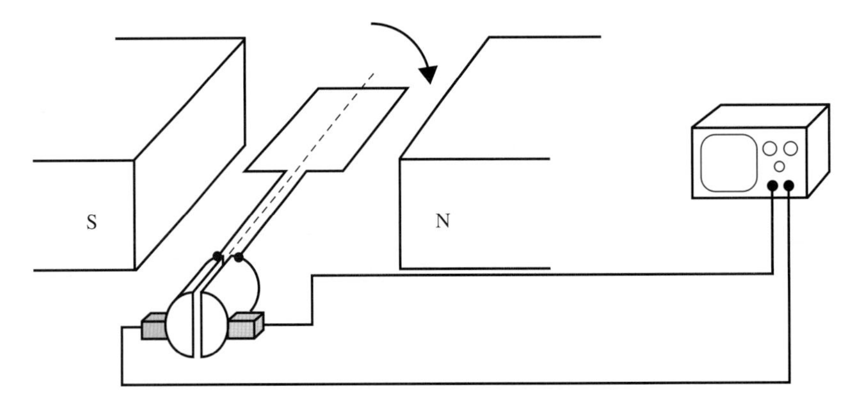

a. In which direction — clockwise or anticlockwise — will the induced current travel through the loop for the first quarter turn as seen from above? **(1 mark)**

b. Calculate the average EMF measured in the loop for the first quarter turn. **(3 marks)**

c. On the axes provided below, sketch the output EMF versus time, t, for the first two rotations. Include a scale on the horizontal axis. **(3 marks)**

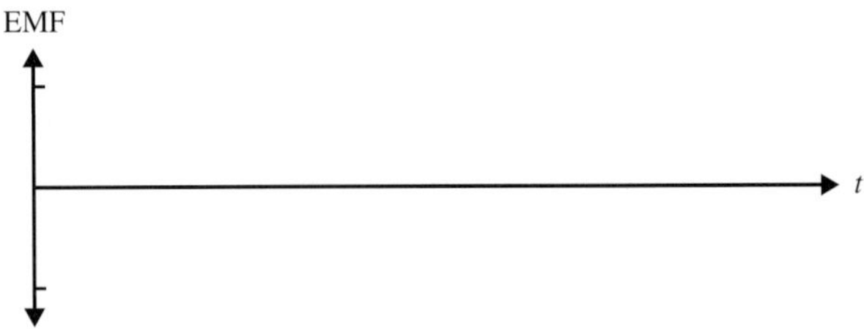

d. Suggest **two** modifications that could be made to the apparatus shown in the figure above that would increase the output EMF of the DC generator. **(2 marks)**

Question 12 (9 marks)

Source: *VCE 2021 Physics Exam, NHT, Section B, Q.5;* © VCAA

Physics students who are investigating the generation of electricity spin a coil at a constant 10 rotations per second in a uniform magnetic field. They observe the output on an oscilloscope. The experimental set-up is shown in Figure 5. The peak voltage produced by the coil is 5 mV.

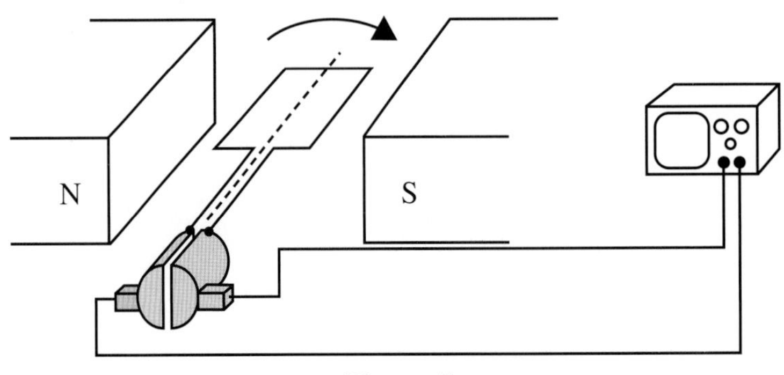

Figure 5

On the axes below, sketch the voltage versus time graph observed on the screen of the oscilloscope for one complete rotation of the coil from the position shown in Figure 5. Include appropriate scales on each axis.

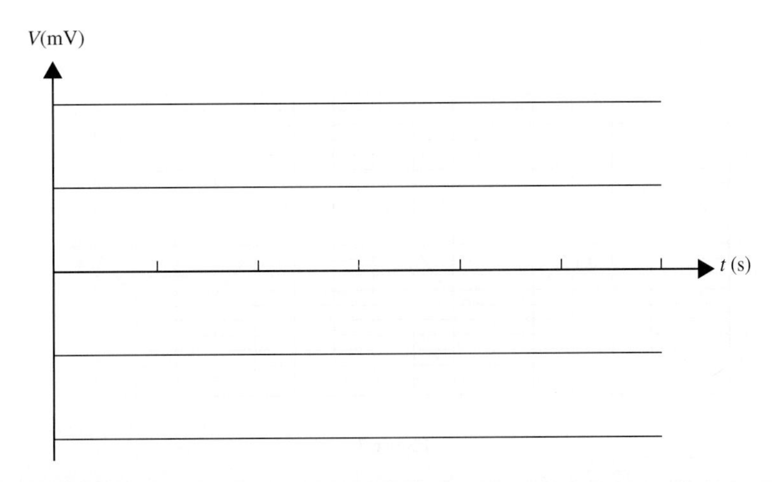

Question 13 (7 marks)

Source: *VCE 2021 Physics Exam, NHT, Section B, Q.6;* © VCAA

Gir and Kau are investigating electromagnetic induction. They have a single wire loop of dimensions XY = 0.030 m long and YZ = 0.020 m wide, which is placed in a uniform magnetic field of strength 0.20 T. The loop is rotated clockwise about an axis, as shown in Figure 6.

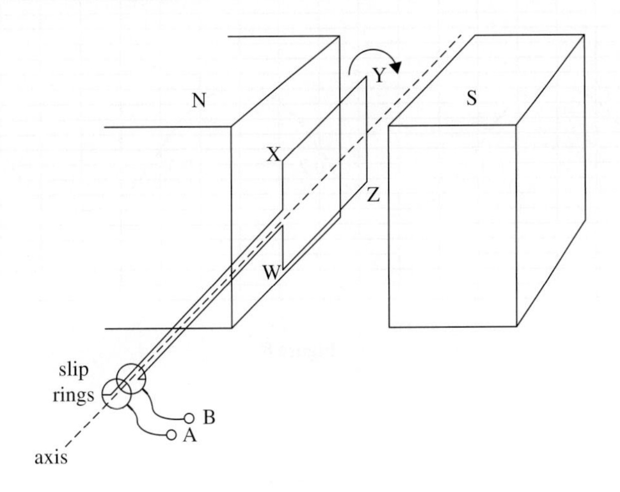

Figure 6

a. Explain the purpose of the slip rings in the apparatus shown in Figure 6. **(2 marks)**

b. Calculate the size of the magnetic flux through the loop when it is oriented as shown in Figure 6. Show your working. **(2 marks)**

The loop is rotated by Kau at a constant frequency, f, and an EMF, ε, is generated. Figure 7 shows the generated EMF versus time trace observed on the screen of an oscilloscope.

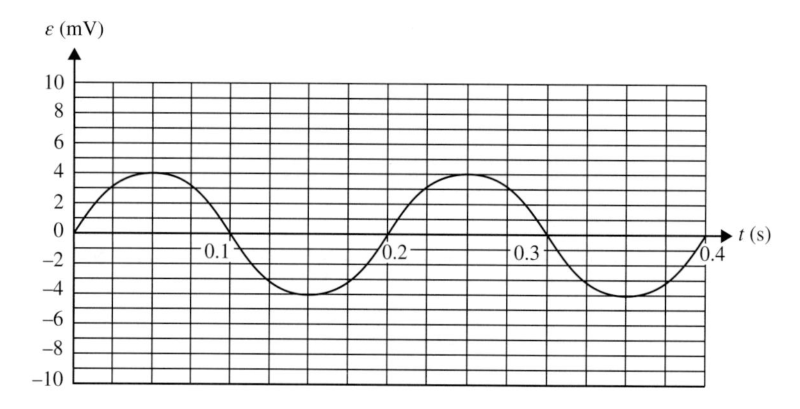

Figure 7

c. Calculate the frequency of the rotation from the oscilloscope trace shown in Figure 7. **(1 mark)**

d. Gir now doubles the number of turns in the loop from one turn to two turns, creating two loops. The loops are again rotated at the same constant frequency, f.
On Figure 8 below, sketch a graph that shows the resulting variation of the EMF with time between points A and B, as labelled in Figure 6. The original output is shown as a dashed line. **(2 marks)**

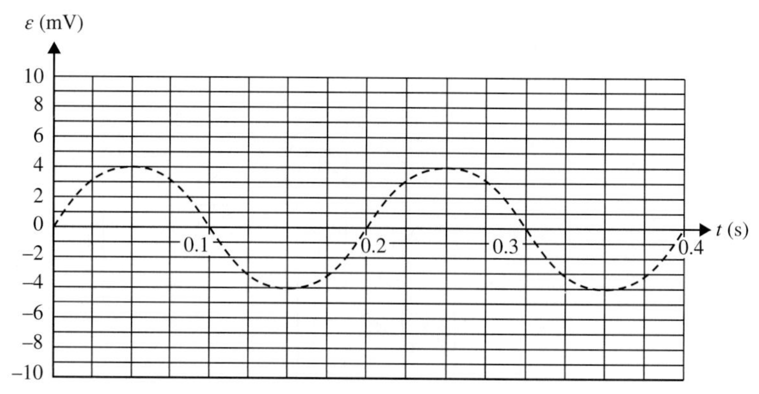

Figure 8

Question 14 (5 marks)

Source: *VCE 2019 Physics Exam, NHT, Section B, Q.2*; © VCAA

A square loop of wire with a cross-sectional area of $0.010\ \text{m}^2$ and 20 turns rotates in a magnetic field of strength 4.0×10^{-2} T. The wires of the loop are connected to two slip rings and an oscilloscope, as shown in Figure 2.

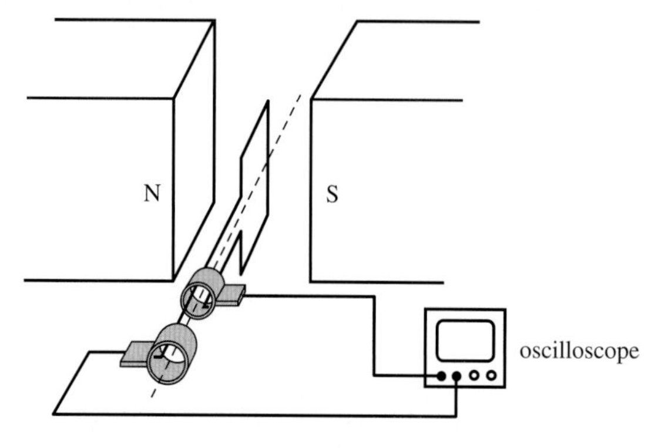

Figure 2

The loop takes 0.10 s to make a quarter rotation (from a position at right angles to the field to a position parallel to the field).

a. Calculate the average magnitude of the induced EMF in the loop as it makes this quarter rotation. Show your working. **(3 marks)**

b. On the axes provided below, sketch the output signal that would be displayed on the oscilloscope over 1.0 s. A value or scale on the *y*-axis is not necessary. Take the position of the loop at $t = 0$ to be that shown in Figure 2. **(2 marks)**

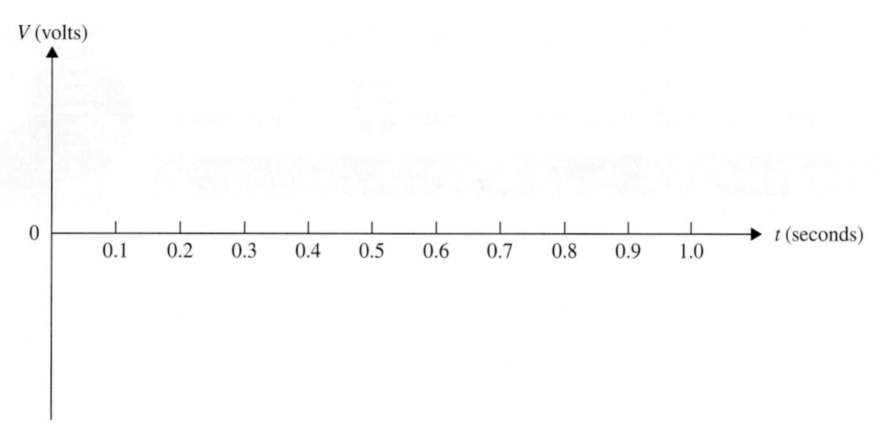

Question 15 (7 marks)

Source: *VCE 2017 Physics Exam, NHT, Section A, Q.13*; © VCAA

Students build a simple electric motor consisting of a single coil and a split-ring commutator, as shown in Figure 21. The magnetic field between the pole pieces is a constant 0.02 T.

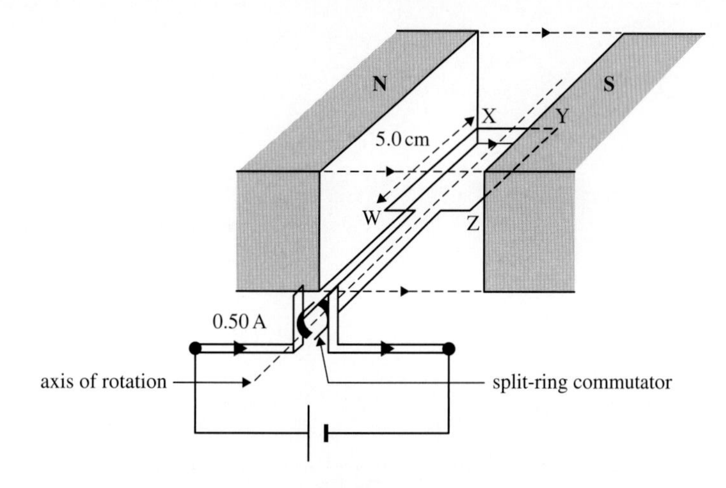

Figure 21

a. Calculate the magnitude of the force on the side WX. **(2 marks)**

b. Will the coil rotate in a clockwise or anticlockwise direction as seen by an observer at the split-ring commutator? Explain your answer. **(3 marks)**

c. Explain the role of the split-ring commutator in the operation of the electric motor. **(2 marks)**

7 Transmission of electricity

KEY KNOWLEDGE

In this topic, you will:

- compare sinusoidal AC voltages produced as a result of the uniform rotation of a loop in a constant magnetic field with reference to frequency, period, amplitude, peak-to-peak voltage (V_{p-p}) and peak-to-peak current (I_{p-p})
- compare alternating voltage expressed as the root-mean-square (rms) to a constant DC voltage developing the same power in a resistive component
- analyse transformer action with reference to electromagnetic induction for an ideal transformer: $\frac{N_1}{N_2} = \frac{V_1}{V_2} = \frac{I_2}{I_1}$
- analyse the supply of power by considering transmission losses across transmission lines.

PRACTICAL WORK AND INVESTIGATIONS

Practical work is a central component of VCE Physics. Experiments and investigations, supported by a **practical investigation eLogbook** and **teacher-led video,** are included in this topic to provide opportunities to undertake investigations and communicate findings.

EXAM PREPARATION

Access past VCAA questions and exam-style questions and their video solutions in every lesson, to ensure you are ready.

7.1 Overview

Hey students! Bring these pages to life online

 Watch videos  Engage with interactivities  Answer questions and check results

Find all this and MORE in jacPLUS

7.1.1 Introduction

Without reliable electricity, everyone's lives would be drastically different. It is easy to take for granted the work that goes into ensuring that households and classrooms safely get a constant supply of electricity. Most electricity is produced at power stations, which are often far away from the towns and cities in which it is consumed. To reduce power loss, electricity must be transmitted sustainably. In this topic you will learn about transformers and their role in limiting power loss in the long-distance transmission of electricity.

FIGURE 7.1 High-voltage power lines are used to transmit electricity over long distances.

LEARNING SEQUENCE

on Resources

Solutions	Solutions — Topic 7 (sol-0821)
Practical investigation eLogbook	Practical investigation eLogbook — Topic 7 (elog-1638)
Digital documents	Key science skills — VCE Physics Units 1–4 (doc-36950) Key terms glossary — Topic 7 (doc-37177) Key ideas summary — Topic 7 (doc-37178)
Exam question booklet	Exam question booklet — Topic 7 (eqb-0104)

7.2 Peak, RMS and peak-to-peak voltages

KEY KNOWLEDGE

- compare sinusoidal AC voltages produced as a result of the uniform rotation of a loop in a constant magnetic field with reference to frequency, period, amplitude, peak-to-peak voltage (V_{p-p}) and peak-to-peak current (I_{p-p})
- compare alternating voltage expressed as the root-mean-square (rms) to a constant DC voltage developing the same power in a resistive component

7.2.1 Peak, RMS and peak-to-peak voltages

The voltage output of an alternator varies with time, producing a sinusoidal graph. This graph, shown in figure 7.2, can be described in terms of the physical quantities listed here.

- The **period**, T, is the time taken for one complete cycle.
- The **frequency**, f, is the number of full cycles completed in one second. The frequency is related to the period by the equation:

$$T = \frac{1}{f}$$

FIGURE 7.2 Sinusoidal graph from voltage output of an alternator

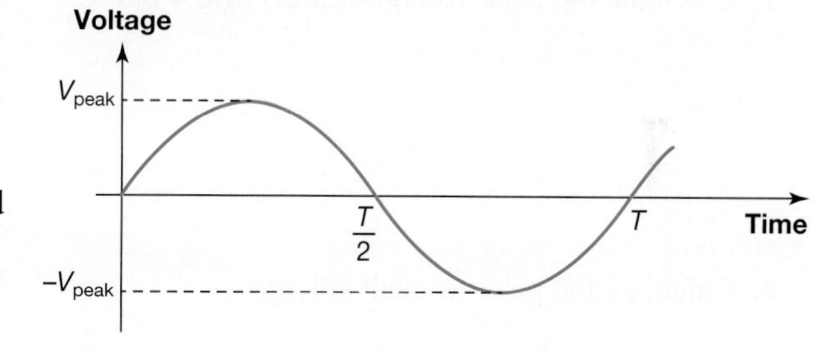

The frequency of the power supplied to consumers in Australia is 50 Hz (1 hertz is one cycle per second). The period is therefore $\frac{1}{50}$ second = 0.02 s.

- The **amplitude** is the maximum variation of the voltage output from zero. It is called the **peak voltage**, V_{peak}. Similarly, the amplitude of the current is called the **peak current**, I_{peak}.
- The **RMS** (root mean square) **voltage**, V_{RMS}, is the value of the constant **direct current** (DC) voltage that would produce the same power as the AC voltage across the same resistance. The RMS voltage is related to the peak voltage by the equation:

$$V_{RMS} = \frac{V_{peak}}{\sqrt{2}}$$

The peak voltage of a 230 V_{RMS} household power supply is 325 V. A 230 V_{RMS} output from a generator delivers the same amount of power as a 230 V DC power supply across the same resistance. Similarly, I_{RMS} is the value of a DC current that generates the same power as an **alternating current** (AC) through the same resistance:

$$I_{RMS} = \frac{I_{peak}}{\sqrt{2}}$$

period the time it takes a source to produce one complete wave (or for a complete wave to pass a given point)

frequency the number of times a wave repeats itself every second

amplitude the maximum variation from zero of a periodic disturbance

peak voltage the amplitude of an alternating voltage

peak current the amplitude of an alternating current

RMS voltage the value of the constant DC voltage that would produce the same power as AC voltage across the same resistance

direct current (DC) an electric current that flows in one direction only

alternating current (AC) an electric current that reverses direction at short, regular intervals

- The **peak-to-peak voltage**, V_{p-p}, is the difference recorded between the maximum and minimum voltages. In the case of a symmetrical AC voltage:

peak-to-peak voltage the difference between the maximum and minimum voltages of a DC voltage

$$V_{p-p} = 2V_{peak}$$

$$I_{p-p} = 2I_{peak}$$

SAMPLE PROBLEM 1 Determining the peak-to-peak voltage from the RMS voltage

A digital multimeter gives a measurement of 6.3 V for the RMS value of an AC voltage. A CRO is used to measure the peak-to-peak voltage. What value do you expect?

THINK	WRITE
1. Calculate the peak voltage when $V_{RMS} = 6.3$ V.	$V_{RMS} = \frac{V_{peak}}{\sqrt{2}}$ $6.3 = \frac{V_{peak}}{\sqrt{2}}$ $V_{peak} = 6.3\sqrt{2}$
2. Calculate the peak-to-peak voltage.	$V_{p-p} = 2V_{peak}$ $= 2 \times 6.3 \times \sqrt{2}$ V $= 18$ V

PRACTICE PROBLEM 1

A toaster is rated at 230 V_{RMS} and 1800 W. (Recall that $P = IV$.) What are the values of the RMS and peak currents?

7.2 Activities

Students, these questions are even better in jacPLUS

Receive immediate feedback and access sample responses

Access additional questions

Track your results and progress

Find all this and MORE in jacPLUS

7.2 Quick quiz  | 7.2 Exercise | 7.2 Exam questions

7.2 Exercise

1. In the past, electronic valves were powered by 6.3 V_{RMS} AC. What was the maximum voltage received by a valve?
2. A CRO shows the following trace. The settings are:
   Y: 10 mV per division
   X: 5 ms per division.

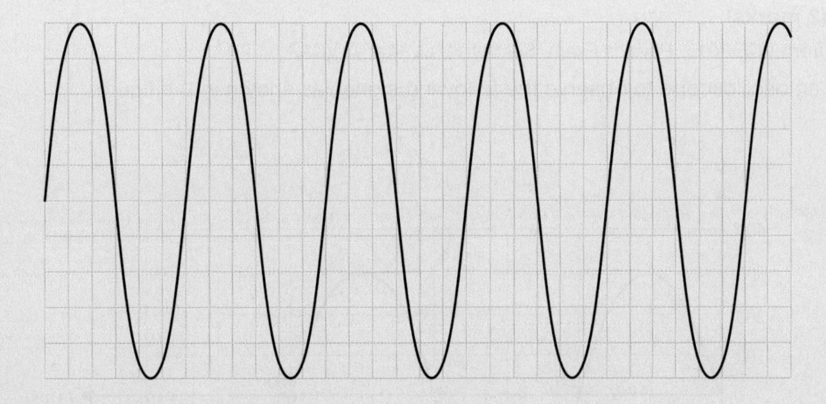

For this AC signal, identify the:

a. period
b. frequency
c. peak voltage
d. peak-to-peak voltage
e. RMS voltage.

3. Some appliances are designed to run off either AC or batteries. The size of the batteries is equivalent to the peak of the AC voltage. If the appliance can run off 9 V DC, what RMS voltage would it also run off?

7.2 Exam questions

Note: Although the power supplied to most Australian consumers has changed from 240 V_{RMS} to 230 V_{RMS}, past exam papers used 240 V_{RMS} in their questions. It is good practice to exercise with both values.

Question 1 (1 mark)

Source: *VCE 2016, Physics Exam, Section A, Q.17.b;* © VCAA

Samira and Mark construct a simple alternator, as shown in the figure.

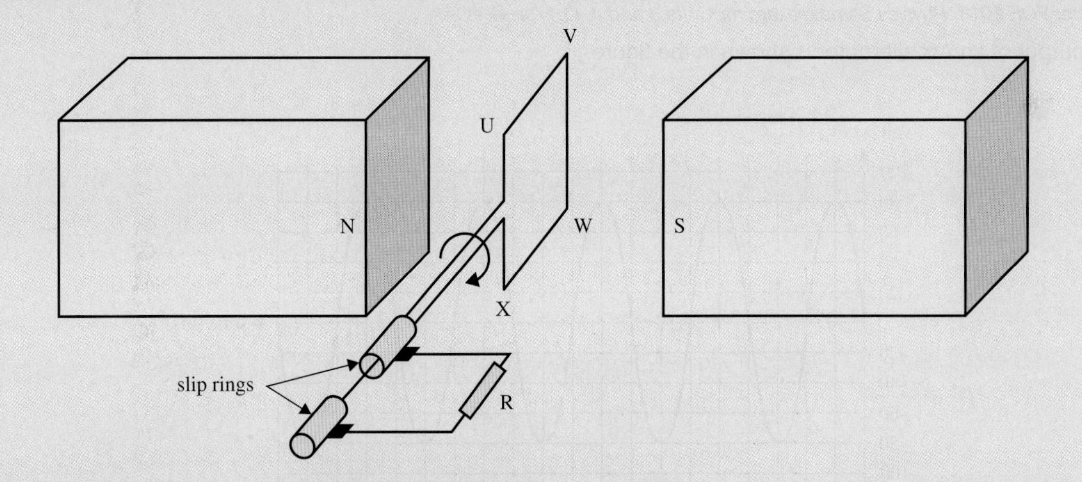

When the coil is rotating steadily, it takes 40 ms for each complete rotation and produces a peak emf of 3.5 V.

Calculate the RMS value of the AC emf.

Question 2 (1 mark)

Source: *VCE 2016, Physics Exam, Section A, Q.17.a;* © VCAA

Refer to the information in question 1.

When the coil is rotating steadily, it takes 40 ms for each complete rotation and produces a peak emf of 3.5 V.

Calculate the frequency of the AC emf.

Question 3 (2 marks)

Source: *Adapted from VCE 2015, Physics Exam, Section A, Q.14.a; © VCAA*

Electricians use an oscilloscope to observe the following signal, as shown in the figure.

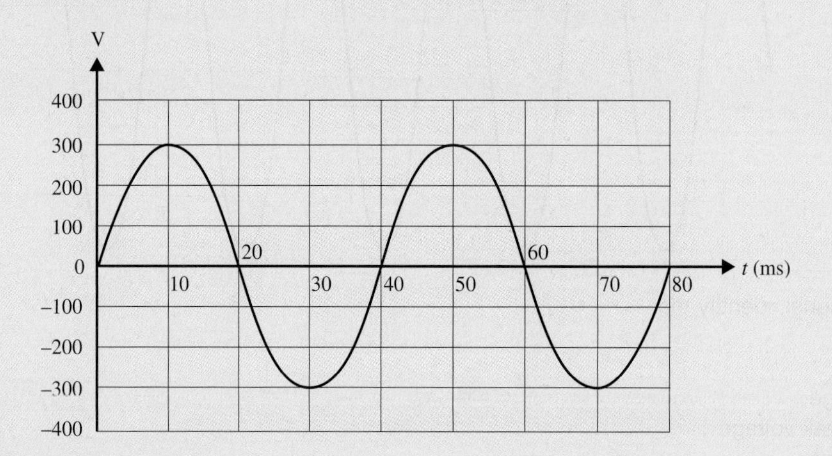

Determine the frequency of the AC (alternating current) observed. Show your working.

Question 4 (1 mark)

MC A CRO is used to measure the peak-to-peak voltage, which is determined to be 24 V. The value for V_{RMS} is:

A. 8.5 V.
B. 12 V.
C. 17 V.
D. 9.6 V.

Question 5 (2 marks)

Source: *VCE 2013, Physics Sample exam for Units 3 and 4, Q.17.a; © VCAA*

The output of an AC alternator is shown in the figure.

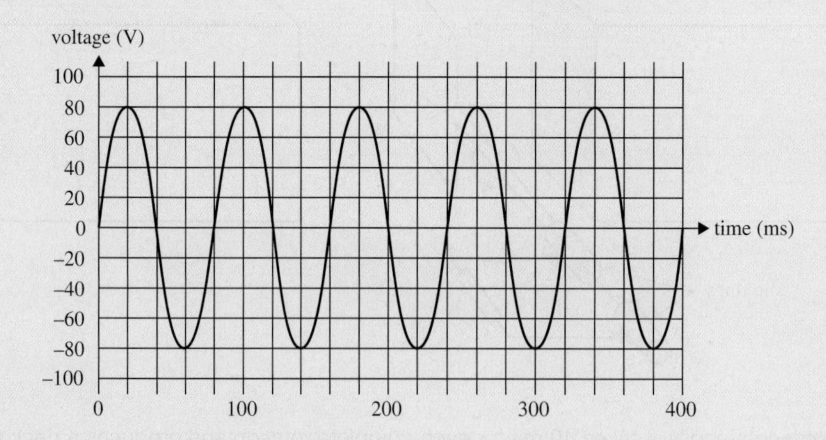

Calculate the frequency of rotation of the alternator.

More exam questions are available in your learnON title.

7.3 Transformers

KEY KNOWLEDGE

- Analyse transformer action with reference to electromagnetic induction for an ideal transformer: $\frac{N_1}{N_2} = \frac{V_1}{V_2} = \frac{I_2}{I_1}$.

Source: VCE Physics Study Design (2024–2027) extracts © VCAA; reproduced by permission.

7.3.1 Electric power

Electric power is generated for a purpose — to provide lighting in streets and homes, and to operate motors in domestic and industrial appliances. But electric power is often generated very far from where it is consumed. This problem *appears* to be simply overcome: make the connecting wires from the generator to the light or motor longer and longer, even stretching to hundreds of kilometres, and you have your basic transmission line.

FIGURE 7.3 Why not just extend the wires from your toaster all the way back to the power plant generator?

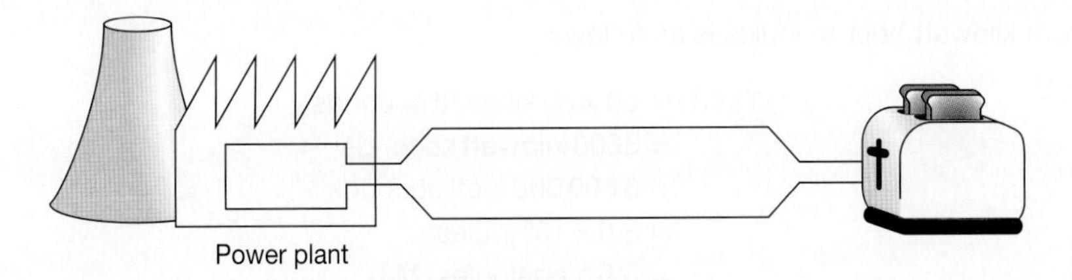

This simple solution might work on the laboratory bench where the connecting wires are so short that their resistance is a very small fraction of the overall resistance in the circuit. However, when the wires extend over kilometres, their resistance becomes significant. So, too, does the power loss in them. The power dissipated in the wires is equal to:

$$P_{\text{loss}} = VI = I^2R$$

where: P_{loss} is the power dissipated in the wires, in W

V is the voltage drop across the wires (not to be confused with the supply voltage), in V

I is the current through the wires, in A

R is the resistance of the wires, in Ω

Hence, the higher the resistance of the wires, the more power is lost in transmission along those wires. In addition, so much of the supply voltage now drops along the wires that the remaining voltage across the devices is insufficient for them to operate properly.

BACKGROUND KNOWLEDGE: Units of power

Electric power is normally measured in watts or megawatts — for example, when comparing electrical generators or deciding between vacuum cleaners. However, the generator supplies *energy*, the vacuum cleaner consumes or transforms *energy*, and it is ultimately people who pay for *energy*. The **power rating**, or wattage, of an electrical appliance indicates the rate at which it uses electricity. The longer the appliance is powered on, the more energy is used and the more the process costs. By definition:

$$1 \text{ watt} = 1 \text{ joule per second}$$
$$1 \text{ joule} = 1 \text{ watt} \times 1 \text{ second} = 1 \text{ watt second}$$

As 1 watt second is equivalent to 1 joule, then

$$1000 \text{ watt seconds} = 1000 \text{ J}$$
$$1 \text{ kilowatt second} = 1 \text{ kJ}.$$

If a 1 kW heater was turned on for 1 s, it would use 1 kilowatt second or 1 kilojoule of electrical energy. If it was turned on for 60 seconds, it would use 60 kilowatt seconds or 60 kilojoules.

The common unit for energy supply and consumption in electricity is the kilowatt hour, which is the amount of energy consumed, for example, by a one-kilowatt heater for one hour. This unit is abbreviated to kWh (e.g. 60 kWh).

The conversion from kilowatt hour to joules is as follows:

$$\begin{aligned} 1 \text{ kWh} &= 60 \times 60 \text{ kilowatt seconds} \\ &= 3600 \text{ kilowatt seconds} \\ &= 3\,600\,000 \text{ watt seconds} \\ &= 3.6 \times 10^6 \text{ joules} \\ &= 3.6 \text{ megajoules (MJ)} \end{aligned}$$

7.3.2 Why use transformers?

The transmission line transfers electrical energy from the generator to the appliance. The electrical energy is generated at a voltage set by the generator. Using Ohm's Law, if the voltage supplied from the generator is constant, then the current drawn from the generator depends on the total resistance in the devices connected to the generator: $I = \frac{V}{R}$. Devices are connected in parallel so that they all have the same voltage. Plugging in additional devices is the same as adding extra resistances in parallel, with each device drawing its own current from the supply. The extra appliances in parallel reduce the total resistance in the circuit.

With more appliances connected, the overall resistance of the circuit is reduced, drawing a larger current from the generator and therefore greater energy is supplied. The amount of energy supplied by the generator every second, or the electrical power supplied, is equal to the product of the voltage supplied by the generator and the current drawn from the generator, $P = VI$.

If the transmission lines are long, the energy wasted due to their resistance becomes a significant fraction of the energy supplied by the generator. If the same amount of energy produced every second (that is, the same power) can be sent along the lines but at a lower current, the energy loss will be less. In fact, since the power loss is given by I^2R (current^2 × resistance of the lines), halving the current through the lines reduces the power loss by a factor of four. This current, and therefore power loss, can be reduced with the use of **transformers**.

power rating the total electrical power required for an appliance or machine to operate normally

transformer a device in which two multi-turn coils may be wound around an iron core. One coil acts as an input while the other acts as an output. The purpose of the transformer is to produce an output AC voltage that is different from the input AC voltage.

tlvd-9002

SAMPLE PROBLEM 2 Determining the total resistance of 100 km of wire to reflect on the usefulness of transformers

A 100-W light globe uses 100 J of energy every second when the voltage across it is 230 V_{RMS}.

a. **Calculate the current through the globe.**

b. **Calculate the resistance of the globe for this current and voltage.**

c. i. **If the globe was connected to a 230-V_{RMS} power supply by 2.00 m of copper wire, what would be the total resistance of the circuit? The wire has a resistance of 0.0220 Ω m^{-1}.**

ii. **What would be the voltage drop across the globe?**

d. i. **If the globe was connected by 100 km of copper wire, what would be the total resistance of the circuit?**

ii. **What would be the voltage drop across the globe now?**

e. **Comment on how the light globe would respond when there is 100 km of wire.**

THINK	WRITE
a. Use the formula $P = VI$ to determine the current through the globe, where $P = 100$ W and $V = 230$ V.	a. $P = VI$ $100 = 230I$ $I = \frac{100}{230}$ $= 0.435\text{A}$
b. Recall and use Ohm's Law, $V = IR$, to determine the resistance of the globe.	b. $V = IR$ $230 = 0.435R$ $R = \frac{230}{0.435}$ $= 529\ \Omega$
c. i. The globe and the wire are connected in series, so add their resistances to determine the total resistance of the circuit.	c. i. $R_{total} = R_{copper} + R_{globe}$ $= (2.00\text{ m} \times 0.0220\ \Omega\text{ m}^{-1}) + 529\ \Omega$ $\approx 529\ \Omega$
ii. 1. Determine the current through the series circuit using Ohm's Law, $V = IR$, where $V = 230$ V (supplied to the circuit) and $R = 529$ Ω (total resistance of the circuit).	ii. $V = IR$ $I = \frac{V}{R}$ $= \frac{230}{529}$ $= 0.435$ A
2. Determine the voltage drop across the globe using Ohm's Law, $V = IR$.	$V = IR$ $= 0.435\text{A} \times 529\ \Omega$ $= 230$ V The voltage drop across the globe is 230 V.
d. i. The globe and the wire are connected in series, so add their resistances to determine the total resistance of the circuit.	d. i. $R_{total} = R_{copper} + R_{globe}$ $= (100 \times 10^3\text{ m} \times 0.0220\ \Omega\text{ m}^{-1}) + 529\ \Omega$ $\approx 2729\ \Omega$ The total resistance of the circuit is 2730 Ω. The resistance of the copper wire is not negligible.

ii. 1. Determine the current through the series circuit using Ohm's Law, $V = IR$, where $V = 230$ V (supplied to the circuit) and $R = 2729\ \Omega$ (total resistance of the circuit)

ii. $V = IR$

$$I = \frac{V}{R} = \frac{230}{2729} = 0.0843 \text{ A}$$

2. Determine the voltage drop across the globe using Ohm's Law, $V = IR$.

$$V = IR = 0.0843 \text{ A} \times 529\ \Omega = 44.6 \text{ V}$$

e. Most of the 230 V supplied to the circuit is dissipated in the copper wire, with only a small amount across the globe.

e. The globe would not light up.

PRACTICE PROBLEM 2

An 800-W toaster uses 800 J of energy every second when the voltage across it is 230 V_{RMS}.

a. **Calculate the current through the toaster.**

b. **Calculate the resistance of the toaster for this current and voltage.**

c. i. **If the toaster was connected to a 230-V power supply by 1.00 m of copper wire, what would be the total resistance of the circuit? The wire has a resistance of 0.0350 $\Omega\ m^{-1}$.**

ii. **What would be the voltage across the toaster?**

d. i. **If the toaster was connected by 20 km of copper wire, what would be the total resistance of the circuit?**

ii. **What would be the voltage across the toaster now?**

e. **Comment on how the toaster would respond.**

7.3.3 How do transformers work?

In 1831 Michael Faraday constructed the first transformer when he demonstrated that an electric current in one circuit had a magnetic effect that could produce an electric current in another circuit.

Faraday's transformer consisted of two sets of wire coils wrapped around a ring of iron. One coil, the primary or input coil, was connected to a battery by a switch, while the other, the secondary coil, was connected to a **galvanometer** (a sensitive current detector). (Faraday's transformer is discussed in section 6.4.1.)

As with other examples of **electromagnetic induction**, the transformer works only when there is a change in magnetic flux in the coils (which is the case with modern transformers).

galvanometer instrument used to detect small electric currents, or to detect the direction of current

electromagnetic induction generation of an electromotive force (emf) in a coil (an electrical conductor) as a result of a changing magnetic field

To sum up, transformers are devices that act to increase or decrease AC voltages, and they are used in many electrical devices and systems.

FIGURE 7.4 Circuit diagram symbol for a transformer

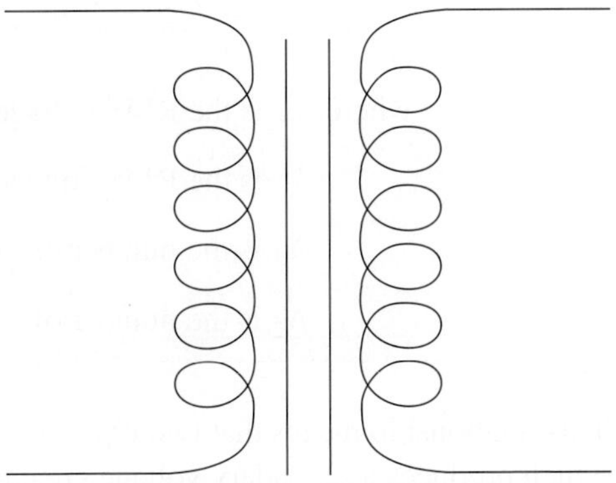

If an AC voltage is applied to the primary coil, an alternating magnetic field will be set up in the iron core. This alternating magnetic field will propagate through the iron core to the secondary coil. Here, the alternating magnetic field will induce an alternating voltage in this coil with the same frequency as the primary AC voltage. Transformers do not work with a constant DC current supply. There will be a brief induced current when the switch is turned off or on, but the constant DC supply does not produce a changing magnetic flux in the primary coil and, hence, no induced current in the secondary coil.

An AC voltage supplied to the primary coil produces an AC voltage at the secondary coil, even though there is no electrical connection between the two coils. How do the magnitudes of the primary and secondary voltages compare? In other words, how do the RMS voltages compare?

FIGURE 7.5 A changing current, I, in the primary coil produces a changing magnetic field, $\boldsymbol{B}$, in the iron core, which is propagated through the iron core to the secondary coil, where the changing magnetic field induces a changing emf in the secondary coil.

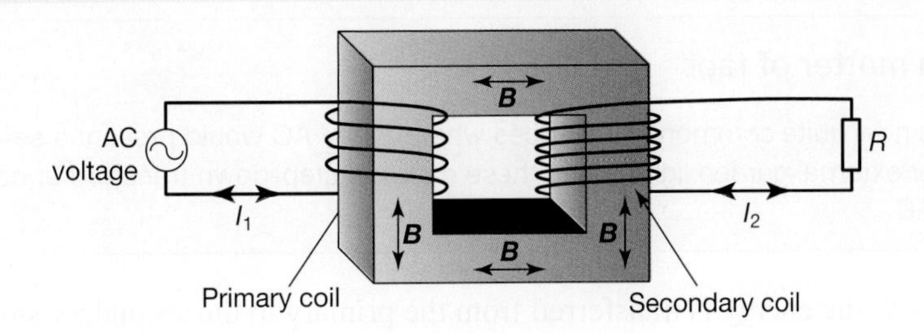

7.3.4 Comparing voltages

When an AC voltage supply, V_1, is connected to the primary coil, it produces a changing magnetic flux in the iron core that will be proportional to the number of turns, N_1, in the coil.

The iron core has constantly changing magnetic flux throughout, which is created by the primary coil. This magnetic flux then passes through the secondary coil, producing an AC voltage at the terminals of the secondary coil. So, applying Faraday's Law to the primary coil gives:

$$V_1 = N_1 \times \frac{\Delta\Phi_B}{\Delta t}$$

Applying it to the secondary coil gives:

$$V_2 = N_2 \times \frac{\Delta\Phi_B}{\Delta t}$$

Combining these equations gives:

$$\frac{\Delta\Phi_B}{\Delta t} = \frac{V_1}{N_1} = \frac{V_2}{N_2}$$

or

$$\frac{V_1}{V_2} = \frac{N_1}{N_2}$$

where: V_1 is the RMS voltage in the primary coil

V_2 is the RMS voltage in the secondary coil

N_1 is the number of turns in the primary coil

N_2 is the number of turns in the secondary coil

This relationship means that two types of transformers can be built. One type, which produces a secondary voltage greater than the primary, is called a **step-up transformer**. In this, the number of secondary turns is greater than the number of primary turns.

The other type is a **step-down transformer**, which features more primary turns than secondary turns. It produces a smaller secondary voltage than the primary voltage. Both types are used in the distribution of electricity from generator to home, and also inside the home.

step-up transformer the output (secondary) voltage produced is greater than the input (primary) voltage

step-down transformer the output (secondary) voltage produced is less than the input (primary) voltage

ideal transformer a transformer that is 100% efficient, meaning its input power is equal to its output power

EXTENSION: As a matter of fact

Low-voltage lighting is now quite common in instances where 230 V AC would present a safety risk (for example, Christmas tree lights or external garden lighting). In these cases, a step-down transformer converts the 230 V AC down to a safer 12 V AC.

If there is no energy loss as the energy is transferred from the primary to the secondary side, then the *power in* to the primary coil will equal the *power out* of the secondary coil, and the transformer is considered an **ideal transformer**. Since power = voltage × current, this can be written as:

$$V_1 \times I_1 = V_2 \times I_2, \text{ and hence:}$$

$$\frac{N_1}{N_2} = \frac{V_1}{V_2} = \frac{I_2}{I_1}$$

where: V_1 is the voltage in the primary coil

V_2 is the voltage in the secondary coil

I_1 is the current in the primary coil

I_2 is the current in the secondary coil

N_1 is the number of turns in the primary coil

N_2 is the number of turns in the secondary coil

Note that the voltages can be RMS voltages, peak voltages or peak-to-peak voltages.

In this relationship, the units used do not matter, as long as they are the same for both the primary and secondary coil. This also applies to the voltage used — it needs to be consistent.

tlvd-9003

SAMPLE PROBLEM 3 Determining the number of turns in a secondary coil of a step-down transformer

A step-down transformer is designed to convert 230 V_{RMS} AC to 12 V_{RMS} AC. If there are 190 turns in the primary coil, how many turns are in the secondary coil?

THINK	WRITE
1. List the known information.	$V_1 = 230$ V; $V_2 = 12$ V; $N_1 = 190$ turns
2. Use the relationship $\frac{V_1}{V_2} = \frac{N_1}{N_2}$ to determine the number of turns in the secondary coil.	$\frac{V_1}{V_2} = \frac{N_1}{N_2}$ $\frac{230}{12} = \frac{190}{N_2}$ $N_2 = \frac{12 \times 190}{230}$ $= 9.9$ The secondary coil consists of approximately 10 turns.

PRACTICE PROBLEM 3

A generator supplies 10 kW of power to a transformer at 1.0 kV. The current in the secondary coil is 0.50 A. What is the turns ratio of the transformer? Is it a step-up or a step-down transformer?

EXTENSION: Household use of transformers

Australian houses are provided with AC electricity that has a value of 230 V_{RMS}. Most electronic circuits are designed to operate at low DC voltages of between 3 V and 12 V. Therefore, household appliances that have electronic circuits in them will either have a 'power-cube' transformer that plugs directly into the power outlet socket, or have transformers built into them.

Power-cube transformers can be found in rechargeable appliances such as mobile phone chargers and other chargers, electric keyboards and laptop computers. You can probably find more in your own home. These transformers also have a rectifier circuit built into them that converts AC to DC.

The RMS value of an AC voltage is a way of describing a voltage that is continuously changing. The voltage actually swings between −325 V and +325 V at a frequency of 50 Hz. This voltage has the same heating effect on a metal conductor as a DC voltage of 230 V; hence, it is usually described as 230 V.

Resources

Weblink Transformer applet

EXTENSION: Ideal versus real transformers

In an ideal transformer, the transformer is assumed to be 100% efficient and energy losses can be ignored. In reality, all transformers lose some energy in transferring electric power from the primary side to the secondary. This energy loss occurs in two areas. The first area is in the wires that make up the primary and secondary coils. This loss is called either *copper loss* (because the wires are usually copper), or *resistive* or I^2R *loss*. This is resistive heat from the wires heating up, so energy and power are lost in the coils. The heat generated from a transformer can be seen in figure 7.6. The loss is usually quite minor. If the transformer is being designed to take large currents, the wires on that side would be made thicker to take the high current and minimise the resistance.

FIGURE 7.6 This infrared photo shows that heat is generated by a transformer. The red areas are the hottest.

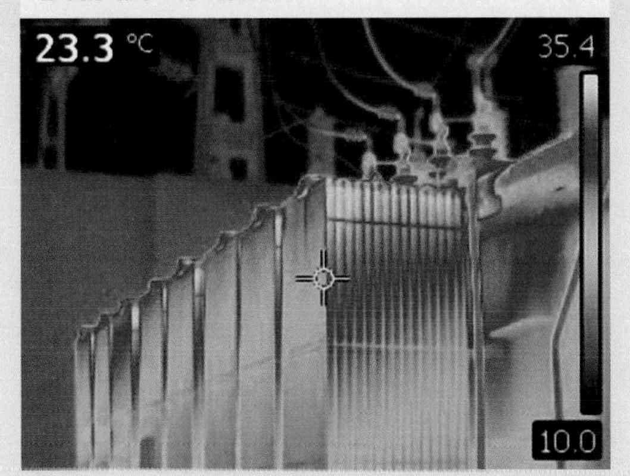

The other area of energy loss in the transformer is in the iron core. This loss is due to induced currents in the iron core. These currents are called **eddy currents**, because they are like the swirls, or eddies, left in the water after a boat has gone by. Eddy currents are an application of Lenz's Law. The magnetic fields set up by eddy currents oppose the changes in the magnetic field acting in the regions of the metal objects.

The changing magnetic flux in the iron core produces a changing voltage in each of the turns of the secondary coils. Iron is an electrical conductor, so it will behave in the same way as the turns of wire. A circular current will be induced in the iron in a plane at right angles to the direction of the changing magnetic flux.

FIGURE 7.7 a. An eddy current induced in an iron core by a changing magnetic field, and **b.** putting the iron core into layers reduces the currents.

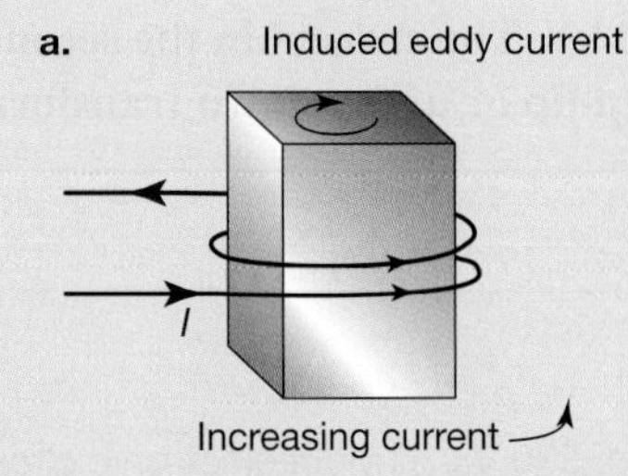

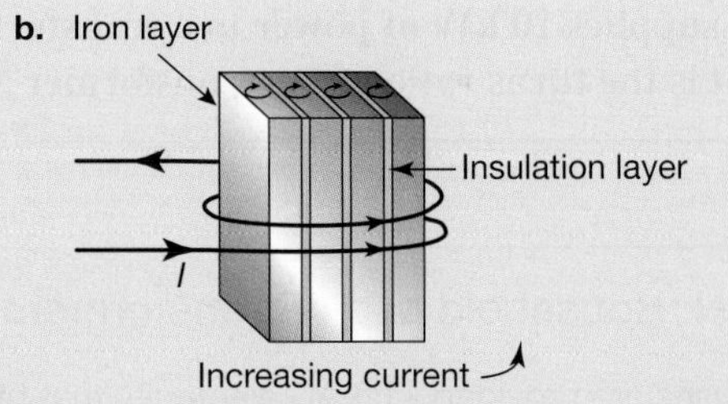

If the iron core was one solid piece of iron, these induced eddy currents would be quite substantial. As iron has a low resistance, it would lead to large energy loss.

To minimise this loss, the iron core is constructed of layers of iron sandwiched between thin layers of insulation. These layers, called laminations, significantly reduce the energy loss. In practice, transformers used to transmit large quantities of energy are about 99% efficient.

eddy current an electric current induced in the iron core of a transformer by changing magnetic fields

elog-1888

tlvd-8789

INVESTIGATION 7.1

online only

Linking coils

Aim

To investigate how the magnetic field of a current-carrying coil can induce a current in another coil

elog-1889

tlvd-8790

INVESTIGATION 7.2

online only

Transformer ins and outs

Aim

To use an AC input voltage across a primary coil to produce an AC output voltage in a secondary coil, and to compare their values

 Resources

 **Interactivity** Transforming voltage and current (int-0118)

7.3 Activities

learn on

Students, these questions are even better in jacPLUS

 Receive immediate feedback and access sample responses

 Access additional questions

 Track your results and progress

Find all this and MORE in jacPLUS

7.3 Quick quiz 	7.3 Exercise	7.3 Exam questions

7.3 Exercise

1. An ideal transformer has 100 turns in the primary coil and 2000 turns in the secondary coil. If the primary coil was connected to 230 V_{RMS} AC, what would be the voltage across the secondary coil?
2. A transformer has 300 turns in the primary coil and six turns in the secondary coil.
   a. If 230 V_{RMS} AC is connected to the primary coil, what will be the voltage across the secondary coil?
   b. If the secondary voltage is 9.0 V_{RMS} AC, what is the voltage across the primary coil?
3. Christmas tree lights need a transformer to convert the 230 V_{RMS} AC to 12 V_{RMS} AC.
   a. If there are 50 coils on the 12 V secondary coil, how many turns are there in the primary coil?
   b. If there are 20 globes connected in parallel to the secondary coil, each of 12 V and 5 W, what is the current in the secondary coil?
   c. What is the current in the primary coil, assuming the transformer is ideal?
4. Explain why a transformer does not work with a constant DC input voltage.
5. Why is the core of transformers made of an alloy of iron that is easy to magnetise?
6. A transformer is used to change 10 000 V_{RMS} to 230 V_{RMS}. There are 2000 turns in the primary coil.
   a. What type of transformer is this?
   b. How many turns are there in the secondary coil?
7. An ideal transformer has 400 turns in the primary coil and 900 turns in the secondary coil. The primary voltage is 60 V_{RMS} and the current in the secondary coil is 0.30 A.
   a. What is the voltage across the secondary turns?
   b. What is the power delivered by the secondary coils?
   c. What is the current in the primary coil?

7.3 Exam questions

Question 1 (1 mark)

Source: *VCE 2021 Physics Exam, Section A, Q.7;* © VCAA

MC A mobile phone charger uses a step-down transformer to transform 240 V AC mains voltage to 5.0 V. The mobile phone draws a current of 3.0 A while charging. Assume that the transformer is ideal and that all readings are RMS.

Which one of the following is closest to the current drawn from the mains during charging?

A. 48 A
B. 16 A
C. 1.2 A
D. 0.06 A

Question 2 (5 marks)

Source: *VCE 2020 Physics Exam, Section B, Q.7; © VCAA*

A rechargeable electric toothbrush uses a transformer circuit, as shown in Figure 7. A secondary coil inside the toothbrush is connected, via an iron core, to a primary coil that is connected to the mains power supply. The mains power is 240 V_{RMS} and the toothbrush recharges at 12 V_{RMS}. The average power delivered by the transformer to the toothbrush is 0.90 W. Assume that the transformer is ideal.

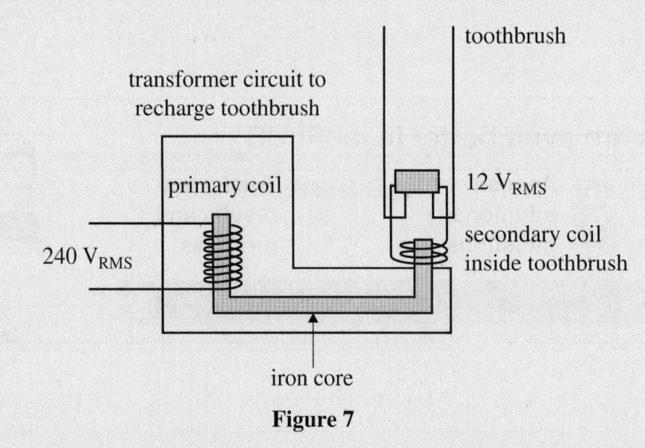

Figure 7

a. Calculate the peak voltage in the secondary coil. Show your working. **(2 marks)**

b. Determine the ratio of the number of turns $\frac{N_p}{N_s}$. **(1 mark)**

c. Calculate the RMS current in the primary coil while the toothbrush is charging. Show your working. **(2 marks)**

Question 3 (1 mark)

Source: *VCE 2019, Physics Exam, Section A, Q.5; © VCAA*

MC A 40 V_{RMS} AC generator and an ideal transformer are used to supply power. The diagram below shows the generator and the transformer supplying 240 V_{RMS} to a resistor with a resistance of 1200 Ω.

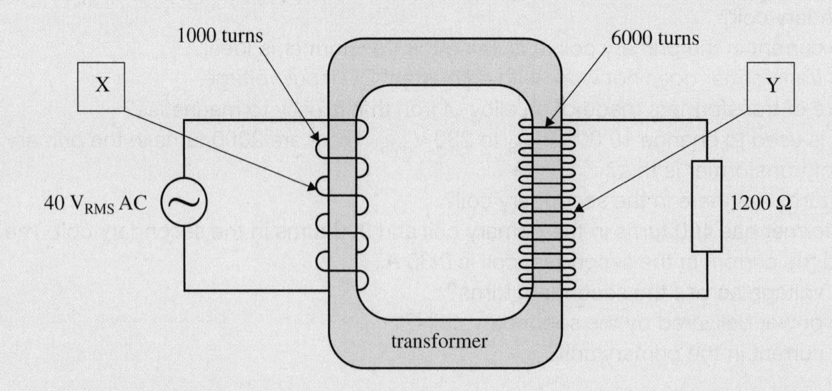

Which of the following correctly identifies the parts labelled X and Y, and the function of the transformer?

	Part X	Part Y	Function of transformer
A.	primary coil	secondary coil	step-down
B.	primary coil	secondary coil	step-up
C.	secondary coil	primary coil	step-down
D.	secondary coil	primary coil	step-up

Question 4 (1 mark)

Source: *VCE 2019, Physics Exam, Section A, Q.6; © VCAA*

MC Refer to the information in question 3.

Which one of the following is closest to the RMS current in the primary circuit?

A. 0.04 A
B. 0.20 A
C. 1.20 A
D. 1.50 A

Question 5 (4 marks)

Source: *Adapted from VCE 2015, Physics Exam, Section A, Q.14.b; © VCAA*

Electricians use an oscilloscope to observe the following signal, as shown in the figure.

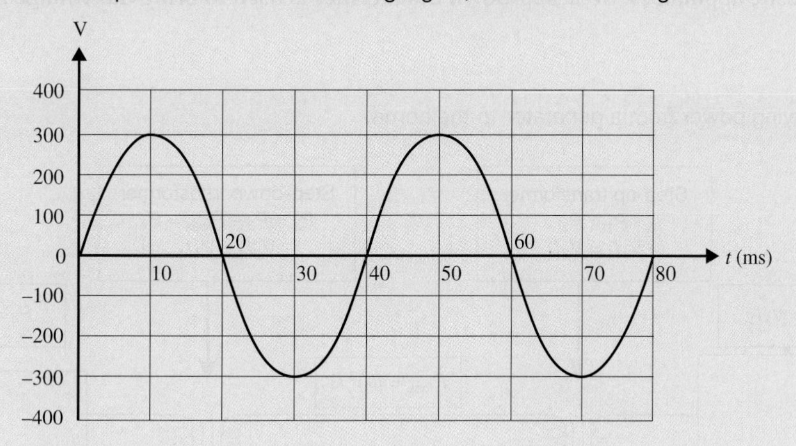

Determine the RMS voltage of the incoming high-voltage input to the transformer. Show your working.

More exam questions are available in your learnON title.

7.4 Power distribution and transmission line losses

KEY KNOWLEDGE

- Analyse the supply of power by considering transmission losses across transmission lines

Source: VCE Physics Study Design (2024–2027) extracts © VCAA; reproduced by permission.

7.4.1 Transmission of power

The power loss in the transmission lines is equal to $P_{loss} = VI = I^2R$, where V is the voltage drop across the lines, I is the current through the lines and R is the resistance of the wires. Be careful not to confuse the voltage drop in the transmission lines with the voltage supplied from the generator. In order to minimise power loss in the transmission wires, large amounts of power produced by the generator need to be transmitted using a very low current. Development of the transformer meant that the AC supply voltage from the generator could be connected to a step-up transformer, allowing transmission lines to increase the voltage supplied by the generator and decrease the current, and so reduce energy loss in the transmission lines.

However, at the other end of the transmission line, the high voltage supplied would be unsuitable, and possibly dangerous, for domestic appliances. So a step-down transformer is used to bring the voltage back down to a safe level for home use.

FIGURE 7.7 Supplying power from a generator to the home

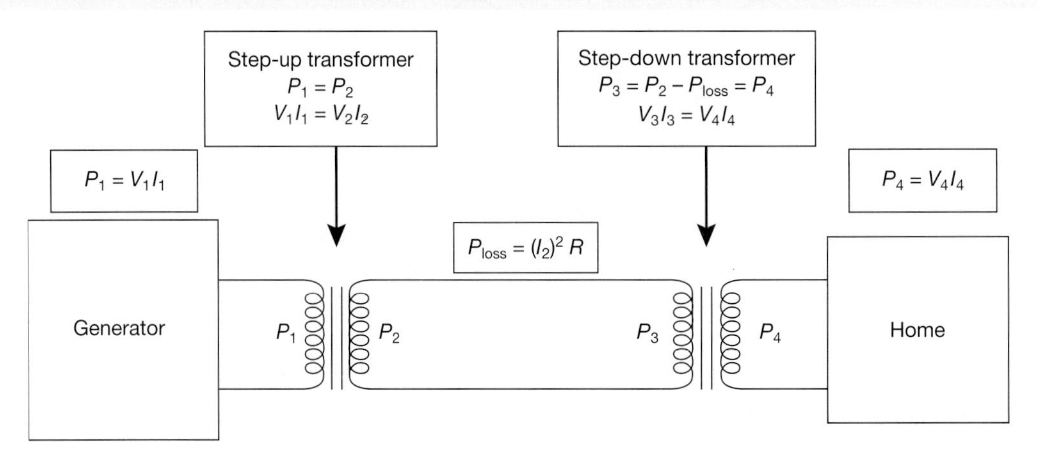

In Victoria, electricity is generated at a variety of voltages. In Yallourn, the voltage is 20 000 V (20 kV). In Newport, the generating voltage is 24 000 V. From the various generators around Victoria, the voltage is stepped up to 500 kV to transmit the electrical energy over the long distances to Melbourne.

When the cables reach the outskirts of Melbourne, the high voltage is stepped down to 66 kV for distribution within the suburban area.

The high-voltage transmission line feeds several outer suburban terminal stations, each of which passes the current to several zone substations, which again, step down the voltage using transformers and split the distribution voltage to go in different directions. These substations can also be disconnected from the transmission grid, allowing for sections to be turned off when required. These substations each connect to hundreds of pole transformers, which then connect to hundreds of homes. As the distribution system spreads further and further down to the domestic consumer, the voltage in the transmission line at each stage gets less and less. Finally, pole transformers step down the voltage to 230 V for domestic use.

Figure 7.8 A transformer at a substation

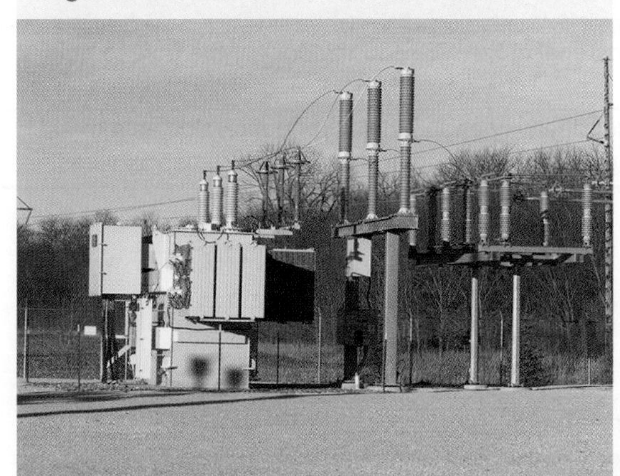

Figure 7.9 A pole transformer

TABLE 7.1 Typical voltages in different sections of the transmission system

Section of transmission system	Voltages
Major power tower or switchyard to terminal substation	220 kV, 330 kV or 500 kV
Terminal substation to zone substation	66 kV
Zone substation to pole-type transformer or underground transformer	22 kV
Pole-type transformer to house	230 V single phase, or 400 V for a three-phase supply

This means that the cables in each section need to be designed to handle the current in that section in a cost-effective way, maximising energy transfer while minimising the cost of doing so. To minimise energy loss, the resistance of the cable needs to be made as small as practicable.

FIGURE 7.10 Victoria's power system — a representation

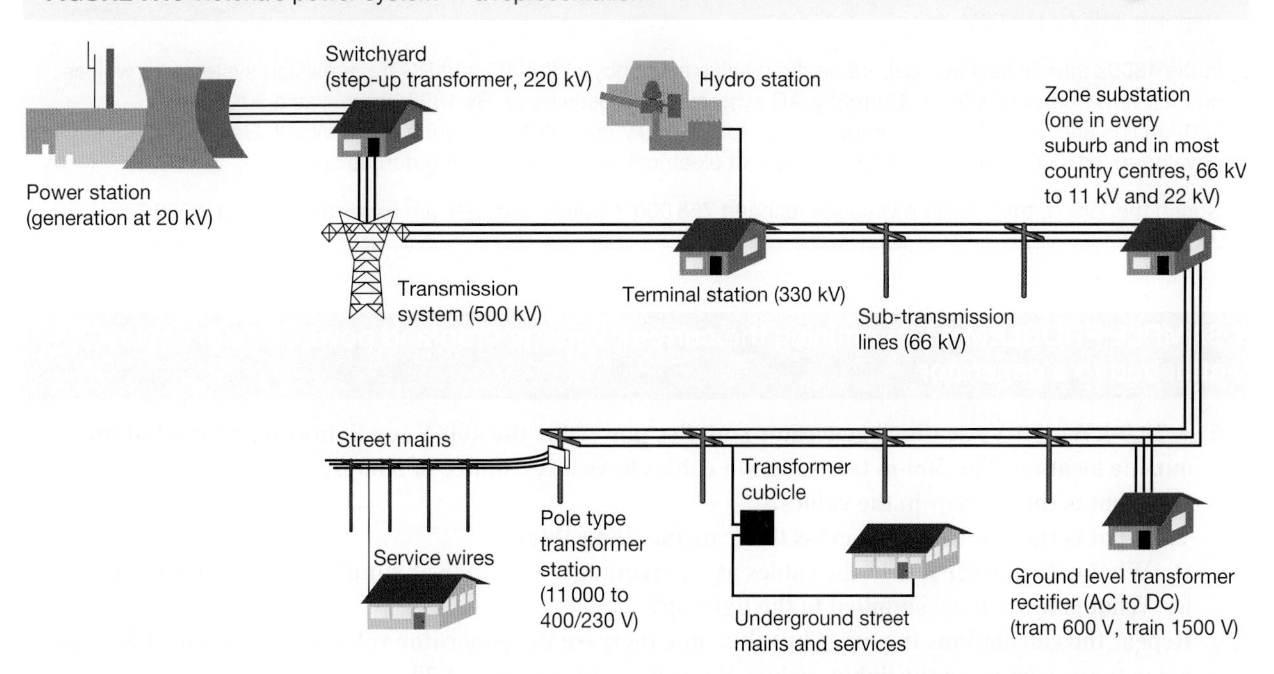

TABLE 7.2 Ways to reduce resistance

Method	Effect
Make the wires thicker	This increases the cost of the material in the wire and the cost of the pole to hold up the heavier wire.
Use a better conductor	Metals differ in their electrical conductivity and in their economic value as a metal. Very good conductors such as gold and silver are too expensive to use as wires.

Imagine that 400 MW of power was available to be transmitted along a transmission line of 4.0 Ω. How would the power losses due to the resistance of the transmission line vary with the voltage across the transmission line? The following table shows some typical values.

TABLE 7.3 Transmission of 400 MW at different voltages in a transmission line with 4.0 Ω resistance

Transmission voltage	1000 kV	500 kV	220 kV	66 kV
Current $\left(I = \frac{P_{tot}}{V}\right)$	400 A	800 A	1800 A	6100 A
Power loss $\left(P_{loss} = I^2R\right)$	640 kW	2.6 MW	13 MW	150 MW
Power loss (%)	0.2%	0.6%	3.3%	37%

EXTENSION: History of power transmission

Electric power was first transmitted in 1882 by Thomas Edison in New York and by St George Lane-Fox in London, both using a DC system (connected to a DC generator) for street lighting. The transmission was at low voltage with considerable transmission power line losses, and so limited to short distances.

Later that decade, George Westinghouse purchased patents for AC generators. His company also improved the design of transformers and developed an AC-based transmission system. In 1886, these new developments allowed power to be transmitted over a distance of a kilometre, stepping up the voltage to 3000 V and then stepping it down to 500 V.

In the 1800s there was much debate on the relative efficiency of the AC and DC transmission systems as well as on their environmental effects. Currently, AC systems are mostly used. By 1898, there was a 30 000-V, 120-km line and, by 1934, the voltage was up to 287 000 V over 430 km. During World War II, German scientists developed 380 000 V and overcame the effect of electrical discharge by using double cables.

During the 1960s, transmission voltages reached 765 000 V. Future voltages are expected to be at 1 000 000 V.

tlvd-9004

SAMPLE PROBLEM 4 Determining the current, voltage drop and power loss in cables attached to a generator

a. A 20.0-kW, 400-V_{RMS} diesel generator supplies power for the 400-V_{RMS} lights on a film set at an outside location. The 500-m transmission cables have a resistance of 5.00 Ω.
- **i. What is the current in the cables?**
- **ii. What is the voltage drop across the transmission cables?**
- **iii. What is the power loss in the cables as a percentage of the power supplied by the generator?**
- **iv. What is the voltage supplied to the lighting?**

b. Repeat the calculations in part a, but this time increase the generator voltage by a factor of 20 and, prior to connection to the lights, reduce the voltage by a factor of 20.

THINK	WRITE
a. i. 1. Draw a diagram showing all known information.	**a. i.** Transmission lines; Generator; $R = 5.00\ \Omega$; Lighting; 500 m; $V = 400$ V; $P = 20.0$ kW
2. The current in the cables is equal to the current coming from the generator. For the generator: $P = 20\ 000$ W; V = 400 V. Use $P = VI$ to determine the current through the cables. (Note: Using $V = IR$ with $V = 400$ V and $R = 5.0\ \Omega$ is incorrect because 400 V is the voltage supplied by the generator, it is not the voltage drop across the cables.)	$P = VI$ $I = \frac{P}{V}$ $= \frac{20\,000}{400}$ $= 50.0$ A The current through the cables is 50.0 A.
ii. In the cables, $I = 50$ A and $R = 5.0\ \Omega$. Use $V = IR$ to determine the voltage drop across the cables.	**ii.** $V = IR$ $= 50.0 \times 5.00$ $= 250$ V The voltage drop across the cables is 250 V.
iii. 1. In the cables, $I = 50$ A and $R = 5.0\ \Omega$. Use $P_{loss} = I^2R$ to determine the power lost in the cables. (Note: This answer could have been obtained by using $P = VI$, with $V = 250$ V from solution ii; however, there is a risk that 400 V may be used by mistake, so it is better to use I^2R.)	**iii.** $P_{loss} = I^2R$ $= 50.0 \times 50.0 \times 5.00$ $= 12\,500$ W
2. Determine the power loss as a percentage of the power supplied by the generator.	$\%P_{loss} = \frac{12\ 500}{20\ 000} \times \frac{100}{1}$ $= 62.5\%$
iv. The voltage supplied by the generator is shared between the transmission cables and the lights: $V_{generator} = V_{cables} + V_{lights}$ where $V_{generator} = 400$ V and $V_{cables} = 250$ V.	**iv.** $V_{generator} = V_{cables} + V_{lights}$ $V_{lights} = V_{generator} - V_{cables}$ $= 400\text{ V} - 250\text{ V}$ $= 150$ V At this distance, the voltage drop across the cables is too much to leave sufficient voltage to operate the lights at their designated voltage. Given the noise of the generators, they cannot be moved closer. Therefore, step-up and step-down transformers with turns ratios of 20 are used to reduce the power loss in the cables and increase the voltage at the lights.

b. i. 1. Draw a diagram showing all known information.

b. i.

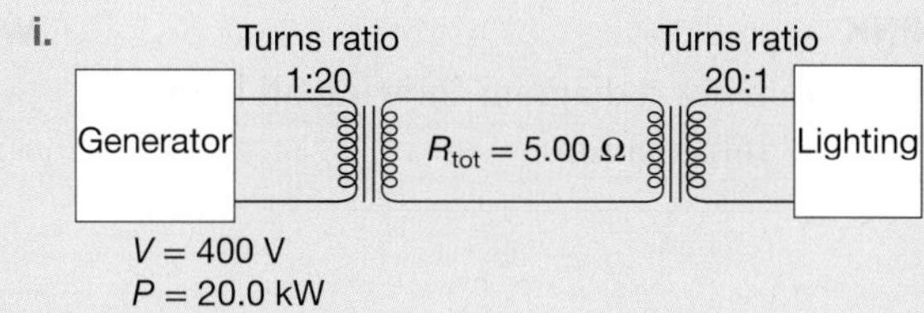

2. Determine the voltage in the secondary coil of the step-up transformer.

$$\frac{V_1}{V_2} = \frac{N_1}{N_2}$$
$$\frac{400}{V_2} = \frac{1}{20}$$
$$V_2 = 20 \times 400$$
$$= 8000 \text{ V}$$

3. The current in the cables is equal to the current coming from the secondary coil of the step-up transformer. Use $P = VI$ to determine the current in the secondary coil of the transformer, and hence the cables, assuming an ideal transformer.

$$P_1 = P_2 = 20\,000 \text{ W}; V_2 = 8000 \text{ V}$$
$$P_1 = P_2 = V_2 I_2$$
$$I_2 = \frac{P_1}{V_2}$$
$$= \frac{20\,000}{8000}$$
$$= 2.50\text{A}$$

ii. In the cables, $I = 2.50$ A and $R = 5.00\ \Omega$. Use $V = IR$ to determine the voltage drop across the cables.

ii. $V = IR$
$$= 2.50 \times 5.00$$
$$= 12.5 \text{ V}$$
The voltage drop across the cables is 12.5 V.

iii. 1. In the cables, $I = 2.50$ A and $R = 5.00\ \Omega$. Use $P_{loss} = I^2R$ to determine the power loss.

iii. $P_{loss} = I^2R$
$$= 2.50 \times 2.50 \times 5.00$$
$$= 31.3 \text{ W}$$

2. Determine the power loss as a percentage of the power supplied by the generator.

$$\%\ P_{loss} = \frac{31.25}{20\,000} \times \frac{100}{1}$$
$$= 0.156\%$$

This is $\left(\frac{1}{20}\right)^2$ or $\frac{1}{400}$ of the original power loss!
This is an impressive reduction.

iv. 1. The voltage supplied to the primary coil of the step-down transformer is the voltage from the secondary coil of the step-up transformer minus the voltage drop across the cables.

iv. Voltage supplied to the step-down transformer is therefore:
8000 V – 12.5 V = 7987.5 V

2. Determine the voltage in the secondary coil of the step-down transformer.

$$\frac{V_1}{V_2} = \frac{N_1}{N_2}$$

$$\frac{7988}{V_2} = \frac{20}{1}$$

$$V_2 = \frac{7987.5}{20}$$

$$= 399.375 \text{ V}$$

$$\approx 400 \text{ V}$$

PRACTICE PROBLEM 4

a. **A remote community uses a 50.0-kW, 250-V_{RMS} generator to supply power to its hospital. The power is delivered by a 100-m cable with total resistance of 0.200 Ω.**
   i. **What is the current in the cables?**
   ii. **What is the voltage drop across the transmission cables?**
   iii. **What is the power loss in the cables as a percentage of the power supplied by the generator?**
   iv. **What is the voltage supplied to the hospital?**

b. **Transformers with a turns ratio of 10:1 are installed. Repeat a with this new ratio.**

EXTENSION: Transmission lines

In transmission lines, the current actually flows through the outer surface of the line to a depth of about 1 mm. This is called the skin effect. It happens because the voltage is applied to the surface of the transmission line and the effect of the voltage decreases exponentially with distance from the surface.

Transmission lines are bare, multi-layered, concentrically stranded aluminium cables with a core of steel or reinforced aluminium for tensile strength. The advantages of having wires in a bundle over a single conductor of the same area include lower resistance to AC currents, lower radio interference and audible noise, and better cooling.

The smaller the sag in a transmission line, the greater will be the tension in the line. As the transmission lines cool, they contract, producing greater tension. High winds also increase tension.

The cost of building a transmission line is very nearly proportional to the input voltage and to the length of the line. The cost to transmit each unit of power is proportional to the length and inversely proportional to the square root of the power. That is, if the power to be transmitted is quadrupled, it can be transmitted twice as far for the same unit cost. It is therefore uneconomical to transmit power over a long distance unless a large quantity of power is involved.

The cost of constructing a line underground rather than above ground ranges from eight times as much (at 69 kV) to 20 times as much (at 500 kV). Underground cables are usually stranded copper, insulated with layers of oil-soaked paper tape. Superconductive cables may make this a more economical proposition.

Basslink uses subsea cables to transmit high-voltage DC between Victoria and Tasmania.

7.4.2 Using Ohm's Law wisely

The relationship $V = IR$ (Ohm's Law) is very useful. It can be applied in many situations in the same problem. This usefulness can lead to error if Ohm's Law is not applied wisely. The errors occur when students assume that, having calculated a value for V, the same value can be used every time $V = IR$ is used.

Rather, $V = IR$ should be remembered as:

Voltage across a section = current through the section × resistance of the section

So, in transmission line problems, the voltage across the output of the generator is different from the voltage across the transmission lines, which is different, in turn, from the voltage across the load at the end of the lines.

A well-labelled diagram can help avoid this confusion.

FIGURE 7.11 An electric circuit with step-up and step-down transformers

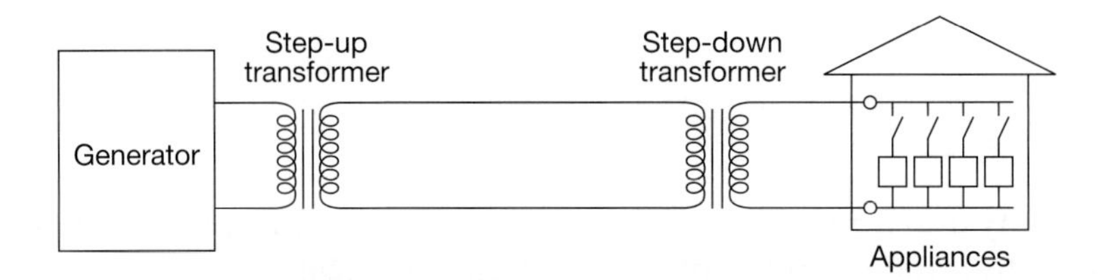

In any electric circuit, the total resistance determines how much current is drawn from the power supply. If there are transformers in the circuit, this statement is still true, but there are different currents and voltages on each side of the transformer.

As more appliances are turned on, there is a larger secondary current, which causes a larger primary current to be drawn from the power supply. This means the secondary current drives the primary current. When all the appliances are turned off, there is no secondary current and, so, no primary current.

However, the primary voltage determines the secondary voltage through the ratio of turns in the transformer.

With more appliances turned on, the current in the transmission lines between the transformers increases. The increased voltage drop across these lines means there will be slightly less voltage across the primary turns of the step-down transformer. This will result in a slight drop in the voltage for each of the appliances. This is a scenario where Ohm's Law can lead to errors in calculations, due to this additional voltage drop.

EXTENSION: Insulating transmission lines

In dry air, sparks can jump a distance of 1 cm for every 10 000 V of potential difference. Therefore, a 330 kV line will spark to a metal tower if it comes within a distance of 33 cm. In high humidity conditions, the distance is smaller. To prevent sparks jumping from transmission lines to the metal support towers, large insulators separate them from each other. It is important that these insulators are strong and have high insulating properties. Suspension insulators, as shown in figure 7.12, are used for all high-voltage power lines operating above 33 kV, where the towers or poles are in a straight line. Note that the individual sections of the insulators are disc shaped. This is because dust and grime collect on the insulators, making them conductive when wet. Many wooden poles catch fire after the first rain following a prolonged dry period because a current flows across wet, dirty insulators. The disc shape of the insulator sections increases the distance that a current has to pass over the surface of the insulator and so decreases the risk. There is also less chance that dirt and grime will collect on the undersides of the sections, and these are also less likely to get wet.

FIGURE 7.12 Suspension insulators are used in high-voltage transmission lines

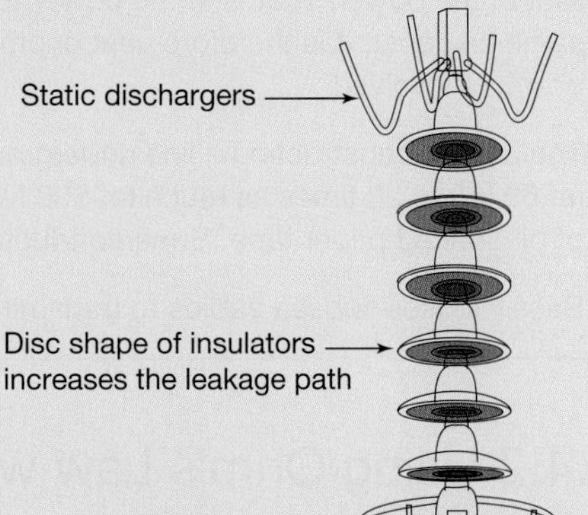

Static dischargers

Disc shape of insulators increases the leakage path

Transmission cable

7.4.3 The advantage of using AC power as a domestic power supply

AC power allows for the use of transformers to increase supply voltages. Transmitting the same power at a higher voltage means less current through the transmission lines and hence less power lost in transmission. Remember that transformers don't work with batteries providing DC voltage (they can work if there is a varying DC power supply), as there is no change in magnetic flux. In order to decrease the power lost in transmission using DC power, the resistance of the transmission would need to be significantly reduced. This can be achieved by increasing the cross-sectional area of the transmission wires, making them heavier and more expensive.

INVESTIGATION 7.3

online only

elog-1890

Modelling power transmission

Aim

a. To investigate power loss through transmission lines
b. To investigate how the use of transformers in power distribution systems can minimise power loss

INVESTIGATION 7.4

online only

elog-1891

Modelling power transmission using a spreadsheet

Aim

a. To investigate power loss through transmission lines
b. To investigate how the use of transformers in power distribution systems can minimise power loss

Resources

Interactivity Losing power (int-0119)

EXTENSION: High-voltage DC power transmission

High-voltage direct current (HVDC) transmission is also being used to transmit power over long distances. An Australian example is Basslink (see figure 7.13), which connects the Tasmanian power grid to the Victorian power grid using a 370-km cable under Bass Strait. The transmission voltage is 400 kV. It has been operating since 2006, and provides power from the Tasmanian grid to Victoria and from there to the national grid when there are high demands on power use, for example during heatwaves.

Because it uses direct current, transformers are not used to increase the voltage and decrease the current. Instead, devices called voltage source converters are used to increase and decrease voltages.

Advantages of HVDC transmission include that it is more stable and that both networks involved do not have to operate on the same frequency nor be synchronised.

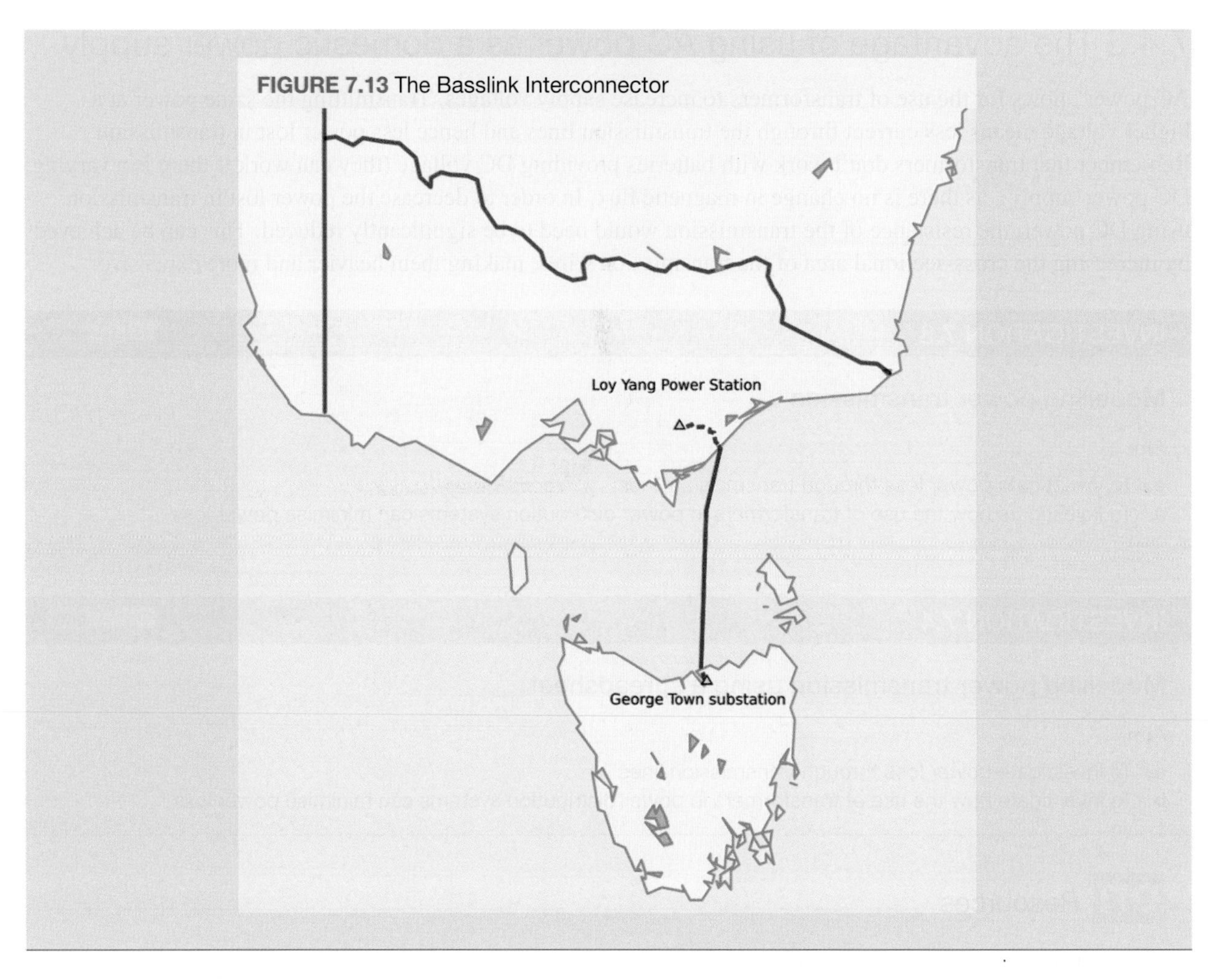

FIGURE 7.13 The Basslink Interconnector

7.4 Activities

learnon

Students, these questions are even better in jacPLUS

Receive immediate feedback and access sample responses

Access additional questions

Track your results and progress

Find all this and MORE in jacPLUS

7.4 Quick quiz	7.4 Exercise	7.4 Exam questions

7.4 Exercise

1. An isolated film set uses a 50-kW generator to produce electricity for lighting and other purposes at 250 V_{RMS}. The generator is connected to lights approximately 100 m away by transmission cables with a combined resistance of 0.30 Ω.
   a. When the generator is operating at full capacity, what current does it supply?

b. What is the power loss in the transmission cables?
c. What is the total drop in voltage across the two cables connected by the generator?
d. What is the voltage supplied to the lights?
e. Two transformers with a turns ratio of 20 are used to first step up the voltage from the generator to the cables, and then to step it down from the cables to the lights. Using this new information, answer parts **b** to **d** again.

2. A house is some distance from the power lines and the connecting cables have a resistance of 0.20 Ω. The appliances in this house would, if all turned on and connected to a power supply of 230 V_{RMS}, draw a current of 40 A.
a. What is the equivalent resistance of the circuit connected to the power lines?
b. What is the equivalent resistance of the appliances in the house when they are all turned on?
c. If the voltage at the power lines is 230 V AC, what is the voltage at the house?
d. A 20 kW workshop that operates off the 230 V supply is installed in the garage in parallel to the house.
i. Calculate the resistance of the workshop.
ii. Calculate the equivalent resistance of the circuit including the house and workshop in parallel.
iii. Calculate the current in the workshop.
iv. Calculate the total current in the transmission lines.
e. Calculate the voltage drop
i. across transmission lines
ii. at the house with the workshop.
The owners now decide to install a step-up transformer and a step-down transformer, each with a turns ratio of 10 : 1, at either end of the transmission lines.
f. If the system draws 120 A from the grid at 230 V, will the voltage at the house and the garage be within 1% of 230 V for the appliances to work properly?
g. At night, the workshop is turned off. Will the voltage at the house increase, decrease or remain unchanged? Give reasons.

3. A generator at a power station produces 220 MW at 23 kV. The voltage is then stepped up to 330 kV. The power passes along transmission lines with a total resistance of 0.40 Ω.
a. What is the current in the transmission lines?
b. What is the power loss in the transmission lines?
c. What is the voltage drop across them?
d. What voltage and power is available to the step-down transformer located at the end of the transmission lines?

4. The maximum electrical power that the generator at a power station can deliver is 500 MW at a voltage of 40 kV. This power is to supply the electricity needs of a distant city. Transmission lines connecting the station to the city have a total resistance of 0.80 Ω. At the city, the transmission lines are connected to a series of step-down transformers that reduce the voltage to 230 V.
a. What percentage of the power delivered by the power station is lost in the transmission lines?
The power loss can be reduced by stepping up the voltage at the generator with a transformer. At the substations on the city's outskirts, the voltage is stepped down. The voltage could be stepped up to 400 kV using a transformer with a turns ratio of 1:10. The same transmission lines could be used, but they would need to be raised higher off the ground and be better insulated at each pole.
b. What would be the effect on the power loss in the transmission lines?

7.4 Exam questions

Question 1 (4 marks)

Source: *VCE 2022 Physics Exam, NHT, Section B, Q.6;* © VCAA

A laptop computer requires a transformer to reduce the voltage to its rechargeable battery while the battery is charging. The power point supplies an RMS voltage of 240 V and delivers an RMS current of 0.35 A. The transformer converts the voltage to an RMS voltage of 8.0 V. Assume that the transformer is ideal.

a. Calculate the ratio of the number of turns $\frac{N_P}{N_S}$. Show your working. **(2 marks)**

b. Calculate the RMS current delivered by the power point while the battery is charging. **(2 marks)**

Question 2 (8 marks)

***Source:** VCE 2019 Physics Exam, NHT, Section B, Q.4; © VCAA*

An electrician is installing a power supply to a yard located 500 m from a farmhouse in order to operate a 240 V_{RMS}, 480 W light globe, as shown in Figure 4. The connecting wires have a total resistance, R_T, of 40 Ω. At the farmhouse, the electrician provides the required input voltage, V_{in}, to the connecting wires for the light globe to operate at 240 V_{RMS} and 480 W.

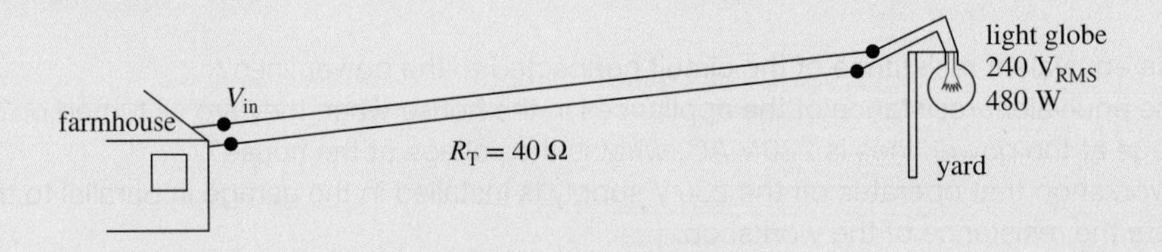

Figure 4

a. When the light globe is operating at 240 V_{RMS} and 480 W, what is the power loss in the connecting wires? Show your working. **(2 marks)**

b. Calculate the RMS voltage of V_{in}. Show your working. **(3 marks)**

c. To reduce the power loss in the connecting wires, the electrician changes the input voltage, V_{in}, and installs an 8 : 1 step-down transformer at the yard. After these changes, the light globe still operates at 240 V_{RMS} and 480 W, as shown in Figure 5.

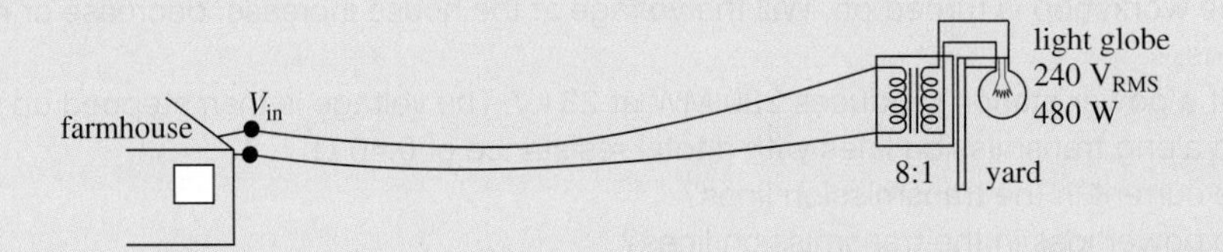

Figure 5

Calculate the RMS power loss in the connecting wires for this new situation. Show your working. **(3 marks)**

Question 3 (12 marks)

***Source:** VCE 2018, Physics Exam, Section B, Q.5; © VCAA*

A Physics class is investigating power loss in transmission lines.

The students construct a model of a transmission system. They first set up the model as shown in Figure 7. The model consists of a variable voltage AC power supply, two transmission lines, each of 4.0 Ω (total resistance = 8.0 Ω), a variable ratio transformer, a light globe and meters as needed. The purpose of the model is to operate the 4.0 V light globe.

A variable ratio transformer is one in which the ratio of turns in primary windings to turns in secondary windings can be varied. The resistance of the connecting wires can be ignored.

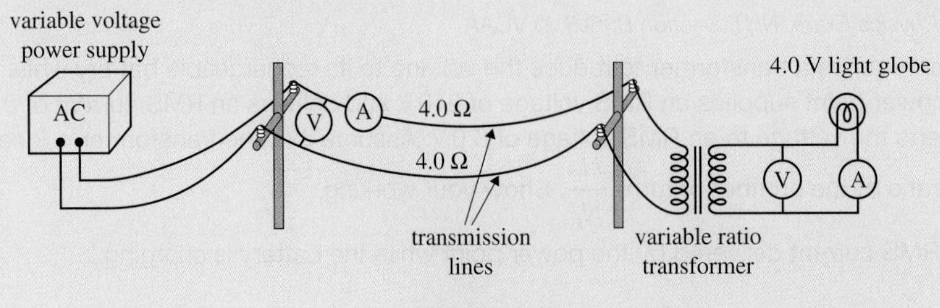

Figure 7

In their first experiment, the transformer is set on a ratio of 4:1 and the current in the transmission lines is measured to be 3.0 A. The light globe is operating correctly, with 4.0 V_{RMS} across it.

a. Calculate the power dissipated in the light globe. Show your working. **(2 marks)**

b. Calculate the voltage output of the power supply. Show your working. **(3 marks)**

c. Calculate the total power loss in the transmission lines. Show your working. **(2 marks)**

d. In a second experiment, the students set the variable ratio of the transformer at 8:1 and adjust the variable voltage power supply so that the light globe operates correctly, with 4.0 V_{RMS} across it.
Calculate the total power loss in the transmission lines in this second experiment. Show your working. **(3 marks)**

e. Suggest **two** reasons why high voltages are often used for the transmission of electric power over long distances. **(2 marks)**

Question 4 (4 marks)

Source: *VCE 2017, Physics Exam, Section B, Q.6; © VCAA*

Figure 5 shows a generator at an electrical power station that generates 100 MW_{RMS} of power at 10 kV_{RMS} AC.

Transformer T_1 steps the voltage up to 500 kV_{RMS} AC for transmission through transmission wires that have a total resistance, R_T, of 3.0 Ω. Transformer T_2 steps the voltage down to 50 kV_{RMS} AC at the substation. Assume that both transformers are ideal.

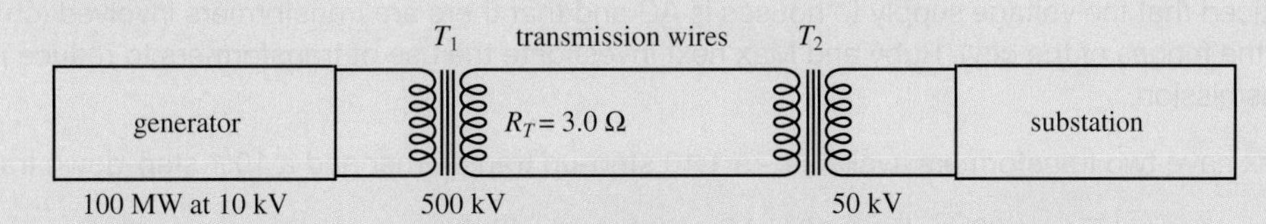

Figure 5

a. The current in the transmission lines is 200 A.
Calculate the total electrical power loss in the transmission wires. **(2 marks)**

b. Transformer T_1 stepped the voltage up to 250 kV_{RMS} AC instead of 500 kV_{RMS} AC.
By what factor would the power loss in the transmission lines increase? **(2 marks)**

Question 5 (12 marks)

Source: *VCE 2016, Physics Exam, Section B, Q.16; © VCAA*

Ruby and Max are investigating the transmission of electric power using a model system, as shown in Figure 20a. The circuit is shown in Figure 20b.

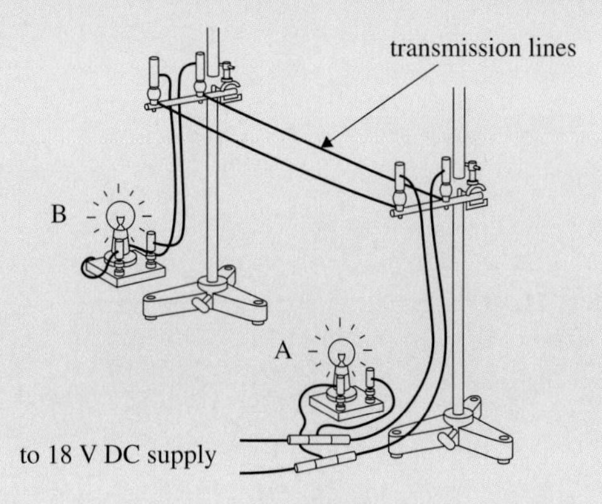

Figure 20a

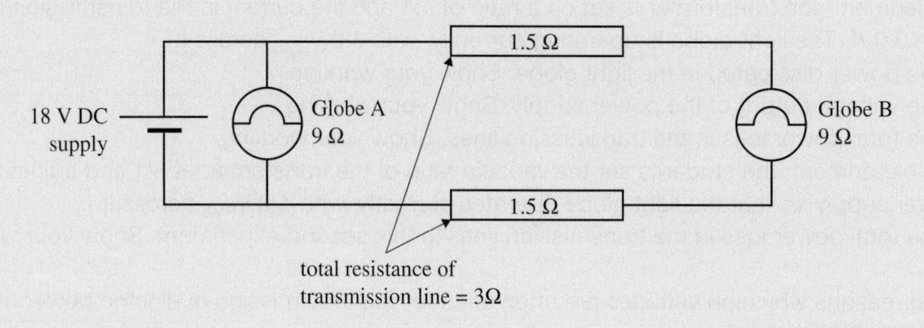

Figure 20b

Ruby and Max use an 18 V DC power supply, as shown in Figure 20b. The two transmission lines have a **total** resistance of 3.0 Ω. Assume that the resistance of the globes is constant at 9.0 Ω and that the other connecting wires have zero resistance.

a. Calculate the power delivered to Globe A. **(2 marks)**

b. Calculate the total voltage drop over the transmission lines. Show your working. **(2 marks)**

c. Calculate the power delivered to Globe B. Show your working. **(3 marks)**

Ruby has noticed that the voltage supply to houses is AC and that there are transformers involved (on street poles and at the fringes of the city). Ruby and Max next investigate the use of transformers to reduce power losses in transmission.

Ruby and Max have two transformers available – a 1:10 step-up transformer and a 10:1 step-down transformer.

d. Redraw the circuit in Figure 20b with an 18 V AC supply and with the transformers correctly connected. Label the transformers as step up and step down. **(2 marks)**

e. Explain why the transformers would reduce the transmission losses. Your answer should include reference to key physics formulas and principles. **(3 marks)**

More exam questions are available in your learnON title.

7.5 Review

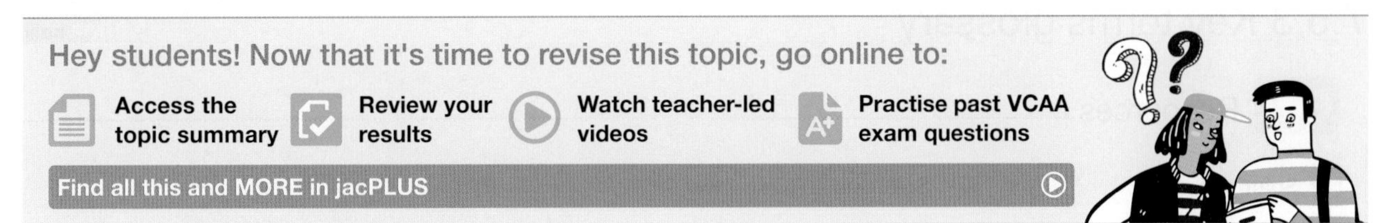

7.5.1 Topic summary

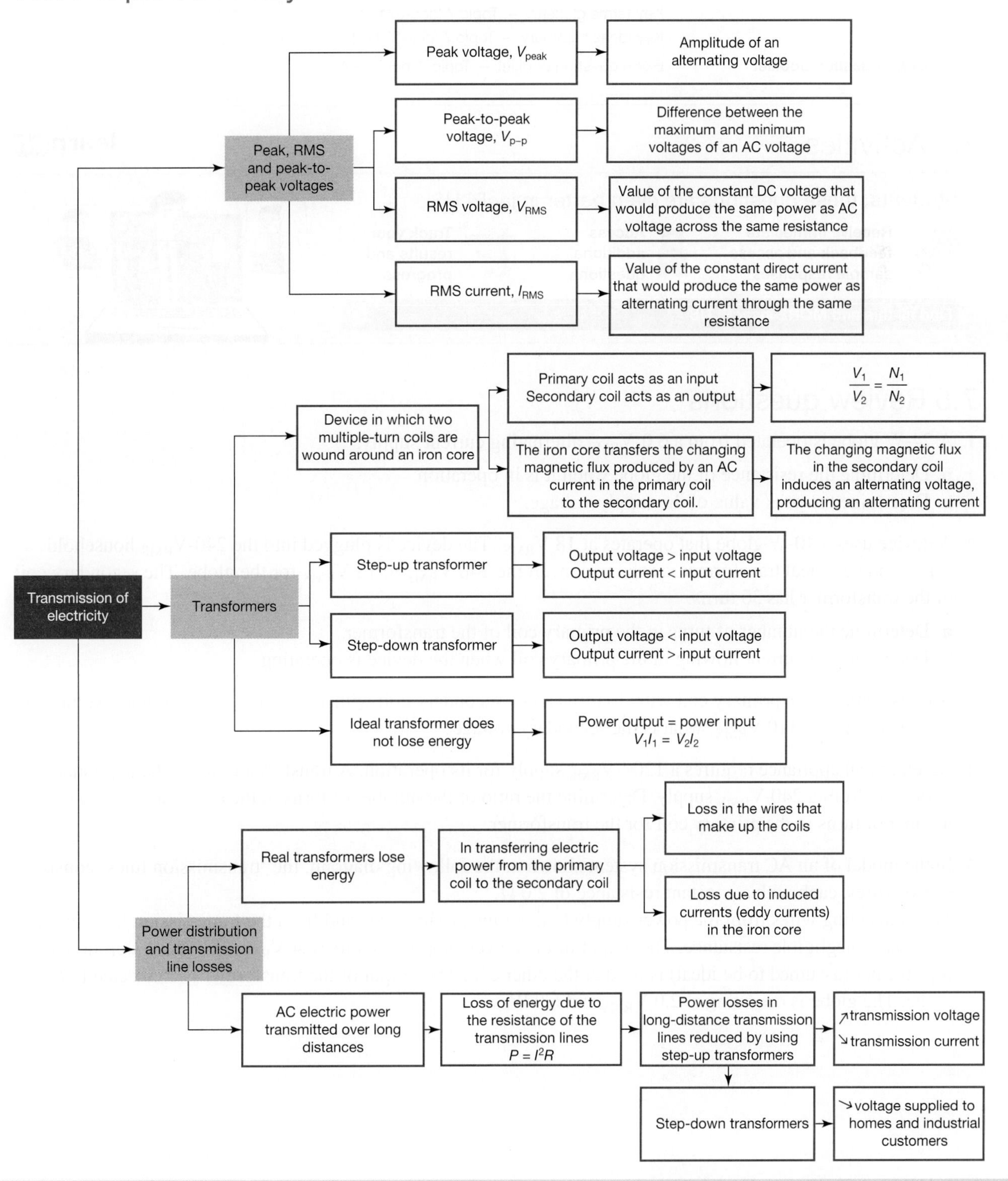

7.5.2 Key ideas summary

7.5.3 Key terms glossary

online only

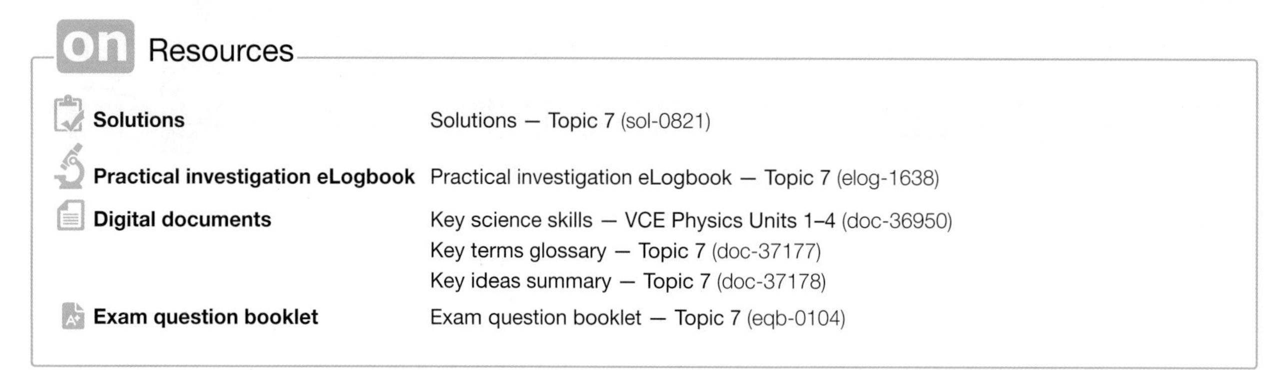

on Resources
Solutions — Solutions — Topic 7 (sol-0821)
Practical investigation eLogbook — Practical investigation eLogbook — Topic 7 (elog-1638)
Digital documents — Key science skills — VCE Physics Units 1–4 (doc-36950)
Key terms glossary — Topic 7 (doc-37177)
Key ideas summary — Topic 7 (doc-37178)
Exam question booklet — Exam question booklet — Topic 7 (eqb-0104)

7.5 Activities

learnon

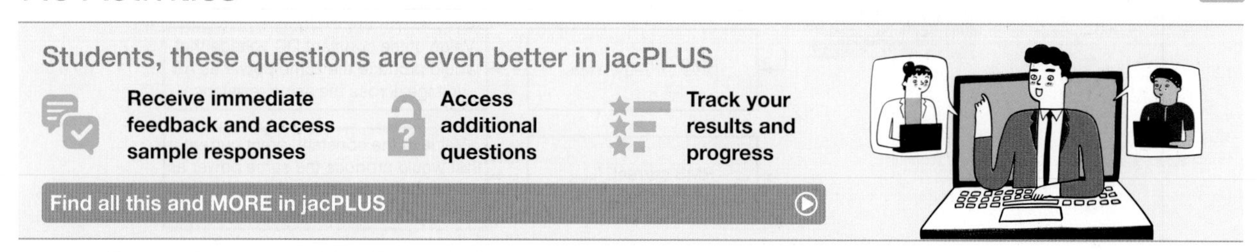

7.5 Review questions

1. A 24-W globe is operated from a 6.0-V_{RMS} alternating current supply.
   a. Determine the resistance of the globe when it is in operation.
   b. Determine the peak value of the supply voltage.

2. A device uses a 40-W globe that operates at 18 V_{RMS}. The device is plugged into the 240-V_{RMS} household supply, and an ideal transformer is used to convert the 240 V_{RMS} to 18 V_{RMS} for the globe. The secondary coil of the transformer has 30 turns.
   a. Determine the number of turns in the primary coil of the transformer.
   b. Determine the current flowing in the primary coil when the device is operating.

3. A transformer has a primary coil with 30 turns and a secondary coil with 1150 turns. If the primary voltage to the transformer is 110 V_{RMS}, what is the secondary voltage?

4. An electrical appliance requires a 1200-V_{RMS} supply for its operation. A transformer allows the appliance to operate from a 240-V_{RMS} supply. Determine the ratio of the number of turns in the secondary coil to the number of turns in the primary coil for the transformer.

5. In the model of an AC transmission system shown in the following diagram, the 'transmission lines' consist of two wires, each with a constant resistance of 2.0 Ω.
   The connecting wires from the power supply to the transmission lines and from the transmission lines to the globe have negligible resistance. The output of the power supply is set to 12.4 V_{RMS} AC. A 5:1 step-down transformer (assumed to be ideal) is used at the other end. The output of the transformer is connected to a globe. The globe is operating at 2.0 V_{RMS} and 6.0 W.

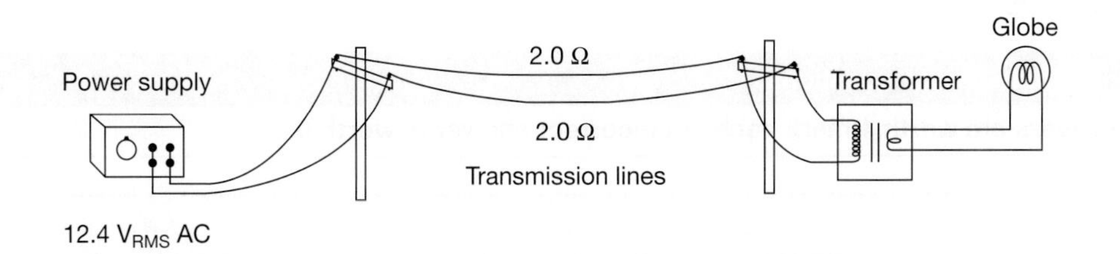

a. The primary coil of the 5:1 step-down transformer has 2000 turns. How many turns does the secondary coil have?
b. With the experiment set up as shown in the diagram, what is the power loss in the transmission lines?
c. Determine the power delivered at each of the following points:
   i. At the input to the primary coil of the transformer
   ii. At the output of the 12-V_{RMS} power supply.
d. What would be observed at the globe if the 18-V_{RMS} AC power supply is replaced by an 18-V battery?

6. Consider the following electrical circuit:

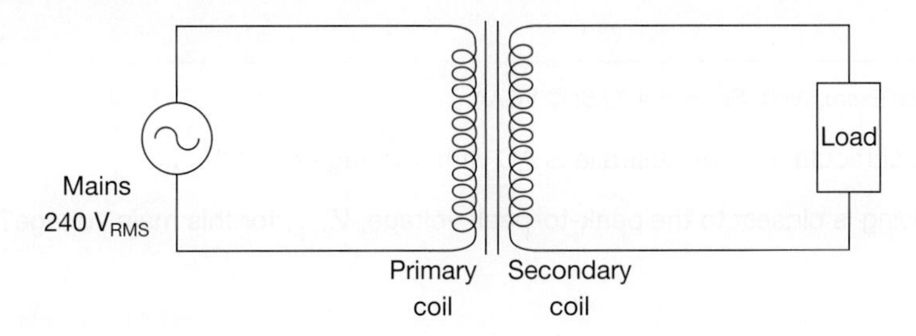

Primary coil: $V_{p\ RMS} = 240\ V_{RMS}$
Secondary coil: $V_{s\ RMS} = 13\ V_{RMS}$
$I_{s\ RMS} = 2.7\ A_{RMS}$
Assuming the transformer is ideal, calculate the RMS current in the primary coil.

7. A transformer is connected to a substation through wires with a total resistance of 4.0 Ω, as shown in the following diagram. This transformer supplies a town with electricity at 240 V_{RMS}, at a rate of 400 kW. The transformer (assumed ideal) has an input voltage of $8.0 \times 10^3\ V_{RMS}$.

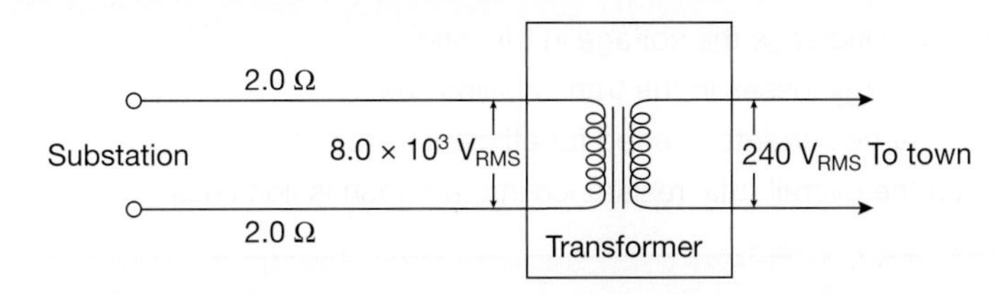

a. If the secondary winding of the transformer has 60 turns, how many turns are there on the primary winding?
b. Determine the RMS current in the wires from the substation to the transformer.
c. What is the power loss in the wires between the substation and the local area transformer?
d. At what voltage is electricity supplied from the substation?

7.5 Exam questions

Section A — Multiple choice questions

All correct answers are worth 1 mark each; an incorrect answer is worth 0.

Question 1

Source: *VCE 2022 Physics Exam, NHT, Section A, Q.5;* © VCAA

The ratio of the number of turns in an ideal step-up transformer is 350 : 1. An alternating RMS current of 30.0 mA is supplied to the primary coil.

The RMS current in the output will be closest to

A. 0 mA
B. 0.086 mA
C. 30.0 mA
D. 1.1×10^4 mA

Question 2

Source: *VCE 2021 Physics Exam, NHT, Section A, Q.6;* © VCAA

The mains voltage in a particular part of Australia is AC with a voltage of $240V_{RMS}$.

Which one of the following is closest to the peak-to-peak voltage, V_{p-p}, for this main voltage?

A. 170 V
B. 340 V
C. 480 V
D. 680 V

Question 3

Source: *VCE 2021 Physics Exam, NHT, Section A, Q.7;* © VCAA

Electrical power stations are often situated far from the cities that require the power that they generate. Which one of the following best describes the reason for the high-voltage transmission of electrical energy?

A. Transformers can be used to increase the voltage in the cities.
B. High voltages reduce the energy losses in the transmission lines.
C. High voltages provide the large currents needed for efficient transmission.
D. High voltages can reduce the overall total resistance in the transmission lines.

Question 4

Source: *VCE 2019 Physics Exam, NHT, Section A, Q.2;* © VCAA

A powerline carries a current of 1000 A DC in the direction east to west. At the point of measurement, Earth's magnetic field is horizontally north and its strength is 5.0×10^{-5} T.

Which one of the following best gives the direction of the electromagnetic force on the powerline?

A. horizontally west
B. horizontally north
C. vertically upwards
D. vertically downwards

Question 5

Source: *VCE 2019 Physics Exam, NHT, Section A, Q.3;* © VCAA

A powerline carries a current of 1000 A DC in the direction east to west. At the point of measurement, Earth's magnetic field is horizontally north and its strength is 5.0×10^{-5} T.

The magnitude of the force on **each metre** of the powerline is best given by

A. 5.0×10^{3} N
B. 5.0×10^{2} N
C. 5.0×10^{-2} N
D. 5.0×10^{-5} N

Question 6

Source: *VCE 2019 Physics Exam, NHT, Section A, Q.5;* © VCAA

A light globe operates at 12 V_{RMS} AC that is supplied by a 240 V to 12 V transformer connected to a 240 V_{RMS} mains supply.

In the transformer, the ratio of turns in the primary (input) to turns in the secondary (output) is

A. 20 : 1
B. 1 : 20
C. 28 : 1
D. 1 : 28

Question 7

Source: *VCE 2019 Physics Exam, NHT, Section A, Q.6;* © VCAA

A light globe operates at 12V_{RMS} AC that is supplied by a 240 V to 12 V transformer connected to a 240 V_{RMS} mains supply.

If the light globe is to be operated using a battery instead of the mains supply, what voltage should the battery have for the light globe to operate correctly?

A. 12 V
B. 17 V
C. 8.5 V
D. 6.0 V

Question 8

Source: *VCE 2018 Physics Exam, NHT, Section A, Q.5;* © VCAA

A step-down transformer is used to convert 240 V_{RMS} AC to 16 V_{RMS} AC.

Assume that the transformer is ideal.

Which one of the following best gives the peak voltage of the input to the transformer?

A. 171 V
B. 240 V
C. 339 V
D. 480 V

Question 9

Source: *VCE 2018 Physics Exam, NHT, Section A, Q.6;* © VCAA

A step-down transformer is used to convert 240 V_{RMS} AC to 16 V_{RMS} AC.

Assume that the transformer is ideal.

The ratio of turns in the primary (input) to turns in the secondary (output) is best given by

A. 15 : 1
B. 1 : 15
C. 24 : 1
D. 1 : 24

Question 10

Source: *VCE 2018 Physics Exam, NHT, Section A, Q.7;* © VCAA

A step-down transformer is used to convert 240 V_{RMS} AC to 16 V_{RMS} AC.

Assume that the transformer is ideal.

The power input to the primary of the transformer is 30 W.

Which one of the following best gives the RMS current in the secondary (output)?

A. 0.50 A
B. 1.9 A
C. 8.0 A
D. 15 A

Section B — Short answer questions

Question 11 (7 marks)

Source: *VCE 2021 Physics Exam, Section B, Q.7;* © VCAA

The generator of an electrical power plant delivers 500 MW to external transmission lines when operating at 25 kV. The generator's voltage is stepped up to 500 kV for transmission and stepped down to 240 V 100 km away (for domestic use). The overhead transmission lines have a total resistance of 30.0 Ω. Assume that all transformers are ideal.

a. Explain why the voltage is stepped up for transmission along the overhead transmission lines. **(2 marks)**
b. Calculate the current in the overhead transmission lines. Show your working. **(2 marks)**
c. Determine the maximum power available for domestic use at 240 V. Show all your working. **(3 marks)**

Question 12 (16 marks)

Source: *VCE 2020 Physics Exam, Section B, Q.18;* © VCAA

Students are modelling the effect of the resistance of electrical cables, *r*, on the transmission of electrical power. They model the cables using the circuit shown in Figure 18.

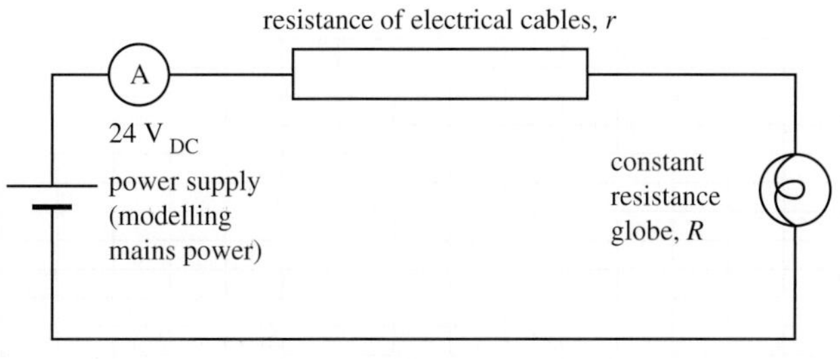

Figure 18

a. The 24 V_{DC} power supply models the mains power.
Describe the effect of increasing the resistance of the electrical cables, *r*, on the brightness of the constant resistance globe, *R*. **(2 marks)**

b. The students investigate the effect of changing *r* by measuring the current in the electrical cables for a range of values. Their results are shown in Table 1 below.

Table 1

Resistance of cables, r (Ω)	Current in cables, i (A)	$\frac{1}{i}\ (A^{-1})$
2.4	2.4	
3.6	2.0	
6.4	1.7	
7.6	1.5	
10.4	1.3	

Identify the dependent and the independent variables in this experiment. Give your reasoning. **(2 marks)**

c. To analyse the data, the students use the following equation to calculate the resistance of the cables for the circuit.

$$r = \frac{24}{i} - R$$

Show that this equation is true for the circuit shown in Figure 18. Show your working. **(2 marks)**

d. Calculate the values of $\frac{1}{i}$ and write them in the spaces provided in the last column of Table 1. **(2 marks)**

e. Plot a graph of *r* on the *y*-axis against $\frac{1}{i}$ on the *x*-axis on the grid provided below. On your graph:

- choose an appropriate scale and numbers for the *x*-axis
- draw a straight line of best fit through the plotted points
- include uncertainty bars (± *x*-direction only) of $\pm 0.02\ A^{-1}$. (Uncertainty bars in the *y*-direction are not required.) **(6 marks)**

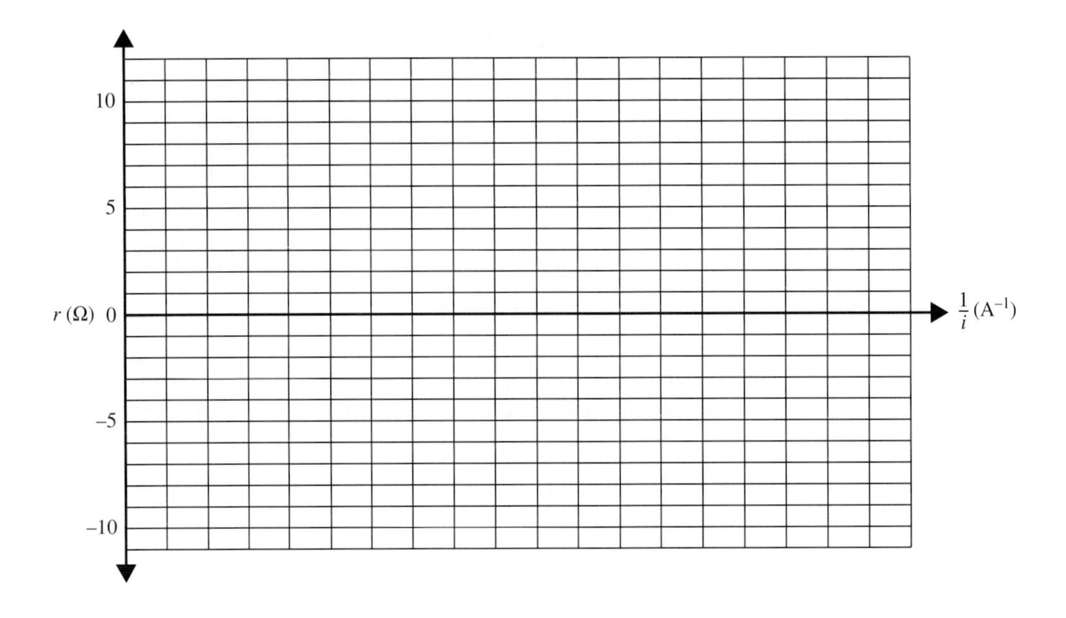

f. Use the straight line of best fit to find the value of the constant resistance globe, R. Give your reasoning. **(2 marks)**

Question 13 (7 marks)

Source: *VCE 2019, Physics Exam, Section B, Q.6*; © VCAA

A home owner on a large property creates a backyard entertainment area. The entertainment area has a low-voltage lighting system. To operate correctly, the lighting system requires a voltage of 12 V_{RMS}. The lighting system has a resistance of 12 Ω.

a. Calculate the power drawn by the lighting system. **(1 mark)**

To operate the lighting system, the home owner installs an ideal transformer at the house to reduce the voltage from 240 V_{RMS} to 12 V_{RMS}. The home owner then runs a 200 m long heavy-duty outdoor extension lead, which has a total resistance of 3 Ω, from the transformer to the entertainment area.

b. The lights are a little dimmer than expected in the entertainment area.
Give **one** possible reason for this and support your answer with calculations. **(4 marks)**

c. Using the same equipment, what changes could the home owner make to improve the brightness of the lights? Explain your answer. **(2 marks)**

Question 14 (10 marks)

Source: *VCE 2021 Physics Exam, NHT, Section B, Q.7*; © VCAA

Angela and Janek are installing two low-voltage lights in their outdoor garden. They have a 240 V_{RMS} AC transformer with an output voltage of 12 V_{RMS}. Each light has a constant resistance of 6.0 Ω. For the purposes of calculations, assume that the transformer is ideal.

a. Describe what is meant by an ideal transformer in terms of the input power and the output power. **(1 mark)**

b. Calculate the ratio of the number of turns of the primary coil to the number of turns of the secondary coil. **(1 mark)**

c. Each light is designed to operate at 12 V_{RMS}.
Calculate the power dissipated in one light when it is operated at 12 V_{RMS}. Show your working.
Angela and Janek now connect the first light, Light 1, to the transformer using two wires, each 12.0 m long, as shown in Figure 9. Each wire has a resistance of 0.05 Ω per metre. **(2 marks)**

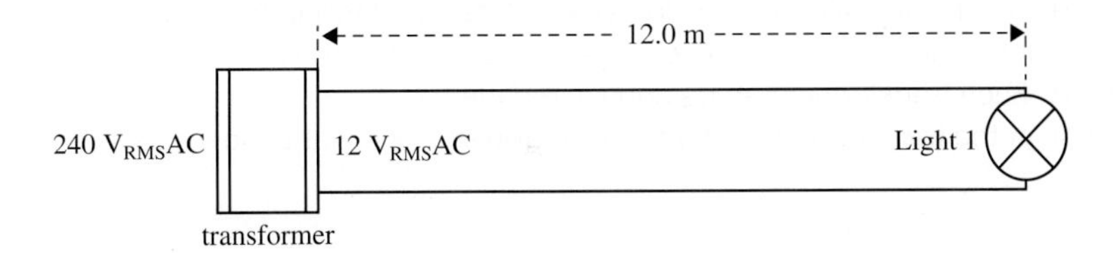

Figure 9

d. Calculate the RMS voltage across Light 1. Show your working. **(3 marks)**

e. Angela and Janek now connect the second light, Light 2, directly across the secondary coil of the transformer, as shown in Figure 10.

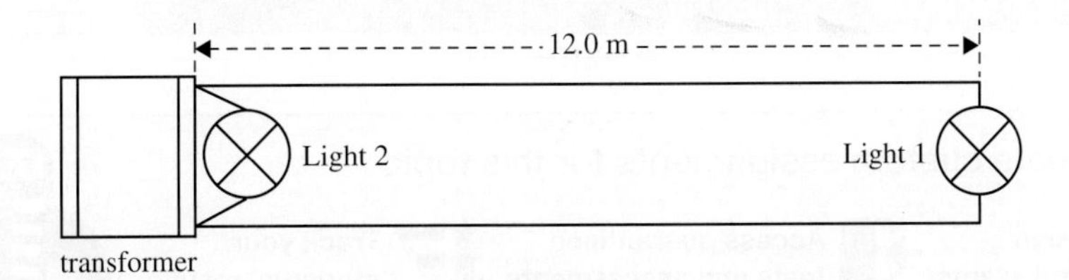

Figure 10

They thought that with the circuit shown in Figure 10, Light 1 and Light 2 would be equally bright. However, they observed that Light 2 was brighter than Light 1.
Explain why Light 2 was observed to be brighter than Light 1. **(3 marks)**

Question 15 (7 marks)

Source: *VCE 2022 Physics Exam, Section B, Q.5*; © VCAA

A wind generator provides power to a factory located 2.00 km away, as shown in Figure 5.

When there is a moderate wind blowing steadily, the generator produces an RMS voltage of 415 V and an RMS current of 100 A.

The total resistance of the transmission wires between the wind generator and the factory is 2.00 Ω.

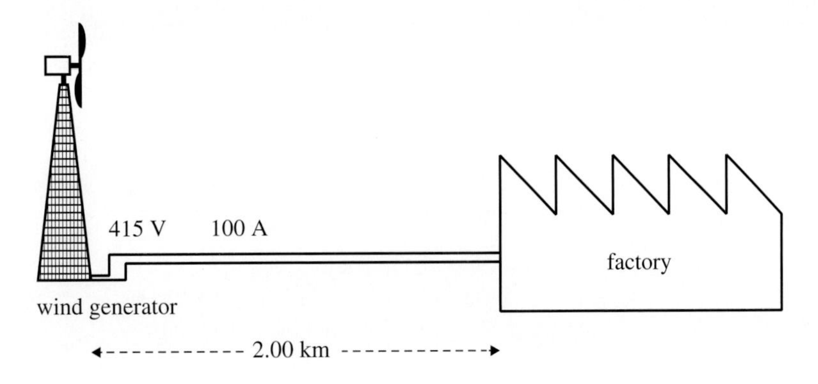

Figure 5

To operate correctly, the factory's machinery requires a power supply of 40 kW.

a. Calculate the power, in kilowatts, produced by the wind generator when there is a moderate wind blowing steadily. **(1 mark)**

b. Determine whether the energy supply system, as shown, will be able to supply power to the factory when the moderate wind is blowing steadily. Justify your answer with calculations. **(3 marks)**

c. The factory's owner decides to limit transmission energy loss by installing two transformers: a step-up transformer with a turns ratio of 1 : 10 at the wind generator and a step-down transformer with a turns ratio of 10 : 1 at the factory. Each transformer can be considered ideal.
With the installation of the transformers, determine the power, in kilowatts, now supplied to the factory. **(3 marks)**

AREA OF STUDY 3 How are fields used in electricity generation?

OUTCOME 3

Analyse and evaluate an electricity generation and distribution system.

PRACTICE EXAMINATION

STRUCTURE OF PRACTICE EXAMINATION		
Section	**Number of questions**	**Number of marks**
A	20	20
B	4	20
	Total	**40**

Duration: 50 minutes
Information:

- This practice examination consists of two parts. You must answer all question sections.
- Pens, pencils, highlighters, erasers, rulers and a scientific calculator are permitted.
- You may use the VCAA Physics formula sheet for this task.

Resources

Weblink VCAA Physics formula sheet

SECTION A — Multiple choice questions

All correct answers are worth 1 mark each; an incorrect answer is worth 0.

Use the following information to answer questions 1 and 2:
A circular conducting coil with 80 loops is placed inside a magnetic field with a field strength of 0.09 T. The area of the circular coil is 38 cm^2.

$\boldsymbol{B} = 0.09$ T

$A = 38$ cm^2

1. What is the magnetic flux through the loop?
   A. 3.4×10^{-4} Wb
   B. 2.7×10^{-2} Wb
   C. 3.4 Wb
   D. 2.7×10^{2} Wb

2. The coil is moved to the left and taken completely out of the magnetic field in a time of 0.35 s. Calculate the magnitude of the EMF induced in the coil.
   A. 9.8×10^{-4} V
   B. 9.6×10^{-3} V
   C. 7.8×10^{-2} V
   D. 2.7×10^{-1} V
3. The magnetic flux through a square loop conductor as it changes over time is plotted on a graph as shown in the following diagram:

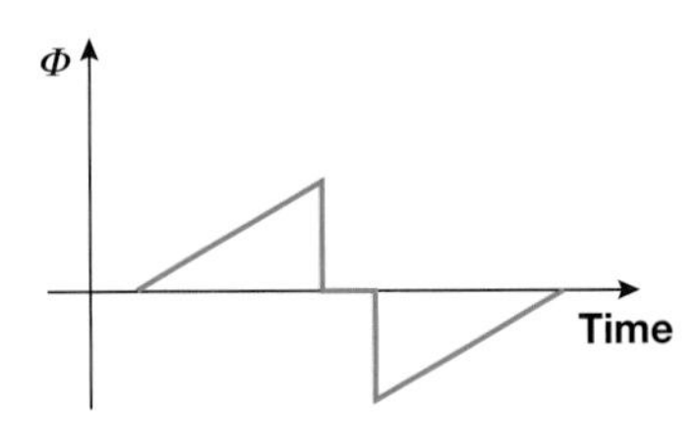

What is the expected graph of the EMF induced in the square loop conductor?

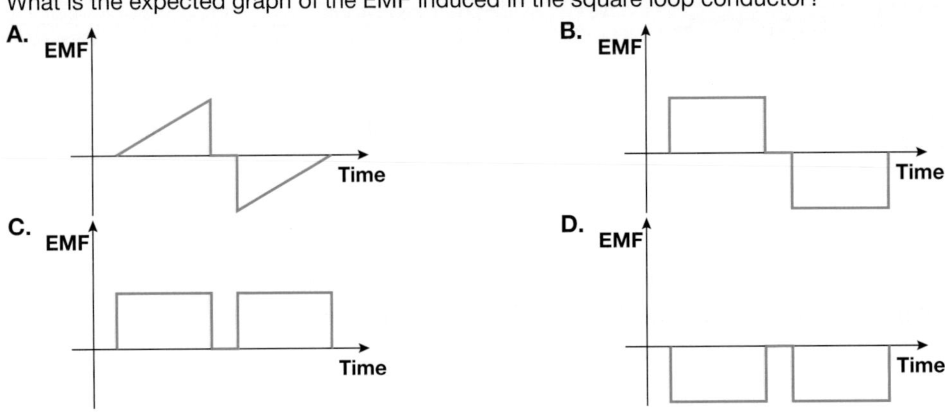

Use the following information to answer questions 4 and 5:
A permanent magnet is allowed to drop, entirely due to gravity, through a circular metal ring with its south pole going down first, as shown in the following diagram:

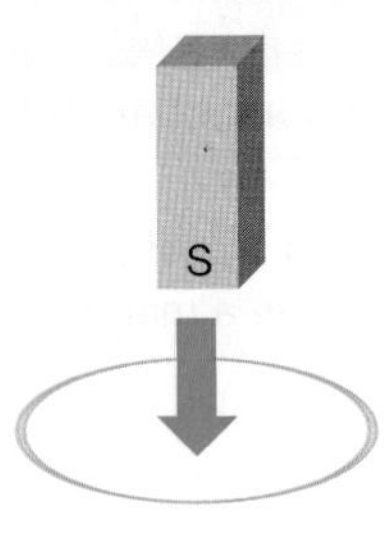

4. All of the following actions will increase the magnitude of the EMF generated in the metal ring except:
   A. increasing the strength of the magnet.
   B. dropping the magnet from a greater height.
   C. using a different metal with lower electrical resistance.
   D. increasing the size of the circle.
5. Viewed from above the ring, what is the direction of the generated current, if any?
   A. Clockwise
   B. Anticlockwise
   C. No current generated, only an EMF
   D. Cannot be determined, as additional information needed; for example, the resistance of the metal ring

6. Which of the following statements is correct?

   A. Slip rings maintain a continuous electrical connection with the spinning loop and are used when a DC output is required.
   B. A split ring commutator is a device used to convert alternating current into direct current.
   C. Motors convert mechanical energy into electrical energy.
   D. A split ring commutator reverses the direction of the coil's current every revolution of the loop.

Use the following information to answer questions 7 and 8:
A solenoid is used to create a magnetic field to the left of a circular metal ring, as shown in the following diagram. When the solenoid circuit is switched on, the solenoid becomes 'active'.

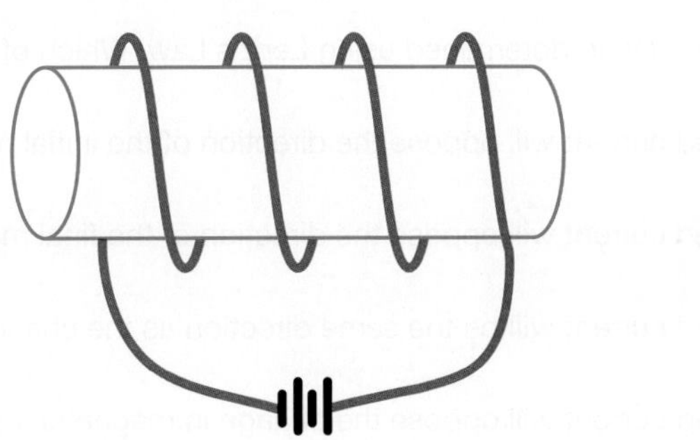

7. The solenoid circuit is initially active and then it is switched off. Viewed from the left of the ring, what is the direction of the generated current, if any?
   A. Clockwise
   B. Anticlockwise
   C. No current generated as the solenoid is no longer active
   D. Cannot be determined, additional information needed, for example, the strength of the solenoid's magnetic field
8. Which of the following actions will *not* induce an EMF in the metal ring?
   A. Switching on an inactive solenoid circuit
   B. Switching off an active solenoid circuit
   C. Moving an active solenoid circuit closer to the metal ring
   D. Moving an inactive solenoid circuit further away from the metal ring
9. A conducting coil is rotating at a constant rate inside a magnetic field with a constant field strength, producing an EMF. The rate of rotation is now reduced to half the rate it was before. Which of the following options best describes the period and the EMF produced by the new rate of rotation?
   A. The period is longer and the EMF produced is higher.
   B. The period is longer and the EMF produced is lower.
   C. The period is shorter and the EMF produced is higher.
   D. The period is shorter and the EMF produced is lower.

Use the following information to answer questions 10 and 11:
A bar magnet is travelling to the right at velocity ***v*** and as it approaches a metal ring, it induces a current in the metal ring.

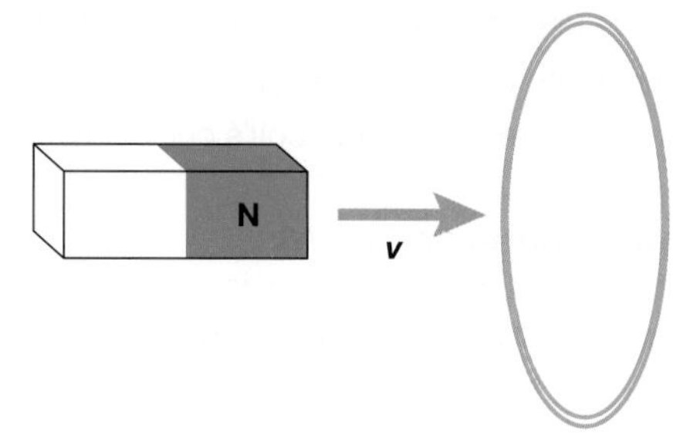

10. The direction of the induced current may be determined using Lenz's Law. Which of the following statements is the best explanation of Lenz's Law?

A. The magnetic field of the induced current will oppose the direction of the initial magnetic flux through the ring.
B. The magnetic field of the induced current will oppose the direction of the final magnetic flux through the ring.
C. The magnetic field of the induced current will be the same direction as the change in magnetic flux through the ring.
D. The magnetic field of the induced current will oppose the change in magnetic flux through the ring.

11. All of the following actions will increase the size of the current in the metal ring except:

A. increasing the strength of the magnet.
B. increasing the velocity of the magnet.
C. increasing the resistance of the metal ring.
D. increasing the diameter of the metal ring.

Use the following information to answer questions 12, 13 and 14:
A mobile phone charger is comprised of a step-down transformer from 240 V_{RMS} AC to 18 V_{RMS} AC. The charger consumes 54 W_{RMS} of power when it is charging the phone. Assume that the phone charger is ideal.

12. If the secondary coil has 360 turns, what is the number of turns in the primary coil?

A. 13 turns
B. 27 turns
C. 4680 turns
D. 4800 turns

13. Calculate the RMS current in the secondary coil.

A. 0.23 A
B. 0.32 A
C. 3.0 A
D. 4.2 A

14. Determine the peak current in the primary coil.

A. 0.23 A
B. 0.32 A
C. 3.0 A
D. 4.2 A

15. Which of the following statements best explains why transformers are AC devices and not DC devices?

A. Direct currents are fixed and cannot be stepped up or stepped down.
B. An alternating current may generate a changing magnetic flux.
C. A direct current generates a weaker magnetic field than an equivalent alternating current.
D. The iron cores of transformers work best with alternating currents.

16. Which of the following statements best describes the role of the iron core in a transformer?
    A. The iron core enables the primary coil to transfer its voltage over to the secondary coil.
    B. The iron core channels the magnetic flux between the primary coil and the secondary coil.
    C. The iron core has a higher resistance than the copper conductors used in the primary and secondary coils.
    D. The iron core focuses the current from the primary coil to reduce the losses in the secondary coil.

Use the following information to answer questions 17 to 20:
A small rural town is powered by a hydroelectric power station 500 m away. The power station produces 84 kW RMS power at 230 V_{RMS}. The voltage is stepped up with a step-up transformer with a turns ratio of 1:20 before transmission to the town. The total resistance of the transmission line is 6 Ω. At the town, the transmission line voltage is stepped down with a step-down transformer with a turns ratio of 20:1 before distribution to individual homes and businesses.

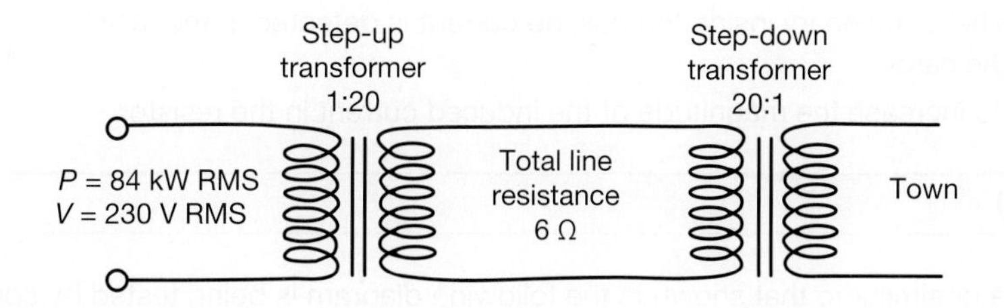

17. Between the power station and the town, which of the following is the expected RMS current in the transmission line closest to?
    A. 7.3 A
    B. 18 A
    C. 3.7×10^2 A
    D. 7.3×10^2 A
18. Which of the following is the expected voltage available to consumers at the town after the step-down transformer?
    A. Equal to 230 V
    B. Less than 230 V
    C. Greater than 230 V
    D. Unknown, since there is insufficient information to determine the voltage
19. If the turns ratio for the step-up transformer is changed to 1:30, and the turns ratio for the step-down transformer is changed to 30:1, what is likely to happen to the transmission line current and the voltage available to consumers at the town?
    A. The transmission line current will increase, while the voltage available will decrease.
    B. The transmission line current will increase, while the voltage available will increase.
    C. The transmission line current will decrease, while the voltage available will decrease.
    D. The transmission line current will decrease, while the voltage available will increase.
20. Which of the following actions will not reduce the transmission line loss?
    A. Reducing the total resistance of the transmission line
    B. Reducing the turns ratio of the step-up transformer
    C. Reducing the transmission line current
    D. Reducing the length of the transmission line

SECTION B — Short answer questions

Question 21 (7 marks)

The diagram below shows a permanent magnet near a coil of wire. The coil is connected to a resistor.

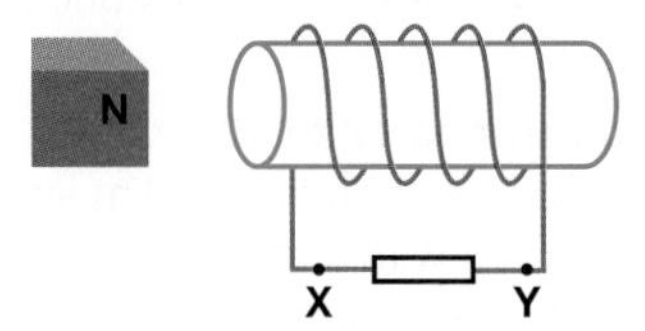

a. Which way will a current flow through the resistor when the magnet moves towards the coil from the position shown? Explain your answer. **(3 marks)**

b. When the magnet is held stationary inside the coil, no current is detected in the resistor. Explain why this is the case. **(2 marks)**

c. Describe two ways to increase the magnitude of the induced current in the resistor. **(2 marks)**

Question 22 (4 marks)

A model electric generator similar to that shown in the following diagram is being tested by some students. The coil consists of 50 turns of wire and has an area of 6.0×10^{-3} m^2. The magnetic field between the poles of the magnets is measured to be 0.15 T.

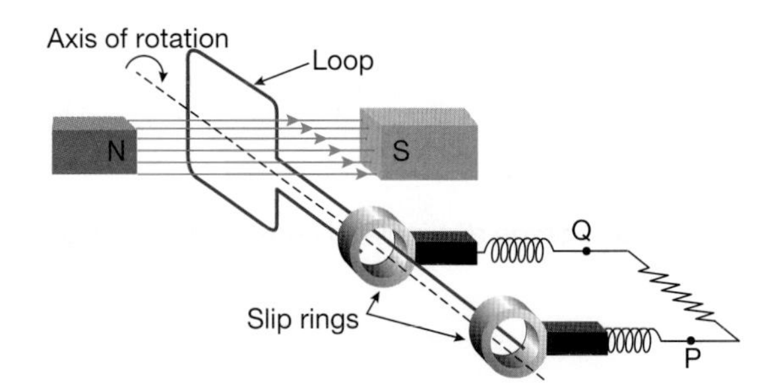

a. Calculate the magnitude of the magnetic flux threading the coil when it is in the position shown. **(1 mark)**

b. The students rotate the coil in a clockwise direction as indicated in the diagram. The coil is rotated a quarter turn (90°). Using the points P and Q, identify which direction the current flows through the external resistor. **(1 mark)**

c. As the students rotate the coil through an angle of 90°, the magnitude of the average induced EMF is 40 mV. What time interval is required for this rotation to produce an average induced EMF of 40 mV during a quarter turn? **(2 marks)**

Question 23 (6 marks)

A step-up transformer has a primary voltage of 240 V_{RMS} and an output RMS voltage of 1200 V. The primary coil of the transformer has 75 turns of wire and carries a current of 4.0 A.

a. How many turns are on the secondary coil of the transformer? **(2 marks)**

b. What is the current in the secondary coil of the transformer? **(2 marks)**

A small 1000-W generator supplies power across some transmission lines. These lines also carry a current of 4.0 A and have a resistance of 3.5 Ω.

c. Calculate the power loss in the cables and express this as a percentage of the power supplied by the generator. **(2 marks)**

Question 24 (3 marks)

A transformer is designed to give 6.0-V output from a 240-V input. A 1.0-A fuse is placed in series with the primary coil.

a. Determine the ratio $\frac{N_{\text{primary}}}{N_{\text{secondary}}}$ where N_{primary} is the number of turns of wire in the primary coil, and $N_{\text{secondary}}$ is the number of turns of wire in the secondary coil. **(1 mark)**

b. What is the maximum power that could be delivered in the primary coil? **(1 mark)**

c. What is the maximum current obtained from the secondary coil? **(1 mark)**

UNIT 3 | AREA OF STUDY 3

PRACTICE SCHOOL-ASSESSED COURSEWORK

ASSESSMENT TASK — COMPARISON AND EVALUATION OF TWO SOLUTIONS TO A PROBLEM, TWO EXPLANATIONS OF A PHYSICS PHENOMENON OR CONCEPT, OR TWO METHODS AND/OR FINDINGS FROM PRACTICAL ACTIVITIES

In this task, you will be required to compare the use of transformers and AC voltages in electricity transmission to the use of copper or silver in transmission lines to reduce energy losses.

- This practice SAC requires you to compare and evaluate two solutions to reduce energy losses during electricity transmission and apply one of the solutions to a real-life situation. A structured set of questions is supplied to assist you.
- You may use the VCAA Physics formula sheet and a scientific calculator for this task.

Total time: 55 minutes (5 minutes reading, 50 minutes writing)

Total marks: 50 marks

TRANSMISSION OF ELECTRICITY — THE IMPORTANCE OF TRANSFORMERS AND AC VOLTAGES TO REDUCE ENERGY LOSSES

A student interested in electricity generation and transmission has gathered several extracts from articles published on websites devoted to physics learning and has made some notes recording useful information.

Read the information gathered by the student and answer the questions that follow to help you compare the use of transformers and AC voltages to the use of copper or silver in transmission lines to reduce energy losses.

The first extract is as follows:

The most common electricity generation processes can be simplified and summarised as using motion-based power (wind or falling-water kinetic energy) or fuel-based power (coal, natural gas, nuclear fuel) plus the motion-based power of steam to power turbines that produce electricity via electromagnetic induction.

In addition, photovoltaic cells and electrochemical cells generate electricity via the photoelectric effect and via redox chemical reactions respectively.

Generally, wind farms, hydroelectric dams, coal-fired or nuclear-fired power stations and even solar farms are not close to where the electrical power is consumed. Electricity needs to be transmitted. Electricity transmission comes at a cost though, as the power loss in transmission lines increases with the resistance of the transmission line, and the intensity of the current in the line.

1. In terms of the resistance, R, of the transmission cable and the current, I, in the transmission line, explain the size of the power dissipated in the cable.
2. Suggest two ways in which the power loss can be decreased.

A second extract is as follows:

High-voltage transmission lines are now commonly made from a central core of galvanised steel wire, to provide strength to the cables, surrounded by strands of lightweight yet durable aluminium to conduct the current. Copper is an even better electrical conductor than aluminium and was the preferred material for transmission lines in earlier days. However, the cost of copper and its high density are major drawbacks.

After reading this extract, the student gathered the following information:

Resistance R (Ω), of a cable of length L metres, cross-section A (m^2) and resistivity ρ (Ω m):

$$R = \frac{\rho L}{A}$$

Metal	Resistivity (Ω m) at 20 °C	Density (kg cm^3) at 20 °C	Cost ($/kg)
Aluminium	2.65×10^{-8}	2.71	0.80
Copper	1.68×10^{-8}	8.96	13.05
Silver	1.59×10^{-8}	10.49	1179.00

3. Copy and complete the table below, prepared by the student, for a 100-km long cable with a 2.0-cm diameter.

Metal	Resistance of cable (Ω) at 20 °C	Mass of cable (kg) at 20 °C	Cost of cable ($)
Aluminium		3405	2724
Copper		11 259	146 936
Silver		13 182	3.21×10^6

4. Assume a 100-km long cable with a 2.0-cm diameter is used to transmit 5.0×10^6 W with 110 kV$_{RMS}$. For both copper and aluminium, calculate the current in the transmission cable.
5. Compare the power loss the aluminium and copper cables for this power and voltage.
6. Write a paragraph to explain why aluminium has overtaken copper as the preferred material for transmission lines.
7. Assume an aluminium cable with a total resistance of 6 Ω is used to transmit 5.0×10^6 W at different voltages. Copy and complete the following table.

Voltage (V$_{RMS}$)	Intensity in cable (A)	Power loss in cable (W)	$\frac{\text{Power loss}}{\text{Power supplied}}$ (%)
240			
33×10^3			
66×10^3			
110×10^3			

8. Explain why high voltages are more suitable than low voltages for the transmission of electricity.

A third extract is as follows:

Transformers can be used in conjunction with AC voltages applied to the primary coil to either increase or decrease the AC potential difference produced at the terminals of the secondary coil. When a step-up transformer is used, the output voltage is larger than the input voltage but with a smaller output current. When a step-down transformer is used, the output voltage is smaller than the input voltage but with a larger output current. Transformers do not work with steady DC voltages applied to the primary coil. A step-up transformer can be used to substantially lower the power loss in a transmission cable without compromising the amount of power delivered to the end user.

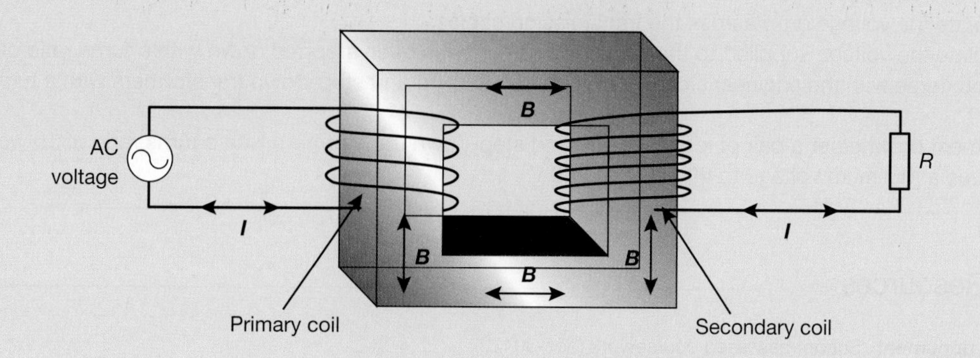

Useful equations:

$$\frac{V_s}{V_p} = \frac{N_s}{N_p}$$

$$V_s I_s = V_p I_p$$

$$V = IR$$

$$P = VI$$

9. The article also states that transformers do not work with steady DC voltages. Explain why this is the case.
10. Comment on the final sentence of the third extract in terms of how a step-up transformer could achieve this outcome.

A week later, the student is reading an article in an engineering magazine about the development of a small independent power supply to a small, remote town in Victoria. The generator uses an array of solar cells connected to an inverter to produce a 240 V_{RMS} AC supply, but at a distance from the small town. The article states the following:

A generator and transmission system is being designed to supply electrical power to a small country town two kilometres away from the generator. The system consists of a generator, two transformers (T1 *located close to the generator and* T2 *located close to the small town) and a transmission line as shown in the following diagram:*

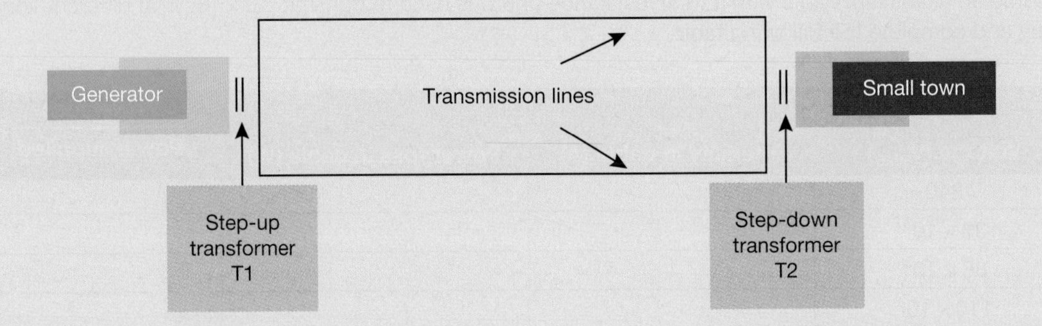

During the day, the generator is expected to supply 6.0×10^5 W *of power with* 240 V_{RMS} *at the primary terminals of* T1. *The power will be transferred via an ideal step-up transformer* T1 *using aluminium transmission lines with a total resistance of* 2.0 Ω *and, finally, an ideal step-down transformer* T2, *with the same turns ratio.*

The town engineer is considering which turns ratio to use for both of the transformers and realises that 220 V is sufficient to run equipment in the city without loss of performance in the appliances used by the residents. The engineer calculates that a turns ratio of 20 would be suitable.

In the following questions, all values calculated are RMS values.

11. Show that the expected current in the primary coil of transformer T1 is 2500 A.
12. By determining the current in the transmission line, calculate the power loss in the transmission line for a turns ratio of 20.
13. Calculate the voltage drop across the transmission cables.
14. Calculate the voltage supplied to the small town after it has been stepped down with a turns ratio of 20.
15. Do you agree with the engineer's choice of a pair of step-up and step-down transformers with a turns ratio of 20?
16. Comment on whether a pair of ideal step-up and step-down transformers with a turns ratio of 15 would produce a minimum voltage to the town of 220 V.

Resources

Digital document School-assessed coursework (doc-39423)

UNIT

4 How have creative ideas and investigation revolutionised thinking in physics?

AREA OF STUDY 1

How has understanding about the physical world changed?

OUTCOME 1

Analyse and apply models that explain the nature of light and matter, and use special relativity to explain observations made when objects are moving at speeds approaching the speed of light.

AREA OF STUDY 2

How is scientific inquiry used to investigate fields, motion or light?

OUTCOME 2

Design and conduct a scientific investigation related to fields, motion or light, and present an aim, methodology and method, results, discussion and a conclusion in a scientific poster.

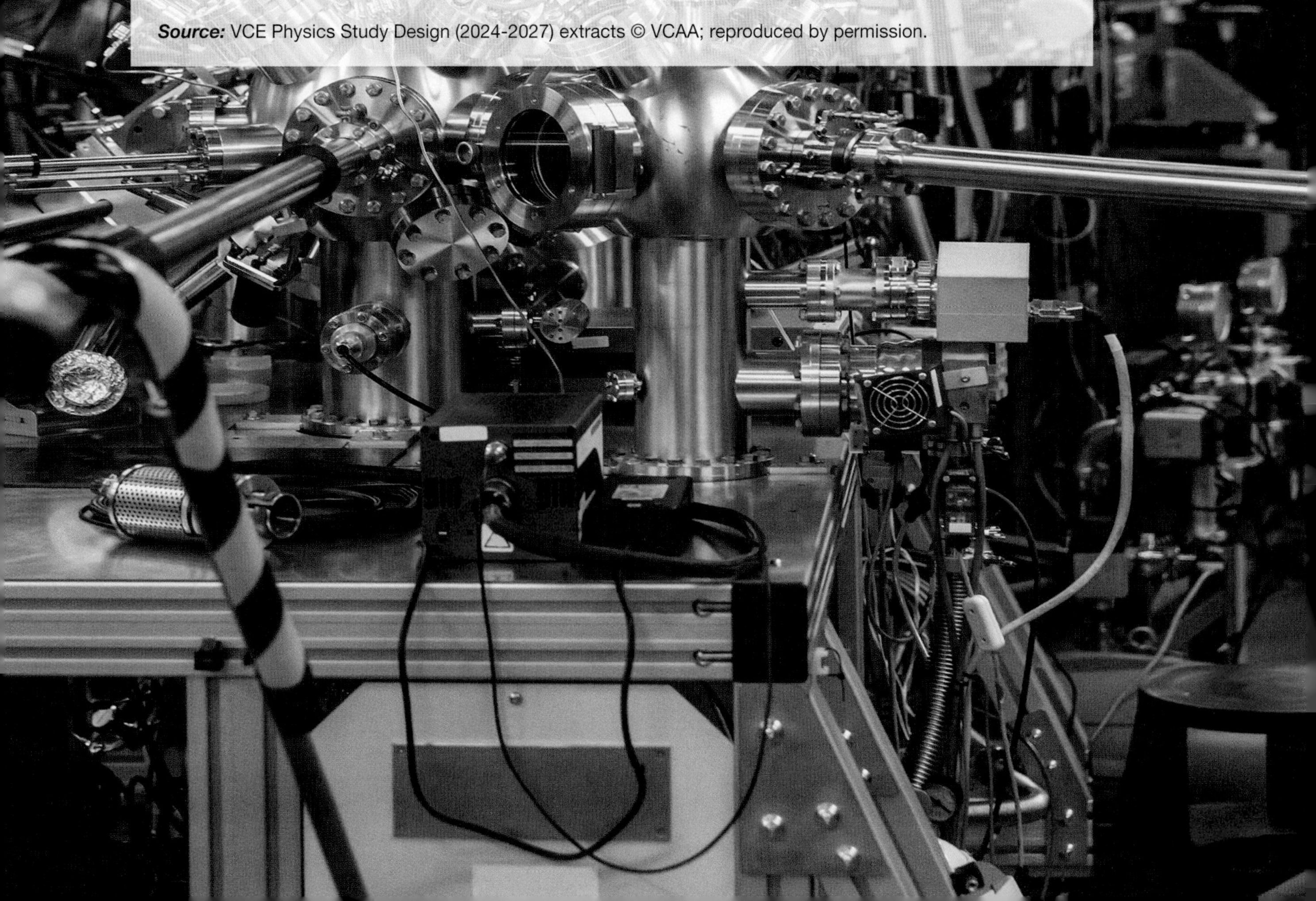

8 Light as a wave

KEY KNOWLEDGE

In this topic, you will:

- describe light as an electromagnetic wave that is produced by the acceleration of charges, which in turn produces changing electric fields and associated changing magnetic fields
- identify that all electromagnetic waves travel at the same speed, c, in a vacuum
- explain the formation of a standing wave resulting from the superposition of a travelling wave and its reflection
- analyse the formation of standing waves (only those with nodes at both ends is required)
- investigate and explain theoretically and practically diffraction as the directional spread of various frequencies with reference to different gap width or obstacle size, including the qualitative effect of changing the $\frac{\lambda}{w}$ ratio and apply this to limitations of imaging using electromagnetic waves
- explain the results of Young's double slit experiment with reference to:
  - evidence for the wave-like nature of light
  - constructive and destructive interference of coherent waves in terms of path differences: $n\lambda$ and $\left(n+\frac{1}{2}\right)\lambda$ respectively, where $n = 0, 1, 2, \ldots$
  - effect of wavelength, distance of screen and slit separation on interference patterns: $\Delta x = \frac{\lambda L}{d}$ when $L \gg d$.

PRACTICAL WORK AND INVESTIGATIONS

Practical work is a central component of VCE Physics. Experiments and investigations, supported by a **practical investigation eLogbook** and **teacher-led video**, are included in this topic to provide opportunities to undertake investigations and communicate findings.

EXAM PREPARATION

Access past VCAA questions and exam-style questions and their video solutions in every lesson, to ensure you are ready.

8.1 Overview

Hey students! Bring these pages to life online

Watch videos

Engage with interactivities

Answer questions and check results

Find all this and MORE in jacPLUS

Light has been considered a mystery from ancient times to the present day. Could light be a type of wave or does it propagate and transfer energy more like a particle? In this topic, you will investigate light as a wave based on its ability to both diffract and interfere. These phenomena, described by a mathematical model, indicate that light can be successfully modelled as a type of wave propagating at a constant speed through a variety of media. James Clerk Maxwell took advantage of new knowledge in the fields of magnetism and electricity, particularly electromagnetic induction, to propose that light was in fact a self-propelled transverse wave consisting of dual electric and magnetic fields oscillating in phase but perpendicular to each other. This mathematical model of light as a wave, combined with the mathematical model of matter as a particle, seemed to make physics complete as a subject area near the end of the nineteenth century. Observations in the twentieth century cast this simplistic notion into doubt resulting in a revolution in the modelling of nature requiring the use of both waves and particles for both light and matter.

FIGURE 8.1 Understanding light as a wave helps to explain many physical phenomena.

LEARNING SEQUENCE

on Resources

Solutions	Solutions — Topic 8 (sol-0822)
Practical investigation eLogbook	Practical investigation eLogbook — Topic 8 (elog-1639)
Digital documents	Key science skills — VCE Physics Units 1–4 (doc-36950) Key terms glossary — Topic 8 (doc-37179) Key ideas summary — Topic 8 (doc-37180)
Exam question booklet	Exam question booklet — Topic 8 (eqb-0105)

8.2 Light as a wave

KEY KNOWLEDGE

- Identify the amplitude, wavelength, period and frequency of waves
- Calculate the wavelength, frequency, period and speed of travel of waves using: $\lambda = \frac{v}{f} = vT$
- Describe light as an electromagnetic wave, which is produced by the acceleration of charges, which in turn produces changing electric fields and associated changing magnetic fields
- Identify that all electromagnetic waves travel at the same speed, c, in a vacuum
- Compare the wavelength and frequencies of different regions of the electromagnetic spectrum, including radio, microwave, infrared, visible, ultraviolet, X-ray and gamma

8.2.1 Review of wave properties

- Light propagates away from its source as a type of wave. The key evidence was that light diffracted, it was capable of interference and slowed down when passing from air into water and other transparent materials.
- A particle model for light was incapable of explaining these phenomena.
- At the end of the century, James Clerk Maxwell produced a highly successful model where light was treated as a transverse wave consisting of both electric and magnetic vibrating fields, requiring no medium to move. Light was also shown that it wasn't a longitudinal wave.

To better appreciate these concepts, a review of wave theory from Unit 1 is essential. (See topics 1 and 2 in *Jacaranda Physics 1 VCE Units 1&2 Fifth Edition* for a more-in depth review.)

The **frequency** of a **periodic wave** is the number of times that it completes a cycle per second. Frequency is measured in hertz (Hz) and 1 Hz = 1 s^{-1}. Frequency is represented by the symbol f.

The **period** of a periodic wave is the time it takes a source to produce a complete **wave**. This is the same as the time taken for a complete wave to pass a given point. The period is measured in seconds and is represented by the symbol T. The frequency f and period T are related.

The period of a wave is the reciprocal of its frequency. For example, if five complete waves pass every second, that is, $f = 5$ Hz, then the period (the time for one complete wavelength to pass) is $\frac{1}{5} = 0.2$ seconds. In other words:

frequency a measure of how many times per second an event happens, such as the number of times a wave repeats itself every second

periodic wave a disturbance that repeats itself at regular intervals

period the time it takes a source to produce one complete wave (or for a complete wave to pass a given point)

wave transfer of energy through a medium without any net movement of matter

transverse wave wave for which the disturbance is at right angles to the direction of propagation

$$f = \frac{1}{T} \Rightarrow T = \frac{1}{f}$$

where: f is the frequency of the wave, in Hz

T is the period of the wave, in s.

A displacement–time graph, as shown in figure 8.2, tracks the movement of a single point on a **transverse periodic wave** as a function of time. It shows how the displacement of a single point on the wave varies. The period of the wave can be identified from this graph.

The **amplitude** of a wave is the size of the maximum disturbance of the medium from its normal state. The units of amplitude vary with wave type. For example, in sound waves the amplitude is measured in the units of pressure, whereas the amplitude of a water wave would normally be measured in centimetres or metres.

In figure 8.2 the wave is the propagation of energy. The diagram shows a single-point particle oscillating up and down while the wave is moving in a perpendicular direction.

FIGURE 8.2 Displacement–time graphs: the movement of a single point on a transverse wave over time

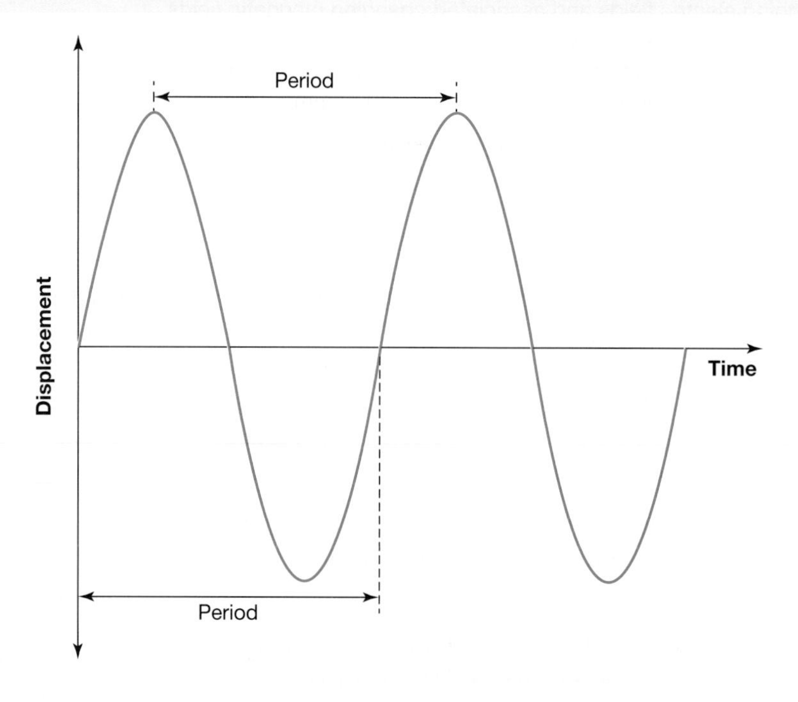

The displacement of all particles along the length of a transverse wave can be represented in a displacement–distance graph as shown in figure 8.3. A displacement–distance graph is like a snapshot of the wave at an instant in time. The amplitude and wavelength of the wave can be easily identified from this graph.

The **wavelength** is the distance between successive corresponding parts of a periodic wave. The wavelength is also the distance travelled by a periodic wave during a time interval of one period. For transverse periodic waves, the wavelength is equal to the distance between successive crests (or troughs).

amplitude size of the maximum disturbance of the medium from its normal state

wavelength distance between successive corresponding parts of a periodic wave

FIGURE 8.3 Displacement–distance graphs: particle displacements along a transverse wave

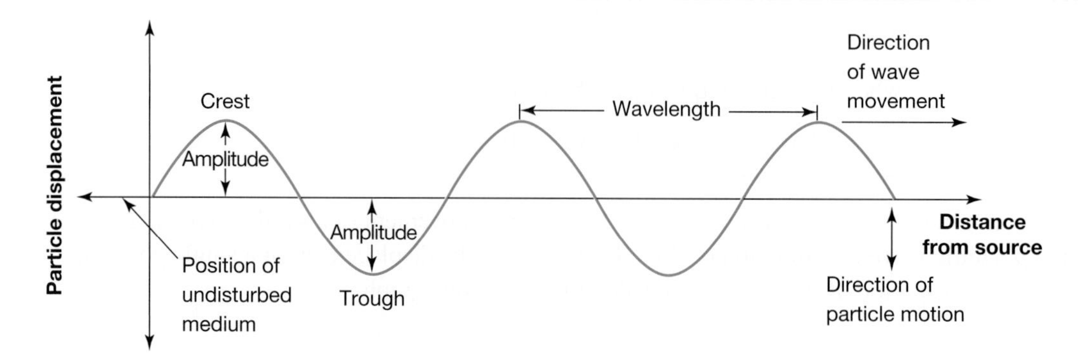

For **longitudinal periodic waves**, such as sound waves (see figure 8.4), the wavelength is equal to the distance between two successive compressions (regions where particles are closest together) or rarefactions (regions where particles are furthest apart). Wavelength is represented by the symbol λ (lambda).

longitudinal periodic wave wave for which the disturbance is parallel to the direction of propagation

FIGURE 8.4 Longitudinal waves in **a.** a slinky **b.** air

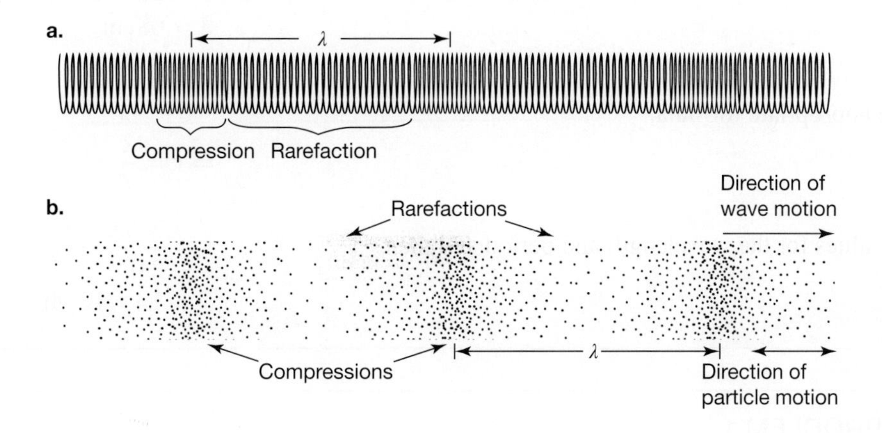

8.2.2 The speed of a wave

Mechanical waves travel away from their source at a speed dictated by the properties of the medium through which they travel and transmit their energy. Light, which is an electromagnetic wave, can travel in a vacuum. For all waves, the speed, v, of a periodic wave is related to the frequency, period and wavelength of the wave. In a time interval of one period, T, the wave travels a distance of one wavelength, λ. Thus:

$$\text{speed} = \frac{\text{distance}}{\text{time}} = \frac{\lambda}{T} = \frac{\lambda}{\frac{1}{f}} = f\lambda$$

This relationship can be written as:

$$v = f\lambda = \frac{\lambda}{T}$$

where: v is the speed of the wave, in m s^{-1}

f is the frequency of the wave, in Hz

λ is the wavelength of the wave, in m

T is the period of the wave, in s.

$v = \lambda \times f$

$\lambda = \frac{v}{f}$

$f = \frac{v}{\lambda}$

This relationship, $v = f\lambda$, is sometimes referred to as the universal wave equation.

The frequency of a periodic wave is determined by the source of the wave. The speed of a periodic wave is determined by the medium through which it is travelling and does not depend on the source frequency or the amplitude of the wave.

tlvd-9005

SAMPLE PROBLEM 1 Calculating the speed of a sound wave

What is the speed of a wave if it has a period of 2.0 ms and a wavelength of 68 cm?

THINK	WRITE
1. Note down the known variables in their appropriate units. Time must be expressed in seconds and length in metres.	$T = 2.0\,\text{ms} = 2.0 \times 10^{-3}\,\text{s}$ $\lambda = 68\,\text{cm} = 0.68\,\text{m}$
2. Choose the appropriate formula.	$v = f\lambda \Rightarrow v = \frac{\lambda}{T}$
3. Substitute values for the wavelength and period and then solve for v.	$v = \frac{0.68\,\text{m}}{2.0 \times 10^{-3}\,\text{s}} = 3.4 \times 10^{2}\,\text{m s}^{-1}$

PRACTICE PROBLEM 1

What is the speed of a wave if it has a period of 1.50 ms and a wavelength of 51.0 cm?

tlvd-9006

SAMPLE PROBLEM 2 Calculating the wavelength of a sound wave

What is the wavelength of a sound of frequency 550 Hz if the speed of sound in air is 335 m s^{-1}?

THINK	WRITE
1. Note down the known variables in their appropriate units. Frequency must be expressed in hertz and speed in m s^{-1}.	$f = 550\text{ Hz}, v = 335\text{ m s}^{-1}$
2. Choose the appropriate formula.	$v = f\lambda \Rightarrow v = \frac{\lambda}{f}$
3. Substitute values for the frequency and speed and then solve for the wavelength.	$= \frac{335\text{ m s}^{-1}}{550\,\text{Hz}} = 0.609\,\text{m}$

PRACTICE PROBLEM 2

A siren produces a sound wave with a frequency of 587 Hz. Calculate the speed of sound if the wavelength of the sound is 0.571 m.

elog-1658

INVESTIGATION 8.1

online only

Investigating waves from a slinky spring

Aim

To observe and investigate the behaviour of waves (or pulses) travelling along a slinky spring

8.2.3 Electromagnetic waves

The question of light being a type of particle, as proposed by Sir Isaac Newton, or a wave, as proposed by Christian Huygens, was settled in the nineteenth century with experiments conducted by Thomas Young and others. The observation that light was both able to diffract and interfere, and studies in the refraction of light, all demonstrated that light is a type of wave and not a particle. Mechanical waves, such as sound and surface waves on water, require a medium to transmit their energy from the wave source to the surrounding environment. Yet light, clearly a type of wave, can pass through a vacuum. No medium is required.

FIGURE 8.5 An electromagnetic wave. The electric (***E***) and magnetic (***B***) fields are uniform in each plane but vary along the direction of the motion of the wave.

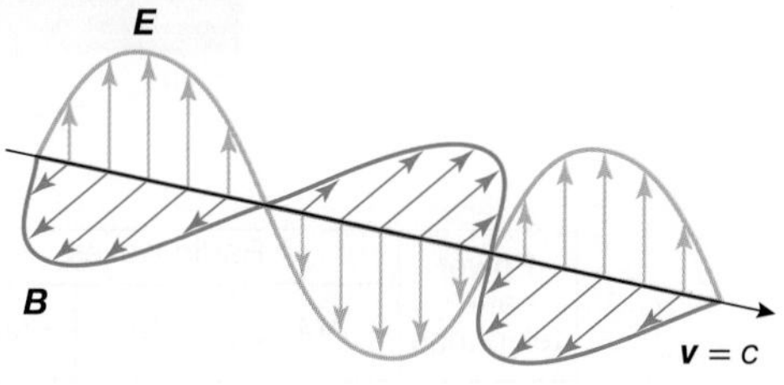

In 1864, James Clerk Maxwell (1831–1879) showed why light doesn't require a medium to propagate. He began with the ideas of electric and magnetic interactions where a changing magnetic flux caused an induced voltage and an associated electric field. It was known that a changing electric flux induced a magnetic field. Maxwell developed a theory predicting that an oscillating electric charge would produce an oscillating electric field, together with a magnetic field oscillating at right angles to the electric field. These inseparable fields would travel together through a vacuum like a wave. The speed of the wave would be constant whether the oscillations were rapid (high frequency and a short wavelength) or very slow (low frequency and a long wavelength). Also, the speed of these electromagnetic waves was independent of the amplitude of the wave. Small oscillations travelled at the same speed as large oscillations. Maxwell remarkably even predicted their speed to be 3×10^8 m s^{-1}, using known electric and magnetic properties of a vacuum. This is the speed of light! Maxwell had formulated a theory that explained how light was produced and travelled through space as electromagnetic waves.

How does an electromagnetic wave form when charges move?

- A charge moving at constant velocity cannot produce electromagnetic waves; the constant motion means there is no change in the electric and magnetic field of the moving charge.
- Electromagnetic waves are produced only when an electric charge undergoes acceleration, deceleration or change in direction.
- When an electrically charged particle vibrates, accelerates or changes direction, this causes a change in the electric field, which in turn creates a change in the magnetic field. This forms an electromagnetic wave.

Maxwell's theory applied not only to visible light (table 8.1) but also to other radiation that people cannot see, such as infrared and ultraviolet radiation. Furthermore, his electromagnetic model of light indicated that light could be described as a transverse wave. His model is used to explain light diffraction and interference phenomena.

TABLE 8.1 Frequency and wavelength of visible light colours

	Red	Orange	Yellow	Green	Cyan	Blue	Violet
Frequency ($\times 10^{12}$ hertz)	430	480	520	570	600	650	730
Wavelength (nanometres)	700	625	580	525	500	460	410

How do the parameters of light relate to your perception of light? It turns out that frequency is associated with your perception of colour in the visible portion of the electromagnetic spectrum, and that brightness is associated with the amplitude. Neither of these parameters influence the speed of light in a vacuum, this being 299 792 458 m s^{-1}.

FIGURE 8.6 Forms of radiation and their place in the electromagnetic spectrum. The visible portion of the spectrum is shown enlarged in the upper part of the diagram.

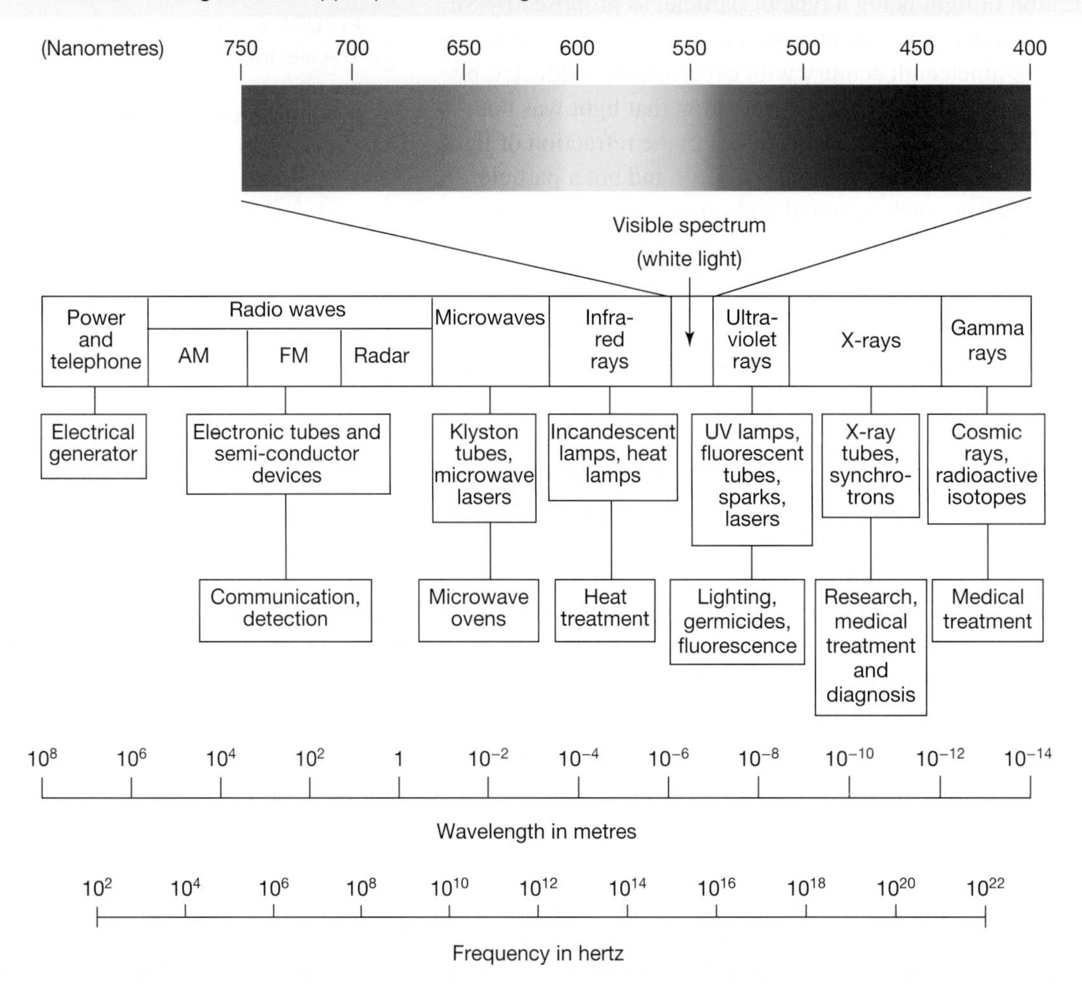

tlvd-9007

SAMPLE PROBLEM 3 Calculating the wavelength and period of light when given frequency

When light with a frequency of 5.60×10^{14} Hz travels through a vacuum, determine the values of:

a. its period

b. its wavelength (in nanometres).

The speed of light in a vacuum is 3.00×10^8 m s^{-1}.

THINK	WRITE
a. The period of a wave is the reciprocal of its frequency.	a. $T = \frac{1}{f}$ $= \frac{1}{5.60 \times 10^{14}}$ $= 1.79 \times 10^{-15}$ s The period of the light is 1.79×10^{-15} s.
b. 1. Use the relationship $\lambda = \frac{c}{f}$ to determine the wavelength.	b. $\lambda = \frac{c}{f}$ $= \frac{3.00 \times 10^8}{5.60 \times 10^{14}}$ $= 5.36 \times 10^{-7}$ m

2. The wavelength of visible light is usually expressed in nanometres (nm) where $1.0\,\text{nm} = 1.0 \times 10^{-9}\text{m}$.

$$\lambda = \frac{5.36 \times 10^{-7}}{1.0 \times 10^{-9}}\text{nm}$$
$$= 5.36 \times 10^{2}\ \text{nm}$$

The wavelength of the light is 536 nanometres.

PRACTICE PROBLEM 3

Determine the frequency and period of light with a wavelength of 450 nm.

8.2 Activities

learnon

Students, these questions are even better in jacPLUS

Receive immediate feedback and access sample responses

Access additional questions

Track your results and progress

Find all this and MORE in jacPLUS

8.2 Quick quiz on	8.2 Exercise	8.2 Exam questions

8.2 Exercise

1. The speed of sound in air is approximately 340 m s^{-1} and one musical note is produced that has a frequency of 256 Hz. Calculate:
   a. its wavelength
   b. its period
   c. how long it takes for the wave to travel 1 km.
2. Copy and complete the following table by using the wave formula.

v (m s^{-1})	f (Hz)	λ (m)
	500	0.67
	12	25
1500		0.30
60		2.5
340	1000	
260	440	

3. In a ripple tank, surface waves are produced by a dipper at a rate of 5 each half-second.
   a. Determine the frequency of the source and its period.
   b. The wavelength is measured to be 2.6 cm. What is the speed of the wave in m s^{-1}?
   c. The frequency is now doubled. Explain how the speed and the wavelength of the wave will change.
4. Take the speed of light in a vacuum as $c = 3.0 \times 10^{8}$ m s^{-1}.
   Calculate the period of orange light that has a frequency of 4.8×10^{14} Hz.
5. Take the speed of light in a vacuum as $c = 3.0 \times 10^{8}$ m s^{-1}.
   Microwaves have a frequency ranging from 1.0×10^{10} through to 1.0×10^{12} Hz. Determine the range of wavelengths associated with microwaves.

6. Take the speed of light in a vacuum as $c = 3.0 \times 10^8$ m s^{-1}.
   X-rays used by dentists have a wavelength of 2.7×10^{-11} m. Determine the frequency, and hence the period of the X-rays produced.
7. Power lines that carry electrical energy use an AC current. These cables emit electromagnetic radiation with a period of 20 ms.
   a. What is the frequency of the radiation emitted by power lines?
   b. What is the wavelength of this radiation?
8. Take the speed of light in a vacuum as $c = 3.0 \times 10^8$ m s^{-1}.
   When blue light of frequency 6.5 × 1014 Hz travelling through the air meets a glass prism, its speed decreases from 3.0×10^8 m s^{-1} to 2.0×10^8 m s^{-1}. Calculate the following:
   a. The wavelength of the blue light in the air
   b. The wavelength of the blue light in the glass
   c. the frequency of the blue light in the glass.
9. Describe how a moving charged particle that changes its motion produces an electromagnetic wave. Your answer should relate to both electric and magnetic fields around a moving charged particle.

8.2 Exam questions

Question 1 (1 mark)

Source: *VCE 2022 Physics Exam, Section A, Q.13; © VCAA*

MC A travelling wave produced at point A is reflected at point B to produce a standing wave on a rope, as represented in the diagram below.

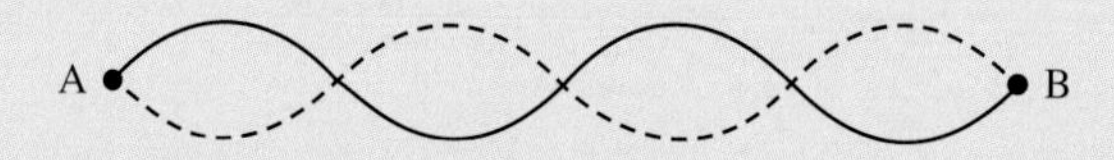

The distance between points A and B is 2.4 m. The period of vibration of the standing wave is 1.6s.
The speed of the travelling wave along the rope is closest to

A. 0.75 m s^{-1}
B. 1.0 m s^{-1}
C. 1.5 m s^{-1}
D. 2.0 m s^{-1}

Question 2 (1 mark)

Source: *VCE 2021 Physics Exam, Section A, Q13; © VCAA*

MC The diagram below shows part of a travelling wave.

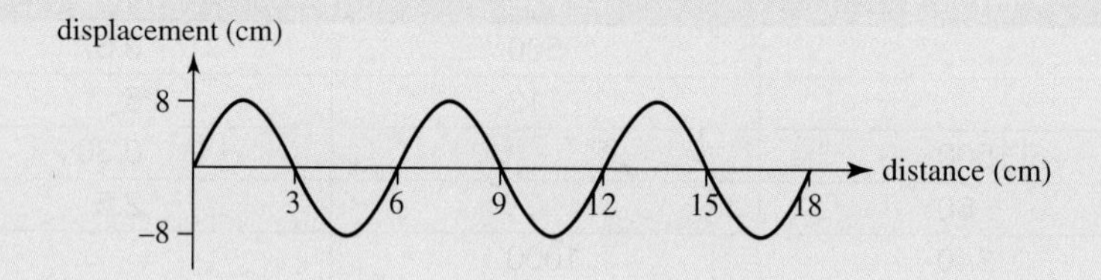

The wave propagates with a speed of 18 m s^{-1}.

Which of the following is closest to the amplitude and frequency of the wave?

A. 8 cm, 3.0 Hz
B. 16 cm, 3.0 Hz
C. 8 cm, 300 Hz
D. 16 cm, 300 Hz

Question 3 (1 mark)

Source: *VCE 2020 Physics Exam, Section B, Q.14;* © VCAA

Figure 13 shows a representation of an electromagnetic wave.

Correctly label Figure 13 using the following symbols.

E – electric field B – magnetic field c – speed of light λ – wavelength

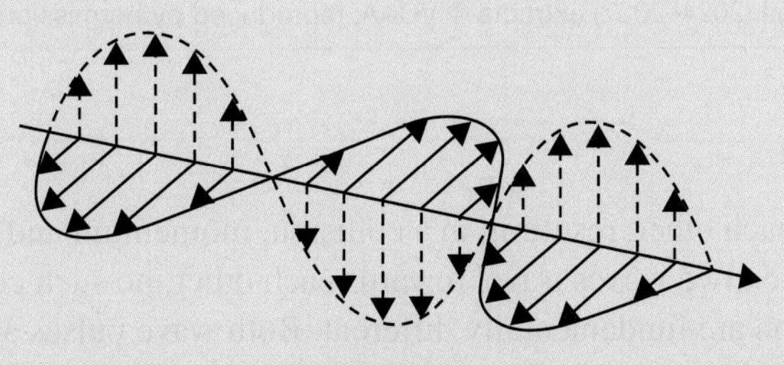

Figure 13

Question 4 (3 marks)

Source: *VCE 2019, Physics Exam, Section B, Q12;* © VCAA

A sinusoidal wave of wavelength 1.40 m is travelling along a stretched string with constant speed v, as shown in the figure below. The time taken point P on the string to move from maximum displacement to zero is 0.120 s.

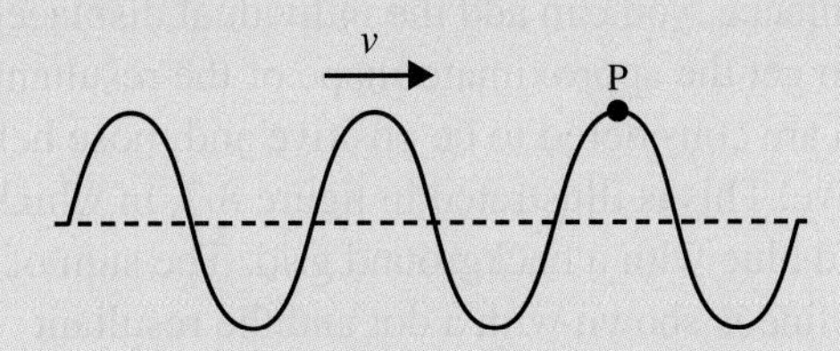

Calculate the speed of the wave, v. Give your answer correct to three significant figures. Show your working.

Question 5 (6 marks)

Source: *VCE 2022 Physics Exam, NHT, Section B, Q.12;* © VCAA

A transverse wave travels to the right along a length of string at a speed of $8.0\,\text{m}\,\text{s}^{-1}$. Figure 8 shows the wave at one instant in time.

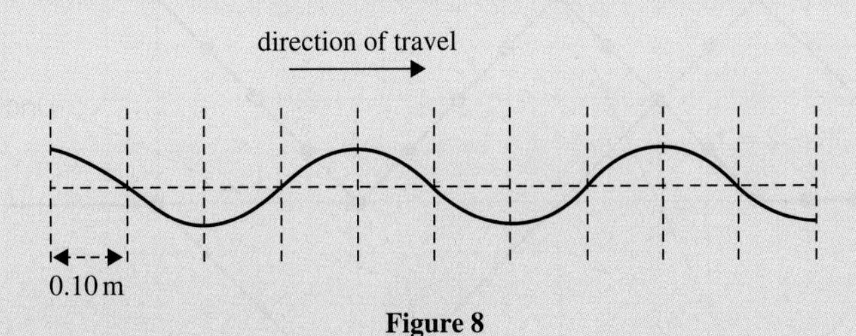

Figure 8

a. Calculate the frequency of the wave shown in Figure 8. **(2 marks)**

b. On Figure 8, draw the wave as it would appear 25 ms after the instant shown. Show any calculations and state any assumptions that you have made. **(2 marks)**

c. The wave source (not shown in Figure 8) is now adjusted to increase the frequency. **(2 marks)**
Explain the effect that this will have on the wavelength, stating any assumptions that you have made.

More exam questions are available in your learnON title.

8.3 Interference, resonance and standing waves

KEY KNOWLEDGE

- Explain the formation of a standing wave resulting from the superposition of a travelling wave and its reflection
- Analyse the formation of standing waves (only those with nodes at both ends is required)

8.3.1 Superposition

When particles travel towards each other, resulting in a collision, momentum and energy are transferred, unlike in the interaction of waves. When two waves travel towards each other, no such collision occurs. In this way, wave motion and particle motion are fundamentally different. Both wave pulses and periodic waves simply pass through each other with neither being modified by each other. This is observed when two pulses pass through each other on a spring. When the pulses are momentarily occupying the same part of the spring, the amplitudes of the individual pulses add together to give the amplitude of the total disturbance of the spring. This effect is known as **superposition** (positioning over). The superposition of waves is responsible for the phenomena known as diffraction and interference. These will be studied in subtopics 8.4 and 8.5 respectively.

The resultant disturbance can be found by applying the superposition principle: '*The resultant wave is the sum of the individual waves*'. For convenience, you can add the individual displacements of the medium at regular intervals where the pulses overlap to get the approximate shape of the resultant wave. Displacements above the position of the undisturbed medium are considered to be positive and those below the position of the undisturbed medium are considered to be negative. This is illustrated in figure 8.7, in which two pulses have been drawn in green and blue with a background grid. The sum of the displacements on each vertical grid line is shown with a dot and the resultant disturbance, drawn in black, is obtained by drawing a line through the dots.

superposition the adding together of amplitudes of two or more waves passing through the same point

FIGURE 8.7 Determining the shape of a resultant disturbance

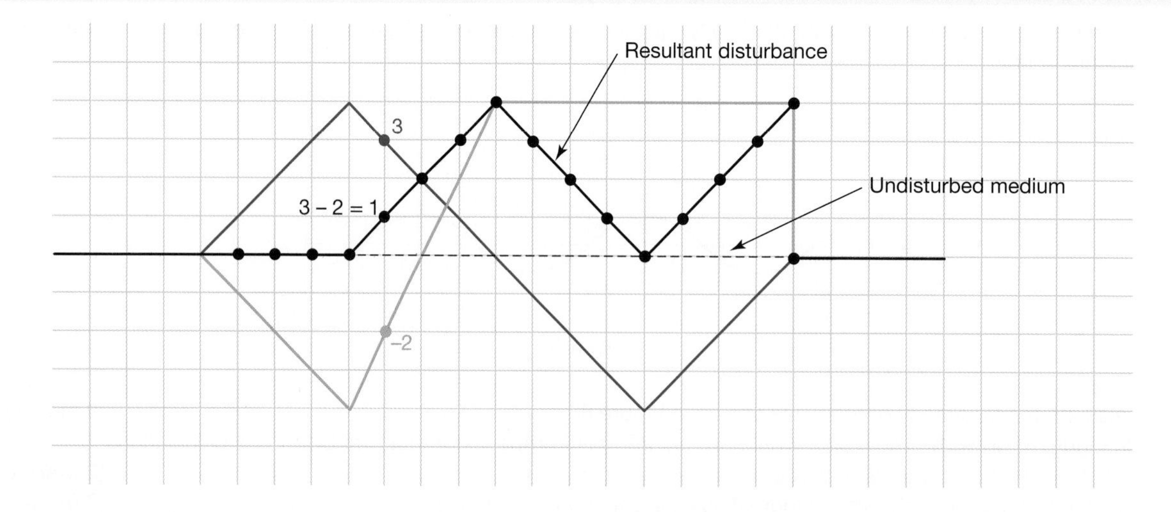

It is possible for a part or whole of a pulse to be 'cancelled out' by another pulse. When this effect occurs, destructive superposition, or **destructive interference**, is said to occur. When two pulses superimpose to give a maximum disturbance of a medium, constructive superposition, or **constructive interference**, is said to occur. This effect is shown in figure 8.8. However, when two pulses superimpose to cause a cancellation, destructive interference is said to occur. The superposition gives a small disturbance.

destructive interference
the addition of two wave disturbances to give an amplitude that is less than either of the two waves

constructive interference
the addition of two wave disturbances to give an amplitude that is greater than either of the two waves

FIGURE 8.8 **a.** Two pulses of different shapes approach each other on a spring. **b.** The pulses begin to pass through each other. **c.** As the pulses pass through each other, the amplitudes of the individual pulses add together to give a resultant disturbance. **d.** After passing through each other, the pulses continue on undisturbed.

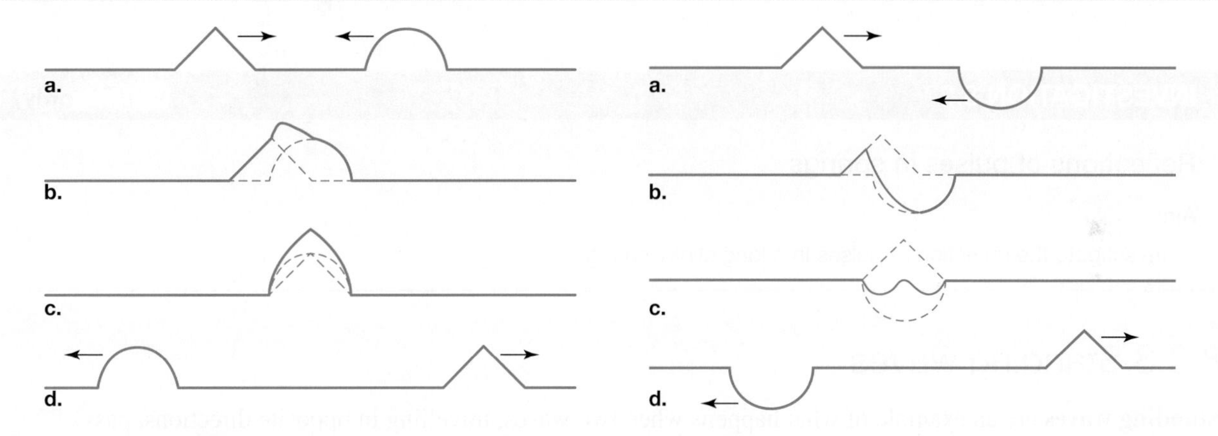

8.3.2 Reflection of waves

When waves arrive at a barrier, reflection occurs. Reflection is the returning of a wave into the medium in which it was originally travelling. When a wave strikes a barrier, or comes to the end of the medium in which it is travelling, part or all of the wave is reflected.

A wave's speed depends only on the medium in which it is travelling through; therefore, the speed of the reflected wave is unchanged. The wavelength and frequency of the reflected wave will also be the same as for the incident (original) wave.

Reflection of transverse waves in strings

When a string has one end fixed so that it is unable to move (e.g. when it is tied to a wall or is held tightly to the 'nut' at the end of a stringed instrument), the reflected wave will be inverted. This is called a change of phase. If the end is free to move, the wave is reflected upright and unchanged, so there is no change of phase. These situations are illustrated in figures 8.9 and 8.10.

FIGURE 8.9 Reflection of a transverse pulse on a string when **a.** and **b.** the end of the string is fixed (as in a guitar). With a fixed end, the reflected wave is inverted; there is a change of phase.

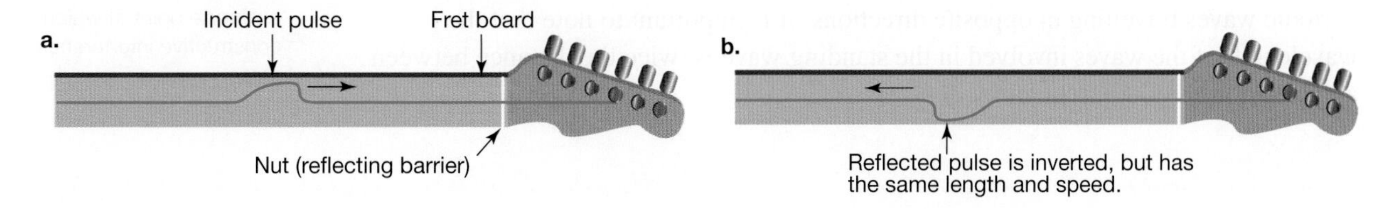

FIGURE 8.10 Reflection of a transverse pulse on a string when a. and b. the end of the string is free to move (as with a loop supported by a retort stand). With an open end, the reflected wave is not inverted; there is no change of phase.

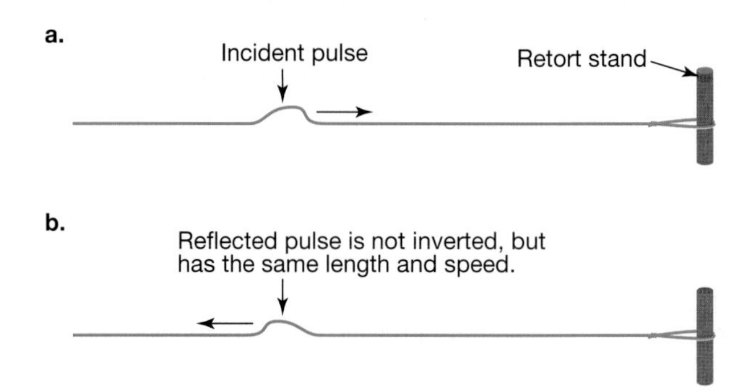

elog-1659
tlvd-10607

INVESTIGATION 8.2 online only

Reflections of pulses in springs

Aim

To investigate the reflection of pulses in a long slinky spring

8.3.3 Standing waves

Standing waves are an example of what happens when two waves, travelling in opposite directions, pass through each other. They can either interfere constructively or destructively. Interference, as applied to light, is explored in detail in subtopics 8.4 and 8.5. The phenomenon of standing waves is an example of interference in a confined space. The restriction may be a guitar string tied down at both ends, or a trumpet closed at the mouthpiece and open at the other end, or even a drum skin stretched tightly and secured at its circumference.

Transverse standing waves

When two periodic waves travelling in opposite directions, and of equal amplitude and frequency (and therefore wavelength), move through an elastic one-dimensional medium, constructive interference and destructive interference occur. In fact, destructive interference occurs at evenly spaced points along the medium and it happens all the time at these points. The medium at these points never moves. Such points in a medium where waves cancel each other at all times are called **nodes**. In between the nodes are points where the waves reinforce each other to give a maximum amplitude of the resultant waveform. This is caused by constructive interference. Such points are called **antinodes**.

When this effect occurs, the individual waves are undetectable. All that is observed are points where the medium is stationary and others where the medium oscillates between two extreme positions. There seems to be a wave but it has no direction of propagation. When this occurs, it is said to be a stationary or standing wave.

Figure 8.11 shows how standing waves are formed in a string by two continuous periodic waves travelling in opposite directions. It is important to note that the wavelength of the waves involved in the standing wave is twice the distance between adjacent nodes (or adjacent antinodes).

standing wave the superposition of two wave trains at the same frequency and amplitude travelling in opposite directions, also known as stationary waves as they do not appear to move through the medium. The nodes and antinodes remain in a fixed position.

node point at which destructive interference takes place

antinode point at which constructive interference takes place

Figure 8.12 shows the motion of a spring as it carries a standing wave. It shows the shape of the spring as it completes one cycle. The time taken to do this is one period (T). Note that:

i. at $t = \frac{T}{4}$ and at $t = \frac{3T}{4}$, the entire medium is momentarily undisturbed at all points

ii. adjacent antinodes are opposite in phase — when one antinode is a crest, those next to it are troughs

iii. there are points labelled N where there is no displacement at all times. These points are evenly spaced at precisely a distance $\frac{\lambda}{2}$.

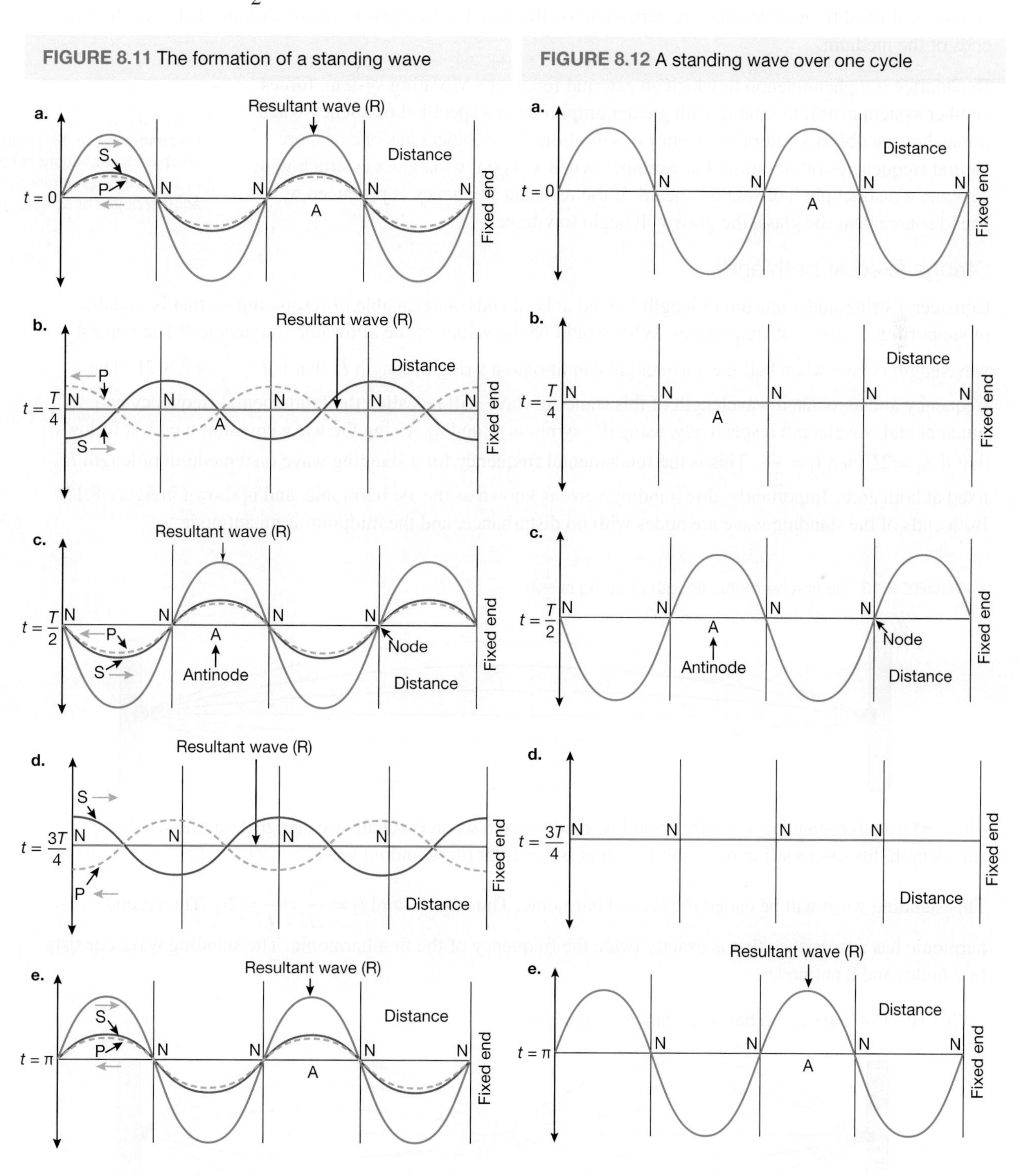

FIGURE 8.11 The formation of a standing wave

FIGURE 8.12 A standing wave over one cycle

8.3.4 Harmonics associated with standing waves

Standing wave theory can be applied to guitar and violin strings, for example, or to water waves reflecting off a barrier, even to suspension bridges with the road fixed to a pylon at both ends. You will learn that this is also vital for understanding the stability of atoms. The important physics here is that when waves are generated, in the case of a guitar by plucking, a violin by using a bow, or by air passing over a bridge, travelling waves in both directions are produced and will reflect off the fixed ends. The reflected waves interfere with each other, and their resultant becomes stationary, vibrating in place: standing waves are created. In all cases, the medium only resonates at fixed frequencies that are dependent on the speed of the waves and the distance between the two ends of the medium.

Resonance is a phenomenon in which an external force, or a vibrating system, forces another system near it to vibrate with greater amplitude at a specified frequency when it matches the object's natural frequency of vibration. Every object has one or more natural frequencies of vibration. For example, when a crystal wine glass is struck with a spoon, a distinct pitch of sound is heard. If the resonant frequency is produced by a sound source near the glass, the glass will begin to vibrate.

resonance when the amplitude of an object's oscillations is increased by the matching vibrations of another object or an external force

Strings fixed at both ends

Consider a string under tension of length L fixed at both ends, and capable of resonating — that is, capable of supporting a variety of frequencies. What would be the values of the allowable frequencies? The longest wavelength occurs when half the wavelength can fit onto a string of length L, that is $L = \frac{\lambda}{2}$ or $\lambda = 2L$. The frequency and associated wavelength of this standing wave will be called the fundamental frequency and fundamental wavelength respectively, using the symbols f_1 and λ_1. Using the wave equation $v = f\lambda$, it follows that if $\lambda_1 = 2L$ then $f_1 = \frac{v}{2L}$. This is the fundamental frequency for a standing wave on a medium of length L fixed at both ends. Importantly, this standing wave is known as the 1st harmonic, and is shown in figure 8.13. Both ends of the standing wave are nodes with no disturbance, and the midpoint is an antinode.

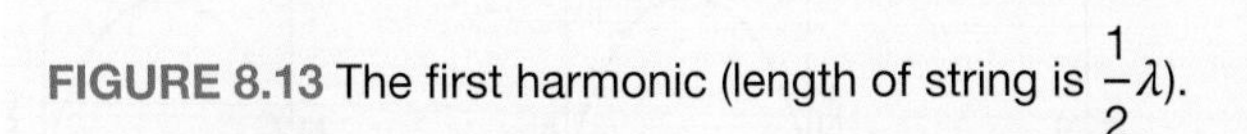

FIGURE 8.13 The first harmonic (length of string is $\frac{1}{2}\lambda$).

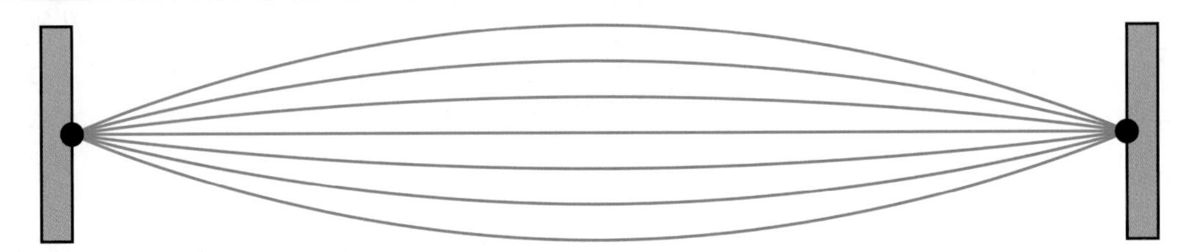

The next possible standing wave that could be made occurs when two half-wavelengths, or one whole wavelength, fits onto a string of length L — that is $\lambda = L$ for this standing wave.

This standing wave will be called the second harmonic. Thus $\lambda_2 = L$ and $f_2 = \frac{2v}{2L} = \frac{v}{L} = 2f_1$. The second harmonic has a frequency that is exactly twice the frequency of the first harmonic. The standing wave consists of 3 nodes and 2 antinodes.

FIGURE 8.14 The second harmonic (length of string is λ).

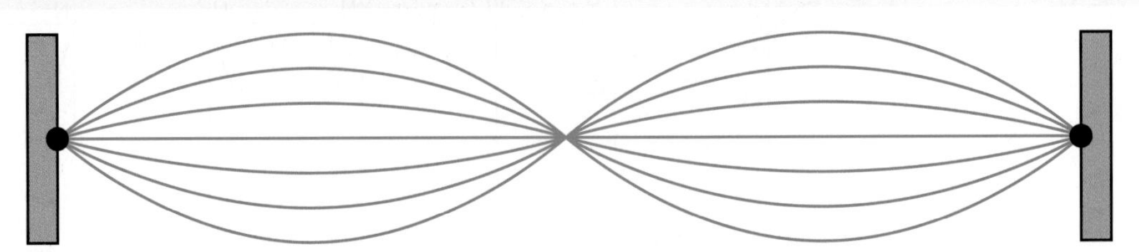

The third harmonic will show that a number pattern emerges for all harmonics. The third harmonic will be again fixed at both ends where 1.5 wavelengths will fit onto a string of length L — that is $L = \frac{3}{2}\lambda$, so $\lambda_3 = \frac{2L}{3}$.

FIGURE 8.15 The third harmonic (length of string is $\frac{3}{2}\lambda$).

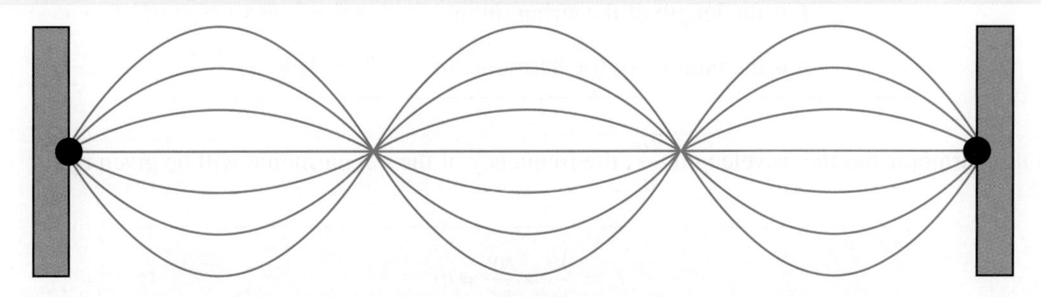

This generalisation for n harmonics is summarised in table 8.2.

TABLE 8.2 Standing waves on a string fixed at both ends

Harmonic	Number of nodes	Number of antinodes	Pattern	Resonant frequency: $f = \frac{v}{\lambda}$
First	2	1		$L = \frac{\lambda_1}{2}$, $f_1 = \frac{v}{2L}$
Second	3	2		$L = \frac{2\lambda_2}{2} = \lambda_2$ $f_2 = \frac{v}{L} = 2f_1$
Third	4	3		$L = \frac{3\lambda_3}{2}$, $f_3 = \frac{3v}{2L} = 3f_1$
Fourth	5	4		$L = \frac{4\lambda_4}{2}$, $f_4 = \frac{4v}{2L} = 4f_1$
***n*th**	$n + 1$	n		$L = \frac{n\lambda_n}{2}$, $f_n = \frac{nv}{2L} = nf_1$ $n = 1, 2, 3, 4, 5...$

In general, for two fixed ends, the *n*th harmonic will have a wavelength given by:

$$\lambda_n = \frac{2L}{n}$$

where: λ_n is the wavelength, in m

L is the length of the string, in m

n is the number of the harmonic ($n = 1, 2, 3, 4, 5 \ldots$)

Since the *n*th harmonic has the wavelength $\frac{2L}{n}$, the frequency of the *n*th harmonic will be given by:

$$f_n = \frac{V}{\lambda_n} = \frac{nv}{2L} = nf_1$$

where: f_n is the frequency of the wave, in Hz

L is the length of the string, in m

n is the number of the harmonic ($n = 1, 2, 3, 4, 5 \ldots$)

v is the velocity of the wave, in m s^{-1}

f_1 is the fundamental frequency, in Hz.

Each of these standing waves or resonances can exist independently of each other and so they give rise to the rich diversity of sound made by musical instruments, even when they are playing the same note. Each instrument will make harmonics of different intensities to produce a spectrum of frequencies unique to that instrument. Nonetheless, the resonant frequencies are governed by simple rules that rely solely on the speed of the travelling wave for the medium, v, and the length of the medium, L, fixed at both ends. This concept applies equally well to pipes used in organs and explains how a musical note can be generated in an air column.

SAMPLE PROBLEM 4 Calculations of standing waves

Two students have created a standing wave using wire under tension attached to two fixed ends as part of an investigation. The distance between the two ends is 1.50 m. At one instant when the amplitude of the standing wave is a maximum, the wave looks like this:

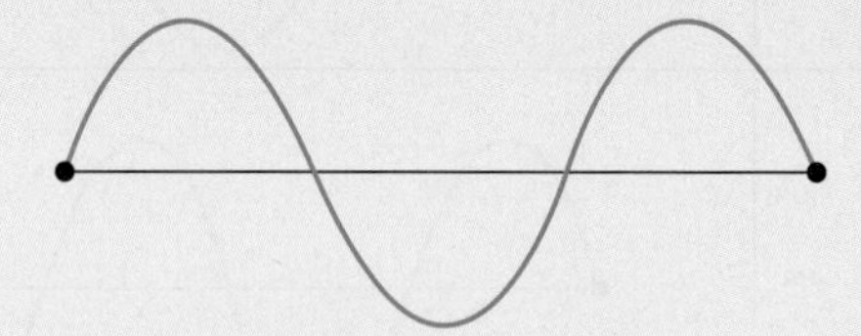

a. **State the number of nodes in this standing wave (including the two fixed ends).**
b. **State the number of antinodes.**
c. **Explain which harmonic of all the possible standing waves this standing wave represents.**
d. **Determine the wavelength of the standing wave.**

The students measure the frequency of the standing wave to be 4.2 Hz.

e. **Calculate the period of the standing wave.**

f. **Determine the speed of the travelling waves on this wire that superimpose to produce this standing wave.**

g. **Determine the fundamental frequency f_1 and wavelength λ_1 for a standing wave on this wire.**

THINK	WRITE
a. A node can be identified as any point where the amplitude of the wave is a minimum. Count the number of nodal points, including both ends.	a. There are four nodes for this standing wave. 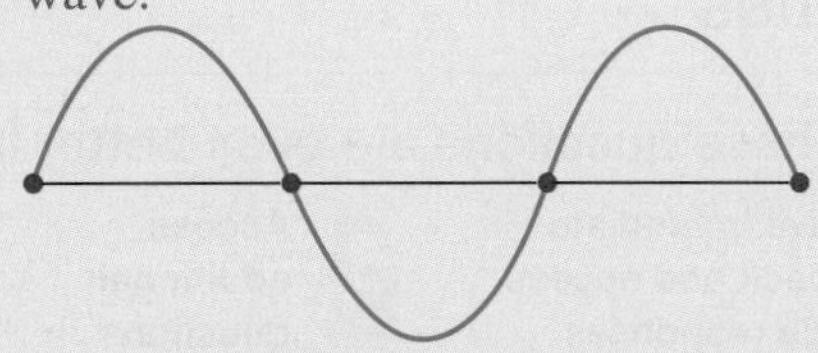
b. An antinode can be identified as any point where the amplitude of the wave is a maximum. Count the number of antinodes.	b. There are three antinodes for this standing wave. 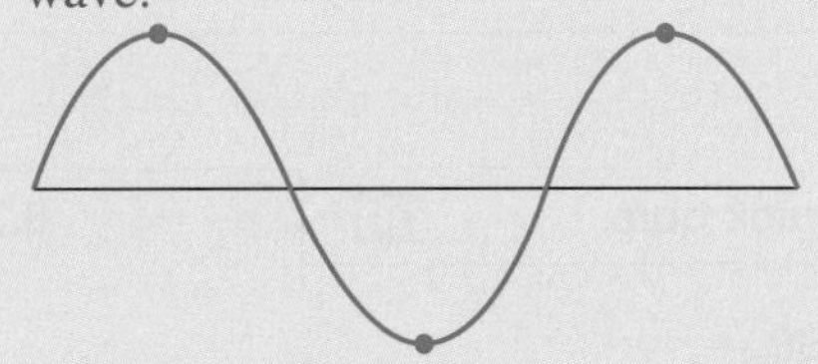
c. The number of antinodes is equal to the harmonic number. There are three antinodes.	c. The wave is the 3rd harmonic for this wire.
d. The length of the wire, 1.5 m is $\frac{3}{2}$ wavelengths.	d. $L=\frac{3}{2}\lambda$ for the 3rd harmonic. Thus: $1.5 = 1.5\lambda$ $\Rightarrow \lambda = 1.0\,\text{m}$
e. The period for the 3rd harmonic is $T=\frac{1}{f_3}$	e. $T=\frac{1}{f_3} = \frac{1}{4.2} = 0.2381 \approx 0.24\ \text{s}$
f. Use the equation $v=f_3\lambda_3$ and solve for v.	f. $v=f_3\lambda_3 = 4.2\times 1.0 = 4.2\ \text{m s}^{-1}$
g. 1. The nth harmonic is given by the expression $f_n=\frac{nv}{2L}=nf_1$ where $f_3=4.2\,\text{Hz}$.	g. 1. $f_3=4.2=3\times f_1$ $f_1=\frac{4.2}{3}=1.4\text{Hz}$ The fundamental frequency for the wire is 1.4 Hz.
2. The fundamental wavelength can be found using the rule $\lambda n=\frac{2L}{n}$	2. $\lambda_1=2L=2\times 1.5=3.0\text{m}$ The fundamental wavelength is 3.0 m.

PRACTICE PROBLEM 4

The tension in the wire is now increased so that the speed of the travelling waves is 6.0 m s^{-1}.

a. **What will be the respective frequencies f_1, f_2 and f_3 of the first, second and third harmonic standing waves now?**

b. **For each of the first three harmonics, what will be the wavelengths λ_1, λ_2 and λ_3?**

8.3 Activities

learnon

Students, these questions are even better in jacPLUS

Receive immediate feedback and access sample responses

Access additional questions

Track your results and progress

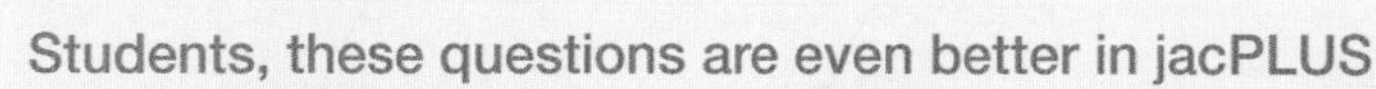

8.3 Quick quiz  | **8.3 Exercise** | **8.3 Exam questions**

8.3 Exercise

1. In each of the following diagrams, two waves move towards each other.

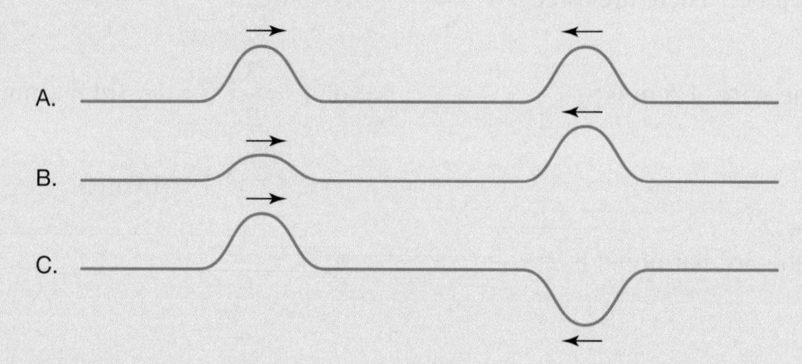

Identify the diagram (or diagrams) showing waves that, as they pass through each other, could experience:
a. only destructive interference
b. only constructive interference.

2. Explain superposition and when it occurs.
3. Explain constructive interference is and when it occurs.
4. The following diagrams show the positions of three sets of two pulses as they pass through each other. Copy the diagrams and sketch the shape of the resultant disturbances.

a. 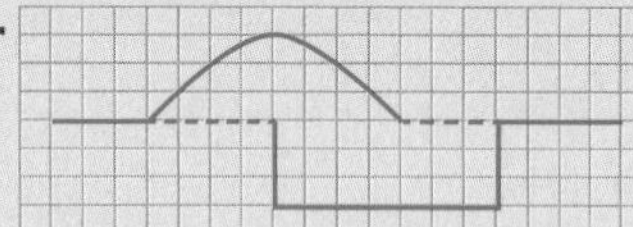
b. 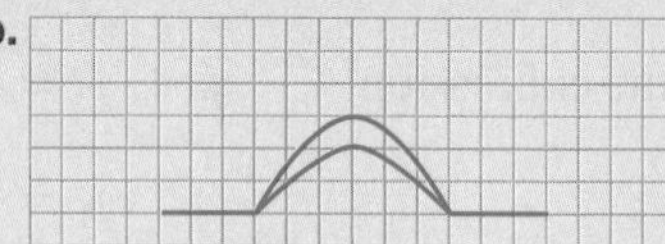
c. 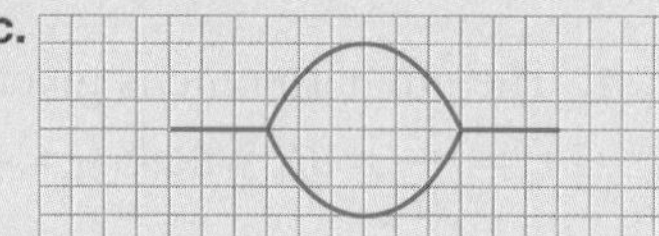

5. Determine the wavelength of a standing wave if the nodes are separated by a distance of 0.75 m.
6. The following figure shows a standing wave in a string. At that instant (t = 0) all points of the string are at their maximum displacement from their rest positions. (The string is fixed at both ends.)

If the period of the standing wave is 0.40 s, sketch diagrams to show the shape of the string at the following times:

a. t = 0.05 s **b.** t = 0.1 s **c.** t = 0.2 s **d.** t = 0.4 s.

7. Kim and Jasmine set up two loudspeakers in the following arrangements:
   - The speakers face each other, so that a standing wave is produced between the speakers.
   - They are 10 m apart.
   - The speakers are in phase and produce a sound of 330 Hz.

   Jasmine uses a microphone connected to a CRO (Cathode Ray Oscilloscope) and detects a series of points between the speakers where the sound intensity is a maximum. These points are at distances of 3.5 m, 4.0 m and 4.5 m from one of the speakers.

   a. Explain what causes the maximum sound intensities at these points.
   b. Determine the wavelength of the sound being used.
   c. Calculate the speed of sound on this occasion.
8. A standing wave is set up by sending continuous waves from opposite ends of a string. The frequency of the waves is 4.0 Hz, the wavelength is 1.2 m and the amplitude is 10 cm.
   a. Calculate the speed of the waves in the string.
   b. Determine the distance between the nodes of the standing wave.
   c. Calculate the maximum displacement of the string from its rest position.
   d. Determine how many times per second the string is straight (flat).
9. A standing wave is set up by sending continuous waves from opposite ends of a string. The frequency of the waves is 2.5 Hz, the wavelength is 2.4 m and the amplitude is 0.040 m.
   a. Calculate the speed of the waves in the string.
   b. Determine the distance between the nodes of the standing wave.
   c. Calculate the maximum displacement of the string from its rest position.
   d. Determine how many times per second the string is straight.
10. A guitar string is capable of supporting many discrete frequencies when plucked. The length of the guitar string is 0.80 m, and travelling waves on the string have a speed of 650 m s^{-1}. Determine the frequency of the first three harmonics.

8.3 Exam questions

Question 1 (4 marks)

Source: *VCE 2020, Physics Exam, Section B, Q.13; © VCAA*

A 0.8m long guitar string is set vibrating at a frequency of 250 Hz. The standing wave envelope created in the guitar string is shown in the following figure.

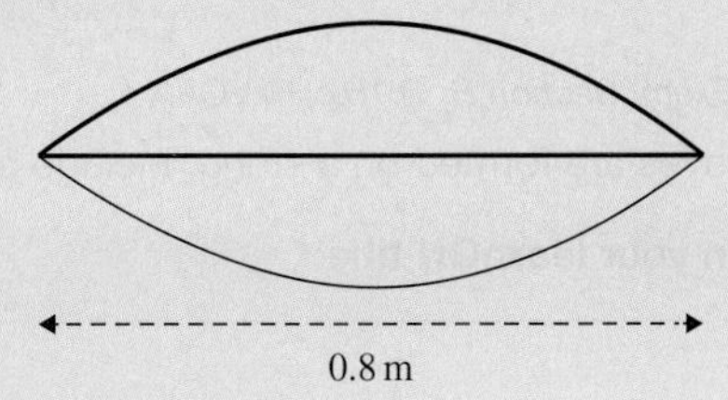

a. Calculate the speed of the wave in the guitar string. **(2 marks)**

b. The frequency of the vibration in the guitar string is tripled to 750 Hz. On the guitar string, draw the shape of the standing wave envelope now created. **(2 marks)**

Question 2 (2 marks)

Source: *VCE 2019, Physics Exam, Section B, Q.13; © VCAA*

In an experimental set-up used to investigate standing waves, a 6.0 m length of string is fixed at both ends, as shown in the following figure. The string is under constant tension, ensuring that the speed of the wave pulses created is a constant 40 m s^{-1}.

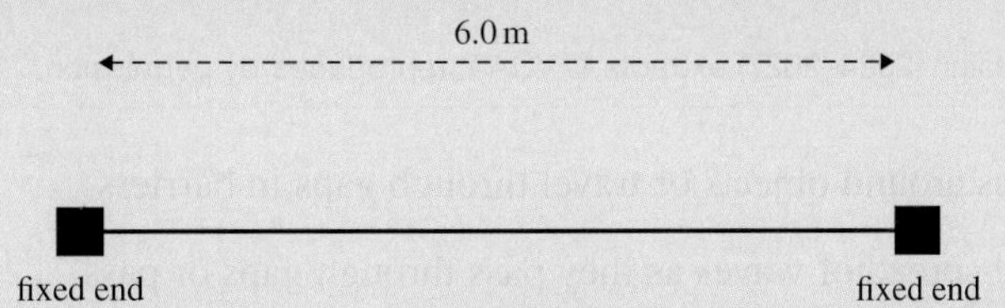

In an initial experiment, a continuous transverse wave of frequency 7.5 Hz is generated along the string.

a. Determine the wavelength of the transverse wave travelling along the string. **(1 mark)**

b. Will a standing wave form? Give a reason for your answer. **(1 mark)**

Question 3 (6 marks)

Source: *VCE 2019 Physics Exam, NHT, Section B, Q.14;* © VCAA

Figure 13 shows a simple apparatus that can be used to determine the frequency of a tuning fork.

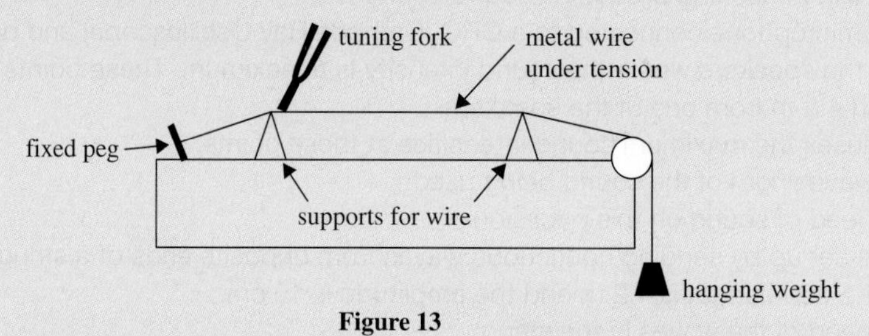

Figure 13

The apparatus consists of two supports and a metal wire that is stretched between a fixed peg and a hanging weight. The wire is under tension.

The tuning fork is set vibrating and is then touched onto the wire close to the left-hand support, which makes the wire vibrate at the same frequency as the tuning fork.

a. Draw a diagram of the simplest standing wave pattern that can exist on the vibrating section of the wire (the fundamental) between the two supports. **(2 marks)**

b. When the distance between the supports is 0.92 m, the fundamental frequency resonates in the wire. Calculate the wavelength of the fundamental. Show your working. **(2 marks)**

c. Calculate the frequency of the tuning fork if the speed of the waves in the wire is 224 m s^{-1}. Show your working. **(2 marks)**

Question 4 (4 marks)

Source: *VCE 2017, Physics Exam, Section B, Q.16.a,b;* © VCAA

Standing waves are formed on a string of length 4.0 m that is fixed at both ends. The speed of the waves is 240 m s^{-1}.

a. Calculate the wavelength of the lowest frequency resonance. **(2 marks)**

b. Calculate the frequency of the second-lowest frequency resonance. **(2 marks)**

Question 5 (3 marks)

Source: *Adapted from VCE 2017, Physics Exam, Section B, Q.16.c;* © VCAA

Explain the physics of how standing waves are formed on a string. Include a diagram in your response.

More exam questions are available in your learnON title.

8.4 Diffraction of light

KEY KNOWLEDGE

- Investigate and explain theoretically and practically diffraction as the directional spread of various frequencies with reference to different gap width or obstacle size, including the qualitative effect of changing the $\frac{\lambda}{w}$ ratio and apply this to limitations of imaging using electromagnetic waves

Source: VCE Physics Study Design (2024–2027) extracts © VCAA; reproduced by permission.

Waves spread out as they pass around objects or travel through gaps in barriers.

Diffraction is the directional spread of waves as they pass through gaps or pass around objects. The amount of diffraction depends on two factors:

i. the wavelength of the wave, λ

ii. the width of the gap or the size of the obstacle, for which w will be used as the variable.

diffraction the spreading out, or bending of, waves as they pass through a small opening or move past the edge of an object

8.4.1 Varying the wavelength and gap width

The diffraction of waves can be modelled with water waves in a ripple tank. Figure 8.16 shows the way that these waves diffract in various situations. The diagrams apply equally well to the diffraction of sound waves and light.

The larger the wavelength λ (in figures 8.16b, d and f), the larger or more significant the amount of diffraction (more spreading).

FIGURE 8.16 Diffraction of water waves: **a.** short wavelength around an object, **b.** long wavelength around the same object, **c.** short wavelength through a gap, **d.** long wavelength through the same gap, **e.** short wavelength around the edge of a barrier and **f.** long wavelength around the edge of the same barrier

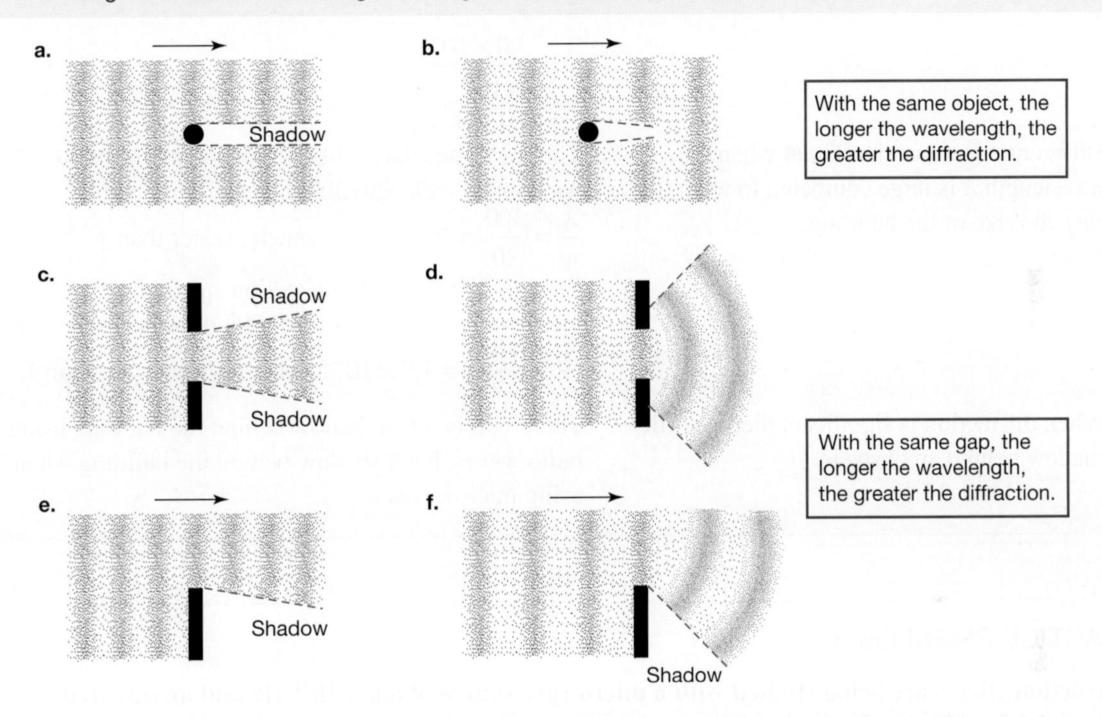

The region where waves do not travel is called a shadow. As a general rule, if the wavelength is less than the size of the object, there will be a significant shadow region.

When waves diffract through a gap of width w in a barrier, the amount of diffraction is proportional to the ratio $\frac{\lambda}{w}$. As the value of this ratio increases, so too does the amount of diffraction that occurs and sharp shadows are less obvious.

tlvd-9009

SAMPLE PROBLEM 5 Describing the amount of diffraction of radio waves and microwaves caused by a building

Electromagnetic radio waves and microwaves are used to transmit data from one tower to another location. However, buildings can act as obstacles. Radio waves have a frequency of the order 10^6 Hz whereas microwaves have a frequency of the order 10^{11} Hz. Describe the amount of diffraction of the two types of waves as they pass around a building. For which wave would there be a significant shadow behind the building? Explain.

Take the speed of light to be 3.0×10^8 m s^{-1} and the building to be an obstacle of size 30 m.

THINK	WRITE
1. First calculate the wavelengths of the two waves using the formula $v = f\lambda$.	For radio waves $v = f\lambda$ $\lambda = \frac{v}{f}$ $= \frac{3.0 \times 10^8}{10^6}$ $= 300\text{m}$ For microwaves $v = f\lambda$ $\lambda = \frac{v}{f}$ $= \frac{3.0 \times 10^8}{10^{11}}$ $= 0.003\text{m} = 3\text{mm}$
2. Diffraction is more significant when the wavelength λ is large compared to w, in this case the size of the building.	There will be a large diffraction spread for radio waves of wavelength 300 m because the ratio $\frac{\lambda}{w} = \frac{300}{30} = 10$, which is much greater than 1. There will be a small diffraction spread for microwaves of wavelength 0.003 m because the ratio $\frac{\lambda}{w} = \frac{0.003}{30} = 1.0 \times 10^{-4}$, which is much less than 1.
3. When diffraction is significant there is little shadow behind an obstacle.	There will be little shadow behind the building using radio waves, but a shadow behind the building when using microwaves.

PRACTICE PROBLEM 5

Diffraction effects are being studied with a microwave source of 6.0×10^{11} Hz and an infrared source of 1.5×10^{13} Hz. Radiation from each source is passed through a narrow slit. The microwaves demonstrate behaviour associated with diffraction, yet the infrared source does not. Take the speed of light to be 3.0×10^8 m s^{-1}.

a. **Use physics to explain why microwaves demonstrate diffraction yet the infrared radiation does not.**
b. **How would you modify the arrangement to observe diffraction effects using infrared radiation?**

8.4.2 Light also diffracts

The word 'diffraction' was coined by Francesco Grimaldi (1618–1663) to describe a specific observation he made of light. He observed that, when sunlight entered a darkened room through a small hole, the spot was larger than would be expected from straight rays of light. He also noted that the border of the spot was fuzzy and included coloured fringes. He observed a similar effect when light passed around a thin wire or a strand of hair.

Newton was aware of Grimaldi's observation of 'diffraction'. He interpreted it using his particle model, arguing that the observed effect was due to light particles interacting with the edges of the hole as a **refraction** effect. He argued that if light was a wave, the bending would be much greater. Newton's conclusions on the particle model were enough for scientists even a hundred years later, in Young's time, to doubt any experimental evidence supporting the wave model.

refraction the bending of light as it passes from one medium into another

FIGURE 8.17 Diffraction of red light through a slit

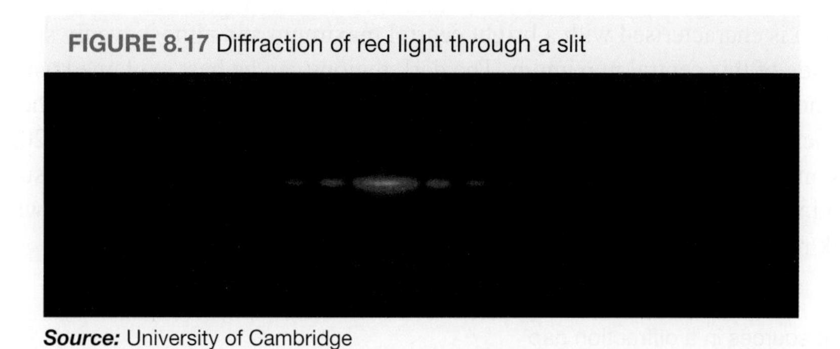

Source: University of Cambridge

As technology improved, the investigation of the diffraction of light revealed more than just the observation of spreading:

- The pattern had a central bright region with narrower and less bright regions either side.
- There was a dark gap between the bright regions.
- The central region was twice as wide as the other regions, which were all about the same size.
- The pattern for red light was more spread out than that for blue light.

FIGURE 8.18 Relative intensity and diffraction patterns through an identical slit for **a.** blue light and **b.** red light. As the wavelength increases, the pattern spreads out more.

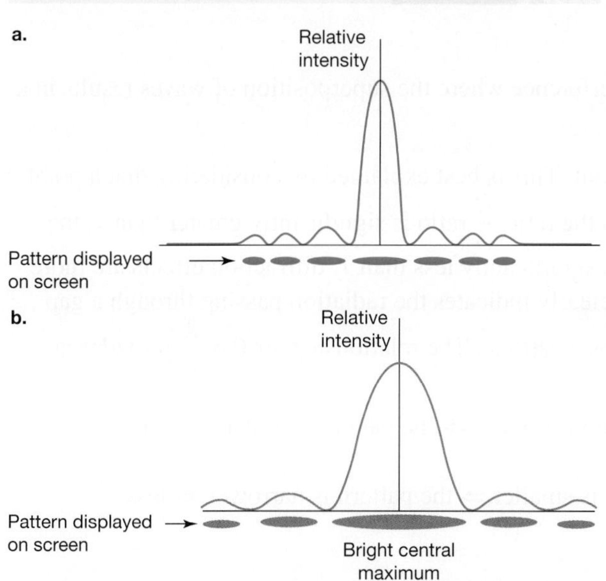

FIGURE 8.19 Diffraction patterns change with gap width. As the gap width gets smaller, the pattern spreads out (more diffraction).

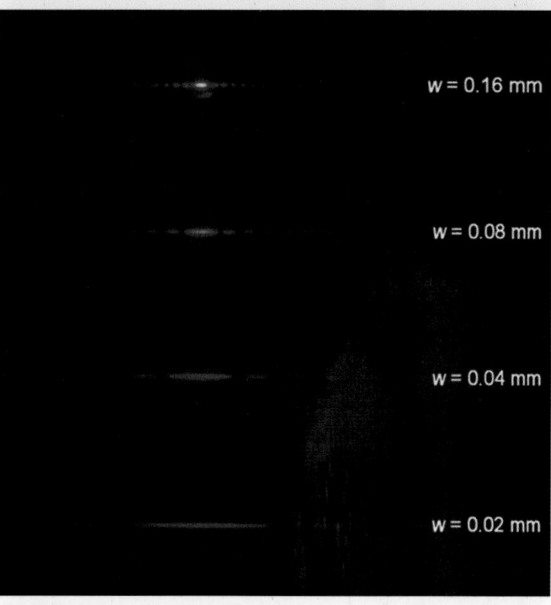

Source: University of Cambridge

Figure 8.18 and figure 8.19 confirm that light satisfies the same relationships as other waves; that is:

- the amount of spreading is proportional to the wavelength, λ
- the amount of spreading is proportional to the inverse of the gap width, $\frac{1}{w}$.

The diffraction pattern is characterised with a bright central maximum and almost evenly spaced bright and dark regions either side of this central maximum. The dark regions can be best explained using the concept of destructive interference. At a dark region on the pattern, part of the wave passing through the gap destructively interferes with another part of the wave passing through the gap. This is shown in figure 8.20, where part of the wave at a_1 combines with part of the wave at b_1. At these dark regions, a crest from a_1 superimposes with a trough from b_1 and vice versa, resulting in wave cancellation or destructive interference, which is seen or interpreted as a dark region.

FIGURE 8.20 Point sources in a diffraction gap

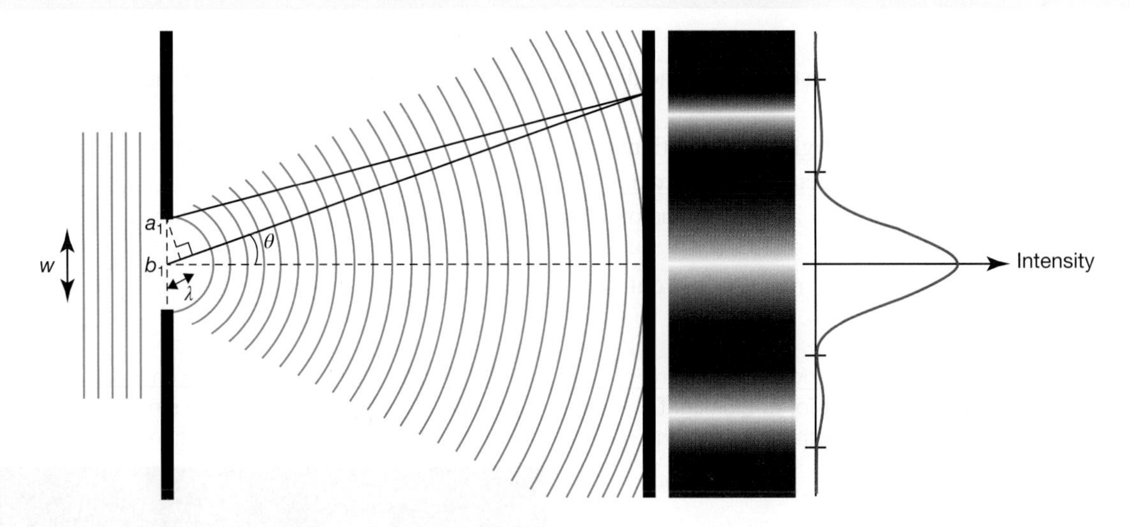

Similarly, bright regions are explained by constructive interference where the superposition of waves results in a larger amplitude.

With decreasing gap width, w, the pattern simply spreads out. This is best explained by considering that a point source of waves spreads out in all directions evenly. When the ratio $\frac{\lambda}{w}$ ratio is significantly greater than 1, the diffraction pattern is readily observable. When this ratio is significantly less than 1, diffraction effects are more difficult to observe. However, if diffraction is observed it clearly indicates the radiation passing through a gap or around an obstacle can be modelled as a wave and not as a particle. The relationship $\sin\theta = \frac{\lambda}{w}$ provides an explanation for the observations of the diffraction of light:

- A longer wavelength ⇒ the angle of the first minimum is greater ⇒ the pattern is wider ⇒ more diffraction.
- A larger gap width ⇒ the angle of the first minimum is smaller ⇒ the pattern is narrower ⇒ less diffraction.

8.4.3 Diffraction and optical instruments

Diffraction limits the usefulness of any optical instrument, whether it be your eye, a microscope or a telescope. It even affects radio telescopes.

The pupil of your eye is the gap through which light enters the eye. The objective lens of a microscope or a telescope determines how much light the instrument captures. These all have a width, so a diffraction effect is unavoidable. Diffraction limits the instrument's capacity to distinguish two objects that are very close to each other.

In the images in figure 8.21, light from two close sources passes an optical device and produces image (a), showing two distinct spots. When the two sources are moved closer together, image (b) is produced, and the spots begin to merge. Moving the two sources even closer together produces image (c); the two spots are now one broad spot. At the separation that produces image (b), the diffraction patterns produced by the optical device begin to overlap so that the central maximum of one pattern sits on the minimum of the other. This separation is the limit of the device to resolve the detail in an image; it is called the diffraction limit or resolution of the device. The higher the resolution, the clearer the image.

The diffraction limit of a device depends on the ratio $\frac{\lambda}{w}$. Thus, a shorter wavelength gives a better resolution, as does a larger aperture for the optical device, such as in the use of electron microscopes.

FIGURE 8.21 Images produced by two point light sources as they get closer, from a. to c.

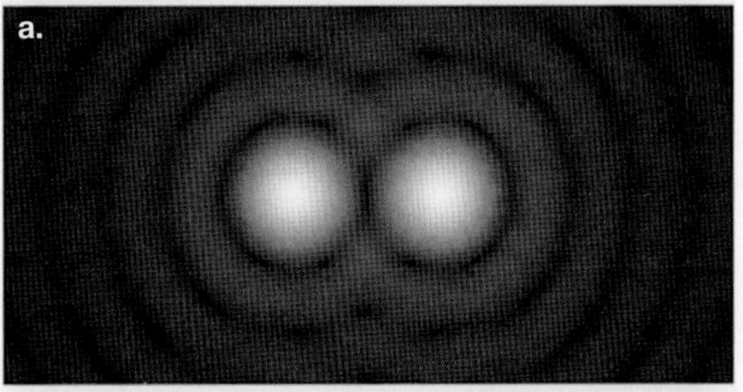

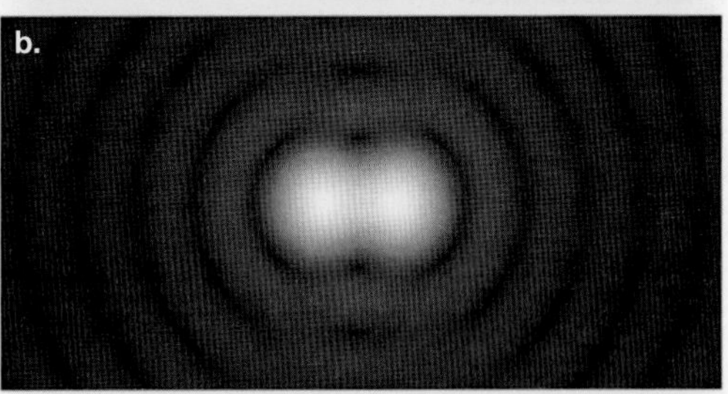

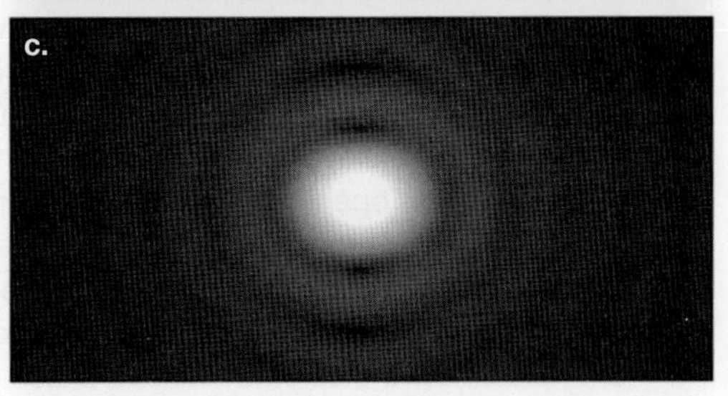

FIGURE 8.22 The diffraction patterns of two point sources overlap as the sources move closer together.

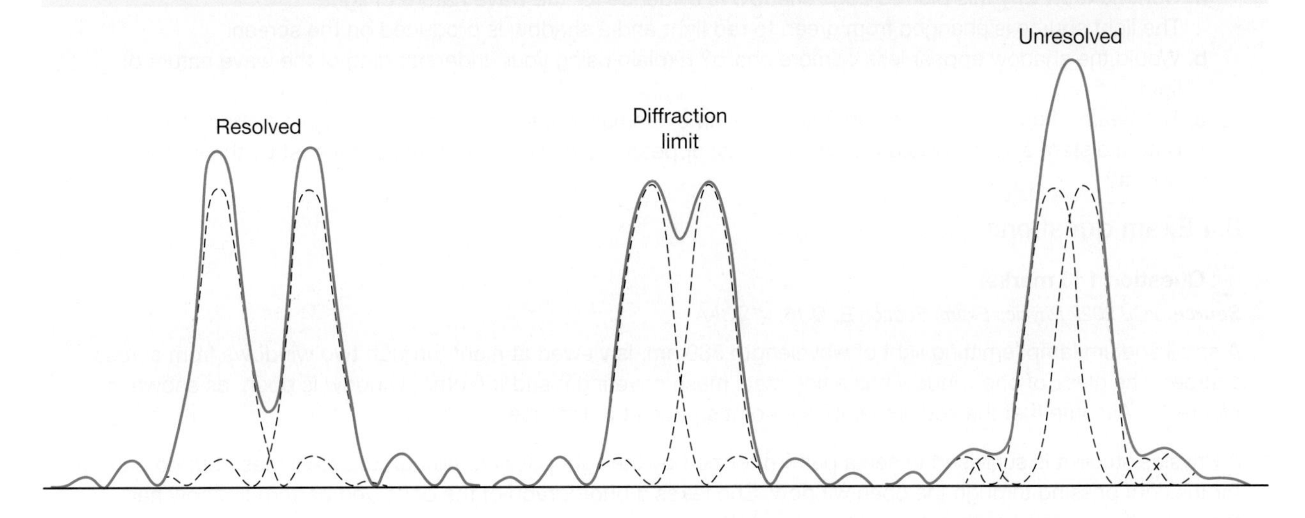

8.4 Activities

Students, these questions are even better in jacPLUS

Receive immediate feedback and access sample responses

Access additional questions

Track your results and progress

Find all this and MORE in jacPLUS

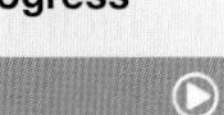

8.4 Quick quiz | 8.4 Exercise | 8.4 Exam questions

8.4 Exercise

1. Light diffracts when it passes through a sufficiently narrow opening. Explain whether this is evidence for light being a type of wave or a stream of particles.
2. A sound of wavelength λ passes through a gap of width w in a barrier. Explain how the following changes will affect the amount of diffraction that occurs.
   a. λ decreases
   b. λ increases
   c. w decreases
   d. w increases
3. Consider the diffraction pattern produced when light passes through a narrow opening.
   a. How does the first minimum from the principal axis in the pattern occur in relation to the interference of waves passing through the narrow opening?
   b. On the same axes, sketch the diffraction pattern produced by blue light and red light passing through the same narrow opening. Be sure to appropriately label.
   c. Repeat part **b** but this time for light passing through an opening that is narrower.
4. White light consisting of all colours passes through a narrow slit and projects onto a distant screen, which shows bright and dark bands with coloured fringes.
   a. Explain how the coloured fringes arise.
   b. Explain why red fringes are observed at the furthermost extent from the central white maximum.
5. A beam of green light is directed at a small obstacle and a shadow is cast onto a distant screen. The shadow is not as sharp as one would expect. Using models for light, predict the characteristics of the shadow.
   a. Comment on why this blurred edge shadow is evidence for the wave nature of light.
   The light source is changed from green to red light and a shadow is produced on the screen.
   b. Would the shadow appear less or more sharp? Explain using your understanding of the wave nature of light.
   c. The beam of green light is restored and, this time, a smaller object is placed in it. Again a shadow is cast onto a distant screen. Would the shadow cast appear less or more sharp than that cast by the larger object?

8.4 Exam questions

Question 1 (3 marks)

Source: *VCE 2022 Physics Exam, Section B, Q.16*; © VCAA

A small sodium lamp, emitting light of wavelength 589 nm, is viewed at night through two windows from across a street. The glass of one window has a fine steel mesh covering it and the other window is open, as shown in Figure 18. Assume that the sodium lamp is a point source at a distance.

A Physics student is surprised to see a pattern formed by the light passing through the steel mesh but no pattern for the light passing through the open window. She takes a photograph of the observed pattern to show her teacher, who assures her that it is a diffraction pattern.

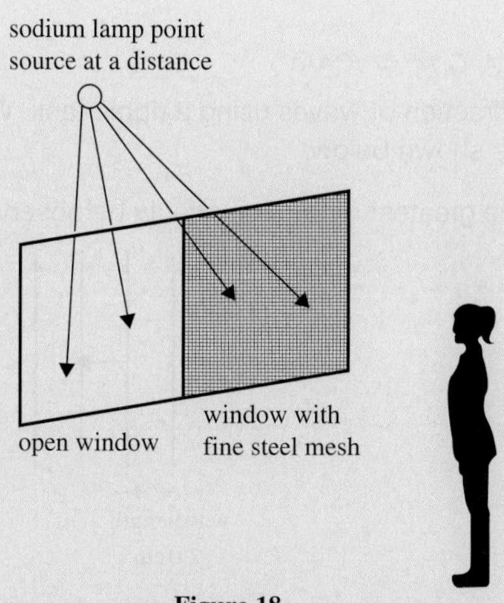

Figure 18

a. State the condition that the fine steel mesh must satisfy for a diffraction pattern to form. **(1 mark)**

b. Explain why the condition stated in part **a** does not apply to the open window. **(2 marks)**

Question 2 (2 marks)

Source: *VCE 2021 Physics Exam, NHT, Section B, Q.14;* © VCAA

To explain different aspects of mechanical waves, a Physics teacher sets up a demonstration in a Physics laboratory using a 0.80 m wide loudspeaker and a microphone. The microphone measures the sound intensity at different positions on a circle around the speaker from position A to position B, as shown in Figure 17.

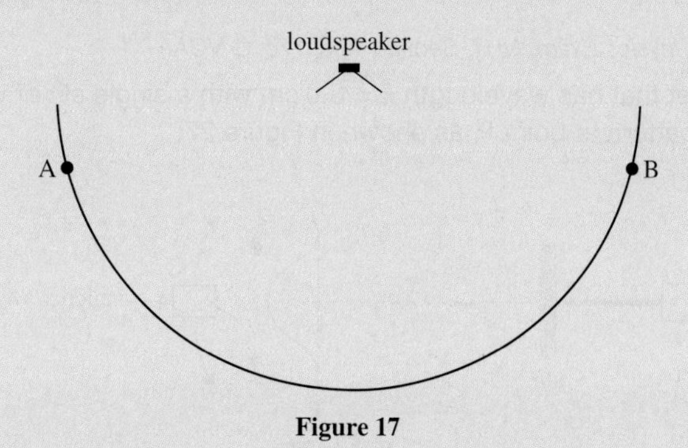

Figure 17

The speed of sound in the Physics laboratory is 334 m s^{-1}. Measurements are made at frequencies of 100 Hz and 10 000 Hz. The loudspeaker emits the 100 Hz and 10 000 Hz frequencies with equal intensity. Figure 18 shows the intensity, I, measured for each frequency at positions on the semicircular line shown in Figure 17 between positions A and B.

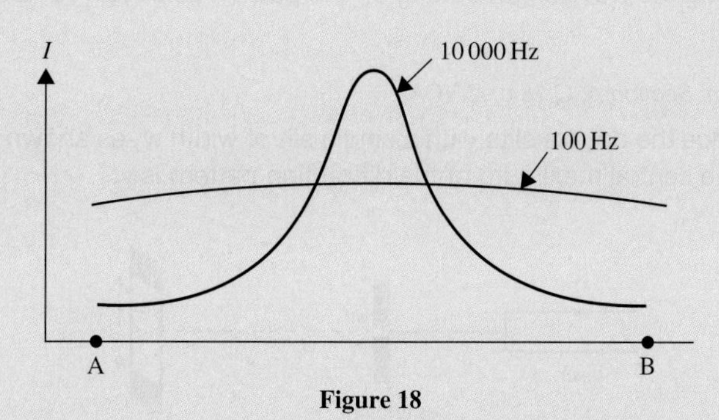

Figure 18

Explain why the response at 10 000 Hz has a greater intensity directly in front of the loudspeaker, while the response at 100 Hz is nearly the same at all positions.

Question 3 (1 mark)

Source: *VCE 2020 Physics Exam, Section A, Q.14; © VCAA*

MC Students are investigating the diffraction of waves using a ripple tank. Water waves are directed towards barriers with gaps of different sizes, as shown below.

In which one of the following would the greatest diffraction effects be observed?

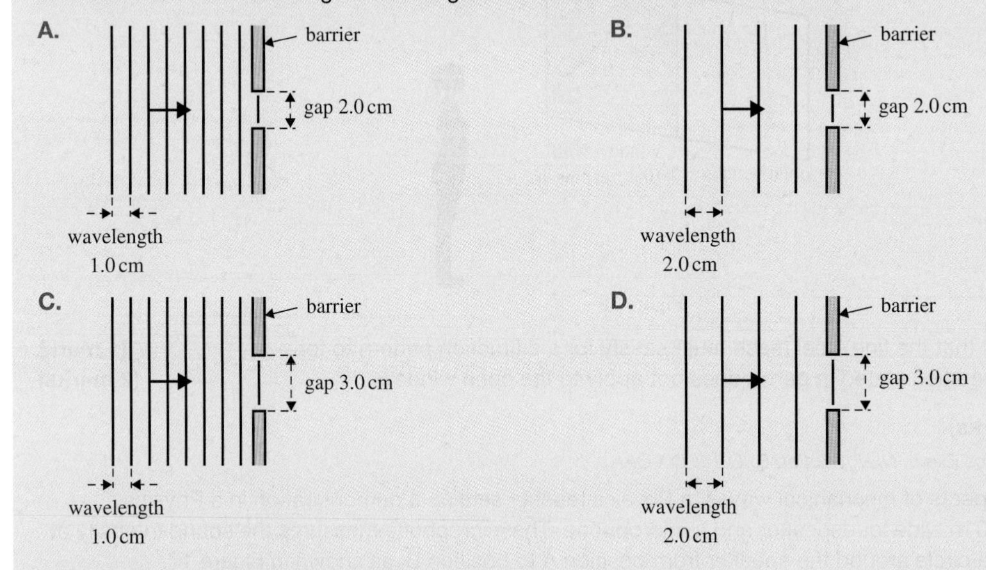

Question 4 (2 marks)

Source: *Adapted from VCE 2017 Physics Exam, NHT, Section A, Q.17c; © VCAA*

A teacher uses a microwave set that has wavelength $\lambda = 3.0$ cm with a single slit of width w and measures the width, y cm, of the diffraction pattern at point P, as shown in Figure 27.

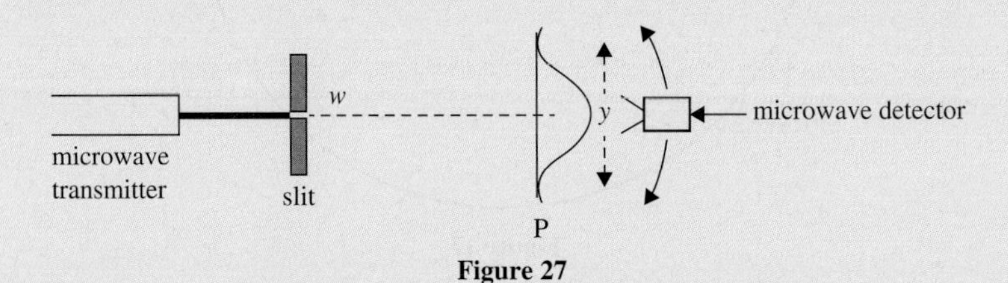

Figure 27

The microwave set has two wavelength settings: 3.0 cm and 6.0 cm. The teacher changes the setting from 3.0 cm and 6.0 cm.

Describe the effect of changing the wavelength setting on the pattern as observed. Explain your answer.

Question 5 (1 mark)

Source: *VCE 2016 Physics Exam, Section A, Q.18.c; © VCAA*

MC Amelia and Rajesh replace the double slits with a single slit of width w, as shown in the following figure. They find that the width of the central maximum of the diffraction pattern is y.

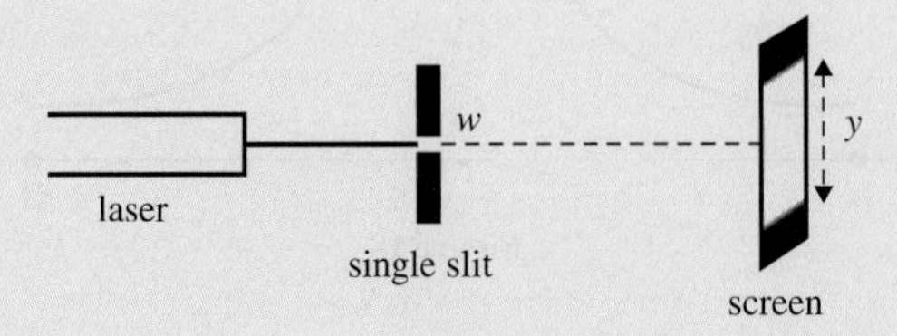

They replace the single slit with another single slit of width $2w$.

Which one of the following (**A–D.**) will Amelia and Rajesh observe in the diffraction pattern?

A. The width will be y, but twice the intensity.
B. The width will be y, but half the intensity.
C. The width will be approximately $2y$.
D. The width will be approximately $\frac{1}{2}y$.

More exam questions are available in your learnON title.

8.5 Interference of light

KEY KNOWLEDGE

- Explain the results of Young's double-slit experiment with reference to:
  - evidence for the wavelike nature of light
  - constructive and destructive interference of coherent waves in terms of path differences: $n\lambda$ and $\left(n+\frac{1}{2}\right)\lambda$ respectively, where $n = 0, 1, 2, \ldots$
  - effect of wavelength, distance of screen and slit separation on interference patterns: $\Delta x = \frac{\lambda L}{d}$ when $L \gg d$

8.5.1 Young's double-slit experiment

Thomas Young (1773–1829) was keenly interested in many things. He had already built a ripple tank to show that the water waves from two point sources with synchronised vibrations show evidence of interference. He was keen to see if he could observe interference with the interaction of two beams of light.

In one of his experiments, Young made a small hole in a window blind. He placed a converging lens behind the hole so that the cone of sunlight became a parallel beam of light. He then allowed light from the small hole to pass through two pinholes that he had punctured close together in a card. On a screen about two metres away from the pinholes, he again noticed coloured bands of light where the light from the two pinholes overlapped. Figure 8.23 shows Young's experimental arrangement.

Young deliberately had just one source, the hole in the blind, because he wanted the light coming though the two pinholes to be **coherent**. Two light sources are coherent when there is a constant phase difference between them, and they have the same frequency. This can also be described as being in phase. If Young had used two separate sources of light, one for each pinhole, their light would have been incoherent, and no discernible pattern on the screen would have been seen. To explain the evenly spaced bright and dark bands on the screen requires light to be treated as a wave. If light were a particle, the two pinholes would result in the light producing two bright spots on the screen instead of a series of bright and darks bands, as observed. Also, if the pinholes were placed closer together, the two bright spots would also be closer together if light behaved like particles. When the pinholes are more closely spaced, the bright and dark bands on the screen are more spread out. This is consistent with properties associated with the interference of coherent waves passing through both openings.

coherent same frequency and waveform (in phase)

FIGURE 8.23 Young's experiment

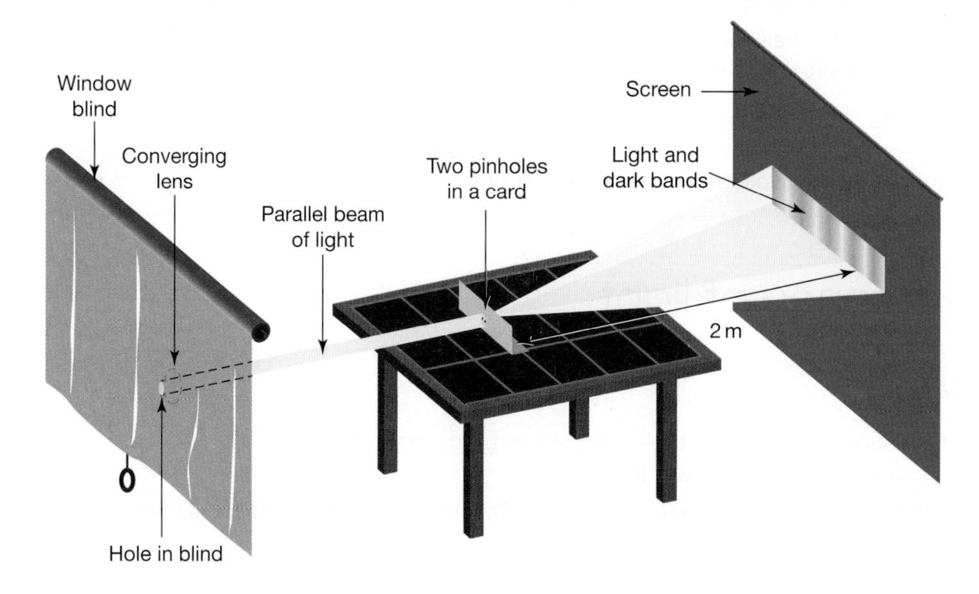

8.5.2 Interference of waves in two dimensions

Interference of waves can be observed in a ripple tank. When two point sources emit continuous waves with the same frequency and amplitude, the waves from each source interfere as they travel away from their respective sources. If the two sources are in phase (producing crests and troughs at the same time as each other), an interference pattern similar to that shown in figure 8.24 is obtained.

FIGURE 8.24 An interference pattern obtained in water by using two point sources that are in phase

Lines are seen on the surface of the water where there is no displacement of the water surface. These lines are called **nodal lines** (see figure 8.25).

FIGURE 8.25 Nodal and antinodal lines in a ripple tank with two sources

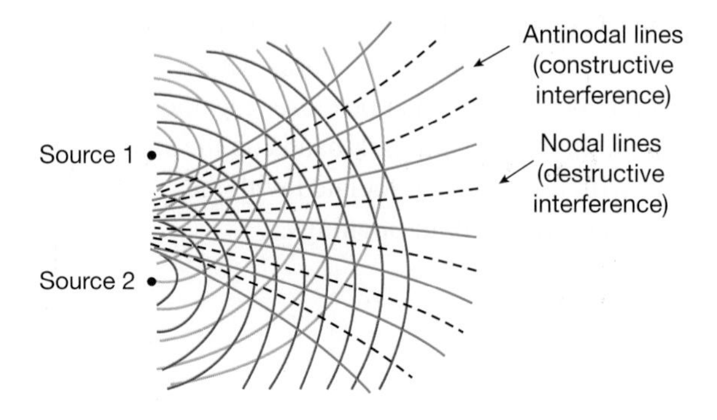

nodal line line where destructive interference occurs on a surface, resulting in no displacement of the surface

They are caused by destructive interference between the two sets of waves. At any point on a nodal line, a crest from one source arrives at the same time as a trough from the other source, and vice versa. Any point on a nodal line is sometimes called a local minimum, because of the minimum disturbance that occurs there, as shown in figure 8.26.

FIGURE 8.26 Destructive interference of waves arriving exactly out of phase

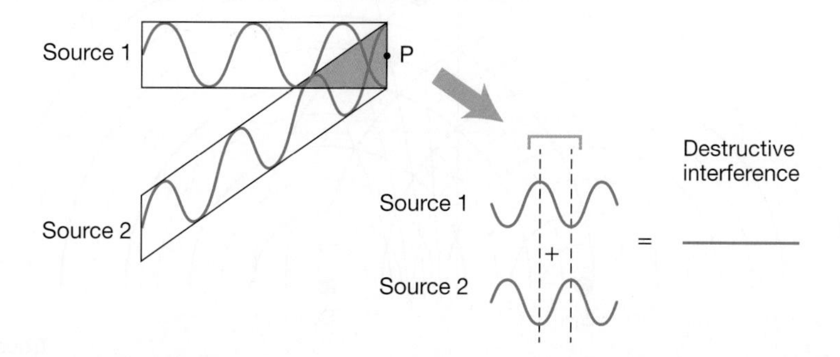

Between the nodal lines are regions where constructive interference occurs. The centres of these regions are called **antinodal lines**. At any point on an antinodal line, a crest from one source arrives at the same time as a crest from the other source, or a trough from each source arrives at the same time. Any point on an antinodal line is sometimes called a local maximum because of the maximum disturbance that occurs there, as shown in figure 8.27.

FIGURE 8.27 Constructive interference of waves arriving in phase

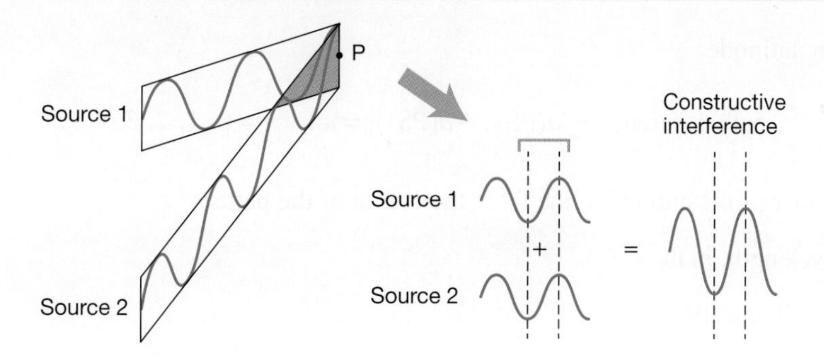

When the two sources are in phase, as shown in figure 8.24, the interference pattern produced is symmetrical with a central antinodal line. Any point on the central antinodal line is an equal distance from each source. Since the sources produce crests at the same time, crests from the two sources will arrive at any point on the central antinodal line at the same time.

Similar analysis will show that, for any point on the first antinodal line on either side of the centre of the pattern, waves from one source have travelled exactly one wavelength further from one source than from the other. This means that crests from one source still coincide with crests from the other, although they were not produced at the same time.

Point P_A in figure 8.28 is on the first antinodal line from the centre of the pattern. It can be seen that point P_A is 4.5 wavelengths from S_1 and 3.5 wavelengths from S_2.

antinodal line line where constructive interference occurs on a surface

FIGURE 8.28 Interference pattern produced by two sources in phase

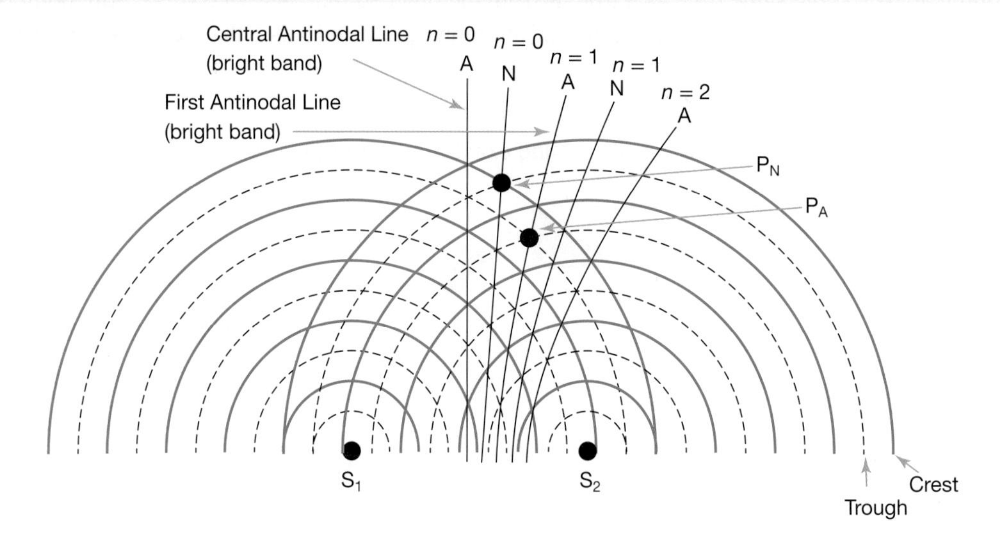

A way to establish whether a point is a local maximum is to look at the distance it is from both sources. The difference between the lengths of the two paths travelled by the waves to reach a given point is called the **path difference**. If the path difference at a point is $n\lambda$, the point is a local maximum (bright band or constructive interference region).

path difference the difference between the lengths of the paths from each of two sources of waves to a point

Constructive interference

For a point to be an antinode:

$$\text{path difference} = |d(\text{PS}_1) - d(\text{PS}_2)| = n\lambda \quad n = 0,\ 1,\ 2,\ 3,\ 4, ...$$

where: n is the number of the antinodal line from the centre of the pattern

λ is the wavelength, in m

P is the point in question

S_1 and S_2 are the sources of the waves

$d(\text{PS}_1)$ is the distance from P to S_1, in m

$d(\text{PS}_2)$ is the distance from P to S_2, in m.

Similar analysis shows that, for a point on a nodal line, the difference in distance from the point to the two sources is $\frac{1}{2}\lambda$ or $1\frac{1}{2}\lambda$ or $2\frac{1}{2}\lambda$ and s0o on. This means that a crest from one source will coincide with a trough from the other source, and vice versa. Point P_N in figure 8.28 is 5 wavelengths from S_1 and 4.5 wavelengths from S_2.

Destructive interference

For a point to be a node:

$$\text{path difference} = |d(PS_1) - d(PS_2)| = \left(n + \frac{1}{2}\right)\lambda \quad n = 0, 1, 2, 3, 4, ...$$

where: n is the number of the nodal line obtained by counting outwards from the central antinodal line

λ is the wavelength, in m

P is the point in question

S_1 and S_2 are the sources of the waves

$d(PS_1)$ is the distance from P to S_1, in m

$d(PS_2)$ is the distance from P to S_2, in m.

tlvd-9010

SAMPLE PROBLEM 6 Determining the path difference on nodal and antinodal lines

Two point sources S1 and S_2 emit waves in phase in a swimming pool. The wavelength of the waves is 1.00 m. P is a point that is 10.00 m from S_1, and P is closer to S_2 than to S_1. How far is P from S_2 if:

a. **P is on the first antinodal line from the central antinodal line**

b. **P is on the first nodal line from the central antinodal line?**

THINK	WRITE
a. 1. $d(PS_1)$ is greater than $d(PS_2)$; If P is on the first antinodal line from the central antinodal line, then: $d(PS_1) - d(PS_2) = \lambda$	a. $d(PS_1) - d(PS_2) = \lambda$
2. $d(PS_1) = 10.00$ m, $\lambda = 1.00$ m	$d(PS_2) = d(PS_1) - \lambda$ $= 10.00 \text{ m} - 1.00 \text{ m}$ $= 9.00 \text{ m}$
b. 1. $d(PS_1)$ is greater than $d(PS_2)$; If P is on the first nodal line from the central antinodal line, then: $d(PS_1) - d(PS_2) = \frac{1}{2}\lambda$	b. $d(PS_1) - d(PS_2) = \frac{1}{2}\lambda$
2. $d(PS_1) = 10.00$ m, $\lambda = 1.00$ m	$d(PS_2) = d(PS_1) - \frac{1}{2}\lambda$ $= 10.00 \text{ m} - 0.50 \text{ m}$ $= 9.50 \text{ m}$

PRACTICE PROBLEM 6

Two point sources S1 and S_2 emit waves in phase in a swimming pool. The wavelength of the waves is 1.00 m. P is a point that is 10.00 m from S_1, and P is closer to S_2 than to S_1. How far is P from S_2 if:

a. **P is on the second antinodal line from the central antinodal line**

b. **P is on the second nodal line from the central antinodal line?**

on Resources

Video eLesson Young's experiment (interference effects with white light) (eles-0027)

Interactivity Young's experiment (interference effects with white light) (int-0051)

8.5.3 Interpreting Young's experiment

Young used the wave model for light to analyse his observations. The hole in the window blind is considered a source of spherical waves. When these waves pass through the pinholes, each pinhole becomes a secondary source of spherical waves (figure 8.29). Waves from the two pinholes overlap on the screen, and their effects add together to produce the pattern (as seen in figure 8.30).

FIGURE 8.29 Light waves in Young's experiment

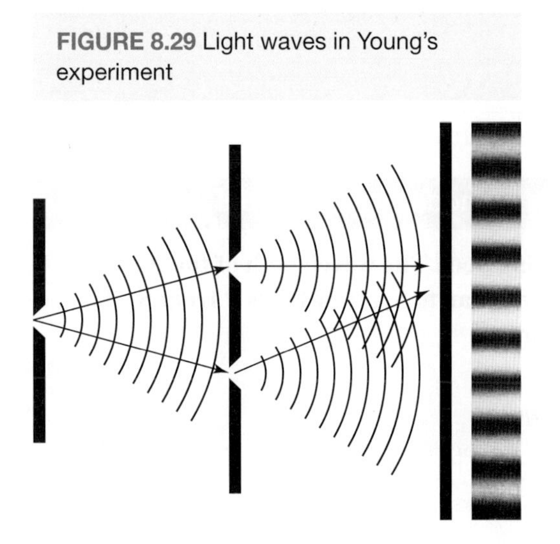

FIGURE 8.30 A light pattern produced by a modern performance of Young's experiment

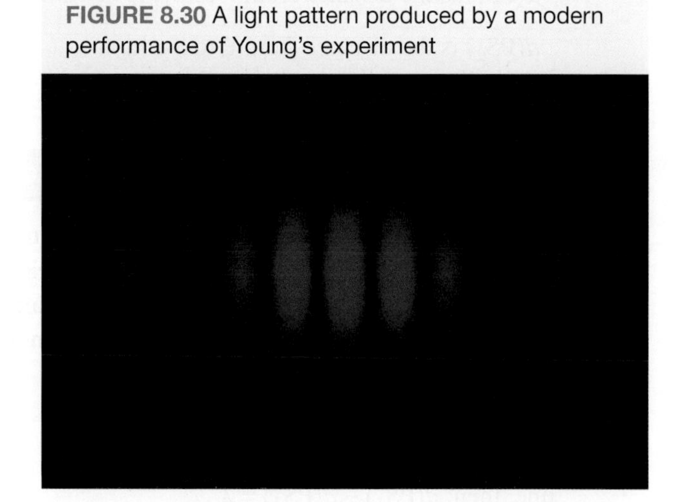

The waves reaching the screen have travelled from the source along two alternative routes, through one pinhole or the other. The difference between the lengths of the two paths is called the path difference. If the path difference results in the crests of the wave from one pinhole always meeting the troughs of the wave from the other pinhole (exactly out of phase) then destructive interference occurs and that place on the screen is a dark band. This links back to the diffraction patterns discussed in subtopic 8.4. Destructive interference occurs when the path difference is a whole number, plus one half, multiplied by the wavelength of the light: $\left(n+\frac{1}{2}\right)\lambda$ where $n = 0, 1, 2, \ldots$. A bright band occurs when the path difference is a whole number of wavelengths, meaning the waves are in phase: crests reinforcing crests and troughs reinforcing troughs.

FIGURE 8.31 Maximum intensity occurs for the maximum amplitude light wave, because of constructive interference. For example, at P, $S_2P - S_1P = \lambda$.

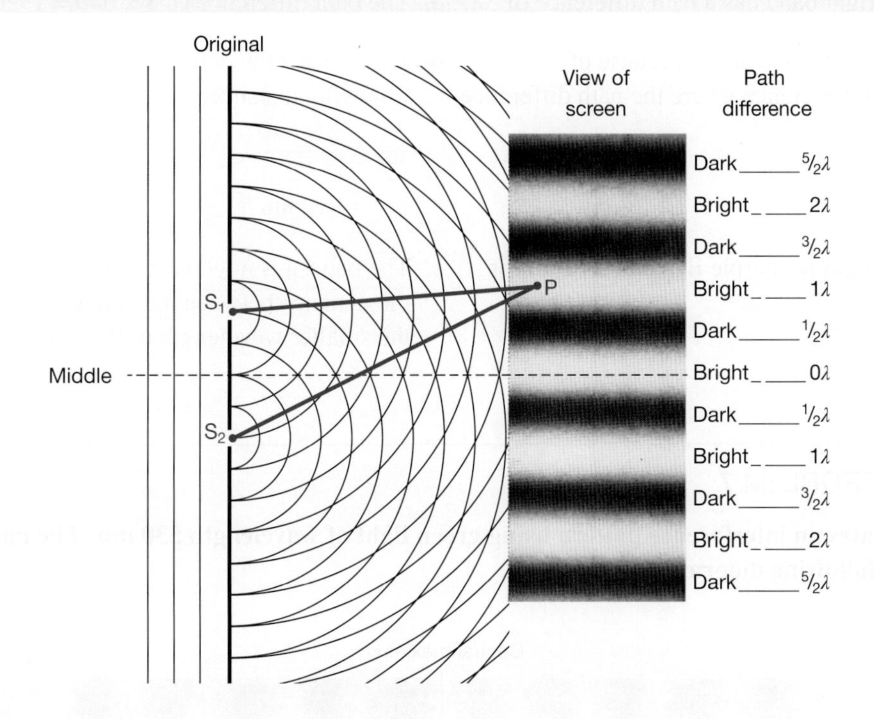

Constructive interference occurs when the path difference is a whole number multiple of the wavelength of the light, n, again where n is the number of bright bands from the central bright band.

This can be summarised as follows:

Constructive interference: path difference = $n\lambda$, $n = 0, 1, 2, 3...$

Destructive interference: path difference = $\left(n + \frac{1}{2}\right)\lambda$, $n = 0, 1, 2, 3...$

where:

λ is the wavelength of the light source

Think about performing Young's experiment with a light source emitting light of only one wavelength, say 600 nm (6×10^{-7} m) in the richly yellow part of the spectrum. Constructive interference will occur if the path difference between the two routes to the screen is 0, 600 nm, 1200 nm, 1800 nm, and so on ($n \times 600$ nm). However, if the path difference is 300 nm, 900 nm, 1500 nm, and so on ($n + 0.5 \times 600$ nm), where n is an integer, then there will be destructive interference.

tlvd-9011

SAMPLE PROBLEM 7 Examining interferences in Young's experiment

Red light of wavelength 640 nm is passed through a pair of slits, S_1 and S_2, to produce an interference pattern, like that shown in figure 8.31.

a. **Determine the path difference for the third bright band from the central bright band.**
b. **Consider the second dark band from the central bright band. Determine how much further S_2 is than S_1 from the second dark band.**
c. **Red light is replaced with purple light. Explain what happens to the interference pattern.**

THINK	WRITE
a. The third bright band has a path difference of 3λ.	a. The path difference is: $3 \times 640 = 1920$ nm.
b. The second dark band arises because of destructive interference where the path difference is $\frac{3\lambda}{2}$.	b. S_2 is further away from this dark band than S_1 by the following distance: $\frac{3\lambda}{2} = \frac{3 \times 640}{2}$ $= 910$ nm
c. The wavelength for purple light is less than for red light.	c. The pattern is now more compact or compressed as the distance between the bands is decreased, due to the smaller wavelength of the purple light.

PRACTICE PROBLEM 7

A student creates an interference pattern using green light of wavelength 530 nm. The pattern is shown in the following diagram.

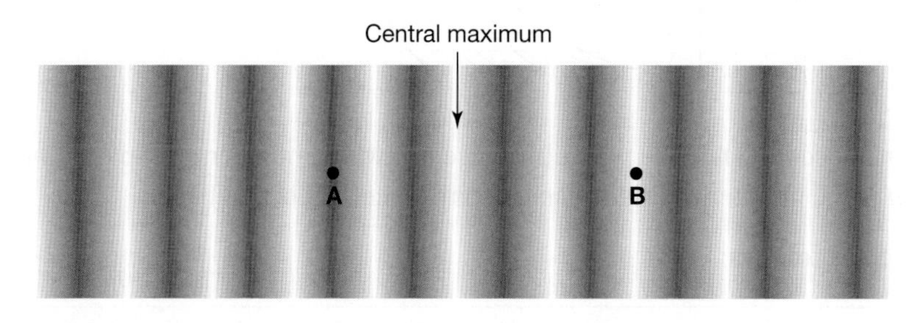

a. **Calculate the path difference for the points marked A and B.**
b. **The student increases the distance between the two slits. Describe what happens to the pattern.**
c. **She now changes the light source from green to red. Describe what happens to the pattern now.**
d. **Explain why the interference pattern is strong evidence for the wave nature of light.**

Resources

Weblink The atomic lab: wave interference

8.5.4 Diffraction and two-slit interference

When light from a point source illuminates a double slit, each slit produces its own diffraction pattern with a wide central maximum and smaller side maxima. If the two slits are close together, these two patterns overlap, and the light coming from each slit interferes with the light coming from the other slit. This produces an interference pattern associated with the two closely separated slits but combined with the effects of diffraction due to the limited width of each slit.

FIGURE 8.32 Interference pattern from two slits that are not extremely narrow. The slits in the bottom section of the image are narrower than those for the top section of the image because the central maximum is wider.

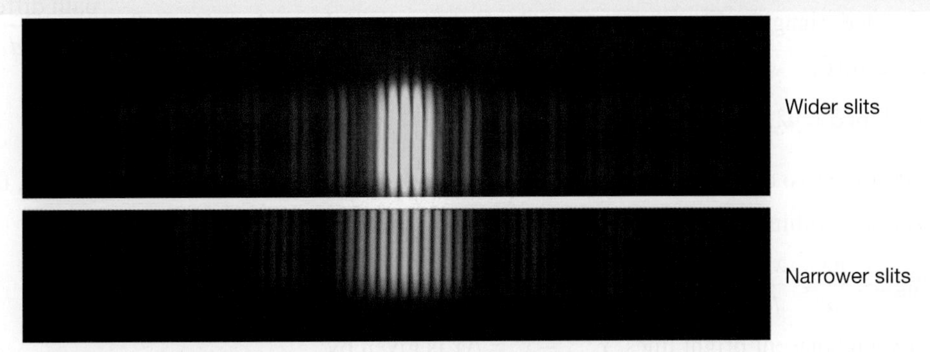

8.5.5 Spacing of bands in an interference pattern

The previous section developed expressions relating the path difference to the light and dark bands in an interference pattern. These expressions are important in understanding Young's experiment, but the path difference cannot be measured. What can be measured in this experiment is:

- the separation of the two slits, d
- the wavelength, λ
- the distance of the screen from the two slits, L
- the spacing between alternate bands in the pattern (either the bright or dark bands), x.

A relationship between these four quantities is $\Delta z = \frac{\lambda L}{d}$ and is useful to calculate any of the unknowns.

If the separation of the two slits, d, is very much less than the distance L, then the two lines S_1P and S_2P are effectively parallel, as shown in figure 8.33. Typically, d is about 1 mm and L is about 2 metres.

FIGURE 8.33 Exploring the relationship between d, λ, L and x

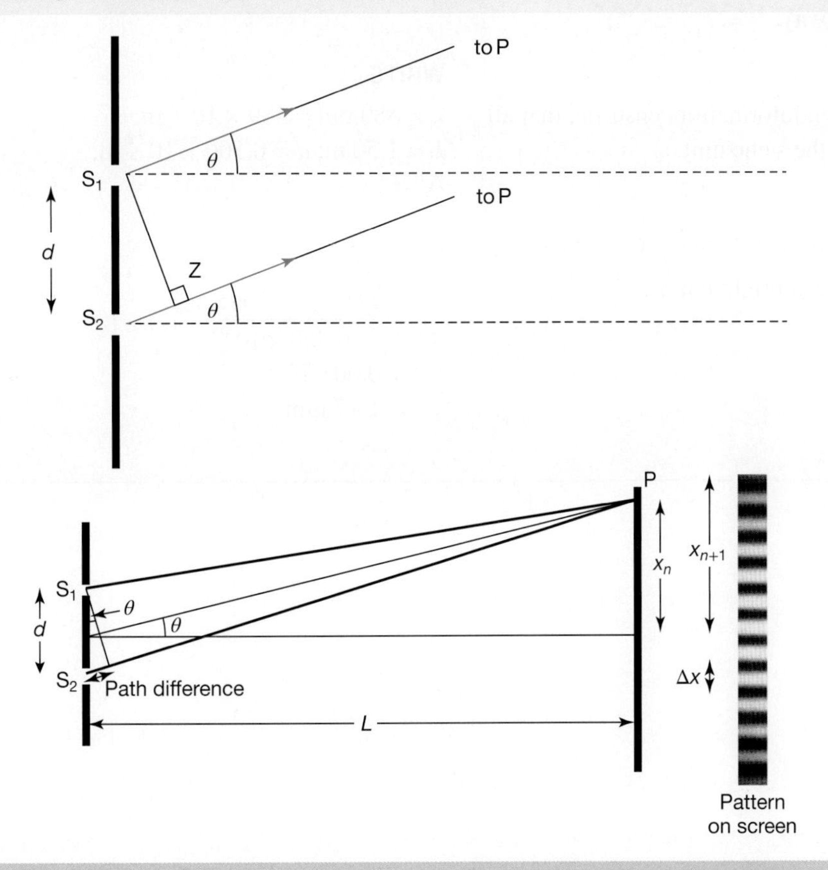

In the first of the two diagrams in figure 8.33, S_1Z is a line drawn across the two light paths at right angles. The distances from S_1 to P and from Z to P will be equal to each other. This means the path difference is S_2Z. From the right-angled triangle with corners at S_1, Z and S_2, and a right angle at Z, $\sin\theta = \frac{\text{path difference}}{d}$, or path difference $= d\sin\theta$.

For bright lines, $d\sin\theta = n\lambda$, where n is a positive integer.

From the second of the two diagrams in figure 8.33, $\tan\theta = \frac{x_n}{L}$, but for small angles less than 10°, $\tan\theta$ and $\sin\theta$ have similar values to within about 1%.

So, for small angles, $\frac{n\lambda}{d} = \frac{x_n}{L}$, giving $x_n = \frac{n\lambda L}{d}$, and for $n+1$, $x_{n+1} = \frac{(n+1)\lambda L}{d}$.

The spacing between adjacent bright lines, $x_{n+1} - x_n = \Delta x$ is given by:

$$\Delta x = \frac{\lambda L}{d} \text{ when } L \gg d$$

where: Δx is the spacing between bands (either light or dark)

λ is the wavelength of the light waves

L is the distance of the screen from the slits

d is the slit separation.

tlvd-9012

SAMPLE PROBLEM 8 Determining the spacing between bright bands in an interference pattern

Sodium light of wavelength 589 nm is directed at a slide containing two slits that are 0.500 mm apart. Determine what the spacing between the bright bands in the interference pattern will be on a screen that is 1.50 m away.

THINK	WRITE
1. List the known information, ensuring that all values are in the same unit.	$\lambda = 589 \text{ nm} = 589 \times 10^{-9}$ m; $L = 1.50$ m; $d = 0.500 \times 10^{-3}$ m; $\Delta x = ?$
2. Use the relationship $\Delta x = \frac{\lambda L}{d}$ to calculate the spacing between bright bands.	$\Delta x = \frac{\lambda L}{d} = \frac{589 \times 10^{-9} \times 1.50}{0.500 \times 10^{-3}} = 0.001\ 77$ m $= 1.77$ mm

PRACTICE PROBLEM 8

Interference bands are formed on a screen that is 2.00 m from a double slit with separation 1.00 mm. The bands are measured to be 1.30 mm apart.

a. **What is the wavelength of the light?**
b. **What is its colour?**
c. **How would the pattern change if blue light was used?**
d. **How could the experimental design be changed to make it easier to measure the line spacing in the pattern?**

How does Young's experiment provide evidence that supports the wave model?

Young postulated that light is a wave and is subject to the superposition principle.

Young's interference experiment showed that lights passing through two slits add together (constructive interference) or cancel each other (destructive interference), creating an interference pattern.

Interference is a property of waves.

This phenomenon cannot be explained unless light is considered as a wave.

Thus Young's experiment provides evidence supporting the wave model of light.

8.5 Activities

learnon

Students, these questions are even better in jacPLUS

Receive immediate feedback and access sample responses

Access additional questions

Track your results and progress

Find all this and MORE in jacPLUS

8.5 Quick quiz on	8.5 Exercise	8.5 Exam questions

8.5 Exercise

1. Young's double-slit experiment was a significant experiment in the development of understanding about light and matter.
   a. Explain what has been learned about the nature of light from Young's experiment.
   b. When light is passed through a pair of narrow closely spaced slits, an interference pattern is formed. Explain what an interference pattern is and how it is formed.

2. Jill and William are studying the effect of passing laser light of wavelength 530 nm through a pair of slits and forming a pattern on a screen several metres away. The following diagram shows the experimental arrangement.

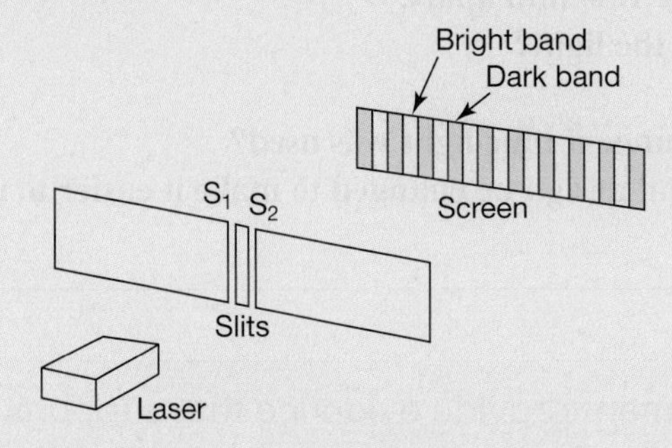

Part of the interference pattern observed is shown in the next diagram. Point C represents the position of the central maximum (bright band), and point W represents the second maximum (bright band) from the centre of the pattern.

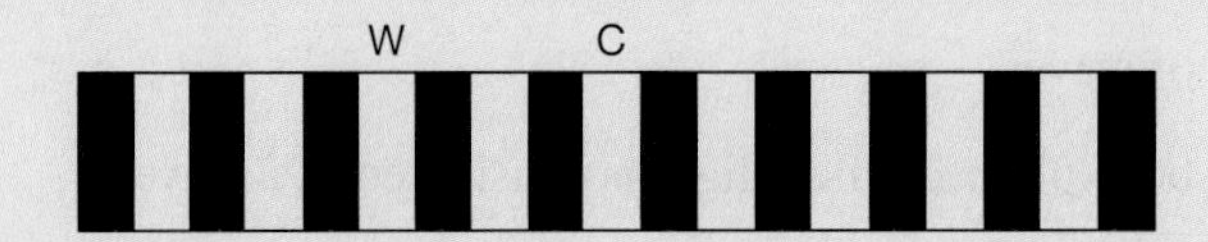

a. Explain whether the pattern on the screen is evidence for the wave nature of light or for the particle nature of light.
b. Determine whether W is a point where constructive or destructive interference is occurring.
c. Determine the path difference $|S_1C - S_2C|$.
d. Determine the path difference $|S_1W - S_2W|$.
e. State the three smallest path differences that would give rise to dark regions on the screen.
f. Determine how many dark regions there are in between C, the central maximum, and W, the second maximum from the central maximum.

3. A student shines a helium–neon laser, which produces light with a wavelength of 633 nm, through two slits and produces a regular pattern of light and dark patches on a screen as shown in the following diagram. The centre of the pattern is the band marked A. Using a wave model, light can be described as having *crests* and *troughs*.

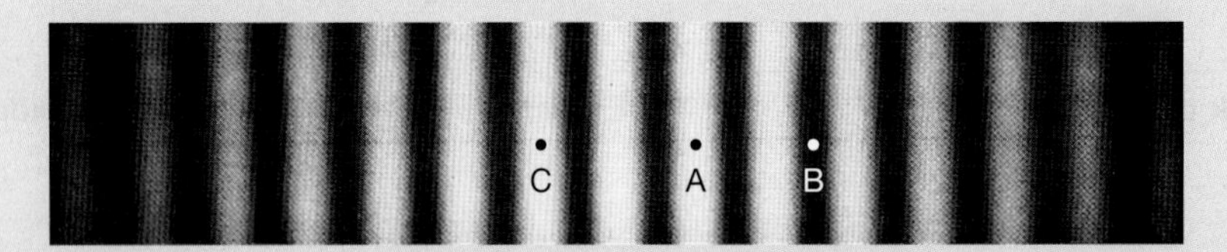

a. Use these terms to explain:
   i. the bright band labelled A
   ii. the dark band labelled B.
b. Determine the path difference in the distance light has travelled from the two slits to:
   i. the bright band labelled A
   ii. the dark band labelled B
   iii. the bright band labelled C.
c. Using the same experimental set-up, but replacing the laser with a green argon ion laser emitting 515 nm light, explain what changes would occur to the interference pattern.
d. The helium–neon laser is set up again. The distance between the two slits is now increased. Explain the changes to the interference pattern shown.
e. The screen on which the interference pattern is projected is moved further away from the slits. Explain the changes to the interference pattern shown.

4. Infrared radiation with wavelength 1.06 μm can be passed through a pair of narrow slits and an interference pattern produced.
   a. List several path differences that would produce constructive interference for the radiation.
   b. Now list several path differences that would produce destructive interference.
5. Light of wavelength 430 nm falls on a double slit of separation 0.500 mm. Determine the distance between the central bright band to the first, second and third bright band respectively in the pattern on a screen placed 1.00 m from the double slit.
6. A group of students measures the wavelength of light emitted from a laser. They do this by producing an interference pattern cast onto the wall of a classroom. The wall is 3.80 m away from the pair of slits, and they locate the central maximum and measure the distance from it to the first bright fringe immediately opposite the central maximum to be 1.90 cm. The slits are separated by a distance of 0.134 mm.
   a. Determine the wavelength of the laser light.
   b. Describe any changes in the pattern if the slits were separated by 0.100 mm.
7. A double slit is illuminated by light of two wavelengths, 600 nm and the other unknown. The two interference patterns overlap with the third dark band of the 600 nm pattern coinciding with the fourth bright band from the central band of the pattern for the light of unknown wavelength. Determine the value of the unknown wavelength.

8.5 Exam questions

Question 1 (1 mark)

Source: *VCE 2021 Physics Exam, NHT, Section A, Q.16; © VCAA*

MC A red laser used in a double-slit experiment creates an interference pattern on a screen, as shown below.

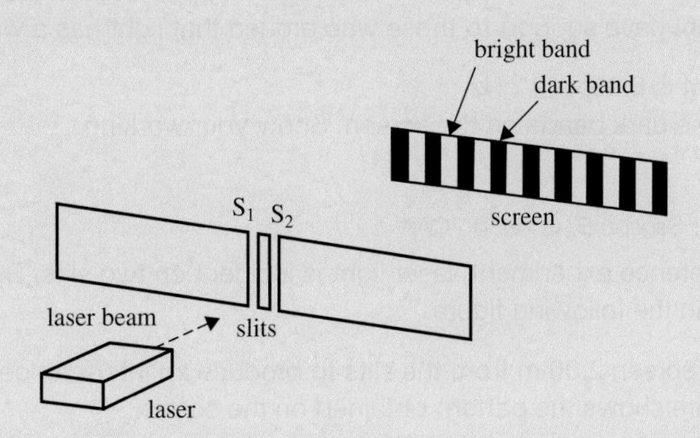

The red laser is replaced with a green laser.

Which one of the following best explains what happens to the spacing between adjacent bright bands when the green laser is used?

A. The spacing increases.
B. The spacing decreases.
C. The spacing stays the same.
D. The spacing cannot be determined from the information given.

Question 2 (6 mark)

Source: *VCE 2022 Physics Exam, Section B, Q.12a-c; © VCAA*

Students conduct an experiment in a Physics laboratory using a laser light source, two narrow slits and a screen, as shown in Figure 10.

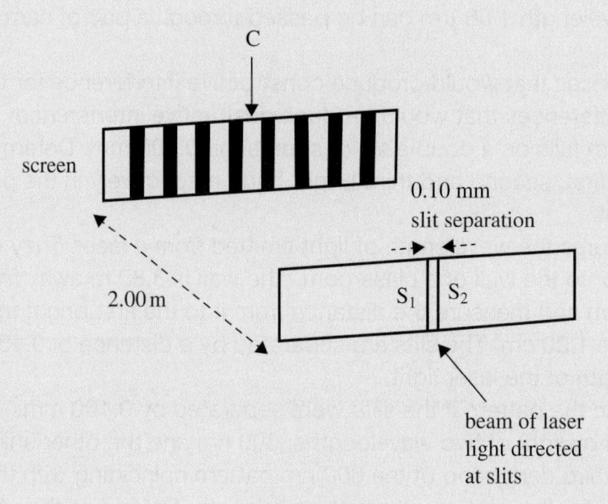

Figure 10

Point C is at the centre of the pattern of light and dark bands on the screen. The slit separation is 0.10 mm and the distance between the two slits and the screen is 2.00 m.

a. The band at point C is a bright band.
Explain why the band at point C is bright and why there is a dark band to the left of the centre. **(2 marks)**

The experiment performed by the students is often described as Young's double-slit experiment.

b. Explain how this experiment gave support to those who argued that light has a wave-like nature. **(2 marks)**

The frequency of the laser light is 6.00×10^{14} Hz.

c. Calculate the spacing of the dark bands on the screen. Show your working. **(2 marks)**

Question 3 (3 marks)

Source: *VCE 2020 Physics Exam, Section B, Q.12; © VCAA*

In a Young's double-slit interference experiment, laser light is incident on two slits, S_1 and S_2, that are 4.0×10^{-4} m apart, as shown in the following figure.

Rays from the slits meet on a screen 2.00 m from the slits to produce an interference pattern. Point C is at the centre of the pattern. The figure shows the pattern obtained on the screen.

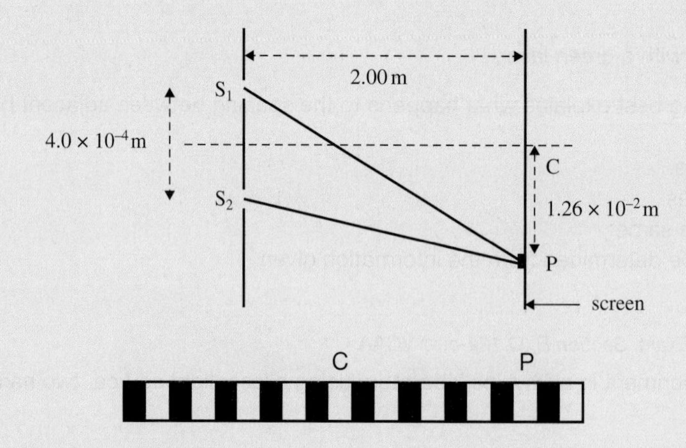

a. There is a bright fringe at point P on the screen.
Explain how this bright fringe is formed. **(2 marks)**

b. The distance from the central bright fringe at point C to the bright fringe at point P is 1.26×10^{-2} m.
Calculate the wavelength of the laser light. Show your working. **(1 mark)**

Question 4 (4 marks)

Source: *VCE 2019 Physics Exam, NHT, Section B, Q.13;* © VCAA

A seawall that is aligned north–south protects a harbour of constant depth from large ocean waves, as shown in Figure 12.

The seawall has two small gaps, S_1 and S_2 which are 60 m apart. Inside the harbour, a small boat sails north parallel to the seawall at a distance of 420 m from the seawall. At point C sits a beacon, equidistant from the two gaps in the seawall.

The boat's captain notices that, at about every 42 m, there is calm water, while there are large waves between those calm points.

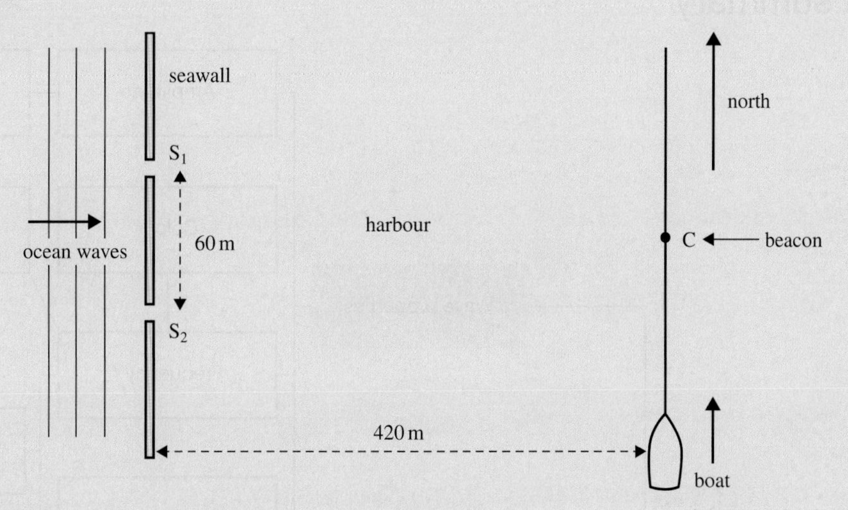

Figure 12

a. Will the beacon at point C be in calm water or large waves? Give a reason for your answer. **(2 marks)**

b. Calculate the wavelength of the ocean waves. Show your working. **(2 marks)**

Question 5 (5 marks)

Source: *VCE 2017 Physics Exam, NHT, Section A, Q.17a&b;* © VCAA

A teacher uses a microwave set that has wavelength $\lambda = 3.0$ cm to demonstrate Young's experiment. The apparatus is shown in Figure 26.

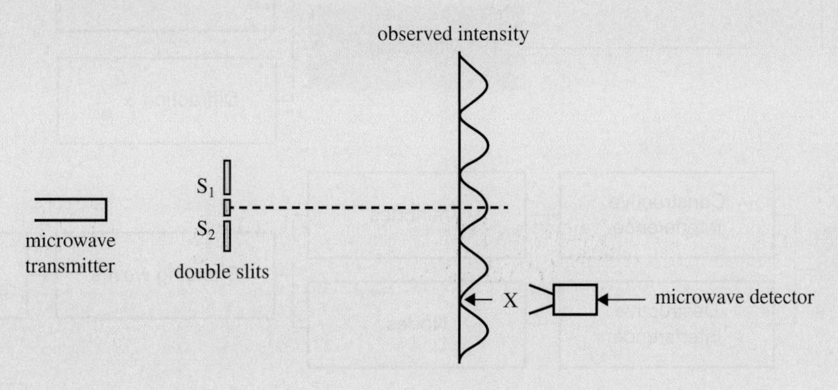

Figure 26

a. The microwave detector is placed at point X (the second nodal line out from the centre). A minimum intensity is observed.
Estimate the path difference $S_1X - S_2X$. **(2 marks)**

b. Explain the importance of Young's experiment in the development of the wave model of light. **(3 marks)**

More exam questions are available in your learnON title.

8.6 Review

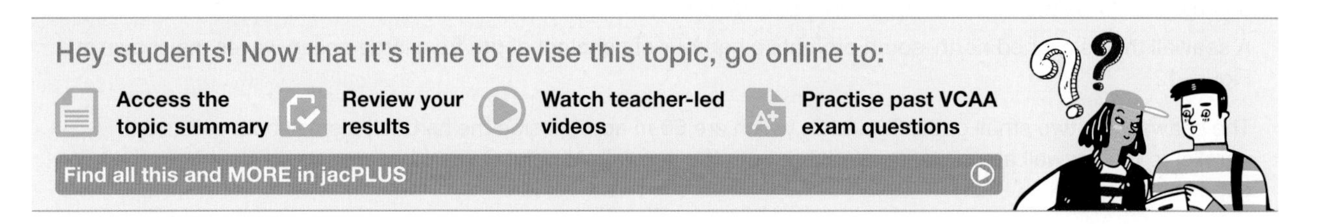

8.6.1 Topic summary

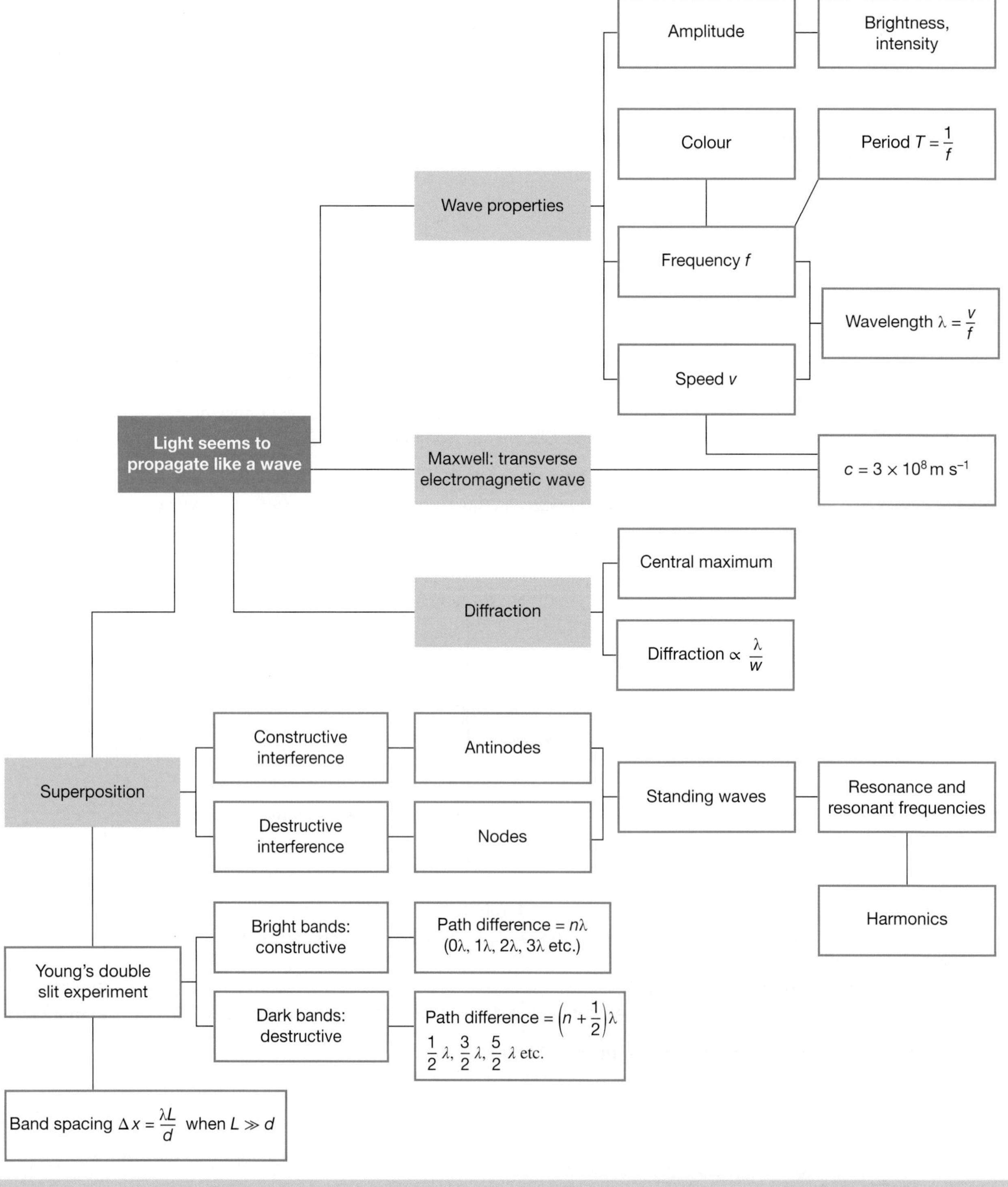

8.6.2 Key ideas summary

8.6.3 Key terms glossary

online only

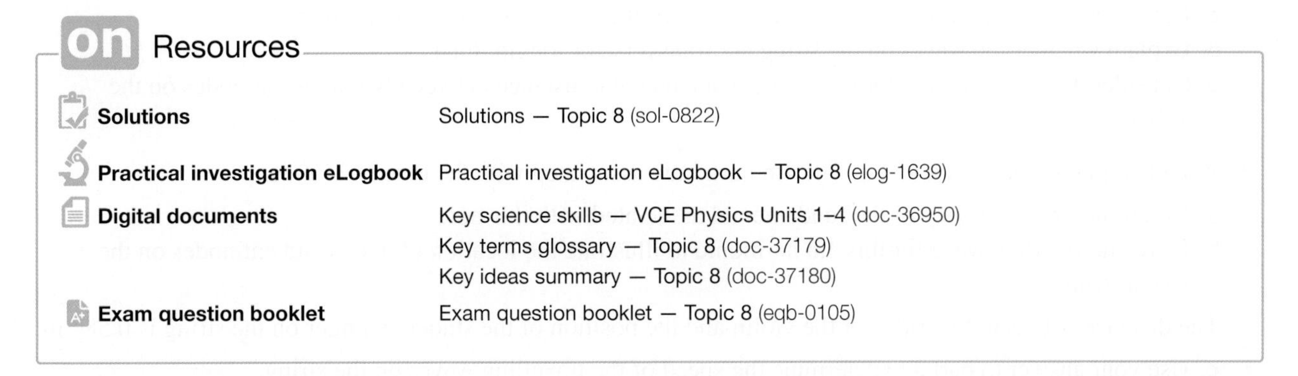

on Resources

Solutions	Solutions — Topic 8 (sol-0822)
Practical investigation eLogbook	Practical investigation eLogbook — Topic 8 (elog-1639)
Digital documents	Key science skills — VCE Physics Units 1–4 (doc-36950) Key terms glossary — Topic 8 (doc-37179) Key ideas summary — Topic 8 (doc-37180)
Exam question booklet	Exam question booklet — Topic 8 (eqb-0105)

8.6 Activities

learnon

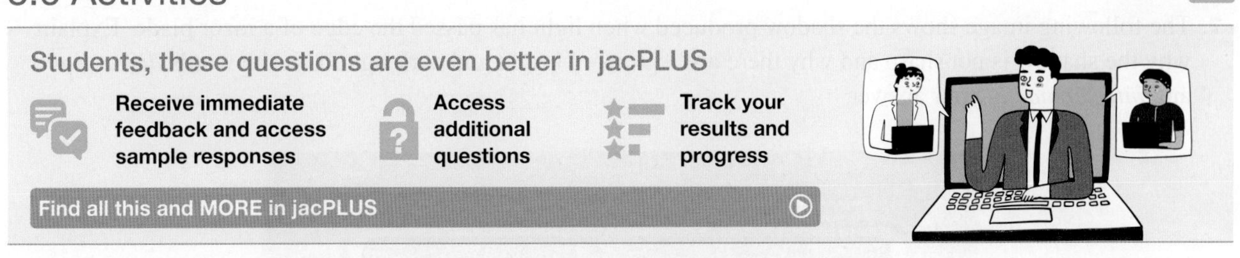

8.6 Review questions

1. A student is studying surface waves using water at a local swimming pool. She makes waves at a rate of two every second and the ripples radiate away from the source.
   a. Calculate the period of the waves.
   b. The waves radiate away from the source at a speed of 2.5 m s^{-1}. Calculate the distance between two adjacent peaks, that is, the wavelength of the waves.
   c. If she increases the rate at which she makes waves, explain what will happen to the wavelength of the waves and what will happen to the speed of the waves.
2. Sound produced by an opera singer has a frequency of 926 Hz.
   a. Calculate the period of the sound wave.
   b. Taking the speed of sound to be 340 m s^{-1}, calculate the wavelength of this sound in air.
3. Blue light has a frequency of 6.5×10^{14} Hz, and yellow light has a frequency of 5.2×10^{14} Hz.
   a. Determine the wavelength of both blue and yellow light in air. Take the speed of light in air to be 3.0×10^{8} m s^{-1}.
   b. Determine the period of both blue and yellow light.
4. Lasers can be rapidly switched on and off to produce a pulse of light. A particular pulse of blue light (6.5×10^{14} Hz) consists of 1.0×10^{6} complete cycles.
   Calculate the distance between the start and end of this pulse. Hint: how much time would it take to produce this pulse?
5. On a guitar, the bottom string is called the E-string. When tuned correctly, the fundamental or first harmonic note has a frequency of 82.41 Hz.
   a. When the E-string is plucked, a series of different frequencies or harmonics is produced. Determine the frequencies of the first four harmonics.

The length of the E-string on a typical acoustic guitar is 0.750 m and the string can be considered to be fixed at both ends.

b. Calculate the wavelength of the fundamental tone on the string.
c. Use your result to part **b** to determine the speed of the travelling waves on the string.
d. Explain whether the waves on the string are transverse or longitudinal.
e. Consider the 3rd harmonic for this string. Calculate the distance between two adjacent nodes on the string.

6. A student plays a note on a violin. The 4th harmonic associated with the note is 2460 Hz.
   a. Determine the frequency of the fundamental associated with this note.
   b. Draw the standing wave for this 4th harmonic to illustrate the location of nodes and antinodes on the violin string.

   The distance between the bridge of the violin and the position of the student's finger on the string is 0.580 m.

   c. Use your answer to part a to determine the speed of the travelling waves on the string.
   d. Determine the distance between two adjacent antinodes on the string for the 4th harmonic.
7. The following image shows the shadow produced when light has passed the edge of a razor blade. Explain why the shadow is not sharp and why there are regions of bright and dark light. Use the words *diffraction* and *interference* in your answer.

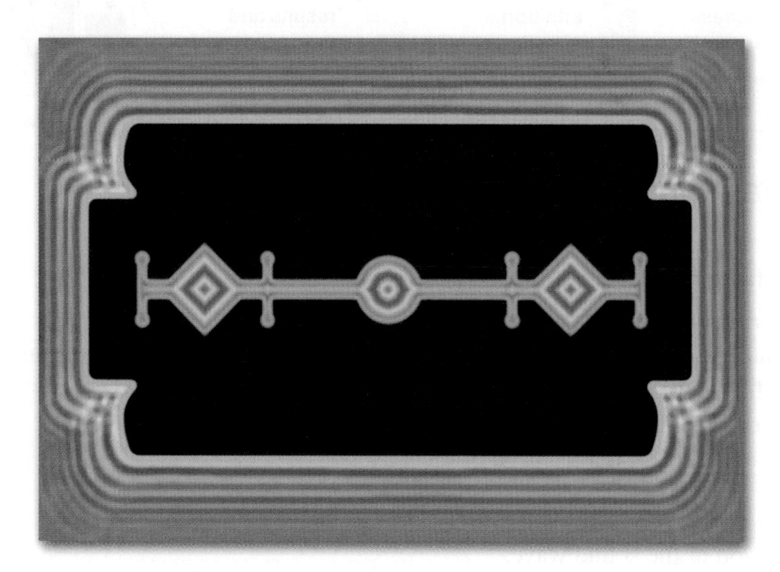

Source: National High Magnetic Field Laboratory

8. Describe what diffraction is and state changes in a diffraction pattern when:
   a. the wavelength of the waves is decreased, keeping the width of the opening constant
   b. the width of the opening is decreased, keeping the wavelength constant.
9. When light passes through a narrow slit, a diffraction pattern is produced and can be displayed on a screen. State two ways in which the amount of diffraction can be increased so that the pattern appears to be more spread out.
10. Using your knowledge of interference, explain why a pattern consisting of many bright and dark regions appears on a screen when light from a single source passes through two closely spaced, narrow slits.
11. A two-source interference pattern is produced in a school pool with two sources S_1 and S_2, each making periodic surface waves with wavelength 0.80 m. This is shown in the following diagram with the horizontal dotted line representing where the path difference is zero for the two sources.

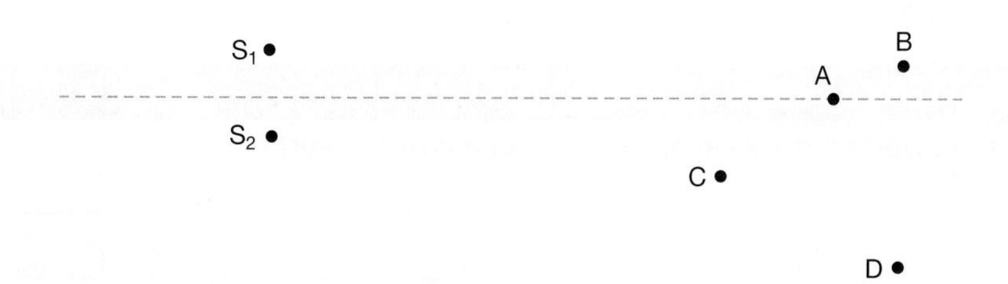

a. The distance $S_1A = 12.0$ m. Calculate the distance S_2A.
b. The point B lies on the first nodal line away from the dotted line. Calculate the path difference $|S_1B - S_2B|$.
c. The path difference $|S_1C - S_2C|$ is 1.6 m. Explain whether waves from S_1 and S_2 are constructively or destructively interfering at point C.
d. The path difference $|S_1D - S_2D|$ is 3.0 m and $S_1D = 14.8$ m. Calculate the distance S_2D.

12. A two-source interference experiment is produced in a ripple tank. A student locates a point of no disturbance X along the first nodal line of the pattern, as shown in the following diagram.

a. In terms of the wavelength λ, calculate the path difference $|S_1X - S_2X|$.
b. The student measures the distances $S_1X = 14.8$ cm and $S_2X = 15.8$ cm. Calculate the wavelength of the surface waves generated in the ripple tank.
c. Waves are generated with frequency 8.0 Hz. Calculate the speed of the surface waves in the ripple tank.

13. The path difference to the third dark fringe in a standard two-slit interference pattern is 1250 nm.
a. Calculate the wavelength of the light used to make the interference pattern.
b. Calculate the path difference for the second bright fringe.

14. A two-slit interference pattern is constructed by a group of students, and a clear pattern like the one shown in the following image is cast onto the wall of a classroom. The distance between two adjacent bright fringes is shown as Δx. State three different ways in which the spacing Δx could be increased by a factor of 2.

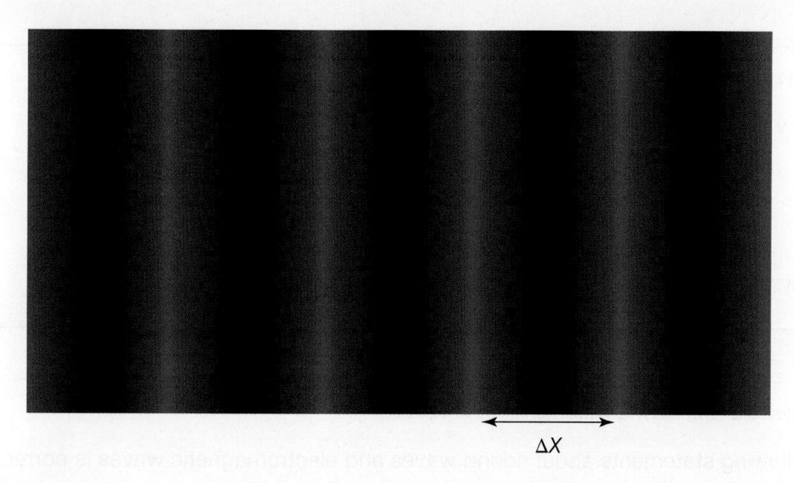

15. A group of students is using a laser that emits light of wavelength 650 nm. They point the beam at a pair of slits that they know has a slit separation of 0.10 mm. They wish to produce a two-slit interference pattern, where the bright fringes are separated by 2.0 cm, as part of a display for an open day at their school. Calculate the distance they must position the screen from the pair of slits to achieve this.

8.6 Exam questions

Section A — Multiple choice questions

All correct answers are worth 1 mark each; an incorrect answer is worth 0.

Question 1

Source: *VCE 2022 Physics Exam, Section A, Q.11;* © VCAA

Which one of the following statements best describes transverse and longitudinal waves?

A. Both transverse waves and longitudinal waves travel in a direction parallel to their vibrations.
B. Both transverse waves and longitudinal waves travel in a direction perpendicular to their vibrations.
C. Transverse waves travel in a direction perpendicular to their vibrations; longitudinal waves travel parallel to their vibrations.
D. Transverse waves travel in a direction parallel to their vibrations; longitudinal waves travel perpendicular to their vibrations.

Question 2

Source: *VCE 2022 Physics Exam, NHT, Section A, Q.12;* © VCAA

The diagram below represents a standing wave on a string fixed at both ends, with a node at the centre. The wave has a frequency of 5.0 Hz and the distance between the two fixed ends is 2.0 m.

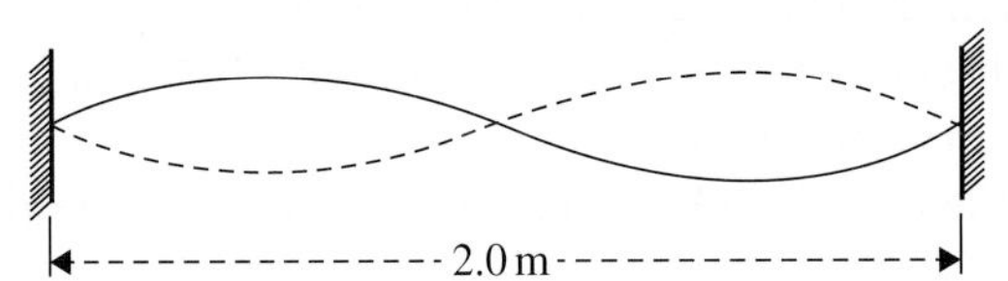

Which one of the following would be closest to the speed of a transverse wave travelling on the string?

A. $0.40\,\text{m}\,\text{s}^{-1}$
B. $2.5\,\text{m}\,\text{s}^{-1}$
C. $5.0\,\text{m}\,\text{s}^{-1}$
D. $10\,\text{m}\,\text{s}^{-1}$

Question 3

Source: *VCE 2019 Physics Exam, NHT, Section A, Q.13;* © VCAA

When a mechanical wave moves through a medium, there is a net transfer of

A. mass.
B. energy.
C. particles.
D. mass and energy.

Question 4

Source: *VCE 2019 Physics Exam, NHT, Section A, Q.14;* © VCAA

Which one of the following statements about sound waves and electromagnetic waves is correct?

A. Both sound waves and electromagnetic waves can travel through a vacuum.
B. Neither sound waves nor electromagnetic waves can travel through a vacuum.
C. Sound waves can travel through a vacuum but electromagnetic waves cannot travel through a vacuum.
D. Sound waves cannot travel through a vacuum but electromagnetic waves can travel through a vacuum.

Question 5

Source: *VCE 2018 Physics Exam, NHT, Section A, Q.14*; © VCAA

Which one of the following best describes electromagnetic waves?

A. They all travel at the same speed in all mediums.
B. They all travel at the same speed in a vacuum.
C. They are not reflected by a surface.
D. They always travel in straight lines.

Question 6

Source: *VCE 2018 Physics Exam, NHT, Section A, Q.15*; © VCAA

Which of the following best gives the different regions of the electromagnetic spectrum in order from longest wavelength to shortest wavelength?

A. ultraviolet, visible light, infra-red, microwaves
B. microwaves, ultraviolet, visible light, infra-red
C. visible light, ultraviolet, infra-red, microwaves
D. microwaves, infra-red, visible light, ultraviolet

Question 7

Source: *VCE 2018, Physics Exam, Section A, Q.12*; © VCAA

A teacher sets up an apparatus to demonstrate Young's double-slit experiment. A pattern of bright and dark bands is observed on the screen, as shown below.

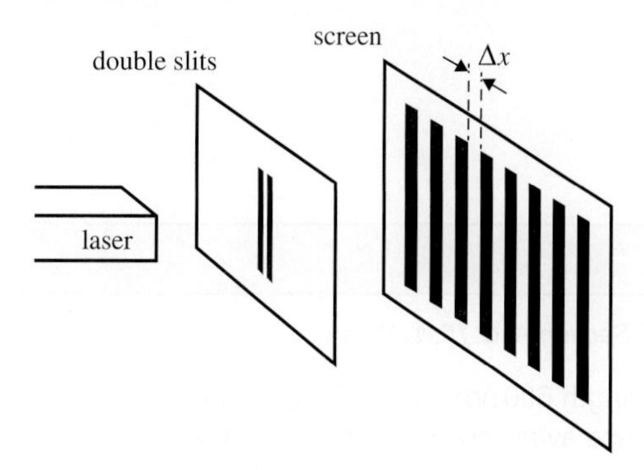

Which one of the following actions will increase the distance, Δx, between the adjacent dark bands in this interference pattern?

A. Decrease the distance between the slits and the screen.
B. Decrease the wavelength of the light.
C. Decrease the slit separation.
D. Decrease the slit width.

Question 8

Source: *VCE 2021 Physics Exam, NHT, Section A, Q.14;* © VCAA

The diagram below represents a standing wave.

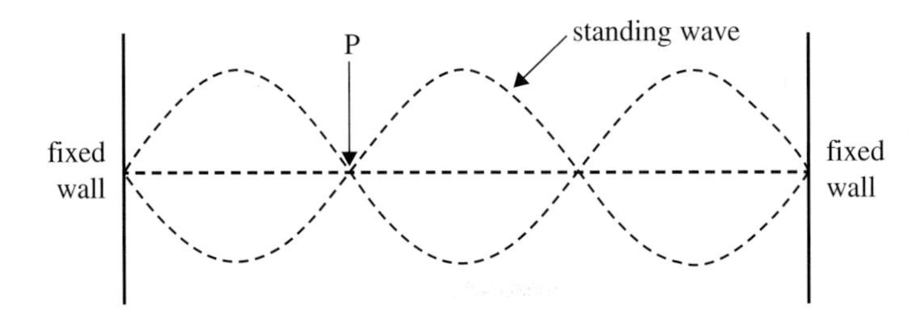

The point P on the standing wave is

A. a node resulting from destructive interference.
B. a node resulting from constructive interference.
C. an antinode resulting from destructive interference.
D. an antinode resulting from constructive interference.

Question 9

Source: *VCE 2015, Physics Exam, Section B, Detailed study 6 Q.1;* © VCAA

A loudspeaker emits a sound of frequency 30 Hz. The speed of sound in air in these conditions is 330 m s^{-1}.

Which one of the following best gives the wavelength of the sound?

A. 30 m
B. 11 m
C. 3.3 m
D. 0.091 m

Question 10

Source: *VCE 2019 Physics Exam, NHT, Section A, Q.15;* © VCAA

Monochromatic laser light of wavelength 600 nm shines through a narrow slit. The intensity of the transmitted light is recorded on a screen some distance away, as shown below in the diagram on the left.

The intensity graph of the pattern seen on the screen is shown below on the right.

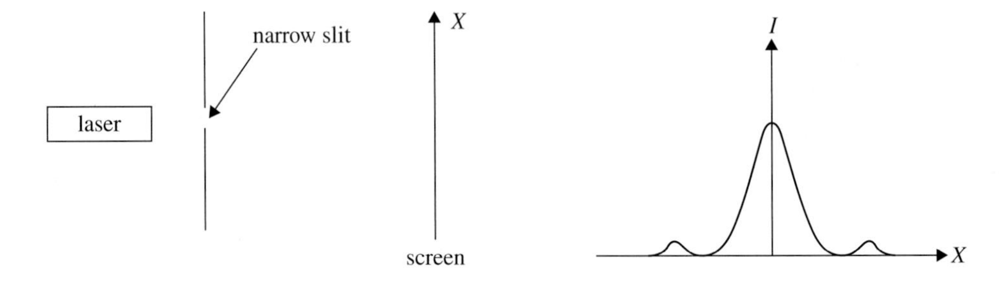

Which one of the following intensity graphs best represents the pattern that would be seen if a slightly wider slit were used?

A.

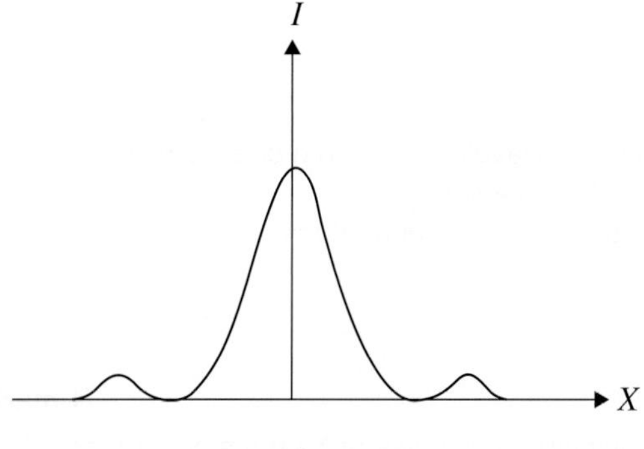

B.

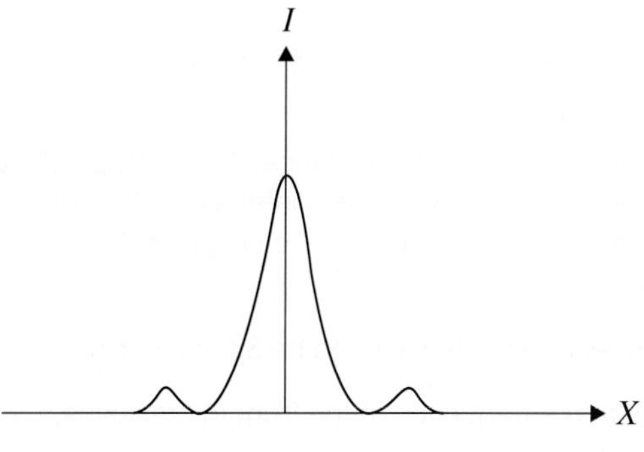

C.

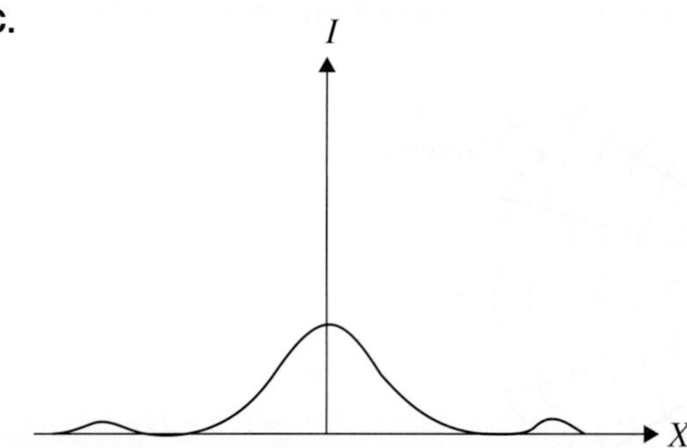

D.

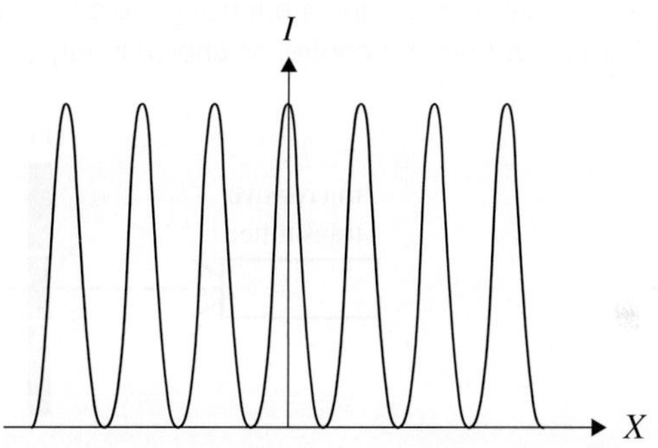

Section B — Short answer questions

Question 11 (4 marks)

Source: *VCE 2021 Physics Exam, NHT, Section B, Q.11a&b*; © VCAA

A transverse wave is travelling through a medium, as shown in Figure 14. The frequency of the source producing the wave is 40 Hz and the wave travels at a speed of 35 m s^{-1}. The amplitude of the wave is 0.50 m.

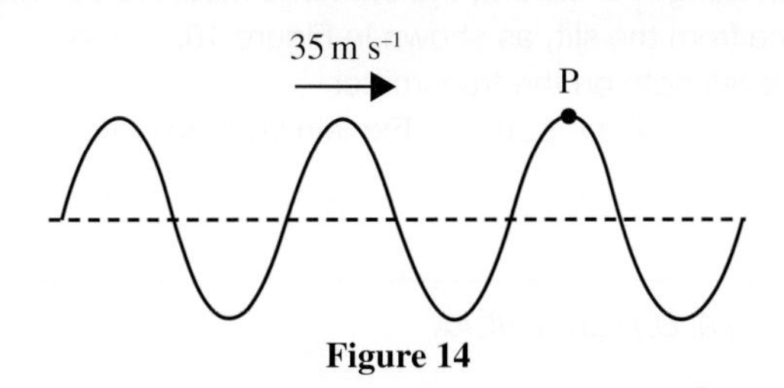

Figure 14

a. What is the period of oscillation for point P in Figure 14? **(1 mark)**

b. On the axes below, sketch the displacement versus time graph for the point P of this transverse wave, showing at least **two** complete cycles. Include scales and units on each axis. **(3 marks)**

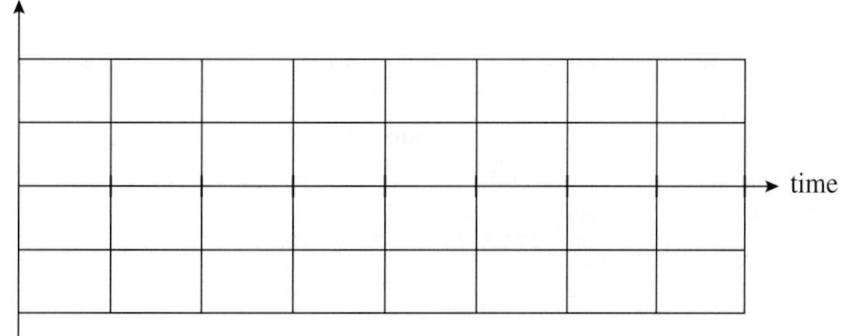

Question 12 (5 marks)

Source: *VCE 2018 Physics Exam, NHT, Section B, Q.11a-c*; © VCAA

Students are using a microwave set to study wave interference.

The set consists of:

- a microwave transmitter that can be set to produce microwaves of wavelength 3.0 cm or 6.0 cm
- a receiver that measures the intensity of the received signal and the wavelength
- plates that can be used to give single or double slits of various widths and separations
- a ruler.

Take the speed of microwaves to be $3.0 \times 10^8 \text{ m s}^{-1}$.

a. Calculate the frequency of the 3 cm microwaves. **(1 mark)**

b. The students set up the equipment using 3.0 cm microwaves, placing the receiver at X on the second nodal line (minima) out from the centre, as shown in Figure 9.

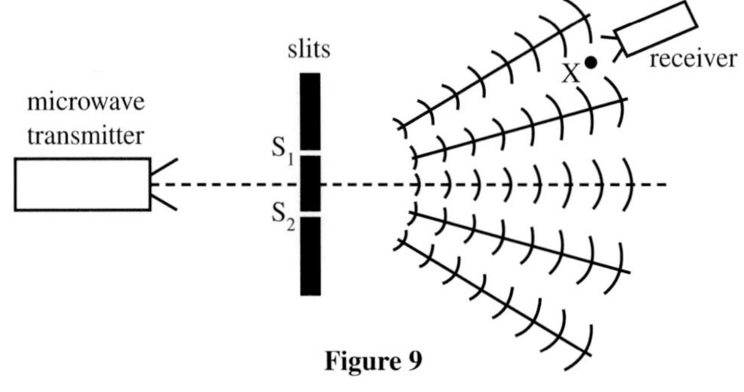

Figure 9

Calculate the path difference $S_2X - S_1X$. Show your working. **(2 marks)**

c. The students now replace the two slits with a slit of width w, as shown in Figure 10.

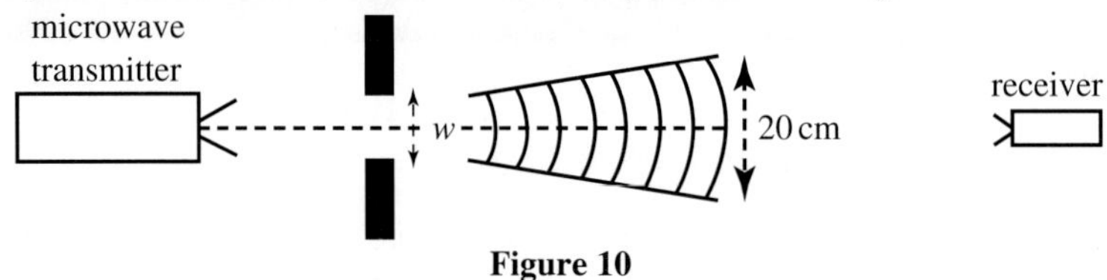

Figure 10

With the transmitter set to a wavelength of 3.0 cm, the students measure the width of the diffraction pattern to be 20 cm at a particular distance from the slit, as shown in Figure 10.
They then switch to a 6.0 cm wavelength on the transmitter.
What effect will this have on the width of the pattern? Explain your answer. **(2 marks)**

Question 13 (4 marks)

Source: *VCE 2018 Physics Exam, Section B, Q.11a,b*; © VCAA

Figure 13 shows two speakers, A and B, facing each other. The speakers are connected to the same signal generator/amplifier and the speakers are simultaneously producing the same 340 Hz sound.

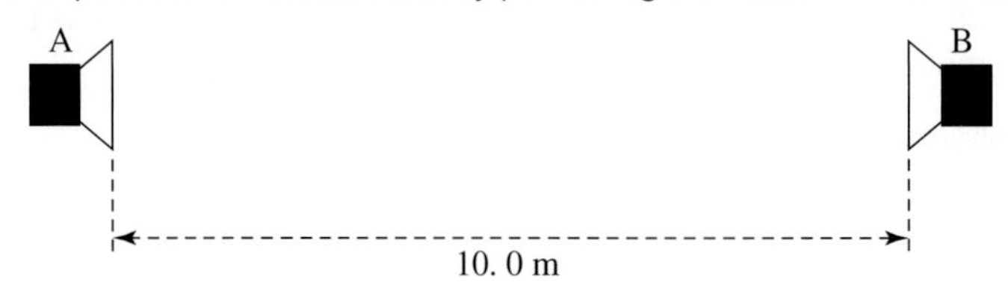

Figure 13

Take the speed of sound to be 340 m s^{-1}.

a. Calculate the wavelength of the sound. **(1 mark)**

b. A student stands in the centre, equidistant from speakers A and B. He then moves towards speaker B and experiences a sequence of loud and quiet regions. He stops at the second region of quietness.
How far has the student moved from the centre? Explain your reasoning. **(3 marks)**

Question 14 (4 marks)

Source: *VCE 2019 Physics Exam, Section B, Q.14a&b; © VCAA*

Students have set up a double-slit experiment using microwaves. The beam of microwaves passes through a metal barrier with two slits, shown as S_1 and S_2 in Figure 13. The students measure the intensity of the resulting beam at points along the line shown. They determine the positions of maximum intensity to be at the points labelled P_0, P_1, P_2 and P_3. Take the speed of electromagnetic radiation to be 3.00×10^8 m s^{-1}.

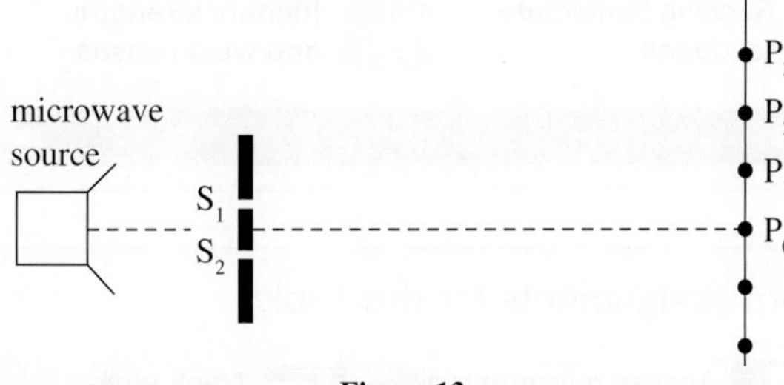

Figure 13

The distance from S_1 to P_3 is 72.3 cm and the distance from S_2 to P_3 is 80.6 cm.

a. What is the frequency of the microwaves transmitted through the slits? Show your working. **(2 marks)**

b. The signal strength is at a minimum approximately midway between points P_0 and P_1.
Explain the reason why the signal strength would be a minimum at this location. **(2 marks)**

Question 15 (4 marks)

Source: *VCE 2018 Physics Exam, Section B, Q.13b&c; © VCAA*

Physics students studying interference set up a double-slit experiment using a 610 nm laser, as shown in Figure 15.

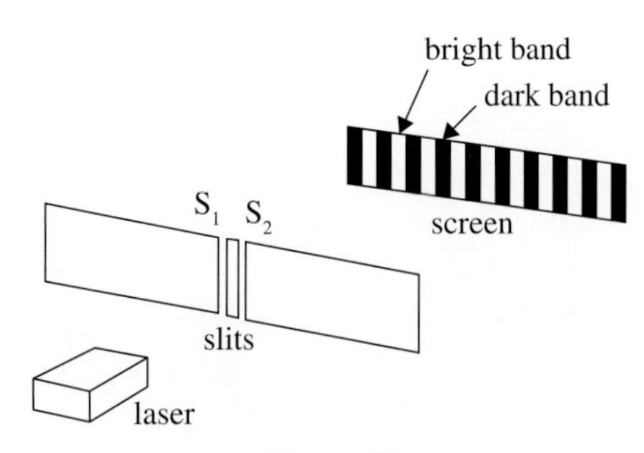

Figure 15

A section of the interference pattern observed by the students is shown in Figure 16. There is a bright band at point C, the centre point of the pattern.

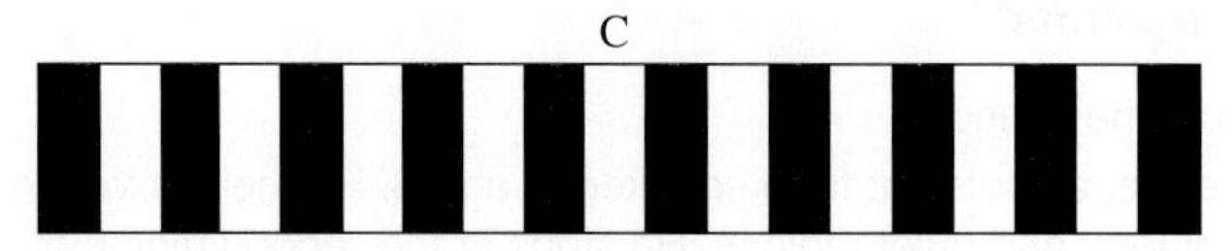

Figure 16

a. Explain why point C is in a bright band rather than in a dark band. **(2 marks)**

b. Another point on the pattern to the right of point C is further from S_1 than S_2 by a distance of 2.14×10^{-6} m. Mark this point on Figure 16 by writing an X above the point. You must use a calculation to justify your answer. **(2 marks)**

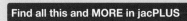

9 Light as a particle

KEY KNOWLEDGE

In this topic, you will:

- apply the quantised energy of photons: $E = hf = \frac{hc}{\lambda}$
- analyse the photoelectric effect with reference to
  - evidence for the particle-like nature of light
  - experimental data in the form of graphs of photocurrent versus electrode potential, and of kinetic energy of electrons versus frequency
  - kinetic energy of emitted photoelectrons: $E_{k\ max} = hf - \phi$, using energy units of joule and electron-volt
  - effects of intensity of incident irradiation on the emission of photoelectrons
- describe the limitation of the wave model of light in explaining experimental results related to the photoelectric effect.

PRACTICAL WORK AND INVESTIGATIONS

Practical work is a central component of VCE Physics. Experiments and investigations, supported by a **practical investigation eLogbook** and **teacher-led video**, are included in this topic to provide opportunities to undertake investigations and communicate findings.

EXAM PREPARATION

Access past VCAA questions and exam-style questions and their video solutions in every lesson, to ensure you are ready.

9.1 Overview

Hey students! Bring these pages to life online

Watch videos

Engage with interactivities

Answer questions and check results

Find all this and MORE in jacPLUS

9.1.1 Introduction

The photoelectric effect, the basis for how solar panels work, shows that light can transfer energy as though it were a particle and not a wave. How is it that the wave model of light works extremely well in some circumstances but not in others? This was the problem facing scientists at the start of the twentieth century, to explain both the photoelectric effect and the way hot bodies emit light (as a wave). It was Albert Einstein who challenged everyone's understanding of the nature of light when he presented his quantum theory of light, which controversially contradicted Charles Maxwell's wave theory. Einstein suggested that light can be modelled as particles in some circumstances. This led to a revolutionary new area of theoretical physics named quantum mechanics.

Today it is accepted that neither a traditional wave model nor a traditional particle model adequately describes the properties of light on its own. Light seems to propagate as a wave but paradoxically transfers energy sometimes as a particle.

FIGURE 9.1 The work of Albert Einstein is central to the present-day understanding of the photoelectric effect.

LEARNING SEQUENCE

Resources

Solutions Solutions — Topic 9 (sol-0823)

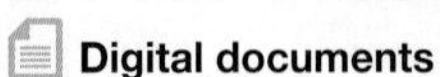

Practical investigation eLogbook Practical investigation eLogbook — Topic 9 (elog-1640)

Digital documents Key science skills — VCE Physics Units 1–4 (doc-36950)
Key terms glossary — Topic 9 (doc-37181)
Key ideas summary — Topic 9 (doc-37182)

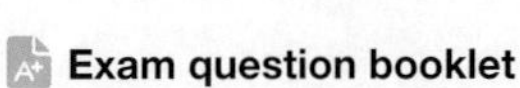

Exam question booklet Exam question booklet — Topic 9 (eqb-0106)

9.2 Could light have particle-like properties as well?

KEY KNOWLEDGE

- Apply the quantised energy of photons: $E = hf = \frac{hc}{\lambda}$

Source: VCE Physics Study Design (2024–2027) extracts © VCAA; reproduced by permission.

9.2.1 Planck's equation

By the latter half of the nineteenth century, the ability of Newtonian mechanics to predict and explain much of the material world was unquestioned. At the same time, discoveries in chemistry showed that the world consisted of different elements (each made up of identical atoms), and compounds (each made up of combinations of atoms in fixed proportions). Most scientists believed that all matter was made up of particles and that the universe was governed by deterministic mechanical laws — that is, they thought the universe was like a big machine. Newtonian mechanics allowed them to explain the workings of the universe in terms of energy transformations, momentum transfer, and the conservation of energy and momentum due to the action of well-understood forces.

Thomas Young had shown that the behaviour of light passing through narrow slits could be explained using the concept of waves and Maxwell produced a model where light was treated as a transverse wave consisting of both electric and magnetic vibrating fields, and requiring no medium for it to move (see topic 8).

Maxwell's theoretical wave model for light was able to show that the energy associated with electromagnetic waves was related to the size or amplitude of the wave. The more intense the light, the greater the amplitude and hence the energy it contained. He was also able to show that an electromagnetic wave had momentum and was thus capable, in principle, of exerting forces on other objects. According to Maxwell's model, the amount of momentum contained in an electromagnetic wave, p, is related to the energy contained in the wave, E, by the following simple equation:

$$p = \frac{E}{c} \text{ or } E = pc$$

where: E is the energy of an electromagnetic wave, in J

p is the magnitude of the momentum of an electromagnetic wave, in N s (or kg m s^{-1})

$c = 3.0 \times 10^8$ m s^{-1} (the speed of light)

At the same time, Max Planck was trying to understand how hot objects emit electromagnetic waves. He studied light emitted by incandescent objects such as the Sun, light bulbs and wood fires. Planck could make his mathematical models fit the available data only if he conceded that the energy associated with the electromagnetic radiation emitted was directly proportional to the frequency of radiation and, importantly, that the energy came in bundles that he called quanta. The word *quanta* is plural for **quantum**, a word meaning a small quantity of a fixed amount. These energy quanta of light are now called photons. It appeared as though light transferred energy to matter and vice versa, behaving more like a localised particle than by a wave spreading out from a source.

What all this meant was not clear — Maxwell's wave model for light worked extremely well and yet understanding incandescent objects required a model that concentrated energy into localised packets called quanta that were more like particles.

A pair of problems existed. One question was how matter could convert some of its kinetic and potential energy into light. Max Planck and other scientists

quantum a small quantity of a fixed amount

were working on this problem as part of their efforts to understand black body radiation (radiation emitted by incandescent objects). The other question was how light could transfer its energy to matter. This process became known as the photoelectric effect.

Planck's conclusion about a particle nature for light did not fit comfortably with the successful wave model that Maxwell proposed. It would remain for Albert Einstein over a decade later to interpret this apparent quandary with other experimental data. In reward for his success, he won the Nobel Prize in Physics in 1921. Einstein's interpretation asserted that light is best thought of as a stream of particles, now called photons, with each photon carrying energy $E_{photon} = hf$ and capable of transferring this energy to other particles such as electrons.

$$E = hf = \frac{hc}{\lambda}$$

where: E is the energy of a quantum of light (quantised energy of photons), in J
f is the frequency of the electromagnetic radiation, in Hz
$h = 6.63 \times 10^{-34}$ J s, which is a constant known as 'Planck's constant'

tlvd-9013

SAMPLE PROBLEM 1 Calculating the energy and momentum of photons

a. Blue light has a frequency of 6.7×10^{14} Hz.
  i. Calculate the energy associated with a photon of blue light.
  ii. Calculate the momentum associated with a quantum of blue light.
b. Calculate the momentum of a quantum of red light of wavelength 650 nm.

THINK	WRITE
a. i. The energy of the blue light E is given by $E = hf$.	a. i. $E = hf$ $= 6.63 \times 10^{-34} \times 6.7 \times 10^{14}$ $= 4.4 \times 10^{-19}$ J
ii. The momentum p is given by $p = \frac{E}{c}$.	ii. $p = \frac{E}{c}$ $= \frac{4.4 \times 10^{-19}}{3.0 \times 10^{8}}$ $= 1.5 \times 10^{-27}$ N s
b. 1. From the wavelength, find the frequency. From the frequency, find the energy.	b. $f = \frac{c}{\lambda}$ and $E = hf \Rightarrow E = \frac{hc}{\lambda}$
2. From the energy, find the momentum.	$p = \frac{E}{c} \Rightarrow p = \frac{hc}{\lambda c} \Rightarrow p = \frac{h}{\lambda}$
3. Substitute the values for h and λ and solve for p, ensuring that the wavelength has been converted to metres (650 nm = 6.50×10^{-7} m).	$p = \frac{h}{\lambda}$ $= \frac{6.63 \times 10^{-34}}{6.50 \times 10^{-7}}$ $= 1.02 \times 10^{-27}$ N s

PRACTICE PROBLEM 1

A quantum of light has a momentum of 9.8×10^{-28} N s. Calculate the frequency of the light.

tlvd-9014

SAMPLE PROBLEM 2 Determining the number of photons emitted each second by a light source

a. What is the energy of each photon emitted by a source of green light having a wavelength of 515 nm?

b. How many photons per second are emitted by a light source emitting a power of 0.3 W as 515 nm light?

THINK

a. The photon energy can be calculated using $E_{photon} = \frac{hc}{\lambda}$
where: $h = 6.63 \times 10^{34}$ J s,
$c = 3.0 \times 10^{8}$ m s^{-1} and
$\lambda = 515$ nm $= 515 \times 10^{-9}$ m

WRITE

a. $$E_{photon} = \frac{hc}{\lambda}$$
$$= \frac{6.63 \times 10^{-34} \times 3.0 \times 10^{8}}{515 \times 10^{-9}}$$
$$= 3.86 \times 10^{-19}\ \text{J}$$

THINK

b. 1. Determine an expression for the power emitted by the globe.

WRITE

b. $$\text{power} = \frac{\text{energy emitted}}{\text{time interval}}$$
$$= \frac{E}{\Delta t}$$
$$= \frac{N E_{photon}}{\Delta t}$$
where: N is the number of photons emitted in the time interval Δt.

2. Rearrange the power equation to make N the subject.

$$N = \frac{\text{power} \times \Delta t}{E_{photon}}$$

3. Calculate the number of photons emitted per second by substituting $\Delta t = 1$s, power $= 0.3$ W, and $E_{photon} = 3.86 \times 10^{-19}$ J

$$N = \frac{0.3 \times 1}{3.86 \times 10^{-19}}$$
$$= 8 \times 10^{17}\ \text{photons s}^{-1}$$

Since each photon carries a tiny amount of energy, huge numbers of photons are emitted from quite ordinary light sources in each second.

PRACTICE PROBLEM 2

A radio station has a 1000-W transmitter and transmits electromagnetic radiation with a frequency 104.6 MHz. Calculate the number of photons emitted per second by the transmitter.

9.2.2 Measuring the energy of light, electrons and photoelectrons

To appreciate the results of the photoelectric effect, it is necessary to be able to calculate both the energy associated with light and the energy associated with a moving particle such as an electron.

Measuring the energy of light

The energy associated with light, E, provided it is treated as a localised object as necessitated by Planck, can be equated to the product of the frequency and Planck's constant: $E = hf$. The speed of light is related to the frequency and wavelength: $c = f\lambda$, in accordance with a wave model for light. For completeness, since $E = pc$, the momentum associated with light, $\boldsymbol{p}$, can be related to the wavelength λ by the equation $p = \frac{h}{\lambda}$. It needs to be mentioned at this stage that both a wave model for light and a particle model for light have been used simultaneously. This usage of two models simultaneously came to be known as the wave–particle duality, and for many years it remained an unresolved component in physics. With the development of quantum mechanics in the 1920s, a consistent mathematical model incorporating both aspects emerged.

Measuring the energy of electrons

Potential differences can be used to accelerate and decelerate charged particles. Kinetic energy of a charged particle is related to the electrical potential difference through which it can be made to move. Understanding this relationship will make understanding the photoelectric effect easier.

The simplest way to accelerate electrons is with two parallel metal plates in an evacuated chamber (figure 9.2). The two plates are connected to a DC power supply (similar to a capacitor connected to a battery). An electron will experience an electric force anywhere in the region between the plates: it will be attracted by the positively charged plate and repelled by the negatively charged plate (see topic 4).

FIGURE 9.2 An electric field set up between two parallel plates connected to a battery

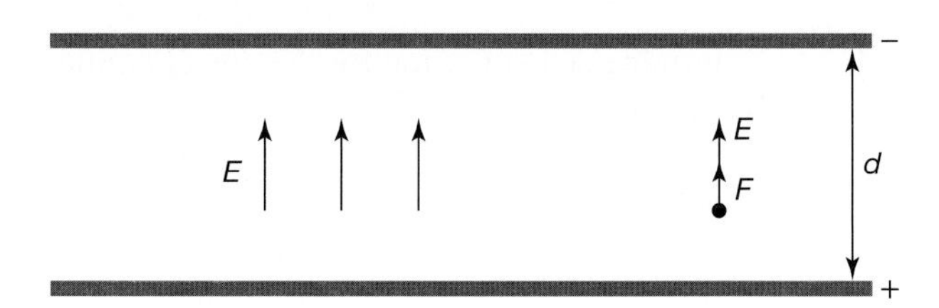

The size of this electric force will be the same throughout this region, as the electric field is uniform. This constant electric force on a charge placed between the plates can be compared to the constant gravitational force on a mass located above the ground. In gravitation, where the force acts on the mass of an object:

Electric force and electric field between parallel charged plates

$$F = qE \text{ and } E = \frac{V}{d}$$

where: F is the electric force on the charged particle, in N

q is the charge of the particle experiencing the force, in C

E is the electric field strength, in N C^{-1} (or V m^{-1})

V is the electric potential difference, in V

d is the distance between the plates

These two relationships for the electric field $\left(E = \frac{F}{q} \text{ and } E = \frac{V}{d}\right)$ give it two equivalent units: newtons per coulomb (N C^{-1}) and volts per metre (V m^{-1}).

These two relationships can also be linked by considering energy. The gain in energy of the electron can be obtained by calculating the work done on the charge to move it from one plate to the other. It can also be obtained by recalling that the voltage across a battery equals the energy gained by one coulomb of charge. So:

$$\text{work done by the potential difference, } W = \text{force} \times \text{distance} = \text{voltage} \times \text{electric charge}$$
$$= F \times d = V \times q$$

The work done by the potential difference, V, on a free electron is equal to the change in the kinetic energy of the electron, ΔE_k. Since kinetic energy is given by the expression $\frac{1}{2}mv^2$, and by further making the assumption that the initial kinetic energy of an electron emitted is zero, a useful non-relativistic equation is obtained.

This equation is interpreted in the following way. For a given potential difference, V, acting on an electron, both the speed of the electron and hence its momentum ($p = mv$), as well as its energy, E_k, can be calculated.

$$E_k = Vq = \frac{1}{2}mv^2$$

where: E_k is the energy of an electron, in J
V is the potential difference an electron passes through, in V
$q = 1.6 \times 10^{-19}$ C is the charge on one electron
$m = 9.1 \times 10^{-31}$ kg is the mass of an electron
v is the velocity of the electron, in m s^{-1}

Thus, an arrangement of negative and positive charged plates can be used to accelerate a charged particle in a straight line. This arrangement came to be known as an **electron gun** (figure 9.3). By reversing the polarity of the charge on the plates, electrons with kinetic energy can be decelerated. The voltage required to achieve this stopping of electrons with energy is known as a stopping voltage and plays a central role in understanding the photoelectric effect.

electron gun a device to provide free electrons for a linear accelerator, usually consists of a hot wire filament with a current supplied by a low-voltage source

FIGURE 9.3 The electrons are attracted across to the positive plate and pass through the hole that is in line with the beam.

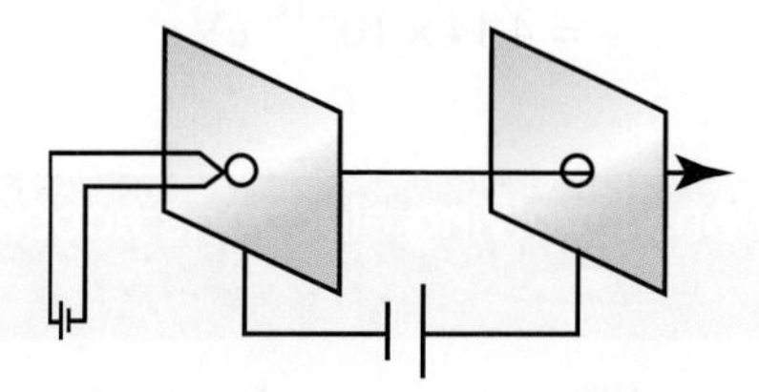

Measuring the energy of photoelectrons

In the photoelectric effect, energy is transferred from light to electrons. Philipp Lenard, a German physicist, was able to measure the maximum kinetic energy of these electrons by applying a stopping voltage to stop them. Recall that the work done on a charge, q, passing through a potential difference, V, is equal to qV. That is, an electron passing through a potential difference of 3.0 V would have 1.6×10^{-19} C $\times$ 3.0 J C^{-1} = 4.8×10^{-19} J or 3.0 eV of work done on it. If the potential difference is arranged so that the emitted electrons leave the positive terminal and are collected at a negative terminal, then electrons lose 4.8×10^{-19} J of energy.

A JOULE AND AN ELECTRON VOLT

Remember that a joule is the electric potential energy change that occurs when one coulomb of charge moves through a potential difference of one volt.

$$1\text{V} = \frac{1\text{J}}{1\text{C}}$$
$$\Rightarrow 1\text{J} = 1\text{C} \times 1\text{V}$$

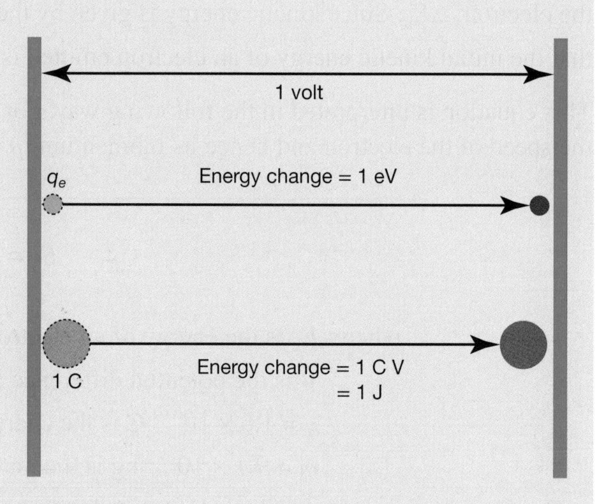

An **electron volt** is defined as the electric potential energy change that occurs when one electronic charge, $q_e = 1.6021 \times 10^{-19}$ C, moves through one volt.

$1 \text{ eV} = 1q_e \times 1 \text{ V}$

Where q_e is the magnitude of charge of an electron:

$$\Rightarrow 1 \text{ eV} = 1.6021 \times 10^{-19} \text{C} \times 1\text{V}$$
$$\Rightarrow 1 \text{ eV} = 1.6021 \times 10^{-19} \text{J}$$

The electron volt is a very useful unit for calculations involving small amounts of energy. To use the equation $E = hf$, Planck's constant (6.63×10^{-34} J s) needs to be converted into electron volts if the energy is given in electron volts (eV).

$$h = 6.63 \times 10^{-34} \text{ J s}$$
$$= \frac{6.63 \times 10^{-34}}{1.6 \times 10^{-19}}$$
$$= 4.14 \times 10^{-15} \text{ eV s}$$

electron volt the quantity of energy acquired by an elementary charge ($q_e = 1.6 \times 10^{-19}$ C) passing through a potential difference of 1 V. Thus, 1.6×10^{-19} J = 1 eV

tlvd-9015

SAMPLE PROBLEM 3 Calculations involving electrons accelerated by an electron gun using potential difference

An electron gun uses a 500-V potential difference to accelerate electrons evaporated from a tungsten filament. Assume that the evaporated electrons have zero kinetic energy.

a. **How much work is done on an electron that has moved across a potential difference of 500 V?**
b. **What type of energy is this work transformed into?**
c. **Calculate the kinetic energy of the electrons in electron volts and joules.**
d. **Using the equation for the kinetic energy, E_k, of a particle with mass m, determine the speed, v, of these electrons.**
e. **Calculate the momentum of these electrons.**

THINK	WRITE
a. Use $W = Vq$ to calculate the work done.	a. $W = Vq$ $= 500 \times 1.6 \times 10^{-19}$ $= 8.0 \times 10^{-17}$ J (or 500 eV)
b. Potential energy available is transformed into the kinetic energy of the electron: $W = Vq = \Delta E_k$	b. kinetic energy

c. Assuming the initial kinetic energy of the electrons evaporated from a tungsten filament is 0, the kinetic energy of the electrons is equal to the work done: $E_k = W$.

c. $E_k = W$

$= 8.0 \times 10^{-17}$ J

(or 500 eV)

d. 1. $E_k = \frac{1}{2}mv^2$, provided the electron speed is sufficiently small to ignore relativistic effects.

d. $E_k = \frac{1}{2}mv^2 = 8.0 \times 10^{-17}$ J

2. Take the mass of an electron to be $m = 9.1 \times 10^{-31}$ kg and solve equation for v.

$$v = \sqrt{\frac{2E_k}{m}}$$

$$= \sqrt{\frac{2 \times 8.0 \times 10^{-17}}{9.1 \times 10^{-31}}}$$

$$= 1.33 \times 10^7 \text{ m s}^{-1}$$

This is substantially slower than the speed of light; therefore, relativistic effects may be ignored.

e. Calculate the momentum using $p = mv$.

e. $p = mv$

$= 9.1 \times 10^{-31} \times 1.33 \times 10^7$

$= 1.2 \times 10^{-23}$ N s

PRACTICE PROBLEM 3

An electron in a beam of electrons generated by an electron gun has energy 1.26×10^{-17} J.

a. **Calculate the energy of this electron in electron volts.**

b. **State the potential difference required to stop electrons with this energy, that is, to remove their kinetic energy and bring them to rest.**

c. **Determine the speed of the electron, assuming that its kinetic energy is given by the equation $E_k = \frac{1}{2}mv^2$.**

d. **Use your answer to (c) to calculate the momentum of this electron.**

Like with the electron, these same principles can be used for photons. A **photon** can be considered a particle of light, a localised packet of energy $E = hf$ and momentum $p = \frac{E}{c} = \frac{h}{\lambda}$.

photon a discrete bundle of electromagnetic radiation. Photons can be thought of as discrete packets of light energy with zero mass and zero electric charge

An electron, modelled as a particle, has a kinetic energy $E_k = \frac{1}{2}mv^2$ and momentum $\boldsymbol{p} = mv$. The kinetic energy can be written as:

$$E_k = \frac{p^2}{2m}$$

9.2 Activities

Students, these questions are even better in jacPLUS

 **Receive immediate feedback and access sample responses**

 **Access additional questions**

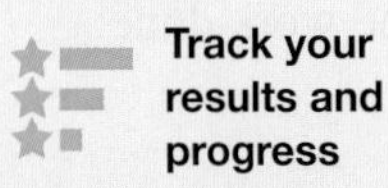 **Track your results and progress**

Find all this and MORE in jacPLUS

9.2 Quick quiz	9.2 Exercise	9.2 Exam questions

9.2 Exercise

1. The light from a red light-emitting diode (LED) has a frequency of 4.59×10^{14} Hz.
   a. What is the wavelength of this light?
   b. What is the period of this light?
2. You can detect light when your eye receives as little as 2.0×10^{-17} J. How many photons of green light is this?
3. Fill in the gaps in the table shown with the missing wavelength, frequency, photon energy and photon momentum values for the five different sources of electromagnetic radiation.

	Source	Wavelength	Frequency	Energy	Momentum
a.	Infrared from CO_2 laser	10.6 μm			
b.	Red helium–neon laser			3.14×10^{-19} J	
c.	Yellow sodium lamp				1.125×10^{-27} kg m s^{-1}
d.	UV from excimer laser		1.55×10^{15} Hz		
e.	X-rays from aluminium			2.01×10^{-16} J	

4. A beam of electrons, each electron having a kinetic energy 3.2×10^{-18} J, is to be stopped by a potential difference. Calculate the stopping voltage (potential difference) required to bring these electrons to rest.
5. The following diagram shows an anode, a cathode and several electrons that have been ejected from the cathode by light. The electrons leaving the cathode surface have been labelled with their kinetic energy and their initial velocity vector. The anode is 5.0 mm from the cathode.

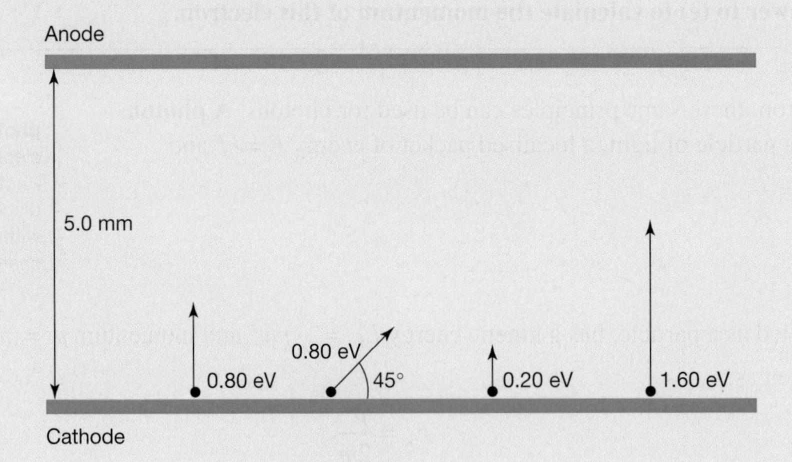

a. Calculate the speed of the electrons that have a kinetic energy of 0.80 eV or alternatively 1.28×10^{-19} J.
b. For each of the following, copy the diagram and sketch the path you would expect each electron to take for each of the potential differences, V. [Hint: think of each electron as a projectile in a uniform electric field, exactly like a ball thrown into the air.]
i. $V = 1.8$ V, with the anode positive relative to the cathode
ii. $V = 1.8$ V, with the anode negative relative to the cathode
iii. $V = 0.8$ V, with the anode negative relative to the cathode.

9.2 Exam questions

Question 1 (1 mark)

Source: *VCE 2022 Physics Exam, Section A, Q.17; © VCAA*

MC Gamma radiation is often used to treat cancerous tumours. The energy of a gamma photon emitted by radioactive cobalt-60 is 1.33 MeV.

Which one of the following is closest to the frequency of the gamma radiation?
A. 1.33×10^{6} Hz
B. 3.21×10^{20} Hz
C. 3.21×10^{21} Hz
D. 2.01×10^{39} Hz

Question 2 (3 marks)

Source: *VCE 2021 Physics Exam, NHT, Section B, Q.16a; © VCAA*

X-rays of wavelength 2.0 nm are emitted from an X-ray source.

Calculate the energy of one photon of these X-rays. Show your working.

Question 3 (1 mark)

Determine the wavelength of a photon with energy 4.52×10^{-34} J.

Question 4 (1 mark)

Determine the momentum of a photon with energy 4.52×10^{-34} J.

Question 5 (3 marks)

a. Determine the kinetic energy of an electron of mass 9.11×10^{-31} kg travelling at 4.50×10^{6} m s^{-1}. **(1 mark)**
b. Determine the speed of an electron of mass 9.11×10^{-31} kg with a kinetic energy of 3.20 eV. **(2 marks)**

More exam questions are available in your learnON title.

9.3 The photoelectric effect and experimental data

KEY KNOWLEDGE

- Analyse the photoelectric effect with reference to:
  - evidence for the particle-like nature of light
  - experimental data in the form of graphs of photocurrent versus electrode potential, and of kinetic energy of electrons versus frequency
  - kinetic energy of emitted photoelectrons: $E_{k\,max} = hf - \phi$, using energy units of joule and electron volt
  - effects of intensity of incident irradiation on the emission of photoelectrons.

The nineteenth-century view of light was developed as a result of the success of the wave model in explaining diffraction and interference. The wave model did a great job!

In 1887, the first signs of behaviour that could not be explained using the wave model almost went unnoticed. Heinrich Hertz was in the middle of experimental work that went on to show radio waves and light were really the same thing — electromagnetic waves. He produced radio waves with a frequency of about 5×10^8 hertz (the unit for frequency was named after him) by creating a spark across the approximately one-centimetre gap between two small metal spheres. The radio waves were detected up to several hundred metres away, by the spark they excited across another air gap, which was between the pointed ends of a circular piece of wire (figure 9.4). Hertz showed that the radio waves travelled at the speed of light. Although Hertz was not aware of it, this was the beginning of radio communication.

During his experiments, Hertz noticed that the spark showing the arrival of the radio waves at the receiver became brighter whenever the gap was simultaneously exposed to ultraviolet radiation. He was puzzled, made note of it but did not follow it up. It is now known that the reason for the brighter spark was because the ultraviolet radiation ejected electrons from the metal points of the detector. The presence of these electrons reduced the electrical resistance of the air gap, so the spark flashed brighter than usual whenever the radio waves were being detected.

This ejection of electrons by light is called the **photoelectric effect**. Following up Hertz's observations of this effect led to a breakthrough in the way the behaviour of light is viewed.

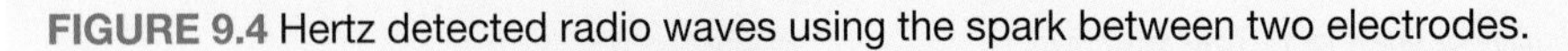

FIGURE 9.4 Hertz detected radio waves using the spark between two electrodes.

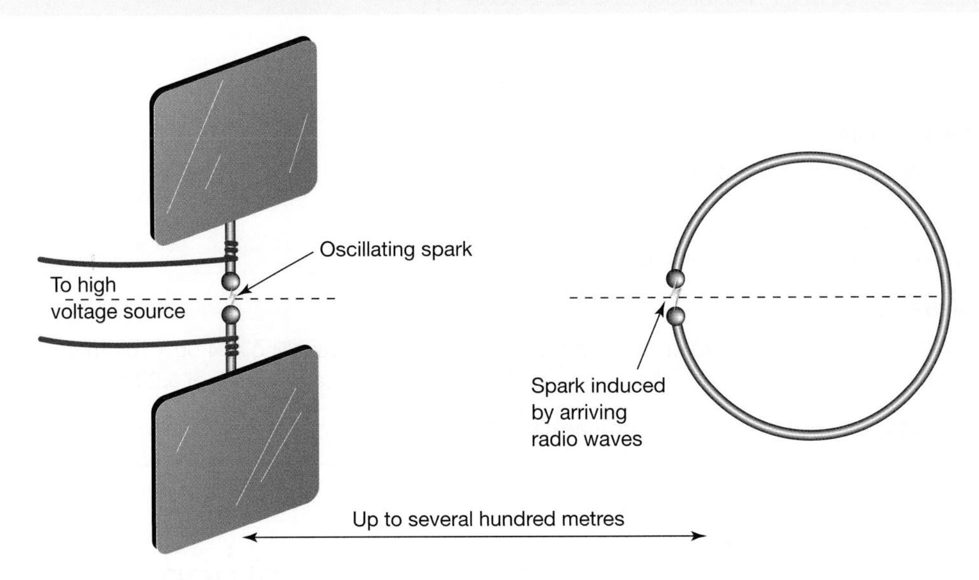

9.3.1 The experiment

Fifteen years passed before Philipp Lenard performed careful experiments to investigate the photoelectric effect. Lenard replaced Hertz's spark gap with two metal electrodes on opposite sides of an evacuated chamber. He investigated the energies of electrons ejected from one of the electrodes when light shone on it. The experimental arrangement used in 1902 by Lenard is shown in figure 9.5. Lenard designed his experiment so that he could vary several features of this arrangement.

- *The frequency and intensity of the light* could be varied. Light sources that emit light of only one frequency are called **monochromatic** light sources. Lenard varied the light intensity either by changing the arc current, or by moving the light source to a different distance from the window.

photoelectric effect the emission of electrons when electromagnetic radiation hits a metal surface

monochromatic light of a single frequency and, hence, very clearly defined colour

- *The potential difference between the electrodes in the chamber* could be varied by changing the position of the slide contact on the coiled resistor. The potential difference could be made either accelerating or retarding for electrons.
- Lenard could vary *the distance between the electrode receiving light, X, and the second electrode, Y.*

FIGURE 9.5 Philipp Lenard's experiment. Note that the point G is earthed, and this earths the electrode *Y*. Electrode *X* could be made either positive or negative relative to electrode *Y*.

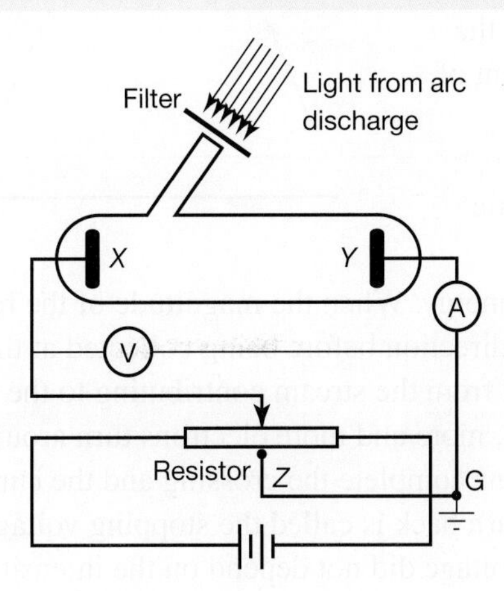

Lenard's findings on the effect of light on the photoelectric effect can be summarised as follows:

- photocurrent increases with the light amplitude
- photocurrent remains constant as the light frequency increases
- applying a retarding voltage decreases the photocurrent
- the stopping voltage is independant of the light amplitude
- the stopping voltage increases with the light frequency
- the kinetic energy of photoelectrons increases with the light frequency
- the emission of photoelectrons is immediate.

The graphs in figure 9.6 illustrate several important parts of Lenard's investigations. The numbers on the diagrams refer to the following numbered points.

① Lenard's results showed that the photocurrent was directly proportional to the light intensity.

FIGURE 9.6 The effect of changing light intensity from I_0 without changing its frequency

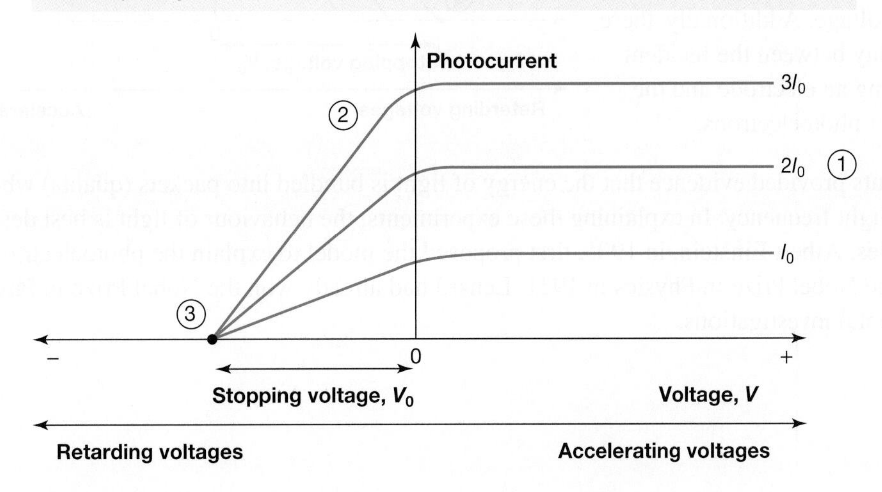

② When a retarding voltage was applied between the electrodes, the current decreased as the magnitude of the voltage increased. If the electric field between the plates exerts a force opposing the motion of the electrons, they would slow down and probably reverse direction before reaching the opposite electrode. This movement of the electron, shown in figure 9.7, is similar to the movement of a ball thrown into the air. The kinetic energy of the electrons would be converted into electric potential energy. When the magnitude of the retarding voltage is low, only the very slow electrons then reverse direction before being collected at the electrode Y and thus only a few electrons would then be removed from the stream contributing to the photocurrent. As the magnitude of the retarding voltage is increased, more and more electrons turn around before reaching the electrode until, at a particular voltage, no electrons complete the crossing and the current drops to zero. This minimum voltage causing all electrons to turn back is called the stopping voltage.

FIGURE 9.7 **a.** An electron in an electric field opposing its motion, and **b.** a ball thrown into the air

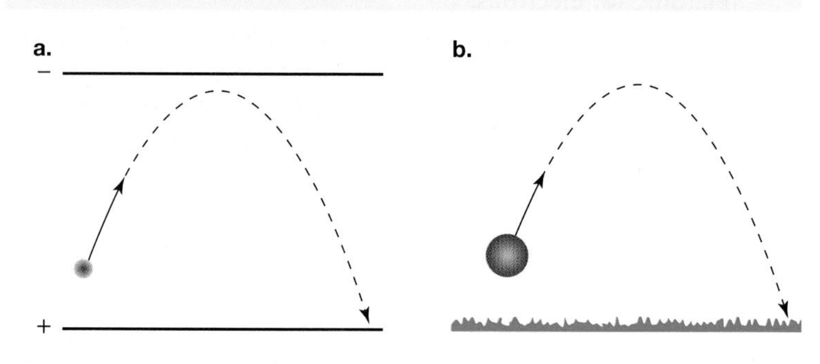

③ Lenard found that the stopping voltage did not depend on the intensity of incident irradiation being used. Brighter light *did not* increase the kinetic energy of the electrons emitted from the cathode. The same potential difference was required to convert all the kinetic energy of the electrons into electric potential energy, no matter how bright the light.

④ Lenard showed that the stopping voltage depends on both the frequency of the light (figure 9.8) and on the material of the electrode. For each material, a minimum frequency (called threshold frequency, f_0) was required for electrons to be ejected. Below this cut-off frequency, no electrons were ever ejected. Above this frequency, a photocurrent could always be detected. The higher the light frequency, the greater the stopping voltage. Additionally, there was no delay between the incident light striking an electrode and the emission of photoelectrons.

FIGURE 9.8 The effect of changing light frequency, without changing its intensity, on the photocurrent of one material

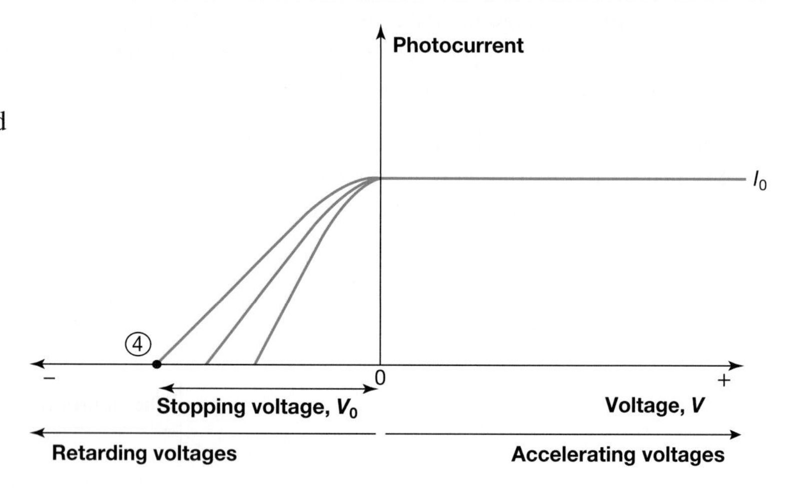

These experiments provided evidence that the energy of light is bundled into packets (quanta) whose energy depends on the light frequency. In explaining these experiments, the behaviour of light is best described as a stream of particles. Albert Einstein, in 1905, first proposed the model to explain the photoelectric effect, for which he won the Nobel Prize in Physics in 1921. Lenard had already won the Nobel Prize in Physics in 1905 for his experimental investigations.

tlvd-9016

SAMPLE PROBLEM 4 Representing the current-versus-stopping voltage curve for a photoelectric cell

The diagram shows the current-versus-stopping voltage curve for a typical photoelectric cell using green light.
The colour is changed to blue, but with a lower intensity.
Sketch the curve that would result from these changes.

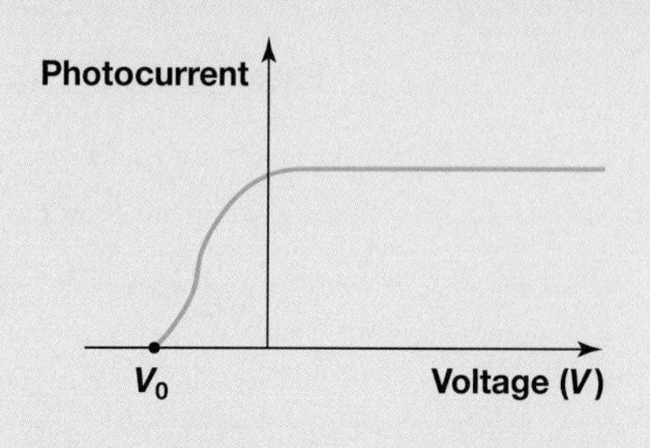

THINK

Because blue light has a higher frequency than green light, the stopping voltage would be greater. The lower intensity would make the photocurrent smaller.

WRITE

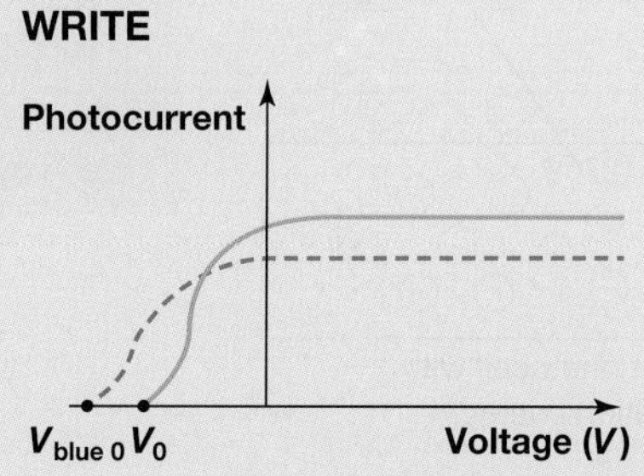

PRACTICE PROBLEM 4

Consider the same arrangement as in sample problem 4 except that, this time, yellow light is used and the photoelectric effect occurs. The intensity of the light is greater than with the green light. Sketch the curve that would result from this change.

9.3.2 Einstein's particle model for light and the photoelectric effect

Let's consider how each of the observations of the photoelectric effect experiment could be explained using a particle model, and why a wave model is not as successful in this situation. A close inspection of the evidence should allow you to decide whether electrons are being hit by particles or waves. The wave model would predict the following:

- a brighter or more intense light should produce photoelectrons with greater energy, therefore requiring a larger stopping voltage
- higher frequency light would not change the kinetic energy of a photoelectron
- photoelectrons would be emitted eventually over the course of irradiation time.

The particle model shows that the entire energy of a *single* photon is transferred to a *single* electron; the photon is gone. Some of the photon energy is required to enable the electron to escape from the electrode. This transferred energy, which enables an electron to escape the attraction of a material, is called its **ionisation energy**. Electrons in the metal have a range of energy levels, so they also have a range of ionisation energies. The minimum ionisation energy is called the **work function**, ϕ, of the material. The photon energy that is 'left over' becomes the kinetic energy of the electron (figure 9.9). Naturally, the electrons requiring the least energy to enable them to escape will leave with the greatest kinetic energy.

ionisation energy the amount of energy required to be transferred to an electron to enable it to escape from a material

work function the minimum energy required to release an electron from the surface of a material

When light of certain frequencies strikes a metal surface, electrons, called photoelectrons in this case, are emitted by the metal. The movement of the freed photoelectrons is the photocurrent.

An electron needs enough energy to be freed from the metal and move away from it.

FIGURE 9.9 A representation of how photon energy is transferred to electrons

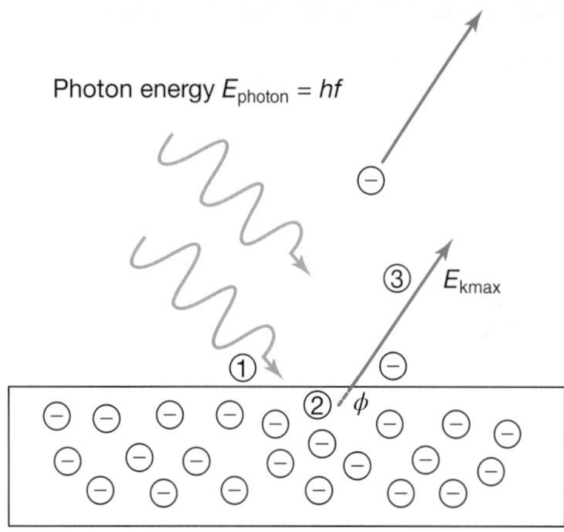

① photon strikes electron in metal with energy $E_{photon} = hf$

② electron absorbs part of photon energy (ϕ) and is ionised

③ electron moves with kinetic energy E_{kmax}

When the light (photon) interacts with an electron, part of the photon energy is used to ionise the electron (release the electron from the metal); this is the work function ϕ. The remaining energy from the photon is converted into kinetic energy (E_{kmax}).

$$E_{photon} = \phi + E_{k\ max}$$
$$hf = \phi + E_{k\ max}$$

where: ϕ is the work function of the metal, in J
h is Planck's constant (6.63×10^{-34} J s or 4.14×10^{-15} eV s)
f is the frequency of the incident photon, in Hz

Note that in some VCAA questions, the work function ϕ is noted W.

Here is how the particle model explains Lenard's experimental observations. The numbering here matches the number of these observations earlier in the topic. (See figures 9.6 and 9.8.)

① *Maximum photocurrent is proportional to incident irradiation (intensity).*
Doubling the intensity without changing frequency doubles the number of photons reaching the electrode each second, but not their individual energy. This doubles the rate of electron emission without changing the energy transferred to each electron, and therefore doubles the maximum photocurrent. That is, increasing the light intensity increases the number of photons striking the electrode, causing more electrons to be released, and hence a higher photocurrent.

② *Retarding voltage reduces photocurrent. A stopping voltage exists above which no electrons reach the second electrode.*
Ejected electrons have a variety of energies, depending on the photon energy and their ionisation energy. A low retarding voltage turns back only the electrons having low kinetic energies. Increasing the retarding voltage will turn back electrons with higher kinetic energies, until at the stopping voltage none can reach the second electrode.

③ *Stopping voltage is independent of light intensity.*
Changing the light intensity does not change its frequency, so the photon energy is not changed. Photoelectrons will have the same range of energies, and so the same stopping voltage is needed to reduce the photocurrent to zero.

④ *Stopping voltage depends on light frequency and material: a cut-off frequency exists (threshold frequency f_0).*
Since the stopping voltage reverses the direction of *all* electrons, it is the voltage required to entirely transform the kinetic energy of the fastest electrons into electric potential energy:

$$E_{\text{k max}} = \text{magnitude of change in electron's electrical potential energy}$$
$$= q_e V_0$$

where q_e here is the *magnitude* of the electronic charge.
The photon model states the following:

$$E_{\text{k max}} = E_{\text{photon}} - \phi$$
$$= hf - \phi$$

So,

$$q_e V_0 = hf - \phi$$

$$V_0 = \frac{1}{q_e}(hf - \phi)$$

Clearly V_0 depends on the light frequency, f, and also on the electrode material through its work function, ϕ. A photon whose energy, hf, is less than the work function, ϕ, cannot supply enough energy for an electron to escape.

The threshold frequency above which photoelectrons are emitted is given by:

$$hf_0 = \phi$$

where: f_0 is the threshold frequency of a metal, in Hz
ϕ is the work function of the metal, in J
h is Planck's constant, in J

tlvd-9017

SAMPLE PROBLEM 5 Determining the size of stopping voltage and kinetic energy of electrons

a. Electrons are emitted from a metal plate with a kinetic energy of 2.6 × 10^{-19} J after being struck by light. What is the size of the stopping voltage required to stop the photocurrent?
b. Calculate the kinetic energy of electrons that a 4.2-V stopping voltage would stop.

THINK		WRITE
a. 1.	Use $E_k = qV_0$ to calculate the stopping voltage. The kinetic energy of each electron is 2.6 × 10^{-19} J. The charge on an electron is 1.6 × 10^{-19} C.	a. $E_k = qV_0$ $2.6\times10^{-19}\text{ J} = 1.6\times10^{-19}\text{ C}\times V_0$ $V_0 = \frac{2.6\times10^{-19}\text{ J}}{1.6\times10^{-19}\text{ C}}$ $= 1.6$ V (accurate to 2 significant figures)
2.	Answer the question.	A stopping voltage of 1.6 V will stop the electrons emitted from the surface.
b. 1.	Use $E_k = qV_0$ to calculate the kinetic energy. The stopping voltage is 4.2 V. The charge of an electron is 1.6 × 10^{-19} C.	b. $E_k = qV_0$ $= 1.6\times10^{-19}\text{ C}\times4.2\text{ V}$ $= 6.72\times10^{-19}$ J $= 6.7\times10^{-19}$ J (accurate to 2 significant figures)
2.	Answer the question.	A stopping voltage of 4.2 V will stop electrons with energy 4.2 eV, which is 4.2 × 1.6 × 10^{-19} = 6.7 × 10^{-19} J.

PRACTICE PROBLEM 5

Electrons are emitted from the surface of a photocell with 4.8 × 10^{-19} J of kinetic energy after being struck by light. What is the size of the stopping voltage that will remove all this kinetic energy from the electrons?

tlvd-9018

SAMPLE PROBLEM 6 Calculations in photocells and the emission of photoelectrons

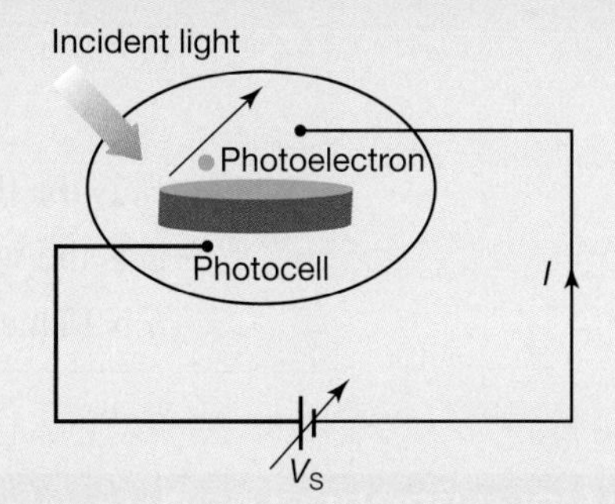

Light with a wavelength of 425 nm strikes a clean metallic surface and photoelectrons are emitted. A voltage of 1.25 V is required to stop the most energetic electrons from being emitted from the photocell.

a. **Calculate the frequency of the photons of light striking the surface.**
b. **Calculate the energy in joules and also in electron volts of a photon.**
c. **State the energy of the emitted electron in both electron volts and joules.**
d. **Calculate the work function ϕ of the metal in electron volts and joules.**
e. **Determine the threshold frequency, f_0, and consequently the maximum wavelength of a photon that will just free a surface electron from the metal.**
f. **Light of wavelength 390 nm strikes the same metal surface. Calculate the stopping voltage.**

THINK	WRITE
a. Use $f = \frac{c}{\lambda}$ to calculate the frequency.	**a.** $f = \frac{c}{\lambda}$ $= \frac{3.0 \times 10^8}{4.25 \times 10^{-7}}$ $= 7.1 \times 10^{14}$ Hz
b. 1. Use $E = hf$ to calculate the energy. *Remember to use your non-rounded figure in your calculations.*	**b.** $E = hf$ $= 6.63 \times 10^{-34} \times 7.1 \times 10^{14}$ $= 4.7 \times 10^{-19}$ J
2. To convert energy in joules into energy in electron volts, divide by 1.6×10^{-19} J eV^{-1}.	$E = \frac{4.7 \times 10^{-19}}{1.6 \times 10^{-19}}$ $= 2.9$ eV
c. Since the stopping voltage is 1.25 V, the energy of the emitted electron is 1.25 eV. The energy in joules can be found by multiplying by 1.6×10^{-19}.	**c.** $1.25 \times 1.6 \times 10^{-19} = 2.0 \times 10^{-19}$ J
d. Using the equation $E_{k\,max} = hf - \phi$, the work function can be found. When the photon energy hf equals 2.92 eV, the electrons have an energy of $1.25 = 2.92 - \phi$.	**d.** $\phi = 2.9 - 1.25$ $= 1.7$ eV $= 2.7 \times 10^{-19}$ J
e. 1. Again use the equation $E_{k\,max} = hf - \phi$. The threshold frequency, f_0, is the frequency below which the photoelectric effect does not occur. At this frequency, electrons are just not able to leave the surface. This model implies $0 = hf_0 - \phi$. Rearrange this equation to give the useful result.	**e.** $f_0 = \frac{\phi}{h}$ $= \frac{2.7 \times 10^{-19}}{6.63 \times 10^{-34}}$ $= 4.0 \times 10^{14}$ Hz
2. The maximum wavelength can be calculated using $\lambda = \frac{c}{f_0}$.	$\lambda = \frac{c}{f_0}$ $= \frac{3.0 \times 10^8}{4.0 \times 10^{14}}$ $= 7.4 \times 10^{-7}$ m or 740 nm
f. 1. Use the equation $E_{k\,max} = h\frac{c}{\lambda} - \phi$ to find the energy of the emitted electrons. It is convenient to use eV here.	**f.** $E_{kmax} = \frac{6.63 \times 10^{-34} \times 3.0 \times 10^8}{3.90 \times 10^{-7}}$ J $- 1.7$ eV $= 5.1 \times 10^{-19}$ J $- 1.7$ eV $= \frac{5.1 \times 10^{-19}}{1.6 \times 10^{-19}} - 1.7$ eV $= 3.19 - 1.7$ eV $= 1.5$ eV
2. Determine the stopping voltage.	A stopping voltage of 1.5 V is required to stop the emitted electrons.

PRACTICE PROBLEM 6

A new photocell with a different metallic surface is used. Again, light of wavelength 425 nm strikes a clean metallic surface and photoelectrons are emitted. This time, a stopping voltage of 0.87 V is required to stop the most energetic electrons from being emitted from the photocell.

a. **State the highest energy of the emitted electrons in both electron volts and joules.**
b. **Calculate the work function, ϕ, of the metal.**
c. **Determine the threshold frequency, f_0, and, consequently, the maximum wavelength of a photon that will just free a surface electron from the photocell.**
d. **Light of wavelength 650 nm strikes the same metal photocell. What will happen? Explain.**

Einstein's insights into using a particle model to explain the photoelectric effect led to his 1905 prediction. He predicted that a graph of stopping voltage versus light frequency would be a straight line whose gradient was independent of the material emitting electrons:

$$V_0 = \frac{1}{|q_e|}(hf - \phi)$$

FIGURE 9.10 V_0 versus f for three different metals

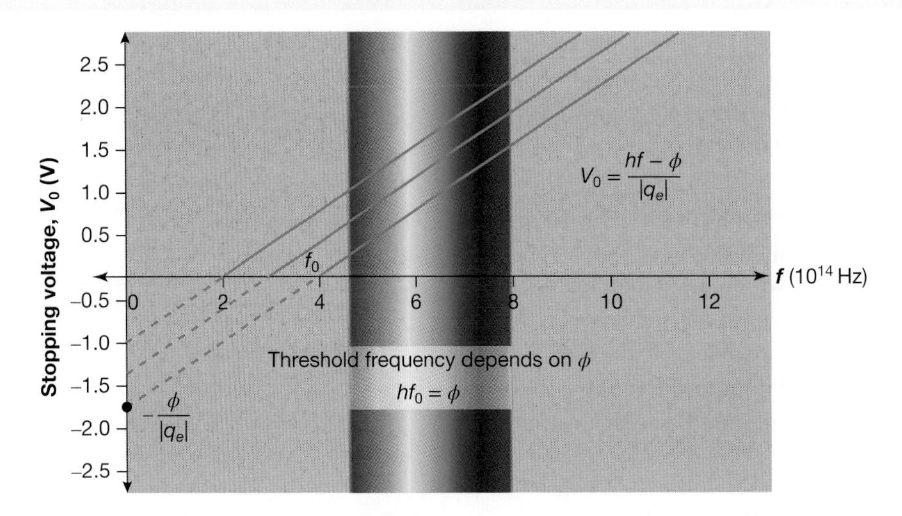

The graphs (figure 9.10) for different materials all have the same slope, $\frac{h}{|q_e|}$, but are displaced up or down, depending on the work function. The threshold frequency, f_0, is where the line meets the frequency axis. Its value is equal to $\frac{\phi}{h}$. No photoelectric effect is observed when light with frequency below f_0 is used.

Einstein said:

> *It seems to me that the observations associated with... the photoelectric effect, and other related phenomena... are more readily understood if one assumes that the energy of light is discontinuously distributed through space... the energy of a light ray spreading out from a point is not continuously spread out over an increasing space, but consists of a finite number of energy quanta which are localised at points in space, which move without dividing, and which can only be produced and absorbed as complete units.*

tlvd-9019

SAMPLE PROBLEM 7 Investigating the photoelectric effect

The following table gives some data collected by students investigating the photoelectric effect using a photocell with a lithium cathode, as shown in the diagram.

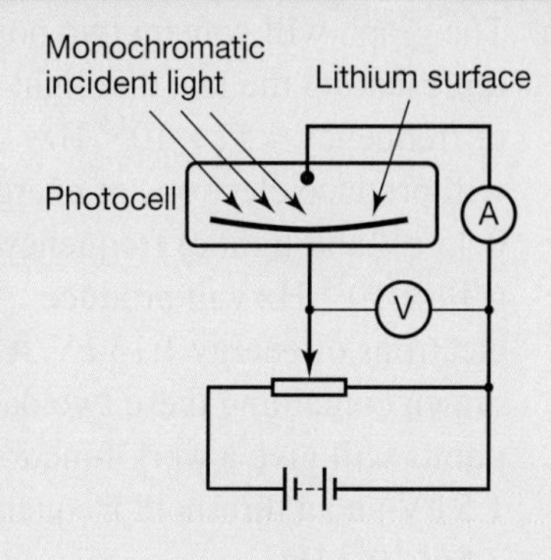

Wavelength of light used (nm)	Frequency of light used (× 10^{14} Hz)	Photon energy of light used (eV)	Stopping voltage readings (V)	Maximum photo-electron energy (J)
663			0.450	
	6.14			1.84×10^{-19}

a. **Complete the table by filling in the missing values.**
b. **Using the data points in the table, plot a graph of maximum photoelectron energy in electron volts versus photon frequency in hertz for the lithium photocell.**
c. **Using the graph, state the values for the following quantities. For each, state what aspect of the graph you used.**
   i. **Planck's constant, h, in the units J s and eV s**
   ii. **The threshold frequency, f_0, for the metal surface in Hz**
   iii. **The work function, ϕ, for the metal surface, in the units J and eV.**
d. **On the same axes, draw and label the graph you would expect to get when using a different photocell, given that it has a work function slightly larger than the one used to collect the data in the table.**

A new photocell is investigated. When light of frequency 9.12×10^{14} Hz is used, a stopping voltage of 1.70 V is required to stop the most energetic electrons.

e. **Calculate the work function of the new photocell, in both joules and electron volts.**
f. **When the battery voltage of the new photocell is set to 0 V, the photocurrent is measured as 48 μA. The intensity of the light is now doubled. Describe what happens now in the electric circuit with the power supply voltage set to 0 V.**
g. **With the intensity still doubled, the voltage is slowly increased from 0 V and the photocurrent slowly reduced to 0 A. State the stopping voltage when the current first equals 0 A.**

THINK

a. Use $c = f\lambda$ to complete columns 1 and 2. Use $E = hf$ to complete column 3, and use the conversion factor for joules to eV to complete columns 4 and 5.

WRITE

a.

Wavelength of light used (nm)	Frequency of light used (× 10^{14} Hz)	Photon energy of light used (eV)	Stopping voltage readings (V)	Maximum photo-electron energy (J)
663	**4.52**	**1.88**	0.450	**7.20×10^{-20}**
489	6.14	**2.54**	**1.15**	1.84×10^{-19}

THINK	WRITE
b. The graph will contain two points representing the fact that light of frequency 4.52×10^{14} Hz will produce electrons of energy 0.45 eV and light of frequency 6.14×10^{14} Hz will produce electrons of energy 1.15 eV. A line drawn containing these two data points will give a work function of 1.5 eV and a threshold frequency of 3.5×10^{14} Hz.	**b.** 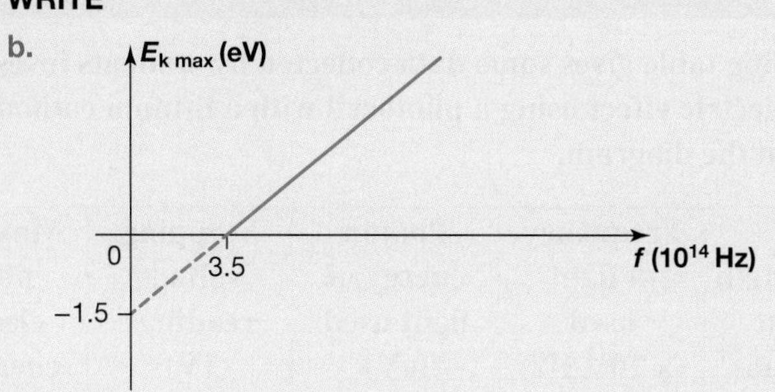
c. i. Planck's constant = gradient of graph	**c. i.** $\text{gradient} = \dfrac{1.84 \times 10^{-19} - 7.20 \times 10^{-20}}{(6.24 - 4.52) \times 10^{14}}$ $= 6.9 \times 10^{-34}$ J s which is close to the accepted value. It also has the value 4.3×10^{-15} eV s.
ii. From the line of best fit in graph (b), the threshold frequency = x-axis intercept.	**ii.** threshold frequency = 3.5×10^{14} Hz.
iii. From the line of best fit in the graph (b), the work function = y-axis intercept.	**iii.** work function = 2.4×10^{-19} J = 1.5 eV.
d. The graph for a photocell with a larger work function will have the same gradient but a lower y-intercept.	**d.** 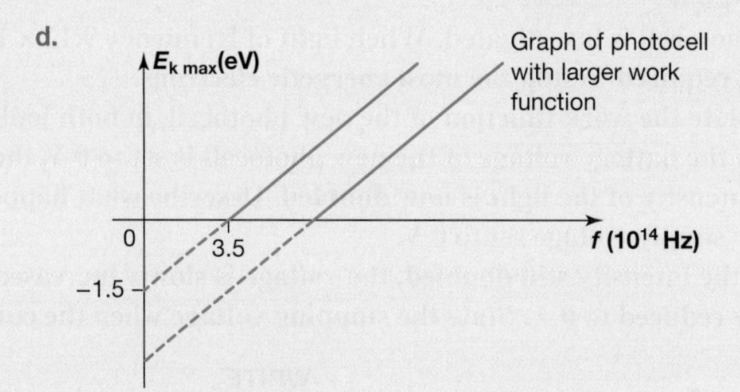
e. Use $E_{kmax} = hf - \phi$ to calculate the work function, ϕ.	**e.** $1.7 \times 1.6 \times 10^{-19} = 6.63 \times 10^{-34} \times 9.12 \times 10^{14} - \phi$ $\phi = 6.05 \times 10^{-19} - 2.72 \times 10^{-19}$ $= 3.33 \times 10^{-19}$ J $= 2.07$ eV
f. With the light intensity doubled, the photocurrent would also double.	**f.** The photocurrent would double.
g. The stopping voltage would remain the same, 1.70 V, as the colour and hence the frequency of the light source is unchanged.	**g.** Stopping voltage = 1.70 V.

PRACTICE PROBLEM 7

The following table gives some data collected by students investigating the photoelectric effect using a photocell with a clean metallic cathode.

Wavelength of light used (nm)	Frequency of light used × 10^{14} (Hz)	Photon energy of light used (eV)	Stopping voltage reading (V)	Maximum photoelectron energy (J)
		3.19		3.78×10^{-19}
524			1.54	

a. **Complete the table by filling in the missing values.**

b. **Using only the data points in the table, plot a graph of maximum photoelectron energy in electron volts versus photon frequency in hertz for the photocell.**

c. **Using the graph, state your values for the following quantities and explain what aspect of the graph was used.**
   i. **Planck's constant, h, in the units J s and eV s**
   ii. **The threshold frequency, f_0, for the metal surface in Hz**
   iii. **The work function, ϕ, for the metal surface, in the units J and eV.**

d. **On the same axes, draw and label the graph that you would expect to get when using a photocell that has a work function slightly larger than the one used to collect the data in the table.**

A new photocell is now investigated. When light of frequency 8.25×10^{14} Hz is used, a stopping voltage of 1.59 V is required to stop the most energetic electrons. In addition, when the battery voltage is set to 0 V, the photocurrent is measured to be 38 μA.

e. **Calculate the work function of the photocell.**

f. **Describe what happens in the electric circuit with the power supply voltage set to 0 V when the light intensity is halved.**

g. **With the intensity still halved, the stopping voltage is now slowly increased from 0 V and the photocurrent slowly reduced to 0 A. State the stopping voltage when the current first equals 0 A.**

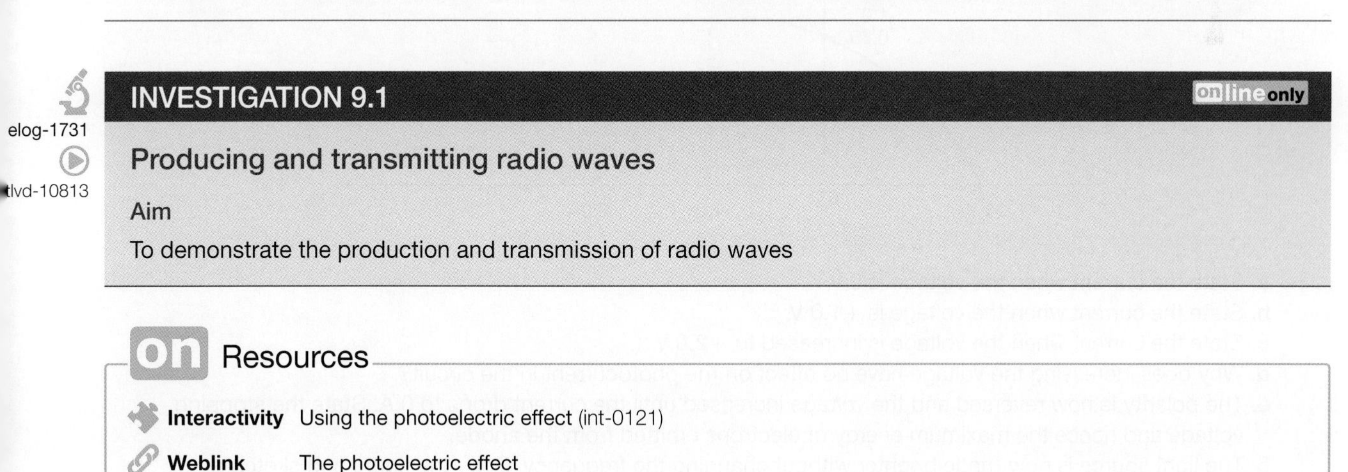

elog-1731

tlvd-10813

INVESTIGATION 9.1

online only

Producing and transmitting radio waves

Aim

To demonstrate the production and transmission of radio waves

on Resources

Interactivity Using the photoelectric effect (int-0121)

Weblink The photoelectric effect

9.3 Activities

learnon

Students, these questions are even better in jacPLUS

Receive immediate feedback and access sample responses

Access additional questions

Track your results and progress

Find all this and MORE in jacPLUS

9.3 Quick quiz	9.3 Exercise	9.3 Exam questions

9.3 Exercise

1. In the diagram shown, the curve shows how the current measured in an experiment involving the photoelectric effect depends on the potential difference between the anode and cathode.
   a. Explain the curve. Why does it reach a constant maximum value at a certain positive voltage, and why does it drop to zero at a certain negative voltage?
   b. If the intensity of the light was increased without changing its frequency, sketch the resulting curve.
   c. If the frequency of the light was increased without changing its intensity, sketch the resulting curve.
   d. If the metal of the cathode was changed, but the light was not changed in any way, sketch the curve that would be obtained.
2. The following curve shows the current in a photoelectric cell versus the potential difference between the anode and the cathode when blue light is shone onto the anode.

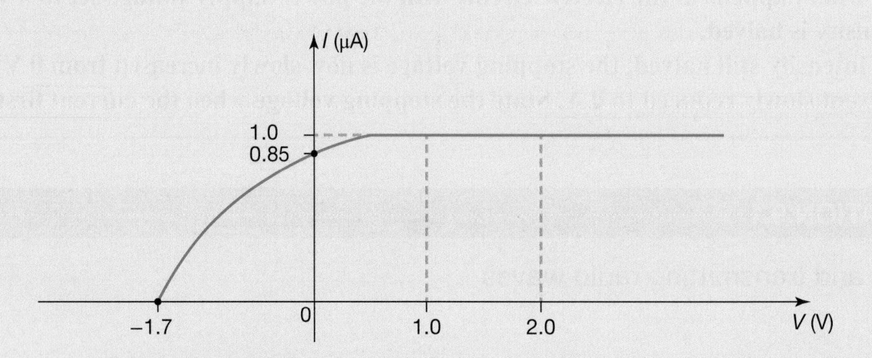

   a. State the current when the voltage is 0 V.
   b. State the current when the voltage is +1.0 V.
   c. State the current when the voltage is increased to +2.0 V.
   d. Why does increasing the voltage have no effect on the photocurrent in the circuit?
   e. The polarity is now reversed and the voltage increased until the current drops to 0 A. State the stopping voltage and hence the maximum energy of electrons emitted from the anode.
   f. The light source is now made brighter without changing the frequency. Copy the figure and sketch a second curve to show the effect of increasing the intensity of the blue light.
   g. The light source is now returned to its original brightness (intensity) and green light is used. A current is still detected. Sketch a third curve to illustrate the effect of using light of this lower frequency.
   h. The apparatus is altered so that the anode consists of a metal with a smaller work function. Again blue light is used. Sketch a fourth curve to illustrate the effect of changing the anode without changing either the brightness or colour of the light.

3. The work function for a particular metal is 3.80 eV. When monochromatic light is shone onto the photocell, electrons with energy 0.670 eV are emitted.
   a. What is the stopping voltage required to stop these electrons?
   b. What frequency of the monochromatic light is used?
   c. What is the threshold frequency of the metallic surface?
4. In a photoelectric effect experiment, the threshold frequency is measured to be 6.2×10^{14} Hz.
   a. Calculate the work function of the metal surface used.
   b. If electrons of maximum kinetic energy 3.4×10^{-19} J are detected when light of a particular frequency is shone onto the apparatus, what is the stopping voltage?
   c. With the same source of light, what is the wavelength and hence the momentum of the photons?
5. One electron ejected from a clean zinc plate by ultraviolet light has a kinetic energy of 4.0×10^{-19} J.
   a. What would be the kinetic energy of this electron when it reached the anode, if a retarding voltage of 1.0 V was applied between the anode and cathode?
   b. What is the minimum retarding voltage that would prevent this electron reaching the anode?
   c. All electrons ejected from the zinc plate are prevented from reaching the anode by a retarding voltage of 4.3 V. What is the maximum kinetic energy of electrons ejected from the zinc?
   d. Sketch a graph of photocurrent versus voltage for this metal surface. Use an arbitrary photocurrent scale.
6. What is the stopping voltage when UV radiation having a wavelength of 200 nm is shone onto a clean gold surface? The work function of gold is 5.10 eV.
7. Robert Millikan performed his photoelectric experiment using a clean potassium surface, with a work function of 2.30 eV. He used a mercury discharge lamp. One wavelength of ultraviolet radiation emitted by the lamp was 254 nm.
   a. What is the maximum kinetic energy of electrons ejected from the potassium surface by this UV radiation?
   b. What voltage would be required to reduce the photocurrent in the cell to zero?
   c. Sketch a graph of maximum electron kinetic energy versus frequency for potassium. Show the point on the graph obtained from the 254 nm UV radiation.
   d. Repeat this sketch for sodium, which has a work function of 2.75 eV.
8. When the surface of a material in a photoelectric effect experiment is illuminated with light from a mercury discharge lamp, the stopping voltages given in the table are measured.
   Plot the stopping voltage versus the light frequency and use the graph to determine its:
   a. threshold frequency
   b. threshold wavelength
   c. work function of the material, in eV
   d. value of Planck's constant.

Wavelength (nm)	Stopping voltage (V)
366	1.48
405	1.15
436	0.93
492	0.62
546	0.36
579	0.24

9.3 Exam questions

Question 1 (4 marks)

Source: *VCE 2020 Physics Exam, Section B, Q.15; © VCAA*

The metal surface in a photoelectric cell is exposed to light of a single frequency and intensity in the apparatus shown in Figure 14.

The voltage of the battery can be varied in value and reversed in direction.

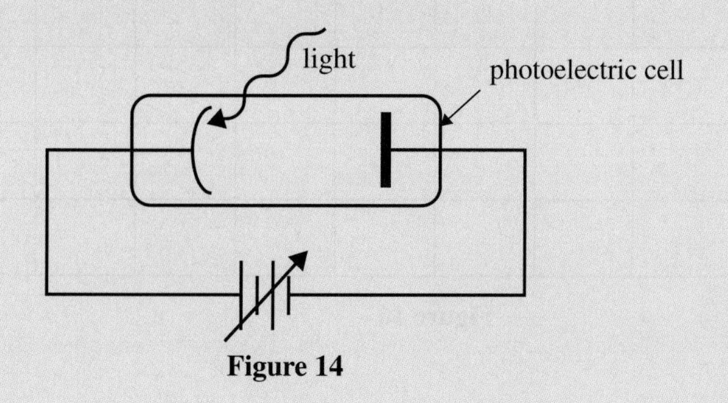

Figure 14

a. A graph of photocurrent versus voltage for one particular experiment is shown in Figure 15. On the figure, draw the trace that would result for another experiment using light of the same frequency but with triple the intensity. **(2 marks)**

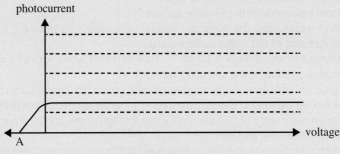

Figure 15

b. What is a name given to the point labelled A on the figure? **(1 mark)**

c. Why does the photocurrent fall to zero at the point labelled A ? **(1 mark)**

Question 2 (6 marks)

Source: *VCE 2019, Physics Exam, Section B, Q.16;* © VCAA

Students are studying the photoelectric effect using the apparatus shown in Figure 15.

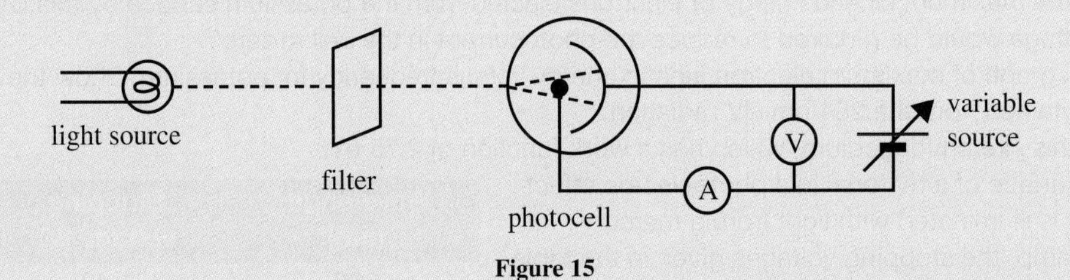

Figure 15

Figure 16 shows the results the students obtained for the maximum kinetic energy ($E_{k\ max}$) of the emitted photoelectrons versus the frequency of the incoming light.

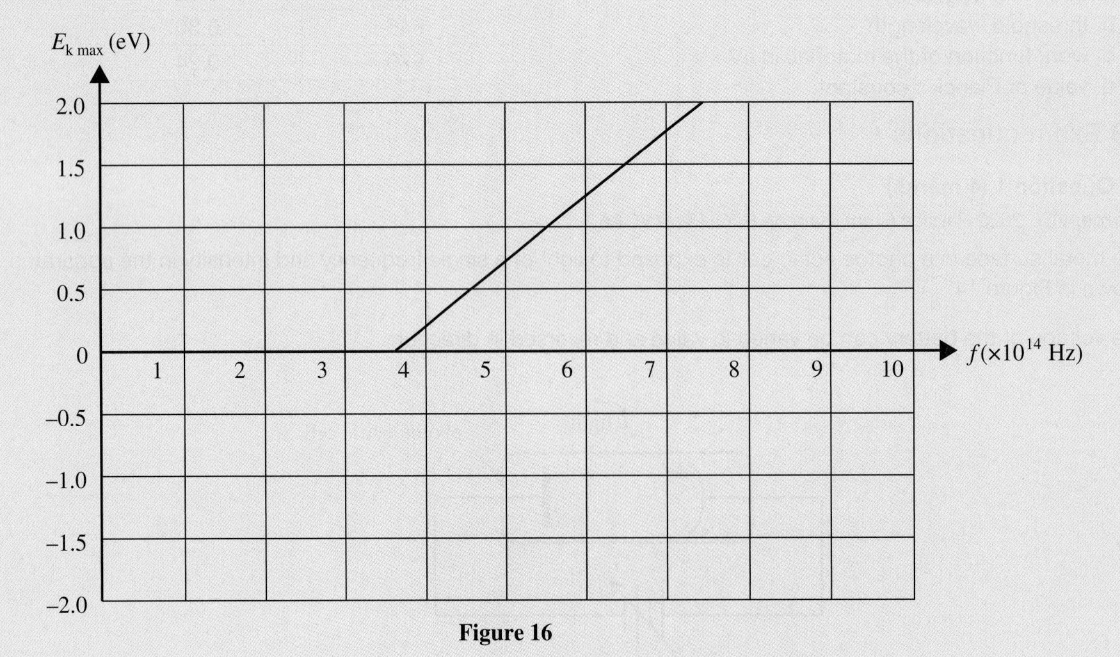

Figure 16

a. Using only data from the graph, determine the values the students would have obtained for
   i. Planck's constant, h. Include a unit in your answer **(2 marks)**
   ii. the maximum wavelength of light that would cause the emission of photoelectrons **(1 mark)**
   iii. the work function of the metal of the photocell. **(1 mark)**

b. The work function for the original metal used in the photocell is ϕ.
   On the graph above, draw the line that would be obtained if a different metal, with a work function of $\frac{1}{2}\phi$, were used in the photocell. **(2 marks)**

Question 3 (4 marks)

Source: *VCE 2017, Physics Exam, Section B, Q.17.a&b;* © VCAA

In an experiment, blue light of frequency 6.25×10^{14} Hz is shone onto the sodium cathode of a photocell. The apparatus is shown in Figure 15.

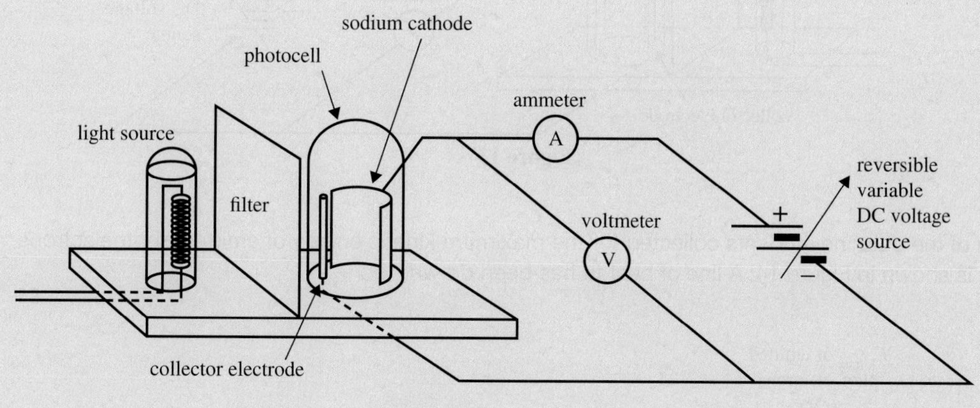

Figure 15

The graph of photoelectric current versus potential difference across the photocell is shown in Figure 16.

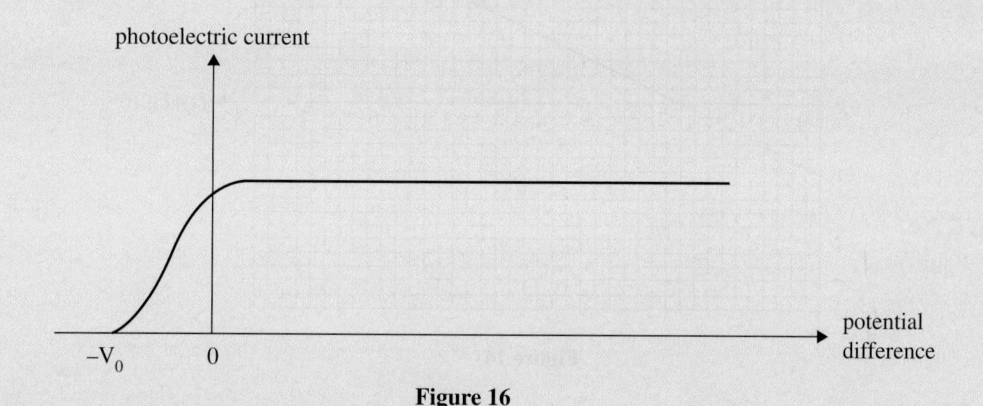

Figure 16

The threshold frequency for sodium is 5.50×10^{14} Hz.

a. What is the cut-off potential, V_o, when blue light of frequency 6.25×10^{14} Hz is shone onto the sodium cathode of the photocell referred to in the diagram and graph? **(2 marks)**

b. On the graph of photoelectric current versus potential difference shown, sketch the curve expected if the light is changed to **ultraviolet** with a **higher intensity** than the original blue light. **(2 marks)**

Question 4 (6 marks)

***Source:** VCE 2018 Physics Exam, NHT, Section B, Q.16; © VCAA*

Students are investigating the photoelectric effect. The apparatus used by the students is shown in Figure 13. A light source shines light through a filter that only allows one frequency of light to pass through. This monochromatic light shines onto a metal plate and photoelectrons are emitted. Different filters allow different frequencies to strike the metal plate. For each frequency, the maximum kinetic energy of the emitted photoelectrons is measured by using a stopping voltage.

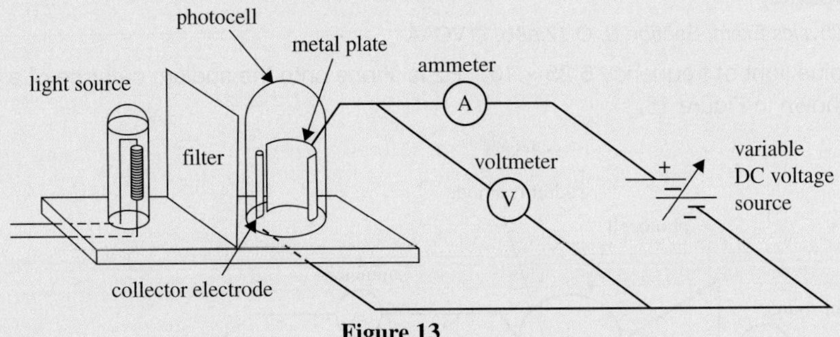

Figure 13

The graph of the data the students collected for the maximum kinetic energy of emitted photoelectrons versus frequency is shown in Figure 14. A line of best fit has been drawn.

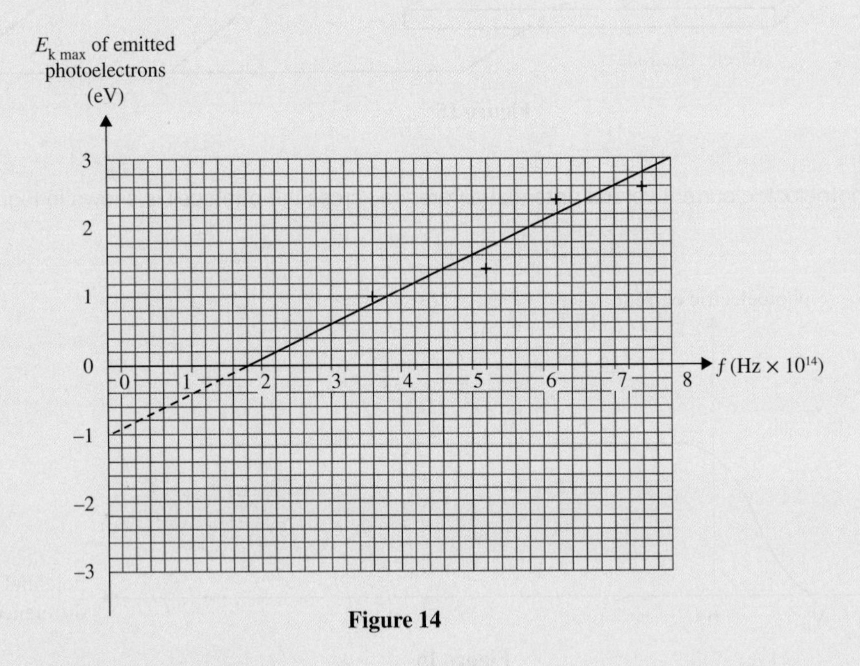

Figure 14

a. Determine the value of Planck's constant, h, that the students would have obtained from this graph. **(2 marks)**

b. Determine the value of the minimum frequency, or cut-off frequency, f_0, that the students would have obtained from this graph. **(1 mark)**

c. Determine the value of the work function of the metal in the plate that the students would have obtained from this graph. **(1 mark)**

d. The students replace the photocell with one that has a different metal plate with a work function of 2.5 eV. On Figure 14, draw in the graph they would now expect. **(2 marks)**

Question 5 (3 marks)

Source: *VCE 2021, Physics Exam, Section B, Q15; © VCAA*

A photoelectric experiment is carried out by students. They measure the threshold frequency of light required for photoemission to be 6.5×10^{14} Hz and the work function to be 3.2×10^{-19} J.

Using the students' measurements, what value would they calculate for Planck's constant? Outline your reasoning and show all your working. Give your answer in joule-seconds.

More exam questions are available in your learnON title.

9.4 Limitations of the wave model

KEY KNOWLEDGE

- Describe the limitation of the wave model of light in explaining experimental results related to the photoelectric effect

Source: VCE Physics Study Design (2024–2027) extracts © VCAA; reproduced by permission.

9.4.1 The photon model

Almost thirty years after the first observation of the photoelectric effect, experimental measurements confirmed the need for a photon model for light. The wave model for light was incapable of explaining the observations of the photoelectric effect, as shown in table 9.2.

TABLE 9.1 Timeline of key discoveries about the photoelectric effect

Date	Event
1887	It all started with Hertz carefully noting the unusual behaviour of sparks across the gaps in his radio wave detector circuit. This was the first observation of the photoelectric effect.
1901	Max Planck solves the black body radiation problem theoretically, paving the way for light to be modelled not only as a wave but also as a localised particle with energy proportional to the frequency of the light, f.
1902	Philipp Lenard carried out experiments to accumulate knowledge about the behaviour of electrons emitted by light. His results revealed several puzzling aspects: electron energies did not depend on the light intensity, and a unique cut-off frequency was found to exist for each material.
1905	The flash of insight was Albert Einstein's, when he realised that all of Lenard's observations could be explained if he changed the way he thought about light — if light energy travelled as particles not waves. He used the particle model to predict that the graph of stopping voltage versus frequency would be straight, with a slope that was the same for all electron emitters.
1915	Robert Millikan sealed the success of Einstein's theory with plots of V_0 versus f for the alkali metals that were straight and parallel to one another. He used the plots to measure Planck's constant. The photon energy was $E_{photon} = hf$.

This need for a photon model to explain the workings of the photoelectric effect fitted very neatly with Planck's black body radiation model, in which a particle model for light was required to make the theory fit with the experimental evidence of light radiated from hot objects. However, both these phenomena contradicted the enormously successful wave model for light summarised by Maxwell's four equations for electromagnetic phenomena. The wave model for light in terms of perpendicular electric and magnetic fields is consistent with observed interference patterns and diffraction patterns, and with the propagation of light at a single, universal speed, c. A wave model for light is also consistent with a large range of electrical and magnetic phenomena, for example electromagnetic induction.

Another chapter in physics was about to begin. The development of quantum mechanics would completely change the way in which scientists viewed the universe. The Newtonian mechanistic world was about to be overthrown. Confusion between particle and wave models for both light and matter would be resolved, but this would take another thirty years to achieve.

TABLE 9.2 Observations made from the photoelectric effect and model predictions

Observation	Wave model prediction (incorrect)	Photon model prediction (correct)
For a given frequency of light, the photocurrent is dependent in a linear fashion on the brightness or intensity of light. • photocurrent increases with light amplitude	✗ The wave model makes no significant prediction other than that brighter light should produce electrons with greater energy, which is not the case.	✓ Intensity of light relates to the number of photons per second striking the photocell. The photocurrent is expected to be dependent on the intensity of light.
The energy of photoelectrons is independent of intensity of light and only linearly dependent on frequency. • kinetic energy of photoelectrons increases with light frequency and amplitude	✗ The energy of electrons is dependent on the intensity of light: the larger the amplitude of the wave, the larger the energy transferred to electrons.	✓ The energy of photoelectrons is linearly dependent on the frequency of light, provided that the energy of a single photon of light is interpreted as equal to hf.
There is no significant time delay between incident light striking a photocell and subsequent emission of electrons, and this observation is independent of the intensity of incident irradiation. • no time lag between incident light and emission of photoelectrons	✗ Time delay is shorter with increasing intensity.	✓ No time delay is expected as individual photons of light strike the photocell and transfer energy to individual electrons.
A threshold frequency exists below which the photoelectric effect does not occur, and this threshold is independent of intensity. • threshold frequency is independent of the intensity of incident radiation	✗ No threshold effect should exist, as energy transfer to electrons from a light source is cumulative and eventually emission will occur.	✓ A threshold frequency is predicted, as photons with energy less than the work function are incapable of freeing electrons from the photocell.

on Resources

Digital document eModelling: Photoelectric effect (doc-0042)

9.4 Activities

learnon

Students, these questions are even better in jacPLUS

Receive immediate feedback and access sample responses

Access additional questions

Track your results and progress

Find all this and MORE in jacPLUS

9.4 Quick quiz	9.4 Exercise	9.4 Exam questions

9.4 Exercise

1. If light behaved like a wave when it transferred energy to electrons, as in the photoelectric effect, what would you expect to happen to the stopping voltage if the intensity of the light were increased?
2. Give two observations that support the particle model for light in the photoelectric effect.
3. Explain why threshold frequency can be adequately explained by the photoelectric effect but not by the wave model of light.
4. Calculating the energy of a 'particle' of light — a photon — by using the frequency associated with an electromagnetic 'wave' of light seems contradictory. Discuss.

9.4 Exam questions

Question 1 (1 mark)

Source: *VCE 2021, Physics Exam, Section A, Q16;* © VCAA

MC The diagram below shows a circuit that is used to study the photoelectric effect.

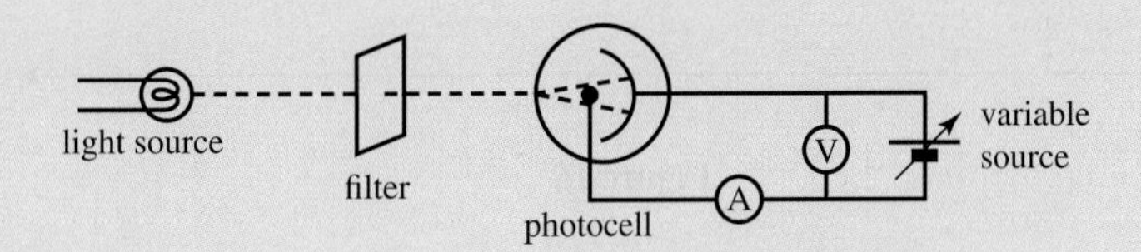

Which one of the following is essential to the measurement of the maximum kinetic energy of the emitted photoelectrons?

A. the level of brightness of the light source
B. the wavelengths that pass through the filter
C. the reading on the voltmeter when the current is at a minimum value
D. the reading on the ammeter when the voltage is at a maximum value

Question 2 (2 marks)

Source: *VCE 2021, Physics Exam, Section B, Q16;* © VCAA

Light can be described by a wave model and also by a particle (or photon) model. The rapid emission of photoelectrons at very low light intensities supports one of these models but not the other.

Identify the model that is supported, giving a reason for your answer.

Question 3 (5 marks)

Source: *VCAA 2017, Physics Exam, Section B, Q17c; © VCAA*

In an experiment, blue light of frequency 6.25×10^{14} Hz is shone onto the sodium cathode of a photocell. The apparatus is shown in Figure 15.

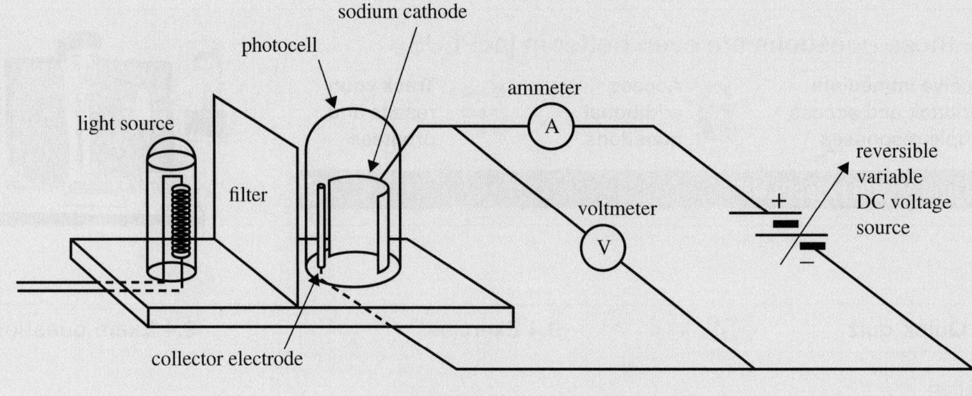

Figure 15

The graph of photoelectric current versus potential difference across the photocell is shown in Figure 16.

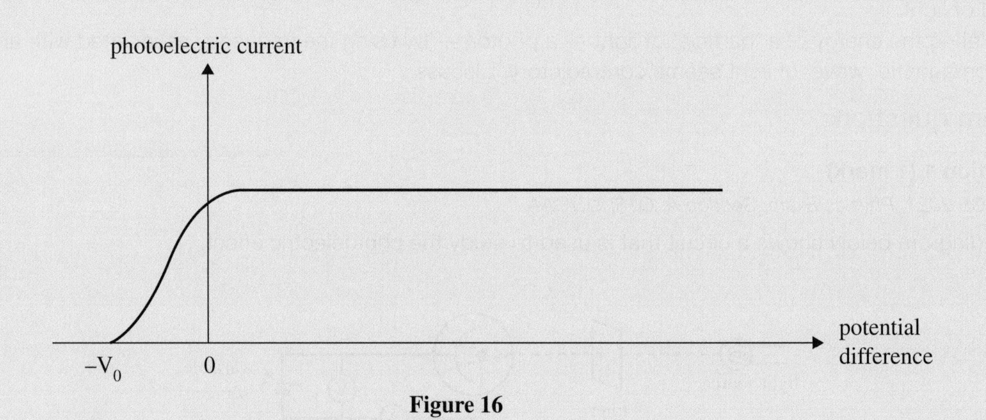

Figure 16

The threshold frequency for sodium is 5.50×10^{14} Hz.

The results of photoelectric effect experiments in general provide strong evidence for the particle-like nature of light.

Outline **two** aspects of these results that provide the strong evidence that is not explained by the wave model of light, and explain why.

Question 4 (3 marks)

Source: *VCE 2018, Physics Exam, NHT, Section B, Q.17; © VCAA*

The results of photoelectric effect experiments provide evidence for the particle-like nature of light. Outline **one** aspect of the results that would provide this evidence. Your response should explain:

- why a wave model of light cannot satisfactorily explain this aspect of the results
- how the photon theory does explain this aspect of the results.

Question 5 (2 marks)

Source: *VCE 2015, Physics Exam, Section A, Q.20.a; © VCAA*

Physicists use the expression 'wave-particle duality' because light sometimes behaves like a particle and electrons sometimes behave like waves.

What evidence do we have that light can behave like a particle? Explain how this evidence supports a particle model of light.

More exam questions are available in your learnON title.

9.5 Review

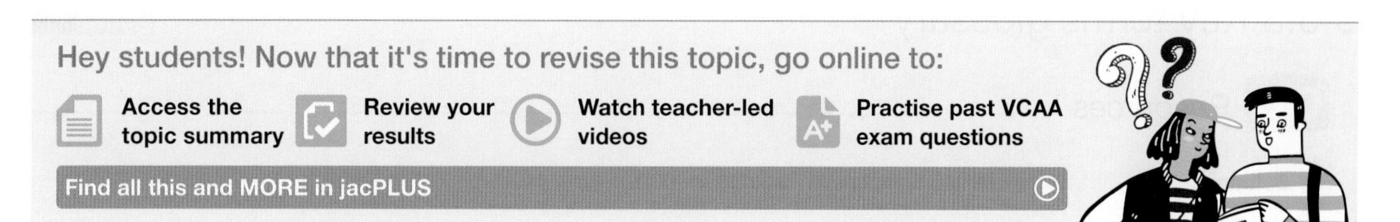

9.5.1 Topic summary

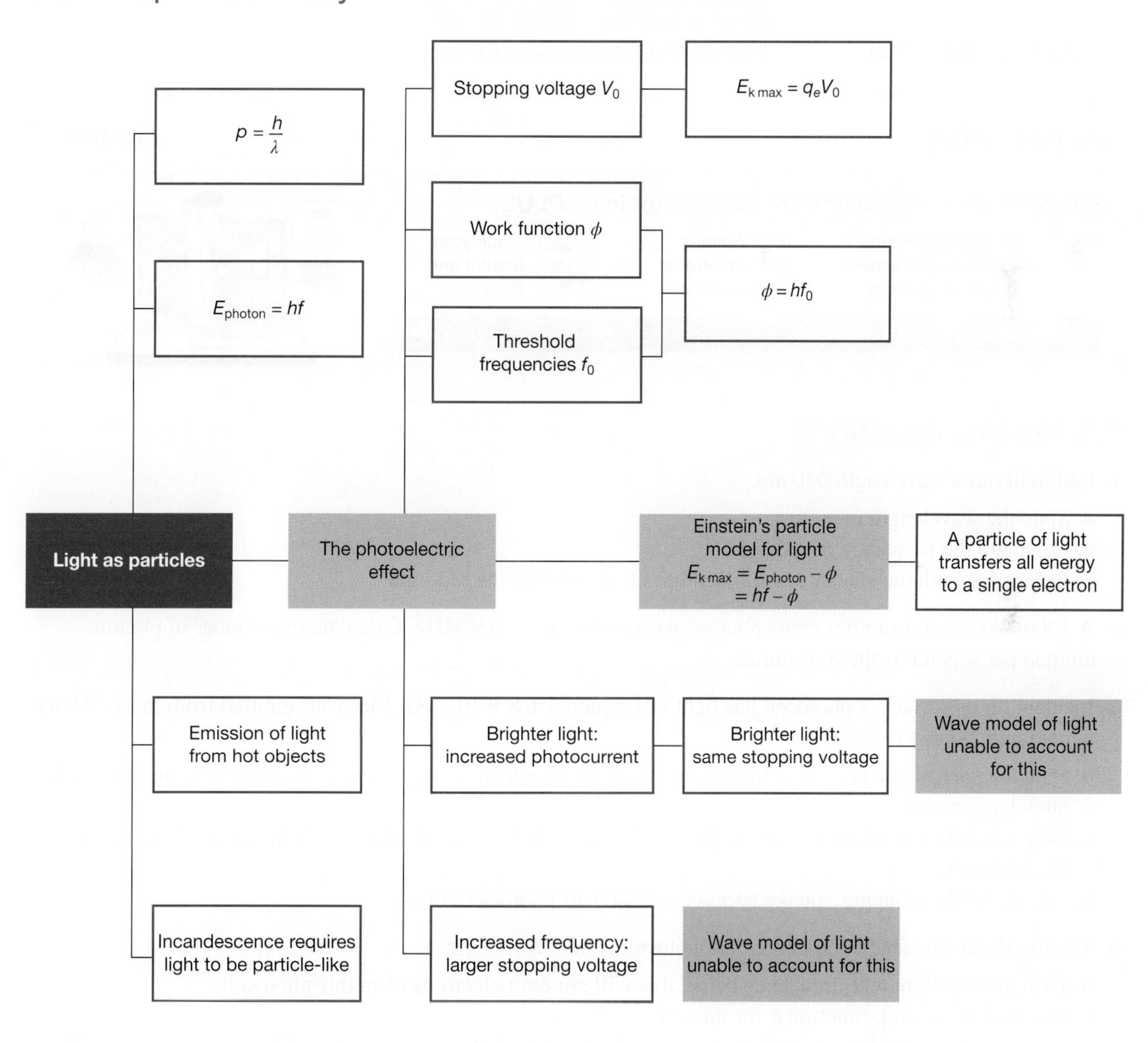

9.5.2 Key ideas summary

9.5.3 Key terms glossary

online only

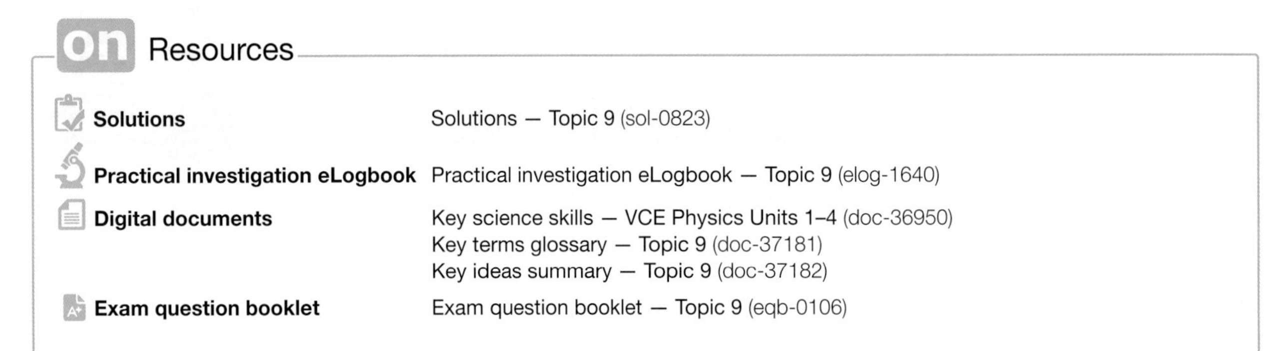

Resources

Solutions	Solutions — Topic 9 (sol-0823)
Practical investigation eLogbook	Practical investigation eLogbook — Topic 9 (elog-1640)
Digital documents	Key science skills — VCE Physics Units 1–4 (doc-36950) Key terms glossary — Topic 9 (doc-37181) Key ideas summary — Topic 9 (doc-37182)
Exam question booklet	Exam question booklet — Topic 9 (eqb-0106)

9.5 Activities

learnon

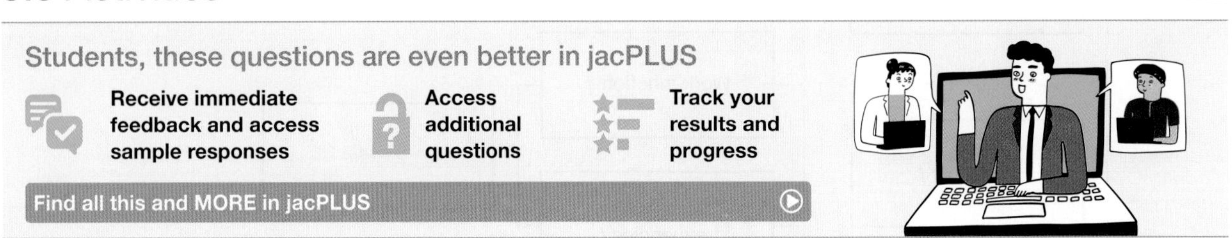

9.5 Review questions

1. Red light has a wavelength 650 nm.
   a. State the wavelength in metres.
   b. Calculate the frequency of red light.
   c. Determine both the energy and momentum of a photon in the red light.
2. A 1000-W radio transmitter emits photons with frequency 105.9 MHz. Calculate the number of photons emitted per second by the transmitter.
3. Incident on its surface, a photocell has light of frequency 6.8×10^{14} Hz. Electrons emitted from this cell have a maximum energy of 2.4×10^{-19} J.
   a. Find the energy of a photon of light incident on the photocell. Give your answer in both the units of joule and electron volt.
   b. Determine the work function of the photocell surface. Give your answer in both the units of joule and electron volt.
   c. Determine the stopping voltage necessary to stop all photoelectrons.
4. The threshold frequency of a particular photocell is 3.50×10^{14} Hz.
   a. Explain why light with frequency below this will not emit electrons from this photocell.
   b. Determine the work function ϕ for this cell.
   c. Calculate the energy of photoelectrons emitted from this photocell when light of frequency 5.10×10^{15} Hz is incident on the cell.
5. Students collect data using a photocell as part of an investigation. They find that, when light of frequency 4.1×10^{14} Hz is used, the stopping voltage is 0.72 V. When they use light of a higher frequency, 6.2×10^{14} Hz, the stopping voltage is increased to 1.60 V.
   a. Use this data to determine a value for Planck's constant in both eV s and J s.
   b. Calculate the work function of this photocell.

6. Students have accurately measured and recorded the following information for a specific photoelectric effect experiment.
incident light frequency: $f = 5.2 \times 10^{12}$ Hz
stopping voltage: $V_0 = 1.10$ V
photocurrent when voltage is 0 V: 24 μA

The following changes are then made:

a. The light is made more intense. State and explain what will happen to the stopping voltage and what will happen to the photocurrent in this situation.
b. The light is returned to its original intensity and a higher frequency is used. State and explain what will happen to the stopping voltage and what will happen to the photocurrent in this situation.
c. The light is returned to its original frequency and intensity, but this time a photocell having a smaller work function is used instead. State and explain what will happen to the stopping voltage and what will happen to the photocurrent in this situation.

7. Consider four observations made concerning the photoelectric effect. Give reasons why a particle model better explains each of these observations and also why a wave model for light is inadequate.

9.5 Exam questions

Section A — Multiple choice questions

All correct answers are worth 1 mark each; an incorrect answer is worth 0.

Question 1

Source: *VCE 2022 Physics Exam, Section A, Q.15; © VCAA*

Which one of the following best provides evidence of light behaving as a particle?

A. photoelectric effect
B. white light passing through a prism
C. diffraction of light through a single slit
D. interference of light passing through a double slit

Question 2

Source: *VCE 2022 Physics Exam, NHT, Section A, Q.16; © VCAA*

When light of a specific frequency strikes a metal surface, photoelectrons are emitted.

If the light intensity is increased but the frequency of the light remains the same, which of the following would be correct?

	Number of photoelectrons emitted	Maximum kinetic energy of photoelectrons
A.	increases	remains the same
B.	remains the same	increases
C.	increases	decreases
D.	remains the same	remains the same

Question 3

Source: *VCE 2021 Physics Exam, NHT, Section A, Q.18;* © VCAA

Experiments on the photoelectric effect involve shining light onto a metal surface. Measurements are made of the number of emitted electrons and their maximum kinetic energy from the metal surface. This is done for different frequencies and intensities of light.

Which one of the following statements would **not** be one of the experimental findings?

A. The ability to eject electrons from this metal depended only on the frequency of light.
B. The stopping potential for the photoelectrons was independent of the light intensity.
C. The maximum kinetic energy of the photoelectrons depended only on the light intensity.
D. At frequencies below the threshold frequency, no electrons were ejected from this metal no matter how high the light intensity was.

Question 4

Source: *VCE 2018 Physics Exam, NHT, Section A, Q.16;* © VCAA

When light of a specific frequency strikes a particular metal surface, photoelectrons are emitted.

If the light intensity is increased but the frequency of the light remains the same, which of the following is correct?

	Number of photoelectrons emitted	Maximum kinetic energy of the photoelectrons
A.	remains the same	remains the same
B.	remains the same	increases
C.	increases	remains the same
D.	increases	increases

Question 5

Source: *VCE 2018 Physics Exam, NHT, Section B, Q.17;* © VCAA

A metal surface has a work function of 2.0 eV.

The minimum energy of an incoming photon required to eject a photoelectron is

A. 3.2×10^{-19} J
B. 1.6×10^{-19} J
C. 8.0×10^{-20} J
D. 4.0×10^{-20} J

Question 6

Source: *VCE 2020 Physics Exam, Section A, Q.16;* © VCAA

The diagram below shows a plot of maximum kinetic energy, $E_{k\ max}$, versus frequency, f, for various metals capable of emitting photoelectrons.

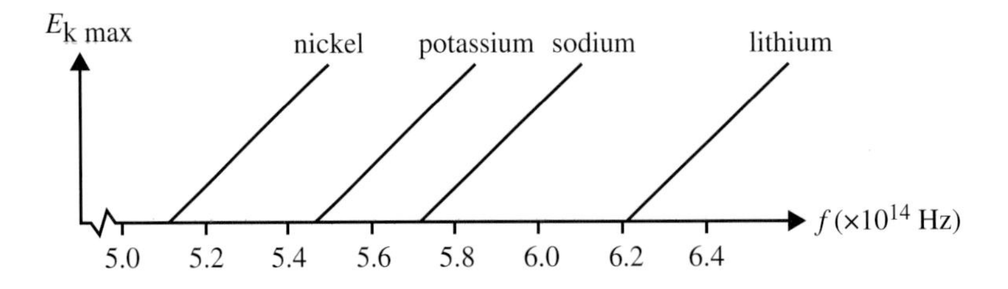

Which one of the following correctly ranks these metals in terms of their work function, from highest to lowest in numerical value?

A. sodium, potassium, lithium, nickel
B. nickel, potassium, sodium, lithium
C. potassium, nickel, lithium, sodium
D. lithium, sodium, potassium, nickel

Question 7

Source: *VCE 2020 Physics Exam, Section A, Q.20;* © VCAA

When photons with energy E strike a metal surface, electrons may be emitted.

The maximum kinetic energy, $E_{k\,max}$, of the emitted electrons is given by $E_{k\,max} = E - W$, where W is the work function of the metal.

Which one of the following graphs best shows the relationship between the maximum kinetic energy of these electrons, $E_{k\,max}$, and the wavelength of the photons, λ?

A.

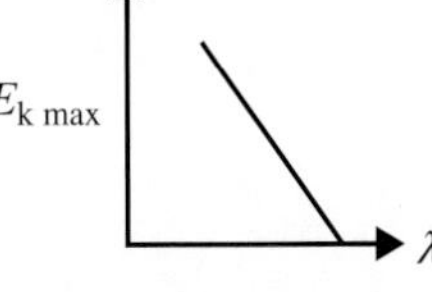

B.

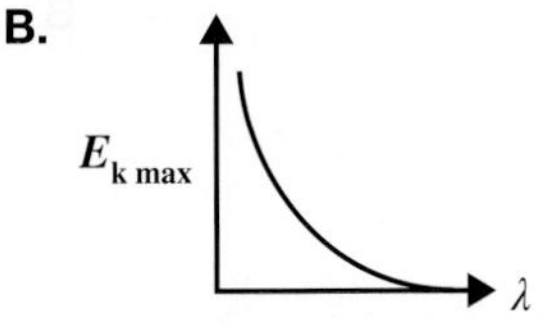

C.

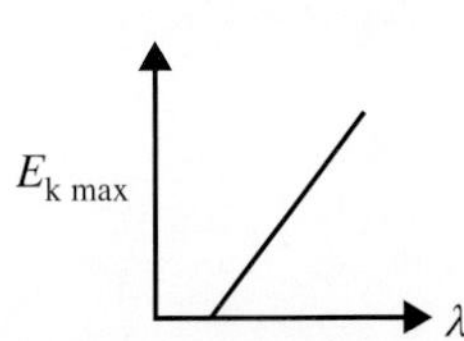

D.

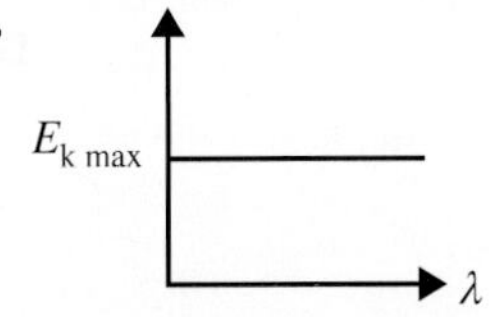

Question 8

Source: *VCE 2019, Physics Exam, Section A, Q.16;* © VCAA

Students are conducting a photoelectric effect experiment. They shine light of known frequency onto a metal and measure the maximum kinetic energy of the emitted photoelectrons.

The students increase the intensity of the incident light.

The effect of this increase would most likely be

A. lower maximum kinetic energy of the emitted photoelectrons.
B. higher maximum kinetic energy of the emitted photoelectrons.
C. fewer emitted photoelectrons but of higher maximum kinetic energy.
D. more emitted photoelectrons but of the same maximum kinetic energy.

Question 9

Source: *VCE 2021, Physics Exam, Section A, Q18;* © VCAA

A monochromatic light source is emitting green light with a wavelength of 550 nm. The light source emits 2.8×10^{16} photons every second.

Which one of the following is closest to the power of the light source?

A. 1.0×10^{-2} W
B. 3.3×10^{-11} W
C. 2.1×10^{9} W
D. 6.3×10^{16} W

Question 10

Source: *VCE 2018, Physics Exam, Q.17;* © VCAA

The results of a photoelectric experiment are displayed in the graph below. The graph shows the maximum kinetic energy ($E_{k\ max}$) of photoelectrons versus the frequency (f) of light falling on the metal surface.

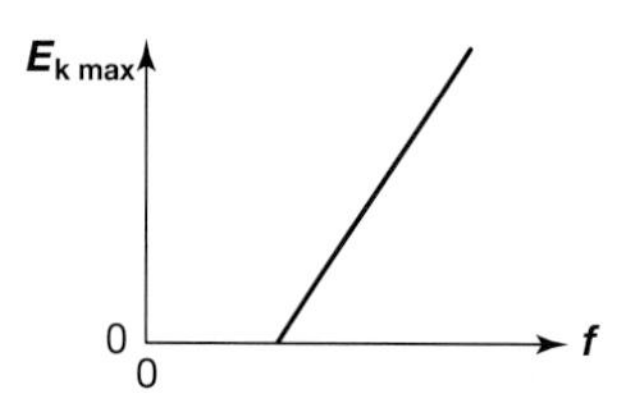

A second experiment is conducted with the original metal surface being replaced by one with a larger work function. The original data is shown with a solid line and the results of the second experiment are shown with a dashed line.

Which one of the following graphs shows the results from the second experiment?

A. 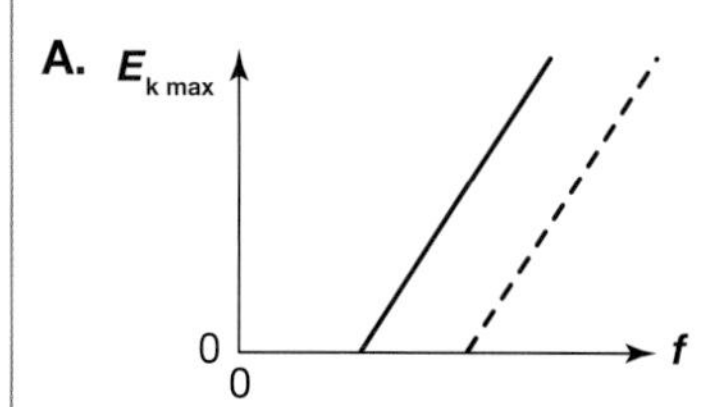

B. 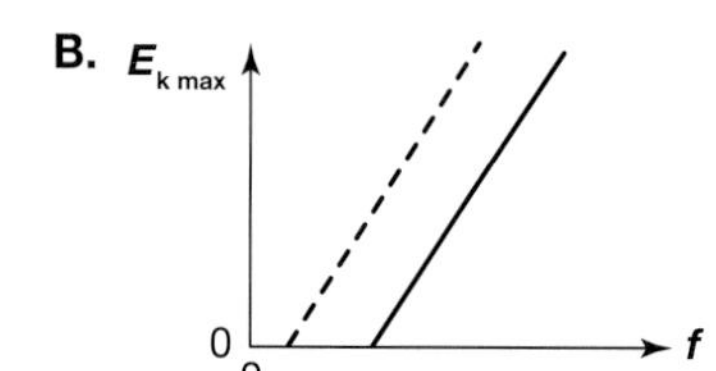

C. 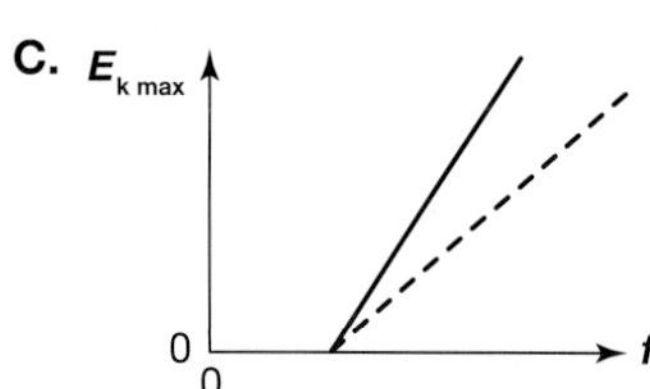

D. 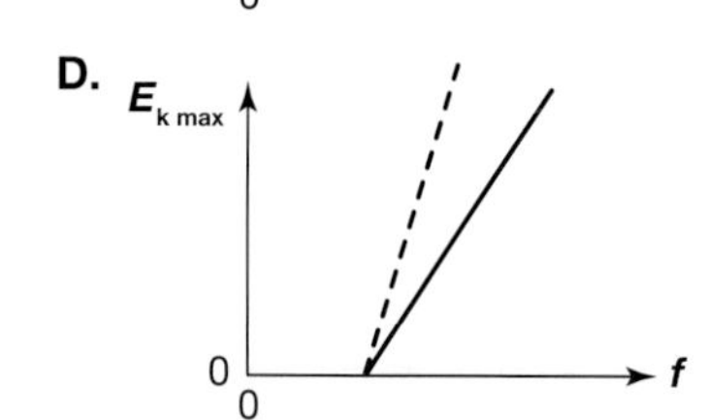

Section B — Short answer questions

Question 11 (9 marks)

Source: *VCE 2022 Physics Exam, Section B, Q.14;* © VCAA

Sam undertakes a photoelectric effect experiment using the apparatus shown in Figure 12. She uses a green filter.

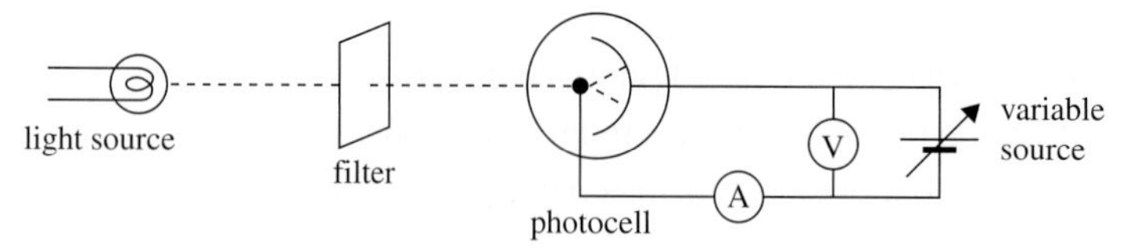

Figure 12

Sam produces a graph of photocurrent, I, in milliamperes, versus voltage, V, in volts, as shown in Figure 13.

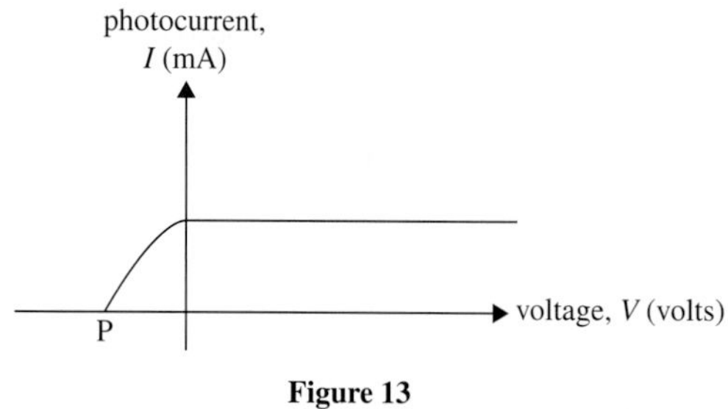

Figure 13

a. Identify what point P represents on the graph in Figure 13. **(1 mark)**

b. Sam then significantly increases the intensity of the light.
Sketch the resulting graph on Figure 14. The dashed line in Figure 14 represents the original data. **(2 marks)**

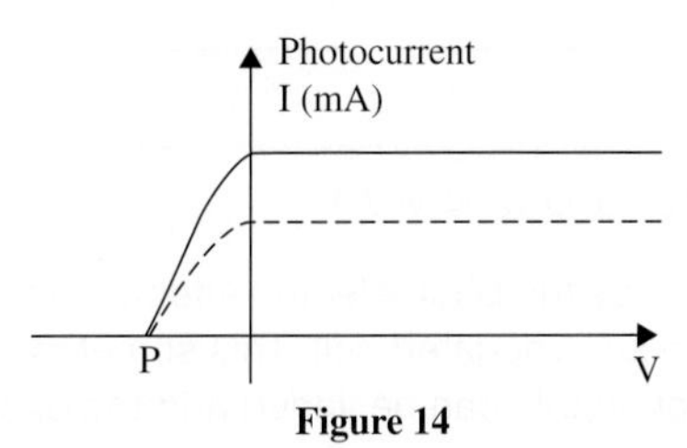

Figure 14

c. Sam replaces the green filter with a violet filter, keeping the light source at the increased intensity.
Sketch the resulting graph on Figure 15. The dashed line in Figure 15 represents the original data. **(2 marks)**

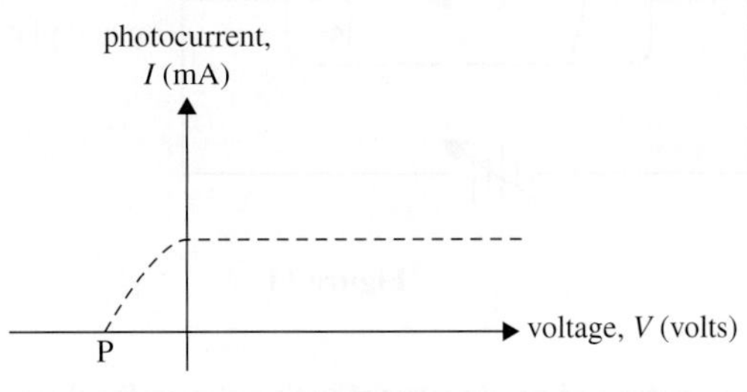

Figure 15

Further experiments produce Figure 16, a graph of maximum kinetic energy, $E_{k\ max}$, of emitted photoelectrons versus frequency, f, of light.

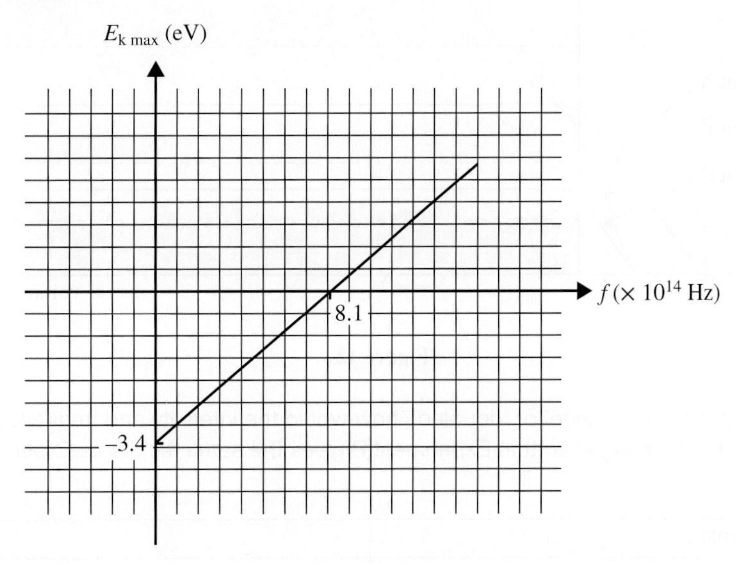

Figure 16

d. Determine the work function, in electron volts, of the metal surface used in the experiment that produced the data shown in Figure 16. **(1 mark)**

e. From the graph shown in Figure 16, calculate, in joule-seconds, the value of Planck's constant. Show your working. **(2 marks)**

f. State **one** limitation of the wave model in explaining the results of the photoelectric effect. **(1 mark)**

Question 12 (6 marks)

Source: *VCE 2022 Physics Exam, NHT, Section B, Q.15; © VCAA*

Figure 11 shows an apparatus used to study the photoelectric effect. Light of various frequencies and intensities can be shone onto the metal plate inside an evacuated cell. This sometimes results in the release of photoelectrons. The voltage of the power supply can be varied and the direction can be reversed.

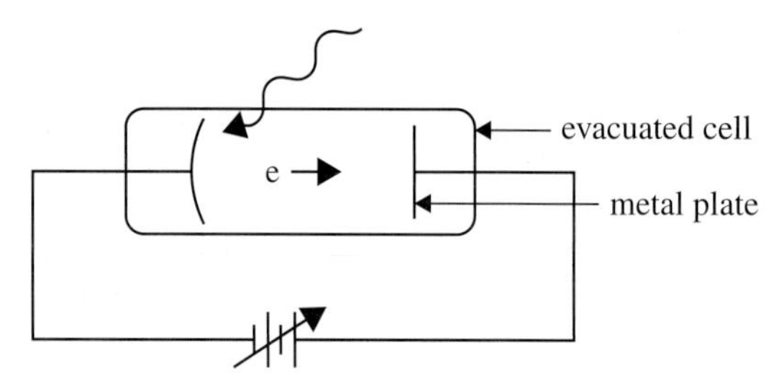

Figure 11

The graph in Figure 12 shows the variation of photocurrent with voltage for three experiments, A, B and C, using light of different frequency and intensity.

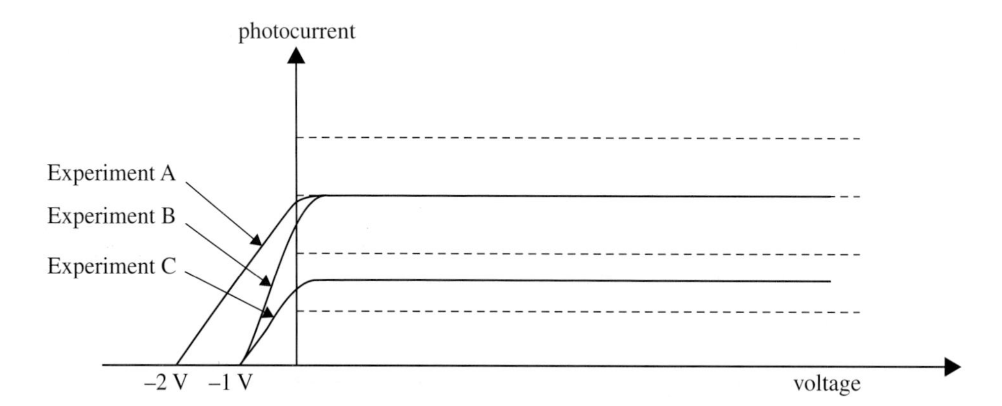

Figure 12

a. Using the terms 'halved', 'no change' or 'doubled', how would the intensity and frequency of the light used in Experiment B need to be changed so that Experiment B gives the same results as Experiment A in Figure 12? **(2 marks)**

Intensity	
Frequency	

b. Using the terms 'halved', 'no change' or 'doubled', how would the intensity and frequency of the light used in Experiment B need to be changed so that Experiment B gives the same results as Experiment C in Figure 12? **(2 marks)**

Intensity	
Frequency	

c. The metal plate is made of a metal that has a work function of 2.93 eV.
Determine whether photoelectrons will be ejected from the metal plate when it is illuminated by light with a wavelength of 700 nm. Show your working. **(2 marks)**

Question 13 (7 marks)

Source: *VCE 2021 Physics Exam, NHT, Section B, Q.15; © VCAA*

The apparatus shown in Figure 19 is used to investigate the photoelectric effect. Light of various wavelengths is shone onto a silver plate (cathode). The work function of silver is 4.9 eV.

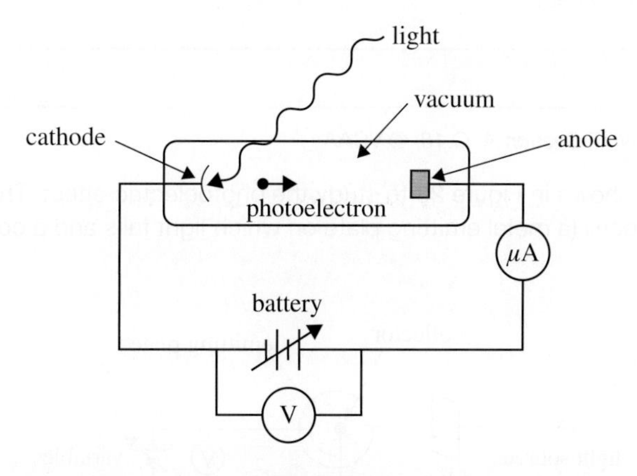

Figure 19

a. Explain what happens when light of wavelength 400 nm hits the silver plate. Use calculations to support your answer. **(2 marks)**

b. Explain what happens when light with a photon energy of 5.4 eV hits the silver plate. **(2 marks)**

c. Which model of light does this photoelectric investigation support? Give two reasons to justify your answer. **(3 marks)**

Question 14 (8 marks)

Source: *VCE 2019 Physics Exam, NHT, Section B, Q.16; © VCAA*

April sets up the apparatus shown in Figure 15 to investigate the photoelectric effect. She can change the frequency of the light incident on the metal plate by changing the filter and she can change the type of metal of which the plate is made.

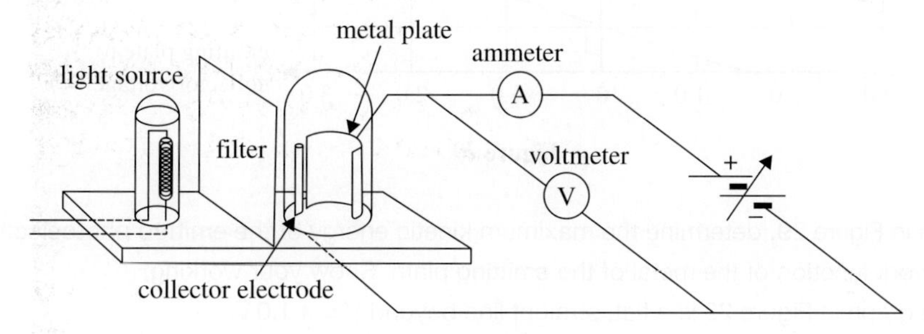

Figure 15

a. For her first experiment, April chooses a filter that gives light of frequency 7.13×10^{14} Hz and a metal plate made of caesium with a work function of 1.95 eV.
April adjusts the voltage of the collector electrode so that the current becomes smaller and smaller.
When the ammeter, A, reaches zero, April records the voltage shown on the voltmeter, V.
Use calculations to determine this voltage. **(3 marks)**

b. For her second experiment, April uses a metal plate made of zinc. Zinc has a threshold frequency for emission of photoelectrons of 1.04×10^{15} Hz. Photoelectrons are emitted.
Calculate the maximum wavelength, in nanometres, of the light for photoelectrons to be emitted from the zinc plate. Show your working. **(2 marks)**

c. For her third experiment, April changes the metal plate from the zinc plate used in the second experiment to a plate made of platinum. Platinum has a threshold frequency of 1.53×10^{15} Hz. April uses light of frequency 7.13×10^{14} Hz but does not make any other changes. Photoelectrons are not emitted.
April observes for a longer time and then increases the intensity of the light beam but still finds that photoelectrons are not emitted.
Explain how April's observations support the particle model of light but do not support the wave model of light in explaining the photoelectric effect. **(3 marks)**

Question 15 (11 marks)

Source: *VCE 2017 Physics Exam, NHT, Section A, Q.18; © VCAA*

Students set up the apparatus shown in Figure 28 to study the photoelectric effect. The apparatus consists of a light source, a filter and a photocell (a metal emitting plate on which light falls and a collecting electrode/collector, all enclosed in a vacuum tube).

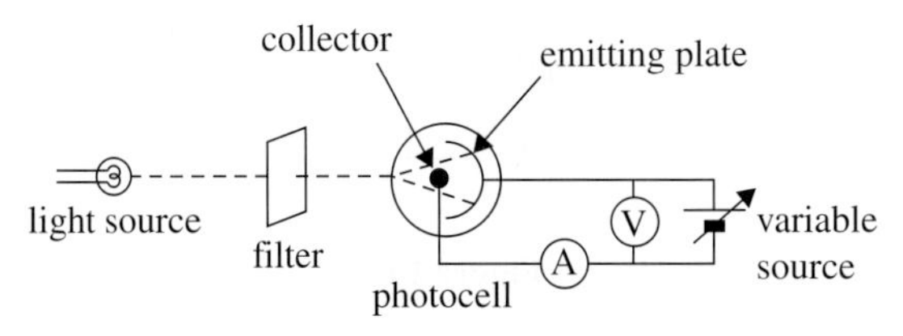

Figure 28

a. The students allow light of 500 nm to shine on the photocell.
Determine the energy of each photon of this light. **(2 marks)**

The students then begin the experiment with the collector negative, with respect to the emitting plate. They gradually reduce the voltage to zero and then increase it to positive values. They measure the current in ammeter A and plot the graph as shown in Figure 29.

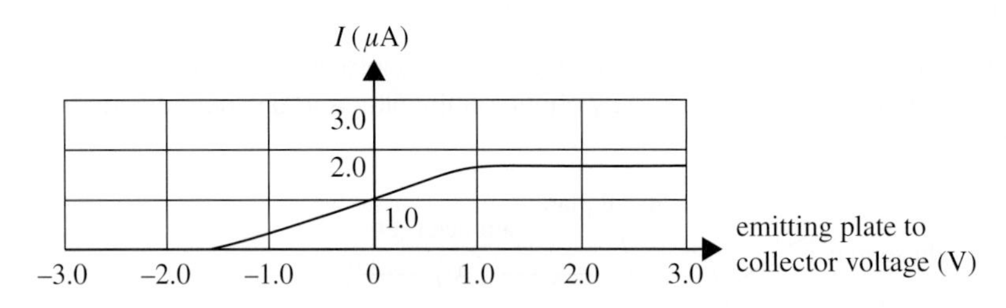

Figure 29

b. Using the graph in Figure 29, determine the maximum kinetic energy of the emitted photoelectrons. **(1 mark)**

c. Determine the work function of the metal of the emitting plate. Show your working. **(2 marks)**

d. Explain why the graph in Figure 29 is a flat, straight line beyond $V = +1.0$ V. **(2 marks)**

e. The students double the intensity of the light, keeping the frequency the same, and plot the results on a graph, with the original current shown as a dashed line.
Which one of the following graphs (**A.–D.**) will they now obtain? **(1 mark)**

A. I (μA)
3.0, 2.0, 1.0
−3.0 −2.0 −1.0 0 1.0 2.0 3.0 emitting plate to collector voltage (V)

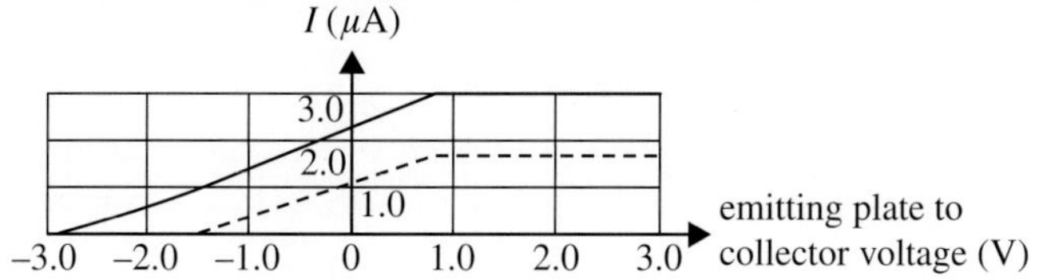

B. I (μA)
3.0, 2.0, 1.0
−3.0 −2.0 −1.0 0 1.0 2.0 3.0 emitting plate to collector voltage (V)

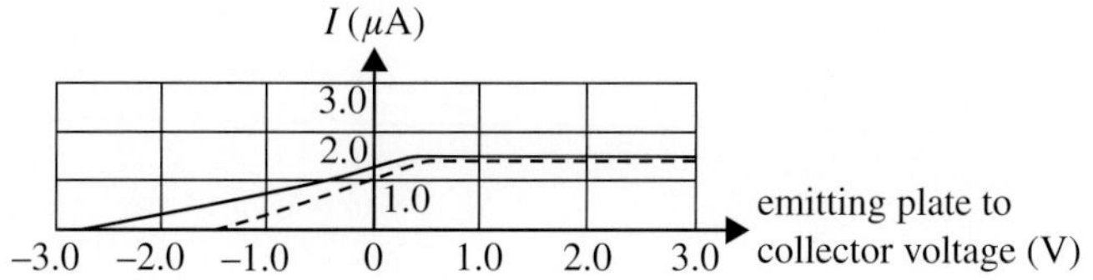

C. I (μA)
3.0, 2.0, 1.0
−3.0 −2.0 −1.0 0 1.0 2.0 3.0 emitting plate to collector voltage (V)

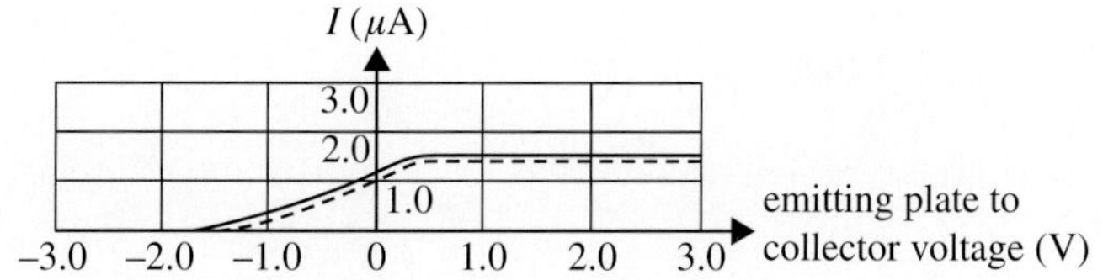

D. I (μA)
3.0, 2.0, 1.0
−3.0 −2.0 −1.0 0 1.0 2.0 3.0 emitting plate to collector voltage (V)

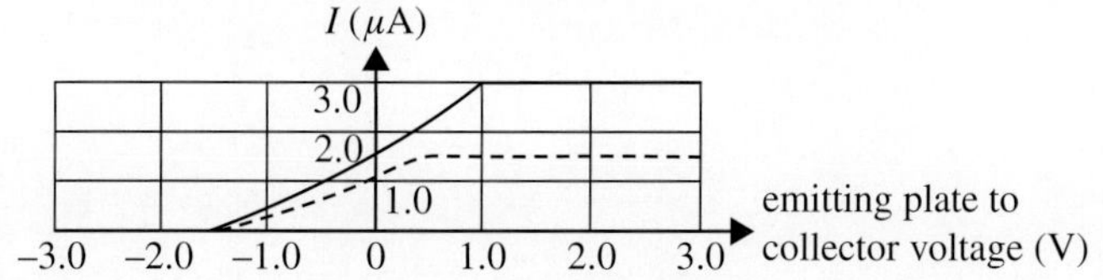

f. Explain your answer to part **e.**, with reference to the particle model and the wave model of light. **(3 marks)**

10 Matter as particles or waves and the similarities between light and matter

KEY KNOWLEDGE

In this topic, you will:

- interpret electron diffraction patterns as evidence for the wave-like nature of matter
- distinguish between the diffraction patterns produced by photons and electrons
- calculate the de Broglie wavelength of matter: $\lambda = \frac{h}{p}$
- compare the momentum of photons and of matter of the same wavelength including calculations using: $p = \frac{h}{\lambda}$
- discuss the importance of the idea of quantisation in the development of knowledge about light and in explaining the nature of atoms
- explain the production of atomic absorption and emission line spectra, including those from metal vapour lamps
- interpret spectra and calculate the energy of absorbed or emitted photons: $E = hf$
- analyse the emission or absorption of a photon by an atom in terms of a change in the electron energy state of the atom, with the difference in the states' energies being equal to the photon energy: $E = hf = \frac{hc}{\lambda}$
- describe the quantised states of the atom with reference to electrons forming standing waves, and explain this as evidence for the dual nature of matter
- interpret the single photon and the electron double slit experiment as evidence for the dual nature of light and matter.

PRACTICAL WORK AND INVESTIGATIONS

Practical work is a central component of VCE Physics. Experiments and investigations, supported by a **practical investigation eLogbook** and **teacher-led video,** are included in this topic to provide opportunities to undertake investigations and communicate findings.

EXAM PREPARATION

Access past VCAA questions and exam-style questions and their video solutions in every lesson, to ensure you are ready.

10.1 Overview

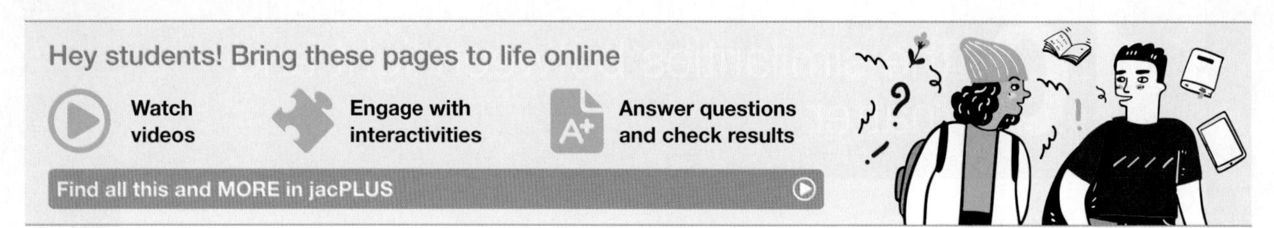

Hey students! Bring these pages to life online
Watch videos
Engage with interactivities
Answer questions and check results
Find all this and MORE in jacPLUS

10.1.1 Introduction

Early in the twentieth century, physicists realised that light can sometimes behave like a wave and at other times like a particle. The question naturally arose: can matter also behave like a particle at certain times and like a wave at other times? In 1920, Louis de Broglie suggested a way to calculate the wavelength of a piece of matter from its momentum. Later that decade, William Bragg showed that individual electrons could diffract when passing through narrow openings — the space between atoms in crystals — and thus have a specified wavelength. A wave model for electrons was developed to explain the diffraction of matter, but was also found to perfectly explain both discrete emission and absorption spectra as well as the inherent stability of atoms themselves. Just as the strings of a guitar emit a spectrum of audio frequencies, an excited atom emits a spectrum of light frequencies, leading us to consider electrons in atoms as a type of standing wave. It seems that a particle–wave model is required for both matter and light. Can light and matter sometimes behave like a wave propagating in all directions at once and sometimes like a particle moving in a single direction only? The answer is: they can!

FIGURE 10.1 Do discoveries related to light behaving like a particle and a wave apply to matter such as electrons?

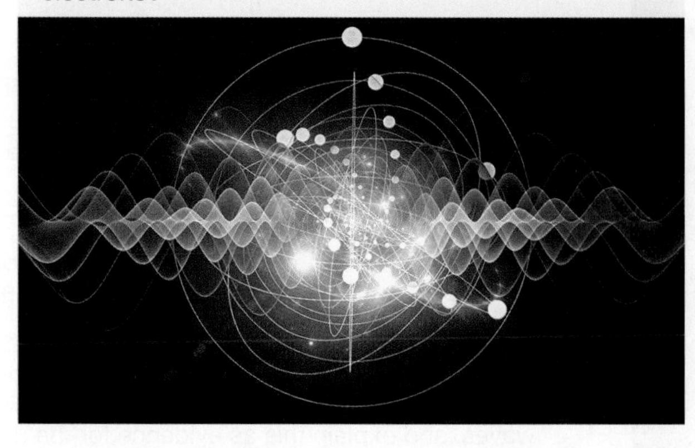

LEARNING SEQUENCE

on Resources

Solutions	Solutions — Topic 10 (sol-0824)
Practical investigation eLogbook	Practical investigation eLogbook — Topic 10 (elog-1641)
Digital documents	Key science skills — VCE Physics Units 1–4 (doc-36950) Key terms glossary — Topic 10 (doc-37183) Key ideas summary — Topic 10 (doc-37184)
Exam question booklet	Exam question booklet — Topic 10 (eqb-0107)

10.2 Matter modelled as a type of wave

KEY KNOWLEDGE

- Interpret electron diffraction patterns as evidence for the wavelike nature of matter
- Calculate the de Broglie wavelength of matter: $\lambda = \frac{h}{p}$

Source: VCE Physics Study Design (2024–2027) extracts © VCAA; reproduced by permission.

10.2.1 The wave behaviour of electrons

matter anything that takes space (has volume) and has a rest mass

By the end of the nineteenth century, it was clear that light exhibited wavelike properties that could be modelled using the wave model of light. At the time, it was also firmly established that **matter** could be modelled as consisting of particles (particle theory). However, as seen in topic 9, the photoelectric effect showed that it was necessary for light to also be modelled as a particle. Could electrons also exhibit wave phenomena as well as demonstrating particle behaviour?

During this time a significant dilemma associated with the stability of atoms also existed. If electrons orbited around the nucleus then they must be accelerating and, according to Maxwell's theory of electromagnetic radiation, they should continuously emit light. Yet, this would cause orbiting electrons to lose energy and spiral inwards into the nucleus. However, atoms are stable so this does not happen. Niels Bohr developed a model in which atomic electrons simply did not emit light, in violation of Maxwell's laws, and had specific discrete amounts of momentum and energy. These orbits were called stationary states. Atoms were known to emit and absorb light at discrete frequencies, a phenomenon not yet understood.

Even though Bohr could calculate electron energies, he could not explain why hydrogen electrons only occupied orbits whose energies were discrete (only taking certain values). Why were these the only possible orbits and how did atoms emit the correct frequency of light to land in another stationary state?

In fact, Rutherford wrote to Bohr:

> *Your ideas are very ingenious and seem to work out well... There seems to me to be one grave difficulty in your hypothesis... namely, how does an electron decide what frequency it is going to vibrate at when it passes from one stationary state into another? It seems to me that you would have to assume that the electron knows beforehand where it is going to stop.*

Light modelled as a wave was known to carry momentum. Using Planck's equation, the amount of momentum was related to energy and to the wavelength. Recall: $p = \frac{E}{c} = \frac{hf}{c} = \frac{h}{\lambda}$.

In 1923 French nobleman Louis de Broglie (1892–1987) suggested that matter also had a wavelength associated with it. He was intrigued by the fact that light exhibited both wavelike and particle-like properties. On this basis, he proposed that matter may also exhibit wavelike properties. De Broglie proposed that the wavelength of a particle, λ, is related to its momentum, p, according to the following equation:

FIGURE 10.2 Louis de Broglie won the Nobel Prize for Physics for his discovery of the wave nature of electrons.

The de Broglie wavelength

$$\lambda = \frac{h}{p} = \frac{h}{mv}$$

where: λ is the **de Broglie wavelength** of the particle, in m

h is Planck's constant (6.63×10^{-34} J s)

p is the momentum of the particle, in kg m s^{-1} (or N s)

m is the mass of the particle, in kg

v is the velocity of the particle, in m s^{-1}

de Broglie wavelength wavelength associated with a particle of matter, in relation to its mass and wavelength

Planck's constant, h, is related to the particle-like behaviour of light and has a value of 6.63×10^{-34} J s. The momentum of matter is given by the product of its mass and velocity.

It can be appreciated why the wave properties of matter are difficult to observe. For example, calculate the de Broglie wavelength of a 70-kg athlete running at a speed of 10 m s^{-1}.

$$\begin{aligned}\lambda &= \frac{h}{mv} \\ &= \frac{6.63 \times 10^{-34}\ \text{J s}}{70\ \text{kg} \times 10\ \text{m s}^{-1}} \\ &= 9.5 \times 10^{-37}\ \text{m}\end{aligned}$$

This wavelength is much too small for any real observation of diffraction effects. The athlete would have to run through an extremely narrow opening for any chance of measurable diffraction. However, for a particle with a small mass, such as an electron travelling at low speed, measurable diffraction is observed. Electrons accelerated through a 100-V potential difference would have a speed of approximately 6.0×10^{6} m s^{-1}, and because the mass of an electron is 9.1×10^{-31} kg, it would have a momentum of:

$$\begin{aligned}p &= mv \\ &= 9.1 \times 10^{-31}\ \text{kg} \times 6.0 \times 10^{6}\ \text{m s}^{-1} \\ &= 5.5 \times 10^{-24}\ \text{kg m s}^{-1}\end{aligned}$$

The de Broglie wavelength for these electrons is:

$$\begin{aligned}\lambda &= \frac{h}{p} \\ &= \frac{6.63 \times 10^{-34}\ \text{J s}}{5.5 \times 10^{-24}\ \text{m s}^{-1}} \\ &= 1.2 \times 10^{-10}\ \text{m}\end{aligned}$$

Note that 10^{-10} m is the order of magnitude of the spacing between atoms in many crystals and recall that if the ratio of wavelength to slit width, $\frac{\lambda}{w}$, is sufficiently large, say greater than $\frac{1}{10}$, then diffraction effects are readily observable.

This observation provides a framework to test whether matter has an associated wavelength, for instance, by using an electron gun to fire a beam of electrons of specific energy, and hence specific momentum and wavelength, at a crystal and observing whether diffraction effects appear. The de Broglie wavelength of a slow-moving electron ($v \ll c$) has the same order of magnitude as the spacing between atoms in a crystal lattice; therefore, diffraction patterns should be observed.

Interpreting electron diffraction

- Diffraction is a property of waves
- Electrons can undergo diffraction when passed through atomic crystals
- Electrons are matter

Electron diffraction is evidence for the wavelike nature of matter.

tlvd-9020

SAMPLE PROBLEM 1 Calculations involving the de Broglie wavelength and the speed of an object

a. Calculate the de Broglie wavelength of a 10.0-g snail whose speed is 0.100 mm s^{-1}.
b. How fast would an electron have to travel to have a de Broglie wavelength of 1.0 μm?

THINK	WRITE
a. The de Broglie wavelength is given by the expression $\lambda = \frac{h}{p} = \frac{h}{mv}$, keeping in mind that mass must be in kilograms and velocity in metres per second.	**a.** $\lambda = \frac{h}{mv}$ $= \frac{6.63 \times 10^{-34}}{10.0 \times 10^{-3} \times 0.100 \times 10^{-3}}$ $= 6.63 \times 10^{-28}$ m
b. The expression $\lambda = \frac{h}{mv}$ can be transposed to make v the subject. The values for Planck's constant, the mass of an electron and the de Broglie wavelength can be substituted in.	**b.** $v = \frac{h}{m\lambda}$ $= \frac{6.63 \times 10^{-34}}{9.1 \times 10^{-31} \times 1.0 \times 10^{-6}}$ $= 7.3 \times 10^{2}$ m s^{-1} The speed of the electron is 7.3×10^{2} m s^{-1}.

PRACTICE PROBLEM 1

Which has the greater de Broglie wavelength: a proton ($m = 1.67 \times 10^{-27}$ kg) travelling at 2.0×10^{4} m s^{-1} or an electron ($m = 9.1 \times 10^{-31}$ kg) travelling at 2.0×10^{5} m s^{-1}?

The de Broglie wavelength for matter is inversely proportional to both the speed and mass. Matter with large wavelengths necessary for wave properties to manifest themselves has to travel slowly and have little mass. Since electrons have a mass that is approximately $\frac{1}{1800}$ that of a proton or neutron, it is easier to detect the wave properties of electrons over those of other fundamental particles such as protons and neutrons.

Resources

Video eLessons De Broglie wavelength (eles-3520)
The momentum of photons and matter (eles-3521)

10.2 Quick quiz	10.2 Exercise	10.2 Exam questions

10.2 Exercise

1. A particle has momentum 3.5×10^{-4} N s. Calculate the de Broglie wavelength of the particle.
2. A particle of mass 3.0 kg has a de Broglie wavelength of 5.0×10^{-36} m. Calculate the speed of the particle.
3. When a particle is accelerated so that its speed increases, state and explain any changes to its de Broglie wavelength.
4. A soccer ball ($m = 420$ g) is kicked towards a goalie defending the net at a speed of 40 m s^{-1}. Calculate the de Broglie wavelength of the soccer ball.
5. Determine what the maximum value of the momentum of an object must be for its de Broglie wavelength to exceed 1.00 nm.
6. Suggest an experiment that in principle could show that matter exhibits wavelike behaviour.

10.2 Exam questions

Question 1 (1 mark)

***Source**: Adapted from VCE 2021 Physics Exam, Section B, Q.17a; © VCAA*

A 'space sail' mounted on a tiny interstellar cylindrical probe relies on the momentum of photons from a nearby star to exert a propulsive force, as shown in the following figure.

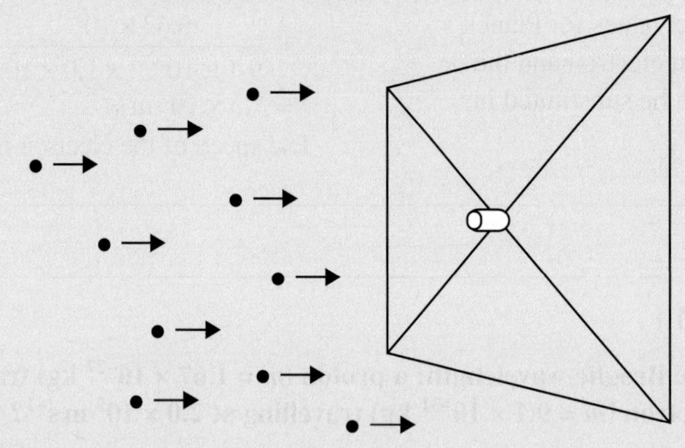

The photons strike the sail at 90° to its surface and reflect elastically. Scientists need to calculate the force exerted by the photons, which have a frequency of 7.0×10^{15} Hz.

Calculate the momentum of a 7.0×10^{15}-Hz photon.

Question 2 (1 mark)

***Source**: VCE 2019 Physics Exam, Section A, Q.14; © VCAA*

MC Electrons of mass 9.1×10^{-31} kg are accelerated in an electron gun to a speed of 1.0×10^{7} m s^{-1}.

The best estimate of the de Broglie wavelength of these electrons is

A. 4.5×10^{-6} m
B. 7.3×10^{-8} m
C. 7.3×10^{-11} m
D. 4.5×10^{-12} m

Question 3 (3 marks)

Source: *VCE 2018 Physics Exam, NHT, Section B, Q.13a; © VCAA*

Electrons are accelerated through a potential difference of 4000 V and then pass through a metallic crystal. The resulting diffraction pattern is observed.

Calculate the de Broglie wavelength of these electrons.

Question 4 (2 marks)

Source: *VCE 2017 Physics Exam, NHT, Section B, Q.19a; © VCAA*

Students use a piece of apparatus to study electron diffraction. The apparatus consists of an evacuated tube, with an electron gun at one end, a crystal in front of the electron gun and a screen at the other end. The apparatus is shown in Figure 30.

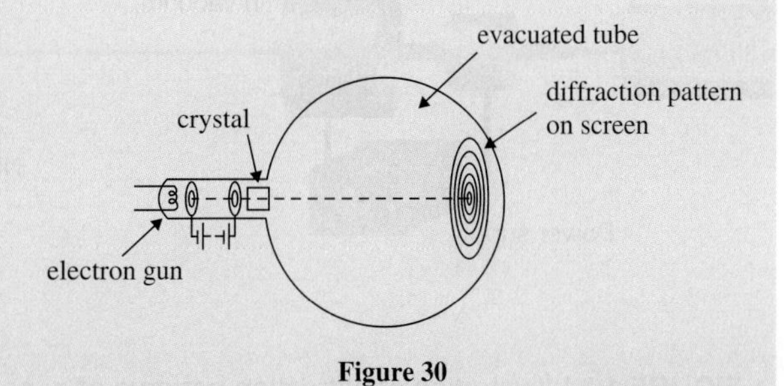

Figure 30

The electrons are accelerated by a voltage to a speed of 2.0×10^{7} m s^{-1}. The mass of an electron is 9.1×10^{-31} kg.

Calculate the de Broglie wavelength of these electrons. Show your working.

Question 5 (2 marks)

Source: *VCE 2015 Physics Exam, Section A, Q.22; © VCAA*

Electrons (of mass 9.1×10^{-31} kg) have a de Broglie wavelength of 1.0×10^{-11} m.

Calculate the speed of these electrons.

More exam questions are available in your learnON title.

10.3 The diffraction of light and matter

KEY KNOWLEDGE

- Distinguish between the diffraction patterns produced by photons and electrons
- Compare the momentum of photons and of matter of the same wavelength including calculations using: $p = \frac{h}{\lambda}$

Source: VCE Physics Study Design (2024–2027) extracts © VCAA; reproduced by permission.

10.3.1 Diffraction of electrons

De Broglie suggested conducting an experiment to confirm whether a beam of electrons could be diffracted from the surface of a crystal. The spaces between atoms could be used as a diffraction grating in much the same way that X-rays were diffracted by thin crystals as suggested by Max von Laue in 1912. Clinton Davisson (1881–1958) and Lester Germer (1896–1971) directed a beam of electrons at a metal crystal in 1927, and the scattered electrons came off in regular peaks as shown in figure 10.3b. This pattern is indicative of diffraction taking place with individual electrons as they scattered off the crystal surface. The diffraction patterns produced by photons and electrons (figure 10.4) are indistinguishable when they are at the same wavelength. In fact, the wavelength determined from the diffraction experiments was exactly as predicted by the de Broglie wavelength formula. In this way, electrons were shown to have wavelike properties. Since then, protons, neutrons and, more recently, atoms have been shown to exhibit wavelike properties, but it begs the question: if matter can exhibit wave characteristics, what is it that is 'waving'? More scientifically, the question is: what physical variable is it that has an amplitude and phase?

FIGURE 10.3 The Davisson and Germer experiment. a. Electrons emitted from a heated filament are accelerated towards the crystal surface. The intensity of reflected electrons is recorded as the angle of the detector is changed. b. Electron intensity as a function of angle.

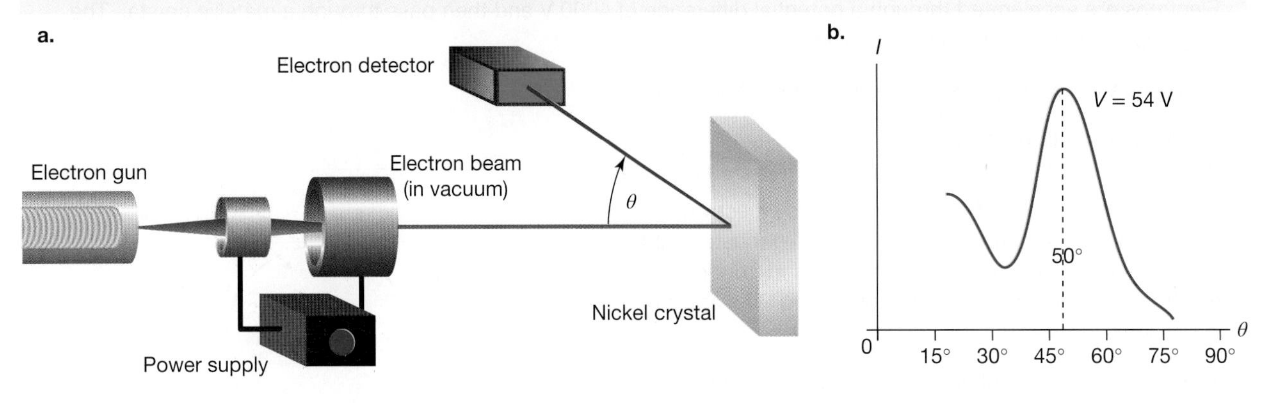

FIGURE 10.4 Illustration of diffraction patterns of a. an X-ray beam and b. an electron beam through a metal lattice

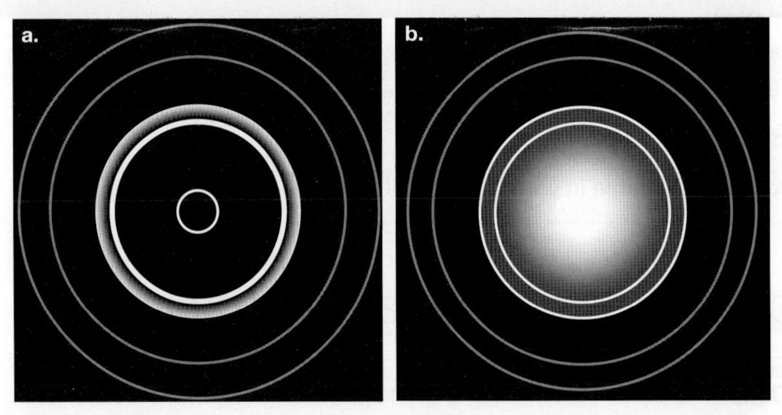

tlvd-9021

SAMPLE PROBLEM 2 Calculating dimensions to observe a diffraction pattern

What dimensions of an array of slits is required to observe diffraction of 60-g tennis balls travelling at 30 m s^{-1}? What about electrons travelling at 3.0 × 10^{6} m s^{-1}?

THINK

1. To observe diffraction effects, the size of the opening needs to be of the same order of magnitude or smaller than the wavelength of the waves.
   Calculate the de Broglie wavelength of the tennis ball and the electron, by substituting $h = 6.63 \times 10^{-34}$ J s, $m = 60$ g $= 0.060$ kg and $v = 30$ m s^{-1} for the tennis ball, and $m = 9.1 \times 10^{-31}$ kg and $v = 3.0 \times 10^{6}$ m s^{-1} for the electrons.

WRITE

The de Broglie wavelength of the tennis ball:

$$\lambda = \frac{h}{mv} = \frac{6.63 \times 10^{-34}}{0.060 \times 30} = 3.7 \times 10^{-34}\,\text{m}$$

The de Broglie wavelength of the electron:

$$\lambda = \frac{h}{mv} = \frac{6.63 \times 10^{-34}}{9.1 \times 10^{-31} \times 3.0 \times 10^{6}} = 2.4 \times 10^{-10}\,\text{m}$$

2. State the order of magnitude of the de Broglie wavelength of the tennis ball and the de Broglie wavelength of the electron.	The de Broglie wavelength of the tennis ball is of the order of 10^{-34} m and that of the electron is of the order of 10^{-10} m.
3. Answer the question.	The distances between atoms in a crystal are of the order of 10^{-10} m, so diffraction and interference can be observed when these electrons are scattered from a crystal. It is not surprising that diffraction and interference effects are never observed with tennis balls, due to the extremely small wavelength, 10^{-34} m, that they have.

PRACTICE PROBLEM 2

At what velocity would neutrons (mass = 1.67×10^{-27} kg) need to move so they demonstrate diffraction effects when passing through an array of slits of width 1.0 µm?

10.3.2 Electrons through foils

Intense, creative interest in fundamental physics ran in the Thomson family. It was J.J. Thomson whose ingenious experiment yielded the measurement of the charge-to-mass ratio of the electron. At that time, there was no doubt that electrons were extremely well modelled as particles. However, G.P. Thomson, son of J.J., continued the exploration of the wave properties of electrons. He fired electrons through a thin polycrystalline metallic foil. The electrons had a much greater momentum than those used by Davisson and Germer. They were able to penetrate the foil and produce a pattern demonstrating diffraction of the electrons by the atoms of the foil — further evidence for wavelike behaviour of electrons. The polycrystalline nature of the foil results in a series of rings of high intensity, as shown in figure 10.5b. Thomson used identical analysis techniques to those used for the diffraction of X-rays through foils, to confirm the de Broglie relationship.

Both Thomsons were awarded Nobel prizes — J.J. in 1897 for measuring a particle-like characteristic of electrons, and G.P. in 1937, together with C.J. Davisson, for demonstrating their wave properties.

FIGURE 10.5 Electrons through foils. a. Diffraction of X-rays and electrons by polycrystalline foils b. Diffraction pattern

a.

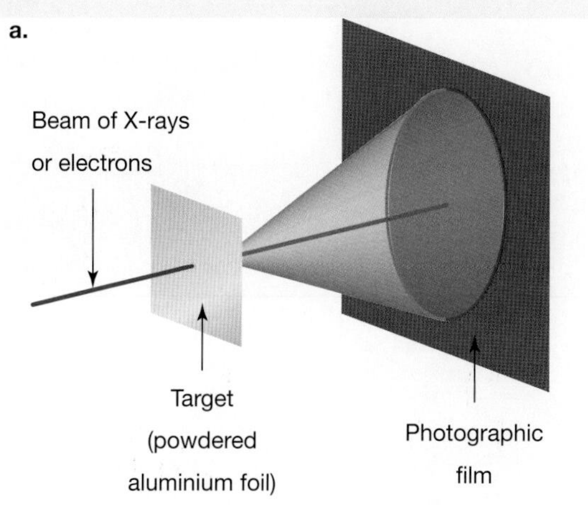

b.

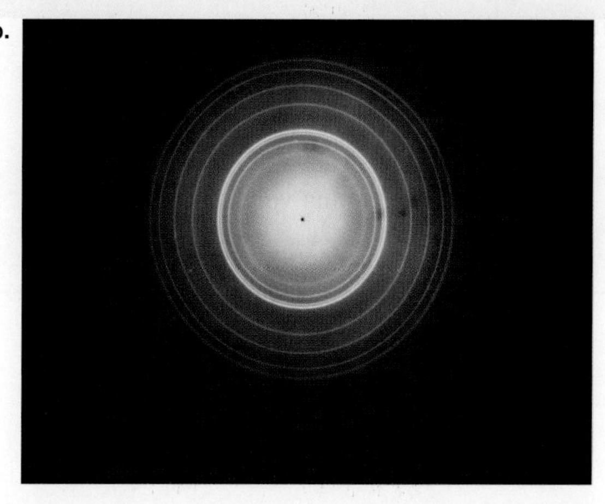

Just as light requires a wave model and a particle model to interpret and explain how it behaves, so too does matter. It behaves like a particle in the sense that work can be done on it to increase its kinetic energy under the action of forces, but matter can also be made to diffract through sufficiently narrow openings and around obstacles. This requires a wave model and the de Broglie wavelength is used to determine the extent of matter's wave behaviour. It appears that both a particle and a wave model are needed for both light and matter. Electrons passed through a voltage, V, acquire a kinetic energy E_k equal to $q_e V$. Since they have kinetic energy, they also possess momentum and, according to de Broglie, a wavelength. A relationship between the de Broglie wavelength of an electron (λ) and the accelerating voltage (V) can be determined.

By equating the kinetic energy of the electron (E_k) to the work done by an accelerating voltage acting on an electron ($q_e V$), the following relationship is obtained

$$E_k = \frac{1}{2} m_e v^2 = q_e V$$
$$m_e v^2 = 2q_e V$$

If we multiply both sides of the equation by the mass of an electron (m_e) seen earlier, we get:

$$(m_e)^2 v^2 = 2m_e q_e V$$

The left-hand side is just the square of the momentum of the electron, and hence by taking the square root of both sides:

$$p = \sqrt{2m_e q_e V}$$

or

$$p = \sqrt{2m_e E_k},$$

remembering that E_k is equal to $q_e V$.

Since the de Broglie wavelength λ is given by $\frac{h}{p}$, it follows that:

$$\lambda = \frac{h}{\sqrt{2m_e q_e V}} = \frac{h}{\sqrt{2m_e E_k}}$$

where: h is Planck's constant (6.63×10^{-34} J s)

m_e is the mass of an electron, in kg

q_e is the charge on one electron, in C

V is the accelerating voltage, in V

E_k is the kinetic energy of the electron, in J

tlvd-9023

SAMPLE PROBLEM 3 Calculating acceleration voltage

What voltage is required to accelerate electrons to a velocity of 3.0×10^6 m s^{-1}?

THINK

To accelerate electrons to a speed of 3.0×10^6 m s^{-1}, the work done by a voltage V needs to be calculated.

$$\Delta E_{\text{k electron}} = \frac{1}{2} m_e v^2$$

$$\Delta E_{\text{k electron}} = -\Delta E_{\text{p electron}}$$

$$= q_e V$$

$$\Rightarrow V = \frac{m_e v^2}{2q_e}$$

where q_e is the *magnitude* of the charge of the electron (1.6×10^{19} C).

WRITE

$$V = \frac{m_e v^2}{2q_e}$$

$$= \frac{9.1 \times 10^{-31} \times (3.0 \times 10^6)^2}{2 \times 1.6 \times 10^{-19}}$$

$$= 26 \text{ V}$$

So, only 26 V is required to accelerate an electron to 3.0×10^6 m s^{-1}.

PRACTICE PROBLEM 3

Calculate the velocity of electrons accelerated from rest by an electron gun whose voltage is set at 13 V.

tlvd-9022

SAMPLE PROBLEM 4 Comparing the wavelength of X-rays and electrons with similar wavelengths

Some of the X-rays used in G.P. Thomson's experiment had a wavelength of 7.1×10^{-11} m. Confirm that the 600-eV electrons have a similar wavelength.

THINK

1. Electrons of energy 600 eV have passed through a voltage equal to 600 V; thus, their energy is $1.6 \times 10^{-19} \times 600$ J. From this, their de Broglie wavelength can be determined. Use the relationship:

$$\lambda = \frac{h}{\sqrt{2m_e E_k}}$$

2. Answer the question.

WRITE

1. $$\lambda = \frac{h}{\sqrt{2m_e E_k}}$$

$$= \frac{6.63 \times 10^{-34}}{\sqrt{2 \times 9.1 \times 10^{-31} \times 1.6 \times 10^{-19} \times 600}}$$

$$= 5.0 \times 10^{-11} \text{ m}$$

2. The de Broglie wavelength of 5.0×10^{-11} m is a similar value to the 7.1×10^{-11} m wavelength of the X-rays.

PRACTICE PROBLEM 4

a. **X-rays of wavelength 0.053 nm are used to investigate the structure of a new plastic. If a beam of electrons is to be used instead of X-rays, what voltage should be used to accelerate these electrons?**
b. **Which has a greater wavelength: a 100-eV photon or a 100-eV electron?**

Comparing the momentum of photons with that of matter of the same wavelength

When examining the energy and momentum of photons and electrons, two useful energy–momentum relationships can be derived. The equations relating energy E and momentum p are different for photons compared to electrons.

Photons

Since $E = hf$ and $f = \frac{c}{\lambda}$, it follows that $E = \frac{hc}{\lambda}$. But since $p = \frac{h}{\lambda}$, it also follows that $E = pc$. The energy of a photon E is directly proportional to the momentum of a photon:

$$E = pc$$

Electrons

Since $E = \frac{1}{2}mv^2 = \frac{1}{2} \times \frac{m^2v^2}{m}$ and $p = mv$, it follows that $E = \frac{p^2}{2m}$. The kinetic energy E of an electron is directly proportional to the square of the electron's momentum:

$$E = \frac{p^2}{2m}$$

Therefore:

- Photons having the same momentum as an electron will have a different amount of energy.
- Photons having the same energy as an electron will have a different amount of momentum.

tlvd-9024

SAMPLE PROBLEM 5 Calculating the momentum and energy of photons and electrons

Consider a photon and an electron that both have a wavelength of 2.0×10^{-10} m.
a. **Calculate and compare the momentum of the photon and the electron.**
b. **Calculate and compare the energy of the photon and the electron.**
c. **Summarise what you have found concerning the momentum and energy of a photon and an electron with the same wavelength.**

THINK	WRITE
a. 1. The momentum of the photon and the electron are governed by the same equation, namely $p = \frac{h}{\lambda}$. Hence, both the photon and the electron will have the same momentum because they have the same associated wavelength.	a. $p = \frac{h}{\lambda} = \frac{6.63 \times 10^{-34}}{2.0 \times 10^{-10}} = 3.3 \times 10^{-24}$ N s

2. Write your observations.

The photon and the electron have the same momentum.

b. 1. To determine the energy of an object from its momentum, the question must be asked whether it is a photon or an object with mass. The relations are different. For the photon, $E = pc$. For the electron, however, $E = \frac{p^2}{2m}$.

b. Energy of the photon:

$$\begin{aligned} E &= pc \\ &= 3.3 \times 10^{-24} \times 3.0 \times 10^8 \\ &= 9.9 \times 10^{-16} \text{ J or } 6.2 \text{ keV} \end{aligned}$$

Energy of the electron:

$$\begin{aligned} E &= \frac{p^2}{2m} \\ &= \frac{(3.3 \times 10^{-24})^2}{2 \times 9.1 \times 10^{-31}} \\ &= 6.0 \times 10^{-18} \text{ J or } 37 \text{ eV} \end{aligned}$$

2. Write your observations.

The electron has substantially less kinetic energy than the photon, even though they have the same momentum.

c. Summarise your observations from parts a and b.

c. Light and matter with the same wavelength will have the same momentum, and vice versa. However, when photons and electrons have the same momentum, they will not necessarily have the same energy. In this problem, the photon has substantially more energy than the electron.

PRACTICE PROBLEM 5

Consider a photon and an electron that both have a wavelength of 1.0×10^{-10} m.

a. **Calculate and compare the momentum of the photon and the electron.**

b. **Calculate and compare the energy of the photon and the electron.**

10.3 Activities

learnon

Students, these questions are even better in jacPLUS

Receive immediate feedback and access sample responses

Access additional questions

Track your results and progress

Find all this and MORE in jacPLUS

10.3 Quick quiz on	10.3 Exercise	10.3 Exam questions

10.3 Exercise

1. Calculate the de Broglie wavelength of the following particles.
   a. A proton (m = 1.67 x 10^{-27} kg) travelling at 3.0 × 10^{7} m s^{-1}
   b. An electron accelerated by a voltage of 54 V, the voltage used by Davisson and Germer in their electron diffraction experiment
   c. A tennis ball (m = 0.20 kg) moving with a speed 50 m s^{-1}
2. In X-ray tubes, the electric potential energy of electrons is transformed into the energy of X-ray photons. Consider a beam of electrons accelerated through 5.0 kV from rest, which rapidly decelerate when they collide with the anode of the tube.
   a. What is the kinetic energy of these electrons as they reach the anode, in joules?
   b. If the entire energy of each electron is transformed into the energy of a single photon, what is the wavelength of the resulting X-rays?
3. Explain what William L. Bragg meant when he said, 'On Mondays, Wednesdays and Fridays light behaves like waves, on Tuesdays, Thursdays and Saturdays like particles, and like nothing at all on Sundays'. Is this a good description of the behaviour of light?
4. A beam consists of electrons with speed 2.5 × 10^{6} m s^{-1} inside an evacuated tube. The beam is directed towards a thin crystal of sodium chloride that can act as a diffraction grating. The spacing between atoms for this crystal is 2.8 × 10^{-10} m.
   a. Calculate the momentum and the de Broglie wavelength for electrons in the beam.
   b. By comparing the wavelength to the atomic spacing, discuss whether or not the electrons would diffract significantly.
5. Electrons may display wave properties and diffract when passed through narrow openings. In an experiment, a scientist uses an electron gun to direct a beam of electrons towards a crystal. The spacing between the atoms in the crystal is about 5.0 × 10^{-10} m. The scientist adjusts the accelerating voltage of the electron gun to 3.0 kV.
   a. Calculate the energy of electrons in the beam in eV and in J.
   b. Calculate the momentum and hence the de Broglie wavelength of the electrons.
   c. Determine whether the scientist should expect to observe significant diffraction effects.
   d. Explain how the scientist should adjust the accelerating voltage to make electrons diffract significantly when passing through the crystal.

   The scientist now decides to use photons to obtain the same diffraction pattern and passes a beam of photons through the same crystal.
   e. What wavelength and, hence, momentum photons should they use?
   f. What is the energy of these photons? Give your answer in eV and in J.
6. Calculate the speed of an electron that has the same de Broglie wavelength as a photon of red light whose frequency is 4.5 × 10^{14} Hz.
7. An electron and a proton are accelerated through the same potential difference.
   a. Determine which will have the greater de Broglie wavelength.
   b. Using a potential difference of 1000 V, calculate the de Broglie wavelength for both an electron and a proton.
      The mass of a proton is 1.67 × 10^{-27} kg.

8. Electrons can be accelerated with a potential difference in an electron gun. To make a beam of electrons whose de Broglie wavelength is 2.0×10^{-10} m, calculate the potential difference that needs to be used.
9. Which has the shorter wavelength, a 10-eV electron or a 10-eV photon?

10.3 Exam questions

Question 1 (7 marks)

***Source:** VCE 2022 Physics Exam, Section B, Q.17; © VCAA*

A materials scientist is studying the diffraction of electrons through a thin metal foil. She uses electrons with an energy of 10.0 keV. The resulting diffraction pattern is shown in Figure 19.

a. Calculate the de Broglie wavelength of the electrons in nanometres. **(4 marks)**

b. The materials scientist then increases the energy of the electrons by a small amount and hence their speed by a small amount. Explain what effect this would have on the de Broglie wavelength of the electrons. Justify your answer. **(3 marks)**

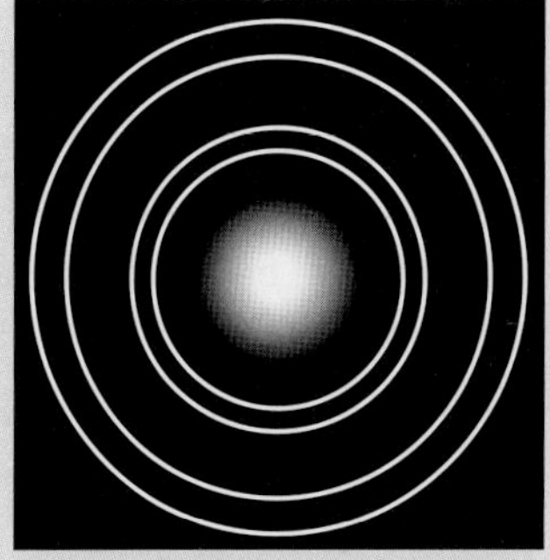

Figure 19

Question 2 (5 marks)

***Source:** VCE 2022 Physics Exam, NHT, Section B, Q.16; © VCAA*

Figure 13 shows the diffraction pattern produced by an X-ray beam consisting of photons of energy 400 eV.

a. Show that the wavelength of an X-ray photon is approximately 3 nm. **(2 marks)**

b. A stream of electrons produces a diffraction pattern with the same spacing as the X-ray diffraction pattern shown in Figure 13. Calculate the speed of an electron in the stream. Take the mass of the electron to be 9.1×10^{-31} kg. **(3 marks)**

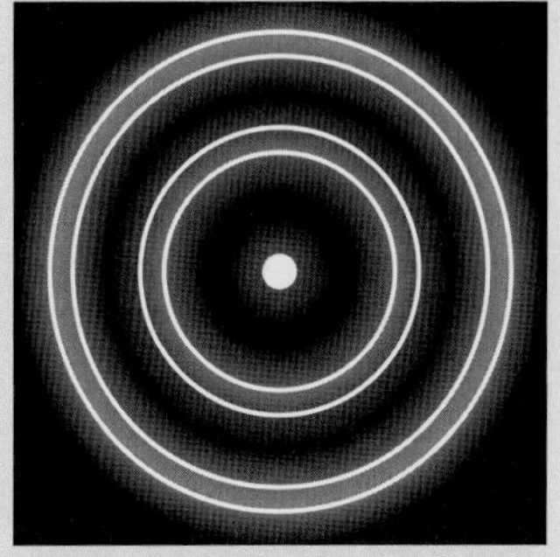

Figure 13

Question 3 (5 marks)

***Source:** VCE 2020 Physics Exam, Section B, Q.16; © VCAA*

A beam of electrons travelling at 1.72×10^{5} m s^{-1} illuminates a crystal, producing a diffraction pattern as shown in Figure 16. Take the mass of an electron to be 9.1×10^{-31} kg. Ignore relativistic effects.

a. Calculate the kinetic energy of one of the electrons. Show your working. **(2 marks)**

b. The electron beam is now replaced by an X-ray beam. The resulting diffraction pattern has the same spacing as that produced by the electron beam. Calculate the energy of one X-ray photon. Show your working. **(3 marks)**

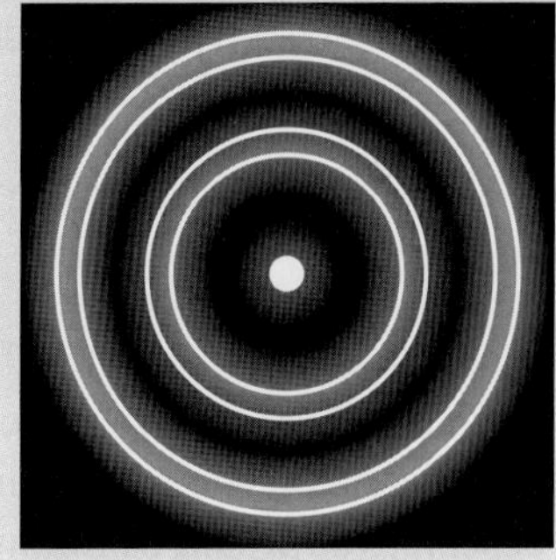

Figure 16

Question 4 (5 marks)

***Source:** VCE 2021 Physics Exam, Section B, Q.18; © VCAA*

Scientists are conducting experiments to compare the circular diffraction patterns formed by X-ray photons and electrons when they pass through small apertures. The X-ray photons have an energy of 100 eV and pass through an aperture of diameter 1.24 μm. The electrons are moving at 5.0×10^{5} m s^{-1}.

a. Show that the de Broglie wavelength of the electrons is equal to 1.46×10^{-9} m. **(1 mark)**

b. The scientists want an aperture for the electrons that forms diffraction patterns with the same spacing as the diffraction patterns formed by the X-ray photons. Calculate the diameter of the aperture that the scientists should choose. Show your working. **(4 marks)**

Question 5 (7 marks)

Source: *VCE 2019 Physics Exam, Section B, Q.17; © VCAA*

Students are comparing the diffraction patterns produced by electrons and X-rays, in which the same spacing of bands is observed in the patterns, as shown schematically in Figure 18. Note that both patterns shown are to the same scale.

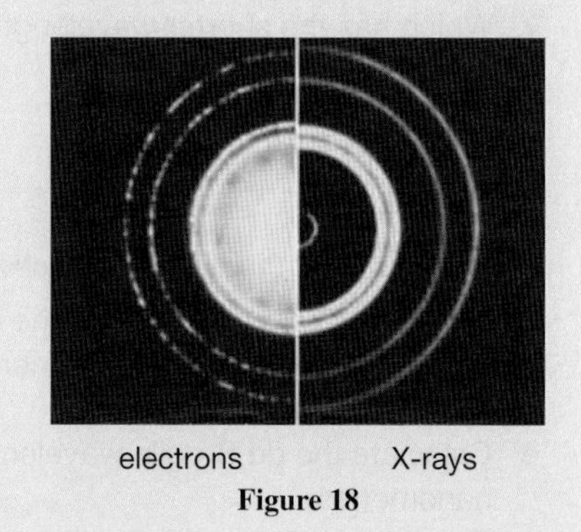

Figure 18

The electron diffraction pattern is produced by 3.0×10^3 eV electrons.

a. Explain why electrons can produce the same spacing of bands in a diffraction pattern as X-rays. **(3 marks)**

b. Calculate the frequency of X-rays that would produce the same spacing of bands in a diffraction pattern as for the electrons. Show your working. **(4 marks)**

More exam questions are available in your learnON title.

10.4 Emission and absorption spectra

KEY KNOWLEDGE

- Discuss the importance of the idea of quantisation in the development of knowledge about light and in explaining the nature of atoms
- Explain the production of atomic absorption and emission line spectra, including those from metal vapour lamps
- Interpret spectra and calculate the energy of absorbed or emitted photons: $E = hf$
- Analyse the emission or absorption of a photon by an atom in terms of a change in the electron energy state of the atom, with the difference in the states' energies being equal to the photon energy: $E = hf = \frac{hc}{\lambda}$

10.4.1 Atoms emit photons

If you dip a loop of wire into a solution of common salt (sodium chloride) in water and then place the loop in the flame of a Bunsen burner, you will observe that the blue flame is transformed into glorious yellow. Placing a slide with two slits to convert the light of the flame into a beam and using a prism to disperse this light, and a telescope to observe, you would be following in the steps of the scientists who developed the field of spectroscopy. You would have constructed a **spectrometer**, which would show spectra produced by the solution of sodium chloride (figure 10.6).

spectrometer device used to disperse light into its spectrum

emission spectra spectra produced when light is emitted from an excited gas and passed through a spectrometer. A spectrum includes a series of bright lines on a dark background. The bright lines correspond to the frequencies of light emitted by the gas.

Sodium atoms in the flame produce the spectrum. This is identical to the spectrum observed when an electric current is passed through a container of sodium gas at low pressure. Spectrometers were used in the early 1860s to identify two new elements, rubidium and caesium, from unidentified colours in the spectrum of the vapour of mineral water.

The colours in the spectrum produced by atoms in this way are known as spectral lines because of the sharp lines they produce on the photographic plate in a spectrometer. These photographs are known as **emission spectra**.

The fact that the spectrum of an element is its 'fingerprint' makes it possible to detect tiny traces of elements in complex mixtures. Astronomers use the light emitted by remote stars to identify elements within the stars. Of more interest here is the contribution that line spectra make to the understanding of both the structure of matter and the behaviour of light. The key is the distinct line nature of the spectra. Sharp lines have precise wavelengths. In the photon model for light, it indicates precise and discrete photon energies. Line spectra indicate that a particular type of atom emits light energy as specific fixed amounts.

FIGURE 10.6 A simple spectrometer

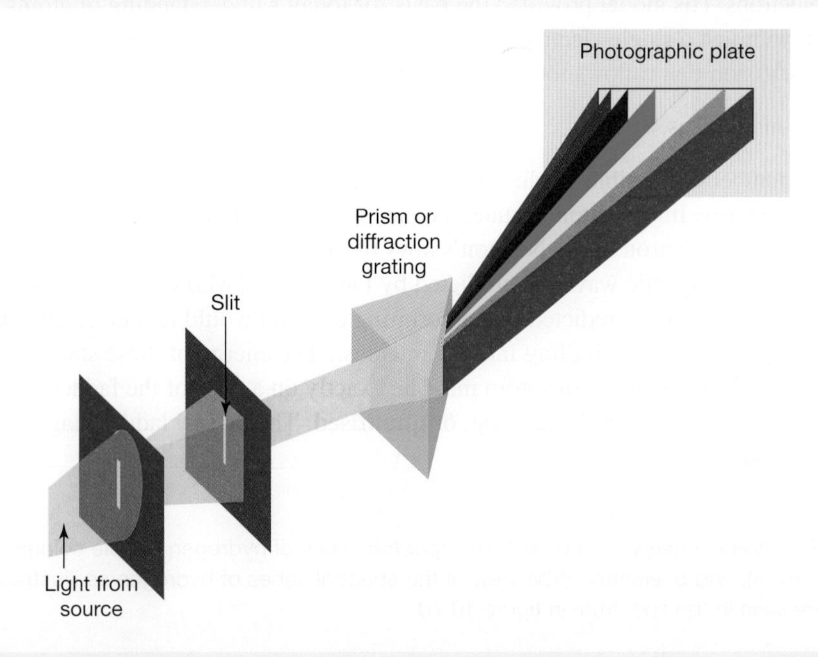

FIGURE 10.7 a. Sodium vapour street light, b. characteristic colour of sodium flame, c. visible line spectrum of sodium atoms, and d. visible line spectrum of hydrogen atoms

This behaviour is remarkably different to that of a hot filament and other incandescent light sources, which emit a continuous spectrum of light and a range of wavelengths, hence energies.

Any model of atomic structure must be able to explain the behaviour of the atom leading to discrete emission spectra. In 1911, Ernest Rutherford, an eminent New Zealander who directed the Cavendish laboratory at Cambridge University, established that electrons must orbit a nucleus with most of the atom being empty space.

Then in 1913, Niels Bohr, a Danish physicist, proposed what was then a revolutionary model for the behaviour of these atoms and electrons. His model provided the basis for today's understanding of atoms. The hydrogen atom — the simplest, with just a single electron revolving around a proton — was the initial testing ground for Bohr's model. The discussion that follows focuses on the hydrogen atom.

Bohr's model had two main ideas.

quantised cannot be divided or broken up into smaller parts

1. Each atom has a number of possible stable states, each state having its own characteristic energy. In each atomic state, the electron is in a stable orbit around the nucleus. The electron obeys Newton's laws of mechanics but does not radiate electromagnetic waves as predicted by Faraday and Maxwell. (This idea of Bohr's was radical. Electromagnetic theory predicted that an orbiting electron would radiate electromagnetic radiation continuously, losing energy and spiralling into the nucleus.) The energy of these states may be imagined as the rungs on a ladder. The energy of the atom must lie exactly on a rung of the ladder, and never between rungs. The energy levels are said to be discrete, or **quantised**. The energy ladder diagram for hydrogen is shown in figure 10.8a.

FIGURE 10.8 a. Atomic energy level view of the spectral series of hydrogen (visible colours are shown as coloured arrows), and b. electron orbit view of the spectral series of hydrogen as illustrated in a. These lines are seen in the spectrum in figure 10.7d.

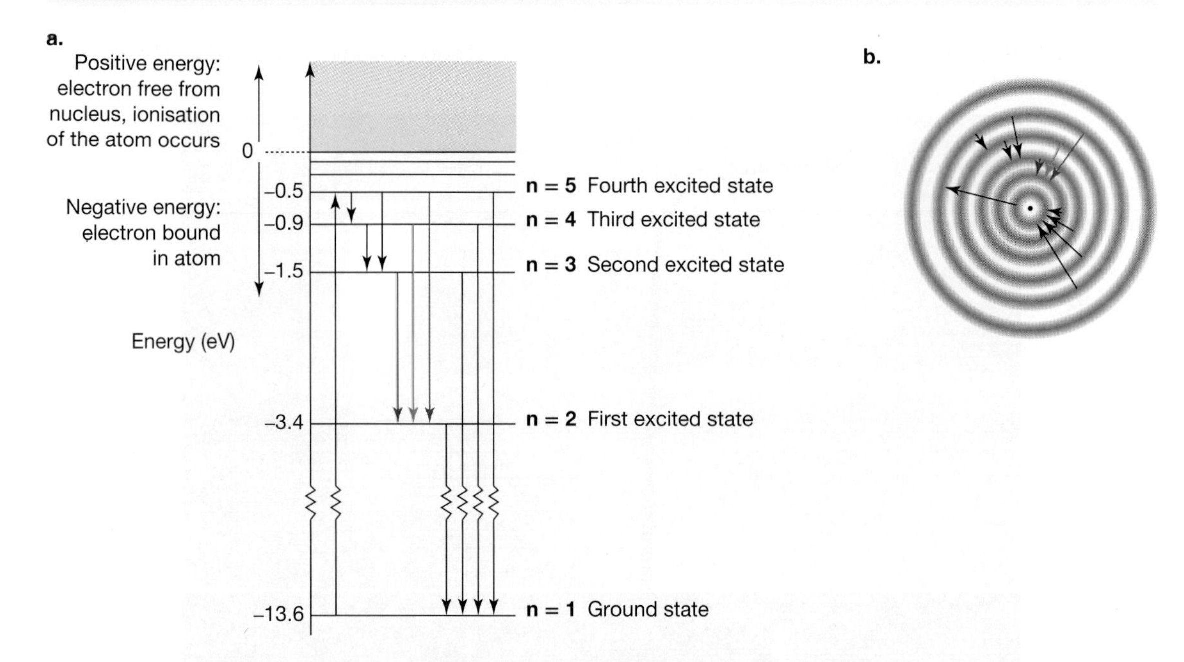

2. An atom can jump up or down from one state to another, corresponding to the transfer of the electron from one orbit to another. When the atom drops to a state having a lower energy, a photon is emitted whose energy is equal to the energy loss of the atom. Alternatively, an atom may absorb a photon, raising the energy of the atom in the process. The energy of the absorbed photon must exactly match the energy difference between the current state of the atom and one of the higher energy states it is allowed to jump to. The energy of the photon absorbed or emitted can be determined using Planck's equation for the energy of a photon:

$$E = hf = \frac{hc}{\lambda}$$

where: E is the energy of the photon produced, in J

h is Planck's constant

f is the frequency of the photon

c is the speed of light (3.0×10^8 m s^{-1})

λ is the wavelength of the photon, in m

Emission and absorption of photons are illustrated in figures 10.9a and 10.9b respectively. The emission of light can be represented as $E_{\text{photon}} = hf = E_{\text{initial}} - E_{\text{final}}$, and the absorption of light can be represented as $E_{\text{photon}} = hf = E_{\text{final}} - E_{\text{initial}}$.

FIGURE 10.9 a. Emission of light **b.** Absorption of light

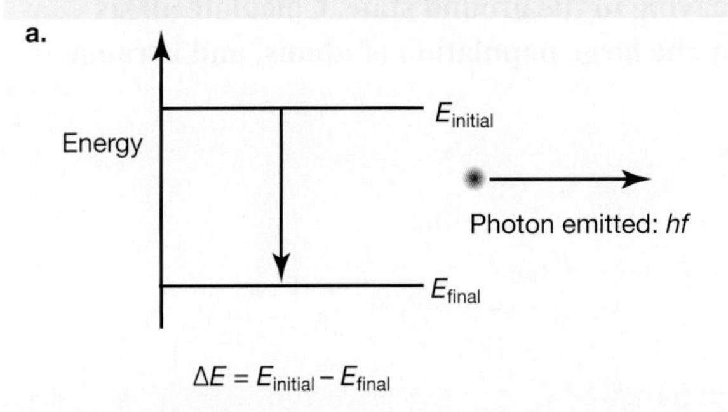

b.

Energy

E_{final}

Photon absorbed: hf

E_{initial}

$\Delta E = E_{\text{final}} - E_{\text{initial}}$

Atoms can gain energy in other ways. Absorption of energy can occur during a collision of an atom with an electron, in a discharge tube, or during a collision with an ion (as occurs in a flame).

States of the atom are commonly labelled using the terms **ground state** for the lowest energy state (or the zero of energy), followed by the **excited states** at higher energy, often labelled first excited state, second excited state, and so on. The most common choice for the zero of energy is the energy of an electron and proton that are completely free from one another — that is, stationary, at an infinite separation. Using this scale, the energy of an electron bound to a proton in a hydrogen atom is negative. The system of a stationary proton and a separated freely moving electron has a positive energy that is equal to the kinetic energy of the electron.

ground state state of an electron in which it has the least possible amount of energy

excited state state of an electron in which it has more energy than its ground state

Let's examine how this explains the emission spectrum of hydrogen (figure 10.7d). Most hydrogen atoms at room temperature are in the ground state. In flames or discharge tubes, atoms are raised to excited states by collisions with other particles. In figure 10.8a, the atom has first been excited into its fourth excited state where it has an energy of −0.5 eV. This is indicated by the upward arrow on the left-hand side of the diagram. Cascading transitions to the ground state are then possible, with a photon emitted during each transition, as indicated by the downward arrows. Some of these downward arrows are shown in black as the emitted photons are not visible light so are not observed in an emission spectrum. Other photons that are emitted as visible colours are those we observe in the spectrum. This model of atomic structure neatly accounted for emission spectra.

It is possible to work backwards to construct an energy level diagram for an atom by collecting an emission spectrum containing many discrete spectral lines and establishing the photon energy associated with each line.

tlvd-9025

SAMPLE PROBLEM 6 Calculating emission energies

Consider an energy level diagram for an atom, as shown.

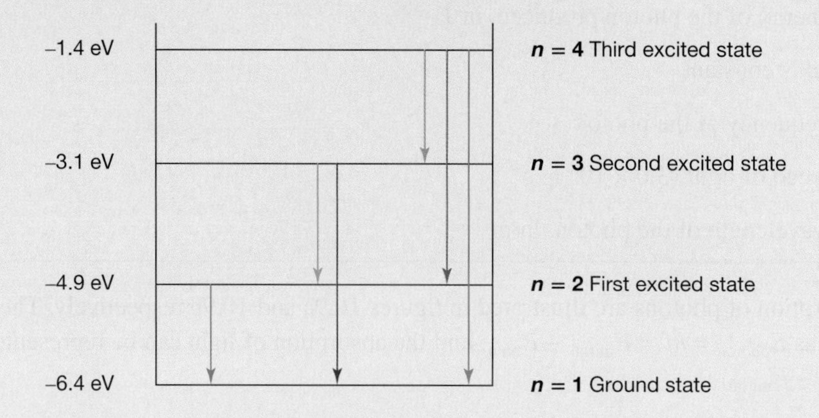

Consider a large population of atoms all excited to the third excited state ($n = 4$), from which an emission spectra is obtainable, resulting in all atoms decaying to the ground state. Calculate all six possible energies in electron volts for photons emitted by the large population of atoms, and arrange them in ascending order.

THINK

1. There are six possible transitions:
   - third excited state to ground state (teal arrow)
   - third to first (red arrow)
   - third to second (orange arrow)
   - second to ground (purple arrow)
   - second to first (green arrow)
   - first to ground (blue arrow)

   The energy of the emitted photon is calculated by finding the difference between the energies of the states for each transition. The highest energy of any photons emitted by this atom occurs when transitioning from the 3rd excited state to the ground state.

WRITE

For the $n = 4$ to $n = 1$ transition:

$$\begin{aligned} E_{\text{photon}} &= E_{\text{initial}} - E_{\text{final}} \\ &= (-1.4\,\text{eV}) - (-6.4\,\text{eV}) \\ &= -1.4\,\text{eV} + 6.4\,\text{eV} \\ &= 5.0\,\text{eV} \end{aligned}$$

THINK

2. The remaining five possible photon energies can be calculated in the same way using $E_{\text{photon}} = E_{\text{initial}} - E_{\text{final}}$.

WRITE

The remaining five calculations give energies of:
- 3.5 eV (third to first)
- 1.7 eV (third to second)
- 3.3 eV (second to ground state)
- 1.8 eV (second to first)
- 1.5 eV (first to ground state).

THINK

3. Arrange the photon energies in ascending order.

WRITE

Arranged in ascending order, the six photon energies are 1.5 eV, 1.7 eV, 1.8 eV, 3.3 eV, 3.5 eV and 5.0 eV.

PRACTICE PROBLEM 6

Consider an energy level diagram for an atom, as shown.

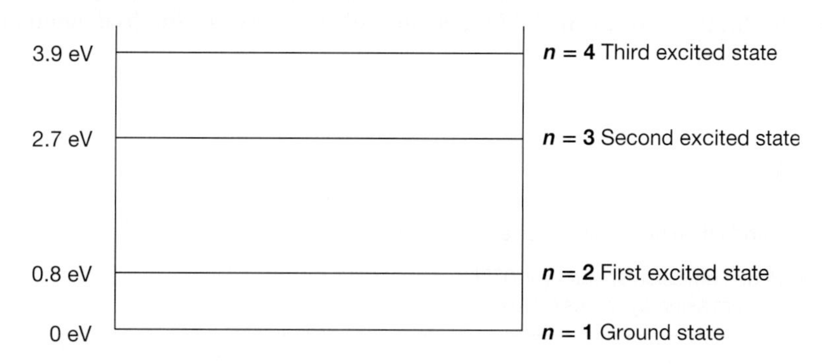

a. **Complete the energy level diagram using arrows to show all the possible transitions.**
b. **Calculate the highest energy and hence the frequency of a photon emitted by this atom in the $n = 4$ state.**
c. **Calculate the lowest energy and hence the frequency of a photon emitted by this atom in the $n = 3$ state.**

tlvd-9026

SAMPLE PROBLEM 7 Calculating emission energies and their corresponding wavelengths

The energies of states of the hydrogen atom are shown in the following table.

H atom state	Energy (eV)
Third excited state $n = 4$	−0.85
Second excited state $n = 3$	−1.51
First excited state $n = 2$	−3.40
Ground state $n = 1$	−13.61

What is the shortest wavelength of light emitted by hydrogen atoms that were initially excited into the third excited state?

THINK

1. Light of the shortest wavelength is emitted when the photons have the greatest energy (when the atoms experience the greatest possible energy change). This will be the transition to the ground state. You need to calculate the energy in eV and convert into joules. Determine the energy of the photon.

2. Determine the wavelength using Planck's constant and the calculated energy change for the photon (in joules).

WRITE

For $n = 4$ to n = 1 transition:

$$E_{\text{photon}} = E_{\text{initial}} - E_{\text{final}}$$
$$= (-0.85\,\text{eV}) - (-13.61\,\text{eV})$$
$$= 12.76\,\text{eV}$$
$$= 12.76\,\text{eV} \times 1.6 \times 10^{-19}\,\text{J}\,\text{eV}^{-1}$$
$$= 2.0 \times 10^{-18}\,\text{J}$$

$$\lambda_{\text{photon}} = \frac{hc}{E_{\text{photon}}}$$
$$= \frac{6.63 \times 10^{-34} \times 3.0 \times 10^{8}}{2.0 \times 10^{-18}}$$
$$= 9.7 \times 10^{-8}\,\text{m}.$$

This is ultraviolet radiation.

PRACTICE PROBLEM 7

What is the lowest frequency and hence longest wavelength light emitted by hydrogen atoms that were initially excited to the third excited state? The energies of the states of the hydrogen atom are found in sample problem 7.

on Resources

Video eLesson The production of atomic spectra (eles-3522)

Interactivities Hydrogen emission spectra (int-0755)
Photon emission by atoms (int-0122)

10.4.2 Atoms absorb photons

Atoms also absorb light. They absorb those particular wavelengths they would emit if the gas was excited. These are the frequencies that correspond to the differences in energy between the energy levels in the atoms. An **absorption spectrum** is a continuous spectrum with a series of dark lines indicating missing frequencies. Absorption spectra are produced by placing a sample of a gas in front of a **continuous spectrum** source, as shown in (figure 10.10).

absorption spectrum spectrum produced when light passes through a cool gas. It includes a series of dark lines that correspond to the frequencies of light absorbed by the gas.

continuous spectrum a spectrum that has no gaps; there are no frequencies or wavelengths missing from the spectrum

The atoms making up the gas absorb particular wavelengths, raising electrons within the atoms into excited states. When the electrons drop back to the ground state, photons are emitted in all directions. This means that the original beam of light has very little of those absorbed colours. The continuous spectrum has dark bands. Generally, the dark bands in an absorption spectrum correspond to the bright lines in an emission spectrum of the same gas if it were hot.

FIGURE 10.10 Absorption spectra are produced by a cool gas absorbing particular wavelengths of light. The absorption spectrum consists of a series of dark lines corresponding to missing wavelengths.

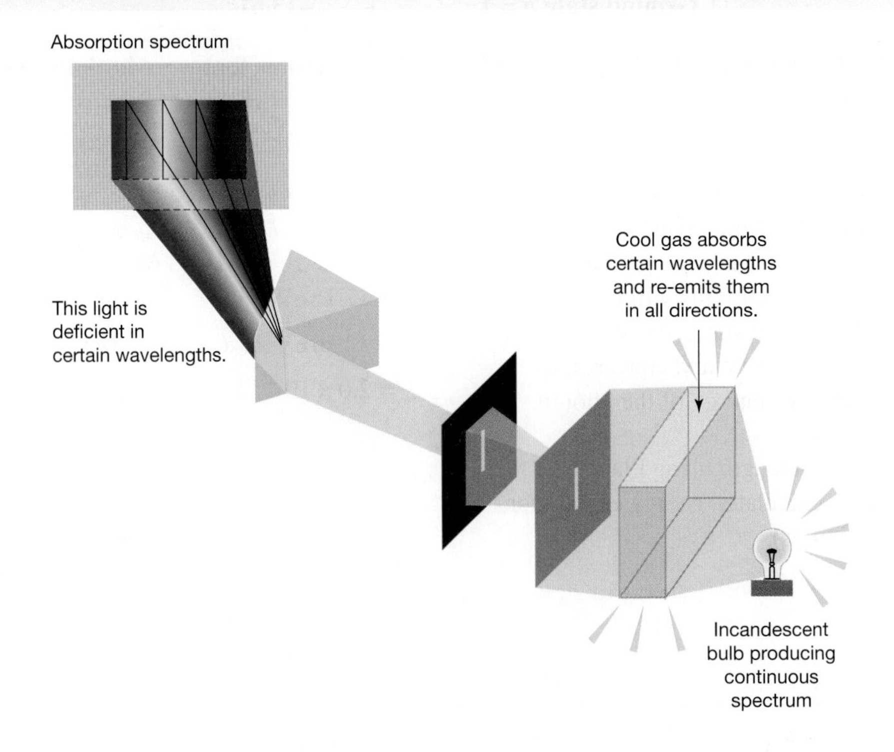

Absorption spectrum
Cool gas absorbs certain wavelengths and re-emits them in all directions.
This light is deficient in certain wavelengths.
Incandescent bulb producing continuous spectrum

Comparing emission and absorption spectra

As they rely on the same energy level structure, the emission and absorption spectra often appear to be negatives of one another. However, there are differences.

The emission spectrum usually includes lines missing from the absorption spectrum of the same element. For example, the emission spectrum of the hydrogen atom includes lines for transitions to all lower states of the atom. However the absorption spectrum of hydrogen atoms at room temperature contains lines in the ultraviolet region only. Each line is linked to a transition beginning at the ground state. This is because virtually all atoms are in the ground state at room temperature — only $\frac{1}{10^{171}}$ are not! The spectrum of hydrogen in the Sun is different. The surface temperature of 6000 K is hot enough for the proportion of atoms in the first excited state to rise to $\frac{1}{10^{8}}$. Not large, but large enough for their absorption spectrum to be detected.

FIGURE 10.11 a. Part of the emission spectrum for hydrogen. The only region of the spectrum shown is that in which transitions to the ground state appear. **b.** The absorption spectrum for hydrogen. The transfer of photon energy to eject an electron is labelled 'ionisation'. All wavelengths less than the limiting value (i.e., photons with energy greater than the limiting value) can be absorbed.

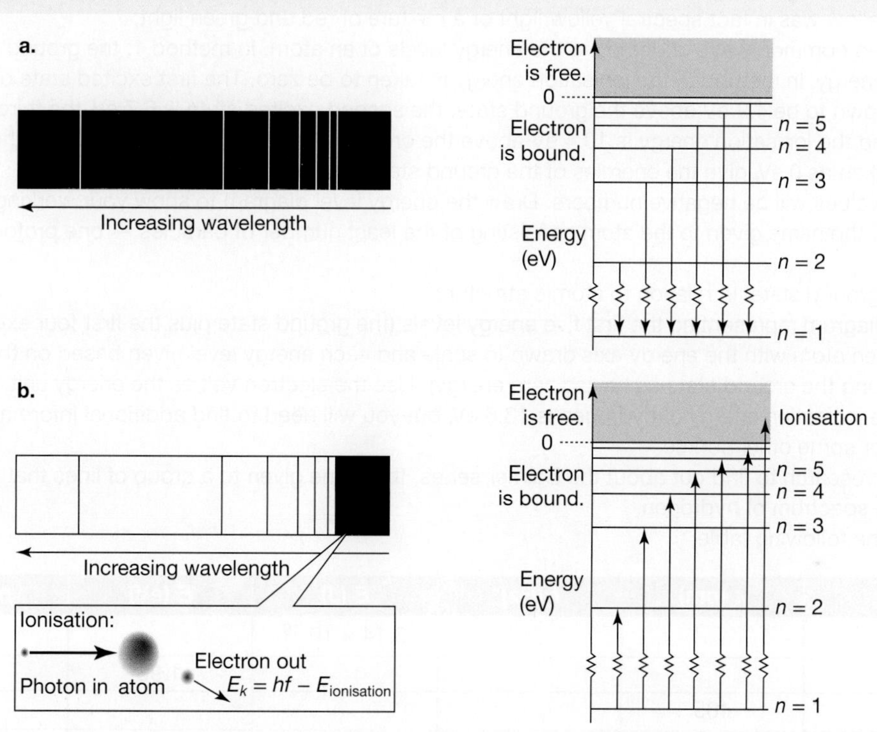

Series of lines in an absorption spectrum converge on a particular wavelength. This wavelength corresponds to the photon energy equal to the ionisation energy of the electron. All light with shorter wavelengths than this limit, and therefore greater photon energy, can be absorbed, removing the electron from the atom completely. This means that all light with wavelengths below the limit may be absorbed. This is an example of the photoelectric effect where a minimum photon energy must be reached before electrons will be ejected.

INVESTIGATION 10.1

online only

elog-1756

vd-10608

Spectroscopes

Aim

To use a spectroscope to observe the spectrum of a light source

10.4 Activities

learnon

Students, these questions are even better in jacPLUS

Receive immediate feedback and access sample responses

Access additional questions

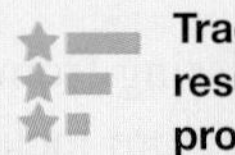
Track your results and progress

Find all this and MORE in jacPLUS

10.4 Quick quiz	10.4 Exercise	10.4 Exam questions

10.4 Exercise

1. How is the reddish glow of light from a dying fire different from the reddish glow of a neon discharge tube?
2. Explain why spectral lines in the emission spectrum of an element correspond to absorption lines in an absorption spectrum for the same element.
3. A beam of red and green light appears yellow. Devise an experiment to decide whether a beam of light that appeared yellow was in fact spectral yellow light or a mixture of red and green light.
4. There are two common ways of depicting the energy levels of an atom. In method 1, the ground state is taken to be zero energy. In method 2, the ionisation energy is taken to be zero. The first excited state of mercury atoms is known to be 4.9 eV above the ground state, the second excited state is 6.7 eV, the third excited state is 8.8 eV, and the ionisation energy is 10.4 eV above the ground state. Using method 2, where the ionisation energy is taken as 0 eV, give the energies of the ground state and the first three excited states.
   Note: Your values will be negative numbers. Draw the energy level diagram to show your working.
5. Hydrogen is the name given to the atom consisting of the least number of particles — one proton and one electron.
   a. Explain 'ground state' in relation to atomic structure.
   b. Draw a diagram representing the first five energy levels (the ground state plus the first four excited states) in a hydrogen atom with the energy axis drawn to scale and each energy level given based on the ground state (taking the ground state as having zero energy). Use the electron volt as the energy unit. As a starting point, the ionisation energy of hydrogen is 13.6 eV, but you will need to find additional information via the internet or some other source.
   c. Conduct research to find out about the Balmer series, the name given to a group of lines that appears in the emission spectrum of hydrogen.
6. Complete the following table.

Element	λ (nm)	f (Hz)	E (J)	E (eV)	p (N s)
Red light			3.14×10^{-19}		
Electron				1.96	
Blue light	405				
Electron	405				

7. The ground state and the first three excited states of hydrogen are shown in the diagram. An emission spectrum of hydrogen gas shows many different spectral lines.
   a. Copy the diagram and label the ground state and first three excited states.
   b. Draw arrows to represent all six possible transitions that may occur when hydrogen atoms in states lower than the fourth excited state emit a photon of light.
   c. Calculate the energy of each of the six possible photons.
   d. Determine the wavelength of the photon having the least and greatest energy in your answer to part **c**.

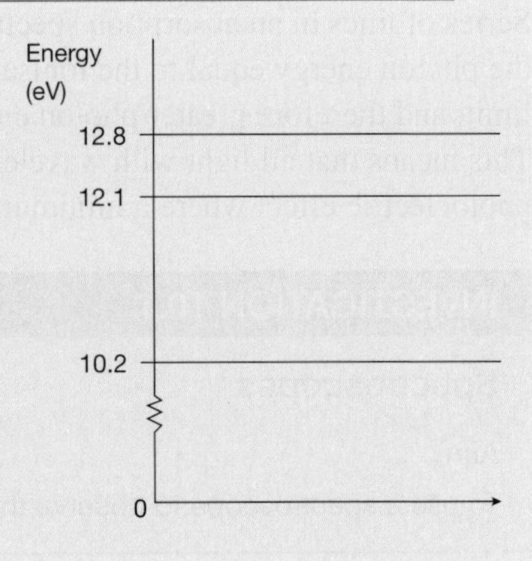

8. Explain why there are dark lines in an absorption spectrum of a gaseous sample. Why are those particular colours missing from the otherwise continuous spectrum of light?
9. When sodium chloride (common salt) is placed in a flame, the flame glows bright gold. The following diagram shows some of the energy levels of a sodium atom.

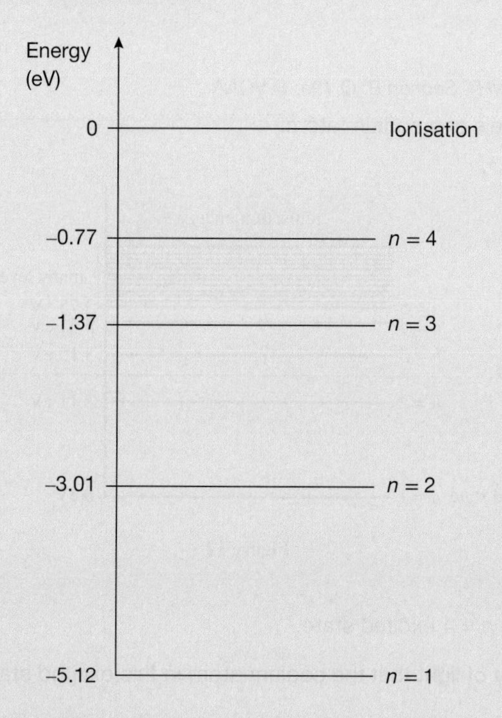

   a. Copy the diagram and label the ground state of the atom, and the first excited state.
   b. Draw arrows to represent the change in energy of atoms in the ground state that absorb energy during collisions with other particles in the flame.
   c. Calculate the wavelength of light emitted by these atoms as they return to the ground state in a single jump. Which energy change is responsible for the yellow glow?

10.4 Exam questions

Question 1 (2 marks)

***Source**: VCE 2022 Physics Exam, Section B, Q.15;* © VCAA

Figure 17 shows some of the energy levels of excited neon atoms. These energy levels are not drawn to scale.

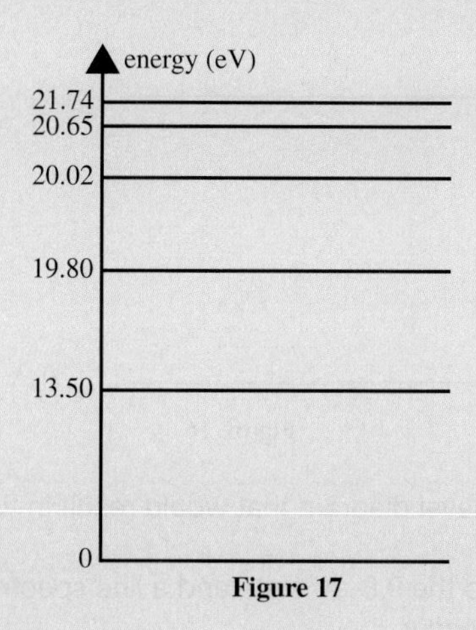

Figure 17

a. Show that the energy transition required for an emitted photon of wavelength 640 nm is 1.94 eV. **(1 mark)**
b. On Figure 17, draw an arrow to show the transition that would emit the photon described in part a. **(1 mark)**

Question 2 (3 marks)

***Source**: VCE 2022 Physics Exam, NHT, Section B, Q.17; © VCAA*

Describe how absorption line spectra are produced and describe their relationship to electron transitions within atoms.

Question 3 (2 marks)

***Source**: VCE 2022 Physics Exam, NHT, Section B, Q.19a; © VCAA*

Figure 14 shows the energy levels of a sodium atom.

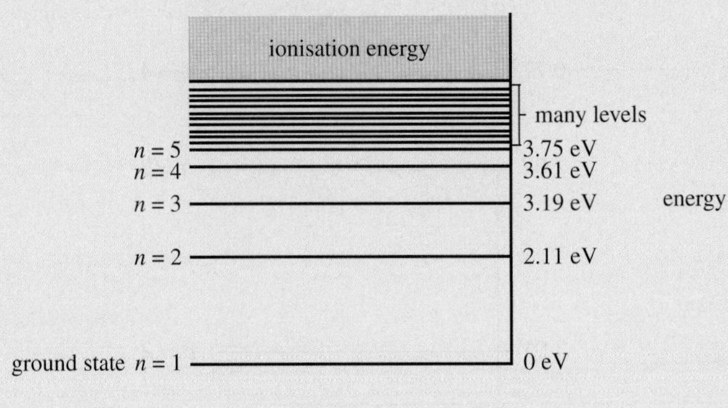

Figure 14

A sodium atom is initially in the $n = 4$ excited state.

Calculate the highest frequency of light that the sodium atom in this excited state could emit.

Question 4 (4 marks)

***Source**: Adapted from VCE 2021 Physics Exam, Section B, Q.19; © VCAA*

A simplified diagram of some of the energy levels of an atom is shown in Figure 16.

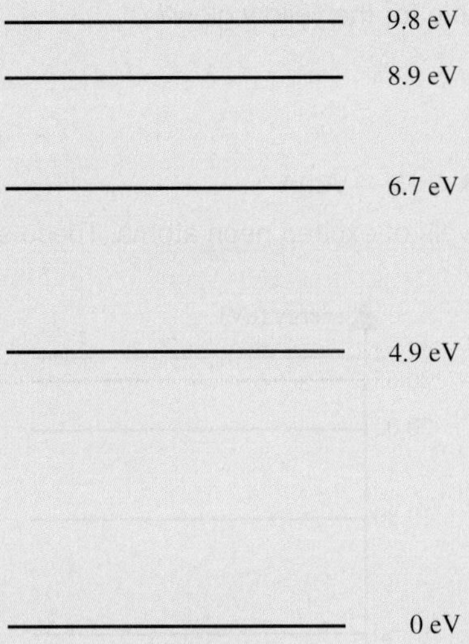

9.8 eV

8.9 eV

6.7 eV

4.9 eV

0 eV

Figure 16

a. Identify the transition on the energy level diagram that would result in the emission of a 565-nm photon. Show your working. **(2 marks)**

b. A sample of the atoms is excited into the 9.8-eV state and a line spectrum is observed as the states decay. Assume that all possible transitions occur.

What is the total number of lines in the spectrum? Explain your answer. You may use Figure 16 to support your answer. **(2 marks)**

Question 5 (5 marks)

Source: *VCE 2020 Physics Exam, Section B, Q.17; © VCAA*

Figure 17 shows the emission spectrum for helium gas.

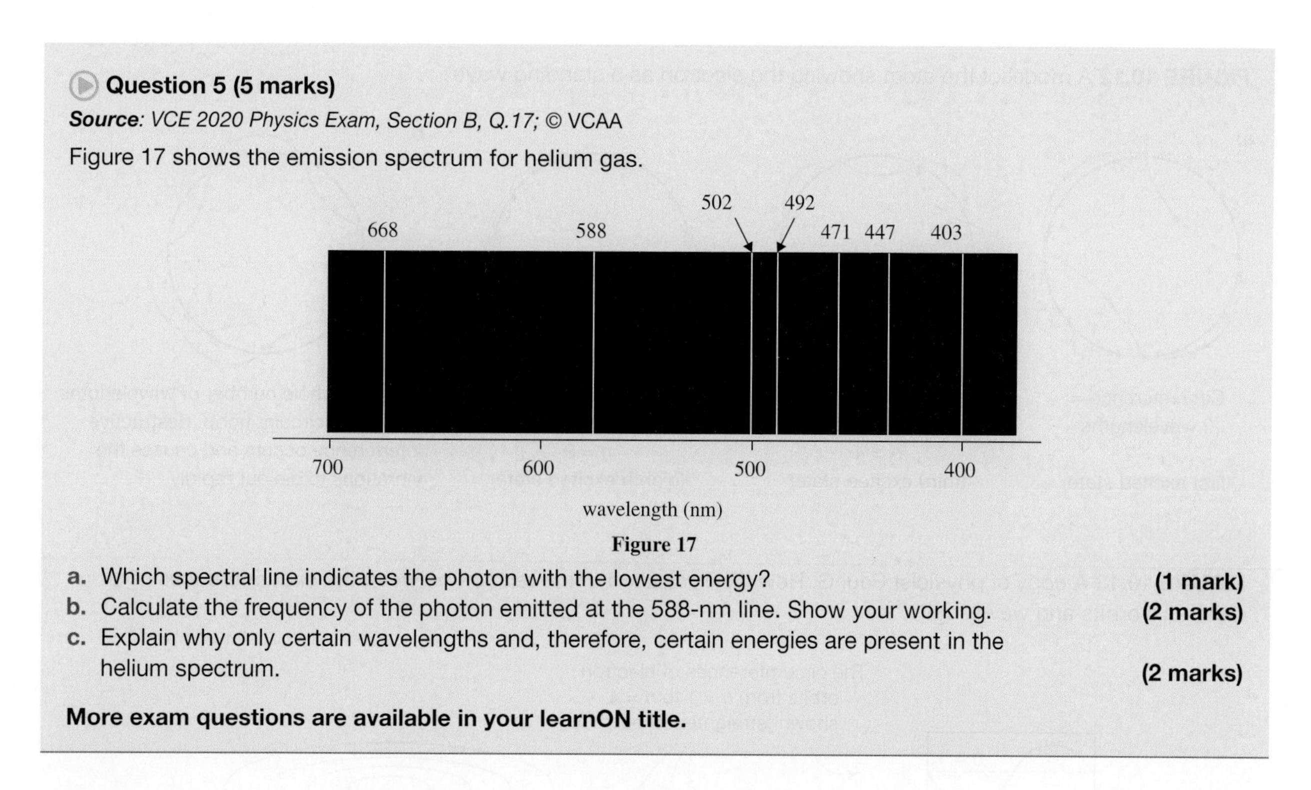

Figure 17

a. Which spectral line indicates the photon with the lowest energy? **(1 mark)**

b. Calculate the frequency of the photon emitted at the 588-nm line. Show your working. **(2 marks)**

c. Explain why only certain wavelengths and, therefore, certain energies are present in the helium spectrum. **(2 marks)**

More exam questions are available in your learnON title.

10.5 Electrons, atoms and standing waves

KEY KNOWLEDGE

- Describe the quantised states of the atom with reference to electrons forming standing waves, and explain this as evidence for the dual nature of matter
- Interpret the single photon and the electron double slit experiment as evidence for the dual nature of light and matter

10.5.1 Electrons modelled as standing waves

Individual electrons act like waves when they are diffracted by atoms in crystals. Do electrons in the atoms also exhibit wavelike properties? They do! Thinking of electrons behaving like waves solved the puzzle of stationary states. This wave model for electrons that are bound within atoms neatly explains why atoms absorb and emit photons of only particular frequencies. This provided the answers to Rutherford's questioning of the Bohr model of the atom. In essence, only waves whose de Broglie wavelength multiplied by an integer ($n\lambda$) set equal to the circumference of a traditional electron orbit are allowed to exist due to these waves being the only ones able to constructively interfere to produce a standing wave. In 1924 de Broglie speculated about the electron in a hydrogen atom displaying wavelike behaviour. A complete description of the hydrogen atom awaited a more sophisticated mathematical treatment called quantum mechanics. The fundamentals of this model were developed by Erwin Schrödinger and Werner Heisenberg later in the 1920s.

Louis de Broglie pictured the electron in a hydrogen atom travelling along one of the allowed orbits around the nucleus, together with its associated wave. In de Broglie's mind, the circumference of each allowed orbit contained a *whole number* of wavelengths of the electron-wave so that it formed a standing wave around the orbit. Thus, $n\lambda = 2\pi r$ or $\lambda = \frac{2\pi r}{n}$ fixes the allowed wavelength. An electron-wave whose wavelength was slightly longer, or shorter, would not join onto itself smoothly. It would quickly collapse due to destructive interference. Only orbits corresponding to standing waves would survive. This is shown in figures 10.12 and 10.13. The concept is identical to the formation of standing waves on stringed instruments.

FIGURE 10.12 A model of the atom showing the electron as a standing wave

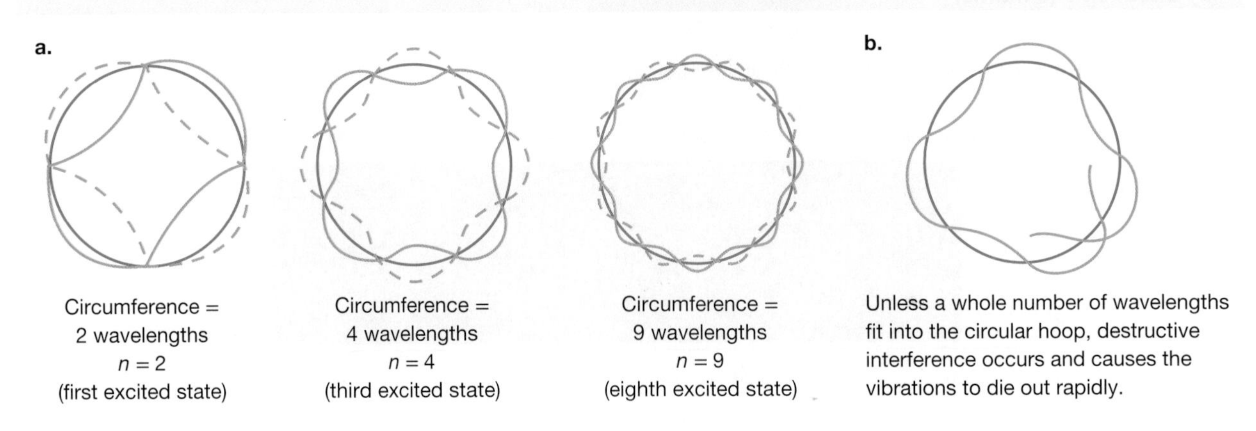

FIGURE 10.13 A copy of physicist Paul G. Hewitt's illustration of the relation between the circumferences of an electron's orbits and wavelength

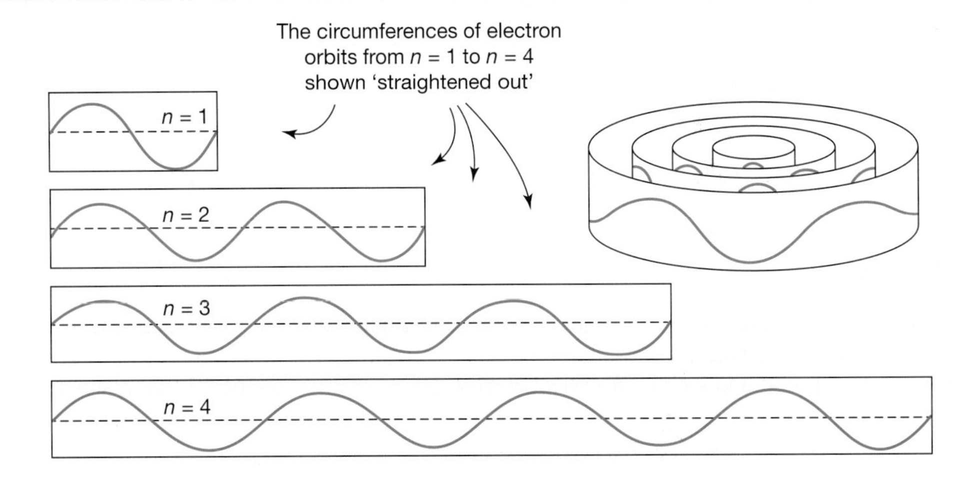

The standing waves produced on a stringed instrument of length l have a series of possible wavelengths $\lambda_n = \frac{2l}{n}$ where n is a positive integer. This series of wavelengths is called a harmonic series. At this level of physics, which is only an introduction to the conceptual nature of quantum mechanics, the harmonic series provides for a series of associated momenta that are discrete in value. This in turn provides for a series of energy states that are also discrete. This connection is in complete agreement with the observation of emission and absorption spectra. When you pluck a guitar string, only certain frequencies are produced. Likewise, when you energise an atom, only certain energy levels are sustainable, resulting in the emission of well-defined frequencies of light in the form of individual photons.

In de Broglie's model of the atom, electrons are viewed as standing waves. It is this interpretation that provides a reasonable explanation for the emission spectra of atoms. It answers Rutherford's remark to Bohr. When a guitar string is plucked, how does it know what frequencies to vibrate at? The answer is the following: the frequencies that equate to the standing waves with wavelengths compatible with the length of the string.

Electrons viewed as standing waves can exist only in stable orbits with precise or discrete wavelengths. This implies that the electrons can only have discrete quantities of momentum. This in turn implies that the electrons can only have discrete amounts of energy. Energy transitions that are made by electrons occur in jumps from one high-energy standing wave to another standing wave of lower energy. In this way, the emission spectra and, hence, absorption spectra can be understood as arising from transitions between quantised energy levels due to electrons having a wavelike character.

It's a consistent story — light displays both wave and particle behaviour and so do electrons and all other forms of matter. The two models are complementary. You observe behaviour consistent with wave properties or particle properties, but not the two simultaneously. Remember how William Bragg expressed it: 'On Mondays, Wednesdays and Fridays light behaves like waves, on Tuesdays, Thursdays and Saturdays like particles, and like nothing at all on Sundays'. This delicate juggling of the two models by both light and matter is known as **wave–particle duality**.

wave–particle duality description of light as having characteristics of both waves and particles. This duality means that neither the wave model nor the particle model adequately explains the properties of light on its own.

There have been many conceptual hurdles for physicists in arriving at this amazingly consistent view of the interaction between light and matter. Their guiding questions always kept them probing for the evidence. Observations and careful analysis gave them the answers. Imagination, creativity and ingenuity were vital in their search for a more complete picture of light and matter.

10.5.2 Single photons exhibit wave properties

It is now known that both light and matter can exhibit both wavelike and particle-like behaviour, depending on the types of experiments performed. For example, when light strikes a material object, it transfers energy as if it is a particle (the photoelectric effect), but when light passes through a narrow opening or a pair of slits, it acts as if it is a wave. Likewise, matter can have work done on it via well-understood forces accelerating it, but matter can also be diffracted when it passes through a crystal, producing diffraction patterns similar to those of X-rays. Also, the behaviour of electrons within atoms can only be understood by treating them as a type of wave phenomena.

The dual nature of light and matter

- Young's experiment (topic 8) supports the wave model of light.
- The photoelectric effect (topic 9) is evidence in support of the particle model of light.
- Electron diffraction is evidence in support of the wave nature of matter.
- The particle properties of matter are well documented.
- Light and matter both have a wave–particle dual nature.

EXTENSION: The Bohr radius and distances between protons and electrons

A hydrogen electron in a stable or stationary state does not move in a circular, or even an elliptical, orbit around the proton — its distance from the proton changes continuously.

In fact the words 'circular', 'elliptical' and 'distance' become inappropriate when an electron is considered not as a particle but as some type of wave phenomenon. To understand and account for emission and absorption spectra, it is better to think of an electron as having no specific location or path; instead, there is only a probability of locating it at various positions. This unpredictability is ultimately related to Heisenberg's uncertainty principle, which asserts that it is not possible to precisely measure both the location and the motion of any object at the same time. This realisation is the cornerstone of contemporary physics and the foundation of quantum mechanics.

The most probable distance from proton to electron is 5.29×10^{-11} m, corresponding to the peak in the probability density curve.

This distance is called the Bohr radius and is often quoted as the 'radius' of the hydrogen atom. It can be useful to think of the hydrogen atom as having a spherical cloud of negative charge surrounding the proton, representing the unpredictable motion of the electron. The diminishing density of the cloud at large distances from the proton indicates that the electron is less likely to be there.

While the proton–electron distance is not fixed or predictable for a particular state of the atom, the total energy of the atom is predictable. As the electron weaves its intricate path around the proton, its kinetic energy changes, being greater when the electron is closer to the nucleus. However, the total energy *does not* change. There is a transformation between the kinetic energy and electric potential energy. Electric potential energy increases as kinetic energy decreases, and vice versa, keeping the total energy constant.

FIGURE 10.14 a. The probability of finding an electron a certain distance from the nucleus, and **b.** the cloud picture of a hydrogen atom

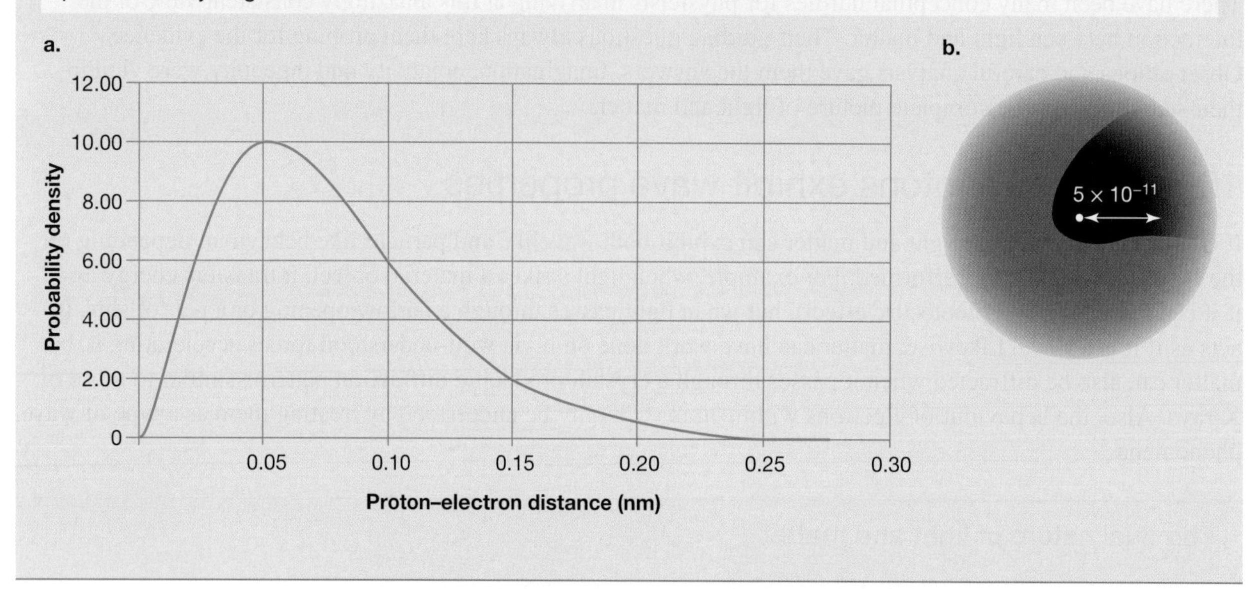

EXTENSION: Taylor's experiment — exploring the duality of waves and particles

A beam of light isn't required to observe wave effects — every single photon has wave properties. Geoffrey Taylor set out to demonstrate this in 1909, while he was a student at the University of Cambridge. Taylor photographed the diffraction pattern in the shadow of a needle, but his photograph took three months of light exposure to produce. He used an extremely dim source, a gas flame, together with several smoked glass screens, to illuminate the entrance slit of a light-tight box. Taylor measured the light intensity entering the box, and estimated that only 10^6 photons entered the box each second. This may not sound like a low intensity, but with a photon speed of 3.0×10^8 m s^{-1}, the average distance between photons was 300 m!

Using a box with a length of 1 m, Taylor could be sure that rarely was there more than one photon travelling through it at any one time, so a vast majority of photons travelled through the box unaccompanied. An image appeared on the photographic plate after three months just as Taylor expected — a pattern of light and dark bands in the shadow of the needle. Taylor compared it to the pattern obtained in a short time with an intense light source and stated, 'In no case was there any diminution in the sharpness of the pattern'. His experiment demonstrated that interference occurred photon by photon, and that the wave of a single photon filled the box, interfering with itself as it diffracted past the edges of the needle.

Taylor's experiment invites one to imagine watching an interference pattern build up on the photographic plate. The first few photons would produce an apparently random sprinkling of spots, each spot due to a single photon changing the chemical state of an ion in the photographic film. As the spots accumulated, they would start to overlap and, gradually, a pattern would emerge from the randomness. During this process, you would never be able to predict precisely where the next photon would strike the plate. The pattern predicted by the wave nature of the light would allow only the prediction of the *probability* of a photon reaching a particular point. This pattern of probabilities would be clear only after many photons had made their mark.

FIGURE 10.15 Taylor's experiment

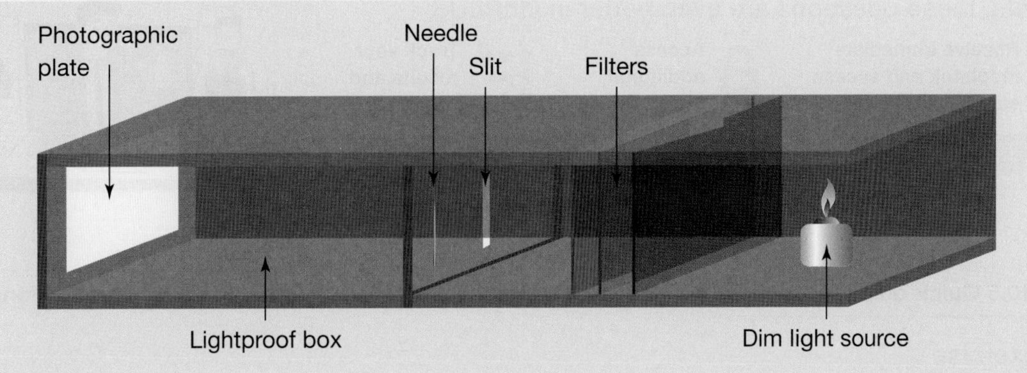

FIGURE 10.16 Imagine the gradual build-up of photon spots into a double-slit interference pattern.

a.

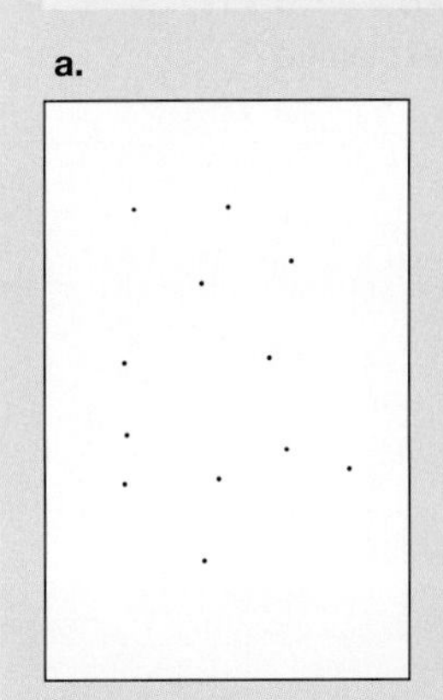

b.

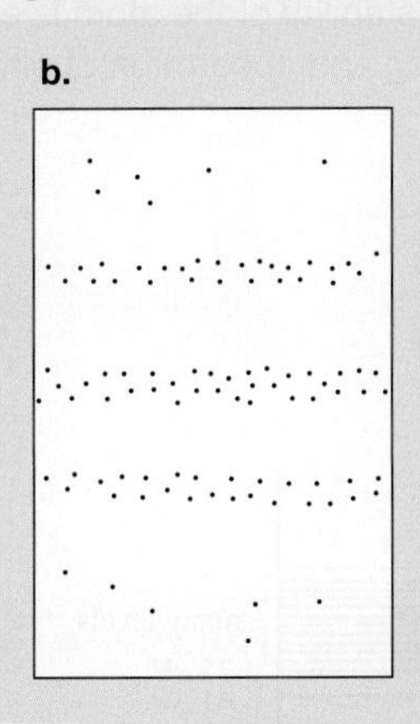

c.

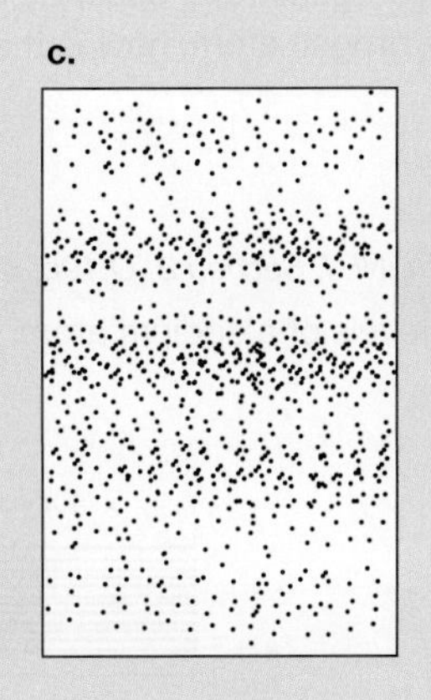

d.

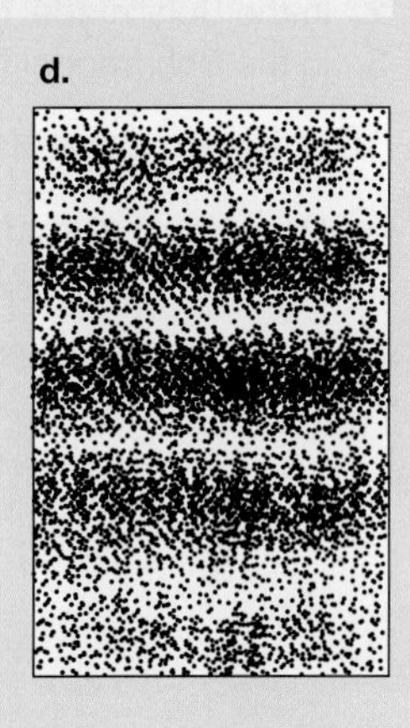

Taylor's experiment is a beautiful demonstration of wave–particle duality. The wave and particle characteristics of light are entangled and cannot be separated. Both models are needed. In fact, even when light is travelling particle by particle, its wave characteristics are there at the same time, determining the outcome.

Similar experiments have been carried out with electrons and neutrons and, more recently, with large molecules. In all cases, the wavelike behaviour of these individual entities when passing through openings has demonstrated wave–particle duality in the form of diffraction effects. It seems that the entities pass through the opening and self-interfere in the process. Importantly, they do this one entity at a time. Over an extended period, a statistical distribution builds up of where these entities go, recorded by where they strike a screen. The distribution is consistent with a wave model analysis for **coherent** waves of the one wavelength passing through an opening, whether it is a single slit, a double slit or any complicated array of openings.

coherent describes light in which all photons are emitted in phase, leading to intense light

Resources

Video eLesson The wave–particle duality of light (eles-0028)

Interactivity The wave–particle duality of light (int-0052)

10.5 Activities

learn**on**

Students, these questions are even better in jacPLUS

Receive immediate feedback and access sample responses

Access additional questions

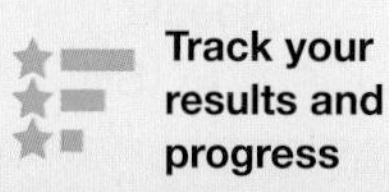
Track your results and progress

Find all this and MORE in jacPLUS

10.5 Quick quiz on	10.5 Exercise	10.5 Exam questions

10.5 Exercise

1. a. Describe the classical model of a hydrogen atom with reference to the electron and proton.
   b. Explain in what way this model predicts that atoms are not stable.
2. In the Bohr model for a hydrogen atom, describe how the stability of the atom is explained.
3. In the Bohr model for a hydrogen atom, how is the emission and absorption of photons explained?

10.5 Exam questions

Question 1 (2 marks)

Source*: VCE 2022 Physics Exam, NHT, Section B, Q.19b; © VCAA*

Figure 14 shows the energy levels of a sodium atom.

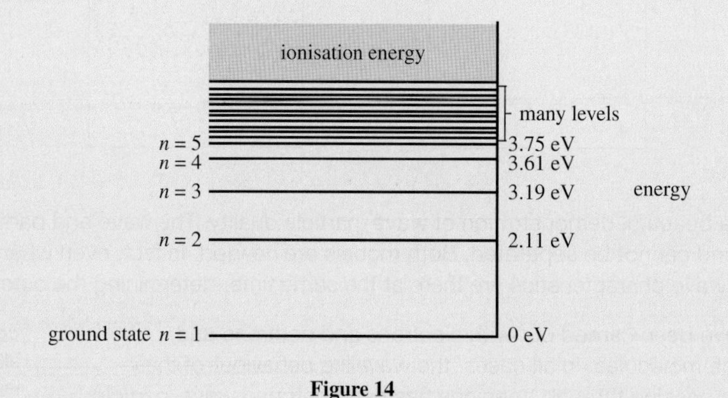

Figure 14

Figure 14 shows some specific energy levels that electrons in a sodium atom can occupy.

Describe how the wave nature of electrons explains the existence of the energy levels shown in Figure 14.

Question 2 (4 marks)

Source*: VCE 2017 Physics Exam, Section B, Q.19; © VCAA*

Roger and Mary are discussing diffraction.

Mary says electrons produce a diffraction pattern.

Roger says this is impossible as diffraction is a wave phenomenon and electrons are particles; diffraction can only be observed with waves, as with electromagnetic waves, such as light and X-rays.

Evaluate Mary's and Roger's statements in light of the current understanding of light and matter. Describe **two** experiments that show the difference between Mary's and Roger's views.

Question 3 (4 marks)

Source: *VCE 2017 Physics Exam, NHT, Section A, Q.21;* © VCAA

De Broglie suggested that the quantised energy states of atoms could be explained in terms of electrons forming standing waves.

Describe how the concept of standing waves can help explain the quantised energy states of an atom. You should include a diagram.

Question 4 (5 marks)

Source: *VCE 2015 Physics Exam, Section A, Q.21;* © VCAA

a. Use the model of quantised states of the atom to explain why only certain energy levels are allowed. **(3 marks)**

b. Illustrate your answer with an appropriate diagram. **(2 marks)**

Question 5 (2 marks)

The following two diagrams illustrate models of the hydrogen atom.

a.

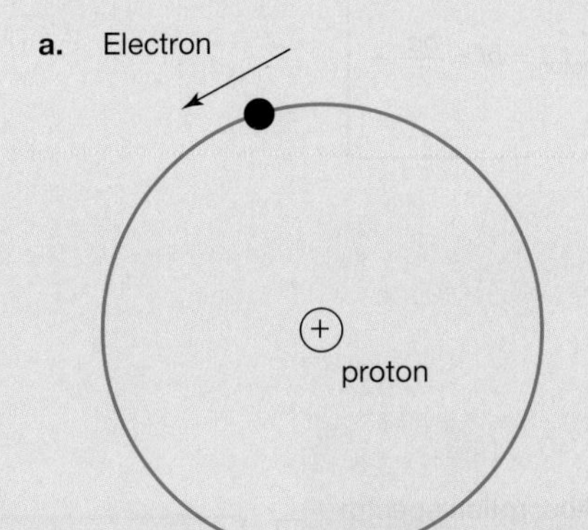

Classical particle model of electron in orbit about a proton

b.

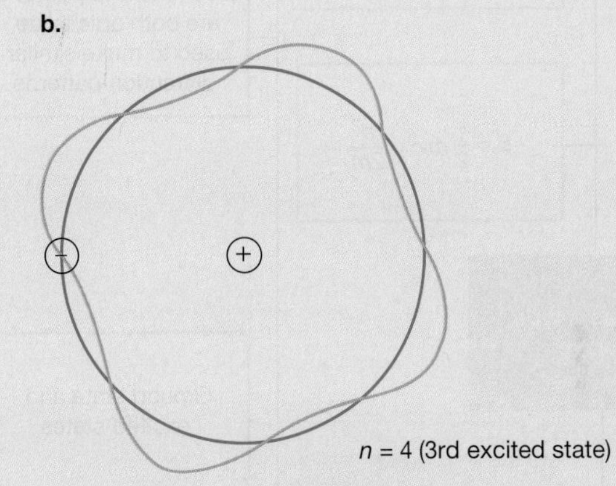

De Broglie wave model of electron in orbit about a proton

The first model is a satellite model where an electron is envisaged as a particle in a stable circular orbit about the nucleus. The second model depicts an electron in the third excited state as a standing wave on the circumference of a circle. Discuss the reason why the first model fails to predict atomic behaviour whereas the second model is consistent with the observation of discrete emission spectra.

More exam questions are available in your learnON title.

10.6 Review

Hey students! Now that it's time to revise this topic, go online to:

Access the topic summary

Review your results

Watch teacher-led videos

Practise past VCAA exam questions

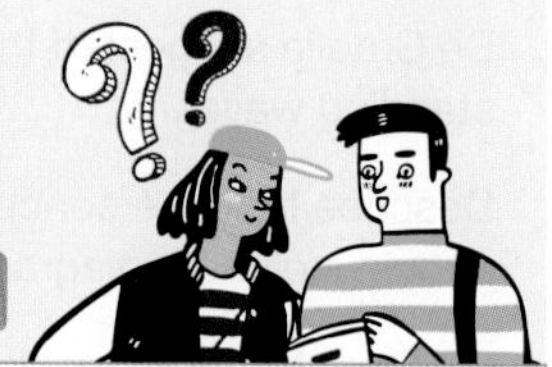

Find all this and MORE in jacPLUS

10.6.1 Topic summary

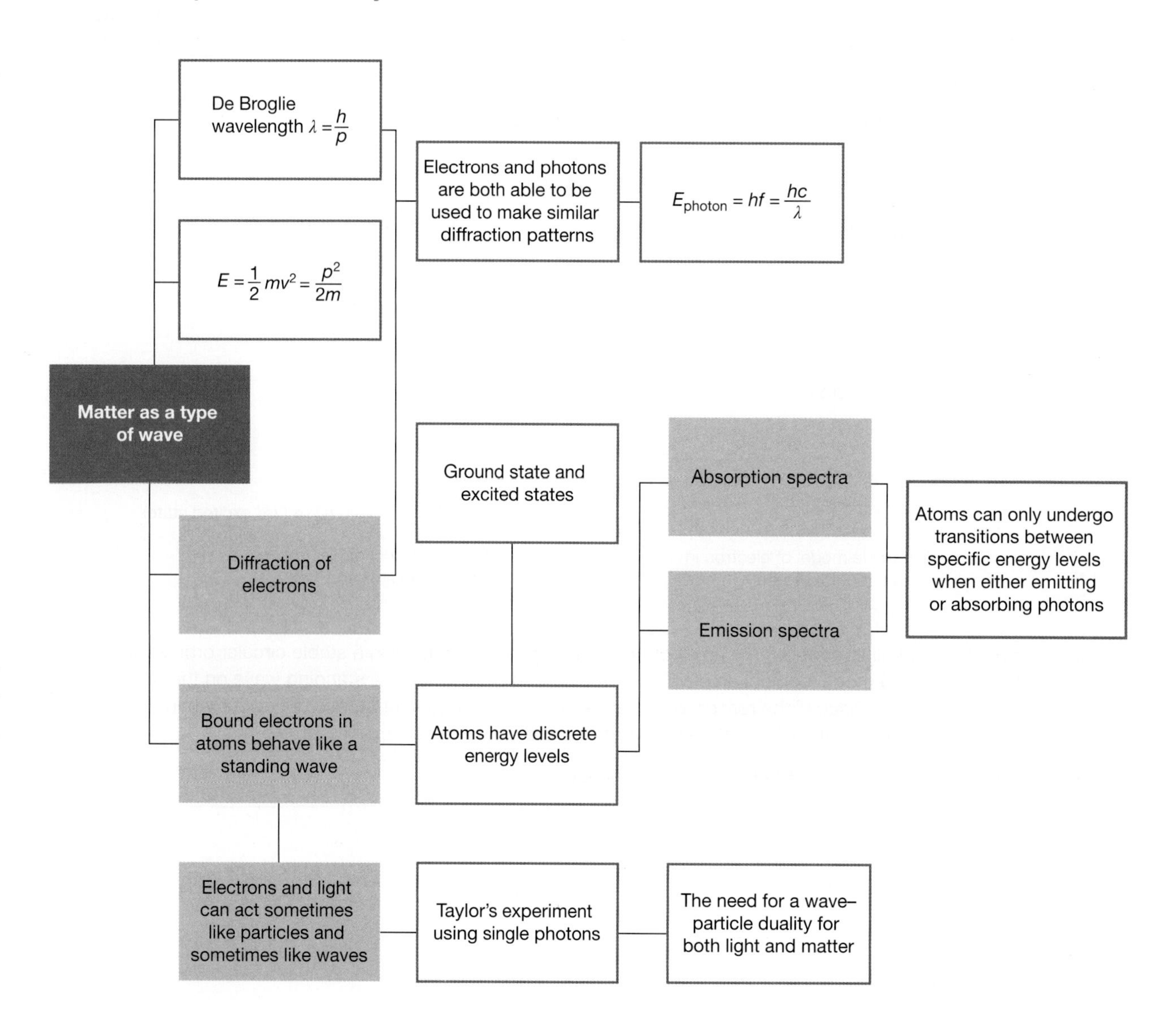

10.6.2 Key ideas summary

10.6.3 Key terms glossary

online only

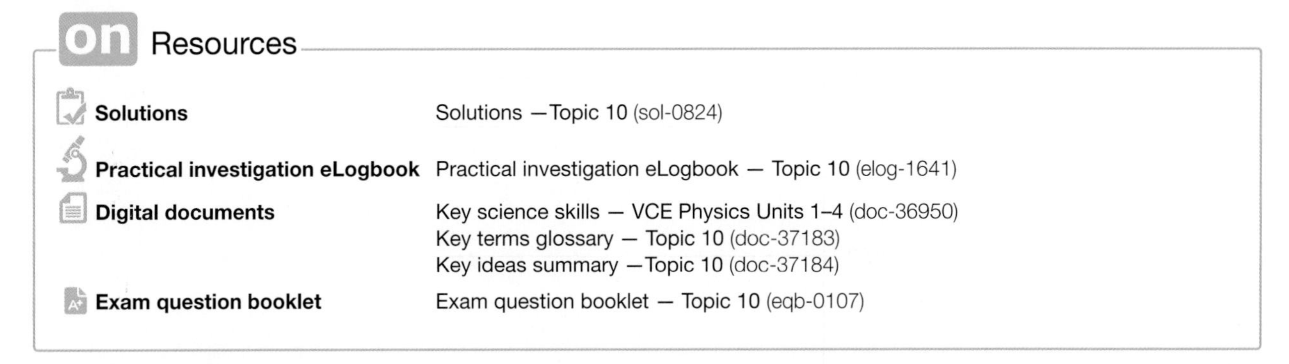

on Resources
Solutions — Solutions — Topic 10 (sol-0824)
Practical investigation eLogbook — Practical investigation eLogbook — Topic 10 (elog-1641)
Digital documents — Key science skills — VCE Physics Units 1–4 (doc-36950)
Key terms glossary — Topic 10 (doc-37183)
Key ideas summary — Topic 10 (doc-37184)
Exam question booklet — Exam question booklet — Topic 10 (eqb-0107)

10.6 Activities

learnon

10.6 Review questions

1. Consider both an electron and a photon. The photon has a frequency of 5.2×10^{14} Hz and the electron has a velocity of 2.8×10^{6} m s^{-1}.
   a. Calculate the wavelength of the photon and the de Broglie wavelength of the electron. Which has the greater value?
   b. Determine which has the greater momentum, the photon or the electron.
   c. Calculate and compare the kinetic energy of the photon and the electron.
2. A photon and an electron both have the same energy, namely 3.0 eV (4.8×10^{-19} J).
   a. Calculate the momentum each. Which is greater in value?
   b. Calculate the wavelength of the photon and the de Broglie wavelength of the electron. Which is greater in magnitude?
   c. If the energy of both the electron and the photon increases, what happens to the momentum of the photon and the electron? What happens to their respective wavelengths?
3. For what important piece of evidence does the diffraction of electrons provide?
4. Explain the difference between an emission and absorption spectrum as it relates to a gas of identical atoms.
5. A beam of electrons accelerated by an electron gun is required for studying the structure of a new molecule. The de Broglie wavelength required is 1.0×10^{-11} m. Calculate the voltage required to accelerate the electrons so they have the required de Broglie wavelength.

6. The following figure is a partial energy level diagram for a hydrogen atom showing the ground state and the first three excited states.

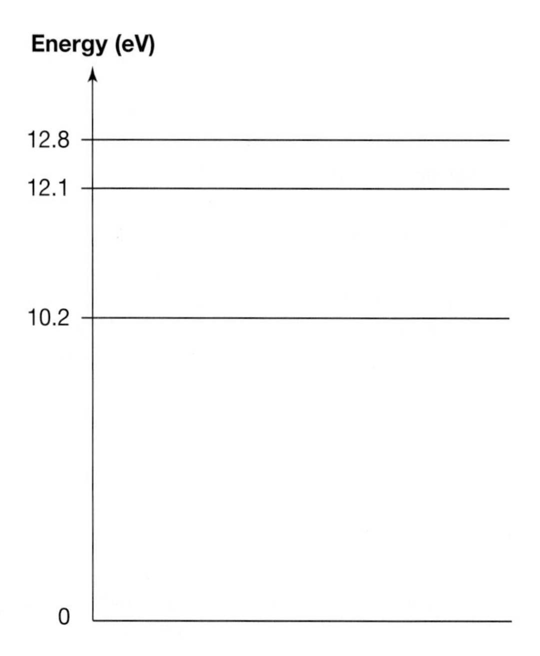

   a. Copy the diagram and draw arrows to represent all six transitions that may occur when a population of hydrogen atoms is excited into the third excited state.
   b. Calculate the energy of each of the six possible photons that may be emitted.
   c. Determine the wavelength of the photon having the least energy.

7. Consider the energy level diagram for a gas Q, shown in the figure.
   a. i. Determine the energy of the photon emitted when an electron in the state $n = 3$ undergoes a transition to the state $n = 2$.
      ii. Determine the frequency and wavelength of this photon.
   b. Determine the wavelength of the photon absorbed by this gas when an electron undergoes a transition from the state $n = 1$ to the state $n = 4$.

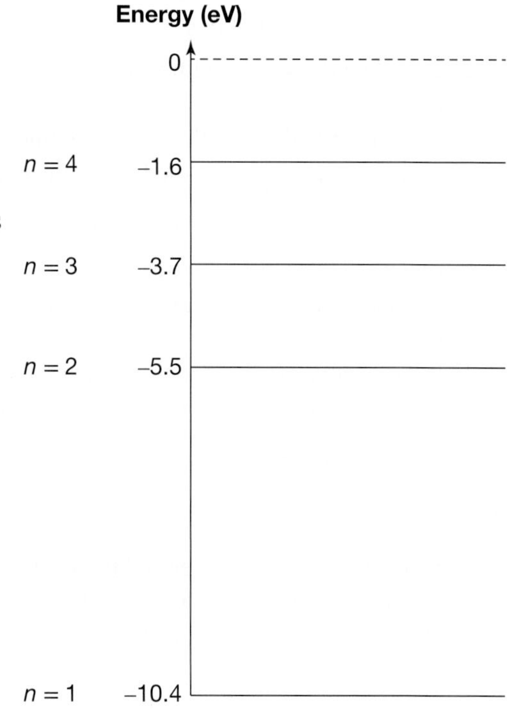

8. In laboratory experiments, the absorption spectrum contains fewer dark lines than the emission spectrum containing bright coloured lines when a gas of the same atoms is used. Explain.

9. What important feature does Taylor's experiment tell us about the dual nature of light?

10.6 Exam questions

Section A — Multiple choice questions

All correct answers are worth 1 mark each; an incorrect answer is worth 0.

Question 1

Source*: VCE 2022 Physics Exam, NHT, Section A, Q.17; © VCAA*

Some of the energy levels of the hydrogen atom are shown in the diagram below. A hydrogen atom has been excited to the 12.8-eV energy level. It returns to the ground state via the three transitions shown.

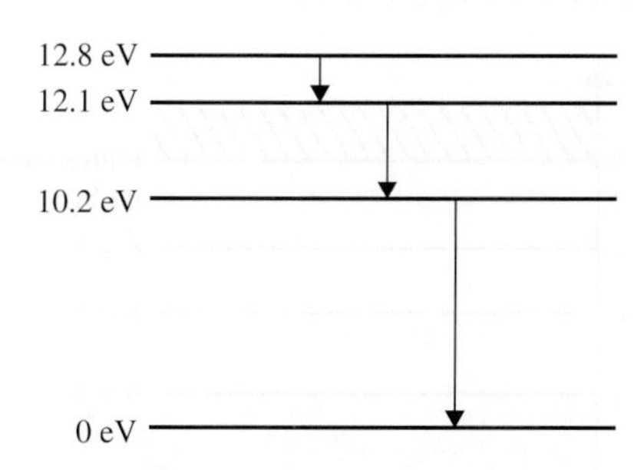

Which of the following indicates the energies of the emitted photons?

A. 0.7 eV, 2.6 eV, 10.2 eV
B. 0.7 eV, 1.9 eV, 10.2 eV
C. 1.9 eV, 2.6 eV, 10.2 eV
D. 10.2 eV, 12.1 eV, 12.8 eV

Question 2

Source*: VCE 2022 Physics Exam, NHT, Section A, Q.19; © VCAA*

Diffraction is a property of waves. Electrons display wave-like properties when producing diffraction patterns.

This is because electrons

A. always carry an electric charge.
B. can move around nuclei in fixed orbits.
C. have a wavelength related to their momentum.
D. can jump between energy levels within an atom.

Question 3

Source*: VCE 2019 Physics Exam, NHT, Section A, Q.19; © VCAA*

Part of the energy-level diagram for an unknown atom is shown.

Which one of the arrows shows a change of energy level corresponding to the absorption of a photon of highest frequency?

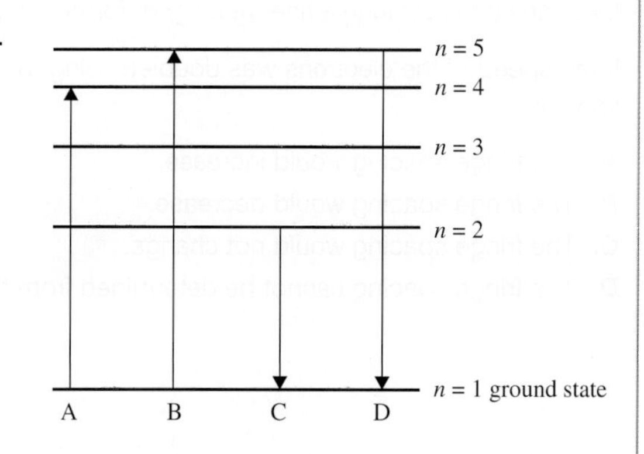

A. A
B. B
C. C
D. D

Question 4

***Source**: Adapted from VCE 2019 Physics Exam, Section A, Q.14; © VCAA*

Electrons of mass 9.1×10^{-31} kg are accelerated in an electron gun to a speed of 4.2×10^{7} m s^{-1}.

The best estimate of the de Broglie wavelength of these electrons is

A. 1.1×10^{-12} m **B.** 1.7×10^{-11} m **C.** 1.7×10^{-8} m **D.** 1.1×10^{-6} m

Question 5

***Source**: Adapted from VCE 2018 Physics Exam, NHT, Section B, Q.12; © VCAA*

Figure 12 shows the energy level diagram for the hydrogen atom.

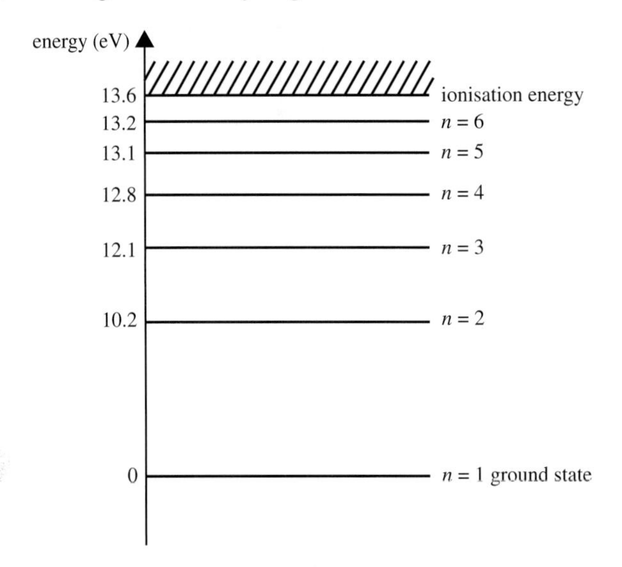

Figure 12

Which of the following indicates the emission energies from the $n = 4$ state?

A. 12.8 eV, 12.1 eV, 10.2 eV
B. 12.1 eV, 10.2 eV, and 1.9 eV
C. 12.8 eV, 12.1 eV, 10.2 eV, 2.6 eV, 1.9 eV, and 0.7 eV
D. 13.1 eV 12.8 eV, 12.1 eV, 10.2 eV, 2.6 eV, 1.9 eV, 0.7 eV and 0.3 eV

Question 6

***Source**: VCE 2019 Physics Exam, Section A, Q.15; © VCAA*

Electrons pass through a fine metal grid, forming a diffraction pattern.

If the speed of the electrons was doubled using the same metal grid, what would be the effect on the fringe spacing?

A. The fringe spacing would increase.
B. The fringe spacing would decrease.
C. The fringe spacing would not change.
D. The fringe spacing cannot be determined from the information given.

Question 7

Source: *VCE 2020 Physics Exam, Section A, Q.17; © VCAA*

The diagram shows some of the energy levels for the electrons within an atom. The arrows labelled A, B, C and D indicate transitions between the energy levels and their lengths indicate the relative size of the energy change.

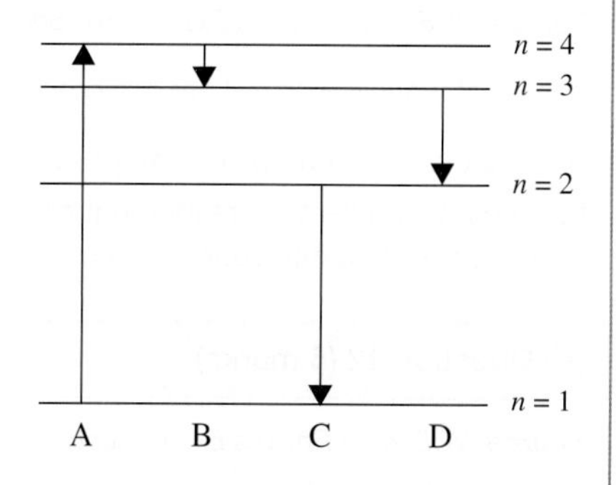

Which transition results in the emission of a photon with the most energy?

A. A

B. B

C. C

D. D

Question 8

Source: *VCE 2021 Physics Exam, Section A, Q.19; © VCAA*

The diagram below shows one representation of a de Broglie standing wave for an electron in orbit around a hydrogen atom.

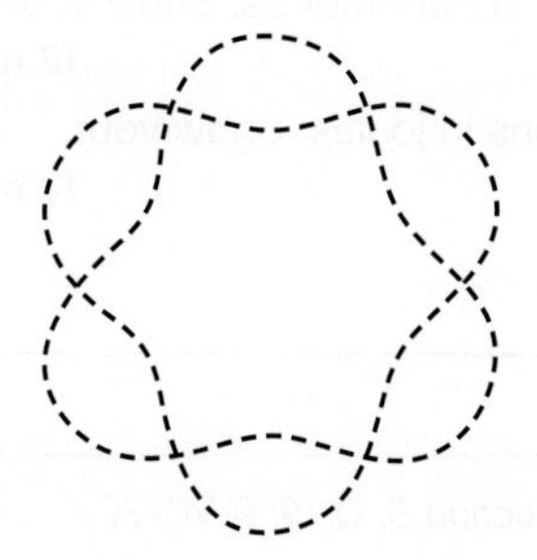

Which one of the following values of n, the number of whole wavelengths, best depicts the de Broglie standing wave pattern shown in the diagram?

A. 2 **B.** 3 **C.** 4 **D.** 6

Question 9

Source: *VCE 2020 Physics Exam, Section A, Q.18; © VCAA*

Quantised energy levels within atoms can best be explained by

A. electrons behaving as individual particles with different energies.

B. electrons behaving as waves, with each energy level representing a diffraction pattern.

C. protons behaving as waves, with only standing waves at particular wavelengths allowed.

D. electrons behaving as waves, with only standing waves at particular wavelengths allowed.

Question 10

Source: *VCE 2017 Physics Exam, Section A, Q.17; © VCAA*

Quantised energy levels within atoms can best be explained by

A. electrons behaving as individual particles with varying energies.

B. atoms having specific energy requirements that can only be satisfied by electrons.

C. electrons behaving as waves, with each energy level representing a diffraction pattern.

D. electrons behaving as waves, with only standing waves at particular wavelengths allowed.

Section B — Short answer questions

Question 11 (6 marks)

Source: *VCE 2021 Physics Exam, NHT, Section B, Q.16;* © VCAA

X-rays of wavelength 2.0 nm are emitted from an X-ray source.

a. Calculate the energy of **one** photon of these X-rays. Show your working. **(3 marks)**

b. The 2.0-nm X-rays are incident on a single narrow slit of width 5×10^{-8} m. Would a diffraction pattern be observed? Justify your answer. **(3 marks)**

Question 12 (5 marks)

Source: *VCE 2018 Physics Exam, Section B, Q.18;* © VCAA

The diffraction patterns for X-rays and electrons through thin polycrystalline aluminium foil have been combined in the diagram in Figure 18, which shows an electron diffraction pattern on the left and an X-ray diffraction pattern on the right. The images are to the same scale.

The X-rays have a photon energy of 8000 eV.

a. Calculate the wavelength of the electrons in nanometres. Show your working. **(2 marks)**

b. Calculate the kinetic energy of the electrons in joules. Show your working. **(3 marks)**

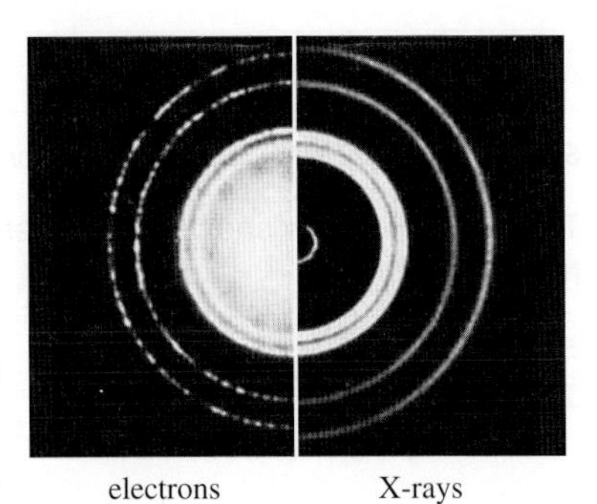

Figure 18

Question 13 (4 marks)

Source: *Adapted from VCE 2018 Physics Exam, Section B, Q.19;* © VCAA

Figure 19 shows the spectrum of light emitted from a hydrogen vapour lamp.

The spectral line, indicated by the arrow on Figure 19, is in the visible region of the spectrum.

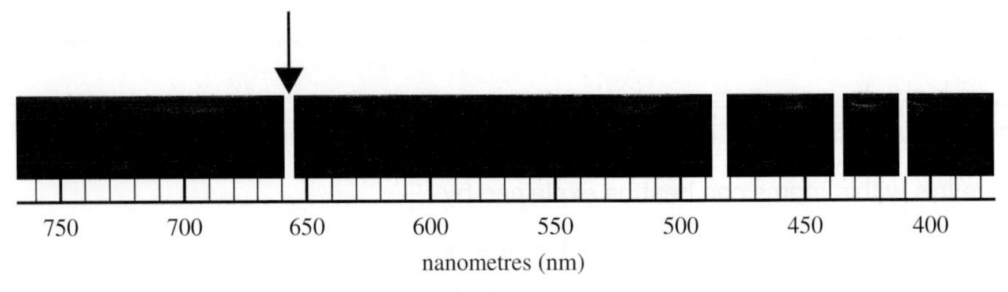

Figure 19

a. The following list gives the four visible colours that are emitted by the hydrogen atom. Select the colour that corresponds to the spectral line indicated by the arrow on Figure 19. **(1 mark)**

A. violet
B. blue-violet
C. blue-green
D. red

b. Explain why the visible spectrum of light emitted from a hydrogen vapour lamp gives **discrete** spectral lines, as shown in Figure 19. **(3 marks)**

Question 14 (6 marks)

***Source**: VCE 2019 Physics Exam, NHT, Section B, Q.11;* © VCAA

Kym and Roger conduct an experiment to observe an electron diffraction pattern. 5000-eV electrons are projected through a diffracting grid and the resulting pattern is observed on a screen.

Kym and Roger want to calculate the wavelength of X-rays that would produce a similarly spaced diffraction pattern.

Kym says that they will need X-rays of 5000 eV.

Roger says that X-rays of a different energy will be needed.

a. Explain why Roger is correct. **(2 marks)**

b. Showing each of the steps involved in your working, calculate the energy of X-rays that would be required to produce the similarly spaced diffraction pattern. **(4 marks)**

Question 15 (5 marks)

***Source**: VCE 2021 Physics Exam, NHT, Section B, Q.17;* © VCAA

Light from a mercury vapour lamp shows a line spectrum related to discrete energy levels. Some of the energy levels for the mercury atom are shown in Figure 20.

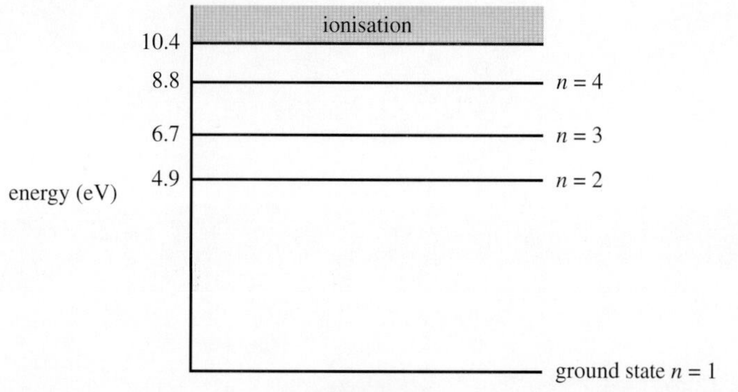

Figure 20

a. Draw an arrow on Figure 20 to indicate the transition between the listed energy states that would produce the lowest frequency of an emitted photon. **(1 mark)**

b. Calculate the energy of the light emitted when the mercury atom makes a transition from the third energy level (n = 3) to its ground state (n = 1). Show your working. **(2 marks)**

c. Explain what happens to a mercury atom in its ground state if a photon of energy 2.1 eV is incident on it. **(2 marks)**

11 Einstein's special theory of relativity and the relationship between energy and mass

KEY KNOWLEDGE

In this topic, you will:

- describe the limitation of classical mechanics when considering motion approaching the speed of light
- describe Einstein's two postulates for his special theory of relativity that:
  - the laws of physics are the same in all inertial (non-accelerated) frames of reference
  - the speed of light has a constant value for all observers regardless of their motion or the motion of the source
- interpret the null result of the Michelson-Morley experiment as evidence in support of Einstein's special theory of relativity
- compare Einstein's special theory of relativity with the principles of classical physics
- describe proper time (t_0) as the time interval between two events in a reference frame where the two events occur at the same point in space
- describe proper length (L_0) as the length that is measured in the frame of reference in which objects are at rest
- model mathematically time dilation and length contraction at speeds approaching c using the equations: $t = \gamma t_0$ and $L = \frac{L_0}{\gamma}$ where $\gamma = \frac{1}{\sqrt{\left(1 - \frac{v^2}{c^2}\right)}}$
- explain and analyse examples of special relativity including that:
  - muons can reach Earth even though their half-lives would suggest that they should decay in the upper atmosphere
  - particle accelerator lengths must be designed to take the effects of special relativity into account
  - time signals from GPS satellites must be corrected for the effects of special relativity due to their orbital velocity
- interpret Einstein's prediction by showing that the total 'mass-energy' of an object is given by:
  $E_{tot} = E_k + E_0 = \gamma mc^2$ where $E_0 = mc^2$, and where kinetic energy can be calculated by:
  $E_k = (\gamma - 1)mc^2$
- apply the energy-mass relationship to mass conversion in the Sun, to positron-electron annihilation and to nuclear transformations in particle accelerators (details of the particular nuclear processes are not required).

EXAM PREPARATION

Access past VCAA questions and exam-style questions and their video solutions in every lesson, to ensure you are ready.

11.1 Overview

Hey students! Bring these pages to life online

 Watch videos

 Engage with interactivities

 Answer questions and check results

Find all this and MORE in jacPLUS

11.1.1 Introduction

Newton's laws of motion revolutionised humanity's understanding of the movement of bodies, gravity and motion. For almost 200 years, they remained unchallenged as some of the most elegant theories the world had ever seen. However, over the course of the nineteenth and twentieth centuries, scientists' understanding of concepts not fully explained by Newton's laws, such as electromagnetism, improved.

Albert Einstein built on the work of many scientists before him and used 'thought experiments' to show that space and time are not absolute but relative to the physical environment. This caused a complete revision of our understanding of time, space and energy. Einstein showed that time intervals and distances between points were not the same for all observers. This led to acceptance that mass is a type of energy.

FIGURE 11.1 The velocity of a yacht can be measured relative to wind, land, water or other yachts, and all of these measurements can be different.

Einstein was named *Time* magazine's 'Person of the Century' in 1999. Not only was he a great humanitarian, having fought for the rights of Jewish and Arab people (after he escaped Nazi Germany) and for the rights of African Americans, but his work on relativity changed the way scientists understand the universe. This topic will introduce you to modern physics and an even better understanding of the workings of the universe.

LEARNING SEQUENCE

 Resources

Solutions	Solutions — Topic 11 (sol-0825)
Digital documents	Key science skills — VCE Physics Units 1–4 (doc-36950)
	Key terms glossary — Topic 11 (doc-37185)
	Key ideas summary — Topic 11 (doc-37186)
Exam question booklet	Exam question booklet — Topic 11 (eqb-0108)

11.2 Einstein's special theory of relativity

KEY KNOWLEDGE

- Describe the limitation of classical mechanics when considering motion approaching the speed of light
- Describe Einstein's two postulates for his special theory of relativity, which state that
  - the laws of physics are the same in all inertial (non-accelerated) frames of reference
  - the speed of light has a constant value for all observers regardless of their motion or the motion of the source.
- Compare Einstein's special theory of relativity with the principles of classical physics.

11.2.1 What is relativity?

Newton's laws held that all observers, regardless of their location and the nature of their motion, were the same in terms of what they observed in the world around them. However, today it is understood that the measurement of the speed of an object depends on the **relative** motion of the observer. So do the measurements of the object's time, kinetic energy, length and mass; that is, these properties are relative rather than fixed. Albert Einstein discovered that some of the physical properties that people assumed to be fixed for all observers actually depend on the observers' motions. But not everything is relative. The laws of physics and the speed of light are the same for all observers. Major developments in physics have come about at times when physicists such as Galileo and Einstein developed a clearer understanding of what is relative and what is not.

FIGURE 11.2 Albert Einstein (1879–1955)

Albert Einstein (1879–1955) is one of the most famous figures in history, largely due to his work on relativity. Einstein did not invent the idea of relativity — it dates back to Galileo — but he brought it into line with nineteenth-century developments in the understanding of light and electricity, leading to some striking changes in how physicists viewed the world. In this topic, you will look at the first revolution in relativity, then explore some of the ideas of Einstein's special theory of relativity.

At this point, it is worth noting that classical kinematics and dynamics will give very accurate predictions for all situations where the speed of an observer relative to another is less than 10% of the speed of light. It is only when dealing with speeds greater than this that it is necessary to use Einstein's model rather than Newton's model, to obtain more accurate calculations. In Newton's model, time intervals between two events and distances between two points are assumed to be the same (i.e. invariant) for all observers regardless of their relative motion, but not so in Einstein's model. This difference comes about because the speed of light is the same for all observers. Moving towards a light source does not make the speed of light greater, and moving away from a light source does not make the speed of light less. This goes against what common sense tells us. Let's start with a theoretical scenario to help distinguish between Newton's (and Galileo's) and Einstein's models for motion.

There is no rest

Consider a motionless police officer pointing her radar gun at an approaching sports car (figure 11.3). She measures the sports car's speed to be 90 km h^{-1}. This agrees with the speed measured by the sports car's speedometer. However, another police car drives towards the sports car in the opposite direction at 60 km h^{-1}. A speed radar is also operating in this car, and it

relative in relation to something else, dependent on the observer

measures the speed of the sports car to be 150 km h^{-1}. So each police officer has a different measurement for the speed of the sports car. Which measurement is correct? The answer is that they are both correct — the speed measured for the car is measured relative to the velocity of the observer.

FIGURE 11.3 Two different measurements of the speed of a car

The sports car is approaching the oncoming police car at the same rate as if the police car was parked and the sports car had a reading of 150 km h^{-1} on its speedometer. The speed of the car is relative to the observer, rather than being an **absolute** quantity agreed on by all observers.

absolute independent of frame of reference, permanent

Newton's First Law of Motion states that an object will remain at rest or in uniform motion in a straight line unless acted on by an unbalanced net force. The speed itself does not matter. In the example shown in figure 11.3, despite the difference in the detected speed, this law works for both of the police officers, as do the other laws of motion.

FIGURE 11.4 A speed limit is the maximum allowed speed relative to the road.

FIGURE 11.5 The radar gun would measure a different speed if it was in a moving vehicle.

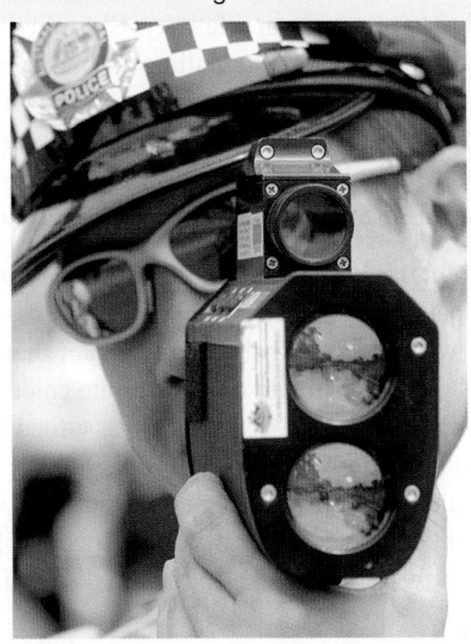

The significance of relative speed becomes all too clear in head-on collisions. For example, you might be driving at only 60 km h^{-1}, but if you collide head-on with someone doing the same speed in the opposite direction, the impact occurs for both cars at 120 km h^{-1}; if one car was stationary, the impact would occur at 60 km h^{-1} for both.

The Italian scientist Galileo Galilei (1564–1642) used sailing ships and cannon balls as examples when considering relative motion, but the physics ideas were the same. In Galileo's time, much of physics was still based on ancient ideas recorded by the Greek philosopher Aristotle (384–322 BC). Aristotle taught that Earth was stationary in the centre of the universe. Motion relative to the centre of Earth was a basis for Aristotelian physics, so a form of relativity was key to physics even before Galileo. But Galileo had to establish a new understanding of relativity before it became widely accepted that Earth moved around the Sun.

FIGURE 11.6 Galileo Galilei (1564–1642), from a nineteenth-century engraving

Galileo's insight helped provide the platform for physics as it is known today, but the idea of a fixed frame of reference persisted. Following on from Galileo, Isaac Newton considered the centre of mass of the solar system to be at absolute rest. James Clerk Maxwell (1831–1879), who put forward the theory of electromagnetism, regarded the medium (luminiferous aether) for electromagnetic waves (light) to be at rest. It was Einstein who let go of the concept of absolute rest, declaring that it was impossible to detect a place at absolute rest and therefore the idea had no consequence. Once again, ideas about relativity were updated to take into account the latest discoveries and enable physics to make enormous leaps of progress.

The measured speed (velocity) of bodies in motion is truly relative to whoever is measuring it. You will return to Einstein's advances shortly, but it's now time to look at some more examples from Galilean relativity.

FIGURE 11.7 What should speed be measured relative to?

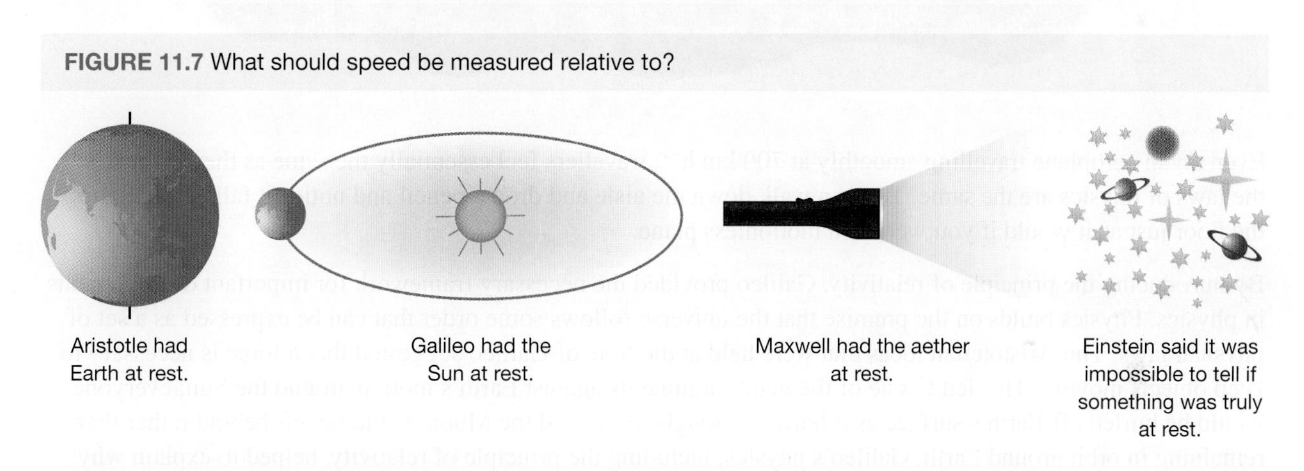

Aristotle had Earth at rest.
Galileo had the Sun at rest.
Maxwell had the aether at rest.
Einstein said it was impossible to tell if something was truly at rest.

Galileo and the principle of relativity

Consider the driver of the sports car discussed earlier. His position relative to features of the landscape he drives through is continuously changing, but inside the car he has the same position relative to the car, the size of the force due to gravity on him is unchanged, and his mass and height have the same measured value as when he was stationary. Everything inside the car behaves just as if it was at rest. On a smooth road at constant speed, his passenger could pour a drink without difficulty. The effect of the bumps in the road would be indistinguishable from a situation in which the car was stationary and someone outside was rocking it. The driver would not experience any difference driving at a constant velocity of 150 km h^{-1} compared to if he was driving at a constant velocity of 90 km h^{-1}.

Nothing inside a vehicle moving with constant velocity can be affected by the magnitude of the velocity. If it was, the following question would need to be asked: which velocity? If a velocity of 90 km h^{-1} caused a passenger to have a mass of 50 kg, but a velocity of 150 km h^{-1} caused the passenger to have a mass of 60 kg, a problem would exist.

This is known as Galilean relativity. Galilean relativity states that the laws of physics do not depend on the velocity of the observer. Galileo played a major role in the development of relativity, and Newton's laws of motion are fully consistent with it. Another way of describing relativity is that there is no way that anyone in the car can measure its velocity without making reference to something external to the car. The sports car driver can measure his speed relative to the two police officers mentioned earlier. He would measure that he is moving relative to each of them at different speeds, but he would not feel any difference. As long as the road is straight and smooth and the car is travelling at a constant speed, there is no way to detect that the car is moving at all! He could be stationary while one police car is approaching him at 90 km h^{-1} and the other at 150 km h^{-1}.

FIGURE 11.8 How can the vehicle that is actually speeding be identified?

Even on an aeroplane travelling smoothly at 700 km h^{-1}, travellers feel essentially the same as they do at rest: the laws of physics are the same. You can walk down the aisle and drop a pencil and notice it fall vertically to the floor just as it would if you were in a motionless plane.

By introducing the principle of relativity, Galileo provided the necessary framework for important developments in physics. Physics builds on the premise that the universe follows some order that can be expressed as a set of physical laws. The Aristotelian ideas that were held at the time of Galileo suggested that a force is necessary to keep objects moving. This led to one of the major arguments against Earth's motion around the Sun: everyone would be hurled off Earth's surface as it hurtled through space, and the Moon would be left behind rather than remaining in orbit around Earth. Galileo's physics, including the principle of relativity, helped to explain why this argument was wrong. Forces are not required to keep objects moving, only to change their motion.

The science of Galileo and Newton was spectacularly successful: it explained the motion of everything from cannon balls to planets. Later, however, as observations and new theories of physics developed in the nineteenth century, physicists faced the challenge of how to make everything fit together. It was not until the early twentieth century that Einstein found a way to make sense of it all.

Examples of Galilean relativity

Here are some examples that support the Galilean principle of relativity:

1. If you are in a car stopped at the lights and another car next to yours slowly rolls past, it is difficult to tell whether you or the other car is moving if nothing but the other car is in view.
2. In IMAX and similar films, viewers can feel as though they are going on a thrilling ride, even though they are actually sitting on a fixed seat in a cinema. Theme parks enhance this effect in virtual reality rides by jolting the chairs in a way that mimics movements you would feel on a real ride. Virtual reality rides are very convincing because what you see and feel corresponds with an expected movement, and your senses do not tell you otherwise. As long as the jolts correspond with the visual effects, there is no way of telling the difference. The motion or lack of motion of the seat is irrelevant.

FIGURE 11.9 A virtual reality ride

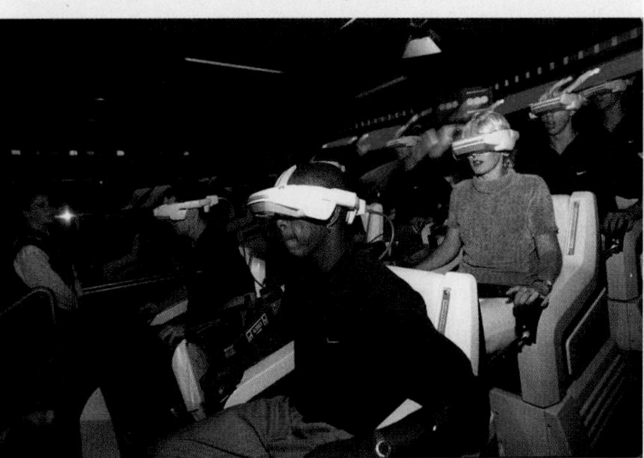

3. Acceleration does not depend on the velocity of the observer. Astronauts in free fall feel weightless, regardless of the magnitude of their velocity. They move along with the same velocity as their spacecraft, as Newton's first law would suggest. When the spacecraft accelerates due to the force of its rocket engines, the force on the person by the rocket is in the forward direction and the effect of the acceleration on the astronaut is noticeable.
4. When you are riding in a car with the window down, most of the wind you feel on your face is due to the motion of the car through the air. It is present even on a still day. Whenever you drive at greater than 60 km h^{-1}, your windscreen is saving you from gale-force winds! Similarly, it is always windy on moving boats. This is because on deck you are not as well protected from the apparent wind as you are in a car.
5. Apparent wind becomes especially significant when sailing, as seen in figure 11.10. As the boat increases its speed, the sailor notices that he is heading more into the wind, even though neither he nor the wind has changed direction relative to the shore. This leads the sailor to change the sail setting to suit the new wind direction.

FIGURE 11.10 The faster the boat moves, the more the wind appears to blow from in front.

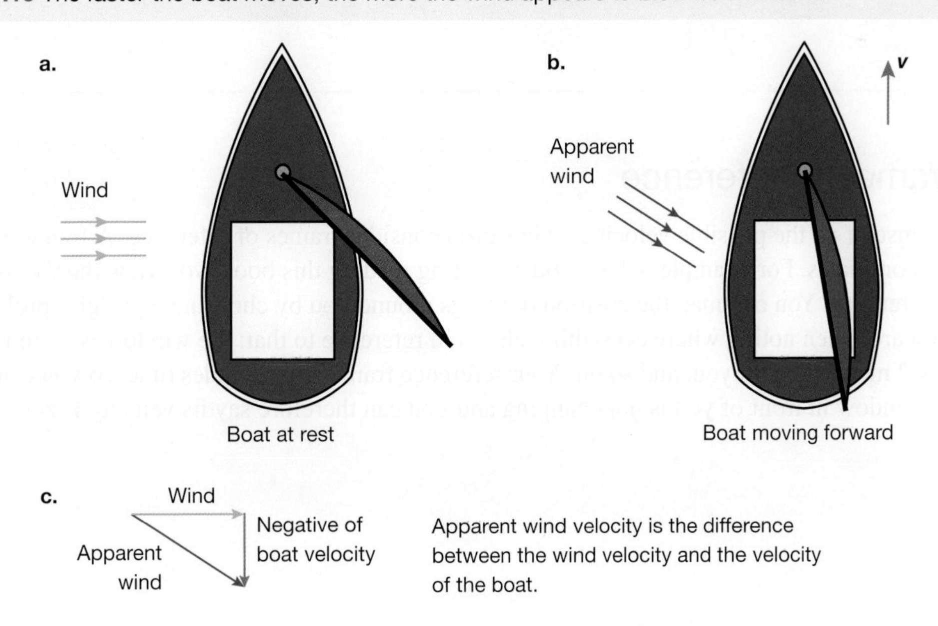

tlvd-9027

SAMPLE PROBLEM 1 Comparing the velocity of cars during collisions

Determine the relative velocity in each of the three scenarios and compare each scenario in terms of velocity.

1. **A car travelling down the highway at 80 km h^{-1} collides with a stationary car.**
2. **A car travelling down the highway at 100 km h^{-1} collides with a car travelling at 20 km h^{-1} in the same direction.**
3. **A car travelling down the highway at 100 km h^{-1} collides with a car travelling at 20 km h^{-1} in the opposite direction.**

THINK	WRITE
1. To determine the relative velocity, you need to consider that the velocity of the first car (car A) relative to the second car (car B) is shown using the formula $$v_{\text{A relative to B}} = v_A - v_B$$ and that the direction of movement is positive.	Scenario 1 $v_A - v_B = 80 - 0 = 80\ \text{km h}^{-1}$ Scenario 2 $v_A - v_B = 100 - 20 = 80\ \text{km h}^{-1}$ Scenario 3 $v_A - v_B = 100 - (-20) = 120\ \text{km h}^{-1}$
2. Provide a comparison of the relative velocities in each scenario.	Although in each case the velocities relative to the road are different, the relative velocities of the cars in scenarios 1 and 2 are the same and will cause similar effects upon colliding. When examining scenarios 2 and 3, the cars are travelling at the same speeds but in opposite directions, and therefore the relative velocities are different and there will be different effects upon the cars colliding.

PRACTICE PROBLEM 1

According to Galilean relativity, which one (or more) of the following variables is relative?

a. **acceleration**
b. **velocity**
c. **time**
d. **mass**

11.2.2 Frames of reference

To help make sense of all the possible velocities, physicists consider frames of reference. A frame of reference is a system of coordinates. For example, where you are sitting reading this book, you view the world through your frame of reference. You can map the position of things around you by choosing an origin (probably the point where you are), then noting where everything else is in reference to that: the window is 1 metre in front of you, the door is 2 metres behind you, and so on. Your reference frame also includes time, so you can see that the position of the window in front of you is not changing and you can therefore say its velocity is zero.

FIGURE 11.11 A reference frame is a set of space and time coordinates that are stationary relative to an observer.

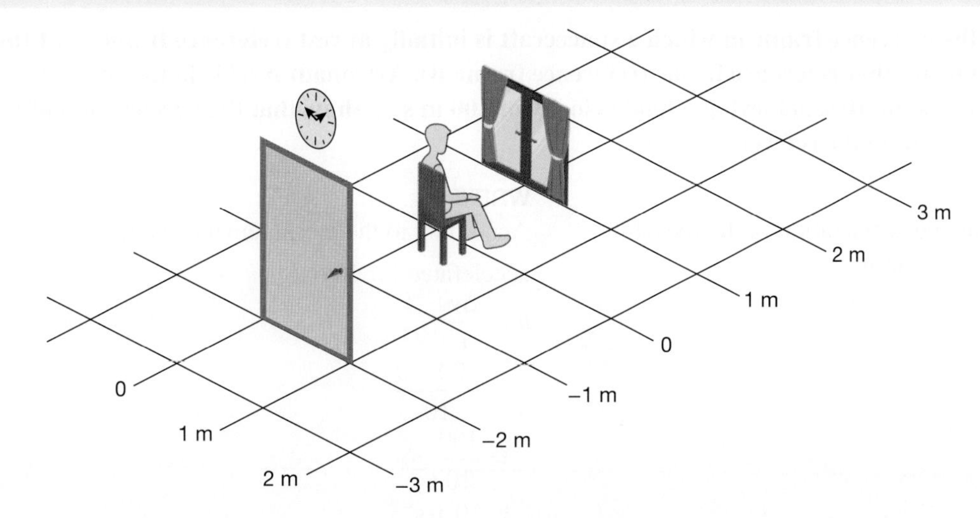

When something is said to be 'at rest', it is at rest in the reference frame in which the observer views the world. In everyday life, there is a tendency to take a somewhat Aristotelian point of view and regard everything from the perspective that Earth is at rest. For example, another student walking behind you has her own reference frame. As she walks, your position in her frame of reference is moving. She could say that she is moving past you while you are stationary, or that she is stationary while you and the rest of the room are on the move!

In many situations, considering Earth to be at rest is a convenient assumption. In more complex examples of motion, such as sports events, car accidents involving two moving vehicles, or the motions of the solar system, it can be useful to choose alternative frames of reference.

In **classical physics**, the differences between frames of reference are their motion and position. In other words, position and speed are relative in classical physics. For example, one person might record an object to have a different position than a second person would (it might be 3 metres in front of the first person but 4 metres behind the second), and the first person might also record it as having a different speed (maybe it is stationary in their frame of reference but approaching the second person at 2 m s^{-1}). The position and speed are dependent on the observer. However, in classical physics both observers can agree on what 3 metres and 2 m s^{-1} are. The rulers are the same and the clocks tick at the same rate in both frames. Time and space are seen as absolute in the classical physics established by Galileo, Newton and the other early physicists.

Frames of reference that are not accelerating are called **inertial reference frames**. An inertial reference frame moves in a straight line at a constant speed relative to other inertial reference frames.

An **invariant quantity** is a quantity that has the same value in all reference frames. In classical physics, mass is the same in all reference frames, so all observers will observe that Newton's second law holds. In sample problem 2, all observers would agree on the forces acting on the astronauts. Unlike velocity, acceleration in Galilean relativity does not depend on the motion of the frame of reference; it is also invariant.

classical physics the physics that predated Einstein's discoveries leading to the laws of relativity and quantum mechanics

inertial reference frame a reference frame that is not accelerating, where Newton's laws hold true

invariant quantity unchanging, regardless of the frame of reference

tlvd-9028

SAMPLE PROBLEM 2 Determining the acceleration in different reference frames

Consider the reference frame in which a spacecraft is initially at rest (reference frame A). Effie is positioned in another reference frame (reference frame B). Astronaut Axel is in the spacecraft and he fires its rockets for 10 s, achieving a final velocity of 100 m s^{-1}. Show that the acceleration of the rocket does not depend on the reference frame.

THINK	WRITE
1. Calculate the acceleration of the rocket in reference frame A.	According to the measurements made in A, the rocket accelerated for 10 s at: $a = \frac{\Delta v}{t} = \frac{100 - 0}{10} = \frac{100}{10} = 10\ \text{m s}^{-2}$ Axel would feel a force in the forward direction.
2. Now choose a different reference frame. Effie is in reference frame B in another spacecraft, moving at 50 m s^{-1} relative to A for instance. Note that you can use any value here, to show that the acceleration will be the same in any frame of reference.	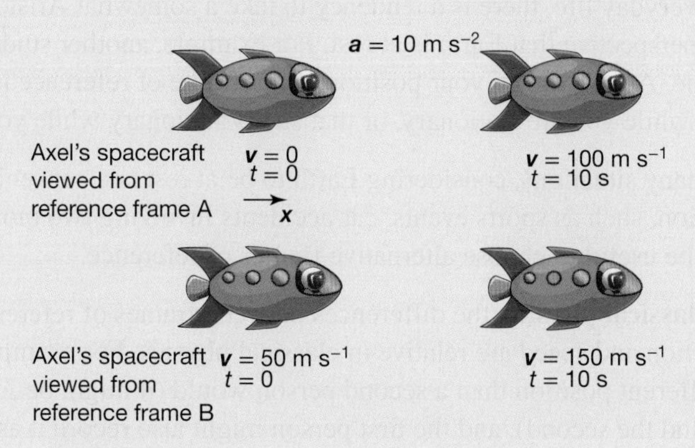
3. Calculate the acceleration of the rocket in reference frame B.	Effie measures the velocity of Axel's spacecraft to change from 50 m s^{-1} to (50 + 100) m s^{-1} in 10 s. From B: $a = \frac{\Delta v}{t} = \frac{150 - 50}{10} = \frac{100}{10} = 10\ \text{m s}^{-2}$
4. Compare the acceleration of the rocket in both reference frames.	The acceleration is the same whether it is measured from frame A or frame B. It will still be 10 m s^{-2} regardless of the speed of the reference frame.

PRACTICE PROBLEM 2

a. **Explain what is meant by the statement 'speed is relative to the frame of reference'.**
b. **By referring to Newton's laws of motion, explain why it is important for acceleration to be invariant but that velocity can be relative.**

In sample problem 2, it is interesting to consider the motion of Axel's spacecraft as viewed by Effie in reference frame B. Reference frame B is in an inertial reference frame as it is not accelerating. Axel, however, looks back at Effie and sees her falling behind at an increasing rate. Is it Axel or Effie who is accelerating? The answer is clear to them: the force experienced by Axel is not felt by Effie. The acceleration can be measured by this force without any reference to the relative motions of other objects; an object's velocity cannot.

11.2.3 Comparing Einstein's theories to earlier principles

Galilean relativity seemed to work well for the motion of massive bodies but, by the nineteenth century, physicists were learning much more about other physical phenomena.

In 1865, James Clerk Maxwell's theory of electromagnetism drew together the key findings of electricity and magnetism to completely describe the behaviour of electric and magnetic fields using a set of four equations. One of the outcomes of this was an understanding of electromagnetic waves. The equations dictated the speed of these waves, and Maxwell noticed that this speed was the same as the speed that had been measured for light. He suggested that light was an electromagnetic wave and predicted the existence of waves with other wavelengths that were soon discovered, such as radio waves. A medium for these fields and waves was proposed, called the **luminiferous aether**. The speed of light, c, was the speed of light relative to this aether.

luminiferous aether hypothesised medium permeating space, supposed to carry electromagnetic waves

The understanding was that if light moves through the aether, then Earth must also be moving through the aether. Changes in the speed of light as Earth orbits the Sun should be detectable. Maxwell predicted that electromagnetic waves would behave like sound and water waves in that the speed of electromagnetic waves in the medium would not depend on the motion of the source or the detector through the medium.

To understand the significance of this aether, consider the sound produced by a jet plane. When the plane is stationary on the runway preparing for takeoff, the sound travels away from the plane at the speed of sound in air, about 340 m s^{-1}. When the plane is flying at a constant speed, say 200 m s^{-1}, the speed of sound is still 340 m s^{-1} in the air. However, to find the speed relative to the reference frame of the plane, the speed of the plane relative to the air must be subtracted. From this it is found that the sound is travelling at:

$340 - 200 = 140$ m s^{-1} in the forward direction relative to the plane

$340 - (-200) = 540$ m s^{-1} in the backward direction relative to the plane.

FIGURE 11.12 Sound moving away from a plane

a. Velocity of sound relative to plane $\boldsymbol{v}$ = 200 m s^{-1}

$\boldsymbol{v}$ = 540 m s^{-1} $\boldsymbol{v}$ = 140 m s^{-1}

b. Velocity of sound relative to air $\boldsymbol{v}$ = 0 m s^{-1}

$\boldsymbol{v}$ = 340 m s^{-1} $\boldsymbol{v}$ = 340 m s^{-1}

In this example, the speed of the plane through the air could be measured by knowing the speed of sound in air (340 m s^{-1}) and measuring the speed of a sound sent from the back of the plane to the front (140 m s^{-1}) in the reference frame of the plane. As long as the plane is flying straight, the speed of the plane relative to the air can be inferred by setting the forward direction as positive and subtracting the velocities:

$$340 - 140 = 200 \text{ m s}^{-1}$$

The speed of the plane has been measured relative to an external reference frame, that of the air, and therefore this example has not violated Galilean relativity. As light had been shown to travel in waves, scientists felt they should be able to measure Earth's speed through the aether in the same way.

tlvd-9029

SAMPLE PROBLEM 3 Explaining Maxwell's concept of electromagnetic waves

Explain how Maxwell's concept of electromagnetic waves such as light challenged the Galilean principle of relativity.

THINK	WRITE
The principle of relativity states that the laws of physics hold true in all inertial reference frames.	Galileo proposed that all inertial frames of reference are equally valid. Maxwell's concept of electromagnetic waves suggested the presence of an absolute frame of reference — the luminiferous aether. That is, observers moving relative to the aether should experience light at different velocities, and therefore not all inertial frames of reference are equally valid, as stated by Galileo.

PRACTICE PROBLEM 3

Assuming that electromagnetic waves travel at *c* relative to the aether, determine the speed of light shining from the rear of a spacecraft that is moving at half the speed of light relative to the aether.

The Michelson–Morley experiment

In 1887, Albert Michelson and Edward Morley devised a method of using interference effects to detect slight changes in the time taken for light to travel through different paths in their apparatus. As with sound travelling from the front and rear of a plane through the air, the light was expected to take different amounts of time to travel in different directions through the luminiferous aether as Earth moved through it. Much to their astonishment, the predicted change in the interference pattern was not observed. It was as though the speed of light was unaffected by the motion of the reference frame of its observer or its source!

FIGURE 11.13 a. The idea behind the Michelson–Morley experiment **b.** A Michelson interferometer

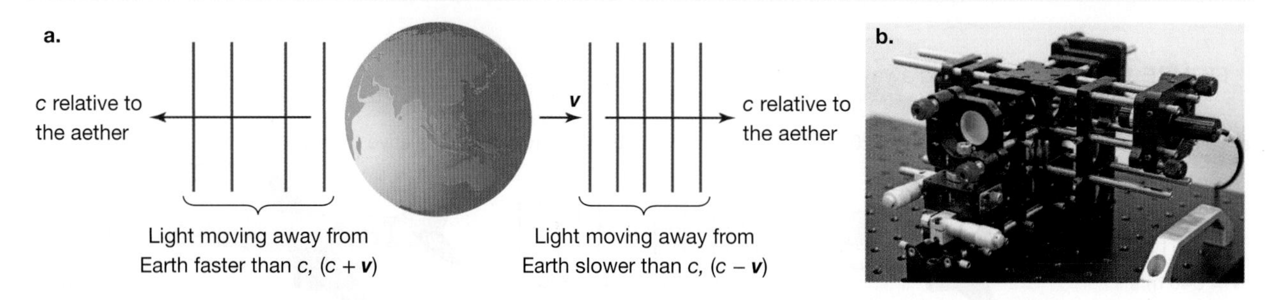

Null-result of the Michelson–Morley experiment

The Michelson–Morley experiment was an experiment to detect the supposed medium (luminiferous aether), through which light propagates, by measuring the speed of light in perpendicular directions. The **null-result** of their experiment (no significant differences in the speed of light were detected) was the first strong evidence against the existence of the luminiferous aether.

The implications from this experiment are that light does not need a medium to propagate, and the speed of light is independent of the motion of the observer.

This is evidence in support of the special theory of relativity.

null-result experimental outcome not showing an expected effect

on Resources

Video eLesson Michelson–Morley experiment (eles-2561)

Developing Einstein's special theory of relativity

Understanding electromagnetic phenomena was the foundation for Einstein's special relativity. In particular, the physicists of the nineteenth century, such as Michael Faraday, knew that they could induce a current in a wire by moving a magnet near the wire. They also knew that if they moved a wire through a magnetic field, a current would be induced in the wire. They saw these as two separate phenomena.

Imagine this: two students are in different physics classes. Annabel has learned in her class that electrons moving in a magnetic field experience a force perpendicular to their direction of motion and in proportion to their speed. Her friend Nicky has learned in her class that a current is induced in a loop of wire when the magnetic flux through the wire changes. Are these two different phenomena? Because they have also learned about the principle of relativity, Annabel and Nicky have doubts. They get together after class to perform experiments. The force depends on the speed. Annabel holds a stationary loop of conducting wire. Nicky moves the north pole of a magnet towards the loop, and they notice that a current is present in the wire as she does this. Nicky says that this is consistent with what she has learned. The conclusion is that a current is induced by a changing magnetic field. Then Nicky holds the magnet still so that the magnetic field is not changing. Annabel moves her loop of wire towards Nicky's magnet. Annabel states that the result agrees with what she learned in class — that electrons and other charged particles experience a force when moving in a magnetic field.

FIGURE 11.14 An experiment in electromagnetism

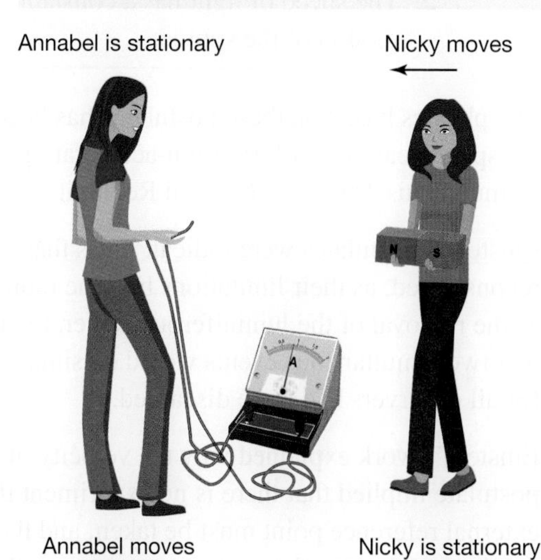

Einstein realised that there was only one phenomenon at work here. Both experiments are doing exactly the same thing, and it is only the relative speeds of the coil and the magnet that are important. This may seem obvious, but to make this jump it was necessary to discard the idea that the electric and magnetic fields depended on the luminiferous aether. It was the relative motion that was important, not whether the magnet or charge was moving through the aether.

11.2.4 Einstein's two postulates of special relativity

Physicists tried all sorts of experiments to detect the motion of Earth through the luminiferous aether, and they attempted to interpret the data in ways that would match the behaviour of light with what they expected would happen. Their attempts were unsuccessful.

Einstein managed to restore order to our understanding of the universe. While others suspected the new theory of electromagnetism to be wrong, Einstein took apart the established theory of Newtonian mechanics, even though its success had given physicists reason to believe in relativity in the first place. Einstein dared to see what would happen if he embraced the results of the Maxwell equations and the experiments with light, and accepted that the speed of light was invariant. The results were surprising and shocking, but this bold insight helped usher in the modern understanding of physics.

Einstein agreed with Galileo that the laws of physics must be the same for all observers, but he added a second requirement: that the speed of light in a vacuum is the same for all observers. The speed of light is invariant and does not differ in different reference frames. He set these two principles down as requirements for the development of theoretical physics. They are known as Einstein's two postulates of special relativity.

Einstein's postulates state that:

1. The laws of physics are the same in all inertial (non-accelerated) frames of reference.
2. The speed of light has a constant value for all observers regardless of their motion or the motion of the source.

The physics based on these postulates has become known as special relativity. It is 'special' because it deals with the special case of uniform (non-accelerating) motion. To deal with gravity and acceleration, Einstein went on to formulate his Theory of General Relativity, but that is beyond the scope of this course.

Einstein's postulates were radical. Ideas that had been taken for granted for centuries had to be completely reconsidered, as their limitations became more obvious with the development of Einstein's postulates. As well as the removal of the luminiferous aether, the intuitive notions that time passed at the same rate for everyone, that two simultaneous events would be simultaneous for all observers, and that distance and mass are the same for all observers had to be discarded.

Einstein's work explained why the velocity of Earth with respect to the aether could not be detected. His first postulate implied that there is no experiment that can be done on Earth to measure the speed of Earth. An external reference point must be taken, and the speed of Earth must be measured relative to that point in order for the speed of Earth to have any meaning. With his second postulate, Einstein also declared that it does not matter in which direction the Michelson–Morley apparatus was pointing in; the light would still travel at the same speed. No change in the interference pattern should be detected when the apparatus was rotated.

tlvd-9030

SAMPLE PROBLEM 4 Comparing Einstein's postulates to earlier physics theories

How do Einstein's postulates differ from the physics that preceded them?

THINK	WRITE
1. Consider Einstein's first postulate, in which the laws of physics are the same in all inertial frames of reference.	Firstly, the principle of relativity is applied to all laws of physics, not just the mechanics of Galileo and Newton.
2. Consider Einstein's second postulate, in which the speed of light has a constant value for all observers regardless of their motion.	Secondly, the speed of light is constant for all observers. Before Einstein, the speed of light was assumed to be relative to its medium, the luminiferous aether.

PRACTICE PROBLEM 4

Einstein realised that measurements that had been regarded as relative (or changing depending on the motion of the observer) were actually invariant. As a result of this, measurements that had been regarded as invariant now had to be regarded as relative. What did he find to be invariant and what did he find to be relative?

Broadening the horizons of understanding

Why did scientists before Einstein (and most people after Einstein) not notice the effects of light speed being invariant? Newton's laws provided a very good approximation for the world experienced by people before the twentieth century. By the beginning of the twentieth century, however, physicists were able to take measurements with incredible accuracy. They were also discovering new particles, such as electrons, that could travel at extremely high speeds. Indeed, these speeds were completely outside the realm of human experience. Light travels at $c = 3.0 \times 10^8$ m s^{-1} or 300 000 km s^{-1}. (To be precise, $c = 299\,792\,458$ m s^{-1}.) At this speed, light covers the distance from Earth to the Moon in roughly 1.3 seconds!

Note: When considering speeds at a significant fraction of the speed of light, it is easier to use the speed of light as the unit. For example, instead of 1.5×10^8 m s^{-1}, a physicist can simply write $0.5c$.

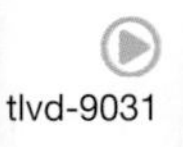
tlvd-9031

SAMPLE PROBLEM 5 Calculating the time taken for an object to accelerate to one-tenth of the speed of light

To get a sense of how fast light travels, Andrei considers how long it would take to accelerate from rest to a tenth of this great speed at the familiar rate of 9.8 m s^{-2} — the acceleration of an object in free fall near the surface of Earth. Calculate how long it would take to reach this speed.

THINK	WRITE
1. List the known information.	$u = 0$ m s^{-1}, $v = 0.1c = 3.0 \times 10^7$ m s^{-1}, $a = 9.8$ m s^{-2}, $t = ?$
2. As acceleration is constant, use the relationship $v = u + at$ to determine the time taken for an object to accelerate from rest to a speed that is a tenth of the speed of light.	$v = u + at$ $t = \frac{v - u}{a}$ $= \frac{3.0 \times 10^7 - 0}{9.8}$ $= 3.06 \times 10^6$ seconds $= 35.4$ days It would take more than 35 days to achieve a speed of $0.1c$! (This is the fastest speed for which use of Newtonian kinematics still gives a reasonable approximation.)

PRACTICE PROBLEM 5

With an acceleration of 9.8 m s^{-2}, occupants of a spacecraft in deep space would reassuringly feel the same force due to gravity that they feel on Earth. What would happen to the astronauts if the acceleration of the spacecraft was much greater, to enable faster space travel?

Light speed really is beyond our normal experience! Maybe Einstein's predictions would not be so surprising if people had more direct experience with objects travelling at great speeds but, as it is, they seem very strange.

AS A MATTER OF FACT

The distance light travels in a year is known as a light-year. Even on Earth, distance is now measured in terms of the speed of light. One metre is defined as the distance light travels in exactly $\frac{1}{c} = \frac{1}{299\,792\,458}$ of a second.

The speed of light is constant

This simple statement of Einstein's second postulate may not seem remarkable. To highlight what it means, light will again be compared with sound. In the nineteenth century, sound and light were thought to have a lot in common, because they both exhibited similar wavelike behaviours, such as diffraction and interference. However, sound is a disturbance of a medium, whereas light does not require any medium at all. Sound has a speed that is relative to its medium. If the source of the sound is moving through the medium, then the speed of the sound relative to the source is different to the speed of sound relative to the medium. Its speed can be different again for an observer depending on the reference frame.

Einstein was saying there is no medium for light, so the concept of the speed of light relative to its medium is not meaningful. Light always moves away from its source at 299 792 458 m s^{-1} and always meets its observer at 299 792 458 m s^{-1}, no matter what the relative speeds of the observer and the source. Even if Earth were hurtling along its orbit at $0.9c$, the result of the Michelson–Morley experiment would have been the same.

As an example, consider a spacecraft in the distant future hurtling towards Earth at $0.5c$. The astronaut sends out a radio message to alert Earth of his impending visit. (Radio waves, as part of the electromagnetic spectrum, have the same speed as visible light.) He notices that, in agreement with the Michelson–Morley measurements of centuries before, the radio waves move away from the spacecraft at speed c. With what speed do they hit Earth? Relative velocity, as treated by Galileo, insists that as the spacecraft already has a speed of $0.5c$ relative to Earth, then the radio waves must strike Earth at $1.5c$. However, this does not happen. The radio waves travel at speed c regardless of the motion of the source and the receiver.

This concept was very difficult for physicists to comprehend, and many resisted Einstein's ideas. But the evidence is irrefutable. Newtonian physics works as a very good approximation only for velocities much less than c. The faster something moves, the more obvious it is that the Newtonian world view does not match reality. It was not until the twentieth century that scientists investigated objects (such as cosmic rays) moving at great speeds. Many technologies and objects (such as satellites in orbit) need to be programmed according to Einsteinian physics rather than Newtonian classical physics.

FIGURE 11.15 A spacecraft approaching Earth at $0.5c$. The radio signal is travelling at c relative to both Earth and the spacecraft.

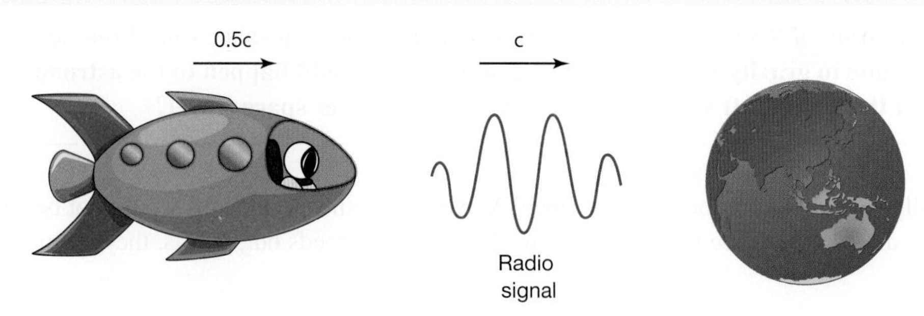

EXTENSION: Space–time diagrams

In 1908, Hermann Minkowski invented a useful method of depicting situations similar to the spacecraft scenario just described. His diagrams are like distance–time graphs with the axes switched around. However, they differ from time–distance graphs in an important way. When reading these diagrams, the markings on the scales for time and position are only correct for the reference frame in which the axes are stationary. Straight lines are drawn diagonally outwards form the source, indicating the constant motion of the light away from the source over time.

Consider this scenario. Light reflecting from planet A radiates in all directions at *c* so that, after one year, the light that left the planet forms a sphere one light-year in radius. Another planet, B, passes planet A at great speed, just missing it. Light from planet B's surface also leaves at *c*, according to the second postulate, forming a sphere around it. However, from planet A's perspective, the light from planet B reaches a distance of one light-year behind planet B before it reaches a distance of one light-year ahead of planet B. Figure 11.16 shows examples of Minkowski diagrams for this scenario. Each diagram is drawn from planet A's reference frame. Diagram (a) shows the situation for planet A, and diagram (b) shows what is happening on planet B according to observers on planet A.

FIGURE 11.16 Minkowski diagrams from planet A's frame of reference

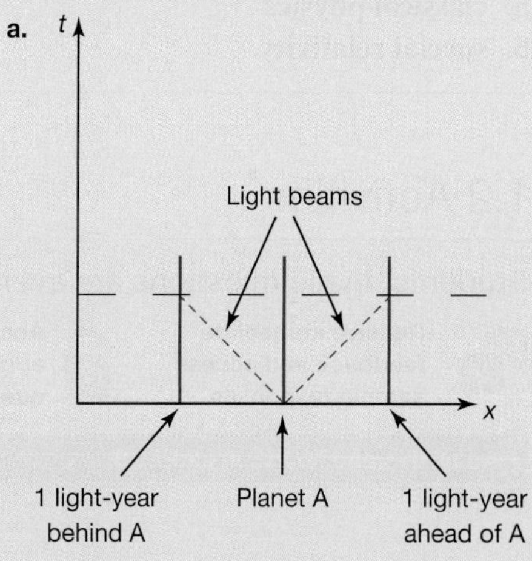

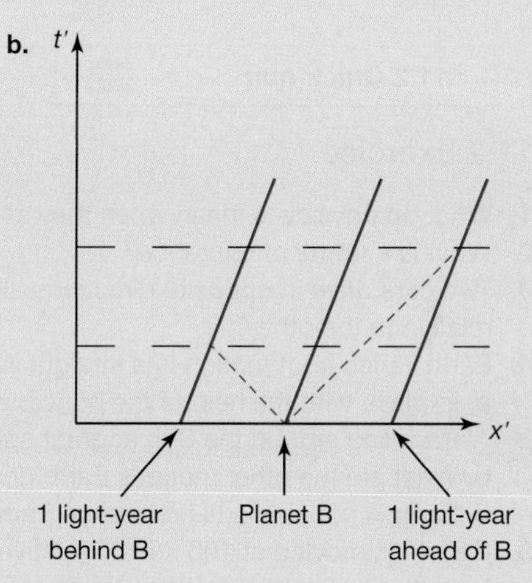

tlvd-9032

SAMPLE PROBLEM 6 Understanding how planets are in the centre of light circles

Using figure 11.16, explain how both planets A and B can be at the centre of their light circles as the postulates demand.

THINK

Use the diagrams to explain how both planets can be at the centre of their light circles as the postulates demand.

WRITE

Diagram (a) shows the situation for planet A. The light radiates in all directions at the same rate, and the diagram shows where the light in one direction and the opposite direction would be after one year.

Diagram (b) shows what is happening on planet B according to observers on planet A. The light moving out behind the moving planet reaches the one-light-year distance sooner than the light moving out from the front! But it is known that planet B is at the centre of this light circle. The way to achieve this is to move away from absolute space and time and understand that these are relative to the observer. When this is done, it is seen as possible for planet B to be at the centre of the light circle. However, this requires that A and B disagree about when two events occur. According to planet A, the different sides of the light circle reach the light-year radius at different times, but from planet B this must occur simultaneously.

PRACTICE PROBLEM 6

State whether the simultaneity of events is invariant or relative in:
a. **classical physics**
b. **special relativity.**

11.2 Activities

learnon

Students, these questions are even better in jacPLUS

Receive immediate feedback and access sample responses

Access additional questions

Track your results and progress

Find all this and MORE in jacPLUS

11.2 Quick quiz 	11.2 Exercise	11.2 Exam questions

11.2 Exercise

1. What do physicists mean when they say that velocity is relative?
2. What is a frame of reference?
3. Two cars drive in opposite directions along a suburban street at 50 km h^{-1}. What is the velocity of one car relative to the other?
4. Earth varies from motion in a straight line by less than 1° each day due to its motion around the Sun.
   a. Explain, with the help of the principle of relativity, why you do not feel Earth moving, even though it is travelling around the Sun at great speed.
   b. What are the other motions Earth undergoes that you cannot feel?
   c. Earth is not an inertial reference frame. Explain why it is often referred to as though it is.
5. If Earth is moving at 100 km s^{-1} relative to the supposed aether, what speed would Michelson have hypothesised for light emitted in the same direction that Earth is travelling?
6. a. Why did Newton's laws seem correct for so long?
   b. Why are Newton's laws still useful?
7. Why is Einstein's second postulate surprising? Give an example to show why Newtonian physicists would think it wrong.
8. A star emits light at speed c. A second star is hurtling towards it at speed $0.3c$. What is the speed of the light when it hits the second star relative to this second star?

11.2 Exam questions

Question 1 (1 mark)

Source: *VCE 2021, Physics Exam, Section A, Q.20; © VCAA*

MC One of Einstein's postulates for special relativity is that the laws of physics are the same in all inertial frames of reference.

Which one of the following best describes a property of an inertial frame of reference?

A. It is travelling at a constant speed.
B. It is travelling at a speed much slower than c.
C. Its movement is consistent with the expansion of the universe.
D. No observer in the frame can detect any acceleration of the frame.

Question 2 (3 marks)

Source: *Adapted from VCE 2019, Physics Exam, Section B, Q.11; © VCAA*

What is the second postulate of Einstein's special theory of relativity regarding the speed of light? Explain how the second postulate differs from the concept of the speed of light in classical physics.

Question 3 (2 marks)

Source: *VCE 2018, Physics Exam, Section B, Q.14; © VCAA*

Jani is stationary in a spaceship travelling at constant speed.

Does this mean that the spaceship must be in an inertial frame of reference? Justify your answer.

Question 4 (1 mark)

Source: *VCE 2017, Physics Exam, Section A, Q.10; © VCAA*

MC A student sits inside a windowless box that has been placed on a smooth-riding train carriage. He conducts a series of motion experiments to investigate frames of reference.

Which one of the following observations is correct?

A. The results when the train accelerates are identical to the results when the train is at rest.
B. The results when the train accelerates differ from the results when the train is in uniform motion in a straight line.
C. The results when the train is at rest differ from the results when the train is in uniform motion in a straight line.
D. The results when the train accelerates are identical to the results when the train is in uniform motion in a straight line.

Question 5 (1 mark)

Source: *VCE 2016, Physics Exam, Section B, Q.2; © VCAA*

MC When Anna is halfway between Earth and the space lab, she sends a radio pulse towards Earth and towards the space lab, as shown in Figure 2.

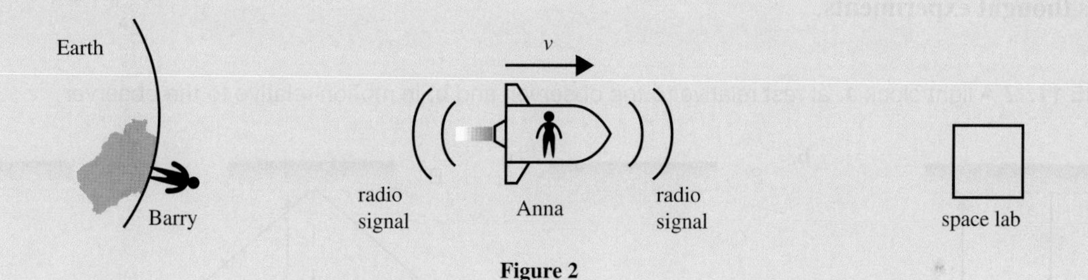

Figure 2

As observed by Anna, which one of the following statements correctly gives the order in which this signal is received by Barry and by the space lab?

A. Barry receives the signal first.
B. The space lab receives the signal first.
C. The signal is received by Barry and the space lab at the same time.
D. It is not possible to predict since special relativity applies to light but not to radio signals.

More exam questions are available in your learnON title.

11.3 Time dilation

KEY KNOWLEDGE

- Describe proper time (t_0) as the time interval between two events in a reference frame where the two events occur at the same point in space
- Model mathematically time dilation and length contraction at speeds approaching *c* using the equations: $t = t_0\gamma$ and $L = \frac{L_0}{\gamma}$ where $\gamma = \frac{1}{\sqrt{\left(1-\frac{v^2}{c^2}\right)}}$

11.3.1 Time is relative

The passing of time can be measured in many ways, including using the position of the Sun in the sky, the position of hands on a watch, the changing of the seasons and the signs of a person ageing. Galileo is known to have made use of the beat of his pulse, the swinging of a pendulum and the dripping of water. As already stated, Newtonian physics assumed that each of these clocks ticked at the same rate regardless of who was observing them. However, Einstein's special theory of relativity shows that this assumption — that time is absolute — is actually wrong. This error becomes apparent when the motion of the clock relative to the observer approaches the speed of light.

Consider a simple clock consisting of two mirrors, A and B, with light reflecting back and forth between them. This is an unusual clock, but it is very useful for illustrating how time is affected by relativity. Experiments that involve pursuing an idea on paper without actually performing them are common in explanations of relativity. They are known as **thought experiments**.

thought experiment an imaginary scenario designed to explore what the laws of physics predict would happen; also known as a gedanker experiment

FIGURE 11.17 A light clock **a.** at rest relative to the observer, and **b.** in motion relative to the observer

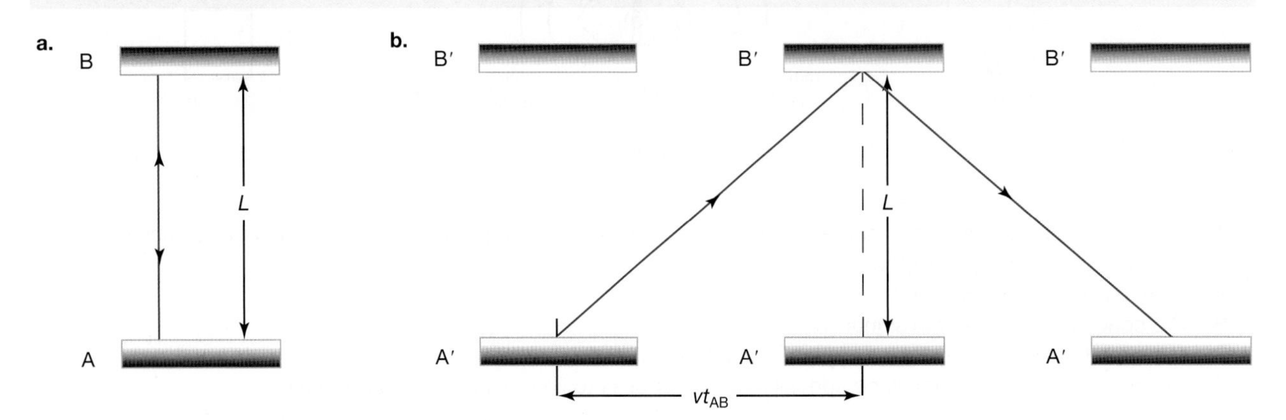

Let the separation of the mirrors be *L*. The time for the pulse of light to pass from mirror A to mirror B and back is calculated in the conventional way:

$$c = \frac{2L}{t_0}$$

$$t_0 = \frac{2L}{c}$$

where t_0 is the time for light to travel from A to B and back, as measured in the frame of reference in which the clock is at rest.

This time, t_0, will be defined as one tick of the clock. In this case, the position of the clock does not change in the frame of reference. The passing of time can be indicated by two events separated by time but not by space — the event of the photon of light first being at A and the event of the photon being back at A.

Imagine an identical clock, with mirrors A′ and B′, moving past this light clock at speed v. At what rate does time pass on this moving clock according to the observer? Label the time interval measured by this clock t to distinguish it from t_0. The light leaves A′ and moves towards B′ at speed c. The speed is still c even though the clock is moving, as stated by Einstein's second postulate. In the time the light makes this journey, the clock moves a distance $d = vt_{AB}$, where t_{AB} is the time the light takes to travel from A′ to B′. Figure 11.17b depicts this situation and shows that the light in the moving frame of reference has further to travel than the light in the rest frame. Using Pythagoras's theorem, the light has travelled a distance of $2\sqrt{L^2 + (vt_{AB})^2}$ from A′ to B′ and back to A′. This is a greater distance than $2L$, given $v \neq 0$ and c is constant. Therefore, the time the light takes to complete the tick must be greater than for the rest clock.

The speed of the light relative to the observer is:

$$c = \frac{d}{t}$$

$$c = \frac{2\sqrt{L^2 + (vt_{AB})^2}}{2t_{AB}}$$

Transpose the equation to make a formula for t:

$$2ct_{AB} = 2\sqrt{L^2 + v^2(t_{AB})^2}$$

$$c^2(t_{AB})^2 = L^2 + v^2(t_{AB})^2$$

But $t_{AB} = \frac{t}{2}$:

$$\frac{c^2t^2}{4} - \frac{v^2t^2}{4} = L^2$$

$$t^2\left(c^2 - v^2\right) = 4L^2$$

$$t = \frac{2L}{\sqrt{c^2 - v^2}}$$

$$= \frac{2L}{c\sqrt{1 - \frac{v^2}{c^2}}}$$

It has already been determined that $t_0 = \frac{2L}{c}$, so

$$t = \frac{t_0}{\sqrt{1 - \frac{v^2}{c^2}}}$$

The expression $\frac{1}{\sqrt{1 - \frac{v^2}{c^2}}}$ appears frequently in special relativity. It is known as the Lorentz factor and is denoted by gamma, γ.

The equation can now be written as:

$$t = t_0\gamma$$

where: $\gamma = \frac{1}{\sqrt{1-\frac{v^2}{c^2}}}$ is the Lorentz factor

v is the speed of the moving frame of reference, in m s^{-1}

c is the speed of light in a vacuum, in m s^{-1}

t is the time measured in the reference frame of the observer, in seconds

t_0 is the time measured in the reference frame of the observed object, or the reference frame where two events occur at the same point in space (proper time), in seconds

The equation $t = t_0\gamma$ is known as the **time dilation** formula. This formula enables the determination of the time interval between two events in a reference frame moving relative to an observer.

Note that gamma is always greater than 1. As a result, t will always be greater than t_0, hence the term 'time dilation'. In a reference frame moving relative to the observer like this, the two events that are being used to mark the time interval (the time between the light being at A) occur at different points in space. The time t_0 is the time measured in a frame of reference where the events occur at the same points in space. It is known as the **proper time**. This is not proper in the sense of correct, but in the sense of property. It is the time in the clock's own reference frame, whatever that clock might be.

Examples:

1. A mechanical clock's large hand moves from the 12 to the 3, showing that 15 minutes have passed. Fifteen minutes is the proper time between the two events of the clock showing the hour and the clock showing quarter past the hour. However, if that clock was moving at great speed relative to an observer, the observer would notice that the time between these two events was longer than 15 minutes. The time is dilated.
2. A candle burns 2 centimetres in 1 hour. One hour is the proper time between the events of the candle being at a particular length and the candle being 2 centimetres shorter. If the candle was moving relative to an observer, the observer would notice that it took longer than 1 hour for the candle to burn down 2 centimetres.
3. A man dies at 89 years of age. His life of 89 years is the time between the events of his birth and his death in his reference frame. To an observer moving past at great speed, the man appears to live longer than 89 years. He does not fit any more into his life; everything he does appears to the observer as if it was slowed down.

time dilation the slowing of time by clocks moving relative to the observer

proper time between two events, the time measured in a frame of reference where the events occur at the same point in space. The proper time of a clock is the time th clock measures in its own referenc frame.

tlvd-9033

SAMPLE PROBLEM 7 Applications of time dilation 1

James observes a clock held by his friend Mabry moving past at 0.5*c*. He notices the hands change from 12 pm to 12.05 pm, indicating that 5 minutes have passed for the clock. How much time has passed for James?

THINK	WRITE
1. Determine the proper time of the clock.	The proper time t_0 is the time interval between the two events: when the clock shows 12 pm and when the clock shows 12.05 pm. The difference is 5 minutes.

2. Calculate the Lorentz factor using $\gamma = \dfrac{1}{\sqrt{1-\dfrac{v^2}{c^2}}}$, where $v = 0.5c$.	$\gamma = \dfrac{1}{\sqrt{1-\dfrac{v^2}{c^2}}}$ $= \dfrac{1}{\sqrt{1-\dfrac{(0.5c)^2}{c^2}}}$ $= \dfrac{1}{\sqrt{1-\dfrac{0.25c^2}{c^2}}}$ $= \dfrac{1}{\sqrt{1-0.25}}$ $= 1.155$
3. Calculate the time taken for the hands to change from 12 pm to 12.05 pm in James's frame of reference, using $t = t_0\gamma$. Ensure you use the non-rounded value for the Lorentz factor in your calculation.	$t = t_0\gamma$ $= 5 \text{ minutes} \times 1.155$ $= 5.774 \text{ minutes}$ James notices that the moving clock takes 5.774 minutes (or 5 minutes, 46.4 seconds) for its hands to move from 12 pm to 12.05 pm.

PRACTICE PROBLEM 7

In another measurement, James looks at his own watch and waits the 5 minutes it takes for the watch to change from 1 pm to 1.05 pm. He then looks at Mabry's clock as she moves past at 0.5*c*. How much time has passed on her clock?

Unlike in Newtonian physics, time intervals in special relativity are not invariant. Rather, they are relative to the observer.

tlvd-9034

SAMPLE PROBLEM 8 Applications of time dilation 2

Mabry is travelling past James at 0.5*c*. She looks at James and sees his watch ticking. How long does she observe it to take for his watch to indicate the passing of 5 minutes?

THINK	WRITE
In this case it is James's watch that is showing the proper time. The situation in sample problem 7 and this situation are symmetrical. Mabry sees James as moving at 0.5*c*, and James sees Mabry moving at 0.5*c*, so Mabry's measurement of time passing on James's watch is the same as James's measurement of time passing on Mabry's watch.	Mabry notices that 5.774 minutes pass when James's watch shows 5 minutes passing.

PRACTICE PROBLEM 8

Aixi listens to a 3-minute song on her phone. As soon as she starts the song, she sees her friend Xiaobo start wrestling with his brother on a spaceship moving by at 0.8*c*. When the song finishes, she sees Xiaobo stop wrestling. How long were the two boys wrestling?

tlvd-9035

SAMPLE PROBLEM 9 Comparing the rates of clocks using the time dilation formula

A car passes Eleanor at 20 m s^{-1}. Compare the rate that a clock in the car ticks with the rate the clock in her hand ticks.

THINK	WRITE
1. Determine the Lorentz factor $\gamma = \frac{1}{\sqrt{1-\frac{v^2}{c^2}}}$ when $v = 20$ m s^{-1}.	$\gamma = \frac{1}{\sqrt{1-\frac{v^2}{c^2}}}$ $= \frac{1}{\sqrt{1-\frac{20^2}{(3\times10^8)^2}}}$ $= 1.000\,000\,000\,000\,0022$
2. Calculate the time that passes on the clock in the car in Eleanor's frame of reference using $t = t_0\gamma$.	$t = t_0\gamma$ $= 1.000\,000\,000\,000\,0022t_0$ The difference between the rates of time in the two perspectives is so small that it is difficult to calculate, much less notice it.

PRACTICE PROBLEM 9

Jonathan observes a clock on a passing spaceship to be ticking at half the rate of his identical clock. What is the relative speed of Jonathan and the passing spaceship?

Newton's assumption that all clocks tick at the same rate, regardless of their inertial reference frame, was very reasonable. Learning the very good approximation of Newton's laws is well justified. They are simpler than Einstein's laws, and they work for all but the highest speeds. A good theory in science has to fit the facts, and Newton's physics fit the data very successfully for 200 years. It was a great theory, but Einstein's is even better.

If Newton knew then what is known now, he would realise that his theories were in trouble. At speeds humans normally experience, time dilation is negligible, but the dilation increases dramatically as objects approach the speed of light. If you passed a planet at 2.9×10^8 m s^{-1}, you would measure the aliens' usual school lessons of 50 minutes as taking 195 minutes. An increase in speed to 2.99×10^8 m s^{-1} would dilate the period to 613 minutes. If you could achieve the speed of light, the period would last forever — time would stop.

Photons do not age, as they do not experience time passing!

on Resources

Weblink Time dilation applet

11.3 Activities

learnon

Students, these questions are even better in jacPLUS

Receive immediate feedback and access sample responses

Access additional questions

Track your results and progress

Find all this and MORE in jacPLUS

11.3 Quick quiz on	11.3 Exercise	11.3 Exam questions

11.3 Exercise

1. What is time dilation? In your explanation, give an example of where time dilation would occur.
2. Which clock runs slow: yours or one in motion relative to you?
3. Draw diagrams of a light clock in motion and at rest to explain why time dilation occurs for moving clocks.
4. Explain the difference between t_0 and t in the time dilation formula.
5. Two spacecraft pass each other with a relative speed of 0.3*c*.
   a. Calculate γ.
   b. A drummer pounds a drum at 100 beats per minute on one of the spacecraft. How many beats per minute would those on the other spacecraft measure as a result of time dilation?

11.3 Exam questions

Question 1 (1 mark)

Source: *VCE 2020 Physics Exam, Section A, Q.12; © VCAA*

MC A high-energy proton is travelling through space at a constant velocity of 2.50×10^8 m s^{-1}.

The Lorentz factor, γ, for this proton would be closest to

A. 1.81
B. 2.44
C. 3.27
D. 3.39

Question 2 (4 marks)

Source: *VCE 2020 Physics Exam, Section B, Q.11; © VCAA*

An astronaut has left Earth and is travelling on a spaceship at 0.800*c* ($\gamma = 1.67$) directly towards the star known as Sirius, which is located 8.61 light-years away from Earth, as measured by observers on Earth.

a. How long will the trip take according to a clock that the astronaut is carrying on his spaceship? Show your working. **(2 marks)**

b. Is the trip time measured by the astronaut in part **a** a proper time? Explain your reasoning. **(2 marks)**

Question 3 (1 mark)

Source: *VCE 2018, Physics Exam, Section A, Q.13; © VCAA*

MC Which one of the following diagrams best represents the graph of γ (the Lorentz factor) versus speed for an electron that is accelerated from rest to near the speed of light, c?

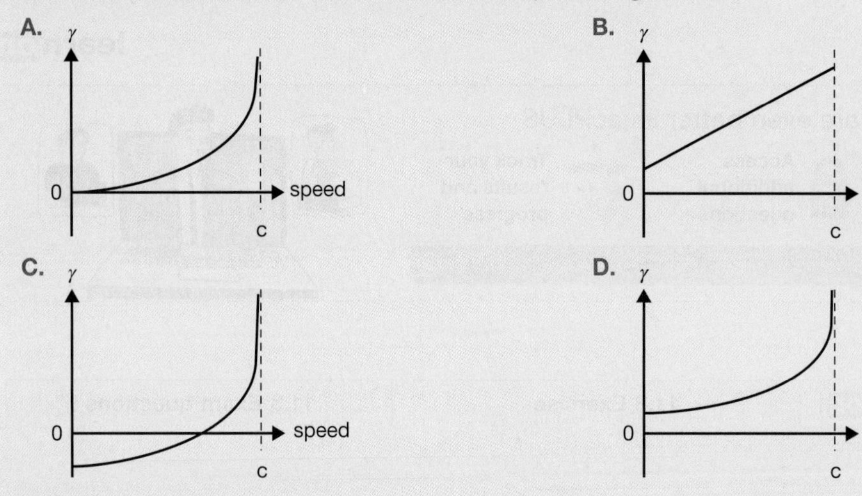

Question 4 (2 marks)

Source: *VCE 2017, Physics Exam, Section B, Q.11.a; © VCAA*

Tests of relativistic time dilation have been made by observing the decay of short-lived particles. A muon, travelling from the edge of the atmosphere to the surface of Earth, is an example of such a particle.

To model this in the laboratory, another elementary particle with a shorter half-life is produced in a particle accelerator. It is travelling at 0.99875c $(\gamma = 20)$. Scientists observe that this particle travels 9.14×10^{-5} m in a straight line from the point where it is made to the point where it decays into other particles. It is not accelerating.

Calculate the lifetime of the particle in the scientists' frame of reference.

Question 5 (1 mark)

Source: *VCE 2016, Physics Exam, Section B, Q.1; © VCAA*

MC Anna and Barry have identical quartz clocks that use the precise period of vibration of quartz crystals to determine time. Barry and his clock are on Earth. Anna accompanies her clock on a rocket travelling at constant high velocity, ***v***, past Earth and towards a space lab (which is stationary relative to Earth), as shown in Figure 1.

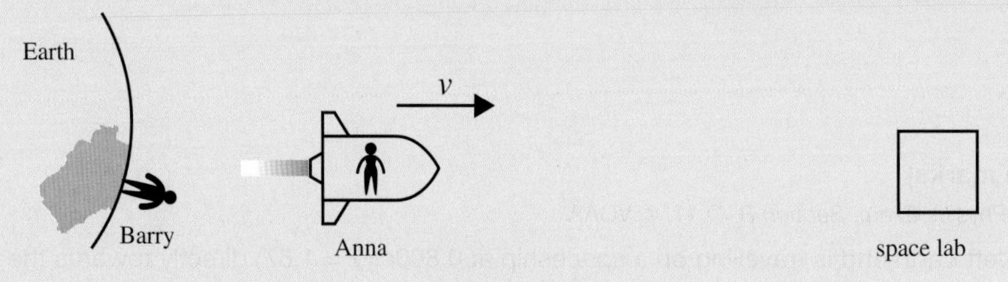

Figure 1

Which one of the following statements correctly describes the behaviour of these two clocks?

A. The period of vibration in Anna's clock (as observed by Anna) will be shorter than the period of vibration in Barry's clock (as observed by Barry).

B. The period of vibration in Anna's clock (as observed by Anna) will be longer than the period of vibration in Barry's clock (as observed by Barry).

C. The period of vibration in Anna's clock (as observed by Anna) will be the same as the period of vibration in Barry's clock (as observed by Barry).

D. Only the time on Barry's clock is reliable because it is in a frame that is not moving.

More exam questions are available in your learnON title.

11.4 Length contraction

KEY KNOWLEDGE

- Describe proper length (L_0) as the length that is measured in the frame of reference in which objects are at rest
- Model mathematically time dilation and length contraction at speeds approaching c using the equations: $t = t_0\gamma$ and $L = \frac{L_0}{\gamma}$ where $\gamma = \frac{1}{\sqrt{\left(1 - \frac{v^2}{c^2}\right)}}$

11.4.1 Length is also relative

Once it is accepted that simultaneity of events and the rate that time passes are dependent on the speed of the observer, it has to be accepted that length must be relative as well. The length of an object is simply the distance between the two ends of the object. To find that distance, the positions of both ends must be noted at the same time. If they were measured at different times, a moving object would have changed position, so the distance between the end that was measured second and the end that was measured first would have changed. The fact that any two inertial reference frames do not agree on which events are simultaneous is going to cause the measurement of length to be different in different reference frames. The speed of light is invariant and time is relative, so there is even more reason to doubt that lengths will be the same for all observers.

A clever thought experiment of Einstein's enables the determination of the effect that the speed of an observer has on a length to be measured. It is essentially the same as the thought experiment used to derive the time dilation equation but with the light clock tipped on its side so that its length is aligned with the direction of its motion.

From the reference frame of the clock, again $t_0 = \frac{2L}{c}$. What about the reference frame of an observer with a speed of v relative to the clock? The distance between the ends of the clock can be measured using the time for light to travel from one end to the other and back, as seen in figure 11.18.

FIGURE 11.18 Light journeys in **a.** a clock at rest and **b.** a clock moving to the right at speed v

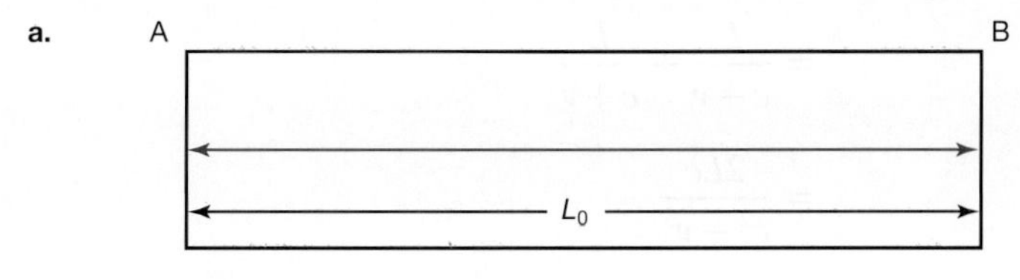

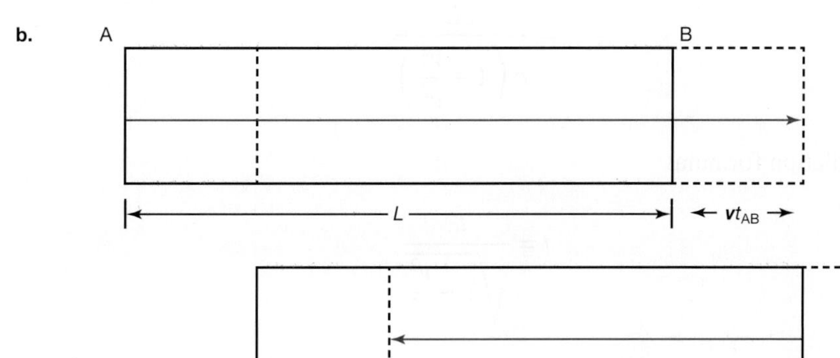

Looking at figure 11.18, the following formulae can be derived.

From A to B:

$$L + vt_{\text{AB}} = ct_{\text{AB}}$$

where:

L = the length of the clock as observed by the moving observer

vt_{AB} = the distance the clock has moved in the time the light passes from A to B

ct_{AB} = the distance the light has travelled passing from A to B.

Transposing the equation to make t_{AB} the subject:

$$t_{\text{AB}} = \frac{L}{c - v}$$

From B to A:

$$L - vt_{\text{BA}} = ct_{\text{BA}}$$

where:

vt_{BA} = the distance the clock has moved in the time the light passes from B back to A

ct_{BA} = the distance the light has travelled passing from B back to A.

Transposing the equation to make t_{BA} the subject:

$$t_{\text{BA}} = \frac{L}{c + v}$$

As A moves to meet the light, the time t_{BA} is less than t_{AB}. The total time is:

$$\begin{aligned} t &= t_{\text{AB}} + t_{\text{BA}} \\ &= \frac{L}{c - v} + \frac{L}{c + v} \\ &= \frac{2Lc}{c^2 - v^2} \\ &= \frac{2L}{c\left(1 - \frac{v^2}{c^2}\right)} \end{aligned}$$

According to the time dilation formula:

$$t = \frac{t_0}{\sqrt{1 - \frac{v^2}{c^2}}}$$

Substituting this for the time in the moving clock gives:

$$\frac{t_0}{\sqrt{1-\frac{v^2}{c^2}}} = \frac{2L}{c\left(1-\frac{v^2}{c^2}\right)}$$

$$t_0 = \frac{2L}{c\sqrt{1-\frac{v^2}{c^2}}}$$

Substituting $t_0 = \frac{2L_0}{c}$ gives:

$$\frac{2L_0}{c} = \frac{2L}{c\sqrt{1-\frac{v^2}{c^2}}}$$

$$\Rightarrow L_0 = \frac{L}{\sqrt{1-\frac{v^2}{c^2}}} \text{ or } L = L_0\sqrt{1-\frac{v^2}{c^2}}$$

The formula $L = L_0\sqrt{1-\frac{v^2}{c^2}}$ is known as the length contraction formula or the Lorentz contraction formula after one of the early pioneers of relativity theory, Hendrik Antoon Lorentz (1853–1928). Length contraction is the shortening of an object in its direction of motion when measured from a reference frame in motion relative to the object.

The **proper length** of an object, L_0, is the length measured in the frame of reference where the object is at rest. L is the length as measured from an inertial reference frame travelling at a velocity v relative to the object. This change in length applies only to the length along the direction of motion. The other dimensions are not affected by this contraction.

proper length the length measured in the rest frame of the object

The proper length is contracted by a factor of $\frac{1}{\gamma}$:

$$L = \frac{L_0}{\gamma}$$

where: $\gamma = \frac{1}{\sqrt{1-\frac{v^2}{c^2}}}$ is the Lorentz factor

v is the speed of the moving frame of reference, in m s^{-1}

c is the speed of light in a vacuum, in m s^{-1}

L is the length measured in the reference frame of the observer, in m

L_0 is the length measured in the reference frame of the observed object, or the reference frame in which objects are at rest (proper length), in m

The Lorentz contraction is negligible at velocities that are commonly experienced. Even at a relative speed of 10% the speed of light, the contraction is less than 1%. As speed increases beyond $0.1c$, however, the contraction increases until at relative speed c, the length becomes zero.

FIGURE 11.19 A spaceship travelling at high speed has its length contracted (in the direction of motion)

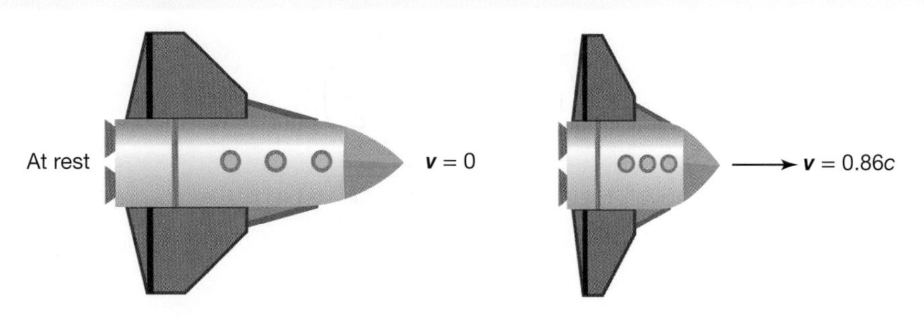

As a matter of fact

George Fitzgerald and Hendrik Lorentz independently proposed an explanation for the result of the Michelson–Morley experiment (in 1889 and 1892 respectively). If the length of the apparatus contracted in the direction of Earth's movement, then the light would take the same time to travel the two paths. This explanation assumed that the aether existed and that light would travel at constant speed through it; therefore, light would travel at different speeds relative to Earth as Earth moved through the aether. This explanation was not completely satisfying as there was no known force that would cause the contraction, and the aether had never been directly detected. The contraction would be measured by those in the reference frame at rest with respect to the aether.

In special relativity, any observer in motion relative to an object measures a contraction. As the contraction is simply a feature of observation from different reference frames, no force is required to cause the contraction. Nothing actually happens to the object in its reference frame.

tlvd-9036

SAMPLE PROBLEM 10 Application of length contraction

Observers on Earth see that the length of a spacecraft travelling at 0.5*c* has contracted. By what percentage of its proper length is the spacecraft contracted according to the observers?

THINK	WRITE
1. Determine the Lorentz factor, $\gamma = \dfrac{1}{\sqrt{1-\dfrac{v^2}{c^2}}}$, for $v = 0.5c$.	$\gamma = \dfrac{1}{\sqrt{1-\dfrac{v^2}{c^2}}} = \dfrac{1}{\sqrt{1-\dfrac{(0.5c)^2}{c^2}}} = \dfrac{1}{\sqrt{1-\dfrac{0.25c^2}{c^2}}} = \dfrac{1}{\sqrt{1-0.25}} = 1.155$
2. Using the Lorentz contraction formula, determine the ratio $\dfrac{L}{L_0}$.	$L = \dfrac{L_0}{\gamma}$, $\dfrac{L}{L_0} = \dfrac{1}{\gamma} = \dfrac{1}{1.155} = 0.866$

3. Interpret the ratio $\frac{L}{L_0}$ as a percentage.	The spacecraft appears to be only 0.866 or 86.6% of its proper length. This is a contraction of 13.4%.

PRACTICE PROBLEM 10

Rebecca and Madeline take measurements of the journey from Melbourne to Sydney. Rebecca stays in Melbourne and stretches a hypothetical tape measure between the two cities. Madeline travels towards Sydney at great speed and measures the distance with her own measuring tape that is in her own reference frame.

a. **How would the two measurements compare, assuming that perfect precision could be achieved?**
b. **Which measurement could be considered to be the proper length of the journey? Explain.**

EXTENSION: The twins paradox

A paradox is a seemingly absurd or contradictory statement. Relativity provides a few paradoxes that are useful in teaching the implications of relativity. The 'twins paradox' is probably the best known. Despite its name, the twins paradox is explained fully by the logic of relativity.

Imagine a spacecraft that starts its journey from Earth. After three years in Earth time, it will turn around and come back, so that those on Earth measure the total time between the events of the launch and the return to take six years. The astronaut, Peter, leaves his twin brother, Mark, on Earth. During this time, Peter and Mark agree that Earth has not moved from its path through space; it is Peter in his spaceship who has gone on a journey and has experienced the effects of acceleration that Mark has not. Mark measures the length of Peter's journey from Earth. His measurement is longer than Peter's due to length contraction, but the speed of Peter is measured relative to Earth. They disagree on distance travelled but not speed, so they must disagree on time taken. This is not just an intellectual dispute — the difference in time will show in their ageing, with Peter actually being younger than Mark on his return to Earth.

People all go on a journey into the future; time cannot be stopped. Relativity shows that the rate at which time progresses depends on the movements made through space on the journey. Coasting along in an inertial reference frame is the longest path to take. Zipping through different reference frames then returning home enables objects to reach the future in a shorter time: they take a longer journey through space but a shorter journey through time.

The twins scenario may sound incredible, but it has been verified experimentally. The most accurate clocks ever built are atomic clocks. They make use of the oscillations of the atoms of particular elements. The period of this oscillation is unaffected even by quite extreme temperatures and accelerations, making the clocks without rival in terms of accuracy. These clocks have been flown around the world on airliners, recording less elapsed time than for similar clocks that remained on the ground. The effect is tiny, but the clocks have more than adequate precision to detect the difference. The difference measured is consistent with the time difference predicted by special relativity.

EXTENSION: The parking spot paradox

Can a long car enter a parking spot that is too short for it by making use of length contraction? The answer is yes and no. To explain, consider another famous paradox of relativity.

Charlotte's car is 8-m long and she proudly drives it at a speed of 0.8*c*. She observes her friend Alexandra, who is stationary on the roadside, and asks her to measure the length of her car. (For the sake of argument, the issues of how a car could travel at such a huge speed, and how Alexandra communicates with Charlotte and measures the car will be ignored.)

Alexandra says that Charlotte must be dreaming if she thinks her car is 8-m long, because she measures it to be only 4.8-m long. She believes her measurements are accurate.

To prove her point, Alexandra marks out a parking spot 4.8-m long. She says that if Charlotte can park her car in the spot, then the car is not as long as she thinks. Charlotte argues that her car will not fit into a parking spot 4.8-m long, but she agrees to the test.

From Charlotte's frame of reference, the parking spot would be merely 2.9-m long. This is because it has a length contraction due to the car's relative motion of 0.8*c*. Alexandra's measuring equipment detects that the front of the car reaches the front of the parking spot at the same instant as the back of the car arrives at the back. However, much to Alexandra's amazement, the stopped car is 8-m long. Charlotte and Alexandra now agree that the stopped car does not fit the 4.8-m parking spot, and that it has a length of 8 m. This may at first seem impossible, which is why it is sometimes called a paradox. Once you consider that Charlotte and Alexandra do not agree on which events are simultaneous, the paradox is resolved. Alexandra measured the front and the back of the car to be within the parking spot at the same time but did not check that the front and back had stopped.

FIGURE 11.20 The parking spot paradox, where **a.** shows Alexandra's view and **b.** shows the view from Charlotte's frame of reference

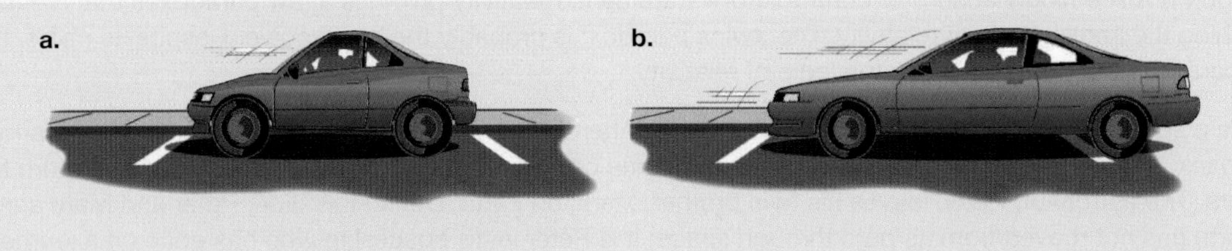

11.4 Activities

learnon

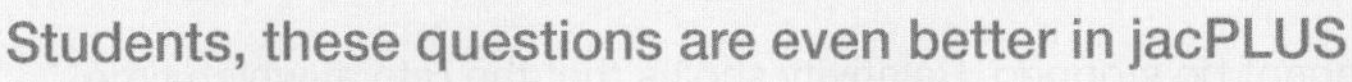

Receive immediate feedback and access sample responses

Access additional questions

Track your results and progress

Find all this and MORE in jacPLUS

11.4 Quick quiz  | **11.4 Exercise** | **11.4 Exam questions**

11.4 Exercise

1. If a box was moving away from you at nearly light speed, which dimensions of the box would undergo length contraction from your perspective: width, height or depth?
2. An alien spacecraft speeds through the solar system at 0.8*c*.
   a. What is the effect of its speed on the length of the spacecraft from the perspective of an alien on board?
   b. What is the effect of its speed on the length of the spacecraft from the perspective of the Sun?
   c. At what speed does light from the Sun reach it?
3. A high-energy physicist detects a particle in a particle accelerator that has a half-life of 20 s when travelling at 0.99*c*.
   a. Calculate the particle's half-life in its rest frame.
   b. The detector is 5-m long. How long would it be in the rest frame of the particle?

4. An astronaut on a spacewalk sees a spacecraft passing at 0.9c. The spacecraft has a proper length of 100 m. What is the length of the spacecraft L due to length contraction according to the astronaut?
5. A rocket of length 12.0 m passes an observer on the Moon. The observer measures the passing rocket to be 8.0-m long. Calculate the velocity of the rocket, in terms of c, in the reference frame of the Moon-based observer.

11.4 Exam questions

Question 1 (4 marks)

***Source:** VCE 2021, Physics Exam, Section B, Q.10; © VCAA*

A new spaceship that can travel at 0.7c has been constructed on Earth. A technician is observing the spaceship travelling past in space at 0.7c, as shown in Figure 10. The technician notices that the length of the spaceship does not match the measurement taken when the spaceship was stationary in a laboratory, but its width matches the measurement taken in the laboratory.

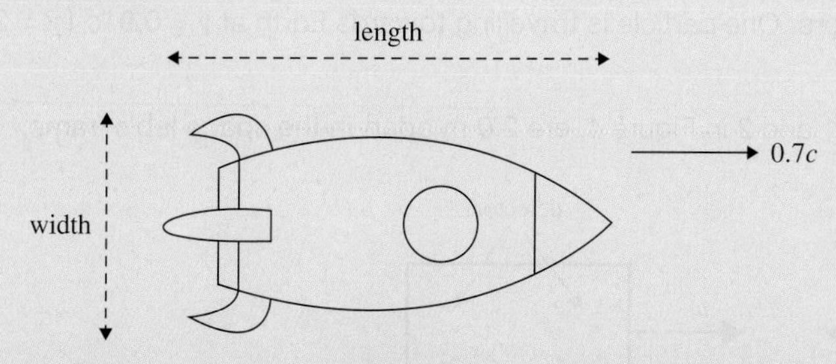

Figure 10

a. Explain, in terms of special relativity, why the technician notices there is a different measurement for the length of the spaceship, but not for the width of the spaceship. **(2 marks)**
b. If the technician measures the spaceship to be 135-m long while travelling at a constant 0.7c, what was the length of the spaceship when it was stationary on Earth? Show your working. **(2 marks)**

Question 2 (1 mark)

***Source:** VCE 2019, Physics Exam, Section A, Q.13; © VCAA*

MC Joanna is an observer in Spaceship A, watching Spaceship B fly past at a relative speed of 0.943c (γ = 3.00). She measures the length of Spaceship B from her frame of reference to be 150 m.

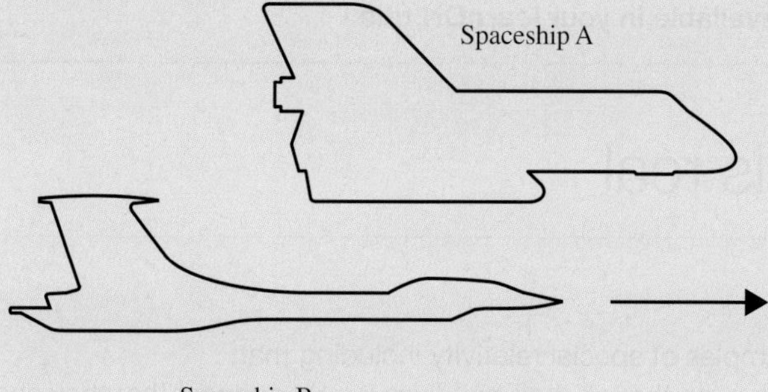

Which one of the following is closest to the proper length of Spaceship B?

A. 50 m
B. 150 m
C. 450 m
D. 900 m

Question 3 (2 marks)

***Source:** VCE 2017, Physics Exam, Section B, Q.10; © VCAA*

The length of a spaceship is measured to be exactly one-third of its rest length as it passes by an observing station. What is the speed of this spaceship, as determined by the observing station, expressed as a multiple of c?

Question 4 (2 marks)

Source: *VCE 2017, Physics Exam, Section B, Q.11.b; © VCAA*

Tests of relativistic time dilation have been made by observing the decay of short-lived particles. A muon, travelling from the edge of the atmosphere to the surface of Earth, is an example of such a particle.

To model this in the laboratory, another elementary particle with a shorter half-life is produced in a particle accelerator. It is travelling at 0.99875*c* $(\gamma = 20)$. Scientists observe that this particle travels 9.14×10^{-5} m in a straight line from the point where it is made to the point where it decays into other particles. It is not accelerating.

Calculate the distance that the particle travels in the laboratory, as measured in the particle's frame of reference.

Question 5 (1 mark)

Source: *VCE 2016, Physics Exam, Section B, Q.7; © VCAA*

MC A space lab travelling at $u = 0.8c$ $(\gamma = 1.67)$ away from Earth can record high-energy charged particles passing through its detectors. One particle is travelling towards Earth at $v = 0.91c$ $(\gamma = 2.4)$ **relative to the space lab.**

Two detectors, numbered 1 and 2 in Figure 4, are 2.0 m apart in the space lab's frame.

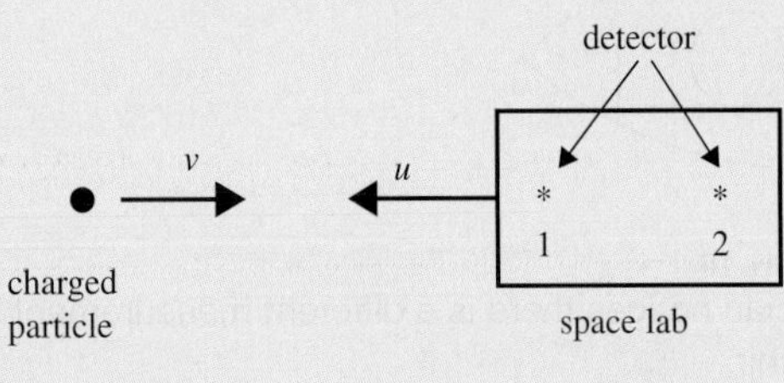

Figure 4

How far apart are the two detectors in this particular particle's frame?

A. 0.83 m
B. 1.2 m
C. 3.3 m
D. 4.8 m

More exam questions are available in your learnON title.

11.5 Relativity is real

KEY KNOWLEDGE

- Explain and analyse examples of special relativity including that:
  - muons can reach Earth even though their half-lives would suggest that they should decay in the upper atmosphere
  - particle accelerator lengths must be designed to take the effects of special relativity into account
  - time signals from GPS satellites must be corrected for the effects of special relativity due to their orbital velocity

Source: VCE Physics Study Design (2024–2027) extracts © VCAA; reproduced by permission.

In this topic, the term *observer* is frequently used. Much of the imagery used in teaching relativity is in principle true but in practicality it is fantasy. Observing in detail anything that is moving at close to the speed of light is not feasible. However, measuring distances and times associated with these objects is reasonable. Images formed of objects moving at speeds approaching *c* will be the result of time dilation, length contraction and other effects including the relativistic Doppler effect and the aberration of light.

Imagine speeding through space in a very fast spacecraft. When you planned your trip on Earth, you forgot to take relativity into account. Everything on board would appear normal throughout the trip, but when you looked out the front window, the effects of relative speed would be obvious. Some examples of what you would see include: the aberration of light, causing the stars to group closer together so that your forward field of vision would be increased; the Doppler effect, causing the colours of stars to change; and the voyage taking much less time than you expected.

11.5.1 The journey of muons

Bruno Rossi and David Hall performed a beautiful experiment in 1941, the results of which are consistent with both time dilation and length contraction.

Earth is constantly bombarded by energetic radiation from space, known as cosmic radiation. These rays collide with the upper atmosphere, producing particles known as muons. Muons are known to have a very short half-life, measured in the laboratory to be 1.56 microseconds. Given the speed at which they travel and the distance they travel through the atmosphere, the vast majority of muons would decay before they hit the ground.

The Rossi–Hall experiment involved measuring the number of muons colliding with a detector on top of a tall mountain and comparing this number with how many muons were detected at a lower point. They found that far more muons survived the journey through the atmosphere than would be predicted without time dilation. The muons were travelling so fast relative to Earth that the muons decayed at a much slower rate for observers on Earth than they would at rest in the laboratory. The journey between the detectors took about 6.5 microseconds according to Earth-based clocks, but the muons decayed as though only 0.7 microseconds had passed. Due to length contraction, the muons did not see the tall mountain but, rather, a small hill. Rossi and Hall were not surprised that the muons survived the journey.

FIGURE 11.21 Muons are a measurable example of special relativistic effects.

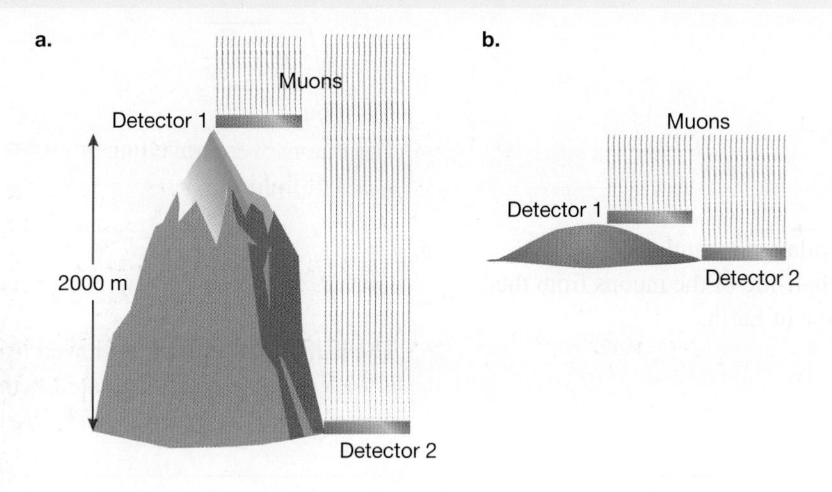

Figure 11.21 shows how the Rossi–Hall experiment was conducted. In figure 11.21a, the number of muons decaying between detectors 1 and 2 implies that less time has passed for the muons than Earth-based clocks suggest. In figure 11.21b, the muons see the distance between the detectors greatly contracted.

tlvd-9037

SAMPLE PROBLEM 11 Exploring the proper time and speed of muons and the value of γ using the Rossi–Hall experiment

Use the description of the Rossi–Hall experiment to answer the following questions.

a. **What is the proper time for the half-life of muons?**
b. **What is the value of gamma as determined from the journey times from the different reference frames?**
c. **How fast were the muons travelling though the atmosphere according to the value for gamma?**
d. **Calculate the half-life of the muons from the reference frame of Earth.**

THINK

a. The proper time for the half-life is in the reference frame of the muon.

WRITE

a. The proper time for the half-life of muons is 1.56 microseconds.

THINK

b. Use the time dilation formula with $t = 6.5$ and $t_0 = 0.7$ to calculate the value of gamma.

WRITE

b. $t = t_0\gamma$

$$\gamma = \frac{t}{t_0}$$
$$= \frac{6.5}{0.7}$$
$$= 9.29$$

THINK

c. Use the expression for the Lorentz factor, $\gamma = \frac{1}{\sqrt{1-\frac{v^2}{c^2}}}$, to calculate the velocity of the muons.

WRITE

c. $\gamma = \frac{1}{\sqrt{1-\frac{v^2}{c^2}}}$

$$v = c\sqrt{1-\frac{1}{\gamma^2}}$$
$$= c\sqrt{1-\frac{1}{9.29^2}}$$
$$= 0.994c$$

The muons were travelling at 99.4% of the speed of light.

THINK

d. Use the time dilation formula, $t = \gamma t_0$, to calculate the half-life of the muons from the reference frame of Earth.

WRITE

d. $t = \gamma t_0$

$$= 9.29 \times 1.56\ \mu s$$
$$= 14.5\ \mu s$$

The half-life of muons as viewed from Earth is 14.5 microseconds compared to the 1.56 microseconds as experienced by the muons.

PRACTICE PROBLEM 11

Use the description of the Rossi–Hall experiment introduced earlier to answer the following questions.

a. **Use the travel time from Earth's reference frame and the speed of the muons to calculate the height of the mountain.**
b. **Use the travel time of the muons to determine how high the mountain appeared to the muons.**

Just as the measured height of a mountain in the muons' rest frame is contracted (smaller) than the height of a mountain in the rest frame of that mountain, so too is the length that an accelerated charged particle travels in an accelerator when viewed in the rest frame of the particle being accelerated.

11.5.2 Particle accelerators and special relativity

Particle accelerators, as the name suggests, are designed to make particles such as electrons and protons move at high velocity, often approaching the speed of light. As a consequence, the time taken for a particle to travel the length of an accelerator can be substantially different in the laboratory frame when compared to the rest frame of the particle. The distance travelled by an accelerated particle can also be fundamentally different due to the high speed of the particles when viewed from both the rest frame of the particle and the rest frame of the laboratory. This is relevant in accelerators designed to make x-rays, such as the synchrotron at Monash University, or in large particle accelerators to study collisions between fundamental particles, such as at CERN, where accurate time and distance measurements are required. If the speed of the particles in the laboratory frame is known, then both time dilation and length contraction effects can be calculated. Compared to the proper length of an accelerator, the length of an accelerator in the rest frame of an accelerated particle would be substantially less. Likewise, the time taken to traverse the length of an accelerator in the rest frame of the particle would be substantially greater when compared to the time taken in the rest frame of the accelerator.

Note that, in the scope of the VCE Physics study design, special relativity is applied only to inertial frames of reference, which are frames not undergoing an acceleration.

Particles moving at a constant speed in circular accelerators are accelerated (their speed is constant, but not their direction) thus circular motion is not an inertial frame of reference and as such is not covered in this topic.

11.5.3 Time dilation and modern technology

Time dilation has great practical significance. A global positioning system (GPS) is able to tell you where you are, anywhere on Earth, in terms of longitude, latitude and altitude, to within a few metres. To achieve this precision, the system has to compensate for time dilation, because it depends on time signals from satellites moving in orbit. Einstein's general relativity also shows that the difference in gravity acting on a satellite in orbit affects the time significantly. Note, however, that general relativity is not included in the VCE Physics study design. Nanosecond accuracy is required for a GPS but, if Newtonian physics was used, the timing would be out by more than 30 microseconds per day. This equates to an error of roughly 11 km per day. GPSs are widely used in satellite navigation, and ships, planes, car drivers and bushwalkers can find their position far more accurately than they ever could using a map.

FIGURE 11.22 With a GPS device you can know your position to within a few metres.

Real-life examples of special relativity

- The time dilation experienced by muons travelling at relativistic speed allows them to reach Earth's surface before they decay.
- In a particle accelerator, in the frame of reference of a particle moving at high speed, the length, in the direction of motion, of the accelerator appears shorter (contracted) to the particle.
- In GPS navigation, relativistic effects have to be taken into account as GPS satellites move at high speed in orbit around Earth.

11.5 Activities

learnon

Students, these questions are even better in jacPLUS

Receive immediate feedback and access sample responses

Access additional questions

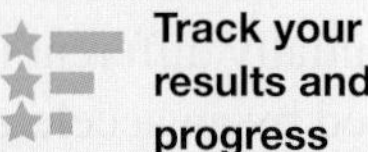
Track your results and progress

Find all this and MORE in jacPLUS

11.5 Quick quiz	11.5 Exercise	11.5 Exam questions

11.5 Exercise

1. Explain why muons reach the surface of Earth in greater numbers than would be predicted by classical physics, given their speed, their half-lives and the distance they need to travel through the atmosphere.
2. A muon forms 30 km above Earth's surface and travels straight down at 0.98*c*. From its frame of reference, what is the distance it has to travel through the atmosphere?
3. The proper time for the half-life of a muon is 1.56 microseconds. If the muon moves at 0.98*c* relative to an observer, what does the observer measure its half-life as?
4. Explain how muons produced by cosmic rays became an early confirmation of special relativity.
5. An anti-proton is travelling in the straight section of a particle accelerator after being accelerated to 0.99*c*. The length of the straight beam-line is 1200 m. Calculate the length of the linear portion of the accelerator in the rest frame of the anti-proton.
6. After being accelerated, electrons travel at a constant speed in a linear accelerator of natural length 140.0 m. From the frame of reference of the electron this length is only 20.0 m.
   a. At what speed are the electrons travelling?
   b. From the rest frame of the laboratory how long does it take for an electron to complete one circuit within the beam-line?

11.5 Exam questions

Question 1 (1 mark)

Source: *VCE 2022 Physics Exam, Section A, Q.19;* © VCAA

MC A particle produced in a linear particle accelerator is travelling at a speed of 2.99×10^8 m s^{-1}.

Take the speed of light to be 3.00×10^8 m s^{-1}. Which one of the following is closest to the Lorentz factor (γ) of the particle?

A. 5.51
B. 7.86
C. 12.3
D. 15.1

Question 2 (2 marks)

Source: *VCE 2022 Physics Exam, Section B, Q.11;* © VCAA

Explain why muons formed in the outer atmosphere can reach the surface of Earth even though their half-lives indicate that they should decay well before reaching Earth's surface.

Question 3 (4 marks)

Source: *VCE 2022 Physics Exam, NHT, Section B, Q.11;* © VCAA

An experiment is set up at a linear accelerator research facility to study muons. The muons created at the research facility are measured to have a speed of 0.950*c* ($\gamma = 3.20$).

a. One muon has a lifetime of 2.3 μs, as measured in the muon's frame of reference.
Calculate this muon's lifetime, as measured by the researchers. Show your working. **(2 marks)**

b. In one observation, a 0.950*c* muon travels 1.5 km, as measured by the researchers.
If measured in the muon's frame of reference, would this length be the same, shorter or longer? Use a calculation to justify your answer. **(2 marks)**

Question 4 (1 mark)

MC An investigator wished to explore muons and found that the proper time experienced by muons travelling between two detectors was 0.32 microseconds. This muon was travelling at 99.4% of the speed of light. The observed time from Earth's reference frame was 4.3 microseconds. The height of the mountain as observed by muons would be:

A. 1282 m.
B. 128 m.
C. 95 m.
D. 950 m.

Question 5 (3 marks)

A GPS satellite has an orbital speed of 14 000 km h^{-1}. Due to its speed, a clock inside the satellite lags by 7 μs each day compared to a clock on Earth. Given that a GPS signal travels at the speed of light, $c = 3.0 \times 10^8$ m s^{-1}, explain why it is necessary for a GPS clock to compensate for the relativistic effect due to its orbital velocity, and to be accurate to the nanosecond, to provide an accurate position on Earth.

More exam questions are available in your learnON title.

11.6 Einstein's relationship between mass and energy

KEY KNOWLEDGE

- Interpret Einstein's prediction by showing that the total 'mass–energy' of an object is given by: $E_{tot} = E_k + E_0 = \gamma mc^2$ where $E_0 = mc^2$, and where kinetic energy can be calculated by: $E_k = (\gamma - 1)\,mc^2$
- Apply the energy-mass relationship to mass conversion in the Sun, to positron-electron annihilation and to nuclear transformations in particle accelerators (details of the particular nuclear processes are not required).

11.6.1 Mass is a type of energy

In Newtonian physics, the work done in accelerating an object is equal to the increase in the kinetic energy of the object, and energy is conserved.

Work done (W) = Force (F) × displacement (s) = Gain in kinetic energy (ΔE_k)

Since $E_k = \frac{1}{2}mv^2$:

$$\Delta E_k = \frac{1}{2}m\left(v^2 - u^2\right) = F \times s$$

Both Newton's laws and special relativity show that energy is conserved. However, the previous equation does not hold in Einstein's special theory of relativity. When work is done on a particle, some of the energy transferred goes towards increasing the **relativistic mass** of the particle. As we will see, in special relativity there is a proper mass, usually referred to as the rest mass, and a relativistic mass dilated by a factor γ.

relativistic mass the mass of a body in motion, relative to an observer, also known as inertial mass

It was in exploring these ideas that Einstein came up with the idea that energy and mass were equivalent and interchangeable, leading to the result of special relativity that people are most familiar with: the equation $E = mc^2$. In fact, it is probably the most well-known equation of all. This formula expresses an equivalence of mass and energy. If work, E, is done on an object, that is, its energy is increased, its mass will increase. Usually, however, this increase in mass is not noticed because it is very small compared to the factor $c^2 = 9.0 \times 10^{16}$ m^2 s^{-2}. According to $\Delta E = \Delta mc^2$, it would take 9.0×10^{16} J of energy to increase a mass by 1 kg.

This is similar to the amount of electrical energy produced in Victoria every year. Conversely, if a 1-kg mass could be converted into electricity, Victoria's electricity needs would be supplied for a year. Nuclear fission reactors produce electricity from the small loss of mass that occurs when large nuclei such as those of uranium-235 undergo fission. The Sun and other stars generate their energy by losing mass to nuclear fusion.

EXTENSION: Deriving the equation $E = mc^2$

A simplified derivation of this equation can help to gain a sense of the physics involved. Consider a box suspended in space, with no external forces acting on it, as shown in figure 11.23. Maxwell found that electromagnetic radiation carries momentum $p = \frac{E}{c}$ where E is the energy transmitted and c is the speed of light. In the context of photons, each photon carries a momentum $p = \frac{E}{c}$. As a result, light exerts pressure on surfaces. This effect can nudge satellites out of orbit over time (orbit decay).

In diagram (a) of figure 11.23, the box begins at rest. The total momentum is zero and its centre of mass is in the centre.

In diagram (b) of figure 11.23, a photon of energy E is emitted from end A travelling to the right, carrying momentum with it. To conserve momentum, the box moves in the direction opposite to the movement of the photon, to the left.

FIGURE 11.23 Einstein's box suspended in space

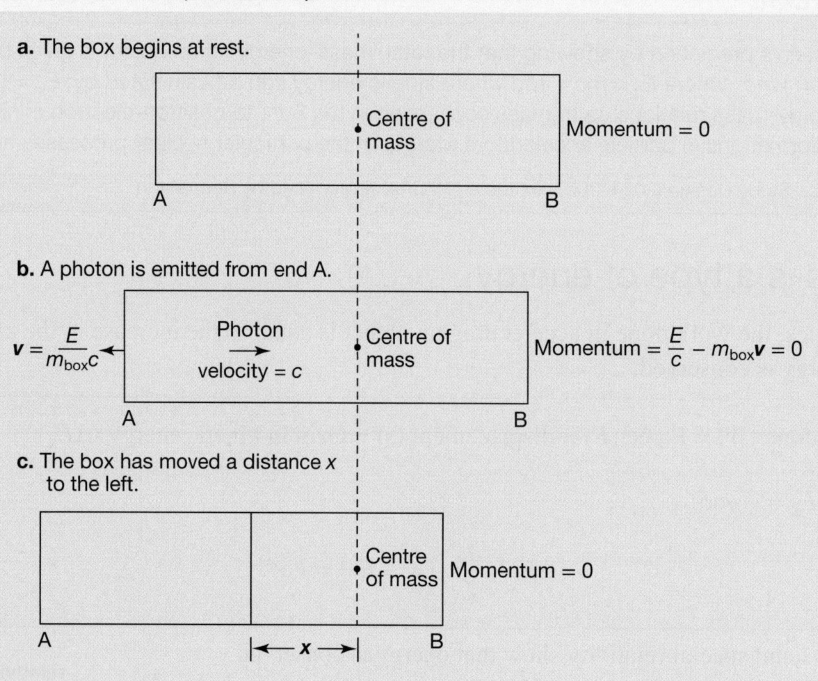

$$p_{photon} + p_{box} = 0$$
$$\Rightarrow \frac{E}{c} - m_{box}v = 0$$

Rearranging gives the velocity of the box in the leftward direction, $v = \frac{E}{m_{box}c}$, a very small number!

In diagram (c) of figure 11.23, after time Δt, the light pulse strikes the other end of the box and is absorbed. The momentum of the photon is also absorbed into the box, bringing the box to a stop. In this process, the box has moved a distance $\boldsymbol{x}$ where $\boldsymbol{x} = \boldsymbol{v}\Delta t$.

Substituting $v = \dfrac{E}{m_{box}c}$ from diagram (b) gives:

$$x = \frac{E\Delta t}{m_{box}c}$$

As $\boldsymbol{v}$ is very small (almost non-existent), it can be assumed that the photon travels the full length of the box.

Substituting $\Delta t = \dfrac{L}{c}$ into $x = \dfrac{E\Delta t}{m_{box}c}$ gives:

$$x = \frac{EL}{m_{box}c^2}$$

$$\text{or } E = \frac{xm_{box}c^2}{L}.$$

There are no external forces acting on the box, so the position of the centre of mass must remain unchanged (see the dotted line in the diagram). The box moved to the left as a result of the transfer of the energy of the photon to the right. Therefore, the transfer of the photon must be the equivalent of a transfer of mass. If you can show that $\dfrac{xm_{box}}{L}$ is the same as the mass equivalent of the transferred energy, you have your answer. To show this, attention must be paid to the shift in the box relative to the centre of mass of the system.

The centre of mass is the point where the box would balance if suspended. This can be determined by balancing moments — the mass times the distance from a reference point. The centre of the box is chosen as the reference point to ensure that the distance x is in the calculations. The moment for the box is $m_{box}\boldsymbol{x}$ anticlockwise, because the mass of the box can be considered to be acting through a point at distance $\boldsymbol{x}$ to the left of the reference point. The photon's equivalent mass is acting at distance $\dfrac{L}{2}$ to the right of the reference point, so its moment is $m\dfrac{L}{2}$ clockwise. However, this moment was acting on the other end of the box before the photon was emitted, so you can consider its absence from that end of the box as an equal moment in the same direction. Therefore:

$$m_{box}x = m\frac{L}{2} + m\frac{L}{2}$$

$$\text{or } m = \frac{m_{box}x}{L} \text{ as required.}$$

Substitute this into $E = \dfrac{xm_{box}c^2}{L}$ and $E = mc^2$ is obtained.

FIGURE 11.24 Balancing Einstein's box

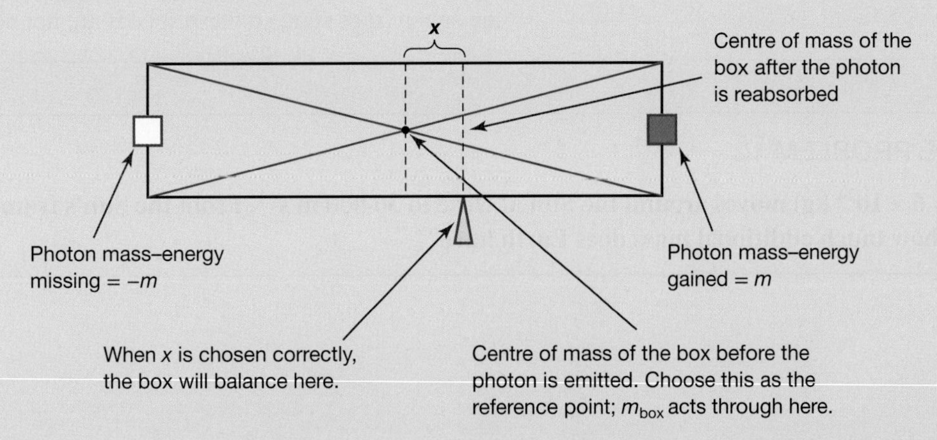
x
Centre of mass of the box after the photon is reabsorbed
Photon mass–energy missing = −m
Photon mass–energy gained = m
When x is chosen correctly, the box will balance here.
Centre of mass of the box before the photon is emitted. Choose this as the reference point; m_box acts through here.

In other words, when the photon carried energy to the other end of the box, it had the same effect as if it had carried mass. In fact, Einstein concluded that energy and mass are equivalent. If you say that some energy has passed from one end of the box to the other, you are equally justified in saying that mass has passed as well. Note the distinction: the photon carries an amount of energy that is equivalent to an amount of mass, but the photon itself does not have mass.

$$\Delta E = \Delta mc^2$$

where: ΔE is the change in energy

Δm is the change in mass

c is the speed of light

One implication of this is that the measurement of mass depends on the relative motion of the observer. The kinetic energy of a body depends on the inertial reference frame from which it is measured. The faster the motion, the greater the kinetic energy. So kinetic energy is relative, and so is mass! Energy is equivalent to mass, so the mass of an object increases as its velocity relative to an observer increases.

The mass of an object that is in the same inertial frame as the observer is called its **rest mass** (m_0). When measured from other reference frames, the mass is given by $m = m_0\gamma$. The derivation of this is complex, so it will not be addressed here.

rest mass mass of an object measured at rest

tlvd-9038

SAMPLE PROBLEM 12 Exploring why a mass cannot exceed the speed of light

Use $m = m_0\gamma$ to show that it is not possible for a mass to exceed the speed of light.

THINK	WRITE
1. Consider the effect on gamma as $v \to c$.	As $v \to c, \gamma \to \dfrac{1}{\sqrt{1-\dfrac{c^2}{c^2}}} \to \infty$ So, as $v \to c$, γ becomes infinitely large.
2. Consider the effect on m as $v \to c$.	$m = m_0\gamma$, so as $v \to c, \gamma \to \infty, m \to \infty$. An object travelling at c would have infinite mass.
3. Consider the effect on gamma if $v > c$.	$\gamma = \dfrac{1}{\sqrt{1-\dfrac{v^2}{c^2}}}$ If $v > c$, then $\dfrac{v^2}{c^2} > 1$ and $1 - \dfrac{v^2}{c^2} < 0$. Speeds larger than c would produce a negative under the square root sign, so these speeds are not possible.

PRACTICE PROBLEM 12

Earth (m = 6 × 10^{24} kg) moves around the Sun at close to 30 000 m s^{-1}. From the Sun's frame of reference, how much additional mass does Earth have?

tlvd-9039

SAMPLE PROBLEM 13 Calculating the mass increase of an accelerated proton

Calculate the mass increase of a proton that is accelerated from rest using 11 GeV of energy, an energy that can be achieved in particle accelerators.

THINK	WRITE
1. Convert 11 GeV into joules.	$\Delta E = 11$ GeV $= 11 \times 10^9 \times 1.6 \times 10^{-19}$ J $= 1.76 \times 10^{-9}$ J
2. Use the equation $\Delta E = \Delta mc^2$ to determine the mass increase that corresponds with a change in energy of 11 GeV.	$\Delta E = \Delta mc^2$ $\Delta m = \dfrac{\Delta E}{c^2}$ $= \dfrac{1.76 \times 10^{-9}\text{ J}}{(3 \times 10^8\text{ m s}^{-1})^2}$ $= 1.96 \times 10^{-26}$ kg Note that the rest mass of a proton is 1.67×10^{-27} kg, so the accelerated proton behaves as though its mass is nearly 12 times its rest mass.

PRACTICE PROBLEM 13

A particle physicist accelerates a proton from rest using 3.45×10^{-10} J of energy. What is the mass increase of the proton if the rest mass of the proton is 1.67×10^{-27} kg?

tlvd-9040

SAMPLE PROBLEM 14 Calculating the speed of a proton using electron volts

In Newtonian physics, if a proton is given 11 GeV of kinetic energy, what would its speed be?

THINK	WRITE
Use the formula $E_k = \frac{1}{2}mv^2$ to determine the speed of the proton, ensuring that the value for kinetic energy used in the formula is in joules. Where the kinetic energy, $E_k = 11$ GeV or 1.76×10^{-9} J, and the mass of a proton, $m = 1.67 \times 10^{-27}$ kg.	$E_k = \dfrac{1}{2}mv^2$ $v = \sqrt{\dfrac{2E_k}{m}}$ $= \sqrt{\dfrac{2 \times 11 \times 10^9 \times 1.6 \times 10^{-19}}{1.67 \times 10^{-27}}}$ $= 1.45 \times 10^9$ m s^{-1} This speed is not possible as the maximum speed attainable is 3.0×10^8 m s^{-1}.

PRACTICE PROBLEM 14

Use Newton's laws to determine the speed of the proton from practice problem 13. What conclusions can you make about Newtonian physics compared to Einstein's physics?

The solution to sample problem 14 is well in excess of the speed of light and is an example of the limitations of Newtonian physics. In relativity, when more energy is given to a particle that is approaching the speed of light, the energy causes a large change in mass and a small change in speed. By doing work on the particle, the particle gains inertia, so the increase in energy has an ever-decreasing effect on the speed. The speed cannot increase beyond the speed of light, no matter how much energy the particle is given.

In particle accelerators, where particles are accelerated to near the speed of light, every tiny increase in the speed of the particles requires huge amounts of energy. Physicists working in this field rely on ever-higher energies to make new discoveries. This costs huge amounts of money. Nonetheless, a number of accelerators have been built that are used by scientists from around the world. This area of research is often called high-energy physics. At these high energies, Newtonian mechanics is hopelessly inadequate and Einstein's relativity is essential.

In particle accelerators, particles such as protons have been accelerated close to the speed of light but, regardless of the amount of energy provided, their speed never exceeds c, the speed of light in a vacuum. With 120 GeV for instance, protons can reach $0.999\,97c$, while with nearly 60 times more energy, they can reach $0.999\,999\,991c$.

As the speed of the particles increases, so does their inertia. It would take an infinite amount of energy for any particle that has mass to reach c.

FIGURE 11.25 Particle accelerators such as the Australian Synchrotron in Melbourne accelerate subatomic particles to near light speeds, where special relativity is essential for understanding the behaviour of the particles. Electrons in the Australian Synchrotron have kinetic energies up to 3 GeV.

11.6.2 Kinetic energy in special relativity

This equivalence of mass and energy has resulted in the term mass–energy. The **mass–energy** of any object is given by $E = mc^2$. With mass–energy, a moving particle has kinetic energy and rest energy. Rest energy is the energy equivalent of the mass at rest given by $E_0 = m_0c^2$.

Therefore:

$$E_{tot} = E_k + E_0$$

mass–energy concept used to describe mass and energy as equivalent, given by $E = mc^2$

Substituting for E_{tot} and E_0:

$$mc^2 = E_k + m_0c^2$$

Rearranging and substituting,

$$\begin{aligned} E_k &= mc^2 - m_0c^2 \\ &= m_0\gamma c^2 - m_0c^2 \\ &= (\gamma - 1)m_0c^2 \end{aligned}$$

The kinetic energy of a particle can be calculated using:

$$E_k = (\gamma - 1)\, m_0c^2$$

where: E_k is the kinetic energy of the particle, in J

γ is the Lorentz factor $\gamma = \dfrac{1}{\sqrt{1 - \dfrac{v^2}{c^2}}}$

m_0 is the rest mass of the particle, in kg

c is the speed of light, in m s^{-1}

v is the speed of the moving frame of reference, in m s^{-1}.

This is the expression we must use for kinetic energy when dealing with high speeds, particularly those exceeding 10% of the speed of light.

tlvd-9041

SAMPLE PROBLEM 15 Calculating kinetic energy using special relativity and classical physics

Calculate the kinetic energy of a 10 000 kg spacecraft travelling at 0.5*c* and compare this with the kinetic energy that you would calculate using classical physics $\left(E_k = \dfrac{1}{2}mv^2\right)$.

THINK	WRITE
1. Use the formula $E_k = (\gamma - 1)\, m_0c^2$ to calculate the kinetic energy of the spacecraft using special relativity, with $v = 0.5c$	$E_k = (\gamma - 1)m_0c^2$ $= \left(\dfrac{1}{\sqrt{1-0.5^2}} - 1\right) \times 10\,000 \times (3 \times 10^8)^2$ $= 1.39 \times 10^{20}$ J
2. Use the formula $E_k = \dfrac{1}{2}mv^2$ to calculate the kinetic energy of the spacecraft using classical physics, with $v = 0.5c = 0.5 \times 3 \times 10^8$ m s^{-1}.	$E_k = \dfrac{1}{2}mv^2$ $= \dfrac{1}{2} \times 10\,000 \times (0.5 \times 3 \times 10^8)^2$ $= 1.13 \times 10^{20}$ J
3. Compare the values for the kinetic energy of the spacecraft using special relativity and classical physics.	The kinetic energy is $\dfrac{1.39}{1.13} = 1.23$ times the value predicted by classical physics.

PRACTICE PROBLEM 15

A particle accelerator is designed to give electrons 3 GeV of kinetic energy. How fast can it make electrons travel?

11.6.3 Mass conversion in the Sun

In Unit 1, you learned about the generation of energy in the core of the Sun and other stars. One of the consequences of Einstein's great contribution to the understanding of relativity is that scientists now understand a great deal about how energy is generated by the Sun. At the centre of it all is the equation $E = mc^2$ (or more precisely, $\Delta E = \Delta mc^2$). The Sun continuously converts mass–energy stored as mass into radiant light and heat. Each second, the Sun radiates enough energy to meet current human requirements for billions of years. It takes the energy generated in the core about 100 000 years to reach the surface. Even if the fusion in the Sun stopped today, it would take tens of thousands of years before there was a significant impact on Earth.

FIGURE 11.26 The Sun's energy comes from nuclear fusion converting mass into energy.

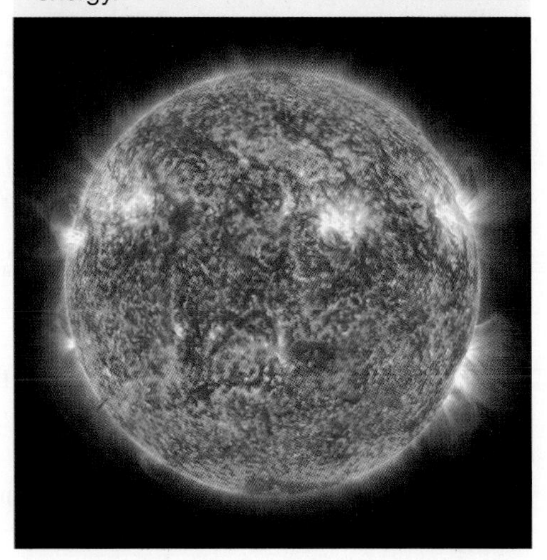

Nuclear fusion

Nuclear fusion is the process of joining smaller nuclei to form one larger, more stable nucleus. This process lowers the potential energy of the combined particles, which in turn lowers the total energy and hence the relativistic mass of these new more stable nuclei, resulting in an output of energy in accordance with $\Delta E = \Delta mc^2$. In essence, nuclear fusion is an exothermic process for isotopes of hydrogen and helium in particular, but all the way up to iron in the periodic table, that produces energy to power stars. Extremely high temperatures are required for nuclear fusion to occur, such as those found on the Sun and other stars.

The Sun is a ball made up mostly of hydrogen plasma and some ionised atoms of lighter elements. The temperatures in the Sun ensure that virtually all of the atoms are ionised. The composition of the Sun is shown in table 11.1.

TABLE 11.1 The composition of the Sun

Element	Percentage of total number of nuclei in the Sun	Percentage of total mass of the Sun
Hydrogen	91.2	71.0
Helium	8.7	27.1
Oxygen	0.078	0.97
Carbon	0.043	0.40
Nitrogen	0.0088	0.096
Silicon	0.0045	0.099
Magnesium	0.0038	0.076
Neon	0.0035	0.058
Iron	0.0030	0.14
Sulfur	0.0015	0.040

At this stage of the Sun's life cycle, ionised hydrogen atoms (i.e. protons) provide the energy. The abundance of protons and the temperatures and pressures in the core of the Sun are sufficient to fuse hydrogen, but not heavier nuclei. The energies of the protons in the Sun have a wide distribution from cool, slow protons to extremely hot, fast protons. Only the most energetic protons, about one in a hundred billion, have the energy required to overcome the electrostatic repulsion and undergo fusion. The Sun is in a very stable phase of fusing hydrogen that is expected to last for billions of years to come.

Fusion in the Sun occurs mainly through the following process:

$$^{1}_{1}\text{H} + ^{1}_{1}\text{H} \rightarrow ^{2}_{1}\text{H} + ^{0}_{1}\beta^{+} + \text{neutrino}$$
$$^{2}_{1}\text{H} + ^{1}_{1}\text{H} \rightarrow ^{3}_{2}\text{He} + \text{gamma photon}$$
$$^{3}_{2}\text{He} + ^{3}_{2}\text{He} \rightarrow ^{4}_{2}\text{He} + 2\,^{1}_{1}\text{H}$$

which can be summed up by the following equation:

$$4\,^{1}_{1}\text{H} \rightarrow ^{4}_{2}\text{He} + 2\,^{0}_{1}\beta^{+} + 2\text{ neutrinos} + 2\text{ gamma photons}$$

The energy is released mainly through the gamma photons and the annihilation of the positrons when they meet free electrons in the Sun. The net result is an enormous release of energy and a corresponding loss of mass. The mass loss has been measured to be 4.4 Tg (4.4×10^{9} kg) per second. As the mass of the Sun is around 2.0×10^{30} kg, even at this incredible rate, there is plenty of hydrogen to sustain it for about twice its age of four and a half billion years.

tlvd-9042

SAMPLE PROBLEM 16 Accounting for the mass differences between the nucleus and individual protons and neutrons

A nucleus of hydrogen-2 made of one proton and one neutron has a smaller mass than the total of an individual proton and an individual neutron. Account for this mass difference.

THINK	WRITE
Consider the mass–energy of the hydrogen-2 nucleus compared to the mass–energy of the separate proton and neutron.	Deuterium (or hydrogen-2) has a mass of 3.3435×10^{-27} kg. A proton has a mass of 1.6726×10^{-27} kg and a neutron has a mass of 1.6759×10^{-27} kg. This leads to a combined mass of 3.3475×10^{-27} kg, showing a mass difference of 0.004×10^{-27} kg or 4.0×10^{-30} kg. Deuterium has a rest-mass and also binding energy that account for this mass deficit. Using $\Delta E = mc^2$, the energy difference from this mass deficit would be 3.6×10^{-13} J. It is clear that the mass of the nucleus is different to the mass of the individual particles, but when the binding energy of the hydrogen-2 nucleus is included, the mass–energy of both is found to be the same. The separate particles have their mass and zero potential energy. The particles bound in the nucleus have a reduced mass and the binding energy of the nucleus. (The binding energy is the energy required to separate the particles. It is released as a combination of increased kinetic energy of the particles and gamma rays.)

PRACTICE PROBLEM 16

Consider the following fusion reaction:

$$^{3}_{2}\text{He} + ^{3}_{2}\text{He} \rightarrow ^{4}_{2}\text{He} + 2^{1}_{1}\text{H}$$

If 30 MeV of energy is released through this reaction, determine how much mass is lost through this reaction. Do not forget to convert MeV to J.

tlvd-9043

SAMPLE PROBLEM 17 Determining the power output of the Sun

At the Sun's core, a huge number of fusion reactions are constantly taking place. As such, the Sun loses approximately 4.4 Tg of mass every second. What is the power output of the Sun?

THINK	WRITE
1. Use the relationship $E = mc^2$ to calculate the energy released by the Sun when its mass loss is 4.4×10^9 kg.	$E = mc^2$ $= 4.4 \times 10^9 \times (3.0 \times 10^8)^2$ J $= 4.0 \times 10^{26}$ J
2. Use the relationship $P = \frac{E}{t}$ to calculate the power output that corresponds to a mass loss of 4.4×10^9 kg s^{-1}.	$P = \frac{E}{t}$ $= \frac{4.0 \times 10^{26}\text{ J}}{1\text{ s}}$ $= 4.0 \times 10^{26}$ W The mass loss of 4.4×10^9 kg s^{-1} equates to a power output of 4.0×10^{26} W.

PRACTICE PROBLEM 17

If the mass of the Sun is approximately 2.0×10^{30} kg and it is estimated that the Sun loses 4.4×10^9 kg s^{-1} during nuclear fusion, how long would you estimate the Sun to be able to continue providing energy for life on Earth?

Particles can also annihilate with their antiparticle and the mass lost is converted into energy, usually in the form of photons. For example, an electron can annihilate with a positron to produce a pair of photons in accordance with $\Delta E = \Delta mc^2$. This process is used to advantage in medical diagnostics with the PET (Positron Emission Tomography) scan. Patients are injected with a radioactive isotope, which beta decays, emitting positrons. These positrons annihilate with electrons, emitting a pair of photons moving away in opposite directions.

tlvd-9044

SAMPLE PROBLEM 18 Calculating the energy of photons produced by the annihilation of an electron and a positron

The rest mass of an electron and a positron is 9.1×10^{-31} kg. An event occurs where an electron annihilates with a nearby positron and a pair of photons is produced. Calculate the energy of each of the photons.

THINK	WRITE
1. The mass of the two particles is converted into energy.	$\Delta E = \Delta mc^2$ and so $\Delta E = 2 \times 9.1 \times 10^{-31} \times (3 \times 10^8)^2$ $= 1.64 \times 10^{-13}$ J
2. The energy is equally shared with the two photons.	Each photon has an energy $\frac{1.64 \times 10^{-13} \text{ J}}{2}$. Thus, the energy is 8.2×10^{-14} J.

PRACTICE PROBLEM 18

At CERN, protons and anti-protons collide and are annihilated to produce energy from which new particles may be formed. Calculate the available energy when a proton annihilates with an anti-proton. Both particles have a mass of 1.6×10^{-27} kg.

11.6 Activities

learnon

11.6 Quick quiz | 11.6 Exercise | 11.6 Exam questions

11.6 Exercise

1. Use your knowledge of relativity to argue that matter cannot travel at the speed of light.
2. a. How much energy would be required to accelerate 1000 kg to:
   i. $0.1c$
   ii. $0.5c$
   iii. $0.8c$
   iv. $0.9c$?
   b. Sketch a graph of energy versus speed using your answers to part a.
3. Explain in words what $E = mc^2$ says about energy and mass.
4. Calculate the rest energy of Earth, which has a rest mass of 5.98×10^{24} kg.
5. Consider Earth to be a mass moving at 30 km s^{-1} relative to a stationary observer. Given that the rest mass of Earth is 5.98×10^{24} kg, what would be the difference between this rest mass and the mass from the point of view of the stationary observer?
6. What is happening to the mass of the Sun over time? Why?

11.6 Exam questions

Question 1 (1 mark)

Source: *VCE 2020 Physics Exam, Section A, Q.13; © VCAA*

MC Matter is converted to energy by nuclear fusion in stars.

If the star Alpha Centauri converts mass to energy at the rate of 6.6×10^{9} kg s^{-1}, then the power generated is closest to

A. 2.0×10^{18} W
B. 2.0×10^{18} J
C. 6.0×10^{26} W
D. 6.0×10^{26} J

Question 2 (1 mark)

Source: *VCE 2018, Physics Exam, Section A, Q.14; © VCAA*

MC Which one of the following statements about the kinetic energy, E_k, of a proton travelling at relativistic speed is the most accurate?

A. The difference between the proton's relativistic E_k and its classical E_k cannot be determined.
B. The proton's relativistic E_k is greater than its classical E_k.
C. The proton's relativistic E_k is the same as its classical E_k.
D. The proton's relativistic E_k is less than its classical E_k.

Question 3 (3 marks)

Source: *VCE 2018, Physics Exam, Section B, Q.15; © VCAA*

A stationary scientist in an inertial frame of reference observes a spaceship moving past her at a constant velocity. She notes that the clocks on the spaceship, which are operating normally, run eight times slower than her clocks, which are also operating normally. The spaceship has a mass of 10 000 kg.

Calculate the kinetic energy of the spaceship in the scientist's frame of reference. Show your working.

Question 4 (2 marks)

Source: *VCE 2018 Physics Exam, NHT, Section B, Q.15; © VCAA*

An unstable subatomic particle, known as a π_0 meson, decays completely into electromagnetic radiation. The rest mass of this π_0 meson is 2.5×10^{-28} kg.

How much energy would be released by this π_0 meson if it decays at rest?

Question 5 (1 mark)

Source: *VCE 2017, Physics Exam, Section A, Q.11; © VCAA*

MC On average, the sun emits 3.8×10^{26} J of energy each second in the form of electromagnetic radiation, which originates from the nuclear fusion reactions taking place in the sun's core.

The corresponding loss in the sun's mass each second would be closest to

A. 2.1×10^{9} kg
B. 4.2×10^{9} kg
C. 8.4×10^{9} kg
D. 2.1×10^{12} kg

More exam questions are available in your learnON title.

11.7 Review

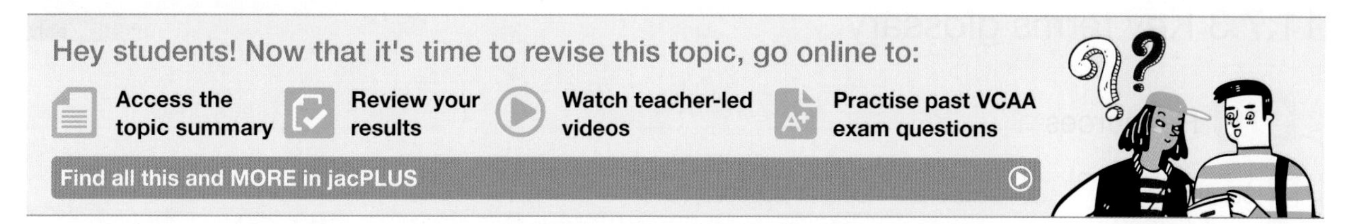

11.7.1 Topic summary

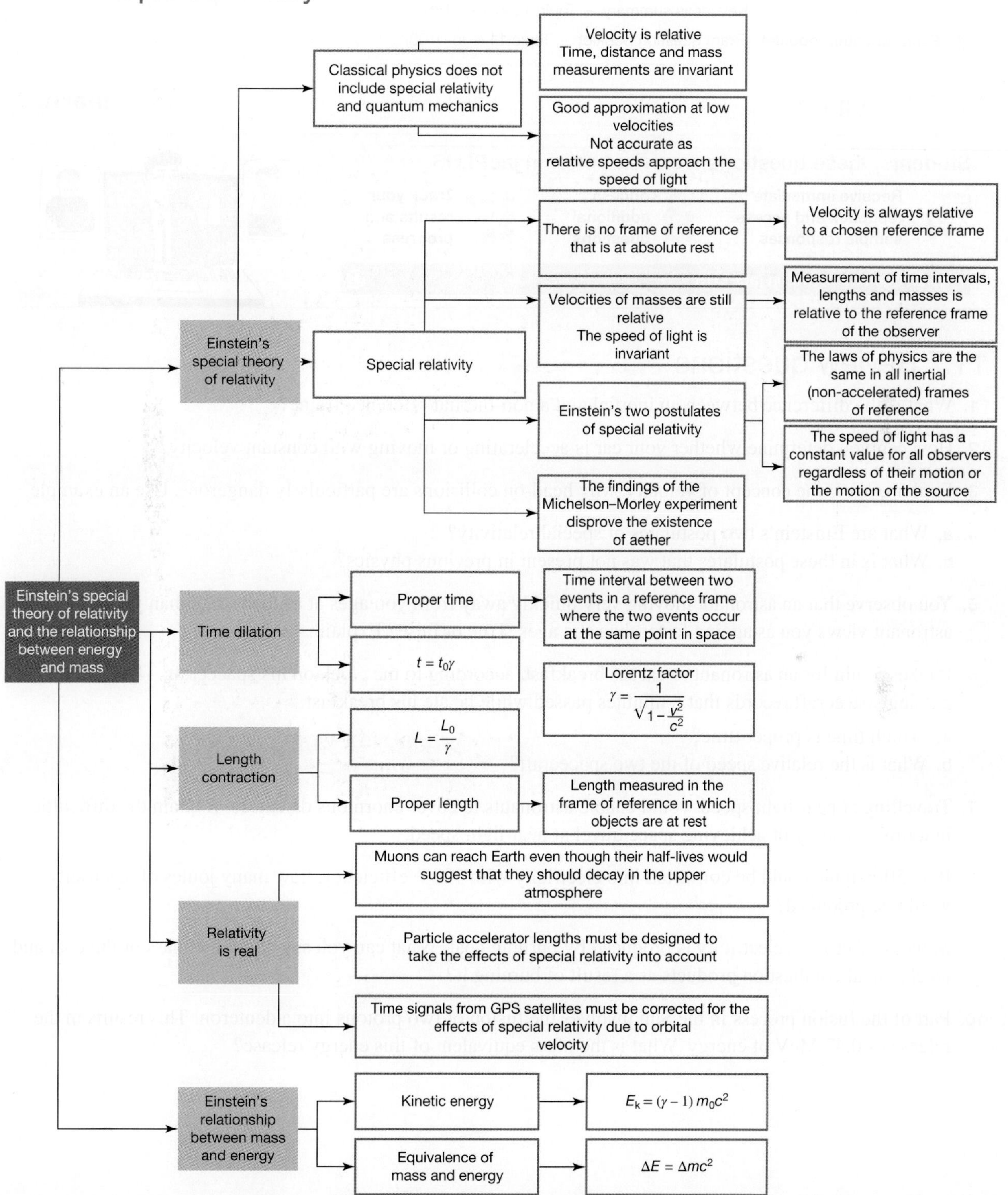

11.7.2 Key ideas summary

11.7.3 Key terms glossary

Resources

Solutions	Solutions — Topic 11 (sol-0825)
Digital documents	Key science skills — VCE Physics Units 1–4 (doc-36950) Key terms glossary — Topic 11 (doc-37185) Key ideas summary — Topic 11 (doc-37186)
Exam question booklet	Exam question booklet — Topic 11 (eqb-0108)

11.7 Activities

learnon

Students, these questions are even better in jacPLUS

Receive immediate feedback and access sample responses

Access additional questions

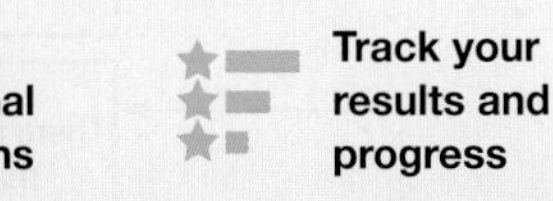
Track your results and progress

Find all this and MORE in jacPLUS

11.7 Review questions

1. What is the difference between an inertial and a non-inertial reference frame?
2. How can you determine whether your car is accelerating or moving with constant velocity?
3. Explain, using the concept of velocity, why head-on collisions are particularly dangerous. Use an example.
4. a. What are Einstein's two postulates of special relativity?
   b. What is in these postulates that was not present in previous physics?
5. You observe that an astronaut moving very quickly away from you ages at a slower rate than you. The astronaut views you as ageing faster than she ages. True or false? Explain.
6. It takes 5 min for an astronaut to eat his breakfast, according to the clock on his spacecraft. The clock on a passing spacecraft records that 8 minutes passed while he ate his breakfast.
   a. Which time is proper time?
   b. What is the relative speed of the two spacecraft?
7. Travelling at near light speed would enable astronauts to cover enormous distances. Explain the difficulties in terms of energy of achieving space travel at near light speed.
8. If a 250-g apple could be converted into electricity with 100% efficiency, how many joules of electricity would be produced?
9. Much of Victoria's electricity is produced by burning coal. What can you say about the mass of the coal and its chemical combustion products as a result of burning it?
10. Part of the fusion process in the Sun involves the fusion of two protons into a deuteron. This results in the release of 0.42 MeV of energy. What is the mass equivalent of this energy release?

11.7 Exam questions

Section A — Multiple choice questions

All correct answers are worth 1 mark each; an incorrect answer is worth 0.

Question 1

Source: *VCE 2022 Physics Exam, Section A, Q.18;* © VCAA

Which one of the following is an example of an inertial frame of reference?

A. a bus travelling at constant velocity
B. an express train that is accelerating
C. a car turning a corner at a constant speed
D. a roller-coaster speeding up while heading down a slope

Question 2

Source: *VCE 2022 Physics Exam, NHT, Section A, Q.10;* © VCAA

Ning travels at 0.67*c* from Earth to the star Proxima Centauri, which is a distance of 4.25 light-years away, as measured by an observer on Earth.

Which one of the following statements is correct?

A. In Ning's frame of reference, the distance to Proxima Centauri is less than 4.25 light-years.
B. In Ning's frame of reference, the distance to Proxima Centauri is more than 4.25 light-years.
C. According to Ning's clock, the trip takes longer than the time measured by Earth-based clocks.
D. In Ning's frame of reference, the distance to Proxima Centauri is exactly equal to 4.25 light-years.

Question 3

Source: *VCE 2022 Physics Exam, NHT, Section A, Q.11;* © VCAA

The star Betelgeuse is classified as a red supergiant. At the core of this star, three stationary helium nuclei fuse to form one carbon nucleus and two gamma-ray photons, as represented by the equation below.

$$^{4}\text{He} + ^{4}\text{He} + ^{4}\text{He} \rightarrow ^{12}\text{C} + \gamma + \gamma$$

The mass of one helium nucleus is 6.645×10^{-27} kg.

The mass of one carbon nucleus is 1.993×10^{-26} kg.

The energy released from the fusion of three helium nuclei is closest to

A. 5.0×10^{-30} J
B. 1.5×10^{-21} J
C. 4.5×10^{-13} J
D. 1.2×10^{-9} J

Question 4

Source: *VCE 2021 Physics Exam, NHT, Section A, Q.13;* © VCAA

Joanna is an observer in Spaceship P and is watching Spaceship Q fly past at a relative speed of 0.943*c* ($\gamma = 3.00$). She observes a stationary clock measuring a time interval of 75.0 s between two events in Spaceship Q. This is a proper time interval.

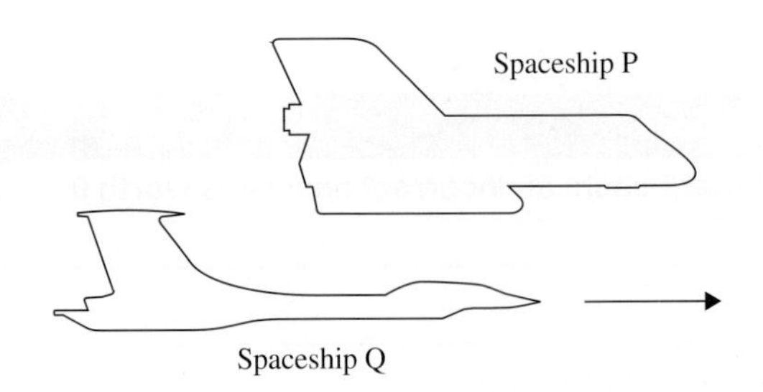

Which one of the following is closest to the time interval observed between the two events in Spaceship P's frame of reference?

A. 15.0 s
B. 25.0 s
C. 125 s
D. 225 s

Question 5

Source: *VCE 2021 Physics Exam, NHT, Section A, Q.20*; © VCAA

A nucleus in an excited energy state emits a gamma ray of energy 3.6×10^{-13} J as it decays to its ground state. The initial mass of the excited nucleus is M_i. The final mass of the nucleus after decay is closest to

A. $M_i - 4 \times 10^{-30}$ kg
B. $M_i - 8 \times 10^{-30}$ kg
C. M_i kg
D. $M_i + 4 \times 10^{-30}$ kg

Question 6

Source: *VCE 2018 Physics Exam, NHT, Section A, Q.11*; © VCAA

An alien spaceship has entered our solar system and is heading directly towards Earth at a speed of 0.6*c*, as shown in the diagram below. When it reaches a distance of 3.0×10^{11} m from Earth (in Earth's frame of reference), the aliens transmit a 'be there soon' signal via a laser beam.

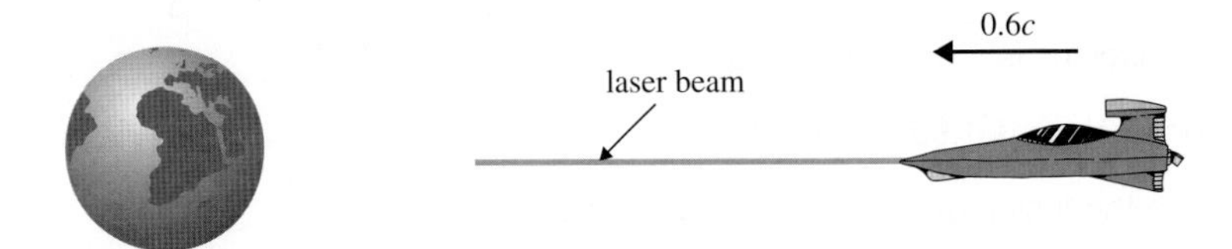

How long will it take for the signal to reach Earth according to an observer on Earth?

A. 1.0 s
B. 1.7 s
C. 625 s
D. 1000 s

Question 7

Source: *VCE 2019 Physics Exam, NHT, Section A, Q.16*; © VCAA

In a particle accelerator, magnesium ions are accelerated to 20.0% of the speed of light. Which one of the following is closest to the Lorentz factor, γ, for the magnesium ions at this speed?

A. 1.02
B. 1.12
C. 1.20
D. 2.24

Question 8

Source: *VCE 2019 Physics Exam, NHT, Section A, Q.17*; © VCAA

The lifetime of stationary muons is measured in a laboratory to be 2.2 μs. The lifetime of relativistic muons produced in Earth's upper atmosphere, as measured by ground-based scientists, is 16 μs. The resulting time dilation observed by the scientists gives a Lorentz factor, γ, of

A. 0.14
B. 1.4
C. 3.5
D. 7.3

Question 9

Source: *VCE 2019 Physics Exam, NHT, Section A, Q.18*; © VCAA

If a particle's kinetic energy is 10 times its rest energy, E_{rest}, then the Lorentz factor, γ, would be closest to

A. 9
B. 10
C. 11
D. 12

Question 10

Source: *VCE 2018 Physics Exam, NHT, Section A, Q.10*; © VCAA

A linear accelerator (linac) accelerates an electron beam to an energy of 100 MeV over a distance of about 10 m. After the first metre of acceleration in the linac, the electrons are travelling at approximately 99.9% of the speed of light. The Lorentz factor, γ, for an electron travelling at this speed would be closest to

A. 22.4
B. 44.8
C. 500
D. 1000

Section B — Short answer questions

Question 11 (2 marks)

Source: *VCE 2022 Physics Exam, Section B, Q.9*; © VCAA

A star is transforming energy at a rate of 2.90×10^{25} W.

Explain the type of transformation involved and what effect, if any, the transformation would have on the mass of the star. No calculations are required.

Question 12 (4 marks)

Source: *VCE 2021 Physics Exam, NHT, Section B, Q.10*; © VCAA

Jacinta is standing still while observing a spaceship passing Earth at a speed of 0.984*c*.

a. Calculate γ for this speed, correct to three significant figures. Show your working. **(2 marks)**

b. The spaceship is travelling to the Alpha Centauri star system in a straight line at this speed. In Jacinta's frame of reference, this distance is measured to be 4.37 light-years (that is, it would take light 4.37 years to travel this distance).

Calculate the time that would be measured by Jacinta for the spaceship's journey, correct to three significant figures. Show your working. **(2 marks)**

Question 13 (3 marks)

Source: *VCE 2019 Physics Exam, NHT, Section B, Q.17*; © VCAA

A spaceship is travelling from Earth to the star system Epsilon Eridani, which is located 10.5 light-years from Earth as measured by Earth-based instruments.

If the spaceship travels at 0.85*c* ($\gamma = 1.90$), determine the duration of the flight as measured by the astronauts on the spaceship travelling to Epsilon Eridani. Take one light-year to be 9.46×10^{15} m. Show your working.

Question 14 (3 marks)

Source: *VCE 2019 Physics Exam, NHT, Section B, Q.18*; © VCAA

Alien astronauts are travelling between star systems aboard a cube-shaped spaceship, as shown in Figure 16. The sides of the cube along the *x*-axis, *y*-axis and *z*-axis measure 3.20×10^3 m in the spaceship's frame of reference.

The spaceship passes Bob, who is on a space station, at speed $v = 0.990c$ ($\gamma = 7.09$).

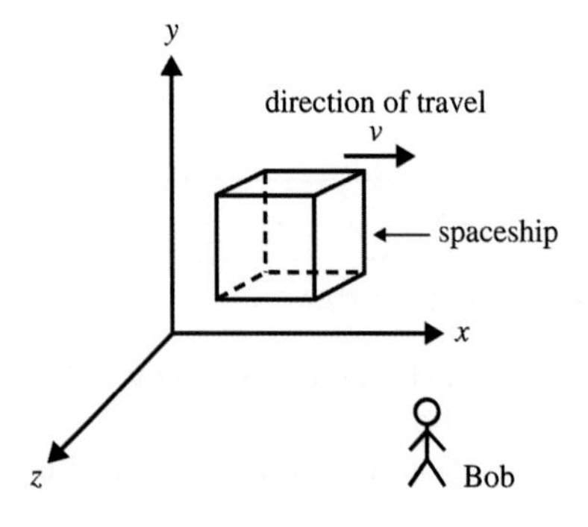

Figure 16

In the table below, determine the dimensions of the cube-shaped spaceship as measured from Bob's frame of reference and explain your reasoning.

length of side along *x*-axis	m
length of side along *y*-axis	m
length of side along *z*-axis	m

Question 15 (3 marks)

Source: *VCE 2018 Physics Exam, NHT, Section B, Q.14*; © VCAA

An Earth-like planet has been discovered orbiting a distant star. A hypothetical mission to this planet is suggested. The planet is 1.0×10^{18} m from Earth. The spaceship suggested for the mission can travel at an average speed of 0.99c. Take $\gamma = 7.1$ for this speed.

Scientists are concerned about the length of time the passengers would have to spend on the spaceship to travel to this planet. Use principles of special relativity to estimate this time, in years, as measured on the spaceship.

UNIT 4 | AREA OF STUDY 1 REVIEW

AREA OF STUDY 1 How has understanding about the physical world changed?

OUTCOME 1

Analyse and apply models that explain the nature of light and matter, and use special relativity to explain observations made when objects are moving at speeds approaching the speed of light.

PRACTICE EXAMINATION

STRUCTURE OF PRACTICE EXAMINATION		
Section	**Number of questions**	**Number of marks**
A	20	20
B	7	30
	Total	**50**

Duration: 50 minutes

Information:

- This practice examination consists of two parts. You must answer all question sections.
- Pens, pencils, highlighters, erasers, rulers and a scientific calculator are permitted.
- You may use the VCAA Physics formula sheet for this task.

Resources

Weblink VCAA Physics formula sheet

SECTION A — Multiple choice questions

All correct answers are worth 1 mark each; an incorrect answer is worth 0.

1. Electrons transition between energy levels with a difference of 2.78 eV. What is the expected frequency of the photons absorbed or emitted during the transition?

A. 1.5×10^{14} Hz

B. 2.1×10^{14} Hz

C. 4.2×10^{14} Hz

D. 6.7×10^{14} Hz

2. A spaceship travels past Earth at a velocity of $v = 0.7c$ and is heading towards the Moon. It transmits an electromagnetic signal that propagates in all directions at the speed of light, c.
To an observer on Earth, what is the speed of the electromagnetic signal just as the spaceship passes Earth?

A. $0.3c$

B. $0.7c$

C. c

D. $1.7c$

3. An experiment with carbon buckyballs $(m = 1.20 \times 10^{-24}\ \text{kg})$ demonstrated diffraction that was consistent with a de Broglie wavelength of 2.5×10^{-12} m. What was the velocity of the buckyballs?
   A. $2.2 \times 10^{1}\ \text{m s}^{-1}$
   B. $2.2 \times 10^{2}\ \text{m s}^{-1}$
   C. $2.2 \times 10^{3}\ \text{m s}^{-1}$
   D. $2.2 \times 10^{4}\ \text{m s}^{-1}$
4. The following two diagrams depict waves:

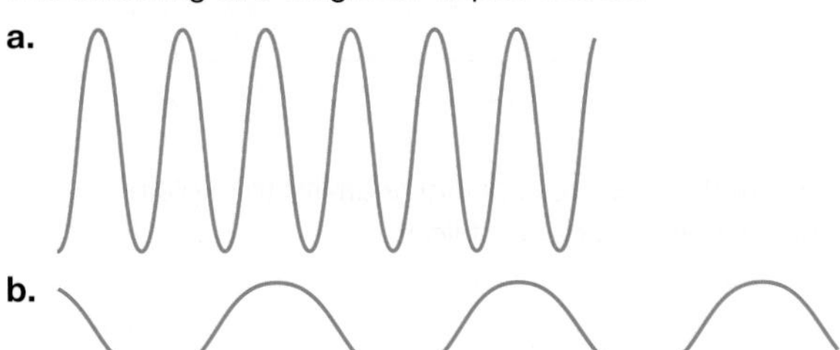

   Diagram (a) represents a bright, green light source. The vertical scale represents the amplitude of the waves and the horizontal scale represents time. The scale is the same in both diagrams.
   What is the best description for the waves in diagram (b)?
   A. Brighter, red light
   B. Dimmer, red light
   C. Brighter, blue light
   D. Dimmer, blue light
5. The brightness of a light source depends on:
   A. the amplitude of the electromagnetic wave.
   B. the speed of the electromagnetic wave.
   C. the frequency of the electromagnetic wave.
   D. the period of the electromagnetic wave.
6. The lifetime of stationary muons is measured in a laboratory to be 2.2 μs. The lifetime of relativistic muons produced in Earth's upper atmosphere, and travelling at 0.995*c* as measured by ground-based scientists, is closest to:
   A. 0.2 s.
   B. 0.7 s.
   C. 7.4 s.
   D. 22 s.
7. The following diagram depicts a standing wave on a string that is fixed at both ends.

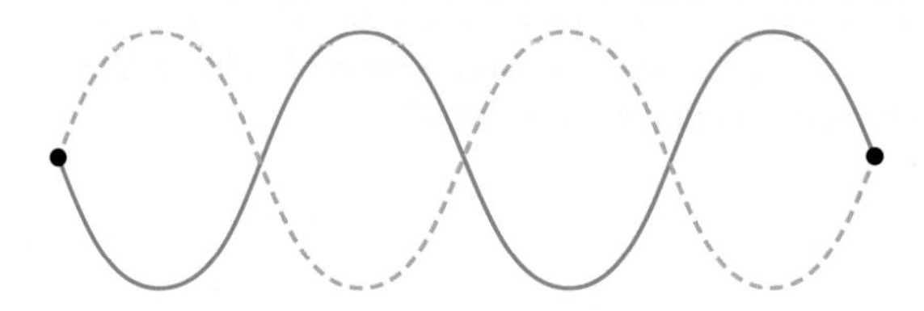

   The fundamental frequency of the wave is 220 Hz. What is the frequency shown in the diagram?
   A. 330 Hz
   B. 440 Hz
   C. 880 Hz
   D. 1760 Hz

8. In the study of the DNA structure, X-rays were used to discern the double-helix configuration of the DNA. DNA molecules are in the range of 3×10^{-9} m.
Which is the best reason for the ability of X-rays to do this?
   A. The wavelengths of X-rays are smaller than the scale of the DNA molecule.
   B. The intensity of the X-rays is sufficient to light up the DNA molecule.
   C. X-ray frequencies are low enough to distinguish the edges of the DNA molecule.
   D. X-ray wavelengths are long enough to distinguish the edges of the DNA molecule.

9. Experiments with photoelectrons provided the key evidence for the particle-like nature of light. If light had behaved in a wave-like manner, which of the following experimental observations would *not* have been made?
   A. A higher intensity light source would produce photoelectrons with higher kinetic energy; thus, a higher stopping voltage would be required.
   B. A lower frequency light source would still produce photoelectrons, but with a time delay.
   C. A higher frequency light source would produce photoelectrons with higher kinetic energy; thus, a higher stopping voltage would be required.
   D. Photoelectrons would be produced at all frequencies.

10. The following graph illustrates photoelectron kinetic energy versus photon frequency. The line labelled Q indicates the relationship for sodium metal. The relationship for another metal, which has a higher work function, Φ, is indicated by which line?

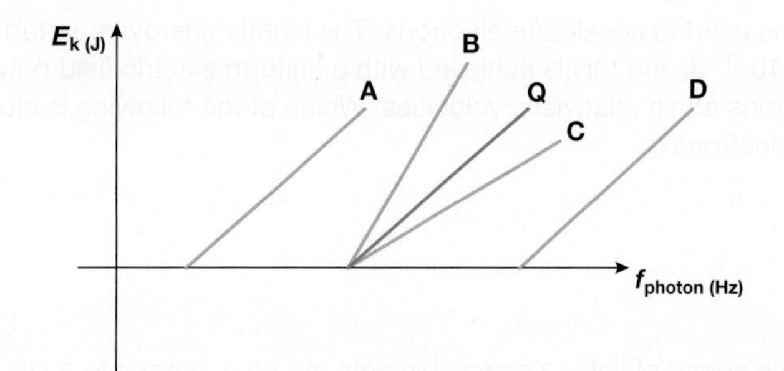

   A. A
   B. B
   C. C
   D. D

11. In the photoelectron experiment, which one of the following would *not* have been an observation supporting the particle-like nature of light?
   A. A photocurrent was produced even at a low frequency.
   B. Increasing the intensity of the light increased the photocurrent.
   C. A higher frequency of light required a higher stopping voltage.
   D. There is a cut-off frequency below which no photocurrent was produced.

12. Light with a frequency of 7.30×10^{14} Hz is incident on a metal with a work function of 1.23 eV. The expected kinetic energy of the photoelectrons is closest to:
   A. 1.79 eV.
   B. 3.02 eV.
   C. 4.25 eV.
   D. 4.87 eV.

13. A photoelectron experiment determined that the cut-off frequency of a metal is 5.56×10^{14} Hz. What is the work function of that metal?
   A. 8.0×10^{-20} J
   B. 1.2×10^{-19} J
   C. 2.3×10^{-19} J
   D. 3.7×10^{-19} J

14. A diffraction pattern produced by X-rays has exactly the same spacing between fringes as a diffraction pattern produced by electrons. Which of the following statements is correct concerning the X-rays and the electrons?
   A. The frequency of the X-rays matches the frequency of the electrons.
   B. The photon energy of the X-rays matches the kinetic energy of the electrons.
   C. The momentum of the X-rays matches the momentum of the electrons.
   D. The intensity of the X-rays matches the intensity of the electrons.
15. The de Broglie wavelength of a golf ball ($m = 0.046$ kg, $d = 0.043$ m) travelling at 70 m s^{-1} is closest to:
   A. 2.1×10^{-34} m.
   B. 3.4×10^{-31} m.
   C. 6.7×10^{-31} m.
   D. There is no de Broglie wavelength as the golf ball cannot diffract.
16. The Michelson–Morley experiment was an experiment to detect the supposed aether through which light propagates, by measuring the speed of light in perpendicular directions.
   The null result of the Michelson–Morley experiment supports:
   A. the existence of a medium through which light propagates.
   B. the independence of the speed of light from the motion of the observer.
   C. the particle model of light.
   D. the wave nature of matter.
17. A particle accelerator is used to accelerate electrons. The kinetic energy imparted to each electron is estimated to be 8.2×10^{-14} J, and this is achieved with a uniform electric field between two parallel plates. The accelerated electrons attain relativistic velocities. Which of the following is closest to the value of the Lorentz factor of the electrons?
   A. 1.0
   B. 1.5
   C. 2.0
   D. 2.5
18. Travelling at 70% of the speed of light, a spaceship sets out on a journey to a star that is 3.9×10^{18} m from Earth, as measured by observers on Earth.
   Which of the following statements is correct?
   A. The Lorentz factor of the spaceship is 1.3.
   B. According to observers on Earth, the trip takes 1.3×10^{10} s.
   C. The distance between Earth and the star, as measured by the occupants of the spaceship, is 2.7×10^{18} m.
   D. According to the astronauts on the spaceship, the trip takes 1.3×10^{10} s.
19. Neutrino particles from the Sun often pass through Earth without interacting at all gravitationally, electrically or magnetically. One such neutrino passes through Earth at a velocity of $v = 0.87c$. Earth's diameter is 1.3×10^{7} m measured in our frame of reference. Measured from the neutrino's frame of reference, what is Earth's diameter?
   A. 3.2×10^{6} m
   B. 6.4×10^{6} m
   C. 1.3×10^{7} m
   D. 2.6×10^{7} m
20. Which of the following statements about producing electromagnetic waves is the most correct?
   A. Electromagnetic waves can only be produced by accelerating electric charges.
   B. Electron energy-level transitions can produce electromagnetic waves.
   C. All electromagnetic waves are produced by electron energy-level transitions.
   D. Electromagnetic waves cannot be produced by accelerating electric charges.

Section B — Short answer questions

Question 21 (3 marks)

A beam of light has a frequency of 4.7×10^{14} Hz.

a. Determine the wavelength for a single photon. **(1 mark)**

b. Calculate the energy for a single photon. Give your answer in electron volts.
Planck's constant: $h = 6.63 \times 10^{-34}\ \text{m}^2\ \text{kg s}^{-1} = 4.14 \times 10^{-15}\ \text{eV s}$ **(1 mark)**

c. Calculate the momentum for a single photon. **(1 mark)**

Question 22 (4 marks)

Coherent light of wavelength 530.0 nm from a laser passes through a pair of slits, and an interference pattern is formed.

a. Calculate the path difference for the third dark band from the central maximum. Give your answer in nm, to 4 significant figures. **(2 marks)**

b. Calculate the path difference for the second bright band from the central maximum. Give your answer in nm, to 4 significant figures. **(2 marks)**

Question 23 (8 marks)

In a photoelectric experiment, a group of students measured the threshold frequency of the metal electrode to be 4.3×10^{14} Hz.

a. Determine the work function for the metal. Give your answer in joules.
Planck's constant: $h = 6.63 \times 10^{-34}\ \text{m}^2\ \text{kg s}^{-1}$ **(1 mark)**

Light of frequency 5.7×10^{14} Hz now strikes the photocell.

b. Calculate the maximum energy of emitted electrons, in joules. **(2 marks)**

c. Calculate the stopping voltage for the photoelectrons.
Charge of an electron: $e = 1.6 \times 10^{-19}$ C **(2 marks)**

d. Discuss the effects on the size of the photocurrent and the stopping voltage when the light intensity is increased. **(3 marks)**

Question 24 (2 marks)

An interference pattern is produced using slits separated by 0.15 mm, and a monochromatic light source casting a pattern onto a screen that is 1.65 m from the slits. The distance between two adjacent bright fringes is measured at 5.8 mm.

Calculate the wavelength of the light source used.

Question 25 (4 marks)

Answer the following.

a. Explain how electron diffraction is evidence in support of the wave nature of matter. **(2 marks)**

b. Explain how the photoelectric effect is evidence in support of the particle model of light. **(2 marks)**

Question 26 (3 marks)

Electrons at the ground state ($n = 1$) of an atom may absorb energy and transition to higher energy levels. They transition back to the ground state, directly or via an intermediate energy level, by emitting photons.

a. In one experiment, an electron transitions from $n = 1$ to $n = 3$. On the diagram provided, draw arrows to represent all the possible transitions for an electron at $n = 3$ back to the ground state. **(1 mark)**

b. An electron at $n = 4$ transitions to $n = 2$. Calculate the expected frequency of the photon emitted in this transition. **(2 marks)**

Question 27 (6 marks)

X-rays with energy of 2.45×10^4 eV generate a diffraction pattern with fringe spacing that matches the diffraction pattern using a stream of accelerated electrons, as shown in the following images.

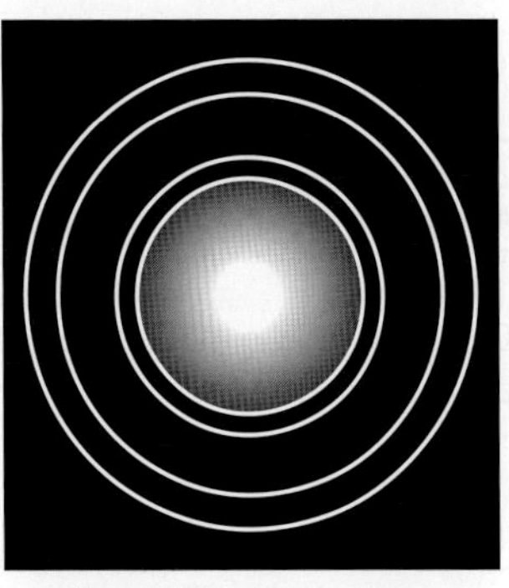

X-ray diffraction pattern

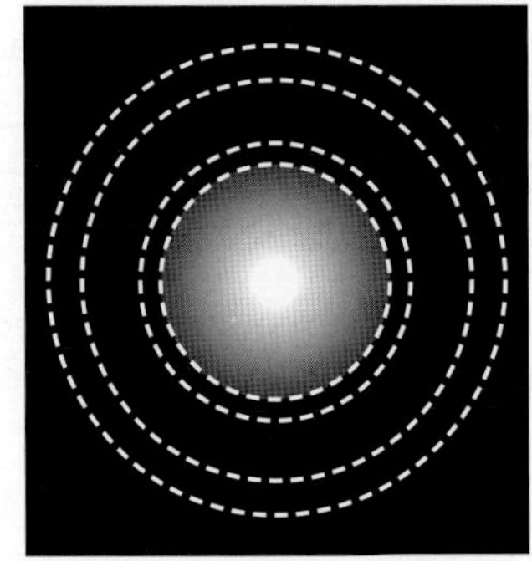

Electron diffraction pattern

a. Determine the wavelength of the X-rays used.
Planck's constant: $h = 6.63 \times 10^{-34}$ m^2 kg s^{-1} $= 4.14 \times 10^{-15}$ eV s **(1 mark)**

b. State the de Broglie wavelength of the electrons used to generate the same diffraction pattern. **(1 mark)**

c. Calculate the kinetic energy imparted to the electrons if they start from rest. **(2 marks)**

d. Calculate the accelerating voltage of the electric field used if the electrons start from rest.
Charge of an electron: $e = 1.6 \times 10^{-19}$ C **(2 marks)**

PRACTICE SCHOOL-ASSESSED COURSEWORK

ASSESSMENT TASK — PROBLEM-SOLVING; APPLYING PHYSICS CONCEPTS AND SKILLS TO REAL-WORLD CONTEXTS

In this task, you will be required to apply physics concepts and skills to the provided stimulus material.

- This practice SAC requires you to apply your problem-solving skills to the provided stimulus material related to the properties of light and matter.
- You may use the VCAA Physics formula sheet and a scientific calculator for this task.

Total time: 55 minutes (5 minutes reading, 50 minutes writing)

Total marks: 35 marks

REAL-WORLD CONTEXT: LIGHT EMISSION SPECTRUM

When a gas of atoms of a single element is struck by white light, the gas emits a series of discrete colours. The emitted light can be passed through a spectrometer to produce what is commonly known as an emission spectrum. The following diagram illustrates how the light emitted by the gas is collected, to produce an emission spectrum.

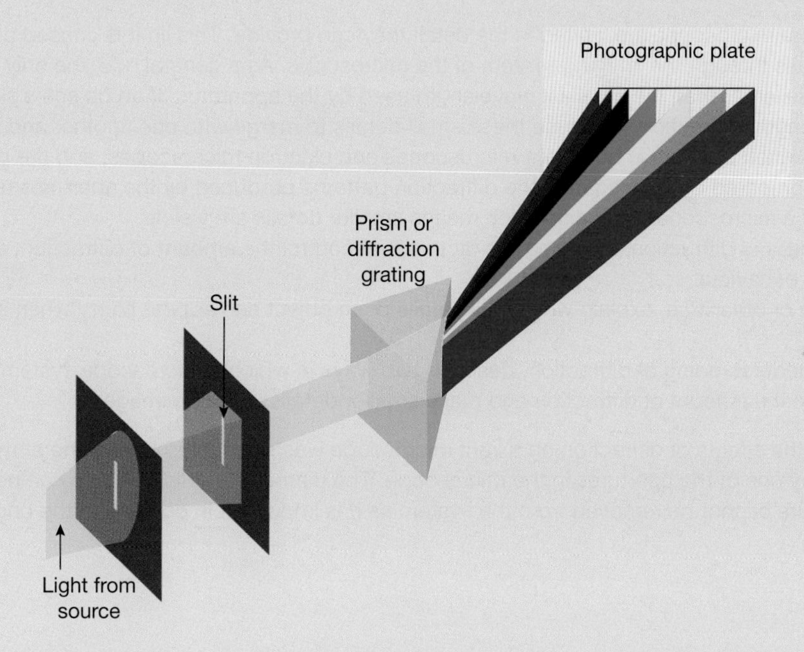

From the emission spectrum, the energy level diagram for the atom can be deduced. Each element produces a unique set of spectral lines, and thus elements can be identified by their line spectra. This is how astronomers can identify the composition of stars, for instance. In one particular experiment, eight different wavelengths were established from an emission spectrum of a monatomic gas. The frequency and energy of the photons for each of the wavelengths can then be determined. The data is shown in the provided table.

Wavelength (nm)	Energy (eV)
185	
254	
365	
405	
436	
546	
578	
1014	

Answer the following questions, using the stimulus material and the data provided.

1. Copy and complete the table by calculating the energy, in eV, corresponding to each wavelength. Show your full working for one of the wavelengths.
2. Explain the emission of a photon by an atom, referencing electron energy states and quantised states.
3. Using the information in the second column of your table, construct an energy-level diagram for the element in the gas where the ground state has an energy of 0 eV. Your energy-level diagram should consist of the ground state and the first three excited states. (The ionisation level for this element is 10.4 eV above the ground state, and should also be shown on your diagram.)
4. In a subsequent experiment, the gas of atoms is illuminated by a beam of photons with energy 2.6 eV. Explain whether there will be a visible line in the spectrum collected this time.

REAL-WORLD CONTEXT: DIFFRACTION IN MICROSCOPY

Images obtained by a microscope are limited in the detail they can provide. This limit is caused by the diffraction of waves as they pass through the imaging system of the microscope. As a general rule, the only objects observable are those larger than the smallest wavelength used by the apparatus. If an object is smaller than the smallest wavelength, diffraction will cause these small details to merge with one another and be impossible to distinguish. This situation appears in both light microscopes and electron microscopes, and the phenomenon can be measured and compared by investigating the diffraction patterns produced by the apertures used within the imaging systems of a microscope. Less diffraction means smaller details are visible.

5. Explain the process of diffraction, including the factors that control the amount of diffraction, and classify it as wave or particle behaviour.
6. Using a diagram or otherwise, explain why small details of an object can appear blurry when seen under a microscope.
7. Based on your understanding of diffraction, describe two ways in which a microscope system could be altered to reduce the amount of diffraction and make smaller details visible in images.

An investigation of the effects of diffraction on a light microscope was conducted by looking at the diffraction pattern produced by one of the apertures in the microscope. The diameter of this aperture can be changed; however, the aperture cannot be removed from the system as it is important in controlling the brightness of the image produced.

Two diffraction patterns were obtained from the same microscope, but under different conditions. These images are shown here.

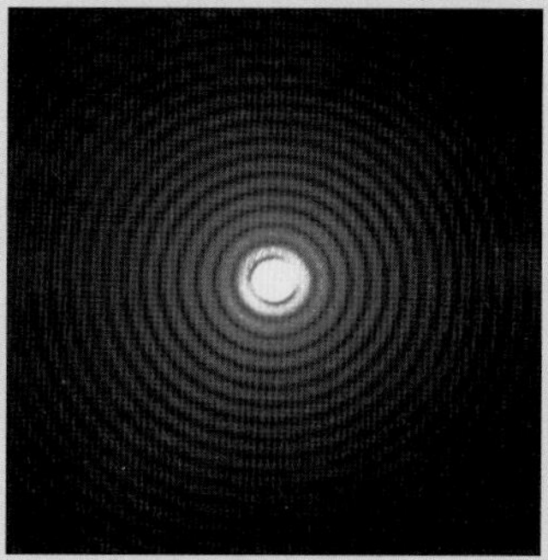
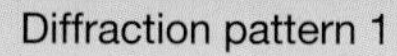
Diffraction pattern 1

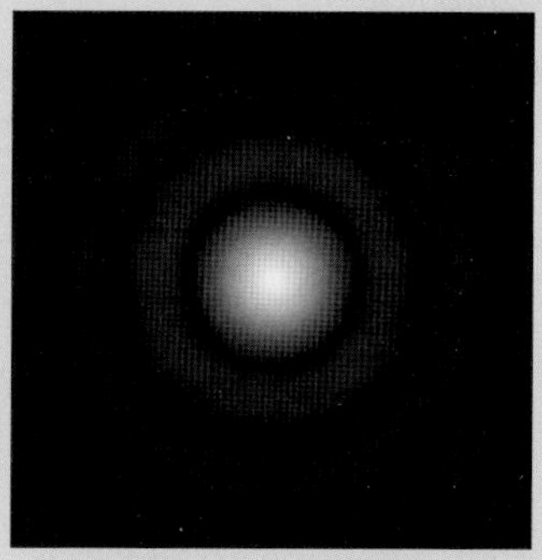
Diffraction pattern 2

8. Draw the diffraction pattern that would be expected if particles or waves of a lower energy compared to diffraction pattern 2 were imaged using the same system.

REAL-WORLD CONTEXT: PHOTOEMISSION ELECTRON MICROSCOPE

The photoelectric effect is the emission of electrons when electromagnetic radiation hits a material. This effect can be used for electron microscopy. Photoemission electron microscopy uses local variations in photoelectrons emitted by a sample to generate a greyscale image consisting of thousands of spots of varying intensity corresponding to the topography of the sample. X-ray or UV photons are directed at a sample surface, and an electron optical column uses the emitted photoelectrons to create a magnified image. In addition, the photoelectrons can be filtered in energy.

Photoemission electron microscopy is used to study the morphology, electronic and chemical properties and the magnetic structures of surfaces and thin film materials with nanometre-scale spatial resolution.

9. Explain how the photoelectric effect is evidence for the particle-like nature of light.
10. Using a diagram or otherwise, explain how, in this context, photon energy is transferred to electrons in the sample, referring to photon energy, kinetic energy of a photoelectron and work function of the sample.

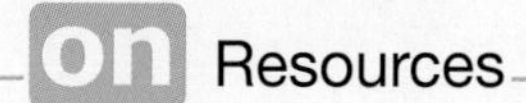
Resources

Digital document School-assessed coursework (doc-39424)

12 Scientific investigations

KEY KNOWLEDGE

online only

In this Area of Study, you will design and conduct a scientific investigation related to fields, motion or light, and present an aim, methodology and method, results, discussion and a conclusion in a scientific poster.

Investigation design

- identify the physics concepts specific to the investigation and explain their significance, including definitions of key terms, and physics representations
- explain the characteristics of the selected scientific methodology and method including: techniques of primary qualitative and quantitative data generation relevant to the selected investigation; and appropriateness of the use of independent, dependent and controlled variables in the selected scientific investigation
- identify and apply concepts of accuracy, precision, repeatability, reproducibility, resolution, and the identification of, and distinction between, error and uncertainty
- identify and apply health, safety and ethical guidelines relevant to the selected investigation

Scientific evidence

- discuss the nature of evidence that supports or refutes a hypothesis, model or theory
- apply methods of organising, analysing and evaluating primary data to identify patterns and relationships including: the physical significance of the gradient of linearised data; causes of uncertainty; use of uncertainty bars; and assumptions and limitations of data, methodologies and methods
- model the scientific practice of using a logbook to authenticate primary data

Science communication

- apply the conventions of science communication: scientific terminology and representations; symbols, equations and formulas; standard abbreviations; significant figures; and units of measurements
- apply the conventions of scientific poster presentation, including succinct communication of the selected scientific investigation, and acknowledgment of references
- explain the key findings and implications of the selected investigation.

KEY SCIENCE SKILLS

- Develop aims and questions, formulate hypotheses and make predictions
- Plan and conduct investigations
- Comply with safety and ethical guidelines
- Generate, collate and record data
- Analyse and evaluate data and investigation methods
- Construct evidence-based arguments and draw conclusions
- Analyse, evaluate and communicate scientific ideas

This topic is available online at **www.jacplus.com.au.**

12.1 Scientific investigations

Hey students! Bring these pages to life online

Watch videos

Engage with interactivities

Answer questions and check results

Find all this and MORE in jacPLUS

LEARNING SEQUENCE

UNIT 4 | AREA OF STUDY 2 REVIEW

AREA OF STUDY 2 How is scientific inquiry used to investigate fields, motion or light?

OUTCOME 2

Design and conduct a scientific investigation related to fields, motion or light, and present an aim, methodology and method, results, discussion and a conclusion in a scientific poster.

The key science skills are a core component of the study of VCE Physics and apply across Units 1 to 4 in all Areas of Study.

- Develop aims and questions, formulate hypotheses and make predictions
- Plan and conduct investigations
- Comply with safety and ethical guidelines
- Generate, collate and record data
- Analyse and evaluate data and investigation methods
- Construct evidence-based arguments and draw conclusions
- Analyse, evaluate and communicate scientific ideas

Information:

- This Area of Study Review is a collation of past VCAA exam questions that focus on key science skills.

APPENDIX 1 Formulae and data

Velocity; acceleration	$\mathbf{v} = \frac{\Delta \mathbf{s}}{\Delta t}$; $\mathbf{a} = \frac{\Delta \mathbf{v}}{\Delta t}$
Equations for constant acceleration	$\mathbf{v} = \mathbf{u} + \mathbf{a}t$ $\mathbf{s} = \mathbf{u}t + \frac{1}{2}\mathbf{a}t^2$ $\mathbf{v}^2 = \mathbf{u}^2 + 2\mathbf{as}$ $\mathbf{s} = \frac{1}{2}(\mathbf{v} + \mathbf{u})t$
Newton's Second Law of Motion	$\Sigma \mathbf{F} = m\mathbf{a}$
Circular motion	$a = \frac{v^2}{r} = \frac{4\pi^2 r}{T^2}$
Hooke's Law	$\mathbf{F} = -k\mathbf{x}$
Elastic potential energy	$E_e = \frac{1}{2}kx^2$
Kinetic energy	$E_k = \frac{1}{2}mv^2$
Newton's Law of Universal Gravitation	$F = G\frac{m_1 m_2}{r^2}$
Gravitational field strength	$g = G\frac{M}{r^2}$
Gravitational potential energy near Earth's surface	$E_g = mg\Delta h$
Impulse	$\mathbf{I} = \mathbf{F}\Delta t = m\Delta \mathbf{v}$
Momentum	$\mathbf{p} = m\mathbf{v}$
Lorentz factor	$\gamma = \frac{1}{\sqrt{1 - \frac{v^2}{c^2}}}$
Time dilation	$t = t_0\gamma$
Length contraction	$L = \frac{L_0}{\gamma}$
Relativistic mass	$m = m_0\gamma$

(continued)

(continued)

Relativistic total energy	$E_{total} = \gamma mc^2$
Rest energy	$E_{rest} = mc^2$
Relativistic kinetic energy	$E_k = (\gamma - 1)\, mc^2$
Magnetic force on a current-carrying conductor	$F = nIlB$
Magnetic force on a moving charge	$\boldsymbol{F} = q\boldsymbol{vB}$
Radius of a charged particle in a magnetic field	$r = \frac{mv}{qB}$
Energy transformation for electrons in an electron gun (< 100 keV)	$\frac{1}{2}m_e v^2 = eV$
Radius of electron path in a magnetic field	$r = \frac{m_e v}{eB}$
Magnetic force on a moving charge, q	$\boldsymbol{F} = q\boldsymbol{vB}$
Electric field between charged plates	$E = \frac{V}{d}$
Energy transformation of charges in an electric field	$\frac{1}{2}mv^2 = qV$
Field of a point charge	$E = \frac{kq}{r^2}$
Force on an electric charge	$\boldsymbol{F} = q\boldsymbol{E}$
Coulomb's Law	$F = \frac{kq_1 q_2}{r^2}$
Voltage; power	$V = RI;\ P = VI = I^2R$
Resistors in series	$R_{equivalent} = R_1 + R_2$
Resistors in parallel	$\frac{1}{R_{equivalent}} = \frac{1}{R_1} + \frac{1}{R_2}$
Transformer action	$\frac{V_1}{V_2} = \frac{N_1}{N_2} = \frac{I_2}{I_1}$
AC voltage and current	$V_{RMS} = \frac{1}{\sqrt{2}}V_{peak};\ I_{RMS} = \frac{1}{\sqrt{2}}I_{peak}$
Electromagnetic induction	EMF : $\varepsilon = \frac{-N\Delta\Phi}{\Delta t}$; flux : $\Phi = B_{\perp}A$

(continued)

(continued)

Transmission losses	$V_{drop} = I_{line}R_{line}$; $P_{loss} = I_{line}^2 R_{line}$
Photoelectric effect	$E_{k\,max} = hf - \Phi$
Photon energy	$E = hf = \frac{hc}{\lambda}$
Photon momentum	$p = \frac{h}{\lambda}$
De Broglie wavelength	$\lambda = \frac{h}{p}$
Wave equation	$v = f\lambda$
Constructive interference	path difference $= n\lambda$
Destructive interference	path difference $= \left(n + \frac{1}{2}\right)\lambda$
Fringe spacing	$\Delta x = \frac{\lambda L}{d}$
Data and constants	
Universal gravitational constant	$G = 6.67 \times 10^{-11}\,\text{N m}^2\text{kg}^{-2}$
Mass of Earth	$M_E = 5.97 \times 10^{24}\,\text{kg}$
Radius of Earth	$R_E = 6.37 \times 10^6\,\text{m}$
Acceleration due to gravity at Earth's surface	$\boldsymbol{g} = 9.8\,\text{m s}^{-2}$ downwards
Mass of an electron	$m_e = 9.1 \times 10^{-31}\,\text{kg}$
Charge of an electron	$e = -1.6 \times 10^{-19}\,\text{C}$
Planck's constant	$h = 6.63 \times 10^{-34}\,\text{J s}$ $= 4.14 \times 10^{-15}\,\text{eV s}$
Speed of light in a vacuum	$c = 3.0 \times 10^8\,\text{m s}^{-1}$
Coulomb constant in air	$k = 8.99 \times 10^9\,\text{N m}^2\text{C}^{-2}$

Prefixes/units

p = pico = 10^{-12}	n = nano = 10^{-9}	μ = micro = 10^{-6}	m = milli = 10^{-3}
k = kilo = 10^3	M = mega = 10^6	G = giga = 10^9	t = tonne = 10^3 kg

Source: VCAA

APPENDIX 2 Periodic table

	Group 1	Group 2	Group 3	Group 4	Group 5	Group 6	Group 7	Group 8	Group 9
Period 1	1 Hydrogen **H** **1.0**								
Period 2	3 Lithium **Li** **6.9**	4 Beryllium **Be** **9.0**							
Period 3	11 Sodium **Na** **23.0**	12 Magnesium **Mg** **24.3**							
Period 4	19 Potassium **K** **39.1**	20 Calcium **Ca** **40.1**	21 Scandium **Sc** **45.0**	22 Titanium **Ti** **47.9**	23 Vanadium **V** **50.9**	24 Chromium **Cr** **52.0**	25 Manganese **Mn** **54.9**	26 Iron **Fe** **55.8**	27 Cobalt **Co** **58.9**
Period 5	37 Rubidium **Rb** **85.5**	38 Strontium **Sr** **87.6**	39 Yttrium **Y** **88.9**	40 Zirconium **Zr** **91.2**	41 Niobium **Nb** **92.9**	42 Molybdenum **Mo** **96.0**	43 Technetium **Tc** **(98)**	44 Ruthenium **Ru** **101.1**	45 Rhodium **Rh** **102.9**
Period 6	55 Caesium **Cs** **132.9**	56 Barium **Ba** **137.3**	57–71 Lanthanoids	72 Hafnium **Hf** **178.5**	73 Tantalum **Ta** **180.9**	74 Tungsten **W** **183.8**	75 Rhenium **Re** **186.2**	76 Osmium **Os** **190.2**	77 Iridium **Ir** **192.2**
Period 7	87 Francium **Fr** **(223)**	88 Radium **Ra** **(226)**	89–103 Actinoids	104 Rutherfordium **Rf** **(261)**	105 Dubnium **Db** **(262)**	106 Seaborgium **Sg** **(266)**	107 Bohrium **Bh** **(264)**	108 Hassium **Hs** **(267)**	109 Meitnerium **Mt** **(268)**

			Key
Period 1	1	2	← Atomic number
	Hydrogen	Helium	← Name
	H	**He**	← Symbol
	1.0	**4.0**	← Relative atomic mass

- Alkali metal
- Alkaline earth metal
- Transition metal
- Lanthanoids
- Actinoids
- Unknown chemical properties
- Post-transition metal
- Metalloid
- Reactive non-metal
- Halide
- Noble gas

Lanthanoids

57 Lanthanum **La** **138.9**	58 Cerium **Ce** **140.1**	59 Praseodymium **Pr** **140.9**	60 Neodymium **Nd** **144.2**	61 Promethium **Pm** **(145)**	62 Samarium **Sm** **150.4**	63 Europium **Eu** **152.0**

Actinoids

89 Actinium **Ac** **(227)**	90 Thorium **Th** **232.0**	91 Protactinium **Pa** **231.0**	92 Uranium **U** **238.0**	93 Neptunium **Np** **(237)**	94 Plutonium **Pu** **(244)**	95 Americium **Am** **(243)**

Group 10	Group 11	Group 12	Group 13	Group 14	Group 15	Group 16	Group 17	Group 18
								2 Helium **He** **4.0**
			5 Boron **B** **10.8**	6 Carbon **C** **12.0**	7 Nitrogen **N** **14.0**	8 Oxygen **O** **16.0**	9 Fluorine **F** **19.0**	10 Neon **Ne** **20.2**
			13 Aluminium **Al** **27.0**	14 Silicon **Si** **28.1**	15 Phosphorus **P** **31.0**	16 Sulfur **S** **32.1**	17 Chlorine **Cl** **35.5**	18 Argon **Ar** **39.9**
28 Nickel **Ni** **58.7**	29 Copper **Cu** **63.5**	30 Zinc **Zn** **65.4**	31 Gallium **Ga** **69.7**	32 Germanium **Ge** **72.6**	33 Arsenic **As** **74.9**	34 Selenium **Se** **79.0**	35 Bromine **Br** **79.9**	36 Krypton **Kr** **83.8**
46 Palladium **Pd** **106.4**	47 Silver **Ag** **107.9**	48 Cadmium **Cd** **112.4**	49 Indium **In** **114.8**	50 Tin **Sn** **118.7**	51 Antimony **Sb** **121.8**	52 Tellurium **Te** **127.6**	53 Iodine **I** **126.9**	54 Xenon **Xe** **131.3**
78 Platinum **Pt** **195.1**	79 Gold **Au** **197.0**	80 Mercury **Hg** **200.6**	81 Thallium **Tl** **204.4**	82 Lead **Pb** **207.2**	83 Bismuth **Bi** **209.0**	84 Polonium **Po** **(210)**	85 Astatine **At** **(210)**	86 Radon **Rn** **(222)**
110 Darmstadtium **Ds** **(271)**	111 Roentgenium **Rg** **(272)**	112 Copernicium **Cn** **(285)**	113 Nihonium **Nh** **(280)**	114 Flerovium **Fl** **(289)**	115 Moscovium **Mc** **(289)**	116 Livermorium **Lv** **(292)**	117 Tennessine **Ts** **(294)**	118 Oganesson **Og** **(294)**

64 Gadolinium **Gd** **157.3**	65 Terbium **Tb** **158.9**	66 Dysprosium **Dy** **162.5**	67 Holmium **Ho** **164.9**	68 Erbium **Er** **167.3**	69 Thulium **Tm** **168.9**	70 Ytterbium **Yb** **173.1**	71 Lutetium **Lu** **175.0**
96 Curium **Cm** **(247)**	97 Berkelium **Bk** **(247)**	98 Californium **Cf** **(251)**	99 Einsteinium **Es** **(252)**	100 Fermium **Fm** **(257)**	101 Mendelevium **Md** **(258)**	102 Nobelium **No** **(259)**	103 Lawrencium **Lr** **(262)**

APPENDIX 3
Astronomical data

	Mean radius of orbit		Orbital period			
	(AU)	(m)	(years)	(seconds)	Equatorial radius (m)	Mass (kg)
Sun					6.96×10^8	1.99×10^{30}
Mercury	0.387	5.79×10^{10}	0.241	7.60×10^6	2.44×10^6	3.29×10^{23}
Venus	0.723	1.08×10^{11}	0.615	1.94×10^7	6.05×10^6	4.87×10^{24}
Earth	1.00	1.50×10^{11}	1.00	3.16×10^7	6.37×10^6	5.97×10^{24}
Moon	2.57×10^{-3}	3.84×10^8	27.3 days*	2.36×10^6*	1.74×10^6	7.35×10^{22}
Mars	1.52	2.28×10^{11}	1.88	5.94×10^7	3.39×10^6	6.39×10^{23}
Jupiter	5.20	7.78×10^{11}	11.9	3.74×10^8	6.99×10^7	1.90×10^{27}
Saturn	9.58	1.43×10^{12}	29.5	9.30×10^8	5.82×10^7	5.68×10^{26}
Titan	8.20×10^{-3}	1.22×10^9	15.9 days*	1.37×10^6*	2.57×10^6	1.35×10^{23}
Uranus	19.2	2.87×10^{12}	84.0	2.65×10^9	2.54×10^7	8.68×10^{25}
Neptune	30.1	4.50×10^{12}	165	5.21×10^9	2.46×10^7	1.02×10^{26}
Pluto	39.48	5.91×10^{12}	248	7.82×10^9	1.19×10^6	1.31×10^{22}

*The orbital period for the Moon and Titan is the time it takes to complete one orbit around Earth and Saturn respectively. All other listed measurements for the orbital period shows the time to orbit the Sun.

The Milky Way	1.50×10^5 light-years across
Alpha Centauri	4.37 light-years away
Andromeda	2.25×10^6 light-years away
Edge of observable universe	4.65×10^{10} light-years away

Source: Data derived from https://solarsystem.nasa.gov.

Answers

1 Newton's laws of motion

1.2 BACKGROUND KNOWLEDGE Motion review

Sample problem 1

a. 3.3 m s^{-2} north
b. 10 m s^{-2} N37°E
c. 2.8 m s^{-2} south-east (or S45°E)

Practice problem 1

a. 3.0 m s^{-2} north
b. 7.1 m s^{-2} south-west
c. 7.1 m s^{-2} 39° south to east

Sample problem 2

a. 10 m south
b. 3.0 m s^{-2} south
c. 6.1 m s^{-1} south

Practice problem 2

a. $\boldsymbol{a} = 3.0 \text{ m s}^{-2}$ south
b. $\boldsymbol{a} = 5.0 \text{ m s}^{-2}$ north
c. $\boldsymbol{v}_{av} = 8.3 \text{ m s}^{-1}$ south

Sample problem 3

a. $\boldsymbol{a} = 3.8 \text{ m s}^{-2}$ down the slope
b. $t = 3.2$ s
c. $s = 4.7$ m
d. $t \approx 1.6$ s

Practice problem 3

a. $\boldsymbol{v} = 30 \text{ m s}^{-1}$
b. $\boldsymbol{a} = 0.63 \text{ m s}^{-2}$
c. $\boldsymbol{v}_{av} = 25 \text{ m s}^{-1}$
d. i. $\boldsymbol{v} = 21 \text{ m s}^{-1}$ ii. $\boldsymbol{v} = 25 \text{ m s}^{-1}$

1.2 Exercise

1. 3.5 m s^{-1}
2. a. $25 \text{ km h}^{-1}\text{s}^{-1}$ S37° E
   b. 6.9 m s^{-2} S37° E
3. a. 25.5 s
   b.

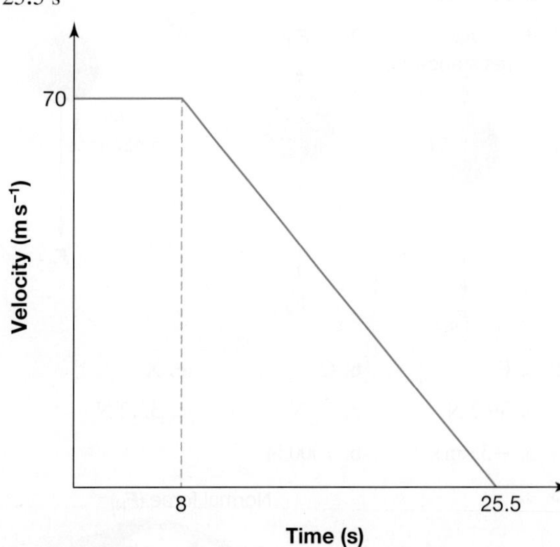

   c. 1172.5 m
4. 632 kg m s^{-1}

1.3 Newton's laws of motion and their application

Sample problem 4

a. $\boldsymbol{a} = 3.00 \text{ m s}^{-2}$ east
b. i. $\boldsymbol{a} = 2.30 \text{ m s}^{-2}$ east
   ii. $\boldsymbol{F}_{tc} = 1.12 \times 10^3$ N

Practice problem 4

a. i. $\boldsymbol{F}_{\text{driving force}} = 500$ N
   ii. $\boldsymbol{F}_T = 100$ N
b. i. $\sum \boldsymbol{F} = 1.2 \times 10^3$ N
   ii. $\boldsymbol{F}_T = 1.3 \times 10^3$ N
   iii. $\boldsymbol{F}_{\text{driving force}} = 4.5 \times 10^3$ N

Sample problem 5

a. $\boldsymbol{F}_N = 663$ N, rounded to 6.6×10^2 N
b. $\boldsymbol{F}_R = 178$ N, rounded to 1.8×10^2 N

Practice problem 5

a. i. $\boldsymbol{F} = 2.5 \times 10^2$ N
   ii. $\boldsymbol{F}_N \approx 8.5 \times 10^2$ N
b. $\boldsymbol{a} \approx 2.0 \text{ m s}^{-2}$

1.3 Exercise

1. a.

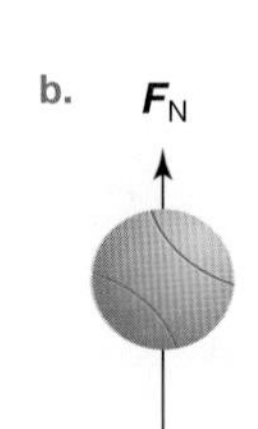

b.

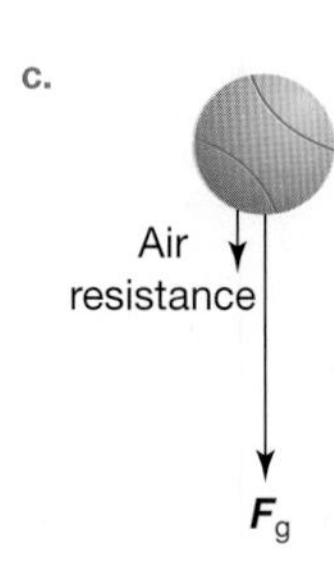

c.

Air resistance

F_g

2. a. F b. C c. X
3. a. 39.2 N b. 10 N c. 32.2 N
4. a. -35 m s^{-2} b. 7000 N
5. a.

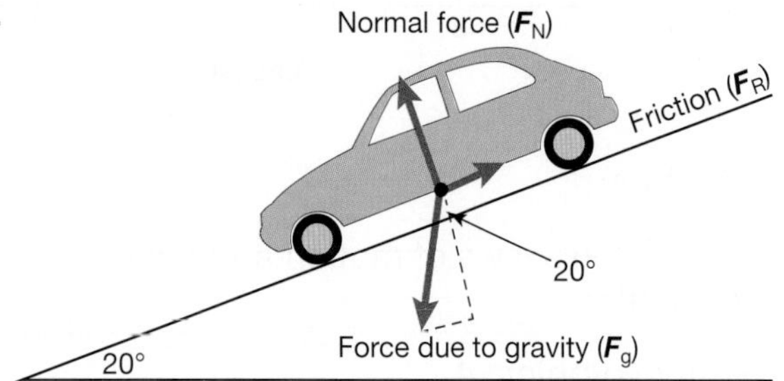

b. 1.4×10^4 N
c. The car is stationary, so the net force on it is zero.
d. 5.0×10^3 N

6. a. As the skier is moving at a constant speed there is no net force.
   b. 2.9×10^2 N
7. a. 3.8×10^2 N north
   b. 1.7×10^3 N north
8. 74 N
9. 8.7°

1.3 Exam questions

1. C
2. B
3. a.

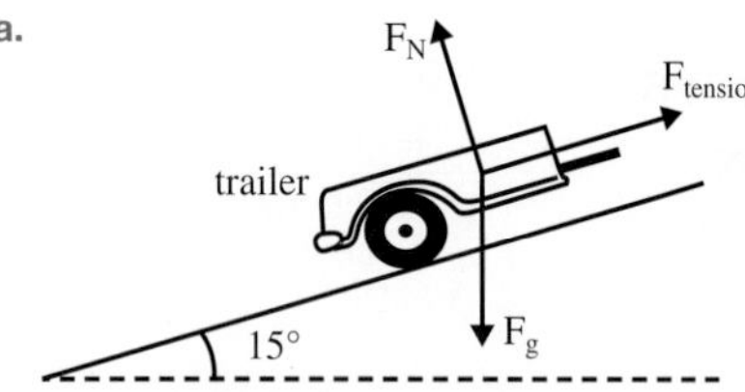

b. $F_{tow} = 5 \times 10^2$ N

4. $v = 2 \text{ m s}^{-1}$
5. $a = 2.0 \text{ m s}^{-2}$

1.4 Projectile motion

Sample problem 6

a. $t = 4.5$ s

b.

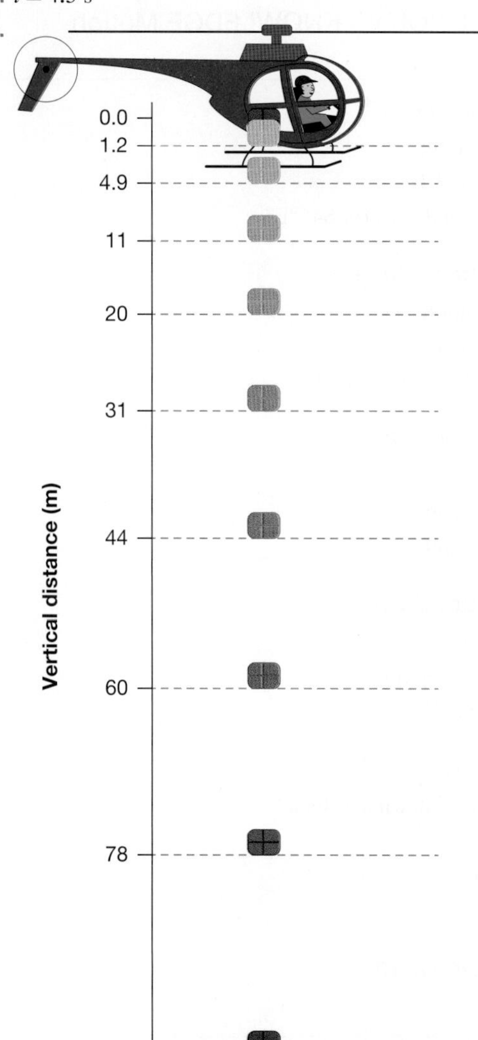

Practice problem 6

a. $s \approx 44$ m

b. $v = 2.9 \times 10^1 \text{ m s}^{-1}$

Sample problem 7

a. $t = 4.5$ s

b. $s = 90$ m

c. See figure at the bottom of the page*

Practice problem 7

a. The cliff is 76 m high

b. $s = 1.6 \times 10^2$ m

c. $v = -39$ m s^{-1}

d. $\theta = 44°$

Sample problem 8

a. $t = 0.41$ s

b. $s = 0.82$ m

c. 9.8 m s^{-2} downwards

d. $v = 4.0$ m s^{-1} downwards

Practice problem 8

a. $v = 3.1$ m s^{-1} downwards

b. $t = 0.64$ s

Sample problem 9

$s = 48$ m, the stunt driver won't make it and will land in the river.

Practice problem 9

a. $u_{vertical} \approx 13.5$ km h^{-1} and $u_{horizontal} \approx 29$ km h^{-1}

b. $t = 0.77$ s

c. $s = 6.2$ m

1.4 Exercise

1.

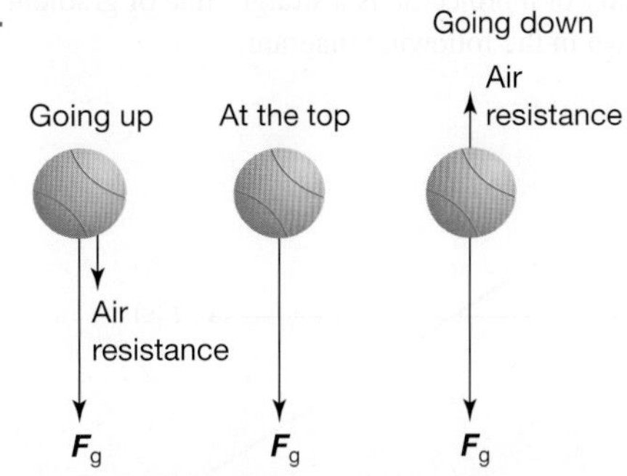

2. The acceleration of a projectile when in motion is due to gravity, which is a constant near Earth's surface (9.8 m s^{-2} downwards). This is displayed by the following acceleration–time graph:

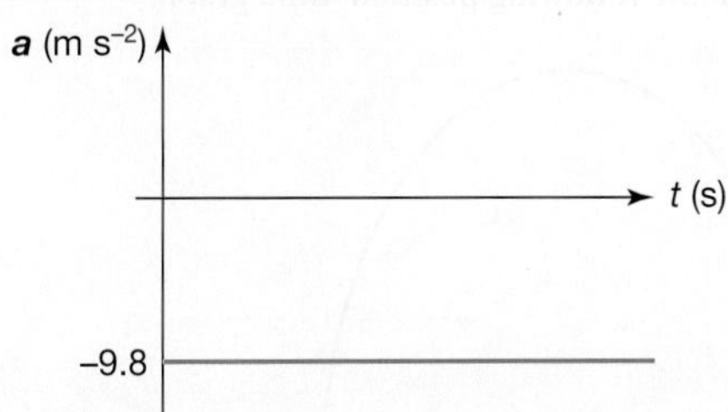

*c.

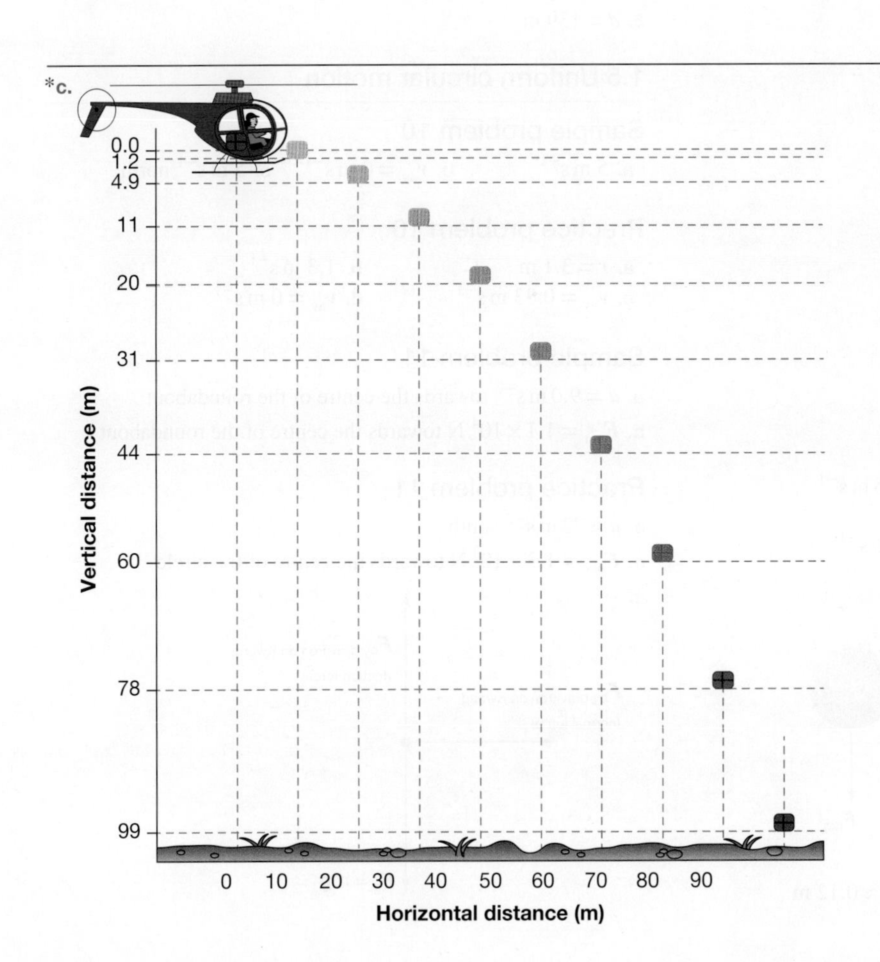

Acceleration is the change in velocity (or the gradient), so the velocity of a projectile is a straight line of gradient –9.8 as shown in the following diagram:

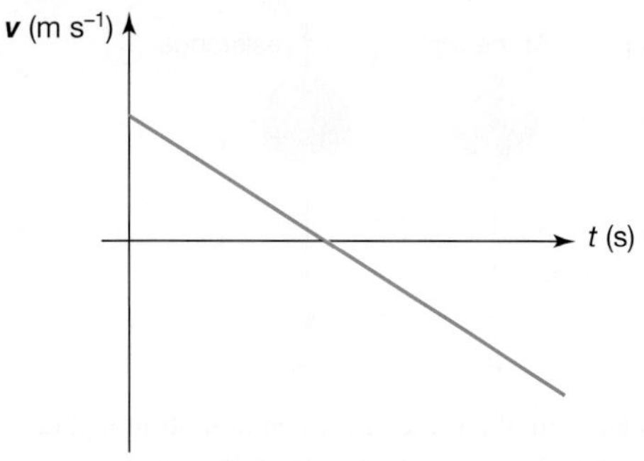

Velocity is the change in position (or the gradient), so the position of a projectile is a negative parabola. This is displayed on the following position–time graph:

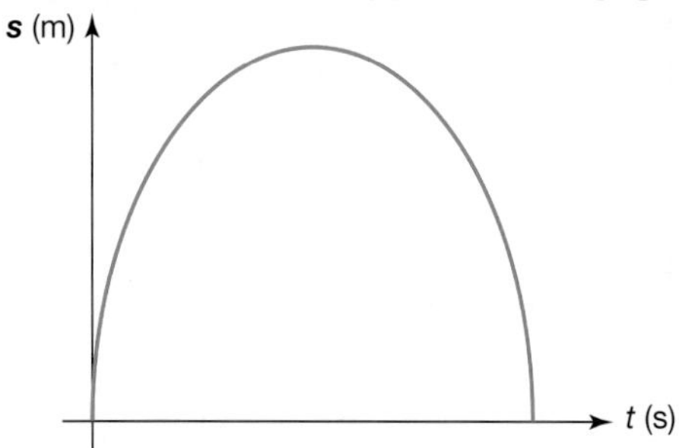

3. a. Vertical component:
   $v_v = 20 \sin 50° \approx 15\ \text{m s}^{-1}$
   Horizontal component:
   $v_h = 20 \cos 50° \approx 13\ \text{m s}^{-1}$
   b. Vertical component:
   $v_v = 11 \cos 23° \approx 10\ \text{m s}^{-1}$
   Horizontal component:
   $v_h = 11 \sin 23° \approx 4.3\ \text{m s}^{-1}$
   c. Vertical component:
   $v_v = 5\ \text{m s}^{-1}$
   Horizontal component:
   $v_h = 5 \sin 0° = 0\ \text{m s}^{-1}$
   d. Vertical component:
   $v_v = 10 \sin 0° = 0\ \text{km h}^{-1}$
   Horizontal component:
   $v_h = 10\ \text{km h}^{-1}$
   e. Vertical component:
   $v_v = 33 \cos 60°$ or $33 \sin 30° = 16.5\ \text{m s}^{-1}$
   Horizontal component:
   $v_h = 33 \sin 60°$ or $33 \cos 30° \approx 29\ \text{m s}^{-1}$
4. a. 3.67 s
   b. 16.5 m
   c. i.

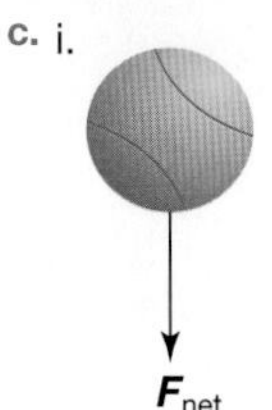

ii.

iii.

5. No
6. a. $\approx 4.76\ \text{m s}^{-1}$ b. $\approx 18°$ c. ≈ 0.12 m
7. a. ≈ 0.71 s
   b. Vertical component ≈ 2.45 m
   Horizontal component ≈ 4.92 m
   c. ≈ 0.30 s
   d. Into the net
8. a. ≈ 0.81 s b. ≈ 9.2 m
   c. ≈ 0.90 s d. ≈ 19.5 m
9. a. The range of the jump is 19.6 metres.
   b. The velocity of the water skier when they hit the water is $15.0\ \text{m s}^{-1}$ in the direction of 37 degrees below the horizontal.
10. $5.4\ \text{m s}^{-1}$
11. $\approx 18.5°$

1.4 Exam questions

1. a. Horizontal component of velocity: $v_H = 7.0 \cos 50° = 4.5\ \text{m s}^{-1}$

$$t = \frac{d}{v} = \frac{3.2}{4.5}$$
$$t = 0.71\ \text{s}$$

   b. The top of the basket is 3.5 m above the ground.
2. a. $d = 1.2$ m b. 0.8 m c. $v_f = 5\ \text{m s}^{-1}$
3. $x = 4.0$ m
4. a. $t = 2.0$ s b. $x = 50$ m c. $Ek_{final} = 1017$ J
5. $d = 139$ m

1.5 Uniform circular motion

Sample problem 10

a. $5\ \text{m s}^{-1}$ b. $v_{av} = 0\ \text{m s}^{-1}$ c. $5\ \text{m s}^{-1}$ north

Practice problem 10

a. $r = 3.1$ m b. $1.3\ \text{m s}^{-1}$
c. $v_{av} = 0.83\ \text{m s}^{-1}$ d. $v_{av} = 0\ \text{m s}^{-1}$

Sample problem 11

a. $a = 9.0\ \text{m s}^{-2}$ towards the centre of the roundabout
b. $F_{net} = 1.1 \times 10^4$ N towards the centre of the roundabout

Practice problem 11

a. $a \approx 22\ \text{m s}^{-2}$ south
b. $F_{net} = 1.3 \times 10^3$ N towards the centre of the circle
c.

$F_{\text{by gravitron on Kwong}}$ (friction force)

$F_{\text{by gravitron on Kwong}}$ (normal force)

$F_g = mg$

Sample problem 12

$v = 12.4\,\text{m s}^{-1}$

Practice problem 12

$\boldsymbol{F}_{\text{net}} = 1.8 \times 10^3\,\text{N}$

Sample problem 13

$v = 13.3\,\text{m s}^{-1}$

Practice problem 13

$\theta \approx 15°$

Sample problem 14

$\theta = 53°$

Practice problem 14

a. $\theta \approx 10°$ **b.** $\boldsymbol{F}_{\text{T}} \approx 5.0 \times 10^2\,\text{N}$

1.5 Exercise

1. **a.** $a = 0.024\,\text{m s}^{-2}$ towards the centre of the circle
   b. ≈ 1.6 N towards the centre of the circle
2. **a.** $\approx 0.050\,\text{m s}^{-2}$
   b. 1.75 N towards the centre of the circle
   c. 75 N towards the centre of the circle
   d. To move along the same path, the child and the train require the same acceleration. As the masses of the child and the train are different, different forces are needed to produce identical accelerations.
3. To go around a bend, a motorcyclist needs a horizontal force acting on the bike towards the centre of the curve of the bend. This is provided by the road acting on the tyres. The force of the road on the tyres needs to act through the centre of mass of the cyclist, otherwise the force will act to tip the bike over. As the horizontal component of this force is acting towards the centre of the curve, the motorcyclist must lean into the curve to avoid falling off.
4. **a.** $\approx 11.7\,\text{m s}^{-2}$
   b. $92.0\,\text{m s}^{-2}$ towards the centre of the circle
   c. 4.60 N towards the centre of the circle
   d. ≈ 4.63 N
5. **a.** ≈ 353 N towards the centre of the circle
   b. 353 N
   c. The radius will increase to 5 metres
6. 82.6°

1.5 Exam questions

1. **a.** $F = 1.6 \times 10^4\,\text{N}$
   b.

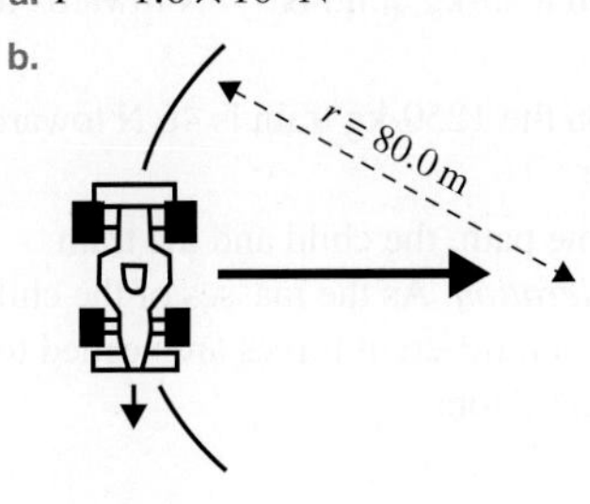

   c. A net horizontal force is required to maintain circular motion. The horizontal force is provided by the friction force on the tyres, by the road.
2. **a.** $r = 3.3 \times 10^3\,\text{m}$
   b. The force of gravity is not zero at the top of the flight. The 'zero gravity experience' is due to the lack of a contact or normal reaction force.
3. **a.** The height of the mass is $2.0\cos 60° = 1.0\,\text{m}$.

$$mgh = \frac{1}{2}mv^2$$
$$2.0 \times 9.8 \times 1.0 = 0.5 \times 2.0 \times v^2$$
$$v = \sqrt{19.6}$$
$$v = 4.4\,\text{m s}^{-1}$$

   b. The mass will reach its maximum velocity at the bottom of the arc, where the maximum amount of gravitational potential energy will have been converted to kinetic energy.
   c. $T = 39\,\text{N}$
4. **a.**

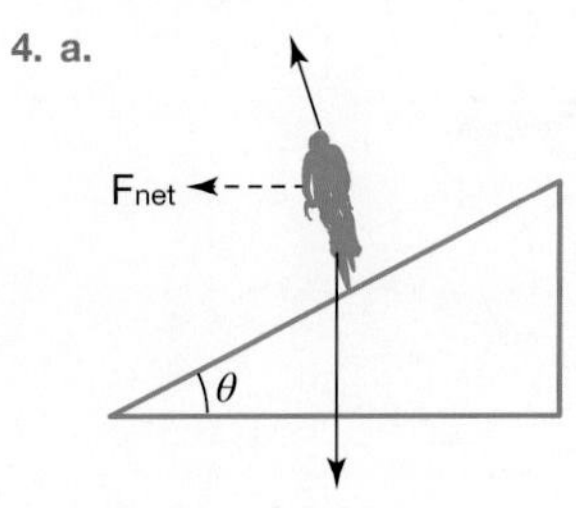

   b. $\theta = 27°$
5. **a.**

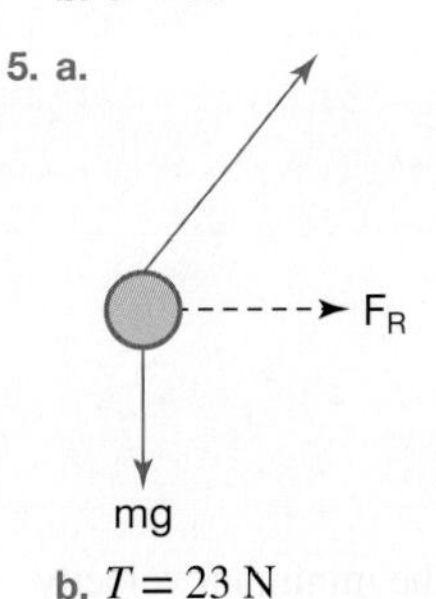

   b. $T = 23\,\text{N}$

1.6 Non-uniform circular motion

Sample problem 15

a. $\boldsymbol{v} \approx 8.9\,\text{m s}^{-1}$

b. $\boldsymbol{F}_{\text{net}} \approx 1.2 \times 10^3\,\text{N}$ upwards

c. $\boldsymbol{F}_{\text{N}} \approx 1800\,\text{N}$

Practice problem 15

a. $\boldsymbol{F}_{\text{net}} \approx 1.1 \times 10^3\,\text{N}$

b. $\boldsymbol{F}_{\text{N}} = 1.7 \times 10^3\,\text{N}$

c. $\boldsymbol{F}_{\text{g}} = 5.6 \times 10^2\,\text{N}$, thus $\boldsymbol{F}_{\text{N}} \approx 2.9 \times \boldsymbol{F}_{\text{g}}$

Sample problem 16

a. $\boldsymbol{v} = 6.6\,\text{m s}^{-1}$

b. $F_N = mg - \dfrac{mv^2}{r}$

The force due to gravity, mg, is constant, so as the speed, $\boldsymbol{v}$, increases, the normal force, $\boldsymbol{F}_{\text{N}}$, gets smaller

c. The passenger will feel lighter.

Practice problem 16

a. $F_N = 2.5 \times 10^3$ N

b. 4.9 m s^{-1}

Sample problem 17

a. $v = 6.26 \text{ m s}^{-1}$

b. $F_{net} = 78.4$ N up

c. $F_N = 80.4$ N up

d. $v = 5.77 \text{ m s}^{-1}$

e. $F_N = 42.7$ N down

Practice problem 17

a. $v \approx 5.0 \text{ m s}^{-1}$

b. $F_{net} = 13$ N up

c. $F_N = 12$ N up

d. $v \approx 4.6 \text{ m s}^{-1}$

e. $F_N \approx 9.8$ N down

1.6 Exercise

1. a. When the ball is at the bottom of the circle
   b. When the ball is at the very top of the circle

2.  a.

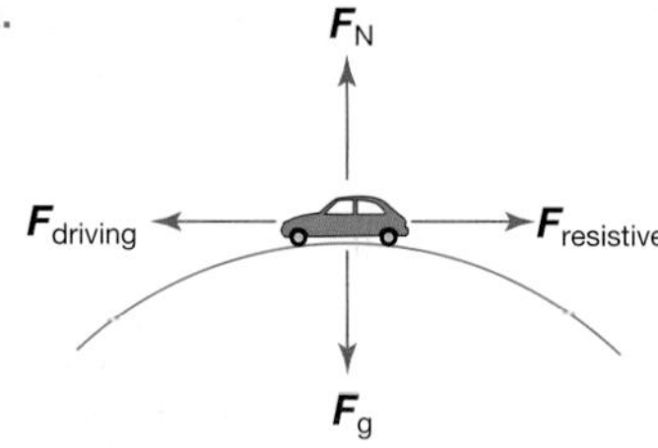

   b. i. 7840 N
      ii. $\approx 6.26 \text{ m s}^{-1}$

3. a. 4.43 m s^{-1} b. ≈ 23.5 N up c. ≈ 22.3 N

4. 1568 N

5. a. 19.6 m s^{-2} b. 2205 N

1.6 Exam questions

1. a. $KE = \frac{1}{2}mv^2$

   $KE = 0.5 \times 0.30 \times 6^2$

   $KE = 5.4$ J

   b. As the velocity at B is greater than the minimum velocity required, the ball will stay on the track.

2. a. $v = 3.9 \text{ m s}^{-1}$ b. $v = 9.8 \text{ m s}^{-1}$

3.  a.

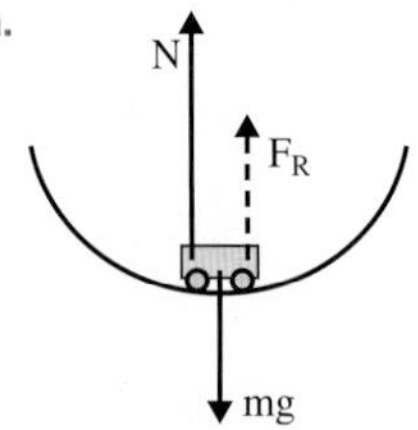

   b. $F = 2900$ N

   c. Roger is incorrect, there is gravity in space. Emily is correct, gravity is still felt at the top of the loop, however, if the reaction force (N) equals zero then a person would feel weightless.

4. $v = 7.9 \text{ m s}^{-1}$

5. $N = 38$ N upwards

1.7 Review

1.7 Review questions

1. The stationary car is pushed forward by the other vehicle. As a result, the seat pushes the body of an occupant forward. This happens almost instantaneously. However, without a headrest, there is nothing to push the occupant's head forward quickly. The head remains at rest until pulled forward by the spine (Newton's First Law of Motion). The head applies an equal and opposite force to the spine (Newton's Third Law of Motion), potentially causing serious injuries.
2. To say that the passenger is thrown forward implies that a force accelerates the passenger. The car slows down rapidly in most collisions as a result of a large external force. The passenger continues to move at the original speed of the car while the car slows down.
3. The matching reaction to the gravitational pull of Earth on you is the gravitational pull of you on Earth.
4. As no forces are acting in the horizontal direction, there can be no horizontal acceleration. Therefore, the horizontal component of velocity must remain constant.
5. The time of a projectile's flight is the time it takes to hit the ground. Therefore, the projectile cannot take longer to complete one part of its motion than the other. Time is the only useful variable that is a scalar and is the same in both the vertical and horizontal directions.
6. When a basketball falls from rest, there is initially no air resistance. As it accelerates downwards due to Earth's gravitational pull, the air resistance increases. The magnitude of the net force, and subsequently its acceleration, decreases. The air resistance continues to increase as the acceleration continues downwards. Because the air resistance is small compared to the basketball's weight, the basketball will not reach its terminal velocity, unless dropped from an aeroplane or helicopter in flight!
7. Newton's First Law states that an object will continue to move in a straight line with constant speed unless an unbalanced force acts on it. Therefore, the mass will continue to move forwards without a propelling force, once in motion. The centripetal force acts to change the direction of the mass, not its speed.
8. a. The braking distance of the train is 200 m.
   b. The speed of the cyclist is 14 m s^{-1}.
   c. The forward force is 2.4×10^4 N.
   d. The additional frictional force is 3.2×10^4 N.
9. a. The centripetal acceleration of the train is $3.7 \times 10^{-2} \text{ m s}^{-2}$.
   b. The net force acting on a 45-kg child is 1.7 N towards the centre of the circle.
   c. The net force acting on the 1250-kg train is 46 N towards the centre of the circle.
   d. To move along the same path, the child and the train require the same *acceleration*. As the masses of the child and the train are different, different forces are needed to produce identical accelerations.

10. a. The speed of the ball is 5.89 m s^{-1}.
    b. The centripetal acceleration of the ball is 30.9 m s^{-2} towards the centre of the circle.
    c. The net force acting on the ball is 3.09 N towards the centre of the circle.
    d. The magnitude of the tension in the string is 3.24 N.
11. The road should be banked at an angle of 84.8°.
12. a. The speed of the gymnast at the point B is 4.43 m s^{-1}.
    b. The centripetal force acting on the gymnast at point B is 3.2×10^2 N upwards.
    c.

1.7 Exam questions

Section A — Multiple choice questions

1. D
2. A
3. C
4. C
5. C
6. C
7. C
8. D
9. D
10. C

Section B — Short answer questions

11. The reaction force to the action of the force due to gravity is the normal force by the floor on Liesel, acting upwards $F_N = 500$ N
12. 9.02 MN
13. The simplest way to start a consideration of forces is to state that at the top of the loop the centripetal force is provided by the gravitational force and the normal force from the track: $\frac{mv^2}{r} = mg + F_N$. For the car to just remain in contact, $F_N = 0$.

    This means that the centripetal force should be provided by gravity alone. That is: $\frac{mv^2}{r} = mg$. Cancelling the ms yields: $v^2 = rg$, as required.

    The most common error was to start with a rearranged form of the required formula (e.g. $v = \sqrt{rg}$) and simply rearrange it. The question required an initial consideration of forces.
14. a.

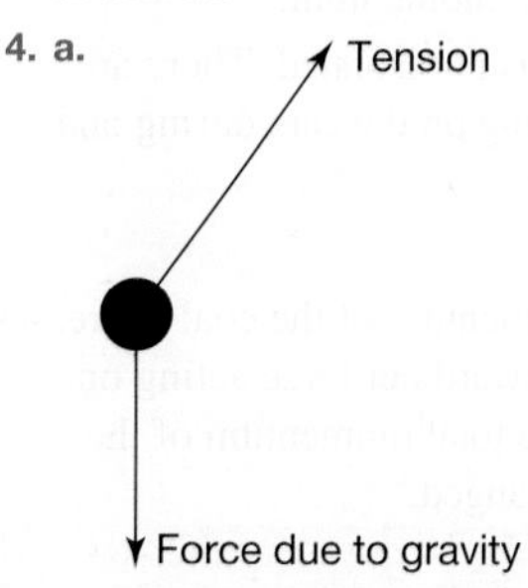

    b. $v = 1.2 \text{ m s}^{-1}$
15. a. $t = \frac{v_{vert} - u_{vert}}{a} = \frac{0 - 25 \sin 39°}{-9.8} = 1.6$ s
    b. Range = 62 m

2 Relationships between force, energy and mass

2.2 Momentum and impulse

Sample problem 1

a. $1.8 \times 10^4 \text{ kg m s}^{-1}$ in a direction opposite to the original direction of the car
b. 1.8×10^4 N s in a direction opposite to the original direction of the car
c. 3.0×10^5 N
d. 3.6×10^5 N

Practice problem 1

a. 2.0×10^3 N s east
b. 2.5×10^3 N east
c. 2.5×10^3 N west

Sample problem 2

12 m s^{-1}

Practice problem 2

4.5 m s^{-1}

Sample problem 3

a. 1500-kg car: $1.80 \times 10^4 \text{ kg m s}^{-1}$
   1200-kg car: $-1.44 \times 10^4 \text{ kg m s}^{-1}$
b. $3.60 \times 10^3 \text{ kg m s}^{-1}$
c. $3.60 \times 10^3 \text{ kg m s}^{-1}$
d. 1.33 m s^{-1}
e. $1.60 \times 10^4 \text{ kg m s}^{-1}$ (or N s) in the direction of motion of the tangled wreck
f. $-1.60 \times 10^4 \text{ kg m s}^{-1}$ (or N s) in the direction opposite that of the 1200-kg car

Practice problem 3

a. 10 m s^{-1}

b. 2.0×10^4 N s north

c. -2.0×10^4 N s (south)

d. 6.0 m s^{-1}

2.2 Exercise

1. Impulse is equal to the change in momentum.
2. No. The system of the two cars is not isolated. There are unbalanced frictional forces acting on the cars during and immediately after the collision.
3. a. 2.0 m s^{-1} south
   b. The vertically downward momentum of the coal decreases to zero because there is an upward net force acting on it when it strikes the cart. The total momentum of the Earth–coal system has not changed.
   c. 2.0 m s^{-1} south
4. a. The forces applied to each car by the other are equal in magnitude and opposite in direction.
   b. The change in velocity of each car is dependent on the mass of the car. Assuming that the sum of the forces other than that applied by the other car is zero, the change in velocity is inversely proportional to the mass of the car.
   c. Assuming that you are properly restrained and that the collision is head-on, your change in velocity (and therefore the deceleration you are subjected to) is less if you are in a heavier car.
   d. The body continues to move in the original direction and at the original speed of your car until an unbalanced force acts on you. If you are not restrained, the unbalanced force will be provided by the windscreen or part of the interior of the car, which has already slowed down. A smaller car will have slowed down more, so the impulse applied to you ($m\Delta v$) will be greater.
5. a. -10 N s
   b. 1.0×10^2 N away from the wall
   c. 10 m s^{-1} away from the wall

2.2 Exam questions

1. a. i. $F = mg \sin\theta$

      $F = 2.0 \times 9.8 \times \sin 25°$

      $F = 8.3$ N

      ii. 1.9 N

   b. i. Using conservation of momentum, $v_f = 2.0 \text{ m s}^{-1}$.

      ii. The collision is inelastic, as there is a decrease in kinetic energy.
2. a.
      - Acceleration at W is greater than zero and less than 9.8 m s^{-2}.
      - Acceleration at X is zero.
      - Acceleration at Y is greater than zero and directed to the left.

   b. 9.2 m

   c. 67 kg m s^{-1}

   d. The momentum is transferred to Earth.
3. a. $F = 8.6$ N upwards
   b. Some of the energy is converted to SPE in the ball. The rest is lost as heat/sound.
   c. The momentum is transferred to Earth.
4. a. 20 N s
   b. $F = 2000$ N
   c. The reduction in kinetic energy indicates that the collision is inelastic.
5. 8.0 N s

2.3 Work done

Sample problem 4

2.53×10^5 J

Practice problem 4

6.26×10^4 J

Sample problem 5

1×10^1 J

Practice problem 5

8 J

2.3 Exercise

1. 200 N
2. 73 J
3. 1.3×10^3 J
4. 4.0 J
5. 3.5×10^2 J

2.3 Exam questions

1. B
2. a. 18 J b. 3.0 m s^{-1}
3. The claim is incorrect; the force is at right angles to the displacement and so zero work is done.
4. 1500 J
5. 18 J

2.4 Kinetic and potential energy

Sample problem 6

a. 4.32×10^4 J b. 1.44×10^5 N

Practice problem 6

a. 8.0×10^4 J b. 4.0×10^3 N

Sample problem 7

2.9 J

Practice problem 7

a. 1.94 J b. 25 cm

Sample problem 8

a. 2.0×10^2 N m^{-1} b. Spring A

a. 2.0 J

Practice problem 8

a. 1.0×10^2 N m^{-1} b. 4.0 J

Sample problem 9

1.0 m s^{-1}

Practice problem 9

a. 0.128 J b. 5.1 cm

Sample problem 10

a. 3.1×10^3 J b. 13 m s^{-1}

Practice problem 10

a. 1.5×10^4 J b. 22 m s^{-1}

Sample problem 11

a. 9.5 m s^{-1}

b. Kinetic energy is not conserved, thus the collision is not elastic.

Practice problem 11

a. i. -0.25 m s^{-1} ii. -1.0 m s^{-1}

b. Both collisions are inelastic as in both cases, the final kinetic energy is less than the initial kinetic energy.

2.4 Exercise

1. a. No. The kinetic energy is not conserved.
   b. Sound, along with some heating of the ball, provides evidence that some of the ball's initial kinetic energy is transformed.
   c. Yes, momentum is conserved, assuming that the ball–ground system is an isolated system.
2. a. 30 m s^{-1} east b. $\frac{1}{26}$
3. a. 9.8 N up b. 38 N m^{-1} c. Spring A
   d. 1.25 J e. Spring B
4. a.

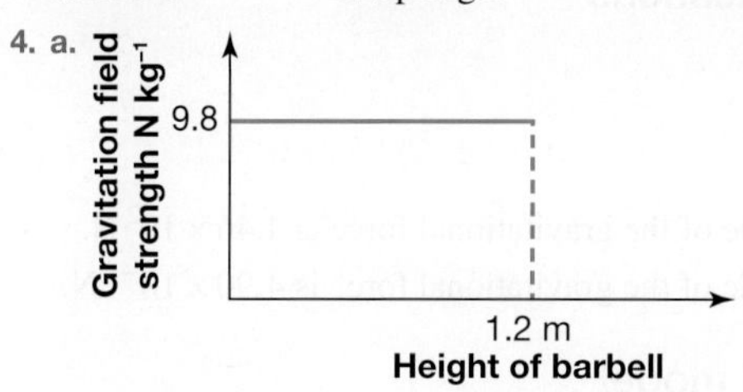

   b. 1.8×10^3 J
   c. 1.8×10^3 J
5. a. 4.1×10^3 J b. 5.5×10^3 J c. 3.1×10^5 J
6. a. 1.8×10^5 J
   b. 1.8×10^5 J
   c. 4.5×10^5 N opposite to the initial direction of motion of the car
7. The depth of penetration would double.

2.4 Exam questions

1. a. $v = 17.3$ m s^{-1}
   b. 12 m
   c. Decrease. Friction will cause the velocity to decrease and since the radius is related to the velocity by $r = \frac{v^2}{g}$, if the velocity decreases the radius will have to decrease as well.
2. a. 0.20 m
   b. i. 0.40 m ii. 1.4 m s^{-1}
3. a. 40 N m^{-1} b. 15 J c. 2.28 m s^{-1}
4. The speed of the car would be 10.7 m s^{-1}.
5. The collision is inelastic.

2.5 Review

2.5 Review questions

1. 0.8 m s^{-1} east
2. 39.2 J
3. a. 3.4×10^6 J b. 3.4×10^6 J
4. a. 1.8×10^4 J b. 2.8×10^4 J
5. a. 140 kg m s^{-1} east b. 2.5 m east of Dean
   c. 1.2 m s^{-1} east d. 1.2 m s^{-1} east
   e. 84 N east
6. a. 9.7×10^3 J b. 1.4×10^4 N c. 2×10^6 N
7. a. 4.4×10^5 J, assuming $m = 4500$ kg and $h = 10$ m.
   b. 1.0×10^3 J, assuming $m = 60$ kg and $h = 1.7$ m.
   c. 15 J, assuming $m = 1.3$ kg and $h = 1.2$ m.
   d. 1.7 J, assuming $m = 58$ *g* and $h = 3$ m.
8. a. 7350 J b. 16 m s^{-1} c. 0 d. 232 N
9. a. 2.9 m s^{-1} east
   b. 3.4×10^4 N s west
   c. The car experiences the greatest change of velocity in magnitude.
   d. They both experience the same change in momentum (in magnitude)
   e. They both experience the same force (in magnitude)
10. a. 7.7×10^{-3} m b. 900 J c. 2.0 m s^{-1}

2.5 Exam questions

Section A — Multiple choice questions

1. D
2. C
3. B
4. C
5. A
6. C
7. C
8. B
9. C
10. B

Section B — Short answer questions

11. $E = 1.54 \times 10^{10}$ J
12. This is an inelastic collision, as there is a decrease in kinetic energy.
13. a. 12 m s^{-1} b. 14 m s^{-1}
14. a. 1.9 m s^{-1} to the left
    b. The kinetic energy is not conserved, thus the collision is inelastic.
    c. i. 358 kN to the right ii. −358 kN to the left
15. 0.86 m

Unit 3 | Area of Study 1 review

Section A — Multiple choice questions

1. A
2. D
3. C
4. A
5. A
6. D
7. D
8. A
9. B
10. C
11. C
12. B
13. B
14. D
15. A
16. B
17. A
18. A
19. B
20. C

Section B — Short answer questions

21. $\boldsymbol{F}_{\text{on bus by wall}} = 6.4 \times 10^4$ N east
22. a. The initial gravitational potential energy of the car is 50 J.
    b. Because of conservation of energy: $E_k = 50$ J.
    c. The maximum speed of the car is 5.8 m s^{-1}.
    d. The work done on the car by the carpet is equal to the kinetic energy of the car at the bottom of the ramp: $W = 50$ J.
    e. The average force exerted on the car by the carpet is 8.3 N.
23. a. The elapsed time is 0.34 s.
    b. The ball left the bench at a speed of 3.7 m s^{-1}.
    c. The speed of the ball just before it strikes the floor is 5.6 m s^{-1}.
24. a. The magnitude of the net force acting on the car is 5.3×10^3 N.
    b. Arrow A
    c. 26°
25. a. The kinetic energy of the pendulum is at its maximum at the bottom of the arc, and the gravitational potential energy is at its maximum when the pendulum is at an angle of 60°. When the pendulum oscillates, its energy is transformed back and forth between gravitational energy and kinetic energy, without loss to the environment.
    b. The maximum speed of the spherical mass is 25.6 m s^{-1}.
    c. The maximum tension in the wire is 548 N.

3 Gravitational fields and their applications

3.2 Newton's Universal Law of Gravitation and the inverse square law

Sample problem 1

a. $\boldsymbol{F}_{\text{on Earth by person}} = 6.9 \times 10^2$ N, towards the centre of Earth.

b. $\boldsymbol{F}_{\text{on Earth by person}} = -6.9 \times 10^2$ N, towards the centre of Earth.

Practice problem 1

$\boldsymbol{F}_{\text{on Earth by Moon}} = 1.98 \times 10^{20}$ N, towards the centre of the Moon.

3.2 Exercise

1. For Mars: $F_g = 2.60 \times 10^2$ N
   For Jupiter: $F_g = 1.98 \times 10^3$ N
2. The magnitude of the force of attraction between Earth and the Sun is 3.60×10^{22} N.
3. The magnitude of the force due to gravity would be $\frac{1}{9}$ of the initial force.
4. $\dfrac{F_{\text{g Mercury}}}{F_{\text{g Earth}}} = 3.76 \times 10^{-1}$
5. a. $\dfrac{F_{g_{S_1}}}{F_{g_{S_2}}} = 1.125$
   b. S_2 should be 3 r_E above Earth's surface.
6. The object must be placed 2.46 r_E from the centre of Earth.

3.2 Exam questions

1. C
2. A
3. D
4. The magnitude of the gravitational force is 4.46×10^2 N.
5. The magnitude of the gravitational force is 4.90×10^{-1} N.

3.3 The field model

Sample problem 2

The gravitational field strength of the Moon at the surface of the Moon is $g = 1.62$ N kg^{-1}.

Practice problem 2

The gravitational field strength of the Sun at the centre of Earth is $g = 5.93 \times 10^{-3}$ N kg^{-1}.

Sample problem 3

The magnitude of the gravitational field on a rocket halfway between Earth and the Moon is $1.06 \times 10^2\,\text{N kg}^{-1}$, towards Earth.

Practice problem 3

$g = \underbrace{3.96 \times 10^{-2}\,\text{N kg}^{-1}}_{\text{from the Sun}} + \underbrace{1.3 \times 10^{-7}\,\text{N kg}^{-1}}_{\text{from Venus}}$

The effect of Venus's gravitational field on Mercury is extremely small.

3.3 Exercise

1. a. $g = 9.81\,\text{N kg}^{-1}$ b. $F_{\text{net}} = 14.7\,\text{N}$

2. a. The ratio of the magnitudes of the forces experienced is $\frac{9}{4}$.

b.

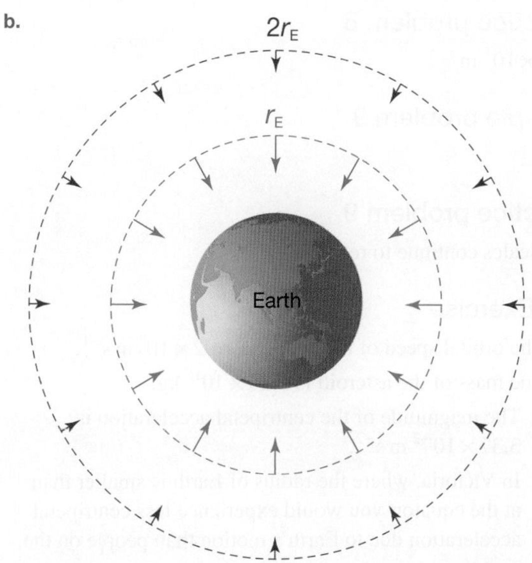

→ Force on object by Earth, at $1r_E$ from the surface

→ Force on object by Earth, at $2r_E$ from the surface

The direction of the force is towards the centre of Earth in both cases.

The ratio of the forces experienced is $\frac{9}{4}$, which should be reflected in the relative length of the force arrows.

c. The field does not have the same magnitude everywhere, thus it is not uniform.

3.

Planet/dwarf planet	Mass (kg)	Radius (m)	$g\,(\text{N kg}^{-1})$
Earth	5.97×10^{24}	6.37×10^{6}	9.81
Mars	6.39×10^{23}	3.39×10^{6}	3.71
Venus	4.87×10^{24}	6.05×10^{6}	8.87
Pluto	1.31×10^{22}	1.19×10^{6}	0.617

4. The gravitational field lines point inwards because masses are only attracted to other masses, and gravity is always attractive, thus an isolated mass will experience an attractive force towards its own centre, as it is the source of the gravitational field.

5. a.

d (km)	$g\,(\text{N kg}^{-1})$
10 000	3.98
20 000	0.995
30 000	4.42×10^{-2}

b. i. Field strength at 20 000 km : Field strength at 10 000 km = 1 : 4

ii. Field strength at 30 000 km : Field strength at 10 000 km = 1 : 9

iii. Field strength at 30 000 km : Field strength at 20 000 km = 4 : 9

c. Comparing ratios demonstrates the inverse square relationship for the field from a point mass.

6. a. $F_g = 97.3\,\text{N}$

b. $F_{\text{up}} = 97.3\,\text{N}$

c. The detector is 6.40×10^6 m from the centre of Earth.

7. a. $g_S = 5.90 \times 10^{-3}\,\text{N kg}^{-1}$

$g_E = 2.70 \times 10^{-3}\,\text{N kg}^{-1}$

b. The gravitational field strength experienced by the Moon from the Sun is approximately twice the gravitational field strength experienced by the Moon from Earth, which is surprising as the Moon orbits Earth. However, in fact both the Moon and Earth orbit the Sun.

8. a. The spacecraft is 3.46×10^8 m from the centre of Earth.

b. $\frac{\text{Distance spacecraft–Earth}}{\text{Distance Earth–Moon}} = \frac{3.46 \times 10^8}{3.84 \times 10^8}$

$\simeq 90\%$

The spacecraft has travelled approximatively 90% of the distance from Earth to the Moon when the net force it experienced is zero.

See figure at the bottom of the page*

*8. b.

3.84 × 10⁸ m

Earth

Moon

x

9. a.

Scenario	dot	$g_c\ (\text{N kg}^{-1})$
(a)	red	6.1104×10^{-3}
(a)	orange	6.1093×10^{-3}
(b)	red	6.1104×10^{-3}
(b)	orange	6.1093×10^{-3}

b. The results support Julian's hypothesis, but not Meredith's hypothesis.

3.3 Exam questions

1. A
2. The mass of Europa is 4.78×10^{22} kg.
3. The gravitational field strength on its surface is 3.40×10^{-5} N kg^{-1}.
4. The gravitational field strength on the surface is 39.2N kg^{-1}.
5. The magnitude of the force due to gravity on it at this altitude 6875 N.
6. 2.72×10^3 N
7. 0.33
8. 4 m s^{-2}
9. 650 N

3.4 Motion in gravitational fields, from projectiles to satellites in space

Sample problem 4

20 m

Practice problem 4

20 m s^{-1}

Sample problem 5

a. 1020 m s^{-1}
b. 0.002 72 m s^{-2}
c. 6.02×10^{24} kg

Practice problem 5

a. 2.14×10^3 m s^{-1}
b. 0.486 m s^{-2}
c. 6.41×10^{23} kg

Sample problem 6

Mercury: 3.36×10^{18}

Venus: 3.35×10^{18}

Mars: 3.36×10^{18}

The values of $\frac{r^3}{T^2}$ for the three planets are approximately the same, confirming Kepler's Third Law.

Practice problem 6

Saturn: 3.38×10^{18}

Uranus: 3.37×10^{18}

Neptune: 3.36×10^{18}

Sample problem 7

6.73×10^5 m

Practice problem 7

2.67×10^7 m

Sample problem 8

3.58×10^7 m

Practice problem 8

1.70×10^7 m

Sample problem 9

0 N

Practice problem 9

The scales continue to read zero.

3.4 Exercise

1. The orbital speed of the Moon is 1.02×10^3 m s^{-1}.
2. The mass of the asteroid is 6.26×10^{16} kg.
3. a. The magnitude of the centripetal acceleration is 3.37×10^{-2} m s^{-2}.
   b. In Victoria, where the radius of Earth is smaller than at the equator, you would experience less centripetal acceleration due to Earth's motion than people on the equator.
4. a. The orbital speed of the space station is 7.65×10^3 m s^{-1}.
   b. The magnitude of the centripetal acceleration of the space station is 8.71 m s^{-2}.
   c. The gravitational field strength that the space station experiences is 8.80 N kg^{-1}.
   d. The answers should be the same, discrepancies may come from rounding off data.
   e. The magnitude of the centripetal force on the space station is 1.06×10^7 N .
   f. The force exerted on the astronaut by the floor of the space station is 0 N.
5. The altitude of the spacecraft is 2.98×10^3m.
6. According to Newton's First Law, an object will move in a straight line, with constant speed, unless an unbalanced force is acting on it. As gravity acts at right angles to the satellite's velocity, it does not change the speed of the satellite; rather, it changes its direction. This causes the satellite to move around Earth. Because the force of gravity and the speed of the satellite remain constant, so must its radius as $r = \frac{mv^2}{F}$; therefore, it cannot move closer to Earth.

7. The period would be increased by a factor of $2\sqrt{2}$.
8. $\frac{\text{distance of Saturn from the Sun}}{\text{distance of Venus from the Sun}} = 13.2$
9. Both the astronaut and the space station are in circular orbit around Earth, they are in free fall. Since they both have the same acceleration, the astronaut's motion is independent of the motion of the space station, so if the astronaut didn't strap herself to the chair, there wouldn't be a normal force by the chair on the astronaut and by the astronaut on the chair, and she would float around.
10. a. One possible answer:

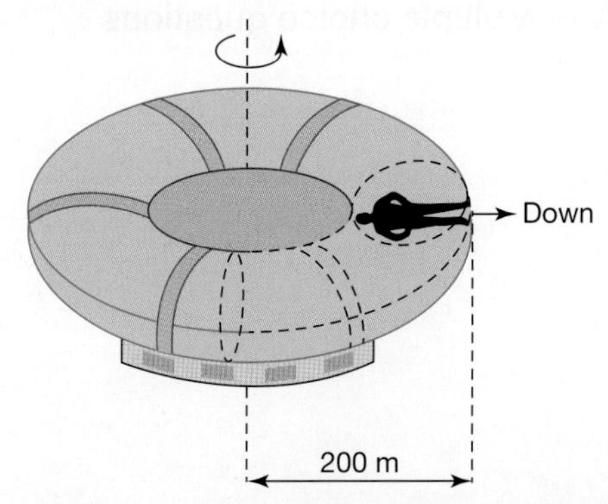

On Earth, a person would feel the ground pushing up on them. In the space station, the outer wall pushes the person in (centripetal force). This means that the person would feel as though the wall is the ground, and the direction opposite to the centripetal force is down.

b. 28 s

3.4 Exam questions

1. a. Titan's orbital speed is 5.6×10^3 m s^{-1}.
   b. The orbital speed increases as the radius decreases.
   c. $T = 4.6$ days
2. a. 1.6×10^{-2} m s^{-2}
   b.

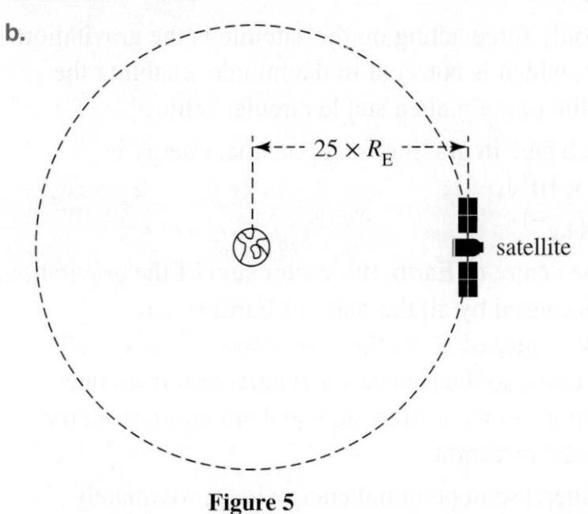

Figure 5

c.

Magnitude of acceleration	Less than
Kinetic energy	Less than
Period	More than

3. The mass of the pulsar is 2.71×10^{30} kg.
4. a. $F = 1.99 \times 10^{20}$ N
   b. The average orbital period of Earth's moon will increase as its orbit radius increases.
5. a. The radius is 3306 m.
   b. The force of gravity is **not** zero at the top of the flight. The 'zero gravity experience' is due to the lack of a contact or normal reaction force.

3.5 Energy changes in gravitational fields

Sample problem 10

4.2×10^8 J

Practice problem 10

a. 1.2×10^{14} J
b. 3.5×10^{14} J

Sample problem 11

7.39×10^8 J

Practice problem 11

2.0×10^8 J

Sample problem 12

a. 4.4×10^4 J
b. 4.4×10^4 J

Practice problem 12

20 J

3.5 Exercise

1. The change in the kinetic energy of the satellite is approximately 6×10^{10} J.
2. The change in in gravitational potential energy is approximately 6×10^9 J.
3. The unit of the area under a force–distance graph is N m = J, which is a unit of energy. The area under a gravitational field strength vs. distance graph has the unit N kg^{-1} = J kg^{-1}, which is a unit of energy per mass.
4. As the space probe gets closer to Planet Q, the gravitational potential energy of the probe decreases, increasing its kinetic energy. As the space probe moves further away from Planet Q, towards Y, kinetic energy is converted into gravitational potential energy. The total energy of the space probe (Planet Q system) is constant because energy is conserved in a closed system.
5. The change in in gravitational potential energy is approximately 10^{10} J.
6. a. The work needed to deploy the shuttle is approximately 10^9 J.
   b. The period of the new satellite is 5.79×10^3 s.
   c. By using $\frac{r^3}{T^2}$ = constant and known radius and period.
   d. The work needed to deploy the satellite would halve and its period would remain the same.
7. a. The magnitude of the gravitational force on the satellite is 1.36×10^4 N.
   b. The loss of gravitational potential energy of the satellite during its fall is approximately 2×10^{10} J.

8. The change in kinetic energy is 2.5 J.
9. The speed of the cannonball is approximately 1.03×10^2 m s^{-1}.

3.5 Exam questions

1. A
2. D
3. a. The magnitude of the thrust force is 9.02 MN and the direction is up.
   b. $E_k = 1.54 \times 10^{10}$ J
   c. $E_g = 1.04 \times 10^9$ J
   d. The gravitational potential energy is converted to kinetic energy.
   The kinetic energy is then converted to heat/light/sound due to friction.
4. $\Delta E_k = 5.7 \times 10^4$ J
5. $\Delta E = 7.2 \times 10^5$ J

3.6 Review

3.6 Review questions

1. The shape of the field is non-uniform, it attracts other masses, and the strength of the field is inversely proportional to the square of the distance from the centre of Earth.
2. $\dfrac{F_{g\ \text{Mars}}}{F_{g\ \text{Moon}}} = 2.29$
3. The rocket experiences a gravitational field that is the sum of the field from the Moon and the field from Earth.
   The two fields point in opposite directions.
   Initially, Earth's field dominates; however, it gets weaker as the rocket moves further away from Earth and the Moon's field becomes stronger until a point is reached where the two fields cancel each other out. From that point on, the field of the Moon dominates the field from Earth.
4. The mass of Jupiter is 1.90×10^{27} kg.
5. $h = 3.0 \times 10^6$ m
6. The period would be decreased by a factor of 64.
7. a. The orbital period of the James Webb Space Telescope is 184 days.
   b. The orbital speed of the James Webb Space Telescope is 3.75×10^2 m s^{-1}.
8. a. Using $g = \dfrac{Gm_E}{(r_E + h)^2}$ and rounding to 3 s.f.:

Altitude h(m)	g (N kg)$^{-1}$
0	9.813
10^3	9.810
10^4	9.783
10^5	9.512
10^6	7.331
6.37×10^6	0.273

   b. Answers will vary. Between sea level and an altitude of 10 km, the variation of the gravitational field strength is approximately 0.3%, which is negligible.
   Thus, using $g = 9.8$ N kg^{-1} as an approximation of Earth's gravitational field strength for altitudes lower than 10 km is acceptable.
9. The change in gravitational potential energy is 4.05×10^2 J.
10. The change in gravitational potential energy is approximately 2×10^8 J.

3.6 Exam questions

Section A — Multiple choice questions

1. B
2. B
3. D
4. A
5. D
6. C
7. D
8. B
9. C
10. D

Section B — Short answer questions

11. a. The satellite must orbit the centre of mass or the gravitational/centripetal force must be directed towards the centre of Earth and the satellite must orbit the same axis or be in the same plane as Earth's rotation.
    b. The altitude of a geostationary satellite must be equal to 3.59×10^7 m.
    c. The speed of an orbiting geostationary satellite is 3.07×10^3 m s^{-1}.
12. a. The orbital radius of the ICON satellite is 6.97×10^6 m.
    b. $T = 5.79 \times 10^3$ s
    c. The only force acting on the satellite is the gravitational force, which is constant in magnitude, enabling the satellite to maintain a stable circular orbit.
    d. The change in gravitational potential energy is 1.56×10^9 J.
13. a. 9.8 N kg^{-1}
    b. At the centre of Earth, the vector sum of the gravitational fields caused by all the mass of Earth is zero.
    At the centre of Earth there are equal masses in all directions, so the gravitational attraction from one direction is balanced by an equal attraction from the opposite direction.
    c. The increase in potential energy is approximately 2.3×10^9 J.
14. a. The force of gravity, which acts towards the centre of Earth, is the only force acting on the satellite.
    b. $T = 4.26 \times 10^4$ seconds
15. a. 4.5×10^3 N
    b. The change in gravitational potential energy of the spacecraft is 1.05×10^{12} J
    c. The period of Europa is 3.06×10^5 s

4 Electric fields and their applications

4.2 Coulomb's Law and electric force

Sample problem 1

The balloons exert a repulsive force on each other (the force value is positive), and the magnitude of the force is 3.6×10^{-6} N.

Practice problem 1

The force between the electron and the proton is attractive (the force value is negative) and its magnitude is 8.1×10^{-8} N.

4.2 Exercise

1. a. The magnitude of the electric force on the 10-μC charge is 1.2×10^2 N.
   b. The direction of the electric force on the 10-μC charge is to the right.
2. a. The magnitude of the electric force on the 5.0-μC charge is 7.9 N.
   b. The direction of the electric force on the 5.0-μC charge is to the left.
3. a. The new magnitude of the electric force is 3.0×10^{-4} N.
   b. The new magnitude of the electric force is 3.6×10^{-3} N.
   c. The new magnitude of the electric force is 2.4×10^{-3} N.
   d. The new magnitude of the electric force is 9.6×10^{-3} N.
4. The two electrons are 9.8×10^{-6} m apart.
5. A test charge would experience a zero net force at a distance 8 cm from A.
6. C

4.2 Exam questions

1. D
2. The magnitude of the electric force is 4.0×10^{10} N.
3. The distance between the two charges is 1.3×10^5 m.
4. The value of the unknown charge is 7.1×10^{14} C.
5. The electron and the proton should be 1.2×10^{-1} m apart.

4.3 The field model for point-like charges

Sample problem 2

The magnitude of the electric field (or electric field strength) is 7.2×10^3 N C^{-1}.

Practice problem 2

The magnitude of the electric field (or electric field strength) is 3.00×10^5 N C^{-1}. The direction of the field is towards the point charge (charged negatively).

Sample problem 3

a.

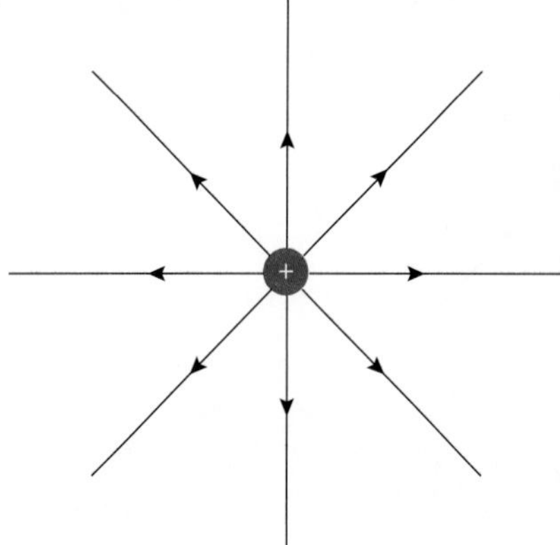

b.

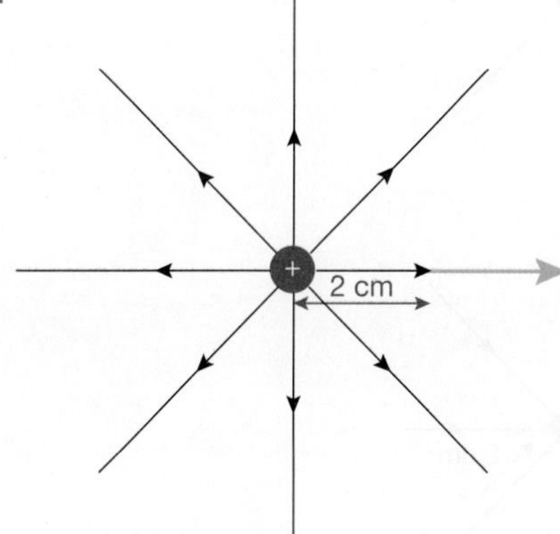

c.

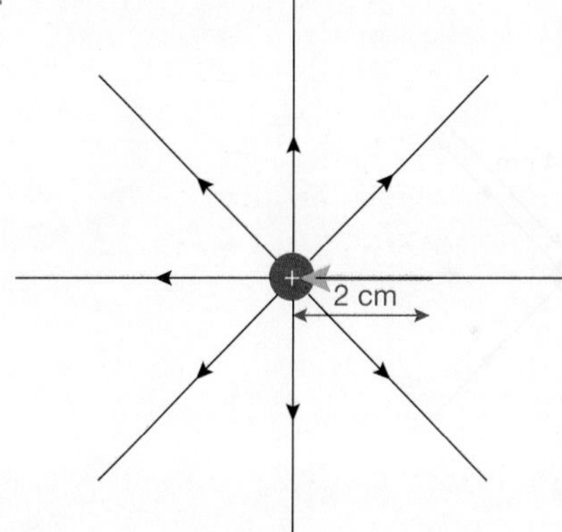

d.

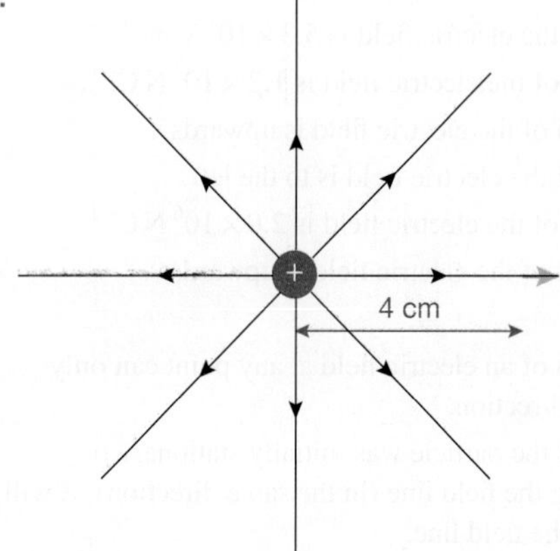

Practice problem 3

a.

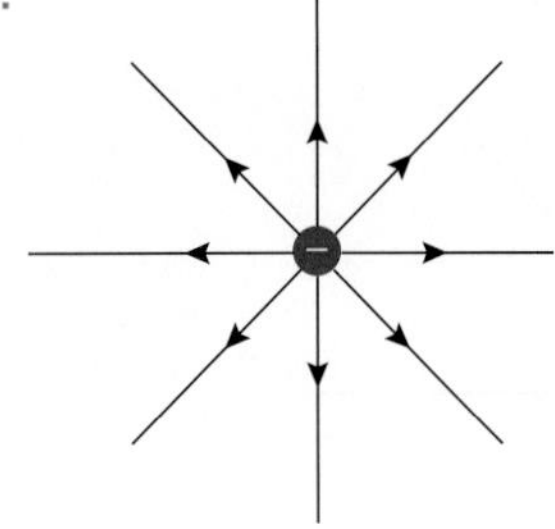

b.

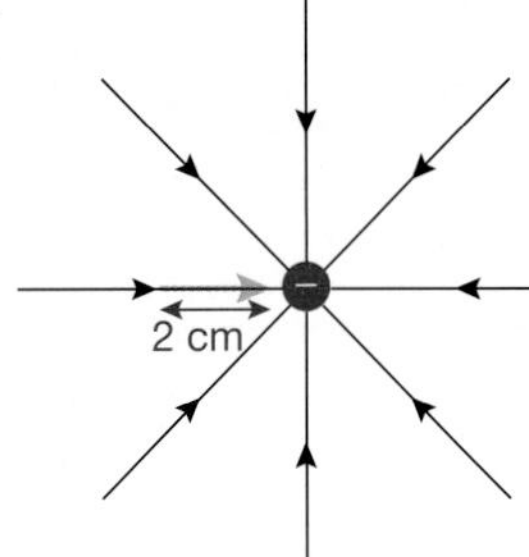

c.

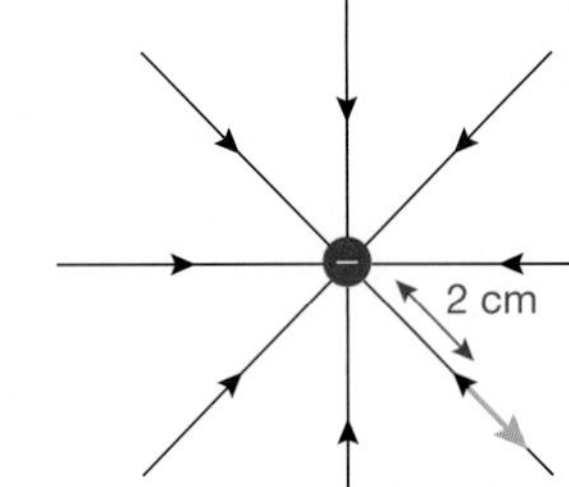

d.

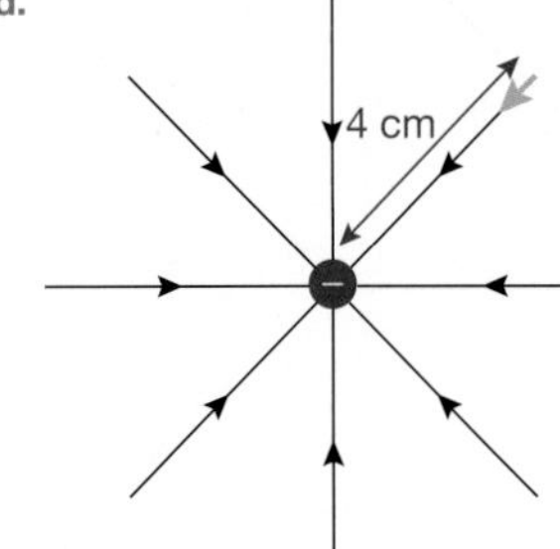

4.3 Exercise

1. The strength of the electric field is 1.2×10^7 V m^{-1}.
2. The strength of the electric field is 5.3×10^1 V m^{-1}.
3. a. The strength of the electric field is 1.2×10^7 N C^{-1}.
   b. The direction of the electric field is upwards.
4. The direction of the electric field is to the left.
5. a. The strength of the electric field is 2.0×10^6 N C^{-1}.
   b. The direction of the electric field is upwards.
6. D
7. a. The direction of an electric field at any point can only point in one direction.
   b. It depends. If the particle was initially stationary or moving along the field line (in the same direction), it will move along the field line.
   If the particle is crossing the field line, it will deviate from its initial direction but it will not follow the field line.

4.3 Exam questions

1. C
2. The direction of the electric field is downwards.
3. The strength of the electric field is 3.7×10^8 V m^{-1}.
4. The unknown charge is 2.0×10^{-10} C.
5. The strength of the electric field 1.0 N C^{-1} at a distance of 1.6×10^{-5} m.

4.4 Electric fields from more than one point-like charge

Sample problem 4

a.

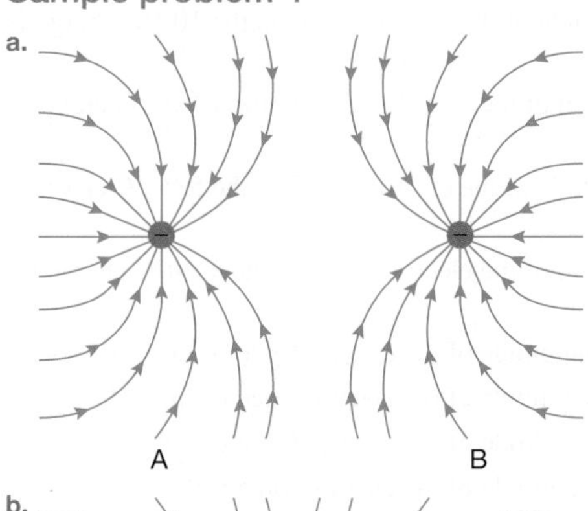

b.

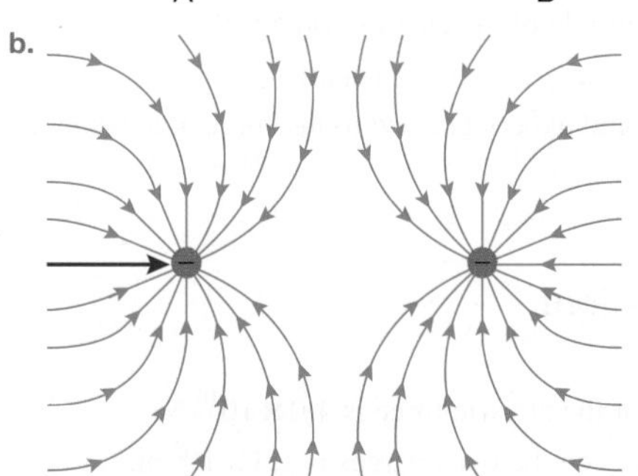

As the charge is positive, it is attracted to the negative particles A and B, which are directly to the right of the charge.

c. When the 1.0-nC charge moves past particle A to the right-hand side, it experiences an attractive force to the left from particle A and also an attractive force to the right from particle B. Because the charge is closer to particle A than particle B, the force acting to the left is stronger than the force to the right, so the net (or overall) force on the 1.0-nC charge at this point is to the left (but it is smaller in magnitude than the force it experienced in part b).
As the charge is moved closer to particle B, the magnitude of the leftward net force decreases to zero at the point halfway between the two particles. When the charge moves closer to particle B than particle A, the net force acts to the right and becomes stronger as the charge moves closer to particle B.

Practice problem 4

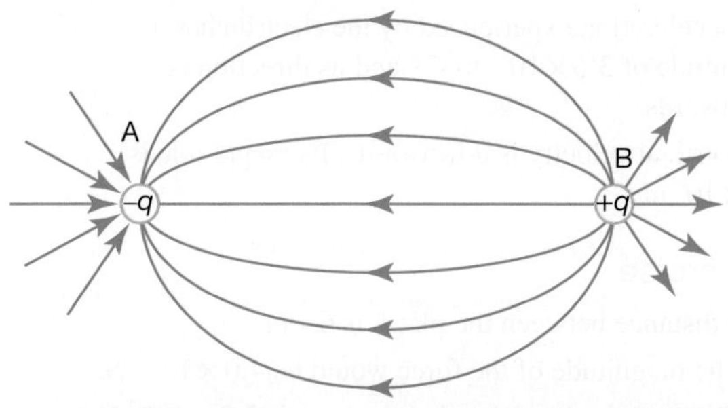

a. A positive charge to the left of A would be attracted to the right towards the negatively charged particle (particle A). A positive charge to the right of A would be both attracted to A by particle A and repelled towards A by particle B, so the force experienced by the charge would be larger, but to the left.

b. As a positive charge moves along an imaginary line separating particles A and B, the direction of the force would always be to the left; however, it would be a minimum at the halfway point. Either side of the midpoint, the inverse decline of the strength of the field from one particle would be less than the increase in the strength of the other particle, causing the total field strength to increase.

4.4 Exercise

1. a. The direction of the electric field is towards B.

b. The direction of the electric field is towards B.

2.

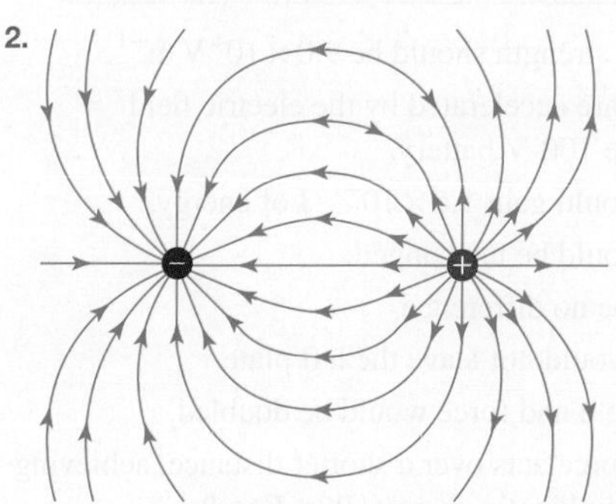

The field lines point towards the negative charge, and away from the positive charge. The field lines should never cross.

3.

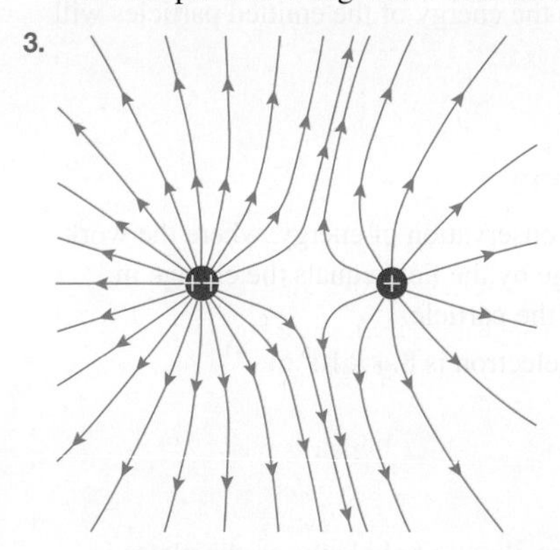

4. D

5. a. See table at the bottom of the page*

b. Josh is correct.

Note that solving $\left(\frac{3.0}{x^2} = \frac{2.0}{(0.15 - x)^2}\right)$ gives $x = 8.1742$ cm.

4.4 Exam questions

1.

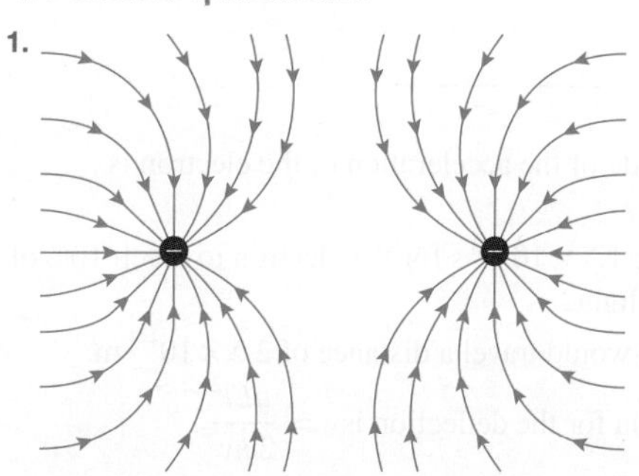

2. A

3. The arrow starts at X and points to the left.

4. The overall field is to the right.

5. a.

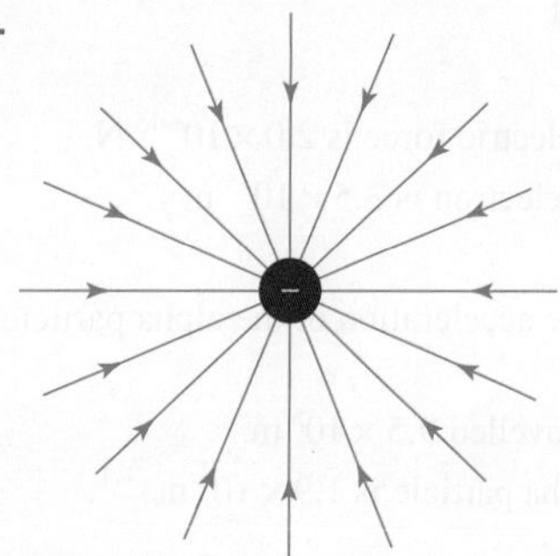

b. The magnitude of the force is 8.4×10^{-5} N.

c. The direction of the electric force is to the right.

d. Yes.

4.5 Uniform electric fields

Sample problem 5

a. The magnitude of the force is 3.2×10^{18} N.

b. The direction of the force is towards the negative plate.

c. The change in velocity is 6.2×10^{3} m s^{-1}.

Practice problem 5

a. The magnitude of the force is 1.6×10^{18} N.

b. The direction of the force is towards the positive plate.

c. The change in velocity is 1.9×10^{5} m s^{-1}.

***5. a.**

x (cm)	1.0	2.5	5.0	7.5	8.0	8.5	9.0	10.0	12.5	14.0
E (N C^{-1})	2.5×10^5	4.2×10^4	9.0×10^3	1.6×10^3	5.4×10^2	-5.2×10^2	-1.7×10^3	-4.5×10^3	-2.7×10^4	-1.8×10^5

4.5 Exercise

1. The top plate is the positively charged plate.
2.

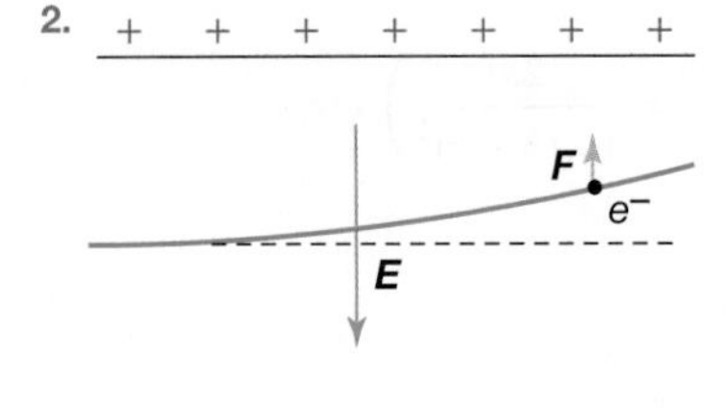

3. a. The magnitude of the acceleration of the electron is 1.8×10^{17} m s^{-2}.
   b. It would take 1.7×10^{-10} s for the electron to reach 10% of the speed of light.
   c. The electron would travel a distance of 2.5×10^{-3} m.
4. a. An expression for the deflection is $s = \dfrac{qEl^2}{2mv^2}$.
   b. The charge required is 8.0×10^{-13} C.
5. A, B

4.5 Exam questions

1. B
2. The magnitude of the electric force is 2.0×10^{-15} N.
3. The acceleration of an electron is 3.5×10^{15} m s^{-2}.
4. C
5. a. The magnitude of the acceleration of the alpha particle is 1.9×10^{14} m s^{-2}.
   b. The alpha particle travelled 9.5×10^5 m.
   c. The speed of the alpha particle is 1.9×10^5 m s^{-1}.

4.6 Energy and motion of charges in electric fields and the linear accelerator

Sample problem 6

The strength of the electric field between the plates is 2.0×10^3 V m^{-1}.

Practice problem 6

The strength of the electric field between the plates is 2.0×10^4 V m^{-1}.

Sample problem 7

a. The change in kinetic energy is 1.6×10^{-17} J.
b. The final speed of the electron is 5.9×10^6 m s^{-1}.

Practice problem 7

a. The change in kinetic energy is 1.6×10^{-16} J.
b. The final speed of the electron is 1.9×10^7 m s^{-1}.

Sample problem 8

a. The acceleration experienced by the electron has a magnitude of 1.8×10^{15} m s^{-2} and its direction is downwards.
b. The vertical velocity is downwards. Its magnitude is 2.1×10^7 m s^{-1}.

Practice problem 8

a. The acceleration experienced by the electron has a magnitude of 3.6×10^{17} m s^{-2} and its direction is downwards.
b. The vertical velocity is downwards. Its magnitude is 1.8×10^8 m s^{-1}.

4.6 Exercise

1. The distance between the plates is 6.7 m.
2. a. The magnitude of the force would be 4.0×10^{-3} N.
   b. The magnitude of the electric force is 5.3×10^{-3} N.
   c. The voltage of the new battery is 180 V.
3. The work done by the electric field is 6.0×10^{-10} J.
4. The change in kinetic energy is 3.0×10^{-8} J.
5. The potential difference is 1.5×10^4 V.
6. a.

Position of the electron	E_{PE} (J)
i. at the negative plate	1.6×10^{-16}
ii. halfway between the plates	8.0×10^{-17}
iii. at the positive plate	0

   b.

Position of the electron	ΔE_k (eV)
i. halfway between the plates	500
ii. at the positive plate	1000

7. The electric field strength should be 9.0×10^4 V m^{-1}.
8. a. The electrons are accelerated by the electric field supplied by the 100-V battery.
   b. An electron would gain 1.6×10^{-17} J of energy.
   c. The answer would be unchanged.
   d. There would be no difference.
   e. The electron would not leave the left plate.
   f. The electric field and force would be doubled.
   g. The stronger force acts over a shorter distance, achieving the same gain in kinetic energy ($W = \boldsymbol{F} \times d$).
9. The overall energy per unit charge available from the field will not change, so the energy of the emitted particles will not change.
10. B, C

4.6 Exam questions

1. a. The principle is conservation of energy, where the work done on the charge by the field equals the change in kinetic energy of the particle.
   b. The speed of the electron is 8.4×10^6 m s^{-1}.
2. $V_0 = 1.1$ kV
3. A
4. A
5. a. The strength of the electric field between the plates is 1.0×10^5 V m^{-1}.
   b. The speed of the electrons is 9.4×10^7 m s^{-1}.

4.7 Review

4.7 Review questions

1. The magnitude of the force is 7.9 N.
2. a. The strength of the electric field is 2.0×10^6 V m^{-1}.
   b. The direction of the electric field is upwards.
3. The strength of the electric field is 1.4×10^{-3} V m^{-1}.
4. The distance between the charges would need to be $2\sqrt{2}r$.
5. a.

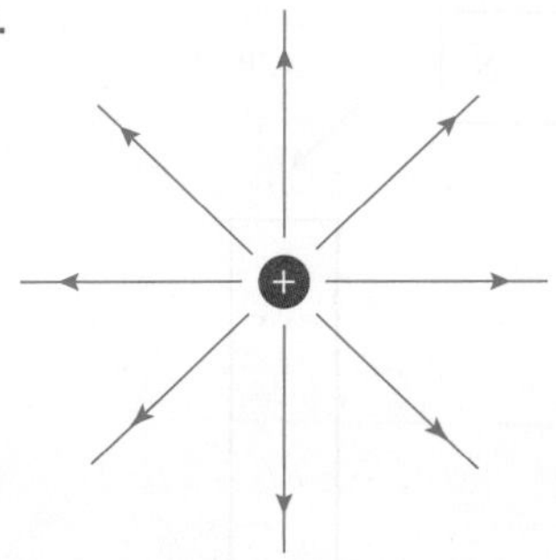

   b. The magnitude of the force is 3.4×10^{-3} N.
   c. Yes.
6.

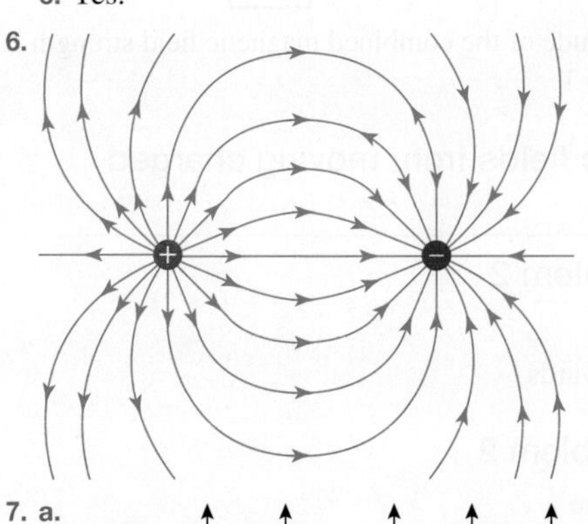

7. a.

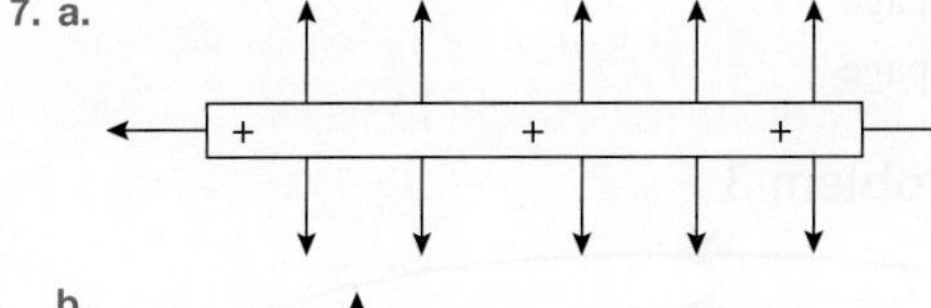

   b.

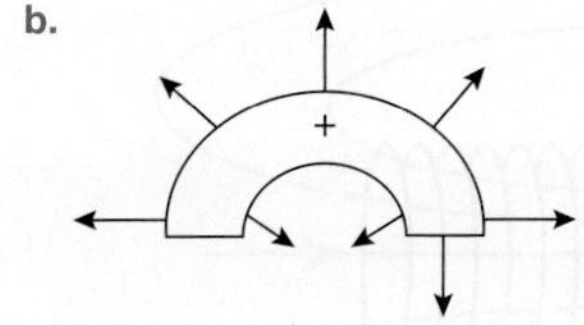

   c. The strength of the electric field in the middle would be 0 V m^{-1}.
8. a. The arrow should point upwards.
   b. The potential difference is 10 V.
   c. The change in kinetic energy is 1.2×10^{-18} J.
9. a. The direction of the electric field is perpendicular to the plates, towards the left plate.
   b. The strength of the electric field is 6.7×10^4 V m^{-1}.
   c. It takes 2.3×10^{-9} s.
   d. The work done is 3.2×10^{-16} J.
   e. The change in kinetic energy is 3.2×10^{-16} J
   f. The change in kinetic energy is 1.6×10^{-16} J
   g. The electron would not leave the positive plate.
10. No, the energy of the emitted particles will not change.

4.7 Exam questions

Section A — Multiple choice questions

1. B
2. D
3. B
4. B
5. B
6. D
7. D
8. C
9. B
10. A

Section B — Short answer questions

11. a. The sphere will move up.
    b. $F = 2.1 \times 10^{-2}$ N
12.

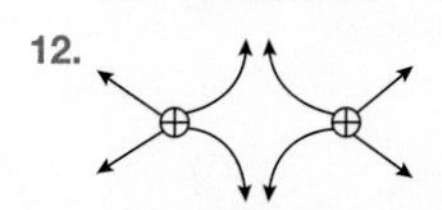

13. a. $E = 5 \times 10^4$ V m^{-1} (or N C^{-1}).
    b. The magnitude of the force is 8×10^{-15} N.
    c. The speed of the electrons is 4.2×10^7 m s^{-1}.
14. The arrow starts at X and points to the left.
15. a. $F = 8.2 \times 10^{-8}$ N
    b. $v = 2.2 \times 10^6$ m s^{-1}

5 Magnetic fields and their applications

5.2 Magnets and magnetic fields

Sample problem 1

a. X: to the right
   Y: to the left
   Z: to the right at an angle of approximately 45° upwards
b. In order of increasing field strength: Y, Z, X

Practice problem 1

a. To the left, along the field lines
b. To the right, along the field lines
c. Between the poles of the horseshoe magnet

5.2 Exercise

1. If a permanent magnet attracts a piece of metal, that metal must be a magnetic material.
2. Both poles can induce a ferromagnetic material, such as an iron nail, to behave like a magnet, thus both ends can attract iron nails.
3. The polarity of Earth's magnetic field at the magnetic pole in the southern hemisphere is north.

4. a. **Loudspeaker**

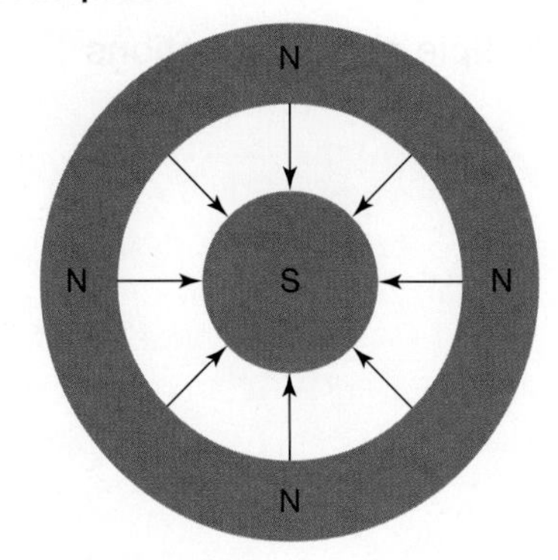

b. **Horseshoe magnet**

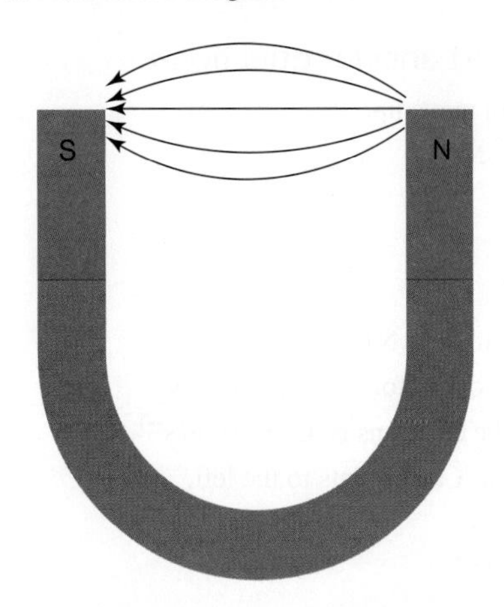

5. As pieces of ferromagnetic material cool, the magnetic domains within the material are aligned by Earth's magnetic field, thus forming natural permanent magnets.

5.2 Exam questions

1. South end of Earth's magnet

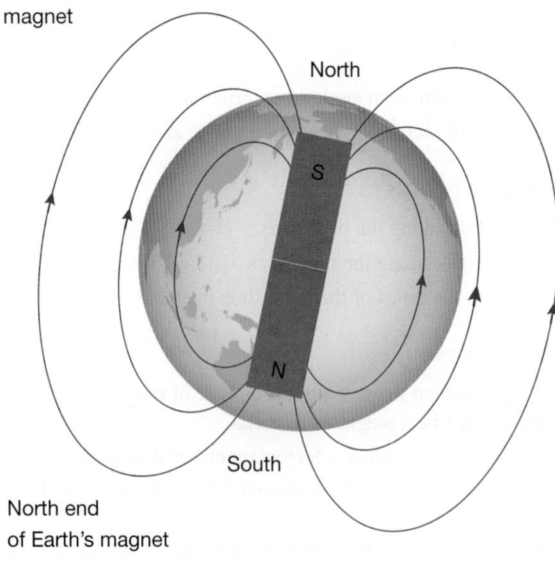

North end of Earth's magnet

2. A
3. a. The force between the magnets is repulsive.

b.

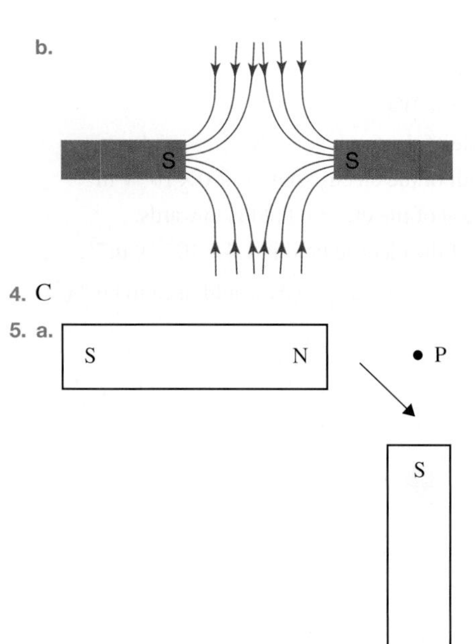

4. C
5. a.

b. The magnitude of the combined magnetic field strength is 1.41×10^{-2} T.

5.3 Magnetic fields from moving charged particles

Sample problem 2

a. Into the page
b. Vertically upwards

Practice problem 2

a. Out of the page
b. Out of the page

Sample problem 3

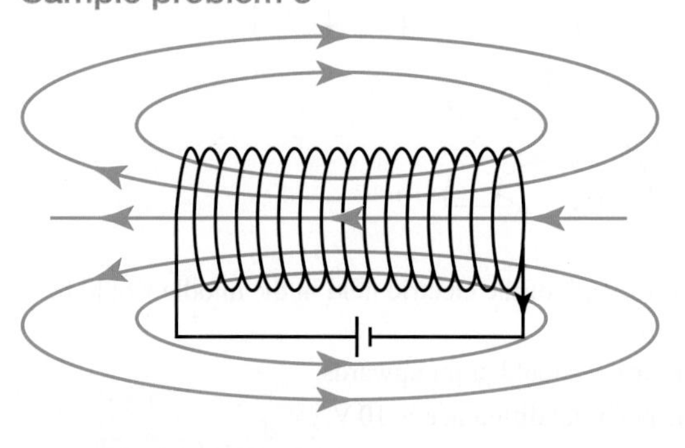

Practice problem 3

The field lines would be the same as in the sample problem, just with the directions reversed.

Sample problem 4

a. North-east direction
b. 35 μT

Practice problem 4

a. 0.014 T

b. The direction of the resultant field depends on the orientation of the two magnets and will be at 45° to the initial magnetic fields (as shown by the green arrows in the following diagrams).

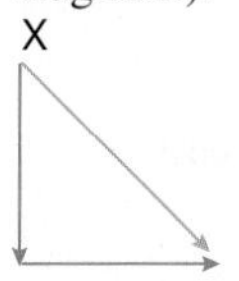

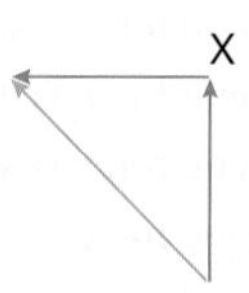

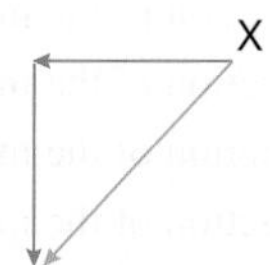

5.3 Exercise

1. a. Out of the page
   b. Out of the page
   c. To the right
2. a. Up the page
   b. To the left
   c. Diagonally down to the right
   d. Into the page
   e. Out of the page
3. a. At W, the magnetic field is coming out of the page. At X, Y and Z, the magnetic field is going into the page.
   b. At W and X, the magnetic field is going from left to right. At Y and Z, the magnetic field is going to the left.
4. Since the iron rods are inside the solenoid, they are induced to become temporary magnets (when the solenoid has a current running through it), both with equivalent orientations of their magnetic poles. Thus, they repel each other because like poles repel.
5. a. The compass needle direction remains the same.
   b. The compass needle points in the opposite direction.

5.3 Exam questions

1. C
2. The current is anti-clockwise.
3.

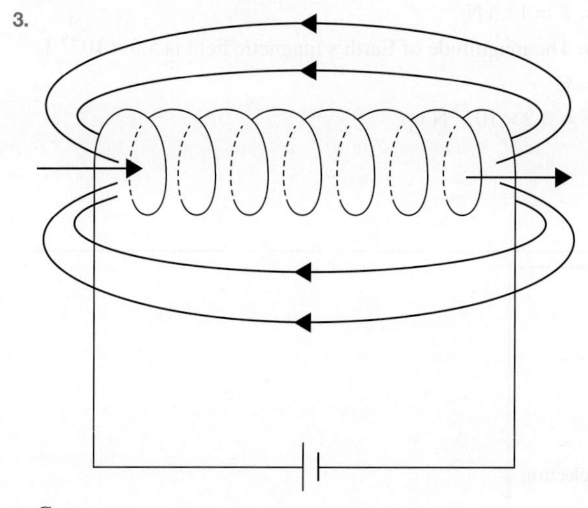

4. C
5. This mnemonic tool is consistent with the right-hand-grip rule, and the fact that field lines exit a solenoid from its north face.
   Thus, it can be used to quickly identify, from the direction of the current when looking at a solenoid's face, whether it is a north face or a south face.

5.4 Using magnetic fields to control charged particles, cyclotrons and mass spectrometers

Sample problem 5

a. 1.4×10^{-15} N

b. The direction of the force is up the page.

Practice problem 5

a. 1.9×10^{-13} N

b. The force is into the page.

c. There is no force on the positron.

Sample problem 6

5.6×10^{-3} m

Practice problem 6

5.3×10^{5} m s^{-1}

Sample problem 7

1.7×10^{5} m s^{-1}

Practice problem 7

1.2×10^{5} N C^{-1}

5.4 Exercise

1. $F = 1.9 \times 10^{-16}$ N
2. a. $F = 0$ N
   b. The particle would pass through the magnetic field undeflected.
3. a. $F = 4.6 \times 10^{-14}$ N
   b. The electron would experience a magnetic force that is perpendicular to the direction of its travel according to the right-hand-slap rule, and thus would be deflected in that direction.
   c. $a = 5.1 \times 10^{16}$ m s^{-2}
4. a. Using the right-hand-slap rule, the charge of the tauon is negative.
   b. The particle will move with the same velocity, in a straight line downwards.
5. a. The magnetic force on the electron is down the page.
   b. The magnetic force on the electron is out of the page
   c. The magnetic force on the proton is down the page and right.
6. If its direction is parallel to the direction of the magnetic field.

7. 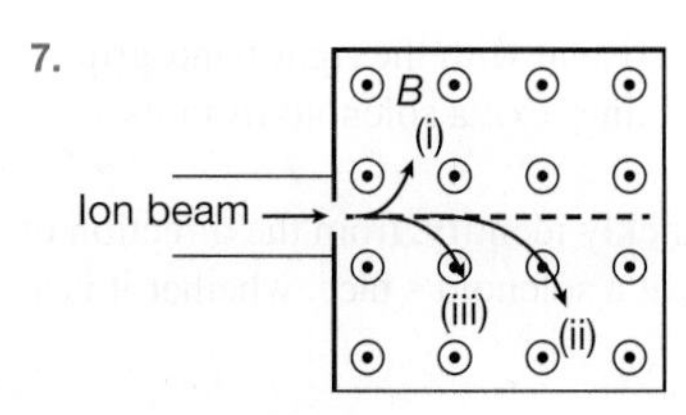

8. a. $r = 0.043$ mm
   b. $r = 78$ mm
   c. $r = 157$ mm
9. Magnetic fields control the movement of particles by separating ions based on their charge and their mass. As different ions (or isotopes) have different mass to charge ratios, they have different path radii as they move. This allows a mass spectrometer to separate ions in a sample with the aid of a magnetic field.
10. The overall trajectory of the positron will be an anti-clockwise helix, with the circular motion perpendicular to the page.
11. $B = 0.63$ mT
12. $v = 1.3 \times 10^6$ m s^{-1}

5.4 Exam questions

1. The diameter of the semicircular path is 1.9×10^{-1} m, or 19 cm.
2. See figure at the bottom of the page*
3. A
4. a. Charge q is negative.
   b. The force is constant in magnitude and always at right angles to the direction of motion, thus the path is circular.
5. The radius of the path is 7.4×10^{-1} m.

5.5 Magnetic forces on current-carrying wires

Sample problem 8

a. 6.0×10^{-3} N   b. 0 N

Practice problem 8

a. 1.25 N   b. 2.3 T

5.5 Exercise

1. $F = 7 \times 10^{-2}$ N
2. $F = 0$ N
3. $F = 1.4$ N
4. a. The direction of the magnetic force is into the page.
   b. The direction of the magnetic force is into the page.
   c. The direction of the magnetic force is south.
   d. The direction of the magnetic force is out of the page.
   e. The direction of the magnetic force is north.
   f. The direction of the magnetic force is east.
   g. The direction of the magnetic force is south-east.
   h. The direction of the magnetic force is south.
5. a. 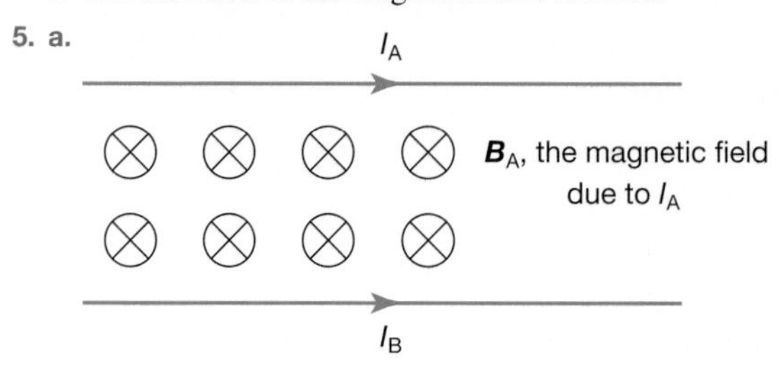

   b. $\boldsymbol{F}_{\text{on B by A}}$ would be up the page according to the right-hand-slap rule.
   c. 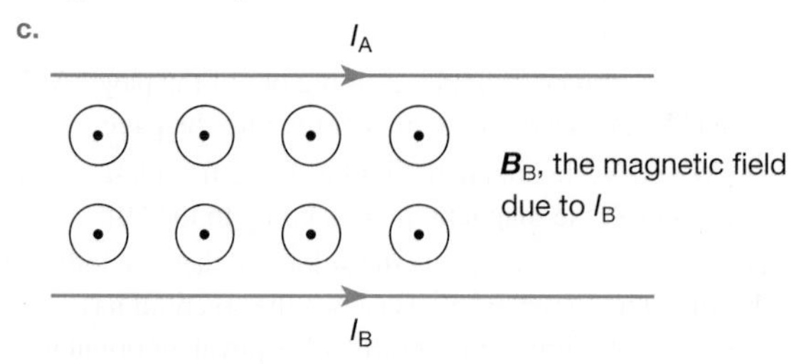

   $\boldsymbol{F}_{\text{on B by A}}$ would be down the page according to the right-hand-slap rule.
   d. Yes, $\boldsymbol{F}_{\text{on B by A}} = -\boldsymbol{F}_{\text{on A by B}}$.

5.5 Exam questions

1. D
2. C
3. a. Draw a downward arrow on side JK.
   b. $F = 13.5$ N
4. a. The magnitude of Earth's magnetic field is 5.3×10^{-5} T.
   b. C
5. $\boldsymbol{F} = 3.2 \times 10^{-3}$ N up

*2. 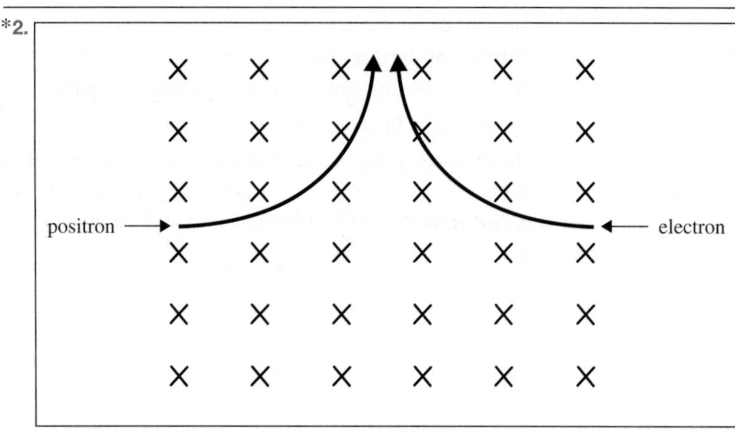

Figure 1

5.6 Applying magnetic forces — the DC motor

Sample problem 9

a. i. 1.5×10^4 N, down

ii. 0 N

iii. 1.5×10^4 N, up

b. Torque is a maximum when the axis of rotation and the line of action of the force are perpendicular to each other, and the distance between them is at a maximum. This occurs when the coil is parallel to the field.

c. Without the commutator, after the loop passes through the vertical position, the direction of the forces on the coil would reverse, causing the loop to slow down and then reverse the direction of rotation. Every time the loop passes through the vertical position, the direction of the forces reverse again, causing the loop to slow and reverse the direction of rotation. The loop would just oscillate around the vertical position.
The commutator enables the direction of current through the loop to be reversed every time the loop passes through the vertical position, reversing the direction of the forces so that the loop can rotate continuously in one direction.

Practice problem 9

a. i. 1.5×10^4 N, up

ii. 0 N

iii. 1.5×10^4 N, down

b. The only difference is that the coil rotates in the opposite direction from the coil in the sample problem.

5.6 Exercise

1. a. Provides a magnetic field
   b. Enables current to flow from stationary conductors
   c. Enables the circuit to switch the direction of the current within the course of one rotation
   d. The more turns there are, the larger the force and the faster the rotation.
2. a. If the coil is perpendicular to the magnetic field, the forces on the sides of the coil are collinear, resulting in no turning force (or torque) on the coil and hence it would not rotate. This can be overcome by having a second coil at right angles to the first coil, each coil having its own section of the split ring commutator.
   b. Yes. This can be achieved either by connecting the motor to an AC power supply or by replacing the split ring commutator with a slip ring commutator.
   c. Yes. This can be achieved by either increasing the voltage of the DC power supply and thereby increasing the current and turning force on the coil.
3. No. The motor would just run backwards and forwards.
4. The two sides of the loop form a parallel circuit supplied by the battery. The current flowing through the loop interacts with the field of the button magnet to create a torque on the loop. As the loop rotates, the direction of the torque is the same, so the loop rotates continuously.
5. a. The field coils (electromagnet) and the armature coils (rotating) may be connected in series to each other or in parallel to each other.
   b. The series-wound motor produces a large starting torque (turning force), making it ideal as motors for trains and trams. The parallel-wound motor can regulate its speed over a range of loads, making it ideal for power tools.
6. D
7. C

5.6 Exam questions

1. a. The current flows from K to L (right-hand-slap rule)
   b. The current is flowing from *L* to *K*.
   c. The magnitude of the current in the coil is 3 A.
2. a. $F = 4.0 \times 10^{-3}$ N
   b. A split ring commutator is a device that reverses the direction of the current flowing through an electric circuit. It is used in a DC motor to reverse the direction of current every half turn, ensuring that the motor continues rotating in the same direction.
3. The coil starts to rotate clockwise.
4. a. $F = 0.90$ N
   The direction is vertically down (D)
   b. The force is 0 N as the current is parallel to the magnetic field.
5. a. A, B
   b. When the coil is horizontal
   When the force is at right-angles to the plane of the coil
   c. Both A and B, as $F = nIlB$

5.7 Similarities and differences between gravitational, electric and magnetic fields

Sample problem 10

Similarities:

1. The gravitational field generated from a point mass and the electric field generated from a positive charge both obey an inverse square law.
2. The gravitational field and electric field are both non-uniform and static.
3. The point mass and positive charge can both be described as monopoles.

Differences:

1. The direction of the gravitational field is inwards towards the point mass, whereas the electric field generated from a positive charge is outwards, away from the point charge.
2. The gravitational field is attractive for all other masses, whereas the electric field generated from a positive charge repels other positive charges and attracts negative charges.

Practice problem 10

The electric field from an electric dipole and the magnetic field from a bar magnet or solenoid have very similar overall shapes. The key differences are as follows:

1. For a bar magnet, there are no field lines inside the magnet between the poles.
2. All the magnetic field lines must form closed loops so that all field lines must eventually loop back to pass through the solenoid from the other side.

3. In the electric dipole, the direction of the field lines change sign as they pass through the poles. This is not the case for magnetic field lines, and it is not possible to isolate points in space as 'the north pole' or 'the south pole'.

5.7 Exercise

1. Repulsion and attraction were both observed.
2. Similarities: the inverse square law exists, proportional to the product of a property of the objects (a consequence of Newton's Third Law).

   Difference: mass cannot be positive or negative, whereas charge can.
3. A uniform magnetic field is found inside a solenoid, whereas the field outside the solenoid is non-uniform, as illustrated below.

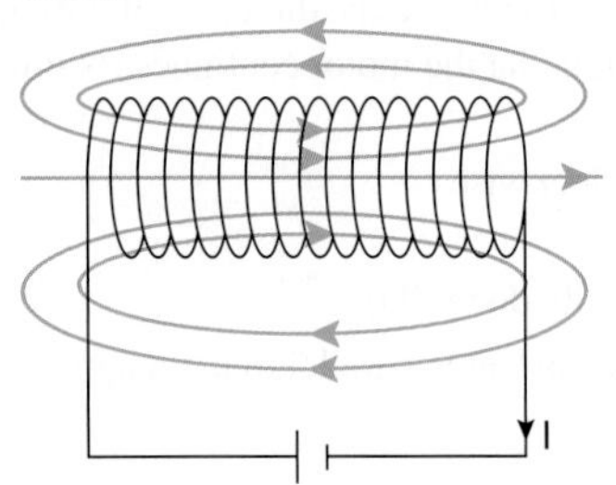

4. The overall field shapes are similar; however, the magnetic field lines all form closed loops, and, unlike the positive charges, the north poles cannot be isolated to single points.

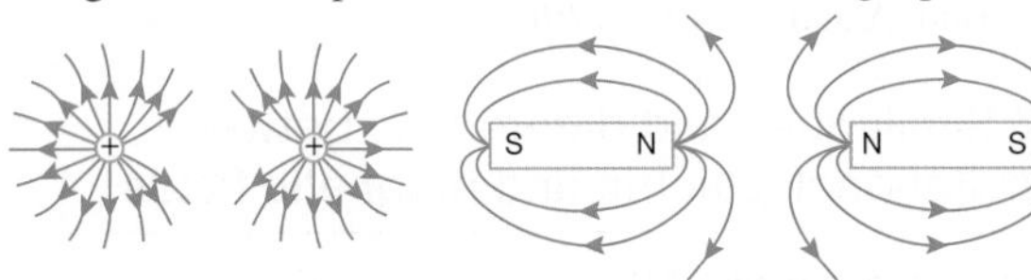

5. $\boldsymbol{F}_g = 3.61 \times 10^{-47}$ N

   $\boldsymbol{F}_e = 8.1 \times 10^{-8}$ N

   The electric force is 39 orders of magnitude greater than the gravitational force; therefore, the gravitational force of attraction has virtually no influence on the motion of the electron and proton in a hydrogen atom.
6. 1.0×10^{-7} N C^{-1}
7. $\boldsymbol{E} = 6.1 \times 10^{10}$ N m^{-1}
8. $q = 5.7 \times 10^{13}$ C
9. $I = 5.0$ A

5.7 Exam questions

1. C
2. D
3. This diagram could correspond to the gravitational field of a point mass, or the electric field of a negatively charged point charge.
4. Noah is correct. This field shape could correspond to the gravitational field near the surface of Earth, it could correspond to the electric field between two positively and negatively charged conducting plates, and it could correspond to the magnetic field inside a large solenoid (or between the arms of a large horseshoe magnet).
5. Magnetic fields can also be created by moving charges, not just by magnetic materials.

5.8 Review

5.8 Review questions

1. a.

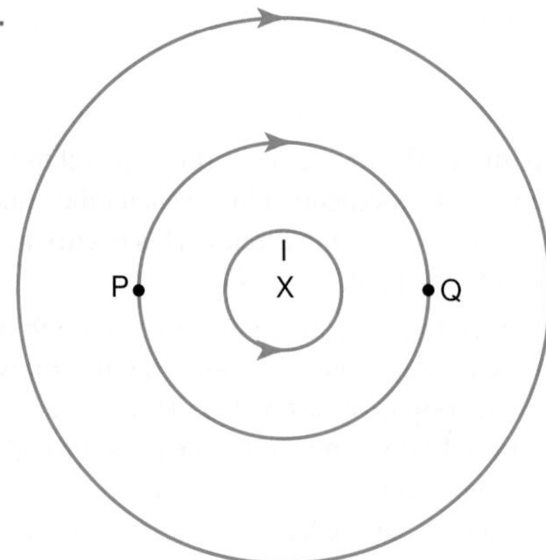

   b. The field from the wire at P is 20 μT, in the same direction as Earth's magnetic field, so the total magnetic field strength is 30 μT pointing up the page. The field from the wire at Q is 20 μT in the opposite direction to Earth's magnetic field, so the total magnetic field strength is 10 μT pointing down the page.
2. The field from an electric dipole and the magnetic field from a solenoid have a very similar overall shape. The key differences are (i) all the magnetic field lines must form closed loops so that all field lines must eventually loop back to pass through the solenoid from the other side.
   (ii) In the electric dipole, the direction of the field lines change sign as they pass through the poles. This is not the case for magnetic field lines, and it is not possible to isolate a point in space as 'the north pole' or 'the south pole'.

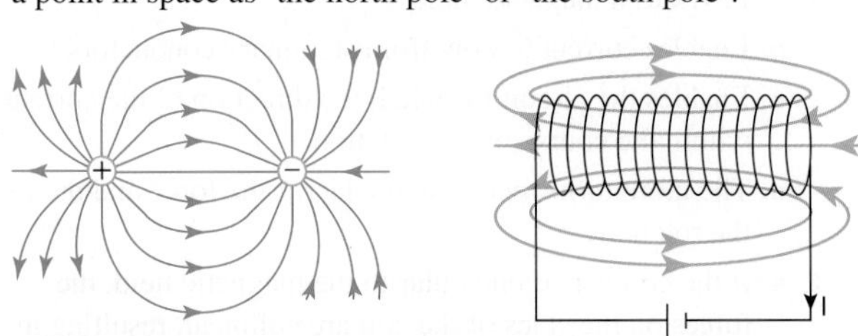

3.

Data set	Gravitational	Electric	Magnetic
Field A	Surface of Earth	Between +ve and −ve charged conducting plate	Inside large solenoid
Field B	Earth	−ve point charge	Not possible
Field C	Not possible	−ve and +ve charge, electric dipole	Not possible (similar to electric dipole, but not identical)

4. a. 2.0×10^{-13} N
   b. 2.2×10^{17} m s^{-2}
   c. 1.2×10^{14} m s^{-2}
5. When the ionisation track appears, the net force on the particle, which is a negative particle, is upwards, so the magnetic field experienced is out of the page.
6. 6.3 mT, perpendicular to the momentum of the electrons
7. a. 3.3×10^5 m s^{-1}
   b. 5.3×10^{-14} N
   c. 0.025 m
8. a. At P, the magnetic field from Q is out of the page, which is indicated by a dot.
   b. At Q, the force on the right-hand wire is to the right.
   c. Forces are in opposite directions, so the wires repel each other.
9. a. $\boldsymbol{F} = nIl\boldsymbol{B}$
   $= 50 \times 3.0 \times 0.15 \times 0.030$
   $= 0.68$ N
   b. The arrow should be pointing vertically upwards.
   c. If the polarity of the DC supply is reversed, then current initially flows in the opposite direction through the loop, so the direction of rotation of the loop is reversed.
   d. For the motor to spin continuously, the split in the ring of the commutator needs to be perpendicular to the direction of the magnetic field.
   This means that the change in current direction in the loop occurs when the loop is perpendicular to the field, or vertical in the diagram shown.
   If the change in current direction does not occur, then as the loop moves past the vertical, the direction of the torque on the loop changes so that the loop is slowed, and it starts to rotate in the opposite direction. This will be the case if the commutator is attached with the split in the ring significantly altered from being perpendicular to the vertical.
10. Yes. Since the direction of the conventional current for electrons is opposite to that for positively charged particles, the force of the magnetic field is directed upwards. This force is balanced by the electric force on the electrons, which is directed downwards.

5.8 Exam questions

Section A — Multiple choice questions

1. A
2. A
3. C
4. B
5. A
6. B
7. A
8. A
9. B
10. D

Section B — Short answer questions

11. a. The magnetic force acting on the electron is always perpendicular to the electron's velocity and has a constant magnitude.
    b. The radius is 4.6×10^{-2} m.
12. a. Draw a downward arrow on the side JK of the coil.
    b. The role of the split ring commutator is to reverse the direction of the current every half turn to maintain a constant direction of rotation.
13. See figure at the bottom of the page*
14.

Field type	Monopoles only	Dipoles only	Both monopoles and dipoles
Gravitation	✓		
Magnetism		✓	
Electricity			✓

15. a. $\dfrac{E}{B}$
    b. $v_0 = 2.0 \times 10^6$ m s^{-1}
    c. i. They would arrive at point Z.
    ii. If v_0 increases, then the radius of the path and the magnitude of the magnetic force on the electron both increase, and the electron is more deviated.
    Using the right-hand-slap rule, the force is upwards for a positively charged particle and downwards for negatively charged particle.
    Thus the electron is more deviated downwards.

Unit 3 | Area of Study 2 review

Section A — Multiple choice questions

1. A
2. A
3. B
4. D
5. C
6. B
7. A
8. C

*13.

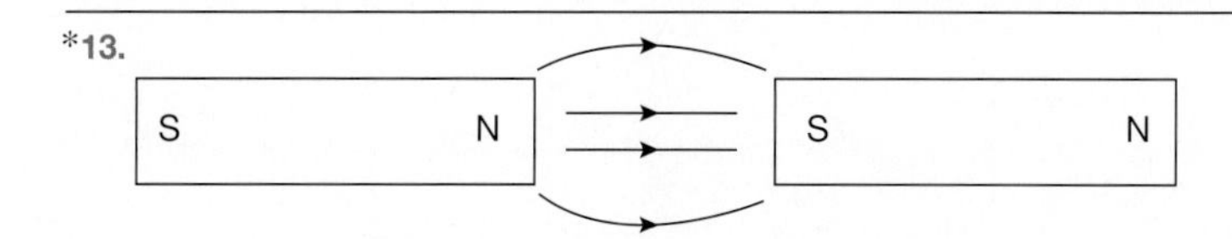

9. B

10. C

11. C

12. D

13. C

14. A

15. B

16. B

17. B

18. B

19. B

20. C

Section B — Short answer questions

21. a. 1.36 N kg^{-1}
    b. $1.64 \times 10^5 \text{ s}$
    c. $5.64 \times 10^{26} \text{ kg}$
    d. $5.55 \times 10^3 \text{ m s}^{-1}$
22. a. 800 V m^{-1}
    b. $1.37 \times 10^{14} \text{ m s}^{-2}$
    c. $1.92 \times 10^{-17} \text{ J}$
    d. $6.50 \times 10^6 \text{ m s}^{-1}$
23. a. Out of the page, or from K to J
    b. 0.32 T
    c. Answers will vary. Sample response:
    The coil will rotate clockwise to a vertical position and stop moving. This is because the magnetic force on the side JK is upward, and the magnetic force on the side LM is downward, and the forces do not impart a twisting motion to the coil to cause it to rotate.
    d. 6.9 cm

6 Generation of electricity

6.2 BACKGROUND KNOWLEDGE Generating voltage and current with a magnetic field

Sample problem 1

5.0 mV

Practice problem 1

80 m s^{-1}

6.2 Exercise

1. 14 mV
2. 0.823 m s^{-1}
3. 0.83 m
4. 0 V
5. 10 T

6.2 Exam questions

1. 0.92 V
2. The energy came from the loss of gravitational potential energy as the rod is falling.
3. When the rod falls faster, a higher voltage difference is created between the two ends of the rod.
4. 0.002 V
5. $5 \times 10^{-4} \text{ V}$
6. a. 6000 m
   b. $3 \times 10^7 \text{ m}^2$

6.3 Magnetic flux

Sample problem 2

a. 0.015 Wb

b. 0.004 Wb

c. 0 Wb

Practice problem 2

0.000 54 Wb

6.3 Exercise

1. Magnetic flux is the amount of magnetic field across an area, and is the product of the magnetic field strength and the perpendicular cross-sectional area.
2. a. 0.15 Wb
   b. $1.8 \times 10^{-4} \text{ Wb}$
   c. $7.1 \times 10^{-3} \text{ Wb}$
3. a. 0 Wb
   b. $7.5 \times 10^{-5} \text{ Wb}$
   c.

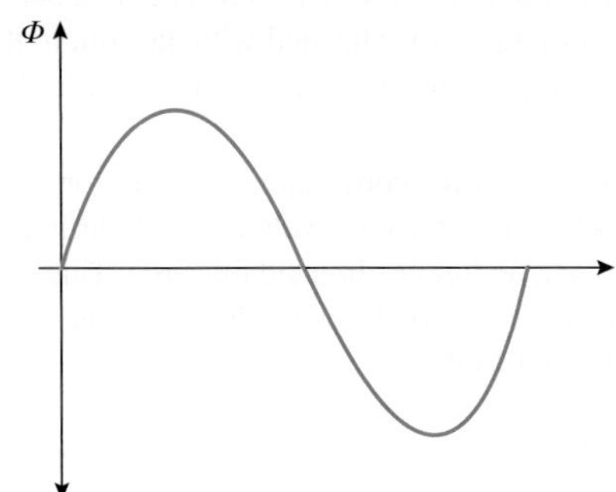

4. 17.5 Wb
5. $3.4 \times 10^{-4} \text{ Wb}$
6. a. 3.0 T
   b. 0.8 m^2
7. $2.0 \times 10^{-5} \text{ Wb}$
8. a. $1.2 \times 10^{-4} \text{ Wb}$
   b. $6.8 \times 10^{-4} \text{ Wb}$
   c. 0 Wb

6.3 Exam questions

1. B
2. 1.7 T
3. 3.0×10^{-5} Wb
4. 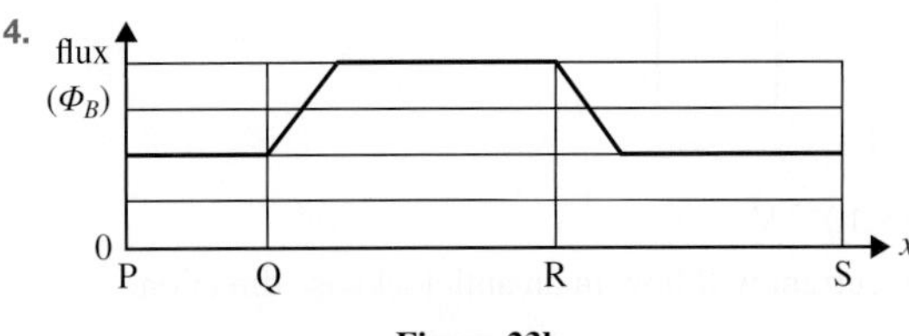

Figure 23b

5. B

6.4 Generating emf from a changing magnetic flux

Sample problem 3

a. −0.025 V

b. The induced current must flow in an anticlockwise direction around the loop.

c. 0.05 A

Practice problem 3

a. 0.0184 V

b. The current must be travelling in the anticlockwise direction.

c. 0.46 A

6.4 Exercise

1. Increasing the number of turns increases the magnetic flux through the coils by increasing the cross-sectional area.
2. Clockwise around the loop
3. a. The flux from the induced current would be out of the page and would be anticlockwise around the top loop.
   b. No
   c. Yes
4. a. An anticlockwise current is produced.
   b. Yes, a clockwise current would be produced.
5. a. Yes, there is an induced current, which is anticlockwise around the loop.
   b. Yes, there is an induced current, which is anticlockwise around the loop.
   c. No, there is no induced current.
   d. No, there is no induced current.
6. a. 0.0157 V
   b. 17 V
   c. 38 V
7. a. 4.0 mA
   b. 0.54 A
   c. 0.2 A
8. 3000 V
9. a. i. Anticlockwise current viewed from above.
   ii. No current in the loop.
   iii. No current in the loop.

   b. 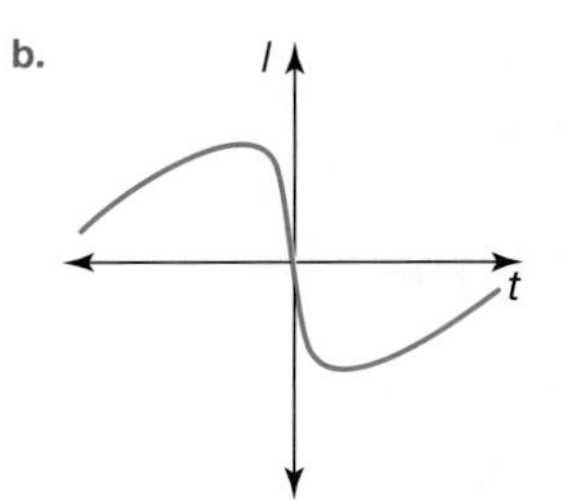

   c. From the loss of gravitational potential energy.
   d. The two areas under the graph — one above the x-axis and the other below the x-axis — would cancel each other out.
10. 0.024 C
11. The passing magnet would induce an emf in both loops; that is, the electrons inside the metal loop as well as the plastic loop would experience a magnetic force from the passing magnet. However, since plastic is an insulator, no current would flow in the plastic loop, in contrast to the metal loop.
12. As the north end of the magnet approaches the loop of conducting wire, it would induce a current in the loop. If Lenz's Law is reversed, the induced current would have a magnetic field that attracts the north end of the magnet. The magnet would accelerate and gain kinetic energy from nothing, hence violating the principle of the conservation of energy.
13. a. A current will flow to induce a N pole at the left end of the coil. Right-hand rule: thumb points to the left, fingers show the direction of current in the coil. Current flows in the direction of X to Y.
    b. There is no change in magnetic flux through the coil so no current is induced.
    c. Any two answers from the following:
    - Move the magnet faster
    - Have more turns on the coil
    - Use a stronger magnet
    - Use a smaller resistance

6.4 Exam questions

1. a. The magnetic flux through the coil will decrease.
   b. The loop experiences a decrease in flux into the page. Lenz's Law states that the induced current will produce an increasing flux into the page. Using the right-hand grip rule, the induced current will flow clockwise around the loop.
   c. From Figure 6a to Figure 6b, the loop is experiencing an increasing flux into the page. Applying the right-hand grip rule, the induced current will flow anticlockwise around the loop. From Figure 6b to Figure 6c, the loop is experiencing a decreasing flux into the page. Applying the right-hand grip rule, the induced current will flow clockwise around the loop.

2. See table at the bottom of the page*

3. a. The current will be clockwise.

   b. $1.5 \times 10^{-4}\ \text{m}^2$

   c. See figure at the bottom of the page**

4. a.

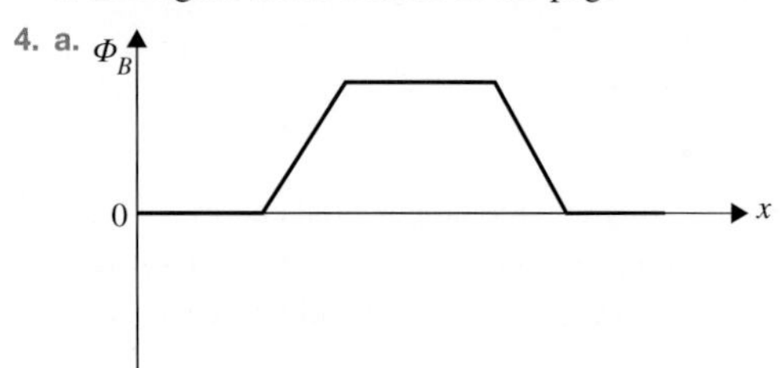

   b.

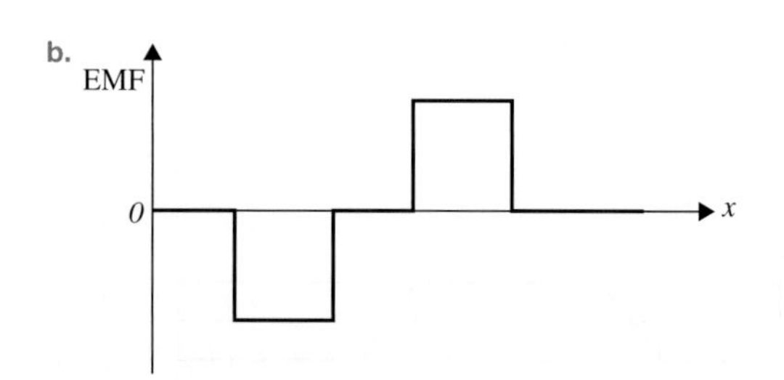

   c. $4.0 \times 10^{-5}\ \text{V}$

   d. The current will flow in an anticlockwise direction.

---

*2.

Movement	Possible EMF	Direction of any induced current (alternating/clockwise/anticlockwise/no current)
A rotation about x-axis	EMF (V) time (s)	Alternating
B moving from Position 1 to Position 2	EMF (V) time (s)	No current
C moving from Position 2 to Position 3	EMF (V) time (s)	Clockwise

**3. c.

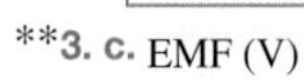

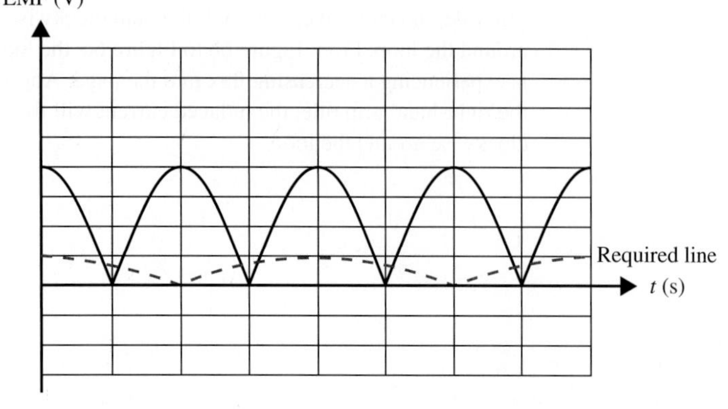

5. a. $\varepsilon = 6.4 \times 10^{-4}$ V

   b. See figure at the bottom of the page*

6.5 Generators and alternators

6.5 Exercise

1. a. $\Phi = 7.5 \times 10^{-2}$ Wb

   b. $\varepsilon = 0.15$ V

   c. The period is doubled and the amplitude is halved. See figure at the bottom of the page**

2. a. 9.0×10^{-4} Wb b. 1.9×10^{2} s

3. a. Current increases

   b. Current decreases

   c. Current increases

4. a. From Q to P (using the right-hand grip rule for coils)

   b.

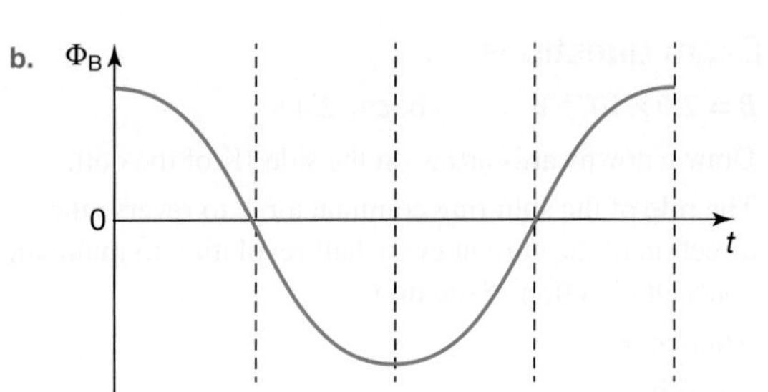

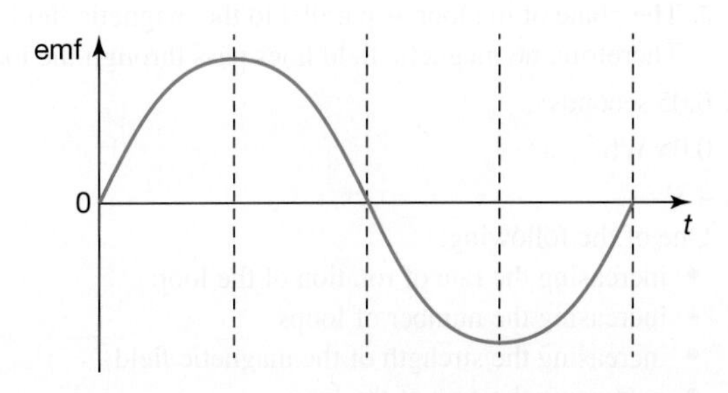

*5. b.

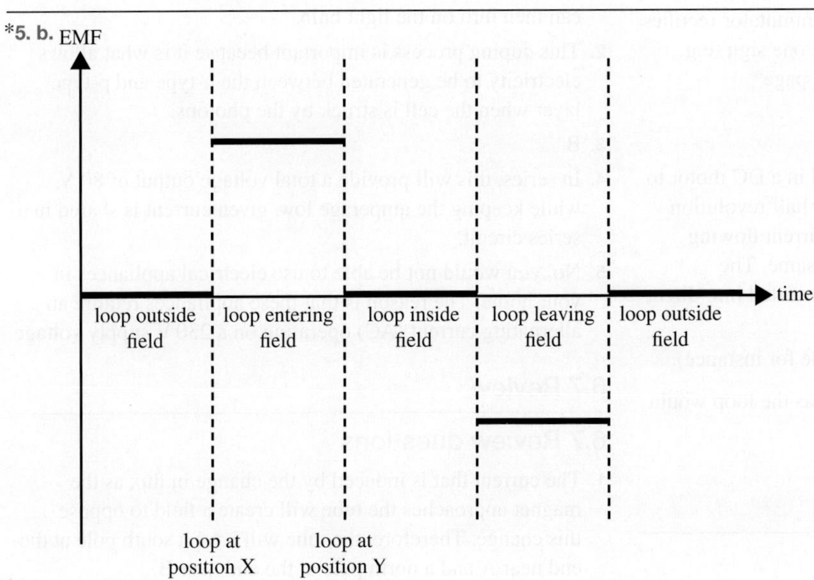

**1. c.

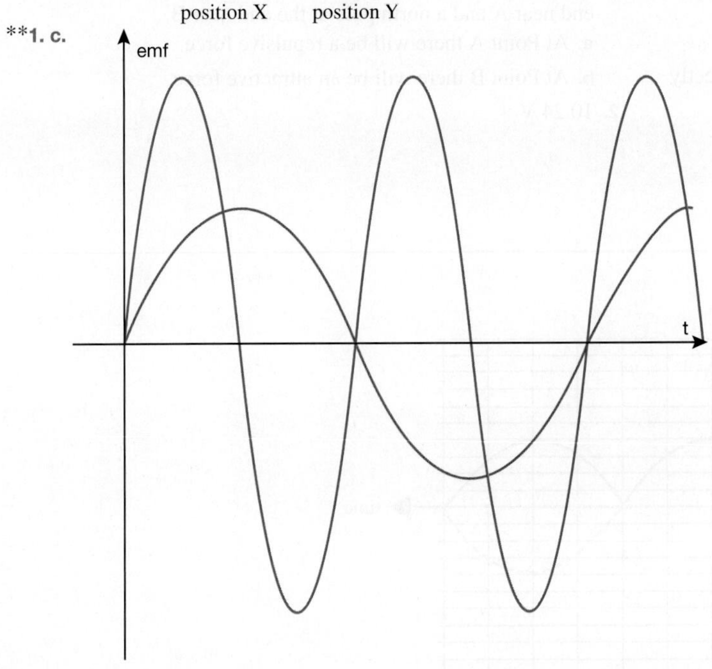

6.5 Exam questions

1. a. $B = 2.0 \times 10^{-2}$ T   b. $\varepsilon = 2.4$ V
2. a. Draw a downwards arrow on the side JK of the coil.
   b. The role of the split ring commutator is to reverse the direction of the current every half revolution to maintain a constant direction of rotation.
3. a. Alternator
   b. i. 0 Wb
      ii. The plane of the loop is parallel to the magnetic field. Therefore, no magnetic field lines pass through the loop.
   c. 0.05 seconds
   d. 0.05 Wb
   e. 4 V
   f. One of the following:
      - increasing the rate of rotation of the loop
      - increasing the number of loops
      - increasing the strength of the magnetic field
      - increasing the area of the loop
   g. Changing the slip rings to a split ring commutator rectifies the output, making the EMF output only one sign (e.g. positive). See figure at the bottom of the page*
4. a. $F = 4.0 \times 10^{-3}$ N
   b. Answers will vary.
      A split ring commutator is a device used in a DC motor to reverse the coil's current direction every half revolution of the loop so that the direction of the current flowing through the external circuit remains the same. The alternating current in the loop is thus converted into direct current.
5. a. Clockwise (using the right-hand-slap rule for instance).
   b. The motor would not function anymore as the loop would only rotate $\frac{1}{4}$ of a turn before stopping.

6.6 Photovoltaic cells

6.6 Exercise

1. A photovoltaic cell is a device that transforms electromagnetic energy, such as light from the Sun, directly into electrical energy.
2. PV panels produce DC voltage. It must be transformed into 230 V AC for use in buildings.
3. If **photons** have enough energy, they cause **electrons** to be removed from atoms in the cell.
   If the photons striking a solar cell **do not** have enough energy, their energy is transformed into **thermal** energy and the solar cell heats up.
4. The n-type layer releases an electron when it absorbs energy from a photon that has enough energy. The electron drifts across the junction to fill in the 'positive holes' of the p-type layer, generating an electric current.
5. B. False.

6.6 Exam questions

1. Photons from the Sun strike the solar cell. If the photons have enough energy, it will remove an electron from the n-type layer. This electron will then travel across the junction (boundary layer) to fill in the positive holes of the p-type layer, resulting in the generation of an electric current, which can then turn on the light bulb.
2. This doping process is important because it is what allows electricity to be generated between the n-type and p-type layer when the cell is struck by the photons.
3. B
4. In series, this will provide a total voltage output of 80 V, while keeping the amperage low, given current is shared in a series circuit.
5. No, you would not be able to use electrical appliances in your home. The reason is that these appliances require an alternating current (AC) operating on a 230 V supply voltage.

6.7 Review

6.7 Review questions

1. The current that is induced by the change in flux as the magnet approaches the tube will create a field to oppose this change. Therefore, the tube will have a south pole at the end near A and a north pole at the end near B.
   a. At Point A there will be a repulsive force.
   b. At Point B there will be an attractive force.
2. 10.24 V

*3. g.

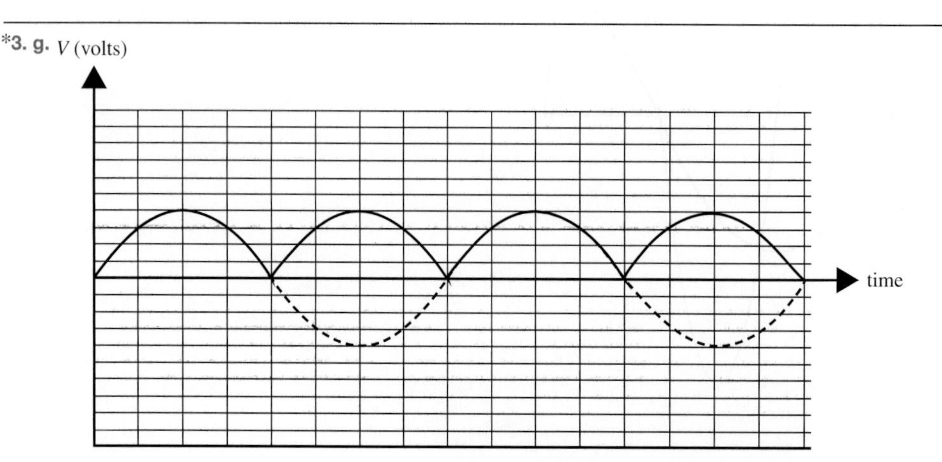

3. 2×10^{-5} Wb

4. $\frac{20\boldsymbol{B}}{3t}$ V

5. 1.5×10^{-3} V

6. An advantage of using photovoltaic cells as an energy source is that:
   - sunlight is free
   - excess energy can be stored in a battery or sold back to the grid
   - no carbon emissions and quiet to run
   - it is not reliant on fossil fuels

   A disadvantage of using photovoltaic cells is that:
   - they can only generate electricity on sunny days
   - to generate the maximum amount of electricity, they must face a certain direction
   - they can be easily damaged
   - they can be costly to install
   - they have low efficiency.

7. a. 5.625×10^{-6} Wb

   b. Ninety degrees about the horizontal axis will render the coil parallel to the magnetic field and thus there will be zero flux threading the coil.

   c. 1.125×10^{-5} V

8. a. 5.71×10^{-3} m^2

   b. 0.16 V

   c. 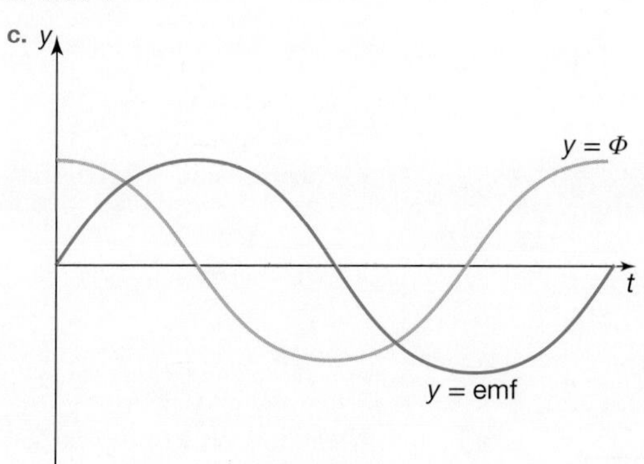

   d. • increase the rate of rotation.
   • Increase the number of coils that create the loop.
   • Increase the magnetic field strength.
   • Increase the size of the coil.

9. a. 0.25 Wb

   b. 1.6 V

   c. 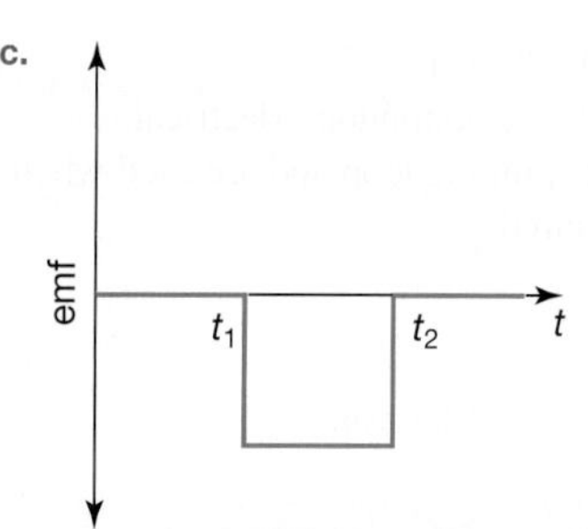

10. a. The induced voltage can be thought of as the negative rate of change of flux.
    Given the flux is being reduced by a constant amount until it reaches zero, it is expected that the voltage will be constant and positive as it seen in the graph in the question.

    b. 3.53×10^{-3} m^2

6.7 Exam questions

Section A — Multiple choice questions

1. C
2. C
3. D
4. A
5. D
6. D
7. C
8. C
9. D
10. A

Section B — Short answer questions

11. a. Anticlockwise.
    The loop is travelling down near the north pole of the magnet and the magnetic force is to the left. Using the right-hand-slap rule, the induced current is into the page, which means the induced current will continue anticlockwise around the loop.

    b. $\varepsilon = -7 \times 10^{-5}$ V

    c. See figure at the bottom of the page*

    d. • increase field strength
    • increase the number of coils
    • increase the area of coil
    • increase the rotation rate or decrease period of rotation.

*11. c. EMF

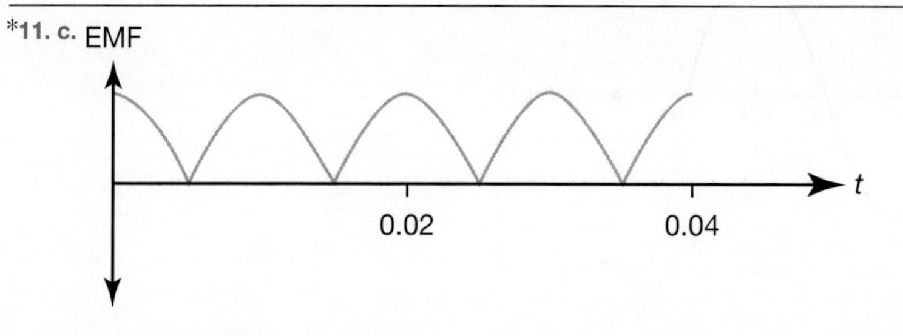

12. See figure at the bottom of the page*
13. a. The slip rings maintain a continuous electrical connection with the spinning loop and are used when an AC output is required.
    b. $\Phi = 1.2 \times 10^{-4}$ Wb
    c. $f = 5$ Hz
    d. See figure at the bottom of the page**
14. a. $\varepsilon = 0.080$ V
    b. See figure at the bottom of the page***
15. a. $F = 5 \times 10^{-4}$ N
    b. Using a rule such as the right-hand-slap rule shows that the side WX will be forced down while the side YZ will be forced upwards resulting in anticlockwise rotation.
    c. The role of a split ring commutator is to reverse the coil's current direction every half revolution of the loop so that the direction of the current flowing through the external circuit remains the same. The alternating current in the loop is thus converted into direct current.

7 Transmission of electricity

7.2 Peak, RMS and peak-to-peak voltages

Sample problem 1

18 V

Practice problem 1

$I_{RMS} = 7.83$ A to 3 s.f.

$I_{peak} = 11.07$ A to 3 s.f.

7.2 Exercise

1. 8.9 V
2. a. 30 ms
   b. 33.3 Hz
   c. 50 mV
   d. 100 mV
   e. 35 mV
3. 0.83 m

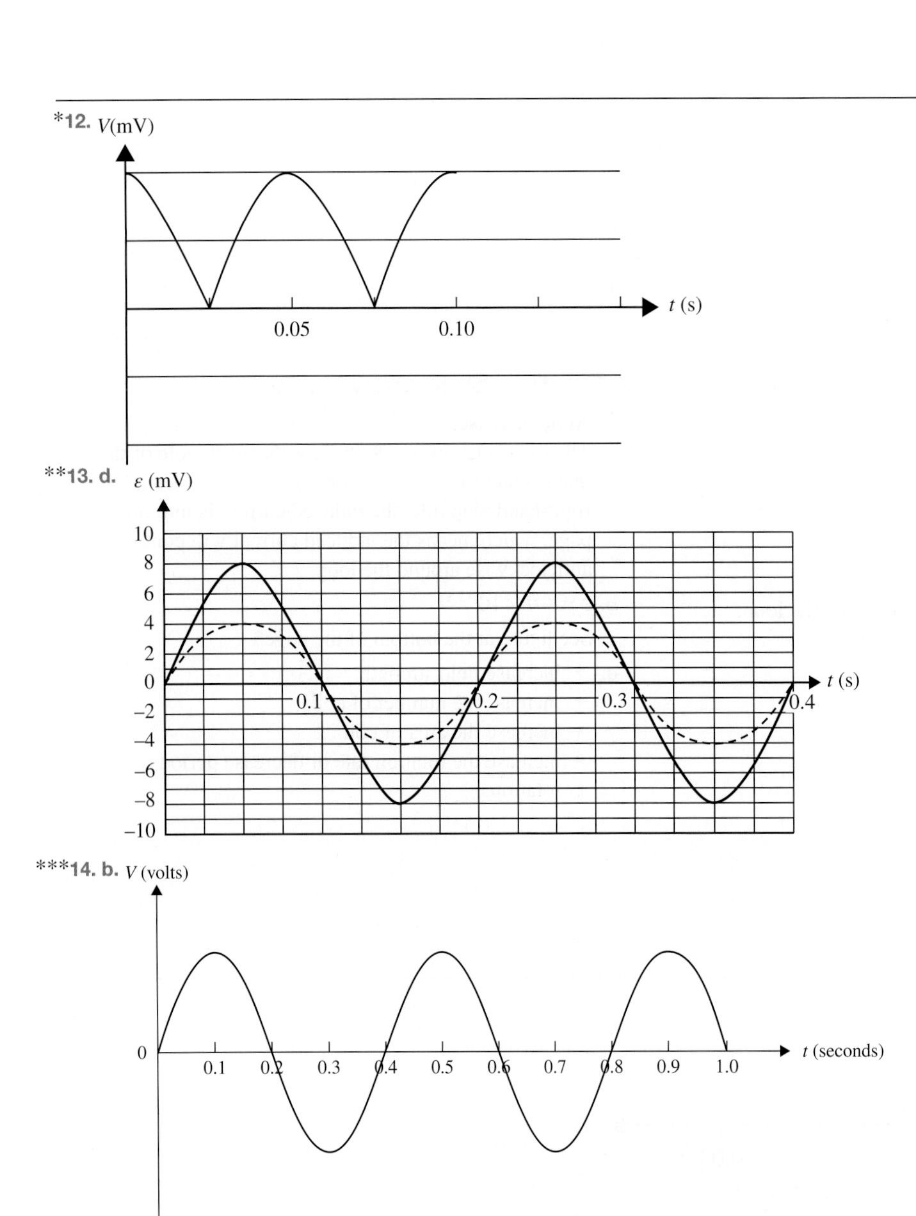

7.2 Exam questions

1. 2.5 V
2. 25 Hz
3. 25 Hz
4. A
5. 12.5 Hz

7.3 Transformers

Sample problem 2

a. $I = 0.435$ A

b. $R = 529\ \Omega$

c. i. $R \approx 529\ \Omega$ ii. 230 V

d. i. $R \approx 2729\ \Omega$ ii. 44.6 V

e. The globe would not light up.

Practice problem 2

a. 3.48 A

b. 66.1 Ω

c. i. 66.1 Ω ii. ≈ 230 V

d. i. 766 Ω ii. 19.8 V

e. The toaster would not operate properly.

Sample problem 3

The secondary coil consists of approximately 10 turns.

Practice problem 3

The turns ratio is 1:20, making this transformer a step-up transformer.

7.3 Exercise

1. 4600 V
2. a. 4.6 V b. 450 V
3. a. 960 turns b. 8 A c. 0.4 A
4. A transformer requires a changing magnetic flux in order to work; thus, a transformer will not work with a DC input voltage, which produces a constant magnetic flux.
5. This allows the magnetic field to change direction quickly as the current changes direction.
6. a. Step-down transformer b. 46 turns
7. a. 135 V
   b. 41 W
   c. 0.68 A

7.3 Exam questions

1. D
2. a. 17 V
   b. 20 : 1
   c. 3.8 mA
3. B
4. C
5. 9.55×10^3 V

7.4 Power distribution and transmission line losses

Sample problem 4

a. i. The current through the cables is 50.0 A.
   ii. The voltage drop across the cables is 250 V.
   iii. 62.5%
   iv. 150 V

b. i. 2.50 A
   ii. 12.5 V
   iii. 0.156%
   iv. 400 V

Practice problem 4

a. i. 200 A
   ii. 40.0 V
   iii. 16.0%
   iv. 210 V

b. i. 20.0 A
   ii. 4.00 V
   iii. 0.160%
   iv. 250 V

7.4 Exercise

1. a. 200 A
   b. 12 kW
   c. 60 V
   d. 190 V
   e. i. 30 W
      ii. 3.0 V
      iii. 250 V
2. a. 5.8 Ω
   b. 5.6 Ω
   c. 222 V
   d. i. 2.7 Ω ii. 1.8 Ω
      iii. 87 A iv. 127 A
   e. i. 25 V ii. 205 V
   f. Yes, the voltage will be within 1% of 230 V for the appliances to work properly.
   g. When the workshop is turned off, it does not draw any current on the transmission line, resulting in a lower transmission line current and therefore a lower voltage drop across the transmission line. This increases the voltage at the primary side of the step-down transformer, and therefore a higher voltage at the house.
3. a. 667 A
   b. 180 kW
   c. 270 V
   d. Voltage = 330 kV
      Power = 220 MW
4. a. 25%
   b. The power loss in transmission would decrease from 25% to 0.25%.

7.4 Exam questions

1. a. $\frac{N_P}{N_S} = 30 : 1$
   b. Due to the error in the wording of the question, either 0.35 A or 10.5 A were accepted.
2. a. $P_{loss} = 160$ W
   b. $V_{in} = 320\ V_{RMS}$
   c. $P_{loss} = 2.5$ W
3. a. 48 W b. 40 V c. 72 W
   d. 18 W e. Answers will vary.
4. a. 120 kW
   b. A factor of 4 increase
5. a. 36 W b. 4.5 V
   c. 20 W d. Answers will vary.
   e. Answers will vary.

7.5 Review

7.5 Review questions

1. a. 1.5 Ω b. 8.5 V
2. a. 400 b. 0.17 A
3. 4200 V
4. $\frac{N_2}{N_1} = \frac{5}{1}$
5. a. 400 turns
   b. 1.4 W
   c. i. The power at the input to the primary coil is the same as the power at the globe: 6.0 W.
      ii. 7.4 W
   d. The globe would not glow.
6. 0.15 A
7. a. 2000 turns b. 50 A
   c. 1.0×10^4 W d. 8.2×10^3 V

7.5 Exam questions

Section A — Multiple choice questions

1. B or D
   Due to the wording of the question, both 1.1×10^4 mA and 0.086 mA were accepted.
2. D
3. B
4. D
5. C
6. A
7. A
8. C
9. A
10. B

Section B — Short answer questions

11. a. Stepping up the voltage allows the current to be reduced while maintaining constant power
    b. 1.0 kA
    c. 30 MW
12. a. The brightness of the globe will be decreased. Students could then refer to reduced current, increased voltage drop in the cables or increased power lost in the cables.
    b. The independent variable is the resistance of the cables. The dependent variable is the current in the cables.
    c. $r = \frac{24}{I} - R$
    d. Answers will vary.
    e. Answers will vary.
    f. The correct value is 7 Ω.
13. a. 12 W
    b. 0.8 A
    c. $P_{loss} = I^2R$, so less current will mean less power loss.
14. a. An ideal transformer is one where no power is lost, that is, $P_{primary} = P_{secondary}$.
    b. $\frac{N_p}{N_s} = \frac{20}{1}$
    c. $P = 24$ W
    d. $V_{globe} = 10$ V
    e.
       - There is less power delivered to Light 1 compared to Light 2.
       - Power is lost along the transmission lines.
       - The observed brightness is proportional to the delivered power.
15. a. $P = 41.5$ kW
    b. No, the power supplied, 21.5 kW, is less than the 40 kW required.
    c. $P_{delivered} = 41.3$ kW

Unit 3 | Area of Study 3 review

Section A — Multiple choice questions

1. A
2. C
3. D
4. C
5. A
6. B
7. B
8. D
9. B
10. D
11. C

12. D
13. C
14. B
15. B
16. B
17. B
18. B
19. D
20. B

Section B — Short answer questions

21. a. X to Y.
 b. When the magnet is stationary there is no change of flux through the coil, thus no EMF and no current is generated.
 c. Answers will vary, see worked solution.
22. a. The magnitude of the flux is 9.0×10^{-4} Wb.
 b. From P to Q
 c. The interval is 1.1 s.
23. a. The secondary coil has 375 turns.
 b. The current is 0.8 A.
 c. The power loss is 5.6%.
24. a. $\frac{N_{primary}}{N_{secondary}} = 1 : 40$
 b. The maximum power is 240 W.
 c. The maximum current is 40 A

8 Light as a wave

8.2 Light as a wave

Sample problem 1

3.4×10^2 m s^{-1}

Practice problem 1

340 m s^{-1}

Sample problem 2

0.609 m

Practice problem 2

335 m s^{-1}

Sample problem 3

a. $T = 1.79 \times 10^{-15}$ s
b. $\lambda = 5.36 \times 10^2$ nm

Practice problem 3

$T = 1.50 \times 10^{-15}$ s

8.2 Exercise

1. a. 1.33 m
 b. 3.91 m s
 c. 2.94 s
2.

v (m s^{-1})	f (Hz)	λ (m)
335	500	0.67
300	12	25
1500	**5000**	0.30
60	**24**	2.5
340	1000	**0.34**
260	440	0.59

3. a. $f = 10$ Hz, $T = 0.1$ s
 b. $v = 0.26$ m s^{-1}
 c. The speed remains the same, the wavelength is halved.
4. 2.1×10^{-15} seconds
5. 3.0 cm to 0.30 mm (fractions of a millimetre to a few centimetres)
6. $f = 1.1 \times 10^{19}$ Hz
 $T = 9.0 \times 10^{-20}$ s
7. a. 50 Hz b. 6.0×10^6 m
8. a. 4.6×10^{-7} m b. 3.1×10^{-7} m c. 4.8×10^7 Hz
9. Both electric and magnetic fields are static about the charge. If the charge changes direction or accelerates in a straight line, a ripple is created in both fields, which is associated with an electromagnetic wave: all accelerated charged particles produce electromagnetic radiation consistent with Maxwell's model for electromagnetic waves.

8.2 Exam questions

1. A
2. C
3. See figure at the bottom of the page*
4. 2.92 m s^{-1}

*3.

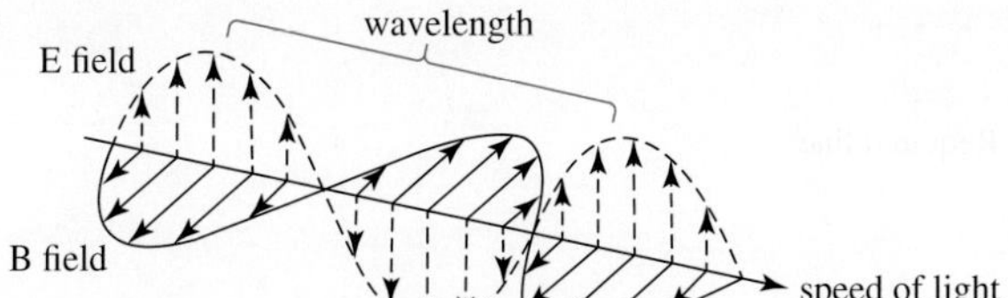

wavelength
E field
B field
speed of light

5. a. 20 Hz
   b. See figure at the bottom of the page*
   c. Assuming that the velocity will remain the same, the wavelength will get shorter.

8.3 Interference, resonance and standing waves

Sample problem 4

a. 4 nodes
b. 3 antinodes
c. 3rd harmonic
d. $\lambda = 1.0$ m
e. $T = \dfrac{1}{4.2} \approx 0.24$ s
f. $4.2\ \text{m s}^{-1}$
g. $f_1 = 1.4$ Hz and $\lambda_1 = 3.0$ m

Practice problem 4

a. $f_1 = 2.0$ Hz, $f_2 = 4.0$ Hz and $f_3 = 6.0$ Hz
b. $\lambda_1 = 3.0$ m, $\lambda_2 = 1.5$ m and $\lambda_3 = 1.0$ m

8.3 Exercise

1. a. Diagram C   b. Diagrams A and B
2. Superposition is the addition of the amplitudes of two or more wave disturbances at a point in space and time. It occurs whenever there is a disturbance arising from two or more sources of waves.
3. Constructive interference occurs when two or more wave disturbances superimpose (add together) to give a resultant amplitude larger than the amplitudes of either wave.
4. a.

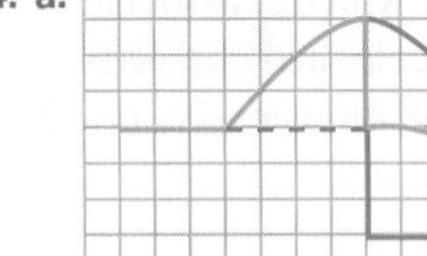

   b.

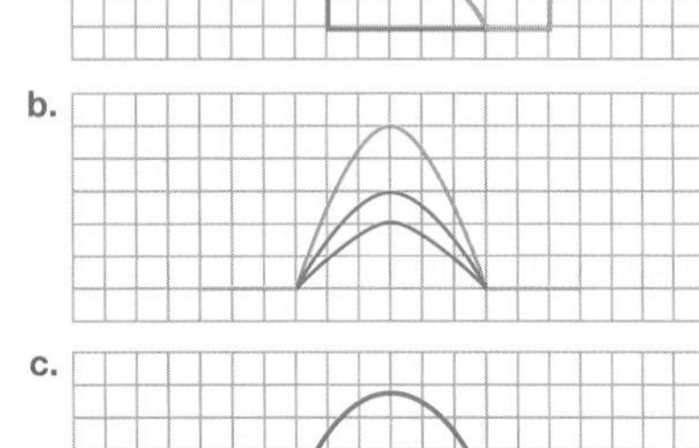

   c.

5. 1.5 m

6. a. $t = 0.05$ s
   b. $t = 0.1$ s
   c. $t = 0.2$ s
   d. $t = 0.4$ s
7. a. Constructive interference
   b. 1.0 m
   c. $330\ \text{m s}^{-1}$
8. a. $4.8\ \text{m s}^{-1}$ b. 0.60 m   c. 0.20 m   d. 8
9. a. $6.0\ \text{m s}^{-1}$ b. 1.2 m   c. 0.08 m   d. 5
10. $f_1 = 406$ Hz, $f_2 = 813$ Hz, $f_3 = 1219$ Hz

8.3 Exam questions

1. a. $400\ \text{m s}^{-1}$
   b.

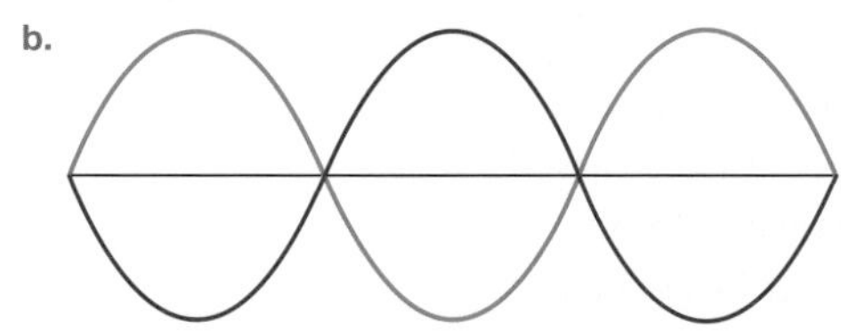

2. a. 5.3 m   b. No
3. a.
   b. $\lambda = 1.84$ m
   c. $f = 122$ Hz
4. a. 8.0 m   b. 60 Hz
5. • Waves travelling along the string are reflected at each end.
   • These waves, travelling in opposite directions, combine constructively and destructively to produce an interference pattern of antinodes and nodes.
   • The diagram below shows two waves at a moment when they destructively interfere to produce a straight horizontal line.

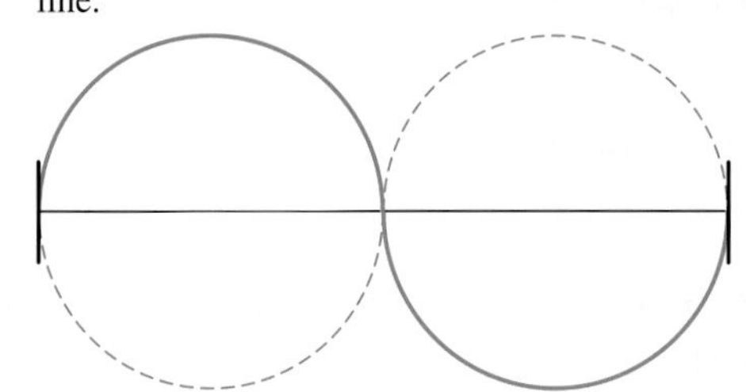

*5. b.

direction of travel

Required line

0.10 m

8.4 Diffraction of light

Sample problem 5

For radio waves, the ratio $\frac{\lambda}{w} = \frac{300}{30} = 10$, much greater than 1

For microwaves, the ratio $\frac{\lambda}{w} = \frac{0.003}{30} = 1.0 \times 10^{-4}$, much lower than 1

There will be little shadow behind the building using radio waves (large diffraction spread) but a shadow behind the building when using microwaves (small diffraction spread).

Practice problem 5

a. $\lambda = \frac{3.0 \times 10^8}{6.0 \times 10^{11}} = 5.0 \times 10^{-4}$ m for the microwaves and $\lambda = \frac{3.0 \times 10^8}{1.5 \times 10^{13}} = 2.0 \times 10^{-5}$ m for the infrared. The microwaves have a much larger wavelength compared to the infrared and would be expected to diffract more readily. If the narrow slit was 5.0×10^{-4} m then $\frac{\lambda}{w} = 1$ for the microwaves whereas $\frac{\lambda}{w} = 0.04$ for the infrared source indicating little significant diffraction.

b. To observe diffraction effects using the infrared source, the slit should be narrowed so that w is comparable to the wavelength 2.0×10^{-5} m.

8.4 Exercise

1. Evidence for light being a type of wave
2. a. As λ decreases, the amount of diffraction decreases.
   b. As λ increases, the amount of diffraction increases.
   c. As w decreases, the amount of diffraction increases.
   d. As w increases the amount of diffraction decreases.
3. a. The first minimum occurs where the path difference for light rays travelling from different places on the opening reaches $\frac{\lambda}{2}$. This will be a region where the intensity of the light will be diminished due to destructive interference.
   b.
   c.

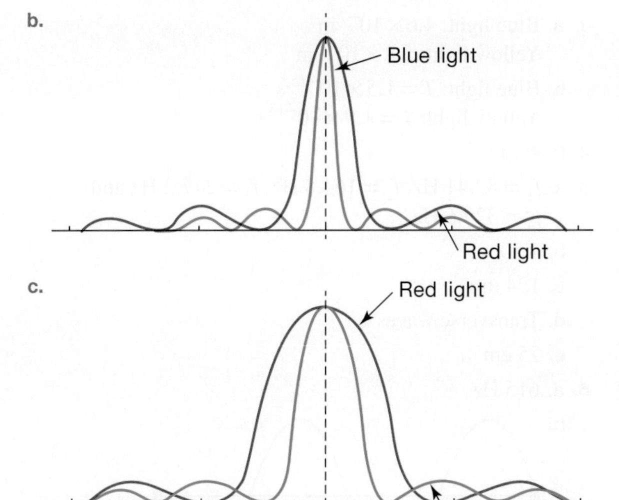

4. a. Diffraction depends on the wavelength or colour of light. White light is a mixture of the different colours in the spectrum, so when it diffracts while passing through a small slit, the colours that make up the white light form slightly different diffraction patterns.
   b. The positions of the minima are given by $\sin\theta = \frac{\lambda}{w}$. The red end of the spectrum has the longest wavelengths; therefore, θ for any minimum occurs at a greater angle than for blue light and other parts of the visible spectrum.
5. a. A blurred edge is evidence of diffraction and hence the wavelike nature of light.
   b. Red light has a longer wavelength compared to green light and so the pattern would reveal more significant diffraction; the edge of the shadow would be more blurry (less sharp).
   c. With a smaller object, the degree of diffraction observed would be more significant than before; the edge of the shadow would be less sharp.

8.4 Exam questions

1. a. To form a diffraction pattern the mesh must have tiny gaps for the light to pass through.
   b. The $\frac{\lambda}{\omega}$ ratio is now much smaller, thus there will be much less diffraction.
2. Diffraction depends on the ratio: $\frac{\lambda}{\omega}$. If this ratio is close to 1 then diffraction is most obvious.

   At 10 000 Hz, $\frac{\lambda}{\omega} << 1$ so diffraction is minimal, whereas at 100 Hz, $\frac{\lambda}{\omega} > 1$, the diffraction is significant.
3. B
4. If the wavelength increases, the width of the pattern also increases.
5. D

8.5 Interference of light

Sample problem 6

a. 9.00 m
b. 9.50 m

Practice problem 6

a. 8.00 m
b. 8.50 m

Sample problem 7

a. 1920 nm
b. 910 nm
c. With purple light (smaller wavelength than red light), the pattern becomes more compact or compressed as the distance between the bands decreases.

Practice problem 7

a. Path difference for point A: 795 nm
   Path difference for point B: 1060 nm
b. The pattern will compress.

c. The pattern will be red and will dilate.

d. The pattern is indicative of superposition of coherent waves and only wavelike sources produce such a pattern.

Sample problem 8

1.77 nm

Practice problem 8

a. 650 nm

b. red light

c. The pattern would be blue and compressed by moving the screen further away from the slits.

d. The line spacing could be made easier to measure if the screen was moved further from the pair of slits.

8.5 Exercise

1. a. Young's experiment clearly demonstrates the interference of light passing through two narrow closely spaced slits. This is strong evidence for the wavelike nature of light.

   b. An interference pattern consists of evenly spaced bright and dark fringes. Bright fringes are produced by the constructive interference of light passing through each slit. This constructive interference occurs when the path difference is 0, λ, 2λ, 3λ and so on. Dark fringes are produced by the destructive interference of light passing through each slit. This destructive interference occurs when the path difference is $\frac{\lambda}{2}, \frac{3\lambda}{2}, \frac{5\lambda}{2}$ and so on.

2. a. Evidence of the wave nature of light

   b. Constructive interference

   c. 0 nm

   d. 1060 nm

   e. 265 nm, 795 nm and 1325 nm

   f. 2 dark fringes

3. a. i. The bright band corresponds to constructive interference, where crests from the light waves coming through the two slits arrive together, and troughs in the light waves arrive together.

   ii. The dark band is where destructive interference occurs. At all times, the sum of the waves from the two slits is zero, with the crests from one slit coinciding with the troughs from the other.

   b. i. 0. The light rays are equidistant from the slits.

   ii. $\frac{\lambda}{2} = 317$ nm

   iii. $2\lambda = 1266$ nm

   c. The wavelength is smaller, so the bright fringes in the interference pattern would be closer together. As the path difference for the bright fringes equals $n\lambda$, a smaller wavelength means that each bright fringe would be closer to the central bright band.

   d. The increase in distance between the slits would result in the bright bands being closer together.

   e. Moving the screen further away would result in the interference pattern spreading out, increasing the distance between the bright bands.

4. a. $n \times 1.06$ µm, $n = 0, 1, 2, 3, ...$

   b. $\left(n - \frac{1}{2}\right) \times 1.06$ µm, $n = 1, 2, 3, ...$

5. 2.58×10^{-3} m

6. a. 670 nm

   b. The distance between adjacent fringes increases to 2.5 cm

7. 450 nm

8.5 Exam questions

1. C

2. a. The point C is bright because the path difference is zero resulting in constructive interference.

   The dark band to the left of C has a path difference of $\frac{\lambda}{2}$, which results in destructive interference.

   b. The experiment demonstrates interference. Interference is a wave phenomenon.

   c. $\omega = 0.01$ m

3. a. The bright fringe in question is the fourth bright fringe so the path difference is four wavelengths. The fact that the fringe is a bright fringe indicates that constructive interference is occurring.

   b. 630 nm

4. a. At point C, the path difference is zero, therefore there will be large waves (constructive interference).

   b. $\lambda = 6$ m

5. a. $S_1X - S_2X = 4.5$ cm

   b. Young's experiment demonstrated interference, interference is a property of waves, therefore, Young's experiment supports the wave model of light.

8.6 Review

8.6 Review questions

1. a. 0.50 s

   b. 1.3 m

   c. The wavelength decreases, the speed is unchanged

2. a. 1.08×10^{-3} s

   b. 0.370 m

3. a. Blue light: 4.6×10^{-7} m
   Yellow light: 5.8×10^{-7} m

   b. Blue light: $T = 1.5 \times 10^{-15}$ s
   Yellow light: $T = 1.9 \times 10^{-15}$ s

4. 0.46 m

5. a. $f_1 = 82.41$ Hz, $f_2 = 164.8$ Hz, $f_3 = 247.2$ Hz and $f_4 = 329.6$ Hz

   b. 1.50 m

   c. 124 m s^{-1}

   d. Transverse waves

   e. 25 cm

6. a. 615 Hz

   b.

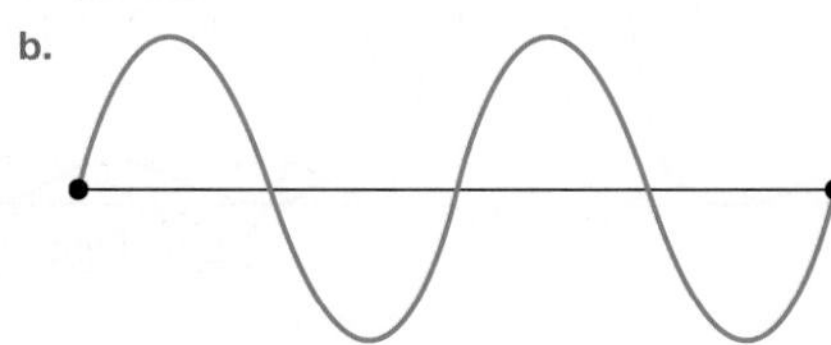

c. 713 m s^{-1}

d. 14.5 cm

7. When light passes around an obstacle or through a sufficiently narrow opening, it displays wavelike behaviour in the form of diffraction. Not only does the direction of light change but, in addition, interference effects, such as those shown in the diagram, produce bright and dark fringes around edges that are well explained by constructive and destructive interference.

8. a. Diffraction is less evident.

   b. Diffraction is more evident.

9. By decreasing the width of the opening or increasing the wavelength of the light

10. Each colour in white light will produce its own diffraction pattern. Each colour will have a slightly different spread, with red spread the most and purple spread the least, due to variations in wavelength. Hence, the pattern has a spectrum of coloured fringes that are consistent with a typical diffraction pattern for waves passing through a narrow opening.

11. a. 12 m

    b. 0.400 m

    c. Constructive interference

    d. 11.8 m

12. a. $\frac{\lambda}{2}$

    b. 2.0 m

    c. 16 m s^{-1}

13. a. 500 nm b. 1000 nm

14. i. increasing the distance of the screen from the slits by a factor of 2

    ii. increasing the wavelength of the light used by a factor of 2

    iii. decreasing the distance between the slits by a factor of 2

15. 3.1 m

8.6 Exam questions

Section A — Multiple choice questions

1. C
2. D
3. B
4. D
5. B
6. D
7. C
8. A
9. B
10. B

Section B — Short answer questions

11. a. $T = 25 \times 10^{-3}$ s

    b. See figure at the bottom of the page*

12. a. $f = 10^{10}$Hz

    b. $S_2X - S_1X = 4.5$ cm

    c. The pattern will widen.

13. a. $\lambda = 1$ m

    b. Thus the student has moved 0.75 m from the centre.

14. a. $f = 1.08 \times 10^{10}$ Hz

    b. The signal strength between P_0 and P_1 is a minimum because the path difference is $\frac{\lambda}{2}$.
    This results in destructive interference.

15. a. The path difference to the point C is zero, and this results in constructive interference at C.

    b.

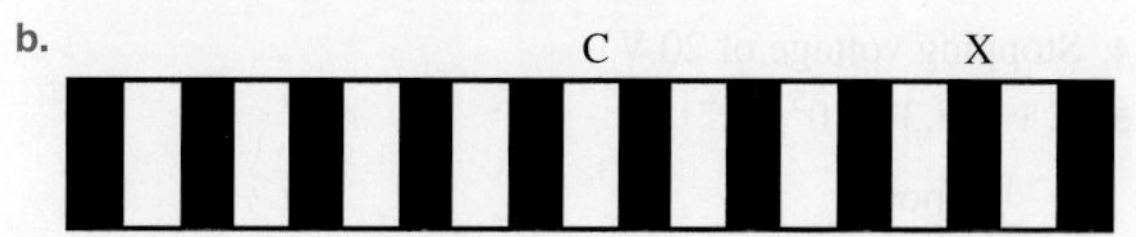

9 Light as a particle

9.2 Could light have particle-like properties as well?

Sample problem 1

a. i. $E = 4.4 \times 10^{-19}$ J

   ii. $p = 1.5 \times 10^{-27}$ N s

b. $p = 1.02 \times 10^{-27}$ N s

Practice problem 1

$f = 4.4 \times 10^{14}$ Hz

Sample problem 2

a. $E_{photon} = 3.86 \times 10^{-19}$ J

b. $N = 8 \times 10^{17}$ photons s^{-1}

*11. b.

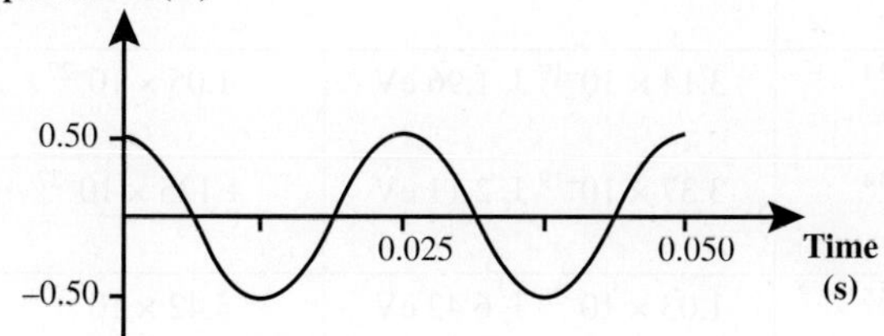

Practice problem 2

$N = 1.4 \times 10^{28}$ photons s^{-1}

Sample problem 3

a. $W = 8.0 \times 10^{-17}$ J

b. Kinetic energy

c. $E_k = 8.0 \times 10^{-17}$ J (or 500 eV)

d. $v = 1.33 \times 10^7$ m s^{-1}

e. $p = 1.2 \times 10^{-23}$ N s

Practice problem 3

a. $E = 79$ eV

b. Potential difference of 79 V

c. $v = 5.3 \times 10^6$ m s^{-1}

d. $p = 4.8 \times 10^{-24}$ N s

9.2 Exercise

1. a. $\lambda = 6.5 \times 10^{-7}$ m
   b. $T = 2.18 \times 10^{-15}$ s
2. 52 photons
3. See table at the bottom of the page*
4. Stopping voltage of 20 V
5. a. $v = 5.3 \times 10^5$ m s^{-1}
   b. i.

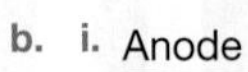

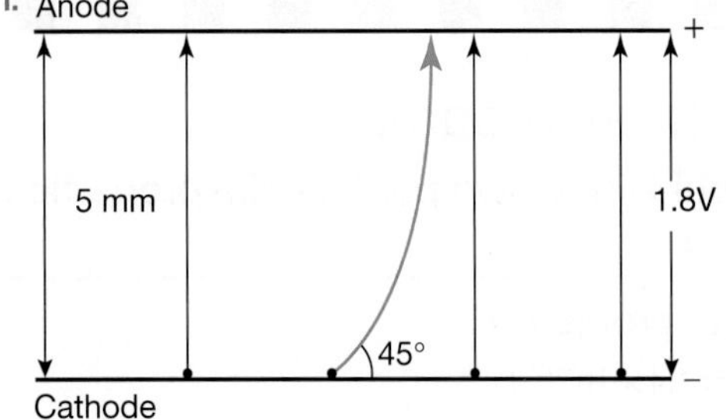

ii.

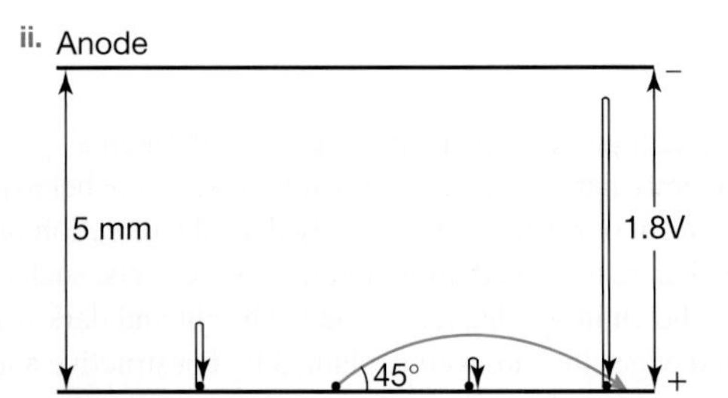

iii.

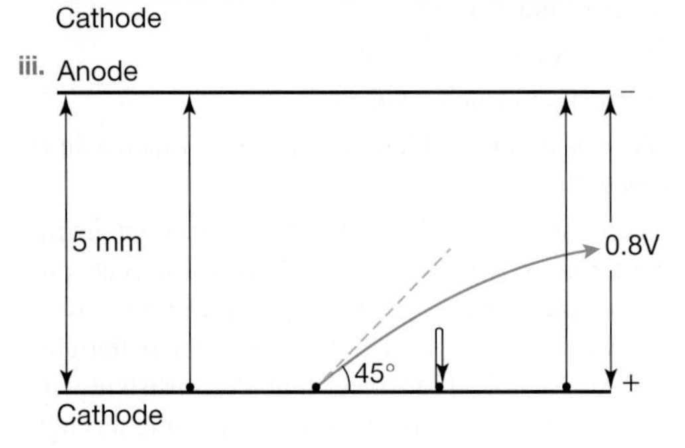

9.2 Exam questions

1. B
2. $E = 6.21$ eV
3. $\lambda = 440$ nm
4. Use $E = pc$ thus $p = \frac{E}{c} = \frac{4.52 \times 10^{-19}}{3.00 \times 10^8} = 1.50 \times 10^{-27}$ N s
5. a. $E_k = 9.22 \times 10^{-18}$ J
   b. 1.06×10^6 m s^{-1}

9.3 The photoelectric effect and experimental data

Sample problem 4

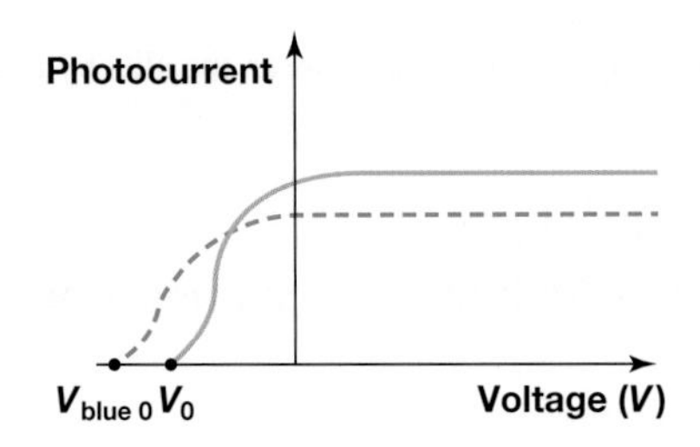

*3.

Source	Wavelength	Frequency (Hz)	Energy	Momentum (N s)
Infrared from CO_2 laser	10.6 μm	2.83×10^{13}	1.88×10^{-20} J, 0.117 eV	6.25×10^{-29}
Red helium–neon laser	633 nm	4.74×10^{14}	3.14×10^{-19} J, 1.96 eV	1.05×10^{-27}
Yellow sodium lamp	589 nm	5.09×10^{14}	3.37×10^{-19} J, 2.11 eV	1.125×10^{-27}
UV from excimer laser	194 nm	1.55×10^{15}	1.03×10^{-18} J, 6.42 eV	3.42×10^{-27}
X-rays from aluminium	0.990 nm	3.03×10^{17}	2.01×10^{-16} J, 1.25 keV	6.69×10^{-25}

Practice problem 4

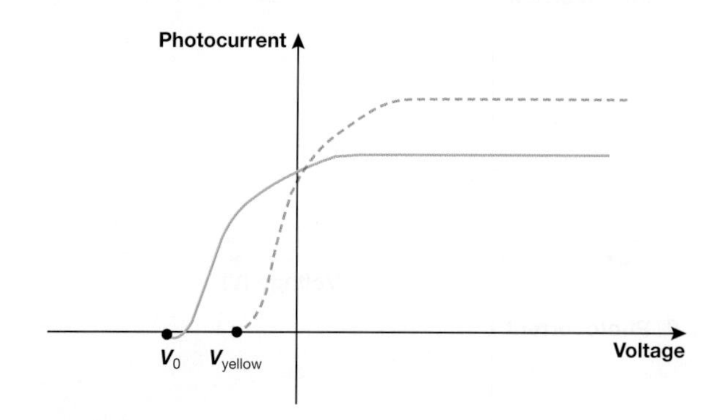

Sample problem 5

a. Stopping voltage of 1.6 V

b. $E_k = 6.7 \times 10^{-19}$ J

Practice problem 5

Stopping voltage of 3.0 V

Sample problem 6

a. $f = 7.1 \times 10^{14}$ Hz

b. $E = 2.9$ eV

c. $1.25 \text{ eV} = 2.0 \times 10^{-19}$ J

d. $\phi = 1.7 \text{ eV} = 2.7 \times 10^{-19}$ J

e. $f_0 = 4.0 \times 10^{14}$ Hz and $\lambda = 7.4 \times 10^{-7} \text{ m} = 740$ nm

f. Stopping voltage of 1.5 V

Practice problem 6

a. 0.87 eV or 1.4×10^{-19} J

b. $\phi = 2.1$ eV

c. $f_0 = 4.96 \times 10^{14}$ Hz and $\lambda = 6.05 \times 10^{-7} \text{ m} = 605$ nm

d. The photoelectric effect will not occur

Sample problem 7

a. See table at the bottom of the page*

b.

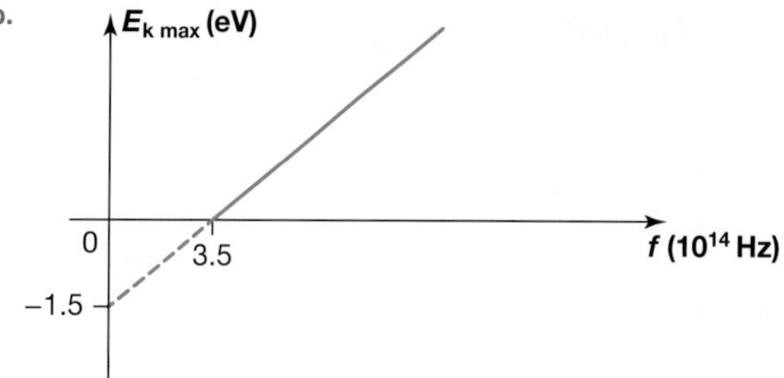

c. i. 6.9×10^{-34} J s $= 4.3 \times 10^{-15}$ eV s using the gradient of the line

ii. $f_0 = 3.5 \times 10^{14}$ Hz using the x-intercept of the line of best fit

iii. $\phi = 2.4 \times 10^{-19} \text{ J} = 1.5$ eV using the y-intercept of the line of best fit

d.

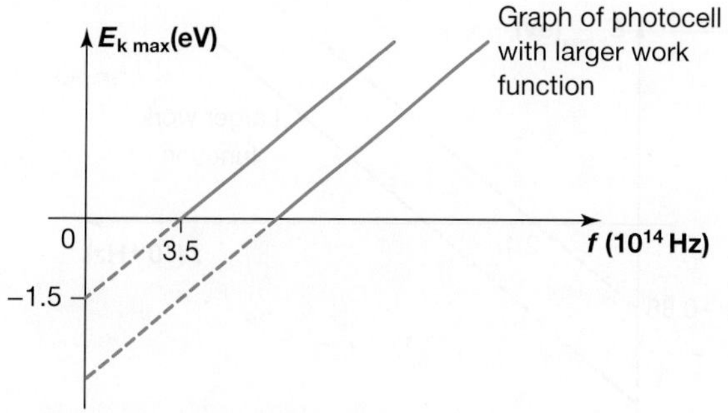

e. $\phi = 3.33 \times 10^{-19} \text{ J} = 2.07$ eV

f. The photocurrent would double

g. The stopping voltage would remain the same.

Practice problem 7

a. See table at the bottom of the page**

---

*a.

Wavelength of light used (nm)	Frequency of light used × 10^{14} (Hz)	Photon energy of light used (eV)	Stopping voltage readings (V)	Maximum photoelectron energy (J)
663	**4.52**	**1.88**	0.450	**7.20×10^{-20}**
489	6.14	**2.54**	**1.15**	1.84×10^{-19}

**a.

Wavelength of light used (nm)	Frequency of light used × 10^{14} (Hz)	Photon energy of light used (eV)	Stopping voltage (V)	Maximum photo-electron energy (J)
390	**7.70**	3.19	**2.36**	3.78×10^{-19}
524	**5.73**	**2.37**	1.54	**2.46×10^{-19}**

b.

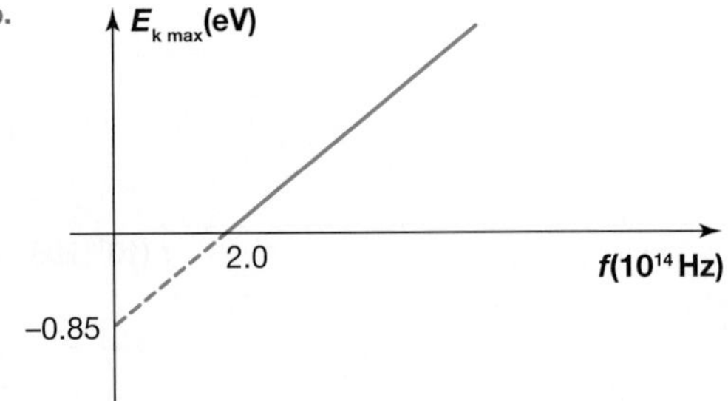

c. i. 6.73×10^{-34} J s $= 4.2 \times 10^{-15}$ eV s using the gradient of the line

ii. $f_0 = 2.0 \times 10^{14}$ Hz using the x-intercept of the line of best fit

iii. $\phi = 1.36 \times 10^{-19}$ J $= 0.85$ eV using the y-intercept of the line of best fit

d.

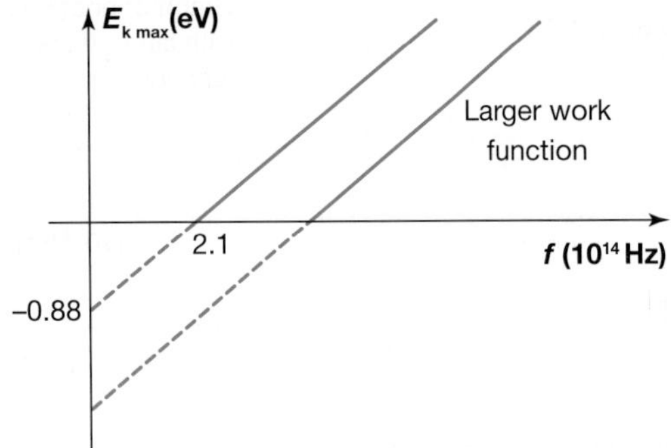

e. $\phi = 1.8$ eV $= 2.88 \times 10^{-19}$ J

f. The current would halve.

g. The stopping voltage would remain the same.

9.3 Exercise

1. a. The maximum current occurs when the accelerating voltage causes all ejected electrons to be collected at the anode. The voltage required for this is greater than zero because some electrons leave at an angle and their parabolic path may miss the anode at lower voltages. When the voltage opposes the motion towards the cathode, electrons travelling towards the anode slow down. When the magnitude of the voltage is large enough, the electrons reverse direction and so do not contribute to the current. At a high enough retarding voltage, *all* electrons turn around before reaching the anode, so the current is zero.

b.

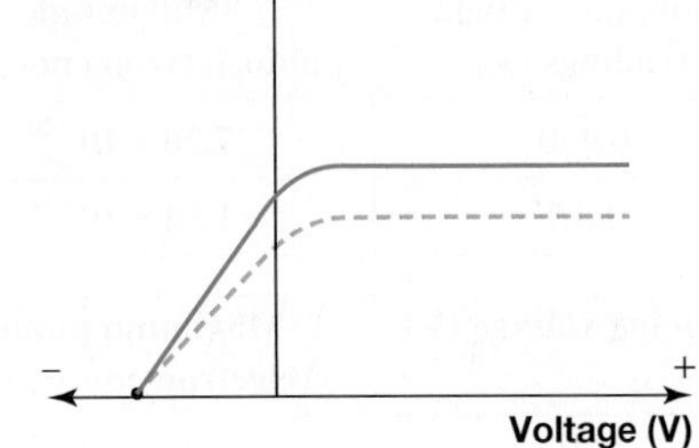

c.

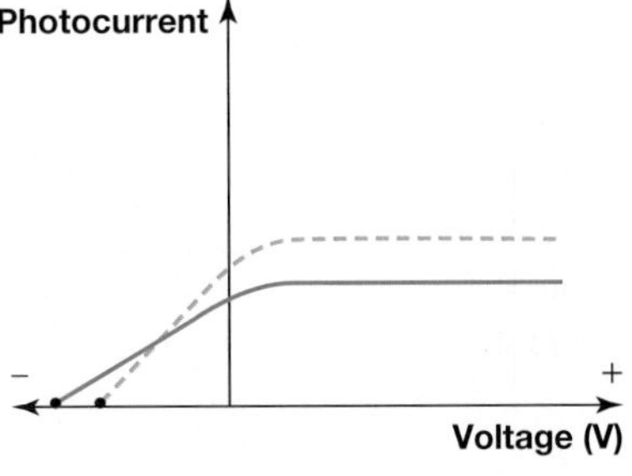

d.

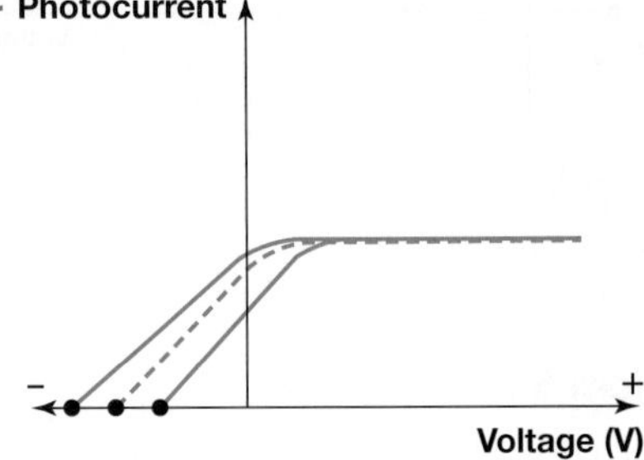

2. a. 0.85 μA

b. 1.0 μA

c. 1.0 μA

d. Once the voltage has reached the point where all electrons are reaching the plate, the current in the photocell is determined solely by the intensity of the light source. There is no increase in light intensity; therefore, there is no increase in the photocurrent.

e. Stopping voltage = 1.7 V or 1.7 eV $= 2.7 \times 10^{-19}$ J

f.

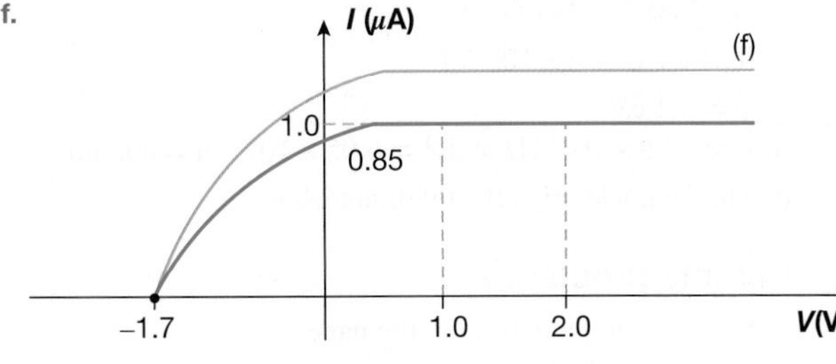

g.

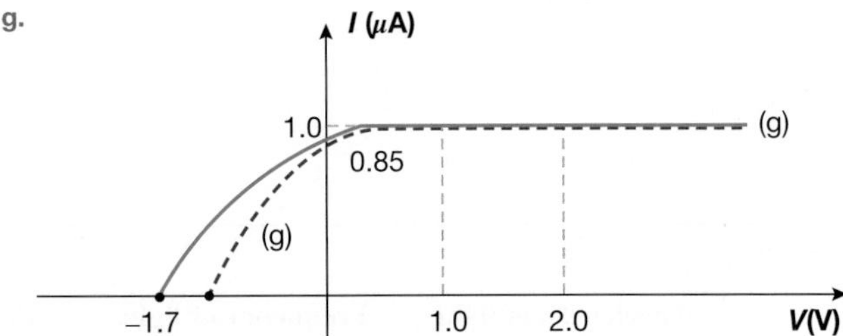

h.

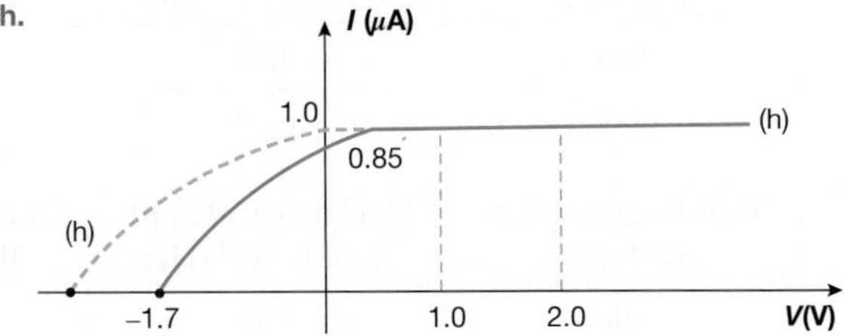

3. a. Stopping voltage = 0.67 V
   b. $f = 1.08 \times 10^{15}$ Hz
   c. $f_0 = 9.17 \times 10^{14}$ Hz
4. a. $\phi = 2.6$ eV $= 4.1 \times 10^{-19}$ J
   b. Stopping voltage = 2.1 V
   c. $\lambda = 2.65 \times 10^{-7}$ m = 265 nm and $p = 2.50 \times 10^{-27}$ N s
5. a. $E_k = 1.5$ eV $= 2.4 \times 10^{-19}$ J
   b. 2.5 V
   c. $E_k = 4.3$ eV $= 6.9 \times 10^{-19}$ J
   d.

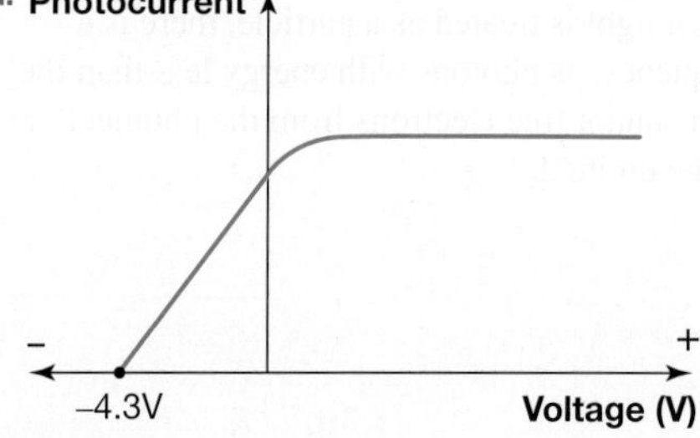

6. Stopping voltage = 1.12 V
7. a. $E_{k\,max} = 2.59$ eV $= 4.14 \times 10^{-19}$ J
   b. 2.59 V
   c.

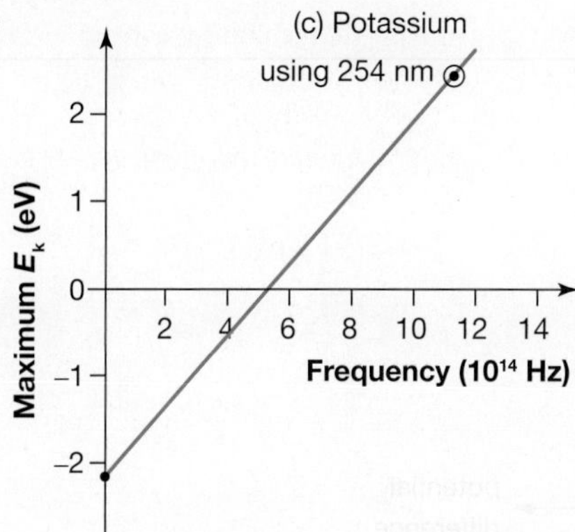

d.

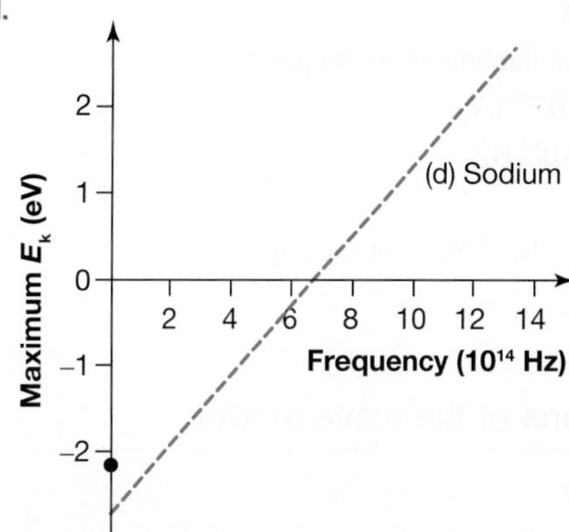

8. a. $f_0 \simeq 4.6 \times 10^{14}$ Hz
   b. $\lambda = 6.5 \times 10^{-7}$ m = 650 nm
   c. $\phi = 1.9$ eV
   d. $h \approx$ gradient $\approx 6.56 \times 10^{-34}$ J s

9.3 Exam questions

1. a.

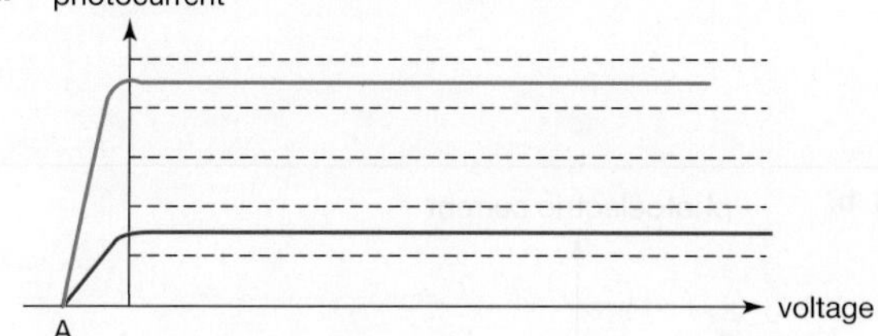

   b. Point A is the stopping voltage.
   c. The stopping voltage is sufficient to turn back even the most energetic photoelectrons.
2. a. i. $h = 5.3. \times 10^{-15}$ eV s
      ii. $\lambda_{max} = 810$ nm
      iii. $\phi \approx 1.9$ eV s
   b. See figure at the bottom of the page*

*2. b.

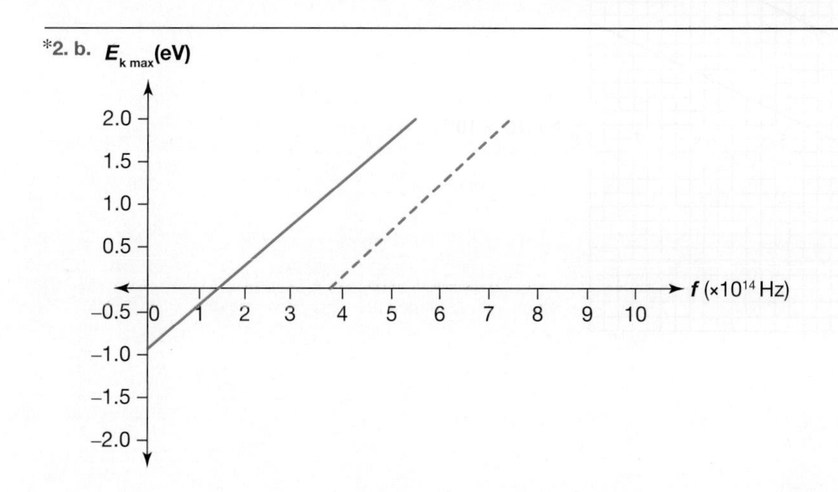

3. a. $V_0 = 0.31\text{ V}$
   b. See figure at the bottom of the page*
4. a. $h = 5.0 \times 10^{-15}\text{ eV s}$
   b. $f_0 = 2.0 \times 10^{14}\text{ Hz}$
   c. $\phi = 1.0\text{ eV}$
   d. See figure at the bottom of the page**
5. $h = \dfrac{\phi}{f_0} = 4.9 \times 10^{-34}\text{ J s}$

9.4 Limitations of the wave model

9.4 Exercise

1. The stopping voltage would increase with the intensity of light.
2. Some examples of observations that support the particle model include:
   - For a given frequency of light, the photocurrent is dependent in a linear fashion on the brightness or intensity of light.
   - The energy of photoelectrons is independent of intensity of light and only linearly dependent on frequency.
   - There is no significant time delay between incident light striking a photocell and the subsequent emission of electrons, and this observation is independent of intensity.
   - A threshold frequency exists below which the photoelectric effect does not occur, and this threshold is independent of intensity.
3. When treating light as a wave, there is no threshold frequency as energy transfer to electrons from a light source is cumulative and, eventually, emission will occur. However, when light is treated as a particle, there is a threshold frequency, as photons with energy less than the work function cannot free electrons from the photocell, and they will not be emitted.

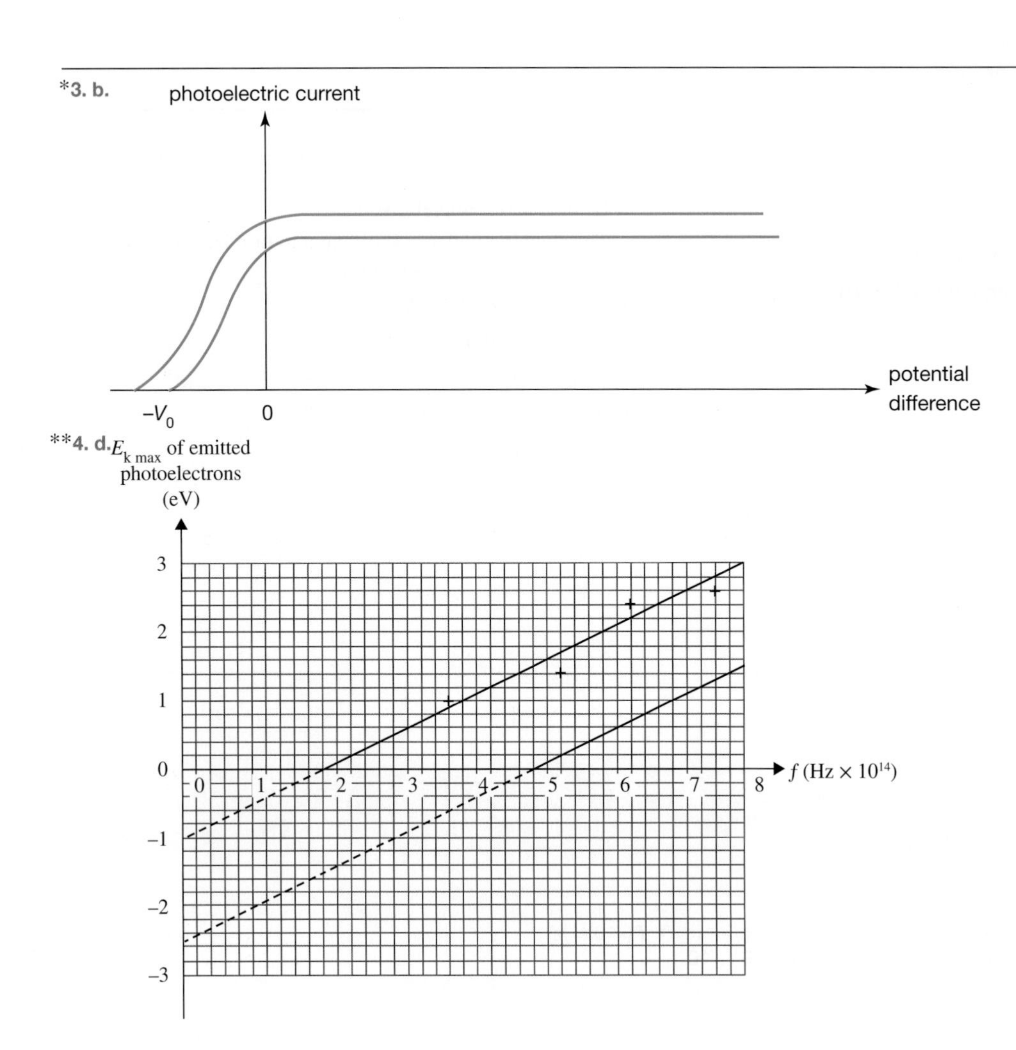

4. Light is thought to behave in two different ways — as both a particle and a wave. Many believe that these are two different models, so it seems strange to use the wave model to complete calculations involving the particle model. However, these two models are not competing models — they exist in duality and work together to explain the behaviour of light. Light behaves as both a 'particle' and a 'wave', so using the frequency from one model to calculate the energy of another model is quite reasonable. Using the wave model, you can relate the frequency and wavelength with the speed of light and, using the particle model, you can relate Planck's constant and frequency with the speed of light. The frequency is in fact useful in both models so, while it seems contradictory, it is incredibly useful for the calculations.

9.4 Exam questions

1. C
2. Particle model
3. • **Absence of time delay**
   - The wave model predicts that no photoelectrons will be emitted until sufficient energy has been delivered to the metal.
   - Experimentally, there is no measurable time delay before electrons are ejected.
   - **The existence of a threshold frequency**
   - The wave model predicts that light of any frequency should be able to eject photoelectrons, since the energy delivered by a wave depends only on the intensity of light.
   - Experiment shows that light below a certain frequency **does not** eject photoelectrons, regardless of its intensity.
   - **The absence of any effect of light intensity on stopping potential**
   - The wave model predicts that greater intensity will deliver more energy to the metal and so increase the maximum KE of the photoelectrons. This should lead to an increase in the stopping potential.
   - Experiment shows that the stopping potential is **not affected** by changing the light intensity.
4. Answers will vary. A sample answer is:
   There should be no threshold frequency. The wave model predicts that all light carries energy proportional to the amplitude of the wave. Therefore, all light should be able to produce photoelectrons. The results show that only light with frequencies above a threshold frequency can produce photoelectrons. The particle model of light predicts that photons have an energy proportional to their frequency and only photons with a high enough frequency will have sufficient energy to release photoelectrons.
5. The evidence is the photoelectric effect.
   The particle model is supported by (any one of):
   - no time delay in electron emission
   - KE of electrons depends on light frequency not intensity
   - the existence of a threshold frequency for light.

9.5 Review

9.5 Review questions

1. a. $\lambda = 6.5 \times 10^{-7}$ m
   b. $f = 4.6 \times 10^{14}$ Hz
   c. $E = 3.1 \times 10^{-19}$ J $= 1.9$ eV and $p = 1.0 \times 10^{-17}$ N s
2. $N = 1.42 \times 10^{28}$ photons s^{-1}
3. a. $E = 2.8$ eV
   b. $\phi = 2.11 \times 10^{-19}$ J $= 1.3$ eV
   c. Stopping voltage = 1.5 V
4. a. Photons with a frequency below the threshold frequency will not have sufficient energy for electrons on the surface, or below the surface, to overcome the work function ϕ. The photoelectric effect will not occur.
   b. $\phi = 2.32 \times 10^{-19}$ J $= 1.45$ eV
   c. $E = 19.7$ eV
5. a. $h = 6.7 \times 10^{-34}$ J s
   b. $\phi = 0.98$ eV $= 1.6 \times 10^{-19}$ J
6. a. The stopping voltage will stay the same, (frequency does not change). The photo electric current will increase with the intensity of light.
   b. The stopping voltage will increase with the frequency, the photocurrent will stay the same as the intensity of light is unchanged.
   c. The stopping voltage will increase and the photocurrent will remain the same.
7. **Reason 1:**
   Since the photocurrent is proportional to the brightness of the light, it lends support to the idea that light is a stream of particles. The greater the number of particles striking a surface per second, the greater the number of electrons emitted per second. A wave model does not make any obvious predictions concerning brightness and current, but incorrectly makes a prediction concerning the brightness of a light source and electron energy.
   Reason 2:
   The energy of the photoelectrons is dependent on the frequency of the light, consistent with the Planck model and consolidated by Einstein; the light consists of individual photons of energy proportional to frequency. A wave model makes no such assertion.
   Reason 3:
   The lack of delay between light striking a surface and the subsequent emission of electrons, both independent of brightness and frequency, indicates that the light arriving as particle-like packets. A wave model would predict a variable delay time due to brightness, which is not observed.
   Reason 4:
   The existence of a threshold frequency below which the photoelectric effect does not occur, regardless of the intensity of light, is not a prediction made using a wave model. It is consistent with a particle model for light where the photons do not have sufficient energy to cause electrons to be emitted regardless of how much time is given for the process. Also implicit is that energy transfers from photons to electrons are in a one-to-one ratio.

9.5 Exam questions

Section A — Multiple choice questions

1. A
2. A
3. C
4. C
5. A
6. D
7. B
8. D
9. A
10. A

Section B — Short answer questions

11. a. Stopping voltage (or cut-off voltage or stopping potential).

b. 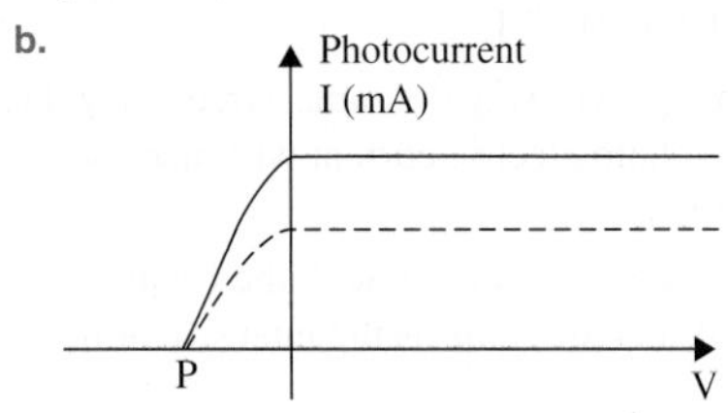

c. 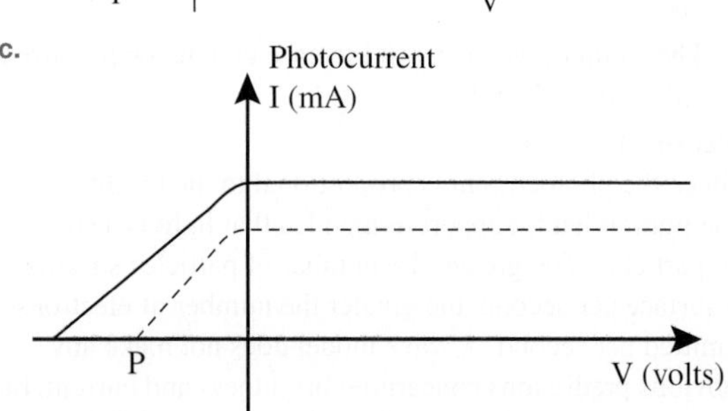

d. 3.4 eV

e. $h = 6.7 \times 10^{-34}$ J s

f. Answers will vary.
The following are possible examples of limitations of the wave model in explaining the results of the photoelectric effect:
- existence of a threshold frequency
- absence of a time delay
- energy of the photoelectrons is independent of the intensity of the light source

12. a.

Intensity	No change
Frequency	Doubled

b.

Intensity	Halved
Frequency	No change

c. $E = 1.77$ eV. This is less than the work function, thus there will be no photoelectrons ejected.

13. a. $E = 4.1$ eV. This is less than the work function, thus there will be no photoelectrons ejected.

b. 5.4 eV is greater than the work function so a photoelectron will be emitted with a kinetic energy of 0.5 eV.

c. The experiment supports the particle model of light.
- The model suggests a threshold frequency and no electrons were emitted by the 400 nm light.
- The model suggests that the energy of the photon is dependent on its frequency and the higher frequency light did produce photoelectrons.

14. a. The stopping voltage is 1.0 V.

b. $\lambda = 288$ nm

c. The particle model predicts a cut-off frequency below which no photoelectrons will be emitted regardless of light intensity. This is supported by April's observations.
The wave model predicts that photoelectron energy can be accumulated over time, this is not supported by April's observations.
The particle model predicts that the energy of a photon is dependent on its frequency and is delivered in a single instant. This is supported by April's observations.

15. a. $E = 2.5$ eV

b. 1.5 eV

c. 1.0 eV

d. At $V = +1.0$ V, there are no more photoelectrons to be collected thus increasing the voltage will not result in an increase in photocurrent.

e. D

f. The wave model predicts that increasing the intensity will increase the kinetic energy of the photoelectrons, which would increase the stopping voltage. This is not what is observed.

10 Matter as particles or waves and the similarities between light and matter

10.2 Matter modelled as a type of wave

Sample problem 1

a. $\lambda = 6.63 \times 10^{-28}$ m

b. $v = 7.3 \times 10^2$ m s^{-1}

Practice problem 1

The electron has a de Broglie wavelength that is approximately 180 times larger than that of the proton.

10.2 Exercise

1. 1.9×10^{-30} m
2. 44.2 m s^{-1}
3. When the speed and hence the momentum of a particle increases, the de Broglie wavelength will decrease with inverse proportion.
4. 3.9×10^{-35} m
5. 6.63×10^{-25} N s
6. A diffraction experiment where a beam of electrons is passed through a narrow opening (or around an obstacle) whose size is smaller than or similar to the size of the de Broglie wavelength associated with the beam of electrons.

10.2 Exam questions

1. $p = 1.5 \times 10^{-26}$ kg m s^{-1}
2. C
3. $\lambda = 2.0 \times 10^{-11}$ m
4. $\lambda = 3.6 \times 10^{-11}$ m
5. $v = 7.3 \times 10^{7}$ m s^{-1}

10.3 The diffraction of light and matter

Sample problem 2

The de Broglie wavelength of the tennis ball is of the order of 10^{-34} m and that of the electron is of the order of 10^{-10} m.

Practice problem 2

$v = 0.40$ m s^{-1}

Sample problem 3

$V = 26$ V

Practice problem 3

$v = 2.1 \times 10^{6}$ m s^{-1}

Sample problem 4

$\lambda_{electrons} = 5.0 \times 10^{-11}$ m, similar to $\lambda_{X\text{-rays}} = 7.1 \times 10^{-11}$ m

Practice problem 4

a. $V = 5.4 \times 10^{2}$ V

b. For the same energy, the photon has a wavelength that is approximately 100 times larger compared to the de Broglie wavelength of the electron.

Sample problem 5

a. $p = 3.3 \times 10^{-24}$ N s
The photon and the electron have the same momentum.

b. $E_{photon} = 9.9 \times 10^{-16}$ J = 6.2 keV and
$E_{electron} = 6.0 \times 10^{-18}$ J = 37 eV
The electron has substantially less kinetic energy than the photon.

c. Light and matter with the same wavelength will have the same momentum but they will not necessarily have the same energy.

Practice problem 5

a. $p = 6.6 \times 10^{-24}$ N s for both the electron and the photon.

b. $E_{photon} = 1.98 \times 10^{-15}$ J = 12.4 keV and
$E_{electron} = 2.4 \times 10^{-17}$ J = 150 eV
The electron has substantially less kinetic energy than the photon.

10.3 Exercise

1. a. $\lambda = 1.3 \times 10^{-14}$ m b. $\lambda = 1.7 \times 10^{-10}$ m
   c. $\lambda = 6.6 \times 10^{-35}$ m
2. a. $E_k = 8.0 \times 10^{-16}$ J b. $\lambda = 2.5 \times 10^{-10}$ m
3. Light exhibits both wave and particle behaviour.
4. a. $p = 2.3 \times 10^{-24}$ kg m s^{-1}
   $\lambda = 2.9 \times 10^{-10}$ m
   b. The electrons will diffract significantly as $\frac{\lambda}{d} \simeq 1$.
5. a. $E = 3000$ eV $= 4.8 \times 10^{-16}$ J
   b. $p = 3.0 \times 10^{-23}$ N s and $\lambda = 2.2 \times 10^{-11}$ m
   c. $\frac{\lambda}{w} = 0.044\< < 1$ thus the scientist should not expect any diffraction effect
   d. The scientist should decrease the accelerating voltage to reduce the energy and the momentum of the electrons, thus increasing the wavelength.
   e. $\lambda = 2.2 \times 10^{-11}$ m and $p = 3.0 \times 10^{-23}$ N s
   f. $E = 9.0 \times 10^{-15}$ J $= 5.6 \times 10^{4}$ eV
6. $v = 1.1 \times 10^{3}$ m s^{-1}
7. a. The electron will have the larger wavelength.
   b. For the electron: $\lambda = 3.9 \times 10^{-11}$ m.
   For the proton: $\lambda = 9.1 \times 10^{-13}$ m.
8. $V = 38$ V
9. The electron has the shorter wavelength of the two.

10.3 Exam questions

1. a. $\lambda = 0.012$ nm
   b. If the speed is increased then the momentum will increase.
   Since $\lambda \propto \frac{1}{v}$, if v increases, λ will decrease.
2. a. $\lambda = 3.11 \times 10^{-9}$ m ≈ 3 nm
   b. $v = 2.34 \times 10^{5}$ m s^{-1}
3. a. $E_k = 1.35 \times 10^{-20}$ J = 0.08 eV
   b. $E_{X\text{-ray}} = 293$ eV
4. a. $\lambda = 1.46 \times 10^{-9}$ m b. $w = 1.46 \times 10^{-7}$ m
5. a. Moving electrons exhibit wave properties with a wavelength known as the de Broglie wavelength.
   Diffraction patterns are dependent on wavelength.
   If the de Broglie wavelength of the electron is the same as the wavelength of the X-ray, then the diffraction patterns will be the same.
   b. $f = 1.34 \times 10^{19}$ Hz

10.4 Emission and absorption spectra

Sample problem 6

$n = 1$ to ground state transition: $E = 1.5$ eV

$n = 3$ to $n = 2$ transition: $E = 1.7$ eV

$n = 2$ to $n = 1$ transition: $E = 1.8$ eV

$n = 2$ to ground state transition: $E = 3.3$ eV

$n = 3$ to $n = 1$ transition: $E = 3.5$ eV

$n = 4$ to $n = 1$ transition: $E = 5.0$ eV

Practice problem 6

a.

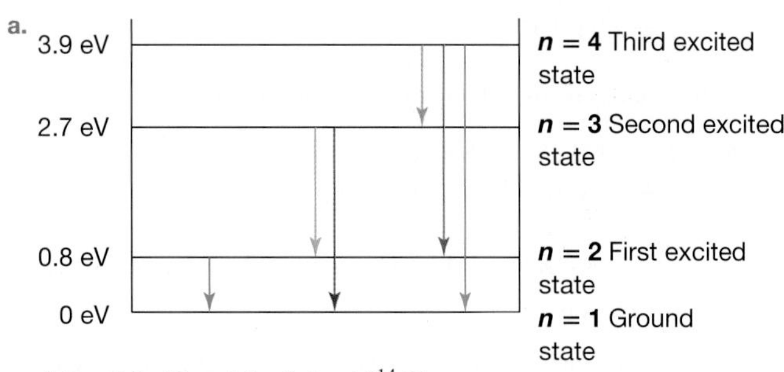

b. $\Delta E = 3.9$ eV and $f = 9.4 \times 10^{14}$ Hz

c. $\Delta E = 1.9$ eV and $f = 4.6 \times 10^{14}$ Hz

Sample problem 7

$\lambda_{photon} = 9.7 \times 10^{-8}$ m, ultraviolet radiation

Practice problem 7

$f = 1.59 \times 10^{14}$ Hz and $\lambda = 1.9 \times 10^{-6}$ m

10.4 Exercise

1. A fire glows with a continuous range of wavelengths, and, in a red fire, the red wavelengths have the greatest intensity. Neon in a discharge tube glows red because the electrons in the neon atoms are excited to specific energies. When the electrons return to the ground state, they produce light of a few fixed wavelengths, mainly in the red part of the spectrum.
2. Emission lines are produced when electrons return from an excited state to a lower energy state. The energy is released in the form of photons of particular frequencies. Absorption lines are produced when light from a continuous spectrum passes through a gas. This light excites some of the electrons in the atoms making up the gas, so photons with the energies allowed by the atoms will be removed from the continuous spectrum. As the energy required to raise an electron to a more excited state is equal to the energy released when the electrons drop back to the lower state, the emission lines and absorption lines for a particular element will be the same.
3. Possible answers include refracting the light through a prism. Spectral yellow will remain yellow, whereas a mixture of green and red light will separate into two beams.
4. Ground state: -10.4 eV
   First excited state: -5.5 eV
   Second excited state: -3.7 eV
   Third excited state: -1.6 eV
5. a. The term 'ground state' defines the lowest energy state of an electron.
   b.

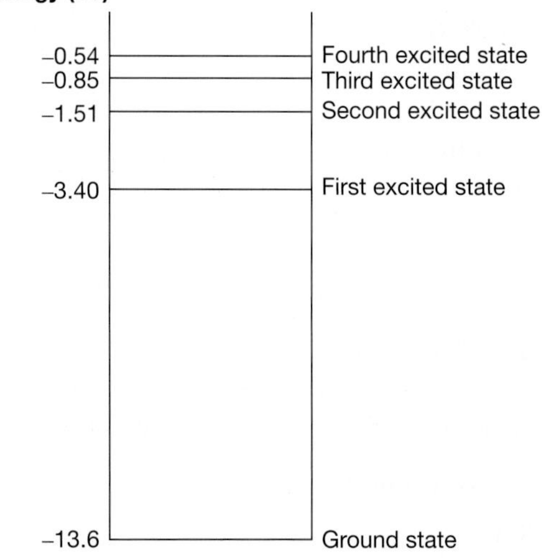

   c. The Balmer series is a collection of transitions from the 3rd, 4th, 5th and 6th excited states to the 2nd excited state in a hydrogen atom. They are grouped because each transition produces a photon in the visible part of the electromagnetic spectrum.
6. See table at the bottom of the page*
7. a.

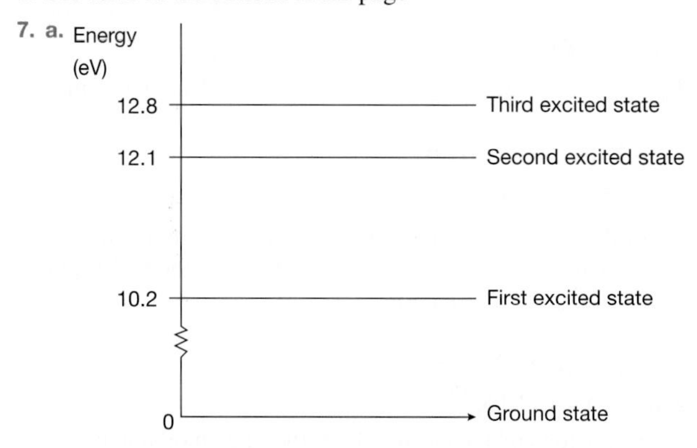

*6.

	λ(nm)	f(Hz)	E(J)	E(eV)	p(N s)
Red light	633	4.74×10^{14}	3.14×10^{-19}	1.96	1.05×10^{-27}
Electron	0.877	—	3.14×10^{-19}	1.96	7.56×10^{-25}
Blue light	405	7.41×10^{14}	4.91×10^{-19}	3.07	1.64×10^{-27}
Electron	405	—	1.48×10^{-24}	9.24×10^{-6}	1.64×10^{-27}

b.

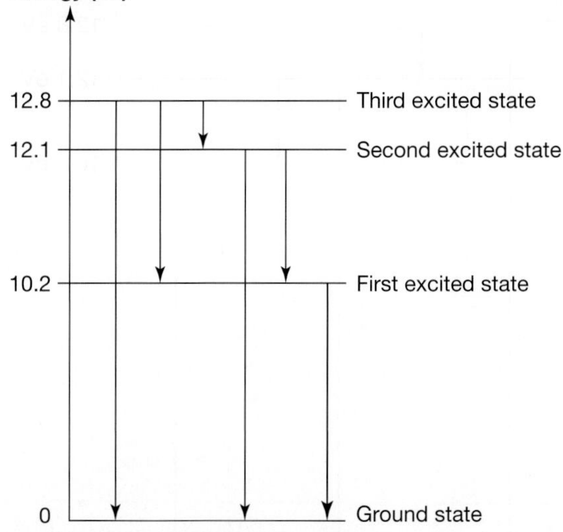

c. $n = 3$ to $n = 2$ transition: $E_1 = 0.7$ eV
$n = 3$ to $n = 1$ transition: $E_2 = 2.6$ eV
$n = 3$ to Ground state transition: $E_3 = 12.8$ eV
$n = 2$ to $n = 1$ transition: $E_4 = 1.9$ eV
$n = 2$ to Ground state transition: $E_5 = 12.1$ eV
$n = 1$ to Ground state transition: $E_6 = 10.2$ eV,

d. Least energetic photon:
$\lambda = 1.8 \times 10^{-6}$ m
Greatest energy photon:
$\lambda = 9.7 \times 10^{-8}$ m

10.4 Exam questions

1. a. $E = 1.94$ eV

b.

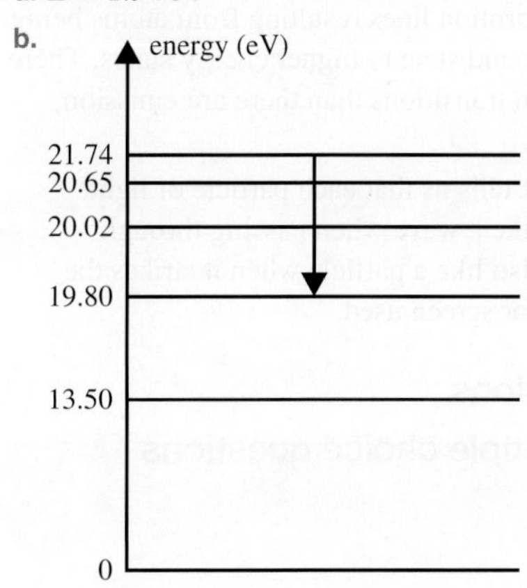

2. Photons with energies matching the transition energies between shells will be absorbed as the electrons are excited to higher energy states. These photons are missing from the spectrum.

3. $f = 8.72 \times 10^{14}$ Hz

4. a. $E = 2.2$ eV
   b. There are 10 transitions (9 different transitions) and 9 spectral lines.

5. a. 668 nm
   b. $f = 5.1 \times 10^{14}$ Hz
   c. Only certain energies are visible because the electrons exist only in certain discrete energy levels. As the electrons transition between these energy levels they can emit only discrete amounts of energy.

10.5 Electrons, atoms and standing waves

10.5 Exercise

1. a. The classical model of a hydrogen atom has an electron orbiting around the proton, which lies at the centre of the atom. This model is often referred to as a planetary model.
   b. An electron orbiting a proton is accelerating and, according to Maxwell, should emit electromagnetic radiation continuously as it orbits. This loss of energy for the electron would cause it to spiral into the nucleus in a very short time. This model does not predict that atoms are stable.

2. In the Bohr model of a hydrogen atom, the electron is modelled as a type of standing wave around the proton. This stationary state is simply asserted not to emit light.

3. In the Bohr model for an atom, light in the form of photons is emitted only when an atom undergoes a transition from one excited state to a lower state. The light emitted is a specific energy (hence frequency, hence colour) accounting for the discrete nature of emission spectra. In a similar fashion, light of specific energies is absorbed when the energies match energy differences in stationary states. Usually, the initial state is the ground state and the final state is an excited state.

10.5 Exam questions

1. Electrons exhibit a wave property. Only orbits with circumferences that are a whole multiple of this wavelength will permit a standing wave to form.

2. Mary is correct and Roger is incorrect.
   - Electrons passed through a crystal will produce a diffraction pattern, just as if X-rays were passed through the crystal.
   - Electrons passed through a single slit will produce a diffraction pattern.

3. Electrons exhibit a wave behaviour.
   Electrons form standing waves in orbits where the circumference is a whole multiple of the electron wavelength.
   This means that only certain discrete energy states can exist.

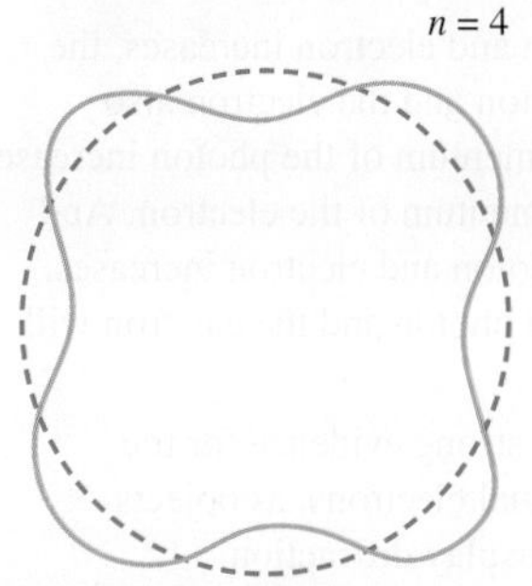

4. a. Answers will vary.
   Electrons have a wavelike nature, with wavelength λ as defined by Louis de Broglie.
   Electrons can be modelled as travelling along one of the allowed orbits around the nucleus, together with their associated wave.
   The circumference of each allowed orbit contains a whole number of wavelengths of the electron-wave: $n\lambda = 2\pi r$.

b. Answers will vary.

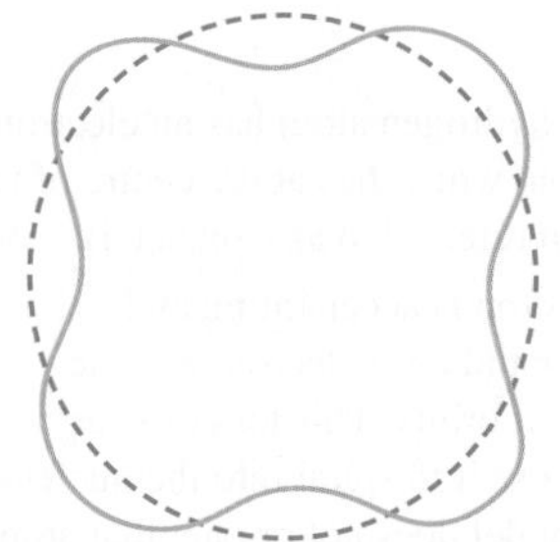

5. In the first model, the electron modelled as a particle is in a circular orbit and thus accelerating. Classical theory predicts that the accelerating electron would continuously emit light and hence spiral into the nucleus. Atoms are inherently unstable using this model and not consistent with observation. The second model treats the electron as a type of standing wave — a resonance. The standing wave is considered stable and does not emit a photon. Only transitions from a higher energy state to a lower energy state are associated with the emission of a single photon. This second model is consistent with observations of emission spectra.

10.6 Review

10.6 Review questions

1. a. $\lambda_{\text{electron}} = 2.6 \times 10^{-10}$ m and $\lambda_{\text{photon}} = 5.8 \times 10^{-7}$ m
   The photon has a longer wavelength than the electron.
   b. $p_{\text{electron}} = 2.6 \times 10^{-24}$ N s and $p_{\text{photon}} = 1.2 \times 10^{-27}$ N s
   The momentum of the electron is greater than the momentum of the photon.
   c. $E_{\text{electron}} = 3.6 \times 10^{-18}$ J and $E_{\text{photon}} = 3.5 \times 10^{-19}$ J
   The energy of the electron is greater than the energy of the photon.
2. a. $p_{\text{electron}} = 9.3 \times 10^{-25}$ N s and $p_{\text{photon}} = 1.6 \times 10^{-27}$ N s
   The momentum of the electron is greater than the momentum of the photon.
   b. $\lambda_{\text{electron}} = 7.1 \times 10^{-10}$ m and $\lambda_{\text{photon}} = 4.1 \times 10^{-7}$ m
   The photon has a longer wavelength than the electron.
   c. As the energy of the photon and electron increases, the momentum of both the photon and the electron also increases, however, the momentum of the photon increases at a faster rate than the momentum of the electron. And as the momentum of the photon and electron increases, the wavelengths of both the photon and the electron will decrease.
3. The diffraction of electrons is strong evidence for the wavelike behaviour of individual electrons, as objects modelled as particles do not display diffraction.
4. In a typical emission spectrum, a series of discrete coloured lines is observed. These lines occur due to atomic transitions from one energy level to a lower energy level within each excited atom. In an absorption spectrum, a continuous spectrum consisting of all colours with black lines is observed. The missing colours (where the black lines are positioned) occur when atoms in their ground state are excited into a higher energy state by the absorption of photons of specific energy consistent with the dark lines in an otherwise continuous spectrum.
5. $V = 1.5 \times 10^4$ V
6. a.

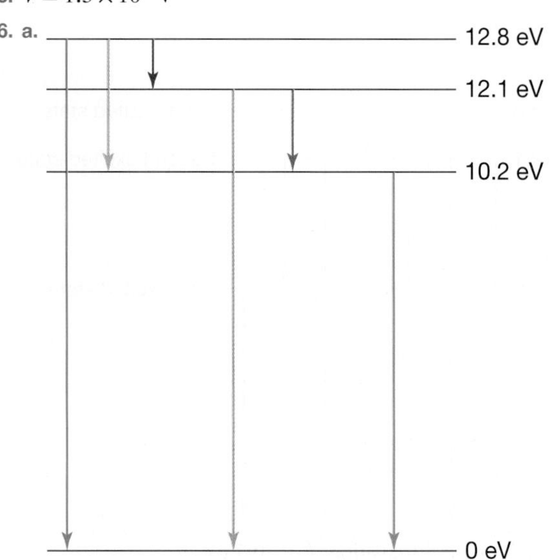

   b. There are six different energy photons. In order from left to right in the diagram in part a, they have energy 12.8 eV, 2.6 eV, 0.70 eV, 12.1 eV, 1.90 eV and 10.2 eV.
   c. $\lambda = 1.8 \times 10^{-6}$ m
7. a. i. $E = 1.8$ eV
   ii. $f = 4.3 \times 10^{14}$ Hz and $\lambda = 6.9 \times 10^{-7}$ m
   b. $\lambda = 1.4 \times 10^{-7}$ m
8. Emission spectra consist of coloured emission lines resulting from photons emitted from atoms when transitions occur between excited states to lower energy states, one of which would be the ground state. Absorption spectra on the other hand consist of absorption lines resulting from atoms being excited from the ground state to higher energy states. There are fewer absorption transitions than there are emission transitions possible.
9. Taylor's experiment tells us that each particle of light, a photon, behaves like a wave when passing through the apparatus and also like a particle when it strikes the photographic plate or screen used.

10.6 Exam questions

Section A — Multiple choice questions

1. B
2. C
3. B
4. B
5. C
6. B
7. C
8. B
9. D
10. D

Section B — Short answer questions

11. a. $E = 6.2 \times 10^2$ eV

b. $\frac{\lambda}{w} = 4 \times 10^{-2}$ thus little to no diffraction will be observed.

12. a. $\lambda = 0.155$ nm

b. $v = 4.7 \times 10^6$ m s^{-1}

13. a. D

b. Answers will vary.
- Electrons orbit nuclei in shells with discrete energy levels.
- Electrons can transition between these shells by absorbing or emitting discrete amounts of energy equal to the difference between the two shells.
- Since transitioning electrons can only release discrete amounts of energy, only discrete spectral lines will be observed.

14. a. Answers will vary. See worked solution.

b. $E = 7.1 \times 10^4$ eV

15. a. See figure at the bottom of the page*

b. $E = 1.07 \times 10^{-18}$ J

c. Nothing will happen.

11 Einstein's special theory of relativity and the relationship between energy and mass

11.2 Einstein's special theory of relativity

Sample problem 1

Scenario 1: 80 km h^{-1}

Scenario 2: 80 km h^{-1}

Scenario 3: 120 km h^{-1}

Practice problem 1

Velocity is relative

Sample problem 2

The acceleration of the rocket is the same whether it is measured from frame A or frame B. It does not depend on the reference frame.

Practice problem 2

a. The measured speed depends entirely on the relative motion of the person measuring the speed. Therefore, it depends on the observer's frame of reference.

b. Newton's three laws of motion do not require the entering of a value of velocity at any point. However, the second law (net force = *ma*) includes a value for acceleration. If the value for acceleration is different, then the net force would be different. By the principle of relativity, physical laws need to give the same answer in all inertial reference frames.

Sample problem 3

Galileo proposed that all inertial frames of reference are equally valid. Maxwell's concept of electromagnetic waves suggested the presence of an absolute frame of reference — the luminiferous aether. That is, observers moving relative to the aether should experience light at different velocities, and therefore not all inertial frames of reference are equally valid, as stated by Galileo.

Practice problem 3

1.5*c*

Sample problem 4

- The principle of relativity is applied to all laws of physics.
- The speed of light is constant for all observers.

Practice problem 4

Einstein found that the speed of light was invariant and that time, distance, simultaneity, mass and kinetic energy were all dependent on the motion or the reference frame of the observer, and hence relative.

Sample problem 5

35 days

Practice problem 5

Even small increases in the acceleration would make the astronauts feel very heavy. Humans cannot withstand large accelerations for long periods of time. If the acceleration was large enough, they would be squashed flat, but, long before that,

***15. a.**

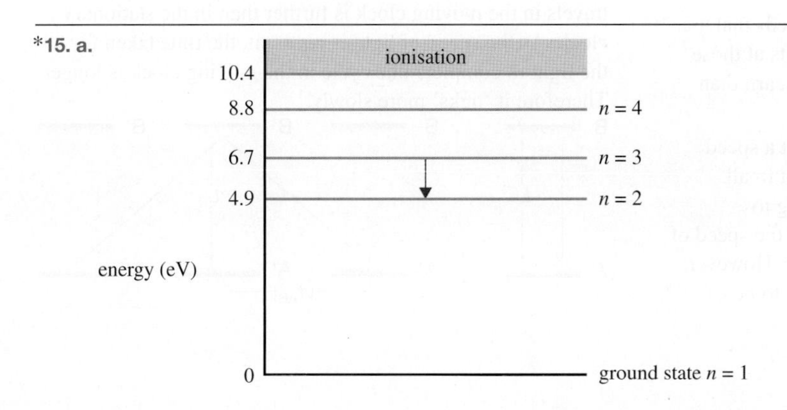

they would be rendered unconscious by the lack of blood flow through the brain.

Sample problem 6

Figure 11.16a in the text shows the situation for planet A. The light radiates in all directions at the same rate, and the diagram shows where the light in one direction and the opposite direction would be after one year.

Figure 11.16b in the text shows what is happening on planet B according to observers on planet A. The light moving out behind the moving planet reaches the one-light-year distance sooner than the light moving out from the front! But it is known that planet B is at the centre of this light circle. The way to achieve this is to move away from absolute space and time and understand that these are relative to the observer. When this is done, it is seen as possible for planet B to be at the centre of the light circle. However, this requires that A and B disagree about when two events occur. According to planet A, the different sides of the light circle reach the light-year radius at different times, but from planet B this must occur simultaneously.

Practice problem 6

a. In classical physics, simultaneity is invariant.

b. In special relativity, simultaneity is relative.

11.2 Exercise

1. Velocity depends on the reference frame in which it is measured.
2. A frame of reference is a set of length and time coordinates that an observer uses to measure an event.
3. 100 km h^{-1}
4. a. The acceleration due to the revolution of Earth around the Sun is about 6×10^{-2} m s^{-2}, which is tiny. According to the principle of relativity, if Earth is not accelerating, the movement of Earth cannot be felt no matter what its speed.
   b. Daily rotation about its axis and precession of the equinoxes (the motion of Earth's axis), which has a period of about 22 000 years.
   c. The accelerations are so tiny that they are difficult to detect.
5. 2.9989×10^8 m s^{-1}
6. a. Newton's laws are an excellent approximation at speeds lower than light speed. For most of human history, there was no way of observing extremely high speeds, meaning that the limitations of Newton's laws were unidentifiable.
   b. Newton's laws are still useful at the lower speeds that are seen in everyday life, providing accurate results at those speeds. They are also much easier to use and learn than Einstein's laws.
7. It would seem impossible. An object travelling at a speed of $0.99c$ towards a stationary observer emits light in all directions. The light travels at speed c. According to Newtonian physics, the observer would measure the speed of the light from the approaching object to be $1.99c$. However, the observer actually measures the speed of light to be c.
8. 3.0×10^8 m s^{-1}

11.2 Exam questions

1. D
2. Einstein's second postulate is 'The speed of light has a constant value for all observers regardless of their motion or the motion of the source.' Classical physics states that if there is relative motion between the source of light and the observer then the measurement of the speed of light would vary.
3. No, the velocity might not be constant (as constant speed is not the same as constant velocity). The ship in question could be travelling in a circular path, or it could be in orbit and still be travelling at a constant speed. Therefore, the spaceship may not be in an inertial frame of reference.
4. B
5. B

11.3 Time dilation

Sample problem 7

5.774 min (of 5 min 46.4 s)

Practice problem 7

4.329 min

Sample problem 8

5.774 min

Practice problem 8

1.8 min (or 1 min 48 s)

Sample problem 9

$t = 1.000\,000\,000\,000\,0022 t_0$

The difference is so small that it will be unnoticeable.

Practice problem 9

$v = 0.866c$

11.3 Exercise

1. Time dilation is the slowing of time for objects in motion relative to the observer. It can be detected with very sensitive instruments when comparing times on board aeroplanes with the time on the ground.
2. The clock in motion relative to you
3. As can be seen in the two figures, the distance the light travels in the moving clock is further than in the stationary clock. As the speed of light is constant, the time taken for the light to complete one cycle in the moving clock is longer. Therefore it 'ticks' more slowly.

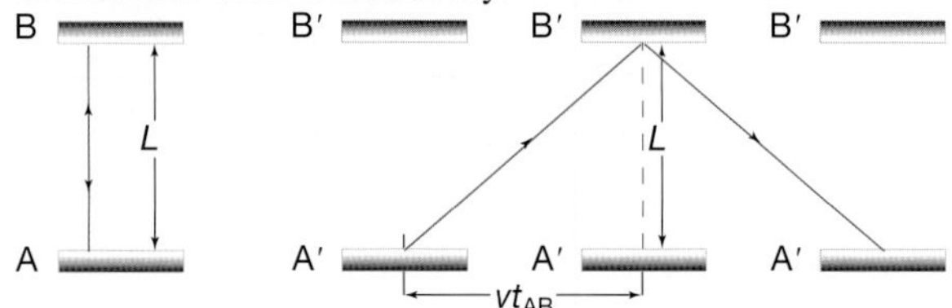

4. Proper time is t_0, which is the time as measured in the reference frame of the event. Conversely, t is the time as measured in a different inertial reference frame.
5. a. 1.048 28
   b. 95.4 beats per minute

11.3 Exam questions

1. A
2. a. 6.44 years
   b. Yes, because the clock is stationary in the astronaut's frame of reference.
3. D
4. 3.05×10^{-13} s
5. C

11.4 Length contraction

Sample problem 10

The spacecraft appears to be only 86.6% of its proper length.

Practice problem 10

a. Madeline will record a distance shorter than the one recorded by Rebecca.
b. Rebecca's measure is the proper length.

11.4 Exercise

1. Depth
2. a. Its length would be unchanged from the perspective of the alien on board.
   b. The length of the spaceship from the perspective of the Sun would be 60% of its proper length.
   c. The speed that light from the Sun reaches it is c.
3. a. 2.82 s b. 0.7053 m
4. 43.59 m
5. $v = 0.75c$

11.4 Exam questions

1. a. The technician is observing length contraction, which only occurs in the axis/direction of motion.
   b. $L_0 = 189$ m
2. C
3. $0.94c$
4. 4.57×10^{-6} m
5. A

11.5 Relativity is real

Sample problem 11

a. 1.56 μs b. $\gamma = 9.29$
c. $v = 0.994c$ d. 14.5 μs

Practice problem 11

a. 1938 m b. 209 m

11.5 Exercise

1. As a result of the muons travelling at very near light speed, their half-lives according to scientists on Earth are much longer than in their rest frame due to time dilation. This means that more of the muons have sufficient time to reach Earth's surface before decaying.
2. 5.97 km
3. 7.839 microseconds
4. In the early days of special relativity, few experiments were available to test whether the phenomena predicted by the theory matched up with measurements. Muons, however, were detectable, were travelling close to the speed of light and had features of their journey that were measurable, in particular the distance travelled and their half-life. The number of muons reaching the ground agreed with the predictions of special relativity but not classical physics.
5. 1.7×10^2 m (to 2 s.f.)
6. a. 2.97×10^8 m s^{-1} b. 4.71×10^{-7} s

11.5 Exam questions

1. C
2. Answers will vay. This is due to time dilation. The half-life of the muon as measured in Earth's frame of reference is longer than the half-life measured in the muon's frame of reference. This explains why muons travel further before they decay.
   OR
   This is due to length contraction. The distance to the surface as measured in the muon's frame of reference is shorter than the distance measured in Earth's frame of reference. This explains why muons can reach the surface before they decay.
3. a. $t = 7.4$ μs
   b. The length would appear shorter. $L = 0.47$ km
4. C
5. With a GPS clock accurate to 7 μs your position on Earth would be accurate to 2 km, whereas with a GPS clock accurate to the nanosecond, your position on Earth would be accurate to 0.3 m.

11.6 Einstein's relationship between mass and energy

Sample problem 12

Speeds larger than c would produce a negative under the square root sign, so these speeds are not possible.

Practice problem 12

3×10^{16} kg

Sample problem 13

$\Delta m = 1.96 \times 10^{-26}\ \text{kg} \simeq 12 \times m_{\text{proton at rest}}$

Practice problem 13

$\Delta m = 3.83 \times 10^{-27}\ \text{kg} \simeq 2 \times m_{\text{proton at rest}}$

Sample problem 14

$v = 1.45 \times 10^9$ m s^{-1}

($v > c$ This speed is not possible.)

Practice problem 14

$v = 6.43 \times 10^8$ m s^{-1}

$v > c$ This speed is not possible, thus Newton's laws are not useful when dealing with objects travelling at high speeds.

Sample problem 15

Einstein's physics: $E_k = 1.39 \times 10^{20}$ J

Newtonian physics: $E_k = 1.13 \times 10^{20}$ J

The kinetic energy is 1.23 times the value predicted by classical physics.

Practice problem 15

$v = 0.999\,999\,985\,4c$

Sample problem 16

The binding energy of the nucleus has not been considered.

Practice problem 16

$\Delta m = 5.33 \times 10^{-29}$ kg

Sample problem 17

$P = 4.0 \times 10^{26}$ W

Practice problem 17

$t = 1.44 \times 10^{13}$years

Sample problem 18

$\Delta E = 8.2 \times 10^{-14}$ J

Practice problem 18

$\Delta E = 2.9 \times 10^{-10}$ J

11.6 Exercise

1. The mass of an object increases with its motion, according to the factor γ. Therefore an object's mass becomes infinite at the speed of light. Also, the length of the object would be zero at the speed of light, and time would cease. Mathematically, speeds beyond the speed of light would require taking the square root of a negative number in the equation for γ.
2. a. i. 4.5×10^{17} J ii. 1.4×10^{19} J iii. 6.0×10^{19} J iv. 1.2×10^{20} J

   b.

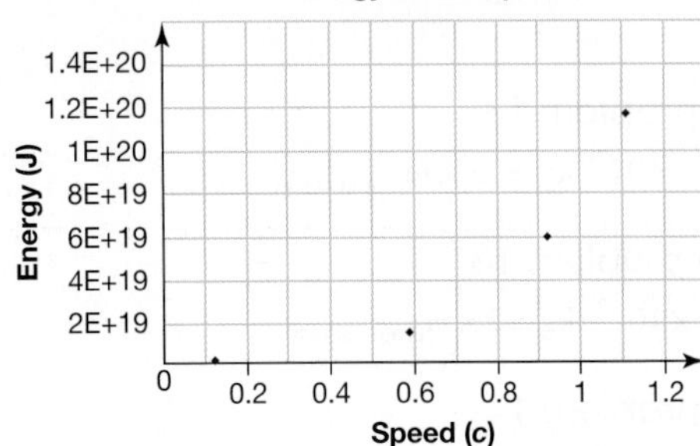

3. Energy and mass are equivalent.
4. 5.4×10^{41} J
5. 2.99×10^{16} kg
6. It decreases as the fusion process converts mass into energy.

11.6 Exam questions

1. C
2. B
3. $E = 6.3 \times 10^{21}$ J
4. $E = 2.3 \times 10^{-11}$ J
5. B

11.7 Review

11.7 Review questions

1. Inertial reference frames are those that are not subject to acceleration. Non-inertial frames are accelerating.
2. There are no indications of movement with constant velocity. While accelerating, however, a force is required and you feel this as a push from your seat.
3. Head-on collisions are particularly dangerous because the velocity of one car relative to the other can be much greater than the velocity of each car individually. For example, two cars, both travelling at 50 km h^{-1} in opposite directions, collide head-on. The velocity of one car relative to the other is 100 km h^{-1}.
4. a. The laws of physics are the same in all inertial reference frames, and light speed is constant in a vacuum, for all observers.
   b. In previous physics, the laws of electromagnetism were not the same in all inertial reference frames — light speed depended on the motion of the source and the receiver, so was not the same for all observers.
5. False. By the principle of relativity, the situation must be symmetrical so that any two inertial reference frames are indistinguishable. Both observers see the other frame as moving and, therefore, both see the other's time running slow.
6. a. 5 minutes b. $0.78c$
7. Accelerating even relatively small masses to near light speed involves enormous amounts of energy. This is not currently feasible. Additionally, humans are not able to sustain large accelerations for extended periods of time. Reaching near light speeds with safe accelerations would take years.
8. 2.25×10^{16} J
9. All the chemical products of combustion must add up to less than the mass of the initial coal because energy has been released.
10. 7.47×10^{-31} kg

11.7 Exam questions

Section A — Multiple choice questions

1. A
2. A
3. C
4. D
5. A

6. D
7. A
8. D
9. C
10. A

Section B — Short answer questions

11. The transformation is a mass–energy transformation due to nuclear fusion. As the energy is radiated away, the mass of the star will decrease.
12. a. $\gamma = 5.61$ b. $t = 4.44$ years
13. $t = 6.5$ years
14.

length of side along x-axis	4.5×10^2 m
length of side along y-axis	3.2×10^3 m
length of side along z-axis	3.2×10^3 m

Length contraction only occurs in the direction of travel, the y and z dimensions would remain unchanged.

15. 15 years

Unit 4 | Area of Study 1 review

Section A — Multiple choice questions

1. D
2. C
3. B
4. B
5. A
6. D
7. C
8. A
9. C
10. D
11. A
12. A
13. D
14. C
15. A
16. B
17. C
18. B
19. B
20. B

Section B — Short answer questions

21. a. 6.4×10^{-7} m b. 1.9 eV c. 1.0×10^{-27} N s
22. a. 1325 nm b. 1060 nm
23. a. 2.9×10^{-19} J
b. 8.8×10^{-20} J
c. 5.5×10^{-1} V
d. Increasing the number of photons will increase the magnitude of the photocurrent, but the stopping voltage will remain the same.
24. 5.3×10^{-7} m
25. a. Diffraction is a property of waves.
Electrons can undergo diffraction when passed through atomic crystals and electrons are matter.
Thus electron diffraction is evidence for the wave-like nature of matter.
b. • The photoelectric effect supports the particle theory of light because the energy required to release photoelectrons from a metal is dependent upon the frequency of light, and not the intensity of light.
• The photoelectric effect proves that energy is quantised. This contradicts the wave model for light with energy arriving continuously.
26. a.

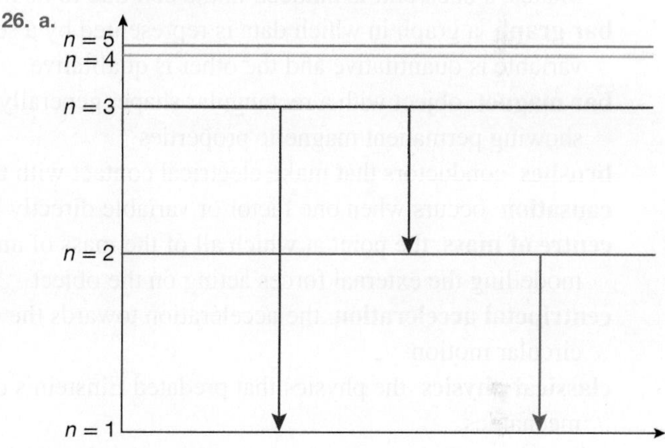

b. 7.7×10^{14} Hz
27. a. 5.07×10^{-11} m b. 5.07×10^{-11} m
c. 9.4×10^{-17} J d. 5.9×10^2 V

GLOSSARY

absolute independent of frame of reference, permanent
absorption spectrum spectrum produced when light passes through a cool gas. It includes a series of dark lines that correspond to the frequencies of light absorbed by the gas.
acceleration rate of change of velocity; a vector quantity
accuracy how close an experimental measurement is to a known or 'true' value
aim a statement outlining the purpose of an investigation, linking the dependent and independent variables
air resistance the force applied to an object opposite to its direction of motion, by the air through which it is moving
alternating current (AC) an electric current that reverses direction at short, regular intervals
alternator device in which the ends of the coil are connected to slip rings, causing the voltage to alternate in direction, inducing an alternating current
amplitude the maximum variation from zero of a periodic disturbance
amplitude size of the maximum disturbance of the medium from its normal state
antinodal line line where constructive interference occurs on a surface
antinode point at which constructive interference takes place
armature frame around which a coil of wire is wound, which rotates in a motor's magnetic field
assumptions ideas that are accepted as true without evidence in order to overcome limitations in experiments
back emf electromagnetic force that opposes the main current flow in a circuit. When the coil of a motor rotates, a back emf is induced in the coil due to its motion in the external magnetic field.
bar graph a graph in which data is represented by a series of bars; bar graphs are usually used when one variable is quantitative and the other is qualitative
bar magnet object with a rectangular shape, generally made up of iron or other ferromagnetic substance, showing permanent magnetic properties
brushes conductors that make electrical contact with the moving split ring commutator in a DC motor
causation occurs when one factor or variable directly influences the results of another factor or variable
centre of mass the point at which all of the mass of an object can be considered to be positioned when modelling the external forces acting on the object
centripetal acceleration the acceleration towards the centre of a circle experienced by an object moving in a circular motion
classical physics the physics that predated Einstein's discoveries leading to the laws of relativity and quantum mechanics
coherent same frequency and waveform (in phase); describes light in which all photons are emitted in phase, leading to intense light
conclusion a section at the end of a report that relates back to the question, sums up key findings and states whether the hypothesis was supported or rejected
constructive interference the addition of two wave disturbances to give an amplitude that is greater than either of the two waves
continuous data quantitative data that can take on any continuous value
continuous spectrum a spectrum that has no gaps; there are no frequencies or wavelengths missing from the spectrum
control group a group that is not affected by the independent variable and is used as a baseline for comparison
controlled variable a variable that is kept constant across different experimental groups
correlation the measure of a relationship between two or more variables
de Broglie wavelength wavelength associated with a particle of matter, in relation to its mass and wavelength
dependent variable the variable that is anticipated to be influenced by the independent variable and is measured by an investigator
design speed speed at which the force due to friction becomes zero, as seen on a banked track

destructive interference the addition of two wave disturbances to give an amplitude that is less than either of the two waves

diffraction the spreading out, or bending of, waves as they pass through a small opening or move past the edge of an object

diode a two-terminal semiconductor device that allows current to pass through it in one direction but not the other

dipole field electric field surrounding a positive charge and a negative charge that are separated by a short distance

direct current (DC) an electric current that flows in one direction only

discrete data quantitative data that can only take on set values

discussion a detailed area of a report in which results are discussed, analysed and evaluated; relationships to concepts are made; errors, limitations and uncertainties are assessed; and suggestions for future improvements are made

displacement measure of the change in position of an object, a vector quantity

distance measure of the full length of the path taken when an object changes position, a scalar quantity

eddy current an electric current induced in the iron core of a transformer by changing magnetic fields

elastic collision collision in which the total kinetic energy is conserved

electric dipole a positive charge and a negative charge that are separated by a short distance

electric field vector field describing the property of the space around a charge that causes a second charge in that space to experience a force due only to the presence of the first charge

electric force force experienced by a charged particle if it is placed within the electric field of a second charged particle

electric monopole single electric point charge, in which all the electric field lines point inward for a net negative electric charge or away for a net positive electric charge

electromagnet temporary magnet produced in the presence of a current-carrying wire

electromagnetic induction generation of an electromotive force (emf) in a coil (an electrical conductor) as a result of a changing magnetic field

electron gun a device to provide free electrons for a linear accelerator, usually consists of a hot wire filament with a current supplied by a low-voltage source

electron volt the quantity of energy acquired by an elementary charge ($q_e = 1.6 \times 10^{-19}$ C) passing through a potential difference of 1 V. Thus, 1.6×10^{-19} J = 1 eV

emf source of voltage that can cause an electric current to flow

emission spectra spectra produced when light is emitted from an excited gas and passed through a spectrometer. A spectrum includes a series of bright lines on a dark background. The bright lines correspond to the frequencies of light emitted by the gas.

errors differences between a measurement taken and the true value that is expected; errors lead to a reduction in the accuracy of an investigation

ethics principles of acceptable and moral conduct determining what is 'right' and what is 'wrong'

excited state state of an electron in which it has more energy than its ground state

experimental bias a type of influence on results in which an investigator either intentionally or unintentionally manipulates results to get a desired outcome

experimental groups test group that is exposed to the independent variable

extrapolation an estimation of a value outside the range of data points tested

falsifiable able to be proven false using evidence

ferromagnetic property of materials, such as iron, cobalt and nickel, that can be easily magnetised (act like a magnet)

frequency a measure of how many times per second an event happens, such as the number of times a wave repeats itself every second or the number of revolutions that an object completes each second

galvanometer instrument used to detect small electric currents, or to detect the direction of current (such as in AC)

generator device in which a rotating coil in a magnetic field is used to induce a voltage

geostationary stationary relative to a point directly below it on Earth's surface. A geostationary orbit has the same period as the rotation of Earth

gravitational field vector field describing the property of space that causes an object with mass to experience a force in a particular direction

gravitational potential energy energy stored in an object as a result of its position relative to another object to which it is attracted by the force of gravity

ground state state of an electron in which it has the least possible amount of energy

histogram a graph in which data is sorted in intervals and frequency is examined, and is used when both pieces of data are quantitative; all columns are connected in a histogram

hypothesis a tentative, testable and falsifiable statement for an observed phenomenon that acts as a prediction for the investigation

ideal transformer a transformer that is 100% efficient, meaning its input power is equal to its output power

impulse product of a force and the time interval during which it acts. Impulse is a vector quantity with SI units of N s.

independent variable the variable that is changed or manipulated by an investigator

induced voltage voltage caused by the separation of charge due to the presence of a magnetic field

inelastic collision collision in which the total kinetic energy is not conserved

inertial reference frame a reference frame that is not accelerating, where Newton's laws hold true

instantaneous speed speed at a particular instant of time

instantaneous velocity velocity at a particular instant of time

interpolation an estimation of a value within the range of data points tested

invariant quantity unchanging, regardless of the frame of reference

inverse square law relationship in which one variable is proportional to the reciprocal of the square of another variable

inverter device that converts direct current electricity produced by solar panels into alternating current electricity, usable in the home

investigation question the focus of a scientific investigation in which experiments act to provide an answer

ionisation energy the amount of energy required to be transferred to an electron to enable it to escape from a material

isolated system system where no external forces act; the only forces acting on objects in the system are those applied by other objects within the system.

kinetic energy the energy associated with the movement of an object. Like all forms of energy, it is a scalar quantity.

left-hand rule rule used to determine the direction of the magnetic force of a magnetic field on a current or moving positive charge

like charges charges with the same type (both positive, or both negative)

limitations factors that affect the interpretation and/or collection of findings in a practical investigation

line graph a graph in which points of data are joined by a connecting line; used when both pieces of data are quantitative (numerical)

linear particle accelerator type of particle accelerator based on the work done by the field in moving a charge from one plate to the other

line of best fit a trend line that is added to a scatterplot to best express the data shown; these are straight lines, and are not required to pass through all points

logbook a record containing all the details of progress through the steps of a scientific investigation

longitudinal periodic wave wave for which the disturbance is parallel to the direction of propagation

luminiferous aether hypothesised medium permeating space, supposed to carry electromagnetic waves

magnet material or object capable of producing a magnetic field and attracting unlike poles and repelling like poles

magnetic field vector field describing the property of the space in which a magnetic object experiences a force

magnetic flux measure of the amount of magnetic field passing through an area; measured in webers (Wb)

magnetic flux density (B) strength of a magnetic field; measured in tesla (T) or weber per square metre (Wb m^{-2})

magnetic induction process by which a substance, such as iron, becomes magnetised by a magnetic field
mass–energy concept used to describe mass and energy as equivalent, given by $E = mc^2$
matter anything that takes space (has volume) and has a rest mass
model a representation of ideas, phenomena or scientific processes; can be a physical model, mathematical model or conceptual model
momentum the product of the mass of an object and its velocity; a vector quantity
monochromatic light of a single frequency and, hence, very clearly defined colour
net force the vector sum of all the forces acting on an object
nodal line line where destructive interference occurs on a surface, resulting in no displacement of the surface
node point at which destructive interference takes place
nominal data qualitative data that has no logical sequence
null-result experimental outcome not showing an expected effect
orbital period the time it takes for a satellite to complete one orbit around a central object
ordinal data qualitative data that can be ordered or ranked
outlier an unusual result that differs from other results
path difference the difference between the lengths of the paths from each of two sources of waves to a point
peak current the amplitude of an alternating current
peak-to-peak voltage the difference between the maximum and minimum voltages of a DC voltage
peak voltage the amplitude of an alternating voltage
period time taken for an object, moving in a circular path and at a constant speed, to complete one revolution; the time it takes a source to produce one complete wave (or for a complete wave to pass a given point)
periodic wave a disturbance that repeats itself at regular intervals
personal errors human errors or mistakes that can affect results but should not be included in analysis
photoelectric effect the emission of electrons when electromagnetic radiation hits a metal surface
photon a discrete bundle of electromagnetic radiation. Photons can be thought of as discrete packets of light energy with zero mass and zero electric charge
photovoltaic cell device that transforms electromagnetic energy, such as light from the Sun, directly into electrical energy. Also known as a PV cell or solar cell.
power rating the total electrical power required for an appliance or machine to operate normally
precision how close multiple measurements of the same investigation are to each other
primary data direct or firsthand evidence obtained from investigations or observations
primary source a document that is a record of direct or firsthand evidence about some phenomenon
proper length the length measured in the rest frame of the object
proper time between two events, the time measured in a frame of reference where the events occur at the same point in space. The proper time of a clock is the time the clock measures in its own reference frame.
qualitative data data with labels or categories rather than a range of numerical quantities; also known as categorical data
quantised cannot be divided or broken up into smaller parts
quantitative data numerical data that examines the quantity of something (e.g. length or time); also known as numerical data
quantum a small quantity of a fixed amount
random errors chance variations in measurements
randomised refers to when the assigning of individuals to an experiment or control group is random and not influenced by external means
refraction the bending of light as it passes from one medium into another
relative in relation to something else, dependent on the observer
relativistic mass the mass of a body in motion, relative to an observer, also known as inertial mass
repeatability how close the results of successive measurements are to each other in the exact same conditions
reproducibility how close results are when the same variable is being measured but under different conditions
resolution smallest change in the quantity being measured that causes a perceptible change in the value indicated on a measuring instrument

resonance when the amplitude of an object's oscillations is increased by the matching vibrations of another object or an external force

rest mass mass of an object measured at rest

restoring force force applied by a spring to resist compression or extension

results a section in a report or poster in which all data obtained is recorded, usually in the form of tables and graphs

right-hand-slap rule rule used to determine the direction of the magnetic force of a magnetic field on a current or moving positive charge

risk assessment a document that examines the different hazards in an investigation and suggests safety precautions

RMS voltage the value of the constant DC voltage that would produce the same power as AC voltage across the same resistance

road friction the force applied by the road surface to the wheels of a vehicle in a direction opposite to the direction of motion of the vehicle

sample size the number of trials in an investigation

satellite an object that is orbiting a larger central mass. Satellites can be natural (such as the Moon) or man-made (such as the International Space Station)

scalar quantity quantity with only a magnitude (size)

scatterplot a graph in which two quantitative variables are plotted as a series of dots

scientific investigation methodology the process of finding the answer to a question through testing and experimentation

scientific method the procedure that must be followed in scientific investigations, consisting of questioning, researching, predicting, observing, experimenting and analysing; also called the scientific process

scientific methodology the type of investigation being conducted to answer a question and resolve a hypothesis

scientific poster a hard-copy or digital poster used to display the key findings from investigations conducted to answer a scientific question or hypothesis

secondary data comments on or summaries and interpretations of primary data

secondary source a document that comments on, summarises or interprets primary data

slip ring an electromechanical device carrying current from a stationary to a rotating structure

solenoid coil of wire wound into a cylindrical shape

spectrometer device used to disperse light into its spectrum

speed the rate at which distance is covered per unit time; a scalar quantity

split ring commutator a device that reverses the direction of the current flowing through an electric circuit every half turn of the loop

standing wave the superposition of two wave trains at the same frequency and amplitude travelling in opposite directions, also known as stationary waves as they do not appear to move through the medium. The nodes and antinodes remain in a fixed position.

step-down transformer the output (secondary) voltage produced is less than the input (primary) voltage

step-up transformer the output (secondary) voltage produced is greater than the input (primary) voltage

strain potential energy energy stored in an object as a result of a reversible change in shape

superposition the adding together of amplitudes of two or more waves passing through the same point

systematic errors errors, usually due to equipment or system errors, that affect the accuracy of a measurement and cannot be improved by repeating an experiment

tentative not fixed or certain; may be changed with new information

terminal velocity velocity reached by a falling object when the upward air resistance becomes equal to the downward force of gravity

testable able to be supported or proven false through the use of observations and investigation

theory a well-supported explanation of phenomena, based on facts that have been obtained through investigations, research and observations

thought experiment an imaginary scenario designed to explore what the laws of physics predict would happen; also known as a gedanken experiment

time dilation the slowing of time by clocks moving relative to the observer

torque the turning effect of the forces on the coil in an electric motor

transformer a device in which two multi-turn coils may be wound around an iron core. One coil acts as an input while the other acts as an output. The purpose of the transformer is to produce an output AC voltage that is different from the input AC voltage.

transverse wave wave for which the disturbance is at right angles to the direction of propagation

uncertainty a limit to the precision of data obtained; a range within which a measurement lies

uniform electric field electrical field in which the strength and direction are constant at every point

United Nations Declaration on the Rights of Indigenous Peoples a universal framework of minimum standards for the survival, dignity and well-being of the indigenous peoples of the world

unlike charges charges with opposite type (one negative, one positive)

validity how accurately an experiment investigates the claim it is intended to investigate

variable any factor that can be changed in an investigation

vector quantity quantity requiring both a direction and a magnitude

velocity the rate of change of position of an object; a vector quantity

velocity selector device that can be used as a velocity filter using a magnetic field to deflect the beam of charged particles in one direction, and an electric field to deflect the beam in the opposite direction

wave transfer of energy through a medium without any net movement of matter

wavelength distance between successive corresponding parts of a periodic wave

wave–particle duality description of light as having characteristics of both waves and particles. This duality means that neither the wave model nor the particle model adequately explains the properties of light on its own.

Wien filter device that can be used as a velocity filter using a magnetic field to deflect the beam of charged particles in one direction, and an electric field to deflect the beam in the opposite direction

work energy transferred to or from another object by the action of a force. Work is a scalar quantity.

work function the minimum energy required to release an electron from the surface of a material

INDEX

E

F

G

H

Q

R

S

T

U

V

W